Spon's Architects' and Builders' Price Book

2009

Spon's Architects' and Builders' Price Book

Edited by

DAVIS LANGDON

2009

One hundred and thirty-fourth edition

First edition 1873
One hundred and thirty-fourth edition published 2009
by Taylor & Francis
2 Park Square, Milton Park, Abingdon, Oxon OX14 4RN

Simultaneously published in the USA and Canada
by Taylor & Francis
270 Madison Avenue, New York, NY 10016

Taylor & Francis is an imprint of the Taylor & Francis Group, an informa business

Printed and bound in Great Britain by
TJ International Ltd, Padstow, Cornwall

Publisher's note
This book has been produced from camera-ready copy supplied by the authors.

British Library Cataloguing in Publication Data
A catalogue record for this book is available from the British Library

ISBN13: 978-0-415-46555-7
Ebook: 978-0-203-88676-2
ISSN: 0306-3046

Contents

Preface to the One Hundred and Thirty Fourth edition

Counter-acting forces are in the ascendancy in the construction industry at this moment in time. Basic fuel and material costs are rising quickly, for example, oil is currently $135 per barrel and Corus have announced new price rises of £80 per tonne as from 29 June and a further £90 per tonne for deliveries from 3 August 2008. Against this, a slow-down in house costs and sales is concerning some economists, whilst liquidity (credit crunch) problems are forcing manufacturers to hold their prices and some developers to temporarily shelve less-promising schemes.

Construction inflation remains difficult to call. Slower growth will ease price pressures but a sharp deterioration in industry prospects could result in significantly lower tender price increases, as tougher market conditions would enforce greater competition than has been seen of late. Material prices are likely to remain elevated, as continuing strong demand from the Asian economies underpins commodity prices and the weak pound makes imports more expensive. Overall tender prices are still forecast to rise between 4% and 7% this year, although some regions may experience lower levels of inflation.

The global economy is still threat; but the IMF has recently revised its forecast for UK growth, up to 1.75% for 2008, from an earlier forecast of 1% growth. The IMF main concerns relate to the need to control inflation, seeing no room for further interest rate cuts in the near future and even suggesting that interest rates may need to rise again.

Prices

The price level of Spon's A&B 2009 has been indexed at 580, an increase of 5.4% over the index of 550 in the 2008 edition. Readers of Spon's A&B are reminded that Spon is the only known price book in which key rates are checked against current tender prices.

From 30 June 2008, the third and final part of a three-year 2006-09 Construction Industry Joint Council agreement on pay and conditions came into effect, raising wage rates from last year by approximately 6%. The revised labour rates provide basic hourly rates of £7.75 for general operatives and £10.30 for the craft rate, and have been taken into account in the preparation of this year's book, along with material prices, generally applicable as at April/May 2008.

Energy costs, but especially oil, have risen further this year, and a number of key material prices are still feeling the effects of high economic demand from China and the Far East, with their prices are rising again. Foremost amongst these are steel, stainless steel (exacerbated by the shortage and volatility of nickel), copper and lead. Readers are urged to check the currency of these particular materials, when using rates from Spon.

Similar care should also be taken with increase to labour costs. This year's national wage award indicates a +6% increase in wages from last year, but, whereas previously there was an influx of Eastern European labour to offset such increases, with the weakening pound, a significant outflow of labour could now boost UK labour costs again.

This Edition

Key additions to this edition include a thoroughly overhauled and expanded section on sustainability and an update of the preliminaries section, for the JCT2005 Standard Building Contract.

Within the Measured Works sections, the 2009 edition includes new items for Patent glazing, Schuco curtain walling, Brise soleil, Velfac aluminium/wood windows, Ecologic roof tiles, WC cubicles, triple glazing etc..

Future measurement changes

The RICS, in consultation with the Construction Confederation, have written a new set of draft rules, with a working title of *New Rules of Measurement*, which will include rules for estimating and cost-planning, as well as simplified rules for trade procurement. It is hoped that the rules for estimating and cost planning will be published at the end of 2008, and, after discussions are complete, that the trade procurement rules will be published in the spring of 2009.

Free updated cost information, between editions, is available on the publisher's web-site to those who register with the publishers. Follow the advice given on the coloured card bound within this volume.

Profits and Overheads

The 2009 edition includes a 5% mark-up for main contractor's overheads and 2½% mark-up for profit on built-up labour rates and material prices in the Measured Major Works section and a 7½% mark-up for overheads and a 5% mark-up for profit in the Measured Minor Works section. For non-analysed sub-contractor prices, only a 2½% mark-up for profit has been included, in both Major and Minor Works sections.

Preliminaries

There are signs that preliminaries costs are beginning to soften, but they still typically range from 15% to 19%, and sometimes, even higher. As last year, we have set our example provision for preliminaries at +16%.

Value Added Tax

Since 1989 most building work, refurbishment and alterations, has been subject VAT, currently at 17½. See Part 1, Value Added Tax, for further details

Prices included within this edition do not include for VAT, which must be added if appropriate.

Part 1: General

This section contains advice on various construction specialisms, legislation, taxes, levies, and sustainability.

Part 2: Approximate Estimating

This section contains the Building Cost and Tender Price Index, information on regional price variations, prices per functional unit and square metre for various types of buildings, building cost models, approximate estimates, cost limits and allowances for 'public sector' building work and a procedure for valuing property for insurance purposes.

Parts 3 & 4: Prices for Measured Work

These sections contain Prices for Measured Work - Major Works, and Prices for Measured Work - Minor Works (on coloured paper). All prices in Parts 3 & 4 exclude the main contractor's preliminaries costs.

Part 5: Professional Fees

This section contains Fees for Professional Services.

Part 6: Rates of Wages

This section includes authorised wage agreements applicable to the Building and associated industries.

Part 7: Daywork

This section contains Daywork and Prime Cost.

Part 8: Tables and Memoranda

This section contains general formulae, weights and quantities of materials, other design criteria and useful memoranda associated with each trade, a list of useful Trade Associations and web-sites with useful costs.

While every effort is made to ensure the accuracy of the information given in this publication, neither the Editors nor Publishers in any way accept liability for loss of any kind resulting from the use of such information

DAVIS LANGDON
MidCity Place
71 High Holborn
London WC1V 6QS

Acknowledgements

Acodrain
ACO Business Park
Hitchin Road
Shefford
Bedfordshire
SG17 5TE
Drainage channels
Tel: 01462 816 666
Fax: 01462 815 895
E-mail: drainsales@aco.co.uk
Website: www.aco.co.uk

Allgood PLC
297 Euston Road
London
NW1 3AQ
Ironmongery
Tel: 020 7387 9951
Fax: 020 7380 1232
E-mail: info@allgood.co.uk
Website: www.allgood.co.uk

Altro Floors
Works Road
Letchworth
Hertfordshire
SG6 1NW
"Altro" sheet, tile flooring and stair nosings
Tel: 01462 480 480
Fax: 01462 480 010
E-mail: info@altro.com
Website: www.altro.com

Alumasc Exterior Building Products Ltd
White House Works
Bold Road
Sutton
St Helens
WA9 4JG
Aluminium rainwater goods
Tel: 01744 648 400
Fax: 01744 648 401
E-mail: info@alumasc-exteriors.co.uk
Website: alumasc-exteriors.co.uk

Amwell Systems Ltd
Buntingford Business Park
Baldock Road
Buntingford
Herfordshire
SG9 9ER
WC cubicles
Tel: 01763 276200
Fax: 01763 276222
E-mail: estimating@amwell-systems.com
Website: www.amwell-systems.com

Andrews Marble & Tiles Ltd
324 - 330, Meanwood Road
Leeds
LS7 2JE
Wall and Floor tiles
Tel: 0113 262 4751
Fax: 0113 262 3337
E-mail: contracts@andrews-tiles.co.uk
Website: www.andrews-tiles.co.uk

Anaco Trading Ltd
PO Box 39
Oundle
Peterborough
PE8 4JT
Stainless steel rebar
Tel: 01832 272 109
Fax: 01832 275759

Armitage Shanks Group Ltd
Armitage
Rugeley
Staffordshire
WS15 4BT
Sanitary fittings
Tel: 0870 122 8822
Fax: 0870 122 8282
E-mail: arm-idealinfo@aseur.com
Website: www.armitage-shanks.co.uk

Armstrong Floor Products Ltd
Hitching Court
Abingdon Business Park
Oxon OX14 1BR
Vinyl flooring products
Tel: 01235 554 848
Fax: 01235 553 583
E-mail: serviceuk@armstrong.com
Website: www.armstrong.com

Astec Projects Ltd
Astec House
187- 189, Kings Road
Reading
Berkshire
RG1 4EX
Suspended ceiling systems
Tel: 0118 958 1333
Fax: 0118 958 1337
E-mail: astec@astecprojects.co.uk
Website: www.astecprojects.co.uk

Bison Concrete Products Ltd
Millennium Court
First Avenue
Centrum 100
Burton Upon Trent
DE14 2WR
Precast concrete flooring systems
Tel: 01283 495 000
Fax: 01283 544 900
E-mail: concrete@bison.co.uk
Website: www.bison.co.uk

Bonar Floors (Nuway)
High Holburn Road
Ripley
Derbyshire
DE5 3NT
Entrance matting and aluminium matwell frames
Tel: 01773 740 688
Fax: 01773 740 640
E-mail: infouk@bonarfloors.com
Website: www.bonarfloors.com

Bostik Findley (Evode Ltd)
Common Road
Stafford
Staffordshire
ST16 3EH
Adhesives etc.
Tel: 01785 272 727
Fax: 01785 241 818
E-mail: sales@bostik-findley.com
Website: www.bostick-findley.com

Bradstone Structural
Northend Works
Ashton Keynes
Swindon
Wiltshire
SN6 6QX
Reconstructed stone walling, kerbs etc.
Tel: 01285 646 844
Fax: 01285 646 891
Website: www.aggregate.com

BRC Building Products
Carver Road
Astonfields Industrial Estate
Staffordshire
ST16 3BP
Brick reinforcement
Tel: 01785 222 288
Fax: 01785 240 029
E-mail: email@brc-special-products.co.uk
Website: www.brc-uk.co.uk

Brewers
122–124, Broadley Street
Edgeware Road
London
NW8 8BB
Sadolin, Sikkens, etc.
Tel: 0207 723 6657
Fax: 0207 706 4662
E-mail: decorating@brewers.co.uk
Website: www.brewers.co.uk

B R Hodgson Ltd
Ability House
121, Brooker Road
Waltham Abbey
Essex
EN9 1JH
Plastering and screeding
Tel: 01992 766777
Fax: 01992 766888
E-mail: brhlondon@aol.com
Website: www.brhodgson.co.uk

Briggs Amasco Roofing & Cladding
Halfords Lane
Smethwick
West Midlands
B66 1BJ
Various types of roofing finishes
Tel: 0121 555 267
Website: www.briggsamasco.co.uk

British Gypsum Ltd
East Leake
Loughborough
Leicestershire
LE12 6HX
Plasterboard and plaster products
Tel: 0115 945 1000
E-mail: BGTechnical.Enquiries@bpb.com
Website: www.british-gypsum.com

Building Innovation
Unit 4, The Cobalt Centre
Kineton Road
Southam
Warwickshire
CV47 0AL
Tapered insulation
Tel: 01926 815 008
Fax: 01926 815 018
E-mail: info@building-innovation.co.uk
Website: www.building-innovation.co.uk

Burlington Slate Ltd
Cavendish House
Kirkby-in-Furness
Cumbria
LA17 7UN
"Westmoreland" slating
Tel: 01229 889 661
Fax: 01229 889 466
E-mail: sales@burlingtonstone.co.uk
Website: www.burlingtonstone.co.uk

Catnic Ltd
Pontypandy Industrial Estate
Caerphilly
Mid Glamorgan
CF83 3GL
Steel lintels
Tel: 02920 337 900
Fax: 02920 867 796
Website: www.catnic.com

Cavity Trays Ltd
Administration Centre
Lufton Trading Estate
Yeovil
Somerset
BA22 8HU
Cavity trays, closers and associated products
Tel: 01935 474 769
Fax: 01935 428 223
E-mail: sales@cavitytrays.co.uk
Website: www.cavitytrays.com

Cementation Foundations Skanska Ltd
Maple Cross House
Denham Way
Maple Cross
Rickmansworth
Hertfordshire
WD3 2SW
Piling
Tel: 01923 423 100
Fax: 01923 777 834
E-mail: cementation.foundations@skanska.co.uk
Wsite:www.cementationfoundations.skanska.co.uk

Cemex UK
Waldorf Way
Denby Dale Road
Wakefield
West Yorkshire
WF2 8DH
Thermabate
Tel: 01924 262 081
Fax: 01924 290 126
E-mail: thermabate.sales@rmc-group.com
Website: www.rmc-group.com

Clement Windows Group Ltd
Clement House
Haslemere
GU27 1HR
Metal windows
Tel: 01428 643393
Fax: 01428 644436
E-mail: info@clementwg.co.uk
Website: www.clementwg.co.uk

Colour Centre
29a Offord Road
London
N1 1EA
Paints, varnishes, stains
Tel: 020 7609 1164
Fax: 020 7700 1345
Website: www.colourcentre.com

Complete Drilling Services Ltd
Unit 3
Norbury Trading Estate
Craignish Avenue
Norbury
London
SW16 4RW
Diamond drilling and sawing
Tel: 020 8679 8833
Fax: 020 8679 7877

Cordek Ltd
Spring Copse Business Park
Slinfold
West Sussex
RH13 7SZ
Trough moulds
Tel: 01403 799 600
Fax: 01403 791 718
E-mail: sales@cordek.com
Website: www.cordek.com

Corus UK Ltd
PO Box 1
Brigg Road
Scunthorpe
North Lincolnshire
DN16 1BP
Structural steel
Tel: 01724 405060
Fax: 01724 404224
Website: www.corusconstructionandindustrial.com

Cox Building Products
Unit 1, Shaw Road
Bishbury
Wolverhampton
WV10 9LA
Rooflights
Tel: 01902 371 800
Fax: 01902 371 810
E-mail: enquiries@coxdome.co.uk
Website: www.coxbp.com

CPM Group Ltd
Mells Road
Mells
Frome
BA11 3PD
Concrete pipes etc.
Tel: 01179 812 791
Fax: 01179 814 511
E-mail: sales@cpm-group.co.uk
Website: www.cpm-group.cp.uk

Custom Metal Fabricators
Central Way
Feltham
Middlesex
TW14 0XJ
Metalwork and balustrading
Tel: 0208 844 0940
Fax: 0208 751 5793
E-mail: dgibbs@cmf.co.uk
Website: www.cmf.co.uk

Decra Roof Systems (UK) Ltd
3, Faraday Centre
Faraday Road
Crawley
West Sussex
RH10 2PX
'Stratos' roofing system and accessories
Tel: 01293 545 058
Fax: 01293 562 709
E-mail: sales@decra.co.uk
Website: www.decra.co.uk

Dennis Ruabon
Hafod Tileries
Ruabon
Wrexham
Clywdd
LL14 6ET
Quarry floor tiles
Tel: 01978 843 484
Fax: 01978 843 276
E-mail: salesenq@dennisruabon.co.uk
Website: www.dennisruabon.co.uk

Donaldson Timber Engineering (SE)
Brunswick Road
Cobbs Wood Industrial Estate
Ashford, Kent
TN23 1EL
Roof trusses
Tel: 01233 895222
Fax: 01233 895220
Website: www.dontim-eng.co.uk

Dow Chemical Company Ltd
Dow Building Solutions
Diamond House
Lotus Park
Kingsbury Crescent
Staines
Tw18 3AG
Insulation products
Tel: 0203 139 4000
Fax: 0203 139 4013
E-mail: Styrofoam-uk@dow.com
Website: www.styrofoameurope.com

Dreadnought Clay Roof Tiles
(Hinton, Perry & Davenhill Ltd)
Dreadnought Works
Pensnett
Brierley Hill
Staffordshire
DY5 4TH
Clay roof tiling
Tel: 01384 774 05
Fax: 01384 745 53
E-mail: sales@dreadnought-tiles.co.uk
Website: www.dreadnought-tiles.co.uk

Essex Insulation Ltd
Essex House
Josselin Road
Burnt Mills
Basildon
Essex
SS13 1EL
Insulation products
Tel: 01268 724455
Fax: 01268 727963
E-mail: info@essexinsulation.co.uk
Website: www.essexinsulation.co.uk

Expamet Building Products
PO Box 52
Longhill Industrial Estate (North)
Hartlepool
Cleveland
TS25 1PR
Expanded metal building products
Tel: 01429 866 611
Fax: 01429 866 633
E-mail: expamet@compuserve.com

Forticrete Architectural Products
Thornhill Works
Calder Road
Dewsbury
West Yorkshire
WF12 9HY
Concrete, stone faced and 'Astra' glazed blocks
Tel: 01924 456 416
Fax: 01924 430 697
E-mail: Info@forticrete.co.uk
Website: www.forticrete.co.uk

Forticrete Ltd
Boss Avenue
Off Grovebury Road
Leighton Buzzard
Bedfordshire
LU7 4SD
"Hardrow" slating
Tel: 01525 244 900
Fax: 01525 850 432
E-mail: roofing@forticrete.co.uk
Website: www.forticrete.co.uk

Franklin Andrews
3 Tythe Barn
Brumstead Road
Stalham
Norwich
NR12 9DH
Griffiths building materials
Price Guides
Tel: 01692 581 545
Fax: 01692 582 893
Website: www.franklinandrews.com

Garador Ltd
Bunford Lane
Yeovil
Somerset
BA20 2YA
Garage doors
Tel: 01935 443 722
Fax: 01935 443 744
Website: www.garador.co.uk

Geberit Sales Ltd
PO Box 87
Aylesford
Kent
ME20 7PJ
Above and below ground drainage goods
Tel: 0800 077 8365
Fax: 01622 710 010
E-mail: technical@geberit-terrain.demon.co.uk
Website: www.geberit.co.uk

Gerfloor Ltd
Wedgnock House, Wedgnock Road
Warwick
Warwickshire
CV34 5AP
Sheet and tile flooring
Tel: 01926 401 500
Fax: 01926 401 647
E-mail: gerflorUK@gerfloor.com
Website: www.gerfloor.com

Grace Construction Products Ltd
Ajax Avenue
Slough
Berkshire
SL1 4BH
Expansion joint fillers and waterbars
Tel: 01753 692 929
Fax: 01753 691 623
E-mail: uksales@grace.com
Website: www.graceconstruction.com

Gradus Wall Protection
Park Green
Macclesfield
Cheshire
SK11 7LZ
Stair edgings and floor trims
Tel: 01625 428922
Fax: 01625 433949
E-mail: sales@gradusworld.com
Website: www.gradusworld.com

Greenberg Glass Ltd
William Moult Street
Liverpool
L5 5AT
Glass and glazing
Tel: 07525 901 400
Fax: 0870 1620 200
E-mail: info@pilkington.com
Website: www.pilkington.co.uk

Halfen Ltd
31, Humphreys Road
Woodside Estate
Dunstable
Bedfordshire
LU5 4TP
Building fixings and fastenings
Tel: 01582 316 300
Fax: 01582 470 304
E-mail: info@halfen.co.uk
Website: www.halfen.co.uk

H&H Celcon Ltd
Celcon House
Ightham
Sevenoaks
Kent
TN15 9HZ
Concrete blocks
Tel: 01732 886 333
Fax: 01732 886 810
E-mail: marketing@celcon.co.uk
Website: www.celcon.co.uk

Hanson Bath and Portland
Garston Road
Frome
Somerset
BA11 1RS
Cotswold stone
Tel: 01373 453 333
Fax: 01373 452 964

Hanson Building Products Ltd
Stewartby
Bedford
MK43 9LZ
Facing bricks etc.
Tel: 0870 525 8258
Fax: 01234 762 040
E-mail: info@hansonbrick.com
Website: www.hanson-brickeurope.com

Hare Structural Engineers
Brandlesholme House
Brandlesholme Road
Bury
Lancashire
BL8 1JJ
Structural steelwork
Tel: 0161 609 0000
Fax: 0161 609 0409
E-mail: http://www.hare.co.uk/
Website: www.hare.co.uk

Hathaway Roofing Ltd
Tindale Crescent
Bishop Auckland
County Durham
DL14 9TL
Sheet wall and roof claddings
Tel: 01388 605 636
Fax: 01388 608 841
E-mail: mark.watson@hathaway-roofing.co.uk
Website: www.hathaway-roofing.co.uk

Hepworth Building Products
Hazelhead
Crow Edge
Sheffield
South Yorkshire
S36 4HG
Clay drainage goods
Tel: 01226 763 561
Fax: 01226 764 827
E-mail: info@hepworthdrainage.co.uk
Website: www.hepworthdrainage.co.uk

Hillaldam Coburn Ltd
Unit 6
Wyvern Estate
Beverly Way
New Malden
Surrey
KT3 4PH
Sliding and folding door gear
Tel: 0208 336 1515
Fax: 0208 336 1414
E-mail: sales@hillaldam.co.uk
Website: www.coburn.co.uk

Hudevad Britain
Bridge House
Bridge Street
Walton-on-Thames
Surrey
KT12 1AL
Hot water radiators and fittings
Tel: 01932 247 835
Fax: 01932 247 694
E-mail: sales@hudevad.co.uk

Hunter Plastics Ltd
Nathan Way
London
SE28 0AE
Plastic rainwater goods
Tel: 0208 855 9851
Fax: 0208 317 7764
E-mail: hunplas@hunterplastics.co.uk
Website: www.hunterplastics.co.uk

Hutchison, VA Flooring Ltd
Units1-3, Building NA
Beeding Close
Southern Cross Trading Estate
Bognor Regis
West Sussex
PO22 9TS
Hardwood flooring
Tel: 01243 841175
Fax: 01243 841173
Website: www.hutchisonflooring.co.uk

Ibstock Building Products Ltd
Leicester Road
Ibstock
Leicestershire
LE67 6HS
Facing bricks
Tel: 01530 261 999
Fax: 01530 261 888
E-mail: marketing@ibstock.co.uk
Website: www.ibstock.co.uk

IMI Yorkshire Copper Tube Ltd
East Lancashire Road
Kirkby
Liverpool
Merseyside
L33 7TU
Copper piping
Tel: 0151 546 2700
Fax: 0151 549 2139

Instafoam and Fibre Ltd
Hogwood Basin
Ivanhoe Road
Finchampstead
Wokingham
RG40 4PZ
Cavity wall Insulation
Tel: 0118 932 8811
Fax: 0118 932 8314
Website: www.instagroup.co.uk

James Latham (Western)
Badminton Road
Yate
Bristol
Avon
BS37 5JX
Softwood, hardwood and panel products
Tel: 01454 315 421
Fax: 01454 323 488
E-mail: plywood.west@lathams.co.uk
Website: www.lathamtimber-co.uk

Jeld-Wen UK Ltd
Snow Hill
Melton Mobray
Leicester
LE131PD
Standard doors and windows
Tel: 0845 122 2890
Fax: 01644 503 403
Website: www.jeld-wen.co.uk

Jewson Ltd
Winterstoke Road
Weston-Super-Mare
North Somerset
BS23 3YB
Carcassing and planed softwood
Tel: 01934 412 822
Fax: 01934 622 276

John Brash and Co Ltd
The Old Shipyard
Gainsborough
Lincolnshire
DN21 1NG
Roofing shingles
Tel: 01427 613 858
Fax: 01427 810 218
E-mail: info@johnbrash.co.uk
Website: www.johnbrash.co.uk

John Guest Speedfit Ltd
Horton Road
West Drayton
Middlesex
UB7 8JL
Push-fit plumbing and heating systems
Tel: 01895 449 233
Fax: 01895 425 319

John Newton & Co Ltd
12 Verney Road
London
SE16 3DH
Waterproofing lathing
Tel: 0207 237 1217
Fax: 0207 252 2769

Junckers Ltd
Unit A
1 Wheaton Road
Whitham
Essex
CM8 3UJ
Hardwood flooring
Tel: 01376 534 700
Fax: 01376 514 401
E-mail: sales@junkers.co.uk
Website: www.junkers.com

Keymer Tiles Ltd
Nye Road
Burgess Hill
West Sussex
RH15 0NG
Hand made clay tiles
Tel: 01444 232 931
Fax: 01444 871 852
E-mail: info@keymer.co.uk
Website: www.keymer.co.uk

Kingspan Access Floor
Burma Drive
Marfleet
Hull
HU9 5SG
Raised and access floors
Tel: 01482 781 701
Fax: 01482 799 272
E-mail: enquiries@kingspanaccessfloors.co.uk
Website: www.kingspanaccessfloors.co.uk

Kingspan Insulation Ltd
Pembridge
Leominster
HR6 9LA
Insulation products
Tel: 0870 850 8555
Fax: 0870 850 8666
E-mail: info.uk@insulation.kingspan.com
Website: www.insulation.kingspan.com

Kingspan Metl-Con Ltd
St Hilda's Street
Sheburn
Malton
YO17 8PQ
Multibeam purlins
Tel: 01944 712 000
Fax: 01944 710 555
E-mail: john.williams@kingspanmetlcon.co.uk
Website: www.kingspanmetlcon.com

Klargester Environmental Engineering Ltd
College Road
Aston Clinton
Aylesbury
Buckinghamshire
HP22 5EW
Petrol interceptors and septic tanks
Tel: 01296 633 000
Fax: 01296 633 001
E-mail: uksales@klargester.co.uk
Website: www.klargester.co.uk

Knauf Insulation Ltd
PO Box 10
Stafford Road
St Helens
WA10 3NS
Crown and other insulation products
Tel: 0844 800 0135
Fax: 01744 612 007
E-mail: john.gaunt@knaufinsulation.co.uk
Website: www.knaufinsulation.co.uk

Lafarge Cement
The Shore
Northfleet
Kent
DA11 9AN
Cement, lime and flintag
Tel: 01474 564 344
Fax: 01474 531 281
Website: www.bluecircle.co.uk

Lafarge Contracting
Pesthouse Lane
Barham
Suffolk
IP6 0PF
Tarmacadam and asphalte roads
Tel: 01473 833146
Fax: 01473 833114
Website: www.lafarge.co.uk

Lamatherm Products Ltd
Forge Industrial Estate
Maesteg
Bridgend
Mid Glamorgan
CF34 0AU
Under foor cavity barriers
Tel: 01656 730 833
Fax: 01656 730 115
E-mail: sales@lamatherm.co.uk
Website: www.lamatherm.co.uk

Light Alloy Ltd
Dales Road
Ipswich
Suffolk
IP1 4R J
"Zig Zag" loft ladders
Tel: 01473 740 445
Fax: 01473 240 002

Lignacite
Meadgate Works
Nazeing
Essex
EN9 2PD
Concrete blocks
Tel: 01992 464441
Fax: 01992 445713
E-mail: info@lignacite.co.uk
Website: www.lignacite.co.uk/

Leaderflush Shapland
PO Box 5404
Nottingham
Nottinghamshire
NG16 4BU
Internal doors
Tel: 0870 240 0666
Fax: 0870 240 0777
E-mail: marketing@lsgroup.co.uk
Website: www.lsgroup.co.uk

Lonsdale Metal Company Ltd
Unit 40, Millmead Industrial Centre
Mill Mead Road
London
N17 9QU
Patent Glazing
Tel: 020 8801 4221
Fax: 020 8801 1287
E-mail: info@lonsdalemetal.co.uk
Website: www.patentglazing.co.uk

Luxaflex UK
Swanscombe Business Centre
17, London Road
Swanscombe
Kent
DA10 0LH
Roller and vertical blinds
Tel: 01322 624 580
Fax: 01322 624 558
E-mail: sales@luxaflex.com
Website: www.luxaflex.com

Maccaferri
7400 The Quadrant, Oxford Business Park North
Garsington Road
Oxford
OX4 2JZ
Gabions
Tel: 01685 770 555
Fax: 01685 774 550
E-mail: marketing@maccaferri.co.uk
Website: www.maccaferri.co.uk

Magrini Ltd
Unit 5, Maybrook Industrial Estate
Brownhills
Walsall
West Midlands
WS8 7DG
Baby equipment
Tel: 01543 375 311
Fax: 01543 361 172
mail:sales@magrini.co.uk
Website: www.magrini.co.uk

Manhole Covers Ltd
Airfield Industrial Estate
Cheddington Lane
Long Marston
Bucks
HP33 4QR
Manhole covers
Tel: 01296 668 850
Fax: 01296 668 080
Website: www.manholecovers.com

Marley Eternit Ltd
Lichfield Road
Burton-on-Trent
DE14 3HD
Roof tiles, sheets and pavings
Tel: 08705 626 400
Fax: 08705 626 450
E-mail:roofingsales@marleyeternit.co.uk
Website: www.marleyeternit.co.uk

Marshalls PLC
Landscape House
Premier Way Lower Business Park
Elland
HX5 9XT
Concrete block paving
Tel: 01422 312 000
Fax: 01422 330 185
Website: www.marshall.co.uk

Metsec Plc
Broadwell Road
Oldbury
Warley
West Midlands
B69 4HE
Lattice beams
Tel: 0121 601 6000
Fax: 0121 601 6119
E-mail: lattice-joists@metsec.com
Website: www.metsec.com

Moelven Laminated Timber Structures Ltd
Unit 10
Vicarage Farm
Winchester Road
Fair Oak
Eastleigh
Hampshire
SO50 7HD
'Gulam' laminated timber beams
Tel: 02380 695 566
Fax: 02380 695 577
E-mail: moelvenlts@aol.com
Website: www.moelven.co.uk

MPG Facades Ltd
Stuart House
Queensgate
Waltham Cross
Hertfordshire
EN8 7TF
Sto rendering and SFS walls
Tel: 01992 807 200
Fax: 01992 807 202
Website: www.mpgcontracts.com

N.D.M Leadwork Specialists Ltd
Mettalum House
Unit 3
89, Manor Farm Road
Alperton
Wembley
HA0 1BA
Metal cladding and roofing
Tel: 020 8991 7310
Fax: 020 8991 7311
E-mail: enquiries@ndmltd.com
Website: www.ndmltd.com

Parker & Highland Joinery Ltd
14a Chartwell Road
Lancing Business Park
Lancing
West Sussex
BN15 8TU
Purpose made joinery
Tel: 01903 756283
Fax: 01903 756599
E-mail: sales@parker-joinery.co.uk
Website: www.parker-joinery.co.uk

Peter Cox Ltd
Chancery House
St. Nicholas Way
Sutton
Surrey
SM1 1JB
Chemical and damp-proofing products
Tel: 0208 661 6600
Fax: 0208 642 0677
E-mail: jim.r.paul@ecolab.com
Website: www.ecolab.com

Polyflor Ltd
PO Box 3
Radcliffe New Road
Whitefield
Manchester
M45 7NR
'Polyflor' contract flooring
Tel: 0161 767 1111
Fax: 0161 767 1128
E-mail: info@polyflor.com
Website: www.polyflor.com

Polypipe Terrain
New Hythe Business Park
Aylesford
ME20 7PJ
PVC soil and waste pipes
Tel: 01622 795 200
Fax: 01622 792 564
Website: www.terraindrainage.com

Pressalit Care Ltd
Potts Marsh Industrial Estate
Westham
East Sussex
BN24 5NH
Grab rails
Tel: 01323 465 800
Fax: 01323 460 248
E-mail: ltd@pressalit.com
Website: www.pressalitcare.com

Proctor Group Ltd
The Haugh
Blairgowrie
Perthshire
PH10 7ER
British 'Sisalkraft' paper and Insulation
Tel: 01250 872261
Fax: 01250 872727
E-mail: insulation@proctorgroup.com
Website: www.proctorgroup.com

Prodema UK & Ireland Ltd
Commerce House
Telford Road
Bicester
OX26 4DL
Timber cladding
Tel: 01869 255 766
Fax: 01869 255 767
Website: www.prodema.com

Profile 22
Stafford Park
Telford
Shropshire
TF3 3AT
uPVC windows
Tel: 01952 290 910
Fax: 01952 290 460
E-mail: mail@profile22.com
Website: www.profile22.co.uk

Promat UK
The Stirling Centre
Eastern Road
Bracknell
Berkshire
RG12 2ST
Fireproofing materials
Tel: 01344 381 301
Fax: 01344 381 300
E-mail: promat@promat.co.uk
Website: www.promat.co.uk

Protim Solignum Ltd
Fieldhouse Lane
Marlow
Bucks
SC7 1LS
Solignum
Tel: 01628 486 644
Fax: 01628 481 276
E-mail: info@osmose.co.uk
Website: www.osmose.co.uk

Quelfire
PO Box 35
Caspian RoadAltrincham
Cheshire
WA14 5QA
Fire protection compound
Tel: 0161 928 7308
Fax: 0161 941 2635

Rawlplug Ltd
Skibo Drive
Thronliebank Industrial Estate
Glasgow
Scotland
G46 8JR
'Rawlbolts' etc.
Tel: 0141 638 2255
Fax: 0141 273 2333
E-mail: info@rawlplug.co.uk
Website: www.rawlplug.co.uk

Redland Head Office, Monier Ltd
Sussex Manor Business Park
Gatwick Road
Crawley, West Sussex
RH10 9NZ
"Redland" roofing tiles
Tel: 0870 601 000
Fax; 0870 642 742
E-mail: roofing@lafarge-roofing.co.uk
Website: www.lafarge-roofing.co.uk

Redpath Buchanan
Unit 2
Jenkins Dale
Chatham
Kent
ME4 5DR
Lightning protection
Tel: 01634 838 454
Fax: 01634 831 002
E-mail: quotations@redpathbuchanan.co.uk
Website: www.redpathbuchanan.co.uk

Rentokil Initial Ltd
Rentokil House
Garland Road
East Grinstead
West Sussex
RH19 1DY
Flame retardant products
Tel: 01342 327 171
Fax: 01342 410 712
E-mail: tstephens@rentokil-initial.co.uk
Website: www.rentokil-initial.com

Rex Bousfield Ltd
Holland Road
Hurst Green
Oxted
Surrey
RH8 9BD
Chipboard products
Tel: 01883 717 033
Fax: 01883 717 890
Website: www.bousfield.com

Richard Lees Steel Decking ltd
Moor Farm Road West
The Airfield, Ashbourne
Derbyshire
DE6 1HD
Structural steel deck flooring
Tel: 01335 300 999
Fax: 01335 300 888
E-mail: rlsd.decks@skanska.co.uk
Website: www.rlsd.com

Richard Potter Timber Merchants
Millstone Lane
Nantwich
Cheshire
CW5 5PN
Carcassing and planed softwoods
Tel: 01270 625 791
Fax: 01270 610 483
E-mail: richardpotter@fortimber.demon.co.uk
Website: www.fortimber.demon.co.uk

RMC Aggregates Ltd
Hazel Old Lane
Hensell
DN14 0QD
Lytag products
Tel: 01977 661 661
Fax: 01977 662 221
E-mail: lytagsales@lytag.co.uk
Website: www.lytag.co.uk

Rock Asphalte Ltd
Latimer House
2, Ravenscourt Road
Hammersmith
London
W6 0UX
Asphalte roofing etc.
Tel: 0208 748 7881
Fax: 0208 748 7225
E-mail: enquiries@rockasphalte.com
Website: www.rockasphalte.com

Rockwool Ltd
Pencoed
Bridgend
Mid Glamorgan
CF35 6NY
Cavity wall batts
Tel: 01656 862 621
Fax: 01656 862 302
E-mail: info@rockwool.co.uk
Website: www.rockwool.co.uk

Roger Wilde Ltd
Chareau House
1 Miles Street
Oldham
OL1 3NW
Glass blocks
Tel: 0161 624 6824
Fax: 0161 627 3770
Website: www.rogerwilde.com

Rom Ltd
Witham Road
Witham
Essex
CM8 3BU
Reinforcement
Tel: 0870 0113 605
Fax: 01376 533 227
E-mail: enquiries@rom.co.uk
Website: www.rom.co.uk

Ryton's Building Products Ltd
Design House
Orion Way
Kettering Business Park
Kettering
Northamptonshire
NN15 6NL
Roof ventilation products
Tel: 01536 511 874
Fax: 01536 310 455
E-mail: admin@rytons.com
Website: www.vents.co.uk

Safeguard Europe
Redkiln Close, Redkiln Way
Horsham
Sussex
RH13 5QL
Vandex 'Super' and 'Premix'
Tel: 01403 210 204
Fax: 01403 217 529
E-mail: info@safeguardeurope.com
Website: www.safeguardeurope.com

Saint-Gobain Pipelines
Lows Lane
Stanton-by-Dale
Ilkeston
Derbyshire
DE7 4QU
Cast iron soil, water and rainwater pipes and fittings
Tel: 0115 930 5000
Fax: 0115 932 9513
E-mail: sales@saint-gobain-pipelines.co.uk
Website: www.saint-gobain-pipelines.co.uk

Sandtoft Roof Tiles
Sandtoft
Doncaster
South Yorkshire
DN8 5SY
Clay roof tiling
Tel: 01427 871200
Fax: 01427 871222
E-mail: info@sandtoft.co.uk
Website: www.sandtoft.co.uk

Schiedel Isokern
14, Haviland Road
Ferndown Industrial Estate
Wimbourne
Dorset
BH21 7RF
Flue pipes and gas blocks
Tel: 01202 861 650
Fax: 01202 861 632
E-mail: sales@isokern.co.uk
Website: www.isokern.co.uk

Screeduct Ltd
Unit 29
Alderminster
Nr Stratford-upon-Avon
Warwickshire
CV37 8NY
Trunking systems and conduits
Tel: 01789 459 211
Fax: 01789 459 255
E-mail: sales@screeduct.com
Website: www.screeduct.com

Sealmaster
Brewery Road
Pampisford
Cambridge
Cambridgeshire
CB22 3HG
Intumescent strips and glazing compound
Tel: 01223 832851
Fax: 01223 837215
E-mail: sales@sealmaster.co.uk
Website: www.sealmaster.co.uk

Sheet Piling (UK) Ltd
Oakfield House
Rough Hey Road
Grimsargh
Preston
PR2 5AR
Sheet piling
Tel: 01772 794141
Fax: 01772 795151
E-mail: enquiries@sheetpilinguk.com
Website: www.sheetpilinguk.com

Sheffield Insulation Ltd
Sanderson Street
Off Scotswood Road
Newcastle upon Tyne
Tyne and Wear
NE4 7LW
Cavity wall insulation
Tel: 0191 2735843
Fax: 0191 2739213

Slate UK
David Wallace International Ltd
Unit 6, Aiffield Approach Business Park
Flookburgh
Grange-over-Sands
Cumbria
LA11 7LS
Spanish roof slates
Tel: 015395 59289
Fax: 015395 58078
Website: www.slate.uk.com

Sterling Hydraulics (Huntley & Sparks) Ltd
Building Products Division
Sterling House
Blacknell Lane
Crewkerne
Somerset
TA18 8LL
'Rigifix' column guards
Tel: 01460 72222
Fax: 01460 76402
E-mail: email@sterling-hydraulics.co.uk
Website: www.sterling-hydraulics.co.uk

Sterling Lloyd Polychem Ltd
Union Bank
Kings Street
Knutsford
Cheshire
WA16 6EF
'Intergitank'
Tel: 01565 633 111
Fax: 01565 633 555
Website: www.sterlinglloyd.com

Swish Building Products
Pioneer House
Lichfield Road Industrial Estate
Tamworth
Staffordshire
B29 7TF
Cladding products
Tel: 01827 371 200
Fax: 01827 371 201
E-mail: marketing@swishbp.co.uk
Website: www.swishbp.co.uk

Szerelmey Ltd
369, Kennington Lane
Vauxhall
London
SE11 5QY
Stonework
Tel: 020 7735 9995
Fax: 020 7793 9800
E-mail: info@szerelmey.com
Website: www.szerelmey.com

Tarkett-Marley Floors Ltd
Dickley Lane
Lenham
Maidstone
Kent
ME17 2QX
Sheet and tile flooring
Tel: 01622 854 000
Fax: 01622 854 500
E-mail: uksales@tarkett.com
Website: www.tarkett.com

Tarmac Central Ltd
Tunstead Quarry
Wormhill
Buxton
SK17 8TG
Portland cement and special mortars
Tel: 01298 768 444
Fax: 01298 768 334
Website: www.tarmac-central.co.uk

Tarmac Ltd
Pudding Mill Lane
Bow
London
E15 2PJ
Ready mixed 'Topmix' concrete
Tel: 0208 555 2415
Fax: 0208 555 2593
E-mail: info@tarmac.co.uk
Website: www.tarmac.co.uk

Tarmac Topblock Ltd
Wergs Hall
Wergs Hall Road
Wolverhampton
West Midlands
WV8 2HZ
Concrete blocks
Tel: 01902 754131
Fax: 01902 743171

Terram Ltd
Mamhillad Park
Pontypool
NP4 0YR
Terram Sheeting
Tel: 01495 757722
Fax: 01495762383
E-mail: shancock@email.msn.com
Website: www.terram.co.uk

Thermalite Hanson Building Products
Stewartby
Bedford
MK43 9LZ
'Thermalite' and 'Conbloc' blocks
Tel: 0870 626 500
Fax: 0870 626 550
E-mail: thermalitesales@hansonplc.com
Website: www.thermalite.co.uk

Tor Coatings Ltd
Portobello Industrial Estate
Birtley
Chester-le-Street
County Durham
DH3 2RE
Solignum etc.
Tel: 0191 410 6611
Fax: 0191 492 0125
Website: www.tor-coatings.com

Uponor Ltd
Hillcote Plant
PO Box 1
Blackwell
Alfreton
Derbyshire
DE55 5JD
MDPE piping and fittings
Tel: 01773 811112
Fax: 01783 812343
E-mail: marketing@uponor.co.uk
Website: www.uponor.com

Velfac Ltd
The Old Livery
Hildersham
Cambridge
CB21 6DR
Aluminium/Wood windows
Tel: 01223 897100
Fax: 01223 897101
E-mail: post@velfac.co.uk
Website: www.velfac.co.uk

Velux Company Ltd
Woodside Way
Glenrothes
Fife
KY7 4ND
'Velux' roof windows & flashings
Tel: 0870 405 7700
Fax: 0870 405 7701
E-mail: enquiries@velux.co.uk
Website: www.velux.co.uk

Visqueen Building Products
Maerdy Industrial Estate
Rhymney
Tredegar
NP22 5PY
'Visqueen' products
Tel: 01685 840 672
Fax: 01685 842 580
E-mail: enquiries@visqueenbuilding.co.uk
Website: www.visqueenbuilding.co.uk

Wavin Plastics Ltd
Parsonage Way
Chippenham
Wiltshire
SN15 5PN
UPVC drainage goods
Tel: 01249 766 600
Fax: 01249 443 286
E-mail: genenqy@wavin.co.uk
Website: www.wavin.co.uk

Welco
Woodgate Business Park
Kettles Wood Drive
Birmingham
B32 3GH
Lockers and shelving systems
Tel: 0121 421 9000
Fax: 0121 421 9888
E-mail: sales@welco.co.uk
Website: www.welco.co.uk

Welsh Slate
Penrhyn Quarry, Bethesda
Bangor
Gwynedd
LL57 4YG
Welsh slates
Tel: 01248 604 202
Fax: 01248 602 447
E-mail: enquiries@welshslate.com
Website: www.welshslate.com

William Blyth Ltd
Hoe Hill
Barton-on-humber
North Lincolnshire
DN18 5ET
Roof tiles
Tel: 01652 632175
Fax: 01652 660966

Yeoman Aggregates Ltd
Stone Terminal
Horn Lane
Acton
London
W3 9EH
Hardcore, gravels etc
Tel: 0208 8966800
Fax: 0208 8966811

Yorkshire Fittings Ltd
PO Box 166
Haigh Park Road
Leeds
LS1 1RD
"Yorkshire" & "Kuterlite" fittings
Tel: 0113 270 6945
Fax: 0113 270 5644
E-mail: info@yorkshirefittings.co.uk
Website: www.yorkshirefittings.co.uk

How to use this Book

First-time users of *Spon's Architects' and Builders' Price Book* and others who may not be familiar with the way in which prices are compiled may find it helpful to read this section before starting to calculate the costs of building work. The level of information on a scheme and availability of detailed specifications will determine which section of the book and which level of prices users should refer to.

APPROXIMATE ESTIMATES (PART 2)

For preliminary estimates/indicative costs before drawings are prepared, refer to the average overall *Building Prices per Functional Units* and multiply this by the proposed number of units to be contained within the building (i.e. number of bedrooms etc.) or *Building Prices per Square Metre* rates and multiply this by the gross internal floor area of the building (the sum of all floor areas measured within external walls) to arrive at an overall preliminary cost. These rates include Preliminaries (the Contractors' costs) but make no allowance for the cost of External Works or VAT.

For budget estimates where preliminary drawings are available, one should be able to measure approximate quantities for all the major components of a building and multiply these by individual rates contained in the *Building Cost Model* or *Approximate Estimating* sections This should produce a more accurate estimate of cost than using overall prices per square metre. Labour and other incidental associated items, although normally measured separately within Bills of Quantities, are deemed included within approximate estimate rates.

MEASURED WORKS (PARTS 3 & 4)

For more detailed estimates or documents such as Bills of Quantities (Quantities of supplied and fixed components in a building, measured from drawings), either use rates from *Prices for Measured Work - Major Works* or *Prices for Measured Work - Minor Works*, depending upon the overall value of the contract. All such prices used may need adjustment for size, site constraints, local conditions location and time, etc., and users are referred to page 46 for an example of how to adjust an estimate for some of these factors. Items within the Measured Works sections are made up of many components: the cost of the material or product; any additional materials needed to carry out the work; the labour involved in unloading and fixing, etc. These components are usually broken down into:

Prime Cost

Commonly known as the "PC", Prime Cost is the actual price of the material such as bricks, blocks, tiles or paint, as sold by suppliers. Prime Cost is given "per square metre", "per 100 bags" or "each" according to the way the supplier sells the product. Unless otherwise stated, prices in *Spon's Architects' and Builders' Price Book* (hereafter referred to as *Spon's A & B*), are deemed to be "delivered to site", (in which case transport costs will be included) and also take account of trade and quantity discounts. Part loads generally cost more than whole loads but, unless otherwise stated, Prime Cost figures are based on average prices for full loads delivered to a hypothetical site in Acton, London W3. Actual prices for "live" tenders will depend on the distance from the supplier, the accessibility of the site, whether the whole quantity ordered is to be supplied in one delivery or at specified dates and market conditions prevailing at the time. Prime Cost figures for commonly-used alternative materials are supplied in listed form at the beginning of some work sections. As stated later, these prices are mainly at "list" prices before deduction of quantity discounts, and therefore require 'discount' adjustment before they can be substituted in place of "PC" figures given for Measured Work items.

Labour

This figure covers the cost of the operation and is calculated on the gang wage rate (skilled or unskilled) and the time needed for the job. A full explanation and build-up is provided on page 171. Particular shortages have been noted in the Bricklaying, Carpentry, Joinery and Shuttering trades. In response to these pressures and corresponding hikes in daily rates/measured prices, basic labour rates for Bricklayers, Carpenters and Joiners have been enhanced by using the plus 10% bonus payment figures for these trades as outlined on page 172. Extras such as highly skilled craft work, difficult access, intermittent working and the need for additional labourers to back up the craftsman (e.g. for handling heavy concrete blocks etc.) all add to the cost. Large regular or continuous areas of work are cheaper to install than smaller areas, since less labour time is wasted moving from one area to another.

Materials

Material prices include the cost of any ancillary materials, nails, screws, waste, etc., which may be needed in association with the main material product/s. If the material being priced varies from a standard measured rate, then identify the difference between the original PC price and the material price and add this to your alternative material price before adding to the labour cost to produce a new overall Total rate. Alternative material prices, where given, are largely based upon "list" prices, before the deduction of quantity discounts etc., and therefore require 'discount' adjustment before they can be substituted in place of "PC" figures given for Measured Work items.

Example:

	PC £	Labour hours	Labour £	Material £	Unit	**Total Rate £**
100 mm Thermalite Turbo block (see page 251)	7.20	0.46	12.04	8.94	m^2	**20.98**
100 mm Toplite standard block (see page 250) - £6.40 including discount	6.40					

Calculation: £8.94/1.075 (O&P) – 7.20 (PC) = £1.12 Take (residue (£1.12) + 6.40 (new PC) x 1.075 = £8.08 (revised material £)

	PC £	Labour hours	Labour £	Material £	Unit	**Total Rate £**
Therefore, 100 mm Toplite block price =	6.40	0.46	12.04	8.08	m^2	**20.12**

Plant

Plant covers the use of machinery ranging from JCB's to shovels and static plant including running costs such as fuel, water supply (which is metered on a construction site), electricity and rubbish disposal. Some items of plant are included within the Measured Works sections e.g. "Groundwork", under a Material/Plant column. Other items are included within the Preliminaries section.

Unit

The Unit is generally based upon measurement guidelines laid out in the Standard Method of Measurement of Building Works – Seventh Edition, published by the Royal Institution of Chartered Surveyors and The Building Employers Confederation.

Total Rate

Prices in the Total Rate column generally include for the supply and fix of items, unless otherwise described.

Overheads and profit

The general overheads of the Main Contractor's business, the head office overheads and any profit sought on capital and turnover employed, is usually covered under a general item of overheads and profit which is applied either to all measured rates as a percentage, or alternatively added to the tender summary or included within Preliminaries (site specific overhead costs). At the present time, we are including an allowance of 5% for overheads and 2½% for profit on built-up labour rates and material prices in the Measured Major Works section and 7½% for overheads and 5% for profit on built-up labour rates and material prices in the Minor Works section. For non-analysed sub-contractor prices, only a 2½% mark-up for profit has been included, in both Major and Minor Works sections.

Preliminaries

Site specific Main Contractor's overheads on a contract, such as insurance, site huts, security, temporary roads and the statutory health and welfare of the labour force, are not directly assignable to individual items so they are generally added as a percentage or calculated allowance after all building component items have been costed and summed. Preliminaries will vary from contract to contract according to the type of construction, difficulties of the site, labour shortage, inclement weather or involvement with other contractors, etc. The overall Preliminary addition for a scheme should be adjusted to allow for these factors. For this edition we have raised Preliminary costs to +16%.

Sub/Specialist-Contractor's costs

For the purpose of this book, these are deemed to include all the above costs, plus 2.5% Main Contractor's discount.

With the exclusion of Main Contractor's preliminaries, the above items combine to form item rates in the **Prices for Measured Works** sections. It will be appreciated that a variation in any one item in any group will affect the final Measured Work price. Any cost variation must be weighed against the total cost of the contract, and a small variation in Prime Cost where the items are ordered in thousands may have more effect on the total cost than a large variation on a few items, while a change in design which introduces the need to use, e.g. earth-moving equipment, which must be brought to the site for that one task, will cause a dramatic rise in the contract cost. Similarly, a small saving on multiple items will provide a useful reserve to cover unforeseen extras.

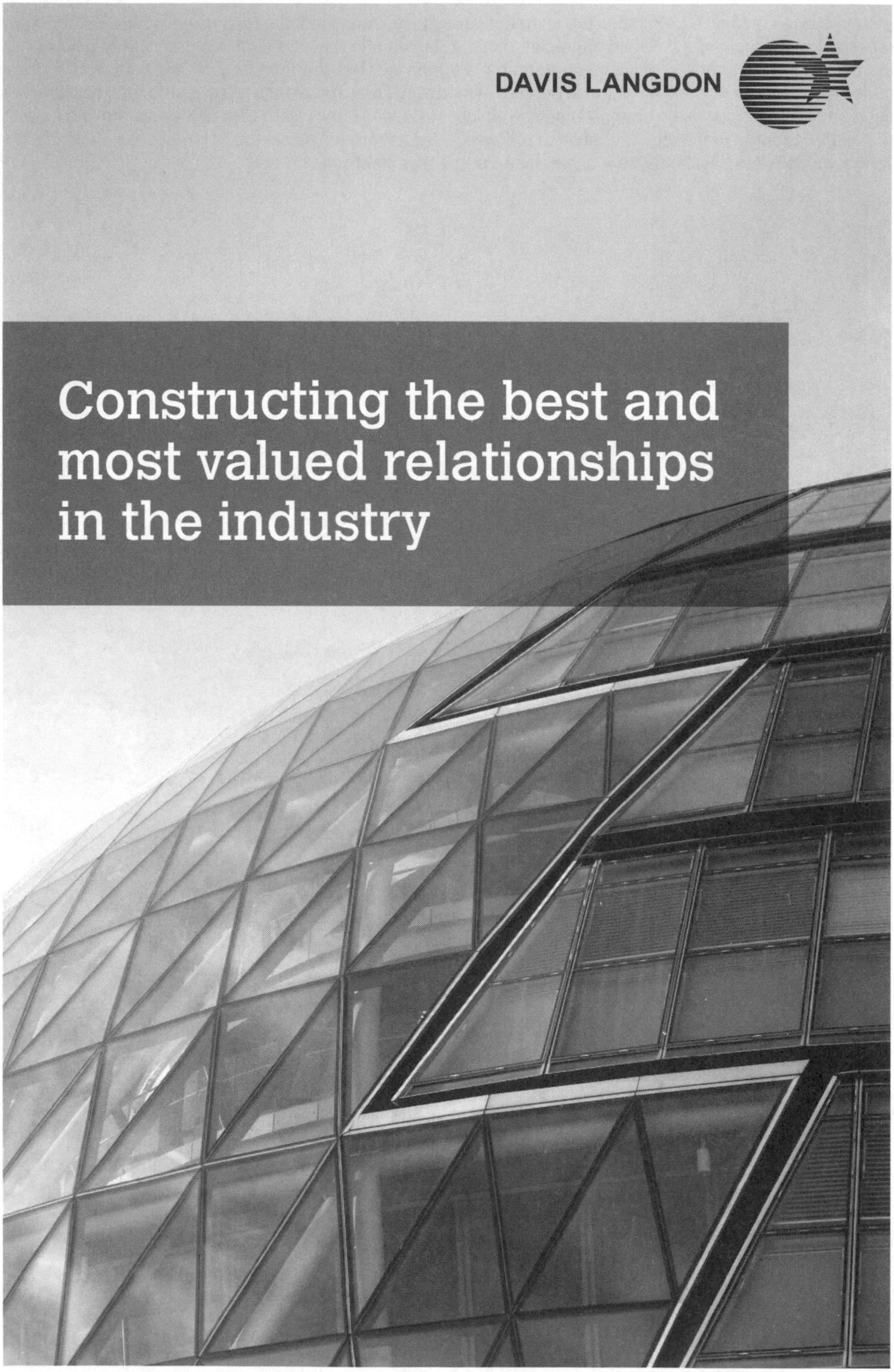

www.davislangdon.com

Offices in Europe & the Middle East, Africa, Asia Pacific, Australasia and the USA

Cost Management | Project Management
Banking Tax and Finance | Building Surveying | Engineering Services | Legal Support Group
Management Consultancy | Specifications and Design Management | VPR

Taylor & Francis
Taylor & Francis Group

Free Updates

with three easy steps…

1. Register today on www.pricebooks.co.uk/updates

2. We'll alert you by email when new updates are posted on our website

3. Then go to www.pricebooks.co.uk/updates and download.

All four Spon Price Books – *Architects' and Builders'*, *Civil Engineering and Highway Works*, *External Works and Landscape* and *Mechanical and Electrical Services* – are supported by an updating service. Three updates are loaded on our website during the year, in November, February and May. Each gives details of changes in prices of materials, wage rates and other significant items, with regional price level adjustments for Northern Ireland, Scotland and Wales and regions of England. The updates terminate with the publication of the next annual edition.

As a purchaser of a Spon Price Book you are entitled to this updating service for the 2009 edition – free of charge. Simply register via the website www.pricebooks.co.uk/updates and we will send you an email when each update becomes available.

If you haven't got internet access we can supply the updates by an alternative means. Please write to us for details: Spon Price Book Updates, Taylor & Francis Marketing Department, 2 Park Square, Milton Park, Abingdon, Oxfordshire, OX14 4RN.

Find out more about Spon books
Visit www.tandfbuiltenvironment.com for more details.

PART 1

General

This part of the book contains the following sections:

Revisions to Part L of the Building Regulations

The following review of the effects of the revised Part L on non-residential buildings is based on updated material originally published in a Building Cost Model on Part L, issued in *Building* Magazine on 5 August 2005. The original article was produced in collaboration with Consultants, Arup.

Introduction

Part L (2006) came into force in April 2006 at a time when awareness of the threat posed by global warming has never been higher. The 2002 revisions signaled an important shift in emphasis from energy conservation to control of emissions of carbon and other greenhouse gases. Revised Part L 2006 is building on this foundation, reducing carbon emissions from new build further, and widening the scope of work required to improve performance in existing buildings. In raising standards, the approach adopted has been to give maximum flexibility to owners and designers in selecting means of compliance.

Changes in the Approach in Part L2

Part L2 deals with non-dwellings. The major headline change is the requirement to achieve savings in carbon emissions compared to the previous 2002 standard. For new build, air-conditioned and mechanically ventilated buildings, the reduction is 28%, whereas for naturally ventilated buildings the reduction is 23.5%.

In addition to the introduction of these challenging targets for carbon emissions reduction, the Revised Part L also involves some significant changes to practice that are having an impact on the work of consultants and contractors.

- The introduction of alternative assessment methodologies for new build and existing buildings. The new build method, based on a national standard assessment methodology is concerned with comparing the carbon emissions of the proposed development with those of a notional building, which complies with the previous 2002 Regulations. The approach for existing buildings is broadly intended to encourage proportional, technically feasible and economic improvements to building performance to ensure that alterations and improvements made to an existing building contribute to a reduction in carbon emissions.

- A move away from prescriptive technical guidance. Users are required to refer to technical details taken from a wide range of sources to identify how to meet the new standards. The new Part L only sets out the required benchmarks. .

- The adoption of a single national calculation methodology. This change, driven by the EPBD, has led to the adoption of a carbon emissions method for buildings other than dwellings. As a result, the simple elemental method of calculation used under previous versions of the regulations is no longer available.

- The Regulations include a provision to include the contribution of low or zero carbon (LZC) technologies, such as solar hot water heating, in the assessment. They also provide flexibility with regard to their application whilst the technologies mature and become more widely available.

The National Calculation Methodology

The new calculation methodology required under involves inputting all aspects of the building design into an assessment tool based on an agreed methodology. The National Calculation Tool, known as iSBEM (Simple Building Evaluation Method), developed by the BRE, is currently the most common application. Others such as CECM are becoming available. The comparative emissions calculation needs to be done fairly early on in the design process and then updated once the design is fixed. This will potentially make it more difficult for the design team to handle changes, and, given the resource requirements of recalculation, this could limit design iteration, although criteria such as perimeter heat gain can be set that will allow some design iterations without recalculation. Furthermore, contractors may be less likely to propose alternative design solutions as a result of the introduction of the new methodology, if they are subsequently required to demonstrate and achieve compliance.

Impact of the Regulations on design

The revised Part L is beginning to affect all aspects of the building design. The transition period was very short, and in effect, all buildings currently being designed will have to meet the new Standard. In addition to insulation levels, proportion and type of glazing, other factors that may be improved are:

- Shading to combat perimeter heat gains, e.g. external shading, double wall facades;
- Boiler efficiencies and chiller coefficients of performance;
- Building envelope air tightness;
- Lighting efficiency, lighting control;
- Fan and pump efficiencies, air and water pressure drops;
- Energy recovery.

A key characteristic of the revisions is, although the carbon emissions target has been cut by some 28% for cooled or mechanically ventilated buildings, none of the minimum acceptable standards to which the design must comply have been changed significantly from those set in 2002. It will continue to be possible to specify, for example, highly glazed buildings, albeit that significant improvement in other aspects of building performance will be necessary. Lighting specification and lamp efficiency are examples of areas where substantial reductions in both energy consumption and cooling loads could be made at a relatively low cost.

Indicative cost implications

Davis Langdon prepared a comparison of comparative office model solutions in Summer 2005 with Consulting Engineers, Arup, in order to show how the 2006 Regulations could be met by using either 40% glazing or different combinations of 100% glazing and solar shading. The cost implications are that using a low proportion of glazing, improvements to the glass specification, services plant efficiency and the introduction of simple heat recovery measures will actually result in a small reduction in capital costs resulting from the reduction in heating and cooling loads and plant sizing.

Option	Envelope solution	Saving/Extra £/m^2	Saving/Extra %
40% glazing	High performance glass	(11.82)	(0.62)
100% glazing	High performance glass, fixed external louvres	61.29	3.19
100% glazing	Clear double glazing with low-e coating, external motorised slatted blinds	96.78	5.06
100% glazing	Double wall facade with externally ventilated cavity, with motorised slatted blinds at close centres	73.12	3.82

The costs in the comparison are at 3rd Quarter 2005 prices based on a central London location. The base building costs include the category A fit-out, preliminaries and contingencies. Costs of demolitions, external works, tenant fit-out, professional fees and VAT are excluded.

Due to the complexity of the changes and the recent introduction of the Regulations, it has not been possible to assess their impact on the rates within the ‘Cost Models’ and the ‘Approximate Estimating’ sections of Spon’s Architects’ and Builders’ Price Book 2008. These sections have not been adjusted to take into account the implications of works required to comply with the revised Part L Regulations. Users should note that the items and their rates in the ‘Prices for Measured Works’ section are compliant.

Capital Allowances

Introduction

Capital Allowances provide tax relief by prescribing a statutory rate of depreciation for tax purposes in place of that used for accounting purposes. They are utilised by government to provide an incentive to invest in capital equipment, including commercial property, by allowing the majority of taxpayers a deduction from taxable profits for certain types of capital expenditure, thereby deferring tax liabilities.

The capital allowances most commonly applicable to real estate are those given for capital expenditure on both new and existing industrial buildings, and plant and machinery in all commercial buildings.

Other types of allowances particularly relevant to property are hotel and enterprise zone allowances, which are in fact variants to industrial buildings allowances code. Enhanced rates of allowances are available on certain types of energy saving and environmentally friendly plant and machinery, whilst reduced rates apply to "integral features" and items with an expected economic life of more than 25 years.

The Act

The primary legislation is contained in the Capital Allowances Act 2001. Amendments to the Act have been made in each subsequent Finance Act. Major changes to the system were announced by the Government in 2007 and the majority of these have now taken effect from April 2008.

The Act is arranged in 12 Parts and was published with an accompanying set of Explanatory Notes.

Plant and Machinery

The Finance Act 1994 introduced major changes to the availability of Capital Allowances on real estate. A definition was introduced which precludes expenditure on the provision of a building from qualifying for plant and machinery, with prescribed exceptions.

List A in Section 21 of the 2001 Act sets out those assets treated as parts of buildings:-

- *Walls, floors, ceilings, doors, gates, shutters, windows and stairs.*
- *Mains services, and systems, for water, electricity and gas.*
- *Waste disposal systems.*
- *Sewerage and drainage systems.*
- *Shafts or other structures in which lifts, hoists, escalators and moving walkways are installed.*
- *Fire safety systems.*

Similarly, List B in Section 22 identifies excluded structures and other assets.

Both sections are, however, subject to Section 23. This section sets out expenditure, which although being part of a building, may still be expenditure on the provision of Plant and Machinery.

List C in Section 23 is reproduced below:

Sections 21 and 22 do not affect the question whether expenditure on any item in List C is expenditure on the provision of Plant or Machinery

1. Machinery (including devices for providing motive power) not within any other item in this list.
2. Electrical systems (including lighting systems) and cold water, gas and sewerage systems provided mainly –
    a. to meet the particular requirements of the qualifying activity, or
    b. to serve particular plant or machinery used for the purposes of the qualifying activity.
3. Space or water heating systems; powered systems of ventilation, air cooling or air purification; and any floor or ceiling comprised in such systems.
4. Manufacturing or processing equipment; storage equipment (including cold rooms); display equipment; and counters, checkouts and similar equipment.
5. Cookers, washing machines, dishwashers, refrigerators and similar equipment; washbasins, sinks, baths, showers, sanitary ware and similar equipment; and furniture and furnishings.
6. Lifts, hoists, escalators and moving walkways.
7. Sound insulation provided mainly to meet the particular requirements of the qualifying activity.
8. Computer, telecommunication and surveillance systems (including their wiring or other links).
9. Refrigeration or cooling equipment.
10. Fire alarm systems; sprinkler and other equipment for extinguishing or containing fires.
11. Burglar alarm systems.
12. Strong rooms in bank or building society premises; safes.
13. Partition walls, where moveable and intended to be moved in the course of the qualifying activity.
14. Decorative assets provided for the enjoyment of the public in hotel, restaurant or similar trades.
15. Advertising hoardings; signs, displays and similar assets.
16. Swimming pools (including diving boards, slides & structures on which such boards or slides are mounted).
17. Any glasshouse constructed so that the required environment (namely, air, heat, light, irrigation and temperature) for the growing of plants is provided automatically by means of devices forming an integral part of its structure.
18. Cold stores.
19. Caravans provided mainly for holiday lettings.
20. Buildings provided for testing aircraft engines run within the buildings.
21. Moveable buildings intended to be moved in the course of the qualifying activity.
22. The alteration of land for the purpose only of installing Plant or Machinery.
23. The provision of dry docks.
24. The provision of any jetty or similar structure provided mainly to carry Plant or Machinery.
25. The provision of pipelines or underground ducts or tunnels with a primary purpose of carrying utility conduits.
26. The provision of towers to support floodlights.
27. The provision of –
    a. any reservoir incorporated into a water treatment works, or
    b. any service reservoir of treated water for supply within any housing estate or other particular locality.
28. The provision of –
    a. silos provided for temporary storage, or
    b. storage tanks.
29. The provision of slurry pits or silage clamps.
30. The provision of fish tanks or fish ponds.
31. The provision of rails, sleepers and ballast for a railway or tramway.
32. The provision of structures and other assets for providing the setting for any ride at an amusement park or exhibition.
33. The provision of fixed zoo cages.

List C is modified from April 2008 by the omission of "Electrical systems (including lighting systems) and cold water from item 2, the omission of item 3, and the omission of "Lifts, hoists, escalators and moving walkways" and their replacement by "Hoists" in item 6.

Capital Allowances on plant and machinery are given in the form of writing down allowances at the rate of 20% per annum on a reducing balance basis. For every £100 of qualifying expenditure £20 is claimable in year 1, £16 in year 2 and so on until either the all the allowances have been claimed or the asset is sold.

Allowances were given at the rate of 25% before April 2008.

Integral Features

A new category of qualifying expenditure on "integral features" has been introduced from April 2008. The following items are integral features:

- An electrical system
- A cold water system
- A space or water heating system, a powered system of ventilation, air cooling or air purification, and any floor or ceiling comprised in such a system
- A lift, an escalator or a moving walkway
- External solar shading

The draft legislation also included active facades but these were subsequently omitted, the explanation given being that allowances are already given on the additional inner skin because it is part of the air-conditioning system.

The reduced writing down allowance of 10% per annum is available on integral features.

The new legislation also includes a rule that prevents a revenue deduction being obtained where expenditure is incurred that is more than 50% of the cost of replacing the integral feature.

Thermal Insulation

For many years the addition of thermal insulation to an existing industrial building has been treated as qualifying for plant and machinery allowances. From April 2008 this has been extended to include all commercial buildings but not residential buildings.

The reduced writing down allowance of 10% per annum is available on thermal insulation

Long Life Assets

The reduced writing down allowance of 10% per annum is available on long-life assets. Allowances were given at the rate of 6% before April 2008.

A long-life asset is defined as plant and machinery that can reasonably be expected to have a useful economic life of at least 25 years. The useful economic life is taken as the period from first use until it is likely to cease to be used as a fixed asset of any business. It is important to note that this likely to be a shorter period than an item's physical life.

Plant and machinery provided for use in a building used wholly or mainly as dwelling house, showroom, hotel, office or retail shop or similar premises, or for purposes ancillary to such use, cannot be long-life assets.

In contrast plant and machinery assets in buildings such as factories, cinemas, hospitals and so on are all potentially long-life assets.

Case Law

The fact that an item appears in List C does not automatically mean that it will qualify for capital allowances. It only means that it may potentially qualify.

Guidance about the meaning of plant has to be found in case law. The cases go back a long way, beginning in 1887. The current state of the law on the meaning of plant derives from the decision in the case of Wimpy International Ltd v Warland and Associated Restaurants Ltd v Warland in the late 1980s.

The Judge in that case said that there were three tests to be applied when considering whether or not an item is plant.

1. Is the item stock in trade? If the answer yes, then the item is not plant.
2. Is the item used for carrying on the business? In order to pass the business use test the item must be employed in carrying on the business; it is not enough for the asset to be simply used in the business. For example, product display lighting in a retail store may be plant but general lighting in a warehouse would fail the test.
3. Is the item the business premises or part of the business premises? An item cannot be plant if it fails the premises test, i.e. if the business use is as the premises (or part of the premises) or place on which the business is conducted. The meaning of part of the premises in this context should not be confused with the law of real property. The Inland Revenue's internal manuals suggest there are four general factors to be considered, each of which is a question of fact and degree:

    - Does the item appear visually to retain a separate identity
    - With what degree of permanence has it been attached to the building
    - To what extent is the structure complete without it
    - To what extent is it intended to be permanent or alternatively is it likely to be replaced within a short period

There is obviously a core list of items that will usually qualify in the majority of cases. However, many other still need to be looked at on a case-by-case basis. For example, decorative assets in a hotel restaurant may be plant but similar assets in an office reception area would almost certainly not be.

One of the benefits of the new integral features rules, apart from simplification, is that items that did not qualify by applying these rules, such as general lighting in an office building, will now qualify albeit at the reduced rate.

Refurbishment Schemes

Building refurbishment projects will typically be a mixture of capital costs and revenue expenses, unless the works are so extensive that they are more appropriately classified a redevelopment. A straightforward repair or a "like for like" replacement of part of an asset would be a revenue expense, meaning that the entire amount can be deducted from taxable profits in the same year.

Where capital expenditure is incurred that is incidental to the installation of plant or machinery then Section 25 of the 2001 Act allows it to be treated as part of the expenditure on the qualifying item. Incidental expenditure will often include parts of the building that would be otherwise disallowed, as shown in the Lists reproduced above. For example, the cost of forming a lift shaft inside an existing building would be deemed to be part of the expenditure on the provision of the lift.

Annual Investment Allowance

The first year allowances previously available to small and medium sized enterprises have been withdrawn from April 2008.

They have been replaced with a new allowance available to all businesses of any size that allows a deduction for the whole of the first £50,000 of qualifying expenditure on plant and machinery.

The Enhanced Capital Allowances Scheme

The scheme is one of a series of measures introduced to ensure that the UK meets its target for reducing greenhouse gases under the Kyoto Protocol. 100% first year allowances are available on products included on the Energy Technology List published on the website at www.eca.gov.uk and other technologies supported by the scheme. All businesses will be able to claim the enhanced allowances, but only investments in new and unused Machinery and Plant can qualify. Leased assets only qualify from 17 April 2002.

There are currently 14 technologies and 55 sub-technologies covered by the scheme.

- Air-to-air energy recovery.
- Automatic monitoring and targeting.
- Boiler equipment.
- Combined heat and power.
- Compact heat exchangers.
- Compressor air equipment.
- Heat pumps.
- HVAC zone controls.
- Lighting.
- Motors and drives.
- Pipe work insulation.
- Refrigeration equipment
- Solar thermal systems.
- Warm air and radiant heaters.

The Finance Act 2003 introduced a new category of environmentally beneficial plant and machinery qualifying for 100% first-year allowances. The Water Technology List includes 14 technologies,

- Cleaning in place equipment.
- Efficient showers
- Efficient taps.
- Efficient toilets.
- Efficient washing machines.
- Flow controllers.
- Leakage detection equipment.
- Meters and monitoring equipment.
- Rainwater harvesting equipment.
- Small scale slurry and sludge dewatering equipment.
- Vehicle wash water reclaim units.
- Water efficient industrial cleaning equipment.
- Water management equipment for mechanical seals.
- Efficient membrane filtration systems.

The list of qualifying technologies will be extended to include waste water recovery and reuse systems, compressed air master controllers, compressed air flow controllers, heat pump dehumidifiers and white LED lighting in 2008.

Buildings and structures long life assets as defined above cannot qualify under the scheme. However, following the introduction of the integral features rules lighting in any non residential building may potentially qualify for enhanced capital allowances if it meets the relevant criteria.

A limited payable ECA tax credit equal to 19% of the loss surrendered was also introduced in April 2008.

Industrial Building Allowances

An industrial building (or structure) is defined in Sections 271 and 274 of the 2001 Act and includes buildings used for the following qualifying purposes:

- Manufacturing
- Processing
- Storage
- Agricultural contracting
- Working foreign plantations
- Fishing
- Mineral extraction

The following undertakings are also qualifying trades:

- Electricity
- Water
- Hydraulic power
- Sewerage
- Transport
- Highway undertakings
- Tunnels
- Bridges
- Inland navigation
- Docks

The definition is extended to include buildings provided for the welfare of workers in a qualifying trade and sports pavilions provided and used for the welfare of workers in any trade. Vehicle repair workshops and roads on industrial estates may also form part of the qualifying expenditure.

Retail shops, showrooms, offices, dwelling houses and buildings used ancillary to a retail purpose are specifically excluded.

The Government announced in 2007 that Industrial Building Allowances (along with Enterprise Zone, Hotel and Agricultural Building Allowances) will be abolished by 2011, with a phased withdrawal beginning in 2008.

Writing-Down Allowances

Allowances are given on qualifying expenditure at the rate of 4% per annum on a straight-line basis over 25 years. The allowance is given if the building is being used for a qualifying purpose on the last day of the accounting period. Where the building is used for a non-qualifying purpose the year's allowance is lost.

The rate will be reduced to 3% for 2008-09, 2% for 2009-10, 1% for 2010-11 and 0% for 2011 onwards.

From 21 March 2007 a balancing adjustment is no longer made on the sale of an industrial building. A purchaser of a used industrial building will be entitled to allowances based on the vendor's tax written down value, rather than the original construction cost adjusted for any periods of non-qualifying use.

The allowances will still be spread equally over the remaining period to the date twenty-five years after first use. However, even if the building was acquired prior to 21 March 2007, whatever the annual allowance given in 2007-08, it will be reduced to ¾ of that amount in 2008-09, ½ in 2009-10, ¼ in 2010-11 and zero from 2011 onwards.

Hotel Allowances

Industrial Building Allowances are also available on capital expenditure incurred on constructing a "qualifying hotel". The building must not only be a "hotel" in the normal sense of the word, but must also be a "qualifying hotel" as defined in Section 279 of the 2001 Act, which means satisfying the following conditions:

- The accommodation is in buildings of a permanent nature
- It is open for at least 4 months in the season (April to October)
- It has 10 or more letting bedrooms
- The sleeping accommodation consists wholly or mainly of letting bedrooms
- The services that it provides include breakfast and an evening meal (i.e. there must be a restaurant), the making of beds and cleaning of rooms.

A hotel may be in more than one building and swimming pools, car parks and similar amenities are included in the definition.

Enterprise Zones

A 100% first year allowance is available on capital expenditure incurred on the construction (or the purchase within two years of first use) of any commercial building within a designated enterprise zone, within ten years of the site being so designated. Like other allowances given under the industrial buildings code the building has a life of twenty-five years for tax purposes.

The majority of enterprise zones had reached the end of their ten-year life by 1993. However, in certain very limited circumstances it may still be possible to claim these allowances up to twenty years after the site was first designated.

Flats Over Shops

Tax relief is available on capital expenditure incurred on or after 11 May 2001 on the renovation or conversion of vacant or underused space above shops and other commercial premises to provide flats for rent.

In order to qualify the property must have been built with 1980 and the expenditure incurred on, or in connection with:

- Converting part of a qualifying building into a qualifying flat.
- Renovating an existing flat in a qualifying building if the flat is, or will be a qualifying flat.
- Repairs incidental to conversion or renovation of a qualifying flat, and
- The cost of providing access to the flat(s).

The property must not have more than 4 storeys above the ground floor and it must appear that, when the property was constructed, the floors above the ground floor were primarily for residential use. The ground floor must be authorised for business use at the time of the conversion work and for the period during which the flat is held for letting. Each new flat must be a self-contained dwelling, with external access separate from the ground-floor premises. It must have no more than 4 rooms, excluding kitchen and bathroom. None flats can be "high value" flats, as defined in the legislation. The new flats must be available for letting as a dwelling for a period of not more than 5 years.

An initial allowance of 100 per cent is available or, alternatively, a lower amount may be claimed, in which case the balance may be claimed at a rate of 25 per cent per annum in subsequent a years. The allowances may be recovered if the flat is sold or ceases to be let within 7 years.

Business Premises Renovation Allowance

The Business Premises Renovation Allowance (BPRA) was first announced in December 2003. The idea behind the scheme is to bring long-term vacant properties back into productive use by providing 100 per cent capital allowances for the cost of renovating and converting unused premises in disadvantaged areas. The legislation was included in Finance Act 2005 and was finally implemented on 11 April 2007 following EU state aid approval.

The legislation is identical in many respects to that for flat conversion allowances. The scheme will apply to properties within one of the areas specified in the Assisted Areas Order 2007 and Northern Ireland.

BPRA will be available to both individuals and companies who own or lease business property that has been unused for 12 months or more. Allowances will be available to a person who incurs qualifying capital expenditure on the renovation of business premises.

Agricultural Building Allowances

Allowances are available on capital expenditure incurred on the construction of buildings and works for the purposes of husbandry on land in the UK. Agricultural building means a building such as a farmhouse or farm building, a fence or other works. A maximum of only one-third of the expenditure on a farmhouse may qualify.

Husbandry includes any method of intensive rearing of livestock or fish on a commercial basis for the production of food for human consumption, and the cultivation of short rotation coppice. Over the years the Courts have held that sheep grazing and poultry farming are husbandry, and that a dairy business and the rearing of pheasants for sport are not. Where the use is partly for other purposes the expenditure can be apportioned.

The rate of allowances available and the way in which the system operates is very similar to that described above for industrial buildings. However, no allowance is ever given if the first use of the building is not for husbandry. A different treatment is also applied following acquisition of a used building unless the parties to the transaction elect otherwise.

Other Capital Allowances

Other types of allowances include those available for capital expenditure on Mineral Extraction, Research and Development, Know-How, Patents, Dredging and Assured Tenancy.

Value Added Tax

Introduction

Value Added Tax (VAT) is a tax on the consumption of goods and services. The UK adopted VAT when it joined the European Community in 1973. The principal source of European law in relation to VAT is Council Directive 2006/112/EC, a recast of Directive 77/388/EEC which is currently restated and consolidated in the UK through the VAT Act 1994 and various Statutory Instruments, as amended by subsequent Finance Acts.

VAT Notice 708: Buildings and construction (June 2007) gives an interpretation of the law in connection with construction works from the point of view of HM Revenue & Customs. VAT tribunals and court decisions since the date of this publication will affect the application of the law in certain instances. The Notice is available on HM Revenue & Customs website at www.hmrc.gov.uk.

The scope of VAT

VAT is payable on:

- Supplies of goods and services made in the UK
- By a taxable person
- In the course or furtherance of business; and
- Which are not specifically exempted or zero-rated.

Rates of VAT

There are three rates of VAT:

- A standard rate, currently 17.5%
- A reduced rate, currently 5%; and
- A zero rate.

Additionally some supplies are exempt from VAT and others are outside the scope of VAT.

Recovery of VAT

When a taxpayer makes taxable supplies he must account for VAT at the appropriate rate of either 17.5% or 5%. This VAT then has to be passed to HM Revenue & Customs and will normally be charged to the taxpayer's customers.

As a VAT registered person, the taxpayer can reclaim from HM Revenue & Customs as much of the VAT incurred on their purchases as relates to the standard-rated, reduced-rated and zero-rated onward supplies they make. A person cannot however reclaim VAT that relates to any non-business activities (but see below) or to any exempt supplies they make.

At predetermined intervals the taxpayer will pay to HM Revenue & Customs the excess of VAT collected over the VAT they can reclaim. However if the VAT reclaimed is more than the VAT collected, the taxpayer can reclaim the difference from HM Revenue & Customs.

Example

X Ltd constructs a block of flats. It sells long leases to buyers for a premium. X Ltd has constructed a new building designed as a dwelling and will have granted a long lease. This sale of a long lease is VAT zero-rated. This means any VAT incurred in connection with the development that which X Ltd will have paid (e.g. payments for consultants and certain preliminary services) will be reclaimable. For reasons detailed below the builder employed by X Ltd will not have charged VAT on his construction services.

Use for Business and Non Business Activities

Where VAT relates partly to business use and partly to non-business use then the basic rule is that it must be apportioned so that only the business element is potentially recoverable. However in some cases VAT on land, buildings and certain construction services purchased for both business and non-business use can be recovered in full by applying what is known as the "Lennartz" mechanism to reclaim the VAT relating to the non-business use and account for VAT on the non business use over a maximum period of 10 years. Legislation regulating the use of the "Lennartz" mechanism was eventually introduced on 1 November 2007.

Taxable Persons

A taxable person is an individual, firm, company etc who is required to be registered for VAT. A person who makes taxable supplies above certain value limits is required to be registered. The current registration limit is £67,000 for 2008-09. The threshold is exceeded if at the end of any month the value of taxable supplies in the period of one year then ending is over the limit, or at any time, if there are reasonable grounds for believing that the value of the taxable supplies in the period of 30 days than beginning will exceed £67,000.

A person who makes taxable supplies below these limits is entitled to be registered on a voluntary basis if they wish, for example in order to recover VAT incurred in relation to those taxable supplies.

In addition, a person who is not registered for VAT in the UK but acquires goods from another EC member state, or make distance sales in the UK, above certain value limits may be required to register for VAT in the UK.

VAT Exempt Supplies

If a supply is exempt from VAT this means that no tax is payable – but equally the person making the exempt supply cannot normally recover any of the VAT on their own costs relating to that supply.

Generally property transactions such as leasing of land and buildings are exempt unless a landlord chooses to standard-rate its supplies by a process known as opting to tax. This means that VAT is added to rental income and also that VAT incurred, on say, an expensive refurbishment, is recoverable.

Supplies outside the scope of VAT

Supplies are outside the scope of VAT if they are:

- Made by someone who is not a taxable person
- Made outside the UK; or
- Not made in the course or furtherance of business

In course or furtherance of business

VAT must be accounted for on all taxable supplies made in the course or furtherance of business with the corresponding recovery of VAT on expenditure incurred.

If a taxpayer also carries out non-business activities then VAT incurred in relation to such supplies is generally not recoverable.

In VAT terms, business means any activity continuously performed which is mainly concerned with making supplies for a consideration. This includes:

- Any one carrying on a trade, vocation or profession;
- The provision of membership benefits by clubs, associations and similar bodies in return for a subscription or other consideration; and
- Admission to premises for a charge.

It may also include the activities of other bodies including charities and non-profit making organisations.

Examples of non-business activities are:

- Providing free services or information;
- Maintaining museums, or particular historic sites;
- Publishing religious or political views.

Construction Services

In general the provision of construction services by a contractor will be VAT standard rated at 17.5%, however, there are a number of exceptions for construction services provided in relation to certain residential and charitable use buildings.

The supply of building materials is VAT standard rated at 17.5%, however, where these materials are supplied as part of the construction services the VAT liability of those materials follows that of the construction services supplied.

Zero-rated construction services

The following construction services are VAT zero-rated including the supply of related building materials.

The construction of new dwellings

The supply of services in the course of the construction of a building designed for use as a dwelling or number of dwellings is zero-rated other than the services of an architect, surveyor or any other person acting as a consultant or in a supervisory capacity.

The following conditions must be satisfied in order for the works to qualify for zero-rating:

1. The work must not amount to the conversion, reconstruction or alteration of an existing building;
2. The work must not be an enlargement of, or extension to, an existing building except to the extent that the enlargement or extension creates an additional dwelling or dwellings;
3. The building must be designed as a dwelling or number of dwellings. Each dwelling must consist of self-contained living accommodation with no provision for direct internal access from the dwelling to any other dwelling or part of a dwelling;
4. Statutory planning consent must have been granted for the construction of the dwelling, and construction carried out in accordance with that consent;
5. Separate use or disposal of the dwelling must not be prohibited by the terms of any covenant, statutory planning consent or similar provision.

The construction of a garage at the same time as the dwelling can also be zero-rated as can the demolition of any existing building on the site of the new dwelling

A building only ceases to be an existing building (see points 1. and 2. above) when it is:

1. Demolished completely to ground level; or when
2. The part remaining above ground level consists of no more than a single façade (or a double façade on a corner site) the retention of which is a condition or requirement of statutory planning consent or similar permission.

The construction of a new building for 'relevant residential or charitable' use

The supply of services in the course of the construction of a building designed for use as a relevant residential or charitable building is zero-rated other than the services of an architect, surveyor or any other person acting as a consultant or in a supervisory capacity.

A 'relevant residential' use building means:

1. A home or other institution providing residential accommodation for children;
2. A home or other institution providing residential accommodation with personal care for persons in need of personal care by reason of old age, disablement, past or present dependence on alcohol or drugs or past or present mental disorder;
3. A hospice;
4. Residential accommodation for students or school pupils
5. Residential accommodation for members of any of the armed forces;
6. A monastery, nunnery, or similar establishment; or
7. An institution which is the sole or main residence of at least 90% of its residents.

The construction of a new building for 'relevant residential or charitable' use – continued

A 'relevant residential' purpose building does not include use as a hospital, a prison or similar institution or as a hotel, inn or similar establishment.

A 'relevant charitable' use means use by a charity:

1. Otherwise than in the course or furtherance of a business; or
2. As a village hall or similarly in providing social or recreational facilities for a local community.

Non qualifying use which is not expected to exceed 10% of the time the building is normally available for use can be ignored. The calculation of business use can be based on time, floor area or head count subject to approval being acquired from HM Revenue & Customs.

The construction services can only be zero-rated if a certificate is given by the end user to the contractor carrying out the works confirming that the building is to be used for a qualifying purpose i.e. for a 'relevant residential or charitable' purpose. It follows that such services can only be zero-rated when supplied to the end user and, unlike supplies relating to dwellings, supplies by sub contractors cannot be zero-rated.

The construction of an annex used for a 'relevant charitable' purpose

Construction services provided in the course of construction of an annexe for use entirely or partly for a 'relevant charitable' purpose can be zero-rated.

In order to qualify the annexe must:

1. Be capable of functioning independently from the existing building;
2. Have its own main entrance; and
3. Be covered by a qualifying use certificate.

The conversion of a non-residential building into dwellings or the conversion of a building from non-residential use to 'relevant residential' use where the supply is to a 'relevant' housing association

The supply to a 'relevant' housing association in the course of conversion of a non-residential building or non-residential part of a building into:

1. A building or part of a building designed as a dwelling or number of dwellings; or
2. A building or part of a building for use solely for a relevant residential purpose,

of any services related to the conversion other than the services of an architect, surveyor or any person acting as a consultant or in a supervisory capacity are zero-rated.

A 'relevant' housing association is defined as:

1. A registered social landlord within the meaning of Part I of the Housing Act 1996
2. A registered housing association within the meaning of the Housing Associations Act 1985 (Scottish registered housing associations), or
3. A registered housing association within the meaning of Part II of the Housing (Northern Ireland) Order 1992 (Northern Irish registered housing associations).

If the building is to be used for a 'relevant residential' purpose the housing association should issue a qualifying use certificate to the contractor completing the works.

The construction of a permanent park for residential caravans

The supply in the course of the construction of any civil engineering work 'necessary for' the development of a permanent park for residential caravans of any services related to the construction can be VAT zero-rated. This includes access roads, paths, drainage, sewerage and the installation of mains water, power and gas supplies.

Certain building alterations for "disabled" persons

Certain goods and services supplied to a "disabled" person, or a charity making these items and services available to "disabled" persons can be zero-rated. The recipient of these goods or services needs to give the supplier an appropriate written declaration that they are entitled to benefit from zero rating.

The following services (amongst others) are zero-rated:

1. the installation of specialist lifts and hoists and their repair and maintenance
2. the construction of ramps, widening doorways or passageways including any preparatory work and making good work
3. the provision, extension and adaptation of a bathroom, washroom or lavatory; and
4. emergency alarm call systems

Approved alterations to protected buildings

A supply in the course of an 'approved alteration' to a 'protected building' of any services other than the services of an architect, surveyor or any person acting as consultant or in a supervisory capacity can be zero-rated.

A 'protected building' is defined as a building that is:

1. designed to remain as or become a dwelling or number of dwellings after the alterations; or
2. is intended for use for a 'relevant residential or charitable purpose' after the alterations; and which is;
3. a listed building or scheduled ancient monument.

A listed building does not include buildings that are in conservation areas, but not on the statutory list, or buildings included in non-statutory local lists.

An 'approved alteration' is an alteration to a 'protected building' that requires and has obtained listed building consent or scheduled monument consent. This consent is necessary for any works that affect the character of a building of special architectural or historic interest.

It is important to note that 'approved alterations' do not include any works of repair or maintenance or any incidental alteration to the fabric of a building that results from the carrying out of repairs or maintenance work.

A 'protected building' that is intended for use for a 'relevant residential or charitable purpose' will require the production of a qualifying use certificate by the end user to the contractor providing the alteration services.

Listed Churches are 'relevant charitable' use buildings and where 'approved alterations' are being carried out zero-rate VAT can be applied. Additionally since April 1 2001, listed places of worship can apply for a grant for repair and maintenance works equal to the difference between the VAT paid at 17.5% on the repair and maintenance works and the amount that would have been charged if VAT had been 5%.

With effect from 1 April 2004, it has been possible to reclaim the full amount of VAT paid on eligible works carried out on or after 1 April 2004. Information relating to the scheme can be obtained from the website at www.lpwscheme.org.uk.

DIY Builders and Converters

Private individuals who decide to construct their own home are able to reclaim VAT they pay on goods they use to construct their home by use of a special refund mechanism made by way of an application to HM Revenue & Customs. This also applies to services provided in the conversion of an existing non-residential building to form a new dwelling.

The scheme is meant to ensure that private individuals do not suffer the burden of VAT if they decide to construct their own home.

Charities may also qualify for a refund on the purchase of materials incorporated into a building used for non-business purposes where they provide their own free labour for the construction of a 'relevant charitable' use building.

Reduced-rated construction services

The following construction services are subject to the reduced rate of VAT of 5%, including the supply of related building materials.

A changed number of dwellings conversion

In order to qualify for the 5% rate there must be a different number of 'single household dwellings' within a building than there were before commencement of the conversion works. A 'single household dwelling' is defined as a dwelling that is designed for occupation by a single household.

These conversions can be from 'relevant residential' purpose buildings, non-residential buildings and houses in multiple occupation.

A house in multiple occupation conversion

This relates to construction services provided in the course of converting a 'single household dwelling', a number of 'single household dwellings', a non-residential building or a 'relevant residential' purpose building into a house for multiple occupation such as a bed sit accommodation.

A special residential conversion

A special residential conversion involves the conversion of a 'single household dwelling', a house in multiple occupation or a non-residential building into a 'relevant residential' purpose building such as student accommodation or a care home.

Renovation of derelict dwellings

The provision of renovation services in connection with a dwelling or 'relevant residential' purpose building that has been empty for two or more years prior to the date of commencement of construction works can be carried out at a reduced rate of VAT of 5%.

Installation of energy saving materials

The supply and installation of certain energy saving materials including insulation, draught stripping, central heating and hot water controls and solar panels in a residential building or a building used for a relevant charitable purpose.

Grant-funded of heating equipment or connection of a gas supply

The grant funded supply and installation of heating appliances, connection of a mains gas supply, supply, installation, maintenance and repair of central heating systems, and supply and installation of renewable source heating systems, to qualifying persons. A qualifying person is someone aged 60 or over or is in receipt of various specified benefits.

Installation of security goods

The grant funded supply and installation of security goods to a qualifying person.

Housing alterations for the elderly

Certain home adaptations that support the needs of elderly people were reduced rated with effect from 1 July 2007.

Building Contracts

Design and build contracts

If a contractor provides a design and build service relating to works to which the reduced or zero rate of VAT is applicable then any design costs incurred by the contractor will follow the VAT liability of the principal supply of construction services.

Management contracts

A management contractor acts as a main contractor for VAT purposes and the VAT liability of his services will follow that of the construction services provided. If the management contractor only provides advice without engaging trade contractors his services will be VAT standard rated.

Construction Management and Project Management

The project manager or construction manager is appointed by the client to plan, manage and co-ordinate a construction project. This will involve establishing competitive bids for all the elements of the work and the appointment of trade contractors. The trade contractors are engaged directly by the client for their services. The VAT liability of the trade contractors will be determined by the nature of the construction services they provide and the building being constructed.

The fees of the construction manager or project manager will be VAT standard rated. If the construction manager also provides some construction services these works may be zero or reduced rated if the works qualify.

Liquidated and Ascertained Damages

Liquidated damages are outside of the scope of VAT as compensation. The employer should not reduce the VAT amount due on a payment under a building contract on account of a deduction of damages. In contrast an agreed reduction in the contract price will reduce the VAT amount.

Similarly, in certain circumstances HM Revenue & Customs may agree that a claim by a contractor under a JCT or other form of contract is also compensation payment and outside the scope of VAT.

The Aggregates Levy

The Aggregates Levy came into operation on 1 April 2002 in the UK, except for Northern Ireland where it has been phased in over five years from 2003.

It was introduced to ensure that the external costs associated with the exploitation of aggregates are reflected in the price of aggregate, and to encourage the use of recycled aggregate. There continues to be strong evidence that the levy is achieving its environmental objectives, with sales of primary aggregate down and production of recycled aggregate up. The Government expects that the rates of the levy will at least keep pace with inflation over time, although it accepts that the levy is still bedding in.

The rate of the levy increased to £1.95 per tonne from 1 April 2008 and is levied on anyone considered to be responsible for commercially exploiting 'virgin' aggregates in the UK and should naturally be passed by price increase to the ultimate user. The rate of levy will increase to £2.00 per tonne from 1 April 2009.

All materials falling within the definition of 'Aggregates' are subject to the levy unless specifically exempted.

It does not apply to clay, soil, vegetable or other organic matter.

The intention is that it will:

- Encourage the use of alternative materials that would otherwise be disposed of to landfill sites.
- Promote development of new recycling processes, such as using waste tyres and glass
- Promote greater efficiency in the use of virgin aggregates
- Reduce noise and vibration, dust and other emissions to air, visual intrusion, loss of amenity and damage to wildlife habitats

The intention is for part of the revenue from the levy to be recycled to business and communities affected by aggregates extraction, through:

- A 0.1 percentage point cut in employers' national insurance contributions
- A new £35 million per annum 'Sustainability Fund' to reduce the need for virgin materials and to limit the effects of extraction on the environmental where it takes place. A number of key priorities were identified with a promise to give effect to them through existing programmes. Following a mid-term review in December 2003, this fund was extended for a further three years.

Definition

'Aggregates' means any rock, gravel or sand which is extracted or dredged in the UK for aggregates use. It includes whatever substances are for the time being incorporated in it or naturally occur mixed with it.

'Exploitation' is defined as involving any one or a combination of any of the following:

- Being removed from its original site
- Becoming subject to a contract or other agreement to supply to any person
- Being used for construction purposes
- Being mixed with any material or substance other than water, except in permitted circumstances

Incidence

It is a tax on primary aggregates production – i.e. 'virgin' aggregates won from a source and used in a location within the UK territorial boundaries (land or sea). The tax is not levied on aggregates which are exported nor on Aggregates imported from outside the UK territorial boundaries.

It is levied at the point of sale.

Exemption from tax

An 'aggregate' is exempt from the levy if it is:

- Material which has previously been used for construction purposes
- Aggregate that has already been subject to a charge to the Aggregates levy
- Aggregate which was previously removed from its originating site before the start date of the levy
- Aggregate which is being returned to the land from which it was won
- Aggregate won from a farm land or forest where used on that farm or forest
- Rock which has not been subjected to an industrial crushing process
- Aggregate won by being removed from the ground on the site of any building or proposed building in the course of excavations carried out in connection with the modification or erection of the building and exclusively for the purpose of laying foundations or of laying any pipe or cable
- Aggregate won by being removed from the bed of any river, canal or watercourse or channel in or approach to any port or harbour (natural or artificial), in the course of carrying out any dredging exclusively for the purpose of creating, restoring, improving or maintaining that body of water
- Aggregate won by being removed from the ground along the line of any highway or proposed highway in the course of excavations for improving, maintaining or constructing the highway + otherwise than purely to extract the aggregate
- Drill cuttings from petroleum operations on land and on the seabed
- Aggregate resulting from works carried out in exercise of powers under the New Road and Street Works Act 1991, the Roads (Northern Ireland) Order 1993 or the Street Works (Northern Ireland) Order 1995
- Aggregate removed for the purpose of cutting of rock to produce dimension stone, or the production of lime or cement from limestone.
- Aggregate arising as a waste material during the processing of the following industrial minerals:
    - ball clay
    - barytes
    - calcite
    - china clay
    - coal, lignite, slate or shale
    - feldspar
    - flint
    - fluorspar
    - fuller's earth
    - gems and semi-precious stones
    - gypsum
    - any metal or the ore of any metal
    - muscovite
    - perlite
    - potash
    - pumice
    - rock phosphates
    - sodium chloride
    - talc
    - vermiculite

However, the levy is still chargeable on any aggregates arising as the spoil or waste from or the by-products of the above exempt processes. This includes quarry overburden.

Anything that consists 'wholly or mainly' of the following is exempt from the levy (note that 'wholly' is defined as 100% but 'mainly' as more than 50%, thus exempting any contained aggregates amounting to less than 50% of the original volumes:

- clay, soil, vegetable or other organic matter
- coal, slate or shale
- china clay waste and ball clay waste

Relief from the levy either in the form of credit or repayment is obtainable where:

- it is subsequently exported from the UK in the form of aggregate
- it is used in an exempt process

- where it is used in a prescribed industrial or agricultural process
- it is waste aggregate disposed of by dumping or otherwise, e.g. sent to landfill or returned to the originating site

A new exemption for aggregate obtained as a by-product of railway, tramway and monorail improvement, maintenance and construction was introduced in 2007.

Discounts

From 1 July 2005 the standard added water percentage discounts listed below can be used. Alternatively a more exact percentage can be agreed and this must be done for dust dampening of aggregates.

- washed sand 7%
- washed gravel 3.5%
- washed rock/aggregate 4%

Impact

The British Aggregates Association suggests that the additional cost imposed by quarries is more likely to be in the order of £2.765 per tonne on mainstream products, applying an above average rate on these in order that by-products and low grade waste products can be held at competitive rates, as well as making some allowance for administration and increased finance charges.

With many gravel aggregates costing in the region of £16.00 to £18.00 per tonne, there is a significant impact on construction costs.

Avoidance

An alternative to using new aggregates in filling operations is to crush and screen rubble which may become available during the process of demolition and site clearance as well as removal of obstacles during the excavation processes.

Example: Assuming that the material would be suitable for fill material under buildings or roads, a simple cost comparison would be as follows (note that for the purpose of the exercise, the material is taken to be 1.80 tonne/m³ and the total quantity involved less than 1,000 m³):

Importing fill material :		**£/m³**	**£/tonne**
Cost of 'new' aggregates delivered to site		31.23	17.35
Addition for Aggregates Tax		3.51	1.95
Total cost of importing fill materials	£	34.74	19.30
Disposing of site material :		**£/m³**	**£/tonne**
Cost of removing materials from site materials	£	21.52	11.95
Crushing site materials :		**£/m³**	**£/tonne**
Transportation of material from excavations or demolition to stockpiles		3.00	1.67
Transportation of material from temporary stockpiles to the crushing plant		4.00	2.22
Establishing plant and equipment on site; removing on completion		2.00	1.11
Maintain and operate plant		9.00	5.00
Crushing hard materials on site		13.00	7.22
Screening material on site		2.00	1.11
Total cost of crushing site materials		33.00	18.33

From the above it can be seen that potentially there is a great benefit in crushing site materials for filling rather than importing fill materials.

Setting the cost of crushing against the import price would produce a saving of £1.74 per m^3. If the site materials were otherwise intended to be removed from the site, then the cost benefit increases by the saved disposal cost to £23.26 per m^3.

Even if there is no call for any or all of the crushed material on site, it ought to be regarded as a useful asset and either sold on in crushed form or else sold with the prospects of crushing elsewhere.

Specimen Unit rates	**unit³**	**£**
Establishing plant and equipment on site; removing on completion		
Crushing plant	trip	1,200.00
Screening plant	trip	600.00
Maintain and operate plant		
Crushing plant	week	7,200.00
Screening plant	week	1,800.00
Transportation of material from excavations or demolition places to temporary stockpiles	m^3	3.00
Transportation of material from temporary stockpiles to the crushing plant	m^3	2.40
Breaking up material on site using impact breakers		
mass concrete	m^3	14.00
reinforced concrete	m^3	16.00
brickwork	m^3	6.00
Crushing material on site		
mass concrete not exceeding 1000m^3	m^3	13.00
mass concrete 1000 - 5000m^3	m^3	12.00
mass concrete over 5000m^3	m^3	11.00
reinforced concrete not exceeding 1000m^3	m^3	15.00
reinforced concrete 1000 - 5000m^3	m^3	14.00
reinforced concrete over 5000m^3	m^3	13.00
brickwork not exceeding 1000m^3	m^3	12.00
brickwork 1000 - 5000m^3	m^3	11.00
brickwork over 5000m^3	m^3	10.00
Screening material on site	m^3	2.00

Land Remediation

The purpose of this section is to review the general background of ground contamination, the cost implications of current legislation and to consider the various remedial measures and to present helpful guidance on the cost of Land Remediation.

It must be emphasised that the cost advice given is an average and that costs can vary considerably from contract to contract depending on individual Contractors, site conditions, type and extent of contamination, methods of working and various other factors as diverse as difficulty of site access and distance from approved tips.

We have structured this Unit Cost section to cover as many aspects of Land Remediation works as possible.

The introduction of the Landfill Directive in July 2004 has had a considerable impact on the cost of Remediation works in general and particularly on the practice of Dig and Dump. The number of Landfill sites licensed to accept Hazardous Waste has drastically reduced and inevitably this has led to increased costs.

Market forces will determine future increases in cost resulting from the introduction of the Landfill Directive and the cost guidance given within this section will require review in light of these factors.

Statutory framework

In July 1999 new contaminated land provisions, contained in Part IIa of the Environmental Protection Act 1990 were introduced. A primary objective of the measures is to encourage the recycling of brownfield land.

Under the Act action to remediate land is required only where there are unacceptable actual or potential risks to health or the environment. Sites that have been polluted from previous land use may not need remediating until the land use is changed. In addition, it may be necessary to take action only where there are appropriate, cost-effective remediation processes that take the use of the site into account.

The Environment Act 1995 amended the Environment Protection Act 1990 by introducing a new regime designed to deal with the remediation of sites which have been seriously contaminated by historic activities. The regime became operational on 1 April 2000. Local authorities and/or the Environment Agency regulate seriously contaminated sites which are known as 'special sites'. The risks involved in the purchase of potentially contaminated sites are high, particularly considering that a transaction can result in the transfer of liability for historic contamination from the vendor to the purchaser.

The contaminated land provisions of the Environmental Protection Act 1990 are only one element of a series of statutory measures dealing with pollution and land remediation that have been and are to be introduced. Others include:

- Groundwater regulations, including pollution prevention measures
- An integrated prevention and control regime for pollution
- Sections of the Water Resources Act 1991, which deals with works notices for site controls, restoration and clean up.

The contaminated land measures incorporate statutory guidance on the inspection, definition, remediation, apportionment of liabilities and recovery of costs of remediation. The measures are to be applied in accordance with the following criteria:

- The standard of remediation should relate to the present use
- The costs of remediation should be reasonable in relation to the seriousness of the potential harm
- The proposals should be practical in relation to the availability of remediation technology, impact of site constraints and the effectiveness of the proposed clean-up method.

Liability for the costs of remediation rests with either the party that 'caused or knowingly permitted' contamination, or with the current owners or occupiers of the land.

Apportionment of liability, where shared, is determined by the local authority. Although owners or occupiers become liable only if the polluter cannot be identified, the liability for contamination is commonly passed on when land is sold.

The ability to forecast the extent and cost of remedial measures is essential for both parties, so that they can be accurately reflected in the price of the land. If neither the polluter nor owner can be found, the clean up is funded from public resources.

The EU Landfill Directive

The Landfill (England and Wales) Regulations 2002 came into force on 15 June 2002 followed by Amendments in 2004 and 2005. These new regulations implement the Landfill Directive (Council Directive 1999/31/EC), which aims to prevent, or to reduce as far as possible, the negative environmental effects of landfill. These regulations have had a major impact on waste regulation and the waste management industry in the UK.

The Scottish Executive and the Northern Ireland Assembly will be bringing forward separate legislation to implement the Directive within their regions.

In summary, the Directive requires that:

- Sites are to be classified into one of three categories: hazardous, non-hazardous or inert, according to the type of waste they will receive.
- Higher engineering and operating standards will be followed.
- Biodegradable waste will be progressively diverted away from landfills.
- Certain hazardous and other wastes, including liquids, explosive waste and tyres will be prohibited from landfills.
- Pre-treatment of wastes prior to landfilling will become a requirement.

On 15 July 2004 the co-disposal of hazardous and non-hazardous waste in the same landfill site ended and in July 2005 new waste acceptance criteria (WAC) were introduced which also prevents the disposal of materials contaminated by coal tar.

The effect of this Directive has been to dramatically reduce the hazardous disposal capacity post July 2004, resulting in a **SIGNIFICANT** increase in remediating costs. There are now less than 20 commercial landfills licensed to accept hazardous waste as a direct result of the implementation of the Directive! There are no sites in Wales and only limited capacity in the South of England. This has significantly increased travelling distance and cost for disposal to landfill. The increase in operating expenses incurred by the landfill operators has also resulted in higher tipping costs.

All hazardous materials designated for disposal off-site are subject to WAC tests. Samples of these materials are taken from site to laboratories in order to classify the nature of the contaminants. These tests, which cost approximately £200 each, have resulted in increased costs for site investigations and as the results may take up to 3 weeks this can have a detrimental effect on programme.

As from 1 July 2008 the WAC derogations which have allowed oil contaminated wastes to be disposed in landfills with other inert substances were withdrawn. As a result the costs of disposing of oil contaminated solids has increased.

There has been a marked slowdown in brownfield development in the UK with higher remediation costs, longer clean-up programmes and a lack of viable treatment options for some wastes.

The UK Government established the Hazardous Waste Forum in December 2002 to bring together key stakeholders to advise on the way forward on the management of hazardous waste.

Effect on Disposal Costs

Although most landfills are reluctant to commit to future tipping prices, tipping costs during the first half of 2008 have generally stabilised. However, there are significant geographical variances, with landfill tip costs in the North of England being substantially less than their counterparts in the Southern regions.

For most projects to remain viable there is an increasing need to treat soil in-situ by bioremediation, soil washing or other alternative long-term remediation measures. Waste untreatable on-site such as coal tar remains a problem.

Development costs and programmes need to reflect this change in methodology.

Types of hazardous waste

- Sludges, acids and contaminated wastes from the oil and gas industry
- Acids and toxic chemicals from chemical and electronics industries
- Pesticides from the agrochemical industry
- Solvents, dyes and sludges from leather and textile industries
- Hazardous compounds from metal industries
- Oil, oil filters and brake fluids from vehicles and machines
- Mercury-contaminated waste from crematoria
- Explosives from old ammunition, fireworks and airbags
- Lead, nickel, cadmium and mercury from batteries
- Asbestos from the building industry
- Amalgam from dentists
- Veterinary medicines

[Source: Sepa]

Foam insulation materials containing ODP (Ozone Depletant Potential) are also considered as hazardous waste under the EC Regulation 2037/2000.

Land remediation techniques

There are two principal approaches to remediation - dealing with the contamination in situ or off site. The selection of the approach will be influenced by factors such as: initial and long term cost, timeframe for remediation, types of contamination present, depth and distribution of contamination, the existing and planned topography, adjacent land uses, patterns of surface drainage, the location of existing on-site services, depth of excavation necessary for foundations and below-ground services, environmental impact and safety, prospects for future changes in land use and long-term monitoring and maintenance of in situ treatment.

In situ techniques

A range of in situ techniques is available for dealing with contaminants, including:

- Clean cover - a layer of clean soil is used to segregate contamination from receptor. This technique is best suited to sites with widely dispersed contamination. Costs will vary according to the need for barrier layers to prevent migration of the contaminant.

- On-site encapsulation - the physical containment of contaminants using barriers such as slurry trench cut-off walls. The cost of on-site encapsulation varies in relation to the type and extent of barriers required.

There are also in situ techniques for treating more specific contaminants, including:

- Bio-remediation - for removal of oily, organic contaminants through natural digestion by micro-organisms. The process is slow, taking from one to three years, and is particularly effective for the long-term improvement of a site, prior to a change of use.

- Soil washing - involving the separation of a contaminated soil fraction or oily residue through a washing process. The dewatered contaminant still requires disposal to landfill. In order to be cost effective, 70 - 90% of soil mass needs to be recovered.

- Vacuum extraction - involving the extraction of liquid and gas contaminants from soil by vacuum.

- Thermal treatment - the incineration of contaminated soils on site. The uncontaminated soil residue can be recycled. By-products of incineration can create air pollution and exhaust air treatment may be necessary.

- Stabilisation - cement or lime, is used to physically or chemically bound oily or metal contaminants to prevent leaching or migration. Stabilisation can be used in both in situ and off-site locations.
- Air sparging – the injection of contaminant-free air into the sub-surface enabling a phase transfer of hydrocarbons from a dissolved state to a vapour phase.
- Chemical oxidisation – the injection of reactive chemical oxidants directly into the soil for the rapid destruction of contaminants.

Off-site techniques

Removal for landfill disposal has, historically, been the most common and cost-effective approach to remediation in the UK, providing a broad spectrum solution by dealing with all contaminants. As discussed above, the implementation of the Landfill Directive has resulted in other techniques becoming more competitive for the disposal of hazardous waste. .

If used in combination with material-handling techniques such as soil washing, the volume of material disposed at landfill sites can be significantly reduced. The disadvantages of the technique include the fact that the contamination is not destroyed, there are risks of pollution during excavation and transfer; road haulage may also cause a local nuisance.

Cost drivers

Cost drivers relate to the selected remediation technique, site conditions and the size and location of a project.

The wide variation of indicative costs of land remediation techniques shown below is largely because of differing site conditions.

Indicative costs of land remediation techniques for 2008 (excluding General Items, testing and landfill tax)		
Remediation technique	**Unit**	**Rate (£/unit)**
Removal – non-hazardous Removal – hazardous Note: excluding any pre-treatment of material	disposed material (m^3) disposed material (m^3)	60 - 120 100 - 200
Clean cover	surface area of site (m^2)	25 - 50
On-site encapsulation	encapsulated material (m^3)	30 - 95
Bio-remediation	treated material (tonne)	15 - 50
Soil washing	treated material (tonne)	50 - 80
Thermal treatment	treated material (tonne)	300 – 1,000

Many other on-site techniques deal with the removal of the contaminant from the soil particles and not the wholesale treatment of bulk volumes. Costs for these alternative techniques are very much Engineer designed and site specific.
Factors that need to be considered include:

- Waste classification of the material
- Underground obstructions, pockets of contamination and live services
- Ground water flows and the requirement for barriers to prevent the migration of contaminants
- Health and safety requirements and environmental protection measures
- Location, ownership and land use of adjoining sites
- Distance from landfill tips, capacity of the tip to accept contaminated materials, and transport restrictions
- The escalating cost of diesel fuel, currently nearing £1.25 per litre (at May 2008 prices)

Other project related variables include size, access to disposal sites and tipping charges; the interaction of these factors can have a substantial impact on overall unit rates.

The tables below set out the costs of remediation using *dig-and-dump* methods for different sizes of project, differentiated by the disposal of non-hazardous and hazardous material. Variation in site establishment and disposal cost accounts for 60 - 70% of the range in cost.

Variation in the costs of land remediation by removal: Non-hazardous Waste			
Item	**Disposal Volume (less than 3000 m³) (£/m³)**	**Disposal Volume (3000 - 10 000 m³) (£/m³)**	**Disposal Volume (more than 10 000 m³) (£/m³)**
General items and site organisation costs	55 - 90	25 - 40	7 - 20
Site investigation and testing	5 - 12	2 - 7	2 - 6
Excavation and backfill	18 - 35	12 - 25	10 - 20
Disposal costs (including tipping charges but not landfill tax)	20 - 35	20 - 35	20 - 35
Haulage	20 - 40	20 - 40	20 - 40
Total (£/m³)	**118 - 212**	**79 - 147**	**59 - 121**
Allowance for site abnormals	*0 - 10 +*	*0 - 15 +*	*0 - 10 +*

Variation in the costs of land remediation by removal: Hazardous Waste			
Item	**Disposal Volume (less than 3000 m³) (£/m³)**	**Disposal Volume (3000 - 10 000 m³) (£/m³)**	**Disposal Volume (more than 10 000 m³) (£/m³)**
General items and site organisation costs	55 - 90	25 - 40	7 - 20
Site investigation and testing	10 - 18	5 - 12	5 - 12
Excavation and backfill	18 - 35	12 - 25	10 - 20
Disposal costs (including tipping charges but not landfill tax)	80 – 170	80 - 170	80 - 170
Haulage	25-120	25 - 120	25 - 120

Total (£/m³)	188 - 433	147 - 367	127 - 342
Allowance for site abnormals	*0 - 10 +*	*0 - 15 +*	*0 - 10 +*

The strict health and safety requirements of remediation can push up the overall costs of site organisation to as much as 50% of the overall project cost. A high proportion of these costs are fixed and, as a result, the unit costs of site organisation increase disproportionally on smaller projects.

Haulage costs are largely determined by the distances to a licensed tip. Current average haulage rates, based on a return journey range from £1.65 to £3.00 per mile. Short journeys to tips, which involve proportionally longer standing times, typically incur higher mileage rates, up to £8. 50 per mile. The volatility of oil prices will also have a major impact on haulage rates.
A further source of cost variation relates to tipping charges. The table below summarises typical tipping charges for 2008, exclusive of landfill tax:

Typical 2008 tipping charges (excluding landfill tax)	
Waste classification	**Charges (£/tonne)**
Non-hazardous wastes	10 - 25
Hazardous wastes	35 - 85
Contaminated liquid	40 - 75
Contaminated sludge	55 - 200

Tipping charges fluctuate in relation to the grades of material a tip can accept at any point in time. This fluctuation is a further source of cost risk. Furthermore, tipping charges in the North of England are generally less than in the rest of the country.

In addition, landfill tips generally charge a tip administration fee of approximately £25 per load, equivalent to £1.25 per tonne. This charge does not apply to non-hazardous wastes.

Landfill tax, increased on 1 April 2008 to £32 a tonne for active waste, is also payable. Exemptions currently available for the disposal of historically contaminated material are being phased out (refer also to *'Landfill Tax'* section).

The government stated in the 2008 Budget that the standard rate of tax will increase by £8 per tonne in subsequent years to a rate of £48 per tonne by 2011. Thereafter, further increases will no doubt occur.

Tax Relief for Remediation of Contaminated Land

The Finance Act 2001 included provisions that allow companies (but not individuals or partnerships) to claim tax relief on capital and revenue expenditure on the "remediation of contaminated land" in the United Kingdom. The relief is available for expenditure incurred on or after 11 May 2001.

From 1 April 2009 Land Remediation relief will be extended to long-term derelict land and will include the removal of Japanese Knotweed (except removal to landfill).

A company is able to claim an additional 50% deduction for "qualifying land remediation expenditure" allowed as a deduction in computing taxable profits, and may elect for the same treatment to be applied to qualifying capital expenditure.

The Relief

Qualifying expenditure may be deducted at 150% of the actual amount expended in computing profits for the year in which it is incurred.

For example, a property trading company may buy contaminated land for redevelopment and incurs £250,000 on qualifying land remediation expenditure that is an allowable for tax purposes. It can claim an additional deduction of £125,000, making a total deduction of £375,000. Similarly, a company incurring qualifying capital expenditure on a fixed asset of the business is able to claim the same deduction provided it makes the relevant election within 2 years.

What is Remediation?

Land remediation is defined as the doing of works including preparatory activities such as condition surveys, to the land in question, any controlled waters affected by the land, or adjoining or adjacent land for the purpose of: -

- Preventing or minimising, or remedying or mitigating the effects of, any harm, or any pollution of controlled waters, by reason of which the land is in a contaminated state; or
- Restoring the land or waters to their former state.

Definitions

Contaminated land is defined as land that, because of substances on or under it, is in such a condition that: -

- Harm is or may possibly be caused; or
- Controlled waters are or likely to be polluted.

Land includes buildings on the land, and expenditure on asbestos removal is expected to qualify for this tax relief. It should be noted that the definition is not the same as that used in the Environmental Protection Act Part 11A.

Harm is defined as meaning: -

- Harm to the health of living organisms;
- Interference with the ecological systems of which any living organisms form part;
- Offence to the senses of human beings;
- Damage to property.

Pollution of controlled waters is defined as the entry into such waters of any poisonous, noxious or polluting matter or any solid waste matter. Nuclear sites are specifically excluded.

Conditions

In order to become qualifying, the expenditure must be in land that was in a contaminated state when it was acquired. The land must not have been contaminated by the company, or by a connected company.

PROFILE22

The Landfill Tax

The Tax

The Landfill tax came into operation on 1 October 1996. It is levied on operators of licensed landfill sites at the following rates with effect from 1 April 2008:

£2.50 per tonne	-	Inactive or inert wastes. Included are soil, stones, brick, plain and reinforced concrete, plaster and glass.
£32 per tonne	-	All other taxable wastes. Included are timber, paint and other organic wastes generally found in demolition work and builders skips.

The rate for "all other taxable wastes" will be increased by £8 per tonne each year at least until 2010/11. The rate for "inactive or inert wastes" will be frozen at £2.50 per tonne to 2009/10.

Mixtures containing wastes not classified as inactive or inert will not qualify for the lower rate of tax unless the amount of non-qualifying material is small and there is no potential for pollution. Water can be ignored and the weight discounted.

Calculating the Weight of Waste

There are two options:

* If licensed sites have a weighbridge, tax will be levied on the actual weight of waste.

* If licensed sites do not have a weighbridge, tax will be levied on the permitted weight of the lorry based on an alternative method of calculation based on volume to weight factors for various categories of waste.

Effect on Prices

The tax is paid by Landfill site operators only. Tipping charges reflect this additional cost.

As an example, Spon's A & B rates for mechanical disposal will be affected as follows:

*	Inactive waste	Spon's A & B 2009 net rate	£16.52 per m^3
		Tax, 2 t per m^3 (un-bulked) @ £2.50	£ 5.00 per m^3
		Spon's rate including tax (page 206)	£21.52 per m^3

Effect on Prices - cont'd

* Active waste — Active waste will normally be disposed of by skip and will probably be mixed with inactive waste. The tax levied will depend on the weight of materials in the skip which can vary significantly.

Exemptions

The following disposals are exempt from Landfill Tax:

* dredgings which arise from the maintenance of inland waterways and harbours.
* naturally occurring materials arising from mining or quarrying operations.
* waste resulting from the cleaning up of historically contaminated land, although to obtain an exemption it is necessary to first obtain a contaminated land certificate from HM Revenue and Customs.
* waste removed from one site to be used on another or to be recycled or incinerated.
* inert waste used to restore landfill sites and to fill working and old quarries where a planning condition or obligation is in existence.

The exemption for waste from contaminated land will be phased out completely by 1 April 2012 and no new applications for landfill tax exemption will be accepted after 1 December 2008.

For further information contact the National Advisory Service, Telephone: 0845 010 9000.

Sustainability

The purpose of this section is to set sustainability into context for the construction industry by describing its evolution, some of the drivers, key issues and the technologies available, together with indicative costs, which enable us to better address this increasingly important issue in the built environment.

BACKGROUND

Over the past two decades governments around the world have begun to recognise that the rate of environmental degradation and current practices of economic development are having significant impacts on the planet and its people and are posing significant challenges to the growth potential of future generations. Bringing this to the fore in 1987, the Brundtland Commission published its report "*Our Common Future*", which alerted the world to the urgency of making progress towards economic development that could be sustained without depleting natural resources or harming the environment and defined the concept as:

"Development that meets the needs of the present without compromising the ability of future generations to meet their own needs"

The report highlighted and described the three fundamental components of sustainable development, namely environmental protection, economic growth and social equity.

Building on the Brundtland Report, the Rio Earth Summit in 1992 represented a significant step in moving the sustainability forward, with international agreements made on climate change and biodiversity. A key outcome of the Summit was Agenda 21 – a framework for tackling social, economic and environmental problems. More importantly though, Agenda 21 required each country to draw up a national strategy for sustainable development, detailing their approach and how it would be delivered.

One of the most significant recent environmental agreements is probably the Kyoto Protocol. It is an international agreement setting targets for industrialised countries to cut their greenhouse gas emissions. The Kyoto Protocol was agreed in 1997, and followed from principles set out in 1992. It finally came into effect in 2005. One hundred and fifty countries have now signed the agreement. However, the USA - the world's biggest producer of greenhouse gases has not yet signed.

Under the protocol, industrialised countries have been set legally binding individual targets to cut their greenhouse gas emissions by the period 2008-2012. The European Union, which consisted of only 15 countries at the time of signing, has been set a reduction target of 8 per cent average across the European Union when compared with 1990 levels.

The UK's own target is 12.5 per cent. Developing countries, including India, China and Brazil, have not been set reduction targets.

The UK looks set to achieve its target, although most of the reductions have been achieved through switching from coal-fired to gas-fired power stations. Carbon dioxide emissions from other sources in the UK continue to rise, however, and it is now considered unlikely the nation will achieve a more ambitious 20 per cent cut which it has volunteered to reach.

LEGISLATION, STRATEGY AND POLICY

Legislation

At a national level in the UK, legislation relating to environmental issues has been in existence for many years. The construction industry will be familiar with issues relating to environmental pollution covered by Acts of Parliament such as:

- The Control of Pollution Act, 1974
- Water Resources Act 1991
- The Environmental Protection Act, 1990 (as amended by the Environment Act 1995)

Since joining the European Union, the UK has been subject to European Union Directives and Regulations on which much UK environmental law is based. The UK legislation process often starts with the issue of a consultation paper (by the government or the environment agencies). The paper explains the proposed legislation, summarises EU legal provisions and gives the public and others an opportunity to comment on the proposal.

Under UK law, Acts form what is known as Primary Legislation. Secondary Legislation is often referred to as Statutory Instruments and made under authority contained in Acts of the UK Parliament They may consists of Orders, Regulations and Rules. Many of the European Directives are enacted in UK law through such statutory instruments and include:

The Building Regulations, 2000 (made under the Building Act, 1984).

The Building Regulations have become very central to the delivery of more sustainable buildings, particularly through Part L which considers energy efficiency and as such is considered key to mitigating the impact of climate change. Part L is currently under revision in order to meet with the requirements of the EU Directive on the energy performance of buildings (EPBD). The Building Regulations were revised in 2002 and again in April 2006. Future revisions are due in 2010 and 2013 and will progressively tighten requirements in line with the Directive. The EPBD has also brought about the recent introduction of energy performance certification for buildings.

Sustainable and Secure Buildings Act 2004

The Sustainable and Secure Buildings Acts gives new powers under the Building Act 1984 to improve the sustainability of buildings, including furthering the conservation of fuel and power.

Waste Management Licensing Regulations 1994 (as amended 1995 and made under the Environment Protection Act 1990 and modified by the Environment Act 1995).

The regulations require that organisations involved in disposing or treating waste must be licensed by the Environment Agency. In most cases, where project site activities only involve the storing of waste which has been produced on that site and this is regularly removed from site, the requirement for licensing will not apply. A waste management licence may be required if, however, certain activities are being undertaken on site.

Contaminated Land (England) Regulations 2000 (made under the Environment Protection Act 1990 and amended by the Environment Act 1995).

This includes land which is actually contaminated, but also land where certain activities have taken place which are likely to have resulted in contamination, for example specified industrial processes. Where land is designated as contaminated, the responsible person (the "Appropriate" person) will be notified and a remediation notice may be served which will require that the land is remediated to a specified standard.

In parallel to these legislative drivers, the response to Agenda 21 and other international agreements has seen the UK government developed its strategy "*A Better Quality of Life*" and more recently *"Securing the Future*" which outlines its vision of sustainable development based around four broad objectives:

- Social progress that recognises the needs of everyone;
- Effective protection of the environment;
- Prudent use of natural resources; and
- Maintenance of high and stable levels of economic growth and employment.

The Climate Change Bill

The Climate Change Bill was introduced in Parliament on 14 November 2007 and completed its passage through the House of Lords on 31 March 2008. It will shortly go to the House of Commons for consideration. The aim is to receive Royal Assent by summer 2008. The proposed Bill provides a clear, credible, long-term framework for the UK to achieve its goals of reducing carbon dioxide emissions, and will ensure that steps are taken towards adapting to the impacts of climate change.

Strategy and policy

In response to the challenge of sustainability, in 2000 the government published its strategy for sustainable construction, *"Building a Better Quality of Life"*, which suggested key themes for action for the construction industry, and in 2006 the *"Sustainable Construction Strategy Report"*, highlighted progress and sector initiatives developed to help deliver the strategy.

Linking into these important strategy documents is the UK Energy Strategy, set out in a White Paper in May 2007, which sets out the key targets for CO_2 reduction, many of which are directly relevant to sustainable design and construction, these include:

- 60% CO_2 reduction by 2050;
- 20% CO_2 reduction by 2010;
- 10% of UK energy supply from renewables by 2010; and
- 10% Good Quality Combined Heat and Power by 2010.

The 2008 budget targets opportunities for achieving zero carbon in new non domestic buildings. It is theoretically possible to reduce carbon emissions from energy use down to zero in the majority of new non-domestic buildings, as long as onsite, near-site and offsite renewable solutions are employed. However, the practicalities of achieving this for all new buildings will provide both a technological and financial challenge. All new buildings should be zero carbon by 2019.

Planning policy guidance notes (PPGs) and their replacements Planning Policy Statements (PPSs) are prepared by the government after public consultation to explain statutory provisions and provide guidance to local authorities and others on planning policy and the operation of the planning system.

They also explain the relationship between planning policies and other policies which have an important bearing on issues of development and land use.

Local authorities must take their contents into account in preparing their development plans. The guidance may also be relevant to decisions on individual planning applications and appeals.

All PPS' are relevant to construction projects to a lesser or greater degree, but those most pertinent to the sustainability agenda are listed below:

Planning Policy Statement 1: Delivering Sustainable Development

This places sustainable development at the core of the planning system and also establishes policy relating to Sustainable Design and Construction.

PPS10 Planning for Sustainable Waste Management

This Planning Policy Statement replaces Planning Policy Guidance Note 10 (Planning and Waste Management) published in 1999 and forms part of the national waste management plan for the UK. They may also be material to decisions on individual planning applications. These policies complement other national planning policies and should be read in conjunction with Government policies for sustainable waste management

Planning Policy Statement 22: Renewable Energy

This sets out the Government's policies for renewable energy, which planning authorities should give regard to when preparing local development frameworks (LDF's) and supporting documents, and when considering planning decisions.

PPS23 Planning and Pollution Control

They are also material to decisions on individual planning applications. Where these policies are not reflected adequately in local development documents, or taken into account in relevant development control decisions.

PPS25 Development and Flood Risk

Planning Policy Statement 25 (PPS25) sets out Government policy on development and flood risk. Its aims are to ensure that flood risk is taken into account at all stages in the planning process to avoid inappropriate development in areas at risk of flooding, and to direct development away from areas of highest risk. Where new development is, exceptionally, necessary in such areas, policy aims to make it safe, without increasing flood risk elsewhere, and, where possible, reducing flood risk overall.

Regional Spatial Strategies, Local Development Frameworks & Supplementary Planning Guidance

The Planning and Compulsory Purchase Act, 2004 set out a two tier planning structure consisting of Regional Spatial Strategies (RSS) and Local Development Frameworks (LDF). The RSS sets out policies relating to the development and use of land for each English region. Importantly the RSS has a statutory duty to contribute to the achievement of sustainable development. In London the RSS is known as the London Plan and notably calls for CO_2 emissions from the total energy needs (heat, cooling and power) of the development to be reduced with at least 10% onsite generation of renewable energy.

Local Authorities set out their own policies on specific issues relevant to sustainable design and construction, such as waste, water and CO_2 emissions in their LDF's and supplementary planning guidance.

Importantly although Planning Policy Statements and Supplementary Planning guidance have no legal standing, compliance or non-compliance against the guidance is a material consideration in granting planning permission and therefore can be used as grounds for refusal.

The Code for Sustainable Homes

On the 27 February 2008 the Government confirmed that from 1 May 2008 it would be mandatory for all new homes to have a rating against the Code. The new regulations for providing for mandatory ratings does not apply to properties (individual or as part of an ongoing development) where the initial notice, full plans or Building Notice have been received by the relevant Local Authority Building Control body prior to 1st May 2008.

The new requirement to have a rating against the Code does not make it mandatory to build a Code home or to have each new home assessed against the Code. It does however mean that all buyers of new homes be given clear information about the sustainability of the new home.

KEY ISSUES

When looking at sustainability in the built environment there are a number of key issues which need to be considered when designing and building new assets or refurbishing existing ones. These span issues relating to social, environmental and economic aspects of sustainability. However, in the context of the physical material resources that are involved in the manifestation of a building project it is possible to characterise the environmental resources into five broad themes:

- Energy/CO_2;
- Materials;
- Resource efficiency;
- Water; and
- Ecology and Biodiversity.

The sustainability of buildings can be assessed using several methodologies that are now available in the market place. In the UK this is predominantly BREEAM, which is available in several variants for different building types. Within each of these themes are a number of measures, activities and technologies which can be utilised to reduce impact and/or increase sustainability performance of an asset. The following table illustrates some of these.

	Energy/CO2	Materials	Resource efficiency	Water
Activity, measure & technology	▪ Maximise heat recovery ▪ Prevent heat losses ▪ Solar gain ▪ Use low energy consumption technologies ▪ Use local micro generation for heating & power from renewable sources ▪ Maximise air tightness	▪ Life cycle analysis ▪ Use of sustainable timber ▪ Low embodied carbon materials ▪ Use of non-toxic materials ▪ Sustainability 'chain of custody' for all materials sought ▪ Use of high recycled content materials	▪ Recycled content analysis ▪ Site waste management planning ▪ Pre-demolition audits ▪ Maximise secondary & recycled materials potential ▪ Designing out waste ▪ Use of standardisation, pre-fabrication and modularisation	▪ Use of efficient fixtures and fittings ▪ Use of recycling and retention systems ▪ Use of natural attenuation systems i.e. SUDS (sustainable urban drainage systems)

Furthermore, other aspects of environmental sustainability should consider impacts such as:

- Land use;
- Archaeology;
- Noise and Vibration; and
- Air Quality.

All major construction projects have a significant impact on the economy and also on society and these impacts should not be disregarded. These impacts can be both negative and positive in nature. A balanced view of sustainability should ensure that these impacts are evaluated and where there is opportunity to provide additional value and benefits these should be realised wherever possible. It is no longer simply a matter of ensuring that contractors are for example, members of Considerate Contractor-type schemes, but that opportunities to train and skill people are identified and implemented. Furthermore, despite what can be very complex supply chains, it is important to begin the process of questioning the ethical as well as environmental considerations of materials and products used in the construction process.

SUSTAINABLE CONSTRUCTION

One of the first questions that should be asked with regard to sustainable construction is whether the building or infrastructure is necessary. For large projects that fall within the scope of the Town and Country Planning (Environmental Impact Assessment) (England and Wales) Regulations 1999 this question is addressed through the "do minimum" analysis. For projects that fall outside this scope it would be considered best practice to make a similar assessment. Sustainable construction should consider land use, and in the case of building related projects it is important to design with the building fabric in mind such that operational impacts are reduced.

Considerations such as building orientation, should take priority over technologies as façade design, lighting, heating and cooling strategies, Even the most sustainably designed buildings may not operate at their optimum if not properly managed. In this respect there is an ever increasing clamour for intelligent buildings. Whether, intelligent or not, a robust handover process with clear operational and maintenance guidance is essential, and so too appropriate monitoring and measurement and post occupancy evaluation. Although the use of management measures and activities is a vital element of increasing sustainability performance it is often difficult to apportion market values to them, making cost appraisals difficult.

Costs associated with the building fabric and any necessary land mitigation is more easily apportioned, although some sites and design options can ensure that these are kept to a minimum. Costs associated with renewable energy technologies are typically the most significant, and therefore, this section will look at the varying sustainable technologies available on the market, their sector applicability and indicative costs.

The costs of energy generating technologies in this section are expressed on the basis of units of energy output rather than the physical dimensions of the system. For example, photovoltaic cells cost on the basis of kWh output as well as £/m2. The reasons for this are that systems should be sized relative to their planned contribution to energy loads, and that this will vary in accordance with site factors such as orientation etc. Furthermore, as most renewable energy systems are designed to provide only a portion of the peak load, it is not practical to establish overall £/m2 benchmarks for renewable technologies. When using the guide prices in this section of the price book, it is recommended that energy outputs are clearly stated in the accompanying assumptions and commentary.

The key factors affecting the selection of a renewable technology are the physical and site constraints, affordability, effectiveness of the systems, and in the case of biomass, availability of a fuel supply chain. Scalability of systems and the requirement for back-up supply sources are particular challenges, as is the pay back period for many technologies.

WIND

In a suitable location, building-integrated wind energy can be an effective source of renewable power generation. Wind turbines are the most widespread wind technology and they convert the force of the wind into electrical energy using rotating blades (mounted horizontally or vertically) that drive a generator to create electricity. They can be connected to the national grid for electricity export, used directly for electricity consumption or used to charge batteries for on-site use. Turbines can range from small domestic turbines producing hundreds of watts of energy to large offshore turbines with a capacity of 5MW and a blade diameter of 10m.

To operate efficiently, micro wind turbines need a clear and uninterrupted source of wind above 5-6 m/s and although the UK has a very good wind resource wind flow is a critical factor in successful energy generation. In urban locations the wind profile will be significantly affected by buildings and trees which create turbulence. To offset this, turbines for urban areas should be specified with caution and where deemed viable erected with the maximum mast height available.

Factors to consider in the adoption of wind turbines
▪ Wind is free and unlimited;
▪ Low cost, relative to output;
▪ No motive power required once installed;
▪ Most of the good wind is in under populated areas such as NW Scotland;
▪ Long operating life and low maintenance requirements.
▪ Problems with visual impact and operational noise;
▪ Requires a long-term wind survey and planning consent;
▪ Substantial pylon structures required; and
▪ Vulnerability to changes to neighbouring development that could disrupt wind flow.
▪ Recent concerns regarding light flicker through rotating blades and potential to trigger epileptic seizure

The micro turbine market is well-served by a number of UK suppliers and specialist installers. The main market is residential, agricultural and light industrial. A few turbines have been fitted to commercial buildings, mainly rooftop installations and electrically integrated. Energy export is technically possible if the site load pattern is suitable, but otherwise the only cost benefit will be the value of each unit of electricity generated. Micro turbines may also be installed on 'folding masts' and simple foundations. Overall project costs, ignoring planning issues costs and project management, are indicated in Figure 1.

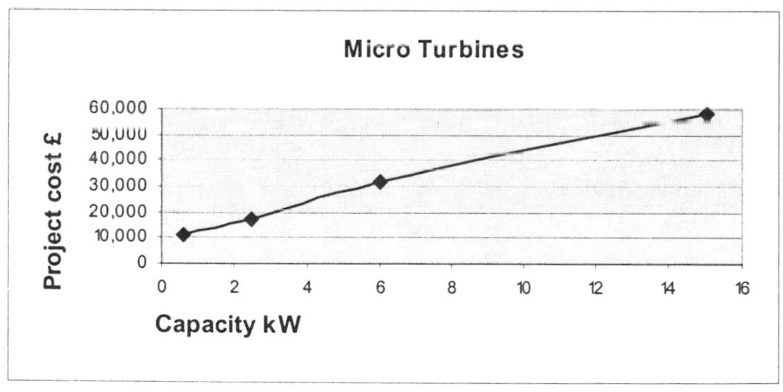

Figure 1 Indicative project costs for micro turbines

Small Turbines

The market for wind turbines in the 10 kW to 100 kW range is currently very 'thin', there being very few suppliers having equipment for sale in this range, and apparently little demand. One reason for this could be that installations larger than, 15 kW rating, require the use of a crane which adds to the cost. Other factors could be larger foundations and taller masts, which may be more difficult in terms of planning.

The BRE report[1] indicates that there is little experience of the operation of such turbines mounted on domestic buildings in urban environments therefore little objective data about their actual performance in terms of power generation, service life and maintenance requirements. This has led to concerns that, in some environments, the installation of micro-wind turbines on housing could increase carbon emissions rather than reduce them.

In addition to the initial embodied carbon and efficiency of the turbine, the payback period is highly sensitive to relatively small changes in one or more of a large number of variable factors, in particular:

- the local wind conditions;
- the size of conurbation and the position within the urban terrain;
- the type of building on which the turbine is mounted and the mounting position;
- the proximity of the surrounding buildings;
- the transport associated with installation and maintenance;
- the maintenance regime; and
- the expected service life of the turbine.

Large Turbines

There is considerable potential at suitable industrial or brownfield sites and a number of installations have been successfully completed. This type of site will often have good access, be remote from residential areas and have substantial electrical demand. In many circumstances it is no more difficult to obtain planning permission for a 50m mast than it is for a 20m one, but a larger, higher turbine will often get better wind above local obstructions. Very few manufacturers offer turbines in the 100-300 kW range, but a market exists in second-hand machines from Europe and these are very much in demand for their potential to reduce capital cost and improve payback.

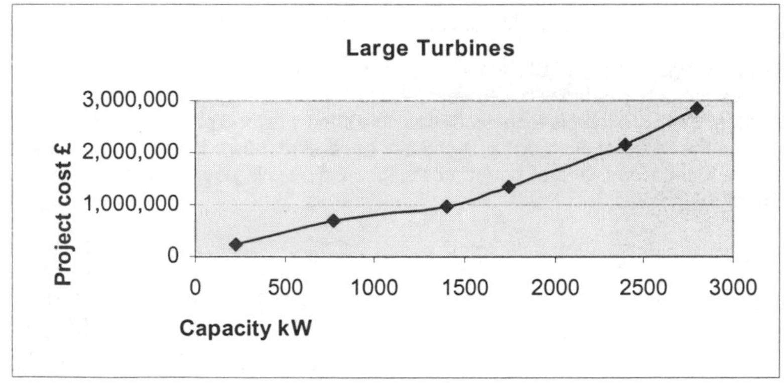

Figure 2 Indicative project costs for large turbines, in optimum conditions.

Exclusions: Professional and statutory fees, funding and VAT

[1] Richard Phillips, Paul Blackmore, Jane Anderson, Michael Clift, Antonio Aguilo-Rullan and Steve Pester. 'Micro-Wind Turbines in Urban Environments, an Assessment', 2007.

Costs

Indicative costs:

Typical 2000 kW turbine project – Breakdown of costs		
	Cost £	**Cost £/kW**
Complete turbine	1,407,000	700
Transportation	56,000	
Foundation	62,000	
Civil works e.g. access roads	34,000*	
Electrical network	45,000*	
Electrical hook-up	51,000*	
Total capital cost	**1,655,000**	**900**
Allowances for warranties & maintenance	£31,000 p.a	£15.00/kW p.a

Exclusions: Professional and statutory fees, funding and VAT
* Allowances are based on an analysis of DTI indicative cost models.

Distance between turbines can affect cable lengths and routing of the cables. In determining the density and spacing of turbines, the costs and losses of the electrical transmission need to be balanced against the aerodynamic losses caused by wakes of closely spaced turbines.

SOLAR PHOTOVOLTAIC CELLS (PV)

Solar photovoltaic's (PV's), sometimes called solar cells generate direct current electrical energy when exposed to light. Solar cells are constructed from semi-conducting materials, such as silicone that absorb solar radiation; upon absorption electrons are displaced within the material, thus starting a flow of current through an external connected circuit, converting solar energy into electrical power. These cells are grouped together to form 'PV modules' that in turn are arranged in 'solar arrays', which are referred to as solar panels.

Although PV's do not face the economy of scale issues of wind technologies, there is a constraint on PV output related to available roof area with a suitable aspect and pitch and the inherent conversion efficiency of the PV. Even for the most efficient monocrystalline panels, $8m^2$ of PV's are required to generate just 1kWp – an output that will typically contribute about a fifth of annual household electricity consumption.

Other aspects affecting the use of PV's include the location and orientation of the panels, shading, temperature control, framing, control systems and inverter efficiency. In addition, once in use, panel cleaning and performance monitoring are essential to ensure the system is operating efficiently. It also has to be taken into account that PV's have a typical lifespan of 25 years so achieving full payback is not guaranteed in a fluctuating energy market.

In utilising PV's it has been found that using them as a direct replacement for an element of a building is currently the best approach – often termed 'building integrated PV's'. This has implications on cost, as the life cycle might be less than the cladding cost. For building integrated PV's specific feasibility and life cycle costing studies should be carried out. An example of this might be using them as roofing or shading elements on south-facing roofs or as an architectural feature integrated within glazing. As result PV's are now available in an increasingly wide variety of forms, such as:

- Façade cladding;
- Roof tiles;
- Roof slates;
- Standing seam and single ply roofing; and
- Freestanding arrays.

Factors to consider in the adoption of Solar Photovoltaic cells
▪ Free energy once the system is installed;
▪ Scalable system based on modular panels;
▪ Potential for substitution of other cladding or roofing materials;
▪ Complementary with wind power;
▪ High initial cost and extended payback;
▪ Must be integrated into the new-build construction programme;
▪ Power generation is not synchronised with peak demand which necessitates export to the National Grid at unattractive unit price. [There is currently a campaign to provide an attractive "feed-in" tariff to stimulate the uptake of PV as practiced in a number of EU member countries]; and
▪ Potential for underperformance, so in-use monitoring required.
▪ Development of Building Integrated PV, which provides glazing function in addition to power generation

Costs

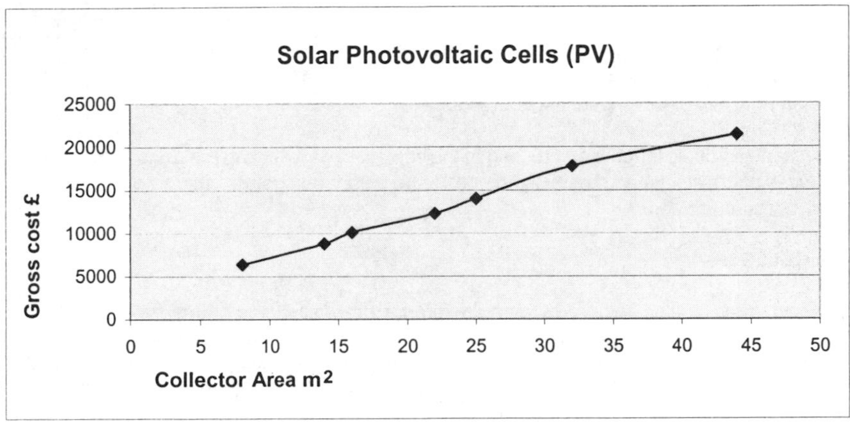

Investment and Returns

Technology	Collector Area m^2	Energy Yield per year (kWh)	Net cost (£)
0.97 kWp Crystalline	8	776	6,000
1.6 kWp Crystalline	14	1,280	8,000
1.94 kWp Crystalline	16	1,552	10,000
2.6 kWp Crystalline	22	2,080	12,000
2.9 kWp Crystalline	25	2,320	13,000
3.9 kWp Crystalline	32	3,120	17,000
4.9 kWp Crystalline	44	3,920	20,000

Energy yields assume a good location for solar i.e. south facing with minimal obstruction and for wind, exposed area with no tall trees or buildings with an average or above UK wind speed (5m/s).

BIOMASS AND CHP SYSTEMS

Biomass heating and combined heat and power (CHP) systems have become a major component of the low-carbon strategy for many projects, as they can provide a large renewable energy component at a relatively low initial cost. Work by the Carbon Trust has demonstrated that both large and small biomass systems were viable even before recent increases in gas and fuel oil prices. These proposals are not without risk, however. Although the technology is well established, few schemes are in operation in the UK and long-term success depends more on the effectiveness of the local supply chain than the quality of the design and installation. Both biomass and CHP is available for both large and small scale use. Biomass can be used independently for producing heat or as a fuel source to power CHP.

Further confidence is required with regard to reliability in the so-called micro CHP market, whilst both large and small scale require a heat and energy consumer which can cause difficulties when there is no heating requirement. Attempts to overcome this imbalance have led to the advocacy of Combined Cooling Heat and Power (also known as Tri-generation), where heat can be turned into a chilled water supply by use of absorption chillers. Typically large scale CHP can be used effectively in developments where there is a high heat requirement and in this context leisure centres and hospitals are often quoted as suitable.

Due to the mechanical complexity of biomass systems, and the need for storage there is a cost premium for the boiler system,, but studies have demonstrated financial viability, particularly in locations which do not have access to mains gas supplies, as alternatives such as fuel oil and LPG are very expensive. The major constraint on the adoption of biomass is the reliable availability of local fuel materials and the practicalities of transporting biomass within urban locations.

Biomass fuels can be waste, residue or energy crops grown specifically for use as wood, or oil fuel. The fuel is transported to the site and stored in a suitable area. Fuel is then delivered by conveyor or pumping systems to the boiler. It is burned to produce hot water in the same way as a coal or oil-fired boilers. It is defined as a 'low carbon' solution rather than zero carbon. Emissions such as NO_X and SO_X[2] have a much greater impact than CO_2.

Factors to consider in choosing biomass, biomass CHP and CHP
▪ Biomass heat output can be controlled but not instantaneously;
▪ Solid biomass used for heat output cannot be throttled back as much as for gas or liquid fuel systems;
▪ Solid biomass feedstock is bulky and needs mechanised feed and extensive storage facilities;
▪ Returns on CHP and electricity generating systems depend heavily on government incentives and under present arrangements large CHP systems provide the best returns;
▪ Heat only systems are very responsive to changes in fuel prices, with small scale heat-only plants producing the best returns because typically the cost of the displaced fuel is more expensive;
▪ Small scale electricity and large scale heat-only systems produce very poor returns;
▪ Biomass is considered as a 'low carbon' solution;
▪ High conversion efficiency of solid biomass to heat energy if based on pellets;
▪ Biomass production has the potential to support local energy industries in rural areas and is arguably more suited to these areas in its use;
▪ Availability of local fuel supply determines carbon output and running costs;
▪ May require secondary heat source for low-season water heating; and
▪ Potential smoke and fumes may increasingly become a concern as uptake of biomass increases.

It is more appropriate for policy to encourage 'off-grid' domestic and industrial users to take first call on the expanding biomass resource rather than commercial schemes. Data shows that 50% of the market potential for industrial applications of CHP could utilise 100% of the UK's available Biomass resource. Given the issues that city centre biomass schemes face in connection with storage, transport, emissions and supply chain management, these might be better addressed by industrial users or their energy suppliers in low-cost locations rather than by developers in prime city centre sites.

---

[2] NO_X and SO_X are Oxides of Nitrogen/Sulphur.

Costs
Indicative costs:

System	Indicative Load kW	Capital Cost (£/KWh)
Gas fired boiler	50 400	90 50
Biomass fired Boiler	50 500	500 250
Biomass fired CHP	1,000	450

Allowance for stand-alone boiler house and fuel store £30,000-60,000 for 50kWh system. Indicative costs exclude flues and plan room installation.

Fuel Costs	Capital Cost (p/KWh)
Wood chip	1.5 to 2.5
Wood pellet	3.0 to 4.5
Fuel oil	4.0 to 5.0
Natural gas	2.5 to 3.0
Bottled LPG	6.5 to 7.0

HEAT AND COOLING PUMPS

In the challenge to reduce building related carbon emissions, heat pumps represent a potentially attractive solution. The principal argument for the wider adoption of heat pumps is their operational efficiency and reliability. Whilst they are more expensive to run than conventional heating and cooling systems, on a pound for pound basis, Heat Pumps provide a similar level of energy output than other low energy/renewable systems such as small scale wind or solar water heating, but are less vulnerable to weather related performance variation. However, as heat pumps use electrical power, they have higher operating costs and contribute less to the reduction of carbon emissions. Several variants are available: closed or open loop, horizontal or vertical. The system can also be used to provide cooling. However, a system's overall output depends on specific ground and aquifer conditions

Ground source heating is a 'low carbon' solution, as electricity is required to run pumps and compressors. However, the pump power conversion rate is typically around 3:1 with potential for a well balanced highly performing system to produce 4:1 or 5:1, making it an effective low carbon system. Ground source heating is best suited to steady background heating loads such as under floor heating and as it cannot be modulated, should be sized to the base load of a building only. Supplementary heating is required for hot water and peak loads, although some of this load could be met with complementary technologies such as solar.

Factors to consider in the adoption of heat pumps.

- Initial costs of the heat pump and its infrastructure;
- Site constraints affecting the location of the heat sink circuits;
- Renewables targets - schemes which are required to invest in low carbon technologies in line with Local Authority targets such as the 'Merton Rule' are likely to adopt explicitly 100% on-site renewable systems such as PVs or biomass boilers;
- The absence of a critical mass of expertise in design, specification and installation – particularly for larger systems. Design standards cannot be transferred from one territory to another due to differences in climate and geology; and
- Lack of independent long term warranties and insurance schemes.

Costs

The table summarises the capital costs of ground source heat pumps and equivalent boiler and chiller systems at a range of scales. The 4kWh system would provide most of the needs of a domestic house, whereas the 400 kWh systems would serve around 5,000 m^2 of net office space in heating mode, or 3,000m^2 in cooling. The costs exclude the wider plant room installation, which is common between the two systems.

The table sets out alternative sources of heating and cooling. The heat pump can provide cooling and can be installed with either a horizontal or vertical network. It illustrates that the cost gap between heat pump systems and conventional installations increases significantly as the size of the installation grows. This reflects the nearly direct relationship between the size of the load and the cost of the energy collector network. Costs can be reduced by integrating the collector network into other aspects of the development such as thermal piles or a car park sub-base. Tapping into an aquifer using an open system will also reduce the potential size of the collector network.

	Small (4 kWh) (£)	**Medium (50 kWh) (£)**	**Large (400 kWh) (£)**
Condensing boiler and flue	1,200	4,200	17,000
Air cooled chiller	1,000	6,000	50,000
Heat pump	3,500 – 4,500	30,000 – 40,000	140,000 – 170,000
Horizontal 'slinky' collector	3,000 – 4,000	40,000 – 50,000	360,000 – 390,000
Vertical borehole collector	2,000 – 3,000	50,000 – 70,000	360,000 – 390,000

Costs exclude distribution pipework from the collection network to the heat pump and plant room installations. Main contractor's preliminaries and overhead and profit are also excluded.

WATER MANAGEMENT SYSTEMS

Water is our most important utility, yet security of supply is almost taken for granted and investment in infrastructure is significantly behind the required rate of replacement, resulting in a significant waste in water through the distribution network. Although the water utilities have increased the rate of replacement of water supply infrastructure, the relatively low cost of water in the UK, capped capital investment and absence of compulsory metering means that there is little incentive for individual developers or occupiers to reduce water consumption.

Water consumption

Practical measures to manage water consumption include metering, low-consumption sanitary fittings and controls, sustainable urban drainage systems (SUDS), rainwater harvesting and greywater recycling. Because of the ready availability of low-cost, high-quality water in the UK, management of water consumption has not demanded the same attention as other sustainability initiatives. For housing development in low-rainfall areas, water management is increasingly finding its way onto the planning agenda.

Rainwater Harvesting Systems

A rainwater harvesting system will collect water via a pipework collection system, typically from roofs but can also include pavings, filter and store it in a collection tank which will typically be located below ground.

The recycled water can be used to flush WCs and urinals, for cleaning and irrigation purposes. Tank size will be dictated by the fact that the water can only be stored for 10-12 days. Mains water systems will still be required to give continuity of supply during periods of low rainfall and drought. This system is most efficient in low rise buildings, compared to city centre blocks and tall buildings.

Grey Water Recycling Systems

An independent waste water collection pipework system collects the water from basins, baths, showers etc, and directs it to a common storage tank which will typically be located below ground. The waste water will be filtered and disinfected to kill bacteria.

On demand the treated water can be pumped back to cisterns for flushing WCs and urinals and also for irrigation purposes. The water can only be stored for 2-3 days dictating tank size and avoiding, where possible, long runs of distribution pipework.

Filtering and treatment are the key issues for greywater systems, particularly in communal systems. The treatments have to be able to cope with worst-case scenarios, such as accidental faecal contamination, and will use chemicals such as bromine, chlorine or ozone or, in some cases, systems based on ultraviolet light. As all systems are designed on a fail-safe basis, it is essential to have effective monitoring systems and easy access for cleaning and maintenance if the full benefit of a greywater system is to be gained.

Reed Bed Drainage

A reed bed drainage system is a highly efficient and cost effective way of removing contaminants from mixed waste water sources, although they are generally used for 'polishing' of water and therefore not suitable for primary treatment. Running costs are low, particularly after the initial growing period. They provide a surface water run-off buffering capacity, especially when combined with balancing ponds and have an increased amenity and biodiversity value, particularly within urban areas.

Reed bed systems are designed bespoke to meet site specific demands and the size of the bed will vary accordingly. Subject to buffering capacity for a 100 year storm event the bed size may need to be increased or a balancing tank installed.

Cistern Reductions

New appliances can be specified to have the minimum flush capacity or dual flush capacities to ensure that water consumption is kept to a minimum. The existing building stock is a far greater challenge and one simple and cost effective solution to reduce the flush capacity of existing WC cisterns is the Hippo Bag or Save-a-Flush solution that involves placing a bag into the cistern that displaces 1 litre of water.

The bag itself costs less than **£5.00** and in a large commercial environment where an efficient programme for the insertion of the bags can be devised the payback period based upon the cost of water saved can be as low as 2 or 3 weeks

Reduced Water Urinals

There are a growing number of urinals available that consume either no or minimal quantities of water by using special liquid air seals to contain odours. Water is still required periodically for cleaning but the saving is significant needing only 2 litres per week for cleaning compared to a typical urinal installation that will use 7.5 litres per hour.

Waterless urinal cost typically in the range **£185.00 – 320.00 (supply only excluding installation)**, replacement cartridges cost circa £37 (supply only) and need to be replaced quarterly

Costs

Indicative costs for water efficient taps:

System Specification	**Indicative Cost (£)**
Sensor tap, battery operated (per unit)	255 – 310
Water-free urinal (per unit)	310 – 360
Aerated tap (per unit)	300 – 1000

Indicative costs for rainwater and grey water harvesting

System Specification	Indicative Cost (£)
Rainwater recycling, 1 below-ground tank, grp with masonry and concrete surround, 12,000l, cover and frame	**11,000**
Rainwater harvesting, 1 subterranean tank, 35,000l, 1 header tank, concrete surround, access hatches, controls, complete	47,000
Rainwater harvesting, 1 subterranean tank, 110,000 litres, 2 header tanks, concrete surround, access hatches, controls, complete	**100,000**
Grey water recycling, overall indicative cost for commercial scale installation, complete	47,000

Drainage Solutions

The following are indicative costs for commercial schemes located in south-east England, current in first quarter 2008.

Infiltration Systems	**Indicative Cost £**
Standard pavings, 150 mm thick sub-base (Not SUDS, given as comparison only)	40-50 (£/m^2)
Permeable pavings, 250 mm crushed rock sub-base, disposal of spoil on-site	45-55(£/m^2)
Permeable pavings, 350 mm crushed rock sub-base, disposal of spoil off-site	**60-65** (£/m^2)
Land drainage, perforated pipes in trenches, granular fill	35-55 (£/m)
Soakaways, crushed rock soakaway, cover and cover slab, geotextile membrane and granular surround, excavation and disposal of soil off-site	**175-315** (£/m^2)
Swales; excavation and formation of swale channel to fall, 1m wide by 750 average deep; geo-textile membrane; vegetation	85-115 (£/m)
Retention systems	
Geocellular storage system; 500 thick storage blocks, perforated land drains in granular surround; sub-base; impermeable membrane; excavation and disposal of spoil off site (surface costs excluded)	115-160 (£/m^2)
Vortex Control Valves: Supply cost of valve and chamber ring only	230 (£/nr)
Vortex Control Valves: Total cost of valve and inspection chamber	2300-3100 £/nr)
Control systems	
Bypass separator–approx 7,000 litre capacity	10,250-11,500 (£/nr)
Full retention separator-approx 7,000 litre capacity	17,000 (£/nr)

Sustainable Urban Drainage Systems

Several of the drainage systems listed in the table above qualify as part of a sustainable drainage solution (SUDS). The principle of SUDS is to mimic natural drainage as closely as possible on sites that have been developed. Fundamentally this is done by providing more porous surfaces and water retention areas, the latter often featuring as an amenity feature. The system can be used to reduce flow from hard impermeable surfaces and in so doing reduce flood risk and through the use of the permeable substrate or vegetation be used as a filtering mechanism for water

GREEN ROOFS

Green roofs are flat or gently pitched roofs overlaid with a growing medium and plants, creating a habitat on what would otherwise be a plain surface. They are sometimes called living roofs or vegetated roof structures. Green roofs are one of the most tangible elements of a sustainability strategy, being visible, natural and distinctive. In many occasions, green roofs are more expensive, subject the building to a heavier load and require regular maintenance. They are 'green' when 1) they are part of a sustainable drainage system collecting rainwater, 2) they contain plant types that form habitats for species, or 3) when there is a sufficient extent of vegetation to mitigate the effects of the 'urban heat island'.

A green roof system will typically comprise a multi-layer composition including insulation boarding, underlay, waterproof membrane, root protection layer, drainage layer, filter mat, growing medium and the vegetation itself. The final appearance of the roof will vary significantly by season and may require a degree of on-going irrigation if site specific factors dictate south facing slopes, high winds or low rainfall. Green roofs can vary considerably in terms of their function from sedum and brown roofs with a minimal amount of plant bearing substrate to those capable of supporting mature shrubs. The design needs to consider local biodiversity needs, and costs will vary significantly dependent on the chosen option. It should be noted that green roofs are compatible with renewable energy technology such as solar thermal or PV provided that this is considered during design.

The sandwich construction, for which there are a wide range of options, is designed to give the maximum water retention for the minimum amount of weight; this is an important consideration as the structure will have to be considered and up-rated to meet the additional load. A typical system will retain approximately 25 litres per m^2 of water which will make a significant contribution to the 20% minimum requirement in Planning Policy Statement 25 for reducing run-off rates.

In addition to providing an attractive, roof-level habitat, living roofs can also contribute to the management of surface water run-off, the mitigation of local air pollutants and the extension of the useful life of the roof membrane. Their contribution to the attenuation of rainwater runoff also contributes to the potential to downsize other "hard" drainage elements within a building.

The following costs are for green roofs with areas ranging from 100 to 1000 m^2.The rates include the specialist contractor's costs, but exclude allowances for main contractors' preliminaries and overheads and profit. Costs are current in April 2008 based on an average UK location. Rates are based on the surface area of the green roof.

Costs

Indicative costs:

Green Roofs	**Indicative Cost** £/m^2
Sedum blanket only	40-45
Sedum blanket with drainage layer and filter fleece	55-75
Sedum blanket on filter fleece and drainage layers, capping layer, and vapour barrier	90-130
Extra for insulation	60
Extra for waterproof membrane and vapour barrier	35-55
250 thick growing medium on drainage board, root membranes and insulation; turf	100-115
225 thick growing medium on filter fleece and LDPE drainage core; plug and hydroseed planting	60-70

OPERATIONAL AND EMBODIED ENERGY/CARBON

Operational Energy

Operational energy can be significant in certain building types, leading to high operational costs and high carbon emissions. Zero and low carbon technologies help reduce the carbon emissions by substituting carbon intensive sources of energy (electricity and gas) with clean energy (such as wind or solar). However priority should be given to reducing operational energy in the first instance, leading to immediate reduction in operational (and life cycle) cost and carbon emissions. Measure to reduce operational energy can centre around the design and orientation of the building, specification for insulation and air tightness, passive or naturally ventilated solutions, energy efficient lighting, heating, cooling and appliances. With operational energy reduced, any zero carbon technologies will then contribute a higher proportion of the energy load required.

Costs

The results of the following table, based on the individual contribution of each reduction measure gives an indication that significant carbon reductions can be made using existing technologies such as lighting control or static cooling, whilst retaining the ability to provide air conditioned space. The aggregate cost of these proposals, inclusive of contractor's on-costs, is likely to be no more than a 5% addition to overall capital costs.

At Davis Langdon we have undertaken extensive research which has been widely published in leading industry publications and journals. The sections on technologies and costs are based on articles previously published by DL in Building Magazine and Building Services Magazine. For further details on our research please see our website **<http://www.davislangdon.com>.**

Embodied Energy

Embodied energy is a significant part of the lifecycle impact of buildings. Every building is a complex combination of many processed materials, each of which contributes to the building's total embodied energy. Renovation and maintenance also add to the embodied energy over a building's life.

Embodied energy can be the equivalent of many years of operational energy. The single most important factor in reducing the impact of embodied energy is to design long life, durable and adaptable buildings. Typical life expectancy of residential buildings, particularly traditional detached, semi-detached and terraced property, is such that embodied energy plays a less significant part than in city centre commercial buildings which are subject to significant refurbishment cycles and have relatively short life spans. Ultimately, both embodied and operational energy need to be considered such that overall energy and carbon intensity can be minimised over the life of the building.

Embodied energy is also expressed as embodied carbon, which reflects the carbon and equivalent emissions resulting from energy use during the life of materials. The embodied carbon emissions of a building are from:

- The CO_2 and equivalent gases (methane, nitrous oxides, etc) produced during the manufacture of materials and products;
- Transport to and assembly on site;
- Maintenance and replacement; and
- Disposal and decomposition.

Factors to consider in reducing embodied energy in buildings
▪ Using less materials for a certain functional unit (e.g. minimum materials for a specific roof span)
▪ Using materials with inherently low embodied energy for a certain functional unit (e.g. wool insulation instead of foam insulation). The Green Guide to Specifications rates various building materials and components by environmental impact including embodied energy.
▪ Using recycled materials or material products with high recycled content (caution should be given towards recycled materials with long transport routes). Specific guides from WRAP are available listing the available products with high recycled content
▪ Minimising waste material on site and recycling as much as possible of the residual waste
▪ Sourcing building materials from local suppliers, reducing transport emissions
▪ Recycling demolition material at disposal
Avoiding or minimizing, where possible, highly manufactured components

Currently, there are no regulations or public policies that call for the reduction of embodied energy or carbon. New EU regulations which require the energy rating and labelling of buildings cover operational energy only. Hence the embodied energy is still low on the sustainability agenda. Recently various tools are becoming available in the construction industry for measuring and mitigating embodied energy, which is driving developers and project teams to utilise them for measuring the carbon impact of developments, although this does not necessarily lead to commitment to mitigation solutions.

IMPLEMENTATION – TAXATION AND GRANTS

Taxation

Many of the technologies discussed within this section attract 100% First Year Tax Allowances called Enhanced Capital Allowances (ECA's), deductible against taxable profit in the year of investment. Along with energy performance, increased tax allowances can be a significant driver for clients to include such technologies within their buildings.

Whilst Capital Allowances are not available to developers seeking to sell the building on completion, or to non-taxpayers, the inclusion of ECA compliant equipment will improve a building's asset rating under the Energy Performance of Buildings Directive discussed earlier in this section. For this reason alone, ECA compliant equipment is likely to become even more important to everybody with an interest in commercial property.

For further information on the Enhanced Capital Allowances scheme and a list of the current technologies attracting these improved tax allowances visit www.eca.gov.uk.

Grants

Regional Development Agencies, local authorities and government organisations, as well as various energy suppliers provide incentives in the form of grants to projects that help to achieve local, regional, national and EU priorities to reduce carbon emissions and reduce waste. Funding is often available to improve energy efficiency of buildings, improve waste management, encourage recycling and promote research and development into new and innovative technologies.

The Department for Business Enterprise and Regulatory Reform is providing grants of up to £1 million until March 2009 to public sector organisations under Phase 2 of the Low Carbon Buildings Programme. This programme awards grants of 50% of eligible costs towards purchasing and installing technologies such as biomass boilers, solar PV or ground source heat pumps. http://www.lowcarbonbuildingsphase2.org.uk/

Grant programmes are constantly changing, so it is advisable to contact a grants advisor at an early stage to see whether your project might be eligible for funding.

PART 2

Approximate Estimating

This part of the book contains the following sections:

Building Costs and Tender Prices

The tables which follow show the changes in building costs and tender prices since 1985. To avoid confusion it is essential that the terms "building costs" and "tender prices" are clearly defined and understood. "Building costs" are the costs incurred by the builder in the course of his business, the principal ones being those for labour and materials. "Tender Price" is the price for which a builder offers to erect a building.

Building costs

This table reflects the fluctuations since 1986 in wages and materials costs to the builder. In compiling the table, the proportion of labour to material has been assumed to be 40:60. The wages element has been assessed from a contract wages sheet revalued for each variation in labour costs, whilst the changes in the costs of materials have been based upon the indices prepared by the Department of Trade and Industry. No allowance has been made for changes in productivity, plus rates or hours worked which may occur in particular conditions and localities.

1976 = 100 (commencing from 1986)

Year	First quarter	Second Quarter	Third quarter	Fourth quarter	Annual Average
1986	256	258	266	267	262
1987	270	272	281	282	276
1988	284	286	299	302	293
1989	305	307	322	323	314
1990	326	329	346	347	337
1991	350	350	360	360	355
1992	361	362	367	368	365
1993	370	371	373	374	372
1994	376	379	385	388	382
1995	392	397	407	407	401
1996	407	408	414	414	411
1997	416	417	423	429	421
1998	430	431	448	447	439
1999	446	443	473	478	460
2000	480	482	497	498	489
2001	498	499	516	515	507
2002	516	522	553	554	536
2003	555	560	578	577	568
2004	579	586	617	618	600
2005	621	623	660	660	641
2006	665	670	694	699	682
2007	703	707	730	732	718
2008	733 (P)	736 (F)	767 (F)	771(F)	752 (F)
2009	774 (F)	776 (F)	802 (F)	805 (F)	789 (F)
2010	808 (F)	810 (F)	834 (F)	837 (F)	822 (F)
2011	842 (F)	846 (F)	870 (F)	872 (F)	857 (F)

Note: P = Provisional F = Forecast

Tender Prices

Tender prices are similar to "building costs" but also take into account market considerations such as the availability of labour and materials, and prevailing economic situation. This means that in "boom" periods, when there is a surfeit of building work to be done, "tender prices" may increase at a greater rate than "building costs", whilst in a period when work is scarce, "tender prices" may actually fall when "building costs" are rising.

This table reflects the changes in tender prices since 1986. It indicates the level of pricing contained in the lowest competitive tenders for new work in the Greater London area (over £3,500,000 in value).

1976 = 100 (commencing from 1986)

Year	First quarter	Second Quarter	Third quarter	Fourth quarter	Annual average
1986	221	226	234	234	229
1987	242	249	265	279	258
1988	289	299	321	328	309
1989	341	335	340	345	340
1990	320	315	312	290	309
1991	272	262	261	254	262
1992	250	248	241	233	243
1993	227	242	233	239	235
1994	239	247	266	256	252
1995	258	265	266	270	265
1996	265	262	270	270	267
1997	275	287	284	287	283
1998	305	312	318	318	313
1999	325	332	330	342	332
2000	348	353	362	375	359
2001	375	383	388	392	384
2002	398	405	423	421	412
2003	425	424	432	434	429
2004	434	436	442	454	442
2005	459	458	466	475	465
2006	480	485	494	506	491
2007	515	520	528	538	525
2008	544 (P)	552 (F)	560 (F)	567 (F)	556 (F)
2009	574 (F)	581 (F)	588 (F)	594 (F)	584 (F)
2010	600 (F)	607 (F)	617 (F)	624 (F)	612 (F)
2011	632 (F)	640 (F)	650 (F)	657 (F)	645 (F)

Note: P = Provisional F = Forecast

Tender prices for the Greater London area during 2007 rose by approximately 5.6% to the first quarter of 2008. With uncertainty about the extent of a UK economic slowdown, a declining housing market, liquidity problems and rising oil and commodity prices, the overall outlook for UK construction looks gloomier than a year ago. However, with a still relatively buoyant market in London, our current prediction is that tender inflation will rise by 4 to 7% over the next year, although other regions may experience lower rates of inflation.

Tender Prices – cont'd

Readers will be kept abreast of tender price movements in the free *Spon's Updates* and also in the *Tender Price Forecast* and *Cost Update* articles, published quarterly in *Building* magazine.

Davis Langdon Tender, Building Cost and Retail Prices Graph May 2008

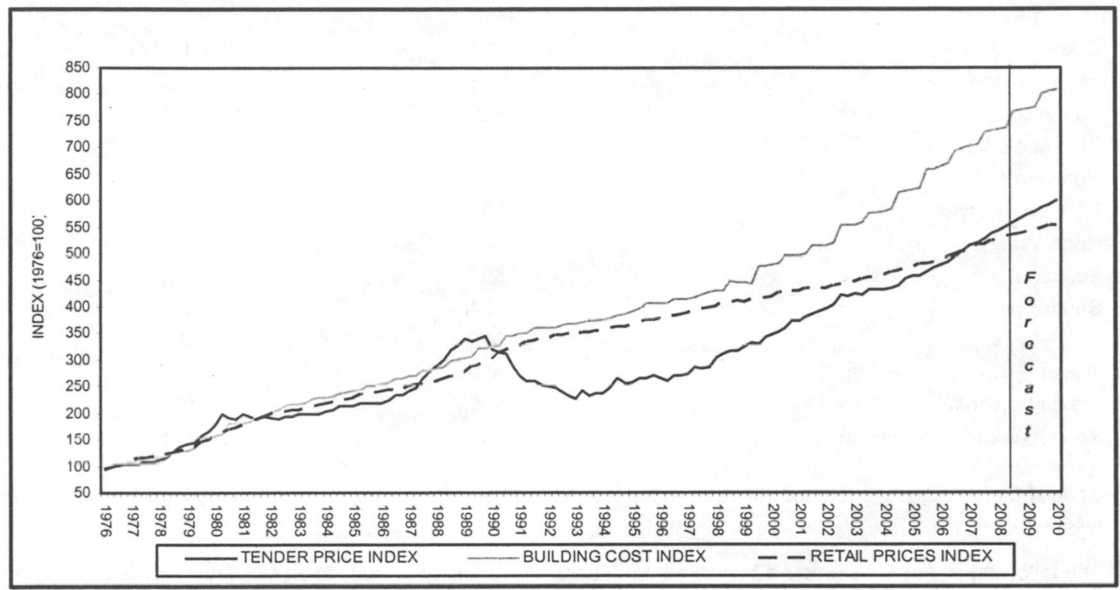

Regional Variations

As well as being aware of inflationary trends when preparing an estimate, it is also important to establish the appropriate price level for the project location.

Prices throughout this book approximately reflect price levels for the first quarter of 2009 in Outer London. Regional variations for certain inner London boroughs can be up to 14% higher while prices in Northern regions can be as much as 13% lower. Broad regional adjustment factors to assist with the preparation of initial estimates are shown in the table on the next page.

Over time, price differentials can change depending on regional workloads and local "hot spots". For the last few years, workloads and prices have been rising faster in regions away from London and South East, but it now appears that this trend has changed. Although London's resurgence in its office sector, may have abated, there is still the workload associated with the Olympics and the regeneration of the East London marshes. Elsewhere, the picture might not be so rosy. *Spon's Updates* and the *Tender Price Forecast* and *Cost Update* featured in *Building* magazine will keep readers informed of the latest regional developments and changes as they occur.

The regional variations shown in the table on the next page are Measured Work location factors based on our forecast of price differentials in each of the economic planning regions in the first quarter 2009. The table shows the forecast first quarter 2009 tender price index for each region plus the recommended percentage adjustments required to the Major Works section of the Prices for Measured Work. (Prices in the book are at a Tender Price Index level of 580 for Outer London).

Measured Work Location Factors

Region	Forecast first quarter 2009 tender price index	Percentage adjustment to *Major Works* section
Outer London	574	- 1.0
Inner London	626	+ 8.0
East Anglia	500	- 13.8
East Midlands	471	- 18.8
Northern	494	- 14.8
Northern Ireland	333	- 42.6
North West	465	- 19.8
Scotland	522	- 10.0
South East	534	- 7.9
South West	505	- 12.9
Wales	500	- 13.8
West Midlands	465	- 19.8
Yorkshire and Humberside	500	- 13.8

Special further adjustment to the above percentages may be necessary when considering city centre or very isolated locations.

The following example illustrates a Measured Work Location Factor adjustment to an estimate prepared using *Spon's* A&*B 2009*, to a price level that reflects the forecast Outer London market conditions for competitive tenders in the first quarter 2009:

Estimate for an Outer London project

		£
A	Value of items priced using Spon's A&B 2009 i.e. Tender Price Index 580	2,025,000
B	Adjustment to increase value of A to forecast price level for first quarter 2009 i.e. Forecast Tender Price Index 574 (574- 580) x 100 = say - 1.0% 580 deduct 1.0% say	- 21,000
		2,004,000
C	Value of items priced using competitive quotations that reflect the market conditions in the first quarter 2009	975,000
		2,979,000
D	Allowance for preliminaries +16% say	476,000
E	Total value of estimate at first quarter 2009 price levels	3,455,000

Alternatively, for a similar estimate for a project in Scotland:

				£
A	Value of items priced using Spon's A&B 2009 i.e. Tender Price Index 580			2,025,000
B	Adjustment to reduce value of A to forecast price level for first quarter 2009 for Scotland (from regional variation table) i.e. Tender Price Index 520 $\frac{(522 - 580)}{580} \times 100 = -10.0\%$	deduct 10.0%	say	- 202,500
				1,822,500
C	Value of items priced using competitive quotations that reflect the market conditions in the first quarter 2009			975,000
				2,797,500
D	Allowance for preliminaries	+16%	say	447,500
E	Total value of estimate at first quarter 2009 price levels			3,245,000

Total Building Cost Location Factors

Total Building Cost Location Factors apply to overall £ per square metre rates, including preliminaries etc.

The following example illustrates the use of the map on page 64 to adjust a scheme for alternative locations in the East Midlands and Scotland.

Light industrial unit located in South East/Outer London region

				£ per Square Metre
A	Original cost of project		say	735.00
B	Cost for project in East Midlands	£735/m² x $\frac{0.87}{1.00}$	say	640.00
C	Cost for project in Scotland	£735/m² x $\frac{0.94}{1.00}$	say	690.00

Scotland 0.94
Northern Ireland 0.72
Northern 0.90
Yorkshire & Humberside 0.90
North West 0.87
East Midlands 0.87
East Anglia 0.91
West Midlands 0.88
Wales 0.90
Southeast 0.95 (excl Lond)
Southwest 0.91
Inner London 1.06
Outer London 1.00
Spons Architects` and Builders` Price Book 2008 Edition
Total Building Cost Location Factors

Building Prices per Functional Unit

Prices given under this heading are average prices, on a *fluctuating basis*, for typical buildings based on a tender price level index of 580(1976 = 100). Prices includes for Preliminaries at 16%, and Overheads and Profit. Unless otherwise stated, prices do not allow for external works, furniture, loose or special equipment and are, of course, exclusive of fees for professional services.

On certain types of buildings there exists a close relationship between its cost and the number of functional units that it accommodates. During the early stages of a project therefore an approximate estimate can be derived by multiplying the proposed unit of accommodation (i.e. hotel bedrooms, car parking spaces etc.) by an appropriate cost.

The following indicative unit areas and costs have been derived from historic data. It is emphasized that the prices must be treated with reserve, as they represent the average of prices from our records and cannot provide more than a rough guide to the cost of a building. There are limitations when using this method of estimating, for example, the functional areas and costs of football stadia are strongly influenced by the extent of front and back of house facilities housed within it, and these areas can vary considerably from scheme to scheme.

The areas may also be used as a "rule of thumb" in order to check on economy of designs. Where we have chosen not to show indicative areas, this is because either ranges are extensive or such figures may be misleading.

Costs have been expressed within a range, although this is not to suggest that figures outside this range will not be encountered, but simply that the calibre of such a type of building can itself vary significantly.

For assistance with the compilation of a closer estimate, or of a Cost Plan, the reader is directed to the *"Building Prices per Square Metre"*, *"Approximate Estimates"* and "*Cost Models*" sections. As elsewhere in this edition, prices do not include V.A.T.

Function		Indicative functional unit area	Indicative functional cost
Utilities, civil engineering facilities (Uniclass D1)			
Car Parking	▪ surface level	20 to 22 m²/car	£1,050 to £1,850 /car
	▪ ground level (under buildings)	22 to 24 m²/car	£2,400 to £4,050 /car
	▪ multi storey	22 to 32 m²/car	£5,300 to £19,000 /car
	▪ semi basement	27 to 30 m²/car	£13,250 to £17,250 /car
	▪ basement	28 to 37 m²/car	£24,000 to £38,000 /car
Administrative, commercial protective service facilities (Uniclass D3)			
Office – air conditioned	▪ low density cellular	15 to 20 m²/person	£25,000 to £30,000 /person
	▪ high density open plan	10 to 15 m²/person	£20,500 to £36,000 /person
Health and welfare facilities (Uniclass D4)			
Hospitals	▪ district general	65 to 85 m²/bed	£95,000 to £170,000 /bed
	▪ teaching	120 + m²/bed	£140,000 to £190,000 /bed
	▪ private	75 to 100 m²/bed	£110,000 to £220,000 /bed
Nursing Homes	▪ residential home	40 to 60 m²/bedroom	£36,000 to £85,000 /bedroom
	▪ nursing homes	40 to 80 m²/bedroom	£49,000 to £130,000 /bedroom

Function		Indicative functional unit area	Indicative functional cost
Recreational facilities (Uniclass 5)			
Football Stadia	▪ basic stand	-	£800 to £900 /seat
	▪ stand plus basic facilities	-	£1,000 to £1,400 /seat
	▪ stand plus extensive facilities	-	£1,350 to £1,750 /seat
	▪ national stadia plus extensive facilities	-	£3,450 to £5,700 /seat
Theatres	▪ theatre refurbishment	-	£9,800 to £18,750 /seat
	▪ workshop (fewer than 500 seats)	-	£10,100 to £16,000 /seat
	▪ more than 500 seats	-	£23,000 to £34,000 /seat
Educational, scientific, information facilities (Uniclass 7)			
Schools	▪ nursery	3 to 5 m²/child	£3,700 to £9,000 /child
	▪ secondary	6 to 10 m²/child	£10,100 to £22,000 /child
	▪ to boarding	10 to 12 m²/child	£11,500 to £20,750 /child
	▪ to special	18 to 20 m²/child	£18,000 to £29,000 /child
Residential facilities (Uniclass 8)			
Housing (private developer)	▪ terraced; two bedroom	55 to 65 m²/gifa	£38,000 to £67,000 /house
	▪ semi-detached; three bedroom	70 to 90 m²/gifa	£56,000 to £99,000 /house
	▪ detached; four bedroom	90 to 100 m²/gifa	£91,000 to £160,000 /house
	▪ low rise flats; two bedroom	55 to 65 m²/gifa	£49,000 to £74,000 /flat
	▪ medium rise flat; two room	55 to 65 m²/gifa	£54,000 to £81,000 /flat
Hotels	▪ luxury city-centre hotel, multi-storey, conference and wet leisure facilities	70 to 120 m²/bedroom	£160,000 to £380,000 /bedroom
	▪ business town centre provincial hotel four to six storeys, conference and wet leisure facilities	70 to 100 m²/bedroom	£101,000 to £230,000 /bedroom
	▪ mid range provincial hotel two to three storeys, conference and leisure facilities	50 to 60 m²/bedroom	£81,000 to £140,000 /bedroom
	▪ city centre aparthotel four to seven storeys, apartments with self-catering facilities	50 to 60 m²/bedroom	£68,000 to £140,000 /bedroom
	▪ budget city-centre hotel four to six storeys, dining bar and facilities	35 to 45 m²/bedroom	£33,000 to £70,000 /bedroom
	▪ mid-range provincial hotel two to three storeys, bedroom extension	33 to 40 m²/bedroom	£53,000 to £81,000 /bedroom
	▪ two to three storey lodge, excluding dining facilities	28 to 35 m²/bedroom	£37,000 to £58,000 /bedroom
	▪ budget roadside hotel	28 to 35 m²/bedroom	£32,000 to £47,000 /bedroom

Function		Indicative functional unit area	Indicative functional cost
Hotel furniture fittings and equipment	▪ budget hotel	-	£4,600 to £9,200 /bedroom
	▪ mid-range hotel	-	£16,250 to £24,000 /bedroom
	▪ luxury hotel	-	£38,000 to £92,000 /bedroom
Students Residences	▪ large turnkey budget schemes (200 + units), simple design, open site; en suite accommodation	18 to 20 m²/bedroom	£19,000 to £32,000 /bedroom
	▪ smaller schemes (40 to 100 units) with mid range specifications, some with en suite bathroom and kitchen facilities	19 to 24 m²/bedroom	£26,000 to £47,000 /bedroom
	▪ smaller high quality courtyard schemes of collegiate style in restricted city centre sites	24 to 28 m²/bedroom	£42,000 to £79,000 /bedroom

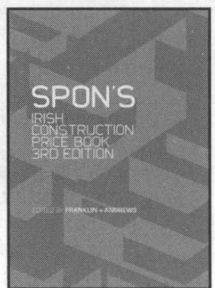

Building Prices per Square Metre

Prices given under this heading are average prices, on a *fluctuating basis*, for typical buildings based on a tender price level index of 580 (1976 = 100). Prices allow for Preliminaries at 16%, and Overheads and Profit. Unless otherwise stated, prices do not allow for external works, furniture, loose or special equipment and are, of course, exclusive of fees for professional services.

Prices are based upon the total floor area of all storeys, measured between external walls and without deduction for internal walls, columns, stairwells, lift wells and the like.

As in previous editions it is emphasized that the prices must be treated with reserve, as they represent the average of prices from our records and cannot provide more than a rough guide to the cost of a building.

In many instances normal commercial pressures together with a limited range of available specifications ensure that a single rate is sufficient to indicate the prevailing average price. However, where such restrictions do not apply a range has been given; this is not to suggest that figures outside this range will not be encountered, but simply that the calibre of such a type of building can itself vary significantly.

For assistance with the compilation of a closer estimate, or of a Cost Plan, the reader is directed to the *"Approximate Estimates"* and "*Cost Models*" sections. As elsewhere in this edition, prices do not include VAT.

	£ per square metre excluding VAT
Utilities, civil engineering facilities (Uniclass D1)	
Surface car parking	55 to 85
Surface car parking; landscaped	80 to 130
Multi storey car parks	
grade & upper level	220 to 300
flat slab	360 to 600
Underground car parks	
partially underground under buildings; naturally ventilated	490 to 580
completely underground under buildings	840 to 1,020
completely underground with landscaped roof	1,030 to 1,220
Railway stations	1,800 to 2,990
Bus and coach stations	840 to 1,800
Bus garages	980 to 1,080
Petrol stations	1,330 to 2,580
Vehicle showrooms with workshops, garages etc	
upto 2,000m²	850 to 1,150
over 2,000m²	800 to 1,030
Vehicle showrooms without workshops, garages etc	
upto 2,000m²	1,010 to 1,120
Vehicle repair and maintenance buildings	
upto 500m²	1,350 to 1,700
over 500m² upto 2,000m²	815 to 1,250
Car wash buildings	690 to 1,110
Garages, domestic	540 to 1,090
Airport facilities (excluding aprons)	
airport terminals	1,850 to 4,050
airport piers/satellites	2,260 to 5,050

	£ per Square metre excluding VAT
Utilities, civil engineering facilities (Uniclass D1) - cont'd	
Airport campus facilities	
cargo handling bases	660 to 1,100
distribution centres	340 to 660
hangars (type C and D aircraft)	1,420 to 1,680
hangars (type E aircraft)	1,690 to 4,210
TV, radio and video studios	1,280 to 2,530
Telephone exchanges	1,020 to 1,560
Telephone engineering centres	850 to 960
Branch Post Offices	1,080 to 1,460
Postal Delivery Offices/Sorting Offices	1,090 to 1,490
Mortuaries	1,950 to 2,710
Substations	1,460 to 2,190
Industrial facilities (Uniclass D2)	
B1 Light industrial/offices buildings	
economical shell, and core with heating only	590 to 1,020
medium shell and core with heating and ventilation	910 to 1,350
high quality shell and core with air conditioning	1,200 to 2,190
developers Category A fit out	490 to 830
tenants Category B fit out	220 to 640
Agricultural storage buildings	520 to 880
Factories	
for letting (incoming services only)	360 to 520
for letting (including lighting, power and heating)	490 to 650
nursery units (including lighting, power and heating)	590 to 880
workshops	560 to 1,090
maintenance/motor transport workshops	660 to 1,150
owner occupation for light industrial use	690 to 880
owner occupation for heavy industrial use	1,140 to 1,330
Factory/office buildings high technology production	
for letting (shell and core only)	640 to 880
for owner occupation (controlled environment, fully finished)	1,380 to 1,850
Laboratory workshops and offices	1,240 to 1,570
High technology laboratory workshop centres, air conditioned	2,910 to 3,720
Warehouse and Distribution centres	
high bay (10-15 m high) for owner occupation (no heating) up to 10,000m²	330 to 410
high bay (10-15 m high) for owner occupation (no heating) 10,000m² to 20,000m²	240 to 330
high bay (16-24 m high) for owner occupation (no heating) over 10,000m² to 20,000m²	350 to 460
high bay (16-24 m high) for owner occupation (no heating) over 20,000m²	260 to 410
Fit out cold stores, refrigerated stores inside warehouse	510 to 950
Industrial Buildings	
Shell with heating to office areas only	
500 - 1,000m²	360 to 840
1,000 - 2,000m²	270 to 750
greater than 2,000m²	360 to 750
Unit including services to production area	
500 - 1,000m²	640 to 1,000
1,000 - 2,000m²	580 to 930
greater than 2,000m²	580 to 930

	£ per Square metre excluding VAT
Administrative, commercial, protective services, facilities (Uniclass D3)	
Embassies	2,010 to 2,910
County Courts	1,790 to 2,230
Magistrates Courts	1,360 to 1,730
Civic offices	
non air conditioned	1,360 to 1,730
fully air conditioned	1,720 to 2,010
Probation/Registrar Offices	990 to 1,420
Offices for letting	
low rise, air conditioned, high quality speculative	1,380 to 1,740
medium rise, air conditioned, high quality speculative, 8-20 storeys	1,690 to 2,310
city fringe deep-plan speculative office tower, air-conditioned	2,190 to 2,660
Offices for owner occupation	
low rise, air conditioned	1,500 to 1,850
medium rise, air conditioned	2,050 to 2,400
high rise, air conditioned	2,310 to 2,890
Offices - City and West End	
high quality, speculative 8-20 storeys, air-conditioned	2,010 to 2,540
high rise, air conditioned, iconic speculative towers	2,660 to 3,570
Business park offices	
functional non air conditioned less than 2,000 m²	840 to 1,090
functional non air conditioned more than 2,000 m²	720 to 950
medium quality non air conditioned less than 2,000 m²	950 to 1,150
medium quality non air conditioned more than 2000 m²	900 to 1,100
medium quality air conditioned less than 2,000m²	1,050 to 1,250
medium quality air conditioned more than 2,000m²	1,000 to 1,200
good quality - naturally ventilated to meet BCO specification (exposed soffits, solar shading) less than 2,000m²	1,140 to 1,250
good quality - naturally ventilated to meet BCO specification (exposed soffits, solar shading) more than 2,000 m²	1,090 to 1,200
high quality air conditioned less than 2,000 m²	1,200 to 1,450
high quality air conditioned more than 2,000 m²	1,150 to 1,400
Large trading floors in medium rise offices	3,180 to 3,810
Two storey ancillary office accommodation to warehouses/factories	1,030 to 1,380
Fitting out offices (nla)	
City/West End	
basic fitting out including carpets, decorations, partitions and services	380 to 400
good quality fitting out including carpets, decorations, partitions and services	490 to 630
and carpets, decorations, partitions, ceilings, furniture, air conditioning and	730 to 1,090
Out-of-town (South East)	
basic fitting out including carpets, decorations, partitions and services	280 to 380
good quality fitting out including carpets, decorations, partitions and services	320 to 460
high quality fitting out including raised floors and carpets, decorations, partitions, ceilings, furniture, air conditioning and electrical services	650 to 1,090
Meeting areas	780 to 1,090
Reception area	1,140 to 1,440

	£ per Square metre excluding VAT
Administrative, commercial, protective services, facilities (Uniclass D3) - cont'd	
Conference suites - City/West End (nia)	1,330 to 1,800
Conference suites - Out-of-town (nia South East)	1,200 to 1,800
Sub equipment room - City/West End (nia)	1,800 to 2,310
Sub equipment room - Out-of-town (nia South East)	1,440 to 2,050
Back of house / storage - City/West End (nia)	630 to 780
Back of house / storage - Out-of-town (nia South East)	340 to 490
Kitchen - City/West End (nia)	2,530 to 3,110
Kitchen - Out-of-town (nia South East)	2,400 to 3,110
Restaurants - City/West End (nia)	1,380 to 2,340
Restaurants - Out-of-town (nia South East)	1,330 to 2,340
Office refurbishment (including developers finish - gifa; central London)	
minor refurbishment	240 to 860
medium refurbishment	840 to 1,330
major refurbishment	1,330 to 1,930
Banks	
local	1,590 to 1,980
city centre/head office	2,280 to 2,930
Building Society Branch Offices	1,440 to 1,880
refurbishment	810 to 1,400
Shop shells	
small	690 to 880
large including department stores and supermarkets	590 to 830
Fitting out shell for small shop (including shop fittings)	
simple store	660 to 810
fashion store	1,270 to 1,560
Fitting out shell for department store or supermarket	1,200 to 2,650
Retail Warehouses	
shell	440 to 630
fitting out, including all display and refrigeration units, check outs and IT systems	300 to 350
Supermarkets	
shell	350 to 700
supermarket fit-out	930 to 1,270
hypermarket fit-out	780 to 1,020
Shopping centres	
malls including fitting out	
comfort cooled	3,240 to 4,990
air conditioned	4,210 to 6,160
food court	4,210 to 6,050
factory outlet centre mall - enclosed	2,830 to 4,570
factory outlet centre mall - open	520 to 960
anchor tenants, capped off services	780 to 1,140
medium/small store units, capped off services	730 to 1,020
centre management	2,050 to 2,530
enclosed surface level service yard	1,560 to 1,930
landlords back of house and service corridors	1,560 to 1,930
Refurbishment	
mall; limited scope	1,050 to 1,620
mall; comprehensive	1,500 to 2,290

	£ per Square metre excluding VAT
Medical health and welfare facilities (Uniclass D4)	
Ambulance stations	910 to 1,360
Ambulance controls centre	1,270 to 2,340
Fire stations	1,370 to 1,920
Police stations	1,380 to 1,980
Prisons	1,560 to 2,290
District hospitals	1,460 to 1,980
refurbishment	660 to 1,370
Hospice	1,540 to 1,890
Private hospitals	1,420 to 2,170
Pharmacies	1,440 to 1,800
Hospital laboratories	1,800 to 2,630
Ward blocks	1,370 to 1,720
refurbishment	630 to 1,020
Geriatric units	1,390 to 1,890
Psychiatric units	1,390 to 1,740
Psycho-geriatric units	1,330 to 1,940
Maternity units	1,390 to 1,890
Operating theatres	1,490 to 2,320
Outpatients/casualty units	1,520 to 2,010
Hospital teaching centres	1,150 to 1,610
Health centres	1,380 to 1,440
Welfare centres	1,460 to 1,730
Day centres	1,250 to 1,730
Group practice surgeries	1,120 to 1,400
Homes for the physically handicapped - houses	1,430 to 1,500
Homes for the mentally handicapped	1,150 to 1,610
Geriatric day hospital	1,250 to 1,620
Accommodation for the elderly	
residential homes	890 to 1,420
nursing homes	1,230 to 1,690
Children's homes	1,030 to 1,620
Homes for the aged	1,080 to 1,430
refurbishment	410 to 1,010
Observation and assessment units	850 to 1,630
Primary Health Care	
doctors surgery - basic	1,100 to 1,480
doctors surgery / medical centre	1,420 to 1,800
Hospitals	
diagnostic and treatment centres	2,770 to 3,290
acute services hospitals	2,430 to 3,460
radiotherapy and oncology units	2,540 to 3,570
community hospitals	2,190 to 2,770
trauma unit	2,310 to 2,590

	£ per Square metre excluding VAT
Recreational facilities (Uniclass D5)	
Public houses	1,220 to 1,690
Dining blocks and canteens in shop and factory	1,200 to 1,690
Restaurants	1,420 to 2,010
Community centres	1,010 to 1,440
General purpose halls	1,010 to 1,560
Visitors' centres	1,360 to 2,340
Youth clubs	950 to 1,400
Arts and drama centres	1,400 to 1,590
Galleries	
refurbishment of historic building to create international standard gallery	3,860 to 6,380
international-standard art gallery	3,210 to 4,400
national-standard art gallery	2,590 to 3,210
independent commercial art gallery	1,390 to 1,800
Arts and drama centres	1,370 to 1,590
Theatres, including seating and stage equipment	
large over 500 seats	4,040 to 5,190
studio/workshop less than 500 seats	2,890 to 4,040
refurbishment	1,730 to 2,890
Concert halls, including seating and stage equipment	2,530 to 4,120
Cinema	
shell	790 to 1,030
multiplex; shell only	1,440 to 2,650
fitting out including all equipment, air-conditioned	860 to 1,590
Exhibition centres	1,570 to 2,100
Swimming pools	
international standard	3,230 to 4,040
local authority standard	1,850 to 2,890
school standard	1,270 to 1,420
leisure pools, including wave making equipment	3,040 to 3,860
Ice rinks	1,400 to 1,670
Rifle ranges	1,110 to 1,390
Leisure centres	
dry	1,380 to 1,960
extension to hotels (shell and fit-out - including pool)	2,100 to 2,940
wet and dry	2,080 to 2,890
Sports halls including changing	790 to 1,400
School gymnasiums	950 to 1,120
Squash courts	950 to 1,400
Indoor bowls halls	590 to 1,130
Bowls pavilions	950 to 1,150
Health and fitness clubs	
out-of-town (shell and fit-out - including pool)	1,330 to 2,170
town centre (fit-out - excluding pool)	
Sports pavilions	990 to 1,520
changing only	1,200 to 1,620
social and changing	1,010 to 1,740
Clubhouses	1,010 to 1,370
Golf clubhouses	960 to 1,690

	£ per Square metre excluding VAT
Religious facilities (Uniclass D6)	
Temples, mosques, synagogues	1,470 to 1,740
Churches	1,220 to 1,750
Mission halls, meeting houses	1,430 to 1,890
Convents	1,340 to 2,070
Crematoria	1,790 to 2,200
Educational, scientific, information facilities (Uniclass D7)	
Nursery Schools	1,220 to 1,800
Primary/junior school	1,440 to 1,930
Secondary/middle schools	1,690 to 2,170
Secondary school and further education college buildings	
classrooms	1,090 to 1,380
laboratories	1,110 to 1,680
craft design and technology	1,100 to 1,680
music	1,020 to 2,100
Extensions to schools	
classrooms	1,200 to 1,690
laboratories	1,520 to 1,670
Sixth form colleges	1,140 to 1,560
Special schools	1,010 to 1,440
Training colleges	980 to 1,470
Management training centres	1,360 to 1,850
Universities	
arts buildings	1,220 to 1,520
science buildings	1,360 to 1,930
College/University Libraries	1,140 to 1,620
Laboratories and offices, low level servicing	1,330 to 2,170
Laboratories (specialist, controlled environment)	1,890 to 3,520
Computer buildings	1,610 to 2,410
Museums and art galleries	
national standard museum	2,990 to 4,210
national standard independent specialist museum (excluding fit-out)	2,190 to 2,700
regional including full air conditioning	2,400 to 3,610
local including full air conditioning	1,800 to 2,650
conversion of existing warehouse to regional standard museum	1,330 to 2,010
conversion of existing warehouse to local standard museum	1,140 to 1,720
Learning resource centre	
economical	1,100 to 1,680
high quality	1,480 to 2,050
Libraries	
branch	1,140 to 1,540
city centre	1,690 to 2,250
collegiate; including fittings	2,650 to 3,360
Conference centres	1,920 to 2,600

	£ per Square metre excluding VAT
Residential facilities (Uniclass D8)	
Local authority and housing association schemes	
Bungalows	
semi detached	900 to 1,080
terraced	770 to 970
Two storey housing	
detached	840 to 1,020
semi detached	780 to 960
terraced	640 to 900
Three storey housing	
semi detached	790 to 1,100
terraced	700 to 1,030
Apartments/Flats	
low rise	900 to 1,140
medium rise	980 to 1,250
Sheltered housing with wardens' accommodation	900 to 1,310
Private developments	
single detached houses	1,020 to 1,590
houses two or three storey	900 to 2,230
high-quality apartments in residential tower - Inner London	2,650 to 2,890
high-quality multi-storey apartments - Inner London	2,050 to 2,340
mid-market apartments in residential tower - Outer London	2,050 to 2,470
prestige-quality apartments multi-storey - Inner London	2,660 to 2,890
development - Inner London	
three-to-four storey villa - Inner London	1,610 to 1,850
residential tower - outer London	2,080 to 2,430
Apartments/Flats - generally	
standard quality, 3-5 storeys	1,030 to 1,380
warehouse conversions to apartments	1,150 to 1,730
Rehabilitation	
Housing	400 to 700
Flats	630 to 980
Hotels (including fittings, furniture and equipment)	
luxury city-centre hotel - multi-storey conference and wet leisure facilities	2,310 to 3,170
business town centre/provincial hotel - 4 to 6 storeys, conference and wet leisure facilities	1,690 to 2,540
mid range provincial hotel - 2 to 3 storeys, conference and leisure facilities	1,610 to 2,100
budget city-centre hotel - 4 to 6 storeys, dining and bar facilities	1,330 to 1,560
budget roadside hotel - 2 to 3 storey lodge, excluding dining facilities	1,090 to 1,500
Hotel accommodation facilities (excluding fittings, furniture and equipment)	
bedroom areas	910 to 1,320
front of house and reception	1,210 to 1,600
restaurant areas	1,350 to 2,010
bar areas	1,180 to 1,850
function rooms/conference facilities	1,010 to 1,850
Students' residences	
large budget schemes with en-suite accommodation	1,150 to 1,730
smaller schemes (40 - 100 units) with mid-range specifications	1,500 to 1,960
smaller high-quality courtyard schemes, college style	1,960 to 2,840

Building Cost Models

Davis Langdon have been producing cost models for publication in *Building* magazine since 1993.

During this 16 year period , over seventy models have been published examining most building types as well as providing detailed coverage of broader issues including sustainability, infrastructure and off site manufacturing.

Trends continue to change. Sustainability, mixed-use developments and the increasing size and complexity of schemes have become evident over the past two years

Although the scope and coverage of the cost models has expanded considerably, the objectives remain constant.

They are:

- To provide detailed elemental cost information derived from a generic building that can be applied to other projects
- To provide a commentary on cost drivers and other design and specification issues
- To compare suitable procurement routes that secure the clients objectives

For this edition of Spon's Davis Langdon have published updated elemental cost data for 12 of the most common building types. All models have been updated to reflect 1st Quarter 2009 tendered rates (tender index 580) Locations do vary for each model, so please make a note of the location and location factor for any adjustments that may need to be made.

Readers may refer to the *Approximate Estimates* section of this book to make adjustments to the models for alternative locations and specifications

DISTRIBUTION CENTRE COST MODEL

This cost model features a detailed cost breakdown of a new build high bay distribution centre with a 15m haunch height. The costs are based on a generic solution with a gross internal floor area of 70,000m², which includes 5% office and ancillary accommodation (3,500m²). Costs of enhancements including the warehouse and office area fit-out and ancillary buildings, together with costs of external works are detailed. Costs of racking and materials handling installations are excluded. The model has been prepared on the assumption that ground conditions are good and that minimal site preparation is required.

- Warehouse: Gross internal floor area 71,700m²
- Office Shell: Net internal floor area 3,500m²
- Warehouse: Net internal floor area 70,000m²

Model location is based on UK average (TPI = 580, LF = 0.93)

This updated cost model is copyright of Davis Langdon LLP, originally published in Building Magazine on 06-Aug-04

Warehouse Shell	Quantity	Unit	Rate	Total (£)	Cost (£/m²)	%
Substructure				**4,181,000**	**58.31**	**26.3%**
· 225mm reinforced concrete ground slab; laser levelled; surface hardener; subbase; perimeter ground beam; lift pits	70,000	m²	50			
· Insitu concrete pad foundations and ground beams	260	nr	1,600			
· Allowance for foundations and retaining walls to dock levellers	100	nr	2,650			
Frame				**3,850,000**	**53.70**	**24.2%**
· Steel propped portal frame, cold rolled purlin sections, surface treatment, including decorations	70,000	m²	55			
Roof				**3,704,000**	**51.66**	**23.3%**
· Composite roof panels; powder coated galvanised steel	70,000	m²	33			
· Extra over for 10% rooflights	7,000	m²	39			
· Roof drainage generally; syphonic system	70,000	m²	8			
· Eaves/valley gutter; galvanised steel; insulation; stop ends	3,100	m²	150			
· Allowance for mansafe system, hatches and access ladders		item	110,000			
External Walls, Windows and Doors				**930,600**	**12.98**	**5.9%**
· Wall cladding system, composite panels and built-up cladding systems; mineral fibre insulation; polyester powder coating	17,750	m²	46			
· Allowance for personnel escape doors	25	m²	975			
· Cladding and details to inside face of parapet walls	1,150	m²	55			
· Level access doors; insulated sectional overhead dock doors	10	nr	2,650			
Dock Leveller Installations				**1,224,600**	**17.08**	**7.7%**
· Insulated sectional overhead dock doors	100	nr	1,900			
· Dock leveller; precast concrete dock pits; wheel guides	100	nr	7,500			
· Dock shelter; heavy duty scissors retracting frame	100	nr	2,150			
· Protection, bollards to door tracks; heavy duty rubber dock buffers		item	42,600			
· Traffic control lights	100	nr	270			
Services Installations				**288,100**	**4.02**	**1.8%**
· Water installations; hot and cold water services		item	44,700			
· Mechanical installations; gas and water connections		item	25,600			
· Electrical installation; general sub-mains and distribution		item	98,000			
· Electrical installation; to dock levellers and access doors		item	110,000			
· Allowance for lightning protection		item	9,800			
Preliminaries and Contingency				**1,701,100**	**23.73**	**10.7%**
· Overheads and profit, site establishment and supervision @		10%				
· Contingency @		2%				
Construction cost (Warehouse shell only, rate based on GIFA)				**15,880,000**	**221.48**	**100%**

DISTRIBUTION CENTRE COST MODEL

Office Shell and Fit-Out	Quantity	Unit	Rate	Total (£)	Cost (£/m²)	%
Substructure				**87,500**	**25.00**	**2.4%**
• Extra for foundations to offices; insitu concrete pad foundations; ground beams; lift pit (upper floor footprint)	1,750	m²	50			
Frame, Upper Floors and Stairs				**419,700**	**119.91**	**11.5%**
• Steel frame, universal sections; surface treatment, fire protection and decoration	3,500	m²	60			
• Upper floors; 200mm thick precast concrete plank and structural screed	1,750	m²	65			
• Allowance for fire stopping to perimeter		item	21,300			
• Precast concrete stairs, mild steel balustrades and handrails; polyester powder coated	2	nr	37,300			
External Walls, Windows and Doors				**497,300**	**142.09**	**13.7%**
• Extra over wall cladding for double glazed ribbon windows	750	m²	460			
• Extra over wall cladding for louvres		item	92,000			
• Allowance for glazed screen		item	33,000			
• Glazed entrance doors; to match glazed screen (per leaf)	14	nr	1,950			
Internal Partitions and Doors				**58,100**	**16.60**	**1.6%**
• 140mm thick blockwork; head restraint; fire stopping	700	m²	70			
• Doors; ironmongery (cost per leaf)	14	nr	650			
Finishes				**332,800**	**95.09**	**9.1%**
• Allowance for wall finishes generally, emulsion paint and ceramic tile (allowance based on floor area)	3,500	m²	14			
• Raised floor to office areas only; 150 cavity; fire barriers	3,200	m²	39			
• Ceramic tiles to reception and WCs	650	m²	60			
• Vinyl sheet and skirtings to corridor areas	750	m²	46			
• Suspended ceiling; mineral fibre tile in exposed lay in grid	3,500	m²	23			
• Extra for moisture resistant tiles	300	m²	13			
Fittings				**743,800**	**212.51**	**20.4%**
• Allowance for open plan office fit-out to category B	3,500	m²	190			
• Allowance for kitchen fittings		item	13,000			
• Allowance for reception fittings and features		item	52,200			
• Allowance for matwells and frames		item	7,100			
• Allowance for WC fittings		item	6,500			
Services Installations				**1,110,100**	**317.17**	**30.5%**
• Sanitary fittings generally	75	nr	520			
• Hot and cold water services. Disposal installations		item	72,000			
• Low temperature hot water heating	3,500	m²	33			
• Mechanical ventilation and comfort cooling	3,500	m²	140			
• Allowance for toilet and Lift Motor Room ventilation		item	33,000			
• Gas and Electrical installation		item	120,000			
• Lighting and emergency lighting and small power	3,500	m²	46			
• Lift installation; 8 person electro-hydraulic		item	26,600			
• Allowance for builder's work in connection say		5%				
Preliminaries and Contingency				**392,200**	**112.06**	**10.8%**
• Overheads and profit, site establishment and supervision @		10%				
• Contingency @		2%				
Construction cost (Office shell and fit-out only, rate based on Office NIA)				**3,642,000**	**1,040.43**	**100%**

DISTRIBUTION CENTRE COST MODEL

Warehouse Fit-Out	Quantity	Unit	Rate	Total (£)	Cost (£/m²)	%
Fixtures and Fittings				**417,000**	**5.82**	**5.4%**
• Allowance for general fixtures and fittings		item	150,000			
• Protection; secondary steelwork, armco barriers, bollards		item	104,000			
• Allowance for internal and external signage		item	59,000			
• Allowance for jockey wheel strips and wheel stops		item	104,000			
Services and Communication Installations				**6,251,500**	**87.19**	**81.5%**
• Gas fired heating to warehouse areas; high level nozzle	68,200	m²	11			
• Roof smoke ventilation system	68,200	m²	8			
• Increased incoming power supply		item	270,000			
• Mains power to mechanical installations; high level lighting	68,200	m²	37			
• Electrical installations; standby generator		item	320,000			
• Roof level sprinklers and storage tanks (category 3)	68,200	m²	16			
• Communications; fire detection and alarm; CCTV; PA	68,200	m²	9			
• Allowance for builder's work in connection		item	110,000			
Preliminaries and Contingency				**999,900**	**13.95**	**13.0%**
• Overheads and profit, site establishment and supervision @		10%				
• Contingency @		5%				
Construction cost (Warehouse fit-out only, rate based on GIFA)				**7,669,000**	**106.96**	**100%**

External Works	Quantity	Unit	Rate	Total (£)	Cost (£/m²)	%
Site Works				**5,335,300**	**74.41**	**64.2%**
▪ Allowance for site preparation	175,000	m²	6			
• Excavation to form ramp to dock levellers		item	65,000			
• Heavy duty access road and service yard	41,700	m²	43			
• Extra for ramped vehicle access		item	65,000			
• Car parking; tarmacadam on subbase	29,500	m²	34			
• Paved areas for pedestrian and maintenance access	3,800	m²	33			
• Allowance for soft landscaping, including reuse of topsoil	30,000	m²	10			
• Boundary fencing; 2.4m high; gates and entrance barriers	1,900	m	85			
• Signage		item	65,000			
• Hardstanding drainage	77,000	m²	10			
External Services				**880,000**	**12.27**	**10.6%**
• Gas, water, electricity and telecommunications connections		item	210,000			
• External lighting installations including BWIC		item	290,000			
• Fire hydrant main; 12 nr hydrants		item	190,000			
• Allowance for builders works in connection with utilities		item	190,000			
Ancillary Buildings				**1,204,000**	**16.79**	**14.5%**
• Vehicle wash; steam clean facility; fuel pump and canopy		item	590,000			
• Sprinkler tank base and housing		item	64,000			
• Gatehouse; transport office; axle weigher; cycle storage		item	550,000			
Preliminaries and Contingency				**890,700**	**12.42**	**10.7%**
• Overheads and profit, site establishment and supervision @		10%				
• Contingency @		2%				
Construction cost (External works only, rate based on GIFA)				**8,310,000**	**115.89**	**100%**
TOTAL DISTRIBUTION CENTRE CONSTRUCTION COST (rate based on GIFA)				**35,501,000**	**495.13**	

SMALL INDUSTRIAL UNIT

A single storey new building with a gross internal floor area of 900m², subdivided into five industrial units. Reinforced concrete ground bearing slab and pads to receive a steel portal frame. Wall and roof cladding is aluminium built up system, with internal blockwork division walls. Each of the five units has a seperate entrance door and one roller shutter door, together with a single WC. Units vary in size from 150m² to 360m²

- Gross internal floor area 900m² Model location is South East England (TPI = 580, LF = 1.00)

This updated cost model is copyright of Davis Langdon LLP and was originally published in Building Magazine on 08-Mar-08

Small Industrial Unit	Quantity	Unit	Rate	Total (£)	Cost (£/m²)	%
Substructure				**138,100**	**153.44**	**20.9%**
• Excavation and disposal off site	190	m³	32			
• Reinforced concrete ground slab, including ground beams and column bases	900	m²	120			
• Power floated and hardener	900	m²	12			
• Strip foundations for party walls	80	m	170			
Frame and Upper Floors				**69,000**	**76.67**	**10.5%**
• Steel propped portal frame, cold rolled purlins, surface treatments (@ 40kg/m²)	36	t	1,700			
• Intumscent paint fire protection to steelwork		item	5,500			
• Allowance for miscellaneous works, protecting columns		item	2,250			
Roof				**73,900**	**82.11**	**11.2%**
• Built up aluminium roof cladding with 180 thick insulation, including all labours	950	m²	55			
• Extra over for Rooflights (10% of total)	95	m²	70			
• Mansafe system	80	m	90			
• Rainwater drainage, aluminium gutters and downpies	120	m	65			
External Wall, Windows and Doors				**140,300**	**155.89**	**21.3%**
• Built up aluminium wall cladding with 130 thick insulation	520	m²	65			
• 2.5m high inner leaf of 140 thick fairface blockwork	380	m²	40			
• 3000 x 4600 high steel sectional overhead doors	5	nr	4,550			
• Aluminium single entrance doors	5	nr	1,350			
• Coated aluminium double glazed window system	150	m²	370			
• Polycarbonate Canopy Entrance - approx 1500 x 1000	5	nr	1,250			
Internal Walls and Partitions and Doors				**42,500**	**47.22**	**6.4%**
• 2 hour fire resistant blockwork party walls	450	m²	75			
• Fireproofing between blockwork and roof		item	2,250			
• Metal stud partitions	50	m²	60			
• Laminated faced internal doorset with softwood frames and ironmongery	5	n	700			
Wall Finishes				**5,300**	**5.89**	**0.8%**
• Emulsion paint to blockwork wall surfaces generally	1,370	m²	3			
• Ceramic wall tiles splashbacks to WC area		item	650			
Floor Finishes				**1,000**	**1.11**	**0.2%**
• Screed and non slip vinyl sheeting to WC areas	15	m²	65			
Ceiling Finishes				**600**	**0.67**	**0.1%**
• Moisture resistant plasterboard to WC with ceiling grid and paint finish	15	m²	42			

SMALL INDUSTRIAL UNIT

Small Industrial Unit (cont'd)	Quantity	Unit	Rate	Total (£)	Cost (£/m²)	%
Sanitary Appliances				**6,300**	**7.00**	**1.0%**
• Mall furniture including bins, benches, bollards	5	nr	1,250			
Disposal Installations				**2,100**	**2.33**	**0.3%**
• Waste, soil and vent installation; uPVC pipework and fittings	900	m²	2			
Hot and Cold Water Installations				**3,600**	**4.00**	**0.5%**
• Hot and cold water supplies to WC's	5	nr	725			
Electrical Installations				**35,700**	**39.67**	**5.4%**
• Small power, basic and emergency lighting	900	m²	23			
• Supply to WC for ventilation, heater etc	5	nr	1,700			
• External lighting generally		item	6,500			
Incoming Services				**16,000**	**17.78**	**2.4%**
• Allowance for incoming, electrical, gas and water services		item	16,000			
Protective Installations				**1,000**	**1.11**	**0.2%**
• Lightning protection		item	1,000			
Communication Installations				**10,400**	**11.56**	**1.6%**
• Fire and intruder alarm	900	m²	12			
Builders Work in Connection				**700**	**0.78**	**0.1%**
• Forming holes and chases etc @		1%				
Preliminaries and Contingency				**113,500**	**126.11**	**17.2%**
• Overheads; profit, site establishment and site supervision @		17%				
• Contingency @		3%				
Construction cost (shell only, rate based on GIFA)				**660,000**	**733.34**	**100%**

CENTRAL LONDON OFFICES

This cost model features a high quality City office scheme arranged over 13 floor and one basement with a gross internal area of 21,300 m². The scheme is steel framed and incorporates an internally ventilated double-wall façade. The wall-floor ratio is 0.46. Air treatment is by a four-pipe fan-coil unit. Costs are based on construction management procurement. Demolitions, site preparation, external works and services beyond Category A, tenant enhancement are excluded.

- Offices: Gross internal floor area 21,300m² — Model location is the City of London
- Offices: Net internal floor area 14,600m² — (TPI = 580, LF = 1.03)

Shell and Core Works	Quantity	Unit	Rate	Total (£)	Cost (£/m²)	%
Substructure				**3,339,500**	**156.78**	**7.5%**
· Break out existing slabs, piles, obstructions and allowance for probing/testing; dewatering		item	500,000			
· Foundations; bored piles with under-ream; ground beams; pile caps	1,940	m²	380			
· Piling platform; mini piles and other works to boundary walls		item	230,000			
· RC basement slab 300mm thick, including waterproofing, excavation and disposal	1,940	m²	190			
· RC mat slab 1200mm thick, including waterproofing, excavation and disposal	200	m²	570			
· Reinforced concrete retaining walls 300mm thick	600	m²	350			
· Reinforced concrete ground floor slab 130mmm thick on profiled metal sheet decking	1,760	m²	75			
· Allowance for car park ramp, slab thickenings to stair foundations, lift/escalator pits, drainage channels, concrete transfer walls etc		item	380,000			
· Allowance for crane base including base piles		item	37,700			
· Attendance on archaeologists and movement monitoring		item	130,000			
· Below slab drainage; other items and sundries		item	500,000			
Frame and Upper Floors				**6,668,300**	**313.07**	**14.9%**
· Structural steel frame including fittings	1,350	t	1,950			
· Extra for built up beams	360	t	310			
· Secondary steelwork, based on an extra 5kg/m²	100	t	2,500			
· Extra for concrete encased beams at ground floor		item	75,000			
· Fire protection to steel frame (90mins intumescent paint)	1,350	t	750			
· Reinforced concrete core walls average	3,300	m²	250			
· Allowance for other structures (e.g. within plant rooms etc)		item	130,000			
· Allowance for expansion joints and other sundries		item	63,000			
· Lightweight reinforced concrete 130mm thick on profiled steel decking; upstands plinths; walkways etc	17,430	m²	90			
Roof				**605,200**	**28.41**	**1.4%**
· Profiled steel decking with 200mm lightweight concrete inc mesh reinforcement; Insulation and acoustics to soffit	1,760	m²	270			
· Proprietary roof; paving slabs; upstands / plinths, hatches/ladders, safety hooks and latchways		item	130,000			
Stairs				**674,800**	**31.68**	**1.5%**
· Steel pan staircases; concrete infills to stair treads; painted mild steel balustrades and handrails (basement to roof; 26	2	nr	240,000			
· Ditto, basement to ground: 2 flights	2	nr	18,900			
· Feature entrance stairs		item	94,000			
· Allowance for stairs/cat ladders and safety rails to plant rooms		item	63,000			

CENTRAL LONDON OFFICES

Shell and Core Works (cont'd)	Quantity	Unit	Rate	Total (£)	Cost (£/m²)	%
External Walls				**9,554,800**	**448.58**	**21.4%**
· Feature wall at ground level		item	630,000			
· Internally-ventilated double wall façade: unitised system incorporating double-glazed outer skin	8,600	m²	875			
· Stainless steel screening to plant enclosures	400	m²	570			
· Glass entrance canopies; cantilevered from building	250	m²	1,250			
· Allowance for stainless steel detailing, articulations, etc		item	570,000			
· Extra for louvres		item	31,500			
· Blockwork walls at roof level, including wind posts	60	m²	130			
· Allowance for visual mock-ups and performance tests		item	250,000			
External Windows and Doors				**251,700**	**11.82**	**0.6%**
· Single and double doors, including disabled pass doors		item	63,000			
· Extra over cladding for revolving doors	2	nr	50,300			
· Extra over screen enclosures for single and double doors		item	12,600			
· Steel roller shutter to loading bay and car park	2	nr	18,900			
· Metal doors in service areas		item	37,700			
Internal Walls and Partitions				**2,025,100**	**95.08**	**4.5%**
· Insitu concrete walls in basement, etc.	540	m²	190			
· Fairfaced blockwork walls at basement, ground and roof levels	3,500	m²	100			
· Curved blockwork entrance feature wall	300	m²	210			
· Drylined core walls	6,950	m²	100			
· Extra for double thickness drylined core walls	1,000	m²	100			
· Other walls/partitions to plant areas, additional walls		item	190,000			
· Glazed screen to shopfronts	70	m²	1,000			
· Veneer faced wc cubicles / doors; access panelling	90	nr	5,050			
Internal Doors				**476,000**	**22.35**	**1.1%**
· Single timber doors	140	nr	1,900			
· Double timber doors	30	nr	3,150			
· Profilex riser doors	35	nr	1,500			
· Other doors: plantrooms; additional access door hatches		item	63,000			
Wall Finishes				**1,299,100**	**60.99**	**2.9%**
· Stone cladding to main entrance lobby	880	m²	440			
· Back-lit glass panelling on steel frame in main entrance lobby	150	m²	1,250			
· Paint to fair face block walls	2,150	m²	6			
· Plaster and paint to blockwork / concrete	3,820	m²	19			
· Skim coat and paint to drylined walls	1,700	m²	10			
· Stone cladding to toilets	450	m²	350			
· Granite cladding to lift lobbies	800	m²	470			
· Lift architraves		item	88,000			
Floor Finishes				**977,900**	**45.91**	**2.2%**
· Granite/stone tiles to main entrance lobby and lift lobbies	1,250	m²	440			
· Stone tiles to toilets including, waterproofing, screed; skirtings	440	m²	310			
· Lightweight screed to circulation and core areas	1,280	m²	38			
· Sealant/hardener to car park, loading bay and plant rooms	1,140	m²	100			
· Vinyl flooring to security areas		item	9,400			
· Entrance mats and matwells		item	50,300			
· Allowance for white lining to carpark and loading bay		item	31,500			
· Allowance for other floor finishes		item	37,700			

CENTRAL LONDON OFFICES

Shell and Core Works (cont'd)	Quantity	Unit	Rate	Total (£)	Cost (£/m²)	%
Ceiling Finishes				**651,200**	**30.57**	**1.5%**
· GRG feature ceiling to main entrance lobby	870	m²	380			
· Feature drylined ceiling to lift lobbies	380	m²	180			
· Metal tile suspended ceilings to toilets	440	m²	95			
· Painted plasterboard on metal framing to corridors etc.	840	m²	65			
· Insulation to car park/loading bay soffits	1,030	m²	25			
· Access panels, bulkheads; detailing; sundry ceiling finishes		item	130,000			
Fittings / Fitting Out (excludes loose furniture)				**671,800**	**31.54**	**1.5%**
· Main entrance reception desk and security desks		item	101,000			
· Stone vanity tops in toilets for basins/taps with mirrors behind	70	m	2,200			
· Soap dispensers/tanks, roll holders, paper towels etc	90	nr	625			
· Extra for fittings to disabled toilets	10	nr	1,900			
· Rubbish compactor		item	31,500			
· Column guards, bollards/crash rails to loading bay/car park, cycle racks, traffic management, statutory signage		item	310,000			
Sanitary Appliances				**105,300**	**4.94**	**0.2%**
· WCs, basins, cleaners sinks, urinals (average rate per point)	200	nr	470			
· Extra for disabled toilets	10	nr	1,125			
Disposal Installations				**293,800**	**13.79**	**0.7%**
· Rainwater disposal system	21,300	m²	4			
· Soil waste and vent installation	21,300	m²	9			
· Extra for drainage to retail areas		item	12,600			
· Condensate drainage	21,300	m²	1			
Water Installations				**293,600**	**13.78**	**0.7%**
· Cold water services: incoming, storage, pumps, distribution	21,300	m²	8			
· Hot water heaters and distribution	21,300	m²	4			
· Water services for vending area	21,300	m²	1			
· Supply to retail areas		item	25,200			
Space Heating and Air Treatment				**1,821,900**	**85.54**	**4.1%**
· Gas installation		item	25,200			
· Boilers		item	82,000			
· Air handling units	21,300	m²	10			
· Chillers	21,300	m²	16			
· LTHW heating installation including pumps and boiler flues	21,300	m²	23			
· Air conditioning installation including fans and ductwork	21,300	m²	14			
· CHW installation including pumps and riser pipework	21,300	m²	18			
Ventilation Installation				**604,000**	**28.36**	**1.4%**
· Toilet and smoke extract ventilation	21,300	m²	9			
· Ventilation to plant room, lift motor rooms, refuse area, etc.		item	56,600			
· Car park and basement ventilation	21,300	m²	9			
· Stair and lobby pressurisation	21,300	m²	8			

CENTRAL LONDON OFFICES

Shell and Core Works (cont'd)	Quantity	Unit	Rate	Total (£)	Cost (£/m²)	%
Electrical Installation				**1,825,400**	**85.70**	**4.1%**
• HV Switchgear and transformer	21,300	m²	10			
• LV distribution; busbars	21,300	m²	29			
• Power to mechanical plant	21,300	m²	3			
• Small power installation	21,300	m²	4			
• Lighting, emergency lighting, including basement and car park	21,300	m²	18			
• Earthing and bonding	21,300	m²	1			
• Enhanced lighting in lobby and other areas		item	63,000			
• External building lighting		item	200,000			
• Standby power installation, including oil system		item	180,000			
Lifts				**1,833,600**	**86.08**	**4.1%**
• Passenger lifts, 21 person serving 10 floors	6	nr	210,000			
• Goods lift serving 10 floors	1	nr	240,000			
• Car park lift	1	nr	63,000			
• Fire fighting lift	1	nr	180,000			
• Enhanced lift car finishes	6	nr	15,100			
Protective Installations				**374,900**	**17.60**	**0.8%**
• Sprinkler Installations; tanks, pumps, risers etc.	21,300	m²	14			
• Dry riser installation	21,300	m²	3			
• Lightning protection	21,300	m²	1			
Communication Installations				**446,900**	**20.98**	**1.0%**
• Fire alarm installations	21,300	m²	15			
• Containment for BMS, security, data, etc	21,300	m²	3			
• Landlord security provisions	21,300	m²	2			
• Disabled alarms		item	31,500			
Special Installations				**789,300**	**37.06**	**1.8%**
• Building management system	21,300	m²	16			
• Allowance for façade cleaning equipment		item	440,000			
Builders Work				**428,400**	**20.11**	**1.0%**
• Builder's work in connection with services installations, including machine bases.	21,300	m²	20			
Preliminaries and Contingency				**8,606,600**	**404.07**	**19.3%**
• Contractor's overheads and profit, site establishment and supervision @		18%				
• Contingency @		5%				
Construction cost (shell and core works only, rate based on GIFA)				**44,619,000**	**2,094.79**	**100%**

CENTRAL LONDON OFFICES

Category A Works	Quantity	Unit	Rate	Total (£)	Cost (£/m²)	%
Wall Finishes				**200,000**	**9.39**	**2.8%**
• Emulsion paint finish to office side of core walls	1,770	m²	6			
• Column casings, including paint, sub-frame, etc	1,180	m²	160			
Floor Finishes				**676,000**	**31.74**	**9.3%**
• Dust sealer to concrete slabs	14,600	m²	1			
• Medium grade fully accessible raised floor, metal faced plycore; 150 nominal depth; including fire barriers	14,600	m²	45			
Ceiling Finishes				**832,200**	**39.07**	**11.5%**
• Concealed grid metal tray suspended ceiling to office areas; acoustic quilt and fire breaks	14,600	m²	57			
Fittings / Fitting Out				**19,000**	**0.89**	**0.3%**
• Statutory signage	14,600	m²	1			
Space Heating and Air Treatment				**2,166,600**	**101.72**	**29.9%**
• Four pipe fancoil units	14,600	m²	23			
• Distribution ductwork, grilles etc	14,600	m²	50			
• CHW installation; insulation	14,600	m²	34			
• LTHW installation; insulation	14,600	m²	28			
• Condensate installation; insulation	14,600	m²	14			
Electrical Installations				**1,114,000**	**52.30**	**15.4%**
• Lighting and emergency lighting installation	14,600	m²	55			
• Distribution boards	14,600	m²	5			
• Earthing and bonding	14,600	m²	3			
• Lighting control	14,600	m²	9			
• Small power to fan coil units	14,600	m²	5			
Protective Installations				**275,900**	**12.95**	**3.8%**
• Sprinkler protection to offices	14,600	m²	19			
Communications Installations				**192,700**	**9.05**	**2.7%**
• Fire alarm installation	14,600	m²	13			
Special Installations				**275,900**	**12.95**	**3.8%**
• Building management system	14,600	m²	19			
Builders Work in Connection				**92,000**	**4.32**	**1.3%**
• Builders work in connection with Category A services	14,600	m²	6			
Preliminaries and Contingency				**1,396,700**	**65.57**	**19.3%**
• Contractor's overheads and profit, site establishment and supervision @		18%				
• Contingency @		5%				
Construction cost (Category A only, rate based on GIFA)				**7,241,000**	**339.95**	**100%**

MIXED USE CITY CENTRE SCHEME

This cost model comprises a scheme with three mixed use retail and residential buildings set upon a shared basement car park and service yard in the West Midlands. Separate cost breakdowns are given for retail, residential and basement car parking. The scheme has three levels of retail with active retail frontage to three sides of each block. Three hundred flats are included in the residential block, of which 100 are developed for the affordable sector. The retail units are left as shells, whereas the residential are fitted to requirements of both open market and affordable sectors. Parking for 100 cars is provided in the basement.

- Apartment: Gross internal floor area 9,500m²
- Retail shell and core: Gross internal floor area 4,000m²
- Car park: Gross internal floor area 6,500m²

Model location is for major city in South-East England (TPI = 580, LF = 0.90)

This updated cost model is copyright of Davis Langdon LLP and was originally published in Building Magazine on 09-Dec-05

Apartment building	Quantity	Unit	Rate	Total (£)	Cost (£/m²)	%
Frame and Upper Floors				**5,058,000**	**259.38**	**13.8%**
• Insitu concrete podium slabs; (columns in retail shell)	4,700	m²	190			
• Insitu reinforced concrete floor slabs and columns, 250mm thick slabs, allowance for forming openings	19,500	m²	170			
• Extra for transfer structure		item	240,000			
• Balconies; bolt on frame, decking and balustrades	100	nr	6,100			
Roof				**1,442,900**	**73.99**	**3.9%**
• Flat roof coverings to roof and podium, single ply membrane, insulation, ballast; drainage	5,330	m²	130			
• Extra for green roof to podium areas	1,500	m²	180			
• Access equipment, latchways, access hatch, balustrade		item	180,000			
• Access equipment; roof cleaning cradle		item	300,000			
Stairs				**594,800**	**30.50**	**1.6%**
• Concrete stairs, stainless steel balustrades, carpet	60	m²	9,700			
• Roof access stairs	3	nr	4,250			
External Walls, Windows and Doors				**5,409,300**	**277.40**	**14.7%**
• Entrance screens and doors at ground floor level	320	m²	650			
• Curtain walling; glazing; polyester powder coated aluminium spandrel panels; double-glazed units; sliding doors	7,350	m²	470			
• Brickwork and reconstituted stone on precast concrete backing walls; sealed double-glazed windows	3,100	m²	410			
• Allowance for solar shading		item	240,000			
• Extra for glazed balustrade in lieu of balconies	180	m²	590			
• Acoustic plant screens	360	m²	360			
Internal Walls, Partitions and Doors				**3,779,200**	**193.81**	**10.3%**
• Core walls, insitu concrete, 250 thick	5,330	m²	170			
• Party walls to apartments and corridors	9,760	m²	75			
• Internal partitions to apartments; acoustic ceiling	14,910	m²	60			
• Apartment entrance doorsets	300	nr	1,075			
• Core area doorsets	120	nr	725			
• Apartment internal doors	1,080	nr	775			
Wall Finishes				**1,387,900**	**71.17**	**3.8%**
• Plaster and emulsion paint	19,500	m²	22			
• Skim coat and emulsion paint	36,500	m²	9			
• Ceramic tiling to bathrooms and kitchen splashbacks	5,500	m²	75			
• Allowance for additional wall finishes		item	240,000			

MIXED USE CITY CENTRE SCHEME

Apartment building (cont'd)	Quantity	Unit	Rate	Total (£)	Cost (£/m²)	%
Floor Finishes				**1,082,800**	**55.53**	**2.9%**
· Acoustic floor, ply on battens	15,500	m²	24			
· Natural wood with skirtings to match; market units	4,800	m²	49			
· Carpet, underlay, skirtings; affordable units	2,400	m²	29			
· Ceramic tiling in bathrooms and kitchens; market units	2,200	m²	75			
· Vinyl sheet in bathrooms and kitchens; affordable units	1,100	m²	36			
· Common areas; carpet on sand cement screed; skirtings	3,800	m²	53			
Ceiling Finishes				**969,700**	**49.73**	**2.6%**
· Ceiling finishes; plasterboard and emulsion paint	19,200	m²	36			
· Allowance for access panels in ceilings	300	nr	725			
· Allowance for enhanced finishes to entrance areas		item	61,000			
Fittings and Furnishings				**1,483,800**	**76.09**	**4.0%**
· Fully fitted quality kitchen to open market units	200	nr	4,850			
· Additional fittings to kitchens to 2 bed apartments	64	nr	1,200			
· Allowance for kitchen fittings and units to RSL specification	100	nr	2,450			
· Allowance for bathroom accessories	300	nr	240			
· Reception area fittings; mailboxes; signage; matwells		item	120,000			
Mechanical and Public Health Installations				**4,259,800**	**218.45**	**11.6%**
· Sanitary fittings to open market units; electric shower	200	nr	1,800			
· Extra for second ensuite bathroom	64	nr	2,200			
· Sanitary fittings to affordable units; electric shower	100	nr	1,700			
· Cleaners sinks, electric water heating; disposal	21	nr	1,450			
· Rainwater disposal	19,500	m²	10			
· Above ground drainage	1,500	nr	240			
· Cold water supply to landlord's areas		item	97,000			
· Hot and cold water supply to apartments	300	nr	3,750			
· Electric heating installation; complete	300	nr	2,050			
· Mechanical ventilation installation; open market flats only	200	nr	1,800			
· Kitchen and bathroom ventilation	300	nr	850			
· Ventilation to plant rooms; smoke extract to staircases		item	220,000			
· Allowance for dry riser installation	12	nr	18,200			
· Building Management System	19,500	m²	6			
Electrical Installations				**2,688,200**	**137.86**	**7.3%**
· Allowance for LV distribution	19,500	m²	4			
· Small power, and lighting to landlord's areas	4,800	m²	70			
· Lighting to open market apartments	200	nr	1,800			
· Lighting to affordable apartments	100	nr	600			
· Small power to apartments; generally	300	m²	1,200			
· Power supply to lifts		item	97,000			
· Containment generally	19,500	m²	12			
· Earthing; bonding and lightning protection		item	120,000			
· Fire alarm installation		item	240,000			
· Communications; TV and radio, Satellite TV, telephone	300	nr	725			
· Security: open market flats; video entry, intruder alarm	200	nr	1,450			
· Security: open market flats; audio entry phone	100	nr	240			
· Access control in landlord's areas	45	nr	2,450			
· CCTV; landlord's areas and external monitoring	15	nr	5,350			
· Emergency communication systems	30	nr	2,900			

MIXED USE CITY CENTRE SCHEME

Apartment building (cont'd)	Quantity	Unit	Rate	Total (£)	Cost (£/m²)	%
Lift Installation				**1,620,000**	**83.08**	**4.4%**
• 17 person fire fighting lifts serving 7 stops	3	nr	160,000			
• 17 person fire fighting lifts serving 15 stops	6	nr	190,000			
Builders Work in Connection				**458,700**	**23.52**	**1.2%**
• Forming holes and chases; firestopping @		5%				
• Extra for additional builders work in retail units		item	30,300			
Preliminaries and Contingency				**6,512,000**	**333.95**	**17.7%**
• Overheads; profit; site establishment and supervision @		18%				
• Contingency @		3%				
Construction cost (Apartment building only, rate based on GIFA)				**36,747,000**	**1,884.46**	**100%**

Retail Unit Shell and Core	Quantity	Unit	Rate	Total (£)	Cost (£/m²)	%
Frame, Upper Floors and Stairs				**3,079,800**	**219.99**	**28.3%**
• Insitu reinforced concrete frame;	14,000	m²	75			
• Insitu concrete upper floor slabs; 325 thick	9,300	m²	190			
• Precast concrete stairs, steel balustrades and handrails	36	nr	7,300			
External Walls, Windows and Doors				**2,641,100**	**188.65**	**24.3%**
• Unitised curtain walling; full height sealed double-glazing	3,180	m²	590			
• Feature solar shading to first floor retail elevations	2,830	m²	170			
• Temporary shopfronts	2,830	m²	85			
• Allowance for soffit cladding	180	m²	240			
Internal Walls, Partitions and Doors				**744,300**	**53.16**	**6.9%**
• Core walls, insitu concrete, 250 thick	900	m²	170			
• 140 thick blockwork; 5m high	3,680	m²	100			
• Allowance for plasterboard linings		item	53,400			
• Single leaf steel doorsets; fire-rated; ironmongery	30	nr	1,200			
• Double leaf steel doorsets; fire-rated; ironmongery	50	nr	1,950			
• Allowance for riser doors etc		item	36,400			
Finishes				**65,900**	**4.71**	**0.6%**
• Wall finishes; emulsion paint finish where required		item	36,400			
• Plant rooms and back of house areas only; floor sealer	850	m²	13			
• Ceiling finishes; sealant or emulsion as required		item	18,200			
Fittings and Furniture				**57,000**	**4.07**	**0.5%**
• Allowance for statutory signage to landlord's areas		item	18,200			
• Bump rails; barriers; edge strips and back of house fittings		item	38,800			
Mechanical and Public Health Installations				**549,900**	**39.28**	**5.1%**
• Cleaners sinks, electric water heating; local disposal	9	nr	1,450			
• Cold water supply; booster pumps		item	9,100			
• Rainwater disposal	14,000	m²	9			
• Above ground drainage		item	18,200			
• Supply and extract to plant rooms and HV/LV rooms only		item	12,100			
• Sprinkler installation; shut off valves in retail units; full installation in back of house		item	220,000			
• Allowance for dry riser installation		item	73,000			
• Building Management System	14,000	m²	6			

MIXED USE CITY CENTRE SCHEME

Retail Unit Shell and Core (cont'd)	Quantity	Unit	Rate	Total (£)	Cost (£/m²)	%
Electrical Installations				**790,200**	**56.44**	**7.3%**
• Allowance for LV distribution	14,000	m²	4			
• Small power to landlord's areas	650	m²	4			
• Lighting and emergency lighting to landlord's areas	650	m²	65			
• Power supply to mechanical plant and lifts	14,000	m²	4			
• Containment generally	14,000	m²	12			
• Earthing and bonding	14,000	m²	4			
• Lightning protection		item	54,600			
• Fire alarm installation; panels; detectors and sounders to landlord's areas; public address and voice alarm system		item	109,000			
• Access control in landlord's areas	15	nr	2,450			
• CCTV installation; landlord's areas and external monitoring	42	nr	5,350			
Lift Installation				**900,000**	**64.29**	**8.3%**
• Lift installation; 26 person goods lifts serving 3 stops	6	nr	150,000			
Builders Work in Connection				**112,000**	**8.00**	**1.0%**
• Forming holes and chases; firestopping @		5%				
Preliminaries and Contingency				**1,924,000**	**137.43**	**17.7%**
• Overheads; profit; site establishment and supervision @		18%				
• Contingency @		3%				
Construction cost (Retail shell and core only, rate based on GIFA)				**10,864,000**	**776.02**	**100%**

Basement Car-park	Quantity	Unit	Rate	Total (£)	Cost (£/m²)	%
Substructure				**4,731,700**	**727.95**	**40.3%**
• Excavation for basement including disposal, obstructions	40,600	m³	22			
• Oversite slab; tensile anchors, waterbars, reinforcement	6,500	m²	240			
• Below slab drainage; gullies and petrol interceptor	6,500	m²	12			
• Sheet piling and concrete retaining wall to perimeter	2,700	m²	490			
• Piled foundations (including pile caps) for buildings above	6,500	m²	90			
• Raised concrete walls to edges of suspended slabs		item	130,000			
• Allowance for sundry concrete works; lift pits etc		item	170,000			
Frame and Stairs				**2,254,000**	**346.77**	**19.2%**
• In situ concrete columns; members of varying sizes	6,500	m²	75			
• Allowance for walls, upstands and movement joints		item	130,000			
• Insitu concrete grade level suspended slab; post tensioned beams; precast infill with structural topping	6,500	m²	240			
• Steps and stairs including finishes, handrails and balusters	18	nr	4,250			
Internal Walls, Partitions and Doors				**411,700**	**63.34**	**3.5%**
• 100mm Blockwork liner wall	1,800	m²	85			
• Blockwork internal walls; average 5.3m high	2,200	m²	100			
• 1 hour fire rated timber doorsets (average rate per leaf)	15	nr	875			
• Steel blast doors to substations (pair)	8	nr	3,200			

MIXED USE CITY CENTRE SCHEME

Basement Car-park (cont'd)	Quantity	Unit	Rate	Total (£)	Cost (£/m²)	%
Finishes				**199,800**	**30.74**	**1.7%**
· Paint finish to blockwork and concrete walls	7,570	m²	6			
· Epoxy paint to floor	6,500	m²	13			
· Extra over for sealing to plant room floors	450	m²	12			
· Allowance for car/ lorry markings and other finishes		item	12,100			
· Paint to soffits of slab; car park only	4,300	m²	7			
· Allowance for additional finishes		item	18,200			
Fixtures and Fittings				**98,200**	**15.11**	**0.8%**
· Allowance for bump rails, etc to landlord areas	575	m²	6			
· Allow for architectural metalwork generally		item	26,700			
· Allowance for car park barriers	2	nr	24,300			
· Allowance for statutory signage		item	19,400			
Mechanical and Public Health Installations				**981,900**	**151.06**	**8.4%**
· Drainage installations, including gullies in plant rooms	6,500	m²	12			
· Cold water system ; landlord's wash down only		item	12,100			
· Car park extract system including impulse fans		item	140,000			
· Allowance for additional exhaust ventilation		item	36,400			
· Basement smoke extract installation	6,500	m²	41			
· Localised plantroom, refuse store, lift shaft and transformer room ventilation		item	61,000			
· Heating / ventilation to security room		item	18,200			
· Sprinkler installation, ordinary hazard, leak detection	6,500	m²	18			
· Allowance for sprinkler tanks and zone valves		item	109,000			
· BMS/ Controls	6,500	m²	22			
Electrical Installations				**871,200**	**134.03**	**7.4%**
· LV and sub mains distribution	6,500	m²	27			
· Lighting and emergency lighting	6,500	m²	30			
· Small power installation	6,500	m²	4			
· Power supply to mechanical plant and lifts	6,500	m²	4			
· Containment generally	6,500	m²	12			
· Earthing and bonding	6,500	m²	4			
· Lightning protection	6,500	m²	4			
· Fire alarm installation to landlord's areas; L3 system; public address and voice alarm system	6,500	m²	24			
· Access control and intruder alarm	15	nr	2,450			
· Emergency communication systems; fire telephones and disabled refuge alarm	10	nr	2,900			
· CCTV; 10 nr cameras and control room installation		item	97,000			
· Gate entry intercom system	2	nr	4,850			
Lift Installation				**29,100**	**4.48**	**0.2%**
· Goods lift		item	29,100			
Builders Work in Connection				**94,100**	**14.48**	**0.8%**
· Forming holes ; chases; firestopping, plant room louvres @		5%				
Preliminaries and Contingency				**2,082,300**	**320.35**	**17.7%**
· Overheads and profit, site establishment and supervision @		18%				
· Contingency @		3%				
Construction cost (basement car-park only, rate based on GIFA)				**11,754,000**	**1,808.31**	**100%**

SUPERMARKET

This cost model features a new build supermarket, together with indicative costs of both store extension and refurbishment projects. The new store has first floor staff accommodation and is built to a value engineered specification. Ventilation and refrigeration installations are based on centralised plant. The extension and refurbishment schemes are based on generic models, the extension is a side extension without the construction of a new entrance.

- Supermarket shell: gross internal floor area 7,530m²
- Supermarket extension: gross internal floor area 7,590m²
- Supermarket refurbishment: sales floor area 1,250m²

Model location is Outer London (TPI = 580, LF = 0.93)

This updated cost model is copyright of Davis Langdon LLP and was originally published in Building Magazine on 13-Jun-03

Supermarket Shell	Quantity	Unit	Rate	Total (£)	Cost (£/m²)	%
Substructure				**439,200**	**58.33**	**15.6%**
· Pad foundations and ground beams	6,960	m²	19			
· Reinforced concrete ground floor slab; powerfloat finish; floor ducts to checkout areas only	6,960	m²	44			
Frame and Upper Floors				**588,200**	**78.11**	**21.0%**
· Steel propped portal frame, cold rolled purlin sections	286	t	1,650			
· Fire casing to columns and beams under first floor	570	m²	36			
· Structural steel frame to form first floor	50	t	1,175			
· Holorib decking and insitu concrete topping to first floor	570	m²	65			
Roof				**923,900**	**122.70**	**32.9%**
· Standing seam aluminium roof, curved, inner liner tray	5,500	m²	115			
· Eaves detail to roof cladding	170	m	170			
· Single layer polymeric built-up roof, including insulation board and inner liner tray to flat roof area	1,640	m²	65			
· Rainwater goods, including syphonic drainage		item	38,100			
· Allowance for mansafe system and hatches to flat roof		item	19,100			
· Profiled metal pvf2 coated cladding to form canopy	210	m²	470			
Stairs				**25,400**	**3.37**	**0.9%**
· Reinforced concrete stairs; steel balustrade and handrails	2	nr	12,700			
External Walls, Windows and Doors				**283,800**	**37.69**	**10.1%**
· Profiled galvanised steel built-up cladding system	890	m²	44			
· Feature cladding band to store front and restaurant	675	m²	25			
· Allowance for louvres, flashings and detailing	65	m²	1,075			
· Allowance for column casings		item	12,700			
· Polyester powder coated aluminium shopfronts	300	m²	390			
· Aluminium windows; sealed double glazed units	25	m²	330			
· Softwood framed, metal lined external doorsets,	17	nr	950			
· Allowance for shutters to loading bay doors		item	3,800			
Disposal Installations				**159,400**	**21.17**	**5.7%**
· Below slab drainage, manholes etc.	6,960	m²	23			
Protective Installations				**5,100**	**0.68**	**0.2%**
· Lightning protection		item	5,100			
Preliminaries and Contingency				**382,000**	**50.73**	**13.6%**
· Contractors site establishment and supervision @		6%				
· Contractors overheads and profit @		4%				
· Contingency @		5%				
Construction cost (Supermarket shell only, rate based on GIFA)				**2,807,000**	**372.78**	**100%**

SUPERMARKET

Supermarket Fit-Out	Quantity	Unit	Rate	Total (£)	Cost (£/m²)	%
Internal Walls and Partitions and Doors				**301,200**	**40.00**	**4.7%**
• Internal metal stud partitions, including sundry metalwork	2,160	m²	70			
• Fire protection / stopping	2,160	m²	25			
• Carpentry and joinery, internal doors and trucking doors		item	64,000			
• Security and fire shutters generally		item	32,000			
Internal Finishes				**426,100**	**56.59**	**6.7%**
• Terrazzo flooring to sales area	5,100	m²	44			
• Checkout duct covers		item	32,000			
• Allowance for aluminium access covers and frames		item	6,000			
• White wall tiles to bakery, including epoxy grout	440	m²	41			
• Ceramic floor tiles to bakery, prep areas, serveries and WC's, including epoxy grout, skirtings, angles, etc	540	m²	55			
• Vinyl sheet flooring to back-up areas, including DPM	425	m²	60			
• Allowance for other miscellaneous wall / floor finishes		item	32,000			
• Suspended ceilings to domestic areas and customer WC's	465	m²	57			
• Allowance for insitu finishes, including screeding, etc		item	32,000			
Furniture and Fittings				**1,657,200**	**220.08**	**25.9%**
• Internal signage		item	23,000			
• Trolley protection rails		item	25,000			
• ATM / cash office		item	38,000			
• Specialist joinery to create front brand wall		item	25,000			
• Shopfitting to Pharmacy		item	38,000			
• Gondolas to sales floor generally	4,645	m²	48			
• Servery, plant and equipment to bakery		item	330,000			
• Shopfitting to specialist areas including hot food, deli, meat and fish, salad bar, etc		item	165,000			
• Checkouts (not including service desk and kiosk)	30	nr	3,800			
• Miscellaneous shopfitting / specialist items, etc		item	191,000			
• Fitting out to customer restaurant (excluding catering equipment)	300	m²	1,050			
• Fitting out to staff WCs		item	20,000			
• Fitting out to staff dining room		item	57,000			
• Fit out to staff offices / meeting room / training room		item	19,000			
• Racking to bulk stock areas		item	64,000			
• Compactor		item	10,200			
Water Installations and Services Equipment				**982,000**	**130.41**	**15.4%**
• Plumbing and water installation		item	38,000			
• Refrigeration distribution installation complete; including pipework and installation		item	191,000			
• Refrigeration plant, including packs and condensers		item	178,000			
• Refrigeration cabinets to shopfloor, including mixture of full height glass door cabinets, open top cabinets, etc	4,645	m²	110			
• Packaged cold stores including controls, etc		item	64,000			
Space Heating and Air Treatment				**597,800**	**79.39**	**9.4%**
• Ventilation ductwork including insulation, grilles, diffusers,	7,530	m²	33			
• Air treatment plant, air handling units and central boiler plant	7,530	m²	27			
• Pipework to heating system		item	121,000			
• Allowance for special attenuation		item	25,000			

SUPERMARKET

Supermarket Fit-Out (cont'd)	Quantity	Unit	Rate	Total (£)	Cost (£/m²)	%
Electrical Installations				**651,900**	**86.57**	**10.2%**
• Mains and sub-mains	7,530	m²	14			
• Trays and trunking	7,530	m²	8			
• Small power	7,530	m²	9			
• General lighting to sales floor (800-1,000 lux)	4,645	m²	36			
• Allowance for feature lighting to sales floor	4,645	m²	13			
• Back of house lighting	2,885	m²	19			
• Power to services installations; containment		item	34,000			
• Works to specific departments / customer restaurant		item	102,000			
Lift Installations				**127,000**	**16.87**	**2.0%**
• 8 person goods lifts to first floor	2	nr	38,000			
• Scissor lift to service yard		item	25,000			
• Dock levellers / shelters	2	nr	13,000			
Protective and Communications Installations				**385,300**	**51.17**	**6.0%**
• Allowance for sprinkler installation	7,530	m²	25			
• Earthing and bonding		item	5,000			
• Fire alarm installation	7,530	nr	10			
• Public address system	7,530	m²	1			
• Telephone installation		item	6,000			
• Structured cabling		item	25,000			
• Allowance for CCTV installation	19	nr	1,800			
• Security tagging		item	44,000			
Special Installations				**343,000**	**45.55**	**5.4%**
• BMS / controls		item	108,000			
• Catering equipment to customer restaurant (115 covers)		item	235,000			
Builders Work				**185,000**	**24.57**	**2.9%**
• Builders work in connection with services @		6%				
Preliminaries and Contingency				**734,500**	**97.54**	**11.5%**
• Contractors site establishment and supervision @		6%				
• Contractors overheads and profit @		4%				
• Contingency @		3%				
Construction cost (Supermarket fit-out only, rate based on GIFA)				**6,391,000**	**848.74**	**100%**

Supermarket - External Works	Quantity	Unit	Rate	Total (£)	Cost (£/m²)	%
Site Works				**997,600**	**132.48**	**28.1%**
• Allowance for site preparation	22,000	m²	11			
• Tarmac surfacing to car park and access road	13,000	m²	25			
• Concrete surfacing to service yard	1,200	m²	44			
• Paving to front of store	800	m²	51			
• Signage and street furniture		item	64,000			
• Trolley bays		item	6,000			
• Allowance for soft landscaping, including topsoiling		item	178,000			
• Boundary fencing		item	89,000			

SUPERMARKET

Supermarket - External Works (cont'd)	Quantity	Unit	Rate	Total (£)	Cost (£/m²)	%
Drainage and External Services				**595,000**	**79.02**	**16.8%**
· Allowance for drainage and sewer connections	22,000	m²	8			
· BWIC underground services		item	78,000			
· External lighting installations including BWIC	15,000	m²	10			
· Allowance for utilities supplies, directs and diversions		item	191,000			
Minor Building Works				**1,481,000**	**196.68**	**41.7%**
· Stand alone substation/pump house; complete		item	19,000			
· 6 pump PFS, including car wash and jet wash		item	572,000			
· Allowance for 1,000ft² kiosk		item	572,000			
· Section 106 / 278 works		item	318,000			
Preliminaries and Contingency				**476,400**	**63.27**	**13.4%**
· Overheads, site establishment and supervision @		10%				
· Contingency @		5%				
Construction cost (External Works only, rate based on GIFA)				**3,550,000**	**471.45**	**100%**

TOTAL SUPERMARKET CONSTRUCTION COST (rate based on GIFA)				**12,748,000**	**1,692.97**

Supermarket - Extension	Quantity	Unit	Rate	Total (£)	Cost (£/m²)	%
Works to Existing				**273,000**	**35.97**	**4.2%**
· Repairs and alterations to existing shell, redecoration		item	273,000			
Extension to Shell				**1,583,000**	**208.56**	**24.4%**
· Extension to rear sales wall including removal of equipment; demolition of existing; pad foundations and floor slab, steel frame, external walls and doors, tiled roof, incorporation of new M&E with existing systems, internal finishes	280	m²	875			
· Ditto to non-entrance side extension	1,115	m²	1,200			
Fitting Out				**3,251,300**	**428.37**	**50.2%**
· Part refurbishment / part new shopfitting to existing sales area and new shopfitting to extended area including M&E	4,365	m²	675			
· Allowance for changes to back-up areas		item	222,000			
· Internal finishes, branding and signage	4,365	m²	19			
External Works				**356,000**	**46.90**	**5.5%**
· External works, groundworks, foul and surface water drainage, services diversions, fencing, and street furniture		item	356,000			
Preliminaries and Contingency				**1,018,700**	**134.22**	**15.7%**
· Overheads, site establishment and supervision @		13%				
· Contingency @		5%				
Construction cost (Extension only, rate based on GIFA)				**6,482,000**	**854.02**	**100%**

SUPERMARKET

Supermarket - Refurbishment	Quantity	Unit	Rate	Total (£)	Cost (£/m²)	%
Remedial Works / Finishes				**1,309,800**	**1,047.84**	**52.0%**
· Removal of existing contaminated ceiling and replacement with metal tile suspended ceiling; to sales floor	1,250	m²	165			
· Repair and make good existing sales area floor finishes	1,250	m²	110			
· Additional lighting to sales floor		item	57,000			
· Refrigeration plant / infrastructure costs		item	197,000			
· Upgrade / remedial works to M & E services		item	489,000			
· Internal finishes, branding and signage		item	76,000			
· Alteration works to existing customer entrance		item	64,000			
· Changes to back-up areas		item	83,000			
Fitting Out				**647,000**	**517.60**	**25.7%**
· Refurbishment of store including new and reused shopfittings; associated M&E works	1,250	m²	400			
· Allowance for new checkouts		item	83,000			
· New staff restaurant and other back of house works		item	64,000			
External Works				**32,000**	**25.60**	**1.3%**
· Allowance for limited external works		item	32,000			
Preliminaries and Contingency				**532,200**	**425.76**	**21.1%**
· Additional cost of nightworks and 24hr working		item	64,000			
· Allowance for special attendance		item	51,000			
· Contractors, site establishment and supervision @		10%				
· Overheads and profit @		4.5%				
· Contingency @		5%				
Construction cost (Refurbishment only, rate based on sales floor area)				**2,521,000**	**2,016.80**	**100%**

HEALTH CENTRE

This cost model details a large joint service centre scheme housing GP, dentistry and social services in a single building located on a tight urban site. The accommodation included 60 consulting rooms. Internal circulation is a key design aspect and the building design is based on two wings of largely cellular space arranged around an enclosed 'street', providing space for reception, cafes and other public facilities. There is extensive service installation which included data infrastructure

- Gross internal floor area, including area of tiers 8,435m² Model location is Outer London (TPI = 580, LF = 1.00)

Health Centre	Quantity	Unit	Rate	Total (£)	Cost (£/m²)	%
Substructure				**1,119,300**	**132.70**	**6.6%**
• 450mm diameter reinforced concrete piling	1,955	m²	200			
• Excavating basement; disposal off site; breaking out obstructions; dewatering	1,900	m³	60			
• Reinforced concrete slab to basement area (565m2) and ground floor with vapour barrier, insulation; underslab ducts and drainage	1,955	m²	200			
• Extra for formation of lift pits		item	20,800			
• Reinforced concrete retaining walls to basement area, including temporary sheet piling; blockwork lining wall	450	m²	450			
Frame and Upper Floors				**1,771,400**	**210.01**	**10.5%**
• Reinforced concrete frame, 7.2 x 7.2m grid; 200mm thick flat slab; in-situ concrete shear walls	8,435	m²	210			
Roof				**266,600**	**31.61**	**1.6%**
• Inverted roof coverings, polymeric roof coverings, insulation; flashings and copings	1,955	m²	100			
• Mansafe system		item	19,000			
• Extra for polycarbonate rooflights; 800 diameter; flashings	15	nr	1,350			
• Allowance for entrance canopies		item	31,800			
Stairs				**433,800**	**51.43**	**2.6%**
• Precast reinforced concrete stair cases; including fins smooth finish to exposed surfaces	6	nr	12,200			
• Balustrades; 1.1m average high; glazed with profiled timber handrails; average rate used	665	m	490			
• Profiled timber handrails	150	m	150			
• Miscellaneous metalwork; cat ladders; open mesh flooring in risers etc		item	12,200			
External Walls, Windows and Doors				**1,864,900**	**221.09**	**11.0%**
• Multi-coloured curtain walling; double-glazed units; composite cladding panels; secondary steel and internal	2,520	m²	390			
• Coloured render system on mesh, insulation, dpm and metsec backing wall.	1,260	m	260			
• Aluminium framed windows	95	m²	440			
• Perforated clay tile rain screen cladding complete including vertical and horizontal support rails	1,300	m²	230			
• Entrance door unit; aluminium framed glazed doors; automatic operation	2	nr	19,600			
• External doors; aluminium framed; polyester coated; to match windows	10	nr	3,800			
• Free standing screens to roof plant; louvre panels as required	650	m	210			

HEALTH CENTRE

Health Centre (cont'd)	Quantity	Unit	Rate	Total (£)	Cost (£/m²)	%
Internal Walls and Partitions				**718,300**	**85.16**	**4.2%**
· Concrete blockwork; head restraints, movement joints	2,000	m²	55			
· Plasterboard partitions; 2 layers of 12.5mm wall board; including 25mm insulation and skim coat finish	7,350	m²	65			
· Internal glazed screens; blinds	225	m²	580			
Internal Doors				**523,900**	**62.11**	**3.1%**
· Flush doors; beech veneer; fire rated; solid core; veneered frame; vision panel and ironmongery	325	nr	1,050			
· Glazed doors and screens; fire-rated; frames, fittings and ironmongery	22	nr	3,650			
· Riser cupboard doors; beech veneer; fire rated, fittings and ironmongery	66	nr	1,550			
Wall Finishes				**212,600**	**25.20**	**1.3%**
· Emulsion paint to wall surfaces generally	17,250	m²	4			
· Ceramic wall tiling 150 x 150mm	900	m²	60			
· Single layer plasterboard dry lining 12.5mm thick	5,150	m²	18			
Floor Finishes				**604,100**	**71.62**	**3.6%**
· Screed; latex self levelling screed 15 to 25mm thick	8,435	m²	15			
· Ceramic floor tiling 600 x 600mm to ground floor atrium	435	m²	120			
· Non slip vinyl sheet flooring; skirtings	315	m²	44			
· Linoleum sheet; skirtings	1,100	m²	50			
· Carpet; softwood skirtings	5,500	m²	51			
· Carpet and nosing to stair cases	4	m²	13,500			
· Allowance for entrance barrier matting		item	24,500			
Ceiling Finishes				**339,000**	**40.19**	**2.0%**
· Suspended ceilings; 600 x 600mm clip in metal tiles	4,170	m²	49			
· Plasterboard MF suspended ceilings and bulkheads	1,280	m²	58			
· Plaster to soffit of reinforced concrete slabs	1,900	m²	20			
· Painting to plasterboard ceilings and bulkheads	3,180	m²	7			
Furniture and Fittings				**689,200**	**81.71**	**4.1%**
Supply and install Group 1 medical equipment (fixed furniture), fit only Group 2 and 3 medical equipment (loose furniture)						
· Reception area, desks and fixed furniture	13	nr	9,800			
· Café, servery counter and fixed furniture	2	nr	11,000			
· Waiting area	13	nr	1,600			
· Consulting room	60	nr	1,950			
· Treatment room	18	nr	1,350			
· Podiatary	3	nr	3,800			
· Dentistry	4	nr	30,600			
· Audiology	2	nr	36,700			
· Family Room	12	nr	370			
· Office room	34	nr	240			
· Classroom	4	nr	1,100			
· Meeting room	14	nr	75			
· WC	27	nr	490			
· Records/store	12	nr	600			
· Utility Room	9	nr	2,450			
· Allowance for additional fittings and furniture		item	110,000			

HEALTH CENTRE

Health Centre (cont'd)	Quantity	Unit	Rate	Total (£)	Cost (£/m²)	%
Sanitary Fittings and Disposal Installations				**325,000**	**38.53**	**1.9%**
• Sanitary fittings generally	200	nr	1,025			
• Waste, soil and vent pipework	200	nr	400			
• Extra for lift sump pumps		item	8,800			
• Rainwater installation	8,435	m²	4			
Water Installations				**409,900**	**48.60**	**2.4%**
• Cold water plant room installation; storage and booster unit	8,435	m²	7			
• Hot water plant room installation; gas fired water heaters, pumps etc	8,435	m²	14			
• Hot and cold water distribution pipework, insulation	8,435	m²	22			
• Allowance for water treatment		item	48,900			
Space Heating and Air Treatment				**1,871,500**	**221.87**	**11.1%**
• Chilled water plant room installation; air cooled chiller; pumps, pipework distribution in plant rooms and risers		item	260,000			
• LTHW plant room installation; boiler and flue; pumps, pipework distribution in plant rooms and risers		item	97,000			
• Air Handling units; total combined capacity 10 m³/s; duct mounted heating and cooling batteries		item	102,000			
• Chilled water distribution; pipework; valves; insulation, trace heating	3,900	m²	65			
• LTHW heating; LST panels and trench heaters; pipework; valves; insulation	8,435	m²	38			
• Air curtains and door heaters; generally		item	13,500			
• Supply and extract ductwork installation serving active chilled beams	3,900	m²	53			
• Extra for enhanced attenuation measures in ductwork		item	19,600			
• Active chilled beam installation	3,900	m²	75			
• Thermal insulation to ductwork	8,435	m²	22			
• Allowance for packaged cooling systems to IT rooms		item	12,200			
• Allowance for dedicated ventilation systems (7nr systems)		item	25,700			
• Dirty extract system	8,435	m²	10			
Electrical and Gas Installations				**1,162,800**	**137.85**	**6.9%**
• Incoming supply and Main LV panel		item	58,700			
• Submains distribution, busbars in risers, distribution boards and isolators; including electrical supplies to main plant	8,435	m²	24			
• Lighting: standard luminaires, wiring, containment,	8,435	m²	46			
• Allowance for emergency lighting	8,435	m²	7			
• Allowance for external lighting		item	34,300			
• General small power, including cabling, trunking and socket outlets	8,435	m²	43			
• Allowance for standby power and UPS		item	41,600			
• Incoming gas supply, including steel pipework, valves etc		item	13,500			
Lift Installations				**166,400**	**19.73**	**1.0%**
• Electric traction lift; 10 person; serving 4nr floors	2	nr	36,700			
• Electric traction lift; 10 person; serving 6nr floors	2	nr	46,500			

HEALTH CENTRE

Health Centre (cont'd)	Quantity	Unit	Rate	Total (£)	Cost (£/m²)	%
Protective, Communications and Special Installations				**1,250,900**	**148.30**	**7.4%**
• Lightning protection; earthing installations:		item	25,200			
• Data and telephone pathway and cable infrastructure	8,435	m²	38			
• Fire detection and alarm system	8,435	m²	15			
• Security system; access control and intruder detection; wiring and equipment	8,434	m²	17			
• Building management and automatic control systems	8,435	m²	51			
• CCTV system		item	37,900			
• Patient call system; LED signboards		item	87,000			
• Public Address system		item	12,200			
• Induction loop system		item	23,200			
• Video/display monitor system in public areas		item	17,100			
• Gas scavenging systems to dental rooms		item	13,500			
• Compressor and vacuum installation to dental rooms		item	15,900			
Builders Work				**129,700**	**15.38**	**0.8%**
• Builder's work in connection with services @		2.5%				
Preliminaries and Contingency				**3,073,700**	**364.40**	**18.2%**
• Testing and commissioning of building services installations		item	86,000			
• Contractors preliminaries, overheads and profit @		18%				
• Contractors contingency @		3%				
Construction cost (Health Centre only, rate based on GIFA)				**16,933,000**	**2,007.49**	**100%**

PRIMARY CARE CENTRE

A single storey, new build, health care centre consisting of six consulting rooms, three nurse rooms, two treatment rooms (one clean and one 'dirty'), a health visitor room, a general purpose room and administration space providing a total of 510m² of accommodation

- Gross internal floor area 8,435m² Model location is Outer London (TPI = 580, LF = 1.00)

This updated cost model is copyright of Davis Langdon LLP and was originally published in Building Magazine on 08-Mar-08

Primary Care Centre	Quantity	Unit	Rate	Total (£)	Cost (£/m²)	%
Substructure				70,800	138.82	8.9%
· Excavation & disposal	140	m²	32			
· Strip foundation with cavity masonry wall	150	m	170			
· 150mm thick ground floor slab with thickenings	510	m²	80			
Roof				105,600	207.06	13.3%
· Precast concrete planks	510	m²	65			
· Single ply membrane, including insulation and labours	510	m²	120			
· Allowance for roof inlets		Item	3,000			
Aluminium rainwater down pipes	30	m	80			
· Sunpipes	8	nr	725			
External Walls				40,800	80.00	5.1%
· Brick cavity wall, facing brick outer skin with cavity, insulation, 140mm inner blockwork leaf	240	m²	170			
Windows and External Doors				58,300	114.31	7.3%
· Aluminium double glazed windows	60	m²	410			
· Single aluminium doors	3	nr	1,700			
· Double aluminium framed glazed doors and screens to entrance	22	m²	600			
· Single steel door to plant room	1	nr	1,050			
· Fixed metal grilles to windows		Item	5,800			
· Electrically operated roller shutter to glazed entrance screens, including controls	22	m²	390			
Internal Walls and Partitions				19,700	38.63	2.5%
· 100 Blockwork	470	m²	37			
· Stud partition		Item	525			
· Security counter grille to reception desk		Item	1,800			
Internal Doors				32,500	63.73	4.1%
· Internal doors, wood veneer, stainless steel ironmongery	20	nr	1,000			
· Internal doors & screen, inner door to lobby	5	m²	600			
· Wrot softwood storage cupboard door, stainless steel ironmongery	3	nr	525			
· Fire rated doors, stainless steel ironmongery	5	nr	1,250			
· Double internal fire rated doors, stainless steel ironmongery	1	nr	1,700			
Wall Finishes				21,000	41.18	2.6%
· Plaster and emulsion paint to wall surfaces generally	900	m²	19			
· Ceramic wall tiling 150 x 150mm	60	m²	65			

PRIMARY CARE CENTRE

Primary Care Centre (cont'd)	Quantity	Unit	Rate	Total (£)	Cost (£/m²)	%
Floor Finishes				**30,100**	**59.02**	**3.8%**
· Screed	510	m²	21			
· Carpet tiles to reception area, staff areas	145	m²	26			
· Vinyl to WC areas & practical areas, including skirtings	320	m²	37			
· Heavy duty vinyl to entrance area, dirty areas	40	m²	47			
· Entrance matting with aluminium matwell	5	m²	370			
Ceiling Finishes				**18,600**	**36.47**	**2.3%**
· Mineral fibre suspended ceilings	80	m²	32			
· Plasterboard ceiling, plaster skim and emulsion paint finish	405	m²	37			
· Moisture resistance plasterboard ceiling, plaster skim and emulsion paint finish	25	m²	42			
Fittings and Fixtures				**35,700**	**70.00**	**4.5%**
Supply and install Group 1 medical equipment (fixed furniture); fit only Group 2 and 3 medical equipment (loose furniture)						
· Reception area, desks and fixed furniture		Item	8,700			
· Waiting room		Item	5,400			
· Consulting rooms	6	nr	1,750			
· Treatment rooms	2	nr	1,200			
· Nurse rooms, including health visitor	4	nr	1,200			
· Records store	1	nr	550			
· Admin rooms	2	nr	650			
· Multi purpose / interview rooms	2	nr	1,000			
Sanitary Fittings and Disposal Installations				**17,000**	**33.33**	**2.1%**
· WCs	6	nr	360			
· Hand basins	6	nr	220			
· Medical handwash basins	12	nr	600			
· Disabled toilet, including WC, wash hand basin, grab rails and other fittings	2	nr	1,350			
· Cleaners sink, 510 x 380	2	nr	850			
· Baby changing unit	1	nr	470			
· Sink & slop hopper	1	nr	1,400			
Disposal Installations				**8,100**	**15.88**	**1.0%**
· Waste, soil and vent installation; uPVC pipework and fittings	510	m²	16			
Water Installations				**16,700**	**32.75**	**2.1%**
· Cold water system	28	nr	320			
· Hot water system	21	nr	370			
Space Heating and Air Treatment				**66,300**	**130.00**	**8.4%**
· Space heating, all costs associated with the supply & installation of the heating system, temperature control and distribution pipework	510	m²	130			
Ventilation Installations				**24,200**	**47.45**	**3.0%**
· Extract ventilation to clean areas		item	14,700			
· Extract ventilation to dirty areas		item	6,300			
· Extract ventilatiom to toilet areas		item	3,150			
Gas Installations				**1,800**	**3.53**	**0.2%**
· Gas Installations, all costs associated with the supply and installation of gas		item	1,800			

PRIMARY CARE CENTRE

Primary Care Centre (cont'd)	Quantity	Unit	Rate	Total (£)	Cost (£/m²)	%
Electrical Installations				**55,100**	**108.04**	**6.9%**
• Mains and sub - mains installation	510	m²	24			
• Small power installation	510	m²	37			
• Lighting and general luminaires; emergency lighting	510	m²	47			
Protective Installations				**2,400**	**4.71**	**0.3%**
• Lightning protection; earthing installations:		item	2,400			
Communications Installations				**28,900**	**56.67**	**3.6%**
• Fire alarm installation; smoke detectors; call points	510	m²	18			
• Telephone and data wireways; internal telephone system	510	m²	6			
• Patient call and induction loop systems		Item	5,800			
• Security installation; intruder detection, panic buttons, disabled toilet alarm, CCTV	510	m²	21			
Specialist Installations				**5,800**	**11.37**	**0.7%**
• BMS system		item	5,800			
Builders Work				**5,700**	**11.18**	**0.7%**
• Builder's work in connection with services @		2.5%				
Preliminaries and Contingency				**128,900**	**252.75**	**16.2%**
• Contractors preliminaries, overheads and profit @		16%				
• Contractors contingency @		3%				
Construction cost (Primary Care Centre, rate based on GIFA)				**794,000**	**1,556.88**	**100%**

50m SWIMMING POOL WITH DRY SPORTS FACILITY COST MODEL

The cost model is based on a regional 50m pool with spectator seating for 350 people. The two storey development includes a creche, café and bar, a climbing wall, a four court multi-purpose sports hall, a health and fitness suite with 50 workstations and a multi-use studio. Works exclude loose FFE, catering equipment and cafe fit-out. The costs include for a movable floor and booms to the 50m pool to allow for mixed use.

- Gross internal floor area 8,600m² Model location is South-East England (TPI = 580, LF = 0.97)

This updated cost model is copyright of Davis Langdon LLP and was originally published in Building Magazine on 24-Feb-06

Swimming Pool	Quantity	Unit	Rate	Total (£)	Cost (£/m²)	%
Substructure				**1,779,400**	**206.91**	**8.9%**
• Excavate and fill site generally to an average depth of 500mm; disposal, allowance for breaking out	5,900	m²	24			
• Ground bearing slabs; reduced levels; blinding; waterproofing, hardcore; reinforcement and lift pits	5,900	m²	110			
• Extra over for forming 300mm reinforced concrete pool tank walls and moveable floor boom pits; waterproofing,	400	m²	240			
• Allowance for formation of dry ducts to pool perimeter	120	m²	950			
• Piling; 600mm CFA piles; 15m deep; testing	5,900	m²	90			
• Excavation and installation of column bases and pile caps; including reinforced concrete; blinding; formwork etc	5,900	m²	42			
Frame and Upper Floors				**1,156,500**	**134.48**	**5.8%**
• In situ upper floor slabs; including profile structural metal decking; concrete; reinforcement; power float finish	2,700	m²	130			
• Precast concrete seating units; supply and erection	300	m²	240			
• Portalised frame to sports hall; structural steel hollow sections, columns, trusses and bracing; intumescent paint	45	t	3,000			
• Hollow section columns supporting Glulam roof beams to pool; intumescent paint	30	t	2,750			
• Main Frame to all other areas; structural steel universal columns, beams, rakers and bracing; intumescent paint	160	t	2,100			
• Concrete shear and core walls; concrete; reinforcement	1,000	m²	180			
Roof				**1,800,500**	**209.36**	**9.0%**
• Glulam beams to pool	400	m	950			
• Steel frame to other areas; universal beams; fireproofing	185	t	2,100			
• Aluminium standing seam construction; including structural liner tray, waterproofing, insulation and roof covering	6,000	m²	120			
• Extra over allowance for rooflights to swimming pool	250	m²	600			
• Roof drainage; rainwater installations generally	6,000	m²	12			
• Roof latchway system	500	m	120			
• Glazed entrance canopy; planar glazing; architectural steelwork		item	30,000			
Stairs				**150,000**	**17.44**	**0.8%**
• Feature staircase; precast concrete steps; carborundum strips; architectural metalwork; glazed balustrade	1	nr	60,000			
• Public access staircase; precast concrete units; carborundum strips; stainless steel handrail	2	nr	30,000			
• Back of house access staircase; precast concrete units; carborundum strips; galvanized steel handrail	1	nr	24,000			
• Roof access cat ladders	2	nr	3,000			

50m SWIMMING POOL WITH DRY SPORTS FACILITY COST MODEL

Swimming Pool (cont'd)	Quantity	Unit	Rate	Total (£)	Cost (£/m²)	%
External Walls, Windows and Doors				**1,278,000**	**148.60**	**6.4%**
• Render finished cavity wall; STO render; blockwork;	1,500	m²	180			
• Aluminium framed double glazed curtain walling	1,200	m²	480			
• Metal flat panel cladding system; secondary steelwork; insulation; internal blockwork	800	m²	300			
• Extra over for louvres	100	m²	420			
• Brise soleil system	400	m²	300			
• Main entrance; single pane laminated glazed screen; ironmongery	1	nr	12,000			
• Escape doors; double escape doors and frames; ironmongery	10	nr	1,800			
Internal Walls and Partitions				**612,000**	**71.16**	**3.1%**
• Blockwork walls; firestopping; head restraints	3,500	m²	60			
• Internal partitions; plasterboard; insulation; studwork	3,000	m²	48			
• Glazed screens	400	m²	420			
• Changing cubicles	50	nr	900			
• WC cubicles	50	nr	900			
Internal Doors				**161,000**	**18.72**	**0.8%**
• Single doors and framework; ironmongery	100	nr	950			
• Double doors and framework; ironmongery	20	nr	1,800			
• Riser cupboard doors; fire rated	50	nr	600			
Wall Finishes				**363,500**	**42.27**	**1.8%**
• Plaster and paint	5,000	m²	24			
• Ceramic tiling (including pool tank walls)	2,500	m²	95			
• Paint finish on concrete walls	1,000	m²	6			
Floor Finishes				**779,000**	**90.58**	**3.9%**
• Vinyl sheeting to dry change and kitchen; skirtings	600	m²	42			
• Carpet tile; skirtings	3,000	m²	42			
• Anti dust sealant to plant and stores	1,200	m²	12			
• Timber sprung floor to sports hall, fitness suite and multi purpose studio; floor battens	1,200	m²	95			
• Tiled ceramic flooring to pool area and wet change (including pool tank floor); skirtings	2,200	m²	95			
• Terrazo stone flooring to reception; skirtings	400	m²	180			
• Levelling screed	8,600	m²	24			
• Allowance for entrance matwell		item	12,000			
Ceiling Finishes				**501,200**	**58.28**	**2.5%**
• Timberboard slat on moisture resistant plasterboard backing to pool hall; timber, plasterboard; hangers	1,520	m²	120			
• Extra for works around rooflights		item	36,000			
• Suspended ceilings; plasterboard; acoustic treatment, paint finish; perimeter trims; hangers	4,200	m²	54			
• Moisture resistant suspended ceilings to wet change and kitchen; plasterboard; paint finish; perimeter trims; hangers	750	m²	65			
• Paint finish on slab soffit	1,200	m²	6			

50m SWIMMING POOL WITH DRY SPORTS FACILITY COST MODEL

Swimming Pool (cont'd)	Quantity	Unit	Rate	Total (£)	Cost (£/m²)	%
Furniture and Fittings				**1,533,600**	**178.33**	**7.7%**
· Pool seating	350	nr	36			
· Moveable floors including 2nr booms to pool	1	nr	780,000			
· Disabled hydraulic access platform	1	nr	60,000			
· Springboards	4	nr	6,000			
· Main reception desk	1	nr	60,000			
· Sundry reception desks	2	nr	12,000			
· Climbing wall	1	nr	120,000			
· Lockers	300	nr	360			
· Balustrades and safety barriers	150	m	600			
· Handrails	100	m	300			
· Bleacher seating	250	nr	300			
· Entrance turnstiles	2	nr	18,000			
· Signage		item	30,000			
· Allowance for mirrors		item	18,000			
· Access ladders to pool		item	30,000			
· Changing room benches		item	36,000			
Sanitary Fittings and Disposal Installations				**445,000**	**51.74**	**2.2%**
· Sanitaryware generally	300	nr	600			
· Below slab foul drainage; area based on building footprint	5,900	m²	12			
· Rainwater disposal; rainwater outlets; gratings; downpipes	8,600	m²	4			
· Soil and waste disposal; foul water and sanitary appliances	8,600	m²	12			
· Enhanced drainage to pool		item	60,000			
Hot and Cold Water Installations				**263,500**	**30.64**	**1.3%**
· Water supply; mains connection; booster set; storage tanks		item	120,000			
· Cold water service; distribution to toilets and changing	8,600	m²	7			
· Hot water services; local electric heating, service to toilets and changing	8,600	m²	6			
· Allowance for cold water drinking points	10	nr	3,000			
Space Heating and Air Treatment				**1,316,000**	**153.02**	**6.6%**
· Boiler installation; flues; pumps; valves; pipework	8,600	m²	24			
· Hot water generators; heat exchangers; buffer vessels	8,600	m²	6			
· Space heating; LTHW; pipework distribution; radiators	8,600	m²	24			
· Under floor heating to changing and reception areas	1,500	m²	36			
· Localised DX cooling to fitness suite and admin areas	1,000	m²	200			
· Ventilation installation ; supply and extract installation	4,200	m²	48			
· Specialist ventilation installation to changing, pools, WCs, kitchen, café and plant areas; air handling and distribution	4,400	m²	90			
Electrical and Gas Installations				**1,380,000**	**160.47**	**6.9%**
· HV/LV mains connection; switchgear; mains switchboard	8,600	m²	12			
· Submains cabling; containment; switchgear; and	8,600	m²	24			
· Small power	8,600	m²	18			
· Power to mechanical services and lifts	8,600	m²	6			
· Lighting installation and luminaires	8,600	m²	48			
· Emergency lighting; exit signs; luminaires	8,600	m²	6			
· Light switching; control system; presence detectors	8,600	m²	6			
· Diesel standby generator		item	180,000			
· Scene lighting control and external feature lighting		item	120,000			
· Gas installation to boilers and kitchen		item	48,000			

50m SWIMMING POOL WITH DRY SPORTS FACILITY COST MODEL

Swimming Pool (cont'd)	Quantity	Unit	Rate	Total (£)	Cost (£/m²)	%
Lift Installations				**192,000**	**22.33**	**1.0%**
· 13 person lift 1nr	1	nr	120,000			
· Goods lift 1nr	1	nr	72,000			
Protective, Communications and Special Installations				**1,778,200**	**206.77**	**8.9%**
· Wet riser installation	8,600	m²	6			
· Lightening protection	8,600	m²	2			
· Handheld firefighting appliance		item	6,000			
· Fire alarm and smoke detector installation	8,600	m²	12			
· Voice alarms	8,600	m²	6			
· Public address system	8,600	m²	6			
· Telephone and data containment	8,600	m²	6			
· Data back-bone wiring	8,600	m²	6			
· Telephone back-bone wiring	8,600	m²	7			
· CCTV installation; wiring, containment; equipment	8,600	m²	18			
· Access control security system; intruder alarm	8,600	m²	12			
· Induction loops to changing, seating, first aid, sports hall, reception and gym	3,000	m²	42			
· Television and FM radio aerial system; satellite system		item	6,000			
· Disabled toilet alarm system		item	12,000			
· Building Management System	8,600	m²	24			
· Pool water treatment; UV treatment		item	720,000			
Builders Work				**300,000**	**34.88**	**1.5%**
· Pads, bases, holes, chases, mortices, supports, walkways painting pipework		item	300,000			
Preliminaries and Contingency				**4,149,600**	**482.51**	**20.8%**
· Contractors preliminaries, overheads and profit @		17.5%				
· Contingency and design reserve @ say		7.5%				
Construction cost (Swimming pool and dry sports only, rate based on GIFA)				**19,939,000**	**2,318.49**	**100%**

REGIONAL STADIUM

The cost model is based on a regional stadium with a total of 25,000 seats. The stadium features a continuous roof enclosing one two-tier stand, with the rest of the seating arranged on a single tier. Back of house areas, hospitality areas and concessions.

- Gross internal floor area, including area of tiers 35,800m² Model location is South-East England (TPI = 580, LF = 0.97)

Regional Stadium	Quantity	Unit	Rate	Total (£)	Cost (£/m²)	%
Substructure				**1,944,200**	**54.31**	**5.1%**
· Excavate and fill site generally to an average depth of 500mm; disposal; allowance for breaking out	20,000	m²	9			
· Ground bearing slabs; blinding; DPM; hardcore; concrete slab with mesh reinforcement; ground beams and lift pits	10,600	m²	60			
· Piling and pile caps: 600 mm diameter piles; 15m deep; complete	10,600	m²	70			
· Column bases; including reinforced concrete; blinding; reinforcement; formwork; etc.	10,600	m²	37			
Frame and Upper Floors				**7,304,800**	**204.04**	**19.0%**
· Main frame, structural steel columns, beams and bracing	1,800	t	2,500			
· Intumescent paint / fireboard and architectural finishes	18,000	m²	14			
· In situ upper floor slabs to concourse areas; waffle construction with perimeter beam strips	14,000	m²	56			
· Precast concrete seating units; stainless steel locating pins; waterproofing	11,100	m²	160			
Roof				**5,098,200**	**142.41**	**13.3%**
· Structural steel main roof structure; high performance paint system	1,080	t	3,100			
· Roof access cat ladders	2	nr	1,850			
· Roof access stairs	2	nr	3,700			
· Latchway systems and walkways	680	m	140			
· Safety balustrades / handrails	560	m	210			
· Roof cladding system to main bowl; aluminium standing seam roofing; clear sections and overhangs	15,800	m²	85			
· Roof drainage: rainwater installations generally	15,800	m²	9			
· Camera Gantries and canopies		item	45,800			
Stairs				**847,100**	**23.66**	**2.2%**
· Reinforced insitu concrete stairs, landings and ramps, with power float finish and non-slip inserts to nosings	1,000	nr	310			
· Precast concrete step units; bolted to precast concrete seating units; forming gangway steps		item	37,100			
· Stair balustrades and handrails	2,000	m	250			
External Walls				**1,654,500**	**46.22**	**4.3%**
· Facing quality blockwork cavity wall to external elevations and bowl elevations	3,100	m²	120			
· Aluminium profiled sheet cladding system including secondary steelwork and insulation	3,300	m	210			
· Extra over for double glazed aluminium framed, facetted cladding system to walls; structural mullions	250	m²	280			
· Extra over sheet cladding for openable single glazed units in metal frames	250	m²	370			
· Glazing and glazed doors to executive boxes	450	m²	420			
· Galvanised steel weld mesh; 8m x 4m panels	1,700	m²	140			

REGIONAL STADIUM

Regional Stadium (cont'd)	Quantity	Unit	Rate	Total (£)	Cost (£/m²)	%
Windows and External Doors				**248,500**	**6.94**	**0.6%**
• Main entrances: laminated glazed screens and doors	30	nr	4,350			
• Escape doors; double escape doors	20	nr	4,350			
• Shutters; allowance for: power operated security shutters		item	31,000			
Internal Walls and Partitions				**1,734,200**	**48.44**	**4.5%**
• Insitu concrete walls; 200mm thick to lift and stair core walls	1,500	m²	140			
• Insitu concrete parapets to seating area	700	m²	140			
• Insitu concrete walls; 200mm thick to vomitories	40	nr	4,350			
• Blockwork division walls	15,000	m²	60			
• Proprietary vandal resistant metal faced toilet cubicles	300	nr	560			
• Glazed screens generally	150	m²	370			
• Front screens and privacy panels to executive suites	26	nr	4,950			
Internal Doors				**585,000**	**16.34**	**1.5%**
• Single doors and framesets; fire resisting; ironmongery	200	nr	1,125			
• Double doors and framesets; fire resisting; ironmongery	100	nr	1,750			
• Fire shutters to concession/bar fronts	20	nr	7,400			
• Rolling shutters generally	10	nr	3,700			
Wall Finishes				**669,100**	**18.69**	**1.7%**
• Render and tiling	4,000	m²	75			
• Plaster and paint	14,400	m²	12			
• Plaster and decorative coverings	100	m²	105			
• Paint finish on concrete or block walls	36,000	m²	5			
Floor Finishes				**544,400**	**15.21**	**1.4%**
• Vinyl sheeting/tiling; levelling screed; skirtings	3,500	m²	37			
• Contract grade carpet, levelling screed; skirtings	4,500	m²	37			
• Stone/ high quality ceramic tile; levelling screed; skirtings	500	m²	140			
• Paint and epoxy finish to concrete slabs; skirtings	16,000	m²	7			
• Tiled ceramic flooring, levelling screed; skirtings	800	m²	75			
Ceiling Finishes				**578,600**	**16.16**	**1.5%**
• Suspended ceilings; mineral fibre	5,100	m²	50			
• Plasterboard ceilings; skim coat and decorations	3,300	m²	37			
• Spray insulation	16,250	m²	12			
Furniture and Fittings				**1,388,300**	**38.78**	**3.6%**
• Padded upholstered seats; fixed units	21,000	nr	25			
• Padded upholstered seats; club seats	4,000	nr	31			
• Safety rails and barriers; to fixed seating bowl	1,750	m²	190			
• Security and crowd control gates; generally	150	m²	925			
• Turnstiles	40	nr	3,700			
• Allowance for signs; generally		item	120,000			
Sanitary Fittings and Disposal Installations				**936,400**	**26.16**	**2.4%**
• Sanitary fittings generally	850	nr	500			
• Below slab foul drainage; complete system,	10,600	m²	31			
• Sanitary fittings; IPS; concession areas, locker rooms etc.	24,700	m²	7			
Water Installations				**411,300**	**11.49**	**1.1%**
• Water supply; mains connection; booster set; storage tanks		item	105,000			
• Hot and cold water services	24,700	m²	12			

REGIONAL STADIUM

Regional Stadium (cont'd)	Quantity	Unit	Rate	Total (£)	Cost (£/m²)	%
Space Heating and Air Treatment				**2,129,300**	**59.48**	**5.6%**
· Space heating; boilers, flues, pumps and pressurisation sets; plant room and riser distribution	24,700	m²	15			
· Space heating; LTHW heating to public areas generally	10,600	m²	25			
· Localised cooling to hospitality areas; DX units		item	200,000			
· Air treatment and ventilation installations	10,600	m²	85			
· Extract installations; extract fans and ductwork to kitchens, toilets etc	2,500	m²	85			
· Smoke Extract Installations:	24,700	m²	7			
Electrical and Gas Installations				**3,253,000**	**90.87**	**8.5%**
· Mains connection; high voltage switchgear / transformers / connections; mains switchboard and busbars	24,700	m²	11			
· Sub mains distribution; switchboards; mains cabling	24,700	m²	6			
· Small power installation	24,700	m²	19			
· Lighting and luminaires	24,700	m²	50			
· Emergency lighting	24,700	m²	7			
· Under Roof Lighting	15,800	m²	7			
· Seating Bowl Lighting	11,100	m²	12			
· Containment installations	24,700	m²	7			
· Power supply to mechanical plant		item	62,000			
· Illuminated signs		item	62,000			
· Allowance for external "feature" lighting		item	210,000			
· Diesel standby generator		item	140,000			
· Gas installation to boilers and kitchen		item	37,100			
Lift Installations				**297,000**	**8.30**	**0.8%**
· 13 Person Lifts	2	nr	105,000			
· Goods Lifts	1	nr	87,000			
Protective, Communications and Special Installations				**2,938,400**	**82.08**	**7.7%**
· Hose Reel Installations		item	24,800			
· Dry Riser Installations		item	24,800			
· Lightning protection; earthing installations:		item	105,000			
· Public address and voice alarm system; main bowl PA	24,700	m²	19			
· Fire alarm system	24,700	m²	15			
· CCTV / security installations	24,700	m²	15			
· Allowance for card access and intruder alarm installations		item	62,000			
· Floodlighting installation		item	370,000			
· Playing surface; Fully heated pitch complete with drainage, irrigation, service ducts etc		item	850,000			
· BMS installation complete	24,700	m²	12			
Builders Work				**210,000**	**5.87**	**0.5%**
· Builder's work in connection with services @		item	210,000			
Preliminaries and Contingency				**5,573,700**	**155.69**	**14.5%**
· Commissioning management		item	62,000			
· Management costs, site establishment and supervision @		8%				
· Contractors overheads and profit @		3%				
· Contingency @		5%				
Construction cost (Regional Stadium only, rate based on GIFA)				**38,346,000**	**1,071.14**	**100%**

PRIMARY SCHOOL EXTENSION

A single storey, three-classroom extension to a primary school. Constructed using traditional masonry cavity walls on concrete strip foundations with a pitched tiled roof. Individual classrooms are formed by load bearing blockwork partitions

- Gross internal floor area 310m² Model location is South East England (TPI = 580, LF = 1.00)

Three classroom extension	Quantity	Unit	Rate	Total (£)	Cost (£/m²)	%
Substructure				**44,600**	**143.87**	**9.1%**
• Excavation & disposal	140	m³	32			
• Concrete strip foundations, masonry work below DPC; blockwork and facing brickwork	90	m	170			
• Reinforced insitu concrete ground slab, including service trench; vapour barrier; hardcore, excavation and disposal	310	m²	80			
Roof				**60,100**	**193.87**	**12.3%**
• Softwood roof trusses	380	m²	47			
• Board insulation to roof	310	m²	21			
• Cement slate roofing including all eaves, ridge tiles and labours; measured on plan	380	m²	65			
• Aluminium rainwater down pipes	40	m	85			
• Aluminium gutters	90	m	75			
• Fire barriers	40	m	21			
External Walls				**37,400**	**120.65**	**7.6%**
• Brick cavity wall, facing brick outer skin with cavity, 140mm inner blockwork leaf	220	m²	170			
Windows and External Doors				**35,600**	**114.84**	**7.3%**
• Proprietary aluminium framed, double glazed windows, doors and solid aluminium faced panels; powder coated finish	55	m²	450			
• Double door steel security doors, including all ironmongery	1	nr	1,800			
• Double door aluminium framed glazed doors and screens to paved areas, including all ironmongery	15	m²	600			
Internal Walls and Partitions				**14,300**	**46.13**	**2.9%**
• Partitions; 100/140 blockwork	285	m²	42			
• WC Cubicle partitions; laminated plastics, including all ironmongery	4	nr	575			
Internal Doors				**11,400**	**36.77**	**2.3%**
• Internal fire doors, Georgian wired glass vision panel, stainless steel ironmongery to classrooms	5	nr	1,050			
• Double fire door, Georgian wired glass, stainless steel ironmongery to corridor	1	nr	1,450			
• Wrot softwood storage cupboard door, stainless steel ironmongery	5	nr	525			
• Non fire rated doors, stainless steel ironmongery to WC	3	nr	675			
Wall Finishes				**17,000**	**54.84**	**3.5%**
• Plaster and 1 mist coat, 3 coats of emulsion paint	760	m²	19			
• Ceramic wall tiles, full height in toilets and selected classroom areas	40	m²	65			

PRIMARY SCHOOL EXTENSION

Three classroom extension (cont'd)	Quantity	Unit	Rate	Total (£)	Cost (£/m²)	%
Floor Finishes				**18,100**	**58.39**	**3.7%**
• Screed cement screed	310	m²	21			
• Carpet tiles	150	m²	26			
• Safety vinyl to WC areas & practical areas, including	90	m²	42			
• Heavy duty vinyl to circulation areas, including skirtings	68	m²	47			
• Entrance matting with aluminium matwell	2	m²	370			
Ceiling Finishes				**11,600**	**37.42**	**2.4%**
• Plasterboard ceiling, plaster skim and emulsion paint finish	280	m²	37			
• Moisture resistance plasterboard ceiling, plaster skim and emulsion paint finish	30	m²	42			
Furniture and Fittings				**22,600**	**72.90**	**4.6%**
• Storage trays containers	3	nr	2,850			
• Storage units - allowance of 1 double cupboard per	3	nr	575			
• Worktops, including cut out for sink, 3000x600	3	nr	150			
• Coat hooks, fixed to masonry	100	nr	32			
• Pinboards, 1000 x 2000	6	nr	85			
• Pinboards, 1000 x 1200	2	nr	60			
• Whiteboards : Interactive	3	nr	1,700			
• Whiteboards : Magnetic	3	nr	120			
• Signage	1	Item	2,300			
• Mirrors 640 x 460	7	nr	47			
Sanitary Fittings				**7,400**	**23.87**	**1.5%**
• WC's	6	nr	360			
• Urinals, including side panel	2	nr	240			
• Hand basins	8	nr	220			
• Disabled toilet, including WC, wash hand basin, grab rails and other fittings	1	nr	1,350			
• Single stainless steel sinks to classrooms, 1200 x 600	3	nr	260			
• Cleaners sink, 510 x 380	1	nr	850			
Disposal Installations				**6,500**	**20.97**	**1.3%**
• Waste, soil and vent installation; uPVC pipework and fittings	310	m²	21			
Water Installations				**11,500**	**37.10**	**2.4%**
• Cold water points	21	nr	320			
• Hot water points	13	nr	370			
Space Heating and Air Treatment				**43,400**	**140.00**	**8.9%**
• Space heating, all costs associated with the supply & installation of the heating system, temperature control and distribution pipework	310	m²	140			
Ventilation Installations				**5,300**	**17.10**	**1.1%**
• Ventilation extraction to toilet areas		item	5,300			
Electrical and Gas Installations				**41,600**	**134.19**	**8.5%**
• Mains and sub - mains installation	310	m²	26			
• Small power installation	310	m²	37			
• Lighting and general luminaires; emergency lighting	310	m²	65			
• Gas installations, all costs associated with the supply and installation of gas	310	m²	6			

PRIMARY SCHOOL EXTENSION

Three classroom extension (cont'd)	Quantity	Unit	Rate	Total (£)	Cost (£/m²)	%
Protective, Communications and Special Installations				**15,900**	**51.29**	**3.3%**
• Lightning protection		item	1,250			
• Fire alarm installation; smoke detectors; call points		item	4,750			
• Telephone and data wireways; internal telephone system		item	2,650			
• Security installation; intruder detection, CCTV to existing control unit etc		item	7,200			
Builders Work				**6,300**	**20.32**	**1.3%**
• Forming holes, chases etc		item	6,300			
Preliminaries and Contingency				**78,400**	**252.90**	**16.0%**
• Management costs; site establishment; site supervision @		16%				
• Contractors contingency @		3%				
Construction cost (Primary School extension, rate based on GIFA)				**489,000**	**1,577.42**	**100%**

APARTMENTS

This cost model is based on a mixed-tenure apartment building in a south-east location, featuring 65 open-market apartments and 35 flats for the affordable sector, in a mix of one and two-bedroom configurations. The scheme also features a 50-place semi-basement car park, providing secure spaces for the open-market element of the scheme. Demolition and site preparation, and external works are excluded

- Apartment block: Gross internal floor area 7,000m²
- Open Market Apartments: Net internal floor area 3,660m²
- Affordable Apartments: Net internal floor area 1,930m²
- Car-Park: Gross internal floor area 1,750m²

Model location is Outer London (TPI = 580, LF = 1.00)

This updated cost model is copyright of Davis Langdon LLP and was published in Building Magazine on 13-Feb-04

Apartment Shell and Core	Quantity	Unit	Rate	Total (£)	Cost (£/m²)	%
Substructure				**784,200**	**112.03**	**8.5%**
• Substructure, piled foundations, pile caps, ground slab	1,000	m²	725			
• Allowance for drainage	1,000	nr	46			
• Allowance for lift pits etc	2	nr	6,600			
Frame and Upper Floors				**1,759,000**	**251.29**	**19.0%**
• Insitu reinforced concrete frame and upper floors	6,650	m²	200			
• Balconies, primary and secondary frame, decking,	65	nr	6,600			
Roof				**160,800**	**22.97**	**1.7%**
• Flat roof coverings, single ply membrane, insulation, ballast; allowance for details to upstands	1,000	m²	100			
• Extra for roof terraces and paving to terraces	350	m²	60			
• Allowance for roof drainage, roof sundries	1,000	m²	20			
• Roof access equipment, latchways, cat ladder, access hatch, safety balustrade		item	19,900			
Stairs				**191,400**	**27.34**	**2.1%**
• RC concrete stairs, mild steel balustrades and handrails	14	m²	9,300			
• Extra over for enhanced finishes to entrance level staircases	2	m²	6,600			
• Balustrade and parapet to terraces; polyester powder coated	120	m	400			
External Walls, Windows and Doors				**2,486,800**	**355.26**	**26.8%**
• Unitised curtain walling; powder coated insulated aluminium spandrel panels; double-glazed tilt and turn windows	4,000	m²	600			
• Extra for doors to balconies, ironmongery	65	nr	1,050			
• Entrance doors, aluminium framed glazed door and screen	1	nr	13,200			
• External fire escape doors, metal, polyester powder coated	2	nr	2,650			
Internal Walls and Partitions				**373,100**	**53.30**	**4.0%**
• Core walls, insitu concrete, 225 thick	420	m²	130			
• Party walls to apartments and corridors; dense concrete block; head restraint, fire stopping	4,900	m²	65			
Internal Doors				**178,000**	**25.43**	**1.9%**
• Fire doors to cores and corridors; hardwood architraves / frames, including basic ironmongery	60	nr	1,000			
• Fire doors to risers; hardwood architraves / frames, including basic ironmongery	20	nr	650			
• Apartment entrance doors; solid core doors, hardwood architraves / frames, including basic quality ironmongery	100	nr	1,050			

APARTMENTS

Apartment Shell and Core (cont'd)	Quantity	Unit	Rate	Total (£)	Cost (£/m²)	%
Wall Finishes				**118,300**	**16.90**	**1.3%**
• Plasterboard to concrete and blockwork, with specialist painted finish; entrance hall	500	nr	65			
• Plasterboard to concrete and blockwork, with emulsion paint finish; lift lobbies and corridors	2,600	nr	33			
Floor Finishes				**124,800**	**17.83**	**1.3%**
• Feature ceramic floor tiles, sand cement screed; entrance hall	50	m²	160			
• Heavy duty carpet, sand cement screed; corridors	1,050	m²	80			
• Ceramic tile, sand cement screed; lift lobbies	170	m²	130			
• Skirtings, surface fixed skirting, painted MDF	810	m	13			
Ceiling Finishes				**55,300**	**7.90**	**0.6%**
• Painted plasterboard with feature bulkheads; reception	50	m²	130			
• Painted plasterboard on battens; lift lobbies and corridors	1,220	m²	40			
Furniture and Fittings				**19,900**	**2.84**	**0.2%**
• Allowance for reception area fittings; mailboxes, signage		item	19,900			
Sanitary Fittings and Disposal Installations				**139,300**	**19.90**	**1.5%**
• Allowance for cleaners sinks	14	nr	650			
• Rainwater disposal	7,000	m²	4			
• Soil, waste and overflow installations; stacks and connections to below ground drainage	7,000	m²	15			
Water Installations				**166,800**	**23.83**	**1.8%**
• Cold water storage tanks, booster pumps, mains distribution pipework, trace heating, water softener/conditioner etc.	7,000	m²	20			
• Hot and cold water services to landlord's areas, including local water storage heaters	1,220	m²	23			
Space Heating and Ventilation				**210,000**	**30.00**	**2.3%**
• Electric panel heaters; landlord's areas	1,220	m²	4			
• Central extract system for bathrooms; ductwork, extract fans	7,000	m²	26			
• Reception area air treatment		item	9,900			
• Supply and extract; plantroom areas		item	13,200			
Electrical Installations				**235,100**	**33.59**	**2.5%**
• Mains switchgear, cabling, containment and landlord's distribution boards	7,000	m²	13			
• Small power; landlord's areas	1,220	m²	8			
• Power supply to mechanical services	7,000	m²	7			
• Lighting and emergency lighting to landlord's areas	1,220	m²	40			
• Feature lighting to entrances		item	19,900			
• Earthing and bonding	7,000	m²	3			
Lift Installation				**238,000**	**34.00**	**2.6%**
• Lift installation; 13 person fire fighting lifts serving 7 storeys	2	nr	119,000			

APARTMENTS

Apartment Shell and Core (cont'd)	Quantity	Unit	Rate	Total (£)	Cost (£/m²)	%
Protective, Communications and Special Installations				**267,500**	**38.21**	**2.9%**
• Allowance for dry riser inlets	1,220	m²	40			
• Lightning protection	7,000	m²	3			
• Fire alarm system to landlord's areas	1,220	m²	46			
• Telephone containment only	7,000	m²	5			
• TV/Satellite system; central aerial and distribution	7,000	m²	5			
• Localised controls for cold water system	7,000	m²	5			
• CCTV and access control to perimeter		item	33,100			
Builders Work				**62,800**	**8.97**	**0.7%**
• Forming holes and chases; firestopping @		5%				
Preliminaries and Contingency				**1,690,900**	**241.56**	**18.3%**
• Testing and commissioning of building services @		2.5%				
• Contractor's overheads and profit, site establishment and supervision @		16%				
• Contingency @		5%				
Construction cost (Apartment shell and core only, rate based on GIFA)				**9,262,000**	**1,323.15**	**100%**

Open Market Apartment Fit-Out	Quantity	Unit	Rate	Total (£)	Cost (£/m²)	%
Internal Walls and Partitions and Doors				**505,400**	**138.09**	**11.6%**
• Metal stud partitions; 1 layer wall board each side; insulation; skim coat	4,575	m²	65			
• Flush doors; non fire rated; single leaf; solid core hardwood veneered; softwood frames; decorations; ironmongery	260	nr	800			
Wall Finishes				**356,400**	**97.38**	**8.2%**
• Plasterboard dry lining; MF framing; to external façade; emulsion paint finish	650	m²	60			
• Plasterboard; to concrete and blockwork walls; emulsion paint	3,730	m²	33			
• Ceramic tiles to kitchens	290	m²	115			
• Ceramic tiles to bathrooms	1,400	m²	115			
Floor Finishes				**405,900**	**110.90**	**9.3%**
• Suspended floor construction; ply on timber battens	3,080	m²	33			
• Edge fixed carpet; PC sum £20/m2; underlay	3,080	m²	46			
• Screed; ceramic tiling; to kitchens and bathrooms	730	m²	140			
• Skirtings; surface fixed skirting, painted MDF	3,700	m	13			
• Skirting; ceramic to match tiling	580	m	20			
Ceiling Finishes				**152,200**	**41.58**	**3.5%**
• Plasterboard suspended ceiling on battens; painting	3,660	m²	37			
• Feature bulkhead to junction with external wall	420	m	20			
• Plasterboard bulkhead for bathroom extract ductwork	65	nr	130			

APARTMENTS

Open Market Apartment Fit-Out (cont'd)	Quantity	Unit	Rate	Total (£)	Cost (£/m²)	%
Furniture and Fittings				**715,600**	**195.52**	**16.4%**
• Fully fitted kitchen to developer's specification with quality laminate worktops; appliances	65	nr	6,600			
• Additional fittings to kitchens to 2 bed apartments	30	nr	2,650			
• Built-in furniture to bedrooms; MDF, softwood frame and doors	95	nr	850			
• Allowance for built-in cloak, meter and airing cupboards	95	nr	400			
• Bathroom furniture, cistern enclosure; shelving	95	nr	530			
• Bathroom accessories, mirrors etc	95	nr	400			
Sanitary Fittings and Disposal Installations				**403,800**	**110.33**	**9.3%**
• Fully fitted bathroom; WC, bidet, washhand basin, pressed steel bath with power shower and screen	65	nr	3,450			
• Fully fitted en-suite shower room; WC, washhand basin, power shower, tray and screen; including all fixtures and fittings	30	nr	2,850			
• Kitchen sink; including all fixtures and fittings	65	nr	420			
• Soil waste and vent installation within apartments; connections to stacks	545	nr	105			
• Allowance for overflow pipework	3,660	nr	3			
Water Installations				**226,700**	**61.94**	**5.2%**
• Cold water supply; connection, meter	65	nr	210			
• Cold water distribution within apartments; final connections with sanitary fittings and appliances	545	nr	140			
• Domestic electric water heaters	65	nr	650			
• Hot water distribution within apartments; final connections with sanitary fittings and appliances	450	nr	210			
Space Heating, Air Treatment and Ventilation				**253,100**	**69.15**	**5.8%**
• Electrical panel heaters; local thermostatic control; power supply measured separately	320	nr	280			
• Electric heated towel rails	95	nr	500			
• Kitchen and bathroom extract, centralised bathroom system; localised kitchen extract with vent to façade, extract fans	160	nr	725			
Electrical Installation				**302,200**	**82.57**	**6.9%**
• Mains and sub-mains; connection; LV distribution boards to apartments; meters	65	nr	500			
• Small power distribution; sockets and fused connection points; wiring	1,685	nr	42			
• Cooker point; wiring	65	nr	130			
• Lighting; pendants, ceiling roses and bulkhead connections, wiring; to general areas	520	nr	36			
• Lighting; low energy fluorescent and low voltage fittings, wiring; to kitchens and bathrooms	580	nr	130			
• Shaving outlet; wiring	95	nr	95			
• Lighting; 5 amp lighting sockets; wiring	390	nr	42			
• Lighting distribution; switches and wiring	640	nr	40			
• Extra for; kitchen pelmet lighting	65	nr	240			
• Extra for; bathroom mirror lighting	95	nr	170			
• Allowance for earthing and bonding	65	nr	210			

APARTMENTS

Open Market Apartment Fit-Out (cont'd)	Quantity	Unit	Rate	Total (£)	Cost (£/m²)	%
Communication Installation				**150,200**	**41.04**	**3.5%**
• Fire alarm; combined detector/sounder; mains supply	130	nr	280			
• Phone points and wiring; 2 nr points	65	nr	150			
• TV sockets and wiring; 2 nr sockets	65	nr	150			
• Video entry phone system	65	nr	1,450			
Builders Work				**66,800**	**18.25**	**1.5%**
• Forming holes and chases; firestopping @		5%				
Preliminaries and Contingency				**813,700**	**222.32**	**18.7%**
• Testing and commissioning of building services @		2.5%				
• Contractor's overheads and profit, site establishment and supervision @		16%				
• Contingency @		5%				
Construction cost (Open Market Apartment fit-out only, rate based on NIA area)				**4,352,000**	**1,189.07**	**100%**

Affordable Market Apartment Fit-Out	Quantity	Unit	Rate	Total (£)	Cost (£/m²)	%
Internal Walls and Partitions and Doors				**227,500**	**117.88**	**15.1%**
• Metal stud partitions; 1 layer wall board each side; insulation; skim coat	2,100	m²	65			
• Flush doors; non fire rated; single leaf; solid core hardwood veneered; softwood frames; decorations; ironmongery	140	nr	650			
Wall Finishes				**144,600**	**74.92**	**9.6%**
• Plasterboard dry lining; MF framing; to external façade; emulsion paint finish	380	m²	60			
• Plasterboard; to concrete and blockwork walls; emulsion paint	2,510	m²	33			
• Ceramic tiles to kitchens	70	m²	95			
• Ceramic tiles to bathrooms	340	m²	95			
Floor Finishes				**152,200**	**78.86**	**10.1%**
• Suspended floor construction; ply on timber battens	1,450	m²	33			
• Edge fixed carpet; PC sum £20/m2; underlay	1,450	m²	33			
• Sand cement screed; ceramic tiling; to kitchens and bathrooms	455	m²	70			
• Skirtings; surface fixed skirting, painted MDF	1,410	m	13			
• Skirting; ceramic to match tiling	460	m	13			
Ceiling Finishes				**70,500**	**36.53**	**4.7%**
• Plasterboard suspended ceiling on battens; painting	1,905	m²	37			
Furniture and Fittings				**183,300**	**94.97**	**12.2%**
• Kitchen fittings to housing association specifications	35	nr	3,300			
• Additional fittings to kitchens to 2 bed apartments	15	nr	650			
• Allowance for built-in furniture to bedrooms; MDF, softwood frame and doors	50	nr	650			
• Allowance for built-in cloak, meter and airing cupboards	35	nr	330			
• Allowance for bathroom furniture, cistern enclosure; shelving	35	nr	200			
• Allowance for bathroom accessories, mirrors etc	35	nr	200			

APARTMENTS

Affordable Market Apartment Fit-Out (cont'd)	Quantity	Unit	Rate	Total (£)	Cost (£/m²)	%
Sanitary Fittings and Disposal Installations				**99,200**	**51.40**	**6.6%**
• Fully fitted bathroom; WC, bidet, washhand basin, pressed steel bath with power shower and screen; including all fixtures and fittings	35	nr	1,700			
• Kitchen sink; including all fixtures and fittings	35	nr	360			
• Soil waste and vent installation within apartments; connections to stacks	210	nr	105			
• Allowance for overflow pipework	1,930	nr	3			
Water Installations				**101,200**	**52.44**	**6.7%**
• Cold water supply; connection, meter	35	nr	210			
• Cold water distribution within apartments; final connections with sanitary fittings and appliances	245	nr	140			
• Domestic electric water heaters	35	nr	650			
• Hot water distribution within apartments; final connections with sanitary fittings and appliances	175	nr	210			
Space Heating, Air Treatment and Ventilation				**111,000**	**57.51**	**7.4%**
• Electrical panel heaters; local thermostatic control; power supply measured separately	170	nr	280			
• Electric heated towel rails; power supply measured	35	nr	360			
• Kitchen and bathroom extract, centralised bathroom system; localised kitchen extract with vent to façade, extract fans	70	nr	725			
Electrical Installation				**90,600**	**46.94**	**6.0%**
• Mains and sub-mains; connection; LV distribution boards to apartments; meters	35	nr	500			
• Small power distribution; sockets and fused connection points; wiring	905	nr	42			
• Cooker point; wiring	35	nr	130			
• Lighting; pendants, ceiling roses and bulkhead connections, wiring; to general areas	280	nr	36			
• Shaving outlet; wiring	35	nr	95			
• Lighting; 5 amp lighting sockets; wiring	245	nr	40			
• Allowance for earthing and bonding	35	nr	210			
Communication Installation				**22,400**	**11.61**	**1.5%**
• Fire alarm; combined detector/sounder; mains supply	35	nr	280			
• Phone points and wiring; 2 nr points	35	nr	80			
• TV sockets and wiring; 2 nr sockets	35	nr	80			
• Audio entry phone system	35	nr	200			
Builders Work				**21,200**	**10.98**	**1.4%**
• Forming holes and chases; firestopping @ say 5%		5%				
Preliminaries and Contingency				**281,000**	**145.60**	**18.7%**
• Testing and commissioning of building services @ say 2.5%		2.5%				
• Contractor's overheads and profit, site establishment and supervision @		16%				
• Contingency @		5%				
Construction cost (Affordable Market Apartment fit-out only, rate based on NIA)				**1,504,700**	**779.64**	**100%**

APARTMENTS

Semi-Basement Car-Park	Quantity	Unit	Rate	Total (£)	Cost (£/m²)	%
Substructure				**523,000**	**298.86**	**31.1%**
· Concrete retaining wall; temporary propping	510	m²	300			
· Excavation and disposal, including dewatering	5,250	m²	64			
· Tie in slab edge to retaining wall	170	m	200			
Frame and Upper Floors				**536,000**	**306.29**	**31.9%**
· Reinforced in-situ concrete columns and suspended slab to ground floor	1,750	m²	240			
· Extra for vehicle ramp		item	53,000			
· Allowance for louvres for natural ventilation	175	m²	360			
Stairs				**21,200**	**12.11**	**1.3%**
· Insitu concrete stairs and half landings; mild steel, polyester coated handrails and balustrades; finishes	2	nr	10,600			
Internal Walls and Partitions				**16,400**	**9.37**	**1.0%**
· Blockwork partitions; facework; 215 average thickness; emulsion paint finish	60	m²	65			
· Reinforced concrete core walls; emulsion paint finish	100	m²	125			
Internal Doors				**11,200**	**6.40**	**0.7%**
· Flush doors; fire rated; double leaf; solid core ply faced; softwood frames; decorations; ironmongery; complete	2	nr	1,300			
· Flush doors; fire rated; single leaf; solid core ply faced; softwood frames; decorations; ironmongery; complete	4	nr	1,050			
· Fire shutters; 120 minutes fire resistance; frame and sub-frame; electric operation	2	nr	2,200			
Finishes				**63,600**	**36.34**	**3.8%**
· Emulsion paint to concrete and blockwork	320	m²	4			
· Allowance for painted floor finish, with car parking demarcation	1,750	m²	9			
· Allowance for insulation to underside of building footprint	1,000	m²	46			
Fittings				**26,500**	**15.14**	**1.6%**
· Car park barriers and operating system		item	19,900			
· Protective bollards; kerbs; barriers; column guards etc		item	0,000			
Electrical Installations				**68,600**	**39.20**	**4.1%**
· Mains and sub - mains installation	1,750	m²	7			
· Lighting and luminaires to car park areas	1,750	m²	26			
· Emergency lighting and luminaires	1,750	m²	7			
Protective and Communications Installations				**103,600**	**59.20**	**6.2%**
· Sprinkler installation; ordinary hazard group 1	1,750	m²	46			
· Fire, smoke detection and alarm system	1,750	m²	13			
Builders Work				**5,200**	**2.97**	**0.3%**
· Forming holes and chases; firestopping @ say 3%		3%				
Preliminaries and Contingency				**304,700**	**174.11**	**18.1%**
· Testing and commissioning of building services @		2.5%				
· Overheads and profit, site establishment and supervision @		16%				
· Contingency @		5%				
Construction cost (Semi-Basement Car-Park only, rate based on GIFA)				**1,680,000**	**959.99**	**100%**

HOTEL

This cost model is for a new build business hotel located in an urban locations in Manchester. The hotel floor area is 8,400m² and utilises a proportion of MMC including bathroom pods and pre-cast structural concrete beams, slabs, crosswalls and external wall panels. Amenities include meeting rooms, bar and restaurants. The costs cover all areas - front of house, back of house and guestrooms. The costs of site preparation, external works and incoming services are excluded.

- Gross internal floor area 8,400m² Model location is Outer London (TPI = 580, LF = 1.00)

This updated cost model is copyright of Davis Langdon LLP and was originally published in Building Magazine on 01-Dec-06

Hotel	Quantity	Unit	Rate	Total (£)	Cost (£/m²)	%
Substructure				**825,000**	**98.21**	**5.3%**
• Excavation, ground beams, filling to levels, lift pits, ground slab	1,500	m²	360			
• Rotary bored piles	1,500	m³	150			
• Underslab drainage	1,500	m²	40			
Frame and Upper Floors				**1,768,000**	**210.48**	**11.4%**
• In-situ concrete frame and flat slab acting as transfer structure; 400 thick slab	1,100	m²	210			
• Precast concrete floor slab, cross walls and stairs (quantity based on floor slab area); self finish quality to cross walls	5,300	m²	290			
Roof				**554,400**	**66.00**	**3.6%**
• Precast concrete roof slab	1,500	m²	170			
• Extra for forming upstands and copings		item	17,300			
• Single ply roof membrane; insulation; rainwater outlets	1,500	m²	140			
• Roof plant room; louvre screens and cladding	100	m²	450			
• Mansafe system		item	9,800			
• Allowance for roof level ancillaries; walkways, plant bases		item	17,300			
Stairs				**176,400**	**21.00**	**1.1%**
• Insitu concrete; ground to first floor (other stairs included in frame and upper floors package)	3	nr	10,400			
• Handrails and balustrades; stainless steel	21	m	4,150			
• Floor and soffit finishes to stairs; nosings		item	58,000			
External Walls, Windows and Doors				**1,708,800**	**203.43**	**11.0%**
• Precast concrete wall panels; insulation and self-coloured render	2,850	m²	320			
• Extra for coated aluminium-framed double glazed windows to guest room floors	670	m²	350			
• Full height glazed window wall; ground floor elevations	500	m²	575			
• Extra over above for double doorsets	9	nr	2,750			
• Extra for glazed entrance lobby		item	120,000			
• Allowance for additional masonry works		item	130,000			
Internal Walls and Partitions				**320,700**	**38.18**	**2.1%**
• Blockwork; 100 and 140 thick	1,800	m²	46			
• Acoustic metal stud partitions	2,500	m²	60			
• Hardwood glazed partitions; fire rated glazing	135	m²	575			
• WC cubicles	10	nr	1,025			

HOTEL

Hotel (cont'd)	Quantity	Unit	Rate	Total (£)	Cost (£/m²)	%
Internal Doors				**541,300**	**64.44**	**3.5%**
• *(Bathroom doors included in pod costs)*						
• Hardwood doors and frames; vision panels; stainless steel ironmongery	80	nr	1,250			
• Bedroom doors; card access control; ironmongery	200	nr	1,550			
• Melamine faced doors and hardwood frames; ironmongery; back of house areas	50	nr	1,025			
• Riser access doors; ironmongery	100	nr	800			
Wall Finishes				**493,500**	**58.75**	**3.2%**
• *(Finishes to bathroom interiors included in bathroom pod costs; some bedroom finishes included in FF&E)*						
• Drylining to bathroom pod	1,800	m²	40			
• Specialist finish to public areas; decorative panels	400	m²	230			
• Applied finish to bedrooms and corridors; vinyl	10,500	m²	23			
• Whiterock cladding to kitchen areas	200	m²	70			
• Ceramic tiling	200	m²	70			
• Timber window boards	400	m	35			
• Corridor corner guards		item	46,000			
Floor Finishes				**508,500**	**60.54**	**3.3%**
• *(Finishes to bathrooms included in bathroom pod cost)*						
• Guest room flooring; edge fixed carpet	6,400	m²	23			
• Screeds generally	7,500	m²	14			
• Wood flooring - front of house	750	m²	170			
• Ceramic tiling	100	m²	120			
• Specialist flooring	225	m²	150			
• Allowance for skirtings and joints generally		item	77,000			
• Entrance matting		item	7,500			
Ceiling Finishes				**262,000**	**31.19**	**1.7%**
• *(Finishes to bathrooms included in bathroom pod cost)*						
• Plasterboard ceiling and bulkheads	6,400	m²	35			
• Extra for acoustic treatment in public areas	900	m²	23			
• Allowance for feature ceilings in front of house areas		item	17,300			
Furniture and Fittings						
(Furniture and fittings to front of house and guestrooms generally included in FF&E budget)						
Sanitary Fittings				**28,800**	**3.43**	**0.2%**
• *(Sanitary ware to guestrooms in bathroom pod cost)*						
• Sanitary ware and fittings to front and back of house only		item	28,800			
Services Equipment				**580,000**	**69.05**	**3.7%**
• Installation of kitchen and servery complete; including catering equipment		item	580,000			
Disposal Installations				**63,500**	**7.56**	**0.4%**
• Waste, soil and vent pipework to guestrooms; stub connections to pods	200	nr	260			
• Rainwater installation		item	11,500			

HOTEL

Hotel (cont'd)	Quantity	Unit	Rate	Total (£)	Cost (£/m²)	%
Hot and Cold Water Installations				**243,600**	**29.00**	**1.6%**
• Hot and cold water installation; incoming main, storage, distribution; valves and accessories in front and back of house; stub connections to pods only	8,400	m²	29			
Space Heating, Air Treatment and Ventilation				**1,137,000**	**135.36**	**7.3%**
• Air conditioning to public areas; main plant; ductwork, pipework, insulation, terminal units, grilles and diffusers	1,000	m²	130			
• Extract ventilation and heating/cooling to guest rooms; main plant; ductwork, pipework, insulation, wall mounted units	200	nr	3,550			
• Extra for supplementary supply ventilation to inboard guest rooms	30	nr	925			
• Staircase pressurisation		item	98,000			
• Toilet extract ventilation; public areas only		item	9,200			
• Supplementary supply and extract ventilation to front and back of house areas; dedicated systems serving restaurant, meeting rooms, lobby etc		item	104,000			
• Kitchen supply and extract		item	58,000			
Electrical and Gas Installations				**1,513,900**	**180.23**	**9.8%**
• *(Lighting to guestroom bathrooms included in bathroom pods, decorative lighting included in FF&E)*						
• Mains and sub-mains distribution	8,400	m²	23			
• Lighting installation to front and back of house; luminaires; emergency lighting	1,500	m²	120			
• Small power to front and back of house	1,500	m²	23			
• Lighting installation to guestrooms; luminaires; emergency lighting	6,400	m²	90			
• Small power to guestrooms	6,400	m²	46			
Fix only allowance for connecting lighting & appliances included in FF&E	200	nr	575			
Electrical supplies to mechanical plant, including guestroom ventilation units		item	46,000			
• External building lighting		item	40,300			
• Incoming gas supply, including steel pipework, valves etc		item	34,500			
Lift Installations				**282,800**	**33.67**	**1.8%**
• Public lifts	2	nr	98,000			
• Service lift	1	nr	58,000			
• Platform lift	1	nr	28,800			
Protective Installations				**40,000**	**4.76**	**0.3%**
• Lightning protection		item	9,200			
• Dry Riser		item	11,500			
• Earthing and bonding	8,400	m²	2			
Communication Installations				**356,500**	**42.44**	**2.3%**
• Fire alarm and smoke detection	8,400	m²	17			
• Disabled WC alarm system		item	9,200			
• Allowance for containment		item	23,000			
• Telephone and data cabling		item	63,000			
• Audio and TV distribution network		item	58,000			
• Security and CCTV systems		item	58,000			

HOTEL

Hotel (cont'd)	Quantity	Unit	Rate	Total (£)	Cost (£/m²)	%
Specialist Installations				**1,305,300**	**155.39**	**8.4%**
· Bathroom pods including protection, sealing duct, door handles and locks	200	nr	5,800			
BMS Controls; Installations and PC	8,400	m²	17			
Builders Work				**40,300**	**4.80**	**0.3%**
· Builder's work in connection with services @		item	40,300			
Preliminaries and Contingency				**2,769,700**	**329.73**	**17.8%**
· Testing and commissioning of building services installations		item	23,000			
· Management costs, site establishment and site supervision. Contractor's preliminaries, overheads and profit @		18%				
· Design reserve @		3%				
Construction cost (Hotel only, rate based on GIFA)				**15,520,000**	**1,847.64**	**100%**

FF&E Items				**2,848,000**	**339.05**	**18.4%**
· Front-of-house and back-of-house items	8,400	m²	120			
· Guestroom fitout, casework, fixed and loose furniture, lighting fittings and apliances	200	nr	9,200			

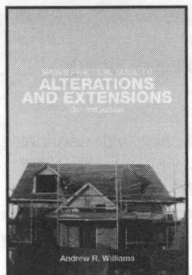

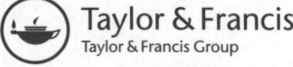

Approximate Estimates

Estimating by means of priced approximate quantities is always more accurate than by using overall prices per square metre. Prices given in this section, which is arranged in elemental order, are derived from "Prices from Measured Works – Major Works" section, but also include for all the incidental items and labours which are normally measured separately in Bills of Quantities. As in other sections, they have been established with a tender price level of 580 (1976 = 100). They include overheads and profit but do not include for preliminaries, details of which are given in Part 3, and which in the current tendering climate amount to approximately 16% of the value of the measured works.

Whilst every effort is made to ensure the accuracy of these figures, they have been prepared for approximate estimating purposes only and on no account should they be used for the preparation of tenders.

Unless otherwise described units denoted as m² refer to appropriate unit areas (rather than gross floor areas).

As elsewhere in this edition prices do not include Value Added Tax, which should be applied at the current rate.

Item	Unit	Range £
1 SUBSTRUCTURE		
Ground floor area (unless otherwise described)		
Strip or base foundations		
Foundations in good ground; reinforced concrete bed: for up to two storey development		
shallow foundations up to 1200 mm deep	m²	66.00 to 135.00
deep foundations up to 2400 mm deep	m²	110.00 to 195.00
extra for		
each additional storey	m²	23.00 to 30.50
Raft foundations		
Raft on poor ground for development up to two storey high	m²	88.00 to 235.00
extra for		
each additional storey	m²	23.00 to 30.50
Piled foundations		
Foundation in poor ground; reinforced concrete slab; for one storey commercial development		
short bore piles to columns only	m²	110.00 to 170.00
short bore piles	m²	145.00 to 205.00
fully piled	m²	195.00 to 290.00
Basements		
Basement floor/wall area (as appropriate)		
Basement (excluding bulk excavation costs)		
Reinforced concrete basement floors		
non – waterproofed	m²	71.00 to 94.00
waterproofed	m²	94.00 to 130.00
Reinforced concrete basement walls		
non – waterproofed	m²	210.00 to 265.00
waterproofed	m²	240.00 to 305.00
sheet piled	m²	485.00 to 600.00
Diaphragm walling	m²	480.00 to 560.00
extra for		
each additional basement level	%	20.00 to 25.00
Underpinning		
In stages not exceeding 1500 mm long from one side of existing wall and foundation; excavate preliminary trench by machine and underpinning pit by hand; partial backfill; partial disposal; earthwork support (open boarded); cutting away projecting foundations; prepare underside of existing; compact base of pit; plain in situ concrete 20.00 N/mm² to 20 mm aggregate (1:2:4); formwork; brickwork in cement mortar (1:3); pitch polymer damp proof course; wedge and pin to underside of existing with slates. Commencing at 1.00m below ground level with common bricks PC £240.00 per 1000; depth of underpinning		
900 mm high; one brick wall	m²	325.00 to 395.00
1500 mm high; one brick wall	m²	465.00 to 565.00
extra for excavating commencing		
2.00 m below ground level	m²	60.00 to 82.00
3.00 m below ground level	m²	120.00 to 160.00
4.00 m below ground level	m²	160.00 to 220.00

Item	Unit	Range £		
Trench fill foundations				
Machine excavation; disposal; plain in-situ concrete 20.00 N/mm² to 20 mm aggregate (1:2:4) trench fill; 300 mm cavity high brickwork in cement mortar (1:3); pitch polymer damp roof course; machine excavation				
With common bricks outer skin PC £240 per 1000				
600 mm x 1000 mm deep	m	115.00	to	130.00
600 mm x 1500 mm deep	m	145.00	to	175.00
extra over for three courses of facing bricks to outer skin				
PC £350.00 per 1000	m	2.35	to	2.75
PC £500.00 per 1000	m	6.85	to	7.98
Strip foundations				
Excavate trench 600 mm wide; partial backfill; partial disposal; earthwork support (risk item); compact base of trench; plain in-situ concrete 20 N/mm² to 20 mm aggregate (1:2:4) 250 mm thick; cavity brickwork/blockwork in cement mortar (1:3); pitch polymer damp proof; course machine excavation; machine excavation				
With common bricks outer skin PC £240 per 1000				
600 mm x 1000 mm deep	m	130.00	to	150.00
600 mm x 1500 mm deep	m	140.00	to	165.00
extra over for three courses of facing bricks to outer skin				
PC £350.00 per 1000	m	2.35	to	2.75
PC £500.00 per 1000	m	6.85	to	7.95
Column bases				
Excavate pit in firm ground by machine; partial backfill; partial disposal; support; compact base of pit; plain in-situ concrete 20.00 N/mm² to 20mm aggregate (1:2:4); formwork; machine excavation; base size				
up to 600 mm x 600 mm x 300 mm; 1000 mm deep pit	m³	740.00	to	860.00
up to 900 mm x 900 mm x 450 mm; 1250 mm deep pit	m³	520.00	to	610.00
up to 1500 mm x 1500 mm x 600 mm; 1500 mm deep pit	m³	385.00	to	450.00
up to 2700 mm x 2700 mm x 1000 mm; 1500 mm deep pit	m³	235.00	to	275.00
extra for				
reinforcement at 50 kg/m³ concrete; base size	m³	51.00	to	62.00
reinforcement at 75 kg/m³ concrete; base size	m³	76.00	to	93.00
reinforcement at 100 kg/m³ concrete; base size	m³	100.00	to	125.00
Concrete Piles				
Total length of pile				
Supply and install concrete Continuous Flight Auger (CFA) piles; set up at each location; cart away spoil				
450 mm reinforced concrete CFA piles	m	42.00	to	67.00
600 mm diameter reinforced concrete CFA piles	m	72.00	to	95.00
mobilisation and de-mobilisation of CFA piling rig	item	9000	to	12000
Steel Piling				
Gross total area of sheets				
Interlocking steel sheet piling to excavation perimeter; Corus LX or similar; extraction on completion	m²	147.00	to	179.00
Concrete Ground Beams				
Reinforced in situ concrete ground beams; bar reinforcement; formwork				
300 mm x 300 mm; reinforcement at 180 kg/m³	m	48.50	to	61.00
450 mm x 450 mm; reinforcement at 200 kg/m³	m	105.00	to	130.00
450mm x 600 mm; reinforcement at 270 kg/m³	m	160.00	to	200.00

Item	Unit	Range £
1 SUBSTRUCTURE – cont'd		
Pile caps		
Excavate pit in firm ground by machine; partial backfill partial disposal; earthwork support; compact base of pit; cut off top of pile and prepare reinforcement; reinforced in-situ concrete 25.00 N/mm² to 20 mm aggregate (1:2:4); formwork		
Reinforcement at 50 kg/m³ concrete; cap size		
up to 900 mm x 900 mm x 1000 mm; one pile	m³	480.00 to 560.00
up to 2100 mm x 2100 mm x 1000 mm; two to three piles	m³	310.00 to 365.00
up to 2700 mm x 2700 mm x 1500 mm; four piles	m³	280.00 to 325.00
extra for		
reinforcement at 75 kg/m³ concrete; cap size	m³	25.00 to 30.50
reinforcement at 100 kg/m³ concrete; cap size	m³	49.00 to 61.00
additional cost of alternative strength concrete		
30.00 N/mm²	m³	1.65 to 2.15
40.00 N/mm²	m³	3.80 to 4.80
Ground slabs		
Mechanical excavation to reduce levels; disposal; level and compact; hardcore bed blinded with sand; 1200 gauge polythene damp proof membrane; in-situ concrete 20.00 N/mm² to 20 mm aggregate (1:2:4)		
150 mm thick concrete slab with 1 layer of A195 fabric reinforcement	m²	62.00 to 78.00
200 mm thick concrete slab with 1 layer of A252 fabric reinforcement	m²	68.00 to 86.00
250 mm thick concrete slab with 1 layer of A393 fabric reinforcement	m²	76.00 to 95.00
extra for every additional 50mm thick concrete	m²	6.50 to 8.15
Add to the foregoing prices for high yield steel bar reinforcement B.S. 4449 straight or bent; at a rate of:-		
25 kg/m³	m³	20.00 to 25.50
50 kg/m³	m³	51.00 to 62.00
75 kg/m³	m³	76.00 to 93.00
100 kg/m³	m³	100.00 to 125.00
additional cost of alternative strength concrete		
30.00 N/mm²	m³	1.60 to 2.05
40.00 N/mm²	m³	3.70 to 4.60
Warehouse Ground Slab		
Steel fibre reinforced floor slab placed using large pour construction techniques providing a finish floor flatness complying with FM2 special +/to 15mm from datum. Excavation; sub-base and damp proof membrane not included		
Nominal 200 mm thick in-situ concrete floor slab; concrete grade C40; reinforced with steel fibres; surface power floated and cured with a spray application of curing and hardening agent	m²	31.50 to 39.00
Temporary Works		
Formation of temporary roads to building perimeter comprising of geotextile membrane and 300 mm MOT type 1	m²	9.50 to 12.60
Installation of wheel wash facility	nr	2900.00 to 3650.00

Item	Unit	Range £
2A FRAME AND 2B UPPER FLOORS		
Comparative Frame and Upper Floors		
Upper floor area (unless otherwise described)		
Reinforced concrete floors and concrete frame		
Suspended slab; no coverings or finishes		
up to six storeys	m²	130.00 to 175.00
seven to twelve storeys	m²	165.00 to 210.00
thirteen to eighteen storeys	m²	255.00 to 330.00
Reinforced concrete floors and steel frame		
Suspended slab; permanent steel shuttering; protected steel frame; no coverings or finishes		
up to six storeys	m²	125.00 to 155.00
seven to twelve storeys	m²	130.00 to 160.00
thirteen to eighteen storeys	m²	165.00 to 235.00
Post-tensioned concrete upper floors		
Reinforced post-tensioned suspended concrete slab 150 mm to 225 mm thick; 40 N/mm2; reinforcement 60 kg/m3; formwork	m²	115.00 to 135.00
Pre-cast concrete floors		
Suspended slab; 75 mm thick screed; no coverings or finishes		
6.00 m span; 5.00 kN/m² loading	m²	54.00 to 76.00
7.50 m span; 5.00 kN/m² loading	m²	56.00 to 84.00
6.00 m span; 8.50 kN/m² loading	m²	72.00 to 81.00
7.50 m span; 8.50 kN/m² loading	m²	73.00 to 84.00
6.00 m span; 12.50 kN/m² loading	m²	79.00 to 89.00
Pre-cast concrete floors; steel frame		
Suspended slabs; unprotected steel frame; no coverings or finishes		
up to three storeys	m²	115.00 to 205.00
Suspended slabs; protected steel frame; no coverings or finishes		
up to six storeys	m²	175.00 to 255.00
seven to twelve storeys	m²	215.00 to 335.00
Softwood floors		
Joisted floor; plasterboard ceiling; skim; emulsion; t&g chipboard; sheet vinyl flooring and painted softwood skirtings	m²	50.00 to 85.00
Reinforced concrete frame		
Generally all formwork assumes four uses		
Reinforced in-situ concrete columns; bar reinforcement; formwork		
Reinforcement rate 180 kg/m³; column size		
225 mm x 225 mm	m	58.00 to 68.00
300 mm x 600 mm	m	145.00 to 170.00
450 mm x 900 mm	m	260.00 to 305.00
Reinforcement rate 240 kg/m³; column size		
225 mm x 225 mm	m	60.00 to 71.00
300 mm x 600 mm	m	155.00 to 180.00
450 mm x 900 mm	m	290.00 to 340.00

Item	Unit	Range £
2A FRAME AND 2B UPPER FLOORS – cont'd		
Reinforced concrete frame – cont'd		
In-situ concrete casing to steel column; formwork; column size		
225 mm x 225 mm	m	47.00 to 55.00
300 mm x 600 mm	m	110.00 to 125.00
450 mm x 900 mm	m	180.00 to 210.00
Reinforced in-situ concrete beams; bar reinforcement; formwork		
Reinforcement rate 200 kg/m³; beam size		
225 mm x 450 mm	m	88.00 to 100.00
300 mm x 600 mm	m	135.00 to 155.00
450 mm x 600 mm	m	175.00 to 205.00
600 mm x 600 mm	m	215.00 to 250.00
Reinforcement rate 240 kg/m³; beam size		
225 mm x 450 mm	m	92.00 to 110.00
300 mm x 600 mm	m	145.00 to 170.00
450 mm x 600 mm	m	190.00 to 220.00
600 mm x 600 mm	m	230.00 to 270.00
In-situ concrete casing to steel beams; formwork; beam size		
225 mm x 450 mm	m	65.00 to 76.00
300 mm x 600 mm	m	92.00 to 110.00
450 mm x 600 mm	m	115.00 to 130.00
600 mm x 600 mm	m	135.00 to 155.00
Steel Frame		
Fabricated steelwork erected on site with bolted connections; primed		
universal beams; grade S275	tonne	1550.00 to 1700.00
universal beams; grade S355	tonne	1600.00 to 1775.00
universal columns; grade S275	tonne	1550.00 to 1775.00
universal columns; grade S355	tonne	1650.00 to 1850.00
composite beams	tonne	1975.00 to 2300.00
lattice beams	tonne	1850.00 to 2175.00
rectangular section columns; grade S355	tonne	2075.00 to 2275.00
composite columns	tonne	1900.00 to 2250.00
roof trusses	tonne	1900.00 to 2250.00
smaller sections	tonne	1550.00 to 1775.00
hollow section circular; square; rectangular	tonne	2075.00 to 2275.00
extra for		
galvanising	tonne	305.00 to 365.00
400 mm Ø x 12 mm thick cells in beams	tonne	260.00 to 315.00
Other floor and frame construction/extras		
Space deck on steel frame; unprotected	m²	315.00 to 365.00
Exposed steel frame for tent/mast structures	m²	185.00 to 440.00
Columns and beams to 18.00 m high bay warehouse unprotected	m²	125.00 to 250.00
Columns and beams to mansard protected	m²	140.00 to 165.00
Feature columns and beams to glazed atrium roof unprotected	m²	140.00 to 220.00
Reinforced concrete cantilevered balcony	nr	2275.00 to 3050.00
Reinforced concrete cantilevered walkways	m²	165.00 to 205.00
Reinforced concrete walkways and supporting frame	m²	190.00 to 280.00
Reinforced concrete core with steel umbrella frame; twelve to twenty four storeys	m²	385.00 to 525.00
extra for		
wrought formwork	m²	4.60 to 12.00
sound reducing quilt in screed	m²	4.85 to 8.10
insulation to avoid cold bridging	m²	4.85 to 11.30

Item	Unit	Range £		
Comparative steel finishes				
primer only	m²	1.60	to	2.45
grit blast and one coat zinc chromate primer	m²	1.70	to	2.80
touch up primer and one coat of two pack epoxy zinc phosphate primer	m²	2.45	to	3.15
Fire Protection				
Gross surface area				
Sprayed mineral fibre				
60 minute protection	m²	10.60	to	16.60
90 minute protection	m²	17.90	to	21.50
Sprayed vermiculite cement				
60 minute protection	m²	12.10	to	17.90
90 minute protection	m²	14.60	to	21.50
Supply and fit fire resistant boarding to steel columns and beams; noggins; brackets angles; intumescent paste. Beamclad or similar				
Board area				
60 minute protection	m²	34.50	to	52.00
60 minute protection	m²	50.00	to	68.00
Intumescent fire protection coating / decoration to exposed steelwork				
Gross surface area				
30 minute protection; on site application	m²	5.25	to	15.70
30 minute protection; on site application	tonne	190.00	to	365.00
60 minute protection; on site application	m²	8.35	to	16.80
60 minute protection; on site application	tonne	210.00	to	365.00
60 minute protection; off site application	m²	15.70	to	26.00
60 minute protection; off site application	tonne	315.00	to	525.00
90 minute protection; off site application	m²	42.00	to	84.00
90 minute protection; off site application	tonne	525.00	to	835.00
2C ROOF				
Roof plan area (unless otherwise described)				
Timber				
Timber roof trusses; insulation; roof coverings; PVC rainwater goods; plasterboard; skim and emulsion to ceilings				
concrete interlocking tile coverings	m²	105.00	to	165.00
clay pan tile coverings	m²	115.00	to	180.00
plain clay tile coverings	m²	140.00	to	205.00
natural slate coverings	m²	155.00	to	220.00
composite slate coverings	m²	120.00	to	190.00
reconstructed stone coverings	m²	120.00	to	225.00
Timber dormer roof trusses; 100 mm thick insulation; roof coverings; PVC rainwater goods; plasterboard; skim and emulsion to ceilings				
concrete interlocking tile coverings	m²	145.00	to	220.00
clay pantile coverings	m²	155.00	to	225.00
plain clay tile coverings	m²	175.00	to	245.00
natural slate coverings	m²	190.00	to	255.00
composite slate coverings	m²	155.00	to	235.00
reconstructed stone coverings	m²	155.00	to	270.00
extra for				
end of terrace semi/detached configuration	m²	36.50	to	41.50
hipped roof configuration	m²	36.50	to	47.50

Item	Unit	Range £
2C ROOF – cont'd		
Steel		
Steel roof trusses and beams; thermal and acoustic insulation		
aluminium profiled composite cladding	m²	235.00 to 310.00
copper roofing or boarding	m²	260.00 to 320.00
Flat roof decking and finishes		
Galvanised steel roof decking; insulation; three layer felt roofing and chippings; 0.70 mm thick steel decking	m²	56.00 to 78.00
Aluminium roof decking; three layer felt roofing and chippings; 0.90 mm thick aluminium decking	m²	87.00 to 87.00
Softwood trussed pitched roofs		
Structure only comprising 75 mm x 50 mm Fink roof trusses at 600 mm centres (measured on plan)	m²	28.00 to 36.50
Structure only comprising 100 mm x 38 mm Fink roof trusses at 600 mm centres (measured on plan)	m²	30.50 to 40.50
Structure only for "Mansard" roof comprising 100 mm x 50 mm roof trusses at 600 mm centres; 70° pitch	m²	30.50 to 43.50
extra for		
forming dormers	m²	535.00 to 750.00
Concrete flat roofs		
Reinforced concrete suspended slab; no coverings or finishes	m²	130.00 to 155.00
Reinforced concrete slabs; on "Holorib" permanent steel shuttering; protected steel frame; no coverings or finishes	m²	185.00 to 315.00
Softwood flat roofs		
Structure only comprising roof joists; 100 mm x 50 mm wall plates; herringbone strutting; no coverings or finishes	m²	43.50 to 61.00
Roof claddings		
Fibre cement sheet profiled cladding; single skin;		
"Profile 6"; single skin; natural grey finish	m²	20.60 to 25.00
"P61 Insulated System"; natural grey finish; metal inner lining panel	m²	36.50 to 48.00
extra for		
coloured fibre cement sheeting	m²	3.45 to 3.95
single skin GRP translucent roof sheets	m²	39.00 to 56.00
double skin GRP translucent roof sheets	m²	50.00 to 83.00
PVF2 coated galvanised steel trapezoidal profile cladding on steel purlins		
single skin trapezoidal	m²	20.50 to 28.50
built up system; insulation; metal inner lining panel	m²	41.50 to 48.50
composite insulated roofing system; 80mm overall panel thickness	m²	60.00 to 68.00
standing seam joints composite insulated roofing system; 80mm overall panel thickness	m²	98.00 to 115.00
Copper roofing with standing seam joints; 80mm insulation breather membrane or vapour barrier	m²	94.00 to 125.00

Item	Unit	Range £
Rooflights/patent glazing and glazed roofs		
Rooflights		
individual polycarbonate rooflights; rectangular	m²	310.00 to 465.00
individual polycarbonate rooflights; circular	m²	425.00 to 845.00
feature/ventilating	m²	265.00 to 490.00
'Velux' style rooflights to traditional roof construction (tiles\slates)	m²	490.00 to 845.00
Patent glazing; including flashings; standard aluminium Georgian wired		
single glazed	m²	325.00 to 435.00
double glazed	m²	380.00 to 515.00
Glazed roof; purpose made polyester powder coated aluminium;		
double glazed low emissivity glass	m²	295.00 to 440.00
feature; to covered walkways	m²	320.00 to 520.00
Screeds/Decks to receive roof coverings		
50 mm thick cement and sand screed	m²	10.90 to 13.00
75 mm thick lightweight bituminous screed and vapour barrier	m²	19.10 to 22.50
18 mm thick external quality plywood boarding	m²	20.50 to 25.00
Comparative tiling and slating finishes/perimeter treatments (including underfelt; battening; eaves courses and ridges)		
Concrete troughed or bold roll interlocking tiles; sloping	m²	27.00 to 43.00
Tudor clay pantiles; sloping	m²	30.00 to 48.00
Natural red pantiles; sloping	m²	35.50 to 54.00
Blue composition (cement fibre) slates; sloping	m²	36.50 to 44.50
Machine made clay plain tiles; sloping	m²	55.00 to 66.00
Welsh natural slates; sloping	m²	80.00 to 170.00
Spanish slates; sloping	m²	64.00 to 86.00
Man made slates; sloping	m²	57.00 to 80.00
Reconstructed stone slates; random slates; sloping	m²	38.00 to 72.00
Handmade sand faced plain tiles; sloping	m²	80.00 to 99.00
Eaves to sloping roof; 200 mm x 25 mm painted softwood fascia; 6 mm thick "Masterboard" soffit lining 225 mm wide		
100mm uPVC gutter	m	24.00 to 32.00
150mm uPVC gutter	m	30.00 to 39.50
100mm cast iron gutter; decorated	m	40.00 to 47.50
150mm cast iron gutter; decorated	m	48.50 to 56.00
Rainwater pipes; fixed to backgrounds; including offsets and shoes		
68mm diameter uPVC	m	8.40 to 11.40
110mm diameter uPVC	m	12.70 to 15.60
75mm diameter cast iron; decorated	m	31.50 to 37.00
100mm diameter cats iron; decorated	m	37.00 to 44.50
Comparative cladding finishes (including boarding; underfelt; labourers; etc.)		
0.91 mm thick aluminium roofing; commercial grade; fixed to boarding		
flat	m²	68.00 to 99.00
sloping	m²	73.00 to 100.00
0.81 mm thick zinc roofing; fixed to boarding		
flat	m²	86.00 to 115.00
sloping	m²	97.00 to 125.00
Copper roofing; fixed to boarding		
0.56 mm thick; flat	m²	76.00 to 105.00
0.56 mm thick; sloping	m²	86.00 to 115.00
Stainless steel sheeting		
0.40 mm thick; sloping	m²	131.00 to 160.00

Item	Unit	Range £
2C ROOF – cont'd		
Lead roofing		
code 5 sheeting; sloping	m²	135.00 to 195.00
code 5 sheeting; vertical to mansard; including insulation	m²	200.00 to 235.00
Flat Roofing Systems		
Includes insulation and vapour control barrier; excludes decking		
Single layer polymer roofing membrane	m²	79.00 to 110.00
Single layer polymer roofing membrane with tapered insulation	m²	96.00 to 160.00
20mm thick Polymer modified asphalt roofing including felt underlay	m²	67.00 to 90.00
High performance bitumen felt roofing system	m²	99.00 to 120.00
High performance polymer modified bitumen membrane	m²	90.00 to 110.00
extra for		
solar reflective paint	m²	2.80 to 3.95
limestone chipping finish	m²	3.40 to 9.05
grip tiles in hot bitumen	m²	38.00 to 47.50
Edges to felt flat roofs; softwood splayed fillet; 280 mm x 25 mm painted softwood fascia; no gutter aluminium		
edge trim	m	37.50 to 42.50
Edges to flat roofs; code 4 lead drip dresses into gutter; 230 mm x 25 mm painted softwood fascia;		
100 mm uPVC gutter	m	33.00 to 51.00
100 mm cast iron gutter; decorated	m	52.00 to 82.00
Landscaped roofs		
Polyester based elastomeric bitumen waterproofing and vapour equalisation layer; copper lined bitumen membrane root barrier and waterproofing layer; separation and slip layers; protection layer; 50 mm thick drainage board; filter fleece; insulation; Sedum vegetation blanket		
intensive (high maintenance to may include trees and shrubs require deeper substrate layers; are generally limited to flat roofs)	m²	150.00 to 195.00
extensive (low maintenance to herbs; grasses; mosses and drought tolerant succulents such as Sedum)	m²	140.00 to 185.00
2D STAIRS		
Reinforced concrete construction		
Escape staircase; granolithic finish; mild steel balustrades and handrails		
3.00 m rise; dogleg	nr	4200.00 to 8500.00
plus or minus for each 300 mm variation in storey height	nr	400.00 to 835.00
Staircase; terrazzo finish; mild steel balustrades and handrails; plastered and decorated soffit;		
3.00 m rise; dogleg	nr	6500.00 to 12500
plus or minus for each 300 mm variation in storey height	nr	645.00 to 1200.00
Staircase; terrazzo finish; stainless steel balustrades and handrails; plastered and decorated soffit		
3.00 m rise; dogleg	nr	8500 to 14500
plus or minus for each 300 mm variation in storey height	nr	835.00 to 1425.00
Staircase; high quality finishes; stainless steel and glass balustrades; plastered and decorated soffit		
3.00 m rise; dogleg	nr	17000 to 21000
plus or minus for each 300 mm variation in storey height	nr	1675.00 to 2700.00

Item	Unit	Range £
Metal construction		
Steel access/fire ladder		
3.00 m high	nr	660.00 to 920.00
4.00 m high; epoxide finished	nr	920.00 to 1550.00
Light duty metal staircase; galvanised finish; perforated treads; no risers; balustrades and handrails; decorated		
3.00 m rise; straight; 900 mm wide	nr	3950.00 to 5000.00
plus or minus for each 300 mm variation in storey height	nr	310.00 to 500.00
Light duty circular metal staircase; galvanised finish; perforated treads; no risers; balustrades and handrails; decorated		
3.00 m rise; straight; 1548 mm diameter	nr	4450.00 to 5500.00
plus or minus for each 300 mm variation in storey height	nr	350.00 to 525.00
Heavy duty cast iron staircase; perforated treads; no risers; balustrades and hand rails; decorated		
3.00 m rise; straight	nr	5100.00 to 6500.00
plus or minus for each 300 mm variation in storey height	nr	515.00 to 665.00
3.00 m rise; spiral; 1548 mm diameter	nr	5500.00 to 7500.00
plus or minus for each 300 mm variation in storey height	nr	555.00 to 750.00
Feature metal staircase; galvanised finish perforated treads; no risers; decorated		
3.00 m rise; spiral balustrades and handrails	nr	7000.00 to 8000.00
3.00 m rise; dogleg; hardwood balustrades and handrails	nr	7500 to 10250
3.00 m rise; dogleg; stainless steel balustrades and handrails	nr	9000 to 14000
plus or minus for each 300 mm variation in storey height	nr	715.00 to 795.00
galvanised steel catwalk; nylon coated balustrading 450mm wide	m	360.00 to 485.00
Timber construction		
per storey (unless otherwise described)		
Softwood staircase; softwood balustrades and hardwood handrail; plasterboard; skim and emulsion to soffit		
2.60 m rise; standard; straight flight	nr	815.00 to 1225.00
2.60 m rise; standard; top three treads winding	nr	995.00 to 1350.00
2.60 m rise; standard; dogleg	nr	1150.00 to 1450.00
Oak staircase; balustrades and handrails; plasterboard; skim and emulsion to soffit		
2.60 m rise; purpose made; dogleg	nr	7500 to 10000
plus or minus for each 300 mm variation in storey height	nr	1050.00 to 1300.00
Comparative finishes/balustrading		
Finishes to treads and risers; including nosings etc.		
vinyl or rubber	m	11.50 to 21.00
carpet (PC sum £25 /m²)	m	28.50 to 43.50
Wall handrails		
Softwood handrail and brackets	m	61.00 to 93.00
Hardwood handrail and brackets	m	81.00 to 135.00
Mild steel handrail and brackets	m	160.00 to 180.00
Stainless steel handrail and brackets	m	155.00 to 205.00
Balustrading and handrails		
Mild steel balustrade and steel or timber handrail	m	285.00 to 365.00
Balustrade and handrail with metal infill panels	m	350.00 to 475.00
Balustrade and handrail with glass infill panels	m	395.00 to 510.00
Stainless steel balustrade and handrail	m	490.00 to 710.00
Stainless steel and glass balustrade	m	625.00 to 890.00

Item	Unit	Range £		
2E EXTERNAL WALLS				
Wall area (unless otherwise described)				
Brick / block walling				
Common brick solid walls; bricks PC £240.00 per 1000				
half brick thick	m²	40.00	to	47.00
one brick thick	m²	73.00	to	88.00
one and a half brick thick	m²	105.00	to	125.00
add or deduct for each variation of £10.00 per 1000 in PC value				
half brick thick	m²	1.10	to	1.75
one brick thick	m²	1.95	to	2.85
one and a half brick thick	m²	2.95	to	3.60
extra for				
fair face one side	m²	1.95	to	2.85
Engineering brick walls; class B; bricks PC £275.00 per 1000				
half brick thick	m²	45.50	to	53.00
one brick thick	m²	86.00	to	96.00
Facing brick walls; sand faced bricks; bricks PC £350.00 per 1000				
half brick thick; pointed one side	m²	57.00	to	68.00
one brick thick; pointed both sides	m²	94.00	to	115.00
Facing bricks solid walls; hand made facings; bricks PC £500.00 per 1000				
half brick thick; pointed one side	m²	83.00	to	95.00
one brick thick; pointed both sides	m²	155.00	to	190.00
add or deduct for each variation of £10.00 per 1000 in PC value				
half brick thick	m²	1.10	to	1.50
one brick thick	m²	1.95	to	2.20
Cavity wall; facing brick outer skin; 50 mm thick insulation; plasterboard on stud inner skin; emulsion				
machine made facings; PC £350.00 per 1000	m²	98.00	to	120.00
hand made facings; PC £475.00 per 1000	m²	115.00	to	140.00
Cavity wall; facing brick outer skin; 50 mm thick insulation; with plaster on lightweight block inner skin; emulsion				
machine made facings; PC £350.00 per 1000	m²	101.00	to	125.00
hand made facings; PC £475.00 per 1000	m²	125.00	to	155.00
add or deduct for				
each variation of £10.00 per 1000 in PC value	m²	1.10	to	1.30
extra for				
heavyweight block inner skin	m²	1.30	to	2.60
insulating block inner skin	m²	2.60	to	6.95
75 mm thick cavity insulation	m²	5.35	to	6.15
100 mm thick cavity insulation	m²	6.85	to	7.65
Aerated lightweight block walls				
100 mm thick	m²	25.00	to	34.50
140 / 150 mm thick	m²	32.50	to	46.50
200 / 215 mm thick	m²	47.50	to	62.00
Dense aggregate block walls				
100 mm thick	m²	23.00	to	32.50
140 mm thick	m²	31.50	to	44.00
Coloured dense aggregate masonry block walls; "Lignacite" or similar				
100 mm thick; hollow	m²	40.00	to	46.50
100 mm thick; solid	m²	47.00	to	58.00
140 mm thick; solid	m²	48.50	to	57.00
140 mm thick; hollow	m²	59.00	to	68.00

Item	Unit	Range £
Cavity wall; coloured masonry block; outer and inner skins; fair faced both sides	m²	76.00 to 94.00
Cavity wall; block outer skin; 50 mm insulation; lightweight block inner skin; outer block rendered	m²	65.00 to 92.00
extra for		
architectural masonry outer block	m²	1.30 to 2.60
75 mm thick cavity insulation	m²	5.35 to 6.05
Reinforced concrete walling		
In situ reinforced concrete 25.00 N/mm²; 13 kg/m² reinforcement; formwork both sides		
150 mm thick	m²	140.00 to 195.00
225 mm thick	m²	165.00 to 210.00
300 mm thick	m²	160.00 to 235.00
Panelled walling		
Pre-cast concrete panels; including insulation; lining and fixings; generally 7.5 m x 150 mm thick x storey height		
standard panels	m²	230.00 to 280.00
standard panels; exposed aggregate finish	m²	255.00 to 305.00
brick clad panels (£350 per 1000 provisional sum for bricks)	m²	360.00 to 460.00
reconstructed stone faced panels	m²	280.00 to 335.00
natural stone faced panels (Portland Stone at £135 / m² provisional sum)	m²	460.00 to 590.00
marble or granite faced panels	m²	515.00 to 870.00
Wall claddings		
Non-asbestos profiled cladding		
"Profile 6"; single skin; natural grey finish	m²	26.00 to 29.50
"P61 Insulated System"; natural grey finish; metal inner lining panel	m²	47.00 to 59.00
extra for		
coloured fibre cement sheeting	m²	3.15 to 3.95
insulated; with 2.8 m high block inner skin; emulsion	m²	38.00 to 44.50
insulated; with 2.8 m high block inner skin plasterboard lining on metal tees; emulsion	m²	43.50 to 63.00
Metal profiled cladding		
coated steel profiled cladding on steel rails; insulated built up system	m²	48.50 to 65.00
coated steel micro-rib profiled cladding on steel rails; composite sandwich panel system	m²	92.00 to 120.00
coated aluminium profiled cladding on steel rails; insulated built up system	m²	45.50 to 69.00
coated aluminium flat panel cladding on steel rails; insulated built up system	m²	140.00 to 185.00
Other cladding systems		
Vitreous enamelled insulated steel sandwich panel system; with insulation board on inner face	m²	175.00 to 215.00
Formalux sandwich panel system; with coloured lining tray; on steel cladding rails	m²	200.00 to 260.00
Aluminium over cladding system rain screen	m²	240.00 to 275.00
High pressure laminate board on a simple metal component framing system	m²	105.00 to 150.00
Timber rainscreen cladding; carrier frame system; bracketry	m²	89.00 to 180.00
Terracotta rainscreen cladding; aluminium support rails; anti-graffiti coating	m²	315.00 to 470.00

Item	Unit	Range £
2E EXTERNAL WALLS – cont'd		
Curtain/glazed walling		
Stick curtain walling with double glazed units; aluminium structural framing and spandrel rails. Standard colour powder coated	m²	430.00 to 540.00
Unitised curtain walling system with double glazed units; aluminium structural framing and spandrel rails. Standard colour powder coated	m²	650.00 to 865.00
Unitised naturally ventilated double curtain walling system with double glazed units; cavity; single opening pane internally; interstitial blinds.		
Standard colour powder coated	m²	970.00 to 1300.00
Fixed Brise Soleil including uni-strut supports	m	305.00 to 410.00
Operable Brise Soleil including uni-strut supports	m	815.00 to 1025.00
Lift surround of double glazed or laminated glass with aluminium or stainless steel framing	m²	810.00 to 1350.00
Patent glazing systems; excluding opening lights and lead flashings etc; 7 mm Georgian wired cast glass; aluminium glazing bars spanning up to 3m at 600 mm spacing	m²	325.00 to 430.00
Patent glazing systems; excluding opening lights and lead flashings etc; 6.8 mm laminate glass; aluminium glazing bars spanning up to 3 m at 600 mm spacing	m²	380.00 to 485.00
Comparative external finishes		
Comparative concrete wall finishes		
wrought formwork one side including rubbing down	m²	3.65 to 7.30
shotblasting to expose aggregate	m²	4.75 to 9.00
bush hammering to expose aggregate	m²	14.60 to 20.50
Comparative in-situ finishes		
two coats "Sandtex Matt" cement paint	m²	8.60 to 12.00
cement and sand plain face rendering	m²	15.20 to 22.50
three coat "Tyrolean" rendering; including backing	m²	36.50 to 50.00
Comparative cladding		
25 mm thick Tongued and grooved "Tanalised" softwood boarding; including battens	m²	36.00 to 45.50
25 mm thick Tongued and grooved western red cedar boarding including battens	m²	42.50 to 54.00
Machine made tiles; including battens	m²	45.50 to 54.00
Best hand made sand faced tiles; including battens	m²	56.00 to 64.00
20 mm x 20 mm thick Mosaic glass or ceramic; in common colours; fixed on prepared surface	m²	110.00 to 125.00
2F WINDOWS AND EXTERNAL DOORS		
Window and external door area (unless otherwise described)		
Softwood windows and external doors		
Standard windows; painted; double glazed	m²	295.00 to 465.00
Purpose made windows; painted; double glazed	m²	355.00 to 560.00
Standard external softwood doors and hardwood frames; doors painted; including ironmongery		
two panelled door; plywood panels	nr	580.00 to 970.00
solid flush door	nr	525.00 to 865.00
two panelled door; glazed panels	nr	990.00 to 1100.00
heavy duty solid flush door		
single leaf	nr	790.00 to 1225.00
double leaf	nr	1400.00 to 2000.00
extra for		
emergency fire exit door	nr	295.00 to 470.00

Item	Unit	Range £
Steel windows and doors		
Standard windows		
double glazed; galvanised; painted	m²	300.00 to 420.00
double glazed; powder coated	m²	300.00 to 430.00
Purpose made windows		
double glazed; powder coated	m²	330.00 to 575.00
Standard doors		
single external steel door; including frame; ironmongery; powder coated finish	nr	745.00 to 1100.00
single external steel security door; including frame; ironmongery; powder coated finish	nr	745.00 to 1250.00
Steel roller shutters		
manual	m²	230.00 to 325.00
electric	m²	280.00 to 430.00
manual; insulated	m²	385.00 to 505.00
electric; insulated	m²	450.00 to 590.00
electric; insulated; fire resistant	m²	450.00 to 785.00
Hardwood windows		
Standard windows; stained		
double glazed	m²	465.00 to 710.00
Purpose made windows; stained		
double glazed	m²	505.00 to 825.00
uPVC windows and external doors		
Purpose made windows		
double glazed	m²	235.00 to 520.00
extra for		
tinted glass	m²	34.00 to 46.50
Aluminium windows; entrance screens and doors		
Standard windows; anodised finish		
single glazed; horizontal sliding sash	m²	275.00 to 355.00
single glazed; vertical sliding sash	m²	430.00 to 525.00
single glazed; casement; in hardwood sub-frame	m²	325.00 to 430.00
double glazed; vertical sliding slash	m²	480.00 to 715.00
double glazed; casement; in hardwood sub-frame	m²	395.00 to 525.00
Purpose made windows		
double glazed	m²	620.00 to 770.00
double glazed; feature; with pre-cast concrete surrounds	m²	1575.00 to 2350.00
purpose made entrance screens and doors double glazed	m²	770.00 to 1250.00
single external aluminium door; frame; ironmongery	nr	1300.00 to 1450.00
Purpose made doors		
revolving door; 2 m diameter; clear laminated glazing; 4 nr wings; glazed curved walls	m²	29000 to 37000
automatic sliding door; bi-parting	m²	1300.00 to 2600.00
Stainless steel entrance screens and doors		
Purpose made screen; double glazed		
with manual doors	m²	1500.00 to 2250.00
with automatic doors	m²	1850.00 to 2650.00
purpose made revolving door 2 m diameter; clear laminated glazing; 4 nr wings; glazed curved walls	m²	43000 to 57000
automatic sliding door; bi-parting	m²	2100.00 to 3400.00

Item	Unit	Range £
2F WINDOWS AND EXTERNAL DOORS – cont'd		
Shop fronts; shutters and grilles		
Flat façade; glass in aluminium framing; manual centre doors only	m	1300.00 to 3050.00
Hardwood and glass; including high enclosed window beds	m	5250.00 to 6250.00
High quality; marble or granite plasters and stair risers; window beds and backings; illuminated signs	m	5750.00 to 7750.00
Temporary timber shop fronts	m	64.00 to 89.00
Grilles or shutters	m	755.00 to 1500.00
Fire shutters; powers operated	m	1250.00 to 1775.00
2G INTERNAL WALLS; PARTITIONS AND DOORS		
Internal partition area (unless otherwise described)		
Timber or metal stud partitions		
Timber stud partitions		
structure only comprising 100 mm x 38 mm softwood studs at 400 mm x 600 mm centres; head and sole plates	m²	15.20 to 19.60
softwood stud and plasterboard partitions; tape and fill joints; emulsion finish	m²	43.50 to 67.00
Metal stud and plasterboard partitions		
90 mm thick partition; 1 layer 13 mm board each side; tape and fill joints; emulsion finish	m²	45.50 to 63.00
30 minute fire resistant partition; 1 layer 13 mm board each side; cavity insulation; tape and fill joints; emulsion finish	m²	45.50 to 68.00
60 minute fire resistant partition; 1 layer 15 mm board each side; cavity insulation; tape and fill joints; emulsion finish	m²	57.00 to 63.00
120 minute fire resistant Shaftwall partition; 2 layers 13 mm Fireline board each side; cavity insulation; tape and fill joints; emulsion finish	m²	68.00 to 78.00
extra for		
vinyl paper in lieu of emulsion	m²	6.10 to 15.40
easy clean finish in lieu of emulsion	m²	14.00 to 22.50
curved work	%	10.00 to 20.00
Glass block wall partition		
Glass block walling; reinforced bars each course; fair faced both sides	m²	205.00 to 270.00
Brick/block partitions		
Common brick half brick thick wall; bricks PC £240.00 per 1000	m²	40.00 to 47.00
Aerated/lightweight block partitions		
100 mm thick	m²	22.50 to 36.00
140 / 150 mm thick	m²	31.50 to 48.50
200 / 215 mm thick	m²	38.00 to 65.00
Dense aggregate block walls		
100 mm thick	m²	23.00 to 32.50
140 / 150 mm thick	m²	31.50 to 44.00
extra for		
fair face both sides	m²	3.15 to 6.95
plaster and emulsion	m²	13.90 to 20.50
curved work	%	10.00 to 20.00

Item	Unit	Range £
Reinforced concrete walls		
150 mm thick	m²	115.00 to 195.00
225 mm thick	m²	115.00 to 210.00
300 mm thick	m²	140.00 to 235.00
extra for		
plaster and emulsion	m²	13.90 to 20.50
Solid partitioning and doors		
Demountable partitioning; aluminium framing; veneer finish doors		
medium quality; 46 mm thick panels factory finish vinyl faced	m²	120.00 to 170.00
high quality; 46 mm thick panels factory finish vinyl faced	m²	155.00 to 235.00
Aluminium internal patent glazing		
single glazed laminated	m²	125.00 to 175.00
double glazed; 1 layer toughened and 1 layer laminated glass	m²	210.00 to 265.00
Demountable aluminium/steel partitioning and doors		
high quality	m²	340.00 to 610.00
high quality; sliding	m²	825.00 to 1000.00
Stainless steel glazed manual doors and screens		
high quality; to inner lobby of malls	m²	485.00 to 1325.00
Special partitioning and doors		
Demountable fire partitions		
enamelled steel; half hour	m²	465.00 to 755.00
stainless steel; half hour	m²	1075.00 to 1375.00
Soundproof partitions; hardwood doors luxury veneered	m²	255.00 to 430.00
WC/Changing cubicles		
WC cubicles; high pressure laminate faced mdf; proprietary system		
back panelling system; including access hatch; frame support	nr	160.00 to 230.00
WC cubicle partition sets; dividing panels; doors and ironmongery	nr	535.00 to 670.00
IPS duct panel including sub-frame and accessories	nr	190.00 to 610.00
Changing cubicles		
aluminium	nr	420.00 to 770.00
aluminium; textured glass and bench seating	nr	695.00 to 920.00
Standard doors		
Standard softwood doors and frames; including lintel; ironmongery; and painting		
flush; hollow core	nr	295.00 to 410.00
flush; hollow core; hardwood faced	nr	280.00 to 400.00
flush; solid core		
single leaf	nr	495.00 to 755.00
double leaf	nr	915.00 to 1450.00
flush; solid core; hardwood faced	nr	475.00 to 1350.00
four panel door	nr	495.00 to 1625.00
Purpose made doors		
Softwood doors and hardwood frames; including lintel; ironmongery; painting and polishing		
solid core; heavy duty		
single leaf	nr	830.00 to 995.00
double leaf	nr	1125.00 to 1450.00
flush solid core; heavy duty; plastic laminate faced		
single leaf	nr	1010.00 to 1425.00
double leaf	nr	1400.00 to 1625.00

Item	Unit	Range £
2G INTERNAL WALLS; PARTITIONS AND DOORS – cont'd		
Purpose made doors – cont'd		
Softwood fire doors and hardwood frames; including lintel; ironmongery; painting and polishing flush; one hour fire resisting		
single leaf	nr	1100.00 to 1250.00
double leaf	nr	1300.00 to 1450.00
flush; one hour fire resisting; plastic laminate faced		
single leaf	nr	1300.00 to 1475.00
double leaf	nr	1625.00 to 1925.00
Softwood doors and pressed steel frames lintel		
flush; half hour fire check; plastic laminate faced	nr	1250.00 to 1525.00
Mahogany doors and frames; including lintel; ironmongery; and polishing		
four panel door	nr	990.00 to 1175.00
3A WALL FINISHES		
Internal wall area (unless otherwise described)		
Sheet/board finishes		
Dry plasterboard lining; taped joints; for direct decoration		
9.50 mm thick Gyproc Wallboard	m²	14.20 to 18.40
12.50 mm thick Gyproc wallboard (half hour fire resisting)	m²	15.20 to 20.00
Dry plasterboard lining; taped joints; for direct decoration; on adhesive dabs		
9.50 mm thick Gyproc Wallboard	m²	15.20 to 20.00
Dry plasterboard lining; taped joints; for direct decoration; on metal channels and adhesive dabs		
9.50 mm thick Gyproc Wallboard	m²	24.00 to 27.00
12.50 mm thick Gyproc Wallboard	m²	25.00 to 29.50
extra for		
22 mm Gyproc Thermaline board	m²	1.95 to 3.45
two layers of 12.50 mm thick Gyproc wallboard (one hour fire resisting)	m²	24.00 to 29.50
9 mm thick Supalux (half hour fire resisting)	m²	25.00 to 37.50
Timber boarding/panelling; on and including battens; plugged to wall		
12 mm thick softwood boarding	m²	28.00 to 47.50
hardwood panelling; t&g & v-jointed	m²	64.00 to 165.00
In-situ wall finishes		
Comparative finishes		
one mist and two coats emulsion paint	m²	3.15 to 4.75
two coats of lightweight plaster	m²	10.70 to 15.60
9.50 mm thick Gyproc Wallboard and skim coat	m²	19.60 to 26.50
12.50 mm thick Gyproc Wallboard and skim coat	m²	20.50 to 28.50
plaster and emulsion	m²	13.90 to 20.50
two coat render and emulsion	m²	23.50 to 34.50
plaster and vinyl wall coverings	m²	19.50 to 29.50
plaster and fabric wall coverings	m²	19.50 to 45.50

Item	Unit	Range £
Rigid tile/panel finishes		
Ceramic wall tiles; including backing		
economical quality	m²	24.00 to 40.00
medium to high quality	m²	35.00 to 55.00
Porcelain mosaic tiling; including backing to swimming pool lining; walls and floors	m²	60.00 to 81.00
"Roman Travertine" marble wall linings; polished	m²	300.00 to 530.00
Metal mirror cladding panels	m²	310.00 to 530.00
Comparative woodwork finishes		
Primer only	m²	1.50 to 1.70
Gloss		
two coats; touch up primer	m²	5.45 to 8.20
three coats; touch up primer	m²	7.60 to 11.40
three coats; touch up primer to small girth n.e. 300 mm	m	3.15 to 4.70
Polyurethane lacquer		
two coats	m²	3.15 to 3.75
three coats	m²	4.85 to 5.70
Flame retardant paint		
three coats	m²	7.30 to 9.00
Polish		
wax polish; seal	m²	7.80 to 16.10
wax polish; stain and body in	m²	13.10 to 15.00
French polish; stain and body in	m²	19.40 to 23.50
3B FLOOR FINISHES		
Internal floor area (unless otherwise described)		
Sheet/board flooring		
Chipboard flooring; t&g joints	m²	10.20 to 14.60
Wrought softwood flooring	m²	23.50 to 35.00
Wrought softwood t&g strip flooring; polished; including fillets	m²	32.00 to 51.00
Wrought hardwood t&g strip flooring; polished; including fillets	m²	53.00 to 97.00
Sprung composition block flooring (sports); court markings; sanding and sealing	m²	83.00 to 110.00
Softwood skirting; gloss paint finish	m	10.40 to 13.00
Hardwood skirting; stained finish	m	17.90 to 25.50
MDF skirting; gloss paint finish	m	7.35 to 12.70
In situ screed and floor finishes		
Latex cement screeds	m²	5.30 to 8.05
Rubber latex non slip solution and epoxy sealant	m²	10.40 to 21.50
Cement and sand (1:3) screeds		
50 mm thick	m²	13.00 to 19.00
75 mm thick	m²	18.40 to 25.00
Granolithic		
20 mm thick	m²	11.80 to 19.30
25 mm thick	m²	16.30 to 24.00
Epoxy floor finish		
1.50 mm to 2.00 mm thick	m²	24.50 to 33.50
5.00 mm to 6.00 mm thick	m²	46.50 to 52.00
Resin floor finish		
5.00 mm to 9.00 mm thick	m²	56.00 to 65.00

Item	Unit	Range £
3B FLOOR FINISHES – cont'd		
Rigid Tile/slab finishes (includes skirtings; excludes screeds)		
Quarry tile flooring	m²	57.00 to 83.00
Glazed ceramic tiled flooring		
standard plain tiles	m²	40.50 to 48.00
anti slip tiles	m²	45.00 to 50.00
designer tiles	m²	88.00 to 105.00
Terrazzo tile flooring 28 mm thick white "Sicilian" marble aggregate tiling	m²	87.00 to 105.00
York stone 50 mm thick paving	m²	115.00 to 160.00
Slate tiles; smooth; straight cut	m²	140.00 to 185.00
Portland stone paving	m²	235.00 to 280.00
Roman "Travertine" marble paving; polished	m²	300.00 to 385.00
Granite paving 20 mm thick paving	m²	415.00 to 530.00
Parquet/wood block finishes		
wrought hardwood block floorings; 25 mm thick; polished; t&g joints	m²	68.00 to 90.00
composition block flooring	m²	81.00 to 95.00
Flexible tiling		
thermoplastic tile flooring	m²	8.80 to 10.40
vinyl floor tiling	m²	9.50 to 15.10
vinyl sheet flooring; heavy duty	m²	16.80 to 43.50
vinyl safety flooring	m²	34.00 to 46.50
linoleum tile flooring	m²	31.00 to 45.50
linoleum sheet flooring	m²	36.00 to 46.50
rubber tile flooring	m²	30.50 to 40.50
rubber sheet flooring	m²	39.50 to 55.00
Carpet tiles; including underlay and fixing		
PC sum £10 / m²	m²	19.10 to 26.50
PC Sum £25 / m²	m²	38.00 to 49.50
Entrance matting and matwell; barrier matting and aluminium trim	m²	270.00 to 420.00
Access floors and finishes		
Raised access floors: excluding 600 mm x 600 mm steel encased particle boards on height adjustable pedestals < 300 mm		
medium grade duty	m²	26.00 to 40.50
heavy grade duty	m²	39.50 to 58.00
battened raft floor with sound insulation fixed to battens; medium quality carpeting	m²	65.00 to 94.00
Common floor coverings bonded to access floor panels		
heavy duty fully flexible vinyl; to BS 3261; type A	m²	10.40 to 31.00
fibre bonded carpet	m²	15.50 to 26.00
high pressure laminate; to BS 2794; class D	m²	11.60 to 29.50
anti static grade fibre bonded carpet	m²	12.00 to 31.00
anti static grade sheet PVC; to BS 3261	m²	16.40 to 26.50
low loop tufted carpet	m²	20.50 to 36.00

Item	Unit	Range £
3C CEILING FINISHES		
Internal ceiling area (unless otherwise described)		
In situ/board finishes		
Decoration only to soffits; one mist and two coats emulsion paint		
to exposed steelwork (surface area)	m²	3.15 to 5.35
to concrete soffits (surface area)	m²	3.15 to 4.75
to plaster / plasterboard	m²	3.15 to 4.75
Plaster to soffits		
lightweight plaster	m²	10.70 to 15.20
plaster and emulsion	m²	14.10 to 25.00
extra for		
gloss paint in lieu of emulsion (surface area)	m²	2.35 to 2.80
Plasterboard to soffits		
12.50 mm Gyproc lath and skim coat	m²	20.50 to 30.50
12.50 mm Gyproc insulating lath and skim coat	m²	23.00 to 34.00
extra for		
"Artex" finish	m²	10.60 to 15.00
Other board finishes; with fire-resisting properties; excluding decoration		
12.50 mm thick Gyproc "Fireline" board	m²	23.00 to 34.50
9 mm thick Supalux	m²	51.00 to 68.00
Specialist plasters; to soffits		
sprayed acoustic plaster; self-finished	m²	31.50 to 43.50
rendering; "Tyrolean" finish	m²	32.50 to 46.50
Other ceiling finishes		
timber boarding	m²	20.50 to 29.50
Suspended and integrated ceilings		
Suspended ceiling		
economical; exposed grid	m²	20.00 to 25.00
medium quality; "Minatone"; concealed grid	m²	42.00 to 47.50
high quality; "Travertone"; concealed grid	m²	49.50 to 70.00
jointless; plasterboard	m²	24.00 to 34.00
Other suspended ceilings		
metal linear strip; "Dampa"/"Luxalon"	m²	34.00 to 67.00
metal tray	m²	46.50 to 60.00
egg-crate	m²	51.00 to 110.00
open grid; "Formalux"/"Dimension"	m²	93.00 to 120.00
Integrated ceilings		
coffered; with steel surfaces	m²	105.00 to 175.00
acoustic suspended ceilings on anti vibration mountings	m²	57.00 to 72.00
Comparative wall and ceiling finishes		
Emulsion paint		
two coats	m²	2.15 to 2.80
one mist and two coats	m²	3.15 to 4.75
Artex plastic compound one coat; textured	m²	10.60 to 15.00
Wall paper	m²	4.85 to 12.60
Hessian wall coverings	m²	10.60 to 19.10
Gloss		
primer and two coats	m²	5.45 to 8.20
primer and three coats	m²	7.60 to 11.40

Item	Unit	Range £
4 FITTINGS AND FURNISHINGS		
Residential fittings (volume housing)		
Kitchen fittings for residential units (not including 'white' goods)		
one person flat/bed-sit	nr	1450.00 to 2275.00
two person flat/house	nr	1700.00 to 2900.00
three person flat/house	nr	2275.00 to 5200.00
four person house	nr	5750.00 to 8750.00
five person house	nr	5750 to 14000
Office furniture and equipment		
Reception desk		
straight counter; 3.5 m long; 2 person	nr	1700.00 to 2275.00
curved counter; 3.5 m long; 2 person	nr	4050.00 to 5750.00
curved counter; 3.5 m long; 2 person; real wood veneer finish	nr	6250 to 11500
Furniture and equipment to general office area		
workstation; 2 m long desk; drawer unit; task chair	nr	570.00 to 870.00
Hotel Bathroom Pods		
Fully fitted out; finished and furnished bathroom pods; installed		
standard pod (4.5m plan area)	nr	4050.00 to 5200.00
accessible pod (4.5m plan area)	nr	5200.00 to 6250.00
5A SANITARY AND DISPOSAL INSTALLATIONS		
Gross internal floor area (unless otherwise described)		
Residential units		
range including WC; wash hand basin; bath	nr	1825.00 to 3050.00
range including WC; wash hand basin; bidet; bath and kitchen sink	nr	2300.00 to 3650.00
range including two WC's; two wash hand basins; bidet bath and kitchen sink	nr	3400.00 to 4700.00
extra for		
rainwater pipe per storey	nr	75.00 to 93.00
soil pipe per storey	nr	165.00 to 185.00
shower over bath	nr	455.00 to 645.00
Industrial buildings		
Warehouse		
minimum provision	m²	12.00 to 16.40
high provision	m²	16.40 to 25.00
Production unit		
minimum provision	m²	17.10 to 24.00
minimum provision; area less than 1000 m²	m²	19.40 to 33.50
high provision	m²	18.20 to 29.50
Retailing outlets		
to superstore	m²	4.20 to 10.30
shopping centre malls; public conveniences; branch connections shop shells	m²	8.20 to 13.30
fitting out public conveniences in shopping mall block	nr	6250.00 to 9750.00
Leisure buildings	m²	14.10 to 18.10
Office and industrial office buildings		
speculative; low rise; area less than 1000 m²	m²	6.05 to 15.50
speculative; low rise	m²	10.80 to 20.50
speculative; medium rise; area less than 1000 m²	m²	12.00 to 22.00
speculative; medium rise	m²	16.10 to 25.00
speculative; high rise	m²	16.10 to 25.00

Item	Unit	Range £
owner-occupied; low rise; area less than 1000 m²	m²	10.30 to 20.50
owner-occupied; low rise	m²	10.30 to 25.00
owner-occupied; medium rise; area less than 1000 m²	m²	16.10 to 25.00
owner-occupied; medium rise	m²	19.40 to 28.50
owner-occupied; high rise	m²	23.00 to 31.50
Hotels		
WC; bath; shower; basin to each bedroom; sanitary accommodation to public areas	m²	28.50 to 71.00
Comparative sanitary fittings/sundries		
Note: Material prices vary considerably; the following composite rates are based on average prices for mid priced fittings:		
Individual sanitary appliances (including fittings)		
Lavatory basins; vitreous china; chromium plated taps; waste; chain and plug; cantilever brackets		
white	nr	215.00 to 280.00
coloured	nr	285.00 to 350.00
Low level WC's; vitreous china pan and cistern; black plastic seat; low pressure ball valve; plastic flush pipe; fixing brackets		
On ground floor		
white	nr	205.00 to 340.00
coloured	nr	275.00 to 340.00
One of a range; on upper floors		
white	nr	400.00 to 500.00
coloured	nr	465.00 to 580.00
Bowl type wall urinal; white glazed vitreous china flushing cistern; chromium plated flush pipes and spreaders; fixing brackets		
white	nr	180.00 to 250.00
Shower tray; glazed fireclay; chromium plated waste; chain and plug; riser pipe; rose and mixing valve		
white	nr	285.00 to 535.00
coloured	nr	340.00 to 565.00
Sink; glazed fireclay; chromium plated waste; chain and plug; fixing		
white	nr	275.00 to 715.00
Sink; stainless steel; chromium plated waste; chain and self coloured		
single drainer	nr	230.00 to 345.00
double drainer	nr	275.00 to 345.00
Bath; reinforced acrylic; chromium plated taps; overflow; waste; chain and plug; "P" trap and overflow connections		
white	nr	315.00 to 590.00
coloured	nr	410.00 to 590.00
Bath; enamelled steel; chromium plated taps; overflow; waste; chain and plug; "P" trap and overflow connections		
white	nr	455.00 to 635.00
coloured	nr	495.00 to 655.00
Soil waste stacks; 3.15 m storey height; branch and connection to drain		
110 mm diameter uPVC	nr	360.00 to 410.00
extra for		
additional floors	nr	180.00 to 215.00
100 mm diameter cast iron; decorated	nr	275.00 to 765.00
extra for		
additional floors	nr	360.00 to 410.00

Item	Unit	Range £
5D WATER INSTALLATIONS		
Gross internal floor area (unless described otherwise)		
Hot and cold water installations		
Complete installations (industrial; leisure shopping malls and the like)	m²	5.65 to 20.50
Complete installations (offices; hotels; residential and the like)	m²	25.00 to 48.00
Mall public conveniences; branch connections to shop shells (gross internal floor area of mall)	m²	5.95 to 10.30
5F HEATING; AIR-CON AND VENTILATION		
Gross internal floor area (unless described otherwise)		
Residential solid fuel radiator heating		
Gas or oil fired hot water service and central heating for		
three radiators	nr	2525.00 to 3550.00
four radiators	nr	3550.00 to 3900.00
five radiators	nr	3850.00 to 4350.00
six radiators	nr	4100.00 to 4650.00
seven radiators	nr	4550.00 to 5250.00
Office space heating and air treatment		
LTHW heating	m²	36.50 to 73.00
Chilled water	m²	26.50 to 41.50
extra for category A fit-out (category A nett area)		
LTHW heating	m²	10.60 to 46.50
chilled water	m²	15.80 to 26.50
Gas or oil-fired convector heating LPHW convector system		
Speculative; area less than 1000 m²	m²	71.00 to 82.00
Speculative	m²	76.00 to 98.00
Owner-occupied; area less than 1000 m²	m²	83.00 to 93.00
Owner-occupied	m²	84.00 to 105.00
Hot air systems		
Warm air heating to sports hall area and the like	m²	15.80 to 21.50
Hot air heating and ventilation to shopping malls; including automatic remote vents in rooflights	m²	110.00 to 140.00
Ventilation system		
Local ventilation to (area to be vented)		
WC's	nr	255.00 to 335.00
Bathroom and toilet areas	m²	21.00 to 25.50
Air extract systems to kitchens; changing rooms etc	m²	26.00 to 42.00
Comfort cooling systems		
2 pipe fan coil for office building up to 3000 m²	m²	52.00 to 67.00
extra for		
category A fit-out (category A nett area)	m²	94.00 to 105.00
2 pipe fan coil for office building over 3000 m² up to 15000 m²	m²	46.50 to 62.00
extra for		
category A fit-out (category A nett area)	m²	84.00 to 94.00

Item	Unit	Range £
Full air-conditioning		
4 pipe fan coil for office building up to 3000 m²	m²	79.00 to 94.00
extra for		
category A fit-out (category A nett area)	m²	135.00 to 155.00
4 pipe fan coil for office building over 3000 m² up to 15000 m²	m²	69.00 to 84.00
extra for		
category A fit-out (category A nett area)	m²	115.00 to 135.00
Variable air volume for office building over 3000 m² up to 15000 m²	m²	63.00 to 74.00
extra for		
category A fit-out (category A nett area)	m²	115.00 to 130.00
Chilled beam exposed services for office building over 3000 m² up to 15000 m²	m²	69.00 to 84.00
extra for		
category A fit-out (category A nett area)	m²	235.00 to 260.00
4 pipe fan coil for hotel to 2 to 5 star	m²	215.00 to 255.00
Variable air volume for hotel to 2 to 5 star	m²	135.00 to 155.00
5H ELECTRICAL INSTALLATIONS		
gross internal area serviced (unless otherwise described)		
Lighting and power installations to		
Residential units		
one person flat/bed-sit	nr	1200.00 to 1850.00
two person flat/house	nr	1400.00 to 2600.00
three person flat/house	nr	1600.00 to 3100.00
four person house	nr	1950.00 to 4750.00
five/six person house	nr	2375.00 to 4750.00
extra for		
intercom	nr	475.00 to 535.00
Industrial buildings		
warehouse area	m²	47.50 to 78.00
production area	m²	54.00 to 86.00
production area; high provision	m²	78.00 to 102.00
office area	m²	115.00 to 140.00
office area; high provision	m²	155.00 to 185.00
Retail outlets		
shopping mall and landlords' areas	m²	78.00 to 130.00
Offices		
buildings up to 3000 m²	m²	15.10 to 25.00
extra for		
category A fit-out (net category A area)	m²	42.00 to 63.00
buildings over 3000 m² up to 15000 m²	m²	18.40 to 27.00
extra for		
category A fit-out (net category A area)	m²	42.00 to 63.00
Hotel		
2 to 3 star	m²	22.00 to 38.00
4 to 5 star	m²	27.00 to 43.00
Lighting installation		
Lighting to		
warehouse area	m²	10.40 to 26.00
factory production / picking area	m²	15.60 to 41.50
leisure and retail buildings	m²	10.40 to 26.00
shopping mall	m²	43.00 to 64.00
emergency lighting	m²	4.75 to 17.20
Standby generators only (life safety only)	m²	3.10 to 12.30

Item	Unit	Range £
5H ELECTRICAL INSTALLATIONS – cont'd		
Mains and sub-mains switchgear and distribution; offices; commercial and retail buildings		
mains intake only	m²	2.05 to 3.80
mains switchgear only	m²	3.10 to 10.70
Mains and sub-mains distribution		
to floors only	m²	5.80 to 43.00
floors; including small power and supplies to equipment	m²	17.30 to 20.50
floors; including lighting and power to landlords areas and supplies to equipment	m²	11.30 to 70.00
floors; including power; communication and supplies to equipment	m²	70.00 to 102.00
shop units; including fire alarms and telephone distribution	m²	6.90 to 17.20
Comparative fittings/rates per point		
Consumer control unit; 63 to 100 Amp 230 volt; switched and insulated; RCDB protection	nr	190.00 to 415.00
Fittings; excluding lamps or light fittings		
lighting point; PVC cables	nr	36.50 to 52.00
lighting point; PVC cables in screwed conduits	nr	41.50 to 62.00
lighting point; MICC cables	nr	52.00 to 83.00
Switch socket outlet; PVC cables		
single	nr	52.00 to 73.00
double	nr	57.00 to 88.00
Switch socket outlet; PVC cables in screwed conduit		
single	nr	73.00 to 94.00
double	nr	78.00 to 115.00
Switch socket outlet; MICC cables		
single	nr	68.00 to 94.00
double	nr	73.00 to 115.00
Immersion heater point (excluding heater)	nr	94.00 to 115.00
Cooker point; including control unit	nr	135.00 to 210.00
5I GAS INSTALLATIONS		
Gross internal floor area (unless described otherwise)		
Gas mains service to plantroom		
shopping mall / supermarket	m²	2.80 to 4.50
warehouse / distribution centre	m²	0.60 to 1.40
office / hotel	m²	1.15 to 2.25
5J LIFT AND CONVEYOR INSTALLATIONS		
Passenger lift 6 to 24 person lifts (standard finish)		
Electric traction passenger lifts		
6-person; 450 kg; 7 stops; 1.6 m/s;	nr	66000 to 98000
8-person; 630 kg; 5 stops; 1.6 m/s;	nr	54000 to 650000
10-person; 800 kg; 8 stops; 1.6 m/s;	nr	68000 to 82000
13-person; 1000 kg; 7 stops; 1.0 m/s;	nr	56000 to 68000
21-person; 1600 kg; 7 stops; 2.0 m/s;	nr	110000 to 130000
24-person; 1800 kg; 4 stops; 2.0 m/s;	nr	95000 to 110000
Electro-hydraulic passenger lifts		
8-person; 630 kg; 4 stops; 0.4 m/s;	nr	41500 to 49500
8-person; 630 kg; 7 stops; 0.6 m/s;	nr	50500 to 61000
8-person; 630 kg; 7 stops; 1.0 m/s;	nr	61000 to 73000
10-person; 800 kg; 3 stops; 0.75 m/s;	nr	48500 to 59000

Item	Unit	Range £		
5J LIFT AND CONVEYOR INSTALLATIONS – cont'd				
Passenger lift 6 to 24 person lifts (standard finish) – cont'd				
16-person; 1250 kg; 6 stops; 0.4 m/s;	nr	57000	to	69000
24-person; 1800 kg; 6 stops; 0.5 m/s;	nr	67000	to	80000
extra for				
lift car LCD TV	nr	5500	to	8000
intelligent group control; 5 cars; 11 stops	nr	26500	to	40500
10-person wall climber lift; 0.50 m/sec; 2 levels	nr	290000	to	370000
Disabled platform lift single wheelchair; 400 kg; 4 stops; 0.16 m/s	nr	6750	to	9250
Escalators				
30° escalator; 0.50 m/sec; enamelled steel glass balustrades				
3.5 m rise; 800 mm step width	nr	66000	to	79000
4.6 m rise; 800 mm step width	nr	70000	to	84000
5.2 m rise; 800 mm step width	nr	73000	to	88000
6 m rise; 800 mm step width	nr	81000	to	97000
extra for				
enhanced finish; enamelled finish; glass balustrade	nr	6500	to	12000
Good lifts				
Hoist	nr	9750	to	35000
Kitchen service hoist 50 kg; 2 levels	nr	8750	to	10500
Electric heavy duty goods lifts				
300 kg; 2 levels; 0.4 m/s	nr	11000	to	13500
1000 kg; 4 levels; 0.6 m/s	nr	35500	to	42500
2000 kg; 3 levels; 0.25 m/s	nr	40000	to	48000
4000 kg; 5 levels; 0.4 m/s	nr	51500	to	62000
Oil hydraulic heavy duty goods lifts				
2000 kg; 3 levels; 0.25 m/s	nr	39000	to	47500
4000 kg; 5 levels; 0.4 m/s	nr	51000	to	62000
Dock levellers				
dock levellers	nr	12000	to	28500
dock leveller and canopy	nr	17000	to	40000
5K PROTECTIVE INSTALLATIONS				
Gross internal floor area (unless described other wise)				
Fire fighting/protective installations				
Fire alarms/appliances				
Offices				
single stage smoke detectors; alarms and controls up to 3000 m²	m²	6.25	to	10.40
single stage smoke detectors; alarms and controls over 3000 m² up to 15000 m²	m²	6.25	to	8.30
Hotels 2 to 5 star	m²	10.40	to	15.60
Shopping mall	m²	7.30	to	21.00
Loose fire fighting equipment	m²	0.20	to	0.45
Hosereels; dry risers and extinguishers	m²	6.80	to	14.60
Sprinkler installations				
landlords areas; supply to shop shells; including fire alarms; appliances etc.	m²	11.30	to	16.90
single level sprinkler systems; alarms and smoke detectors; low hazard	m²	14.60	to	20.50
single level sprinkler systems; alarms and smoke detectors; ordinary hazard	m²	16.90	to	22.50
double level sprinkler systems; alarms and smoke detectors; high hazard	m²	31.50	to	41.00
Smoke vents				
automatic smoke vents over glazed shopping mall	m²	36.00	to	70.00
smoke control ventilation to atria	m²	70.00	to	88.00
lightning protection	m²	1.05	to	4.20

Item	Unit	Range £		
5L COMMUNICATION INSTALLATIONS				
Clock installation	m²	0.50	to	1.55
Security alarm system	m²	2.25	to	3.40
Telephone system	m²	1.15	to	2.25
Public address; television aerial and clocks	m²	3.40	to	6.80
Closed-circuit television	m²	4.50	to	5.65
Public address system	m²	12.40	to	14.60
5M SPECIAL INSTALLATIONS				
Photo-voltaics	m²	730.00	to	835.00
Window cleaning equipment				
twin track	m	160.00	to	215.00
manual trolley/cradle	nr	11500	to	14000
automatic trolley/cradle	nr	26000	to	33500
Laundry chute	nr	15000	to	20000
Sauna	nr	15000	to	20000
Jacuzzi installation	nr	9750	to	24000
Wave machine; four chamber wave generation equipment	nr	60000	to	84000
Swimming pool including structure; finishing's; ventilation; heating and filtration	m²	1200.00	to	1375.00
5N GENERAL BWIC WITH SERVICES				
Gross internal floor area				
Warehouses; sports halls and shopping malls				
main supplies; lighting and power to landlord areas	m²	1.35	to	5.40
central heating and electrical installation	m²	4.20	to	16.00
central heating; electrical and lift installation	m²	6.25	to	16.20
air conditioning; electrical and ventilation installations	m²	19.40	to	32.00
Offices and hotels				
main supplies; lighting and power to landlord areas	m²	4.30	to	15.10
central heating and electrical installation	m²	11.90	to	16.20
central heating; electrical and lift installation	m²	14.00	to	19.40
air conditioning; electrical and ventilation installations	m²	28.00	to	34.50
6A SITE WORK				
Preparatory excavation and sub-bases				
Excavating				
spread and lightly consolidate top soil form spoil 150 mm thick; by machine	m²	2.10	to	3.20
spread and lightly consolidate top soil form spoil 150 mm thick; by hand	m²	5.30	to	15.90
Seeded and planted areas				
Plant supply; planting; maintenance and 12 months guarantee				
seeded areas	m²	3.85	to	7.65
turfed areas	m²	5.00	to	10.00
Planted areas (per m² of surface area)				
herbaceous plants	m²	4.25	to	5.70
climbing plants	m²	5.70	to	10.00
general planting	m²	12.70	to	25.50
woodland	m²	19.10	to	38.50
shrubbed planting	m²	25.50	to	71.00

Item	Unit	Range £
dense planting	m²	32.00 to 64.00
shrubbed area including allowance for small trees	m²	38.50 to 89.00
Trees		
advanced nursery stock trees (12 to 200 cm girth)	nr	155.00 to 190.00
semi-mature trees; 5 to 8 m high		
coniferous	nr	510.00 to 1275.00
deciduous	nr	770.00 to 2125.00
Parklands		
Surface area (unless otherwise described)		
NOTE: Work on parklands will involve different techniques of earth shifting and cultivation. The following rates include for normal surface excavation; they include for the provision of any land drainage.		
Parklands; including cultivating ground; applying fertiliser; etc and seeding with parks type grass	ha	17500 to 20500
Lakes including excavation average 10 m deep; laying 1.5 mm thick butyl rubber sheet and spreading top soil evenly on top 300 mm deep		
between 1 and 5 hectare in area	ha	370000 to 410000
Land drainage		
NOTE: If land drainage is required on a project; the propensity of the land to flood will decide the spacing of the land drains. Costs include for excavation and backfilling of trenches and laying agricultural clay drain pipes with 75 mm diameter lateral runs average 600 mm deep; and 100 mm diameter mains runs average 750 mm deep.		
Land drainage to parkland with laterals at 30 m centres and main runs at 100 m centres	ha	3650 to 10500
Paved areas		
Gravel paving rolled to falls and chambers paving on sub-base; including excavation	m²	10.50 to 1.40
Tarmacadam paving; two layers; limestone or igneous chipping finish paving on Sub-base; including excavation	m²	20.10 to 29.50
Pre-cast concrete paving slabs on sub-base; including excavation	m²	35.50 to 46.50
extra for		
tactile slabs	m²	12.30 to 17.80
Pre-cast concrete block paviours to footways including excavation; sub-base; edgings	m²	35.50 to 46.50
Brick paviours on sub-base; including excavation	m²	69.00 to 80.00
Granite setts on sub base; including excavation	m²	93.00 to 115.00
York stone slab paving on sub-base; including excavation	m²	105.00 to 175.00
Cobblestone paving cobblestones on sub-base; including excavation	m²	80.00 to 105.00
Car Parking alternatives		
Surface level parking; including lighting and drainage	car	1250.00 to 1850.00
Surface landscaped	car	1850.00 to 3150.00
At ground level with deck or building over	car	6250.00 to 8750.00
Multi-storey parking; including lighting and drainage		
multi-storey flat slab	car	8750 to 12500
multi-storey warped slab	car	10000 to 14500

Item	Unit	Range £
6A SITE WORK – cont'd		
All purpose roads		
Tarmacadam or reinforced concrete roads; including all earthworks; drainage; pavements; lighting; signs; fencing and safety barriers		
Single 7.30 m wide carriageway	m	1175.00 to 1400.00
Wide single 10.00 m wide carriageway	m	1300.00 to 1550.00
Dual two lane road 7.30 m wide carriageway	m	2250.00 to 1775.00
Dual three lane road 11.00 m wide carriageway	m	2250.00 to 2700.00
Road crossings		
NOTE: Costs include road markings; beacons; lights; signs; advance danger signs etc.		
Zebra crossing	nr	5200.00 to 5750.00
Pelican crossing	nr	19000 to 21500
Footbridges		
Footbridge of either pre-cast concrete or steel construction up to 6.00 m wide; 6.00 m high including deck; access stairs and ramp; parapets etc.		
5 m span between piers or abutments	m²	1175.00 to 1800.00
2 0m span between piers or abutments	m²	1150.00 to 3000.00
Footbridge of timber (stress graded with concrete piers)		
12 m span between piers or abutments	m²	1015.00 to 1175.00
Roadbridges		
Roadbridges including all excavation; reinforcement; formwork; concrete; bearings; expansion joints; deck water proofing and finishing's; parapets etc. deck area		
Reinforced concrete bridge with pre-cast beams		
10 m span	m²	1250.00 to 1800.00
15 m span	m²	1150.00 to 3000.00
Reinforced concrete bridge with prefabricated steel beams		
20 m span	m²	1200.00 to 1625.00
30 m span	m²	1150.00 to 1500.00
Underpass		
Provision of underpasses to new roads; constructed as part of a road building programme		
Pre-cast concrete pedestrian underpass		
3 m wide x 2.5 m high	m	3600.00 to 4750.00
Pre-cast concrete vehicle underpass		
7 m wide x 5 m high	m	19500 to 23500
14 m wide x 5 m high	m	33500 to 47000
Roundabouts		
Roundabout on existing dual carriageway; including perimeter road; drainage and lighting; signs and disruption while under construction	nr	410000 to 610000
Guard rails and parking bollards etc.		
Open metal post and rail fencing 1 m high	m	150.00 to 175.00
Galvanised steel post and rail fencing 2 m high	m	165.00 to 220.00
Steel guard rails and vehicle barriers	m	53.00 to 81.00
Parking bollards pre-cast concrete or steel	nr	120.00 to 255.00
Vehicle control barrier; manual pole	nr	940.00 to 1150.00
Galvanised steel cycle stand	nr	44.50 to 58.00
Galvanised steel flag staff	nr	1175.00 to 1500.00

Item	Unit	Range £
Street Furniture		
Reflected traffic signs 0.25m² area on steel post	nr	110.00 to 205.00
Internally illuminated traffic signs dependent on area	nr	215.00 to 300.00
Externally illuminated traffic signs dependent on area	nr	540.00 to 1450.00
Lighting to pedestrian areas and estates roads on 4.00 m to 6.00 m columns with up to 70 W lamps	nr	235.00 to 355.00
Lighting to main roads		
10 m to 12 m columns with 250 W lamps	nr	560.00 to 690.00
12 m to 15 m columns with 400 W high pressure sodium lighting	nr	720.00 to 870.00
benches to hardwood and pre-cast concrete	nr	215.00 to 290.00
Litter bins		
pre-cast concrete	nr	215.00 to 245.00
hardwood slatted	nr	89.00 to 120.00
cast iron	nr	365.00 to 400.00
large aluminium	nr	625.00 to 730.00
Bus stops	nr	405.00 to 595.00
Bus stops including basic shelter	nr	930.00 to 1475.00
Pillar box	nr	335.00 to 475.00
Telephone box	nr	3650.00 to 4200.00
Playground equipment		
Modern swings with flat rubber safety seats: four seats; two bays	nr	1500.00 to 1950.00
Stainless steel slide; 3.4 m long	nr	1775.00 to 2275.00
Climbing frame to igloo type 3.2 m x 3.75 m on plan x 2 m high	nr	1725.00 to 2250.00
See-saw comprising timber plank on sealed ball bearings; 3.96 m x 230 mm x 70 mm thick board	nr	1200.00 to 1550.00
Wickstead "Tumbleguard" type safety surfacing around play equipment	m²	101.00 to 130.00
Bark particles type safety surfacing 150 mm thick on hardcore bed	m²	13.60 to 18.10
Fencing and screen walls; ancillary building etc		
Chain link fencing; plastic coated		
1.2 m high	m	19.70 to 23.00
1.8 m high	m	28.00 to 31.00
Timber fencing		
1.2 m high chestnut pale facing	m	22.00 to 25.00
1.8 m high cross-boarded fencing	m	56.00 to 69.00
Screen walls; one brick thick; including foundations etc.		
1.8 m high facing brick screen wall	m	280.00 to 345.00
1.8 m high coloured masonry block boundary wall	m	310.00 to 395.00
6B DRAINAGE		
Overall £/m² allowances		
Site drainage (per m² of paved area)	m²	8.00 to 21.50
Building drainage (per m² of gross floor area)	m²	8.00 to 18.10
Machine excavation; grade bottom; earthwork support; laying and jointing pipes and accessories; backfill and compact; disposal of surplus soil		
Vitrified clay pipes and fittings; "Hepseal" socketted; with push fit flexible joints		
up to 1.5 m deep; nominal size		
up to 150 mm diameter	m	50.00 to 65.00
up to 300 mm diameter	m	62.00 to 84.00
over 1.5m not exceeding 3 m deep; nominal size		
up to 150 mm diameter	m	70.00 to 88.00
up to 300 mm diameter	m	120.00 to 135.00

Item	Unit	Range £
6B DRAINAGE – cont'd		
Class M tested concrete centrifugally spun pipes and fittings; flexible joints		
up to 1.5 m deep; nominal size		
300 mm diameter	m	47.50 to 60.00
up to 600 mm diameter	m	89.00 to 101.00
over 1.5 m not exceeding 3m deep; nominal size		
up to 600 mm diameter	m	110.00 to 140.00
900 mm diameter	m	185.00 to 225.00
1200 mm diameter	m	275.00 to 325.00
Cast iron "Timesaver" drain pipes and fittings; mechanical coupling joints		
up to 1.5 m deep; nominal size		
100 mm diameter	m	63.00 to 73.00
150 mm diameter	m	95.00 to 110.00
over 1.5 m not exceeding 3 m deep; nominal size		
100 mm diameter	m	73.00 to 115.00
150 mm diameter	m	110.00 to 145.00
uPVC pipes and fittings; lip seal coupling joints		
up to 1.5 m deep; nominal size		
100 mm diameter	m	20.50 to 26.00
160 mm diameter	m	26.00 to 33.00
over 1.5 m not exceeding 3 m deep; nominal size		
100 mm diameter	m	34.00 to 49.50
160 mm diameter	m	41.00 to 56.00
uPVC "UltratoRib" ribbed pipes and fittings; sealed ring push fit joints		
up to 1.5 m deep; nominal size		
150 mm diameter	m	22.50 to 34.00
300 mm diameter	m	48.50 to 61.00
over 1.5 m not exceeding 3 m deep; nominal size		
150 mm diameter	m	46.50 to 74.00
225 mm diameter	m	59.00 to 71.00
300 mm diameter	m	71.00 to 88.00
Brick Manholes		
Excavate pit in firm ground; partial backfill; partial disposal; earthwork support; compact base of pit; plain in situ concrete 20 N/mm² to 20 mm aggregate (1:2:4) base; formwork; one brick wall of engineering bricks PC £275.00 per 1000 in cement mortar (1:3) finished fair face; vitrified clay channels; plain in-situ concrete 25 N/mm² to 20 mm aggregate (1:2:4) cover and reducing slabs; fabric reinforcement; formwork step irons; medium duty cover and frame		
Internal size of manhole		
600 mm x 450 mm; cover to invert		
not exceeding 1 m	nr	475.00 to 565.00
over 1 m not exceeding 1 m	nr	615.00 to 710.00
over 1 m not exceeding 2 m	nr	655.00 to 805.00
900 x 600 mm; cover to invert		
not exceeding 1 m	nr	520.00 to 625.00
over 1 m not exceeding 1.5 m	nr	750.00 to 855.00
over 1.5 m not exceeding 2 m	nr	750.00 to 970.00
900 x 900 mm; cover to invert		
not exceeding 1 m	nr	595.00 to 800.00
over 1 m not exceeding 1.5 m	nr	725.00 to 965.00
over 1 m not exceeding 2 m	nr	865.00 to 1125.00

Item	Unit	Range £
1200 mm x 1800 mm; cover to invert		
not exceeding 1 m	nr	1010.00 to 1275.00
over 1 m not exceeding 1 m	nr	1200.00 to 1600.00
over 1 m not exceeding 2 m	nr	1400.00 to 1775.00
with reducing slab and brick shaft internal size 600 mm x 450 mm; depth from cover to invert		
over 2 m not exceeding 3 m	nr	1250.00 to 1675.00
over 3 m not exceeding 4 m	nr	2250.00 to 2900.00
Concrete manholes		
Excavate pit in firm ground; disposal; earthwork support; compact base of pit; plain in-situ concrete 20 N/mm² to 20 mm aggregate (1:2:4) base; formwork; reinforced pre-cast concrete chamber and shaft rings; taper pieces and cover slabs bedded jointed and pointed in cement mortar (1:3) weak mix concrete filling to working space; vitrified clay channels; plain in-situ concrete 25 N/mm² to 20 mm aggregate (1:1:5:3) benchings; step irons; medium duty cover and frame		
Internal diameter of manhole		
1350mm diameter; cover to invert		
up to 1.5 m	nr	905.00 to 1125.00
over 1.5 m not exceeding 2 m	nr	1010.00 to 1250.00
over 2 m not exceeding 3 m	nr	1450.00 to 1775.00
1500mm diameter; cover to invert		
up to 1.50 m	nr	1050.00 to 1350.00
over 1 m not exceeding 2 m	nr	1125.00 to 1500.00
over 2 m not exceeding 3 m	nr	1550.00 to 1925.00
1800mm diameter; cover to invert		
up to 1.5 m	nr	1350.00 to 1750.00
over 1.5 m not exceeding 2 m	nr	1475.00 to 1900.00
over 2 m not exceeding 3 m	nr	1775.00 to 2250.00
with taper piece and shaft 675 mm diameter; depth from cover to invert depth from cover to invert		
up to 2 m	nr	1250.00 to 1500.00
over 2 m not exceeding 3 m	nr	1400.00 to 2250.00
over 3 m not exceeding 4 m	nr	1575.00 to 2650.00
6C EXTERNAL SERVICES		
Gross internal area (unless otherwise described)		
Service runs		
All laid in trenches including excavation and backfill with excavated material		
Water main		
75 mm uPVC main	m	58.00 to 73.00
Electric main		
600 / 1000 volt cables. Two core 25 mm diameter cable including 100 mm diameter clayware duct	m	37.00 to 47.00
Gas main		
150 mm diameter gas pipe	m	63.00 to 84.00
Telephone		
100 mm diameter uPVC duct	m	26.00 to 42.00

Item	Unit	Range £
6C EXTERNAL SERVICES – cont'd		
Connection areas		
The privatisation of telephone; water; gas and electricity has complicated the assessment of service connection charges. Typically; service connection charges will include the actual cost of the direct connection plus an assessment of distribution costs from the main. The latter cost is difficult to estimate as it depends on the type of scheme and the distance from the mains. In addition; service charges are complicated by discounts that maybe offered. For instance; the electricity boards will charge less for housing connections if the house is all electric. However; typical charges for an estate of 200 houses might be as follows		
Water	house	500.00 to 1000.00
Electric		
all electric	house	300.00 to 400.00
gas/electric	house	275.00 to 800.00
pre-packaged sub-station housing	nr	22500 to 30000
Gas		
gas connection to house	house	600.00 to 800.00
governing station	nr	15000 to 2500
Telephone	house	200.00 to 275.00
Sewerage	house	500.00 to 600.00

Cost Limits and Allowances

Information given under this heading is based upon the cost targets currently in force for buildings financed out of public funds, i.e., hospitals, schools and public authority housing. The information enables the cost limit for a scheme to be calculated and is not intended to be a substitute for estimates prepared from drawings and specifications.

The cost limits are generally set as target costs based upon the user accommodation. However ad-hoc additions can be agreed with the relevant Authority in exceptional circumstances.

The documents setting out cost targets are almost invariably complex and cover a range of differing circumstances. They should be studied carefully before being applied to any scheme; this study should preferably be undertaken in consultation with a Chartered Quantity Surveyor.

HOSPITAL BUILDINGS

The index level for all business' cases for all health buildings is now set at a Median Index of Public Sector Building Tender Prices (MIPS) Variation of Price (VOP) index level of 480 with a Firm Price (FP) differential allowance of 2.08% increase on the VOP level, that is, a Firm Price (FP) index level of 490. The effective date for these increased allowances was 1 August 2007 (see letter from Rob Smith, Head of the Gateway Team and Director of Estates and Facilities Department of Health)

Both the Cost Allowances (DCAGs) and the Equipment Cost Allowance Guides (ECAGs) are now included in version 1.0 of the Health Capital Investment (HCI) document, issued in March 1997. This supersedes previous publications including version 13 of the Concise 4D/DCAG database.

"Quarterly Briefing" is now the official document for the notification of all new/revised DCAGs and ECAGs. Users should check Briefings issued from and including volume 13 no. 3 to ensure that all revisions to the HCI have been identified.

The NHS Estates has recently completed research into alternative indices for its Equipment Price Index, this was due to concerns that the index did not reflect market conditions as it had not been effectively maintained over the past few years. The result has been the development of a new equipment price index which reflects more accurately the market indicators for equipment.

The new index is a combination of various sources of information. This includes taking into consideration equipment-specific indices issued by the NHS Executive, namely the Health Service Cost Index, the retail price index and a general construction materials index. The combination of all this produces a new index series which reflects price movement and inflationary effects specifically associated with equipment.

The new equipment price index is set at 1997 = 100.

The ECAGs in the HCI document result from a complete revision of existing cost guidance and therefore supersede all previous equipment cost guidance.

During the calculation of capital costs for business cases, consideration must be given to the date when the equipment is likely to be purchased in relation to the construction tender index base date. This is the base date that the capital costs are adjusted to once a business case has been approved. Most equipment is likely to be purchased near the end of a building contract and will therefore have a later base date than the construction date.

Further advice can be obtained from NHS Estates (see Useful Addresses for Further Information on page 955).

EDUCATIONAL BUILDINGS (Procedures and Cost Guidance)

The former Department of Education and science is now called the Department of Children, Schools and Families.

The Department of Children, Schools and Families no longer publish guidance costs, as the data is no longer collected.

Capital funding mainly is devolved to schools, local authority and associated sectors under three main programmes. Current funding allocations relate to 2008/11. Most funding outside Building Schools for the Future is allocated on a formulaic basis related to pupil numbers

1. Devolved funding for local authorities and schools (2008/9 funding of £2,619m)

Devolved Formula capital (DFC) gives maintained schools direct funding to invest in their buildings and equipment. It is initially allocated to local education authorities (LEAs) who are then required to allocate the funding directly to schools using a simple national formula. In general, this funding should be invested in priorities agreed locally and identified in the local Asset Management Plan. It can be rolled forward to enable larger projects to be planned. In April 2007,the then DfES published research into how schools use this funding, which showed strong support from schools for its contribution to raising standard.

This direct investment initiative is aimed directly at schools, to provide each school with it's own capital, to address it's own priorities. Local Authorities, who deal with more strategic matters, expect schools to contribute DFC to any new projects associated with the school. It is mainly used to ensure that locally decided improvements can be made to unmodernised schools, and that modernised schools are properly maintained. There will be a higher rate for schools that are unmodernised and a lower rate (50 per cent) for those that have had 80 per cent or more of their floor areas modernised over the last ten years. These programmes also provide investment to every local authority in the country to focus on national and local priorities for modernisation, access and pupil places. Funding is allocated on the basis of relative need, the amount of modernised or unmodernised schools and pupil numbers.

2. Long-term funding programmes (Strategic) (2008/9 funding of £3,004m)

Building Schools for the future is the Governments flagship programme, launched in July 2003, with the aim of rebuilding or refurbishing all secondary schools in England over the next 15 years, at a total cost of £45 billion.. This investment id distributed in a series of 'waves' and will enable the continued rolling-out of the long-term Building Schools for the Future (BSF) (including academies) and the new Primary Capital Programmes which will rebuild or refurbish all secondary schools and at least half of all primary schools over the next 15 years.

The body responsible for delivering the Government's secondary school renewal programme is Partnership for Schools, and this includes contributing up to 50% of the salary for a dedicated Project Manager, to help deliver projects on time and to budget.

3. Targeted funding programmes (Targeted) (2008/9 funding of £1,046m)

These will support Government priorities in local authorities with the highest need. The programmes will provide capital funding to support higher standards and diversity of provision including promoter projects, fresh starts, new federations and applications for capital to expand successful and popular schools. There is also £150 million earmarked to support the introduction of kitchens in areas of greatest need. And there is £8 million each for 76 local authorities yet to be involved in BSF to: support Diploma provision for 14-19-year-olds; and to improve buildings for children with special educational needs and disabilities.

Formulaic funding excluding Building Schools for the Future.

The bulk of schools capital funding is allocated by formula to authorities and schools so that they can address their local priorities. Where allocations are for specific projects, funding is according to guidance for the relevant programme.

Cost multipliers

Cost Multipliers are costs per pupil for the construction of accommodation to provide for additional pupil places. The Multipliers are compiled to inform Basic Need funding allocations. It is likely that the Multipliers will in future be updated every two or three years, to suit the allocation process.

The 2008-9 Multipliers, based on project pricing levels at 4Q 2008, are as follows:

Primary	£ 12,257
Secondary	£ 18,469
Post 16	£ 20,030

These Multipliers are the averages of Multipliers for new schools and extensions to existing schools, weighted to reflect the national balance of such projects.

Each Multiplier has an area-per-place factor, derived from the BB98 or BB99 area standards. This is multiplied by a cost-per-m² factor. Allowances are added for external works, furniture and equipment and professional fees. The Multipliers exclude ICT equipment, site abnormals, site acquisition costs, VAT and the effect of regional variations in prices. Adjustments for location factors are downloadable from the Department's web-site.

Further Information for DCF and targeted funding is available at http://www.teachernet.gov.uk/management/resourcesfinanceandbuilding/FSP/, and for BSF, at http://www.p4s.org.uk/library/bsf_guidance.jsp/FundingGuidance

HOUSING ASSOCIATION SCHEMES

Information and other current literature is available from one of the offices of **The Housing Corporation**, whose headquarters are at 149 Tottenham Court Road, London, W1P 7BN, and who can be accessed via a national telephone number of **0845 230 7000** (please use for all telephone enquiries)

Other **Housing Corporation** offices are located at:

Central

- Cambridge: Westbrook Centre, Block A Suite 1, Milton Road, Cambridge, CB4 1YG; Fax: 01223 272531
- Leicester: Attenborough House, 109/119 Charles Street, Leicester LE1 1FQ; Fax: 0116 242 4801
- Wolverhampton: 31 Waterloo Road, Wolverhampton WV1 4DJ Fax;: 01902 795001

London

- London: Waverley House, 7–12 Noel Street, London W1F 8BA; Fax: 020 7292 4401

North

- Leeds: 1 Park Lane, Leeds, LS3 1EP; Fax: 0113 233 7101
- Manchester: 4th Floor, One Piccadilly Gardens, Manchester, M1 1RG;Fax: 0161 242 5901
- Gateshead: St George's House, Kingsway, Team Valley, Gateshead NE11 0NA Fax: 0191 482 7666

South East

- Croydon: Leon House, High Street, Croydon, Surrey CR9 1UH ;Fax: 020 8253 1444

South West

- Exeter: Beaufort House, 51 New North Road, Exeter EX4 4EP;Fax: 01392 428201

The Housing Corporation no longer uses **Total Cost Indicators (TCIs)** to assess bids for grant funding, although these are still available through the Housing Corporation website. Instead they invite bids under '**The National Affordable Housing Programme (NAHP) 2008 – 11 Prospectus**', issued in September 2007.

An Invitation to Bid (ITB) has been prepared for organisation, and details the information requirements with which each bid must comply. There are two associated documents which Bidders need to use, and these are:

(i) The NAHP Prospectus, which introduces the programme and products that the Corporation are looking to find, and

(ii)The Bid Information requirements, which takes Bidders through the process of submitting a Bid on line via the Corporations Information Management System (IMS).

There are two ways to apply for the NAHP 2008 – 11:

(i) through the Partnering Programme (open to all Housing Associations and unregistered bodies), or

(ii) through the specialist Programme (open to Housing Associations only),

but for the 2008 – 11 programmes, initial bids must have been submitted by 2 November 2007.

Since the new 'National Affordable Housing Programme 2006 – 08 Prospectus' was introduced, dwellings having to be designed to achieve an Ecohomes rating of Very Good; this is an increase from the requirement of good for schemes commenced prior to April 2006.

A new Capital Funding Guide was issued in 2008. For copies of all recent information and access to the IMS, go onto the Housing Corporations web-site on www.housingcorp.gov.uk

For internal purposes, some Registered Social Landlords (RSLs) still use the earlier TCIs.

Property Insurance

The problem of adequately covering by insurance the loss and damage caused to buildings by fire and other perils has been highlighted in recent years by the increasing rate of inflation.

There are a number of schemes available to the building owner wishing to insure his property against the usual risk. Traditionally the insured value must be sufficient to cover the actual cost of reinstating the building. This means that in addition to assessing the current value an estimate has also to be made of the increases likely to occur during the period of the policy and of rebuilding which, for a moderate size building, could amount to a total of three years. Obviously such an estimate is difficult to make with any degree of accuracy, if it is too low the insured may be penalized under the terms of the policy and if too high will result in the payment of unnecessary premiums.

There are variations on the traditional method of insuring which aim to reduce the effects of over estimating and details of these are available from the appropriate offices. For the convenience of readers who may wish to make use of the information contained in this publication in calculating insurance cover required the following may be of interest.

1 PRESENT COST

The current rebuilding costs may be ascertained in a number of ways:

(a) where the actual building cost is known this may be updated by reference to tender prices (page 60);
(b) by reference to average published prices per square metre of floor area (page 59). In this case it is important to understand clearly the method of measurement used to calculate the total floor area on which the rates have been based;
(c) by professional valuation;
(d) by comparison with the known cost of another similar building.

Whichever of these methods is adopted regard must be paid to any special conditions that may apply, i.e., a confined site, complexity of design, or any demolition and site clearance that may be required.

2 ALLOWANCE FOR INFLATION

The "Present Cost" when established will usually, under the conditions of the policy, be the rebuilding cost on the first day of the policy period. To this must be added a sum to cover future increases. For this purpose, using the historical indices on pages 59 to 60, as a base and taking account of the likely change in building costs and tender climate the following annual average indices are predicted for the future.

Year	Cost Index	Tender Index
1997	421	283
1998	439	313
1999	460	332
2000	489	359
2001	507	384
2002	536	412
2003	567	429
2004	600	442
2005	641	465
2006	682	491
2007	718	525
2008	752 (F)	556 (F)
2009	789 (F)	584 (F)
2010	822(F)	612 (F)
2011	857 (F)	645 (F

3 FEES

To the total of 1 and 2 above must be added an allowance for fees.

4 VALUE ADDED TAX (V.A.T.)

To the total of 1 to 3 above must be added Value Added Tax. Historically, relief may have been given to total reconstruction following fire damage etc. Since the 1989 Finance Act, such work, except for self-contained dwellings and other residential buildings and certain non-business charity buildings, has attracted V.A.T. and the limit of insurance cover should be raised to follow this.

5 EXAMPLE

An assessment for insurance cover is required in the fourth quarter of 2008 for a property which cost £200,000 when completed in 1976.

Present Cost

Known cost at mid 1976	=		£200,000
Predicted tender index fourth quarter 2008	=	567	
Tender index fourth quarter 1976	=	100	
Increase in tender index	=	467%	
applied to known cost	=		£934,000
Present cost (excluding any allowance for demolition)	=		£1,134,000

Allowance for inflation

Present cost at day one of policy	=		£1,134,000
Predicted tender index fourth quarter 2009	=	594	
Predicted tender index fourth quarter 2008	=	567	
Increase in tender index	=	4.76%	
applied to present cost	=	say	£ 54,000
Anticipated cost at expiry of policy	=		£1,188,000

Assuming that total damage is suffered on the last day of the currency of the policy and that planning and documentation would require a period of twelve months before re-building could commence, then a further similar allowance must be made.

Predicted tender index fourth quarter 2010	=	624	
Predicted tender index fourth quarter 2009	=	594	
Increase in tender index	=	5.05%	
applied to cost at expiry of policy	=	say	£ 60,000
Anticipated cost at tender date	=		£1,248,000

Assuming that reconstruction would take one year, allowance must be made for the increases In costs which would directly or indirectly be met under a building contract.

Predicted cost index fourth quarter 2011	=	872	
Predicted cost index fourth quarter 2010	=	837	
Increase in cost index	=	4.18%	

This is the total increase at the end of the one year period.			
The amount applicable to the contract would be about half,		say	£ 26,000.
Estimated cost of reinstatement	=		£1,274,000

SUMMARY OF EXAMPLE

Estimated cost of reinstatement	=	£ 1,274,000
Add professional fees at, say 16%	=	£ 204,000
Sub-total		£ 1,478,000
Add for V.A.T., currently at 17½% say	=	£ 259,000
Total insurance cover required	=	**£ 1,737,000**

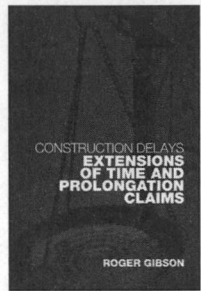

PART 3

Prices for Measured Works - Major Works

INTRODUCTION

The rates contained in "Prices for Measured Works - Major Works" are intended to apply to a project in the outer London area costing about £3,500,000 (including Preliminaries) and assume that reasonable quantities of all types of work are required. Similarly it has been necessary to assume that the size of the project warrants the sub-letting of all types of work normally sub-let. Adjustments should be made to standard rates for time, location, local conditions, site constraints and any other factors likely to affect costs of a specific scheme.

The distinction between builders' work and work normally sub-let is stressed because prices for work which can be sub-let may well be quite inadequate for the contractor who is called upon to carry out relatively small quantities of such work himself.

As explained in more detail later, Measured Works prices are generally based upon wage rates which came into force on 30 June 2008, and material costs from April/May 2008. Built-up prices include an allowance of 5% for overheads and 2½% for profit, whereas non-analysed sub-contractor prices only include mark-up of 2½% for profit. They do not allow for preliminary items that are dealt with under a separate heading (see page 175) or for any Value Added Tax.

The format of this section is so arranged that, in the case of work normally undertaken by the Main Contractor, the constituent parts of the total rate are shown enabling the reader to make such adjustments as may be required in particular circumstances. Similar details have also been given for work normally sub-let although it has not been possible to provide this in all instances.

As explained in the Preface, there is a facility available to readers, which enables a comparison to be made between the level of prices in this section and current tenders by means of a tender index. The tender index for this Major Works section of Spons is 580 (as shown on the front cover) which is close to our current forecast tender price index for the outer London region for the first quarter of 2008.

To adjust prices for other regions/times, the reader is recommended to refer to the explanations and examples on how to apply these tender indices, given on page 62.

There follow explanations and definitions of the basis of costs in the "Prices for Measured Work" section under the following headings:

- Overhead charges and profit
- Labour hours and Labour £ column
- Material £ column
- Material/Plant £ columns
- Total rate £ column

OVERHEAD CHARGES AND PROFIT

Rates checked against winning tenders include overhead charges and profit at current levels.

LABOUR HOURS AND LABOUR £ COLUMNS

"Labour rates" are based upon typical gang costs divided by the number of primary working operatives for the trade concerned, and for general building work include an allowance for trade supervision (see below). "Labour hours" multiplied by "Labour rate" with the appropriate addition for overhead charges and profit gives "Labour £". In some instances, due to variations in gangs used, "Labour rate" figures have not been indicated, but can be calculated by dividing "Labour £" by "Labour hours".

Building craft operatives and labourers

From 30 June 2008 guaranteed minimum weekly earnings in the London area for craft operatives and general operatives are £401.70 and £302.25 respectively; to these rates have been added allowances for the items below in accordance with the recommended procedure of the Institute of Building in its "Code of Estimating Practice". The resultant hourly rates on which the "Prices for Measured Work" have generally been based are £14.27 and £10.69 for craft operatives and labourers, respectively.

- Lost time
- Construction Industry Training Board Levy
- Holidays with pay
- Accidental injury, retirement and death benefits scheme
- Sick pay
- National Insurance
- Severance pay and sundry costs
- Employer's liability and third party insurance

NOTE: For travelling allowances and site supervision see "Preliminaries" section.

The table which follows illustrates how the "all-in" hourly rates referred to on page170 have been calculated. Productive time has been based on a total of 1954 hours worked per year.

		Craft Operatives		General Operatives	
		£	£	£	£
Wages at standard basic rate					
productive time	44.30wks	401.70	17,795.31	302.25	13,389.68
Lost time allowance	0.9 wks	401.70	361.53	302.25	272.03
Non-productive overtime	5.80 wks	602.55	3,494.79	453.38	2,629.58
			21,651.63		16,291.28
Extra payments under National Working Rules	45.2 wks		-		-
Sick Pay	1 wk		-		-
CITB Allowance (0.50% of payroll)	1 year		122.15		91.91
Holiday pay	4.20 wks	479.02	2,011.88	360.43	1,513.81
Public Holiday pay	1.60 wks	479.02	766.43	360.43	576.69
Employer's contribution to retirement cover scheme (death and accident cover is provided free)	52.0 wks	5.00	260.00	5.00	260.00
National Insurance (average weekly payment)	48 wks	44.30	2,126.40	30.00	1,440
			26,938.49		20,173.68
Severance pay and sundry costs	Plus	1.5%	404.08	1.5%	302.61
			27,342.57		20,476.29
Employer's Liability and Third Party Insurance	Plus	2.0%	546.85	2.0%	409.53
Total cost per annum			**£27,889.42**		**£20,885.82**
Total cost per hour			**£14.27**		**£10.69**

NOTES:

1 Absence due to sickness has been assumed to be for periods not exceeding 3 days for which no payment is due (Working Rule 20.7.3).

2 EasyBuild Stakeholder Pension effective from 1 July 2002. Death and accident benefit cover is provided free of charge. Taken as £5.00/week average as range increased for 2006/09 wage award

3 All N.I. Payments are at not-contracted out rates applicable from April 2008. National Insurance is paid for 48 complete weeks (52wks-4.2wks) is based on employer making regular monthly payments into the Template holiday pay scheme and by doing so the employer achieves National Insurance savings on holiday wages.

The “labour rates” used in the Measured Work sections have been based on the following gang calculations which generally include an allowance for supervision by a foreman or ganger. Alternative labour rates are given showing the effect of various degrees of bonus.

Gang	Total Gang rate £/hour	Productive unit rate £/hour		Alternative labour rates £/hour			
				Normal	+10%	+20%	+30%
Groundwork Gang							
1 Ganger	1 x 11.53 =	11.53					
6 Labourers	6 x 10.69 =	64.14					
		75.67	÷ 6.5 =	11.64	12.83	14.02	15.20
Concreting gang							
1 Foreman	1 x 15.12 =	15.12					
4 Skilled Labourers	4 x 11.53 =	46.12					
		61.24	÷ 4.5 =	13.61	15.00	16.38	17.76
Steelfixing Gang							
1 Foreman	1 x 15.12 =	15.12					
4 Steelfixers	4 x 14.27 =	57.08					
		72.20	÷ 4.5 =	16.04	17.67	19.30	20.93
Formwork Gang							
1 Foreman	1 x 15.12 =	15.12					
10 Carpenters	10 x 14.27 =	142.70					
1 Labourer	1 x 10.69 =	10.69					
		168.51	÷ 10.5 =	16.05	17.68	19.31	20.89
Bricklaying/Light Blockwork Gang							
1 Foreman	1 x 15.12 =	15.12					
6 Bricklayers	6 x 14.27 =	85.62					
4 Labourers	4 x 10.69 =	42.76					
		143.50	÷ 6.5 =	22.08	24.32	26.57	28.81
Dense Blockwork Gang							
1 Foreman	1 x 15.12 =	15.12					
6 Bricklayers	6 x 14.27 =	85.62					
6 Labourers	6 x 10.69 =	64.14					
		164.88	÷ 6.5 =	25.37	27.95	30.53	33.11
Carpentry/Joinery Gang							
1 Foreman	1 x 15.12 =	15.12					
5 Carpenters	5 x 14.27 =	71.35					
1 Labourer	1 x 10.69 =	10.69					
		97.16	÷ 5.5 =	17.67	19.46	21.25	23.05
Craft Operative (Painter, Slater, etc.)	1 x 14.27 =	14.27	÷ 1 =	14.27	15.72	17.17	18.62
1 and 1 Gang							
1 Craft Operative	1 x 14.27 =	14.27					
1 Skilled Labourer	1 x 11.53 =	11.53					
		25.80	÷ 1 =	25.80	28.43	31.05	33.67
2 and 1 Gang							
2 Craft Operatives	2 x 14.27 =	28.54					
1 Skilled Labourer	1 x 11.53 =	11.53					
		40.07	÷ 2 =	20.04	22.08	24.11	26.15
Small Labouring Gang (making good)							
1 Foreman	1 x 15.12 =	15.12					
4 Skilled Labourers	4 x 11.53 =	46.12					
		61.24	÷ 4.5 =	13.61	15.00	16.38	17.76
Drain Laying Gang/Clayware							
2 Skilled Labourers	2 x 11.53 =	23.06	÷ 2 =	11.53	12.71	13.88	15.05

Sub-Contractor's operatives
Similar labour rates are shown in respect of sub-let trades where applicable.

Plumbing operatives
From 7 January 2008 the hourly earnings for technical and trained plumbers are £13.46 and £10.39, respectively; to these rates have been added allowances similar to those added for building operatives (see below). The resultant average hourly rate on which the "Prices for Measured Work" have been based is £16.83. The items referred to above for which allowance has been made are:

Tool allowance
Plumbers' welding supplement
Holidays with pay
Pension and welfare stamp
National Insurance "contracted out"
Severance pay and sundry costs
Employer's liability and third party insurance

No allowance has been made for supervision as we have assumed the use of a team of technical or trained plumbers who are able to undertake such relatively straightforward plumbing works, e.g. on housing schemes, without supervision.

The table which follows shows how the average hourly rate referred to above has been calculated. Productive time has been based on a total of 1912.50 hours worked per year.

		Technical Plumber		Trained Plumber	
		£	£	£	£
Wages at standard basic rate					
productive time	1687.5hrs	13.46	22,713.75	10.39	17,533.13
Overtime (paid at standard basic rate)	157.5hrs	13.46	2,119.95	10.39	1,636.43
Overtime	67.5hrs	20.19	1,362.83	15.59	1,051.99
Plumber's welding supplement (gas and arc)	1912.4hrs	0.46	879.75		0.00
			27,076.28		20,221.54
Employer's contribution to					
holiday credit/welfare					
stamps (to provide for 29 days)	60 credits	60.30	3,618.00	46.40	2,784.00
pension (6.5% of earnings)	46.0wks	43.48	2,000.08	32.59	1,499.14
Holiday top-up funding	60 credits	1.26	75.60	0.99	59.40
(Provided by employer)					
National Insurance	46.0wks	62.11	2,857.06	42.99	1,977.54
			35,627.02		26,541.62
Severance pay and sundry costs	Plus	1.5%	534.41	1.5%	398.12
			36,161.43		26,939.74
Employer's Liability and Third Party Insurance	Plus	2.0%	723.23	2.0%	538.79
Total cost per annum			**£36,884.66**		**£27,478.53**
Total cost per hour			**£19.29**		**£14.37**
Average all-in rate per hour				**£16.83**	

SUB/SPECIALIST-CONTRACTOR'S COSTS

Where Sub/Specialist-Contractor's figures have been provided, we have not been able to show build-ups, as these are not widely available. Any prices from such companies are deemed to include all their costs to a main contractor, including their own overheads, profit, preliminaries and a 2.5% Main Contractor's discount.

MATERIAL £ COLUMN

Many items have reference to a "PC" value. This indicates the prime cost of the principal material delivered to site in the outer London area assuming appropriate discounts for large quantities. When obtaining material prices from other sources, it is important to identify any discounts that may apply. Some manufacturers only offer 5 to 10% discount for the largest of orders; or "firm" orders (as distinct from quotations). For other materials, discounts of 30% to 40% may be obtained, depending on value of order, preferential position of the purchaser or state of the market. The "Material £" column indicates the total materials cost including delivery, waste, sundry materials and an allowance (currently 7½%) for overhead charges and profit for the unit of work concerned. Alternative material prices are listed, excluding discount, at the beginning of many sections. If these are used they require 'discount' adjustment before they can be substituted in place of "PC" figures given for Measured Work items. All material prices quoted are exclusive of Value Added Tax.

If material only or alternative material prices are indicated, they have not been extended into the TOTAL RATE £ COLUMN, and their values exclude overheads and profit.

PLANT COSTS (included in the MATERIAL £ COLUMN)

Plant costs have been based on current weekly hire charges and estimated weekly cost of oil, grease, ropes (where necessary), site servicing and cartage charges. The total amount is divided by 30 (assuming 25% idle time) to arrive at a cost per working hour of plant. To this hourly rate is added one hour fuel consumption and one hour for an operator where indicated; the rate to be calculated in accordance with the principles set out earlier in this section, i.e. with an allowance for plus rates, etc.
For convenience the all-in rates per hour used in the calculations of "Prices for Measured Work" are shown below and where included in this book, are included in the MATERIAL £ COLUMN.

Plant	Labour	"All-in" rate per hour £
Excavator (4 wheeled - 0.76 m³ shovel, 0.24 m³ bucket)	Driver	27.50
Excavator (JCB 3C - 0.24 m³ bucket)	Driver	25.00
Excavator (JCB 3C off centre - 0.24 m³ bucket)	Driver	27.50
Excavator (Hitachi EX120 - 0.53 m³ bucket)	Driver	32.50
Dumper (2.30 m³)	Driver	17.50
Two tool portable compressor (125 cfm)*		
per breaking tool		2.33
per punner foot and stem rammer		2.00
Roller		
Bomag BW75S - pedestrian double drum		5.00
Bomag BW120AD - tandem		6.00
5/3.50 cement mixer		2.00
Kango heavy duty breaker		1.00
Power float		1.67
Light percussion drill		0.83

* Operation of compressor by tool operator

TOTAL RATE £ COLUMN

"Total rate £" column is the sum of "Labour £" and "Material £" columns. This column excludes any allowance for "Preliminaries" which must be taken into account if one is concerned with the total cost of work.
The example of "Preliminaries" in the following section indicates that in the absence of detailed calculations currently 16% should be added to all Main Contractor's prices for measured work to arrive at total cost for the project (excluding VAT).

A PRELIMINARIES

The number of items priced in the “Preliminaries” section of Bills of Quantities and the manner in which they are priced vary considerably between Contractors. Some Contractors, by modifying their percentage factor for overheads and profit, attempt to cover the costs of “Preliminary” items in their “Prices for Measured Work”. However, the cost of “Preliminaries” will vary widely according to job size and complexity, site location, accessibility, degree of mechanisation practicable, position of the Contractor's head office and relationships with local labour/domestic Sub-Contractors. It is therefore usually far safer to price “Preliminary” items separately on their merits according to the job.

In amending the Preliminaries/General Conditions section for SMM7, the Joint Committee stressed that the preliminaries section of a bill should contain two types of cost significant item:

1. Items which are not specific to work sections but which have an identifiable cost which is useful to consider separately in tendering e.g. contractual requirements for insurances, site facilities for the employer's representative and payments to the local authority.
2. Items for fixed and time-related costs which derive from the contractor's expected method of carrying out the work, e.g. bringing plant to and from site, providing temporary works and supervision.

A fixed charge is for work the cost of which is to be considered as independent of duration. A time related charge is for work the cost of which is to be considered as dependent on duration. The fixed and time-related subdivision given for a number of preliminaries items will enable tenderers to price the elements separately should they so desire. Tenderers also have the facility at their discretion to extend the list of fixed and time-related cost items to suit their particular methods of construction.

The opportunity for Tenderers to price fixed and time-related items in A30-A37, A40-A44 and A51-A52 has been noted against the following appropriate items although we have not always provided guidance as costs can only be assessed in the light of circumstances of a particular job.
Works of a temporary nature are deemed to include rates, fees and charges related thereto in Sections A36, A41, A42, and A44, all of which will probably be dealt with as fixed charges.
In addition to the cost significant items required by the method, other preliminaries items which are important from other points of view, e.g. quality control requirements, administrative procedures, may need to be included to complete the Preliminaries/General conditions as a comprehensive statement of the employer's requirements.

Typical clause descriptions from a “Preliminaries/General Conditions” section are given below together with details of those items that are most likely to be priced in detail here when submitting tenders.
An example in pricing “Preliminaries” follows, and this assumes the form of contract used is the JCT 2005 Standard Building Contract With Quantities SBC/Q and the value, including “Preliminaries”, is approximately £3,500,000 The contract is estimated to take 60 weeks to complete and the value is built up as follows:

		£
Labour value		1,100,000
Material value		925,000
Provisional sums and all Sub-Contractors		975,000
	£	3,000,000

At the end of the section the example is summarised to give a total value of “Preliminaries” for the project.

A PRELIMINARIES

A PRELIMINARIES/GENERAL CONDITIONS

(NOTE The term "Not priced" or "Generally not priced" where used throughout this section means either that the cost implication is negligible or that it is usually included elsewhere in the tender.)

Preliminary particulars

A10 **Project particulars** - Not priced

A11 **Tender and Contract Drawings** - Not priced

A12 **The Site/Existing buildings -** Generally not priced

The reference to the site and existing buildings relates only to access and those buildings that could have an influence on cost. This could arise from their close proximity making access difficult, their heights relative to the possible use of tower cranes or the fragility of, for example, an historic building necessitating special care.

A13 **Description of the work** - Generally not priced

A20 **The Contract/Sub-contract -** Generally not priced (except where indicated)

(The JCT2005 Standard Building Contract is assumed)

(Note: Most of the contract particulars tend to be either priced elsewhere in specific preliminaries clauses or in general allowances for overheads etc. In a number of instances the cost implication is negligible and included elsewhere in the tender. Where prices are included against listed contract particulars they tend to be of a specialist nature with measurable risk attributed to the specific contract obligation)

Section 1: Definitions and Interpretation – Not priced
Definitions
Interpretation

Section 2: Carrying out the Works – Generally not priced (except where marked with an *)
Contractors Obligations
Possession
Supply of Documents, Setting Out etc.
The contract conditions may require a master programme to be prepared. This will normally form part of head office overheads and therefore is not priced separately here.
Errors, Discrepencies and Divergences
CDP Design Work
Where there is a Contractor's Design Portion, Design liabilities and limitation are identified here. Design costs are usually included with the related work section. See also note on Professional Indemnity Insurance in section 6 below.
Fees, Royalties and Patent Rights
Unfixed Materials and Goods; property, risk etc
Adjustment of Completion Date
Practical Completion, Lateness and Liquidated Damages
Partial Possession by Employer
Defects *
Inevitably some defects will arise after practical completion and an allowance will often be made to cover this. An allowance of say 0.25 to 0.50% should be sufficient, e.g. Example

Defects after completion

Based on 0.25% of the contract sum
£3,500,000 x 0.25%, say £ 9,000

A PRELIMINARIES

Contractors Design Documents
Contractor is to supply as built drawings for works included in Contractor Design Portion. Costs are likely to be included in Specialist Sub-contract work package or general overheads unless design works is extensive and of a special nature

Section 3: Control of the Works – Not priced
Access for Employers Agent
Sub-Letting
Architect/ Contract Administrators's Instructions
Antiquities
CDM Regulations

Section 4: Payment – Not priced
Contract Sum and Adjustments
Certificates and Payments
Gross Valuation
Retention
Fluctuations
Loss and Expense

Section 5: Variations – Not priced
General
The Valuation Rules

Section 6: Injury, Damage and Insurance – Generally not priced (except where marked with an *)
Injury to Persons and Property
Sets out liability of Contractor against personal injury or death of any person arising out of or in the course of or caused by the carrying out of the Works
Insurance against Personal Injury and Property Damage
The Contractor's Employer's Liability and Public Liability policies (which would both be involved under this heading) are often in the region of 0.50 to 0.60% on the value of his own contract work (excluding provisional sums and work by sub-contractors whose prices should allow for these insurances). This is normally included in the Contractor's overheads therefore not included here.
No requirement is made upon the Contractor to insure the liability of the Employer unless it is stated in the contract particulars that insurance may be required and the Architect/CA instructs the Contractor to take out a joint names policy for the sum as stated in the particulars. If instructed the amount expended by the Contractor to take out and maintain insurance is added to the Contract Sum
Insurance of the Works *
If at the Contractor's risk, the insurance cover must be sufficient to include the full cost of reinstatement, all increases in cost, professional fees and any consequential costs such as demolition. The average provision for fire risk is 0.10% of the value of the work after adding for increased costs and professional fees, e.g. Example

Contractor's Liability - Insurance of works against fire, etc.		£
Contract value (including "Preliminaries"), say		3,500,000
Estimated increased costs during contract period, say 3%		105,000
		3,605,000
Estimated increased costs incurred during period of reinstatement, say 5%		180,000
		3,785,000
Professional fees @ 16%		605,000
		4,390,000
Allow 0.1% say	£	4,400

A PRELIMINARIES

A20 **The Contract/Sub-contract** – cont'd

Section 6: Injury, Damage and Insurance – cont'd

NOTE:Insurance premiums are liable to considerable variation, depending on the Contractor, the nature of the work and the market in which the insurance is placed.

CDP Professional Indemnity Insurance
When the works include a Contractor Designed Portion, Professional indemnity Insurance is now a contract condition. Inclusion of the premium here will depend on who is carrying out the design i.e. if it is specialist work it would be normal for the specialists rates to include the premium. The Contractor is still liable for Professional Negligence of his sub-contractor's which may carry a premium in its self. This is likely to be included in the Contractor's overheads unless the design responsibility from the project is of a particularly high risk.
Joint Fire Code: Compliance

Section 7: Assignment, Third Party Rights and Collateral Warranties
– Generally not priced (except occasionally where marked with an *)
Assignment
Clauses 7A to 7F: Preliminary
This section reflects the JCT policy of introducing Third Party Rights into its contracts while maintaining the option of using alternative collateral warranties
Third Party Rights from Contractor
The form of Third Party Rights to be granted (where applicable) are set out in Schedule 5 and are substantially identical to those in the corresponding Collateral Warranty
Collateral Warranties *
Agreement of contractor and sub-contractor third party warranties is often complex and can involve legal input. This cost can vary depending on the size of project, number of third parties involved, together with the scope of contractor design responsibility. Costs are normally part of the Contractor's overheads and therefore not priced here.

Section 8: Termination – Not priced
General
Termination by Employer
Termination by Contractor
Termination by Either Party
Consequences of Termination under Clauses 8-9 to 8-11, etc.

Section 9: Settlement of Disputes – Generally not priced (except where marked with an *)
Mediation
Adjudication
Arbitration

General Note If the Contractor is required to provide sureties for the fulfilment of the work the usual method of providing this is by a bond provided by one or more insurance companies. The cost of a performance bond depends largely on the financial standing of the applying Contractor. Figures tend to range from 0.15 to 0.25% of the net tender sum (tender sum – preliminaries = net tender sum).

A30-A37 EMPLOYERS' REQUIREMENTS

These include the following items but costs can only be assessed in the light of circumstances on a particular job. Details should be given for each item and the opportunity for the Tenderer to separately price items related to fixed charges and time related charges.

A30 **Tendering/Sub-letting/Supply**

A31 **Provision, content and use of documents**

A PRELIMINARIES

A32 Management of the works

This includes Client's specific requirements for management of the works including supervision, management schemes, such as the Considerate Constructors, specific programming requirements, site meetings, progress reports , control of cost etc. The Contractor should allow for costs here where not already included in his general management and Staff cost section A40.

A33 Quality standards/control

Most Contractor's undertaking major projects will have accredited Quality Assurance schemes such as ISO 9001. The cost of running these schemes is normally accounted for in overhead charges however, specific Employers Requirements for specic or special Quality Control activity may be priced here.

A34 Security/Safety/Protection

This includes allowing for specific execution or product hazards, site security requirements, constraints due to working on occupied buildings or hazardous areas. Also, protection against or control of noise, pollution, fire, waste, adjoining buildings, public and private roads, live services, deterioration, security and work in all sections.

(i) Control of noise, pollution and other obligations
The Local Authority, Landlord or Management Company may impose restrictions on the timing of certain operations, particularly noisy of dust-producing operations, which may necessitate the carrying out of these works outside normal working hours or using special tools and equipment. The situation is most likely to occur in built-up areas such as city centres, shopping malls etc., where the site is likely to be in close proximity to offices, commercial or residential property.

(ii) Maintenance of public and private roads
Some additional value or allowance may be required against this item to insure/protect against damage to entrance gates, kerbs or bridges caused by extraordinary traffic in the execution of the works.

(iii) The requirements of the SIte Waste Management Plans Regulations 2008 came into force on 6 April 2008. This requires the preparation of project specific waste management plans together with additional management resource particularly on projects in excess of £0.5m in value. It is expected that the cost of meeting these requirements will be included in General management and staff costs. Additional costs for treatment of waste will depend on the project location etc and may incur additional costs within work sections or within A42 Contractor's General Cost Items - Services and facilities .

A35 Specific limitations on method/sequence/timing

This includes design constraints, method and sequence of work, access, possession and use of the site, use or disposal of materials found, start of work, working hours, employment of labour and sectional possession or partial possession etc.

A36 Facilities/Temporary work/Services

This includes offices, sanitary accommodation, temporary fences, hoardings, screens and roofs, name boards, technical and surveying equipment, temperature and humidity, telephone/facsimile installation and rental/maintenance, special lighting and other general requirements, etc. The attainment and maintenance of suitable levels of facilities and services necessary for satisfactory completion of the work including the installation of joinery, suspended ceilings, lift machinery, etc. is the responsibility of the contractor.

The installation of telephones or facsimiles for the use of the Employer, and all related charges therewith, shall be given as a provisional sum.

A PRELIMINARIES

A37 Operation/Maintenance of the finished building

Requirements for spares and replacement parts are usually identified in work sections and therefore priced elsewhere. However, the Employer often requires the Contractor to provide significant documentation and training on completion of the works including a Building Manual (incorporating the Health and Safety File) together with a separate Building Log book. The cost is likely to be included either of part of the overhead or within site management and staff costs

A40-A44 CONTRACTORS GENERAL COST ITEMS

For items A41-A44 it shall be clearly indicated whether such items are to be "Provided by the Contractor" or "Made available (in any part) by the Employer".

A40 Management and staff (Provided by the Contractor)

Includes management, trades supervision, engineering, programming and production, quantity surveying, support staff and the like.

Typical allowance for Management and Staff could be 5% to 8% of the net tender sum (tender sum excluding preliminaries).

Based on 6.25% of £3,000,000 - say	£	187,500

A41 Site accommodation (Provided by the Contractor or made available by the Employer)

This includes all temporary offices, laboratories, cabins, stores, compounds, canteens, sanitary facilities and the like for the Contractor's and his domestic sub-contractors' use (temporary office for a Clerk of Works is covered under obligations and restrictions imposed by the Employer).

Typical costs for jack-type timber or steel vandal-proof offices are as follows, based upon a twelve months minimum hire period they exclude furniture which could add a further £15.00 - £20.00 per week.

Typical rates for other units are as follows:

Size	Rate £/Week
Office Cabins - 24 ft x 10 ft (23 m²)	30 - 35
Office Cabins - 32 ft x 10 ft (30 m²)	45 - 55
Fire rated Cabins - 24 ft x 10 ft (23 m²)	85 - 95
Fire rated Cabin - 32 ft x 10 ft (30 m²)	105 - 115
Meeting Room - 24 ft x 10 ft (23 m²)	30 - 35
Meeting Room - 32 ft x 10 ft (30 m²)	45 - 55
Mess Cabins (incl Furniture)	45 - 55
Drying Rooms (incl Furniture)	55 - 60
Safestore-20ft x 8ft (15 m²)	15 - 20
Container with Padlock-20ft x 8ft (15 m²)	10 - 15
Toilets	50 - 60
Fire Signs and notices	35 - 40
Fire Extingusihers	5 - 10
Haulage to and from site – Site offices, Storage sheds, Toilets	£800

A PRELIMINARIES

Typical allowance for site accommodation is 0.40% to 0.60% of the net tender sum (tender sum excluding preliminaries).

Based on 0.50% of £3,000,000, say £ 15,000

A42 **Services and facilities** (Provided by the Contractor or made available by the Employer)

This generally includes the provision of all of the Contractor's own services, power, lighting, fuels, water, telephone and administration, safety, health and welfare, storage of materials, rubbish disposal, cleaning, drying out, protection of work in all sections, security, maintaining public and private roads, small plant and tools and general attendance on nominated sub-contractors.
However, this section does not cover fuel for testing and commissioning permanent installations, which would be measured under Sections Y51 and Y81. Examples of build-ups/allowances for some of the major items are provided below:

A42/110 -120 Lighting and power for the works

The Contractor is usually responsible for providing all temporary lighting and power for the works and all charges involved. On large sites this could be expensive and involve sub-stations and the like, but on smaller sites it is often limited to general lighting (depending upon time of year) and power for power operated tools for which a small diesel generator and some transformers usually proves adequate.

Typical costs are:

	Rate £/week
Power and Lighting	
Cost of locating existing services	
3 Kva transformer	£10 - £15
5 Kva transformer	£15 - £25
10 Kva transformer	£25 - £35
4 Way Distribution Box	£10 - £15
50ft extension lead	£3.50 - £7.50
2 x 500W floodlights	£15 - £30
2 x 500W Floodlights on stands	£15 - £30
Generators	
4 Kva	£55
8 Kva	£155
12.5 Kva silenced	£195
25 Kva silenced	£240
50 Kva silenced	£360
100 Kva silenced	£450
200 Kva silenced	£705

Typical allowance for lighting and power could be 1.5% to 2.0% of the net tender sum (tender sum excluding preliminaries)

Based on 1.50% of £3,000,000, say £ 45,000

A PRELIMINARIES

A42 Services and facilities - cont'd

A42/140 Water for the works

Charges should properly be ascertained from the local Water Authority. If these are not readily available, an allowance of 0.10 to 0.15% of the value of the net tender sum (tender sum excluding preliminaries) is probably adequate, providing water can be obtained directly from the mains. Failing this, each case must be dealt with on its merits. In all cases an allowance should also be made for temporary plumbing including site storage of water if required.

Useful rates for temporary plumbing include:

Water for the works	
Connection main works	£500
Piping	£15 per metre
Standpipe	£150-£250
Water charges	£10 per week
Allow for hoses etc	£250

Based on 0.15% of £3,000,000, say £ 4,500

A42/150 Temporary telephones for the use of the Contractor

Communications
Useful rates are:

Line installation costs/line	£100
Line rental	£50 per quarter
Mobile call charges	£50 per week
Server	£5,000
External ringing device	£100
Radiophone	£15 per week
Walkie/talkie	£15 per week
Broad band	£20 per month
Fax machine	£400
Connection of fax line	£100
Rental of fax line	£50 per quarter
Photocopier	£35 per week
Call Charges per week	£35 per week per line

Typical allowance for communications could be 0.20% to 0.30% of the net tender sum (tender sum excluding preliminaries)

Based on 0.20% of £3,000,000, say £ 6,000

A42/160 Safety, health and welfare of workpeople

The Contractor is required to comply with the Code of Welfare Conditions for the Building Industry which sets out welfare requirements for the following: Shelter from inclement weather; Accommodation for clothing; Accommodation and provision for meals; Provision of drinking water; Sanitary conveniences; Washing facilities; First aid; Site conditions

A PRELIMINARIES

A variety of self-contained mobile or jack-type units are available for hire and a selection of rates is given below:

Kitchen with cooker, fridge, sink unit, water heater and basin		
32 ft x 10 ft jack-type	£162.50	per week
Mess room with water heater, wash basin and seating		
16 ft x 7 ft 6 in mobile	£90.00	per week
16 ft x 9 ft jack-type	£80.00	per week
Welfare unit with drying rack, lockers, tables, seating, cooker, heater, sink and basin		
22 ft x 7 ft 6 in mobile	£130.00	per week
Toilets (mains type)		
One pan unit	£47.50	per week
Three pan unit mobile (one fire-rated)	£140.00	per week
Four pan unit jack-type (one fire-rated)	£180.00	per week

Allowance must be made in addition for transport costs to and from site, setting up costs, connection to mains, fuel supplies and attendance

Site first aid kit £ 10.00 per week

A general provision to comply with the above code is often 0.1 to 0.15% of the measured work value. The costs of safety supervisors (required for firms employing more than 20 people) are usually part of head office overhead costs.

Example
Safety, health and welfare

Combined charge
The fixed charge would normally represent a proportion of the following allowance, with the majority allocated to time related charges.

Based on 0.12% of £3,000,000- say £ 3,600

A42/180 Removing rubbish, protective casings and coverings and cleaning the works on completion. This includes removing surplus materials and final cleaning of the site prior to handover. Allow for sufficient skips for the site throughout the contract duration and for some operatives time at the end of the contract for final clearing and cleaning ready for handover.

Cost of skips - approx. £200.00 each
A general allowance of 0.3% to 0.5% of measured work value is probably sufficient.

Example
Removing rubbish, etc., and cleaning
Combined charge
The fixed charge would normally represent an allowance for final clearing of the works on completion with the residue for cleaning throughout the contract period

Based on 0.3% of £3,000,000 – say £ 9,000

A PRELIMINARIES

A42 Services and facilities - cont'd

A42/200a Drying the works

Use or otherwise of an installed heating system will probably determine the value to be placed against this item. Dependent upon the time of year, say allow 0.03% to 0.05% of the contract value to cover this item

Example

	Rate £/week
Dehumidifier	
Small	£25
Large	£50
Turbo dryer	£35
JetAir heaters (gas included in rate)	£35

Typical allowance for drying the works could be 0.03% to 0.05% of the net tender sum (tender sum excluding preliminaries)

Based on 0.05% of £3,000,000, say £ 1,500

A42/210 Protecting the works from inclement weather

In areas likely to suffer particularly inclement weather, some nominal allowance should be included for tarpaulins, polythene sheeting, battening, etc., and the effect of any delays in concreting or brickwork by such weather.

Typical allowance for protection of works could be 0.12% to 0.15% of the net tender sum (tender sum excluding preliminaries)

Based on 0.12% of £3,000,000, say £3,600

A42/220 Security

This includes watchman, electronic surveillance and protection of scaffolds

Typical allowance for protection of works could be 0.10% to 0.15% of the net tender sum (tender sum excluding preliminaries)

Based on 0.12% of £3,000,000 say £3,600

A42/240 Small plant and tools

Small plant and hand tools are usually assessed as between 0.10% and 0.12% of total labour value.

	Rate£/week
Useful rates are:	
Mixers	£15 - £20
Compressor & tools	£100 - £130
Small tools	£50 - £60
Concrete test tubes	£10 - £15

Typical allowance for small plant and tools could be 0.10% to 0.12% of the net tender sum (tender sum excluding preliminaries)

Based on 0.10% of £3,000,000 say	£3,000	
Testing and commissioning: Water, fuel, gas, electricity and other	£3,000	
Allowance of say		£6,000

A PRELIMINARIES

A43 Mechanical plant

This includes for cranes, hoists, personnel transport, transport, earthmoving plant, concrete plant, piling plant, paving and surfacing plant, etc. SMM6 required that items for protection or for plant be given in each section, whereas SMM7 provides for these items to be covered under A34, A42 and A43, as appropriate.

A43/110 Plant

Quite often, the Contractors own plant and plant employed by sub-contractors are included in measured rates, (e.g. for earthmoving, concrete or piling plant) and the Editors have adopted this method of pricing where they believe it to be appropriate. As for other items of plant e.g. cranes, hoists, site vans etc., these tend to be used by a variety of trades and are therefore often priced in the preliminaries section. A Typical allowance is 1.5%-2.5% of the net tender sum (tender sum excluding preliminaries) An example of such an item follows:. e.g. Example

Tower Crane - Luffing jib 30m radius-4/5t max. load

Fixed Costs	£
Erection	8,000
Dismantling	8,000
Base and base Angles	9,500
Radios Chain	3,000
Signage	1,500
Flood Lights	1,000
Time related charge	
Hire of crane say 16 weeks @ £1,750	28,000
Banksman Cost say 16 weeks @ £775/week	12,400
Power Consumed	3,750
Bonus	3,250
	£78,400

A43/130 Personnel transport

The labour rates per hour on which "Prices for Measured Work" have been based do not cover travel and lodging allowances, which must be assessed according to the appropriate working rule agreement.

Example
Personnel transport

Assuming all labour can be found within the London region, the labour value of £1,100,000 represents approximately 750 man weeks. Assume each man receives an allowance of £7.80 per day or £39.00 per week of five days.

Combined/time related charge

750 man weeks @ £39.00, say	£	30,000

A43/140 Plant transport

Allowance of say	£	2,000

A PRELIMINARIES

A44 **Temporary works** (Provided by the Contractor or made available by the Employer)

This includes for temporary roads, temporary walkways, access scaffolding, support scaffolding and propping, hoardings, fans, fencing etc., hardstanding and traffic regulations etc. The Contractor should include maintaining any temporary works in connection with the items, adapting, clearing away and making good, and all notices and fees to Local Authorities and public undertakings. On fluctuating contracts, i.e. where Option A and B is incorporated there is no allowance for fluctuations in respect of plant and temporary works and in such instances allowances must be made for any increases likely to occur over the contract period.

Examples of build-ups/allowances for some items are provided below against the relevant preliminaries reference:

A44/110 Temporary roads, crossings and similar items. Quite often consolidated bases of eventual site roads are used throughout a contract to facilitate movement of materials around the site. However, during the initial setting up of a site, with drainage works outstanding, this is not always possible and occasionally temporary roadways have to be formed and ground levels later reinstated.

Typical costs are:

3.5m wide @ £8-12/m²	£11,500

Useful costs are:

Type 1 fill say 150mm thick	£7.00/m²
Terram	£1.25/m²
Dig	£4.50/m³
Cartaway	£25/m³

A44/120 Temporary Walkways

Typical cost	£1,000

A44/130 Access Scaffolding

The General Contractor's standing scaffolding is usually undertaken by specialist Sub-Contractors who will submit quotations based on the specific requirements of the works

Access Scaffolding	(Typical allowance)		£45,500
Scissor lifts (SL30 Flying Carpet		£175 - £200	
Scissor lifts (Gennie boom 245)		£275-£325	
Cherry picker :		£275-£325	

A44/140 Support scaffolding and propping	£1,000

A44/150 Temporary fencing, hoarding, screens, fans, planked footways, guardrails, gantries, and similar items. This item must be considered in some detail as it is dependant on site perimeter, phasing of the work, work within existing buildings, etc.

A PRELIMINARIES

Useful rates include:

Hoarding 2.30 m high of 18 mm thick plywood with 50 mm x 100 mm sawn softwood studding, rails and posts, including later dismantling		
	undecorated	£85.00/m (£37.00/m²)
	decorated one side	100.00/m (£43.50/m²)
Pair of gates for hoarding		extra £500.00 per pair
Cleft Chestnut fencing 1.20 m high including dismantling		£10.50/
Morarflex "T-Plus" scaffold sheeting		£5.000/m²

Example
Temporary hoarding
Combined fixed charge

		£
Decorated plywood hoarding		
100m @ £100.00		10,000
extra for one pair of gates		500
	£	10,500

A44/160 Temporary Hardstandings

£8-£12/m²	£	1,200

A44/170 Traffic regulations

Waiting and unloading restrictions can occasionally add considerably to costs, resulting in forced overtime or additional weekend working. Any such restrictions must be carefully assessed for the job in hand.

Typical Allowance £ 1,500

A44/200 Additional Temporary Works Items

Insert below further cost items as may be required, with fixed charges and time related charge

	£	
Setting Out Equipment	250	
Sign Boards	1200	
Photographs	500	
Considerate Construstors Scheme	1000	
Sundries	4000	
Programme	5000	
Towers,Trestles et	750	
Sub total		£ 12,700

A50 Work/Materials by the Employer

A description shall be given of works by others directly engaged by the Employer and any attendance that is required shall be priced.

A PRELIMINARIES

A51 Nominated sub-contractors

Not applicable with JCT 2005 Contracts

A52 Nominated suppliers

Not applicable with JCT 2005 Contracts

A53 Work by statutory authorities

Works which are to be carried out by a Local Authority or statutory undertakings shall be given as provisional sums.

A54 Provisional work

SMM7 requires the identification of provisional sums as being for either defined or undefined work. The rules require that each sum for defined work should be accompanied in the bills of quantities by a description of the work sufficiently detailed for the tenderer to make allowance for its effect in the pricing of relevant preliminaries. The information should also enable the length of time required for execution of the work to be estimated and its position in the sequence of construction to be determined and incorporated into the programme. Where Provisional Sums are given for undefined work the Contractor will be deemed not to have made any allowance in programming, planning and pricing preliminaries. Any provision for Contingencies shall be given as a provisional sum for undefined work.

A55 Dayworks

To include provisional sums for: Labour, Materials and Goods and Plant

A PRELIMINARIES

Summary of Preliminaries costs included in previous pages

Items		£
A20 sec 2	Defects after completion	9,000
A20 sec 6	Insurance of the works against fire, etc.	4,400
A40	Management and staff	187,500
A41	Contractor's accommodation	15,000
A42/110-120	Lighting and power for the works	45,000
A42/140	Water for the works	4,500
A42/150	Temporary telephones	6,000
A42/160	Safety, health and welfare	3,600
A42/180	Removing rubbish, etc., and cleaning	9,000
A42/200a	Drying the works	1,500
A42/210	Protection of the works	3,600
A42/220	Security	3,600
A42/240	Small plant and tools	6,000
A43/110	Mechanical plant	78,400
A43/130	Personnel transport	30,000
A43/140	Plant transport	2,000
A44/110	Temporary roads	11,500
A44/110	Temporary walkways	1,000
A44/130	Access scaffolding	45,500
A44/140	Support scaffolding and propping	1,000
A44/150	Hoardings, fans, fencing, etc.	10,500
A44/160	Temporary hardstandings	1,200
A44/170	Traffic regulations	1,500
A44/200	Additional temporary works	12,700
	TOTAL £	494,000

It is emphasized that the above is an example only of the way in which Preliminaries may be priced and it is essential that for any particular contract or project the items set out in Preliminaries should be assessed on their respective values. The value of the Preliminaries items in recent tenders received by the editors varies from a 15% to a 19% addition to all other costs. The above example represents approximately a 16% addition to the value of measured work.

A PRELIMINARIES

NEW ITEMS

The One Hundred and Thirty-fourth edition of Spon's Architects' and Builders' Price Book introduces a number of new items into both the Measured Works sections. These new items, as part of a rolling programme of improvements, compliment and expand the existing information that makes Spon's the most detailed, professionally relevant source of construction price data. Pages 191 to 199 and 505 to 511 separately identify the new items also introduced into the relevant Major and Minor Works sections of the book, respectively. Any suggestions for further suitable new items in future editions should be addressed to the authors.

B NEW ITEMS

Item	PC £	Labour hours	Labour £	Material £	Unit	Total rate £
New items are also included in their appropriate work section						
E10 IN SITU CONCRETE CONSTRUCTION						
Proprietary voided Bubbledeck, Cobiax or other equal and approved slab; concrete mix RC35; to achieve design loadings of 5.0 kN/m² live and 3.0 kN/m² dead; with trowelled finish						
Beds						
360 mm overall thickness	-	-	-	-	m²	**112.75**
Extra for						
Additional concrete 600 mm wide at edges where formers omitted at junctions with walls etc.	-	-	-	-	m	**46.13**
E30 REINFORCEMENT FOR IN SITU CONCRETE						
Bars; stainless steel; to LDX2101® (EN 1.4362)						
NOTE: LDX2101® (EN 1.4362) is a new low Ni, Mn bearing stainless steel alloy, which offers greater price stability and cost effectiveness, and is expected to be adopted into the British Standard in the near future.						
32 mm diameter nominal size						
straight	3023.75	17.00	323.29	3403.52	tonne	**3726.81**
bent	3280.00	21.00	390.62	3686.20	tonne	**4076.82**
25 mm diameter nominal size						
straight	3023.75	18.00	342.31	3399.99	tonne	**3742.30**
bent	3280.00	18.00	342.31	3682.67	tonne	**4024.98**
20 mm diameter nominal size						
straight	3023.75	20.00	380.35	3404.63	tonne	**3784.98**
bent	3228.75	20.00	380.35	3630.78	tonne	**4011.13**
16 mm diameter nominal size						
straight	3023.75	22.00	418.38	3413.25	tonne	**3831.64**
bent	3228.75	22.00	418.38	3639.40	tonne	**4057.78**
12 mm diameter nominal size						
straight	3023.75	24.00	456.42	3421.88	tonne	**3878.29**
bent	3177.50	24.00	456.42	3591.49	tonne	**4047.90**
10 mm diameter nominal size						
straight	3023.75	26.00	494.45	3431.61	tonne	**3926.06**
bent	3177.50	26.00	494.45	3601.22	tonne	**4095.67**
8 mm diameter nominal size						
straight	3023.75	28.00	528.11	3440.23	tonne	**3968.35**
bent	3177.50	28.00	528.11	3609.84	tonne	**4137.96**
F30 ACCESSORIES/SUNDRY ITEMS FOR BRICK/BLOCK WALLING						
Damp proof courses						
"Engerseal" polymer elastomeric damp proof course or other equal and approved; 200 mm laps; in gauged morter (1:1:6)						
width exceeding 225 mm; horizontal	2.37	0.23	6.02	2.74	m²	**8.76**
width exceeding 225 mm; forming cavity gutters in hollow walls; horizontal	-	0.37	9.68	2.74	m²	**12.43**
width not exceeding 225 mm; horizontal	-	0.46	12.04	2.74	m²	**14.78**
width not exceeding 225 mm; vertical	-	0.69	18.06	2.74	m²	**20.80**

B NEW ITEMS

Item	PC £	Labour hours	Labour £	Material £	Unit	Total rate £
F30 ACCESSORIES/SUNDRY ITEMS FOR BRICK/BLOCK WALLING – cont'd						
"Zedex CPT" (Co-Polymer Thermoplastic) damp proof course or other equal and approved; 200 mm laps; in gauged mortar (1:1:6)						
width exceeding 225 mm; horizontal	3.26	0.23	6.02	3.77	m²	**9.79**
width exceeding 225 mm wide; forming cavity gutters in hollow walls; horizontal	-	0.37	9.68	3.77	m²	**13.45**
width not exceeding 225 mm; horizontal	-	0.46	12.04	3.77	m²	**15.81**
width not exceeding 225 mm; vertical	-	0.69	18.06	3.77	m²	**21.83**
"Alumite" aluminium cored bitumen gas retardent damp proof course or other equal and approved; 200 mm laps; in gauged mortar (1:1;6)						
width exceeding 225 mm; horizontal	4.80	0.31	8.11	5.55	m²	**13.67**
width exceeding 225 mm; forming cavity gutters in hollow walls; horizontal	-	0.49	12.83	5.55	m²	**18.38**
width not exceeding 225 mm; horizontal	-	0.60	15.70	5.55	m²	**21.26**
width not exceeding 225 mm; vertical	-	0.83	21.72	5.55	m²	**27.28**
Galvanised steel lintels; "Catnic" or other equal and approved; built into brickwork or blockwork						
90/125 range "CG" open back lintel for cavity wall						
750 mm long	36.85	0.23	6.02	39.70	nr	**45.72**
900 mm long	44.22	0.28	7.33	47.63	nr	**54.96**
1200 mm long	58.06	0.32	8.38	62.53	nr	**70.91**
1500 mm long	72.41	0.37	9.68	77.97	nr	**87.65**
1800 mm long	91.62	0.42	10.99	98.64	nr	**109.63**
2100 mm long	108.58	0.46	12.04	116.89	nr	**128.94**
2400 mm long	153.38	0.56	14.66	165.11	nr	**179.77**
90/125 range "CUB" open back lintel for cavity wall						
2700 mm long	242.82	0.65	17.01	261.38	nr	**278.39**
3000 mm long	385.77	0.74	19.37	415.22	nr	**434.59**
90/125 range "CU" open back lintel for cavity wall						
3300 mm long	293.17	0.83	21.72	315.56	nr	**337.28**
3600 mm long	319.82	0.93	24.34	344.25	nr	**368.59**
3900 mm long	346.23	1.02	26.70	372.67	nr	**399.36**
4200 mm long	363.35	0.46	12.04	391.09	nr	**403.13**
"CN100" single lintel for 75 mm internal walls						
1050 mm long	17.30	0.28	7.33	18.64	nr	**25.96**
1200 mm long	21.50	0.32	8.38	23.16	nr	**31.53**
"CN5XA" single lintel for 100 mm internal walls						
1050 mm long	32.67	0.28	7.33	35.18	nr	**42.51**
1200 mm long	33.52	0.32	8.38	36.09	nr	**44.47**

B NEW ITEMS

Item	PC £	Labour hours	Labour £	Material £	Unit	Total rate £
H10 PATENT GLAZING						
Patent glazing; aluminium alloy bars 2.55 m long at 622 mm centres; fixed to supports						
Roof cladding						
single glazed with 6.4 mm laminated glass	-	-	-	-	m²	**125.00**
thermally broken and double glazed with low-e clear toughened and laminated double glazed units; aluminium finished RAL matt colour	-	-	-	-	m²	**350.00**
Extra for opening roof vents						
600 mm x 900 mm top hung opening roof vent; manually operated	-	-	-	-	nr	**400.00**
600 mm x 900 mm top hung opening roof vent; electrically operated	-	-	-	-	nr	**500.00**
Skylight						
Self-supporting hipped or gable ended lantern/skylight thermally broken and double glazed with low-e clear toughened and laminated double glazed units; aluminium finished RAL matt colour	-	-	-	-	m²	**700.00**
Wall cladding						
single glazed with 6.4 mm laminated glass	-	-	-	-	m²	**130.00**
thermally broken and double glazed with low-e clear toughened and laminated double glazed units; aluminium finished RAL matt colour	-	-	-	-	m²	**368.00**
H11 CURTAIN WALLING						
Stick curtain walling system; Schuco FW50+ proprietary system or other equal and approved						
Polyester powder coated solid colour matt finish or natural anodised curtain walling with mullions spaced 1.5m apart and spanning typical storey height of 3.8m. Floor to ceiling glass sealed units with 8.8mm low E coated laminated inner pane, air filled cavity and 8mm clear annealed outer pane, retained by external pressure plates and caps. Rates to include 0.8m deep solid spandrel panels, all brackets, membranes, fire stopping between floors and external access equipment						
Flat system; drilling and screwing; to metal	-	-	-	-	m²	**450.00**
Extra over for						
neutral selective high performance coating in lieu of low E, for assisting in solar control	-	-	-	-	m²	**40.00**
outer glass pane to be toughened and heat soak tested or heat strengthened in lieu of annealed	-	-	-	-	m²	**20.00**
inner laminated glass to be toughened and heat soak tested laminated, or heat strengthened laminated	-	-	-	-	m²	**40.00**
flush glass finish without external face caps, achieved by concealed toggle fixings locating within perimeter channels within sealed units including silicone sealing between glass panes	-	-	-	-	m²	**50.00**
typical coping detail, including pressed aluminium profiles, membranes, seals, etc	-	-	-	-	m	**250.00**
typical cill detail, including pressed aluminium profiles, membranes, seals, etc	-	-	-	-	m	**200.00**
intermediate transoms (per transom)	-	-	-	-	m	**40.00**

B NEW ITEMS

Item	PC £	Labour hours	Labour £	Material £	Unit	Total rate £
H11 CURTAIN WALLING – cont'd						
Unitised curtain walling system; Schuco Skyline 65 proprietary system or other equal and approved						
Polyester powder coated solid colour matt finish or natural anodised curtain walling with mullions spaced 1.5m apart and spanning typical storey height of 3.8m. Floor to ceiling glass sealed units with 8.8mm low E coated laminated inner pane, air filled cavity and 8mm clear annealed outer pane, retained by external beading system. Rates to include 0.8m deep solid spandrel panels, all brackets, membranes, fire stopping between floors & external access equipment						
Flat system; drilling and screwing; to metal	-	-	-	-	m²	**650.00**
Extra over for						
neutral selective high performance coating in lieu of low E, for assisting in solar control	-	-	-	-	m²	**40.00**
outer glass pane to be toughened and heat soak tested or heat strengthened in lieu of annealed	-	-	-	-	m²	**20.00**
inner laminated glass to be toughened and heat soak tested laminated, or heat strengthened laminated	-	-	-	-	m²	**40.00**
flush glass finish without external face caps, achieved by carrier frames with glass sealed units factory silicone bonded; often referred to as SSG (Structural Silicone Glazing)	-	-	-	-	m²	**100.00**
typical coping detail, including pressed aluminium profiles, membranes, seals, etc	-	-	-	-	m	**250.00**
typical cill detail, including pressed aluminium profiles, membranes, seals, etc	-	-	-	-	m	**200.00**
Other curtain walling systems/costs						
Unitised curtain walling system; bespoke solution via specialist façade contractor based in mainland Europe. Generally as described in 1J but comprising a project specific solution, thus additional design development. Note: These rates are subject to currency fluctuations between £ and €. The rate opposite assumes £1 = €1.40	-	-	-	-	m²	**750.00**
Project specific performance testing for bespoke unitised curtain walling system	-	-	-	-	nr	**75000.00**
Visual mock-ups are often required for bespoke curtain walling solutions and in cases for proprietary unitised and stick curtain walling projects	-	-	-	-	nr	**25000.00**
All curtain walling projects should be site hose tested. The rate depends upon the quantum of joints to be tested, generally 5%. Assume 5 days @ £1000	-	-	-	-	nr	**5000.00**
Brise soleil, to mitigate the effects of solar gain and enable compliance with Part L of the Building regulations. There are a variety of material types which can be adopted for the purpose of solar shading, including but not limited to; Aluminium, Glass, Timber. South elevations require horizontal shading to combat high sun angles, whereas east and west elevations require vertical fins to accommodate low angle sun paths. The rate opposite assumes a single natural anodised extruded aluminium fin, with brackets and orientated either horizontally or vertically						
300 mm deep	-	-	-	-	m	**125.00**

B NEW ITEMS

Item	PC £	Labour hours	Labour £	Material £	Unit	**Total rate £**
H60 PLAIN ROOF TILING						
Concrete interlocking tiles; Marley Eternit "Ecologic Ludlow Major" granule finish tiles or other equal and approved; 420 mm x 330 mm; to 75 mm lap; on 25 mm x 38 mm battens and type 1F reinforced underlay						
Roof coverings (PC £ per 1000)	795.60	0.32	9.51	11.65	m²	**21.15**
Extra over coverings for						
fixing every tile	-	0.02	0.59	0.78	m²	**1.37**
eaves; eaves filler	-	0.04	1.19	0.59	m	**1.78**
verges; 150 mm wide asbestos free strip undercloak	-	0.21	6.24	1.91	m	**8.15**
dry verge system; extruded white pvc	-	0.14	4.16	11.36	m	**15.51**
segmental ridge cap to dry verge	-	0.02	0.59	3.63	m	**4.22**
valley trough tiles; cutting both sides	-	0.51	15.15	25.02	m	**40.16**
segmental ridge tiles	-	0.46	13.66	1.05	m	**14.72**
segmental hip tiles; cutting both sides	-	0.60	17.82	3.75	m	**21.57**
dry ridge tiles; segmental including batten sections; unions and filler pieces	-	0.28	8.32	11.54	m	**19.86**
segmental mono-ridge tiles	-	0.46	13.66	17.12	m	**30.78**
gas ridge terminal	-	0.46	13.66	68.59	nr	**82.26**
holes for pipes and the like	-	0.19	5.64	-	nr	**5.64**
H92 RAINSCREEN CLADDING						
Reynobond rainscreen cladding; aluminium composite material cassettes with thermoplastic cores, back ventilated, including insulation, vapour control membrane and aluminium support system						
4 mm thick cladding; fixed to walls	-	-	-	-	m²	**164.00**
Terracotta clay rainscreen cladding; including insulation, vapour control membrane and aluminium support system						
400 x 200 x 30 mm tile cladding; fixed to walls	-	-	-	-	m²	**297.25**
K32 FRAMED PANEL CUBICLE PARTITIONS						
Toilet cubicle partitions; Amwells or other equal and approved; standard colours and ironmongery; assembling and screwing to floor and wall						
"Axis" standard cubicle set; 800 mm x 1500 mm x 1980 mm high per cubicle, with polished aluminium framing; 19 mm melamine-faced chipboard divisions and doors						
One cubicle set; 2 nr panels; 1 nr door	-	3.25	130.00	280.00	nr	**410.00**
range of 3 cubicle sets; 4 nr panels; 3 nr doors	-	9.75	390.00	800.00	nr	**1190.00**
range of 6 cubicle sets; 7 nr panels; 6 nr doors	-	19.50	780.00	1580.00	nr	**2360.00**
Reduction of 1 nr panel for end unit adjoining side wall	-	-	-	-110.00	nr	**-**
"Minima" designer cubicle set; 800 mm x 1500 mm x 2100 mm high per cubicle, with satin polished stainless steel framing; 18 mm high pressure laminated (HPL) chipboard divisions and doors						
One cubicle set; 2 nr panels; 1 nr door	-	3.25	130.00	590.00	nr	**720.00**
range of 3 cubicle sets; 4 nr panels; 3 nr doors	-	9.75	390.00	1530.00	nr	**1920.00**
range of 6 cubicle sets; 7 nr panels; 6 nr doors	-	19.50	780.00	2940.00	nr	**3720.00**
Reduction of 1 nr panel for end unit adjoining side wall	-	-	-	-160.00	nr	**-**

B NEW ITEMS

Item	PC £	Labour hours	Labour £	Material £	Unit	Total rate £
K32 FRAMED PANEL CUBICLE PARTITIONS – cont'd						
Toilet cubicle partitions; Amwells – cont'd						
"Sylan " corporate cubicle set; 800 mm x 1500 mm x 2400 mm high per cubicle, with sating finished stainless steel ironmongery; 30 mm high pressure laminated (HPL) chipboard divisions and 44 mm solid cored real wood veneered doors and pilasters						
One cubicle set; 2 nr panels; 1 nr door	-	5.00	200.00	1745.00	nr	**1945.00**
range of 3 cubicle sets; 4 nr panels; 3 nr doors	-	15.00	600.00	4715.00	nr	**5315.00**
range of 6 cubicle sets; 7 nr panels; 6 nr doors	-	30.00	1200.00	9165.00	nr	**10365.00**
Reduction of 1 nr panel for end unit adjoining side wall	-	-	-	-355.00	nr	-
L10 WINDOWS/ROOFLIGHTS/SCREENS/ LOUVRES						
SUPPLY ONLY PRICES						
Thermally broken composite double glazed aluminium/ timber windows; 'Velfac 200'; with a maximum glazing U value of 1.5 W/m²K; low e glazing with laminated glass unless otherwise specified; including multi point espagnolette locking mechanisms and other ironmongery						
NOTE: The following supply only prices are for standard windows, to which fixings, sealants etc. labour and overheads and profit need to be added, before they may be used to arrive at a guide price for a complete unit.						
Outward opening standard fixed sash casement windows						
1200 mm x 1200 mm single fixed pane; low-e glass 4/16/4	-	-	-	260.00	nr	-
2200 mm x 2200 mm single fixed pane; low-e glass 6/12/6	-	-	-	660.00	nr	-
1200 mm x 2200 mm three fixed panes; low-e glass 4/16/4	-	-	-	520.00	nr	-
Outward opening standard sash casement windows						
1600 mm x 1600 mm with two sidehung sashes; low-e glass 4/16/4	-	-	-	530.00	nr	-
1600 mm x 1600 mm with two sidehung projecting sashes; low-e glass 4/16/4	-	-	-	590.00	nr	-
2000 mm x 1600 mm with one sidehung sash next to a tophung projecting sash over a fixed sash; low-e glass 4/16/4	-	-	-	650.00	nr	-
1200 mm x 2200 mm with fixed lower sash and tophung projecting upper sash; lower low-e upper low-e glass 4 toughened/16/6.4; upper low-e glass 4/16/4	-	-	-	515.00	nr	-
1200 mm x 2200 mm with fixed lower sash and fully reversible upper sash; lower low-e upper low-e glass 4 toughened/16/6.4; upper low-e glass 4/16/4	-	-	-	560.00	nr	-

B NEW ITEMS

Item	PC £	Labour hours	Labour £	Material £	Unit	Total rate £
Outward opening standard doors						
2200 mm x 2200 mm French casement patio door; low-e toughened glass 4/16/4	-	-	-	995.00	nr	-
2200 mm x 2200 mm Sliding patio door; low-e glass 4 toughened/16/4 laminated	-	-	-	1240.00	nr	-
Guide price for installation of preceding windows	-	1.00	60.00	-	m²	**60.00**
SUPPLY AND FIX PRICES						
Aluminium windows; Schuco AWS 50 proprietary system or equal and approved						
Polyester powder coated solid colour matt finish or natural anodised window system of glass sealed units with 6.4mm low E coated laminated inner pane, air fillled cavity and 6mm clear annealed outer pane. Rates to include all brackets, membranes, cills, silicone seals, trade contractor preliminaries, including external access equipment						
Ribbon construction windows 1.5m high	-	-	-	-	m²	**450.00**
Punched hole windows fixing into prepared apertures by others	-	-	-	-	m²	**500.00**
Extra over for						
1.25 m x 1.5m opening vents, assuming tilt and turn operation	-	-	-	-	m²	**150.00**
neutral selective high performance coating in lieu of low E, for assisting in solar control	-	-	-	-	m²	**40.00**
outer glass pane to be toughened and heat soak tested or heat strengthened in lieu of annealed	-	-	-	-	m²	**20.00**
inner laminated glass to be toughened and heat soak tested laminated, or heat strengthened laminated	-	-	-	-	m²	**40.00**
Louvres, Brise Soleils and frames; polyester powder coated aluminium; fixing in position including brackets						
Louvre; Levolux or other equal and approved; 5 rows of 400 aerofins set in steel plate frame						
6700 mm x 2200 mm (14.75 m² overall)	-	-	-	-	m²	**300.00**
Brise Soleil; Levolux or other equal and approved; on galvanised steel cantilever beams and runners						
1000 mm deep	-	-	-	-	m	**400.01**
L40 GENERAL GLAZING						
Factory made double hermetically sealed units; with inner pane of Pilkington's K low emissivity coated glass; to wood or metal with screwed or clipped beads						
Two panes; BS EN 14449; clear float glass; 4 mm thick; 6 mm air space						
0.35 m² - 2.00 m²	-	-	-	-	m²	**94.26**
Two panes; BS EN 14449; clear float glass; 6 mm thick; 6 mm air space						
0.35 m² - 2.0 m²	-	-	-	-	m²	**109.74**
2.00 m² - 4.00 m²	-	-	-	-	m²	**165.05**

B NEW ITEMS

Item	PC £	Labour hours	Labour £	Material £	Unit	Total rate £
L40 GENERAL GLAZING – cont'd						
Factory made triple hermetically sealed units; with inner pane of Pilkington's K low emissivity coated glass; to wood or metal with screwed or clipped beads						
Three panes; BS EN 14449; clear float glass; 4 mm thick; 6 mm air spaces						
0.35 m² - 2.00 m²	-	-	-	-	m²	**151.76**
Three panes; BS EN 14449; clear float glass; 6 mm thick; 6 mm air spaces						
0.35 m² - 2.0 m²	-	-	-	-	m²	**176.72**
2.00 m² - 4.00 m²	-	-	-	-	m²	**265.74**
R12 DRAINAGE BELOW GROUND						
Accessories; grates and covers						
Polypropylene access covers and frames; supplied by Manhole Covers Ltd or other equal and approved; to suit PPIC inspection chambers; bedding and pointing in frame.						
450 mm dia; class A15	15.30	1.30	16.13	18.36	nr	**34.49**
450 mm dia; classB125; kite-marked	39.60	1.30	16.13	45.16	nr	**61.30**
Ductile iron heavy duty road gratings and frame; Manhole Covers Ltd or other equal and approved; bedding and pointing in cement and sand (1:3); one course half brick thick wall in semi-engineering bricks in cement mortar (1:3)						
420 mm x 420 mm x 75 mm hinged road grating and frame; ref C250; kite-marked	35.55	2.25	27.92	42.16	nr	**70.08**
445 mm x 445 mm x 75 mm double triangular road grating and frame; ref C250; kite-marked	37.80	2.25	27.92	44.64	nr	**72.56**
435 mm x 435 mm x 100 mm pedestrian mesh road grating and frame; ref D400	66.60	2.25	27.92	76.41	nr	**104.33**
MANHOLES						
Coated cast or ductile iron access covers and frames; to BS EN124; supplied by Manhole Covers Ltd or other equal and approved; bedding frame in cement and sand (1:3); cover in grease and sand						
Light duty; cast iron; rectangular single seal solid top						
750 mm x 600 mm; class A15	98.50	1.50	18.61	110.40	nr	**129.01**
Medium duty; ductile iron; rectangular single seal solid top						
450 mm x 450 mm x 40 mm; class C250; kite-marked	57.60	2.00	24.82	65.27	nr	**90.09**
600 mm x 450 mm x 40 mm; slide-out; class C250; kite-marked	71.55	2.00	24.82	80.66	nr	**105.48**
600 mm x 600 mm x 40 mm; slide-out; class C250; kite-marked	76.50	2.00	24.82	86.12	nr	**110.94**
760 mm x 600 mm x 40 mm; slide-out; class C250; kite-marked	114.30	2.00	24.82	127.82	nr	**152.64**
Heavy duty; ductile iron; solid top						
450 mm x 450 mm x 75 mm; single seal; class C250; kite-marked	80.55	2.50	31.02	90.59	nr	**121.61**
600 mm x 450 mm x 75 mm; single seal; class C250; kite-marked	85.50	2.50	31.02	96.05	nr	**127.08**

B NEW ITEMS

Item	PC £	Labour hours	Labour £	Material £	Unit	Total rate £
600 mm x 600 mm x 75 mm; single seal; class C250; kite-marked	96.30	2.50	31.02	107.97	nr	**138.99**
450 mm x 450 mm x 100 mm; double triangular; class D400; kite-marked	89.10	2.50	31.02	100.02	nr	**131.05**
600 mm x 450 mm x 100 mm; double triangular; class D400; kite-marked	80.10	2.50	31.02	90.09	nr	**121.12**
750 mm x 600 mm x 100 mm; double triangular; class D400; kite-marked	169.78	2.50	31.02	189.02	nr	**220.04**
1220 mm x 675 mm x 100 mm; double triangular; class D400; kite-marked	202.50	3.50	43.43	225.12	nr	**268.55**
T31 LOW TEMPERATURE HOT WATER HEATING						
Radiators; Hudevad Heat Emitters or other equal and approved						
Plan Fiona double panel convector; 600 mm high; front, back plates and convector fins with intergrated top grille; wheelhead and lockshield valves						
500 mm long x 68 mm deep; 584 watts output	76.61	1.85	38.41	95.60	nr	**134.01**
1400 mm long x 68 mm deep; 1634 watts output	185.47	2.15	44.64	212.77	nr	**257.40**
1400 mm long x 98 mm deep; 2022 watts output	207.52	2.15	44.64	236.50	nr	**281.14**
P5K horizontal single panel convector; 600 mm high; wheelhead and lockshield valves						
500 mm long; 412 watts output	59.37	1.75	36.33	77.05	nr	**113.38**
1400 mm long; 1154 watts output	121.67	2.15	44.64	144.10	nr	**188.73**
2000 mm long; 1648 watts output	163.32	2.40	49.83	188.92	nr	**238.75**
P5KV vertical single panel convector; 600 mm long; wheelhead and lockshield valves						
1400 mm high; 960 watts output	139.87	2.40	49.83	163.69	nr	**213.52**
2200 mm high; 1492 watts output	195.31	2.60	53.98	223.36	nr	**277.34**

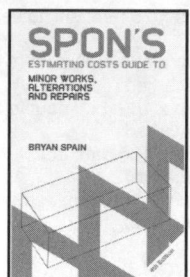

D GROUNDWORK

Item	PC £	Labour hours	Labour £	Material /Plant £	Unit	Total rate £
D20 EXCAVATING AND FILLING						
Prices are applicable to excavation in firm soil. Multiplying factors for other soils are as follows:						
Mechanical / Hand						
Clay x 2.00 / x 1.20						
Compact gravel x 3.00 / x 1.50						
Soft chalk x 4.00 / x 2.00						
Hard rock x 5.00 / x 6.00						
Running sand or silt x 6.00 / x 2.00						
Site preparation						
Removing trees						
girth 600 mm - 1.50 m	-	18.50	231.76	-	nr	**231.76**
girth 1.50 - 3.00 m	-	32.50	407.15	-	nr	**407.15**
girth exceeding 3.00 m	-	46.50	582.53	-	nr	**582.53**
Removing tree stumps						
girth 600 mm - 1.50 m	-	0.93	11.65	45.97	nr	**57.63**
girth 1.50 - 3.00 m	-	0.93	11.65	67.23	nr	**78.88**
girth exceeding 3.00 m	-	0.93	11.65	92.05	nr	**103.71**
Clearing site vegetation						
bushes, scrub, undergrowth, hedges and trees and tree stumps not exceeding 600 mm girth	-	0.03	0.38	-	m²	**0.38**
Lifting turf for preservation						
stacking	-	0.32	4.01	-	m²	**4.01**
Excavating; by machine						
Topsoil for preservation						
average depth 150 mm	-	0.02	0.25	1.15	m²	**1.40**
add or deduct for each 25 mm variation in average depth	-	0.01	0.13	0.27	m²	**0.39**
To reduce levels						
maximum depth not exceeding 0.25 m	-	0.05	0.63	1.51	m³	**2.14**
maximum depth not exceeding 1.00 m	-	0.03	0.38	1.07	m³	**1.44**
maximum depth not exceeding 2.00 m	-	0.05	0.63	1.51	m³	**2.14**
maximum depth not exceeding 4.00 m	-	0.07	0.88	1.86	m³	**2.74**
Basements and the like; commencing level exceeding 0.25 m below existing ground level						
maximum depth not exceeding 1.00 m	-	0.07	0.88	1.36	m³	**2.24**
maximum depth not exceeding 2.00 m	-	0.05	0.63	1.05	m³	**1.68**
maximum depth not exceeding 4.00 m	-	0.07	0.88	1.36	m³	**2.24**
maximum depth not exceeding 6.00 m	-	0.09	1.13	1.78	m³	**2.91**
maximum depth not exceeding 8.00 m	-	0.12	1.50	2.10	m³	**3.60**
Pits						
maximum depth not exceeding 0.25 m	-	0.31	3.88	5.56	m³	**9.45**
maximum depth not exceeding 1.00 m	-	0.33	4.13	4.93	m³	**9.07**
maximum depth not exceeding 2.00 m	-	0.39	4.89	5.56	m³	**10.45**
maximum depth not exceeding 4.00 m	-	0.47	5.89	6.30	m³	**12.18**
maximum depth not exceeding 6.00 m	-	0.49	6.14	6.61	m³	**12.75**
Extra over pit excavating for commencing level exceeding 0.25 m below existing ground level						
1.00 m below	-	0.03	0.38	0.73	m³	**1.11**
2.00 m below	-	0.05	0.63	1.05	m³	**1.68**
3.00 m below	-	0.06	0.75	1.36	m³	**2.12**
4.00 m below	-	0.09	1.13	1.78	m³	**2.91**

D GROUNDWORK

Item	PC £	Labour hours	Labour £	Material /Plant £	Unit	Total rate £
D20 EXCAVATING AND FILLING – cont'd						
Excavating; by machine – cont'd						
Trenches; width not exceeding 0.30 m						
maximum depth not exceeding 0.25 m	-	0.26	3.26	4.51	m³	**7.77**
maximum depth not exceeding 1.00 m	-	0.28	3.51	3.88	m³	**7.39**
maximum depth not exceeding 2.00 m	-	0.33	4.13	4.51	m³	**8.65**
maximum depth not exceeding 4.00 m	-	0.40	5.01	5.56	m³	**10.57**
maximum depth not exceeding 6.00 m	-	0.46	5.76	6.61	m³	**12.37**
Trenches; width exceeding 0.30 m						
maximum depth not exceeding 0.25 m	-	0.23	2.88	4.20	m³	**7.08**
maximum depth not exceeding 1.00 m	-	0.25	3.13	3.46	m³	**6.59**
maximum depth not exceeding 2.00 m	-	0.30	3.76	4.20	m³	**7.96**
maximum depth not exceeding 4.00 m	-	0.35	4.38	4.93	m³	**9.32**
maximum depth not exceeding 6.00 m	-	0.43	5.39	6.30	m³	**11.68**
Extra over trench excavating for commencing level exceeding 0.25 m below existing ground level						
1.00 m below	-	0.03	0.38	0.73	m³	**1.11**
2.00 m below	-	0.05	0.63	1.05	m³	**1.68**
3.00 m below	-	0.06	0.75	1.36	m³	**2.12**
4.00 m below	-	0.09	1.13	1.78	m³	**2.91**
For pile caps and ground beams between piles						
maximum depth not exceeding 0.25 m	-	0.39	4.89	7.35	m³	**12.23**
maximum depth not exceeding 1.00 m	-	0.35	4.38	6.61	m³	**11.00**
maximum depth not exceeding 2.00 m	-	0.39	4.89	7.35	m³	**12.23**
To bench sloping ground to receive filling						
maximum depth not exceeding 0.25 m	-	0.09	1.13	1.78	m³	**2.91**
maximum depth not exceeding 1.00 m	-	0.07	0.88	2.10	m³	**2.98**
maximum depth not exceeding 2.00 m	-	0.09	1.13	1.78	m³	**2.91**
Extra over any types of excavating irrespective of depth						
excavating below ground water level	-	0.13	1.63	2.41	m³	**4.04**
next existing services	-	2.55	31.95	1.36	m³	**33.31**
around existing services crossing excavation	-	5.80	72.66	3.88	m³	**76.54**
Extra over any types of excavating irrespective of depth for breaking out existing materials						
rock	-	2.95	36.96	15.41	m³	**52.37**
concrete	-	2.55	31.95	11.23	m³	**43.17**
reinforced concrete	-	3.60	45.10	16.66	m³	**61.76**
brickwork, blockwork or stonework	-	1.85	23.18	8.47	m³	**31.65**
Extra over any types of excavating irrespective of depth for breaking out existing hard pavings, 75 mm thick						
coated macadam or asphalt	-	0.19	2.38	0.59	m²	**2.97**
Extra over any types of excavating irrespective of depth for breaking out existing hard pavings, 150 mm thick						
concrete	-	0.39	4.89	1.78	m²	**6.67**
reinforced concrete	-	0.58	7.27	2.39	m²	**9.66**
coated macadam or asphalt and hardcore	-	0.26	3.26	0.64	m²	**3.90**
Working space allowance to excavations						
reduce levels, basements and the like	-	0.07	0.88	1.36	m²	**2.24**
pits	-	0.19	2.38	3.88	m²	**6.26**
trenches	-	0.18	2.25	3.46	m²	**5.72**
pile caps and ground beams between piles	-	0.20	2.51	3.88	m²	**6.39**
Extra over excavating for working space for backfilling in with special materials						
hardcore	-	0.13	1.63	15.34	m²	**16.97**
sand	-	0.13	1.63	23.47	m²	**25.10**

D GROUNDWORK

Item	PC £	Labour hours	Labour £	Material /Plant £	Unit	**Total rate £**
40 mm - 20 mm gravel	-	0.13	1.63	26.08	m²	**27.71**
plain in situ ready mixed designated concrete C7.5 - 40 mm aggregate	-	0.93	15.01	52.84	m²	**67.86**
Excavating; by hand						
Topsoil for preservation						
average depth 150 mm	-	0.23	2.88	-	m²	**2.88**
add or deduct for each 25 mm variation in average depth	-	0.03	0.38	-	m²	**0.38**
To reduce levels						
maximum depth not exceeding 0.25 m	-	1.44	18.04	-	m³	**18.04**
maximum depth not exceeding 1.00 m	-	1.63	20.42	-	m³	**20.42**
maximum depth not exceeding 2.00 m	-	1.80	22.55	-	m³	**22.55**
maximum depth not exceeding 4.00 m	-	1.99	24.93	-	m³	**24.93**
Basements and the like; commencing level exceeding 0.25 m below existing ground level						
maximum depth not exceeding 1.00 m	-	1.90	23.80	-	m³	**23.80**
maximum depth not exceeding 2.00 m	-	2.04	25.56	-	m³	**25.56**
maximum depth not exceeding 4.00 m	-	2.73	34.20	-	m³	**34.20**
maximum depth not exceeding 6.00 m	-	3.33	41.72	-	m³	**41.72**
maximum depth not exceeding 8.00 m	-	4.02	50.36	-	m³	**50.36**
Pits						
maximum depth not exceeding 0.25 m	-	2.13	26.68	-	m³	**26.68**
maximum depth not exceeding 1.00 m	-	2.75	34.45	-	m³	**34.45**
maximum depth not exceeding 2.00 m	-	3.30	41.34	-	m³	**41.34**
maximum depth not exceeding 4.00 m	-	4.18	52.37	-	m³	**52.37**
maximum depth not exceeding 6.00 m	-	5.17	64.77	-	m³	**64.77**
Extra over pit excavating for commencing level exceeding 0.25 m below existing ground level						
1.00 m below	-	0.42	5.26	-	m³	**5.26**
2.00 m below	-	0.88	11.02	-	m³	**11.02**
3.00 m below	-	1.30	16.29	-	m³	**16.29**
4.00 m below	-	1.71	21.42	-	m³	**21.42**
Trenches; width not exceeding 0.30 m						
maximum depth not exceeding 0.25 m	-	1.85	23.18	-	m³	**23.18**
maximum depth not exceeding 1.00 m	-	2.76	34.58	-	m³	**34.58**
maximum depth not exceeding 2.00 m	-	3.24	40.59	-	m³	**40.59**
maximum depth not exceeding 4.00 m	-	3.96	49.61	-	m³	**49.61**
maximum depth not exceeding 6.00 m	-	5.10	63.89	-	m³	**63.89**
Trenches; width exceeding 0.30 m						
maximum depth not exceeding 0.25 m	-	1.80	22.55	-	m³	**22.55**
maximum depth not exceeding 1.00 m	-	2.46	30.82	-	m³	**30.82**
maximum depth not exceeding 2.00 m	-	2.88	36.08	-	m³	**36.08**
maximum depth not exceeding 4.00 m	-	3.66	45.85	-	m³	**45.85**
maximum depth not exceeding 6.00 m	-	4.68	58.63	-	m³	**58.63**
Extra over trench excavating for commencing level exceeding 0.25 m below existing ground level						
1.00 m below	-	0.42	5.26	-	m³	**5.26**
2.00 m below	-	0.88	11.02	-	m³	**11.02**
3.00 m below	-	1.30	16.29	-	m³	**16.29**
4.00 m below	-	1.71	21.42	-	m³	**21.42**
For pile caps and ground beams between piles						
maximum depth not exceeding 0.25 m	-	2.78	34.83	-	m³	**34.83**
maximum depth not exceeding 1.00 m	-	2.96	37.08	-	m³	**37.08**
maximum depth not exceeding 2.00 m	-	3.52	44.10	-	m³	**44.10**
To bench sloping ground to receive filling						
maximum depth not exceeding 0.25 m	-	1.30	16.29	-	m³	**16.29**
maximum depth not exceeding 1.00 m	-	1.48	18.54	-	m³	**18.54**
maximum depth not exceeding 2.00 m	-	1.67	20.92	-	m³	**20.92**

D GROUNDWORK

Item	PC £	Labour hours	Labour £	Material /Plant £	Unit	Total rate £
D20 EXCAVATING AND FILLING – cont'd						
Excavating; by hand – cont'd						
Extra over any types of excavating irrespective of depth						
excavating below ground water level	-	0.32	4.01	-	m³	**4.01**
next existing services	-	0.93	11.65	-	m³	**11.65**
around existing services crossing excavation	-	1.85	23.18	-	m³	**23.18**
Extra over any types of excavating irrespective of depth for breaking out existing materials						
rock	-	4.63	58.00	7.53	m³	**65.54**
concrete	-	4.16	52.11	6.28	m³	**58.39**
reinforced concrete	-	5.55	69.53	8.79	m³	**78.32**
brickwork, blockwork or stonework	-	2.78	34.83	3.77	m³	**38.59**
Extra over any types of excavating irrespective of depth for breaking out existing hard pavings, 60 mm thick						
precast concrete paving slabs	-	0.28	3.51	-	m²	**3.51**
Extra over any types of excavating irrespective of depth for breaking out existing hard pavings, 75 mm thick						
coated macadam or asphalt	-	0.37	4.64	0.50	m²	**5.14**
Extra over any types of excavating irrespective of depth for breaking out existing hard pavings, 150 mm thick						
concrete	-	0.65	8.14	0.88	m²	**9.02**
reinforced concrete	-	0.83	10.40	1.26	m²	**11.66**
coated macadam or asphalt and hardcore	-	0.46	5.76	0.63	m²	**6.39**
Working space allowance to excavations						
reduce levels, basements and the like	-	2.13	26.68	-	m²	**26.68**
pits	-	2.22	27.81	-	m²	**27.81**
trenches	-	1.94	24.30	-	m²	**24.30**
pile caps and ground beams between piles	-	2.31	28.94	-	m²	**28.94**
Extra over excavation for working space for backfilling with special materials						
hardcore	-	0.74	9.27	13.29	m²	**22.57**
sand	-	0.74	9.27	21.43	m²	**30.70**
40 mm - 20 mm gravel	-	0.74	9.27	24.04	m²	**33.31**
plain in situ concrete ready mixed designated concrete; C7.5 - 40 mm aggregate	-	1.02	16.47	50.80	m²	**67.27**
Earthwork support (average "risk" prices)						
Maximum depth not exceeding 1.00 m						
distance between opposing faces not exceeding 2.00 m	-	0.10	1.25	0.41	m²	**1.66**
distance between opposing faces 2.00 - 4.00 m	-	0.11	1.38	0.48	m²	**1.86**
distance between opposing faces exceeding 4.00 m	-	0.12	1.50	0.60	m²	**2.11**
Maximum depth not exceeding 2.00 m						
distance between opposing faces not exceeding 2.00 m	-	0.12	1.50	0.48	m²	**1.98**
distance between opposing faces 2.00 - 4.00 m	-	0.13	1.63	0.60	m²	**2.23**
distance between opposing faces exceeding 4.00 m	-	0.14	1.75	0.76	m²	**2.51**
Maximum depth not exceeding 4.00 m						
distance between opposing faces not exceeding 2.00 m	-	0.16	2.00	0.60	m²	**2.61**
distance between opposing faces 2.00 - 4.00 m	-	0.16	2.00	0.76	m²	**2.76**
distance between opposing faces exceeding 4.00 m	-	0.18	2.25	0.96	m²	**3.21**

D GROUNDWORK

Item	PC £	Labour hours	Labour £	Material /Plant £	Unit	Total rate £
Maximum depth not exceeding 6.00 m						
distance between opposing faces not exceeding 2.00 m	-	0.18	2.25	0.72	m²	**2.97**
distance between opposing faces 2.00 - 4.00 m	-	0.19	2.38	0.96	m²	**3.34**
distance between opposing faces exceeding 4.00 m	-	0.22	2.76	1.20	m²	**3.95**
Maximum depth not exceeding 8.00 m						
distance between opposing faces not exceeding 2.00 m	-	0.23	2.88	0.96	m²	**3.84**
distance between opposing faces 2.00 - 4.00 m	-	0.28	3.51	1.20	m²	**4.70**
distance between opposing faces exceeding 4.00 m	-	0.33	4.13	1.43	m²	**5.57**
Earthwork support (open boarded)						
Maximum depth not exceeding 1.00 m						
distance between opposing faces not exceeding 2.00 m	-	0.28	3.51	0.84	m²	**4.35**
distance between opposing faces 2.00 - 4.00 m	-	0.31	3.88	0.96	m²	**4.84**
distance between opposing faces exceeding 4.00 m	-	0.35	4.38	1.20	m²	**5.58**
Maximum depth not exceeding 2.00 m						
distance between opposing faces not exceeding 2.00 m	-	0.35	4.38	0.96	m²	**5.34**
distance between opposing faces 2.00 - 4.00 m	-	0.39	4.89	1.15	m²	**6.04**
distance between opposing faces exceeding 4.00 m	-	0.44	5.51	1.43	m²	**6.95**
Maximum depth not exceeding 4.00 m						
distance between opposing faces not exceeding 2.00 m	-	0.44	5.51	1.08	m²	**6.60**
distance between opposing faces 2.00 - 4.00 m	-	0.50	6.26	1.34	m²	**7.60**
distance between opposing faces exceeding 4.00 m	-	0.56	7.02	1.67	m²	**8.69**
Maximum depth not exceeding 6.00 m						
distance between opposing faces not exceeding 2.00 m	-	0.56	7.02	1.20	m²	**8.21**
distance between opposing faces 2.00 - 4.00 m	-	0.61	7.64	1.50	m²	**9.15**
distance between opposing faces exceeding 4.00 m	-	0.70	8.77	1.91	m²	**10.68**
Maximum depth not exceeding 8.00 m						
distance between opposing faces not exceeding 2.00 m	-	0.74	9.27	1.56	m²	**10.83**
distance between opposing faces 2.00 - 4.00 m	-	0.83	10.40	1.80	m²	**12.20**
distance between opposing faces exceeding 4.00 m	-	0.97	12.15	2.39	m²	**14.54**
Earthwork support (close boarded)						
Maximum depth not exceeding 1.00 m						
distance between opposing faces not exceeding 2.00 m	-	0.74	9.27	1.67	m²	**10.94**
distance between opposing faces 2.00 - 4.00 m	-	0.81	10.15	1.91	m²	**12.06**
distance between opposing faces exceeding 4.00 m	-	0.90	11.27	2.39	m²	**13.67**
Maximum depth not exceeding 2.00 m						
distance between opposing faces not exceeding 2.00 m	-	0.93	11.65	1.91	m²	**13.56**
distance between opposing faces 2.00 - 4.00 m	-	1.02	12.78	2.29	m²	**15.07**
distance between opposing faces exceeding 4.00 m	-	1.11	13.91	2.87	m²	**16.77**
Maximum depth not exceeding 4.00 m						
distance between opposing faces not exceeding 2.00 m	-	1.16	14.53	2.15	m²	**16.68**
distance between opposing faces 2.00 - 4.00 m	-	1.30	16.29	2.67	m²	**18.96**
distance between opposing faces exceeding 4.00 m	-	1.43	17.91	3.35	m²	**21.26**
Maximum depth not exceeding 6.00 m						
distance between opposing faces not exceeding 2.00 m	-	1.44	18.04	2.39	m²	**20.43**
distance between opposing faces 2.00 - 4.00 m	-	1.57	19.67	3.01	m²	**22.68**
distance between opposing faces exceeding 4.00 m	-	1.76	22.05	3.83	m²	**25.87**

D GROUNDWORK

Item	PC £	Labour hours	Labour £	Material /Plant £	Unit	Total rate £
D20 EXCAVATING AND FILLING – cont'd						
Earthwork support (close boarded) – cont'd						
Maximum depth not exceeding 8.00 m						
distance between opposing faces not exceeding 2.00 m	-	1.76	22.05	3.11	m²	**25.16**
distance between opposing faces 2.00 - 4.00 m	-	1.94	24.30	3.59	m²	**27.89**
distance between opposing faces exceeding 4.00 m	-	2.22	27.81	4.30	m²	**32.11**
Extra over earthwork support for						
Curved	-	0.02	0.25	0.41	m²	**0.66**
Below ground water level	-	0.28	3.51	0.37	m²	**3.87**
Unstable ground	-	0.46	5.76	0.72	m²	**6.48**
Next to roadways	-	0.37	4.64	0.60	m²	**5.24**
Left in	-	0.60	7.52	16.74	m²	**24.25**
Earthwork support (average "risk" prices - inside existing buildings)						
Maximum depth not exceeding 1.00 m						
distance between opposing faces not exceeding 2.00 m	-	0.18	2.25	0.60	m²	**2.86**
distance between opposing faces 2.00 - 4.00 m	-	0.19	2.38	0.69	m²	**3.07**
distance between opposing faces exceeding 4.00 m	-	0.22	2.76	0.84	m²	**3.60**
Maximum depth not exceeding 2.00 m						
distance between opposing faces not exceeding 2.00 m	-	0.22	2.76	0.69	m²	**3.45**
distance between opposing faces 2.00 - 4.00 m	-	0.24	3.01	0.91	m²	**3.92**
distance between opposing faces exceeding 4.00 m	-	0.32	4.01	1.01	m²	**5.02**
Maximum depth not exceeding 4.00 m						
distance between opposing faces not exceeding 2.00 m	-	0.28	3.51	0.91	m²	**4.42**
distance between opposing faces 2.00 - 4.00 m	-	0.31	3.88	1.08	m²	**4.97**
distance between opposing faces exceeding 4.00 m	-	0.34	4.26	1.27	m²	**5.53**
Maximum depth not exceeding 6.00 m						
distance between opposing faces not exceeding 2.00 m	-	0.34	4.26	1.03	m²	**5.29**
distance between opposing faces 2.00 - 4.00 m	-	0.38	4.76	1.27	m²	**6.03**
distance between opposing faces exceeding 4.00 m	-	0.43	5.39	1.50	m²	**6.89**
Disposal; by machine						
Excavated material						
inactive waste off site; to tip not exceeding 13 km (using lorries); including Landfill Tax	-	-	-	21.52	m³	**21.52**
active non-hazardous waste off site; to tip not exceeding 13 km (using lorries); including Landfill Tax	-	-	-	95.00	m³	**95.00**
inactive waste on site; depositing in spoil heaps; average 25 m distance	-	-	-	0.89	m³	**0.89**
on site; spreading; average 25 m distance	-	0.20	2.51	0.62	m³	**3.13**
on site; depositing in spoil heaps; average 50 m distance	-	-	-	1.51	m³	**1.51**
on site; spreading; average 50 m distance	-	0.20	2.51	1.15	m³	**3.66**
on site; depositing in spoil heaps; average 100 m distance	-	-	-	2.66	m³	**2.66**
on site; spreading; average 100 m distance	-	0.20	2.51	1.78	m³	**4.28**
on site; depositing in spoil heaps; average 200 m distance	-	-	-	3.37	m³	**3.37**
on site; spreading; average 200 m distance	-	0.20	2.51	2.40	m³	**4.90**

D GROUNDWORK

Item	PC £	Labour hours	Labour £	Material /Plant £	Unit	Total rate £
Disposal; by hand						
Excavated material						
inactive waste; off site; to tip not exceeding 13 km (using lorries); including Landfill Tax	-	0.74	9.27	29.60	m³	**38.87**
active non-hazardous waste; off site; to tip not exceeding 13 km (using lorries); including Landfill Tax	-	1.25	15.66	99.88	m³	**115.54**
inactive waste on site; depositing in spoil heaps; average 25 m distance	-	1.02	12.78	-	m³	**12.78**
on site; spreading; average 25 m distance	-	1.34	16.79	-	m³	**16.79**
on site; depositing in spoil heaps; average 50 m	-	1.34	16.79	-	m³	**16.79**
on site; spreading; average 50 m distance	-	1.62	20.29	-	m³	**20.29**
on site; depositing in spoil heaps; average 100 m distance	-	1.94	24.30	-	m³	**24.30**
on site; spreading; average 100 m distance	-	2.22	27.81	-	m³	**27.81**
on site; depositing in spoil heaps; average 200 m distance	-	2.87	35.95	-	m³	**35.95**
on site; spreading; average 200 m distance	-	3.15	39.46	-	m³	**39.46**
Filling to excavations; by machine						
Average thickness not exceeding 0.25 m						
arising from the excavations	-	0.17	2.13	2.40	m³	**4.53**
obtained off site; hardcore	22.88	0.19	2.38	28.87	m³	**31.25**
obtained off site; granular fill type one	28.41	0.19	2.38	37.31	m³	**39.69**
obtained off site; granular fill type two	26.73	0.19	2.38	35.37	m³	**37.75**
Average thickness exceeding 0.25 m						
arising from the excavations	-	0.14	1.75	1.78	m³	**3.53**
obtained off site; hardcore	19.61	0.16	2.00	24.47	m³	**26.47**
obtained off site; granular fill type one	28.41	0.16	2.00	36.42	m³	**38.43**
obtained off site; granular fill type two	26.73	0.16	2.00	34.48	m³	**36.48**
Filling to make up levels; by machine						
Average thickness not exceeding 0.25 m						
arising from the excavations	-	0.24	3.01	3.18	m³	**6.19**
obtained off site; imported topsoil	17.01	0.24	3.01	22.11	m³	**25.12**
obtained off site; hardcore	22.88	0.28	3.51	29.60	m³	**33.10**
obtained off site; granular fill type one	28.41	0.28	3.51	38.04	m³	**41.55**
obtained off site; granular fill type two	26.73	0.28	3.51	36.09	m³	**39.60**
obtained off site; sand	32.08	0.28	3.51	42.28	m³	**45.78**
Average thickness exceeding 0.25 m						
arising from the excavations	-	0.20	2.51	2.33	m³	**4.84**
obtained off site; imported topsoil	17.01	0.20	2.51	21.26	m³	**23.77**
obtained off site; hardcore	19.61	0.24	3.01	24.98	m³	**27.99**
obtained off site; granular fill type one	28.41	0.24	3.01	36.94	m³	**39.95**
obtained off site; granular fill type two	26.73	0.24	3.01	34.99	m³	**38.00**
obtained off site; sand	32.08	0.24	3.01	41.18	m³	**44.19**
Filling to excavations; by hand						
Average thickness not exceeding 0.25 m						
arising from the excavations	-	1.16	14.53	-	m³	**14.53**
obtained off site; hardcore	22.88	1.25	15.66	26.47	m³	**42.13**
obtained off site; granular fill type one	28.41	1.48	18.54	32.87	m³	**51.41**
obtained off site; granular fill type two	26.73	1.48	18.54	30.93	m³	**49.47**
obtained off site; sand	32.08	1.48	18.54	37.11	m³	**55.65**
Average thickness exceeding 0.25 m						
arising from the excavations	-	0.93	11.65	-	m³	**11.65**
obtained off site; hardcore	19.61	1.02	12.78	22.69	m³	**35.47**
obtained off site; granular fill type one	28.41	1.20	15.03	32.87	m³	**47.91**
obtained off site; granular fill type two	26.73	1.20	15.03	30.93	m³	**45.96**
obtained off site; sand	32.08	1.20	15.03	37.11	m³	**52.14**

D GROUNDWORK

Item	PC £	Labour hours	Labour £	Material /Plant £	Unit	Total rate £
D20 EXCAVATING AND FILLING – cont'd						
Filling to make up levels; by hand						
Average thickness not exceeding 0.25 m						
arising from the excavations	-	1.25	15.66	7.26	m³	**22.92**
obtained off site; imported topsoil	17.01	1.25	15.66	25.57	m³	**41.23**
obtained off site; hardcore	22.88	1.39	17.41	34.55	m³	**51.96**
obtained off site; granular fill type one	28.41	1.54	19.29	41.80	m³	**61.09**
obtained off site; granular fill type two	26.73	1.54	19.29	39.85	m³	**59.15**
obtained off site; sand	32.08	1.54	19.29	46.04	m³	**65.33**
Average thickness exceeding 0.25 m						
arising from the excavations	-	1.02	12.78	5.92	m³	**18.70**
arising from on site spoil heaps; average 25 m distance; multiple handling	-	2.22	27.81	12.91	m³	**40.73**
obtained off site; imported topsoil	17.01	1.02	12.78	24.23	m³	**37.01**
obtained off site; hardcore	19.61	1.34	16.79	30.49	m³	**47.28**
obtained off site; granular fill type one	28.41	1.43	17.91	41.22	m³	**59.13**
obtained off site; granular fill type two	26.73	1.43	17.91	39.27	m³	**57.19**
obtained off site; sand	32.08	1.43	17.91	45.46	m³	**63.37**
Surface packing to filling						
To vertical or battered faces	-	0.17	2.13	0.12	m²	**2.25**
Surface treatments						
Compacting						
filling; blinding with sand	-	0.04	0.50	2.02	m²	**2.52**
bottoms of excavations	-	0.04	0.50	0.02	m²	**0.52**
Trimming						
sloping surfaces	-	0.17	2.13	-	m²	**2.13**
sloping surfaces; in rock	-	0.93	10.70	1.76	m²	**12.46**
Filter membrane; one layer; laid on earth to receive granular material						
"Terram 500" filter membrane or other equal and approved; one layer; laid on earth	-	0.04	0.50	0.27	m²	**0.77**
"Terram 700" filter membrane or other equal and approved; one layer; laid on earth	-	0.04	0.50	0.29	m²	**0.79**
"Terram 1000"; filter membrane or other equal and approved; one layer; laid on earth	-	0.04	0.50	0.28	m²	**0.78**
"Terram 2000"; filter membrane or other equal and approved; one layer; laid on earth	-	0.04	0.50	0.74	m²	**1.24**

D GROUNDWORK

Item	PC £	Labour hours	Labour £	Material /Plant £	Unit	Total rate £
D30 CAST IN PLACE PILING						
NOTE: The following approximate prices, for the quantities of piling quoted, are for work on clear open sites with reasonable access. They are based on 500 mm nominal diameter piles, normal concrete mix 20.00 N/mm² reinforced for loading up to 40,000 kg depending upon ground conditions and include up to 0.16 m of projecting reinforcement at top of pile. The prices do not allow for removal of spoil.						
*** indicates work normally carried out by the Main Contractor**						
Minipile cast-in-place concrete piles						
Provision of all plant (2 nr rigs) ; including bringing to and removing from site; maintenance, erection and dismantling at each pile position for 100 nr piles	-	-	-	-	item	**23000.00**
Bored piles						
450 mm diameter piles; reinforced; 10 m long	-	-	-	-	nr	**1400.00**
add for additional piles length up to 15 m	-	-	-	-	m	**140.00**
deduct for reduction in pile length	-	-	-	-	m	**14.00**
Cutting off tops of piles*	-	1.20	25.88	-	m	**25.88**
Blind bored piles						
500 mm diameter	-	-	-	-	m	**125.00**
Delays						
rig standing time	-	-	-	-	hour	**275.00**
Extra over piling						
breaking through obstructions	-	-	-	-	hour	**320.00**
Pile tests						
working to 600 kN/t; using tension piles as reaction; first pile	-	-	-	-	nr	**5535.00**
working to 600 kN/t; using tension piles as reaction; subsequent piles	-	-	-	-	nr	**5305.00**
Rotary bored cast-in-place concrete piles						
Provision of all plant (1 nr rig); including bringing to and removing from site; maintenance, erection and dismantling at each pile position for 100 nr piles	-	-	-	-	item	**13850.00**
Bored piles						
500 mm diameter piles; reinforced; 10 m long	-	-	-	-	nr	**460.00**
add for additional piles length up to 15 m	-	-	-	-	m	**46.00**
deduct for reduction in pile length	-	-	-	-	m	**24.00**
Cutting off tops of piles*	-	1.20	25.88	-	m	**25.88**
Blind bored piles						
500 mm diameter	-	-	-	-	m	**22.00**
Delays						
rig standing time	-	-	-	-	hour	**230.00**
Extra over piling						
breaking through obstructions	-	-	-	-	hour	**276.00**
Pile tests						
working to 600 kN/t; using tension piles as reaction; first pile	-	-	-	-	nr	**4380.00**
working to 600 kN/t; using tension piles as reaction; subsequent piles	-	-	-	-	nr	**4250.00**

D GROUNDWORK

Item	PC £	Labour hours	Labour £	Material /Plant £	Unit	Total rate £
D32 STEEL PILING						
"Arcelor" 600 mm wide 'U' shaped steel sheeting piling; or other equal and approved; pitched and driven						
Provision of all plant for sheet pile installation; including bringing to and removing from site; maintenance, erection and dismantling; assuming one rig for 1500 m² of piling						
Leader rig with vibratory hammer	-	-	-	-	item	**3550.00**
Conventional rig	-	-	-	-	item	**4675.00**
Silent vibrationless rig	-	-	-	-	item	**5350.00**
Supply only of standard sheet pile sections						
PU12	-	-	-	-	m²	**90.00**
PU18-1	-	-	-	-	m²	**100.00**
PU22-1	-	-	-	-	m²	**113.00**
PU25	-	-	-	-	m²	**130.00**
PU32	-	-	-	-	m²	**157.50**
Pitching and driving of sheet piles; using the following plant						
Leader rig with vibratory hammer	-	-	-	-	m²	**19.50**
Conventional rig	-	-	-	-	m²	**27.50**
Silent vibrationless rig	-	-	-	-	m²	**41.00**
Provision of all plant for sheet pile extraction; including bringing to and removing from site; maintenance, erection and dismantling; assuming one rig for 1500 m² of piling						
Leader rig with vibratory hammer	-	-	-	-	item	**3550.00**
Conventional rig	-	-	-	-	item	**4675.00**
Silent vibrationless rig	-	-	-	-	item	**4625.00**
Extraction of sheet piles; using the following plant						
Leader rig with vibratory hammer	-	-	-	-	m²	**14.60**
Conventional rig	-	-	-	-	m²	**17.00**
Silent vibrationless rig	-	-	-	-	m²	**25.30**
Credit on extracted piles; recovered in re-usable lenghts; for sheet pile sections						
PU12	-	-	-	-	m²	**64.00**
PU18-1	-	-	-	-	m²	**70.00**
PU22-1	-	-	-	-	m²	**80.00**
PU25	-	-	-	-	m²	**90.00**
PU32	-	-	-	-	m²	**110.00**

D GROUNDWORK

Item	PC £	Labour hours	Labour £	Material /Plant £	Unit	Total rate £
D40 EMBEDDED RETAINING WALLING						
Diaphragm walls; contiguous panel construction; panel lengths not exceeding 5 m						
Provision of all plant; including bringing to and removing from site; maintenance, erection and dismantling; assuming one rig for 1000 m² of walling	-	-	-	-	item	**138000.00**
Excavation for diaphragm wall; excavated material removed from site; Bentonite slurry supplied and disposed of						
600 mm thick walls	-	-	-	-	m³	**275.00**
1000 mm thick walls	-	-	-	-	m³	**275.00**
Ready mixed reinforced in situ concrete; normal portland cement; C30 - 10 mm aggregate in walls	-	-	-	-	m³	**117.50**
Reinforcement bar; BS 4449 cold rolled deformed square high yield steel bars; straight or bent						
25 mm - 40 mm diameter	-	-	-	-	tonne	**925.00**
20 mm diameter	-	-	-	-	tonne	**925.00**
16 mm diameter	-	-	-	-	tonne	**925.00**
Formwork 75 mm thick to form chases	-	-	-	-	m²	**59.96**
Construct twin guide walls in reinforced concrete; together with reinforcement and formwork along the axis of the diaphragm wall	-	-	-	-	m	**260.00**
Delays						
rig standing	-	-	-	-	hour	**830.00**
D41 CRIB WALLS/GABIONS/REINFORCED EARTHWORKS						
Gabion baskets						
Wire mesh gabion baskets; Maccaferri Ltd or other equal and approved; galvanised mesh 80 mm x 100 mm; filling with broken stones 125 mm - 200 mm size						
2.00 x 1.00 x 0.50 m	16.32	1.00	21.57	101.74	nr	**123.31**
2.00 x 1.00 x 0.50 m; pvc coated	20.78	1.00	21.57	106.79	nr	**128.36**
2.00 x 1.00 x 1.00 m	22.88	2.00	43.14	192.47	nr	**235.60**
2.00 x 1.00 x 1.00 m; pvc coated	29.30	2.00	43.14	199.72	nr	**242.85**
"Reno" mattress gabion baskets or other equal and approved; Maccaferri Ltd; filling with broken stones 125 mm - 200 mm size						
6.00 x 2.00 x 0.17 m	63.16	2.00	43.14	230.17	nr	**273.31**
6.00 x 2.00 x 0.23 m	68.76	2.50	53.92	293.02	nr	**346.94**
6.00 x 2.00 x 0.30 m	79.39	3.00	64.70	365.97	nr	**430.68**
D50 UNDERPINNING						
Excavating; by machine						
Preliminary trenches						
maximum depth not exceeding 1.00 m	-	0.23	2.88	7.37	m³	**10.25**
maximum depth not exceeding 2.00 m	-	0.28	3.51	8.88	m³	**12.39**
maximum depth not exceeding 4.00 m	-	0.32	4.01	10.39	m³	**14.40**
Extra over preliminary trench excavating for breaking out existing hard pavings, 150 mm thick						
concrete	-	0.65	8.14	0.88	m²	**9.02**

D GROUNDWORK

Item	PC £	Labour hours	Labour £	Material /Plant £	Unit	Total rate £
D50 UNDERPINNING – cont'd						
Excavating; by hand						
Preliminary trenches						
maximum depth not exceeding 1.00 m	-	2.68	33.57	-	m³	**33.57**
maximum depth not exceeding 2.00 m	-	3.05	38.21	-	m³	**38.21**
maximum depth not exceeding 4.00 m	-	3.93	49.23	-	m³	**49.23**
Extra over preliminary trench excavating for breaking out existing hard pavings, 150 mm thick						
concrete	-	0.28	3.51	2.13	m²	**5.64**
Underpinning pits; commencing from 1.00 m below existing ground level						
maximum depth not exceeding 0.25 m	-	4.07	50.99	-	m³	**50.99**
maximum depth not exceeding 1.00 m	-	4.44	55.62	-	m³	**55.62**
maximum depth not exceeding 2.00 m	-	5.32	66.65	-	m³	**66.65**
Underpinning pits; commencing from 2.00 m below existing ground level						
maximum depth not exceeding 0.25 m	-	5.00	62.64	-	m³	**62.64**
maximum depth not exceeding 1.00 m	-	5.37	67.27	-	m³	**67.27**
maximum depth not exceeding 2.00 m	-	6.24	78.17	-	m³	**78.17**
Underpinning pits; commencing from 4.00 m below existing ground level						
maximum depth not exceeding 0.25 m	-	5.92	74.16	-	m³	**74.16**
maximum depth not exceeding 1.00 m	-	6.29	78.80	-	m³	**78.80**
maximum depth not exceeding 2.00 m	-	7.17	89.82	-	m³	**89.82**
Extra over any types of excavating irrespective of depth						
excavating below ground water level	-	0.32	4.01	-	m³	**4.01**
Earthwork support to preliminary trenches (open boarded - in 3.00 m lengths)						
Maximum depth not exceeding 1.00 m						
distance between opposing faces not exceeding 2.00 m	-	0.37	4.64	1.56	m²	**6.20**
Maximum depth not exceeding 2.00 m						
distance between opposing faces not exceeding 2.00 m	-	0.46	5.76	1.91	m²	**7.68**
Maximum depth not exceeding 4.00 m						
distance between opposing faces not exceeding 2.00 m	-	0.59	7.39	2.39	m²	**9.78**
Earthwork support to underpinning pits (open boarded - in 3.00 m lengths)						
Maximum depth not exceeding 1.00 m						
distance between opposing faces not exceeding 2.00 m	-	0.41	5.14	1.67	m²	**6.81**
Maximum depth not exceeding 2.00 m						
distance between opposing faces not exceeding 2.00 m	-	0.51	6.39	2.15	m²	**8.54**
Maximum depth not exceeding 4.00 m						
distance between opposing faces not exceeding 2.00 m	-	0.65	8.14	2.63	m²	**10.77**

D GROUNDWORK

Item	PC £	Labour hours	Labour £	Material /Plant £	Unit	Total rate £
Earthwork support to preliminary trenches (closed boarded - in 3.00 m lengths)						
Maximum depth not exceeding 1.00 m						
1.00 m deep	-	0.93	11.65	2.63	m²	**14.28**
Maximum depth not exceeding 2.00 m						
distance between opposing faces not exceeding 2.00 m	-	1.16	14.53	3.35	m²	**17.88**
Maximum depth not exceeding 4.00 m						
distance between opposing faces not exceeding 2.00 m	-	1.43	17.91	4.06	m²	**21.98**
Earthwork support to underpinning pits (closed boarded - in 3.00 m lengths)						
Maximum depth not exceeding 1.00 m						
distance between opposing faces not exceeding 2.00 m	-	1.02	12.78	2.87	m²	**15.65**
Maximum depth not exceeding 2.00 m						
distance between opposing faces not exceeding 2.00 m	-	1.28	16.04	3.59	m²	**19.62**
Maximum depth not exceeding 4.00 m						
distance between opposing faces not exceeding 2.00 m	-	1.57	19.67	4.54	m²	**24.21**
Extra over earthwork support for						
Left in	-	0.69	8.64	16.74	m²	**25.38**
Cutting away existing projecting foundations						
Concrete						
maximum width 150 mm; maximum depth 150 mm	-	0.15	1.88	0.17	m	**2.05**
maximum width 150 mm; maximum depth 225 mm	-	0.22	2.76	0.26	m	**3.01**
maximum width 150 mm; maximum depth 300 mm	-	0.30	3.76	0.35	m	**4.10**
maximum width 300 mm; maximum depth 300 mm	-	0.58	7.27	0.68	m	**7.94**
Masonry						
maximum width one brick thick; maximum depth one course high	-	0.04	0.50	0.05	m	**0.56**
maximum width one brick thick; maximum depth two courses high	-	0.13	1.63	0.15	m	**1.78**
maximum width one brick thick; maximum depth three courses high	-	0.25	3.13	0.29	m	**3.42**
maximum width one brick thick; maximum depth four courses high	-	0.42	5.26	0.48	m	**5.75**
Preparing the underside of existing work to receive the pinning up of the new work						
Width of existing work						
380 mm wide	-	0.56	7.02	-	m	**7.02**
600 mm wide	-	0.74	9.27	-	m	**9.27**
900 mm wide	-	0.93	11.65	-	m	**11.65**
1200 mm wide	-	1.11	13.91	-	m	**13.91**
Disposal; by hand						
Excavated material						
off site; to tip not exceeding 13 km (using lorries); including Landfill Tax based on inactive waste	-	0.74	9.27	37.00	m³	**46.27**
Filling to excavations; by hand						
Average thickness exceeding 0.25 m						
arising from the excavations	-	0.93	11.65	-	m³	**11.65**

D GROUNDWORK

Item	PC £	Labour hours	Labour £	Material /Plant £	Unit	Total rate £
D50 UNDERPINNING – cont'd						
Surface treatments						
Compacting						
bottoms of excavations	-	0.04	0.50	0.02	m²	**0.52**
Plain in situ ready mixed designated concrete; C10 - 40 mm aggregate; poured against faces of excavation						
Underpinning						
thickness not exceeding 150 mm	-	3.42	55.21	94.57	m³	**149.78**
thickness 150 - 450 mm	-	2.87	46.33	94.57	m³	**140.90**
thickness exceeding 450 mm	-	2.50	40.36	94.57	m³	**134.93**
Plain in situ ready mixed designated concrete; C20 - 20 mm aggregate; poured against faces of excavation						
Underpinning						
thickness not exceeding 150 mm	-	3.42	55.21	98.80	m³	**154.01**
thickness 150 - 450 mm	-	2.87	46.33	98.80	m³	**145.14**
thickness exceeding 450 mm	-	2.50	40.36	98.80	m³	**139.16**
Extra for working around reinforcement	-	0.28	4.52	-	m³	**4.52**
Sawn formwork; sides of foundations in underpinning						
Plain vertical						
height exceeding 1.00 m	-	1.48	28.16	5.98	m²	**34.14**
height not exceeding 250 mm	-	0.51	9.70	1.72	m²	**11.42**
height 250 - 500 mm	-	0.79	15.03	3.19	m²	**18.22**
height 500 mm - 1.00 m	-	1.20	22.83	5.98	m²	**28.81**
Reinforcement bar; BS 4449 hot rolled deformed square high yield steel bars						
20 mm diameter nominal size						
bent	643.50	24.00	447.68	778.85	tonne	**1226.52**
16 mm diameter nominal size						
bent	643.50	26.00	485.71	787.47	tonne	**1273.18**
12 mm diameter nominal size						
bent	682.50	28.00	523.75	839.11	tonne	**1362.86**
10 mm diameter nominal size						
bent	699.40	30.00	561.78	867.49	tonne	**1429.27**
8 mm diameter nominal size						
bent	734.50	32.00	595.45	914.83	tonne	**1510.28**
Common bricks; in cement mortar (1:3)						
Walls in underpinning						
one brick thick (PC £ per 1000)	240.00	2.22	58.11	37.07	m²	**95.18**
one and a half brick thick	-	3.05	79.83	55.25	m²	**135.08**
two brick thick	-	3.79	99.20	76.40	m²	**175.60**
Class A engineering bricks; in cement mortar (1:3)						
Walls in underpinning						
one brick thick (PC £ per 1000)	356.00	2.22	58.11	53.17	m²	**111.28**
one and a half brick thick	-	3.05	79.83	79.40	m²	**159.23**
two brick thick	-	3.79	99.20	108.59	m²	**207.79**

D GROUNDWORK

Item	PC £	Labour hours	Labour £	Material /Plant £	Unit	Total rate £
Class B engineering bricks; in cement mortar (1:3)						
Walls in underpinning						
one brick thick (PC £ per 1000)	288.00	2.22	58.11	43.74	m²	**101.85**
one and a half brick thick	-	3.05	79.83	65.26	m²	**145.09**
two brick thick	-	3.79	99.20	89.73	m²	**188.94**
Add or deduct for variation of £10.00/1000 in PC of bricks						
one brick thick	-	-	-	1.39	m²	-
one and a half bricks thick	-	-	-	2.09	m²	-
two bricks thick	-	-	-	2.79	m²	-
"Zedex CPT" (Co-Polymer Thermoplastic) damp proof course or other equal and approved; 200 mm laps; in gauged mortar (1:1:6)						
Horizontal						
width exceeding 225 mm	3.26	0.23	6.02	3.77	m²	**9.79**
width not exceeding 225 mm	-	0.46	12.04	3.77	m²	**15.81**
"Hyload" (pitch polymer) damp proof course or similar; 150 mm laps; in cement mortar (1:3)						
Horizontal						
width exceeding 225 mm	3.89	0.23	6.02	4.50	m²	**10.52**
width not exceeding 225 mm	3.98	0.46	12.04	4.60	m²	**16.64**
"Alumite" aluminium cored bitumen gas retardent damp proof course or other equal and approved; 200 mm laps; in gauged mortar (1:1;6)						
Horizontal						
width exceeding 225 mm	4.80	0.31	8.11	5.55	m²	**13.67**
width not exceeding 225 mm	-	0.60	15.70	5.55	m²	**21.26**
Two courses of slates in cement mortar (1:3)						
Horizontal						
width exceeding 225 mm	-	1.39	36.38	24.96	m²	**61.35**
width not exceeding 225 mm	-	2.31	60.46	25.52	m²	**85.99**
Wedging and pinning						
To underside of existing construction with slates in cement mortar (1:3)						
width of wall - half brick thick	-	1.02	26.70	5.92	m	**32.62**
width of wall - one brick thick	-	1.20	31.41	11.85	m	**43.26**
width of wall - one and a half brick thick	-	1.39	36.38	17.77	m	**54.15**

E IN SITU CONCRETE/LARGE PRECAST CONCRETE

Item	PC £	Labour hours	Labour £	Material £	Unit	Total rate £
BASIC MIXED CONCRETE PRICES						
DESIGNED MIXES						
Definition: "Mix for which the purchaser is responsible for specifying the required performances and the producer is responsible for selecting the mix proportions to produce the required performance".						
NOTE: The following prices are for designed mix concrete ready for placing excluding any allowance for waste, discount or overheads and profit. Prices are based upon delivery to site within a 5 mile (8 km) radius of concrete mixing plant, using full loads.						
Grade C7.5; cement to BS12; 10 mm aggregate	-	-	-	75.64	m³	-
Grade C7.5; cement to BS12; 20 mm aggregate	-	-	-	74.10	m³	-
Grade C7.5; cement to BS12; 40 mm aggregate	-	-	-	73.61	m³	-
Grade C7.5; sulphate resistant cement; 10 mm aggregate	-	-	-	82.92	m³	-
Grade C7.5; sulphate resistant cement; 20 mm aggregate	-	-	-	82.29	m³	-
Grade C7.5; sulphate resistant cement; 40 mm aggregate	-	-	-	80.89	m³	-
Grade C10; cement to BS12; 10 mm aggregate	-	-	-	77.24	m³	-
Grade C10; cement to BS12; 20 mm aggregate	-	-	-	75.67	m³	-
Grade C10; cement to BS12; 40 mm aggregate	-	-	-	74.80	m³	-
Grade C10; sulphate resistant cement; 10 mm aggregate	-	-	-	84.60	m³	-
Grade C10; sulphate resistant cement; 20 mm aggregate	-	-	-	83.03	m³	-
Grade C10; sulphate resistant cement; 40 mm aggregate	-	-	-	82.16	m³	-
Grade C15; cement to BS12; 10 mm aggregate	-	-	-	78.49	m³	-
Grade C15; cement to BS12; 20 mm aggregate	-	-	-	76.89	m³	-
Grade C15; cement to BS12; 40 mm aggregate	-	-	-	75.98	m³	-
Grade C15; sulphate resistant cement; 10 mm aggregate	-	-	-	85.93	m³	-
Grade C15; sulphate resistant cement; 20 mm aggregate	-	-	-	84.33	m³	-
Grade C15; sulphate resistant cement; 40 mm aggregate	-	-	-	83.42	m³	-
Grade C20; cement to BS12; 10 mm aggregate	-	-	-	79.74	m³	-
Grade C20; cement to BS12; 20 mm aggregate	-	-	-	78.10	m³	-
Grade C20; cement to BS12; 40 mm aggregate	-	-	-	77.19	m³	-
Grade C20; sulphate resistant cement; 10 mm aggregate	-	-	-	87.26	m³	-
Grade C20; sulphate resistant cement; 20 mm aggregate	-	-	-	85.62	m³	-
Grade C20; sulphate resistant cement; 40 mm aggregate	-	-	-	84.71	m³	-
Grade C25; cement to BS12; 10 mm aggregate	-	-	-	81.00	m³	-
Grade C25; cement to BS12; 20 mm aggregate	-	-	-	79.33	m³	-
Grade C25; cement to BS12; 40 mm aggregate	-	-	-	78.39	m³	-
Grade C25; sulphate resistant cement; 10 mm aggregate	-	-	-	89.55	m³	-
Grade C25; sulphate resistant cement; 20 mm aggregate	-	-	-	87.88	m³	-
Grade C25; sulphate resistant cement; 40 mm aggregate	-	-	-	86.94	m³	-

E IN SITU CONCRETE/LARGE PRECAST CONCRETE

Item	PC £	Labour hours	Labour £	Material £	Unit	Total rate £
Grade C30; cement to BS12; 10 mm aggregate	-	-	-	81.41	m³	-
Grade C30; cement to BS12; 20 mm aggregate	-	-	-	79.72	m³	-
Grade C30; cement to BS12; 40 mm aggregate	-	-	-	78.80	m³	-
Grade C30; sulphate resistant cement; 10 mm aggregate	-	-	-	89.96	m³	-
Grade C30; sulphate resistant cement; 20 mm aggregate	-	-	-	88.27	m³	-
Grade C30; sulphate resistant cement; 40 mm aggregate	-	-	-	87.35	m³	-
Grade C40; cement to BS12; 10 mm aggregate	-	-	-	85.80	m³	-
Grade C40; cement to BS12; 20 mm aggregate	-	-	-	84.23	m³	-
Grade C40; sulphate resistant cement; 10 mm aggregate	-	-	-	95.30	m³	-
Grade C40; sulphate resistant cement; 20 mm aggregate	-	-	-	93.73	m³	-
Grade C50; cement to BS12; 10 mm aggregate	-	-	-	86.06	m³	-
Grade C50; cement to BS12; 20 mm aggregate	-	-	-	84.24	m³	-
Grade C50; sulphate resistant cement; 10 mm aggregate	-	-	-	95.56	m³	-
Grade C50; sulphate resistant cement; 20 mm aggregate	-	-	-	93.74	m³	-
STANDARD MIXES						
Definition "Mix selected from the restricted list given in section 4 of BS 5328 : 2 : 1991 and made with a restricted range of materials".						
NOTE: The following prices are for standard mix concrete ready for placing excluding any allowance for waste, discount or overheads and profit. Prices are based upon delivery to site within a 5 mile (8 km) radius of concrete mixing plant, using full loads.						
Designated concrete mix; GEN0	-	-	-	71.11	m³	-
Designated concrete mix; GEN1	-	-	-	72.92	m³	-
Designated concrete mix; GEN2	-	-	-	74.77	m³	-
Designated concrete mix; GEN3	-	-	-	76.62	m³	-
Designated concrete mix; RC20/25	-	-	-	78.87	m³	-
Designated concrete mix; RC25/30	-	-	-	79.33	m³	-
Designated concrete mix; RC30/37	-	-	-	79.72	m³	-
Designated concrete mix; RC35/45	-	-	-	82.84	m³	-
Designated concrete mix; RC40/50	-	-	-	84.23	m³	-
Designated concrete mix; FND3; sulphate-resisting	-	-	-	87.40	m³	-
Designed concrete mix; ST 1	-	-	-	75.56	m³	-
Designed concrete mix; ST 2	-	-	-	76.86	m³	-
Designed concrete mix; ST 3	-	-	-	78.14	m³	-
Designed concrete mix; ST 4	-	-	-	79.41	m³	-
Designed concrete mix; ST 5	-	-	-	81.12	m³	-
LIGHTWEIGHT CONCRETE						
Grade 25; pumped; Lytag medium and natural sand	-	-	-	127.40	m³	-
Grade 30; pumped; Lytag medium and natural sand	-	-	-	132.30	m³	-
Grade 35; pumped; Lytag medium and natural sand	-	-	-	137.20	m³	-
Reduction for un-pumped concrete	-	-	-	-1.96	m³	-

E IN SITU CONCRETE/LARGE PRECAST CONCRETE

Item	PC £	Labour hours	Labour £	Material £	Unit	Total rate £
BASIC MIXED CONCRETE PRICES – cont'd						
SITE MIXED CONCRETE						
Mix 7.50 N/mm²; cement to BS12 (1:8); 40 mm aggregate	-	-	-	80.08	m³	-
Mix 7.50 N/mm²; sulphate resisting cement (1:8): 40 mm aggregate	-	-	-	89.18	m³	-
Mix 10.00 N/mm²; cement to BS12 (1:8): 40 mm aggregate	-	-	-	82.80	m³	-
Mix 10.00 N/mm²; sulphate resisting cement (1:8): 40 mm aggregate.	-	-	-	92.00	m³	-
Mix 20.00 N/mm²; cement to BS12 (1:2:4); 20 mm aggregate	-	-	-	88.36	m³	-
Mix 20.00 N/mm²; sulphate resisting cement (1:2:4); 20 mm aggregate	-	-	-	97.76	m³	-
Mix 25.00 N/mm²; cement to BS12 (1:1:5:3); 20 mm aggregate	-	-	-	90.25	m³	-
Mix 25.00 N/mm²; sulphate resisting cement (1:1:5:3); 20 mm aggregate	-	-	-	99.75	m³	-
ADD TO THE PRECEDING PRICES FOR:						
Rapid-hardening cement to BS 12	-	-	-	8.55	m³	-
Polypropylene fibre additive	-	-	-	4.75	m³	-
Air entraned concrtete	-	-	-	4.18	m³	-
Water repellant additive	-	-	-	4.46	m³	-
Distance per mile in excess of 5 miles (8 km)	-	-	-	0.50	m³	-
Part loads per m³ below full load	-	-	-	23.75	m³	-
OTHER MATERIAL PRICES						
CEMENTS						
Ordinary portland to BS12	-	-	-	106.55	tonne	-
Lighting high alumina	-	-	-	435.88	tonne	-
Sulfacrete sulphate resisting	-	-	-	174.35	tonne	-
Ferrocrete rapid hardening	-	-	-	271.21	tonne	-
Snowcrete white cement	-	-	-	242.16	tonne	-
CEMENT ADMIXTURES						
Febtone colourant - red, marigold, yellow, brown, black	-	-	-	9.45	kg	-
Febproof waterproof	-	-	-	9.97	5 ltrs	-
Febond PVA bonding agent	-	-	-	17.23	5 ltrs	-
Febspeed frostproofer and hardener	-	-	-	6.12	5 ltrs	-
SUPPLY AND FIX PRICES						
E10 IN SITU CONCRETE CONSTRUCTION						
NOTE: The following concrete material prices include an allowance for shrinkage factors at plus 2½% (or 5% when poured against earth or unblinded hardcore) and waste at plus 5%, together with overheads and profit at plus 5% and 2.5% respectively.						
PC Sums are designated basic mixed concrete supply only prices						

E IN SITU CONCRETE/LARGE PRECAST CONCRETE

Item	PC £	Labour hours	Labour £	Material £	Unit	**Total rate £**
Plain in situ ready mixed designated concrete; C7.5 - 40 mm aggregate						
Foundations	73.18	1.20	19.37	84.76	m³	**104.14**
Isolated foundations	-	1.39	22.44	84.76	m³	**107.20**
Beds						
thickness not exceeding 150 mm	-	1.62	26.15	84.76	m³	**110.92**
thickness 150 - 450 mm	-	1.16	18.73	84.76	m³	**103.49**
thickness exceeding 450 mm	-	0.93	15.01	84.76	m³	**99.78**
Screeded beds; protection to compressible formwork						
50 mm thick	-	0.10	1.61	4.13	m²	**5.75**
75 mm thick	-	0.15	2.42	6.21	m²	**8.63**
100 mm thick	-	0.20	3.23	8.27	m²	**11.50**
Filling hollow walls						
thickness not exceeding 150 mm	-	3.15	50.85	84.76	m³	**135.62**
Column casings						
stub columns beneath suspended ground slabs	-	4.50	72.65	84.76	m³	**157.41**
Plain in situ ready mixed designated concrete; C10 - 40 mm aggregate						
Foundations	74.31	1.20	19.37	86.07	m³	**105.44**
Isolated foundations	-	1.39	22.44	86.07	m³	**108.51**
Beds						
thickness not exceeding 150 mm	-	1.62	26.15	86.07	m³	**112.22**
thickness 150 - 450 mm	-	1.16	18.73	86.07	m³	**104.80**
thickness exceeding 450 mm	-	0.93	15.01	86.07	m³	**101.08**
Filling hollow walls						
thickness not exceeding 150 mm	-	3.15	50.85	86.07	m³	**136.92**
Plain in situ ready mixed designated concrete; C10 - 40 mm aggregate; poured on or against earth or unblinded hardcore						
Foundations	74.31	1.25	20.18	88.17	m³	**108.35**
Isolated foundations	-	1.48	23.89	88.17	m³	**112.06**
Beds						
thickness not exceeding 150 mm	-	1.71	27.61	88.17	m³	**115.78**
thickness 150 - 450 mm	-	1.25	20.18	88.17	m³	**108.35**
thickness exceeding 450 mm	-	0.97	15.66	88.17	m³	**103.83**
Plain in situ ready mixed designated concrete; C20 - 20 mm aggregate						
Foundations	77.63	1.20	19.37	89.92	m³	**109.29**
Isolated foundations	-	1.39	22.44	89.92	m³	**112.36**
Beds						
thickness not exceeding 150 mm	-	1.76	28.41	89.92	m³	**118.33**
thickness 150 - 450 mm	-	1.20	19.37	89.92	m³	**109.29**
thickness exceeding 450 mm	-	0.93	15.01	89.92	m³	**104.93**
Filling hollow walls						
thickness not exceeding 150 mm	-	3.15	50.85	89.92	m³	**140.77**
Plain in situ ready mixed designated concrete; C20 - 20 mm aggregate; poured on or against earth or unblinded hardcore						
Foundations	77.63	1.25	20.18	92.12	m³	**112.30**
Isolated foundations	-	1.48	23.89	92.12	m³	**116.01**
Beds						
thickness not exceeding 150 mm	-	1.85	29.87	92.12	m³	**121.98**
thickness 150 - 450 mm	-	1.30	20.99	92.12	m³	**113.11**
thickness exceeding 450 mm	-	0.97	15.66	92.12	m³	**107.78**

Item	PC £	Labour hours	Labour £	Material £	Unit	Total rate £
E10 IN SITU CONCRETE CONSTRUCTION – cont'd						
Reinforced in situ ready mixed designated concrete; C25 - 20 mm aggregate						
Foundations	78.85	1.30	20.99	91.33	m³	**112.32**
Ground beams	-	2.59	41.81	91.33	m³	**133.14**
Isolated foundations	-	1.57	25.35	91.33	m³	**116.67**
Beds						
thickness not exceeding 150 mm	-	2.04	32.93	91.33	m³	**124.26**
thickness 150 - 450 mm	-	1.48	23.89	91.33	m³	**115.22**
thickness exceeding 450 mm	-	1.20	19.37	91.33	m³	**110.70**
Slabs						
thickness not exceeding 150 mm	-	3.24	52.31	91.33	m³	**143.63**
thickness 150 - 450 mm	-	2.59	41.81	91.33	m³	**133.14**
thickness exceeding 450 mm	-	2.31	37.29	91.33	m³	**128.62**
Coffered and troughed slabs						
thickness 150 - 450 mm	-	2.96	47.79	91.33	m³	**139.11**
thickness exceeding 450 mm	-	2.59	41.81	91.33	m³	**133.14**
Extra over for sloping						
not exceeding 15 degrees	-	0.23	3.71	-	m³	**3.71**
over 15 degrees	-	0.46	7.43	-	m³	**7.43**
Walls						
thickness not exceeding 150 mm	-	3.42	55.21	91.33	m³	**146.54**
thickness 150 - 450 mm	-	2.73	44.07	91.33	m³	**135.40**
thickness exceeding 450 mm	-	2.41	38.91	91.33	m³	**130.24**
Beams						
isolated	-	3.70	59.73	91.33	m³	**151.06**
isolated deep	-	4.07	65.71	91.33	m³	**157.03**
attached deep	-	3.70	59.73	91.33	m³	**151.06**
Beam casings						
isolated	-	4.07	65.71	91.33	m³	**157.03**
isolated deep	-	4.44	71.68	91.33	m³	**163.01**
attached deep	-	4.07	65.71	91.33	m³	**157.03**
Columns	-	4.44	71.68	91.33	m³	**163.01**
Column casings	-	4.90	79.10	91.33	m³	**170.43**
Staircases	-	5.55	89.60	91.33	m³	**180.93**
Upstands	-	3.56	57.47	91.33	m³	**148.80**
Reinforced in situ ready mixed designated concrete; C35 -20 mm aggregate						
Foundations	82.84	1.30	20.99	95.95	m³	**116.94**
Ground beams	-	2.59	41.81	95.95	m³	**137.76**
Isolated foundations	-	1.57	25.35	95.95	m³	**121.30**
Beds						
thickness not exceeding 150 mm	-	2.04	32.93	95.95	m³	**128.88**
thickness 150 - 450 mm	-	1.48	23.89	95.95	m³	**119.84**
thickness exceeding 450 mm	-	1.20	19.37	95.95	m³	**115.32**
Slabs						
thickness not exceeding 150 mm	-	3.24	52.31	95.95	m³	**148.26**
thickness 150 - 450 mm	-	2.59	41.81	95.95	m³	**137.76**
thickness exceeding 450 mm	-	2.31	37.29	95.95	m³	**133.24**
Coffered and troughed slabs						
thickness 150 - 450 mm	-	2.96	47.79	95.95	m³	**143.74**
thickness exceeding 450 mm	-	2.59	41.81	95.95	m³	**137.76**
Extra over for sloping						
not exceeding 15 degrees	-	0.23	3.71	-	m³	**3.71**
over 15 degrees	-	0.46	7.43	-	m³	**7.43**

E IN SITU CONCRETE/LARGE PRECAST CONCRETE

Item	PC £	Labour hours	Labour £	Material £	Unit	Total rate £
Walls						
thickness not exceeding 150 mm	-	3.42	55.21	95.95	m³	**151.16**
thickness 150 - 450 mm	-	2.73	44.07	95.95	m³	**140.02**
thickness exceeding 450 mm	-	2.41	38.91	95.95	m³	**134.86**
Beams						
isolated	-	3.70	59.73	95.95	m³	**155.68**
isolated deep	-	4.07	65.71	95.95	m³	**161.66**
attached deep	-	3.70	59.73	95.95	m³	**155.68**
Beam casings						
isolated	-	4.07	65.71	95.95	m³	**161.66**
isolated deep	-	4.44	71.68	95.95	m³	**167.63**
attached deep	-	4.07	65.71	95.95	m³	**161.66**
Columns	-	4.44	71.68	95.95	m³	**167.63**
Column casings	-	4.90	79.10	95.95	m³	**175.05**
Staircases	-	5.55	89.60	95.95	m³	**185.55**
Upstands	-	3.56	57.47	95.95	m³	**153.42**
Reinforced in situ ready mixed designated concrete; C40 -20 mm aggregate						
Foundations	83.31	1.30	20.99	96.49	m³	**117.48**
Ground beams	-	2.59	41.81	96.49	m³	**138.30**
Isolated foundations	-	1.57	25.35	96.49	m³	**121.84**
Beds						
thickness not exceeding 150 mm	-	2.04	32.93	96.49	m³	**129.42**
thickness 150 - 450 mm	-	1.48	23.89	96.49	m³	**120.38**
thickness exceeding 450 mm	-	1.20	19.37	96.49	m³	**115.86**
Slabs						
thickness not exceeding 150 mm	-	3.24	52.31	96.49	m³	**148.80**
thickness 150 - 450 mm	-	2.59	41.81	96.49	m³	**138.30**
thickness exceeding 450 mm	-	2.31	37.29	96.49	m³	**133.78**
Coffered and troughed slabs						
thickness 150 - 450 mm	-	2.96	47.79	96.49	m³	**144.28**
thickness exceeding 450 mm	-	2.59	41.81	96.49	m³	**138.30**
Extra over for sloping						
not exceeding 15 degrees	-	0.23	3.71	-	m³	**3.71**
over 15 degrees	-	0.46	7.43	-	m³	**7.43**
Walls						
thickness not exceeding 150 mm	-	3.42	55.21	96.49	m³	**151.70**
thickness 150 - 450 mm	-	2.73	44.07	96.49	m³	**140.56**
thickness exceeding 450 mm	-	2.41	38.91	96.49	m³	**135.40**
Beams						
isolated	-	3.70	59.73	96.49	m³	**156.22**
isolated deep	-	4.07	65.71	96.49	m³	**162.19**
attached deep	-	3.70	59.73	96.49	m³	**156.22**
Beam casings						
isolated	-	4.07	65.71	96.49	m³	**162.19**
isolated deep	-	4.44	71.68	96.49	m³	**168.17**
attached deep	-	4.07	65.71	96.49	m³	**162.19**
Columns	-	4.44	71.68	96.49	m³	**168.17**
Column casings	-	4.90	79.10	96.49	m³	**175.59**
Staircases	-	5.55	89.60	96.49	m³	**186.09**
Upstands	-	3.56	57.47	96.49	m³	**153.96**

E IN SITU CONCRETE/LARGE PRECAST CONCRETE

Item	PC £	Labour hours	Labour £	Material £	Unit	Total rate £
E10 IN SITU CONCRETE CONSTRUCTION – cont'd						
Proprietary voided Bubbledeck, Cobiax or other equal and approved slab; concrete mix RC35; to achieve design loadings of 5.0 kN/m² live and 3.0 kN/m² dead; with trowelled finish						
Beds						
360 mm overall thickness	-	-	-	-	m²	**112.75**
Extra for						
Additional concrete 600 mm wide at edges where formers omitted at junctions with walls etc.	-	-	-	-	m	**46.13**
Extra over vibrated concrete for						
Reinforcement content over 5%	-	0.51	8.23	-	m³	**8.23**
Grouting with cement mortar (1:1)						
Stanchion bases						
10 mm thick	-	0.93	15.01	0.12	nr	**15.14**
25 mm thick	-	1.16	18.73	0.31	nr	**19.04**
Grouting with epoxy resin						
Stanchion bases						
10 mm thick	-	1.16	18.73	8.69	nr	**27.41**
25 mm thick	-	1.39	22.44	22.20	nr	**44.64**
Grouting with "Conbextra GP" cementitious grout						
Stanchion bases						
10 mm thick	-	1.16	18.73	1.57	nr	**20.30**
25 mm thick	-	1.39	22.44	4.02	nr	**26.46**
Grouting with "Conbextra HF" flowable cementitious grout						
Stanchion bases						
10 mm thick	-	1.16	18.73	1.94	nr	**20.67**
25 mm thick	-	1.39	22.44	4.96	nr	**27.40**
Filling; plain in situ designated concrete; C20 - 20 mm aggregate						
Mortices	-	0.09	1.45	0.48	nr	**1.93**
Holes	-	0.23	3.71	100.91	m³	**104.63**
Chases exceeding 0.01 m²	-	0.19	3.07	100.91	m³	**103.98**
Chases not exceeding 0.01 m²	-	0.14	2.26	1.01	m	**3.27**
Sheeting to prevent moisture loss						
Building paper; lapped joints						
subsoil grade 410; horizontal on foundations	-	0.02	0.32	0.45	m²	**0.77**
standard grade 420; horizontal on slabs	-	0.04	0.65	0.67	m²	**1.31**
Polythene sheeting; lapped joints; horizontal on slabs						
250 microns; 0.25 mm thick	-	0.04	0.65	0.50	m²	**1.14**
"Visqueen" sheeting or other equal and approved; lapped joints; horizontal on slabs						
250 microns; 0.25 mm thick	-	0.04	0.65	0.41	m²	**1.06**
300 microns; 0.30 mm thick	-	0.05	0.81	0.47	m²	**1.28**

E IN SITU CONCRETE/LARGE PRECAST CONCRETE

Item	PC £	Labour hours	Labour £	Material £	Unit	Total rate £
E20 FORMWORK FOR IN SITU CONCRETE						
NOTE: Generally all formwork based on four uses unless otherwise stated.						
Sides of foundations; basic finish						
Plain vertical						
height exceeding 1.00 m	-	1.48	28.16	8.18	m²	**36.34**
height exceeding 1.00 m; left in	-	1.30	24.74	19.65	m²	**44.39**
height not exceeding 250 mm	-	0.42	7.99	3.22	m	**11.22**
height not exceeding 250 mm; left in	-	0.42	7.99	5.70	m	**13.69**
height 250 - 500 mm	-	0.79	15.03	6.78	m	**21.81**
height 250 - 500 mm; left in	-	0.69	13.13	13.62	m	**26.75**
height 500 mm - 1.00 m	-	1.11	21.12	8.18	m	**29.30**
height 500 mm - 1.00 m; left in	-	1.06	20.17	19.65	m	**39.82**
Sides of foundations; polystyrene sheet formwork; Cordek "Claymaster" or other equal and approved; 50 mm thick						
Plain vertical						
height exceeding 1.00 m; left in	-	0.30	5.71	9.84	m²	**15.55**
height not exceeding 250 mm; left in	-	0.10	1.90	2.46	m	**4.36**
height 250 - 500 mm; left in	-	0.16	3.04	4.92	m	**7.96**
height 500 mm - 1.00 m; left in	-	0.24	4.57	9.84	m	**14.40**
Sides of foundations; polystyrene sheet formwork; Cordek "Claymaster" or other equal and approved; 75 mm thick						
Plain vertical						
height exceeding 1.00 m; left in	-	0.30	5.71	14.74	m²	**20.45**
height not exceeding 250 mm; left in	-	0.10	1.90	3.68	m	**5.59**
height 250 - 500 mm; left in	-	0.16	3.04	7.37	m	**10.41**
height 500 mm - 1.00 m; left in	-	0.24	4.57	14.74	m	**19.31**
Sides of foundations; polystyrene sheet formwork; Cordek "Claymaster" or other equal and approved; 100 mm thick						
Plain vertical						
height exceeding 1.00 m; left in	-	0.30	5.71	19.63	m²	**25.34**
height not exceeding 250 mm; left in	-	0.10	1.90	4.91	m	**6.81**
height 250 - 500 mm; left in	-	0.16	3.04	9.82	m	**12.86**
height 500 mm - 1.00 m; left in	-	0.24	4.57	19.63	m	**24.20**
Combined heave presssure relief insulation and compressible board substructure formwork; Cordeck "Cellcore CP" or other equal and approved; butt joints; securely fixed in place						
Plain horizontal						
200 mm thick; beneath slabs; left in	-	0.60	11.42	20.08	m²	**31.50**
250 mm thick; beneath slabs; left in	-	0.65	12.37	22.14	m²	**34.51**
300 mm thick; beneath slabs; left in	-	0.70	13.32	24.08	m²	**37.40**
Sides of ground beams and edges of beds; basic finish						
Plain vertical						
height exceeding 1.00 m	-	1.53	29.11	8.13	m²	**37.24**
height not exceeding 250 mm	-	0.46	8.75	3.17	m	**11.92**
height 250 - 500 mm	-	0.83	15.79	6.72	m	**22.52**
height 500 mm - 1.00 m	-	1.16	22.07	8.13	m	**30.20**

E IN SITU CONCRETE/LARGE PRECAST CONCRETE

Item	PC £	Labour hours	Labour £	Material £	Unit	Total rate £
E20 FORMWORK FOR IN SITU CONCRETE – cont'd						
Edges of suspended slabs; basic finish						
Plain vertical						
height not exceeding 250 mm	-	0.69	13.13	3.28	m	**16.41**
height 250 - 500 mm	-	1.02	19.41	5.40	m	**24.81**
height 500 mm - 1.00 m	-	1.62	30.83	8.24	m	**39.06**
Sides of upstands; basic finish						
Plain vertical						
height exceeding 1.00 m	-	1.85	35.20	10.24	m²	**45.44**
height not exceeding 250 mm	-	0.58	11.04	3.38	m	**14.42**
height 250 - 500 mm	-	0.93	17.70	6.93	m	**24.63**
height 500 mm - 1.00 m	-	1.62	30.83	10.24	m	**41.06**
Steps in top surfaces; basic finish						
Plain vertical						
height not exceeding 250 mm	-	0.46	8.75	3.43	m	**12.19**
height 250 - 500 mm	-	0.74	14.08	6.99	m	**21.07**
Steps in soffits; basic finish						
Plain vertical						
height not exceeding 250 mm	-	0.51	9.70	2.70	m	**12.40**
height 250 - 500 mm	-	0.81	15.41	4.88	m	**20.29**
Machine bases and plinths; basic finish						
Plain vertical						
height exceeding 1.00 m	-	1.48	28.16	8.13	m²	**36.29**
height not exceeding 250 mm	-	0.46	8.75	3.17	m	**11.92**
height 250 - 500 mm	-	0.79	15.03	6.72	m	**21.76**
height 500 mm - 1.00 m	-	1.16	22.07	8.13	m	**30.20**
Soffits of slabs; basic finish						
Slab thickness not exceeding 200 mm						
horizontal; height to soffit not exceeding 1.50 m	-	1.67	31.78	7.56	m²	**39.33**
horizontal; height to soffit 1.50 - 3.00 m	-	1.62	30.83	7.66	m²	**38.49**
horizontal; height to soffit 1.50 - 3.00 m (based on 5 uses)	-	1.53	29.11	6.34	m²	**35.45**
horizontal; height to soffit 1.50 - 3.00 m (based on 6 uses)	-	1.48	28.16	5.46	m²	**33.62**
horizontal; height to soffit 3.00 - 4.50 m	-	1.57	29.87	7.92	m²	**37.80**
horizontal; height to soffit 4.50 - 6.00 m	-	1.67	31.78	8.18	m²	**39.96**
Slab thickness 200 - 300 mm						
horizontal; height to soffit 1.50 - 3.00 m	-	1.67	31.78	10.08	m²	**41.86**
Slab thickness 300 - 400 mm						
horizontal; height to soffit 1.50 - 3.00 m	-	1.71	32.54	11.29	m²	**43.83**
Slab thickness 400 - 500 mm						
horizontal; height to soffit 1.50 - 3.00 m	-	1.80	34.25	12.50	m²	**46.75**
Slab thickness 500 - 600 mm						
horizontal; height to soffit 1.50 - 3.00 m	-	1.94	36.91	12.50	m²	**49.42**
Extra over soffits of slabs for						
sloping not exceeding 15 degrees	-	0.19	3.62	-	m²	**3.62**
sloping exceeding 15 degrees	-	0.37	7.04	-	m²	**7.04**

E IN SITU CONCRETE/LARGE PRECAST CONCRETE

Item	PC £	Labour hours	Labour £	Material £	Unit	Total rate £
Soffits of slabs; Richard Lees galvanised steel permanent shuttering; or other equal and approved						
Slab thickness not exceeding 200 mm						
0.9 mm S350 'Holorib' decking; height to soffit 1.50 - 3.00 m	18.90	0.28	5.72	20.95	m2	**26.66**
0.9 mm S350 'Holorib' decking; height to soffit 3.00 - 4.50 m	18.90	0.30	6.23	20.95	m2	**27.18**
1.2 mm S350 'Holorib' decking; height to soffit 3.00 - 4.50 m	22.83	0.30	6.23	25.31	m2	**31.55**
0.9 mm S350 'Ribdeck E60' decking; height to soffit 3.00 - 4.50 m	15.01	0.30	6.23	16.63	m2	**22.87**
1.2 mm S350 'Ribdeck E60' decking; height to soffit 3.00 - 4.50 m	17.94	0.30	6.23	19.89	m2	**26.12**
0.9 mm S350 'Ribdeck AL' decking; height to soffit 3.00 - 4.50 m	15.44	0.30	6.23	17.11	m2	**23.35**
1.2 mm S350 'Ribdeck AL' decking; height to soffit 3.00 - 4.50 m	18.52	0.30	6.23	20.53	m2	**26.77**
0.9 mm S350 'Ribdeck 80' decking; height to soffit 3.00 - 4.50 m	17.15	0.30	6.23	19.01	m2	**25.24**
1.2 mm S350 'Ribdeck 80' decking; height to soffit 3.00 - 4.50 m	20.74	0.30	6.23	22.99	m2	**29.22**
Edge trim and restraints to decking						
Edge trim 1.2 mm x 300 mm girth	-	0.23	4.68	6.12	m	**10.80**
Edge trim 1.2 mm x 350 mm girth	-	0.23	4.68	6.96	m	**11.64**
Edge trim 1.2 mm x 400 mm girth	-	0.23	4.68	7.73	m	**12.40**
Bearings to decking; connection to steel work with 'thru-deck' welded shear studs						
19 mm dia x 95 mm high studs at 100 mm centres	-	-	-	14.48	m	**-**
19 mm dia x 95 mm high studs at 200 mm centres	-	-	-	7.24	m	**-**
19 mm dia x 95 mm high studs at 300 mm centres	-	-	-	4.82	m	**-**
19 mm dia x 120 mm high studs at 100 mm centres	-	-	-	16.72	m	**-**
19 mm dia x 120 mm high studs at 200 mm centres	-	-	-	8.36	m	**-**
19 mm dia x 120 mm high studs at 300 mm centres	-	-	-	5.57	m	**-**
Soffits of landings; basic finish						
Slab thickness not exceeding 200 mm						
horizontal; height to soffit 1.50 - 3.00 m	-	1.67	31.78	8.14	m²	**39.92**
Slab thickness 200 - 300 mm						
horizontal; height to soffit 1.50 - 3.00 m	-	1.76	33.49	10.81	m²	**44.30**
Slab thickness 300 - 400 mm						
horizontal; height to soffit 1.50 - 3.00 m	-	1.80	34.25	12.14	m²	**46.39**
Slab thickness 400 - 500 mm						
horizontal; height to soffit 1.50 - 3.00 m	-	1.90	36.15	13.47	m²	**49.63**
Slab thickness 500 - 600 mm						
horizontal; height to soffit 1.50 - 3.00 m	-	2.04	38.82	13.47	m²	**52.29**
Extra over soffits of landings for						
sloping not exceeding 15 degrees	-	0.19	3.62	-	m²	**3.62**
sloping exceeding 15 degrees	-	0.37	7.04	-	m²	**7.04**
Soffits of coffered or troughed slabs; basic finish						
Cordek "Correx" trough mould or other equal and approved; 300 mm deep; ribs of mould at 600 mm centres and cross ribs at centres of bay; slab thickness 300 - 400 mm						
horizontal; height to soffit 1.50 - 3.00 m	-	2.31	43.95	13.12	m²	**57.08**
horizontal; height to soffit 3.00 - 4.50 m	-	2.41	45.86	13.39	m²	**59.24**
horizontal; height to soffit 4.50 - 6.00 m	-	2.50	47.57	13.54	m²	**61.11**
Top formwork; basic finish						
Sloping exceeding 15 degrees	-	1.39	26.45	5.50	m²	**31.95**

E IN SITU CONCRETE/LARGE PRECAST CONCRETE

Item	PC £	Labour hours	Labour £	Material £	Unit	Total rate £
E20 FORMWORK FOR IN SITU CONCRETE – cont'd						
Walls; basic finish						
Vertical	-	1.67	31.78	10.08	m²	**41.86**
Vertical; height exceeding 3.00 m above floor level	-	2.04	38.82	10.34	m²	**49.16**
Vertical; interrupted	-	1.94	36.91	10.34	m²	**47.26**
Vertical; to one side only	-	3.24	61.65	13.03	m²	**74.68**
Battered	-	2.59	49.28	10.81	m²	**60.10**
Beams; basic finish						
Attached to slabs						
regular shaped; square or rectangular; height to soffit 1.50 - 3.00 m	-	2.04	38.82	9.77	m²	**48.59**
regular shaped; square or rectangular; height to soffit 3.00 - 4.50 m	-	2.13	40.53	10.08	m²	**50.61**
regular shaped; square or rectangular; height to soffit 4.50 - 6.00 m	-	2.22	42.24	10.34	m²	**52.59**
Attached to walls						
regular shaped; square or rectangular; height to soffit 1.50 - 3.00 m	-	2.13	40.53	9.77	m²	**50.30**
Isolated						
regular shaped; square or rectangular; height to soffit 1.50 - 3.00 m	-	2.22	42.24	9.77	m²	**52.01**
regular shaped; square or rectangular; height to soffit 3.00 - 4.50 m	-	2.31	43.95	10.08	m²	**54.04**
regular shaped; square or rectangular; height to soffit 4.50 - 6.00 m	-	2.41	45.86	10.34	m²	**56.20**
Extra over beams for						
regular shaped; sloping not exceeding 15 degrees	-	0.28	5.33	1.16	m²	**6.49**
regular shaped; sloping exceeding 15 degrees	-	0.56	10.66	2.32	m²	**12.97**
Beam casings; basic finish						
Attached to slabs						
regular shaped; square or rectangular; height to soffit 1.50 - 3.00 m	-	2.13	40.53	9.77	m²	**50.30**
regular shaped; square or rectangular; height to soffit 3.00 - 4.50 m	-	2.22	42.24	10.08	m²	**52.32**
Attached to walls						
regular shaped; square or rectangular; height to soffit 1.50 - 3.00 m	-	2.22	42.24	9.77	m²	**52.01**
Isolated						
regular shaped; square or rectangular; height to soffit 1.50 - 3.00 m	-	2.31	43.95	9.77	m²	**53.72**
regular shaped; square or rectangular; height to soffit 3.00 - 4.50 m	-	2.41	45.86	10.08	m²	**55.94**
Extra over beam casings for						
regular shaped; sloping not exceeding 15 degrees	-	0.28	5.33	1.16	m²	**6.49**
regular shaped; sloping exceeding 15 degrees	-	0.56	10.66	2.32	m²	**12.97**
Columns; basic finish						
Attached to walls						
regular shaped; square or rectangular; height to soffit 1.50 - 3.00 m	-	2.04	38.82	8.18	m²	**47.00**
Isolated						
regular shaped; square or rectangular; height to soffit 1.50 - 3.00 m	-	2.13	40.53	8.18	m²	**48.71**
regular shaped; circular; not exceeding 300 mm diameter; height to soffit 1.50 - 3.00 m	-	3.70	70.40	14.40	m²	**84.81**
regular shaped; circular; 300 - 600 mm diameter; height to soffit 1.50 - 3.00 m	-	3.47	66.03	12.50	m²	**78.53**

E IN SITU CONCRETE/LARGE PRECAST CONCRETE

Item	PC £	Labour hours	Labour £	Material £	Unit	Total rate £
regular shaped; circular; 600 - 900 mm diameter; height to soffit 1.50 - 3.00 m	-	3.24	61.65	12.24	m²	**73.89**
Column casings; basic finish						
Attached to walls						
regular shaped; square or rectangular; height to soffit 1.50 - 3.00 m	-	2.13	40.53	8.18	m²	**48.71**
Isolated						
regular shaped; square or rectangular; height to soffit 1.50 - 3.00 m	-	2.22	42.24	8.18	m²	**50.43**
Recesses or rebates						
12 x 12 mm	-	0.06	1.14	0.26	m	**1.40**
25 x 25 mm	-	0.06	1.14	0.47	m	**1.62**
25 x 50 mm	-	0.06	1.14	0.62	m	**1.76**
50 x 50 mm	-	0.06	1.14	0.86	m	**2.00**
Nibs						
50 x 50 mm	-	0.51	9.70	1.11	m	**10.81**
100 x 100 mm	-	0.72	13.70	1.35	m	**15.06**
100 x 200 mm	-	0.96	18.27	8.76	m	**27.03**
Extra over a basic finish for fine formed finishes						
Slabs	-	0.30	5.71	-	m²	**5.71**
Walls	-	0.30	5.71	-	m²	**5.71**
Beams	-	0.30	5.71	-	m²	**5.71**
Columns	-	0.30	5.71	-	m²	**5.71**
Add to prices for basic formwork for						
Curved radius 6.00 m - 50%						
Curved radius 2.00 m - 100%						
Coating with retardant agent	-	0.01	0.19	0.35	m²	**0.54**
Wall kickers; basic finish						
Height 150 mm	-	0.46	8.75	2.32	m	**11.07**
Height 225 mm	-	0.60	11.42	2.74	m	**14.16**
Suspended wall kickers; basic finish						
Height 150 mm	-	0.58	11.04	2.57	m	**13.61**
Wall ends, soffits and steps in walls; basic finish						
Plain						
width exceeding 1.00 m	-	1.76	33.49	10.08	m²	**43.57**
width not exceeding 250 mm	-	0.56	10.66	2.44	m	**13.09**
width 250 - 500 mm	-	0.88	16.74	5.56	m	**22.31**
width 500 mm - 1.00 m	-	1.39	26.45	10.08	m	**36.53**
Openings in walls						
Plain						
width exceeding 1.00 m	-	1.94	36.91	10.08	m²	**47.00**
width not exceeding 250 mm	-	0.60	11.42	2.44	m	**13.86**
width 250 - 500 mm	-	1.02	19.41	5.56	m	**24.97**
width 500 mm - 1.00 m	-	1.57	29.87	10.08	m	**39.96**
Stairflights						
Width 1.00 m; 150 mm waist; 150 mm undercut risers string, width 300 mm	-	4.63	88.10	21.40	m	**109.50**
Width 2.00 m; 200 mm waist; 150 mm undercut risers string, width 350 mm	-	8.33	158.50	60.34	m	**218.85**

E IN SITU CONCRETE/LARGE PRECAST CONCRETE

Item	PC £	Labour hours	Labour £	Material £	Unit	Total rate £
E20 FORMWORK FOR IN SITU CONCRETE – cont'd						
Mortices						
Girth not exceeding 500 mm						
depth not exceeding 250 mm; circular	-	0.14	2.66	0.67	nr	**3.33**
Holes						
Girth not exceeding 500 mm						
depth not exceeding 250 mm; circular	-	0.19	3.62	0.99	nr	**4.61**
depth 250 - 500 mm; circular	-	0.28	5.33	2.74	nr	**8.07**
Girth 500 mm - 1.00 m						
depth not exceeding 250 mm; circular	-	0.23	4.38	1.56	nr	**5.93**
depth 250 - 500 mm; circular	-	0.35	6.66	4.82	nr	**11.48**
Girth 1.00 - 2.00 m						
depth not exceeding 250 mm; circular	-	0.42	7.99	4.82	nr	**12.81**
depth 250 - 500 mm; circular	-	0.62	11.80	10.14	nr	**21.94**
Girth 2.00 - 3.00 m						
depth not exceeding 250 mm; circular	-	0.56	10.66	9.47	nr	**20.13**
depth 250 - 500 mm; circular	-	0.83	15.79	59.48	nr	**75.27**
E30 REINFORCEMENT FOR IN SITU CONCRETE						
NOTE: Prices of steel and fabric reinforcement are particularly volatile at the time of going to press, so Readers are encouraged to contact their suppliers and check prices for currency based on anticipated delivery dates						
Bars; BS 4449; hot rolled deformed high steel bars; grade 500C						
40 mm diameter nominal size						
straight	653.90	16.00	304.28	736.42	tonne	**1040.69**
bent	678.60	16.00	304.28	763.67	tonne	**1067.94**
32 mm diameter nominal size						
straight	648.70	17.00	323.29	731.20	tonne	**1054.49**
bent	673.40	17.00	323.29	758.45	tonne	**1081.74**
25 mm diameter nominal size						
straight	643.50	18.00	342.31	726.20	tonne	**1068.51**
bent	669.50	18.00	342.31	754.88	tonne	**1097.20**
20 mm diameter nominal size						
straight	643.50	20.00	380.35	727.61	tonne	**1107.96**
bent	669.50	18.00	342.31	756.30	tonne	**1098.61**
16 mm diameter nominal size						
straight	643.50	22.00	418.38	729.25	tonne	**1147.63**
bent	669.50	22.00	418.38	757.93	tonne	**1176.31**
12 mm diameter nominal size						
straight	682.50	24.00	456.42	774.42	tonne	**1230.84**
bent	725.40	24.00	456.42	821.75	tonne	**1278.16**
10 mm diameter nominal size						
straight	699.40	26.00	494.45	796.34	tonne	**1290.79**
bent	743.60	26.00	494.45	845.10	tonne	**1339.55**
8 mm diameter nominal size						
straight	734.50	28.00	528.11	837.21	tonne	**1365.32**
links	734.50	31.00	585.17	839.28	tonne	**1424.44**
bent	734.50	28.00	528.11	837.21	tonne	**1365.32**

E IN SITU CONCRETE/LARGE PRECAST CONCRETE

Item	PC £	Labour hours	Labour £	Material £	Unit	Total rate £
Bars; stainless steel; to EN 1.4301						
32 mm diameter nominal size						
straight	3075.00	17.00	323.29	3460.05	tonne	**3783.35**
bent	3382.50	21.00	390.62	3799.27	tonne	**4189.90**
25 mm diameter nominal size						
straight	3075.00	18.00	342.31	3456.53	tonne	**3798.84**
bent	3382.50	18.00	342.31	3795.75	tonne	**4138.06**
20 mm diameter nominal size						
straight	3075.00	20.00	380.35	3461.17	tonne	**3841.52**
bent	3382.50	20.00	380.35	3800.39	tonne	**4180.74**
16 mm diameter nominal size						
straight	3075.00	22.00	418.38	3469.79	tonne	**3888.17**
bent	3280.00	22.00	418.38	3695.94	tonne	**4114.32**
12 mm diameter nominal size						
straight	3075.00	24.00	456.42	3478.41	tonne	**3934.83**
bent	3280.00	24.00	456.42	3704.56	tonne	**4160.97**
10 mm diameter nominal size						
straight	3075.00	26.00	494.45	3488.15	tonne	**3982.60**
bent	3280.00	26.00	494.45	3714.30	tonne	**4208.75**
8 mm diameter nominal size						
straight	3075.00	28.00	528.11	3496.77	tonne	**4024.89**
bent	3280.00	28.00	528.11	3722.92	tonne	**4251.03**
Bars; stainless steel; to EN 1.4462						
32 mm diameter nominal size						
straight	4920.00	17.00	323.29	5495.38	tonne	**5818.67**
bent	5073.75	21.00	390.62	5664.99	tonne	**6055.61**
25 mm diameter nominal size						
straight	4920.00	18.00	342.31	5491.85	tonne	**5834.16**
bent	5073.75	18.00	342.31	5661.46	tonne	**6003.77**
20 mm diameter nominal size						
straight	4920.00	20.00	380.35	5496.49	tonne	**5876.84**
bent	5073.75	20.00	380.35	5666.10	tonne	**6046.45**
16 mm diameter nominal size						
straight	4920.00	22.00	418.38	5505.11	tonne	**5923.50**
bent	5073.75	22.00	418.38	5674.73	tonne	**6093.11**
12 mm diameter nominal size						
straight	4920.00	24.00	456.42	5513.74	tonne	**5970.15**
bent	5073.75	24.00	456.42	5683.35	tonne	**6139.76**
10 mm diameter nominal size						
straight	4920.00	26.00	494.45	5523.47	tonne	**6017.92**
bent	5073.75	26.00	494.45	5693.08	tonne	**6187.53**
8 mm diameter nominal size						
straight	4920.00	28.00	528.11	5532.09	tonne	**6060.21**
bent	5073.75	28.00	528.11	5701.70	tonne	**6229.82**

E IN SITU CONCRETE/LARGE PRECAST CONCRETE

Item	PC £	Labour hours	Labour £	Material £	Unit	Total rate £
E30 REINFORCEMENT FOR IN SITU CONCRETE – cont'd						
Bars; stainless steel; to LDX2101® (EN 1.4362)						
Note: LDX2101® (EN 1.4362) is a new low Ni, Mn bearing stainless steel alloy, which offers greater price stability and cost effectiveness, and is expected to be adopted into the British Standard in the near future.						
32 mm diameter nominal size						
straight	3023.75	17.00	323.29	3403.52	tonne	**3726.81**
bent	3280.00	21.00	390.62	3686.20	tonne	**4076.82**
25 mm diameter nominal size						
straight	3023.75	18.00	342.31	3399.99	tonne	**3742.30**
bent	3280.00	18.00	342.31	3682.67	tonne	**4024.98**
20 mm diameter nominal size						
straight	3023.75	20.00	380.35	3404.63	tonne	**3784.98**
bent	3228.75	20.00	380.35	3630.78	tonne	**4011.13**
16 mm diameter nominal size						
straight	3023.75	22.00	418.38	3413.25	tonne	**3831.64**
bent	3228.75	22.00	418.38	3639.40	tonne	**4057.78**
12 mm diameter nominal size						
straight	3023.75	24.00	456.42	3421.88	tonne	**3878.29**
bent	3177.50	24.00	456.42	3591.49	tonne	**4047.90**
10 mm diameter nominal size						
straight	3023.75	26.00	494.45	3431.61	tonne	**3926.06**
bent	3177.50	26.00	494.45	3601.22	tonne	**4095.67**
8 mm diameter nominal size						
straight	3023.75	28.00	528.11	3440.23	tonne	**3968.35**
bent	3177.50	28.00	528.11	3609.84	tonne	**4137.96**
Fabric; BS 4449						
Ref A98 (1.54 kg/m²)						
400 mm minimum laps	1.77	0.12	2.28	2.09	m²	**4.37**
strips in one width; 600 mm width	1.77	0.15	2.85	2.09	m²	**4.94**
strips in one width; 900 mm width	1.77	0.14	2.66	2.09	m²	**4.75**
strips in one width; 1200 mm width	1.77	0.13	2.47	2.09	m²	**4.56**
Ref A142 (2.22 kg/m²)						
400 mm minimum laps	1.68	0.12	2.28	1.99	m²	**4.27**
strips in one width; 600 mm width	1.68	0.15	2.85	1.99	m²	**4.84**
strips in one width; 900 mm width	1.68	0.14	2.66	1.99	m²	**4.65**
strips in one width; 1200 mm width	1.68	0.13	2.47	1.99	m²	**4.46**
Ref A193 (3.02 kg/m²)						
400 mm minimum laps	2.29	0.12	2.28	2.71	m²	**4.99**
strips in one width; 600 mm width	2.29	0.15	2.85	2.71	m²	**5.56**
strips in one width; 900 mm width	2.29	0.14	2.66	2.71	m²	**5.37**
strips in one width; 1200 mm width	2.29	0.13	2.47	2.71	m²	**5.18**
Ref A252 (3.95 kg/m²)						
400 mm minimum laps	3.00	0.13	2.47	3.55	m²	**6.03**
strips in one width; 600 mm width	3.00	0.16	3.04	3.55	m²	**6.60**
strips in one width; 900 mm width	3.00	0.15	2.85	3.55	m²	**6.41**
strips in one width; 1200 mm width	3.00	0.14	2.66	3.55	m²	**6.22**

E IN SITU CONCRETE/LARGE PRECAST CONCRETE

Item	PC £	Labour hours	Labour £	Material £	Unit	**Total rate £**
Ref A393 (6.16 kg/m²)						
400 mm minimum laps	4.67	0.15	2.85	5.53	m²	**8.39**
strips in one width; 600 mm width	4.67	0.18	3.42	5.53	m²	**8.96**
strips in one width; 900 mm width	4.67	0.17	3.23	5.53	m²	**8.77**
strips in one width; 1200 mm width	4.67	0.16	3.04	5.53	m²	**8.58**
Ref B196 (3.05 kg/m²)						
400 mm minimum laps	4.27	0.12	2.28	5.05	m²	**7.33**
strips in one width; 600 mm width	4.27	0.15	2.85	5.05	m²	**7.90**
strips in one width; 900 mm width	4.27	0.14	2.66	5.05	m²	**7.71**
strips in one width; 1200 mm width	4.27	0.13	2.47	5.05	m²	**7.52**
Ref B283 (3.73 kg/m²)						
400 mm minimum laps	2.83	0.12	2.28	3.35	m²	**5.63**
strips in one width; 600 mm width	2.83	0.15	2.85	3.35	m²	**6.20**
strips in one width; 900 mm width	2.83	0.14	2.66	3.35	m²	**6.01**
strips in one width; 1200 mm width	2.83	0.13	2.47	3.35	m²	**5.82**
Ref B385 (4.53 kg/m²)						
400 mm minimum laps	3.44	0.13	2.47	4.07	m²	**6.54**
strips in one width; 600 mm width	3.44	0.16	3.04	4.07	m²	**7.11**
strips in one width; 900 mm width	3.44	0.15	2.85	4.07	m²	**6.92**
strips in one width; 1200 mm width	3.44	0.14	2.66	4.07	m²	**6.73**
Ref B503 (5.93 kg/m²)						
400 mm minimum laps	4.50	0.15	2.85	5.33	m²	**8.18**
strips in one width; 600 mm width	4.50	0.18	3.42	5.33	m²	**8.75**
strips in one width; 900 mm width	4.50	0.17	3.23	5.33	m²	**8.56**
strips in one width; 1200 mm width	4.50	0.16	3.04	5.33	m²	**8.37**
Ref B785 (8.14 kg/m²)						
400 mm minimum laps	6.18	0.17	3.23	7.32	m²	**10.55**
strips in one width; 600 mm width	6.18	0.20	3.80	7.32	m²	**11.13**
strips in one width; 900 mm width	6.18	0.19	3.61	7.32	m²	**10.94**
strips in one width; 1200 mm width	6.18	0.18	3.42	7.32	m²	**10.74**
Ref B1131 (10.90 kg/m²)						
400 mm minimum laps	8.28	0.18	3.42	9.81	m²	**13.23**
strips in one width; 600 mm width	8.28	0.24	4.56	9.81	m²	**14.37**
strips in one width; 900 mm width	8.28	0.22	4.18	9.81	m²	**13.99**
strips in one width; 1200 mm width	8.28	0.20	3.80	9.81	m²	**13.61**
Ref D49 (0.77 kg/m²)						
100 mm minimum laps; bent	1.58	0.24	4.56	1.87	m²	**6.43**

E IN SITU CONCRETE/LARGE PRECAST CONCRETE

Item	PC £	Labour hours	Labour £	Material £	Unit	Total rate £
E40 DESIGNED JOINTS IN IN SITU CONCRETE						
Formed; Fosroc impregnated fibreboard joint filler or other equal and approved						
Width not exceeding 150 mm						
12.50 mm thick	-	0.14	2.66	1.46	m	**4.13**
20 mm thick	-	0.19	3.62	2.21	m	**5.83**
25 mm thick	-	0.23	4.38	2.30	m	**6.68**
Width 150 - 300 mm						
12.50 mm thick	-	0.23	4.38	2.31	m	**6.68**
20 mm thick	-	0.23	4.38	3.57	m	**7.94**
25 mm thick	-	0.23	4.38	4.21	m	**8.59**
Width 300 - 450 mm						
12.50 mm thick	-	0.28	5.33	3.24	m	**8.57**
20 mm thick	-	0.28	5.33	4.97	m	**10.30**
25 mm thick	-	0.28	5.33	5.86	m	**11.19**
Formed; Grace Servicised "Kork-pak" waterproof bonded cork joint filler board or other equal and approved						
Width not exceeding 150 mm						
10 mm thick	-	0.14	2.66	3.05	m	**5.72**
13 mm thick	-	0.14	2.66	3.10	m	**5.76**
19 mm thick	-	0.14	2.66	4.02	m	**6.69**
25 mm thick	-	0.14	2.66	4.60	m	**7.26**
Width 150 - 300 mm						
10 mm thick	-	0.19	3.62	5.61	m	**9.22**
13 mm thick	-	0.19	3.62	5.70	m	**9.32**
19 mm thick	-	0.19	3.62	7.55	m	**11.17**
25 mm thick	-	0.19	3.62	8.70	m	**12.32**
Width 300 - 450 mm						
10 mm thick	-	0.23	4.38	8.56	m	**12.93**
13 mm thick	-	0.23	4.38	8.70	m	**13.08**
19 mm thick	-	0.23	4.38	11.47	m	**15.85**
25 mm thick	-	0.23	4.38	13.20	m	**17.57**
Sealants; Fosroc"Pliastic 77" hot poured rubberized bituminous compound or other equal and approved						
Width 10 mm						
25 mm depth	-	0.17	3.23	1.01	m	**4.25**
Width 12.50 mm						
25 mm depth	-	0.18	3.43	1.24	m	**4.67**
Width 20 mm						
25 mm depth	-	0.19	3.62	2.02	m	**5.64**
Width 25 mm						
25 mm depth	-	0.20	3.81	2.49	m	**6.29**
Sealants; Fosroc "Thioflex 600" gun grade two part polysulphide or other equal and approved						
Width 10 mm						
25 mm depth	-	0.05	0.95	3.85	m	**4.80**
Width 12.50 mm						
25 mm depth	-	0.06	1.14	4.81	m	**5.95**
Width 20 mm						
25 mm depth	-	0.07	1.33	7.69	m	**9.03**
Width 25 mm						
25 mm depth	-	0.08	1.52	9.62	m	**11.14**

E IN SITU CONCRETE/LARGE PRECAST CONCRETE

Item	PC £	Labour hours	Labour £	Material £	Unit	Total rate £
Sealants; Grace Servicised "Paraseal" polysulphide compound or other equal and approved; priming with Grace Servicised "Primer P"						
Width 10 mm						
25 mm depth	-	0.19	2.78	2.81	m	**5.59**
Width 13 mm						
25 mm depth	-	0.19	2.78	3.60	m	**6.38**
Width 19 mm						
25 mm depth	-	0.23	3.37	5.19	m	**8.56**
Width 25 mm						
25 mm depth	-	0.23	3.37	6.77	m	**10.14**
Waterstops; Grace Servicised or other equal and approved;						
Hydrophilic strip water stop; lapped joints; cast into concrete						
5 x 20 mm "Servistrip AH 205"	-	0.30	4.39	5.31	m	**9.70**
50 x 20 mm "Adcor 500S"	-	0.30	4.39	7.02	m	**11.41**
"Servitite" Internal 10mm thick pvc water stop; flat dumbell type; heat welded joints; cast into concrete						
Servitite 150; 150mm wide	7.27	0.23	4.37	9.01	m	**13.38**
flat angle	13.15	0.28	5.32	18.18	nr	**23.51**
vertical angle	13.07	0.28	5.32	18.10	nr	**23.42**
flat three way intersection	18.96	0.37	7.04	26.47	nr	**33.51**
vertical three way intersection	21.97	0.37	7.04	29.79	nr	**36.83**
four way intersection	23.24	0.46	8.75	33.00	nr	**41.75**
Servitite 230; 230mm wide	10.67	0.23	4.37	12.84	m	**17.22**
flat angle	15.59	0.28	5.32	22.41	nr	**27.74**
vertical angle	19.08	0.28	5.32	26.27	nr	**31.59**
flat three way intersection	22.61	0.37	7.04	32.81	nr	**39.84**
vertical three way intersection	43.14	0.37	7.04	55.45	nr	**62.49**
four way intersection	28.10	0.46	8.75	41.43	nr	**50.18**
Servitite AT200; 200 mm wide	14.35	0.23	4.37	17.01	m	**21.38**
flat angle	17.88	0.28	5.32	26.60	nr	**31.93**
vertical angle	19.74	0.28	5.32	28.66	nr	**33.98**
flat three way intersection	31.98	0.37	7.04	45.64	nr	**52.68**
vertical three way intersection	23.67	0.37	7.04	36.47	nr	**43.51**
four way intersection	36.99	0.46	8.75	54.57	nr	**63.32**
Servitite K305; 305 mm wide	17.61	0.28	5.32	20.69	m	**26.01**
flat angle	25.52	0.32	6.09	38.50	nr	**44.58**
vertical angle	28.07	0.32	6.09	41.31	nr	**47.39**
flat three way intersection	35.86	0.42	7.99	55.11	nr	**63.10**
vertical three way intersection	43.81	0.42	7.99	63.89	nr	**71.87**
four way intersection	48.85	0.51	9.70	74.58	nr	**84.28**
"Serviiseal" External pvc water stop; PVC water stop; centre bulb type; heat welded joints; cast into concrete						
Serviseal 195; 195mm wide	4.66	0.23	4.37	6.06	m	**10.43**
flat angle	8.15	0.28	5.32	11.49	nr	**16.81**
vertical angle	14.63	0.28	5.32	18.64	nr	**23.97**
flat three way intersection	13.97	0.37	7.04	19.20	nr	**26.24**
four way intersection	21.08	0.46	8.75	28.25	nr	**37.00**
Serviseal 240; 240 mm wide	5.93	0.23	4.37	7.49	m	**11.86**
flat angle	9.27	0.28	5.32	13.30	nr	**18.62**
vertical angle	15.31	0.28	5.32	19.96	nr	**25.29**
flat three way intersection	15.67	0.37	7.04	21.93	nr	**28.97**
four way intersection	23.33	0.46	8.75	31.88	nr	**40.63**

E IN SITU CONCRETE/LARGE PRECAST CONCRETE

Item	PC £	Labour hours	Labour £	Material £	Unit	Total rate £
E40 DESIGNED JOINTS IN IN SITU CONCRETE – cont'd						
Waterstops; Grace Servicised – cont'd						
Serviseal AT240; 240 mm wide	16.94	0.23	4.37	19.93	m	**24.30**
flat angle	16.65	0.28	5.32	26.42	nr	**31.75**
vertical angle	15.69	0.28	5.32	25.36	nr	**30.69**
flat three way intersection	25.64	0.37	7.04	40.40	nr	**47.44**
four way intersection	38.61	0.46	8.75	58.70	nr	**67.44**
Serviseal K320; 320 mm wide	7.95	0.28	5.32	9.78	m	**15.10**
flat angle	20.87	0.32	6.09	27.91	nr	**33.99**
vertical angle	4.77	0.32	6.09	15.59	nr	**21.67**
flat three way intersection	30.55	0.42	7.99	41.07	nr	**49.06**
four way intersection	38.05	0.51	9.70	51.76	nr	**61.45**
E41 WORKED FINISHES/CUTTING TO IN SITU CONCRETE						
Worked finishes						
Tamping by mechanical means	-	0.02	0.32	0.09	m²	**0.41**
Power floating	-	0.16	2.58	0.31	m²	**2.89**
Trowelling	-	0.31	5.00	-	m²	**5.00**
Hacking						
by mechanical means	-	0.31	5.00	0.36	m²	**5.36**
by hand	-	0.65	10.49	-	m²	**10.49**
Lightly shot blasting surface of concrete	-	0.37	5.97	-	m²	**5.97**
Blasting surface of concrete	-	0.65	10.49	2.58	m²	**13.08**
to produce textured finish	-	0.65	10.49	0.75	m²	**11.25**
Sand blasting (blast and vac method)	-	-	-	-	m²	**32.29**
Wood float finish	-	0.12	1.94	-	m²	**1.94**
Tamped finish						
level or to falls	-	0.06	0.97	-	m²	**0.97**
to falls	-	0.09	1.45	-	m²	**1.45**
Spade finish	-	0.14	2.26	-	m²	**2.26**
Cutting chases						
Depth not exceeding 50 mm						
width 10 mm	-	0.31	5.00	0.64	m	**5.64**
width 50 mm	-	0.46	7.43	0.82	m	**8.25**
width 75 mm	-	0.61	9.85	0.99	m	**10.84**
Depth 50 - 100 mm						
width 75 mm	-	0.83	13.40	1.57	m	**14.97**
width 100 mm	-	0.93	15.01	1.67	m	**16.69**
width 100 mm; in reinforced concrete	-	1.39	22.44	2.64	m	**25.08**
Depth 100 - 150 mm						
width 100 mm	-	1.20	19.37	1.95	m	**21.32**
width 100 mm; in reinforced concrete	-	1.85	29.87	3.38	m	**33.24**
width 150 mm	-	1.48	23.89	2.27	m	**26.17**
width 150 mm; in reinforced concrete	-	2.22	35.84	3.78	m	**39.62**
Cutting rebates						
Depth not exceeding 50 mm						
width 50 mm	-	0.46	7.43	0.82	m	**8.25**
Depth 50 - 100 mm						
width 100 mm	-	0.93	15.01	1.67	m	**16.69**

E IN SITU CONCRETE/LARGE PRECAST CONCRETE

Item	PC £	Labour hours	Labour £	Material £	Unit	Total rate £
NOTE: The following rates for cutting holes and mortices in concrete allow for diamond drilling.						
Diamond drilling						
Cutting holes and mortices in concrete; per 25 mm depth						
25 mm diameter	-	-	-	-	nr	**1.84**
32 mm diameter	-	-	-	-	nr	**1.43**
52 mm diameter	-	-	-	-	nr	**1.75**
78 mm diameter	-	-	-	-	nr	**2.03**
107 mm diameter	-	-	-	-	nr	**2.21**
127 mm diameter	-	-	-	-	nr	**2.44**
152 mm diameter	-	-	-	-	nr	**2.91**
200 mm diameter	-	-	-	-	nr	**3.87**
250 mm diameter	-	-	-	-	nr	**5.67**
300 mm diameter	-	-	-	-	nr	**7.52**
Cutting holes and mortices in reinforced concrete; per 25 mm depth						
25 mm diameter	-	-	-	-	nr	**2.40**
32 mm diameter	-	-	-	-	nr	**2.12**
52 mm diameter	-	-	-	-	nr	**2.03**
78 mm diameter	-	-	-	-	nr	**2.12**
107 mm diameter	-	-	-	-	nr	**2.49**
127 mm diameter	-	-	-	-	nr	**2.91**
152 mm diameter	-	-	-	-	nr	**3.41**
200 mm diameter	-	-	-	-	nr	**4.94**
250 mm diameter	-	-	-	-	nr	**7.47**
300 mm diameter	-	-	-	-	nr	**9.64**
Other items in reinforced concrete						
diamond chasing; per 25 x 25 mm section	-	-	-	-	m	**11.07**
forming box; per 25 mm depth (per m of perimeter)	-	-	-	-	m	**4.43**
diamond floor sawing; per 25 mm depth	-	-	-	-	m	**2.54**
diamond track mount or ring sawing; per 25 mm depth	-	-	-	-	m	**9.22**
stitch drilling 107 mm diameter hole; per25 mm depth	-	-	-	-	nr	**2.08**

E IN SITU CONCRETE/LARGE PRECAST CONCRETE

Item	PC £	Labour hours	Labour £	Material £	Unit	Total rate £
E42 ACCESSORIES CAST INTO IN SITU CONCRETE						
Foundation bolt boxes						
Temporary plywood; for group of 4 nr bolts						
75 x 75 x 150 mm	-	0.42	7.99	1.01	nr	**9.00**
75 x 75 x 250 mm	-	0.42	7.99	1.28	nr	**9.28**
Expanded metal; Expamet Building Products Ltd or other equal and approved						
75 mm diameter x 150 mm long	-	0.28	5.33	2.16	nr	**7.49**
75 mm diameter x 300 mm long	-	0.28	5.33	2.22	nr	**7.54**
100 mm diameter x 450 mm long	-	0.28	5.33	2.36	nr	**7.68**
Foundation bolts and nuts						
Black hexagon						
10 mm diameter x 100 mm long	-	0.23	4.38	0.70	nr	**5.08**
12 mm diameter x 120 mm long	-	0.23	4.38	1.07	nr	**5.45**
16 mm diameter x 160 mm long	-	0.28	5.33	2.97	nr	**8.30**
20 mm diameter x 180 mm long	-	0.28	5.33	3.47	nr	**8.80**
Masonry slots						
Galvanised steel; dovetail slots; 1.20 mm thick; 18G						
1000 mm long	-	0.25	4.76	1.99	m	**6.75**
100 mm long	-	0.07	1.33	0.24	nr	**1.57**
Galvanised steel; metal insert slots; Halfen Ltd or other equal and approved; 2.50 mm thick; end caps and foam filling						
41 x 41 mm; ref P3270	-	0.37	7.04	6.62	m	**13.66**
41 x 41 x 100 mm; ref P3250	-	0.09	1.71	0.90	nr	**2.61**
41 x 41 x 150 mm; ref P3251	-	0.09	1.71	1.18	nr	**2.90**
Cramps						
Mild steel; once bent; one end shot fired into concrete; other end fanged and built into brickwork joint						
200 mm girth	-	0.14	2.93	0.55	nr	**3.48**
Column guards						
White nylon coated steel; "Rigifix" or other equal and approved; Huntley and Sparks Ltd; plugging; screwing to concrete; 1.50 mm thick						
75 x 75 x 1000 mm	-	0.74	14.08	20.40	nr	**34.48**
Galvanised steel; "Rigifix" or other equal and approved; Huntley and Sparks Ltd; 3 mm thick						
75 x 75 x 1000 mm	-	0.56	10.66	14.50	nr	**25.16**
Galvanised steel; "Rigifix" or other equal and approved; Huntley and Sparks Ltd; 4.50 mm thick						
75 x 75 x 1000 mm	-	0.56	10.66	19.50	nr	**30.15**
Stainless steel; "HKW" or other equal and approved; Halfen Ltd; 5 mm thick						
50 x 50 x 1200 mm	-	0.93	17.70	27.81	nr	**45.50**
50 x 50 x 2000 mm	-	1.11	21.12	55.91	nr	**77.03**
Channels						
Stainless steel; Halfen Ltd or other equal and approved						
ref 38/17/HTA	-	0.32	6.09	24.79	m	**30.88**
ref 41/22/HZA; 80 mm long; including "T" headed bolts and plate washers	-	0.09	1.71	15.57	nr	**17.28**

E IN SITU CONCRETE/LARGE PRECAST CONCRETE

Item	PC £	Labour hours	Labour £	Material £	Unit	Total rate £
Channel ties						
Stainless steel; Halfen Ltd or other equal and approved						
ref HTS - B12; 150 mm projection; including insulation retainer	-	0.03	0.79	0.35	nr	**1.14**
ref HTS - B12; 200 mm projection; including insulation retainer	-	0.03	0.79	0.43	nr	**1.22**
E50 PRECAST CONCRETE LARGE UNITS						
Contractor designed precast concrete staircases and landings; including all associated steel supports and fixing in position						
Straight staicases; 280 mm treads; 170 mm undercut risers						
1200 mm wide; 2750 mm rise	-	-	-	-	nr	**1652.81**
1200 mm wide; 3750 mm rise	-	-	-	-	nr	**2203.75**
Dogleg staicases						
1200 mm wide; one full width half landing; 2750 mm rise	-	-	-	-	nr	**2534.31**
1200 mm wide; one full width half landing; 3750 mm rise	-	-	-	-	nr	**3305.63**
Extra over for 200 mm concrete landing support walls	-	-	-	-	nr	**771.31**
1800 mm wide; one full width half landing; 2750 mm rise	-	-	-	-	nr	**3581.09**
1800 mm wide; one full width half landing; 3750 mm rise	-	-	-	-	nr	**4682.97**
Extra over for 200 mm concrete landing support walls	-	-	-	-	nr	**1190.03**
E60 PRECAST/COMPOSITE CONCRETE DECKING						
Prestressed precast concrete structural suspended floors; Bison "Hollowcore" or other equal and approved; supplied and fixed on hard level bearings, to areas of 500 m² per site visit; top surface screeding and ceiling finishes by others						
Floors to dwellings, offices, car parks, shop retail floors, hospitals, school teaching rooms, staff rooms and the like; superimposed load of 5.00 kN/m²						
floor spans up to 3.00 m; 1200 mm x 150 mm	-	-	-	-	m²	**43.97**
floor spans 3.00 m - 6.00 m; 1200 mm x 150 mm	-	-	-	-	m²	**44.77**
floor spans 6.00 m - 7.50 m; 1200 mm x 200 mm	-	-	-	-	m²	**45.16**
floor spans 7.50 m - 9.50 m; 1200 mm x 250 mm	-	-	-	-	m²	**51.08**
floor spans 9.50 m - 12.00 m; 1200 mm x 300 mm	-	-	-	-	m²	**51.96**
floor spans 12.00 m - 12.50 m; 1200 mm x 350 mm	-	-	-	-	m²	**54.46**
floor spans 12.50 m - 14.00 m; 1200 mm x 400 mm	-	-	-	-	m²	**59.64**
floor spans 14.00 m - 15.00 m; 1200 mm x 450 mm	-	-	-	-	m²	**60.58**
Floors to shop stockrooms, light warehousing, schools, churches or similar places of assembly, light factory accommodation, laboratories and the like; superimposed load of 8.50 kN/m²						
floor spans up to 3.00 m; 1200 mm x 150 mm	-	-	-	-	m²	**44.23**
floor spans 3.00 m - 6.00 m; 1200 mm x 200 mm	-	-	-	-	m²	**44.96**
floor spans 6.00 m - 7.50 m; 1200 mm x 250 mm	-	-	-	-	m²	**51.60**
Floors to heavy warehousing, factories, stores and the like; superimposed load of 12.50 kN/m²						
floor spans up to 3.00 m; 1200 mm x 150 mm	-	-	-	-	m²	**44.51**
floor spans 3.00 m - 6.00 m; 1200 mm x 250 mm	-	-	-	-	m²	**51.35**

E IN SITU CONCRETE/LARGE PRECAST CONCRETE

Item	PC £	Labour hours	Labour £	Material £	Unit	Total rate £
E60 PRECAST/COMPOSITE CONCRETE DECKING – cont'd						
Prestressed precast concrete structural suspended floors; Bison "Hollowcore – cont'd						
Prestressed precast concrete staircase, supplied and fixed in conjunction with Bison "Hollowcore" flooring system or similar; comprising 2 nr 1100 mm wide flights with 7 nr 275 mm treads, 8 nr 185 mm risers and 150 mm waist; 1 nr 2200 mm x 1400 mm x 150 mm half landing and 1 nr top landing						
3.00 m storey height	-	-	-	-	nr	**1998.13**
Composite floor comprising reinforced in situ ready-mixed concrete 30.00 N/mm²; on and including 1.20 mm thick "Holorib" steel deck permanent shutting; complete with reinforcement to support imposed loading and A142 anti-crack mesh						
150 mm thick suspended slab; 5.00 kN/m² loading						
1.50 m - 3.00 m high to soffit	-	1.43	25.31	43.82	m²	**69.12**
3.00 m - 4.50 m high to soffit	-	1.43	25.31	45.22	m²	**70.53**
4.50 m - 6.00 m high to soffit	-	1.67	29.87	45.76	m²	**75.63**
200 mm thick suspended slab; 7.50 kN/m² loading						
1.50 m - 3.00 m high to soffit	-	1.47	25.95	48.40	m²	**74.35**
3.00 m - 4.50 m high to soffit	-	1.47	25.95	49.80	m²	**75.75**
4.50 m - 6.00 m high to soffit	-	1.70	30.33	50.34	m²	**80.66**

F MASONRY

Item	PC £	Labour hours	Labour £	Material £	Unit	Total rate £
BASIC MORTAR PRICES						
Coloured mortar materials (£/tonne); (excluding cement)						
light	-	-	-	58.37	tonne	-
medium	-	-	-	60.72	tonne	-
dark	-	-	-	73.45	tonne	-
extra dark	-	-	-	73.45	tonne	-
Mortar materials (£/tonne)						
cement	-	-	-	113.76	tonne	-
sand	-	-	-	20.26	tonne	-
lime	-	-	-	180.83	tonne	-
white cement	-	-	-	210.96	tonne	-
Mortar materials (£/5 litres)						
"Cemplas Super" mortar plasticiser	-	-	-	6.21	5 litre	-
SUPPLY AND FIX PRICES						
F10 BRICK/BLOCK WALLING						
Common bricks; PC £240.00 per 1000; in gauged mortar (1:1:6)						
Walls						
half brick thick	-	0.93	24.34	18.60	m²	**42.94**
half brick thick; building against other work; concrete	-	1.02	26.70	19.89	m²	**46.59**
half brick thick; building overhand	-	1.16	30.36	18.60	m²	**48.96**
half brick thick; curved; 6.00 m radii	-	1.20	31.41	18.60	m²	**50.01**
half brick thick; curved; 1.50 m radii	-	1.57	41.09	21.38	m²	**62.47**
one brick thick	-	1.57	41.09	37.20	m²	**78.30**
one brick thick; curved; 6.00 m radii	-	2.04	53.40	39.98	m²	**93.37**
one brick thick; curved; 1.50 m radii	-	2.54	66.48	40.63	m²	**107.11**
one and a half brick thick	-	2.13	55.75	55.80	m²	**111.55**
one and a half brick thick; battering	-	2.45	64.13	55.80	m²	**119.93**
two brick thick	-	2.59	67.79	74.40	m²	**142.19**
two brick thick; battering	-	3.05	79.83	74.40	m²	**154.24**
337 average thick; tapering, one side	-	2.68	70.15	55.80	m²	**125.95**
450 average thick; tapering, one side	-	3.47	90.83	74.40	m²	**165.23**
337 average thick; tapering, both sides	-	3.10	81.14	55.80	m²	**136.94**
450 average thick; tapering, both sides	-	3.89	101.82	75.05	m²	**176.07**
facework one side, half brick thick	-	1.02	26.70	18.60	m²	**45.30**
facework one side, one brick thick	-	1.67	43.71	37.20	m²	**80.91**
facework one side, one and a half brick thick	-	2.22	58.11	55.80	m²	**113.91**
facework one side, two brick thick	-	2.68	70.15	74.40	m²	**144.55**
facework both sides, half brick thick	-	1.11	29.05	18.60	m²	**47.65**
facework both sides, one brick thick	-	1.76	46.07	37.20	m²	**83.27**
facework both sides, one and a half brick thick	-	2.31	60.46	55.80	m²	**116.27**
facework both sides, two brick thick	-	2.78	72.76	74.40	m²	**147.17**
Isolated piers						
one brick thick	-	2.36	61.77	37.20	m²	**98.97**
two brick thick	-	3.70	96.85	75.05	m²	**171.90**
three brick thick	-	4.67	122.23	112.90	m²	**235.13**
Isolated casings						
half brick thick	-	1.20	31.41	18.60	m²	**50.01**
one brick thick	-	2.04	53.40	37.20	m²	**90.60**
Chimney stacks						
one brick thick	-	2.36	61.77	37.20	m²	**98.97**
two brick thick	-	3.70	96.85	75.05	m²	**171.90**
three brick thick	-	4.67	122.23	112.90	m²	**235.13**

F MASONRY

Item	PC £	Labour hours	Labour £	Material £	Unit	Total rate £
F10 BRICK/BLOCK WALLING – cont'd						
Common bricks; PC £240.00 per 1000; in gauged mortar (1:1:6) – cont'd						
Projections						
225 mm width; 112 mm depth; vertical	-	0.28	7.33	3.93	m	**11.26**
225 mm width; 225 mm depth; vertical	-	0.56	14.66	7.86	m	**22.52**
337 mm width; 225 mm depth; vertical	-	0.83	21.72	11.79	m	**33.51**
440 mm width; 225 mm depth; vertical	-	0.93	24.34	15.71	m	**40.06**
Closing cavities						
width of cavity 50 mm, closing with common brickwork half brick thick; vertical	-	0.28	7.33	0.94	m	**8.27**
width of cavity 50 mm, closing with common brickwork half brick thick; horizontal	-	0.28	7.33	2.91	m	**10.24**
width of cavity 50 mm, closing with common brickwork half brick thick; including damp proof course; vertical	-	0.37	9.68	1.83	m	**11.51**
width of cavity 50 mm, closing with common brickwork half brick thick; including damp proof course; horizontal	-	0.32	8.38	3.55	m	**11.93**
width of cavity 75 mm, closing with common brickwork half brick thick; vertical	-	0.28	7.33	1.39	m	**8.71**
width of cavity 75 mm, closing with common brickwork half brick thick; horizontal	-	0.28	7.33	4.30	m	**11.63**
width of cavity 75 mm, closing with common brickwork half brick thick; including damp proof course; vertical	-	0.37	9.68	2.27	m	**11.95**
width of cavity 75 mm, closing with common brickwork half brick thick; including damp proof course; horizontal	-	0.32	8.38	4.94	m	**13.32**
Bonding to existing						
half brick thick	-	0.28	7.33	1.02	m	**8.35**
one brick thick	-	0.42	10.99	2.05	m	**13.04**
one and a half brick thick	-	0.65	17.01	3.07	m	**20.08**
two brick thick	-	0.88	23.03	4.09	m	**27.12**
Arches						
height on face 102 mm, width of exposed soffit 102 mm, shape of arch - segmental, one ring	-	1.57	31.36	6.83	m	**38.19**
height on face 102 mm, width of exposed soffit 215 mm, shape of arch - segmental, one ring	-	2.04	43.66	8.99	m	**52.66**
height on face 102 mm, width of exposed soffit 102 mm, shape of arch - semi-circular, one ring	-	1.99	42.35	6.83	m	**49.18**
height on face 102 mm, width of exposed soffit 215 mm, shape of arch - semi-circular, one ring	-	2.50	55.70	8.99	m	**64.70**
height on face 215 mm, width of exposed soffit 102 mm, shape of arch - segmental, two ring	-	1.99	42.35	8.81	m	**51.16**
height on face 215 mm, width of exposed soffit 215 mm, shape of arch - segmental, two ring	-	2.45	54.39	12.96	m	**67.35**
height on face 215 mm, width of exposed soffit 102 mm, shape of arch - semi-circular, two ring	-	2.68	60.41	8.81	m	**69.22**
height on face 215 mm, width of exposed soffit 215 mm, shape of arch - semi-circular, two ring	-	3.05	70.10	12.96	m	**83.05**
ADD or DEDUCT to walls for variation of £10.00/1000 in PC of common bricks						
half brick thick	-	-	-	0.70	m²	-
one brick thick	-	-	-	1.39	m²	-
one and a half brick thick	-	-	-	2.09	m²	-
two brick thick	-	-	-	2.79	m²	-

F MASONRY

Item	PC £	Labour hours	Labour £	Material £	Unit	Total rate £
Class B engineering bricks; PC £288.00 per 1000; in cement mortar (1:3)						
Walls						
half brick thick	-	1.02	26.70	22.16	m²	**48.86**
one brick thick	-	1.67	43.71	44.32	m²	**88.03**
one brick thick; building against other work	-	1.99	52.09	46.49	m²	**98.58**
one brick thick; curved; 6.00 m radii	-	2.22	58.11	44.32	m²	**102.43**
one and a half brick thick	-	2.22	58.11	66.48	m²	**124.59**
one and a half brick thick; building against other work	-	2.68	70.15	66.48	m²	**136.63**
two brick thick	-	2.78	72.76	88.64	m²	**161.41**
337 mm thick; tapering, one side	-	2.87	75.12	66.48	m²	**141.60**
450 mm thick; tapering, one side	-	3.70	96.85	88.64	m²	**185.49**
337 mm thick; tapering, both sides	-	3.33	87.16	66.48	m²	**153.64**
450 mm thick; tapering, both sides	-	4.21	110.19	89.37	m²	**199.56**
facework one side, half brick thick	-	1.11	29.05	22.16	m²	**51.21**
facework one side, one brick thick	-	1.76	46.07	44.32	m²	**90.39**
facework one side, one and a half brick thick	-	2.31	60.46	66.48	m²	**126.95**
facework one side, two brick thick	-	2.87	75.12	88.64	m²	**163.76**
facework both sides, half brick thick	-	1.20	31.41	22.16	m²	**53.57**
facework both sides, one brick thick	-	1.85	48.42	44.32	m²	**92.74**
facework both sides, one and a half brick thick	-	2.41	63.08	66.48	m²	**129.56**
facework both sides, two brick thick	-	2.96	77.48	88.64	m²	**166.12**
Isolated piers						
one brick thick	-	2.59	67.79	44.32	m²	**112.11**
two brick thick	-	4.07	106.53	89.37	m²	**195.90**
three brick thick	-	5.00	130.87	134.41	m²	**265.28**
Isolated casings						
half brick thick	-	1.30	34.03	22.16	m²	**56.19**
one brick thick	-	2.22	58.11	44.32	m²	**102.43**
Projections						
225 mm width; 112 mm depth; vertical	-	0.32	8.38	4.70	m	**13.07**
225 mm width; 225 mm depth; vertical	-	0.60	15.70	9.39	m	**25.10**
337 mm width; 225 mm depth; vertical	-	0.88	23.03	14.09	m	**37.12**
440 mm width; 225 mm depth; vertical	-	1.02	26.70	18.78	m	**45.48**
Bonding to existing						
half brick thick	-	0.32	8.38	1.22	m	**9.59**
one brick thick	-	0.46	12.04	2.44	m	**14.48**
one and a half brick thick	-	0.65	17.01	3.66	m	**20.67**
two brick thick	-	0.97	25.39	4.88	m	**30.27**
ADD or DEDUCT to walls for variation of £10.00/1000 in PC of bricks						
half brick thick	-	-	-	0.70	m²	-
one brick thick	-	-	-	1.39	m²	-
one and a half brick thick	-	-	-	2.09	m²	-
two brick thick	-	-	-	2.79	m²	-

F MASONRY

Item	PC £	Labour hours	Labour £	Material £	Unit	Total rate £
F10 BRICK/BLOCK WALLING – cont'd						
ALTERNATIVE FACING BRICK PRICES (PC £ per 1000)						
Discounts of 10 - 20% available depending on quantity / status.						
Ibstock facing bricks; 215 x 102.5 x 65 mm						
Aldridge Brown Blend	-	-	-	390.00	1000	-
Aldridge Leicester Anglican Red Rustic	-	-	-	319.00	1000	-
Ashdown Cottage Mixture	-	-	-	322.00	1000	-
Ashdown Crowborough Multi	-	-	-	417.00	1000	-
Ashdown Pevensey Multi	-	-	-	390.00	1000	-
Cattybrook Bristol Gold	-	-	-	326.00	1000	-
Chailey Stock	-	-	-	390.00	1000	-
Dorking Multi	-	-	-	316.00	1000	-
Funton Second Hard Stock	-	-	-	430.00	1000	-
Holbook Smooth Red	-	-	-	340.00	1000	-
Leicester Red Stock	-	-	-	343.00	1000	-
Roughdales Red Multi Rustic	-	-	-	317.00	1000	-
Roughdales Trafford Multi Rustic	-	-	-	350.00	1000	-
Stourbridge Himley Mixed Russet	-	-	-	476.00	1000	-
Stourbridge Kenilworth Multi	-	-	-	304.00	1000	-
Stourbridge Pennine Pastone	-	-	-	371.00	1000	-
Strattford Red Rustic	-	-	-	298.00	1000	-
Swanage Handmade Restoration	-	-	-	686.00	1000	-
Tonbridge Handmade Multi	-	-	-	668.00	1000	-
Hanson Brick Limited, London brand; 215 x 102.5 x 65 mm						
Autumn leaf	-	-	-	413.00	1000	-
Brecken Grey	-	-	-	319.00	1000	-
Burghley Red Rustic	-	-	-	319.00	1000	-
Chiltern	-	-	-	378.00	1000	-
Claydon Red Multi	-	-	-	330.00	1000	-
Dawn Red	-	-	-	326.00	1000	-
Georgian	-	-	-	326.00	1000	-
Hathaway Brindled	-	-	-	380.00	1000	-
Hereward Light	-	-	-	339.00	1000	-
Heather	-	-	-	378.00	1000	-
Honey Buff	-	-	-	313.00	1000	-
Ironstone	-	-	-	326.00	1000	-
Milton Buff`	-	-	-	346.00	1000	-
Mixed Brown Brindle Rustic	-	-	-	414.00	1000	-
Old English Brindled Red	-	-	-	316.00	1000	-
Orient Gold	-	-	-	426.00	1000	-
Regency	-	-	-	355.00	1000	-
Sandfaced	-	-	-	364.00	1000	-
Saxon Gold	-	-	-	363.00	1000	-
Selected Regrades	-	-	-	207.00	1000	-
Sunset Red	-	-	-	335.00	1000	-
Tudor	-	-	-	373.00	1000	-
Windsor	-	-	-	329.00	1000	-
Sherbourne Red Pavers	-	-	-	19.83	m²	-
Coxmoor Rose Multi Pavers	-	-	-	21.61	m²	-

F MASONRY

Item	PC £	Labour hours	Labour £	Material £	Unit	Total rate £
SUPPLY AND FIX PRICES						
Facing bricks; machine made facings; PC £350.00 per 1000 (unless otherwise stated); in gauged mortar (1:1:6)						
Walls						
facework one side, half brick thick; stretcher bond	-	1.20	31.41	26.19	m²	**57.60**
facework one side, half brick thick, flemish bond with snapped headers	-	1.39	36.38	26.24	m²	**62.62**
facework one side, half brick thick, stretcher bond; building against other work; concrete	-	1.30	34.03	27.53	m²	**61.56**
facework one side, half brick thick; flemish bond with snapped headers; building against other work; concrete	-	1.48	38.74	27.53	m²	**66.27**
facework one side, half brick thick, stretcher bond; building overhand	-	1.48	38.74	26.24	m²	**64.98**
facework one side, half brick thick; flemish bond with snapped headers; building overhand	-	1.67	43.71	26.24	m²	**69.95**
facework one side, half brick thick; stretcher bond; curved; 6.00 m radii	-	1.76	46.07	26.24	m²	**72.30**
facework one side, half brick thick; flemish bond with snapped headers; curved; 6.00 m radii	-	1.99	52.09	26.24	m²	**78.32**
facework one side, half brick thick; stretcher bond; curved; 1.50 m radii	-	2.22	58.11	30.29	m²	**88.39**
facework one side, half brick thick; flemish bond with snapped headers; curved; 1.50 m radii	-	2.59	67.79	30.29	m²	**98.08**
facework both sides, one brick thick; two stretcher skins tied together	-	2.08	54.44	54.42	m²	**108.86**
facework both sides, one brick thick; flemish bond	-	2.13	55.75	52.48	m²	**108.23**
facework both sides, one brick thick; two stretcher skins tied together; curved; 6.00 m radii	-	2.87	75.12	58.47	m²	**133.59**
facework both sides, one brick thick; flemish bond; curved; 6.00 m radii	-	2.96	77.48	56.52	m²	**134.00**
facework both sides, one brick thick; two stretcher skins tied together; curved; 1.50 m radii	-	3.56	93.18	63.16	m²	**156.34**
facework both sides, one brick thick; flemish bond; curved; 1.50 m radii	-	3.70	96.85	61.22	m²	**158.07**
Isolated piers						
facework both sides, one brick thick; two stretcher skins tied together	-	2.45	64.13	55.52	m²	**119.65**
facework both sides, one brick thick; flemish bond	-	2.50	65.44	55.52	m²	**120.96**
Isolated casings						
facework one side, half brick thick; stretcher bond	-	1.85	48.42	26.24	m²	**74.66**
facework one side, half brick thick; flemish bond with snapped headers	-	2.04	53.40	26.24	m²	**79.63**
Projections						
225 mm width; 112 mm depth; stretcher bond; vertical	-	0.28	7.33	5.63	m	**12.95**
225 mm width; 112 mm depth; flemish bond with snapped headers; vertical	-	0.37	9.68	5.63	m	**15.31**
225 mm width; 225 mm depth; flemish bond; vertical	-	0.60	15.70	11.25	m	**26.96**
328 mm width; 112 mm depth; stretcher bond; vertical	-	0.56	14.66	8.44	m	**23.10**
328 mm width; 112 mm depth; flemish bond with snapped headers; vertical	-	0.65	17.01	8.44	m	**25.46**
328 mm width; 225 mm depth; flemish bond; vertical	-	1.11	29.05	16.84	m	**45.90**
440 mm width; 112 mm depth; stretcher bond; vertical	-	0.83	21.72	11.25	m	**32.98**

F MASONRY

Item	PC £	Labour hours	Labour £	Material £	Unit	Total rate £
F10 BRICK/BLOCK WALLING – cont'd						
Facing bricks; machine made facings; PC £350.00 per 1000 – cont'd						
Projections – cont'd						
440 mm width; 112 mm depth; flemish bond with snapped headers; vertical	-	0.88	23.03	11.25	m	**34.28**
440 mm width; 225 mm depth; flemish bond; vertical	-	1.62	42.40	22.50	m	**64.91**
Arches						
height on face 215 mm, width of exposed soffit 102 mm, shape of arch - flat	-	0.93	20.02	7.70	m	**27.72**
height on face 215 mm, width of exposed soffit 215 mm, shape of arch - flat	-	1.39	32.06	13.53	m	**45.58**
height on face 215 mm, width of exposed soffit 102 mm, shape of arch - segmental, one ring	-	1.76	35.25	10.51	m	**45.76**
height on face 215 mm, width of exposed soffit 215 mm, shape of arch segmental, one ring	-	2.13	44.94	16.09	m	**61.03**
height on face 215 mm, width of exposed soffit 102 mm, shape of arch - semi-circular, one ring	-	2.68	59.33	10.51	m	**69.84**
height on face 215 mm, width of exposed soffit 215 mm, shape of arch - semi-circular, one ring	-	3.61	83.67	16.09	m	**99.76**
height on face 215 mm, width of exposed soffit 102 mm, shape of arch - segmental, two ring	-	2.17	45.98	10.51	m	**56.49**
height on face 215 mm, width of exposed soffit 215 mm, shape of arch - segmental; two ring	-	2.82	63.00	16.09	m	**79.09**
height on face 215 mm, width of exposed soffit 102 mm, shape of arch - semi-circular, two ring	-	3.61	83.67	10.51	m	**94.18**
height on face 215 mm, width of exposed soffit 215 mm, shape of arch - semi-circular, two ring	-	5.00	120.06	16.09	m	**136.15**
Arches; cut voussoirs (PC £ per 1000)	3500.00					
height on face 215 mm, width of exposed soffit 102 mm, shape of arch - segmental, one ring	-	1.80	36.30	59.10	m	**95.40**
height on face 215 mm, width of exposed soffit 215 mm, shape of arch - segmental, one ring	-	2.27	48.60	113.28	m	**161.88**
height on face 215 mm, width of exposed soffit 102 mm, shape of arch - semi-circular, one ring	-	2.04	42.58	59.10	m	**101.68**
height on face 215 mm, width of exposed soffit 215 mm, shape of arch - semi-circular, one ring	-	2.59	56.98	113.28	m	**170.25**
height on face 320 mm, width of exposed soffit 102 mm, shape of arch - segmental, one and a half ring	-	2.41	52.26	113.22	m	**165.48**
height on face 320 mm, width of exposed soffit 215 mm, shape of arch - segmental, one and a half ring	-	3.15	71.63	228.76	m	**300.40**
Arches; bullnosed specials (PC £ per 1000)	2000.00					
height on face 215 mm, width of exposed soffit 102 mm, shape of arch - flat	-	0.97	21.06	33.15	m	**54.22**
height on face 215 mm, width of exposed soffit 215 mm, shape of arch - flat	-	1.43	33.10	65.07	m	**98.17**
Bullseye windows; 600 mm diameter						
height on face 215 mm, width of exposed soffit 102 mm, two rings	-	4.63	110.37	13.78	nr	**124.15**
height on face 215 mm, width of exposed soffit 215 mm, two rings	-	6.48	158.79	25.89	nr	**184.68**
Bullseye windows; 600 mm; cut voussoirs (PC £ per 1000)	3500.00					
height on face 215 mm, width of exposed soffit 102 mm, one ring	-	3.89	91.00	144.17	nr	**235.17**
height on face 215 mm, width of exposed soffit 215 mm, one ring	-	5.37	129.74	286.67	nr	**416.41**

F MASONRY

Item	PC £	Labour hours	Labour £	Material £	Unit	Total rate £
Bullseye windows; 1200 mm diameter						
height on face 215 mm, width of exposed soffit 102 mm, two rings	-	7.22	178.16	30.16	nr	**208.32**
height on face 215 mm, width of exposed soffit 215 mm, two rings	-	10.36	260.35	54.96	nr	**315.31**
Bullseye windows; 1200 mm diameter; cut voussoirs (PC £ per 1000)	3500.00					
height on face 215 mm, width of exposed soffit 102 mm, one ring	-	6.11	149.11	251.09	nr	**400.20**
height on face 215 mm, width of exposed soffit 215 mm, one ring	-	8.70	216.90	495.54	nr	**712.44**
ADD or DEDUCT for variation of £10.00 per 1000 in PC of facing bricks in 102 mm high arches with 215 mm soffit	-	-	-	0.31	m	**-**
Facework sills						
150 mm x 102 mm; headers on edge; pointing top and one side; set weathering; horizontal	-	0.51	13.35	5.63	m	**18.97**
150 mm x 102 mm; cant headers on edge; pointing top and one side; set weathering; horizontal (PC £ per 1000)	2000.00	0.56	14.66	31.08	m	**45.74**
150 mm x 102 mm; bullnosed specials; headers on flat; pointing top and one side; horizontal (PC £ per 1000)	2000.00	0.46	12.04	31.08	m	**43.12**
Facework copings						
215 mm x 102 mm; headers on edge; pointing top and both sides; horizontal	-	0.42	10.99	5.72	m	**16.72**
260 mm x 102 mm; headers on edge; pointing top and both sides; horizontal	-	0.65	17.01	8.51	m	**25.52**
215 mm x 102 mm; double bullnose specials; headers on edge; pointing top and both sides; horizontal (PC £ per 1000)	2000.00	0.46	12.04	31.18	m	**43.22**
260 mm x 102 mm; single bullnose specials; headers on edge; pointing top and both sides; horizontal (PC £ per 1000)	2000.00	0.65	17.01	62.11	m	**79.13**
ADD or DEDUCT for variation of £10.00 per 1000 in PC of facing bricks in copings 215 mm wide, 102 mm high	-	-	-	0.15	m	**-**
Extra over facing bricks for; facework ornamental bands and the like, plain bands						
flush; horizontal; 225 mm width; entirely of stretchers (PC £ per 1000)	400.00	0.19	4.97	0.62	m	**5.59**
Extra over facing brick for; facework quoins						
flush; mean girth 320 mm (PC £ per 1000)	400.00	0.28	7.33	0.62	m	**7.95**
Bonding to existing						
facework one side, half brick thick; stretcher bond	-	0.46	12.04	1.45	m	**13.49**
facework one side, half brick thick; flemish bond with snapped headers	-	0.46	12.04	1.45	m	**13.49**
facework both sides, one brick thick; two stretcher skins tied together	-	0.65	17.01	2.89	m	**19.91**
facework both sides, one brick thick; flemish bond	-	0.65	17.01	2.89	m	**19.91**
ADD or DEDUCT for variation of £10.00 per 1000 in PC of facing bricks; in walls built entirely of facings; in stretcher or flemish bond						
half brick thick	-	-	-	0.70	m²	**-**
one brick thick	-	-	-	1.39	m²	**-**

F MASONRY

Item	PC £	Labour hours	Labour £	Material £	Unit	Total rate £
F10 BRICK/BLOCK WALLING – cont'd						
Facing bricks; hand made; PC £500.00 per 1000 (unless otherwise stated); in gauged mortar (1:1:6)						
Walls						
facework one side, half brick thick; stretcher bond	-	1.20	31.41	36.65	m²	**68.06**
facework one side, half brick thick; flemish bond with snapped headers	-	1.39	36.38	36.65	m²	**73.03**
facework one side; half brick thick; stretcher bond; building against other work; concrete	-	1.30	34.03	37.94	m²	**71.97**
facework one side, half brick thick; flemish bond with snapped headers; building against other work; concrete	-	1.48	38.74	37.94	m²	**76.68**
facework one side, half brick thick; stretcher bond; building overhand	-	1.48	38.74	36.65	m²	**75.39**
facework one side, half brick thick; flemish bond with snapped headers; building overhand	-	1.67	43.71	36.65	m²	**80.36**
facework one side, half brick thick; stretcher bond; curved; 6.00 m radii	-	1.76	46.07	36.65	m²	**82.71**
facework one side, half brick thick; flemish bond with snapped headers; curved; 6.00 m radii	-	1.99	52.09	40.99	m²	**93.07**
facework one side, half brick thick; stretcher bond; curved 1.50 m radii	-	2.22	58.11	36.65	m²	**94.76**
facework one side, half brick thick; flemish bond with snapped headers; curved; 1.50 m radii	-	2.59	67.79	43.88	m²	**111.67**
facework both sides, one brick thick; two stretcher skins tied together	-	2.08	54.44	75.24	m²	**129.68**
facework both sides, one brick thick; flemish bond	-	2.13	55.75	73.30	m²	**129.05**
facework both sides; one brick thick; two stretcher skins tied together; curved; 6.00 m radii	-	2.87	75.12	81.02	m²	**156.14**
facework both sides, one brick thick; flemish bond; curved; 6.00 m radii	-	2.96	77.48	79.08	m²	**156.56**
facework both sides, one brick thick; two stretcher skins tied together; curved; 1.50 m radii	-	3.56	93.18	87.45	m²	**180.63**
facework both sides, one brick thick; flemish bond; curved; 1.50 m radii	-	3.70	96.85	85.51	m²	**182.36**
Isolated piers						
facework both sides, one brick thick; two stretcher skins tied together	-	2.45	64.13	76.35	m²	**140.47**
facework both sides, one brick thick; flemish bond	-	2.50	65.44	76.35	m²	**141.78**
Isolated casings						
facework one side, half brick thick; stretcher bond	-	1.85	48.42	36.65	m²	**85.07**
facework one side, half brick thick; flemish bond with snapped headers	-	2.04	53.40	36.65	m²	**90.04**
Projections						
225 mm width; 112 mm depth; stretcher bond; vertical	-	0.28	7.33	7.94	m	**15.27**
225 mm width; 112 mm depth; flemish bond with snapped headers; vertical	-	0.37	9.68	7.94	m	**17.62**
225 mm width; 225 mm depth; flemish bond; vertical	-	0.60	15.70	15.88	m	**31.58**
328 mm width; 112 mm depth; stretcher bond; vertical	-	0.56	14.66	11.92	m	**26.58**
328 mm width; 112 mm depth; flemish bond with snapped headers; vertical	-	0.65	17.01	11.92	m	**28.93**
328 mm width; 225 mm depth; flemish bond; vertical	-	1.11	29.05	23.79	m	**52.84**
440 mm width; 112 mm depth; stretcher bond; vertical	-	0.83	21.72	15.88	m	**37.60**
440 mm width; 112 mm depth; flemish bond with snapped headers; vertical	-	0.88	23.03	15.88	m	**38.91**
440 mm width; 225 mm depth; flemish bond; vertical	-	1.62	42.40	31.76	m	**74.16**

F MASONRY

Item	PC £	Labour hours	Labour £	Material £	Unit	Total rate £
Arches						
height on face 215 mm, width of exposed soffit 102 mm, shape of arch - flat	-	0.93	20.02	10.01	m	**30.03**
height on face 215 mm, width of exposed soffit 215 mm, shape of arch - flat	-	1.39	32.06	18.21	m	**50.27**
height on face 215 mm, width of exposed soffit 102 mm, shape of arch - segmental, one ring	-	1.76	35.25	12.82	m	**48.07**
height on face 215 mm, width of exposed soffit 215 mm, shape of arch - segmental, one ring	-	2.13	44.94	20.72	m	**65.65**
height on face 215 mm, width of exposed soffit 102 mm, shape of arch - semi-circular, one ring	-	2.68	59.33	12.82	m	**72.15**
height on face 215 mm, width of exposed soffit 215 mm, shape of arch - semi-circular, one ring	-	3.61	83.67	20.72	m	**104.39**
height on face 215 mm, width of exposed soffit 102 mm, shape of arch - segmental, two ring	-	2.17	45.98	12.82	m	**58.80**
height on face 215 mm, width of exposed soffit 215 mm, shape of arch - segmental, two ring	-	2.82	63.00	20.72	m	**83.71**
height on face 215 mm, width of exposed soffit 102 mm, shape of arch - semi-circular, two ring	-	3.61	83.67	12.82	m	**96.49**
height on face 215 mm, width of exposed soffit 215 mm, shape of arch - semi-circular, two ring	-	5.00	120.06	20.72	m	**140.77**
Arches; cut voussoirs (PC £ per 1000)	3700.00					
height on face 215 mm, width of exposed soffit 102 mm, shape of arch - segmental, one ring	-	1.80	36.30	62.18	m	**98.48**
height on face 215 mm, width of exposed soffit 215 mm, shape of arch - segmental, one ring	-	2.27	48.60	119.45	m	**168.05**
height on face 215 mm, width of exposed soffit 102 mm, shape of arch - semi-circular, one ring	-	2.04	42.58	62.18	m	**104.76**
height one face 215 mm, width of exposed soffit 215 mm, shape of - arch semi-circular, one ring	-	2.59	56.98	119.45	m	**176.42**
height on face 320 mm, width of exposed soffit 102 mm, shape of arch - segmental, one and a half ring	-	2.41	52.26	119.39	m	**171.65**
height on face 320 mm, width of exposed soffit 215 mm, shape of arch - segmental, one and a half ring	-	3.15	71.63	241.10	m	**312.74**
Arches; bullnosed specials (PC £ per 1000)	2100.00					
height on face 215 mm, width of exposed soffit 102 mm, shape of arch - flat	-	0.97	21.06	34.70	m	**55.76**
height on face 215 mm, width of exposed soffit 215 mm, shape of arch - flat	-	1.43	33.10	68.19	m	**101.29**
Bullseye windows; 600 mm diameter						
height on face 215 mm, width of exposed soffit 102 mm, two ring	-	4.63	110.37	18.64	nr	**129.01**
height on face 215 mm, width of exposed soffit 215 mm, two ring	-	6.48	158.79	47.25	nr	**206.04**
Bullseye windows; 600 mm diameter; cut voussoirs (PC £ per 1000)	3700.00					
height on face 215 mm, width of exposed soffit 102 mm, one ring	-	3.89	91.00	152.26	nr	**243.27**
height on face 215 mm, width of exposed soffit 215 mm, one ring	-	5.37	129.74	302.86	nr	**432.60**
Bullseye windows; 1200 mm diameter						
height on face 215 mm, width of exposed soffit 102 mm, two ring	-	7.22	178.16	39.88	nr	**218.04**
height on face 215 mm, width of exposed soffit 215 mm, two ring	-	10.36	260.35	74.40	nr	**334.75**

F MASONRY

Item	PC £	Labour hours	Labour £	Material £	Unit	Total rate £
F10 BRICK/BLOCK WALLING – cont'd						
Facing bricks; hand made PC £500.00 per 1000 – cont'd						
Bullseye windows; 1200 mm diameter; cut voussoirs (PC £ per 1000)	3700.00					
height on face 215 mm, width of exposed soffit 102 mm, one ring	-	6.11	149.11	264.97	nr	**414.08**
height on face 215 mm, width of exposed soffit 215 mm, one ring	-	8.70	216.90	523.30	nr	**740.20**
ADD or DEDUCT for variation of £10.00 per 1000 in PC of facing bricks in 102 mm high arches with 215 mm soffit	-	-	-	0.31	m	-
Facework sills						
150 mm x 102 mm; headers on edge; pointing top and one side; set weathering; horizontal	-	0.51	13.35	7.94	m	**21.29**
150 mm x 102 mm; cant headers on edge; pointing top and one side; set weathering; horizontal (PC £ per 1000)	2100.00	0.56	14.66	32.62	m	**47.28**
150 mm x 102 mm; bullnosed specials; headers on edge; pointing top and one side; horizontal (PC £ per 1000)	2100.00	0.46	12.04	32.62	m	**44.66**
Facework copings						
215 mm x 102 mm; headers on edge; pointing top and both sides; horizontal	-	0.42	10.99	8.04	m	**19.03**
260 mm x 102 mm; headers on edge; pointing top and both sides; horizontal	-	0.65	17.01	11.98	m	**28.99**
215 mm x 102 mm; double bullnose specials; headers on edge; pointing top and both sides; horizontal (PC £ per 1000)	2100.00	0.46	12.04	32.72	m	**44.76**
260 mm x 102 mm; single bullnose specials; headers on edge; pointing top and both sides; horizontal (PC £ per 1000)	2100.00	0.65	17.01	65.20	m	**82.21**
ADD or DEDUCT for variation of £10.00 per 1000 in PC of facing bricks in copings 215 mm wide, 102 mm high	-	-	-	0.15	m	-
Extra over facing bricks for; facework ornamental bands and the like, plain bands						
flush; horizontal; 225 mm width; entirely of stretchers (PC £ per 1000)	540.00	0.19	4.97	0.62	m	**5.59**
Extra over facing bricks for; facework quoins						
flush mean girth 320 mm (PC £ per 1000)	540.00	0.28	7.33	0.62	m	**7.95**
Bonding ends to existing						
facework one side, half brick thick; stretcher bond	-	0.46	12.04	2.03	m	**14.07**
facework one side, half brick thick; flemish bond with snapped headers	-	0.46	12.04	2.03	m	**14.07**
facework both sides, one brick thick; two stretcher skins tied together	-	0.65	17.01	4.05	m	**21.06**
facework both sides, one brick thick; flemish bond	-	0.65	17.01	4.05	m	**21.06**
ADD or DEDUCT for variation of £10.00/1000 in PC of facing bricks; in walls built entirely of facings; in stretcher or flemish bond						
half brick thick	-	-	-	0.70	m²	-
one brick thick	-	-	-	1.39	m²	-

F MASONRY

Item	PC £	Labour hours	Labour £	Material £	Unit	Total rate £
Facing bricks; slips 50 mm thick; in gauged mortar (1:1:6) built up against concrete including flushing up at back (ties not included)						
Walls (PC £ per 1000)	1200.00	1.85	48.42	85.24	m²	**133.66**
Edges of suspended slabs; 200 mm wide	-	0.56	14.66	17.05	m	**31.71**
Columns; 400 mm wide	-	1.11	29.05	34.10	m	**63.15**
Engineering bricks; and bullnosed specials; in cement mortar (1:3)						
Facework steps						
215 mm x 102 mm; all headers-on-edge; edges set with bullnosed specials; pointing top and one side; set weathering; horizontal (specials PC £ per 1000)	2000.00	0.51	13.35	31.11	m	**44.45**
returned ends pointed	-	0.14	3.66	5.84	nr	**9.51**
430 mm x 102 mm; all headers-on-edge; edges set with bullnosed specials; pointing top and one side; set weathering; horizontal (engineering bricks PC £ per 1000)	356.00	0.74	19.37	36.85	m	**56.22**
returned ends pointed	-	0.19	4.97	7.47	nr	**12.44**
Facing tile bricks; Ibstock "Tilebrick" or other equal and approved; in gauged mortar (1:1:6)						
Walls						
facework one side; half brick thick; stretcher bond (PC £ per 1000)	1125.00	0.87	22.77	58.55	m²	**81.32**
Extra over facing tile bricks for						
fair ends; 79 mm long	-	0.28	7.33	25.93	m	**33.26**
fair ends; 163 mm long	-	0.28	7.33	25.93	m	**33.26**
90° x 1/2 exernal return	-	0.28	7.33	55.14	m	**62.47**
90° internal return	-	0.28	7.33	65.16	m	**72.49**
45° external return	-	0.28	7.33	55.14	m	**62.47**
45° internal return	-	0.28	7.33	55.14	m	**62.47**
angled verge	-	0.28	7.33	31.16	m	**38.49**
ALTERNATIVE BLOCK PRICES						
Discounts of 0 - 20% available depending on quantity / status.						
Aerated concrete Durox "Supablocs"; 630 mm x 225 mm						
100 mm	-	-	-	12.05	m²	-
130 mm	-	-	-	17.01	m²	-
140 mm	-	-	-	19.66	m²	-
150 mm	-	-	-	20.13	m²	-
215 mm	-	-	-	28.76	m²	-
Hanson Conbloc blocks: 450 x 225 mm						
Cream fair faced						
100 mm hollow	-	-	-	8.61	m²	-
100 mm solid	-	-	-	9.33	m²	-
140mm hollow	-	-	-	12.77	m²	-
140 mm solid	-	-	-	15.19	m²	-
190 mm hollow	-	-	-	18.19	m²	-
190 mm solid	-	-	-	20.04	m²	-
215 mm hollow	-	-	-	18.15	m²	-

F MASONRY

Item	PC £	Labour hours	Labour £	Material £	Unit	Total rate £
F10 BRICK/BLOCK WALLING – cont'd						
ALTERNATIVE BLOCK PRICES – cont'd						
Fenlite						
100 mm solid; 3.50 N/mm²	-	-	-	7.18	m²	-
100 mm solid; 7.00 N/mm²	-	-	-	7.56	m²	-
140 mm solid; 3.50 N/mm²	-	-	-	10.72	m²	-
Standard Dense						
100 mm solid	-	-	-	6.69	m²	-
140 mm solid	-	-	-	10.01	m²	-
140 mm hollow	-	-	-	9.95	m²	-
190 mm hollow	-	-	-	14.31	m²	-
190 mm solid	-	-	-	14.02	m²	-
215 mm hollow	-	-	-	15.00	m²	-
Celcon blocks; 450 mm x 225 mm						
75 mm Standard	-	-	-	8.10	m²	-
100 mm Standard	-	-	-	10.48	m²	-
150 mm Standard	-	-	-	15.73	m²	-
265 mm Standard footing	-	-	-	27.80	m²	-
100 mm solid	-	-	-	12.35	m²	-
215 mm hollow	-	-	-	22.48	m²	-
100 mm Solar	-	-	-	10.48	m²	-
125 mm Solar	-	-	-	13.11	m²	-
150 mm Solar	-	-	-	15.73	m²	-
215 mm Solar	-	-	-	22.55	m²	-
265 mm Solar	-	-	-	27.80	m²	-
Forticrete painting quality blocks; 450 mm x 225 mm						
100 mm hollow	-	-	-	11.52	m²	-
100 mm solid	-	-	-	12.83	m²	-
140 mm hollow	-	-	-	15.91	m²	-
140 mm solid	-	-	-	18.56	m²	-
190 mm hollow	-	-	-	21.11	m²	-
190 mm solid	-	-	-	24.07	m²	-
215 mm hollow	-	-	-	22.15	m²	-
215 mm solid	-	-	-	26.63	m²	-
Lignacite "Lignacrete" standard blocks; 450 mm x 225 mm; 7.3 N/mm²						
100 mm	-	-	-	7.01	m²	-
140 mm	-	-	-	10.01	m²	-
150 mm	-	-	-	11.86	m²	-
190 mm	-	-	-	14.47	m²	-
215 mm	-	-	-	15.73	m²	-
Tarmac "Hemelite"; 450 mm x 225 mm						
100 mm solid; 3.50 N/mm²	-	-	-	6.23	m²	-
100 mm solid; 7.00 N/mm²	-	-	-	6.45	m²	-
140 mm solid; 7.00 N/mm²	-	-	-	9.17	m²	-
190 mm solid; 7.00 N/mm²	-	-	-	13.24	m²	-
215 mm solid; 7.00 N/mm²	-	-	-	16.32	m²	-
Tarmac "Toplite" standard blocks; 450 mm x 225 mm						
100 mm	-	-	-	6.40	m²	-
140 mm	-	-	-	8.96	m²	-
150 mm	-	-	-	9.60	m²	-
215 mm	-	-	-	13.76	m²	-
Tarmac "Toplite" GTI (thermal) blocks; 450 x 225 mm						
115 mm	-	-	-	7.36	m²	-
125 mm	-	-	-	8.00	m²	-
130 mm	-	-	-	8.32	m²	-
140 mm	-	-	-	8.96	m²	-
150 mm	-	-	-	9.60	m²	-
215 mm	-	-	-	13.76	m²	-

F MASONRY

Item	PC £	Labour hours	Labour £	Material £	Unit	Total rate £
SUPPLY AND FIX PRICES						
Lightweight aerated concrete blocks; Thermalite "Turbo" blocks or other equal and approved; in gauged mortar (1:2:9)						
Walls						
100 mm thick	7.20	0.46	12.04	8.94	m²	**20.98**
115 mm thick	8.28	0.46	12.04	10.28	m²	**22.32**
125 mm thick	9.00	0.46	12.04	11.17	m²	**23.21**
130 mm thick	9.36	0.46	12.04	11.62	m²	**23.66**
140 mm thick	10.08	0.51	13.35	12.52	m²	**25.86**
150 mm thick	10.80	0.51	13.35	13.41	m²	**26.76**
190 mm thick	13.68	0.56	14.66	16.98	m²	**31.64**
200 mm thick	14.40	0.56	14.66	17.88	m²	**32.54**
215 mm thick	15.48	0.56	14.66	19.22	m²	**33.88**
Isolated piers or chimney stacks						
190 mm thick	-	0.83	21.72	16.98	m²	**38.71**
215 mm thick	-	0.83	21.72	19.22	m²	**40.95**
Isolated casings						
100 mm thick	-	0.51	13.35	8.94	m²	**22.29**
115 mm thick	-	0.51	13.35	10.28	m²	**23.63**
125 mm thick	-	0.51	13.35	11.17	m²	**24.52**
140 mm thick	-	0.56	14.66	12.52	m²	**27.17**
Extra over for fair face; flush pointing						
walls; one side	-	0.04	1.05	-	m²	**1.05**
walls; both sides	-	0.09	2.36	-	m²	**2.36**
Closing cavities						
width of cavity 50 mm, closing with lightweight blockwork 100 mm thick; vertical	-	0.23	6.02	0.50	m	**6.52**
width of cavity 50 mm, closing with lightweight blockork 100 mm thick; including damp proof course; vertical	-	0.28	7.33	1.38	m	**8.71**
width of cavity 75 mm, closing with lightweight blockwork 100 mm thick; vertical	-	0.23	6.02	0.72	m	**6.74**
width of cavity 75 mm, closing with lightweight blockwork 100 mm thick; including damp proof course; vertical	-	0.28	7.33	1.60	m	**8.93**
Bonding ends to common brickwork						
100 mm thick	-	0.14	3.66	1.03	m	**4.69**
115 mm thick	-	0.14	3.66	1.18	m	**4.85**
125 mm thick	-	0.23	6.02	1.29	m	**7.31**
130 mm thick	-	0.23	6.02	1.34	m	**7.36**
140 mm thick	-	0.23	6.02	1.45	m	**7.47**
150 mm thick	-	0.23	6.02	1.55	m	**7.57**
190 mm thick	-	0.28	7.33	1.96	m	**9.28**
200 mm thick	-	0.28	7.33	2.06	m	**9.39**
215 mm thick	-	0.32	8.38	2.22	m	**10.60**

F MASONRY

Item	PC £	Labour hours	Labour £	Material £	Unit	Total rate £
F10 BRICK/BLOCK WALLING – cont'd						
Lightweight aerated concrete blocks; Thermalite "Shield" blocks; in thin joint mortar						
Walls						
75 mm thick	5.95	0.28	7.33	8.10	m²	**15.42**
90 mm thick	6.30	0.30	7.85	20.08	m²	**27.93**
100 mm thick	7.00	0.32	8.38	9.70	m²	**18.07**
140 mm thick	9.80	0.34	8.90	13.58	m²	**22.48**
150 mm thick	10.50	0.36	9.42	14.56	m²	**23.98**
190 mm thick	13.30	0.40	10.47	18.42	m²	**28.89**
200 mm thick	14.00	0.42	10.99	19.39	m²	**30.39**
Isolated piers or chimney stacks						
190 mm thick	-	0.60	15.70	18.42	m²	**34.12**
Isolated casings						
75 mm thick	-	0.35	9.16	8.45	m²	**17.61**
90 mm thick	-	0.35	9.16	8.87	m²	**18.03**
100 mm thick	-	0.35	9.16	9.70	m²	**18.86**
140 mm thick	-	0.38	9.95	13.58	m²	**23.53**
Lightweight aerated concrete blocks; Thermalite "Shield"; in gauged mortar (1:2:9)						
Walls						
75 mm thick	5.95	0.42	10.99	6.72	m²	**17.71**
90 mm thick	6.30	0.42	10.99	7.16	m²	**18.15**
100 mm thick	7.00	0.46	12.04	7.95	m²	**19.99**
140 mm thick	9.80	0.51	13.35	11.13	m²	**24.48**
150 mm thick	10.50	0.51	13.35	11.93	m²	**25.28**
190 mm thick	13.30	0.56	14.66	15.10	m²	**29.76**
200 mm thick	14.00	0.56	14.66	15.90	m²	**30.56**
Isolated piers or chimney stacks						
190 mm thick	-	0.83	21.72	15.10	m²	**36.83**
Isolated casings						
75 mm thick	-	0.51	13.35	6.72	m²	**20.07**
90 mm thick	-	0.51	13.35	7.16	m²	**20.50**
100 mm thick	-	0.51	13.35	7.95	m²	**21.30**
140 mm thick	-	0.56	14.66	11.13	m²	**25.79**
Extra over for fair face; flush pointing						
walls; one side	-	0.04	1.05	-	m²	**1.05**
walls; both sides	-	0.09	2.36	-	m²	**2.36**
Closing cavities						
width of cavity 50 mm, closing with lightweight blockwork 100 mm thick; vertical	-	0.23	6.02	0.45	m	**6.47**
width of cavity 50 mm, closing with lightweight blockork 100 mm thick; including damp proof course; vertical	-	0.28	7.33	1.33	m	**8.66**
width of cavity 75 mm, closing with lightweight blockwork 100 mm thick; vertical	-	0.23	6.02	0.65	m	**6.67**
width of cavity 75 mm, closing with lightweight blockwork 100 mm thick; including damp proof course; vertical	-	0.28	7.33	1.53	m	**8.86**
Bonding ends to common brickwork						
75 mm thick	-	0.09	2.36	0.77	m	**3.13**
90 mm thick	-	0.09	2.36	0.83	m	**3.18**
100 mm thick	-	0.14	3.66	0.92	m	**4.58**
140 mm thick	-	0.23	6.02	1.30	m	**7.32**
150 mm thick	-	0.23	6.02	1.38	m	**7.40**
190 mm thick	-	0.28	7.33	1.75	m	**9.08**
200 mm thick	-	0.28	7.33	1.84	m	**9.17**

F MASONRY

Item	PC £	Labour hours	Labour £	Material £	Unit	Total rate £
Lightweight smooth face aerated concrete blocks; Thermalite "Smooth Face" blocks or other equal and approved; in gauged mortar (1:2:9); flush pointing one side						
Walls						
100 mm thick	9.50	0.56	14.66	11.67	m²	**26.32**
140 mm thick	13.30	0.65	17.01	16.33	m²	**33.34**
150 mm thick	14.25	0.65	17.01	17.50	m²	**34.51**
190 mm thick	18.05	0.74	19.37	22.16	m²	**41.53**
200 mm thick	20.43	0.74	19.37	25.02	m²	**44.39**
215 mm thick	19.00	0.74	19.37	23.39	m²	**42.76**
Isolated piers or chimney stacks						
190 mm thick	-	0.93	24.34	22.16	m²	**46.50**
200 mm thick	-	0.93	24.34	25.02	m²	**49.36**
215 mm thick	-	0.93	24.34	23.39	m²	**47.73**
Isolated casings						
100 mm thick	-	0.69	18.06	11.67	m²	**29.73**
140 mm thick	-	0.74	19.37	16.33	m²	**35.70**
Extra over for fair face flush pointing walls; both sides	-	0.04	1.05	-	m²	**1.05**
Bonding ends to common brickwork						
100 mm thick	-	0.23	6.02	1.33	m	**7.35**
140 mm thick	-	0.23	6.02	1.87	m	**7.89**
150 mm thick	-	0.28	7.33	2.00	m	**9.33**
190 mm thick	-	0.32	8.38	2.53	m	**10.91**
200 mm thick	-	0.32	8.38	2.86	m	**11.23**
215 mm thick	-	0.32	8.38	2.69	m	**11.06**
Lightweight smooth face aerated concrete blocks; Thermalite "Party Wall" blocks or other equal and approved; in gauged mortar (1:2:9); flush pointing one side						
Walls						
100 mm thick	7.00	0.56	14.66	7.95	m²	**22.61**
215 mm thick	15.05	0.74	19.37	17.09	m²	**36.46**
Isolated piers or chimney stacks						
215 mm thick	-	0.93	24.34	17.09	m²	**41.44**
Isolated casings						
100 mm thick	-	0.69	18.06	7.95	m²	**26.01**
Extra over for fair face flush pointing walls; both sides	-	0.04	1.05	-	m²	**1.05**
Bonding ends to common brickwork						
100 mm thick	-	0.23	6.02	0.92	m	**6.94**
215 mm thick	-	0.32	8.38	1.99	m	**10.36**
Lightweight aerated high strength concrete blocks (7.00 N/mm²); Thermalite "High Strength" blocks or other equal and approved; in cement mortar (1:3)						
Walls						
100 mm thick	9.00	0.46	12.04	11.14	m²	**23.18**
140 mm thick	12.60	0.51	13.35	15.59	m²	**28.94**
150 mm thick	13.50	0.51	13.35	16.70	m²	**30.05**
190 mm thick	17.10	0.56	14.66	21.15	m²	**35.81**
200 mm thick	18.00	0.56	14.66	22.27	m²	**36.93**
215 mm thick	19.35	0.56	14.66	23.94	m²	**38.60**
Isolated piers or chimney stacks						
190 mm thick	-	0.83	21.72	21.15	m²	**42.88**
200 mm thick	-	0.83	21.72	22.27	m²	**43.99**
215 mm thick	-	0.83	21.72	23.94	m²	**45.67**

F MASONRY

Item	PC £	Labour hours	Labour £	Material £	Unit	Total rate £
F10 BRICK/BLOCK WALLING – cont'd						
Lightweight aerated high strength concrete blocks (7.00 N/mm²); Thermalite "High Strength" blocks – cont'd						
Isolated casings						
100 mm thick	-	0.51	13.35	11.14	m²	**24.49**
140 mm thick	-	0.56	14.66	15.59	m²	**30.25**
150 mm thick	-	0.56	14.66	16.70	m²	**31.36**
190 mm thick	-	0.69	18.06	21.15	m²	**39.21**
200 mm thick	-	0.69	18.06	22.27	m²	**40.33**
215 mm thick	-	0.69	18.06	23.94	m²	**42.00**
Extra over for flush pointing						
walls; one side	-	0.04	1.05	-	m²	**1.05**
walls; both sides	-	0.09	2.36	-	m²	**2.36**
Bonding ends to common brickwork						
100 mm thick	-	0.23	6.02	1.28	m	**7.30**
140 mm thick	-	0.23	6.02	1.80	m	**7.82**
150 mm thick	-	0.28	7.33	1.92	m	**9.25**
190 mm thick	-	0.32	8.38	2.43	m	**10.81**
200 mm thick	-	0.32	8.38	2.56	m	**10.94**
215 mm thick	-	0.32	8.38	2.76	m	**11.14**
Lightweight concrete blocks; Thermalite "Trenchblock" or other equal and approved; with tongued and grooved joints; in cement mortar (1:4)						
Walls						
255 mm thick	18.74	0.60	15.70	23.10	m²	**38.81**
275 mm thick	20.21	0.65	17.01	24.91	m²	**41.92**
305 mm thick	22.05	0.70	18.32	27.15	m²	**45.47**
355 mm thick	26.09	0.75	19.63	32.00	m²	**51.63**
Concrete blocks; Thermalite "Trenchblock" 7.00N/mm² or other equal and approved; with tongued and grooved joints; in cement mortar (1:4)						
Walls						
255 mm thick	25.50	0.70	18.32	31.20	m²	**49.52**
275 mm thick	27.50	0.75	19.63	33.64	m²	**53.27**
305 mm thick	30.00	0.80	20.94	36.67	m²	**57.61**
355 mm thick	35.50	0.85	22.25	43.25	m²	**65.50**
Medium dense smooth faced concrete blocks; Lignacite standard and paint grade 3.60N/mm² blocks or other equal and approved; in gauged mortar (1:2:9); flush pointing one side						
Walls						
100 mm thick	7.04	0.62	16.23	8.75	m²	**24.98**
140 mm thick	10.30	0.72	18.85	12.78	m²	**31.63**
150 mm thick	10.97	0.74	19.37	13.61	m²	**32.98**
190 mm thick	14.00	0.86	22.51	17.37	m²	**39.88**
215 mm thick	15.01	0.95	24.87	18.67	m²	**43.53**
Isolated piers or chimney stacks						
190 mm thick	-	1.14	29.84	17.37	m²	**47.21**
215 mm thick	-	1.26	32.98	18.67	m²	**51.65**

F MASONRY

Item	PC £	Labour hours	Labour £	Material £	Unit	Total rate £
Isolated casings						
100 mm thick	-	0.78	20.42	8.75	m²	**29.17**
140 mm thick	-	0.90	23.56	12.78	m²	**36.34**
Extra over for fair face flush pointing						
walls; both sides	-	0.04	1.05	-	m²	**1.05**
Bonding ends to common brickwork						
100 mm thick	-	0.23	6.02	1.01	m	**7.03**
140 mm thick	-	0.23	6.02	1.48	m	**7.50**
150 mm thick	-	0.28	7.33	1.57	m	**8.90**
190 mm thick	-	0.32	8.38	2.00	m	**10.37**
215 mm thick	-	0.32	8.38	2.16	m	**10.54**
Medium dense smooth faced concrete blocks; Lignacite standard and paint grade 7.30N/mm² blocks or other equal and approved; in gauged mortar (1:2:9); flush pointing one side						
Walls						
100 mm thick	7.16	0.62	16.23	8.89	m²	**25.12**
140 mm thick	10.39	0.72	18.85	12.89	m²	**31.73**
150 mm thick	11.54	0.74	19.37	14.29	m²	**33.65**
190 mm thick	14.45	0.86	22.51	17.90	m²	**40.41**
215 mm thick	16.01	0.95	24.87	19.85	m²	**44.72**
Isolated piers or chimney stacks						
190 mm thick	-	1.14	29.84	17.90	m²	**47.74**
215 mm thick	-	1.26	32.98	19.85	m²	**52.83**
Isolated casings						
100 mm thick	-	0.78	20.42	8.89	m²	**29.31**
140 mm thick	-	0.90	23.56	12.89	m²	**36.45**
Dense aggregate concrete blocks; Hanson "Conbloc" or other equal and approved; in cement mortar (1:2:9)						
Walls or partitions or skins of hollow walls						
75 mm thick; solid	5.32	0.56	16.85	6.61	m²	**23.45**
100 mm thick; solid	5.88	0.69	20.76	7.38	m²	**28.13**
140 mm thick; solid	11.56	0.83	24.97	14.27	m²	**39.24**
140 mm thick; hollow	11.10	0.74	22.26	13.72	m²	**35.98**
190 mm thick; hollow	13.10	0.93	27.98	16.30	m²	**44.28**
215 mm thick; hollow	13.67	1.02	30.68	17.07	m²	**47.76**
Isolated piers or chimney stacks						
140 mm thick; hollow	-	1.02	30.68	13.72	m²	**44.40**
190 mm thick; hollow	-	1.34	40.31	16.30	m²	**56.61**
215 mm thick; hollow	-	1.53	46.02	17.07	m²	**63.10**
Isolated casings						
75 mm thick; solid	-	0.69	20.76	6.61	m²	**27.36**
100 mm thick; solid	-	0.74	22.26	7.38	m²	**29.64**
140 mm thick; solid	-	0.93	27.98	14.27	m²	**42.25**
Extra over for fair face; flush pointing						
walls; one side	-	0.09	2.71	-	m²	**2.71**
walls; both sides	-	0.14	4.21	-	m²	**4.21**
Bonding ends to common brickwork						
75 mm thick solid	-	0.14	4.21	0.76	m	**4.97**
100 mm thick solid	-	0.23	6.92	0.85	m	**7.77**
140 mm thick solid	-	0.28	8.42	1.65	m	**10.07**
140 mm thick hollow	-	0.28	8.42	1.58	m	**10.01**
190 mm thick hollow	-	0.32	9.63	1.88	m	**11.51**
215 mm thick hollow	-	0.37	11.13	1.99	m	**13.12**

F MASONRY

Item	PC £	Labour hours	Labour £	Material £	Unit	Total rate £
F10 BRICK/BLOCK WALLING – cont'd						
Dense aggregate concrete blocks; (7.00 N/mm²) Forticrete "Shepton Mallet Common" blocks or other equal and approved; in cement mortar (1:3)						
Walls						
75 mm thick; solid	7.93	0.56	16.85	9.75	m²	**26.59**
100 mm thick; hollow	6.47	0.69	20.76	8.14	m²	**28.90**
100 mm thick; solid	6.04	0.69	20.76	7.63	m²	**28.39**
140 mm thick; hollow	9.55	0.74	22.26	11.98	m²	**34.24**
140 mm thick; solid	9.55	0.83	24.97	11.98	m²	**36.94**
190 mm thick; hollow	12.92	0.93	27.98	16.21	m²	**44.19**
190 mm thick; solid	12.91	1.02	30.68	16.19	m²	**46.87**
215 mm thick; hollow	11.72	1.02	30.68	14.91	m²	**45.59**
215 mm thick; solid	14.14	1.16	34.89	17.77	m²	**52.67**
Dwarf support wall						
140 mm thick; solid	-	1.16	34.89	11.98	m²	**46.87**
190 mm thick; solid	-	1.34	40.31	16.19	m²	**56.50**
215 mm thick; solid	-	1.53	46.02	17.77	m²	**63.80**
Isolated piers or chimney stacks						
140 mm thick; hollow	-	1.02	30.68	11.98	m²	**42.66**
190 mm thick; hollow	-	1.34	40.31	16.21	m²	**56.52**
215 mm thick; hollow	-	1.53	46.02	14.91	m²	**60.93**
Isolated casings						
75 mm thick; solid	-	0.69	20.76	9.75	m²	**30.50**
100 mm thick; solid	-	0.74	22.26	7.63	m²	**29.89**
140 mm thick; solid	-	0.93	27.98	11.98	m²	**39.95**
Extra over for fair face; flush pointing						
walls; one side	-	0.09	2.71	-	m²	**2.71**
walls; both sides	-	0.14	4.21	-	m²	**4.21**
Bonding ends to common brickwork						
75 mm thick solid	-	0.14	4.21	1.11	m	**5.32**
100 mm thick solid	-	0.23	6.92	0.88	m	**7.79**
140 mm thick solid	-	0.28	8.42	1.38	m	**9.80**
190 mm thick solid	-	0.32	9.63	1.85	m	**11.48**
215 mm thick solid	-	0.37	11.13	2.05	m	**13.18**
Dense aggregate coloured concrete blocks; Forticrete "Shepton Mallet Bathstone" or other equal and approved; in coloured gauged mortar (1:1:6); flush pointing one side						
Walls						
100 mm thick hollow	24.44	0.74	22.26	28.16	m²	**50.42**
100 mm thick solid	24.44	0.74	22.26	28.16	m²	**50.42**
140 mm thick hollow	35.41	0.83	24.97	40.76	m²	**65.72**
140 mm thick solid	35.41	0.93	27.98	40.76	m²	**68.73**
215 mm thick hollow	40.47	1.16	34.89	46.88	m²	**81.78**
Isolated piers or chimney stacks						
140 mm thick solid	-	1.25	37.60	40.76	m²	**78.36**
215 mm thick hollow	-	1.57	47.23	46.88	m²	**94.11**
Extra over blocks for						
100 mm thick half lintel blocks; ref D14	-	0.23	6.92	21.75	m	**28.67**
140 mm thick half lintel blocks; ref H14	-	0.28	8.42	38.90	m	**47.32**
140 mm thick quoin blocks; ref H16	-	0.32	9.63	32.87	m	**42.49**
140 mm thick cavity closer blocks; ref H17	-	0.32	9.63	35.25	m	**44.87**
140 mm thick cill blocks; ref H21	-	0.28	8.42	25.92	m	**34.34**

F MASONRY

Item	PC £	Labour hours	Labour £	Material £	Unit	Total rate £
"Astra-Glaze" satin-gloss glazed finish blocks or other equal and approved; Forticrete Ltd; standard colours; in gauged mortar (1:1:6); joints raked out; gun applied latex grout to joints						
Walls or partitions or skins of hollow walls						
100 mm thick; glazed one side	94.85	0.93	27.98	107.62	m²	**135.60**
extra; glazed square end return	59.50	0.37	11.13	34.18	m	**45.31**
100 mm thick; glazed both sides	124.94	1.11	33.39	141.62	m²	**175.01**
100 mm thick lintel 200 mm high; glazed one side	-	0.83	17.61	28.55	m	**46.16**
"Fireborn" terracotta blocks or other equal and approved; Ibstock Brick Ltd; in coloured gauged mortar (1:1:6); flush pointing one side						
Walls or partitions or skins of hollow walls						
102.50 mm thick; stretcher bond	47.25	0.33	9.93	54.09	m²	**64.02**
102.50 mm thick; stack bond	47.25	0.35	10.53	54.05	m²	**64.58**
F11 GLASS BLOCK WALLING						
NOTE: The following specialist prices for glass block walling; supplied by Roger Wilde Ltd; assume standard blocks in panels of 50 m²; no fire rating; work in straight walls at ground level; and all necessary ancillary fixing; strengthening; easy access; pointing and expansion materials etc.						
Hollow glass block walling; Pittsburgh Corning sealed "Thinline" or other equal and approved; in cement mortar joints; reinforced with 6 mm diameter stainless steel rods; pointed both sides with mastic or other equal and approved						
Walls; facework both sides						
115 mm x 115 mm x 80 mm flemish blocks	-	-	-	-	m²	**555.43**
190 mm x 190 mm x 80 mm flemish; cross reeded or clear blocks	-	-	-	-	m²	**257.56**
240 mm x 240 mm x 80 mm flemish; cross reeded or clear blocks	-	-	-	-	m²	**275.47**
240 mm x 115 mm x 80 mm flemish or clear blocks	-	-	-	-	m²	**360.58**

F MASONRY

Item	PC £	Labour hours	Labour £	Material £	Unit	Total rate £
F20 NATURAL STONE RUBBLE WALLING						
Cotswold Guiting limestone or other equal and approved ; laid dry						
Uncoursed random rubble walling						
275 mm thick	-	2.07	51.71	75.69	m²	**127.40**
350 mm thick	-	2.46	60.96	96.31	m²	**157.27**
425 mm thick	-	2.81	69.10	116.96	m²	**186.06**
500 mm thick	-	3.15	76.97	137.61	m²	**214.58**
Cotswold Guiting limestone or other equal and approved; bedded; jointed and pointed in cement:lime mortar (1:2:9)						
Uncoursed random rubble walling; faced and pointed; both sides						
275 mm thick	-	1.98	49.21	79.48	m²	**128.69**
350 mm thick	-	2.18	53.19	101.13	m²	**154.32**
425 mm thick	-	2.39	57.44	122.81	m²	**180.26**
500 mm thick	-	2.59	61.42	144.49	m²	**205.92**
Coursed random rubble walling; rough dressed; faced and pointed one side						
114 mm thick	-	1.48	37.35	49.31	m²	**86.66**
150 mm thick	-	1.76	46.77	49.80	m²	**96.57**
Fair returns on walling						
114 mm wide	-	0.02	0.52	-	m	**0.52**
150 mm wide	-	0.03	0.79	-	m	**0.79**
275 mm wide	-	0.06	1.57	-	m	**1.57**
350 mm wide	-	0.08	2.09	-	m	**2.09**
425 mm wide	-	0.10	2.62	-	m	**2.62**
500 mm wide	-	0.12	3.14	-	m	**3.14**
Fair raking cutting or circular cutting						
114 mm wide	-	0.20	5.27	7.16	m	**12.43**
150 mm wide	-	0.25	6.59	7.16	m	**13.75**
Level uncoursed rubble walling for damp proof courses and the like						
275 mm wide	-	0.19	5.28	8.55	m	**13.82**
350 mm wide	-	0.20	5.55	10.79	m	**16.35**
425 mm wide	-	0.21	5.83	13.10	m	**18.93**
500 mm wide	-	0.22	6.11	15.38	m	**21.49**
Copings formed of rough stones; faced and pointed all round						
275 mm x 200 mm (average) high	-	0.56	14.37	31.22	m	**45.59**
350 mm x 250 mm (average) high	-	0.75	18.99	43.83	m	**62.82**
425 mm x 300 mm (average) high	-	0.97	24.31	61.74	m	**86.05**
500 mm x 300 mm (average) high	-	1.23	30.48	83.95	m	**114.44**

F MASONRY

Item	PC £	Labour hours	Labour £	Material £	Unit	Total rate £
F22 CAST STONE WALLING/DRESSINGS						
Reconstructed limestone walling; Bradstone 100 bed weathered "Cotswold" or "North Cerney" masonry blocks or other equal and approved; laid to pattern or course recommended; bedded; jointed and pointed in approved coloured cement:lime mortar (1:2:9)						
Walls; facing and pointing one side						
Rebastone Split	-	1.00	26.17	21.00	m²	**47.17**
Rebastone Rustic	-	1.00	26.17	21.00	m²	**47.17**
masonry blocks; random uncoursed	-	1.04	27.22	36.97	m²	**64.19**
extra; returned ends	-	0.37	9.68	29.10	m	**38.78**
extra; plain L shaped quoins	-	0.12	3.14	36.25	m	**39.39**
traditional walling; coursed squared	-	1.30	34.03	36.97	m²	**71.00**
squared coursed rubble	-	1.25	32.72	38.11	m²	**70.83**
squared random rubble	-	1.30	34.03	37.95	m²	**71.97**
squared and pitched rock faced walling; coursed	-	1.34	35.07	37.95	m²	**73.02**
ashlar; 440 x 215 x 100 mm thick	-	1.10	28.79	38.74	m²	**67.53**
rough hewn rockfaced walling; random	-	1.39	36.38	37.78	m²	**74.17**
extra; returned ends	-	0.15	3.93	-	m	**3.93**
Isolated piers or chimney stacks; facing and pointing one side						
Rebastone Split	-	1.40	36.64	21.00	m²	**57.64**
Rebastone Rustic	-	1.40	36.64	21.00	m²	**57.64**
masonry blocks; random uncoursed	-	1.43	37.43	36.97	m²	**74.40**
traditional walling; coursed squared	-	1.80	47.11	36.97	m²	**84.09**
squared coursed rubble	-	1.76	46.07	38.11	m²	**84.18**
squared random rubble	-	1.80	47.11	37.95	m²	**85.06**
squared and pitched rock faced walling; coursed	-	1.90	49.73	37.95	m²	**87.68**
ashlar; 440 x 215 x 100 mm thick	-	1.54	40.31	38.74	m²	**79.05**
rough hewn rockfaced walling; random	-	1.94	50.78	37.78	m²	**88.56**
Isolated casings; facing and pointing one side						
Rebastone Split	-	1.20	31.41	21.00	m²	**52.41**
Rebastone Rustic	-	1.20	31.41	21.00	m²	**52.41**
masonry blocks; random uncoursed	-	1.25	32.72	36.97	m²	**69.69**
traditional walling; coursed squared	-	1.57	41.09	36.97	m²	**78.07**
squared coursed rubble	-	1.53	40.05	38.11	m²	**70.10**
squared random rubble	-	1.57	41.09	37.95	m²	**79.04**
squared and pitched rock faced walling; coursed	-	1.62	42.40	37.95	m²	**80.35**
ashlar; 440 x 215 x 100 mm thick	-	1.32	34.55	38.74	m²	**73.29**
rough hewn rockfaced walling; random	-	1.67	43.71	37.78	m²	**81.50**
Fair returns 100 mm wide						
Rebastone Split	-	0.10	2.62	-	m²	**2.62**
Rebastone Rustic	-	0.10	2.62	-	m²	**2.62**
masonry blocks; random uncoursed	-	0.11	2.88	-	m²	**2.88**
traditional walling; coursed squared	-	0.14	3.66	-	m²	**3.66**
squared coursed rubble	-	0.13	3.40	-	m²	**3.40**
squared random rubble	-	0.14	3.66	-	m²	**3.66**
squared and pitched rock faced walling; coursed	-	0.14	3.66	-	m²	**3.66**
ashlar; 440 x 215 x 100 mm thick	-	0.14	3.66	-	m²	**3.66**
rough hewn rockfaced walling; random	-	0.15	3.93	-	m²	**3.93**
Fair raking cutting or circular cutting						
100 mm wide	-	0.17	4.45	-	m	**4.45**
Quoin						
ashlar; 440 x 215 x 215 x 100 mm thick	-	0.75	19.63	74.30	m	**93.93**

F MASONRY

Item	PC £	Labour hours	Labour £	Material £	Unit	Total rate £
F22 CAST STONE WALLING/DRESSINGS – cont'd						
Reconstructed limestone dressings; "Bradstone Architectural" dressings in weathered "Cotswold" or "North Cerney" shades or other equal and approved; bedded, jointed and pointed in approved coloured cement:lime mortar (1:2:9)						
Copings; twice weathered and throated						
305 mm x 76 mm; type A	-	0.37	9.68	21.45	m	**31.13**
Extra for						
fair end	-	-	-	10.64	nr	**-**
returned mitred fair end	-	-	-	10.64	nr	**-**
Copings; once weathered and throated						
305 mm x 76 mm	-	0.37	9.68	21.13	m	**30.82**
356 mm x 76 mm	-	0.37	9.68	19.59	m	**29.27**
Extra for						
fair end	-	-	-	10.64	nr	**-**
returned mitred fair end	-	-	-	10.64	nr	**-**
Pier caps; four times weathered and throated						
305 mm x 305 mm	-	0.23	6.02	12.64	nr	**18.66**
381 mm x 381 mm	-	0.23	6.02	18.74	nr	**24.77**
457 mm x 457 mm	-	0.28	7.33	25.65	nr	**32.98**
533 mm x 533 mm	-	0.28	7.33	35.60	nr	**42.92**
Splayed corbels						
479 mm x 100 mm x 215 mm	-	0.14	3.66	21.01	nr	**24.67**
665 mm x 100 mm x 215 mm	-	0.19	4.97	29.05	nr	**34.03**
100 mm x 140 mm lintels; rectangular; reinforced with mild steel bars						
all lengths to 2.07 m	-	0.26	6.81	33.46	m	**40.26**
100 mm x 215 mm lintels; rectangular; reinforced with mild steel bars						
all lengths to 2.85 m	-	0.30	7.85	35.73	m	**43.58**
Sills to suit standard windows; stooled 100 mm at ends						
197 mm x140 mm; not exceeding 1.97 m long	-	0.28	7.33	45.18	m	**52.51**
197 mm x140 mm; not exceeding 1.97 m long	-	0.28	7.33	50.73	m	**58.06**
Window surround; traditional with label moulding; for single light; sill 146 mm x 133 mm; jambs 146 mm x 146 mm; head 146 mm x 105 mm; including all dowels and anchors						
overall size 508 mm x 1479 mm	143.44	0.83	21.72	156.00	nr	**177.73**
Window surround; traditional with label moulding; three light; for windows 508 mm x 1219 mm; sill 146 mm x 133 mm; jambs 146 mm x 146 mm; head 146 mm x 103 mm; mullions 146 mm x 108 mm; including all dowels and anchors						
overall size 1975 mm x 1479 mm	337.38	2.17	56.80	367.16	nr	**423.96**
Door surround; moulded continuous jambs and head with label moulding; including all dowels and anchors						
door 839 mm x 1981 mm in 102 mm x 64 mm frame	307.85	1.53	40.05	332.13	nr	**372.18**

F MASONRY

Item	PC £	Labour hours	Labour £	Material £	Unit	Total rate £
F30 ACCESSORIES/SUNDRY ITEMS FOR BRICK/BLOCK/STONE WALLING						
Forming cavities						
In hollow walls						
width of cavity 50 mm; polypropylene ties; three wall ties per m²	-	0.05	1.31	0.21	m²	**1.52**
width of cavity 50 mm; galvanised steel twisted wall ties; three wall ties per m²	-	0.05	1.31	0.51	m²	**1.81**
width of cavity 50 mm; stainless steel butterfly wall ties; three wall ties per m²	-	0.05	1.31	0.17	m²	**1.48**
width of cavity 50 mm; stainless steel twisted wall ties; three wall ties per m²	-	0.05	1.31	0.46	m²	**1.77**
width of cavity 75 mm; polypropylene ties; three wall ties per m²	-	0.05	1.31	0.21	m²	**1.52**
width of cavity 75 mm; galvanised steel twisted wall ties; three wall ties per m²	-	0.05	1.31	0.54	m²	**1.85**
width of cavity 75 mm; stainless steel butterfly wall ties; three wall ties per m²	-	0.05	1.31	0.18	m²	**1.48**
width of cavity 75 mm; stainless steel twisted wall ties; three wall ties per m²	-	0.05	1.31	1.02	m²	**2.33**
Damp proof courses						
Polythene damp proof course or other equal and approved; 200 mm laps; in gauged mortar (1:1:6)						
width exceeding 225 mm; horizontal	0.56	0.23	6.02	0.65	m²	**6.67**
width exceeding 225 mm; forming cavity gutters in hollow walls; horizontal	-	0.37	9.68	0.65	m²	**10.34**
width not exceeding 225 mm; horizontal	-	0.46	12.04	0.65	m²	**12.69**
width not exceeding 225 mm; vertical	-	0.69	18.06	0.65	m²	**18.71**
"Engerseal" polymer elastomeric damp proof course or other equal and approved; 200 mm laps; in gauged morter (1:1:6)						
width exceeding 225 mm; horizontal	2.37	0.23	6.02	2.74	m²	**8.76**
width exceeding 225 mm; forming cavity gutters in hollow walls; horizontal	-	0.37	9.68	2.74	m²	**12.43**
width not exceeding 225 mm; horizontal	-	0.46	12.04	2.74	m²	**14.78**
width not exceeding 225 mm; vertical	-	0.69	18.06	2.74	m²	**20.80**
"Zedex CPT" (Co-Polymer Thermoplastic) damp proof course or other equal and approved; 200 mm laps; in gauged mortar (1:1:6)						
width exceeding 225 mm; horizontal	3.26	0.23	6.02	3.77	m²	**9.79**
width exceeding 225 mm wide; forming cavity gutters in hollow walls; horizontal	-	0.37	9.68	3.77	m²	**13.45**
width not exceeding 225 mm; horizontal	-	0.46	12.04	3.77	m²	**15.81**
width not exceeding 225 mm; vertical	-	0.69	18.06	3.77	m²	**21.83**
"Hyload" (pitch polymer) damp proof course or other equal and approved; 150 mm laps; in gauged mortar (1:1:6)						
width exceeding 225 mm; horizontal	3.62	0.23	6.02	4.19	m²	**10.21**
width exceeding 225 mm; forming cavity gutters in hollow walls; horizontal	-	0.37	9.68	4.19	m²	**13.87**
width not exceeding 225 mm; horizontal	-	0.46	12.04	4.19	m²	**16.23**
width not exceeding 225 mm	-	0.69	18.06	4.19	m²	**22.25**

F MASONRY

Item	PC £	Labour hours	Labour £	Material £	Unit	Total rate £
F30 ACCESSORIES/SUNDRY ITEMS FOR BRICK/BLOCK/STONE WALLING – cont'd						
Damp proof courses – cont'd						
"Nubit" bitumen and polyester based damp proof course or other equal and approved; 200 mm laps; in gauged mortar (1:1:6)						
width exceeding 225 mm; horizontal	4.47	0.23	6.02	5.17	m²	**11.19**
width exceeding 225 mm wide; forming cavity gutters in hollow walls; horizontal	-	0.37	9.68	5.17	m²	**14.86**
width not exceeding 225 mm; horizontal	-	0.46	12.04	5.17	m²	**17.21**
width not exceeding 225 mm; vertical	-	0.69	18.06	5.17	m²	**23.23**
"Permabit" bitumen polymer damp proof course or other equal and approved; 150 mm laps; in gauged mortar (1:1:6)						
width exceeding 225 mm; horizontal	7.35	0.23	6.02	8.50	m²	**14.52**
width exceeding 225 mm; forming cavity gutters in hollow walls; horizontal	-	0.37	9.68	8.50	m²	**18.19**
width not exceeding 225 mm; horizontal	-	0.46	12.04	8.50	m²	**20.54**
width not exceeding 225 mm; vertical	-	0.69	18.06	8.50	m²	**26.56**
"Alumite" aluminium cored bitumen gas retardant damp proof course or other equal and approved; 200 mm laps; in gauged mortar (1:1;6)						
width exceeding 225 mm; horizontal	4.80	0.31	8.11	5.55	m²	**13.67**
width exceeding 225 mm; forming cavity gutters in hollow walls; horizontal	-	0.49	12.83	5.55	m²	**18.38**
width not exceeding 225 mm; horizontal	-	0.60	15.70	5.55	m²	**21.26**
width not exceeding 225 mm; verticall	-	0.83	21.72	5.55	m²	**27.28**
Milled lead damp proof course; BS 1178; 1.80 mm thick (code 4), 175 mm laps; in cement:lime mortar (1:2:9)						
width exceeding 225 mm; horizontal (PC £/kg)	1.57	1.85	48.42	43.76	m²	**92.18**
width not exceeding 225 mm; horizontal	-	2.78	72.76	43.76	m²	**116.52**
Two courses slates in cement:mortar (1:3)						
width exceeding 225 mm; horizontal	-	1.39	36.38	11.13	m²	**47.51**
width exceeding 225 mm; vertical	-	2.08	54.44	11.13	m²	**65.57**
"Synthaprufe" damp proof membrane or other equal and approved; PC £45.65/25 litres; three coats brushed on						
width not exceeding 150 mm; vertical	-	0.31	3.57	4.89	m²	**8.45**
width 150 mm - 225 mm; vertical	-	0.30	3.45	4.89	m²	**8.34**
width 225 mm - 300 mm; vertical	-	0.28	3.22	4.89	m²	**8.11**
width exceeding 300 mm wide; vertical	-	0.26	2.99	4.89	m²	**7.88**
Joint reinforcement						
"Brickforce" galvanised steel joint reinforcement						
width 60 mm; ref GBF40W60B25	-	0.05	1.31	0.64	m	**1.95**
width 100 mm; ref GBF40W100B25	-	0.07	1.83	0.74	m	**2.57**
width 150 mm; ref GBF40W150B25	-	0.09	2.22	0.92	m	**3.15**
width 175 mm; ref GBF40W175B25	-	0.10	2.62	1.10	m	**3.72**
"Brickforce" stainless steel joint reinforcement						
width 60 mm; ref SBF35W60BSC	-	0.05	1.31	1.39	m	**2.70**
width 100 mm; ref SBF35W100BSC	-	0.07	1.83	1.43	m	**3.26**
width 150 mm; ref SBF35W150BSC	-	0.09	2.22	1.46	m	**3.68**
width 175 mm; ref SBF35W175BSC	-	0.10	2.62	1.59	m	**4.21**
width 60 mm; ref SBF40W60BSC	-	0.06	1.44	1.74	m	**3.18**
width 100 mm; ref SBF40W100BSC	-	0.07	1.83	1.80	m	**3.63**
width 150 mm; ref SBF40W150BSC	-	0.09	2.22	1.82	m	**4.05**
width 175 mm; ref SBF40W175BSC	-	0.10	2.62	2.00	m	**4.62**

F MASONRY

Item	PC £	Labour hours	Labour £	Material £	Unit	Total rate £
"Wallforce" stainless steel joint reinforcement						
width 240 mm; ref SWF35W240	-	0.12	3.14	5.50	m	**8.64**
width 260 mm; ref SWF35W260	-	0.13	3.40	5.74	m	**9.14**
width 275 mm; ref SWF35W275	-	0.14	3.66	5.80	m	**9.47**
Weather fillets						
Weather fillets in cement:mortar (1:3)						
50 mm face width	-	0.11	2.88	0.04	m	**2.92**
100 mm face width	-	0.19	4.97	0.18	m	**5.15**
Angle fillets						
Angle fillets in cement:mortar (1:3)						
50 mm face width	-	0.11	2.88	0.04	m	**2.92**
100 mm face width	-	0.19	4.97	0.18	m	**5.15**
Pointing in						
Pointing with mastic						
wood frames or sills	-	0.09	1.38	0.56	m	**1.94**
Pointing with polysulphide sealant						
wood frames or sills	-	0.09	1.38	1.77	m	**3.15**
Wedging and pinning						
To underside of existing construction with slates in cement mortar (1:3)						
width of wall - one brick thick	-	0.74	19.37	2.38	m	**21.75**
width of wall - one and a half brick thick	-	0.93	24.34	4.77	m	**29.11**
width of wall - two brick thick	-	1.11	29.05	7.15	m	**36.21**
Joints						
Hacking joints and faces of brickwork or blockwork						
to form key for plaster	-	0.24	2.76	-	m²	**2.76**
Raking out joint in brickwork or blockwork for turned-in edge of flashing						
horizontal	-	0.14	3.66	-	m	**3.66**
stepped	-	0.19	4.97	-	m	**4.97**
Raking out and enlarging joint in brickwork or blockwork for nib of asphalt						
horizontal	-	0.19	4.97	-	m	**4.97**
Cutting grooves in brickwork or blockwork						
for water bars and the like	-	0.23	2.65	0.39	m	**3.04**
for nib of asphalt; horizontal	-	0.23	2.65	0.39	m	**3.04**
Preparing to receive new walls						
top existing 215 mm wall	-	0.19	4.97	-	m	**4.97**
Cleaning and priming both faces; filling with pre-formed closed cell joint filler and pointing one side with polysulphide sealant; 12 mm deep						
expansion joints; 12 mm wide	-	0.23	4.94	3.52	m	**8.46**
expansion joints; 20 mm wide	-	0.28	5.71	5.16	m	**10.86**
expansion joints; 25 mm wide	-	0.32	6.21	6.23	m	**12.44**
Fire resisting horizontal expansion joints; filling with joint filler; fixed with high temperature slip adhesive; between top of wall and soffit						
wall not exceeding 215 mm wide; 10 mm wide joint with 30 mm deep filler (one hour fire seal)	-	0.23	6.02	6.29	m	**12.31**
wall not exceeding 215 mm wide; 10 mm wide joint with 30 mm deep filler (two hour fire seal)	-	0.23	6.02	6.29	m	**12.31**
wall not exceeding 215 mm wide; 20 mm wide joint with 45 mm deep filler (two hour fire seal)	-	0.28	7.33	9.45	m	**16.77**
wall not exceeding 215 mm wide; 30 mm wide joint with 75 mm deep filler (three hour fire seal)	-	0.32	8.38	23.52	m	**31.89**

F MASONRY

Item	PC £	Labour hours	Labour £	Material £	Unit	Total rate £
F30 ACCESSORIES/SUNDRY ITEMS FOR BRICK/BLOCK/STONE WALLING – cont'd						
Joints – cont'd						
Fire resisting vertical expansiojn joints; filling with joint filler; fixed with high temperature slip adhesive; with polysulphide sealant one side; between end of wall and concrete						
wall not exceeding 215 mm wide; 20 mm wide joint with 45 mm deep filler (two hour fire seal)	-	0.37	8.60	12.99	m	**21.59**
Slate and tile sills						
Sills; two courses of machine made plain roofing tiles set weathering; bedded and pointed	-	0.56	14.66	5.13	m	**19.79**
Sundries						
Weep holes						
Perpend units; plastic	-	0.02	0.52	0.13	nr	**0.65**
Chimney pots; red terracotta; plain or cannon-head; setting and flaunching in cement mortar (1:3)						
185 mm diameter x 300 mm long	15.12	1.67	43.71	17.72	nr	**61.43**
185 mm diameter x 600 mm long	26.80	1.85	48.42	30.29	nr	**78.71**
185 mm diameter x 900 mm long	48.72	1.85	48.42	53.88	nr	**102.30**
Air bricks						
Air bricks; red terracotta; building into prepared openings						
215 mm x 65 mm	-	0.07	1.83	1.57	nr	**3.41**
215 mm x 140 mm	-	0.07	1.83	2.18	nr	**4.01**
215 mm x 215 mm	-	0.07	1.83	6.08	nr	**7.91**
Gas flue blocks						
Gas flue system; Schiedel "HP"or other equal and approved; concrete blocks built in; in flue joint mortar mix; cutting brickwork or blockwork around						
recess.unit; ref HP1	-	0.09	2.36	2.51	nr	**4.86**
cover block; ref HP2	-	0.09	2.36	5.51	nr	**7.86**
222 mm standard block with nib; ref HP3	-	0.09	2.36	4.05	nr	**6.40**
112 mm standard block with nib; ref HP3112	-	0.09	2.36	3.19	nr	**5.54**
72 mm standard block with nib; ref HP372	-	0.09	2.36	3.19	nr	**5.54**
222 mm standard block without nib; ref HP4	-	0.09	2.36	4.05	nr	**6.40**
112 mm standard block without nib; ref HP4112	-	0.09	2.36	3.19	nr	**5.54**
72 mm standard block without nib; ref HP472	-	0.09	2.36	3.19	nr	**5.54**
120 mm side offset block; ref HP5	-	0.09	2.36	4.27	nr	**6.63**
70 mm back offset block; ref HP6	-	0.09	2.36	13.71	nr	**16.07**
vertical exit block; ref HP7	-	0.09	2.36	8.15	nr	**10.51**
angled entry/exit block; ref HP8	-	0.09	2.36	8.11	nr	**10.47**
reverse rebate block; ref HP9	-	0.09	2.36	5.95	nr	**8.30**
corbel block; ref HP10	-	0.09	2.36	7.98	nr	**10.33**
lintel unit; ref HP11	-	0.09	2.36	7.45	nr	**9.80**
Proprietary items						
External Door and window cavity closers; "Thermabate" or equivalent; inclusive of flange clips; jointing strips; wall fixing ties and adhesive tape						
closing cavities; width of cavity 50 mm	-	0.14	3.66	3.89	m	**7.56**
closing cavities; width of cavity 75 mm	-	0.14	3.66	4.38	m	**8.05**
closing cavities; width of cavity 100 mm	-	0.14	3.66	4.38	m	**8.05**

F MASONRY

Item	PC £	Labour hours	Labour £	Material £	Unit	Total rate £
"Westbrick" cavity closers or other equal and approved; Manthorpe Building Products Ltd						
closing cavities; width of cavity 50 mm	-	0.14	3.66	4.83	m	**8.49**
"Type H cavicloser" or other equal and approved; uPVC universal cavity closer, insulator and damp proof course by Cavity Trays Ltd; built into cavity wall as work proceeds, complete with face closer and ties						
closing cavities; width of cavity 50 mm - 100 mm	-	0.07	1.83	5.43	m	**7.26**
"Type L" durropolyethelene lintel stop ends or other equal and approved; Cavity Trays Ltd; fixing with butyl anchoring strip; building in as the work proceeds						
adjusted to lintel as required	-	0.04	1.05	0.55	nr	**1.59**
"Type W" polypropylene weeps/vents or other equal and approved; Cavity Trays Ltd; built into cavity wall as work proceeds						
100/115 mm x 65 mm x 10 mm including lock fit wedges	-	0.04	1.05	0.41	nr	**1.46**
extra; extension duct 200/225 mm x 65 mm x 10 mm	-	0.07	1.83	0.71	nr	**2.54**
"Type X" polypropylene abutment cavity tray or other equal and approved; Cavity Trays Ltd; built into facing brickwork as the work proceeds; complete with Code 4 flashing; intermediate/catchment tray with short leads (requiring soakers); to suit roof of						
17 - 20 degree pitch	-	0.05	1.31	6.58	nr	**7.89**
21 - 25 degree pitch	-	0.05	1.31	6.12	nr	**7.43**
26 - 45 degree pitch	-	0.05	1.31	5.85	nr	**7.16**
"Type X" polypropylene abutment cavity tray or other equal and approved; Cavity Trays Ltd; built into facing brickwork as the work proceeds; complete with Code 4 flashing; intermediate/catchment tray with long leads (suitable only for corrugated roof tiles); to suit roof of						
17 - 20 degree pitch	-	0.05	1.31	8.90	nr	**10.20**
21 - 25 degree pitch	-	0.05	1.31	8.18	nr	**9.49**
26 - 45 degree pitch	-	0.05	1.31	7.55	nr	**8.85**
"Type X" polypropylene abutment cavity tray or other equal and approved; Cavity Trays Ltd; built into facing brickwork as the work proceeds; complete with Code 4 flashing; ridge tray with short/long leads; to suit roof of						
17 - 20 degree pitch	-	0.05	1.31	14.98	nr	**16.29**
21 - 25 degree pitch	-	0.05	1.31	13.88	nr	**15.19**
26 - 45 degree pitch	-	0.05	1.31	12.36	nr	**13.66**
Servicised "Bituthene MR" aluminium faced gas resistant cavity flashing or other equal and approved; sealed at joints with Servitape 30mm; in gauged mortar (1:1:6)						
width exceeding 225 mm wide	-	0.79	20.68	13.27	m²	**33.95**
"Expamet" stainless steel wall starters or other equal and approved; plugged and screwed						
to suit walls 60 mm - 75 mm thick	-	0.23	2.65	13.38	m	**16.03**
to suit walls 100 mm - 115 mm thick	-	0.23	2.65	14.72	m	**17.37**
to suit walls 125 mm - 180 mm thick	-	0.37	4.26	19.70	m	**23.95**
to suit walls 190 mm - 260 mm thick	-	0.46	5.29	25.16	m	**30.45**

F MASONRY

Item	PC £	Labour hours	Labour £	Material £	Unit	Total rate £
F30 ACCESSORIES/SUNDRY ITEMS FOR BRICK/BLOCK/STONE WALLING – cont'd						
Stainless steel posts, channels and ties						
Windposts; 130 x 70 x 6 mm; including one piece through ties						
1200 mm overall long	-	-	-	-	nr	**137.44**
3000 mm overall long	-	-	-	-	nr	**281.46**
4800 mm overall long	-	-	-	-	nr	**410.46**
Wall restraint channel ties; vertical channels; welded to steelwork. with lateral resatrint ties						
channel reference 28/15; tie reference HTS-B9; 200 mm long; one end of tie secured to channel; other end and debonding sleeve built into horizontal joint of masonry at 250 mm centres	-	0.12	3.14	10.78	m	**13.92**
Brickwork support angle welded to bracket reference HC6C or other equal and approved; Halfen Ltd; to suit 75 mm cavity, support to brickwork 6000 mm high						
6 mm thick; bolting with M12 x 50 mm T head bolts to cast in channel (not included)	-	0.32	6.70	91.05	m	**97.75**
Wall restraint individually fixed ties; fixed to steelwork						
ties reference HTS-B9; 200 mm long; one end of tie secured to channel; other end and debonding sleeve built into horizontal joint of masonry at 250 mm centres	-	0.02	0.52	0.50	nr	**1.02**
Head restraints; sliding brick anchors reference SBA/L at 900 mm horizontal centres; 500 mm deep tying into two courses of blockwork; fixed to steelwork						
2 nr ties reference HTS-B12; 200 mm long; built into horizontal joint of masonry	-	0.05	1.31	15.91	nr	**17.22**
Head restraint fixings; sliding brick anchors with 500 mm long stem; 2 nr 100 mm projection HST brick anchor ties or other equal and approved; Halfen Ltd; fixing with bolts to concrete soffit (bolts not included)						
ref. SBA/L	-	0.19	3.98	6.77	nr	**10.75**
Ties in walls; 200 mm long butterfly type; building into joints of brickwork or blockwork						
galvanised steel or polypropylene	-	0.02	0.52	0.08	nr	**0.61**
stainless steel	-	0.02	0.52	0.07	nr	**0.59**
Ties in walls; 20 mm x 3 mm x 200 mm long twisted wall type; building into joints of brickwork or blockwork						
galvanised steel	-	0.02	0.52	0.20	nr	**0.73**
stainless steel	-	0.02	0.52	0.19	nr	**0.71**
Anchors in walls; 25 mm x 3 mm x 100 mm long; one end dovetailed; other end building into joints of brickwork or blockwork						
galvanised steel	-	0.05	1.31	0.21	nr	**1.52**
stainless steel	-	0.05	1.31	0.53	nr	**1.84**
Slotted frame cramp; Halfen Ltd or other equal and approved; fixing by bolting (bolts measured elsewhere)						
ref. HTS - FH12; 150 mm projection	-	0.07	1.03	0.41	nr	**1.44**
Single expansion bolt; Halfen Ltd or other equal and approved: including washer						
8 mm diameter; ref. SEB 8	-	0.11	2.30	0.67	nr	**2.97**
Fixing cramps; 25 mm x 3 mm x 250 mm long; once bent; fixed to back of frame; other end building into joints of brickwork or blockwork						
galvanised steel	-	0.05	1.31	0.20	nr	**1.51**

F MASONRY

Item	PC £	Labour hours	Labour £	Material £	Unit	Total rate £
Galvanised steel lintels; "Catnic" or other equal and approved; built into brickwork or blockwork						
70/125 Range "CG" open back lintel for cavity wall						
750 mm long	33.13	0.23	6.02	35.70	nr	**41.72**
900 mm long	39.57	0.28	7.33	42.63	nr	**49.96**
1200 mm long	52.00	0.32	8.38	56.01	nr	**64.38**
1500 mm long	65.44	0.37	9.68	70.47	nr	**80.16**
1800 mm long	89.82	0.42	10.99	96.70	nr	**107.70**
2100 mm long	105.76	0.46	12.04	113.87	nr	**125.91**
2400 mm long	146.03	0.56	14.66	157.20	nr	**171.86**
70/125 Range "CUB" open back lintel for cavity wall						
2700 mm long	215.36	0.65	17.01	231.82	nr	**248.84**
3000 mm long	349.37	0.74	19.37	376.05	nr	**395.42**
70/125 Range "CU" open back lintel for cavity wall						
3300 mm long	269.07	0.83	21.72	289.63	nr	**311.35**
3600 mm long	293.51	0.93	24.34	315.93	nr	**340.27**
3900 mm long	317.75	1.02	26.70	342.02	nr	**368.72**
4200 mm long	354.84	0.46	12.04	381.93	nr	**393.97**
90/125 Range "CG" open back lintel for cavity wall						
750 mm long	36.85	0.23	6.02	39.70	nr	**45.72**
900 mm long	44.22	0.28	7.33	47.63	nr	**54.96**
1200 mm long	58.06	0.32	8.38	62.53	nr	**70.91**
1500 mm long	72.41	0.37	9.68	77.97	nr	**87.65**
1800 mm long	91.62	0.42	10.99	98.64	nr	**109.63**
2100 mm long	108.58	0.46	12.04	116.89	nr	**128.94**
2400 mm long	153.38	0.56	14.66	165.11	nr	**179.77**
90/125 Range "CUB" open back lintel for cavity wall						
2700 mm long	242.82	0.65	17.01	261.38	nr	**278.39**
3000 mm long	385.77	0.74	19.37	415.22	nr	**434.59**
90/125 Range "CU" open back lintel for cavity wall						
3300 mm long	293.17	0.83	21.72	315.56	nr	**337.28**
3600 mm long	319.82	0.93	24.34	344.25	nr	**368.59**
3900 mm long	346.23	1.02	26.70	372.67	nr	**399.36**
4200 mm long	363.35	0.46	12.04	391.09	nr	**403.13**
"CN92" single lintel; for 75 mm internal wall						
1050 mm long	5.60	0.28	7.33	6.05	nr	**13.38**
1200 mm long	6.33	0.32	8.38	6.84	nr	**15.21**
"CN102" single lintel; for 100 mm internal wall						
1050 mm long	7.08	0.28	7.33	7.64	nr	**14.97**
1200 mm long	7.82	0.32	8.38	8.43	nr	**16.81**
"CN100" single lintel; for 75 mm internal wall						
1050 mm long	17.30	0.28	7.33	18.64	nr	**25.96**
1200 mm long	21.50	0.32	8.38	23.16	nr	**31.53**
"CN5XA" single lintel; for 100 mm internal wall						
1050 mm long	32.67	0.28	7.33	35.18	nr	**42.51**
1200 mm long	33.52	0.32	8.38	36.09	nr	**44.47**

F MASONRY

Item	PC £	Labour hours	Labour £	Material £	Unit	Total rate £
F31 PRECAST CONCRETE SILLS/LINTELS/ COPING FEATURES						
Mix 20.00 N/mm² - 20 mm aggregate (1:2:4)						
Lintels; plate; prestressed bedded						
100 mm x 70 mm x 750 mm long	3.57	0.37	9.68	3.86	nr	**13.55**
100 mm x 70 mm x 900 mm long	4.22	0.37	9.68	4.57	nr	**14.25**
100 mm x 70 mm x 1050 mm long	5.02	0.37	9.68	5.42	nr	**15.10**
100 mm x 70 mm x 1200 mm long	5.68	0.37	9.68	6.14	nr	**15.82**
150 mm x 70 mm x 900 mm long	5.80	0.46	12.04	6.28	nr	**18.32**
150 mm x 70 mm x 1050 mm long	6.77	0.46	12.04	7.32	nr	**19.36**
150 mm x 70 mm x 1200 mm long	7.73	0.46	12.04	8.36	nr	**20.40**
220 mm x 70 mm x 900 mm long	8.94	0.56	14.66	9.68	nr	**24.33**
220 mm x 70 mm x 1200 mm long	11.89	0.56	14.66	12.85	nr	**27.51**
220 mm x 70 mm x 1500 mm long	14.91	0.65	17.01	16.10	nr	**33.11**
265 mm x 70 mm x 900 mm long	9.55	0.56	14.66	10.33	nr	**24.99**
265 mm x 70 mm x 1200 mm long	12.75	0.56	14.66	13.78	nr	**28.43**
265 mm x 70 mm x 1500 mm long	15.88	0.65	17.01	17.15	nr	**34.16**
265 mm x 70 mm x 1800 mm long	19.02	0.74	19.37	20.53	nr	**39.90**
Lintels; rectangular; reinforced with mild steel bars; bedded						
100 mm x 145 mm x 900 mm long	8.64	0.56	14.66	9.32	nr	**23.98**
100 mm x 145 mm x 1050 mm long	10.15	0.56	14.66	10.94	nr	**25.60**
100 mm x 145 mm x 1200 mm long	11.48	0.56	14.66	12.38	nr	**27.03**
225 mm x 145 mm x 1200 mm long	14.56	0.74	19.37	15.72	nr	**35.09**
225 mm x 225 mm x 1800 mm long	28.22	1.39	36.38	30.42	nr	**66.80**
Lintels; boot; reinforced with mild steel bars; bedded						
250 mm x 225 mm x 1200 mm long	23.38	1.11	29.05	25.21	nr	**54.27**
275 mm x 225 mm x 1800 mm long	31.34	1.67	43.71	33.78	nr	**77.49**
Padstones						
300 mm x 100 mm x 75 mm	4.77	0.28	7.33	5.16	nr	**12.49**
225 mm x 225 mm x 150 mm	7.24	0.37	9.68	7.83	nr	**17.52**
450 mm x 450 mm x 150 mm	19.15	0.56	14.66	20.75	nr	**35.41**
Mix 30.00 N/mm² - 20 mm aggregate (1:1:2)						
Copings; once weathered; once throated; bedded and pointed						
152 mm x 76 mm	5.80	0.65	17.01	6.62	m	**23.64**
178 mm x 64 mm	6.40	0.65	17.01	7.30	m	**24.32**
305 mm x 76 mm	10.80	0.74	19.37	12.37	m	**31.74**
extra for fair ends	-	-	-	5.33	nr	-
extra for angles	-	-	-	6.05	nr	-
Copings; twice weathered; twice throated; bedded and pointed						
152 mm x 76 mm	5.80	0.65	17.01	6.62	m	**23.64**
178 mm x 64 mm	6.35	0.65	17.01	7.24	m	**24.26**
305 mm x 76 mm	10.80	0.74	19.37	12.37	m	**31.74**
extra for fair ends	-	-	-	5.33	nr	-
extra for angles	-	-	-	6.05	nr	-
Sills; splayed top edge, stooled ends; bedded and pointed						
140 mm x 85 mm	20.27	0.75	19.63	22.98	m	**42.61**
180 mm x 85 mm	25.34	0.75	19.63	28.71	m	**48.34**

G STRUCTURAL/CARCASSING METAL/TIMBER

Item	PC £	Labour hours	Labour £	Material £	Unit	**Total rate £**
BASIC STEEL PRICES						
UNIVERSAL BEAMS AND COLUMNS						
NOTE: The following July 2008 basis prices from Corus Construction & Industrial, are for basic specifications and quantities of BS EN10025 grade 275JR steel (over 10 tonnes of one quantity, one serial size and one thickness in lengths between 6 m and 18½ m, for delivery to outer London). They include typical current end user discounts.						
The additional cost of grade 355JR is				30.00	tonne	
Readers should contact Corus to check the currency of these prices.						
Universal beams (kg/m)						
1016 x 305 mm (222, 249, 272, 314, 349, 393, 438, 487)	-	-	-	928.00	tonne	-
914 x 419 mm (343, 388)	-	-	-	918.00	tonne	-
914 x 305 mm (201, 224, 253, 289)	-	-	-	913.00	tonne	-
838 x 292 mm (176, 194, 226)	-	-	-	908.00	tonne	-
762 x 267 mm (134, 147, 173, 197)	-	-	-	908.00	tonne	-
686 x 254 mm (125, 140, 152, 170)	-	-	-	908.00	tonne	-
610 x 305 mm (149, 179, 238)	-	-	-	898.00	tonne	-
610 x 229 mm (101, 113, 125, 140)	-	-	-	883.00	tonne	-
610 x 178 mm (82, 92, 100)	-	-	-	883.00	tonne	-
533 x 312 mm (150, 182, 219, 272)	-	-	-	868.00	tonne	-
533 x 210 mm (82, 92, 101, 109, 122)	-	-	-	868.00	tonne	-
533 x 165 mm (66, 74, 85)	-	-	-	868.00	tonne	-
457 x 191 mm (67, 74, 82, 89, 98)	-	-	-	858.00	tonne	-
457 x 152 mm (52, 60, 67, 74, 82)	-	-	-	858.00	tonne	-
406 x 178 mm (54, 60, 67, 74)	-	-	-	863.00	tonne	-
406 x 140 mm (39, 46)	-	-	-	863.00	tonne	-
356 x 171 mm (45, 51, 57, 67)	-	-	-	863.00	tonne	-
356 x 127 mm (33, 39)	-	-	-	863.00	tonne	-
305 x 165 mm (40, 46, 54)	-	-	-	858.00	tonne	-
305 x 127 mm (37, 42, 48)	-	-	-	858.00	tonne	-
305 x 102 mm (25, 28, 33)	-	-	-	858.00	tonne	-
254 x 102 mm (22, 25, 28)	-	-	-	873.00	tonne	-
203 x 133 mm (25, 30)	-	-	-	873.00	tonne	-
203 x 102 mm (23)	-	-	-	873.00	tonne	-
178 x 102 mm (19)	-	-	-	873.00	tonne	-
152 x 89 mm (16)	-	-	-	883.00	tonne	-
127 x 76 mm (13)	-	-	-	883.00	tonne	-
Universal columns (kg/m)						
356 x 406 mm (235, 287, 340, 393, 467, 551, 634)	-	-	-	918.00	tonne	-
356 x 368 mm (129, 153, 177, 202)	-	-	-	918.00	tonne	-
305 x 305 mm (97, 118, 137, 158, 198, 240, 283)	-	-	-	913.00	tonne	-
254 x 254 mm (73, 89, 107, 132, 167)	-	-	-	863.00	tonne	-
203 x 203 mm (46, 52, 60, 71, 86)	-	-	-	863.00	tonne	-
152 x 152 mm (23, 30, 37)	-	-	-	873.00	tonne	-
Channels (kg/m)						
430 x 100 mm (64.4)	-	-	-	913.00	tonne	-
380 x 100 mm (54.0)	-	-	-	913.00	tonne	-
300 x 100 mm (45.5)	-	-	-	883.00	tonne	-
300 x 90 mm (41.4)	-	-	-	883.00	tonne	-
260 x 90 mm (34.8)	-	-	-	883.00	tonne	-
260 x 75 mm (27.6)	-	-	-	883.00	tonne	-

G STRUCTURAL/CARCASSING METAL/TIMBER

Item	PC £	Labour hours	Labour £	Material £	Unit	Total rate £
BASIC STEEL PRICES – cont'd						
UNIVERSAL BEAMS AND COLUMNS – cont'd						
Channels (kg/m) – cont'd						
230 x 90 mm (32.2)	-	-	-	883.00	tonne	-
230 x 75 mm (25.7)	-	-	-	883.00	tonne	-
200 x 90 mm (29.7)	-	-	-	883.00	tonne	-
200 x 75 mm (23.4)	-	-	-	848.00	tonne	-
180 x 90 mm (26.1)	-	-	-	883.00	tonne	-
180 x 75 mm (20.3)	-	-	-	848.00	tonne	-
150 x 90 mm (23.9)	-	-	-	883.00	tonne	-
150 x 75 mm (17.9)	-	-	-	848.00	tonne	-
125 x 65 mm (14.8)	-	-	-	848.00	tonne	-
100 x 50 mm (10.2)	-	-	-	848.00	tonne	-
Equal angles (mm)						
200 x 200 mm (16,18,20,24)	-	-	-	838.00	tonne	-
150 x 150 mm (10,12,15,18)	-	-	-	833.00	tonne	-
120 x 120 mm (8, 10, 12, 15)	-	-	-	838.00	tonne	-
100 x 100 mm (8, 10, 12, 15)	-	-	-	823.00	tonne	-
90 x 90 mm (6, 7, 8, 10, 12)	-	-	-	823.00	tonne	-
Unequal angles (mm)						
200 x 150 mm (12,15, 18)	-	-	-	853.00	tonne	-
200 x 100 mm (10, 12, 15)	-	-	-	848.00	tonne	-
150 x 90 mm (10, 12, 15)	-	-	-	833.00	tonne	-
150 x 75 mm (10, 12, 15)	-	-	-	823.00	tonne	-
125 x 75 mm (8, 10, 12)	-	-	-	823.00	tonne	-
100 x 75 mm (8, 10, 12)	-	-	-	823.00	tonne	-
100 x 65 mm (7, 8, 10)	-	-	-	823.00	tonne	-
Please refer to the Corus Price List for other extras to basis prices						
HOLLOW SECTIONS						
NOTE: The following basis March 2008 prices from Corus Tubes, are for for basic specifications and quantities of BS EN10025 grade 355J2H steel in one size, thickness, length, steelgrade and surface finish and include for delivery to outer London. They include typical current end user discounts.						
Readers should contact Corus to check the currency of these prices.						
A further increase was expected to be announced in May 2008, but details were not available at the time of going to press.						
Hot formed structural circular hollow section; S355J2H Grade 50D (kg/m)						
26.90 x 3.20 mm (1.87); (approximately 535.00 metres per tonne)	-	-	-	161.23	100 m	-
33.70 x 2.60 mm (1.99); (approximately 503.00 metres per tonne)	-	-	-	177.08	100 m	-
33.70 x 3.20 mm (2.41); (approximately 415.00 metres per tonne)	-	-	-	214.45	100 m	-
33.70 x 4.00 mm (2.93); (approximately 342.00 metres per tonne)	-	-	-	263.19	100 m	-
42.40 x 2.60 mm (2.55); (approximately 392.00 metres per tonne)	-	-	-	219.83	100 m	-

G STRUCTURAL/CARCASSING METAL/TIMBER

Item	PC £	Labour hours	Labour £	Material £	Unit	Total rate £
42.40 x 3.20 mm (3.09); (approximately 324.00 metres per tonne)	-	-	-	266.39	100 m	-
42.40 x 4.00 mm (3.79); (approximately 264.00 metres per tonne)	-	-	-	338.62	100 m	-
42.40 x 5.00 mm (2.61); (approximately 217.00 metres per tonne)	-	-	-	414.10	100 m	-
48.30 x 3.20 mm (3.56); (approximately 281.00 metres per tonne)	-	-	-	306.91	100 m	-
48.30 x 4.00 mm (4.37); (approximately 229.00 metres per tonne)	-	-	-	379.60	100 m	-
48.30 x 5.00 mm (5.34); (approximately 188.00 metres per tonne)	-	-	-	464.23	100 m	-
60.30 x 3.20 mm (4.51); (approximately 222.00 metres per tonne)	-	-	-	388.81	100 m	-
60.30 x 4.00 mm (5.55); (approximately 181.00 metres per tonne)	-	-	-	498.54	100 m	-
60.30 x 5.00 mm (6.82); (approximately 147.00 metres per tonne)	-	-	-	612.62	100 m	-
76.10 x 2.90 mm (5.24); (approximately 191.00 metres per tonne)	-	-	-	464.38	100 m	-
76.10 x 3.20 mm (5.75); (approximately 174.00 metres per tonne)	-	-	-	495.71	100 m	-
76.10 x 4.00 mm (7.11); (approximately 141.00 metres per tonne)	-	-	-	638.67	100 m	-
76.10 x 5.00 mm (8.77); (approximately 115.00 metres per tonne)	-	-	-	787.78	100 m	-
88.90 x 3.20 mm (6.76); (approximately 148.00 metres per tonne)	-	-	-	582.78	100 m	-
88.90 x 4.00 mm (8.38); (approximately 120.00 metres per tonne)	-	-	-	722.44	100 m	-
88.90 x 5.00 mm (10.30); (approximately 97.10 metres per tonne)	-	-	-	925.21	100 m	-
88.90 x 6.30 mm (12.80); (approximately 78.00 metres per tonne)	-	-	-	1149.78	100 m	-
114.30 x 3.20 mm (8.76); (approximately 114.00 metres per tonne)	-	-	-	786.89	100 m	-
114.30 x 3.60 mm (9.83); (approximately 102.00 metres per tonne)	-	-	-	883.00	100 m	-
114.30 x 4.00 mm (10.90); (approximately 92.00 metres per tonne)	-	-	-	1003.56	100 m	-
114.30 x 5.00 mm (13.50); (approximately 74.10 metres per tonne)	-	-	-	1212.66	100 m	-
114.30 x 6.30 mm (16.80); (approximately 59.60 metres per tonne)	-	-	-	1785.00	100 m	-
139.70 x 5.00 mm (16.60); (approximately 60.30 metres per tonne)	-	-	-	1431.45	100 m	-
139.70 x 6.30 mm (20.70); (approximately 48.40 metres per tonne)	-	-	-	1785.00	100 m	-
139.70 x 8.00 mm (26.00); (approximately 38.50 metres per tonne)	-	-	-	2242.03	100 m	-
139.70 x 10.00 mm (32.00); (approximately 31.30 metres per tonne)	-	-	-	2838.98	100 m	-
168.30 x 5.00 mm (20.10); (approximately 49.80 metres per tonne)	-	-	-	1733.26	100 m	-
168.30 x 6.30 mm (25.20); (approximately 39.70 metres per tonne)	-	-	-	2173.04	100 m	-
168.30 x 8.00 mm (31.60); (approximately 31.70 metres per tonne)	-	-	-	2724.93	100 m	-
168.30 x 10.00 mm (39.00); (approximately 25.70 metres per tonne)	-	-	-	3460.01	100 m	-

G STRUCTURAL/CARCASSING METAL/TIMBER

Item	PC £	Labour hours	Labour £	Material £	Unit	Total rate £
BASIC STEEL PRICES – cont'd						
HOLLOW SECTIONS – cont'd						
Hot formed structural circular hollow section; – cont'd						
168.30 x 12.50 mm (48.00); (approximately 20.80 metres per tonne)	-	-	-	4389.72	100 m	-
193.70 x 5.00 mm (23.30); (approximately 42.90 metres per tonne)	-	-	-	2009.20	100 m	-
193.70 x 6.30 mm (29.10); (approximately 34.40 metres per tonne)	-	-	-	2509.35	100 m	-
193.70 x 8.00 mm (36.60); (approximately 27.30 metres per tonne)	-	-	-	3156.08	100 m	-
193.70 x 10.00 mm (45.30); (approximately 22.10 metres per tonne)	-	-	-	4018.94	100 m	-
193.70 x 12.50 mm (55.90); (approximately 17.90 metres per tonne)	-	-	-	5472.74	100 m	-
219.10 x 5.00 mm (26.40); (approximately 37.90 metres per tonne)	-	-	-	2374.55	100 m	-
219.10 x 6.30 mm (33.10); (approximately 30.20 metres per tonne)	-	-	-	2977.18	100 m	-
219.10 x 8.00 mm (41.60); (approximately 24.10 metres per tonne)	-	-	-	3741.71	100 m	-
219.10 x 10.00 mm (51.60); (approximately 19.40 metres per tonne)	-	-	-	4641.16	100 m	-
219.10 x 12.50 mm (63.70); (approximately 15.70 metres per tonne)	-	-	-	5975.26	100 m	-
219.10 x 16.00 mm (80.10); (approximately 12.50 metres per tonne)	-	-	-	7204.59	100 m	-
244.50 x 8.00 mm (46.70); (approximately 21.50 metres per tonne)	-	-	-	4200.43	100 m	-
244.50 x 10.00 mm (57.80); (approximately 17.40 metres per tonne)	-	-	-	5198.82	100 m	-
244.50 x 12.50 mm (71.50); (approximately 14.00 metres per tonne)	-	-	-	6706.93	100 m	-
244.50 x 16.00 mm (90.20); (approximately 11.10 metres per tonne)	-	-	-	8113.04	100 m	-
273.00 x 6.30 mm (41.40); (approximately 24.20 metres per 0tonne)	-	-	-	3723.72	100 m	-
273.00 x 8.00 mm (52.30); (approximately 19.10 metres per tonne)	-	-	-	4704.12	100 m	-
273.00 x 10.00 mm (64.90); (approximately 15.40 metres per tonne)	-	-	-	5837.43	100 m	-
273.00 x 12.50 mm (80.30); (approximately 12.50 metres per tonne)	-	-	-	7532.39	100 m	-
273.00 x 16.00 mm (101.00); (approximately 9.91 metres per tonne)	-	-	-	9084.44	100 m	-
323.90 x 6.30 mm (49.30); (approximately 20.30 metres per tonne)	-	-	-	4434.29	100 m	-
323.90 x 8.00 mm (62.30); (approximately 16.10 metres per tonne)	-	-	-	5603.57	100 m	-
323.90 x 10.00 mm (77.40); (approximately 12.90 metres per tonne)	-	-	-	6961.74	100 m	-
323.90 x 12.50 mm (96.00); (approximately 10.40 metres per tonne)	-	-	-	9005.10	100 m	-
323.90 x 16.00 mm (121.00); (approximately 8.27 metres per tonne)	-	-	-	10883.34	100 m	-
355.60 x 16.00 mm (134.00); (approximately 7.47metres per tonne)	-	-	-	12052.63	100 m	-
406.40 x 6.30 mm (62.20); (approximately 16.10 metres per tonne)	-	-	-	5594.57	100 m	-

G STRUCTURAL/CARCASSING METAL/TIMBER

Item	PC £	Labour hours	Labour £	Material £	Unit	Total rate £
406.40 x 8.00 mm (78.60); (approximately 12.70 metres per tonne)	-	-	-	7069.67	100 m	-
406.40 x 10.00 mm (97.80); (approximately 10.20 metres per tonne)	-	-	-	8796.62	100 m	-
406.40 x 12.50 mm (121.00); (approximately 8.27 metres per 0tonne)	-	-	-	11350.18	100 m	-
406.40 x 16.00 mm (154.00); (approximately 6.50 metres per tonne)	-	-	-	13851.52	100 m	-
457.00 x 8.00 mm (88.60); (approximately 11.30 metres per tonne)	-	-	-	7969.12	100 m	-
457.00 x 10.00 mm (110.00); (approximately 9.09 metres per tonne)	-	-	-	9893.95	100 m	-
457.00 x 12.50 mm (137.00); (approximately 7.30 metres per tonne)	-	-	-	12851.03	100 m	-
457.00 x 16.00 mm (174.00); (approximately 5.75 metres per tonne)	-	-	-	15650.42	100 m	-
508.00 x 10.00 mm (123.00); (approximately 8.13 metres per tonne)	-	-	-	11063.23	100 m	-
508.00 x 12.50 mm (153.00); (approximately 6.54 metres per tonne)	-	-	-	14351.88	100 m	-
508.00 x 16.00 mm (194.00); (approximately 5.16 metres per tonne)	-	-	-	17449.32	100 m	-
Hot formed structural square hollow section; S355J2H Grade 50D (kg/m)						
40 x 40 x 3.00 mm (3.41); (approximately 293.30 metres per tonne)	-	-	-	289.31	100 m	-
40 x 40 x 3.20 mm (3.61); (approximately 277.00 metres per tonne)	-	-	-	306.28	100 m	-
40 x 40 x 4.00 mm (4.39); (approximately 227.80 metres per tonne)	-	-	-	372.45	100 m	-
40 x 40 x 5.00 mm (5.28); (approximately 189.40 metres per tonne)	-	-	-	447.96	100 m	-
50 x 50 x 3.00 mm (4.35); (approximately 229.90 metres per tonne)	-	-	-	369.06	100 m	-
50 x 50 x 3.20 mm (4.62); (approximately 216.50 metres per tonne)	-	-	-	391.96	100 m	-
50 x 50 x 4.00 mm (5.64); (approximately 216.50 metres per tonne)	-	-	-	478.50	100 m	-
50 x 50 x 5.00 mm (6.85); (approximately 177.30 metres per tonne)	-	-	-	581.16	100 m	-
50 x 50 x 6.30 mm (8.31); (approximately 120.30 metres per tonne)	-	-	-	705.03	100 m	-
60 x 60 x 3.00 mm (5.29); (approximately 189.00 metres per tonne)	-	-	-	448.81	100 m	-
60 x 60 x 3.20 mm (5.62); (approximately 177.90 metres per tonne)	-	-	-	476.80	100 m	-
60 x 60 x 4.00 mm (6.90); (approximately 145.00 metres per tonne)	-	-	-	609.10	100 m	-
60 x 60 x 5.00 mm (8.42); (approximately 118.80 metres per tonne)	-	-	-	743.28	100 m	-
60 x 60 x 6.30 mm (10.30); (approximately 97.10 metres per tonne)	-	-	-	909.24	100 m	-
60 x 60 x 8.00 mm (12.50); (approximately 80.00 metres per tonne)	-	-	-	1103.45	100 m	-
70 x 70 x 3.60 mm (7.40); (approximately 135.10 metres per tonne)	-	-	-	632.91	100 m	-
70 x 70 x 5.00 mm (9.99); (approximately 100.10 metres per tonne)	-	-	-	881.87	100 m	-
70 x 70 x 6.30 mm (12.30); (approximately 81.30 metres per tonne)	-	-	-	1085.79	100 m	-

G STRUCTURAL/CARCASSING METAL/TIMBER

Item	PC £	Labour hours	Labour £	Material £	Unit	Total rate £
BASIC STEEL PRICES – cont'd						
HOLLOW SECTIONS – cont'd						
Hot formed structural square hollow section – cont'd						
70 x 70 x 8.00 mm (15.00); (approximately 66.70 metres per tonne)	-	-	-	1324.14	100 m	-
80 x 80 x 3.60 mm (8.53); (approximately 117.20 metres per tonne)	-	-	-	729.56	100 m	-
80 x 80 x 4.00 mm (9.41); (approximately 106.30 metres per tonne)	-	-	-	818.83	100 m	-
80 x 80 x 5.00 mm (11.60); (approximately 86.20 metres per tonne)	-	-	-	1024.00	100 m	-
80 x 80 x 6.30 mm (14.20); (approximately 70.40 metres per tonne)	-	-	-	1253.51	100 m	-
80 x 80 x 8.00 mm (17.50); (approximately 57.10 metres per tonne)	-	-	-	1544.82	100 m	-
90 x 90 x 3.60 mm (9.66); (approximately 103.50 metres per tonne)	-	-	-	826.20	100 m	-
90 x 90 x 4.00 mm (10.70); (approximately 93.50 metres per tonne)	-	-	-	931.08	100 m	-
90 x 90 x 5.00 mm (13.10); (approximately 76.30 metres per tonne)	-	-	-	1156.41	100 m	-
90 x 90 x 6.30 mm (16.20); (approximately 61.70 metres per tonne)	-	-	-	1430.07	100 m	-
90 x 90 x 8.00 mm (20.10); (approximately 49.80 metres per tonne)	-	-	-	1774.35	100 m	-
100 x 100 x 4.00 mm (11.90); (approximately 84.00 metres per tonne)	-	-	-	995.99	100 m	-
100 x 100 x 5.00 mm (14.70); (approximately 68.00 metres per tonne)	-	-	-	1297.65	100 m	-
100 x 100 x 6.30 mm (18.20); (approximately 54.90 metres per tonne)	-	-	-	1606.62	100 m	-
100 x 100 x 8.00 mm (22.60); (approximately 44.20 metres per tonne)	-	-	-	2067.49	100 m	-
100 x 100 x 10.00 mm (27.40); (approximately 36.50 metres per tonne)	-	-	-	2355.13	100 m	-
120 x 120 x 5.00 mm (17.80); (approximately 56.20 metres per tonne)	-	-	-	1482.46	100 m	-
120 x 120 x 6.30 mm (22.20); (approximately 45.00 metres per tonne)	-	-	-	1848.92	100 m	-
120 x 120 x 8.00 mm (27.60); (approximately 36.20 metres per tonne)	-	-	-	2298.65	100 m	-
120 x 120 x 10.00 mm (33.70); (approximately 29.70 metres per tonne)	-	-	-	2896.63	100 m	-
120 x 120 x 12.50 mm (40.90); (approximately 24.40 metres per tonne)	-	-	-	3857.26	100 m	-
140 x 140 x 5.00 mm (21.00); (approximately 17.60 metres per tonne)	-	-	-	1924.47	100 m	-
140 x 140 x 6.30 mm (26.10); (approximately 38.30 metres per tonne)	-	-	-	2173.72	100 m	-
140 x 140 x 8.00 mm (32.60); (approximately 30.70 metres per tonne)	-	-	-	2715.08	100 m	-
140 x 140 x 10.00 mm (40.00); (approximately 25.00 metres per tonne)	-	-	-	3438.14	100 m	-
140 x 140 x 12.50 mm (48.70); (approximately 20.50 metres per tonne)	-	-	-	4593.65	100 m	-
150 x 150 x 5.00 mm (21.00); (approximately 44.20 metres per tonne)	-	-	-	1882.23	100 m	-
150 x 150 x 6.30 mm (28.10); (approximately 35.60 metres per tonne)	-	-	-	2340.29	100 m	-

G STRUCTURAL/CARCASSING METAL/TIMBER

Item	PC £	Labour hours	Labour £	Material £	Unit	**Total rate £**
150 x 150 x 8.00 mm (35.10); (approximately 28.50 metres per tonne)	-	-	-	2923.28	100 m	-
150 x 150 x 10.00 mm (43.10); (approximately 23.20 metres per tonne)	-	-	-	3704.59	100 m	-
150 x 150 x 12.50 mm (52.70); (approximately 19.00 metres per tonne)	-	-	-	4970.95	100 m	-
160 x 160 x 5.00 mm (24.10); (approximately 41.50 metres per tonne)	-	-	-	2149.50	100 m	-
160 x 160 x 6.30 mm (30.10); (approximately 33.20 metres per tonne)	-	-	-	2684.64	100 m	-
160 x 160 x 8.00 mm (37.60); (approximately 26.60 metres per tonne)	-	-	-	3353.57	100 m	-
160 x 160 x 10.00 mm (46.30); (approximately 21.60 metres per tonne)	-	-	-	4129.53	100 m	-
160 x 160 x 12.50 mm (56.60); (approximately 17.70 metres per tonne)	-	-	-	5253.23	100 m	-
180 x 180 x 6.30 mm (34.00); (approximately 29.40 metres per tonne)	-	-	-	3032.49	100 m	-
180 x 180 x 8.00 mm (42.70); (approximately 23.40 metres per tonne)	-	-	-	3808.45	100 m	-
180 x 180 x 10.00 mm (52.50); (approximately 19.00 metres per tonne)	-	-	-	4682.52	100 m	-
180 x 180 x 12.50 mm (64.40); (approximately 15.50 metres per tonne)	-	-	-	5977.18	100 m	-
180 x 180 x 16.00 mm (80.20); (approximately 12.50 metres per tonne)	-	-	-	7641.18	100 m	-
200 x 200 x 5.00 mm (30.40); (approximately 32.90 metres per tonne)	-	-	-	2711.40	100 m	-
200 x 200 x 6.30 mm (38.00); (approximately 26.30 metres per tonne)	-	-	-	3389.25	100 m	-
200 x 200 x 8.00 mm (47.70); (approximately 21.00 metres per tonne)	-	-	-	4253.21	100 m	-
200 x 200 x 10.00 mm (58.80); (approximately 17.00 metres per tonne)	-	-	-	5244.42	100 m	-
200 x 200 x 12.50 mm (72.30); (approximately 13.80 metres per tonne)	-	-	-	6710.40	100 m	-
200 x 200 x 16.00 mm (90.30); (approximately 11.10 metres per tonne)	-	-	-	8603.47	100 m	-
250 x 250 x 6.30 mm (47.90); (approximately 20.90 metres per tonne)	-	-	-	4272.24	100 m	-
250 x 250 x 8.00 mm (60.30); (approximately 16.60 metres per tonne)	-	-	-	5378.20	100 m	-
250 x 250 x 10.00 mm (74.50); (approximately 13.40 metres per tonne)	-	-	-	6644.71	100 m	-
250 x 250 x 12.50 mm (91.90); (approximately 10.90 metres per tonne)	-	-	-	8529.54	100 m	-
250 x 250 x 16.00 mm (115.00); (approximately 8.70 metres per tonne)	-	-	-	10956.80	100 m	-
300 x 300 x 6.30 mm (57.80); (approximately 17.30 metres per tonne)	-	-	-	5155.23	100 m	-
300 x 300 x 8.00 mm (72.80); (approximately 13.70 metres per tonne)	-	-	-	6493.09	100 m	-
300 x 300 x 10.00 mm (90.20); (approximately 11.10 metres per tonne)	-	-	-	7985.41	100 m	-
300 x 300 x 12.50 mm (112.00); (approximately 8.93 metres per tonne)	-	-	-	10395.09	100 m	-
300 x 300 x 16.00 mm (141.00); (approximately 7.09 metres per tonne)	-	-	-	13434.00	100 m	-
350 x 350 x 8.00 mm (85.40); (approximately 11.70 metres per tonne)	-	-	-	7616.89	100 m	-

G STRUCTURAL/CARCASSING METAL/TIMBER

Item	PC £	Labour hours	Labour £	Material £	Unit	Total rate £
BASIC STEEL PRICES – cont'd						
HOLLOW SECTIONS – cont'd						
Hot formed structural square hollow section – cont'd						
350 x 350 x 10.00 mm (106.00); (approximately 9.43 metres per tonne)	-	-	-	9454.22	100 m	-
350 x 350 x 12.50 mm (131.00); (approximately 7.63 metres per onne)	-	-	-	12158.54	100 m	-
350 x 350 x 16.00 mm (166.00); (approximately 6.02 metres per tonne)	-	-	-	15815.91	100 m	-
400 x 400 x 10.00 mm (122.00); (approximately 8.20 metres per tonne)	-	-	-	10881.28	100 m	-
400 x 400 x 12.50 mm (151.00); (approximately 6.62 metres per tonne)	-	-	-	14014.81	100 m	-
400 x 400 x 16.00 mm (191.00); (approximately 5.24 metres per tonne)	-	-	-	18197.82	100 m	-
Hot formed structural rectangular hollow section; S355J2H Grade 50D (kg/m)						
50 x 30 x 3.20 mm (3.61); (approximately 277.00 metres per tonne)	-	-	-	306.28	100 m	-
60 x 40 x 3.00 mm (4.35); (approximately 229.90 metres per tonne)	-	-	-	3854.06	100 m	-
60 x 40 x 4.00 mm (5.64); (approximately 177.30 metres per tonne)	-	-	-	477.81	100 m	-
60 x 40 x 5.00 mm (6.85); (approximately 146.00 metres per tonne)	-	-	-	581.16	100 m	-
80 x 40 x 3.20 mm (5.62); (approximately 177.90 metres per tonne)	-	-	-	476.80	100 m	-
80 x 40 x 4.00 mm (6.90); (approximately 144.90 metres per tonne)	-	-	-	609.10	100 m	-
80 x 40 x 5.00 mm (8.42); (approximately 118.80 metres per tonne)	-	-	-	1440.28	100 m	-
80 x 40 x 6.30 mm (10.30); (approximately 97.10 metres per tonne)	-	-	-	909.24	100 m	-
80 x 40 x 8.00 mm (12.50); (approximately 80.00 metres per tonne)	-	-	-	1103.45	100 m	-
90 x 50 x 3.60 mm (7.40); (approximately 135.10 metres per tonne)	-	-	-	632.91	100 m	-
90 x 50 x 5.00 mm (9.99); (approximately 100.10 metres per tonne)	-	-	-	881.87	100 m	-
90 x 50 x 6.30 mm (12.30); (approximately 81.30 metres per tonne)	-	-	-	1085.79	100 m	-
100 x 50 x 3.00 mm (6.71); (approximately 149.00 metres per tonne)	-	-	-	551.61	100 m	-
100 x 50 x 3.20 mm (7.13); (approximately 140.30 metres per tonne)	-	-	-	586.14	100 m	-
100 x 50 x 4.00 mm (8.78); (approximately 113.90 metres per tonne)	-	-	-	750.94	100 m	-
100 x 50 x 5.00 mm (10.80); (approximately 92.60 metres per tonne)	-	-	-	953.38	100 m	-
100 x 50 x 6.30 mm (13.30); (approximately 75.20 metres per tonne)	-	-	-	1174.07	100 m	-
100 x 50 x 8.00 mm (16.30); (approximately 61.30 metres per tonne)	-	-	-	1438.89	100 m	-
100 x 60 x 3.60 mm (8.53); (approximately 117.20 metres per tonne)	-	-	-	729.35	100 m	-
100 x 60 x 5.00 mm (11.60); (approximately 86.20 metres per tonne)	-	-	-	1024.00	100 m	-
100 x 60 x 6.30 mm (14.20); (approximately 70.40 metres per tonne)	-	-	-	1253.51	100 m	-

G STRUCTURAL/CARCASSING METAL/TIMBER

Item	PC £	Labour hours	Labour £	Material £	Unit	**Total rate £**
100 x 60 x 8.00 mm (17.50); (approximately 57.10 metres per tonne)	-	-	-	1544.82	100 m	-
120 x 60 x 3.60 mm (9.66); (approximately 103.50 metres per tonne)	-	-	-	829.63	100 m	-
120 x 60 x 5.00 mm (13.10); (approximately 76.30 metres per tonne)	-	-	-	1156.41	100 m	-
120 x 60 x 6.30 mm (16.20); (approximately 61.70 metres per tonne)	-	-	-	1430.07	100 m	-
120 x 60 x 8.00 mm (20.10); (approximately 49.80 metres per tonne)	-	-	-	1774.35	100 m	-
120 x 80 x 5.00 mm (14.70); (approximately 68.00 metres per tonne)	-	-	-	1297.65	100 m	-
120 x 80 x 6.30 mm (18.20); (approximately 54.90 metres per tonne)	-	-	-	1606.62	100 m	-
120 x 80 x 8.00 mm (22.60); (approximately 44.20 metres per tonne)	-	-	-	2067.49	100 m	-
120 x 80 x 10.00 mm (27.40); (approximately 36.50 metres per tonne)	-	-	-	2355.13	100 m	-
150 x 100 x 5.00 mm (18.60); (approximately 53.80 metres per tonne)	-	-	-	1549.09	100 m	-
150 x 100 x 6.30 mm (23.10); (approximately 43.30 metres per tonne)	-	-	-	1923.87	100 m	-
150 x 100 x 8.00 mm (28.90); (approximately 34.60 metres per tonne)	-	-	-	2406.92	100 m	-
150 x 100 x 10.00 mm (35.30); (approximately 28.30 metres per tonne)	-	-	-	3034.15	100 m	-
150 x 100 x 12.50 mm (42.80); (approximately 23.40 metres per tonne)	-	-	-	4037.12	100 m	-
160 x 80 x 4.00 mm (14.40); (approximately 69.40 metres per tonne)	-	-	-	1199.30	100 m	-
160 x 80 x 5.00 mm (17.80); (approximately 56.20 metres per tonne)	-	-	-	1482.46	100 m	-
160 x 80 x 6.30 mm (22.20); (approximately 45.00 metres per tonne)	-	-	-	1848.92	100 m	-
160 x 80 x 8.00 mm (27.60); (approximately 36.20 metres per tonne)	-	-	-	2298.65	100 m	-
160 x 80 x 10.00 mm (33.70); (approximately 29.70 metres per tonne)	-	-	-	2896.63	100 m	-
200 x 100 x 5.00 mm (22.60); (approximately 44.20 metres per tonne)	-	-	-	1882.23	100 m	-
200 x 100 x 6.30 mm (28.10); (approximately 35.60 metres per tonne)	-	-	-	2340.29	100 m	-
200 x 100 x 8.00 mm (35.10); (approximately 28.50 metres per tonne)	-	-	-	2923.28	100 m	-
200 x 100 x 10.00 mm (43.10); (approximately 23.20 metres per tonne)	-	-	-	3704.59	100 m	-
200 x 100 x 12.50 mm (52.70); (approximately 19.00 metres per tonne)	-	-	-	4970.95	100 m	-
200 x 120 x 5.00 mm (24.10); (approximately 41.50 metres per tonne)	-	-	-	2149.50	100 m	-
200 x 120 x 6.30 mm (30.10); (approximately 33.20 metres per tonne)	-	-	-	2684.64	100 m	-
200 x 120 x 8.00 mm (37.60); (approximately 26.60 metres per tonne)	-	-	-	3353.57	100 m	-
200 x 120 x 10.00 mm (46.30); (approximately 21.60 metres per tonne)	-	-	-	4129.53	100 m	-
200 x 150 x 8.00 mm (41.40); (approximately 24.15 metres per tonne)	-	-	-	3692.50	100 m	-
200 x 150 x 10.00 mm (51.00); (approximately 19.60 metres per tonne)	-	-	-	4548.73	100 m	-

G STRUCTURAL/CARCASSING METAL/TIMBER

Item	PC £	Labour hours	Labour £	Material £	Unit	Total rate £
BASIC STEEL PRICES – cont'd						
HOLLOW SECTIONS – cont'd						
Hot formed structural rectangular hollow section – cont'd						
250 x 100 x 10.00 mm (51.00); (approximately 19.60 metres per tonne)	-	-	-	4548.73	100 m	-
250 x 100 x 12.50 mm (62.50); (approximately 16.00 metres per tonne)	-	-	-	5800.83	100 m	-
250 x 150 x 5.00 mm (30.40); (approximately 32.90 metres per tonne)	-	-	-	2711.40	100 m	-
250 x 150 x 6.30 mm (38.00); (approximately 26.30 metres per tonne)	-	-	-	3389.25	100 m	-
250 x 150 x 8.00 mm (47.70); (approximately 21.00 metres per tonne)	-	-	-	4254.40	100 m	-
250 x 150 x 10.00 mm (58.80); (approximately 17.00 metres per tonne)	-	-	-	5244.42	100 m	-
250 x 150 x 12.50 mm (72.30); (approximately 13.80 metres per tonne)	-	-	-	6710.40	100 m	-
250 x 150 x 16.00 mm (90.30); (approximately 11.10 metres per tonne)	-	-	-	8603.47	100 m	-
300 x 100 x 8.00 mm (47.70); (approximately 21.00 metres per tonne)	-	-	-	4254.40	100 m	-
300 x 100 x 10.00 mm (58.80); (approximately 17.00 metres per tonne)	-	-	-	5244.42	100 m	-
300 x 200 x 6.30 mm (47.90); (approximately 20.90 metres per tonne)	-	-	-	4272.24	100 m	-
300 x 200 x 8.00 mm (60.30); (approximately 16.60 metres per tonne)	-	-	-	5378.20	100 m	-
300 x 200 x 10.00 mm (74.50); (approximately 13.40 metres per tonne)	-	-	-	6644.71	100 m	-
300 x 200 x 12.50 mm (91.90); (approximately 10.90 metres per tonne)	-	-	-	8529.54	100 m	-
300 x 200 x 16.00 mm (115.00); (approximately 8.70 metres per tonne)	-	-	-	10956.80	100 m	-
400 x 200 x 8.00 mm (72.80); (approximately 13.70 metres per tonne)	-	-	-	6493.09	100 m	-
400 x 200 x 10.00 mm (90.20); (approximately 11.10 metres per tonne)	-	-	-	8045.01	100 m	-
400 x 200 x 12.50 mm (112.00); (approximately 8.93 metres per tonne)	-	-	-	10395.09	100 m	-
400 x 200 x 16.00 mm (141.00); (approximately 7.09 metres per tonne)	-	-	-	13434.00	100 m	-
450 x 250 x 8.00 mm (85.40); (approximately 11.70 metres per tonne)	-	-	-	7616.89	100 m	-
450 x 250 x 10.00 mm (106.00); (approximately 9.43 metres per tonne)	-	-	-	9454.22	100 m	-
450 x 250 x 12.50 mm (131.00); (approximately 7.63 metres per tonne)	-	-	-	12158.54	100 m	-
450 x 250 x 16.00 mm (166.00); (approximately 6.02 metres per tonne)	-	-	-	15815.91	100 m	-
500 x 300 x 8.00 mm (98.00); (approximately 10.20 metres per tonne)	-	-	-	8740.70	100 m	-
500 x 300 x 10.00 mm (122.00); (approximately 8.20 metres per tonne)	-	-	-	10881.28	100 m	-
500 x 300 x 12.50 mm (151.00); (approximately 6.62 metres per tonne)	-	-	-	14014.81	100 m	-
500 x 300 x 16.00 mm (191.00); (approximately 5.24 metres per tonne)	-	-	-	18197.82	100 m	-

G STRUCTURAL/CARCASSING METAL/TIMBER

Item	PC £	Labour hours	Labour £	Material £	Unit	Total rate £
SUPPLY AND FIX PRICES						
G10 STRUCTURAL STEEL FRAMING						
Framing, fabrication; weldable steel; BS EN 10025: 2004 Grade S275; hot rolled structural steel sections; welded fabrication						
Columns						
weight not exceeding 40 kg/m	-	-	-	-	tonne	**1538.73**
weight not exceeding 40 kg/m; castellated	-	-	-	-	tonne	**2024.89**
weight not exceeding 40 kg/m; curved	-	-	-	-	tonne	**2019.35**
weight not exceeding 40 kg/m; square hollow section	-	-	-	-	tonne	**1989.83**
weight not exceeding 40 kg/m; circular hollow section	-	-	-	-	tonne	**2097.76**
weight 40 - 100 kg/m	-	-	-	-	tonne	**1342.24**
weight 40 - 100 kg/m; castellated	-	-	-	-	tonne	**1745.37**
weight 40 - 100 kg/m; curved	-	-	-	-	tonne	**1716.77**
weight 40 - 100 kg/m; square hollow section	-	-	-	-	tonne	**1709.39**
weight 40 - 100 kg/m; circular hollow section	-	-	-	-	tonne	**1799.80**
weight exceeding 100 kg/m	-	-	-	-	tonne	**1214.93**
weight exceeding 100 kg/m; castellated	-	-	-	-	tonne	**1542.42**
weight exceeding 100 kg/m; curved	-	-	-	-	tonne	**1535.04**
weight exceeding 100 kg/m; square hollow section	-	-	-	-	tonne	**1649.43**
weight exceeding 100 kg/m; circular hollow section	-	-	-	-	tonne	**1757.36**
Beams						
weight not exceeding 40 kg/m	-	-	-	-	tonne	**1585.78**
weight not exceeding 40 kg/m; castellated	-	-	-	-	tonne	**2073.78**
weight not exceeding 40 kg/m; curved	-	-	-	-	tonne	**2060.86**
weight not exceeding 40 kg/m; square hollow section	-	-	-	-	tonne	**2160.49**
weight not exceeding 40 kg/m; circular hollow section	-	-	-	-	tonne	**2543.33**
weight 40 - 100 kg/m	-	-	-	-	tonne	**1312.72**
weight 40 - 100 kg/m; castellated	-	-	-	-	tonne	**1672.49**
weight 40 - 100 kg/m; curved	-	-	-	-	tonne	**1649.43**
weight 40 - 100 kg/m; square hollow section	-	-	-	-	tonne	**2031.35**
weight 40 - 100 kg/m; circular hollow section	-	-	-	-	tonne	**2361.60**
weight exceeding 100 kg/m	-	-	-	-	tonne	**1211.24**
weight exceeding 100 kg/m; castellated	-	-	-	-	tonne	**1546.11**
weight exceeding 100 kg/m; curved	-	-	-	-	tonne	**1538.73**
weight exceeding 100 kg/m; square hollow section	-	-	-	-	tonne	**1972.31**
weight exceeding 100 kg/m; circular hollow section	-	-	-	-	tonne	**2239.83**
Bracings						
weight not exceeding 40 kg/m	-	-	-	-	tonne	**1882.82**
weight not exceeding 40 kg/m; square hollow section	-	-	-	-	tonne	**2384.66**
weight not exceeding 40 kg/m; circular hollow section	-	-	-	-	tonne	**2384.66**
weight 40 - 100 kg/m	-	-	-	-	tonne	**1752.75**
weight 40 - 100 kg/m; square hollow section	-	-	-	-	tonne	**2243.52**
weight 40 - 100 kg/m; circular hollow section	-	-	-	-	tonne	**2243.52**
weight exceeding 100 kg/m	-	-	-	-	tonne	**1660.50**
weight exceeding 100 kg/m; square hollow section	-	-	-	-	tonne	**2142.05**
weight exceeding 100 kg/m; circular hollow section	-	-	-	-	tonne	**2142.05**

G STRUCTURAL/CARCASSING METAL/TIMBER

Item	PC £	Labour hours	Labour £	Material £	Unit	Total rate £
G10 STRUCTURAL STEEL FRAMING – cont'd						
Framing, fabrication; weldable steel; BS EN 10025: 2004 Grade S275 – cont'd						
Purlins and cladding rails						
weight not exceeding 40 kg/m	-	-	-	-	tonne	**1452.94**
weight not exceeding 40 kg/m; square hollow section	-	-	-	-	tonne	**2416.03**
weight not exceeding 40 kg/m; circular hollow section	-	-	-	-	tonne	**2416.03**
weight 40 - 100 kg/m	-	-	-	-	tonne	**1297.04**
weight 40 - 100 kg/m; square hollow section	-	-	-	-	tonne	**2178.02**
weight 40 - 100 kg/m; circular hollow section	-	-	-	-	tonne	**2178.02**
weight exceeding 100 kg/m	-	-	-	-	tonne	**1199.25**
weight exceeding 100 kg/m; square hollow section	-	-	-	-	tonne	**2107.91**
weight exceeding 100 kg/m; circular hollow section	-	-	-	-	tonne	**2107.91**
Grillages						
weight not exceeding 40 kg/m	-	-	-	-	tonne	**1636.52**
weight 40 - 100 kg/m	-	-	-	-	tonne	**1320.10**
weight exceeding 100 kg/m	-	-	-	-	tonne	**1260.13**
Trestles, towers and built up columns						
straight	-	-	-	-	tonne	**2201.09**
Trusses and built up girders						
straight	-	-	-	-	tonne	**2201.09**
curved	-	-	-	-	tonne	**2702.00**
Fittings	-	-	-	-	tonne	**2162.34**
Add to the aforementioned prices for:						
grade 355 steelwork	-	-	-	-	%	**6.92**
Framing, erection						
Trial erection	-	-	-	-	tonne	**368.08**
Permanent erection on site	-	-	-	-	tonne	**368.08**
Surface preparation						
At works						
blast cleaning	-	-	-	-	m²	**2.58**
Surface treatment						
At works						
galvanising	-	-	-	-	m²	**12.64**
shotblasting and priming to SA 2.5	-	-	-	-	m²	**6.79**
touch up primer and one coat of two pack epoxy zinc phosphate primer	-	-	-	-	m²	**4.71**
intumescent paint fire protection (30 minutes); spray applied	-	-	-	-	m²	**10.69**
intumescent paint fire protection (60 minutes); spray applied	-	-	-	-	m²	**16.04**
Extra over for; separate decorative sealer top coat	-	-	-	-	m²	**3.21**
On site						
intumescent paint fire protection (30 minutes); spray applied	-	-	-	-	m²	**8.02**
intumescent paint fire protection (30 minutes) to circular columns etc.; spray applied	-	-	-	-	m²	**13.44**
intumescent paint fire protection (60 minutes) to UBs etc.; spray applied	-	-	-	-	m²	**10.16**
intumescent paint fire protection (60 minutes) to circular columns etc.; spray applied	-	-	-	-	m²	**17.05**
Extra over for; separate decorative sealer top coat	-	-	-	-	m²	**2.67**

G STRUCTURAL/CARCASSING METAL/TIMBER

Item	PC £	Labour hours	Labour £	Material £	Unit	Total rate £
Metsec Lightweight Steel Framing System; or other equal and approved; as inner leaf to external wall; studs typically at 600 mm centres; including provision for all openings, abutments, junctions and head details etc.						
Inner leaf; with supports and perimeter sections for external metal cladding (measured separately)						
100 mm thick steel walling	-	-	-	-	m²	**62.33**
150 mm thick steel walling	-	-	-	-	m²	**66.35**
200 mm thick steel walling	-	-	-	-	m²	**76.51**
Inner leaf; with 16 mm Pyroc sheething board						
100 mm thick steel walling	-	-	-	-	m²	**79.28**
150 mm thick steel walling	-	-	-	-	m²	**83.29**
200 mm thick steel walling	-	-	-	-	m²	**93.46**
16 mm Pyroc sheething board fixed to slab perimeter not exceeding 300 mm	-	-	-	-	m	**8.79**
Inner leaf; with 16 mm Pyroc sheething board and 40 mm Thermawall TW55 insulation supported by halfen channels type 28/15 fixed to studs at 450 mm centres.						
100 mm thick steel walling	-	-	-	-	m²	**88.69**
150 mm thick steel walling	-	-	-	-	m²	**92.71**
200 mm thick steel walling	-	-	-	-	m²	**102.88**
16 mm Pyroc sheething board and 40 mm Thermawall TW55 insulation fixed to slab perimeter not exceeding 300 mm	-	-	-	-	m	**10.04**
Cold formed galvanised steel; Kingspan "Multibeam" or other equal and approved						
Cold rolled purlins and cladding rails						
175 x 65 x 1.40 mm gauge purlins or rails; fixed to steelwork	-	0.04	0.73	7.60	m	**8.32**
175 x 65 x 1.60 mm gauge purlins or rails; fixed to steelwork	-	0.04	0.73	8.09	m	**8.81**
175 x 65 x 2.00 mm gauge purlins or rails; fixed to steelwork	-	0.04	0.73	9.77	m	**10.50**
205 x 65 x 1.40 mm gauge purlins or rails; fixed to steelwork	-	0.04	0.73	8.41	m	**9.14**
205 x 65 x 1.60 mm gauge purlins or rails; fixed to steelwork	-	0.04	0.73	9.13	m	**9.86**
205 x 65 x 2.00 mm gauge purlins or rails; fixed to steelwork	-	0.04	0.73	10.40	m	**11.13**
Heavy duty Zed section spacers						
vertically; across cladding rails; fixed to steelwork	-	0.05	0.91	6.87	m	**7.78**
Cleats						
weld-on for 175 mm purlin or rail	-	0.10	1.82	2.18	nr	**4.00**
bolt-on for 175 mm purlin or rail; including fixing bolts	-	0.02	0.36	4.44	m	**4.80**
weld-on for 205 mm purlin or rail	-	0.10	1.82	2.48	nr	**4.30**
bolt-on for 205 mm purlin or rail; including fixing bolts	-	0.02	0.36	4.79	m	**5.16**
Tubular ties						
1500 mm long; bolted diagonally across purlins or cladding rails	-	0.02	0.36	4.78	m	**5.15**

G STRUCTURAL/CARCASSING METAL/TIMBER

Item	PC £	Labour hours	Labour £	Material £	Unit	Total rate £
G12 ISOLATED STRUCTURAL METAL MEMBERS						
Isolated structural member; weldable steel; BS EN 10025: 2004 Grade S275; hot rolled structural steel sections						
Plain member; beams						
weight not exceeding 40 kg/m	-	-	-	-	tonne	**1182.64**
weight 40 - 100 kg/m	-	-	-	-	tonne	**1151.28**
weight exceeding 100 kg/m	-	-	-	-	tonne	**1121.76**
Metsec open web steel lattice beams or other equal and approved; in single members; raised 3.50 m above ground; ends built in						
Beams; one coat zinc phosphate primer at works						
220 mm deep; to span 6.00 m (11.50 kg/m); ref B22	-	0.19	5.28	37.17	m	**42.44**
270 mm deep; to span 7.00 m (11.50 kg/m); ref B27	-	0.19	5.28	37.17	m	**42.44**
300 mm deep; to span 8.00 m (12.50 kg/m); ref B30	-	0.23	6.39	40.07	m	**46.46**
350 mm deep; to span 9.00 m (14.00 kg/m); ref B35	-	0.23	6.39	44.92	m	**51.30**
350 mm deep; to span 10.00 m (20.00 kg/m); ref D35	-	0.28	7.77	64.29	m	**72.07**
450 mm deep; to span 11.00 m (21.00 kg/m); ref D45	-	0.32	8.89	67.20	m	**76.08**
450 mm deep; to span 12.00 m (32.50 kg/m); ref G45	-	0.46	12.77	104.00	m	**116.78**
Beams; galvanised						
220 mm deep; to span 6.00 m (11.50 kg/m); ref B22	-	0.19	5.28	39.11	m	**44.38**
270 mm deep; to span 7.00 m (11.50 kg/m); ref B27	-	0.19	5.28	39.11	m	**44.38**
300 mm deep; to span 8.00 m (12.50 kg/m); ref B30	-	0.23	6.39	42.98	m	**49.37**
350 mm deep; to span 9.00 m (14.00 kg/m); ref B35	-	0.23	6.39	47.82	m	**54.21**
350 mm deep; to span 10.00 m (20.00 kg/m); ref D35	-	0.28	7.77	68.16	m	**75.94**
450 mm deep; to span 11.00 m (21.00 kg/m); ref D45	-	0.32	8.89	72.04	m	**80.92**
450 mm deep; to span 12.00 m (32.50 kg/m); ref G45	-	0.46	12.77	110.78	m	**123.56**
G20 CARPENTRY/TIMBER FRAMING/FIRST FIXING						
BASIC TIMBER PRICES						
Hardwood; Joinery quality (£/m³)						
American Cherry	-	-	-	1700.47	m³	-
American White Ash	-	-	-	624.23	m³	-
American White Oak	-	-	-	1119.30	m³	-
Beech	-	-	-	823.33	m³	-
Douglas Fir	-	-	-	742.61	m³	-
European Oak	-	-	-	1969.54	m³	-
Idigbo	-	-	-	796.42	m³	-
Iroko	-	-	-	925.58	m³	-
Maple	-	-	-	1194.64	m³	-
Poplar	-	-	-	575.79	m³	-
Sapele	-	-	-	930.96	m³	-
Utile	-	-	-	1060.11	m³	-
Red Meranti	-	-	-	791.04	m³	-

G STRUCTURAL/CARCASSING METAL/TIMBER

Item	PC £	Labour hours	Labour £	Material £	Unit	**Total rate £**
Softwood; Carcassing quality (£/m³)						
2.00 m - 4.80 m lengths	-	-	-	285.00	m³	-
4.80 m - 6.00 m lengths	-	-	-	270.00	m³	-
6.00 m - 9.00 m lengths	-	-	-	300.00	m³	-
G.S. Grade	-	-	-	18.00	m³	-
S.S. Grade	-	-	-	36.00	m³	-
Softwood; Joinery quality (£/m³)	-	-	-	339.02	m³	-
Timber Treatment (£/m³)						
Pre-treatment of timber by vacuum/pressure impregnation, excluding transport costs and any subsequent seasoning:						
interior work; minimum salt retention 4.00 kg/m³	-	-	-	45.60	m³	-
exterior work; minimum salt retention 5.30 kg/m³	-	-	-	52.50	m³	-
Pre-treatment of timber including flame proofing all purposes; minimum salt retention 36.00 kg/m³	-	-	-	139.47	m³	-
"Aquaseal" timber treatments - (£/25 litres)						
"Timbershield"	-	-	-	109.61	25 litre	-
"Longlife Wood Protector"	-	-	-	55.24	25 litre	-
SUPPLY AND FIX PRICES						
Sawn softwood; untreated						
Floor members						
38 mm x 100 mm	-	0.11	2.30	1.10	m	**3.40**
38 mm x 150 mm	-	0.13	2.72	1.52	m	**4.24**
47 mm x 75 mm	-	0.11	2.30	1.17	m	**3.47**
47 mm x 100 mm	-	0.13	2.72	1.45	m	**4.17**
47 mm x 125 mm	-	0.13	2.72	1.77	m	**4.50**
47 mm x 150 mm	-	0.14	2.93	2.10	m	**5.03**
47 mm x 175 mm	-	0.14	2.93	2.56	m	**5.49**
47 mm x 200 mm	-	0.15	3.14	2.86	m	**6.00**
47 mm x 225 mm	-	0.15	3.14	3.32	m	**6.46**
47 mm x 250 mm	-	0.16	3.35	3.74	m	**7.09**
75 mm x 125 mm	-	0.15	3.14	2.66	m	**5.80**
75 mm x 150 mm	-	0.15	3.14	3.15	m	**6.29**
75 mm x 175 mm	-	0.15	3.14	3.77	m	**6.91**
75 mm x 200 mm	-	0.16	3.35	4.39	m	**7.74**
75 mm x 225 mm	-	0.16	3.35	4.97	m	**8.32**
75 mm x 250 mm	-	0.17	3.56	6.06	m	**9.62**
100 mm x 150 mm	-	0.20	4.19	4.25	m	**8.44**
100 mm x 200 mm	-	0.21	4.40	6.36	m	**10.76**
100 mm x 250 mm	-	0.23	4.82	8.10	m	**12.92**
100 mm x 300 mm	-	0.25	5.24	7.93	m	**13.16**
Wall or partition members						
25 mm x 25 mm	-	0.06	1.26	0.44	m	**1.70**
25 mm x 38 mm	-	0.06	1.26	0.50	m	**1.75**
25 mm x 75 mm	-	0.08	1.68	0.61	m	**2.29**
38 mm x 38 mm	-	0.08	1.68	0.58	m	**2.25**
38 mm x 50 mm	-	0.08	1.68	0.72	m	**2.40**
38 mm x 75 mm	-	0.11	2.30	0.89	m	**3.19**
38 mm x 100 mm	-	0.14	2.93	1.10	m	**4.03**
47 mm x 50 mm	-	0.11	2.30	0.78	m	**3.08**
47 mm x 75 mm	-	0.14	2.93	1.20	m	**4.13**
47 mm x 100 mm	-	0.17	3.56	1.48	m	**5.04**
47 mm x 125 mm	-	0.18	3.77	1.80	m	**5.57**
75 mm x 75 mm	-	0.17	3.56	1.62	m	**5.18**
75 mm x 100 mm	-	0.19	3.98	2.27	m	**6.25**
100 mm x 100 mm	-	0.19	3.98	2.95	m	**6.93**

G STRUCTURAL/CARCASSING METAL/TIMBER

Item	PC £	Labour hours	Labour £	Material £	Unit	Total rate £
G20 CARPENTRY/TIMBER FRAMING/FIRST FIXING – cont'd						
Sawn softwood; untreated – cont'd						
Joist strutting; herringbone						
47 mm x 50 mm; depth of joist 150 mm	-	0.46	9.63	1.86	m	**11.49**
47 mm x 50 mm; depth of joist 175 mm	-	0.46	9.63	1.89	m	**11.53**
47 mm x 50 mm; depth of joist 200 mm	-	0.46	9.63	1.93	m	**11.56**
47 mm x 50 mm; depth of joist 225 mm	-	0.46	9.63	1.96	m	**11.59**
47 mm x 50 mm; depth of joist 250 mm	-	0.46	9.63	2.00	m	**11.63**
Joist strutting; block						
47 mm x 150 mm; depth of joist 150 mm	-	0.28	5.86	2.47	m	**8.34**
47 mm x 175 mm; depth of joist 175 mm	-	0.28	5.86	2.93	m	**8.80**
47 mm x 200 mm; depth of joist 200 mm	-	0.28	5.86	3.24	m	**9.10**
47 mm x 225 mm; depth of joist 225 mm	-	0.28	5.86	3.70	m	**9.56**
47 mm x 250 mm; depth of joist 250 mm	-	0.28	5.86	4.11	m	**9.98**
Cleats						
225 mm x 100 mm x 75 mm	-	0.19	3.98	0.56	nr	**4.54**
Extra for stress grading to above timbers						
general structural (GS) grade	-	-	-	20.83	m^3	-
special structural (SS) grade	-	-	-	41.65	m^3	-
Extra for protecting and flameproofing timber with "Celgard CF" protection or other equal and approved						
small sections	-	-	-	114.53	m^3	-
large sections	-	-	-	109.95	m^3	-
Wrot surfaces						
plain; 50 mm wide	-	0.02	0.42	-	m	**0.42**
plain; 100 mm wide	-	0.03	0.63	-	m	**0.63**
plain; 150 mm wide	-	0.04	0.84	-	m	**0.84**
Sawn softwood; "Tanalised"						
Floor members						
38 mm x 75 mm	-	0.11	2.30	1.00	m	**3.30**
38 mm x 100 mm	-	0.11	2.30	1.24	m	**3.54**
38 mm x 150 mm	-	0.13	2.72	1.72	m	**4.45**
47 mm x 75 mm	-	0.11	2.30	1.31	m	**3.61**
47 mm x 100 mm	-	0.13	2.72	1.63	m	**4.35**
47 mm x 125 mm	-	0.13	2.72	2.00	m	**4.73**
47 mm x 150 mm	-	0.14	2.93	2.37	m	**5.31**
47 mm x 175 mm	-	0.14	2.93	2.88	m	**5.81**
47 mm x 200 mm	-	0.15	3.14	3.23	m	**6.37**
47 mm x 225 mm	-	0.15	3.14	3.73	m	**6.87**
47 mm x 250 mm	-	0.16	3.35	4.19	m	**7.54**
75 mm x 125 mm	-	0.15	3.14	3.00	m	**6.14**
75 mm x 150 mm	-	0.15	3.14	3.56	m	**6.70**
75 mm x 175 mm	-	0.15	3.14	4.24	m	**7.39**
75 mm x 200 mm	-	0.16	3.35	4.94	m	**8.29**
75 mm x 225 mm	-	0.16	3.35	5.59	m	**8.94**
75 mm x 250 mm	-	0.17	3.56	6.74	m	**10.30**
100 mm x 150 mm	-	0.20	4.19	4.80	m	**8.99**
100 mm x 200 mm	-	0.21	4.40	7.09	m	**11.49**
100 mm x 250 mm	-	0.23	4.82	9.01	m	**13.83**
100 mm x 300 mm	-	0.25	5.24	9.02	m	**14.26**
Wall or partition members						
25 mm x 25 mm	-	0.06	1.26	0.46	m	**1.72**
25 mm x 38 mm	-	0.06	1.26	0.53	m	**1.79**
25 mm x 75 mm	-	0.08	1.68	0.68	m	**2.36**
38 mm x 38 mm	-	0.08	1.68	0.63	m	**2.30**
38 mm x 50 mm	-	0.08	1.68	0.79	m	**2.47**
38 mm x 75 mm	-	0.11	2.30	1.00	m	**3.30**

G STRUCTURAL/CARCASSING METAL/TIMBER

Item	PC £	Labour hours	Labour £	Material £	Unit	Total rate £
38 mm x 100 mm	-	0.14	2.93	1.24	m	**4.17**
47 mm x 50 mm	-	0.11	2.30	0.87	m	**3.18**
47 mm x 75 mm	-	0.14	2.93	1.34	m	**4.27**
47 mm x 100 mm	-	0.17	3.56	1.66	m	**5.22**
47 mm x 125 mm	-	0.18	3.77	2.03	m	**5.80**
75 mm x 75 mm	-	0.17	3.56	1.82	m	**5.38**
75 mm x 100 mm	-	0.19	3.98	2.54	m	**6.52**
100 mm x 100 mm	-	0.19	3.98	3.31	m	**7.29**
Roof members; flat						
38 mm x 75 mm	-	0.13	2.72	1.00	m	**3.72**
38 mm x 100 mm	-	0.13	2.72	1.24	m	**3.96**
38 mm x 125 mm	-	0.13	2.72	1.48	m	**4.21**
38 mm x 150 mm	-	0.13	2.72	1.72	m	**4.45**
47 mm x 100 mm	-	0.13	2.72	1.63	m	**4.35**
47 mm x 125 mm	-	0.13	2.72	2.00	m	**4.73**
47 mm x 150 mm	-	0.14	2.93	2.37	m	**5.31**
47 mm x 175 mm	-	0.14	2.93	2.88	m	**5.81**
47 mm x 200 mm	-	0.15	3.14	3.23	m	**6.37**
47 mm x 225 mm	-	0.15	3.14	3.73	m	**6.87**
47 mm x 250 mm	-	0.16	3.35	4.19	m	**7.54**
75 mm x 150 mm	-	0.15	3.14	3.56	m	**6.70**
75 mm x 175 mm	-	0.15	3.14	4.24	m	**7.39**
75 mm x 200 mm	-	0.16	3.35	4.94	m	**8.29**
75 mm x 225 mm	-	0.16	3.35	5.59	m	**8.94**
75 mm x 250 mm	-	0.17	3.56	6.74	m	**10.30**
Roof members; pitched						
25 mm x 100 mm	-	0.11	2.30	0.98	m	**3.29**
25 mm x 125 mm	-	0.11	2.30	1.32	m	**3.63**
25 mm x 150 mm	-	0.14	2.93	1.59	m	**4.52**
25 mm x 175 mm	-	0.16	3.35	1.86	m	**5.21**
25 mm x 200 mm	-	0.17	3.56	2.13	m	**5.69**
38 mm x 100 mm	-	0.14	2.93	1.24	m	**4.17**
38 mm x 125 mm	-	0.14	2.93	1.48	m	**4.42**
38 mm x 150 mm	-	0.14	2.93	1.72	m	**4.65**
38 mm x 175 mm	-	0.16	3.35	2.04	m	**5.39**
38 mm x 200 mm	-	0.17	3.56	2.35	m	**5.91**
47 mm x 50 mm	-	0.11	2.30	0.84	m	**3.15**
47 mm x 75 mm	-	0.14	2.93	1.31	m	**4.24**
47 mm x 100 mm	-	0.17	3.56	1.63	m	**5.19**
47 mm x 125 mm	-	0.17	3.56	2.00	m	**5.56**
47 mm x 150 mm	-	0.19	3.98	2.37	m	**6.35**
47 mm x 175 mm	-	0.19	3.98	2.88	m	**6.86**
47 mm x 200 mm	-	0.19	3.98	3.23	m	**7.21**
47 mm x 225 mm	-	0.19	3.98	3.73	m	**7.71**
75 mm x 100 mm	-	0.23	4.82	2.48	m	**7.30**
75 mm x 125 mm	-	0.23	4.82	3.00	m	**7.81**
75 mm x 150 mm	-	0.23	4.82	3.56	m	**8.38**
100 mm x 150 mm	-	0.28	5.86	4.83	m	**10.69**
100 mm x 175 mm	-	0.28	5.86	5.61	m	**11.48**
100 mm x 200 mm	-	0.28	5.86	7.09	m	**12.96**
100 mm x 225 mm	-	0.31	6.49	8.05	m	**14.54**
100 mm x 250 mm	-	0.31	6.49	9.01	m	**15.50**
Plates						
38 mm x 75 mm	-	0.11	2.30	1.03	m	**3.34**
38 mm x 100 mm	-	0.14	2.93	1.24	m	**4.17**
47 mm x 75 mm	-	0.14	2.93	1.31	m	**4.24**
47 mm x 100 mm	-	0.17	3.56	1.63	m	**5.19**
75 mm x 100 mm	-	0.19	3.98	2.48	m	**6.46**
75 mm x 125 mm	-	0.22	4.61	2.97	m	**7.58**
75 mm x 150 mm	-	0.25	5.24	3.53	m	**8.77**

G STRUCTURAL/CARCASSING METAL/TIMBER

Item	PC £	Labour hours	Labour £	Material £	Unit	Total rate £
G20 CARPENTRY/TIMBER FRAMING/FIRST FIXING – cont'd						
Sawn softwood; "Tanalised" – cont'd						
Plates; fixing by bolting						
38 mm x 75 mm	-	0.20	4.19	1.00	m	**5.18**
38 mm x 100 mm	-	0.23	4.82	1.24	m	**6.05**
47 mm x 75 mm	-	0.23	4.82	1.31	m	**6.12**
47 mm x 100 mm	-	0.26	5.45	1.63	m	**7.07**
75 mm x 100 mm	-	0.29	6.07	2.48	m	**8.56**
75 mm x 125 mm	-	0.31	6.49	2.97	m	**9.46**
75 mm x 150 mm	-	0.34	7.12	3.53	m	**10.65**
Joist strutting; herringbone						
47 mm x 50 mm; depth of joist 150 mm	-	0.46	9.63	2.04	m	**11.68**
47 mm x 50 mm; depth of joist 175 mm	-	0.46	9.63	2.08	m	**11.72**
47 mm x 50 mm; depth of joist 200 mm	-	0.46	9.63	2.12	m	**11.76**
47 mm x 50 mm; depth of joist 225 mm	-	0.46	9.63	2.16	m	**11.79**
47 mm x 50 mm; depth of joist 250 mm	-	0.46	9.63	2.20	m	**11.83**
Joist strutting; block						
47 mm x 150 mm; depth of joist 150 mm	-	0.28	5.86	2.75	m	**8.61**
47 mm x 175 mm; depth of joist 175 mm	-	0.28	5.86	3.25	m	**9.12**
47 mm x 200 mm; depth of joist 200 mm	-	0.28	5.86	3.60	m	**9.47**
47 mm x 225 mm; depth of joist 225 mm	-	0.28	5.86	4.11	m	**9.97**
47 mm x 250 mm; depth of joist 250 mm	-	0.28	5.86	4.57	m	**10.43**
Cleats						
225 mm x 100 mm x 75 mm	-	0.19	3.98	0.62	nr	**4.60**
Extra for stress grading to above timbers						
general structural (GS) grade	-	-	-	20.83	m³	**-**
special structural (SS) grade	-	-	-	41.65	m³	**-**
Extra for protecting and flameproofing timber with "Celgard CF" protection or other equal and approved						
small sections	-	-	-	114.53	m³	**-**
large sections	-	-	-	109.95	m³	**-**
Wrot surfaces						
plain; 50 mm wide	-	0.02	0.42	-	m	**0.42**
plain; 100 mm wide	-	0.03	0.63	-	m	**0.63**
plain; 150 mm wide	-	0.04	0.84	-	m	**0.84**
Trussed rafters, stress graded sawn softwood pressure impregnated; raised through two storeys and fixed in position						
"W" type truss (Fink); 22.5 degree pitch; 450 mm eaves overhang						
5.00 m span	-	1.48	31.00	25.79	nr	**56.79**
7.60 m span	-	1.62	33.93	32.47	nr	**66.40**
10.00 m span	-	1.85	38.75	53.28	nr	**92.02**
"W" type truss (Fink); 30 degree pitch; 450 mm eaves overhang						
5.00 m span	-	1.48	31.00	26.03	nr	**57.03**
7.60 m span	-	1.62	33.93	34.31	nr	**68.24**
10.00 m span	-	1.85	38.75	56.17	nr	**94.92**
"W" type truss (Fink); 45 degree pitch; 450 mm eaves overhang						
4.60 m span	-	1.48	31.00	27.47	nr	**58.47**
7.00 m span	-	1.62	33.93	40.82	nr	**74.75**

G STRUCTURAL/CARCASSING METAL/TIMBER

Item	PC £	Labour hours	Labour £	Material £	Unit	**Total rate £**
"Mono" type truss; 17.5 degree pitch; 450 mm eaves overhang						
3.30 m span	-	1.30	27.23	20.40	nr	**47.63**
5.60 m span	-	1.48	31.00	30.59	nr	**61.59**
7.00 m span	-	1.71	35.81	38.25	nr	**74.07**
"Attic" type truss; 45 degree pitch; 450 mm eaves overhang						
5.00 m span	-	2.91	60.95	56.86	nr	**117.81**
7.60 m span	-	3.05	63.88	100.96	nr	**164.84**
9.00 m span	-	3.24	67.86	128.99	nr	**196.84**
"Moelven Toreboda" glulam timber beams or other equal and approved; Moelven Laminated Timber Structures; LB grade whitewood; pressure impregnated; phenbol resorcinal adhesive; clean planed finish; fixed						
Laminated roof beams						
56 mm x 225 mm	-	0.51	10.68	6.13	m	**16.81**
66 mm x 315 mm	-	0.65	13.61	10.11	m	**23.72**
90 mm x 315 mm	-	0.83	17.38	13.78	m	**31.17**
90 mm x 405 mm	-	1.06	22.20	17.72	m	**39.92**
115 mm x 405 mm	-	1.34	28.06	22.65	m	**50.71**
115 mm x 495 mm	-	1.67	34.98	27.68	m	**62.66**
115 mm x 630 mm	-	2.04	42.73	35.23	m	**77.95**
"Masterboard" or other equal and approved; 6 mm thick						
Eaves, verge soffit boards, fascia boards and the like						
over 300 mm wide	6.60	0.65	13.61	8.21	m²	**21.83**
75 mm wide	-	0.19	3.98	0.63	m	**4.61**
150 mm wide	-	0.22	4.61	1.23	m	**5.84**
225 mm wide	-	0.26	5.45	1.83	m	**7.28**
300 mm wide	-	0.28	5.86	2.43	m	**8.30**
Plywood; external quality; 12 mm thick						
Eaves, verge soffit boards, fascia boards and the like						
over 300 mm wide	8.01	0.76	15.92	9.85	m²	**25.76**
75 mm wide	-	0.23	4.82	0.75	m	**5.57**
150 mm wide	-	0.27	5.65	1.48	m	**7.13**
225 mm wide	-	0.31	6.49	2.20	m	**8.69**
300 mm wide	-	0.34	7.12	2.92	m	**10.05**
Plywood; external quality; 15 mm thick						
Eaves, verge soffit boards, fascia boards and the like						
over 300 mm wide	10.06	0.76	15.92	12.21	m²	**28.13**
75 mm wide	-	0.23	4.82	0.93	m	**5.75**
150 mm wide	-	0.27	5.65	1.83	m	**7.49**
225 mm wide	-	0.31	6.49	2.73	m	**9.23**
300 mm wide	-	0.34	7.12	3.63	m	**10.76**
Plywood; external quality; 18 mm thick						
Eaves, verge soffit boards, fascia boards and the like						
over 300 mm wide	12.01	0.76	15.92	14.47	m²	**30.39**
75 mm wide	-	0.23	4.82	1.10	m	**5.92**
150 mm wide	-	0.27	5.65	2.17	m	**7.83**
225 mm wide	-	0.31	6.49	3.24	m	**9.73**
300 mm wide	-	0.34	7.12	4.31	m	**11.43**

G STRUCTURAL/CARCASSING METAL/TIMBER

Item	PC £	Labour hours	Labour £	Material £	Unit	Total rate £
G20 CARPENTRY/TIMBER FRAMING/FIRST FIXING – cont'd						
Plywood; marine quality; 18 mm thick						
Gutter boards; butt joints						
over 300 mm wide	9.30	0.86	18.01	11.34	m²	**29.35**
150 mm wide	-	0.31	6.49	1.70	m	**8.19**
225 mm wide	-	0.34	7.12	2.57	m	**9.69**
300 mm wide	-	0.38	7.96	3.40	m	**11.36**
Eaves, verge soffit boards, fascias boards and the like						
over 300 mm wide	-	0.76	15.92	11.34	m²	**27.26**
75 mm wide	-	0.23	4.82	0.86	m	**5.68**
150 mm wide	-	0.27	5.65	1.70	m	**7.36**
225 mm wide	-	0.31	6.49	2.54	m	**9.03**
300 mm wide	-	0.34	7.12	3.37	m	**10.49**
Plywood; marine quality; 25 mm thick						
Gutter boards; butt joints						
over 300 mm wide	12.93	0.93	19.48	15.53	m²	**35.01**
150 mm wide	-	0.32	6.70	2.33	m	**9.03**
225 mm wide	-	0.37	7.75	3.51	m	**11.26**
300 mm wide	-	0.42	8.80	4.66	m	**13.46**
Eaves, verge soffit boards, fascia baords and the like						
over 300 mm wide	-	0.81	16.96	15.53	m²	**32.50**
75 mm wide	-	0.24	5.03	1.18	m	**6.21**
150 mm wide	-	0.29	6.07	2.33	m	**8.40**
225 mm wide	-	0.29	6.07	3.48	m	**9.55**
300 mm wide	-	0.37	7.75	4.63	m	**12.38**
Sawn softwood; untreated						
Gutter boards; butt joints						
19 mm thick; sloping	-	1.16	24.29	6.56	m²	**30.86**
19 mm thick; 75 mm wide	-	0.32	6.70	0.50	m	**7.21**
19 mm thick; 150 mm wide	-	0.37	7.75	0.95	m	**8.70**
19 mm thick; 225 mm wide	-	0.42	8.80	1.69	m	**10.49**
25 mm thick; sloping	-	1.16	24.29	10.27	m²	**34.56**
25 mm thick; 75 mm wide	-	0.32	6.70	0.64	m	**7.34**
25 mm thick; 150 mm wide	-	0.37	7.75	1.50	m	**9.25**
25 mm thick; 225 mm wide	-	0.42	8.80	2.37	m	**11.16**
Cesspools with 25 mm thick sides and bottom						
225 mm x 225 mm x 150 mm	-	1.11	23.25	1.97	nr	**25.22**
300 mm x 300 mm x 150 mm	-	1.30	27.23	2.59	nr	**29.81**
Individual supports; firrings						
50 mm wide x 36 mm average depth	-	0.14	2.93	1.36	m	**4.30**
50 mm wide x 50 mm average depth	-	0.14	2.93	2.06	m	**4.99**
50 mm wide x 75 mm average depth	-	0.14	2.93	2.65	m	**5.58**
Individual supports; bearers						
25 mm x 50 mm	-	0.09	1.88	0.62	m	**2.50**
38 mm x 50 mm	-	0.09	1.88	0.78	m	**2.67**
50 mm x 50 mm	-	0.09	1.88	0.81	m	**2.69**
50 mm x 75 mm	-	0.09	1.88	1.23	m	**3.11**
Individual supports; angle fillets						
38 mm x 38 mm	-	0.09	1.88	0.55	m	**2.43**
50 mm x 50 mm	-	0.09	1.88	0.69	m	**2.58**
75 mm x 75 mm	-	0.11	2.30	1.40	m	**3.70**

G STRUCTURAL/CARCASSING METAL/TIMBER

Item	PC £	Labour hours	Labour £	Material £	Unit	Total rate £
Individual supports; tilting fillets						
19 mm x 38 mm	-	0.09	1.88	0.34	m	**2.23**
25 mm x 50 mm	-	0.09	1.88	0.53	m	**2.42**
38 mm x 75 mm	-	0.09	1.88	0.81	m	**2.69**
50 mm x 75 mm	-	0.09	1.88	1.03	m	**2.92**
75 mm x 100 mm	-	0.14	2.93	1.90	m	**4.83**
Individual supports; grounds or battens						
13 mm x 19 mm	-	0.04	0.84	0.27	m	**1.10**
13 mm x 32 mm	-	0.04	0.84	0.27	m	**1.10**
25 mm x 50 mm	-	0.04	0.84	0.56	m	**1.40**
Individual supports; grounds or battens; plugged and screwed						
13 mm x 19 mm	-	0.14	2.93	0.31	m	**3.25**
13 mm x 32 mm	-	0.14	2.93	0.31	m	**3.25**
25 mm x 50 mm	-	0.14	2.93	0.61	m	**3.54**
Framed supports; open-spaced grounds or battens; at 300 mm centres one way						
25 mm x 50 mm	-	0.14	2.93	1.84	m²	**4.77**
25 mm x 50 mm; plugged and screwed	-	0.42	8.80	2.02	m²	**10.82**
Framed supports; at 300 mm centres one way and 600 mm centres the other way						
25 mm x 50 mm	-	0.69	14.45	2.76	m²	**17.21**
38 mm x 50 mm	-	0.69	14.45	3.59	m²	**18.04**
50 mm x 50 mm	-	0.69	14.45	3.73	m²	**18.18**
50 mm x 75 mm	-	0.69	14.45	5.81	m²	**20.26**
75 mm x 75 mm	-	0.69	14.45	7.93	m²	**22.38**
Framed supports; at 300 mm centres one way and 600 mm centres the other way; plugged and screwed						
25 mm x 50 mm	-	1.16	24.29	3.26	m²	**27.56**
38 mm x 50 mm	-	1.16	24.29	4.10	m²	**28.39**
50 mm x 50 mm	-	1.16	24.29	4.24	m²	**28.53**
50 mm x 75 mm	-	1.16	24.29	6.32	m²	**30.61**
75 mm x 75 mm	-	1.16	24.29	8.44	m²	**32.73**
Framed supports; at 500 mm centres both ways						
25 mm x 50 mm; to bath panels	-	0.83	17.38	3.59	m²	**20.98**
Framed supports; as bracketing and cradling around steelwork						
25 mm x 50 mm	-	1.30	27.23	3.90	m²	**31.13**
50 mm x 50 mm	-	1.39	29.11	5.26	m²	**34.38**
50 mm x 75 mm	-	1.48	31.00	8.18	m²	**39.18**
Sawn softwood; "Tanalised"						
Gutter boards; butt joints						
19 mm thick; sloping	-	1.16	24.29	7.26	m²	**31.55**
19 mm thick; 75 mm wide	-	0.32	6.70	0.55	m	**7.26**
19 mm thick; 150 mm wide	-	0.37	7.75	1.05	m	**8.80**
19 mm thick; 225 mm wide	-	0.42	8.80	1.85	m	**10.64**
25 mm thick; sloping	-	1.16	24.29	11.18	m²	**35.47**
25 mm thick; 75 mm wide	-	0.32	6.70	0.71	m	**7.41**
25 mm thick; 150 mm wide	-	0.37	7.75	1.64	m	**9.39**
25 mm thick; 225 mm wide	-	0.42	8.80	2.57	m	**11.37**
Cesspools with 25 mm thick sides and bottom						
225 mm x 225 mm x 150 mm	-	1.11	23.25	2.16	nr	**25.41**
300 mm x 300 mm x 150 mm	-	1.30	27.23	2.83	nr	**30.06**
Individual supports; firrings						
50 mm wide x 36 mm average depth	-	0.14	2.93	1.43	m	**4.36**
50 mm wide x 50 mm average depth	-	0.14	2.93	2.15	m	**5.08**
50 mm wide x 75 mm average depth	-	0.14	2.93	2.79	m	**5.72**

G STRUCTURAL/CARCASSING METAL/TIMBER

Item	PC £	Labour hours	Labour £	Material £	Unit	Total rate £
G20 CARPENTRY/TIMBER FRAMING/FIRST FIXING – cont'd						
Sawn softwood; "Tanalised" – cont'd						
Individual supports; bearers						
25 mm x 50 mm	-	0.09	1.88	0.66	m	**2.55**
38 mm x 50 mm	-	0.09	1.88	0.85	m	**2.73**
50 mm x 50 mm	-	0.09	1.88	0.90	m	**2.79**
50 mm x 75 mm	-	0.09	1.88	1.37	m	**3.25**
Individual supports; angle fillets						
38 mm x 38 mm	-	0.09	1.88	0.57	m	**2.46**
50 mm x 50 mm	-	0.09	1.88	0.74	m	**2.62**
75 mm x 75 mm	-	0.11	2.30	1.50	m	**3.81**
Individual supports; tilting fillets						
19 mm x 38 mm	-	0.09	1.88	0.36	m	**2.24**
25 mm x 50 mm	-	0.09	1.88	0.55	m	**2.44**
38 mm x 75 mm	-	0.09	1.88	0.86	m	**2.75**
50 mm x 75 mm	-	0.09	1.88	1.10	m	**2.98**
75 mm x 100 mm	-	0.14	2.93	2.04	m	**4.97**
Individual supports; grounds or battens						
13 mm x 19 mm	-	0.04	0.84	0.27	m	**1.11**
13 mm x 32 mm	-	0.04	0.84	0.28	m	**1.12**
25 mm x 50 mm	-	0.04	0.84	0.60	m	**1.44**
Individual supports; grounds or battens; plugged and screwed						
13 mm x 19 mm	-	0.14	2.93	0.32	m	**3.25**
13 mm x 32 mm	-	0.14	2.93	0.33	m	**3.26**
25 mm x 50 mm	-	0.14	2.93	0.65	m	**3.59**
Framed supports; open-spaced grounds or battens; at 300 mm centres one way						
25 mm x 50 mm	-	0.14	2.93	1.99	m²	**4.92**
25 mm x 50 mm; plugged and screwed	-	0.42	8.80	2.17	m²	**10.97**
Framed supports; at 300 mm centres one way and 600 mm centres the other way						
25 mm x 50 mm	-	0.69	14.45	2.99	m²	**17.44**
38 mm x 50 mm	-	0.69	14.45	3.94	m²	**18.39**
50 mm x 50 mm	-	0.69	14.45	4.19	m²	**18.64**
50 mm x 75 mm	-	0.69	14.45	6.50	m²	**20.95**
75 mm x 75 mm	-	0.69	14.45	8.95	m²	**23.41**
Framed supports; at 300 mm centres one way and 600 mm centres the other way; plugged and screwed						
25 mm x 50 mm	-	1.16	24.29	3.49	m²	**27.79**
38 mm x 50 mm	-	1.16	24.29	4.44	m²	**28.74**
50 mm x 50 mm	-	1.16	24.29	4.69	m²	**28.99**
50 mm x 75 mm	-	1.16	24.29	7.00	m²	**31.30**
75 mm x 75 mm	-	1.16	24.29	9.46	m²	**33.75**
Framed supports; at 500 mm centres both ways						
25 mm x 50 mm; to bath panels	-	0.83	17.38	3.89	m²	**21.27**
Framed supports; as bracketing and cradling around steelwork						
25 mm x 50 mm	-	1.30	27.23	4.22	m²	**31.45**
50 mm x 50 mm	-	1.39	29.11	5.90	m²	**35.01**
50 mm x 75 mm	-	1.48	31.00	9.14	m²	**40.14**
Wrought softwood						
Gutter boards; tongued and grooved joints						
19 mm thick; sloping	-	1.39	29.11	7.67	m²	**36.79**
19 mm thick; 75 mm wide	-	0.37	7.75	0.73	m	**8.47**
19 mm thick; 150 mm wide	-	0.42	8.80	1.12	m	**9.91**
19 mm thick; 225 mm wide	-	0.46	9.63	1.78	m	**11.42**

G STRUCTURAL/CARCASSING METAL/TIMBER

Item	PC £	Labour hours	Labour £	Material £	Unit	Total rate £
25 mm thick; sloping	-	1.39	29.11	9.86	m²	**38.97**
25 mm thick; 75 mm wide	-	0.37	7.75	0.81	m	**8.56**
25 mm thick; 150 mm wide	-	0.42	8.80	1.37	m	**10.17**
25 mm thick; 225 mm wide	-	0.46	9.63	2.09	m	**11.73**
Eaves, verge soffit boards, fascia boards and the like						
19 mm thick; over 300 mm wide	-	1.15	24.09	9.95	m²	**34.03**
19 mm thick; 150 mm wide; once grooved	-	0.19	3.98	1.68	m	**5.66**
25 mm thick; 150 mm wide; once grooved	-	0.19	3.98	1.47	m	**5.45**
25 mm thick; 175 mm wide; once grooved	-	0.19	3.98	2.02	m	**6.00**
32 mm thick; 225 mm wide; once grooved	-	0.23	4.82	2.60	m	**7.42**
Wrought softwood; "Tanalised"						
Gutter boards; tongued and grooved joints						
19 mm thick; sloping	-	1.39	29.11	8.37	m²	**37.48**
19 mm thick; 75 mm wide	-	0.37	7.75	0.78	m	**8.53**
19 mm thick; 150 mm wide	-	0.42	8.80	1.22	m	**10.02**
19 mm thick; 225 mm wide	-	0.46	9.63	1.94	m	**11.57**
25 mm thick; sloping	-	1.39	29.11	10.77	m²	**39.88**
25 mm thick; 75 mm wide	-	0.37	7.75	0.88	m	**8.63**
25 mm thick; 150 mm wide	-	0.42	8.80	1.51	m	**10.31**
25 mm thick; 225 mm wide	-	0.46	9.63	2.30	m	**11.93**
Eaves, verge soffit boards, fascia boards and the like						
19 mm thick; over 300 mm wide	-	1.15	24.09	10.64	m²	**34.73**
19 mm thick; 150 mm wide; once grooved	-	0.19	3.98	1.78	m	**5.76**
25 mm thick; 150 mm wide; once grooved	-	0.19	3.98	1.61	m	**5.59**
25 mm thick; 175 mm wide; once grooved	-	0.20	4.19	2.18	m	**6.37**
32 mm thick; 225 mm wide; once grooved	-	0.23	4.82	2.86	m	**7.68**
Straps; mild steel; galvanised						
Standard twisted vertical restraint; fixing to softwood and brick or blockwork						
27.5 mm x 2.5 mm x 400 mm girth	-	0.23	4.82	1.09	nr	**5.90**
27.5 mm x 2.5 mm x 600 mm girth	-	0.24	5.03	1.52	nr	**6.54**
27.5 mm x 2.5 mm x 800 mm girth	-	0.25	5.24	2.16	nr	**7.39**
27.5 mm x 2.5 mm x 1000 mm girth	-	0.28	5.86	2.80	nr	**8.66**
27.5 mm x 2.5 mm x 1200 mm girth	-	0.29	6.07	3.36	nr	**9.43**
Hangers; mild steel; galvanised						
Joist hangers 0.90 mm thick; The Expanded Metal Company Ltd "Speedy" or other equal and approved; for fixing to softwood; joist sizes						
50 mm wide; all sizes to 225 mm deep	1.25	0.11	2.30	1.56	nr	**3.86**
75 mm wide; all sizes to 225 mm deep	1.31	0.14	2.93	1.70	nr	**4.63**
100 mm wide; all sizes to 225 mm deep	1.40	0.17	3.56	1.88	nr	**5.44**
Joist hangers 2.50 mm thick; for building in; joist sizes						
50 mm x 100 mm	2.42	0.07	1.57	2.84	nr	**4.41**
50 mm x 125 mm	2.43	0.07	1.57	2.85	nr	**4.42**
50 mm x 150 mm	2.28	0.09	1.99	2.73	nr	**4.72**
50 mm x 175 mm	2.39	0.09	1.99	2.85	nr	**4.84**
50 mm x 200 mm	2.65	0.11	2.41	3.19	nr	**5.59**
50 mm x 225 mm	2.81	0.11	2.41	3.37	nr	**5.78**
75 mm x 150 mm	3.52	0.09	1.99	4.12	nr	**6.11**
75 mm x 175 mm	3.30	0.09	1.99	3.88	nr	**5.87**
75 mm x 200 mm	3.52	0.11	2.41	4.17	nr	**6.58**
75 mm x 225 mm	3.77	0.11	2.41	4.46	nr	**6.87**
75 mm x 250 mm	3.99	0.13	2.83	4.76	nr	**7.58**
100 mm x 200 mm	4.37	0.11	2.41	5.14	nr	**7.55**

G STRUCTURAL/CARCASSING METAL/TIMBER

Item	PC £	Labour hours	Labour £	Material £	Unit	Total rate £
G20 CARPENTRY/TIMBER FRAMING/FIRST FIXING – cont'd						
Metal connectors; mild steel; galvanised						
Round toothed plate; for 10 mm or 12 mm diameter bolts						
38 mm diameter; single sided	-	0.01	0.21	0.38	nr	**0.59**
38 mm diameter; double sided	-	0.01	0.21	0.42	nr	**0.63**
50 mm diameter; single sided	-	0.01	0.21	0.41	nr	**0.61**
50 mm diameter; double sided	-	0.01	0.21	0.45	nr	**0.66**
63 mm diameter; single sided	-	0.01	0.21	0.59	nr	**0.80**
63 mm diameter; double sided	-	0.01	0.21	0.66	nr	**0.86**
75 mm diameter; single sided	-	0.01	0.21	0.87	nr	**1.08**
75 mm diameter; double sided	-	0.01	0.21	0.91	nr	**1.12**
framing anchor	-	0.14	2.93	0.71	nr	**3.65**
Bolts; mild steel; galvanised						
Fixing only bolts; 50 mm - 200 mm long						
6 mm diameter	-	0.03	0.63	-	nr	**0.63**
8 mm diameter	-	0.03	0.63	-	nr	**0.63**
10 mm diameter	-	0.04	0.84	-	nr	**0.84**
12 mm diameter	-	0.04	0.84	-	nr	**0.84**
16 mm diameter	-	0.05	1.05	-	nr	**1.05**
20 mm diameter	-	0.05	1.05	-	nr	**1.05**
Bolts						
Expanding bolts; "Rawlbolt" projecting type or other equal and approved; Rawl Fixings; plated; one nut; one washer						
6 mm diameter; ref M6 10P	-	0.09	1.88	0.54	nr	**2.42**
6 mm diameter; ref M6 25P	-	0.09	1.88	0.61	nr	**2.49**
6 mm diameter; ref M6 60P	-	0.09	1.88	0.77	nr	**2.66**
8 mm diameter; ref M8 25P	-	0.09	1.88	0.91	nr	**2.79**
8 mm diameter; ref M8 60P	-	0.09	1.88	0.93	nr	**2.82**
10 mm diameter; ref M10 15P	-	0.09	1.88	1.15	nr	**3.03**
10 mm diameter; ref M10 30P	-	0.09	1.88	1.20	nr	**3.08**
10 mm diameter; ref M10 60P	-	0.09	1.88	1.25	nr	**3.14**
12 mm diameter; ref M12 15P	-	0.09	1.88	1.86	nr	**3.75**
12 mm diameter; ref M12 30P	-	0.10	2.09	0.18	nr	**2.27**
12 mm diameter; ref M12 75P	-	0.09	1.88	2.42	nr	**4.30**
16 mm diameter; ref M16 35P	-	0.09	1.88	4.48	nr	**6.37**
16 mm diameter; ref M16 75P	-	0.09	1.88	4.96	nr	**6.84**
Expanding bolts; "Rawlbolt" loose bolt type or other equal and approved; Rawl Fixings; plated; one bolt; one washer						
6 mm diameter; ref M6 10L	-	0.09	1.88	0.57	nr	**2.46**
6 mm diameter; ref M6 25L	-	0.09	1.88	0.71	nr	**2.59**
6 mm diameter; ref M6 40L	-	0.09	1.88	0.72	nr	**2.60**
8 mm diameter; ref M8 25L	-	0.09	1.88	0.90	nr	**2.78**
8 mm diameter; ref M8 40L	-	0.09	1.88	1.02	nr	**2.91**
10 mm diameter; ref M10 10L	-	0.09	1.88	1.06	nr	**2.94**
10 mm diameter; ref M10 25L	-	0.09	1.88	1.17	nr	**3.05**
10 mm diameter; ref M10 50L	-	0.09	1.88	1.24	nr	**3.12**
10 mm diameter; ref M10 75L	-	0.09	1.88	1.43	nr	**3.32**
12 mm diameter; ref M12 10L	-	0.09	1.88	1.83	nr	**3.71**
12 mm diameter; ref M12 25L	-	0.09	1.88	1.94	nr	**3.82**
12 mm diameter; ref M12 40L	-	0.09	1.88	2.12	nr	**4.00**
12 mm diameter; ref M12 60L	-	0.09	1.88	2.22	nr	**4.11**
16 mm diameter; ref M16 30L	-	0.09	1.88	4.52	nr	**6.40**
16 mm diameter; ref M16 60L	-	0.09	1.88	4.85	nr	**6.73**

G STRUCTURAL/CARCASSING METAL/TIMBER

Item	PC £	Labour hours	Labour £	Material £	Unit	Total rate £
Truss clips						
Truss clips; fixing to softwood; joist size						
38 mm wide	0.25	0.14	2.93	0.57	nr	**3.50**
50 mm wide	0.23	0.14	2.93	0.56	nr	**3.49**
Sole plate angles; mild steel galvanised						
Sole plate angle; fixing to softwood and concrete						
112 mm x 40 mm x 76 mm	0.56	0.19	3.98	1.62	nr	**5.60**
Chemical anchors						
R-CAS Spin-in epoxy acrylate capsules and standard studs or other equal and approved; Rawl Fixings; with nuts and washers; drilling masonry						
capsule ref 60-408; stud ref 60-448	-	0.25	5.24	1.45	nr	**6.68**
capsule ref 60-410; stud ref 60-454	-	0.28	5.86	1.56	nr	**7.43**
capsule ref 60-412; stud ref 60-460	-	0.31	6.49	1.86	nr	**8.36**
capsule ref 60-416; stud ref 60-472	-	0.34	7.12	2.70	nr	**9.82**
capsule ref 60-420; stud ref 60-478	-	0.36	7.54	4.91	nr	**12.45**
capsule ref 60-424; stud ref 60-484	-	0.40	8.38	5.77	nr	**14.14**
R-CAS Spin-in epoxy acrylate capsules and stainless steel studs or other equal and approved; Rawl Fixings; with nuts and washers; drilling masonry						
capsule ref 60-408; stud ref 60-905	-	0.25	5.24	2.46	nr	**7.70**
capsule ref 60-410; stud ref 60-910	-	0.28	5.86	3.22	nr	**9.08**
capsule ref 60-412; stud ref 60-915	-	0.31	6.49	4.34	nr	**10.83**
capsule ref 60-416; stud ref 60-920	-	0.34	7.12	7.14	nr	**14.26**
capsule ref 60-420; stud ref 60-925	-	0.36	7.54	11.99	nr	**19.53**
capsule ref 60-424; stud ref 60-930	-	0.40	8.38	19.31	nr	**27.69**
R-CAS Spin-in epoxy acrylate capsules and standard internal threaded sockets or other equal and approved; Rawl Fixings; drilling masonry						
capsule ref 60-408; socket ref 60-650	-	0.25	5.24	1.70	nr	**6.94**
capsule ref 60-410; socket ref 60-656	-	0.28	5.86	1.74	nr	**7.60**
capsule ref 60-412; socket ref 60-662	-	0.31	6.49	2.09	nr	**8.59**
capsule ref 60-416; socket ref 60-668	-	0.34	7.12	2.63	nr	**9.75**
capsule ref 60-420; socket ref 60-674	-	0.36	7.54	4.05	nr	**11.59**
capsule ref 60-424; socket ref 60-676	-	0.40	8.38	6.50	nr	**14.88**
R-CAS Spin-in epoxy acrylate capsules and stainless steel internal threaded sockets or other equal and approved; Rawl Fixings; drilling masonry						
capsule ref 60-408; socket ref 60-943	-	0.25	5.24	2.98	nr	**8.22**
capsule ref 60-410; socket ref 60-945	-	0.28	5.86	3.02	nr	**8.88**
capsule ref 60-412; socket ref 60-947	-	0.31	6.49	3.42	nr	**9.91**
capsule ref 60-416; socket ref 60-949	-	0.34	7.12	4.61	nr	**11.73**
capsule ref 60-420; socket ref 60-951	-	0.36	7.54	6.45	nr	**13.99**
capsule ref 60-424; socket ref 60-955	-	0.40	8.38	11.61	nr	**19.99**
R-CAS Spin-in epoxy acrylate capsules, perforated sleeves and standard studs or other equal and approved; Rawl Fixings; in low density material; with nuts and washers; drilling masonry						
capsule ref 60-408; sleeve ref 60-538; stud ref 60-448	-	0.25	5.24	3.38	nr	**8.62**
capsule ref 60-410; sleeve ref 60-544; stud ref 60-454	-	0.28	5.86	3.73	nr	**9.59**
capsule ref 60-412; sleeve ref 60-550; stud ref 60-460	-	0.31	6.49	4.25	nr	**10.74**
capsule ref 60-416; sleeve ref 60-562; stud ref 60-472	-	0.34	7.12	5.13	nr	**12.26**

Item	PC £	Labour hours	Labour £	Material £	Unit	Total rate £
G20 CARPENTRY/TIMBER FRAMING/FIRST FIXING – cont'd						
Chemical anchors – cont'd						
R-CAS Spin-in epoxy acrylate capsules, perforated sleeves and stainless steel studs or other equal and approved; Rawl Fixings; in low density material; with nuts and washers; drilling masonry						
capsule ref 60-408; sleeve ref 60-538; stud ref 60-905	-	0.25	5.24	4.40	nr	**9.64**
capsule ref 60-410; sleeve ref 60-544; stud ref 60-910	-	0.28	5.86	5.38	nr	**11.25**
capsule ref 60-412; sleeve ref 60-550; stud ref 60-915	-	0.31	6.49	6.73	nr	**13.22**
capsule ref 60-416; sleeve ref 60-562; stud ref 60-920	-	0.34	7.12	9.58	nr	**16.70**
R-CAS Spin-in epoxy acrylate capsules, perforated sleeves and standard internal threaded sockets or other equal and approved; Rawl Fixings;in low density material; with nuts and washers; drilling masonry						
capsule ref 60-408; sleeve ref 60-538; socket ref 60-650	-	0.25	5.24	3.64	nr	**8.88**
capsule ref 60-410; sleeve ref 60-544; socket ref 60-656	-	0.28	5.86	3.90	nr	**9.77**
capsule ref 60-412; sleeve ref 60-550; socket ref 60-662	-	0.31	6.49	4.48	nr	**10.97**
R-CAS Spin-in epoxy acrylate capsules, perforated sleeves and stainless steel internal threaded sockets or other equal and approved; Rawl Fixings; in low density material; drilling masonry						
capsule ref 60-416; sleeve ref 60-562; socket ref 60-668	-	0.34	7.12	5.06	nr	**12.18**
capsule ref 60-408; sleeve ref 60-538; socket ref 60-943	-	0.25	5.24	4.92	nr	**10.16**
capsule ref 60-410; sleeve ref 60-544; socket ref 60-945	-	0.28	5.86	5.18	nr	**11.05**
capsule ref 60-412; sleeve ref 60-550; socket ref 60-947	-	0.31	6.49	5.81	nr	**12.30**
capsule ref 60-416; sleeve ref 60-562; socket ref 60-949	-	0.34	7.12	7.05	nr	**14.17**

H CLADDING/COVERING

Item	PC £	Labour hours	Labour £	Material £	Unit	Total rate £
H10 PATENT GLAZING						
Patent glazing; aluminium alloy bars 2.55 m long at 622 mm centres; fixed to supports						
Roof cladding						
single glazed with 6.4 mm laminated glass	-	-	-	-	m²	**125.00**
single glazed with 7 mm thick Georgian wired cast glass	-	-	-	-	m²	**135.00**
thermally broken and double glazed with low-e clear toughened and laminated double glazed units; aluminium finished RAL matt colour	-	-	-	-	m²	**350.00**
Extra for opening roof vents						
600 mm x 900 mm top hung opening roof vent; manually operated	-	-	-	-	nr	**400.00**
600 mm x 900 mm top hung opening roof vent; electrically operated	-	-	-	-	nr	**500.00**
Skylight						
Self-supporting hipped or gable ended lantern/skylight thermally broken and double glazed with low-e clear toughened and laminated double glazed units; aluminium finished RAL matt colour	-	-	-	-	m²	**700.00**
Associated code 4 lead flashings						
top flashing; 210 mm girth	-	-	-	-	m	**55.00**
bottom flashing; 240 mm girth	-	-	-	-	m	**63.00**
end flashing; 300 mm girth	-	-	-	-	m	**68.00**
Wall cladding						
single glazed with 6.4 mm laminated glass	-	-	-	-	m²	**130.00**
single glazed with 7 mm thick Georgian wired cast glass	-	-	-	-	m²	**140.00**
thermally broken and double glazed with low-e clear toughened and laminated double glazed units; aluminium finished RAL matt colour	-	-	-	-	m²	**368.00**
Extra for aluminium alloy perimeter members						
38 mm x 38 mm x 3 mm angle jamb	-	-	-	-	m	**19.00**
pressed cill member	-	-	-	-	m	**38.00**
pressed channel head and PVC case	-	-	-	-	m	**38.00**
H11 CURTAIN WALLING						
Stick curtain walling system; Schuco FW50+ proprietary system or other equal and approved						
Polyester powder coated solid colour matt finish or natural anodised curtain walling with mullions spaced 1.5m apart and spanning typical storey height of 3.8m. Floor to ceiling glass sealed units with 8.8mm low E coated laminated inner pane, air filled cavity and 8mm clear annealed outer pane, retained by external pressure plates and caps. Rates to include 0.8m deep solid spandrel panels, all brackets, membranes, fire stopping between floors and external access equipment						
Flat system; drilling and screwing; to metal	-	-	-	-	m²	**450.00**
Extra over for						
neutral selective high performance coating in lieu of low E, for assisting in solar control	-	-	-	-	m²	**40.00**
outer glass pane to be toughened and heat soak tested or heat strengthened in lieu of annealed	-	-	-	-	m²	**20.00**

H CLADDING/COVERING

Item	PC £	Labour hours	Labour £	Material £	Unit	Total rate £
H11 CURTAIN WALLING – cont'd						
Stick curtain walling system; Schuco FW50+ – cont'd						
Extra over flat system for						
inner laminated glass to be toughened and heat soak tested laminated, or heat strengthened laminated	-	-	-	-	m²	**40.00**
flush glass finish without external face caps, achieved by concealed toggle fixings locating within perimeter channels within sealed units including silicone sealing between glass panes	-	-	-	-	m²	**50.00**
typical coping detail, including pressed aluminium profiles, membranes, seals, etc	-	-	-	-	m	**250.00**
typical cill detail, including pressed aluminium profiles, membranes, seals, etc	-	-	-	-	m	**200.00**
intermediate transoms (per transom)	-	-	-	-	m	**40.00**
Unitised curtain walling system; Schuco Skyline 65 proprietary system or other equal and approved						
Polyester powder coated solid colour matt finish or natural anodised curtain walling with mullions spaced 1.5m apart and spanning typical storey height of 3.8m. Floor to ceiling glass sealed units with 8.8mm low E coated laminated inner pane, air filled cavity and 8mm clear annealed outer pane, retained by external beading system. Rates to include 0.8m deep solid spandrel panels, all brackets, membranes, fire stopping between floors and external access equipment						
Flat system; drilling and screwing; to metal	-	-	-	-	m²	**650.00**
Extra over for						
neutral selective high performance coating in lieu of low E, for assisting in solar control	-	-	-	-	m²	**40.00**
outer glass pane to be toughened and heat soak tested or heat strengthened in lieu of annealed	-	-	-	-	m²	**20.00**
inner laminated glass to be toughened and heat soak tested laminated, or heat strengthened laminated	-	-	-	-	m²	**40.00**
flush glass finish without external face caps, achieved by carrier frames with glass sealed units factory silicone bonded; often referred to as SSG (Structural Silicone Glazing)	-	-	-	-	m²	**100.00**
typical coping detail, including pressed aluminium profiles, membranes, seals, etc	-	-	-	-	m	**250.00**
typical cill detail, including pressed aluminium profiles, membranes, seals, etc	-	-	-	-	m	**200.00**
Other curtain walling systems/costs						
Unitised curtain walling system; bespoke solution via specialist façade contractor based in mainland Europe. Generally as described in 1J but comprising a project specific solution, thus additional design development. Note: These rates are subject to currency fluctuations between £ and €. The rate opposite assumes £1 = €1.40	-	-	-	-	m²	**750.00**
Project specific performance testing for bespoke unitised curtain walling system	-	-	-	-	nr	**75000.00**

H CLADDING/COVERING

Item	PC £	Labour hours	Labour £	Material £	Unit	Total rate £
Visual mock-ups are often required for bespoke curtain walling solutions and in cases for proprietary unitised and stick curtain walling projects	-	-	-	-	nr	**25000.00**
All curtain walling projects should be site hose tested. The rate depends upon the quantum of joints to be tested, generally 5%. Assume 5 days @ £1000	-	-	-	-	nr	**5000.00**
Brise soleil, to mitigate the effects of solar gain and enable compliance with Part L of the Building regulations. There are a variety of material types which can be adopted for the purpose of solar shading, including but not limited to; Aluminium, Glass, Timber. South elevations require horizontal shading to combat high sun angles, whereas east and west elevations require vertical fins to accommodate low angle sun paths. The rate opposite assumes a single natural anodised extruded aluminium fin, with brackets and orientated either horizontally or vertically						
300 mm deep	-	-	-	-	m	**125.00**
H20 RIGID SHEET CLADDING						
"Resoplan" sheet or other equal and approved; Eternit UK Ltd; flexible neoprene gasket joints; fixing with stainless steel screws and coloured caps						
6 mm thick cladding to walls						
over 300 mm wide	-	1.94	40.63	57.39	m²	**98.02**
not exceeding 300 mm wide	-	0.65	13.61	21.00	m	**34.61**
Eternit 2000 "Glasal" sheet or other equal and approved; Eternit UK Ltd; flexible neoprene gasket joints; fixing with stainless steel screws and coloured caps						
7.50 mm thick cladding to walls						
over 300 mm wide	-	1.94	40.63	50.56	m²	**91.19**
not exceeding 300 mm wide	-	0.65	13.61	18.95	m	**32.56**
external angle trim	-	0.09	1.88	9.78	m	**11.66**
7.50 mm thick cladding to eaves, verge soffit boards, fascia boards or the like						
100 mm wide	-	0.46	9.63	9.68	m	**19.32**
200 mm wide	-	0.56	11.73	14.32	m	**26.05**
300 mm wide	-	0.65	13.61	18.95	m	**32.56**
Prodema ProdEX high density resin-bonded cellulose fibre weatherboarding panels; including secondary supports and fixing						
Walls						
8 mm Panels face fixed on to timber battens	-	-	-	-	m²	**151.70**
8 mm Panels face fixed on to aluminium rails	-	-	-	-	m²	**170.66**
8 mm Panels adhesive fixed on to timber battens or aluminium rails	-	-	-	-	m²	**180.14**
10 mm Panels secret fixed on to helping hand aluminium system	-	-	-	-	m²	**208.59**

H CLADDING/COVERING

Item	PC £	Labour hours	Labour £	Material £	Unit	Total rate £
H30 FIBRE CEMENT PROFILED SHEET CLADDING						
Asbestos-free corrugated sheets; Eternit "2000" or other equal and approved						
Roof cladding; sloping not exceeding 50 degrees; fixing to steel purlins with hook bolts						
"Profile 3"; natural grey	-	0.23	6.83	18.21	m²	**25.04**
"Profile 3"; coloured	-	0.23	6.83	20.92	m²	**27.75**
"Profile 6"; natural grey	-	0.28	8.32	14.24	m²	**22.55**
"Profile 6"; coloured	-	0.28	8.32	16.18	m²	**24.50**
"Profile 6"; natural grey; insulated 80 glass fibre infill; lining panel	-	0.46	13.66	30.44	m²	**44.11**
"Profile 6"; coloured; insulated 80 glass fibre infill; lining panel	-	0.46	13.66	35.01	m²	**48.67**
Accessories; to "Profile 3" cladding; natural grey						
eaves filler	-	0.09	2.67	11.68	m	**14.36**
external corner piece	-	0.11	3.27	8.63	m	**11.90**
apron flashing	-	0.11	3.27	11.68	m	**14.95**
plain wing or close fitting two piece adjustable capping to ridge	-	0.16	4.75	10.95	m	**15.70**
ventilating two piece adjustable capping to ridge	-	0.16	4.75	16.85	m	**21.60**
Accessories; to "Profile 6" cladding; natural grey						
eaves filler	-	0.09	2.67	7.04	m	**9.72**
external corner piece	-	0.11	3.27	8.02	m	**11.28**
apron flashing	-	0.11	3.27	7.83	m	**11.09**
underglazing flashing	-	0.11	3.27	10.32	m	**13.58**
plain cranked crown to ridge	-	0.16	4.75	15.49	m	**20.24**
plain wing or close fitting two piece adjustable capping to ridge	-	0.16	4.75	14.08	m	**18.83**
ventilating two piece adjustable capping to ridge	-	0.16	4.75	18.00	m	**22.75**
H31 METAL PROFILED/FLAT SHEET CLADDING/ COVERING/SIDING						
Lightweight galvanised steel roof tiles; Decra Roof Systems 'Stratos'; or other equal and approved; coated finish						
Roof coverings	-	0.23	6.83	21.62	m²	**28.45**
Accessories for roof cladding						
pitched "D" ridge	-	0.09	2.67	9.86	m	**12.53**
barge cover (handed)	-	0.09	2.67	10.60	m	**13.28**
in line air vent	-	0.09	2.67	53.21	nr	**55.88**
in line soil vent	-	0.09	2.67	76.77	nr	**79.44**
gas flue terminal	-	0.19	5.64	99.64	nr	**105.29**
Galvanised steel strip troughed sheets; Corus Products or other equal and approved						
Roof cladding or decking; sloping not exceeding 50 degrees; fixing to steel purlins with plastic headed self-tapping screws						
0.7 mm thick; 46 profile	-	-	-	-	m²	**13.44**
0.7 mm thick; 60 profile	-	-	-	-	m²	**14.72**
0.7 mm thick; 100 profile	-	-	-	-	m²	**16.01**

H CLADDING/COVERING

Item	PC £	Labour hours	Labour £	Material £	Unit	Total rate £
Galvanised steel strip troughed sheets; PMF Strip Mill Products or other equal and approved						
Roof cladding; sloping not exceeding 50 degrees; fixing to steel purlins with plastic headed self-tapping screws						
0.7 mm thick type HPS200 13.5/3 corrugated	-	-	-	-	m²	**15.17**
0.7 mm thick type HPS200 R32/1000	-	-	-	-	m²	**13.77**
0.7 mm thick type Arcline 40; plasticol finished	-	-	-	-	m²	**20.06**
Extra over last for aluminium roof cladding or decking	-	-	-	-	m²	**7.78**
Accessories for roof cladding						
HPS200 Drip flashing; 250 mm girth	-	-	-	-	m	**4.54**
HPS200 Ridge flashing; 375 mm girth	-	-	-	-	m	**5.94**
HPS200 Gable flashing; 500 mm girth	-	-	-	-	m	**7.55**
HPS200 Internal angle; 625 mm girth	-	-	-	-	m	**8.71**
Zalutite coated steel flat composite panel cladding; Kingspan or other equal and approved; outer panel 0.7 mm gauge HPS200 colourcoated; HCFC free LPCB FM/FW core and 0.4 mm stucco embossed lining panel with bright white ployester paint finish						
Roof cladding; vertical fixing to steel rails (measured elsewhere)						
80 mm wall panel; ref. KS1000RW	-	-	-	-	m²	**40.33**
Wall cladding; vertical fixing to steel rails (measured elsewhere)						
60 mm wall panel; ref. KS1000RW	-	-	-	-	m²	**38.80**
70 mm wall panel; ref. KS1000RW	-	-	-	-	m²	**39.69**
80 mm wall panel; ref. KS1000RW	-	-	-	-	m²	**40.59**
70 mm wall panel; ref. KS1000MR	-	-	-	-	m²	**55.57**
80 mm wall panel; ref. KS1000MR	-	-	-	-	m²	**56.47**
70 mm wall panel; ref. KS900MR	-	-	-	-	m²	**59.16**
80 mm wall panel; ref. KS900MR	-	-	-	-	m²	**60.05**
70 mm wall panel ; ref. KS600MR	-	-	-	-	m²	**81.69**
80 mm wall panel ; ref. KS600MR	-	-	-	-	m²	**82.46**
Extra over for						
raking cutting to 60 mm KS1000RW panel including waste	-	-	-	-	m	**19.18**
raking cutting to 70 mm KS1000RW panel including waste	-	-	-	-	m	**19.58**
raking cutting to 80 mm KS1000RW panel including waste	-	-	-	-	m	**19.99**
raking cutting to 70 mm KS1000MR panel including waste	-	-	-	-	m	**27.22**
raking cutting to 80 mm KS1000MR panel including waste	-	-	-	-	m	**27.58**
raking cutting to 70 mm KS900MR panel including waste	-	-	-	-	m	**28.77**
raking cutting to 80 mm KS900MR panel including waste	-	-	-	-	m	**29.17**
raking cutting to 70 mm KS600MR panel including waste	-	-	-	-	m	**38.37**
raking cutting to 80 mm KS600MR panel including waste	-	-	-	-	m	**38.76**

H CLADDING/COVERING

Item	PC £	Labour hours	Labour £	Material £	Unit	Total rate £
H31 METAL PROFILED/FLAT SHEET CLADDING/ COVERING/SIDING – cont'd						
Zalutite coated steel flat composite panel cladding; Kingspan – cont'd						
Extra over for – cont'd						
panel bearers ' 1500 mm centres	-	-	-	-	m	**6.76**
vertical tophat joint in HPS200	-	-	-	-	m	**11.90**
vertical tophat joint with cap in HPS200	-	-	-	-	m	**15.33**
cranked KS1000MR panel	-	-	-	-	m	**110.33**
cranked KS900MR panel	-	-	-	-	m	**122.58**
cranked KS600MR panel	-	-	-	-	m	**183.88**
roof penetration; 150 mm dia. opening; with top hat flashing and collar 150 mm high; and silicone joint to roofsheet	-	-	-	-	nr	**58.19**
roof penetration; 250 mm dia. opening; with top hat flashing and collar 150 mm high; and silicone joint to roofsheet	-	-	-	-	nr	**82.81**
GRP Transluscent rooflights; factory assembled						
Rooflight; vertical fixing to steel purlins (measured elsewhere)						
double skin; class 3 over 1	-	-	-	-	m²	**58.25**
triple skin; class 3 over 1	-	-	-	-	m²	**59.84**
Wall cladding; Gasell Profiles Ltd or equal and approved; steel GA50-30 profiled sheeting to outer face; steel; GA600 lining to inner face; including profile fillers; sealing						
Coverings; fixing to and including vertical and horizontal secondary supports						
250 mm girth	-	-	-	-	m²	**69.10**
H32 PLASTICS PROFILED SHEET CLADDING/ COVERING/SIDING						
Extended, hard skinned, foamed PVC-UE profiled sections; Swish Celuka or other equal and approved; Class 1 fire rated to BS 476; Part 7; in white finish						
Wall cladding; vertical; fixing to timber						
100 mm shiplap profiles; Code 001	-	0.35	7.33	45.65	m²	**52.98**
150 mm shiplap profiles; Code 002	-	0.32	6.70	40.51	m²	**47.21**
125 mm feather-edged profiles; Code C208	-	0.34	7.12	44.10	m²	**51.22**
Vertical angles	-	0.19	3.98	4.82	m	**8.80**
Raking cutting	-	0.14	2.93	-	m	**2.93**
Holes for pipes and the like	-	0.03	0.63	-	nr	**0.63**

H CLADDING/COVERING

Item	PC £	Labour hours	Labour £	Material £	Unit	Total rate £
H41 GLASS REINFORCED PLASTICS PANEL CLADDING FEATURES						
Glass fibre translucent sheeting grade AB class 3						
Roof cladding; sloping not exceeding 50 degrees; fixing to timber purlins with drive screws; to suit						
"Profile 3" or other equal and approved	17.17	0.18	5.35	21.59	m²	**26.94**
"Profile 6" or other equal and approved	16.09	0.23	6.83	20.38	m²	**27.21**
Roof cladding; sloping not exceeding 50 degrees; fixing to timber purlins with hook bolts; to suit						
"Profile 3" or other equal and approved	17.17	0.23	6.83	23.27	m²	**30.11**
"Profile 6" or other equal and approved	16.09	0.28	8.32	22.06	m²	**30.38**
"Longrib 1000" or other equal and approved	15.82	0.28	8.32	21.75	m²	**30.07**
H51 NATURAL STONE SLAB CLADDING FEATURES						
Portland Whitbed limestone bedded and jointed in cement - lime - mortar (1:2:9); slurrying with weak lime and stone dust mortar; flush pointing and cleaning on completion (cramps etc. not included)						
Facework; one face plain and rubbed; bedded against backing						
50 mm thick stones	-	-	-	-	m²	**317.13**
63 mm thick stones	-	-	-	-	m²	**361.20**
75 mm thick stones	-	-	-	-	m²	**408.50**
100 mm thick stones	-	-	-	-	m²	**435.38**
Fair returns on facework						
50 mm wide	-	-	-	-	m	**4.30**
63 mm wide	-	-	-	-	m	**5.38**
75 mm wide	-	-	-	-	m	**7.53**
100 mm wide	-	-	-	-	m	**9.68**
Fair raking cutting on facework						
50 mm thick	-	-	-	-	m	**19.35**
63 mm thick	-	-	-	-	m	**21.50**
75 mm thick	-	-	-	-	m	**25.80**
100 mm thick	-	-	-	-	m	**27.95**
Copings; once weathered, and throated; rubbed; set horizontal or raking						
250 mm x 50 mm	-	-	-	-	m	**150.50**
extra for external angle	-	-	-	-	nr	**26.88**
extra for internal angle	-	-	-	-	nr	**26.88**
300 mm x 50 mm	-	-	-	-	m	**159.10**
extra for external angle	-	-	-	-	nr	**26.88**
extra for internal angle	-	-	-	-	nr	**32.25**
350 mm x 75 mm	-	-	-	-	m	**177.38**
extra for external angle	-	-	-	-	nr	**26.88**
extra for internal angle	-	-	-	-	nr	**34.40**
400 mm x 100 mm	-	-	-	-	m	**212.85**
extra for external angle	-	-	-	-	nr	**32.25**
extra for internal angle	-	-	-	-	nr	**45.15**
450 mm x 100 mm	-	-	-	-	m	**258.00**
extra for external angle	-	-	-	-	nr	**40.85**
extra for internal angle	-	-	-	-	nr	**55.90**
500 mm x 125 mm	-	-	-	-	m	**392.38**
extra for external angle	-	-	-	-	nr	**55.90**
extra for internal angle	-	-	-	-	nr	**69.88**

H CLADDING/COVERING

Item	PC £	Labour hours	Labour £	Material £	Unit	Total rate £
H51 NATURAL STONE SLAB CLADDING FEATURES – cont'd						
Portland Whitbed limestone – cont'd						
Band courses; plain; rubbed; horizontal						
225 mm x 112 mm	-	-	-	-	m	**118.25**
300 mm x 112 mm	-	-	-	-	m	**158.03**
extra for stopped ends	-	-	-	-	nr	**6.45**
extra for external angles	-	-	-	-	nr	**6.45**
Band courses; moulded 100 mm girth on face; rubbed; horizontal						
125 mm x 75 mm	-	-	-	-	m	**134.38**
extra for stopped ends	-	-	-	-	nr	**21.50**
extra for external angles	-	-	-	-	nr	**26.88**
extra for internal angles	-	-	-	-	nr	**53.75**
150 mm x 75 mm	-	-	-	-	m	**155.88**
extra for stopped ends	-	-	-	-	nr	**21.50**
extra for external angles	-	-	-	-	nr	**37.63**
extra for internal angles	-	-	-	-	nr	**64.50**
200 mm x 100 mm	-	-	-	-	m	**177.38**
extra for stopped ends	-	-	-	-	nr	**21.50**
extra for external angles	-	-	-	-	nr	**53.75**
extra for internal angles	-	-	-	-	nr	**96.75**
250 mm x 150 mm	-	-	-	-	m	**258.00**
extra for stopped ends	-	-	-	-	nr	**21.50**
extra for external angles	-	-	-	-	nr	**53.75**
extra for internal angles	-	-	-	-	nr	**107.50**
300 mm x 250 mm	-	-	-	-	m	**419.25**
extra for stopped ends	-	-	-	-	nr	**21.50**
extra for external angles	-	-	-	-	nr	**75.25**
extra for internal angles	-	-	-	-	nr	**129.00**
Coping apex block; two sunk faces; rubbed						
650 mm x 450 mm x 225 mm	-	-	-	-	nr	**526.75**
Coping kneeler block; three sunk faces; rubbed						
350 mm x 350 mm x 375 mm	-	-	-	-	nr	**419.25**
450 mm x 450 mm x 375 mm	-	-	-	-	nr	**483.75**
Corbel; turned and moulded; rubbed						
225 mm x 225 mm x 375 mm	-	-	-	-	nr	**344.00**
Slab surrounds to openings; one face splayed; rubbed						
75 mm x 100 mm	-	-	-	-	m	**75.25**
75 mm x 200 mm	-	-	-	-	m	**102.13**
100 mm x 100 mm	-	-	-	-	m	**91.38**
125 mm x 100 mm	-	-	-	-	m	**102.13**
125 mm x 150 mm	-	-	-	-	m	**123.63**
175 mm x 175 mm	-	-	-	-	m	**150.50**
225 mm x 175 mm	-	-	-	-	m	**177.38**
300 mm x 175 mm	-	-	-	-	m	**215.00**
300 mm x 225 mm	-	-	-	-	m	**279.50**
Slab surrounds to openings; one face sunk splayed; rubbed						
75 mm x 100 mm	-	-	-	-	m	**96.75**
75 mm x 200 mm	-	-	-	-	m	**123.63**
100 mm x 100 mm	-	-	-	-	m	**112.88**
125 mm x 100 mm	-	-	-	-	m	**123.63**
125 mm x 150 mm	-	-	-	-	m	**145.13**
175 mm x 175 mm	-	-	-	-	m	**172.00**
225 mm x 175 mm	-	-	-	-	m	**198.88**
300 mm x 175 mm	-	-	-	-	m	**236.50**
300 mm x 225 mm	-	-	-	-	m	**301.00**

H CLADDING/COVERING

Item	PC £	Labour hours	Labour £	Material £	Unit	Total rate £
extra for throating	-	-	-	-	m	**10.75**
extra for rebates and grooves	-	-	-	-	m	**23.65**
extra for stooling	-	-	-	-	m	**40.85**
Sundries - stone walling						
Coating backs of stones with brush applied cold bitumen solution; two coats						
limestone facework	-	0.19	2.78	1.63	m²	**4.41**
Cutting grooves in limestone masonry for						
water bars or the like	-	-	-	-	m	**10.75**
Mortices in limestone masonry for						
metal dowel	-	-	-	-	nr	**2.15**
metal cramp	-	-	-	-	nr	**4.30**
"Eurobrick" insulated brick cladding systems or other equal and approved ; extruded polystyrene foam insulation; brick slips bonded to insulation panels with "Eurobrick" gun applied adhesive or other equal and approved; pointing with formulated mortar grout						
25 mm insulation to walls						
over 300 mm wide; fixing with proprietary screws and plates to timber	-	1.39	33.77	49.11	m²	**82.88**
50 mm insulation to walls						
over 300 mm wide; fixing with proprietary screws and plates; to timber	-	1.39	33.77	53.64	m²	**87.41**
Stainless steel cramps and dowels; Halfen-Deha or other equal and approved; one end built into brickwork or set in slot in concrete						
Dowel						
8 mm diameter x 75 mm long	0.19	0.04	1.05	0.22	nr	**1.27**
10 mm diameter x 150 mm long	0.56	0.04	1.05	0.64	nr	**1.68**
Pattern "J" tie						
25 mm x 3 mm x 100 mm	0.42	0.06	1.57	0.47	nr	**2.04**
Pattern "S" cramp; with two 20 mm turndowns (190 mm girth)						
25 mm x 3 mm x 150 mm	0.59	0.06	1.57	0.66	nr	**2.23**
Pattern "B" anchor; with 8 mm x 75 mm loose dowel						
25 mm x 3 mm x 150 mm	0.72	0.09	2.36	0.82	nr	**3.17**
Pattern "Q" tie						
25 mm x 3 mm x 200 mm	0.73	0.06	1.57	0.82	nr	**2.39**
38 mm x 3 mm x 250 mm	1.47	0.06	1.57	1.67	nr	**3.24**
Pattern "P" half twist tie						
25 mm x 3 mm x 200 mm	0.79	0.06	1.57	0.89	nr	**2.46**
38 mm x 3 mm x 250 mm	1.24	0.06	1.57	1.40	nr	**2.97**
H53 CLAY SLAB/CLADDING/FEATURES						
Terracotta cladding and panels; LockClad or equal and approved; 240 mm x 390 mm x 14½ mm thick terracotta panels; including horizontal rails, clips and vertical spacers, insulation, structural liner trays and fixings						
Walls						
over 300 mm wide	-	-	-	-	m²	**199.95**

H CLADDING/COVERING

Item	PC £	Labour hours	Labour £	Material £	Unit	Total rate £
H60 PLAIN ROOF TILING						
ALTERNATIVE TILE PRICES (£/1000)						
Discounts of 5 - 25% available depending on quantity / status.						
Clay tiles; plain, interlocking and pantiles						
Dreadnought						
Red smooth/sandfaced	-	-	-	335.00	1000	-
Country brown smooth/sandfaced	-	-	-	375.00	1000	-
Brown Antique smooth/sandfaced	-	-	-	385.00	1000	-
Blue / Dark Heather	-	-	-	410.00	1000	-
Sandtoft pantiles						
Bridgewater Double Roman	-	-	-	4580.00	1000	-
"Gaelic"	-	-	-	2020.00	1000	-
Arcadia	-	-	-	1308.00	1000	-
William Blyth pantiles						
"Barco bold Roll"	-	-	-	889.00	1000	-
"Celtic (french)"	-	-	-	1005.00	1000	-
Concrete tiles; plain and interlocking						
Marley Eternit roof tiles						
"Anglia"	-	-	-	602.00	1000	-
"Ashmore"	-	-	-	725.00	1000	-
"Duo Modern"	-	-	-	908.00	1000	-
"Pewter Mendip"	-	-	-	1045.00	1000	-
"Malvern	-	-	-	976.00	1000	-
"Plain"	-	-	-	347.00	1000	-
Redland roof tiles						
"Redland 49"	-	-	-	638.00	1000	-
"50 Double Roman"	-	-	-	912.00	1000	-
"Mini Stoneworld"	-	-	-	970.00	1000	-
"Grovebury"	-	-	-	924.00	1000	-
SUPPLY AND FIX PRICES						
NOTE: The following items of tile roofing unless otherwise described, include for conventional fixing assuming normal exposure with appropriate nails and/or rivets or clips to pressure impregnated softwood battens fixed with galvanised nails; prices also include for all bedding and pointing at verges, beneath ridge tiles, etc.						
Clay interlocking plain tiles; Sandtoft "20/20" natural red faced or other equal and approved; 370 mm x 223 mm; to 75 mm lap; on 25 mm x 38 mm battens and type 1F reinforced underlay						
Roof coverings (PC £ per 1000)	881.92	0.42	12.48	18.49	m²	**30.96**
Extra over coverings for						
fixing every tile	-	0.02	0.59	1.86	m²	**2.46**
double course at eaves	-	0.28	8.32	13.17	m	**21.48**
verges; extra single undercloak course of plain tiles	-	0.28	8.32	5.10	m	**13.41**
open valleys; cutting both sides	-	0.17	5.05	3.99	m	**9.04**
dry ridge tiles	-	0.56	16.63	13.77	m	**30.41**
dry hips; cutting both sides	-	0.69	20.50	12.23	m	**32.72**
holes for pipes and the like	-	0.19	5.64	-	nr	**5.64**

H CLADDING/COVERING

Item	PC £	Labour hours	Labour £	Material £	Unit	Total rate £
Clay pantiles; Sandtoft Goxhill "Old English"; red sand faced or other equal and approved; 342 mm x 241 mm; to 75 mm lap; on 25 mm x 38 mm battens and type 1F reinforced underlay						
Roof coverings (PC £ per 1000)	983.68	0.42	12.48	22.26	m²	**34.73**
Extra over coverings for						
fixing every tile	-	0.02	0.59	4.66	m²	**5.26**
other colours	-	-	-	1.15	m²	**-**
double course at eaves	-	0.31	9.21	6.00	m	**15.21**
verges; extra single undercloak course of plain tiles	-	0.28	8.32	12.46	m	**20.78**
open valleys; cutting both sides	-	0.17	5.05	4.45	m	**9.50**
ridge tiles; tile slips	-	0.56	16.63	31.49	m	**48.12**
hips; cutting both sides	-	0.69	20.50	35.93	m	**56.43**
holes for pipes and the like	-	0.19	5.64	-	nr	**5.64**
Clay pantiles; William Blyth's "Lincoln" natural or other equal and approved; 343 mm x 280 mm; to 75 mm lap; on 19 mm x 38 mm battens and type 1F reinforced underlay						
Roof coverings (PC £ per 1000)	1020.00	0.42	12.48	22.25	m²	**34.73**
Extra over coverings for						
fixing every tile	-	0.02	0.59	4.66	m²	**5.26**
other colours	-	-	-	1.48	m²	**-**
double course at eaves	-	0.31	9.21	6.14	m	**15.35**
verges; extra single undercloak course of plain tiles	-	0.28	8.32	12.47	m	**20.78**
open valleys; cutting both sides	-	0.17	5.05	4.59	m	**9.64**
ridge tiles; tile slips	-	0.56	16.63	24.85	m	**41.49**
hips; cutting both sides	-	0.69	20.50	29.44	m	**49.94**
holes for pipes and the like	-	0.19	5.64	-	nr	**5.64**
Clay plain tiles; Hinton, Perry and Davenhill "Dreadnought" smooth red machine-made or other equal and approved; 265 mm x 165 mm; on 19 mm x 38 mm battens and type 1F reinforced underlay						
Roof coverings; to 64 mm lap (PC £ per 1000)	301.50	0.97	28.81	28.23	m²	**57.04**
Wall coverings; to 38 mm lap	-	1.16	34.46	23.90	m²	**58.35**
Extra over coverings for						
other colours	-	-	-	2.56	m²	**-**
ornamental tiles	-	-	-	21.47	m²	**-**
double course at eaves	-	0.23	6.83	3.22	m	**10.05**
verges	-	0.28	8.32	1.27	m	**9.59**
swept valleys; cutting both sides	-	0.60	17.82	36.62	m	**54.45**
bonnet hips; cutting both sides	-	0.74	21.98	36.70	m	**58.68**
external vertical angle tiles; supplementary nail fixings	-	0.37	10.99	41.67	m	**52.66**
half round ridge tiles	-	0.56	16.63	12.57	m	**29.20**
holes for pipes and the like	-	0.19	5.64	-	nr	**5.64**

H CLADDING/COVERING

Item	PC £	Labour hours	Labour £	Material £	Unit	Total rate £
H60 PLAIN ROOF TILING – cont'd						
Concrete interlocking tiles; Marley Eternit "Anglia" granule finish tiles or other equal and approved; 387 mm x 230 mm; to 75 mm lap; on 25 mm x 38 mm battens and type 1F reinforced underlay						
Roof coverings (PC £ per 1000)	511.70	0.42	12.48	12.41	m²	**24.89**
Extra over coverings for						
fixing every tile	-	0.02	0.59	0.78	m²	**1.37**
eaves; eaves filler	-	0.04	1.19	11.15	m	**12.34**
verges; 150 mm wide asbestos free strip undercloak	-	0.21	6.24	1.91	m	**8.15**
valley trough tiles; cutting both sides	-	0.51	15.15	24.37	m	**39.52**
segmental ridge tiles; tile slips	-	0.51	15.15	5.67	m	**20.82**
segmental hip tiles; tile slips; cutting both sides	-	0.65	19.31	7.40	m	**26.71**
dry ridge tiles; segmental including batten sections; unions and filler pieces	-	0.28	8.32	11.54	m	**19.86**
segmental mono-ridge tiles	-	0.51	15.15	19.81	m	**34.96**
gas ridge terminal	-	0.46	13.66	68.59	nr	**82.26**
holes for pipes and the like	-	0.19	5.64	-	nr	**5.64**
Concrete interlocking tiles; Marley Eternit "Ludlow Major" granule finish tiles; 420 mm x 330 mm; to 75 mm lap; on 25 mm x 38 mm battens and type 1F reinforced underlay						
Roof coverings (PC £ per 1000)	741.20	0.32	9.51	11.04	m²	**20.55**
Extra over coverings for						
fixing every tile	-	0.02	0.59	0.78	m²	**1.37**
eaves; eaves filler	-	0.04	1.19	0.59	m	**1.78**
verges; 150 mm wide asbestos free strip undercloak	-	0.21	6.24	1.91	m	**8.15**
dry verge system; extruded white pvc	-	0.14	4.16	11.36	m	**15.51**
segmental ridge cap to dry verge	-	0.02	0.59	3.63	m	**4.22**
valley trough tiles; cutting both sides	-	0.51	15.15	24.89	m	**40.04**
segmental ridge tiles	-	0.46	13.66	1.05	m	**14.72**
segmental hip tiles; cutting both sides	-	0.60	17.82	3.57	m	**21.39**
dry ridge tiles; segmental including batten sections; unions and filler pieces	-	0.28	8.32	11.54	m	**19.86**
segmental mono-ridge tiles	-	0.46	13.66	17.12	m	**30.78**
gas ridge terminal	-	0.46	13.66	68.59	nr	**82.26**
holes for pipes and the like	-	0.19	5.64	-	nr	**5.64**
Concrete interlocking tiles; Marley Eternit "Ecologic Ludlow Major" granule finish tiles; 420 mm x 330 mm; to 75 mm lap; on 25 mm x 38 mm battens and type 1F reinforced underlay						
Roof coverings (PC £ per 1000)	795.60	0.32	9.51	11.65	m²	**21.15**
Extra over coverings for						
fixing every tile	-	0.02	0.59	0.78	m²	**1.37**
eaves; eaves filler	-	0.04	1.19	0.59	m	**1.78**
verges; 150 mm wide asbestos free strip undercloak	-	0.21	6.24	1.91	m	**8.15**
dry verge system; extruded white pvc	-	0.14	4.16	11.36	m	**15.51**
segmental ridge cap to dry verge	-	0.02	0.59	3.63	m	**4.22**
valley trough tiles; cutting both sides	-	0.51	15.15	25.02	m	**40.16**
segmental ridge tiles	-	0.46	13.66	1.05	m	**14.72**
segmental hip tiles; cutting both sides	-	0.60	17.82	3.75	m	**21.57**
dry ridge tiles; segmental including batten sections; unions and filler pieces	-	0.28	8.32	11.54	m	**19.86**
segmental mono-ridge tiles	-	0.46	13.66	17.12	m	**30.78**
gas ridge terminal	-	0.46	13.66	68.59	nr	**82.26**
holes for pipes and the like	-	0.19	5.64	-	nr	**5.64**

H CLADDING/COVERING

Item	PC £	Labour hours	Labour £	Material £	Unit	Total rate £
Concrete interlocking tiles; Marley Eternit "Mendip" granule finish double pantiles or other equal and approved; 420 mm x 330 mm; to 75 mm lap; on 22 mm x 38 mm battens and type 1F reinforced underlay						
Roof coverings (PC £ per 1000)	754.80	0.32	9.51	11.11	m²	**20.62**
Extra over coverings for						
fixing every tile	-	0.02	0.59	0.78	m²	**1.37**
eaves; eaves filler	-	0.02	0.59	10.70	m	**11.29**
verges; 150 mm wide asbestos free strip undercloak	-	0.21	6.24	1.91	m	**8.15**
dry verge system; extruded white pvc	-	0.14	4.16	11.36	m	**15.51**
segmental ridge cap to dry verge	-	0.02	0.59	3.63	m	**4.22**
valley trough tiles; cutting both sides	-	0.51	15.15	24.92	m	**40.07**
segmental ridge tiles	-	0.51	15.15	5.67	m	**20.82**
segmental hip tiles; cutting both sides	-	0.65	19.31	8.23	m	**27.53**
dry ridge tiles; segmental including batten sections; unions and filler pieces	-	0.28	8.32	11.54	m	**19.86**
segmental mono-ridge tiles	-	0.46	13.66	19.42	m	**33.09**
gas ridge terminal	-	0.46	13.66	68.59	nr	**82.26**
holes for pipes and the like	-	0.19	5.64	-	nr	**5.64**
Concrete interlocking tiles; Marley Eternit "Modern" smooth finish tiles or other equal and approved; 420 mm x 220 mm; to 75 mm lap; on 25 mm x 38 mm battens and type 1F reinforced underlay						
Roof coverings (PC £ per 1000)	771.80	0.32	9.51	11.63	m²	**21.14**
Extra over coverings for						
fixing every tile	-	0.02	0.59	0.78	m²	**1.37**
verges; 150 wide asbestos free strip undercloak	-	0.21	6.24	1.91	m	**8.15**
dry verge system; extruded white pvc	-	0.19	5.64	11.36	m	**17.00**
"Modern" ridge cap to dry verge	-	0.02	0.59	3.63	m	**4.22**
valley trough tiles; cutting both sides	-	0.51	15.15	24.96	m	**40.11**
"Modern" ridge tiles	-	0.46	13.66	10.01	m	**23.67**
"Modern" hip tiles; cutting both sides	-	0.60	17.82	12.63	m	**30.45**
dry ridge tiles; "Modern"; including batten sections; unions and filler pieces	-	0.28	8.32	20.50	m	**28.82**
"Modern" mono-ridge tiles	-	0.46	13.66	17.12	m	**30.78**
gas ridge terminal	-	0.46	13.66	68.59	nr	**82.26**
holes for pipes and the like	-	0.19	5.64	-	nr	**5.64**
Concrete interlocking tiles; Marley Eternit "Wessex" smooth finish tiles or other equal and approved; 413 mm x 330 mm; to 75 mm lap; on 25 mm x 38 mm battens and type 1F reinforced underlay						
Roof coverings (PC £ per 1000)	1168.75	0.32	9.51	16.07	m²	**25.58**
Extra over coverings for						
fixing every tile	-	0.02	0.59	0.78	m²	**1.37**
verges; 150 mm wide asbestos free strip undercloak	-	0.21	6.24	1.91	m	**8.15**
dry verge system; extruded white pvc	-	0.19	5.64	11.36	m	**17.00**
"Modern" ridge cap to dry verge	-	0.02	0.59	3.63	m	**4.22**
valley trough tiles; cutting both sides	-	0.51	15.15	25.86	m	**41.01**
"Modern" ridge tiles	-	0.46	13.66	10.01	m	**23.67**
"Modern" hip tiles; cutting both sides	-	0.60	17.82	13.97	m	**31.80**
dry ridge tiles; "Modern"; including batten sections; unions and filler pieces	-	0.28	8.32	20.50	m	**28.82**
"Modern" mono-ridge tiles	-	0.46	13.66	17.12	m	**30.78**
gas ridge terminal	-	0.46	13.66	68.59	nr	**82.26**
holes for pipes and the like	-	0.19	5.64	-	nr	**5.64**

H CLADDING/COVERING

Item	PC £	Labour hours	Labour £	Material £	Unit	Total rate £
H60 PLAIN ROOF TILING – cont'd						
Concrete interlocking slates; Redland "Richmond" smooth finish tiles; 430 x 380; to 75 mm lap; on 25 mm x 38 mm battens and type 1F reinforced underlay						
Roof coverings (PC £ per 1000)	991.95	0.32	9.51	12.02	m²	**21.53**
Extra over coverings for						
fixing every tile	-	0.02	0.59	0.78	m²	**1.37**
eaves; eaves filler	-	0.02	0.59	5.57	m	**6.16**
verges; extra single undercloak course of plain tiles	-	0.23	6.83	3.76	m	**10.59**
ambi-dry verge system	-	0.19	5.64	10.88	m	**16.52**
ambi-dry verge eave/ridge end piece	-	0.02	0.59	3.89	m	**4.48**
universal valley trough tiles; cutting both sides	-	0.56	16.63	34.87	m	**51.51**
universal hip tiles; cutting both sides	-	0.60	17.82	12.86	m	**30.68**
universal angle ridge tiles	-	0.46	13.66	9.50	m	**23.16**
dry ridge system; universal angle ridge tiles	-	0.23	6.83	25.30	m	**32.13**
universal mono-pitch angle ridge tiles	-	0.51	15.15	18.25	m	**33.40**
gas ridge terminal	-	0.46	13.66	71.18	nr	**84.84**
ridge vent with 110 mm diameter flexible adaptor	-	0.46	13.66	84.48	nr	**98.14**
holes for pipes and the like	-	0.19	5.64	-	nr	**5.64**
Concrete interlockingslates; Redland "Stonewold II" smooth finish tiles; 430 mm x 380 mm; to 75 mm lap; on 25 mm x 38 mm battens and type 1F reinforced underlay						
Roof coverings (PC £ per 1000)	1518.10	0.32	9.51	19.77	m²	**29.28**
Extra over coverings for						
fixing every tile	-	0.02	0.59	0.98	m²	**1.58**
verges; extra single undercloak course of plain tiles	-	0.28	8.32	3.76	m	**12.07**
ambi-dry verge system	-	0.19	5.64	10.88	m	**16.52**
ambi-dry verge eave/ridge end piece	-	0.02	0.59	3.89	m	**4.48**
valley trough tiles; cutting both sides	-	0.51	15.15	35.42	m	**50.57**
universal angle ridge tiles	-	0.46	13.66	9.50	m	**23.16**
universal hip tiles; cutting both sides	-	0.60	17.82	14.64	m	**32.47**
dry ridge system; universal angle ridge tiles	-	0.23	6.83	25.30	m	**32.13**
universal mono-pitch angle ridge tiles	-	0.51	15.15	18.25	m	**33.40**
universal gas flue angle ridge tile	-	0.46	13.66	71.90	nr	**85.56**
universal angle ridge vent tile with 110 mm diameter adaptor	-	0.46	13.66	72.75	nr	**86.42**
holes for pipes and the like	-	0.19	5.64	-	nr	**5.64**
Concrete interlocking tiles; Redland "Norfolk" smooth finish pantiles; 381 mm x 229 mm; to 75 mm lap; on 25 mm x 38 mm battens and type 1F reinforced underlay						
Roof coverings (PC £ per 1000)	563.55	0.42	12.48	14.65	m²	**27.13**
Extra over coverings for						
fixing every tile	-	0.04	1.19	0.34	m²	**1.53**
eaves; eaves filler	-	0.04	1.19	1.21	m	**2.40**
verges; extra single undercloak course of plain tiles	-	0.28	8.32	6.63	m	**14.95**
valley trough tiles; cutting both sides	-	0.56	16.63	33.46	m	**50.09**
universal ridge tiles	-	0.46	13.66	12.77	m	**26.44**
universal hip tiles; cutting both sides	-	0.60	17.82	15.96	m	**33.78**
universal gas flue ridge tile	-	0.46	13.66	71.93	nr	**85.59**
universal ridge vent tile with 110 mm diameter adaptor	-	0.50	14.85	84.22	nr	**99.07**
holes for pipes and the like	-	0.19	5.64	-	nr	**5.64**

H CLADDING/COVERING

Item	PC £	Labour hours	Labour £	Material £	Unit	Total rate £
Concrete interlocking tiles; Redland "Regent" granule finish bold roll tiles or other equal and approved; 418 mm x 332 mm; to 75 mm lap; on 25 mm x 38 mm battens and type 1F reinforced underlay						
Roof coverings (PC £ per 1000)	799.00	0.32	9.51	11.82	m²	**21.32**
Extra over coverings for						
fixing every tile	-	0.03	0.89	0.76	m²	**1.65**
eaves; eaves filler	-	0.04	1.19	0.91	m	**2.10**
verges; extra single undercloak course of plain tiles	-	0.23	6.83	3.23	m	**10.06**
cloaked verge system	-	0.14	4.16	7.71	m	**11.87**
valley trough tiles; cutting both sides	-	0.51	15.15	32.98	m	**48.13**
universal ridge tiles	-	0.46	13.66	12.77	m	**26.44**
universal hip tiles; cutting both sides	-	0.60	17.82	15.48	m	**33.30**
dry ridge system; universal ridge tiles	-	0.23	6.83	43.64	m	**50.47**
universal half round mono-pitch ridge tiles	-	0.51	15.15	26.89	m	**42.04**
universal gas flue ridge tile	-	0.46	13.66	71.93	nr	**85.59**
universal ridge vent tile with 110 mm diameter adaptor	-	0.46	13.66	84.22	nr	**97.88**
holes for pipes and the like	-	0.19	5.64	-	nr	**5.64**
Concrete interlocking tiles; Redland "Renown" granule finish tiles or other equal and approved; 418 mm x 330 mm; to 75 mm lap; on 25 mm x 38 mm battens and type 1F reinforced underlay						
Roof coverings (PC £ per 1000)	771.80	0.32	9.51	11.52	m²	**21.03**
Extra over coverings for						
fixing every tile	-	0.02	0.59	0.41	m²	**1.00**
verges; extra single undercloak course of plain tiles	-	0.23	6.83	3.87	m	**10.70**
cloaked verge system	-	0.14	4.16	7.80	m	**11.96**
valley trough tiles; cutting both sides	-	0.51	15.15	32.89	m	**48.04**
universal ridge tiles	-	0.46	13.66	12.77	m	**26.44**
universal hip tiles; cutting both sides	-	0.60	17.82	15.39	m	**33.21**
dry ridge system; universal ridge tiles	-	0.23	6.83	42.57	m	**49.40**
universal half round mono-pitch ridge tiles	-	0.51	15.15	26.89	m	**42.04**
universal gas flue ridge tile	-	0.46	13.66	71.93	nr	**85.59**
universal ridge vent tile with 110 mm diameter adaptor	-	0.46	13.66	84.22	nr	**97.88**
holes for pipes and the like	-	0.19	5.64	-	nr	**5.64**
Concrete plain tiles; BS EN 490 group A; 267 mm x 165 mm; on 25 mm x 38 mm battens and type 1F reinforced underlay						
Roof coverings; to 64 mm lap (PC £ per 1000)	325.55	0.97	28.81	29.86	m²	**58.67**
Wall coverings; to 38 mm lap	-	1.16	34.46	25.33	m²	**59.79**
Extra over coverings for						
ornamental tiles	-	-	-	18.90	m²	**-**
double course at eaves	-	0.23	6.83	3.38	m	**10.22**
verges	-	0.31	9.21	1.51	m	**10.72**
swept valleys; cutting both sides	-	0.60	17.82	32.36	m	**50.19**
bonnet hips; cutting both sides	-	0.74	21.98	32.44	m	**54.42**
external vertical angle tiles; supplementary nail fixings	-	0.37	10.99	23.07	m	**34.06**
half round ridge tiles	-	0.46	13.66	8.24	m	**21.90**
third round hip tiles; cutting both sides	-	0.46	13.66	10.45	m	**24.11**
holes for pipes and the like	-	0.19	5.64	-	nr	**5.64**

H CLADDING/COVERING

Item	PC £	Labour hours	Labour £	Material £	Unit	Total rate £
H60 PLAIN ROOF TILING – cont'd						
Sundries						
Hip irons						
galvanised mild steel; fixing with screws	-	0.09	2.67	2.04	nr	**4.71**
"Rytons Clip strip" or other equal and approved; continuous soffit ventilator						
51 mm wide; plastic; code CS351	-	0.28	8.32	0.65	m	**8.97**
"Rytons over fascia ventilator" or other equal and approved; continuous eaves ventilator						
40 mm wide; plastic; code OFV890	-	0.09	2.67	1.34	m	**4.01**
"Rytons roof ventilator" or other equal and approved; to suit rafters at 600 mm centres						
250 mm deep x 43 mm high; plastic; code TV600	-	0.09	2.67	1.10	m	**3.78**
"Rytons push and lock ventilators" or other equal and approved; circular						
83 mm diameter; plastic; code PL235	-	0.04	0.84	0.17	nr	**1.01**
Fixing only						
lead soakers (supply cost not included)	-	0.07	1.45	-	nr	**1.45**
Pressure impregnated softwood counter battens; 25 mm x 50 mm						
450 mm centres	-	0.06	1.78	1.41	m²	**3.19**
600 mm centres	-	0.04	1.19	1.07	m²	**2.26**
Underlay; BS EN 13707 type 1B; bitumen felt weighing 14 kg/10 m²; 75 mm laps						
To sloping or vertical surfaces	0.40	0.02	0.59	1.14	m²	**1.73**
Underlay; BS EN 13707 type 1F; reinforced bitumen felt weighing 22.50 kg/10 m²; 75 mm laps						
To sloping or vertical surfaces	0.75	0.02	0.59	1.53	m²	**2.12**
Underlay; Visqueen "Tilene 200P" or other equal and approved; micro-perforated sheet; 75 mm laps						
To sloping or vertical surfaces	0.53	0.02	0.59	1.30	m²	**1.90**
Underlay; "Powerlon 250 BM" or other equal and approved; reinforced breather membrane; 75 mm laps						
To sloping or vertical surfaces	0.56	0.02	0.59	1.35	m²	**1.94**
Underlay; "Anticon" or other equal and approved sarking membrane; Euroroof Ltd; polyethylene; 75 mm laps						
To sloping or vertical surfaces	-	0.02	0.59	1.77	m²	**2.37**
H61 FIBRE CEMENT SLATING						
Asbestos-free artificial slates; Eternit "Garsdale/E2000T" or other equal and approved; to 75 mm lap; on 19 mm x 50 mm battens and type 1F reinforced underlay						
Coverings; 500 mm x 250 mm slates						
roof coverings	-	0.60	17.82	21.99	m²	**39.82**
wall coverings	-	0.74	21.98	21.99	m²	**43.98**
Coverings; 600 mm x 300 mm slates						
roof coverings	-	0.46	13.66	17.91	m²	**31.58**
wall coverings	-	0.60	17.82	17.91	m²	**35.73**

H CLADDING/COVERING

Item	PC £	Labour hours	Labour £	Material £	Unit	Total rate £
Extra over slate coverings for						
double course at eaves	-	0.23	6.83	4.49	m	**11.32**
verges; extra single undercloak course	-	0.31	9.21	0.96	m	**10.17**
open valleys; cutting both sides	-	0.19	5.64	3.83	m	**9.47**
stop end	-	0.09	2.67	9.36	nr	**12.03**
roll top ridge tiles	-	0.56	16.63	31.34	m	**47.98**
stop end	-	0.09	2.67	17.37	nr	**20.04**
mono-pitch ridge tiles	-	0.46	13.66	36.63	m	**50.30**
stop end	-	0.09	2.67	39.82	nr	**42.50**
duo-pitch ridge tiles	-	0.46	13.66	29.67	m	**43.34**
stop end	-	0.09	2.67	29.20	nr	**31.87**
half round hip tiles; cutting both sides	-	0.19	5.64	62.84	m	**68.49**
holes for pipes and the like	-	0.19	5.64	-	nr	**5.64**
H62 NATURAL SLATING						
NOTE: The following items of slate roofing unless otherwise described, include for conventional fixing assuming "normal exposure" with appropriate nails and/or rivets or clips to pressure impregnated softwood battens fixed with galvanised nails; prices also include for all bedding and pointing at verges; beneath verge tiles etc.						
Natural slates; BS EN 12326 Part 2; Spanish blue grey; uniform size; to 75 mm lap; on 25 mm x 50 mm battens and type 1F reinforced underlay						
Coverings; 400 mm x 250 mm slates						
roof coverings (PC £ per 1000)	560.00	0.83	24.65	22.31	m²	**46.97**
wall coverings	-	1.06	31.49	22.31	m²	**53.80**
Coverings; 500 mm x 250 mm slates						
roof coverings (PC £ per 1000)	918.75	0.73	21.68	24.85	m²	**46.53**
wall coverings	-	0.88	26.14	24.85	m²	**50.99**
Coverings; 600 mm x 300 mm slates						
roof coverings (PC £ per 1000)	1465.63	0.56	16.63	25.29	m²	**41.93**
wall coverings	-	0.69	20.50	25.29	m²	**45.79**
Extra over coverings for						
double course at eaves	-	0.28	8.32	6.41	m	**14.73**
verges; extra single undercloak course	-	0.39	11.58	3.47	m	**15.05**
open valleys; cutting both sides	-	0.20	5.94	13.25	m	**19.19**
blue/black glass reinforced concrete 152 mm half round ridge tiles	-	0.46	13.66	14.91	m	**28.57**
blue/black glass reinforced concrete 125 mm x 125 mm plain angle ridge tiles	-	0.46	13.66	14.91	m	**28.57**
mitred hips; cutting both sides	-	0.20	5.94	13.25	m	**19.19**
blue/black glass reinforced concrete 152 mm half round hip tiles; cutting both sides	-	0.65	19.31	28.16	m	**47.47**
blue/black glass reinforced concrete 125 mm x 125 mm plain angle hip tiles; cutting both sides	-	0.65	19.31	28.14	m	**47.45**
holes for pipes and the like	-	0.19	5.64	-	nr	**5.64**

H CLADDING/COVERING

Item	PC £	Labour hours	Labour £	Material £	Unit	Total rate £
H62 NATURAL SLATING – cont'd						
Natural slates; BS EN 12326 Part 2; Welsh blue grey; uniform size; to 75 mm lap; on 25 mm x 50 mm battens and type 1F reinforced underlay						
Coverings; 400 mm x 250 mm slates						
roof coverings (PC £ per 1000)	1355.75	0.83	24.65	44.34	m²	**69.00**
wall coverings	-	1.06	31.49	44.34	m²	**75.83**
Coverings; 500 mm x 250 mm slates						
roof coverings (PC £ per 1000)	2337.50	0.73	21.68	54.83	m²	**76.51**
wall coverings	-	0.88	26.14	54.83	m²	**80.97**
Coverings; 500 mm x 300 mm slates						
roof coverings (PC £ per 1000)	2618.00	0.67	19.90	51.59	m²	**71.49**
wall coverings	-	0.80	23.76	51.59	m²	**75.35**
Coverings; 600 mm x 300 mm slates						
roof coverings (PC £ per 1000)	5469.75	0.56	16.63	82.31	m²	**98.94**
wall coverings	-	0.69	20.50	82.31	m²	**102.80**
Extra over coverings for						
double course at eaves	-	0.28	8.32	21.34	m	**29.66**
verges; extra single undercloak course	-	0.39	11.58	12.52	m	**24.10**
open valleys; cutting both sides	-	0.20	5.94	49.45	m	**55.39**
blue/black glazed ware 152 mm half round ridge tiles	-	0.46	13.66	9.54	m	**23.20**
blue/black glazed ware 125 mm x 125 mm plain angle ridge tiles	-	0.46	13.66	26.94	m	**40.61**
mitred hips; cutting both sides	-	0.20	5.94	49.45	m	**55.39**
blue/black glazed ware 152 mm half round hip tiles; cutting both sides	-	0.65	19.31	58.99	m	**78.30**
blue/black glazed ware 125 mm x 125 mm plain angle hip tiles; cutting both sides	-	0.65	19.31	76.39	m	**95.70**
holes for pipes and the like	-	0.19	5.64	-	nr	**5.64**
Natural slates; Westmoreland green; random lengths; 457 mm - 229 mm proportionate widths to 75 mm lap; in diminishing courses; on 25 mm x 50 mm battens and type 1F underlay						
Roof coverings (PC £ per tonne)	1890.00	1.06	31.49	121.49	m²	**152.98**
Wall coverings	-	1.34	39.80	121.49	m²	**161.30**
Extra over coverings for						
double course at eaves	-	0.61	18.12	22.53	m	**40.65**
verges; extra single undercloak course slates 152 mm wide	-	0.69	20.50	19.42	m	**39.92**
holes for pipes and the like	-	0.28	8.32	-	nr	**8.32**
H63 RECONSTRUCTED STONE SLATING/TILING						
Reconstructed stone slates; "Hardrow Slates"; standard colours; or similar; 75 mm lap; on 25 mm x 50 mm battens and type 1F reinforced underlay						
Coverings; 457 mm x 305 mm slates						
roof coverings	15.52	0.74	21.98	23.87	m²	**45.85**
wall coverings	-	0.93	27.63	23.87	m²	**51.50**
Coverings; 457 mm x 457 mm slates						
roof coverings	15.57	0.60	17.82	23.46	m²	**41.29**
wall coverings	-	0.79	23.47	23.46	m²	**46.93**
Extra over 457 mm x 305 mm coverings for						
double course at eaves	-	0.28	8.32	4.40	m	**12.71**
verges; pointed	-	0.39	11.58	0.07	m	**11.66**
open valleys; cutting both sides	-	0.20	5.94	10.20	m	**16.14**
ridge tiles	-	0.46	13.66	26.18	m	**39.84**
hip tiles; cutting both sides	-	0.65	19.31	22.17	m	**41.48**

H CLADDING/COVERING

Item	PC £	Labour hours	Labour £	Material £	Unit	Total rate £
Reconstructed stone slates; Bradstone "Cotswold" style or other equal and approved; random lengths 550 mm - 300 mm; proportional widths; to 80 mm lap; in diminishing courses; on 25 mm x 50 mm battens and type 1F reinforced underlay						
Roof coverings (all-in rate inclusive of eaves and verges)	27.86	0.97	28.81	37.83	m²	**66.64**
Extra over coverings for						
open valleys/mitred hips; cutting both sides	-	0.42	12.48	14.17	m²	**26.65**
ridge tiles	-	0.61	18.12	18.81	m	**36.93**
hip tiles; cutting both sides	-	0.97	28.81	32.02	m	**60.83**
holes for pipes and the like	-	0.28	8.32	-	nr	**8.32**
Reconstructed stone slates; Bradstone "Moordale" style or other equal and approved; random lengths 550 mm - 450 mm; proportional widths; to 80 mm lap; in diminishing course; on 25 mm x 50 mm battens and type 1F reinforced underlay						
Roof coverings (all-in rate inclusive of eaves and verges)	26.22	0.97	28.81	35.96	m²	**64.78**
Extra over coverings for						
open valleys/mitred hips; cutting both sides	-	0.42	12.48	13.33	m²	**25.81**
ridge tiles	-	0.61	18.12	18.81	m	**36.93**
holes for pipes and the like	-	0.28	8.32	-	nr	**8.32**
H64 TIMBER SHINGLING						
Red cedar sawn shingles preservative treated; uniform length 400 mm; to 125 mm gauge; on 25 mm x 38 mm battens and type 1F reinforced underlay						
Roof coverings; 125 mm gauge, 2.28 m²/bundle (PC £ per bundle)	56.25	0.97	28.81	33.91	m²	**62.72**
Wall coverings; 190 mm gauge, 3.47 m²/bundle	-	0.74	21.98	22.65	m²	**44.63**
Extra over coverings for						
double course at eaves	-	0.19	5.64	3.58	m	**9.22**
open valleys; cutting both sides	-	0.19	5.64	6.83	m	**12.47**
pre-formed ridge capping	-	0.28	8.32	9.46	m	**17.77**
pre-formed hip capping; cutting both sides	-	0.46	13.66	16.29	m	**29.95**
double starter course to cappings	-	0.09	2.67	0.99	m	**3.66**
holes for pipes and the like	-	0.14	4.16	-	nr	**4.16**

H CLADDING/COVERING

Item	PC £	Labour hours	Labour £	Material £	Unit	Total rate £
H71 LEAD SHEET COVERINGS/FLASHINGS						
Milled Lead; BS EN 12588; on and including Geotec underlay						
The following rates are based upon the measurement allowances and the coverage rules of SMM7 clause M2(a-f)						
Roof and dormer coverings						
1.80 mm thick (code 4) roof coverings (20.41 kg/m²)						
flat (in wood roll construction (PC £ per kg)	2.05	0.90	27.68	65.81	m²	**93.49**
pitched (in wood roll construction)	-	1.00	30.75	66.10	m²	**96.85**
pitched (in welded seam construction)	-	0.90	27.68	65.81	m²	**93.49**
vertical (in welded seam construction)	-	1.00	30.75	63.28	m²	**94.03**
1.80 mm thick (code 4) dormer coverings						
flat (in wood roll construction)	-	0.68	20.76	65.18	m²	**85.94**
pitched (in wood roll construction)	-	0.75	23.06	65.39	m²	**88.45**
pitched (in welded seam construction)	-	0.68	20.76	65.18	m²	**85.94**
vertical (in welded seam construction)	-	1.50	46.13	63.28	m²	**109.40**
2.24 mm thick (code 5) roof coverings (25.40 kg/m²)						
flat (in wood roll construction)	-	0.94	29.06	80.03	m²	**109.09**
pitched (in wood roll construction)	-	1.05	32.29	80.33	m²	**112.61**
pitched (in welded seam construction)	-	0.94	29.06	80.03	m²	**109.09**
vertical (in welded seam construction)	-	1.05	32.29	77.37	m²	**109.65**
2.24 mm thick (code 5) dormer coverings						
flat (in wood roll construction)	-	0.71	21.80	79.37	m²	**101.17**
pitched (in wood roll construction)	-	0.79	24.23	79.59	m²	**103.82**
pitched (in welded seam construction)	-	0.71	21.80	79.37	m²	**101.17**
vertical (in welded seam construction)	-	1.57	48.43	77.37	m²	**125.80**
2.65 mm thick (code 6) roof coverings (30.05 kg/m²)						
flat (in wood roll construction)	-	0.99	30.44	93.29	m²	**123.73**
pitched (in wood roll construction)	-	1.10	33.83	93.60	m²	**127.42**
pitched (in welded seam construction)	-	0.99	30.44	93.29	m²	**123.73**
vertical (in welded seam construction)	-	1.10	33.83	90.50	m²	**124.32**
2.65 mm thick (code 6) dormer coverings						
flat (in wood roll construction)	-	0.74	22.85	92.60	m²	**115.44**
pitched (in wood roll construction)	-	0.82	25.37	92.82	m²	**118.19**
pitched (in welded seam construction)	-	0.74	22.85	92.60	m²	**115.44**
vertical (in welded seam construction)	-	1.65	50.74	90.50	m²	**141.23**
3.15 mm thick (code 7) roof coverings (35.72 kg/m²)						
flat (in wood roll construction)	-	1.06	32.53	109.49	m²	**142.03**
pitched (in wood roll construction)	-	1.18	36.13	109.82	m²	**145.95**
pitched (in welded seam construction)	-	1.06	32.53	109.49	m²	**142.03**
vertical (in welded seam construction)	-	1.18	36.13	106.51	m²	**142.64**
3.15 mm thick (code 7) dormer coverings						
flat (in wood roll construction)	-	0.79	24.38	108.75	m²	**133.13**
pitched (in wood roll construction)	-	0.88	27.09	108.99	m²	**136.08**
pitched (in welded seam construction)	-	0.79	24.38	108.75	m²	**133.13**
vertical (in welded seam construction)	-	1.76	54.21	106.51	m²	**160.72**
3.55 mm thick (code 8) roof coverings (40.26 kg/m²)						
flat (in wood roll construction)	-	1.15	35.30	122.57	m²	**157.87**
pitched (in wood roll construction)	-	1.27	39.21	122.92	m²	**162.13**
pitched (in welded seam construction)	-	1.15	35.30	122.57	m²	**157.87**
vertical (in welded seam construction)	-	1.27	39.21	119.32	m²	**158.53**
3.55 mm thick (code 8) dormer coverings						
flat (in wood roll construction)	-	0.86	26.48	121.75	m²	**148.22**
pitched (in wood roll construction)	-	0.96	29.40	122.02	m²	**151.41**
pitched (in welded seam construction)	-	0.86	26.48	121.75	m²	**148.22**
vertical (in welded seam construction)	-	1.91	58.82	119.32	m²	**178.15**

H CLADDING/COVERING

Item	PC £	Labour hours	Labour £	Material £	Unit	Total rate £
Sundries						
patination oil to finished work surfaces	-	0.03	0.77	0.28	m²	**1.04**
chalk slurry to underside of panels	-	0.33	10.24	2.48	m²	**12.72**
provision of 45 x 45 mm wood rolls at 600 mm centres (per m)	-	0.10	3.08	1.17	m	**4.25**
dressing over glazing bars and glass	-	0.25	7.69	0.85	m	**8.53**
soldered nail head	-	0.01	0.25	0.07	nr	**0.32**
1.32 mm thick (code 3) lead flashings, etc.						
Soakers (14.18 kg per m²)						
200 x 200 mm	-	0.02	0.46	1.71	nr	**2.17**
300 x 300 mm	-	0.02	0.46	3.85	nr	**4.32**
1.80 mm thick (code 4) lead flashings, etc.						
Flashings; wedging into grooves (20.41 kg per m²)						
150 mm girth	-	0.25	7.69	8.65	m	**16.33**
200 mm girth	-	0.25	7.69	11.53	m	**19.21**
240 mm girth	-	0.25	7.69	13.83	m	**21.52**
300 mm girth	-	0.25	7.69	17.29	m	**24.98**
Stepped flashings; wedging into grooves						
180 mm girth	-	0.50	15.38	10.37	m	**25.75**
270 mm girth	-	0.50	15.38	15.56	m	**30.94**
Linings to sloping gutters						
390 mm girth	-	0.40	12.30	22.48	m	**34.78**
450 mm girth	-	0.45	13.84	25.94	m	**39.77**
600 mm girth	-	0.55	16.91	34.58	m	**51.49**
Cappings to hips or ridges						
450 mm girth	-	0.50	15.38	25.94	m	**41.31**
600 mm girth	-	0.60	18.45	34.58	m	**53.03**
Saddle flashings; dressing and bossing						
450 x 450 mm	-	0.50	15.38	13.08	nr	**28.45**
600 x 200 mm	-	0.50	15.38	22.16	nr	**37.53**
Slates; with 150 mm high collar						
450 x 450 mm; to suit 50 mm diameter pipe	-	0.75	23.06	15.15	nr	**38.21**
450 x 450 mm; to suit 100 mm diameter pipe	-	0.75	23.06	16.50	nr	**39.56**
450 x 450 mm; to suit 150 mm diameter pipe	-	0.75	23.06	17.86	nr	**40.92**
2.24 mm thick (code 5) lead flashings, etc.						
Flashings; wedging into grooves (25.40 kg per m²)						
150 mm girth	-	0.25	7.69	10.76	m	**18.45**
200 mm girth	-	0.25	7.69	14.34	m	**22.03**
240 mm girth	-	0.25	7.69	17.21	m	**24.90**
300 mm girth	-	0.25	7.69	21.52	m	**29.20**
Stepped flashings; wedging into grooves						
180 mm girth	-	0.50	15.38	12.91	m	**28.28**
270 mm girth	-	0.50	15.38	19.37	m	**34.74**
Linings to sloping gutters						
390 mm girth	-	0.40	12.30	27.97	m	**40.27**
450 mm girth	-	0.45	13.84	32.28	m	**46.11**
600 mm girth	-	0.55	16.91	43.03	m	**59.94**
Cappings to hips or ridges						
450 mm girth	-	0.50	15.38	32.28	m	**47.65**
600 mm girth	-	0.60	18.45	43.03	m	**61.48**
Saddle flashings; dressing and bossing						
450 x 450 mm	-	0.50	15.38	15.93	nr	**31.31**
600 x 200 mm	-	0.50	15.38	27.23	nr	**42.60**
Slates; with 150 mm high collar						
450 x 450 mm; to suit 50 mm diameter pipe	-	0.75	23.06	18.33	nr	**41.39**
450 x 450 mm; to suit 100 mm diameter pipe	-	0.75	23.06	20.01	nr	**43.08**
450 x 450 mm; to suit 150 mm diameter pipe	-	0.75	23.06	21.71	nr	**44.77**

H CLADDING/COVERING

Item	PC £	Labour hours	Labour £	Material £	Unit	Total rate £
H72 ALUMINIUM SHEET COVERINGS/FLASHINGS						
Aluminium roofing; commercial grade; on and including Geotec underlay						
The following rates are based upon nett 'deck' or 'wall' areas, and depart from SMM7 coverage rules						
Roof, dormer and wall coverings						
0.7 mm thick roof coverings; mill finish						
flat (in wood roll construction) (PC per kg)	4.50	1.00	30.75	23.68	m²	**54.43**
eaves detail ED1	-	0.20	6.15	3.03	m	**9.18**
abutment upstands at perimeters	-	0.33	10.15	1.21	m	**11.36**
pitched over 3 degrees (in standing seam construction)	-	0.75	23.06	20.05	m²	**43.11**
vertical (in angled or flat seam construction)	-	0.80	24.60	20.05	m²	**44.65**
0.7 mm thick dormer coverings; mill finish						
flat (in wood roll construction)	-	1.50	46.13	23.68	m²	**69.80**
eaves detail ED1	-	0.20	6.15	3.03	m	**9.18**
pitched over 3 degrees (in standing seam construction)	-	1.25	38.44	20.05	m²	**58.48**
vertical (in angled or flat seam construction)	-	1.35	41.51	20.05	m²	**61.56**
0.7 mm thick roof coverings; Pvf2 finish						
flat (in wood roll construction) (PC per kg)	5.60	1.00	30.75	28.12	m²	**58.87**
eaves detail ED1	-	0.20	6.15	3.77	m	**9.92**
abutment upstands at perimeters	-	0.33	10.15	1.51	m	**11.65**
pitched over 3 degrees (in standing seam construction)	-	0.75	23.06	23.60	m²	**46.66**
vertical (in angled or flat seam construction)	-	0.80	24.60	23.60	m²	**48.20**
0.7 mm thick dormer coverings; Pvf2 finish						
flat (in wood roll construction)	-	1.50	46.13	28.12	m²	**74.24**
eaves detail ED1	-	0.20	6.15	3.77	m	**9.92**
pitched over 3 degrees (in standing seam construction)	-	1.25	38.44	23.60	m²	**62.03**
vertical (in angled or flat seam construction)	-	1.35	41.51	23.60	m²	**65.11**
0.7 mm thick aluminium flashings, etc.						
Flashings; wedging into grooves; mill finish						
150 mm girth (PC per kg)	4.50	0.25	7.69	1.82	m	**9.50**
240 mm girth	-	0.25	7.69	2.91	m	**10.59**
300 mm girth	-	0.25	7.69	3.63	m	**11.32**
Stepped flashings; wedging into grooves; mill finish						
180 mm girth	-	0.50	15.38	2.18	m	**17.55**
270 mm girth	-	0.50	15.38	3.27	m	**18.64**
Flashings; wedging into grooves; Pvf2 finish						
150 mm girth (PC per kg)	5.60	0.25	7.69	2.26	m	**9.95**
240 mm girth	-	0.25	7.69	3.62	m	**11.30**
300 mm girth	-	0.25	7.69	4.52	m	**12.21**
Stepped flashings; wedging into grooves; Pvf2 finish						
180 mm girth	-	0.50	15.38	2.71	m	**18.09**
270 mm girth	-	0.50	15.38	4.07	m	**19.44**
Sundries						
provision of square batten roll at 500 mm centres (per m)	-	0.10	3.08	1.35	m	**4.42**

H CLADDING/COVERING

Item	PC £	Labour hours	Labour £	Material £	Unit	Total rate £
Standing seam aluminium roof cladding; "Kal-zip" Hoogovens Aluminium Building Systems Ltd or other equal and approved; ref AA 3004 A1 Mn1 Mg1; standard natural aluminium stucco embossed finish						
Roof coverings (lining sheets not included); sloping not exceeding 50 degrees; 305 mm wide units						
0.90 mm thick	-	-	-	-	m²	**80.64**
1.00 mm thick	-	-	-	-	m²	**91.91**
1.20 mm thick	-	-	-	-	m²	**114.51**
Extra over for						
polyester coating	-	-	-	-	m²	**6.65**
PVF2 coating	-	-	-	-	m²	**8.21**
smooth curved	-	-	-	-	m²	**39.86**
factory formed tapered sheets	-	-	-	-	m²	**33.88**
raking cutting	-	-	-	-	m	**19.28**
Accessories for roof coverings						
thermal insulation quilt; 60 mm thick	-	-	-	-	m²	**1.73**
thermal insulation quilt; 80 mm thick	-	-	-	-	m²	**3.94**
thermal insulation quilt; 100 mm thick	-	-	-	-	m²	**4.35**
thermal insulation quilt; 150 mm thick	-	-	-	-	m²	**7.88**
semi-rigid insulation slab; 30 mm thick; tissue faced	-	-	-	-	m²	**9.46**
semi-rigid insulation slab; 50 mm thick	-	-	-	-	m²	**10.69**
"Kal-Foil" vapour check	-	-	-	-	m²	**3.71**
Lining sheets; "Kal-bau" aluminium ref TR 30/150; natural aluminium stucco embossed finish; fixed to steel purlins with stainless steel screws						
0.90 mm gauge	-	-	-	-	m²	**19.73**
Lining sheets; profiled steel; 1000 mm cover; bright white polyester paint finish; fixed to steel purlins with stainless steel screws						
0.70 mm gauge; ref TRS30/150S	-	-	-	-	m²	**12.85**
0.70 mm gauge; ref TRS35/200S	-	-	-	-	m²	**11.35**
extra over for						
crimped curve	-	-	-	-	m²	**23.17**
perforated sheet	-	-	-	-	m²	**4.39**
tapered 0.70 mm plain galvanised steel liner	-	-	-	-	m²	**20.08**
plastisol coating on external liners	-	-	-	-	m²	**2.68**
Eaves detail for 305 mm wide "Kal-zip" roof cladding units; including high density polythylene foam fillers; 2 mm extruded alloy drip angle; fixed to Kalzip sheet using stainless steel blind rivets						
40 mm x 20 mm angle: single skin	-	-	-	-	m	**17.48**
70 mm x 30 mm angle: single skin	-	-	-	-	m	**18.97**
40 mm x 20 mm angle: double skin	-	-	-	-	m	**26.90**
70 mm x 30 mm angle: double skin	-	-	-	-	m	**29.01**
Verge detail for 305 mm wide "Kal-zip" roof cladding units						
2 mm extruded aluminium alloy gable closure section; fixed with stainless steel blind sealed rivets; gable end hook/verge clip fixed to ST clip with stainless steel screws and 2 mm extruded aluminium tolerance clip	-	-	-	-	m	**15.09**

Item	PC £	Labour hours	Labour £	Material £	Unit	Total rate £
H72 ALUMINIUM SHEET COVERINGS/FLASHINGS – cont'd						
Standing seam aluminium roof cladding; "Kal-zip – cont'd						
Ridge detail for 305 mm wide "Kal-zip" roof cladding units						
abutment ridge including natural aluminium stucco embossed "U" type ridge closures fixed with stainless steel blind steel rivets; "U" type polyethylene ridge fillers and 2 mm extruded aluminium alloy support Zed fixed with stainless steel blind sealed rivets; fixing with rivets through small seam of "Kal-zip" into ST clip using stainless steel blind sealed rivets	-	-	-	-	m	**54.19**
duo ridge including natural aluminium stucco embossed "U" type ridge closures fixed with stainless steel blind sealed rivets; "U" type polythylene ridge fillers and 2 mm extruded aluminium alloy support Zed fixed with stainless steel blind sealed rivets; fixing with rivets through small seam of "Kal-zip" into ST clip using stainless steel blind steel rivets	-	-	-	-	m	**81.40**
0.9 mm Roof accessories						
Eaves closure; 200 mm girth	-	-	-	-	m	**8.05**
Ridge; 600 mm girth	-	-	-	-	m	**17.45**
Verge; 700 mm girth	-	-	-	-	m	**22.86**
1.0 mm Flashings etc.; fixing/wedging into grooves						
Flashing; 500 mm girth	-	-	-	-	m	**17.45**
Flashing; 750 mm girth	-	-	-	-	m	**23.06**
Flashing; 1000 mm girth	-	-	-	-	m	**32.43**
1.2 mm Flashings etc.; fixing/wedging into grooves						
Flashing; 500 mm girth	-	-	-	-	m	**19.20**
Flashing; 750 mm girth	-	-	-	-	m	**25.36**
Flashing; 1000 mm girth	-	-	-	-	m	**32.04**
1.4 mm Flashings etc.; fixing/wedging into grooves						
Flashing; 500 mm girth	-	-	-	-	m	**22.08**
Flashing; 750 mm girth	-	-	-	-	m	**29.16**
Flashing; 1000 mm girth	-	-	-	-	m	**43.81**
Aluminium Alumasc "Eavesline" coping system; polyester powder coated						
Coping; fixing straps plugged and screwed to brickwork						
362 mm wide; for parapet wall 271 - 300 mm wide	-	0.50	10.38	24.55	m	**34.93**
Extra for						
90 degree angle	-	0.25	5.19	58.49	nr	**63.68**
90 degree tee junction	-	0.35	7.27	62.75	nr	**70.02**
stop end	-	0.15	3.11	34.79	nr	**37.90**
stop end upstand	-	0.20	4.15	37.52	nr	**41.67**

H CLADDING/COVERING

Item	PC £	Labour hours	Labour £	Material £	Unit	Total rate £
H73 COPPER STRIP SHEET COVERINGS/ FLASHINGS						
Copper roofing; BS EN 504; on and including Geotec underlay						
The following rates are based upon nett 'deck' or 'wall' areas, and depart from SMM7 coverage rules						
Roof and dormer coverings						
0.6 mm thick roof coverings; mill finish						
flat (in wood roll construction) (PC per kg)	6.25	1.00	30.75	74.88	m²	**105.63**
eaves detail ED1	-	0.20	6.15	9.25	m	**15.40**
abutment upstands at perimeters	-	0.33	10.15	4.62	m	**14.77**
pitched over 3 degrees (in standing seam construction)	-	0.75	23.06	61.01	m²	**84.07**
vertical (in angled or flat seam construction)	-	0.80	24.60	61.01	m²	**85.61**
0.6 mm thick dormer coverings; mill finish						
flat (in wood roll construction)	-	1.50	46.13	74.88	m²	**121.01**
eaves detail ED1	-	0.20	6.15	9.25	m	**15.40**
pitched over 3 degrees (in standing seam construction)	-	1.25	38.44	61.01	m²	**99.45**
vertical (in angled or flat seam construction)	-	1.35	41.51	61.01	m²	**102.52**
0.6 mm thick roof coverings; oxid finish						
flat (in wood roll construction) (PC per kg)	7.30	1.00	30.75	86.54	m²	**117.29**
eaves detail ED1	-	0.20	6.15	10.80	m	**16.95**
abutment upstands at perimeters	-	0.33	10.15	5.40	m	**15.55**
pitched over 3 degrees (in standing seam construction)	-	0.75	23.06	70.33	m²	**93.40**
vertical (in angled or flat seam construction)	-	0.80	24.60	70.33	m²	**94.93**
0.6 mm thick dormer coverings; oxid finish						
flat (in wood roll construction)	-	1.50	46.13	86.54	m²	**132.66**
eaves detail ED1	-	0.20	6.15	10.80	m	**16.95**
pitched over 3 degrees (in standing seam construction)	-	1.25	38.44	70.33	m²	**108.77**
vertical (in angled or flat seam construction)	-	1.35	41.51	70.33	m²	**111.85**
0.6 mm thick roof coverings; KME pre-patinated finish						
flat (in wood roll construction)	56.50	1.10	33.83	119.53	m²	**153.36**
eaves detail ED1	-	0.20	6.15	15.20	m	**21.35**
abutment upstands at perimeters	-	0.33	10.15	7.60	m	**17.75**
pitched over 3 degrees (in standing seam construction)	-	0.85	26.14	96.73	m²	**122.87**
vertical (in angled or flat seam construction)	-	0.90	27.68	96.73	m²	**124.40**
0.6 mm thick dormer coverings; KME pre-patinated finish						
flat (in wood roll construction)	-	1.50	46.13	119.53	m²	**165.66**
eaves detail ED1	-	0.20	6.15	15.20	m	**21.35**
pitched over 3 degrees (in standing seam construction)	-	1.25	38.44	96.73	m²	**135.17**
vertical (in angled or flat seam construction)	-	1.35	41.51	96.73	m²	**138.24**
0.7 mm thick roof coverings; mill finish						
flat (in wood roll construction) (PC per kg)	6.25	1.00	30.75	84.97	m²	**115.72**
eaves detail ED1	-	0.20	6.15	10.59	m	**16.74**
abutment upstands at perimeters	-	0.33	10.15	5.30	m	**15.44**
pitched over 3 degrees (in standing seam construction)	-	0.75	23.06	69.08	m²	**92.14**
vertical (in angled or flat seam construction)	-	0.80	24.60	69.08	m²	**93.68**

H CLADDING/COVERING

Item	PC £	Labour hours	Labour £	Material £	Unit	Total rate £
H73 COPPER STRIP SHEET COVERINGS/ FLASHINGS – cont'd						
Roof and dormer coverings – cont'd						
0.7 mm thick dormer coverings; mill finish						
flat (in wood roll construction)	-	1.50	46.13	84.97	m²	**131.10**
eaves detail ED1	-	0.20	6.15	10.59	m	**16.74**
pitched over 3 degrees (in standing seam construction)	-	1.25	38.44	69.08	m²	**107.52**
vertical (in angled or flat seam construction)	-	1.35	41.51	69.08	m²	**110.59**
0.7 mm thick roof coverings; oxid finish						
flat (in wood roll construction) (PC per kg)	7.30	1.00	30.75	98.32	m²	**129.07**
eaves detail ED1	-	0.20	6.15	12.37	m	**18.52**
abutment upstands at perimeters	-	0.33	10.15	6.19	m	**16.33**
pitched over 3 degrees (in standing seam construction)	-	0.75	23.06	79.76	m²	**102.82**
vertical (in angled or flat seam construction)	-	0.80	24.60	79.76	m²	**104.36**
0.7 mm thick dormer coverings; oxid finish						
flat (in wood roll construction)	-	1.50	46.13	98.32	m²	**144.45**
eaves detail ED1	-	0.20	6.15	12.37	m	**18.52**
pitched over 3 degrees (in standing seam construction)	-	1.25	38.44	79.76	m²	**118.20**
vertical (in angled or flat seam construction)	-	1.35	41.51	79.76	m²	**121.27**
0.7 mm thick roof coverings; KME pre-patinated finish						
flat (in wood roll construction)	65.00	1.10	33.83	136.68	m²	**170.51**
eaves detail ED1	-	0.20	6.15	17.49	m	**23.64**
abutment upstands at perimeters	-	0.33	10.15	8.74	m	**18.89**
pitched over 3 degrees (in standing seam construction)	-	0.85	26.14	110.45	m²	**136.59**
vertical (in angled or flat seam construction)	-	0.90	27.68	110.45	m²	**138.13**
0.7 mm thick dormer coverings; KME pre-patinated finish						
flat (in wood roll construction)	-	1.50	46.13	136.68	m²	**182.81**
eaves detail ED1	-	0.20	6.15	17.49	m	**23.64**
pitched over 3 degrees (in standing seam construction)	-	1.25	38.44	110.45	m²	**148.89**
vertical (in angled or flat seam construction)	-	1.35	41.51	110.45	m²	**151.96**
0.6 mm thick copper flashings, etc.						
Flashings; wedging into grooves; mill finish						
150 mm girth (PC per kg)	6.25	0.25	7.69	6.94	m	**14.63**
240 mm girth	-	0.25	7.69	11.10	m	**18.79**
300 mm girth	-	0.25	7.69	13.88	m	**21.56**
Stepped flashings; wedging into grooves; mill finish						
180 mm girth	-	0.50	15.38	8.33	m	**23.70**
270 mm girth	-	0.50	15.38	12.49	m	**27.86**
Flashings; wedging into grooves; oxid finish						
150 mm girth (PC per kg)	7.30	0.25	7.69	8.10	m	**15.79**
240 mm girth	-	0.25	7.69	12.96	m	**20.65**
300 mm girth	-	0.25	7.69	16.20	m	**23.89**
Stepped flashings; wedging into grooves; oxid finish						
180 mm girth	-	0.50	15.38	9.72	m	**25.10**
270 mm girth	-	0.50	15.38	14.58	m	**29.96**
Flashings; wedging into grooves; KME pre-patinated finish						
150 mm girth (PC per m²)	56.50	0.25	7.69	11.40	m	**19.09**
240 mm girth	-	0.25	7.69	18.24	m	**25.93**
300 mm girth	-	0.25	7.69	22.80	m	**30.49**

H CLADDING/COVERING

Item	PC £	Labour hours	Labour £	Material £	Unit	Total rate £
Stepped flashings; wedging into grooves; KME pre-patinated finish						
180 mm girth	-	0.50	15.38	13.68	m	**29.06**
270 mm girth	-	0.50	15.38	20.52	m	**35.90**
0.7 mm thick copper flashings, etc.						
Flashings; wedging into grooves; mill finish						
150 mm girth (PC per kg)	6.25	0.25	7.69	7.95	m	**15.63**
240 mm girth	-	0.25	7.69	12.71	m	**20.40**
300 mm girth	-	0.25	7.69	15.89	m	**23.58**
Stepped flashings; wedging into grooves; mill finish						
180 mm girth	-	0.50	15.38	9.54	m	**24.91**
270 mm girth	-	0.50	15.38	14.30	m	**29.68**
Flashings; wedging into grooves; oxid finish						
150 mm girth (PC per kg)	7.30	0.25	7.69	9.31	m	**17.00**
240 mm girth	-	0.25	7.69	14.90	m	**22.59**
300 mm girth	-	0.25	7.69	18.62	m	**26.31**
Stepped flashings; wedging into grooves; oxid finish						
180 mm girth	-	0.50	15.38	11.17	m	**26.55**
270 mm girth	-	0.50	15.38	16.76	m	**32.13**
Flashings; wedging into grooves; KME pre-patinated finish						
150 mm girth (PC per m²)	65.00	0.25	7.69	13.12	m	**20.80**
240 mm girth	-	0.25	7.69	20.99	m	**28.67**
300 mm girth	-	0.25	7.69	26.23	m	**33.92**
Stepped flashings; wedging into grooves; KME pre-patinated finish						
180 mm girth	-	0.50	15.38	15.74	m	**31.12**
270 mm girth	-	0.50	15.38	23.61	m	**38.99**
Sundries						
provision of square batten roll at 500 mm centres (per m)	-	0.10	3.08	1.35	m	**4.42**
H74 ZINC STRIP SHEET COVERINGS/FLASHINGS						
Zinc roofing; BS EN 506; on and including Delta Trella underlay						
The following rates are based upon nett 'deck' or 'wall' areas, and depart from SMM7 coverage rules						
Roof, dormer and wall coverings						
0.7 mm thick roof coverings; pre-weathered Rheinzink						
flat (in wood roll construction) (PC per kg)	3.60	1.00	30.75	43.34	m²	**74.09**
eaves detail ED1	-	0.20	6.15	7.31	m	**13.46**
abutment upstands at perimeters	-	0.33	10.15	3.66	m	**13.80**
pitched over 3 degrees (in standing seam construction)	-	0.75	23.06	36.03	m²	**59.09**
0.7 mm thick dormer coverings; pre-weathered Rheinzink						
flat (in wood roll construction)	-	1.50	46.13	43.34	m²	**89.47**
eaves detail ED1	-	0.20	6.15	7.31	m	**13.46**
pitched over 3 degrees (in standing seam construction)	-	1.25	38.44	36.03	m²	**74.47**
0.8 mm thick wall coverings; pre-weathered Rheinzink						
vertical (in angled or flat seam construction)	-	0.80	24.60	40.34	m²	**64.94**
0.8 mm thick dormer coverings; pre-weathered Rheinzink						
vertical (in angled or flat seam construction)	-	1.35	41.51	40.34	m²	**81.85**

H CLADDING/COVERING

Item	PC £	Labour hours	Labour £	Material £	Unit	Total rate £
H74 ZINC STRIP SHEET COVERINGS/FLASHINGS – cont'd						
0.7 mm thick zinc flashings, etc.; pre-weathered Rheinzink						
Flashings; wedging into grooves						
150 mm girth (PC per kg)	3.60	0.25	7.69	4.20	m	**11.88**
240 mm girth	-	0.25	7.69	6.72	m	**14.40**
300 mm girth	-	0.25	7.69	8.39	m	**16.08**
Stepped flashings; wedging into grooves						
180 mm girth	-	0.50	15.38	5.04	m	**20.41**
270 mm girth	-	0.50	15.38	7.56	m	**22.93**
Integral box gutter						
900 mm girth; 2 x bent; 2 x welted	-	1.00	30.75	34.70	m	**65.45**
Valley gutter						
600 mm girth; 2 x bent; 2 x welted	-	0.75	23.06	20.99	m	**44.05**
Hips and ridges						
450 mm girth; 2 x bent; 2 x welted	-	1.00	30.75	12.59	m	**43.34**
Sundries						
provision of trapezoidal batten roll at 500 mm centres (per m)	-	0.10	3.08	1.35	m	**4.42**
H75 STAINLESS STEEL SHEET COVERINGS/ FLASHINGS						
Terne-coated stainless steel roofing; Associated Lead Mills Ltd; or other equal and approved: on and including Metmatt underlay						
The following rates are based upon nett 'deck' or 'wall' areas, and depart from SMM7 coverage rules						
Roof, dormer and wall coverings in 'Uginox' grade 316; marine						
0.4 mm thick roof coverings						
flat (in wood roll construction) (PC per kg)	9.00	1.00	30.75	60.54	m²	**91.29**
eaves detail ED1	-	0.20	6.15	7.26	m	**13.41**
abutment upstands at perimeters	-	0.33	10.15	3.63	m	**13.78**
pitched over 3 degrees (in standing seam construction)	-	0.75	23.06	49.64	m²	**72.70**
0.5mm thick dormer coverings						
flat (in wood roll construction)	-	1.50	46.13	75.07	m²	**121.19**
eaves detail ED1	-	0.20	6.15	7.26	m	**13.41**
pitched over 3 degrees (in standing seam construction)	-	1.25	38.44	61.27	m²	**99.70**
0.5 mm thick wall coverings						
vertical (in angled or flat seam construction)	-	0.80	24.60	61.27	m²	**85.87**
vertical (with Coulisseau joint construction)	-	1.25	38.44	63.30	m²	**101.73**
0.5 mm thick 'Uginox' grade 316 flashings, etc.						
Flashings; wedging into grooves						
150 mm girth (PC per kg)	9.00	0.25	7.69	6.90	m	**14.59**
240 mm girth	-	0.25	7.69	11.04	m	**18.73**
300 mm girth	-	0.25	7.69	13.80	m	**21.49**
Stepped flashings; wedging into grooves						
180 mm girth	-	0.50	15.38	8.28	m	**23.66**
270 mm girth	-	0.50	15.38	12.42	m	**27.80**

H CLADDING/COVERING

Item	PC £	Labour hours	Labour £	Material £	Unit	Total rate £
Fan apron						
250 mm girth	-	0.25	7.69	11.50	m	**19.19**
Integral box gutter						
900 mm girth; 2 x bent; 2 x welted	-	1.00	30.75	47.39	m	**78.14**
Valley gutter						
600 mm girth; 2 x bent; 2 x welted	-	0.75	23.06	33.68	m	**56.74**
Hips and ridges						
450 mm girth; 2 x bent; 2 x welted	-	1.00	30.75	20.70	m	**51.45**
Sundries						
provision of square batten roll at 500 mm centres (per m)	-	0.10	3.08	1.35	m	**4.42**
H76 FIBRE BITUMEN THERMOPLASTIC SHEET COVERINGS/FLASHINGS						
Glass fibre reinforced bitumen strip slates; "Ruberglas 105" or other equal and approved; 1000 mm x 336 mm mineral finish; to external quality plywood boarding (boarding not included)						
Roof coverings	7.94	0.23	6.83	9.49	m²	**16.32**
Wall coverings	-	0.37	10.99	9.49	m²	**20.48**
Extra over coverings for						
double course at eaves; felt soaker	-	0.19	5.64	6.30	m	**11.94**
verges; felt soaker	-	0.14	4.16	5.22	m	**9.38**
valley slate; cut to shape; felt soaker and cutting both sides	-	0.42	12.48	8.32	m	**20.79**
ridge slate; cut to shape	-	0.28	8.32	5.22	m	**13.54**
hip slate; cut to shape; felt soaker and cutting both sides	-	0.42	12.48	8.18	m	**20.66**
holes for pipes and the like	-	0.48	14.26	-	nr	**14.26**
Bostik Findley "Flashband Plus" sealing strips and flashings or other equal and approved; special grey finish						
Flashings; wedging at top if required; pressure bonded; to walls						
100 mm girth	-	0.23	4.78	0.55	m	**5.33**
150 mm girth	-	0.31	6.44	0.94	m	**7.38**
225 mm girth	-	0.37	7.68	1.45	m	**9.14**
300 mm girth	-	0.42	8.72	1.76	m	**10.48**
H92 RAINSCREEN CLADDING						
Western Red Cedar tongued and grooved wall cladding on and including treated softwrood battens on breather mambrane, 10 mm Eternit Blueclad board and 50 mm insulation board; the whole fixed to Metsec frame system; including sealing all joints etc.						
26 mm thick cladding to walls; boards laid horizontally	-	-	-	-	m²	**99.17**
Reynobond rainscreen cladding; aluminium composite material cassettes with thermoplastic cores, back ventilated, including insulation, vapour control membrane and aluminium support system						
4 mm thick cladding; fixed to walls	-	-	-	-	m²	**164.00**
Terracotta clay rainscreen cladding; including insulation, vapour control membrane and aluminium support system						
400 x 200 x 30 mm tile cladding; fixed to walls	-	-	-	-	m²	**297.25**

J WATERPROOFING

Item	PC £	Labour hours	Labour £	Material £	Unit	Total rate £
J10 SPECIALIST WATERPROOF RENDERING						
"Sika" waterproof rendering or other equal and approved; steel trowelled						
20 mm work to walls; three coat; to concrete base						
over 300 mm wide	-	-	-	-	m²	**45.13**
not exceeding 300 mm wide	-	-	-	-	m²	**68.38**
25 mm work to walls; three coat; to concrete base						
over 300 mm wide	-	-	-	-	m²	**53.33**
not exceeding 300 mm wide	-	-	-	-	m²	**82.05**
40 mm work to walls; four coat; to concrete base						
over 300 mm wide	-	-	-	-	m²	**78.63**
not exceeding 300 mm wide	-	-	-	-	m²	**123.08**
J20 MASTIC ASPHALT TANKING/DAMP PROOF MEMBRANES						
Mastic asphalt to BS 6925 Type T 1097						
13 mm thick one coat coverings to concrete base; flat; subsequently covered						
over 300 mm wide	-	-	-	-	m²	**13.19**
225 mm - 300 mm wide	-	-	-	-	m²	**37.91**
150 mm - 225 mm wide	-	-	-	-	m²	**41.57**
not exceeding 150 mm wide	-	-	-	-	m²	**51.93**
20 mm thick two coat coverings to concrete base; flat; subsequently covered						
over 300 mm wide	-	-	-	-	m²	**16.61**
225 mm - 300 mm wide	-	-	-	-	m²	**34.22**
150 mm - 225 mm wide	-	-	-	-	m²	**47.86**
not exceeding 150 mm wide	-	-	-	-	m²	**55.92**
30 mm thick three coat coverings to concrete base; flat; subsequently covered						
over 300 mm wide	-	-	-	-	m²	**26.65**
225 mm - 300 mm wide	-	-	-	-	m²	**54.93**
150 mm - 225 mm wide	-	-	-	-	m²	**59.59**
not exceeding 150 mm wide	-	-	-	-	m²	**72.62**
13 mm thick two coat coverings to brickwork base; vertical; subsequently covered						
over 300 mm wide	-	-	-	-	m²	**36.67**
225 mm - 300 mm wide	-	-	-	-	m²	**52.71**
150 mm - 225 mm wide	-	-	-	-	m²	**56.91**
not exceeding 150 mm wide	-	-	-	-	m²	**74.38**
20 mm thick three coat coverings to brickwork base; vertical; subsequently covered						
over 300 mm wide	-	-	-	-	m²	**59.32**
225 mm - 300 mm wide	-	-	-	-	m²	**71.01**
150 mm - 225 mm wide	-	-	-	-	m²	**77.91**
not exceeding 150 mm wide	-	-	-	-	m²	**101.03**
Turning into groove 20 mm deep	-	-	-	-	m	**0.70**
Internal angle fillets; subsequently covered	-	-	-	-	m	**4.11**

J WATERPROOFING

Item	PC £	Labour hours	Labour £	Material £	Unit	Total rate £
J21 MASTIC ASPHALT ROOFING/INSULATION/ FINISHES						
Mastic asphalt to BS 6925 Type R 988						
20 mm thick two coat coverings; felt isolating membrane; to concrete (or timber) base; flat or to falls or slopes not exceeding 10 degrees from horizontal						
over 300 mm wide	-	-	-	-	m²	**17.56**
225 mm - 300 mm wide	-	-	-	-	m²	**27.12**
150 mm - 225 mm wide	-	-	-	-	m²	**31.69**
not exceeding 150 mm wide	-	-	-	-	m²	**40.78**
Add to the above for covering with:						
10 mm thick limestone chippings in hot bitumen	-	-	-	-	m²	**2.86**
coverings with solar reflective paint	-	-	-	-	m²	**3.22**
300 mm x 300 mm x 8 mm g.r.p. tiles in hot bitumen	-	-	-	-	m²	**48.57**
Cutting to line; jointing to old asphalt	-	-	-	-	m	**5.56**
13 mm thick two coat skirtings to brickwork base						
not exceeding 150 mm girth	-	-	-	-	m	**11.97**
150 mm - 225 mm girth	-	-	-	-	m	**13.74**
225 mm - 300 mm girth	-	-	-	-	m	**16.82**
13 mm thick three coat skirtings; expanded metal lathing reinforcement nailed to timber base						
not exceeding 150 mm girth	-	-	-	-	m	**20.12**
150 mm - 225 mm girth	-	-	-	-	m	**23.98**
225 mm - 300 mm girth	-	-	-	-	m	**28.04**
13 mm thick two coat fascias to concrete base						
not exceeding 150 mm girth	-	-	-	-	m	**11.97**
150 mm - 225 mm girth	-	-	-	-	m	**13.74**
20 mm thick two coat linings to channels to concrete base						
not exceeding 150 mm girth	-	-	-	-	m	**26.32**
150 mm - 225 mm girth	-	-	-	-	m	**29.93**
225 mm - 300 mm girth	-	-	-	-	m	**30.80**
20 mm thick two coat lining to cesspools						
250 mm x 150 mm x 150 mm deep	-	-	-	-	nr	**25.79**
Collars around pipes, standards and like members	-	-	-	-	nr	**18.45**
Accessories						
Eaves trim; extruded aluminium alloy; working asphalt into trim						
"Alutrim"; type A roof edging or other equal and approved	-	-	-	-	m	**11.01**
extra; angle	-	-	-	-	nr	**6.14**
Roof screed ventilator - aluminium alloy						
"Extr-aqua-vent" or other equal and approved; set on screed over and including dished sinking; working collar around ventilator	-	-	-	-	nr	**21.31**
J30 LIQUID APPLIED TANKING/DAMP PROOF MEMBRANES						
Tanking and damp proofing						
"Synthaprufe" or other equal and approved; blinding with sand; horizontal on slabs						
two coats	-	0.19	2.78	2.31	m²	**5.10**
three coats	-	0.26	3.81	3.40	m²	**7.21**
"Tretolastex 202T" or other equal and approved; on vertical surfaces of concrete						
two coats	-	0.19	2.78	0.51	m²	**3.29**
three coats	-	0.26	3.81	0.76	m²	**4.57**

J WATERPROOFING

Item	PC £	Labour hours	Labour £	Material £	Unit	Total rate £
J30 LIQUID APPLIED TANKING/DAMP PROOF MEMBRANES – cont'd						
Tanking and damp proofing – cont'd						
One coat Vandex "Super" 0.75 kg/m² slurry or other equal and approved; one consolidating coat of Vandex "Premix" 1 kg/m² slurry; horizontal on beds						
over 225 mm wide	-	0.32	4.69	3.69	m²	**8.37**
"Intergritank" MMA (Methyl Methacrylate) resin elastomeric structural waterproffing membrane; in two separate 1mm colour coded coats; on primed substrate						
over 225 mm wide	-	-	-	-	m²	**32.80**
J40 FLEXIBLE SHEET TANKING/DAMP PROOF MEMBRANES						
Tanking and damp proofing						
Visqueen self-adhesive damp proof membrane						
over 300 mm wide; horizontal	-	-	-	-	m²	**7.75**
not exceeding 300 mm wide; horizontal	-	-	-	-	m	**2.98**
Tanking primer for self-adhesive dpm						
over 300 mm wide; horizontal	-	-	-	-	m²	**5.36**
not exceeding 300 mm wide; horizontal	-	-	-	-	m	**2.45**
"Bituthene" sheeting or other equal and approved; lapped joints; horizontal on slabs						
3000 grade	-	0.09	1.32	4.72	m²	**6.03**
8000 grade	-	0.10	1.46	6.56	m²	**8.02**
5000HD heavy duty grade	-	0.12	1.76	6.24	m²	**8.00**
"Bituthene" sheeting or other equal and approved; lapped joints; dressed up vertical face of concrete						
8000 grade	-	0.17	2.49	6.56	m²	**9.05**
RIW "Structureseal" tanking and damp proof membrane; or other equal and approved						
over 300 mm wide; horizontal	-	-	-	-	m²	**6.27**
"Structureseal" Fillet						
40mm x 40mm	-	-	-	-	m	**4.59**
Ruberoid "Plasfrufe 2000SA" self-adhesive damp proof membrane						
over 300 mm wide; horizontal	-	-	-	-	m²	**14.76**
not exceeding 300 mm wide; horizontal	-	-	-	-	m	**5.66**
Extra for 50 mm thick sand blinding	-	-	-	-	m²	**2.58**
"Servi-pak" protection board or other equal and approved; butt jointed; taped joints; to horizontal surfaces;						
3 mm thick	-	0.14	2.05	5.07	m²	**7.12**
6 mm thick	-	0.14	2.05	7.58	m²	**9.63**
12 mm thick	-	0.19	2.78	13.38	m²	**16.16**
"Servi-pak" protection board or other equal and approved; butt jointed; taped joints; to vertical surfaces						
3 mm thick	-	0.19	2.78	5.07	m²	**7.85**
6 mm thick	-	0.19	2.78	7.58	m²	**10.37**
12 mm thick	-	0.23	3.37	13.38	m²	**16.75**
"Bituthene" reinforcing strip or other equal and approved; 70 mm wide						
Bitutape 4000	-	0.09	1.32	0.50	m	**1.82**
Expandite "Famflex" hot bitumen bonded waterproof tanking or other equal and approved; 150 mm laps						
horizontal; over 300 mm wide	-	0.37	5.42	15.66	m²	**21.07**
vertical; over 300 mm wide	-	0.60	8.79	15.66	m²	**24.44**

J WATERPROOFING

Item	PC £	Labour hours	Labour £	Material £	Unit	Total rate £
J41 BUILT UP FELT ROOF COVERINGS						
NOTE: The following items of felt roofing, unless otherwise described, include for conventional lapping, laying and bonding between layers and to base; and laying flat or to falls, crossfalls or to slopes not exceeding 10 degrees - but exclude any insulation etc.						
Felt roofing; BS EN 13707; suitable for flat roofs						
Three layer coverings first layer type 3G; subsequent layers type 3B bitumen glass fibre based felt	-	-	-	-	m²	**14.08**
Extra over felt for covering with and bedding in hot bitumen						
13 mm thick stone chippings	-	-	-	-	m²	**4.08**
300 mm x 300 mm x 8 mm g.r.p. tiles	-	-	-	-	m²	**43.61**
working into outlet pipes and the like	-	-	-	-	m²	**11.05**
Skirtings; three layer; top layer mineral surfaced; dressed over tilting fillet; turned into groove						
not exceeding 200 mm girth	-	-	-	-	m	**10.48**
200 mm - 400 mm girth	-	-	-	-	m	**12.95**
Coverings to kerbs; three layer						
400 mm - 600 mm girth	-	-	-	-	m	**16.77**
Linings to gutters; three layer						
400 mm - 600 mm girth	-	-	-	-	m	**20.37**
Collars around pipes and the like; three layer mineral surface; 150 mm high						
not exceeding 55 mm nominal size	-	-	-	-	nr	**11.13**
55 mm - 110 mm nominal size	-	-	-	-	nr	**11.13**
Three layer coverings; two base layers type 5U bitumen polyester based felt; top layer type 5B polyester based mineral surfaced felt; 10 mm stone chipping covering; bitumen bonded	-	-	-	-	m²	**23.98**
Coverings to kerbs						
not exceeding 200 mm girth	-	-	-	-	m	**10.13**
200 mm - 400 mm girth	-	-	-	-	m	**13.25**
Outlets and dishing to gullies						
300 mm diameter	-	-	-	-	nr	**12.07**
"Andersons" high performance polyester-based roofing system or other equal and approved						
Two layer coverings; first layer HT 125 underlay; second layer HT 350; fully bonded to wood/fibre base	-	-	-	-	m²	**19.88**
Extra over for						
top layer mineral surfaced	-	-	-	-	m²	**1.69**
13 mm thick stone chippings	-	-	-	-	m²	**4.08**
third layer of type 3B as underlay for concrete or screeded base	-	-	-	-	m²	**5.20**
working into outlet pipes and the like	-	-	-	-	nr	**12.06**
Skirtings; two layer; top layer mineral surfaced; dressed over tilting fillet; turned into groove						
not exceeding 200 mm girth	-	-	-	-	m	**10.13**
200 mm - 400 mm girth	-	-	-	-	m	**13.25**
Coverings to kerbs; two layer						
400 mm - 600 mm girth	-	-	-	-	m	**17.17**
Linings to gutters; three layer						
400 mm - 600 mm girth	-	-	-	-	m	**18.46**
Collars around pipes and the like; two layer; 150 mm high						
not exceeding 55 mm nominal size	-	-	-	-	nr	**12.06**
55 mm - 110 mm nominal size	-	-	-	-	nr	**12.06**

J WATERPROOFING

Item	PC £	Labour hours	Labour £	Material £	Unit	Total rate £
J41 BUILT UP FELT ROOF COVERINGS – cont'd						
"Ruberoid Challenger SBS" high performance roofing or other equal and approved (10 year guarantee specification)						
Two layer coverings; first and second layers Ruberglas 120 GP; fully bonded to wood, fibre or cork base	-	-	-	-	m²	**13.08**
Extra over for						
top layer mineral surfaced	-	-	-	-	m²	**4.62**
13 mm thick stone chippings	-	-	-	-	m²	**4.08**
third layer of "Rubervent 3G" as underlay for concrete or screeded base	-	-	-	-	m²	**5.18**
working into outlet pipes and the like	-	-	-	-	nr	**11.97**
Skirtings; two layer; top layer mineral surfaced; dressed over tilting fillet; turned into groove						
not exceeding 200 mm girth	-	-	-	-	m	**9.98**
200 mm - 400 mm girth	-	-	-	-	m	**13.06**
Coverings to kerbs; two layer						
400 mm - 600 mm girth	-	-	-	-	m	**16.93**
Linings to gutters; three layer						
400 mm - 600 mm girth	-	-	-	-	m	**18.13**
Collars around pipes and the like; two layer, 150 mm high						
not exceeding 55 mm nominal size	-	-	-	-	nr	**11.97**
55 mm - 110 mm nominal size	-	-	-	-	nr	**11.97**
"Ruberfort HP 350" high performance roofing or other equal and approved						
Two layer coverings; first layer Ruberfort HP 180; second layer Ruberfort HP 350; fully bonded; to wood; fibre or cork base	-	-	-	-	m²	**15.41**
Extra over for						
top layer mineral surfaced	-	-	-	-	m²	**6.38**
13 mm thick stone chippings	-	-	-	-	m²	**4.08**
third layer of "Rubervent 3G"; as underlay for concrete or screeded base	-	-	-	-	m²	**5.18**
working into outlet pipes and the like	-	-	-	-	nr	**12.11**
Skirtings; two layer; top layer mineral surface; dressed over tilting fillet; turned into groove						
not exceeding 200 mm girth	-	-	-	-	m	**10.18**
200 mm - 400 mm girth	-	-	-	-	m	**13.32**
Coverings to kerbs; two layer						
400 mm - 600 mm girth	-	-	-	-	m	**17.28**
Linings to gutters; three layer						
400 mm - 600 mm girth	-	-	-	-	m	**22.31**
Collars around pipes and the like; two layer; 150 mm high						
not exceeding 55 mm nominal size	-	-	-	-	nr	**12.11**
55 mm - 110 mm nominal size	-	-	-	-	nr	**12.11**
"Ruberoid Superflex Firebloc" high performance roofing or other equal and approved (15 year guarantee specification)						
Two layer coverings; first layer Superflex 180; second layer Superflex 250; fully bonded to wood; fibre or cork base	-	-	-	-	m²	**19.06**

J WATERPROOFING

Item	PC £	Labour hours	Labour £	Material £	Unit	Total rate £
Extra over for						
top layer mineral surfaced	-	-	-	-	m²	**4.54**
13 mm thick stone chippings	-	-	-	-	m²	**4.08**
third layer of "Rubervent 3G" as underlay for concrete or screeded base	-	-	-	-	m²	**5.18**
working into outlet pipes and the like	-	-	-	-	nr	**13.75**
Skirtings; two layer; top layer mineral surfaced; dressed over tilting fillet; turned into groove						
not exceeding 200 mm girth	-	-	-	-	m	**11.91**
200 mm - 400 mm girth	-	-	-	-	m	**15.71**
Coverings to kerbs; two layer						
400 mm - 600 mm girth	-	-	-	-	m	**20.97**
Linings to gutters; three layer						
400 mm - 600 mm girth	-	-	-	-	m	**22.71**
Collars around pipes and the like; two layer; 150 mm high						
not exceeding 55 mm nominal size	-	-	-	-	nr	**13.75**
55 mm - 110 mm nominal size	-	-	-	-	nr	**13.75**
"Ruberoid Ultra prevENt" high performance roofing or other equal and approved (20 year guarantee specification)						
Two layer coverings; first layer Ultra prevENt underlay; second layer Ultra prevENt mineral surface cap sheet.	-	-	-	-	m²	**35.06**
Extra over for						
third layer of "Rubervent 3G" as underlay for concrete or screeded base	-	-	-	-	m²	**5.18**
working into outlet pipes and the like	-	-	-	-	nr	**16.63**
Skirtings; two layer; dressed over tilting fillet; turned into groove						
not exceeding 200 mm girth	-	-	-	-	m	**14.90**
200 mm - 400 mm girth	-	-	-	-	m	**19.85**
Coverings to kerbs; two layer						
400 mm - 600 mm girth	-	-	-	-	m	**27.40**
Linings to gutters; three layer						
400 mm - 600 mm girth	-	-	-	-	m	**28.63**
Collars around pipes and the like; two layer; 150 mm high						
not exceeding 55 mm nominal size	-	-	-	-	nr	**16.61**
55 mm - 110 mm nominal size	-	-	-	-	nr	**16.61**
Accessories						
Eaves trim; extruded aluminium alloy; working felt into trim						
Rubertrim; type FL/G; 65 mm face	-	-	-	-	m	**12.20**
extra over for; external angle	-	-	-	-	nr	**12.29**
Roof screed ventilator - aluminium alloy						
"Extr-aqua-vent" or other equal and approved - set on screed over and including dished sinking and collar	-	-	-	-	nr	**38.61**

J WATERPROOFING

Item	PC £	Labour hours	Labour £	Material £	Unit	Total rate £
J41 BUILT UP FELT ROOF COVERINGS – cont'd						
Insulation board underlays						
Vapour barrier						
reinforced; metal lined	-	-	-	-	m²	**11.67**
Rockwool; Duorock flat insulation board (0.25 U-value)						
140 mm thick	-	-	-	-	m²	**31.97**
Kingspan Thermaroof TR21 zero OPD urethene insulation board						
50 mm thick	-	-	-	-	m²	**19.66**
90 mm thick	-	-	-	-	m²	**33.86**
100 mm thick (0.25 U-value)	-	-	-	-	m²	**37.61**
Wood fibre boards; impregnated; density 220 - 350 kg/m³						
12.70 mm thick	-	-	-	-	m²	**5.30**
Tapered insulation board underlays						
Tapered PIR (Polyisocyanurate) boards; bedded in hot bitumen						
average thickness achieving 0.25W/m²K	21.00	-	-	-	m²	**52.70**
minimum thickness achieving 0.25W/m²K	27.00	-	-	-	m²	**58.51**
Tapered PIR boards; mechanically fastened						
average thickness achieving 0.25W/m²K	-	-	-	-	m²	**55.03**
minimum thickness achieving 0.25W/m²K	-	-	-	-	m²	**60.84**
Tapered Rockwool boards; bedded in hot bitumen						
average thickness achieving 0.25W/m²K	21.00	-	-	-	m²	**77.54**
minimum thickness achieving 0.25W/m²K	27.00	-	-	-	m²	**84.32**
Tapered Rockwool boards; mechanically fastened						
average thickness achieving 0.25W/m²K	-	-	-	-	m²	**79.86**
minimum thickness achieving 0.25W/m²K	-	-	-	-	m²	**86.64**
Insulation board overlays						
Dow "Roofmate SL" extruded polystyrene foam boards or other equal and approved						
50 mm thick	-	-	-	-	m²	**13.01**
140 mm thick (0.25 U-value)	-	-	-	-	m²	**22.70**
Dow "Roofmate LG" extruded polystyrene foam boards or other equal and approved						
80 mm thick	-	-	-	-	m²	**46.32**
100 mm thick	-	-	-	-	m²	**49.66**
120 mm thick	-	-	-	-	m²	**53.02**
J42 SINGLE LAYER PLASTICS ROOF COVERINGS						
"Trocal S" PVC roofing or other equal and approved						
Coverings	-	-	-	-	m²	**19.86**
Skirtings; dressed over metal upstands						
not exceeding 200 mm girth	-	-	-	-	m	**15.43**
200 mm - 400 mm girth	-	-	-	-	m	**18.97**
Coverings to kerbs						
400 mm - 600 mm girth	-	-	-	-	m	**34.72**
Collars around pipes and the like; 150 mm high						
not exceeding 55 mm nominal size	-	-	-	-	nr	**10.61**
55 mm - 110 mm nominal size	-	-	-	-	nr	**10.61**
"Trocal" metal upstands or other equal and approved						
not exceeding 200 mm girth	-	-	-	-	m	**11.25**
200 mm - 400 mm girth	-	-	-	-	m	**14.60**

J WATERPROOFING

Item	PC £	Labour hours	Labour £	Material £	Unit	Total rate £
Sarnafil polymeric waterproofing membrane; ref. S327-12EL; Sarnabar mechanically fastened system; 85mm thick Sarnaform G CFC & HCFC free (0.25 U-value) rigid urethene insulation board mechanially fastened; Sarnavap 1000E vapour control layer loose laid all laps sealed.						
Roof coverings						
Pitch not exceeding 5°; to metal decking or the like	-	-	-	-	m²	**55.69**
Sarnafil polymeric waterproofing membrane; ref. G410 - 12ELF fleece backed membrane; fully adhered system; 90mm thick Sarnaform G CFC & HCFC free (0.25 U-value) insulation board bedded in hot bitumen; BS 747 type 5U felt vapour control layer in hot bitumen; prime concrete with spirit priming solution.						
Roof coverings						
Pitch not exceeding 5°; to concrete base or the like	-	-	-	-	m²	**67.55**
Coverings to kerbs; parapet flashing; Sarnatrim 50 mm deep on face 100 mm fixing arm; standard Sarnafil detail 1.1						
not exceeding 200 mm girth	-	-	-	-	m	**36.57**
200 mm - 400 mm girth	-	-	-	-	m	**38.57**
400 mm - 600 mm girth	-	-	-	-	m	**42.84**
Eaves detail; Sarnatrmetal drip edge to gutter; standard Sarnafil detail 1.3						
not exceeding 200 mm girth	-	-	-	-	m	**19.85**
Skirtings/Upstands; skirting to brickwork with galvanised steel counter flashing to top edge; standard Sarnafil detail 2.3						
not exceeding 200 mm girth	-	-	-	-	m	**29.91**
200 mm - 400 mm girth	-	-	-	-	m	**31.91**
400 mm - 600 mm girth	-	-	-	-	m	**36.17**
Skirtings/Upstands; skirting to brickwork with Sarnametal Raglet to chase; standard Sarnafil detail 2.8						
not exceeding 200 mm girth	-	-	-	-	m	**41.48**
200 mm - 400 mm girth	-	-	-	-	m	**43.49**
400 mm - 600 mm girth	-	-	-	-	m	**47.76**
Collars around pipe standards, and the like						
50 mm diameter x 150 mm high	-	-	-	-	nr	**34.52**
100 mm diameter x 150 mm high	-	-	-	-	nr	**34.52**
Outlets and dishing to gullies						
Fix Sarnadrain pvc rainwater outlet; 110 mm diameter; weld membrane to same; fit plastic leafguard	-	-	-	-	nr	**111.18**

J WATERPROOFING

Item	PC £	Labour hours	Labour £	Material £	Unit	Total rate £
J43 PROPRIETARY ROOF DECKING WITH FELT FINISH						
"Bitumetal" flat roof construction or other equal and approved; fixing to timber, steel or concrete; flat or sloping; vapour check; 32 mm thick polyurethane insulation; 3G perforated felt underlay; two layers of glass fibre base felt roofing; stone chipping finish						
0.70 mm thick galvanized steel						
35 mm deep profiled decking; 2.38 m span	-	-	-	-	m²	**69.34**
46 mm deep profiled decking; 2.96 m span	-	-	-	-	m²	**69.73**
60 mm deep profiled decking; 3.74 m span	-	-	-	-	m²	**70.89**
100 mm deep profiled decking; 5.13 m span	-	-	-	-	m²	**72.13**
0.90 mm thick aluminium; mill finish						
35 mm deep profiled decking; 1.79 m span	-	-	-	-	m²	**74.36**
60 mm deep profiled decking; 2.34 m span	-	-	-	-	m²	**74.78**
"Bitumetal" flat roof construction or other equal and approved; fixing to timber, steel or concrete; flat or sloping; vapour check; 32 mm polyurethane insulation; 3G perforated felt underlay; two layers of polyester based roofing; stone chipping finish						
0.70 mm thick galvanised steel						
35 mm deep profiled decking; 2.38 m span	-	-	-	-	m²	**75.14**
46 mm deep profiled decking; 2.96 m span	-	-	-	-	m²	**75.54**
60 mm deep profiled decking; 3.74 m span	-	-	-	-	m²	**76.69**
100 mm deep profiled decking; 5.13 m span	-	-	-	-	m²	**77.94**
0.90 mm thick aluminium; mill finish						
35 mm deep profiled decking; 1.79 m span	-	-	-	-	m²	**80.16**
60 mm deep profiled decking; 2.34 m span	-	-	-	-	m²	**80.58**

K LININGS/SHEATHING/DRY PARTITIONING

Item	PC £	Labour hours	Labour £	Material £	Unit	**Total rate £**
ALTERNATIVE SHEET LINING MATERIAL PRICES						
Discounts of 0 - 20% available depending on quantity/status						
Fibreboard; 19 mm Decorative faced (£/10 m²)						
Ash	-	-	-	111.35	10 m²	-
Beech	-	-	-	106.45	10 m²	-
Oak	-	-	-	107.35	10 m²	-
Edgings; self adhesive (£/50 m roll)						
22 mm Ash	-	-	-	17.50	50 m	-
22 mm Beech	-	-	-	17.50	50 m	-
22 mm Oak	-	-	-	17.50	50 m	-
Chipboard Standard Grade (£/10m²)						
12 mm	-	-	-	26.85	10 m²	-
18 mm	-	-	-	37.93	10 m²	-
22 mm	-	-	-	46.40	10 m²	-
25 mm	-	-	-	53.16	10 m²	-
Chipboard; melamine faced (£/10m²)						
12 mm	-	-	-	45.60	10 m²	-
18 mm	-	-	-	52.58	10 m²	-
Medium density fibreboard; external quality (£/10m²)						
6 mm	-	-	-	50.47	10 m²	-
9 mm	-	-	-	67.03	10 m²	-
19 mm	-	-	-	108.90	10 m²	-
25 mm	-	-	-	151.70	10 m²	-
Plasterboards						
Wallboard plank (£/100m²)						
9.5 mm	-	-	-	188.00	100 m²	-
12.5 mm	-	-	-	188.00	100 m²	-
15 mm	-	-	-	226.00	100 m²	-
Moisture resistant board (£/100 m²)						
9.5 mm	-	-	-	296.00	100 m²	-
Fireline board (£/100 m²)						
12.5 mm	-	-	-	233.00	100m²	-
15 mm	-	-	-	280.00	100 m²	-

K LININGS/SHEATHING/DRY PARTITIONING

Item	PC £	Labour hours	Labour £	Material £	Unit	Total rate £
SUPPLY AND FIX PRICES						
K10 PLASTERBOARD DRY LINING/PARTITIONS/ CEILINGS						
Linings; "Gyproc GypLyner" metal framed wall lining system; or other equal and approved; floor and ceiling channels plugged and screwed to concrete						
Tapered edge panels; joints filled with joint filler and joint tape to receive direct decoration; one layer of 12.5 mm thick Gyproc Wallboard; or other equal and approved						
height 2.10 m - 2.40 m	-	1.05	22.18	18.51	m	**40.69**
height 2.40 m - 2.70 m	-	1.15	24.30	20.44	m	**44.75**
height 2.70 m - 3.00 m	-	1.27	26.86	22.32	m	**49.18**
height 3.00 m - 3.30 m	-	1.45	30.68	24.20	m	**54.88**
height 3.30 m - 3.60 m	-	1.63	34.50	25.74	m	**60.24**
height 3.60 m - 3.90 m	-	1.88	39.80	27.95	m	**67.75**
height 3.90 m - 4.20 m	-	2.11	44.67	29.83	m	**74.49**
Linings; "Gyproc GypLyner IWL" independent walling system or other equal and approved; comprising 48 mm wide metal I stud frame; 50 mmm wide metal C stud floor and ceiling channels; plugged and screwed to concrete						
62.5 mm partition; outer skin of 12.50 mm thick tapered edge wallboard one side; joints filled with joint filler and joint tape to receive direct decoration						
height 2.10 m - 2.40 m	-	3.05	64.38	12.52	m	**76.90**
height 2.40 m - 2.70 m	-	3.56	75.31	14.92	m	**90.23**
height 2.70 m - 3.00 m	-	3.94	83.39	16.33	m	**99.72**
height 3.00 m - 3.30 m	-	4.58	96.92	17.78	m	**114.70**
height 3.30 m - 3.60 m	-	4.95	104.80	19.26	m	**124.06**
height 3.60 m - 3.90 m	-	5.65	119.58	20.75	m	**140.33**
height 3.90 m - 4.20 m	-	6.20	131.22	22.24	m	**153.47**
62.5 mm partition; outer skin of 12.50 mm thick tapered edge wallboard one side; filling cavity with "Isowool high performance slab (2405); wallboard joints filled with joint filler and joint tape to receive direct decoration						
height 2.10 m - 2.40 m	-	3.05	64.38	21.07	m	**85.45**
height 2.40 m - 2.70 m	-	3.56	75.31	24.53	m	**99.84**
height 2.70 m - 3.00 m	-	3.94	83.39	27.01	m	**110.40**
height 3.00 m - 3.30 m	-	4.58	96.92	29.53	m	**126.45**
height 3.30 m - 3.60 m	-	4.95	104.80	32.09	m	**136.88**
height 3.60 m - 3.90 m	-	5.65	119.58	34.64	m	**154.22**
height 3.90 m - 4.20 m	-	6.20	131.22	37.20	m	**168.43**

K LININGS/SHEATHING/DRY PARTITIONING

Item	PC £	Labour hours	Labour £	Material £	Unit	Total rate £
"Gypwall Rapid/db Plus" metal stud housing partitioning system; or other equal and approved; floor and ceiling channels plugged and screwed to concrete						
75 mm partition; 43/44 mm studs and channels; one layer of 15 mm SoundBloc Rapid each side; joints filled with joint filler and joint tape to receive direct decoration						
height 2.10 m - 2.40 m; studs at 900 mm centres	-	2.70	57.30	27.32	m	**84.62**
height 2.10 m - 2.40 m; studs at 900 mm centres; with 25 mm Isowool 1200 insulation within the stud cavity	-	2.70	57.30	29.15	m	**86.45**
height 2.10 m - 2.40 m; studs at 450 mm centres	-	3.70	78.24	29.98	m	**108.22**
height 2.10 m - 2.40 m; studs at 450 mm centres; with 25 mm Isowool 1200 insulation within the stud cavity	-	3.70	78.24	31.81	m	**110.05**
height 2.40 m - 2.70 m; studs at 450 mm centres	-	4.07	86.07	33.34	m	**119.40**
height 2.40 m - 2.70 m; studs at 450 mm centres; with 25 mm Isowool 1200 insulation within the stud cavity	-	4.07	86.07	35.17	m	**121.24**
102 mm partition; 70/72 mm studs and channels; one layer of 15 mm SoundBloc Rapid each side; joints filled with joint filler and joint tape to receive direct decoration						
height 2.10 m - 2.40 m; studs at 900 mm centres	-	3.00	63.58	30.31	m	**93.89**
height 2.10 m - 2.40 m; studs at 900 mm centres; with 25 mm Isowool 1200 insulation within the stud cavity	-	3.00	63.58	32.14	m	**95.72**
height 2.10 m - 2.40 m; studs at 450 mm centres	-	4.00	84.52	34.69	m	**119.21**
height 2.10 m - 2.40 m; studs at 450 mm centres; with 25 mm Isowool 1200 insulation within the stud cavity	-	4.00	84.52	36.52	m	**121.05**
height 2.40 m - 2.70 m; studs at 900 mm centres	-	3.32	70.36	33.37	m	**103.73**
height 2.40 m - 2.70 m; studs at 900 mm centres; with 25 mm Isowool 1200 insulation within the stud cavity	-	3.32	70.36	35.21	m	**105.56**
height 2.40 m - 2.70 m; studs at 450 mm centres	-	4.32	91.30	38.55	m	**129.85**
height 2.40 m - 2.70 m; studs at 450 mm centres; with 25 mm Isowool 1200 insulation within the stud cavity	-	4.32	91.30	40.38	m	**131.68**
"Gyproc" metal stud proprietary partitions or other equal and approved; comprising 48 mm wide metal stud frame; 50 mm wide floor channel plugged and screwed to concrete through 38 mm x 48 mm tanalised softwood sole plate						
Tapered edge panels; joints filled with joint filler and joint tape to receive direct decoration; 80 mm thick partition; one hour; one layer of 15 mm thick "Fireline" board or other equal and approved each side						
height 2.10 m - 2.40 m	-	3.89	81.97	24.60	m	**106.57**
height 2.40 m - 2.70 m	-	4.49	94.79	28.41	m	**123.19**
height 2.70 m - 3.00 m	-	5.00	105.59	31.23	m	**136.83**
height 3.00 m - 3.30 m	-	5.78	122.05	34.35	m	**156.41**
height 3.30 m - 3.60 m	-	6.34	133.91	37.01	m	**170.91**
height 3.60 m - 3.90 m	-	7.59	160.21	39.91	m	**200.12**
height 3.90 m - 4.20 m	-	8.14	171.86	42.82	m	**214.67**
angles	-	0.19	4.04	1.57	m	**5.61**
T-junctions	-	0.09	1.88	-	m	**1.88**
fair ends	-	0.19	4.04	0.55	m	**4.59**

K LININGS/SHEATHING/DRY PARTITIONING

Item	PC £	Labour hours	Labour £	Material £	Unit	Total rate £
K10 PLASTERBOARD DRY LINING/PARTITIONS/ CEILINGS – cont'd						
"Gyproc" metal stud proprietary partitions – cont'd						
Tapered edge panels; joints filled with joint filler and joint tape to receive direct decoration; 100 mm thick partition; two hour; two layers of 12.50 mm thick "Fireline" board or other equal and approved both sides						
height 2.10 m - 2.40 m	-	4.81	101.24	34.03	m	**135.26**
height 2.40 m - 2.70 m	-	5.53	116.57	39.03	m	**155.59**
height 2.70 m - 3.00 m	-	6.15	129.68	43.03	m	**172.71**
height 3.00 m - 3.30 m	-	6.13	129.38	47.33	m	**176.71**
height 3.30 m - 3.60 m	-	7.72	162.81	51.15	m	**213.96**
height 3.60 m - 3.90 m	-	7.59	160.21	55.27	m	**215.48**
height 3.90 m - 4.20 m	-	9.76	205.78	59.32	m	**265.10**
angles	-	0.28	5.93	1.68	m	**7.60**
T-junctions	-	0.09	1.88	-	m	**1.88**
fair ends	-	0.28	5.93	0.66	m	**6.58**
Gypsum plasterboard; BS EN 520; plain grade tapered edge wallboard; fixing on dabs or with nails; joints left open to receive "Artex" finish or other equal and approved; to softwood base						
9.50 mm board to ceilings						
over 300 mm wide	-	0.23	4.82	2.07	m²	**6.88**
9.50 mm board to beams						
girth not exceeding 600 mm	-	0.28	5.86	1.26	m²	**7.12**
girth 600 mm - 1200 mm	-	0.37	7.75	2.50	m²	**10.25**
12.50 mm board to ceilings						
over 300 mm wide	-	0.31	6.49	2.15	m²	**8.64**
12.50 mm board to beams						
girth not exceeding 600 mm	-	0.28	5.86	1.32	m²	**7.18**
girth 600 mm - 1200 mm	-	0.37	7.75	2.58	m²	**10.33**
Gypsum plasterboard to BS EN 520; fixing on dabs or with nails; joints filled with joint filler and joint tape to receive direct decoration; to softwood base						
Plain grade tapered edge wallboard						
9.50 mm board to walls						
wall height 2.40 m - 2.70 m	-	0.93	19.85	6.98	m	**26.83**
wall height 2.70 m - 3.00 m	-	1.06	22.64	7.77	m	**30.40**
wall height 3.00 m - 3.30 m	-	1.20	25.63	8.54	m	**34.18**
wall height 3.30 m - 3.60 m	-	1.39	29.67	9.37	m	**39.05**
9.50 mm board to reveals and soffits of openings and recesses						
not exceeding 300 mm wide	-	0.19	4.04	1.42	m	**5.47**
300 mm - 600 mm wide	-	0.37	7.87	2.03	m	**9.91**
9.50 mm board to faces of columns - 4 nr						
not exceeding 600 mm total girth	-	0.46	9.79	2.89	m	**12.68**
600 mm - 1200 mm total girth	-	0.93	19.82	4.11	m	**23.93**
1200 mm - 1800 mm total girth	-	1.20	25.63	5.33	m	**30.96**
9.50 mm board to ceilings						
over 300 mm wide	-	0.39	8.29	2.59	m²	**10.88**
9.50 mm board to faces of beams - 3 nr						
not exceeding 600 mm total girth	-	0.56	11.92	2.85	m	**14.77**
600 mm - 1200 mm total girth	-	1.02	21.74	4.07	m	**25.81**
1200 mm - 1800 mm total girth	-	1.30	27.76	5.29	m	**33.05**

K LININGS/SHEATHING/DRY PARTITIONING

Item	PC £	Labour hours	Labour £	Material £	Unit	Total rate £
12.50 mm board to walls						
wall height 2.40 m - 2.70 m	-	0.97	20.69	7.12	m	**27.81**
wall height 2.70 m - 3.00 m	-	1.11	23.68	7.92	m	**31.60**
wall height 3.00 m - 3.30 m	-	1.25	26.68	8.71	m	**35.39**
wall height 3.30 m - 3.60 m	-	1.43	30.51	9.55	m	**40.06**
12.50 mm board to reveals and soffits of openings and recesses						
not exceeding 300 mm wide	-	0.19	4.04	1.46	m	**5.51**
300 mm - 600 mm wide	-	0.37	7.87	2.07	m	**9.95**
12.50 mm board to faces of columns - 4 nr						
not exceeding 600 mm total girth	-	0.46	9.79	2.97	m	**12.76**
600 mm - 1200 mm total girth	-	0.93	19.82	4.21	m	**24.03**
1200 mm - 1800 mm total girth	-	1.20	25.63	5.45	m	**31.08**
12.50 mm board to ceilings						
over 300 mm wide	-	0.41	8.71	2.64	m²	**11.35**
12.50 mm board to faces of beams - 3 nr						
not exceeding 600 mm total girth	-	0.56	11.92	2.91	m	**14.83**
600 mm - 1200 mm total girth	-	1.02	21.74	4.15	m	**25.89**
1200 mm - 1800 mm total girth	-	1.30	27.76	5.39	m	**33.15**
external angle; with joint tape bedded and covered with "Jointex" or other equal and approved	-	0.11	2.37	0.39	m	**2.76**
Tapered edge wallboard TEN						
12.50 mm board to walls						
wall height 2.40 m - 2.70 m	-	0.97	20.69	8.19	m	**28.88**
wall height 2.70 m - 3.00 m	-	1.11	23.68	9.11	m	**32.79**
wall height 3.00 m - 3.30 m	-	1.25	26.68	10.03	m	**36.70**
wall height 3.30 m - 3.60 m	-	1.43	30.51	10.98	m	**41.49**
12.50 mm board to reveals and soffits of openings and recesses						
not exceeding 300 mm wide	-	0.19	4.04	1.58	m	**5.63**
300 mm - 600 mm wide	-	0.37	7.87	2.31	m	**10.19**
12.50 mm board to faces of columns - 4 nr						
not exceeding 600 mm total girth	-	0.46	9.79	3.21	m	**13.00**
600 mm - 1200 mm total girth	-	0.93	19.82	4.69	m	**24.51**
1200 mm - 1800 mm total girth	-	1.20	25.63	6.16	m	**31.80**
12.50 mm board to ceilings						
over 300 mm wide	-	0.41	8.71	3.04	m²	**11.75**
12.50 mm board to faces of beams - 3 nr						
not exceeding 600 mm total girth	-	0.56	11.92	3.15	m	**15.06**
600 mm - 1200 mm total girth	-	1.02	21.74	4.63	m	**26.36**
1200 mm - 1800 mm total girth	-	1.30	27.76	6.10	m	**33.86**
external angle; with joint tape bedded and covered with "Jointex" or other equal and approved	-	0.11	2.37	0.39	m	**2.76**
Tapered edge plank						
19 mm plank to walls						
wall height 2.40 m - 2.70 m	-	1.02	21.74	13.02	m	**34.76**
wall height 2.70 m - 3.00 m	-	1.20	25.57	14.47	m	**40.04**
wall height 3.00 m - 3.30 m	-	1.30	27.73	15.92	m	**43.65**
wall height 3.30 m - 3.60 m	-	1.53	32.61	17.42	m	**50.02**
19 mm plank to reveals and soffits of openings and recesses						
not exceeding 300 mm wide	-	0.20	4.25	2.12	m	**6.37**
300 mm - 600 mm wide	-	0.42	8.92	3.39	m	**12.31**
19 mm plank to faces of columns - 4 nr						
not exceeding 600 mm total girth	-	0.51	10.84	4.28	m	**15.12**
600 mm - 1200 mm total girth	-	0.97	20.66	6.83	m	**27.49**
1200 mm - 1800 mm total girth	-	1.25	26.68	9.38	m	**36.06**

K LININGS/SHEATHING/DRY PARTITIONING

Item	PC £	Labour hours	Labour £	Material £	Unit	Total rate £
K10 PLASTERBOARD DRY LINING/PARTITIONS/ CEILINGS – cont'd						
Gypsum plasterboard to BS EN 520 – cont'd						
Tapered edge plank – cont'd						
19 mm plank to ceilings						
over 300 mm wide	-	0.43	9.13	4.82	m²	**13.95**
19 mm plank to faces of beams - 3 nr						
not exceeding 600 mm total girth	-	0.60	12.75	4.22	m	**16.97**
600 mm - 1200 mm total girth	-	1.06	22.57	6.77	m	**29.35**
1200 mm - 1800 mm total girth	-	1.34	28.60	9.32	m	**37.92**
Thermal Board						
27 mm board to walls						
wall height 2.40 m - 2.70 m	-	1.06	22.57	20.36	m	**42.93**
wall height 2.70 m - 3.00 m	-	1.23	26.20	22.62	m	**48.82**
wall height 3.00 m - 3.30 m	-	1.34	28.53	24.89	m	**53.43**
wall height 3.30 m - 3.60 m	-	1.62	34.49	27.20	m	**61.69**
27 mm board to reveals and soffits of openings and recesses						
not exceeding 300 mm wide	-	0.21	4.46	2.94	m	**7.40**
300 mm - 600 mm wide	-	0.43	9.13	5.02	m	**14.15**
27 mm board to faces of columns - 4 nr						
not exceeding 600 mm total girth	-	0.52	11.05	5.91	m	**16.96**
600 mm - 1200 mm total girth	-	1.02	21.71	10.09	m	**31.80**
1200 mm - 1800 mm total girth	-	1.30	27.73	14.27	m	**42.00**
27 mm board to ceilings						
over 300 mm wide	-	0.46	9.76	7.54	m²	**17.30**
27 mm board to faces of beams - 3 nr						
not exceeding 600 mm total girth	-	0.56	11.88	5.85	m	**17.74**
600 mm - 1200 mm total girth	-	1.06	22.54	10.03	m	**32.58**
1200 mm - 1800 mm total girth	-	1.43	30.48	14.21	m	**44.69**
50 mm board to walls						
wall height 2.40 m - 2.70 m	-	1.06	22.57	21.78	m	**44.35**
wall height 2.70 m - 3.00 m	-	1.30	27.66	24.22	m	**51.88**
wall height 3.00 m - 3.30 m	-	1.43	30.42	26.64	m	**57.06**
wall height 3.30 m - 3.60 m	-	1.71	36.38	29.11	m	**65.49**
50 mm board to reveals and soffits of openings and recesses						
not exceeding 300 mm wide	-	0.23	4.88	3.10	m	**7.98**
300 mm - 600 mm wide	-	0.46	9.76	5.34	m	**15.10**
50 mm board to faces of columns - 4 nr						
not exceeding 600 mm total girth	-	0.56	11.88	6.28	m	**18.16**
600 mm - 1200 mm total girth	-	1.11	23.59	10.77	m	**34.36**
1200 mm - 1800 mm total girth	-	1.43	30.45	15.26	m	**45.71**
50 mm board to ceilings						
over 300 mm wide	-	0.49	10.39	8.07	m²	**18.46**
50 mm board to faces of beams - 3 nr						
not exceeding 600 mm total girth	-	0.58	12.30	6.32	m	**18.62**
600 mm - 1200 mm total girth	-	1.17	24.85	10.85	m	**35.70**
1200 mm - 1800 mm total girth	-	1.57	33.41	15.38	m	**48.79**

K LININGS/SHEATHING/DRY PARTITIONING

Item	PC £	Labour hours	Labour £	Material £	Unit	Total rate £
White plastic faced gypsum plasterboard to BS EN 520; industrial grade square edge wallboard; fixing on dabs or with screws; butt joints; to softwood base						
12.50 mm board to walls						
wall height 2.40 m - 2.70 m	-	0.69	14.45	15.36	m	**29.82**
wall height 2.70 m - 3.00 m	-	0.83	17.38	17.07	m	**34.45**
wall height 3.00 m - 3.30 m	-	0.97	20.32	18.77	m	**39.08**
wall height 3.30 m - 3.60 m	-	1.11	23.25	20.47	m	**43.72**
12.50 mm board to reveals and soffits of openings and recesses						
not exceeding 300 mm wide	-	0.15	3.14	1.72	m	**4.86**
300 mm - 600 mm wide	-	0.30	6.28	3.41	m	**9.70**
12.50 mm board to faces of columns - 4 nr						
not exceeding 600 mm total girth	-	0.39	8.17	3.50	m	**11.67**
600 mm - 1200 mm total girth	-	0.78	16.34	6.95	m	**23.29**
1200 mm - 1800 mm total girth	-	1.02	21.36	10.37	m	**31.73**
Plasterboard jointing system; filling joint with jointing compounds						
To ceilings						
to suit 9.50 mm or 12.50 mm thick boards	-	0.09	1.88	2.25	m	**4.14**
Angle trim; plasterboard edge support system						
To ceilings						
to suit 9.50 mm or 12.50 mm thick boards	-	0.09	1.88	2.10	m	**3.99**
Gyproc SoundBloc plasterboard with higher density core; fixing on dabs or with nails; joints filled with joint filler and joint tape to receive direct decoration; to softwood base						
Tapered edge board						
12.50 mm board to walls						
wall height 2.40 m - 2.70 m	-	0.97	20.69	9.27	m	**29.96**
wall height 2.70 m - 3.00 m	-	1.11	23.68	10.31	m	**33.99**
wall height 3.00 m - 3.30 m	-	1.24	26.47	11.34	m	**37.81**
wall height 3.30 m - 3.60 m	-	1.43	30.51	12.42	m	**42.93**
12.50 mm board to ceilings						
over 300 mm wide	-	0.41	8.71	3.44	m²	**12.15**
15.00 mm board to walls						
wall height 2.40 m - 2.70 m	-	1.00	21.32	10.93	m	**32.25**
wall height 2.70 m - 3.00 m	-	1.14	24.31	12.15	m	**36.46**
wall height 3.00 m - 3.30 m	-	1.27	27.10	13.37	m	**40.47**
wall height 3.30 m - 3.60 m	-	1.46	31.14	14.63	m	**45.77**
15.00 mm board to reveals and soffits of openings and recesses						
not exceeding 300 mm wide	-	0.20	4.25	1.89	m	**6.14**
300 mm - 600 mm wide	-	0.38	8.08	2.92	m	**11.00**
15.00 mm board to ceilings						
over 300 mm wide	-	0.43	9.13	4.05	m²	**13.18**

K LININGS/SHEATHING/DRY PARTITIONING

Item	PC £	Labour hours	Labour £	Material £	Unit	Total rate £
K10 PLASTERBOARD DRY LINING/PARTITIONS/ CEILINGS – cont'd						
Two layers of gypsum plasterboard to BS 1230; plain grade square and tapered edge wallboard; fixing on dabs or with nails; joints filled with joint filler and joint tape; top layer to receive direct decoration; to softwood base						
19 mm two layer board to walls						
wall height 2.40 m - 2.70 m	-	1.30	27.60	12.69	m	**40.29**
wall height 2.70 m - 3.00 m	-	1.48	31.43	14.11	m	**45.55**
wall height 3.00 m - 3.30m	-	1.67	35.48	15.53	m	**51.01**
wall height 3.30 m - 3.60m	-	1.94	41.19	16.99	m	**58.18**
19 mm two layer board to reveals and soffits of openings and recesses						
not exceeding 300 mm wide	-	0.28	5.93	2.11	m	**8.04**
300 mm - 600 mm wide	-	0.56	11.85	3.33	m	**15.19**
19 mm two layer board to faces of columns - 4 nr						
not exceeding 600 mm total girth	-	0.69	14.61	4.31	m	**18.92**
600 mm - 1200 mm total girth	-	1.34	28.41	6.77	m	**35.18**
1200 mm - 1800 mm total girth	-	1.67	35.48	9.23	m	**44.70**
25 mm two layer board to walls						
wall height 2.40 m - 2.70 m	-	1.39	29.49	12.83	m	**42.31**
wall height 2.70 m - 3.00 m	-	1.57	33.32	14.26	m	**47.58**
wall height 3.00 m - 3.30 m	-	1.76	37.36	15.70	m	**53.06**
wall height 3.30 m - 3.60 m	-	2.04	43.29	17.17	m	**60.46**
25 mm two layer board to reveals and soffits of openings and recesses						
not exceeding 300 mm wide	-	0.28	5.93	2.15	m	**8.08**
300 mm - 600 mm wide	-	0.56	11.85	3.37	m	**15.23**
25 mm two layer board to faces of columns - 4 nr						
not exceeding 600 mm total girth	-	0.69	14.61	4.39	m	**19.00**
600 mm - 1200 mm total girth	-	1.34	28.41	6.87	m	**35.28**
1200 mm - 1800 mm total girth	-	1.67	35.48	9.35	m	**44.82**
Gyproc Dri-Wall dry lining system or other equal or approved; plain grade tapered edge wallboard; fixed to walls with adhesive; joints filled with joint filler and joint tape; to receive direct decoration						
9.50 mm board to walls						
wall height 2.40 m - 2.70 m	-	1.11	23.62	9.46	m	**33.08**
wall height 2.70 m - 3.00 m	-	1.28	27.24	10.49	m	**37.74**
wall height 3.00 m - 3.30 m	-	1.43	30.45	11.53	m	**41.98**
wall height 3.30 m - 3.60 m	-	1.67	35.54	12.62	m	**48.16**
9.50 mm board to reveals and soffits of openings and recesses						
not exceeding 300 mm wide	-	0.23	4.88	1.65	m	**6.53**
300 mm - 600 mm wide	-	0.46	9.76	2.53	m	**12.29**
9.50 mm board to faces of columns - 4 nr						
not exceeding 600 mm total girth	-	0.58	12.30	3.31	m	**15.61**
600 mm - 1200 mm total girth	-	1.14	24.22	5.25	m	**29.47**
1200 mm - 1800 mm total girth	-	1.43	30.45	6.82	m	**37.27**
Angle; with joint tape bedded and covered with "Jointex" or other equal and approved						
internal	-	0.05	1.08	0.39	m	**1.47**
external	-	0.11	2.37	0.39	m	**2.76**

K LININGS/SHEATHING/DRY PARTITIONING

Item	PC £	Labour hours	Labour £	Material £	Unit	Total rate £
Gyproc Dri-Wall M/F dry lining system or other equal or approved; mild steel furrings fixed to walls with adhesive; tapered edge wallboard screwed to furrings; joints filled with joint filler and joint tape						
12.50 mm board to walls						
wall height 2.40 m - 2.70 m	-	1.48	31.37	14.09	m	**45.46**
wall height 2.70 m - 3.00 m	-	1.69	35.83	15.64	m	**51.47**
wall height 3.00 m - 3.30 m	-	1.90	40.29	17.21	m	**57.50**
wall height 3.30 m - 3.60 m	-	2.22	47.06	18.83	m	**65.89**
12.50 mm board to reveals and soffits of openings and recesses						
not exceeding 300 mm wide	-	0.23	4.88	1.44	m	**6.32**
300 mm - 600 mm wide	-	0.46	9.76	2.10	m	**11.86**
Lafarge plasterboard to BS 1230; fixing on dabs or with screws; joints filled with joint filler and joint tape to receive direct decoration; to softwood Megadeco wallboard						
12.50 mm board to walls						
wall height 2.40 m - 2.70 m	-	0.97	20.69	11.57	m	**32.26**
wall height 2.70 m - 3.00 m	-	1.11	23.68	12.86	m	**36.55**
wall height 3.00 m - 3.30 m	-	1.25	26.68	14.15	m	**40.83**
wall height 3.30 m - 3.60 m	-	1.43	30.51	15.43	m	**45.94**
12.50 mm board to ceilings						
over 300 mm wide	-	0.41	8.71	4.29	m²	**13.00**
Gypsum cladding; Glasroc "Firecase s" board or other equal and approved; fixed with adhesive; joints pointed in adhesive						
25 mm thick column linings, faces - 4; 2 hour fire protection rating						
not exceeding 600 mm girth	-	0.30	6.28	17.63	m	**23.91**
600 mm - 1200 mm girth	-	0.45	9.42	26.96	m	**36.39**
1200 mm - 1800 mm girth	-	0.60	12.57	36.30	m	**48.86**
30 mm thick beam linings, faces - 3; 2 hour fire protection rating						
not exceeding 600 mm girth	-	0.60	12.57	17.02	m	**29.59**
600 mm - 1200 mm girth	-	0.90	18.85	27.57	m	**46.42**
1200 mm - 1800 mm girth	-	1.20	25.13	38.11	m	**63.24**
Vermiculite gypsum cladding; "Vermiculux" board or other equal and approved; fixed with adhesive; joints pointed in adhesive						
25 mm thick column linings, faces - 4; 2 hour fire protection rating						
not exceeding 600 mm girth	-	0.30	6.28	15.58	m	**21.87**
600 mm - 1200 mm girth	-	0.45	9.42	30.88	m	**40.30**
1200 mm - 1800 mm girth	-	0.60	12.57	46.17	m	**58.74**
30 mm thick beam linings, faces - 3; 2 hour fire protection rating						
not exceeding 600 mm girth	-	0.60	12.57	20.47	m	**33.04**
600 mm - 1200 mm girth	-	0.90	18.85	40.66	m	**59.51**
1200 mm - 1800 mm girth	-	1.20	25.13	60.84	m	**85.97**
55 mm thick column linings, faces - 4 ; 4 hour fire protection rating						
not exceeding 600 mm girth	-	0.35	7.33	43.53	m	**50.86**
600 mm - 1200 mm girth	-	0.50	10.47	86.77	m	**97.24**
1200 mm - 1800 mm girth	-	0.65	13.61	130.01	m	**143.62**

K LININGS/SHEATHING/DRY PARTITIONING

Item	PC £	Labour hours	Labour £	Material £	Unit	Total rate £
K10 PLASTERBOARD DRY LINING/PARTITIONS/ CEILINGS – cont'd						
Vermiculite gypsum cladding – cont'd						
60 mm thick beam linings, faces - 3; 4 hour fire protection rating						
not exceeding 600 mm girth	-	0.70	14.66	47.03	m	**61.70**
600 mm - 1200 mm girth	-	1.00	20.94	93.78	m	**114.72**
1200 mm - 1800 mm girth	-	1.02	21.36	138.80	m	**160.16**
Add to the above for						
plus 3% for work 3.50 m - 5.00 m high						
plus 6% for work 5.00 m - 6.50 m high						
plus 12% for work 6.50 m - 8.00 m high						
plus 18% for work over 8.00 m high						
Cutting and fitting around steel joints, angles, trunking, ducting, ventilators, pipes, tubes, etc						
over 2 m girth	-	0.42	8.80	-	m	**8.80**
not exceeding 0.30 m girth	-	0.28	5.86	-	nr	**5.86**
0.30 m - 1 m girth	-	0.37	7.75	-	nr	**7.75**
1 m - 2 m girth	-	0.51	10.68	-	nr	**10.68**
K11 RIGID SHEET FLOORING/SHEATHING/ LININGS/CASINGS						
Blockboard (Birch faced)						
Lining to walls 18 mm thick						
over 300 wide	5.14	0.46	9.63	6.17	m²	**15.81**
not exceeding 300 wide	-	0.30	6.28	1.87	m	**8.15**
holes for pipes and the like	-	0.04	0.84	-	nr	**0.84**
Chipboard (plain)						
Lining to walls 12 mm thick						
over 300 mm wide	2.04	0.35	7.33	2.59	m²	**9.92**
not exceeding 300 mm wide	-	0.20	4.19	0.79	m	**4.98**
holes for pipes and the like	-	0.02	0.42	-	nr	**0.42**
Lining to walls 15 mm thick						
over 300 mm wide	2.39	0.37	7.75	2.99	m²	**10.74**
not exceeding 300 mm wide	-	0.22	4.61	0.92	m	**5.52**
holes for pipes and the like	-	0.03	0.63	-	nr	**0.63**
Two-sided 15 mm thick pipe casing; to softwood framing (not included)						
300 mm girth	-	0.56	11.73	1.00	m	**12.73**
600 mm girth	-	0.65	13.61	1.83	m	**15.44**
Three-sided 15 mm thick pipe casing; to softwood framing (not included)						
450 mm girth	-	1.16	24.29	1.50	m	**25.80**
900 mm girth	-	1.39	29.11	2.77	m	**31.89**
extra for 400 mm x 400 mm removable access panel; brass cups and screws; additional framing	-	0.93	19.48	1.04	nr	**20.51**
Lining to walls 18 mm thick						
over 300 mm wide	2.86	0.39	8.17	3.60	m²	**11.77**
not exceeding 300 mm wide	-	0.25	5.24	1.08	m	**6.32**
holes for pipes and the like	-	0.04	0.84	-	nr	**0.84**

K LININGS/SHEATHING/DRY PARTITIONING

Item	PC £	Labour hours	Labour £	Material £	Unit	Total rate £
Fire-retardent chipboard; Antivlam or other equal and approved; Class 1 spread of flame						
Lining to walls 12 mm thick						
over 300 mm wide	-	0.35	7.33	8.64	m²	**15.97**
not exceeding 300 mm wide	-	0.20	4.19	2.61	m	**6.80**
holes for pipes and the like	-	0.02	0.42	-	nr	**0.42**
Lining to walls 18 mm thick						
over 300 mm wide	-	0.39	8.17	11.21	m²	**19.38**
not exceeding 300 mm wide	-	0.25	5.24	3.38	m	**8.62**
holes for pipes and the like	-	0.04	0.84	-	nr	**0.84**
Lining to walls 22 mm thick						
over 300 mm wide	-	0.41	8.59	14.49	m²	**23.07**
not exceeding 300 mm wide	-	0.28	5.86	4.36	m	**10.23**
holes for pipes and the like	-	0.05	1.05	-	nr	**1.05**
Chipboard Melamine faced; white matt finish; laminated masking strips						
Lining to walls 15 mm thick						
over 300 mm wide	3.00	0.97	20.32	3.95	m²	**24.27**
not exceeding 300 mm wide	-	0.63	13.19	1.29	m	**14.49**
holes for pipes and the like	-	0.06	1.26	-	nr	**1.26**
Chipboard boarding and flooring						
Boarding to floors; butt joints						
18 mm thick	3.74	0.28	5.86	4.61	m²	**10.48**
Boarding to floors; tongued and grooved joints						
18 mm thick	3.64	0.30	6.28	4.50	m²	**10.79**
22 mm thick	4.69	0.32	6.70	5.71	m²	**12.41**
Acoustic Chipboard flooring						
Boarding to floors; tongued and grooved joints						
chipboard on blue bat bearers	-	-	-	-	m²	**19.67**
chipboard on New Era levelling system	-	-	-	-	m²	**26.95**
Laminated engineered board flooring; 180 or 240 mm face widths; with 6 mm wear surface down to tongue; pre-finished laquered, oiled or untreated.						
Boarding to floors; micro bevel or square edge						
Country laquered; on 10 mm Pro Foam	-	-	-	-	m²	**46.73**
Rustic laquered; on 10 mm Pro Foam	-	-	-	-	m²	**48.91**
Plywood flooring						
Boarding to floors; tongued and grooved joints						
18 mm thick	6.73	0.41	8.59	8.07	m²	**16.66**
22 mm thick	8.38	0.45	9.42	9.99	m²	**19.41**
Plywood; external quality; 18 mm thick						
Boarding to roofs; butt joints						
flat to falls	12.01	0.37	7.75	14.18	m²	**21.93**
sloping	12.01	0.40	8.38	14.18	m²	**22.56**
vertical	12.01	0.53	11.10	14.18	m²	**25.28**
Plywood; external quality; 12 mm thick						
Boarding to roofs; butt joints						
flat to falls	8.01	0.37	7.75	9.56	m²	**17.31**
sloping	8.01	0.40	8.38	9.56	m²	**17.93**
vertical	8.01	0.53	11.10	9.56	m²	**20.66**

K LININGS/SHEATHING/DRY PARTITIONING

Item	PC £	Labour hours	Labour £	Material £	Unit	Total rate £
K11 RIGID SHEET FLOORING/SHEATHING/ LININGS/CASINGS – cont'd						
Glazed hardboard to BS EN 622; on and including 38 mm x 38 mm sawn softwood framing						
3.20 mm thick panel						
to side of bath	-	1.67	34.98	5.10	nr	**40.07**
to end of bath	-	0.65	13.61	1.48	nr	**15.10**
Insulation board to BS EN 622						
Lining to walls 12 mm thick						
over 300 mm wide	1.61	0.22	4.61	2.09	m²	**6.70**
not exceeding 300 mm wide	-	0.13	2.72	0.65	m	**3.37**
holes for pipes and the like	-	0.01	0.21	-	nr	**0.21**
Non-asbestos board; "Masterboard" or other equal and approved; sanded finish						
Lining to walls 6 mm thick						
over 300 mm wide	6.60	0.31	6.49	7.81	m²	**14.30**
not exceeding 300 mm wide	-	0.19	3.98	2.35	m	**6.33**
Lining to ceilings 6 mm thick						
over 300 mm wide	6.60	0.41	8.59	7.81	m²	**16.40**
not exceeding 300 mm wide	-	0.25	5.24	2.35	m	**7.58**
holes for pipes and the like	-	0.02	0.42	-	nr	**0.42**
Lining to walls 9 mm thick						
over 300 mm wide	13.95	0.33	6.91	16.31	m²	**23.22**
not exceeding 300 mm wide	-	0.19	3.98	4.90	m	**8.88**
Lining to ceilings 9 mm thick						
over 300 mm wide	13.95	0.42	8.80	16.31	m²	**25.11**
not exceeding 300 mm wide	-	0.27	5.65	4.90	m	**10.55**
holes for pipes and the like	-	0.03	0.63	-	nr	**0.63**
Non-asbestos board; "Supalux" or other equal and approved; sanded finish						
Lining to walls 6 mm thick						
over 300 mm wide	10.93	0.31	6.49	12.82	m²	**19.31**
not exceeding 300 mm wide	-	0.19	3.98	3.85	m	**7.83**
Lining to ceilings 6 mm thick						
over 300 mm wide	10.93	0.41	8.59	12.82	m²	**21.40**
not exceeding 300 mm wide	-	0.25	5.24	3.85	m	**9.09**
holes for pipes and the like	-	0.03	0.63	-	nr	**0.63**
Lining to walls 9 mm thick						
over 300 mm wide	16.25	0.33	6.91	18.98	m²	**25.89**
not exceeding 300 mm wide	-	0.19	3.98	5.70	m	**9.68**
Lining to ceilings 9 mm thick						
over 300 mm wide	16.25	0.42	8.80	18.98	m²	**27.77**
not exceeding 300 mm wide	-	0.27	5.65	5.70	m	**11.35**
holes for pipes and the like	-	0.03	0.63	-	nr	**0.63**
Lining to walls 12 mm thick						
over 300 mm wide	21.53	0.37	7.75	25.08	m²	**32.83**
not exceeding 300 mm wide	-	0.22	4.61	7.53	m	**12.14**
Lining to ceilings 12 mm thick						
over 300 mm wide	21.53	0.49	10.26	25.08	m²	**35.34**
not exceeding 300 mm wide	-	0.30	6.28	7.53	m	**13.81**
holes for pipes and the like	-	0.04	0.84	-	nr	**0.84**

K LININGS/SHEATHING/DRY PARTITIONING

Item	PC £	Labour hours	Labour £	Material £	Unit	Total rate £
Non-asbestos board; "Monolux 40" or other equal and approved; 6 mm x 50 mm "Supalux" cover fillets or other equal and approved one side						
Lining to walls 19 mm thick						
over 300 mm wide	43.60	0.65	13.61	52.58	m²	**66.20**
not exceeding 300 mm wide	-	0.46	9.63	17.64	m	**27.27**
Lining to walls 25 mm thick						
over 300 mm wide	52.27	0.69	14.45	62.62	m²	**77.07**
not exceeding 300 mm wide	15.68	0.49	10.26	20.65	m	**30.91**
Plywood (Eastern European); internal quality						
Lining to walls 4 mm thick						
over 300 mm wide	2.72	0.34	7.12	3.38	m²	**10.50**
not exceeding 300 mm wide	-	0.22	4.61	1.03	m	**5.64**
Lining to ceilings 4 mm thick						
over 300 mm wide	2.72	0.46	9.63	3.38	m²	**13.01**
not exceeding 300 mm wide	-	0.30	6.28	1.03	m	**7.31**
holes for pipes and the like	-	0.02	0.42	-	nr	**0.42**
Lining to walls 6 mm thick						
over 300 mm wide	3.55	0.37	7.75	4.33	m²	**12.08**
not exceeding 300 mm wide	-	0.24	5.03	1.32	m	**6.34**
Lining to ceilings 6 mm thick						
over 300 mm wide	3.55	0.49	10.26	4.33	m²	**14.60**
not exceeding 300 mm wide	-	0.32	6.70	1.32	m	**8.02**
holes for pipes and the like	-	0.02	0.42	-	nr	**0.42**
Two-sided 6 mm thick pipe casings; to softwood framing (not included)						
300 mm girth	-	0.74	15.50	1.40	m	**16.90**
600 mm girth	-	0.93	19.48	2.63	m	**22.11**
Three-sided 6 mm thick pipe casing; to softwood framing (not included)						
450 mm girth	-	1.06	22.20	2.11	m	**24.31**
900 mm girth	-	1.25	26.18	3.98	m	**30.16**
Lining to walls 12 mm thick						
over 300 mm wide	5.86	0.43	9.01	7.01	m²	**16.01**
not exceeding 300 mm wide	-	0.28	5.86	2.12	m	**7.98**
Lining to ceilings 12 mm thick						
over 300 mm wide	5.86	0.56	11.73	7.01	m²	**18.74**
not exceeding 300 mm wide	-	0.37	7.75	2.12	m	**9.87**
holes for pipes and the like	-	0.03	0.63	-	nr	**0.63**
Lining to walls 18 mm thick						
over 300 mm wide	8.57	0.46	9.63	10.14	m²	**19.78**
not exceeding 300 mm wide	-	0.30	6.28	3.06	m	**9.34**
Lining to ceilings 18 mm thick						
over 300 mm wide	8.57	0.60	12.57	10.14	m²	**22.71**
not exceeding 300 mm wide	-	0.40	8.38	3.06	m	**11.44**
holes for pipes and the like	-	0.03	0.63	-	nr	**0.63**
Plywood (Eastern European); external quality						
Lining to walls 4 mm thick						
over 300 mm wide	3.91	0.34	7.12	4.76	m²	**11.88**
not exceeding 300 mm wide	-	0.22	4.61	1.44	m	**6.05**
Lining to ceilings 4 mm thick						
over 300 mm wide	3.91	0.46	9.63	4.76	m²	**14.39**
not exceeding 300 mm wide	-	0.30	6.28	1.44	m	**7.73**
holes for pipes and the like	-	0.02	0.42	-	nr	**0.42**

Item	PC £	Labour hours	Labour £	Material £	Unit	Total rate £
K11 RIGID SHEET FLOORING/SHEATHING/ LININGS/CASINGS – cont'd						
Plywood (Eastern European) – cont'd						
Lining to walls 6.5 mm thick						
over 300 mm wide	3.91	0.37	7.75	4.75	m²	**12.50**
not exceeding 300 mm wide	-	0.24	5.03	1.44	m	**6.47**
Lining to ceilings 6.5 mm thick						
over 300 mm wide	3.91	0.49	10.26	4.75	m²	**15.01**
not exceeding 300 mm wide	-	0.32	6.70	1.44	m	**8.14**
holes for pipes and the like	-	0.02	0.42	-	nr	**0.42**
Two-sided 6.5 mm thick pipe casings; to softwood framing (not included)						
300 mm girth	-	0.74	15.50	1.53	m	**17.03**
600 mm girth	-	0.93	19.48	2.89	m	**22.36**
Three-sided 6.5 mm thick pipe casing; to softwood framing (not included)						
450 mm girth	-	1.06	22.20	2.29	m	**24.50**
900 mm girth	-	1.25	26.18	4.36	m	**30.54**
Lining to walls 9 mm thick						
over 300 mm wide	5.18	0.40	8.38	6.23	m²	**14.60**
not exceeding 300 mm wide	-	0.26	5.45	1.89	m	**7.33**
Lining to ceilings 9 mm thick						
over 300 mm wide	5.18	0.53	11.10	6.23	m²	**17.33**
not exceeding 300 mm wide	-	0.34	7.12	1.89	m	**9.01**
holes for pipes and the like	-	0.03	0.63	-	nr	**0.63**
Lining to walls 12 mm thick						
over 300 mm wide	6.00	0.43	9.01	7.18	m²	**16.18**
not exceeding 300 mm wide	-	0.28	5.86	2.17	m	**8.03**
holes for pipes and the like	-	0.03	0.63	-	nr	**0.63**
Two-sided 12 mm thick pipe casing; to softwood framing (not included)						
300 mm girth	-	0.69	14.45	2.26	m	**16.71**
600 mm girth	-	0.83	17.38	4.34	m	**21.72**
Three-sided 12 mm thick pipe casing; to softwood framing (not included)						
450 mm girth	-	0.93	19.48	3.39	m	**22.86**
900 mm girth	-	1.11	23.25	6.54	m	**29.79**
extra for 400 mm x 400 mm removable access panel; brass cups and screws; additional framing	-	1.00	20.94	1.04	nr	**21.98**
Lining to ceilings 12 mm thick						
over 300 mm wide	6.00	0.56	11.73	7.18	m²	**18.91**
not exceeding 300 mm wide	-	0.37	7.75	2.17	m	**9.92**
holes for pipes and the like	-	0.03	0.63	-	nr	**0.63**
Extra over wall linings fixed with nails for screwing	-	-	-	-	m²	**1.92**
Preformed white melamine faced plywood casings; Pendock Profiles Ltd or other equal and approved; to softwood battens (not included)						
Skirting trunking profile; plain butt joints in the running length						
45 mm x 150 mm; ref TK150	-	0.11	2.30	44.31	m	**46.62**
extra for stop end	-	0.04	0.84	5.42	nr	**6.26**
extra for external corner	-	0.09	1.88	40.43	nr	**42.31**
extra for internal corner	-	0.09	1.88	40.43	nr	**42.31**
Casing profiles						
150 mm x 150 mm; ref MX150/150; 5 mm thick	-	0.11	2.30	6.22	m	**8.53**
extra for stop end	-	0.04	0.84	10.80	nr	**11.64**
extra for external corner	-	0.09	1.88	53.88	nr	**55.76**
extra for internal corner	-	0.09	1.88	40.43	nr	**42.31**

K LININGS/SHEATHING/DRY PARTITIONING

Item	PC £	Labour hours	Labour £	Material £	Unit	Total rate £
Internal quality American Cherry veneered plywood; 6 mm thick						
Lining to walls						
over 300 mm wide	5.11	0.41	8.59	6.08	m²	**14.67**
not exceeding 300 mm wide	-	0.27	5.65	1.86	m	**7.51**
"Tacboard" or other equal and approved; Eternit UK Ltd; fire resisting boards; butt joints; to softwood base						
Lining to walls; 6 mm thick						
over 300 mm wide	-	0.31	6.49	8.39	m²	**14.88**
not exceeding 300 mm wide	-	0.19	3.98	2.55	m	**6.53**
Lining to walls; 9 mm thick						
over 300 mm wide	-	0.33	6.91	15.31	m²	**22.22**
not exceeding 300 mm wide	-	0.20	4.19	4.63	m	**8.82**
Lining to walls; 12 mm thick						
over 300 wide	-	0.37	7.75	19.88	m²	**27.63**
not exceeding 300 mm wide	-	0.22	4.61	6.00	m	**10.61**
"Tacfire" or other equal and approved; Eternit UK Ltd; fire resisting boards						
Lining to walls; 6 mm thick						
over 300 mm wide	-	0.31	6.49	11.21	m²	**17.70**
not exceeding 300 mm wide	-	0.19	3.98	3.40	m	**7.38**
Lining to walls; 9 mm thick						
over 300 mm wide	-	0.33	6.91	17.08	m²	**23.99**
not exceeding 300 mm wide	-	0.20	4.19	5.16	m	**9.35**
Lining to walls; 12 mm thick						
over 300 mm wide	-	0.37	7.75	22.48	m²	**30.23**
not exceeding 300 mm wide	-	0.22	4.61	6.78	m	**11.39**
K13 RIGID SHEET FINE LININGS/PANELLING						
Perforated steel acoustic wall panels; Eckel type HD EFP or other equal and approved; polyurethene enamel finish; fibrous glass acoustic insulation						
Walls						
over 300 mm wide; fixed to timber or masonry	-	-	-	-	m²	**185.45**
K14 GLASS REINFORCED GYPSUM LININGS/ PANELLING						
Glass reinforced gypsum Glasroc Multi-board or other equal and approved; fixing with nails; joints filled with joint filler and joint tape; finishing with "Jointex" or other equal and approved to receive decoration; to softwood base						
10 mm board to walls						
wall height 2.40 m - 2.70 m	-	0.93	19.85	50.39	m	**70.24**
wall height 2.70 m - 3.00 m	-	1.06	22.64	56.00	m	**78.63**
wall height 3.00 m - 3.30 m	-	1.20	25.63	61.60	m	**87.23**
wall height 3.30 m - 3.60 m	-	1.39	29.67	67.25	m	**96.92**
12.50 mm board to walls						
wall height 2.40 m - 2.70 m	-	0.97	20.69	65.88	m	**86.57**
wall height 2.70 m - 3.00 m	-	1.11	23.68	73.21	m	**96.89**
wall height 3.00 m - 3.30 m	-	1.25	26.68	80.53	m	**107.21**
wall height 3.30 m - 3.60 m	-	1.43	30.51	87.90	m	**118.41**

K LININGS/SHEATHING/DRY PARTITIONING

Item	PC £	Labour hours	Labour £	Material £	Unit	Total rate £
K20 TIMBER BOARD FLOORING/SHEATHING/ LININGS/CASINGS						
Sawn softwood; untreated						
Boarding to roofs; 150 mm wide boards; butt joints						
19 mm thick; flat; over 300 mm wide	-	0.42	8.80	6.13	m²	**14.93**
19 mm thick; flat; not exceeding 300 mm wide	-	0.28	5.86	1.87	m	**7.73**
19 mm thick; sloping; over 300 mm wide	-	0.46	9.63	6.13	m²	**15.77**
19 mm thick; sloping; not exceeding 300 mm wide	-	0.31	6.49	1.87	m	**8.36**
19 mm thick; sloping; laid diagonally; over 300 mm wide	-	0.58	12.15	6.13	m²	**18.28**
19 mm thick; sloping; laid diagonally; not exceeding 300 mm wide	-	0.37	7.75	1.87	m	**9.62**
25 mm thick; flat; over 300 mm wide	-	0.42	8.80	9.83	m²	**18.63**
25 mm thick; flat; not exceeding 300 mm wide	-	0.28	5.86	2.98	m	**8.84**
25 mm thick; sloping; over 300 mm wide	-	0.46	9.63	9.83	m²	**19.47**
25 mm thick; sloping; not exceeding 300 mm wide	-	0.31	6.49	2.98	m	**9.47**
25 mm thick; sloping; laid diagonally; over 300 mm wide	-	0.58	12.15	9.83	m²	**21.98**
25 mm thick; sloping; laid diagonally; not exceeding 300 mm wide	-	0.37	7.75	2.98	m	**10.73**
Boarding to tops or cheeks of dormers; 150 mm wide boards; butt joints						
19 mm thick; laid diagonally; over 300 mm wide	-	0.74	15.50	6.13	m²	**21.63**
19 mm thick; laid diagonally; not exceeding 300 mm wide	-	0.46	9.63	1.87	m	**11.50**
19 mm thick; laid diagonally; area not exceeding 1.00 m² irrespective of width	-	0.93	19.48	5.76	nr	**25.23**
Sawn softwood; "Tanalised"						
Boarding to roofs; 150 wide boards; butt joints						
19 mm thick; flat; over 300 mm wide	-	0.42	8.80	6.82	m²	**15.62**
19 mm thick; flat; not exceeding 300 mm wide	-	0.28	5.86	2.07	m	**7.94**
19 mm thick; sloping; over 300 mm wide	-	0.46	9.63	6.82	m²	**16.46**
19 mm thick; sloping; not exceeding 300 mm wide	-	0.31	6.49	2.07	m	**8.57**
19 mm thick; sloping; laid diagonally; over 300 mm wide	-	0.58	12.15	6.82	m²	**18.97**
19 mm thick; sloping; laid diagonally; not exceeding 300 mm wide	-	0.37	7.75	2.07	m	**9.82**
25 mm thick; flat; over 300 mm wide	-	0.42	8.80	10.75	m²	**19.54**
25 mm thick; flat; not exceeding 300 mm wide	-	0.28	5.86	3.25	m	**9.12**
25 mm thick; sloping; over 300 mm wide	-	0.46	9.63	10.75	m²	**20.38**
25 mm thick; sloping; not exceeding 300 mm wide	-	0.31	6.49	3.25	m	**9.75**
25 mm thick; sloping; laid diagonally; over 300 mm wide	-	0.58	12.15	10.75	m²	**22.89**
25 mm thick; sloping; laid diagonally; not exceeding 300 mm wide	-	0.37	7.75	3.25	m	**11.00**
Boarding to tops or cheeks of dormers; 150 mm wide boards; butt joints						
19 mm thick; laid diagonally; over 300 mm wide	-	0.74	15.50	6.82	m²	**22.32**
19 mm thick; laid diagonally; not exceeding 300 mm wide	-	0.46	9.63	2.07	m	**11.71**
19 mm thick; laid diagonally; area not exceeding 1.00 m² irrespective of width	-	0.93	19.48	6.45	nr	**25.92**
Wrought softwood						
Boarding to floors; butt joints						
19 mm x 75 mm boards	-	0.56	11.73	8.34	m²	**20.06**
19 mm x 125 mm boards	-	0.51	10.68	6.38	m²	**17.06**
22 mm x 150 mm boards	-	0.46	9.63	7.03	m²	**16.67**

K LININGS/SHEATHING/DRY PARTITIONING

Item	PC £	Labour hours	Labour £	Material £	Unit	Total rate £
25 mm x 100 mm boards	-	0.51	10.68	7.68	m²	**18.36**
25 mm x 150 mm boards	-	0.46	9.63	7.81	m²	**17.44**
Boarding to floors; tongued and grooved joints						
19 mm x 75 mm boards	-	0.65	13.61	8.96	m²	**22.58**
19 mm x 125 mm boards	-	0.60	12.57	7.15	m²	**19.72**
22 mm x 150 mm boards	-	0.56	11.73	7.24	m²	**18.97**
25 mm x 100 mm boards	-	0.60	12.57	9.10	m²	**21.67**
25 mm x 150 mm boards	-	0.56	11.73	8.53	m²	**20.26**
Boarding to internal walls; tongued and grooved and V-jointed						
12 mm x 100 mm boards	-	0.74	15.50	6.59	m²	**22.09**
16 mm x 100 mm boards	-	0.74	15.50	7.43	m²	**22.92**
19 mm x 100 mm boards	-	0.74	15.50	7.84	m²	**23.34**
19 mm x 125 mm boards	-	0.69	14.45	8.47	m²	**22.92**
19 mm x 125 mm boards; chevron pattern	-	1.11	23.25	8.47	m²	**31.72**
25 mm x 125 mm boards	-	0.69	14.45	9.36	m²	**23.81**
12 mm x 100 mm boards; knotty pine	-	0.74	15.50	5.59	m²	**21.09**
Boarding to internal ceilings						
12 mm x 100 mm boards	-	0.93	19.48	6.59	m²	**26.07**
16 mm x 100 mm boards	-	0.93	19.48	7.43	m²	**26.90**
19 mm x 100 mm boards	-	0.93	19.48	7.84	m²	**27.32**
19 mm x 125 mm boards	-	0.88	18.43	8.47	m²	**26.90**
19 mm x 125 mm boards; chevron pattern	-	1.30	27.23	8.47	m²	**35.70**
25 mm x 125 mm boards	-	0.88	18.43	9.36	m²	**27.79**
12 mm x 100 mm boards; knotty pine	-	0.93	19.48	5.59	m²	**25.07**
Boarding to roofs; tongued and grooved joints						
19 mm thick; flat to falls	-	0.51	10.68	7.24	m²	**17.92**
19 mm thick; sloping	-	0.56	11.73	7.24	m²	**18.97**
19 mm thick; sloping; laid diagonally	-	0.72	15.08	7.24	m²	**22.32**
25 mm thick; flat to falls	-	0.51	10.68	9.14	m²	**19.82**
25 mm thick; sloping	-	0.56	11.73	9.14	m²	**20.87**
Boarding to tops or cheeks of dormers; tongued and grooved joints						
19 mm thick; laid diagonally	-	0.93	19.48	7.24	m²	**26.72**
Wrought softwood; "Tanalised"						
Boarding to roofs; tongued and grooved joints						
19 mm thick; flat to falls	-	0.51	10.68	7.93	m²	**18.61**
19 mm thick; sloping	-	0.56	11.73	7.93	m²	**19.66**
19 mm thick; sloping; laid diagonally	-	0.72	15.08	7.93	m²	**23.01**
25 mm thick; flat to falls	-	0.51	10.68	10.05	m²	**20.73**
25 mm thick; sloping	-	0.56	11.73	10.05	m²	**21.78**
Boarding to tops or cheeks of dormers; tongued and grooved joints						
19 mm thick; laid diagonally	-	0.93	19.48	7.93	m²	**27.41**
Wood strip; 22 mm thick; Junckers All in Beech Sylva Sport Premium pre-treated or other equal and approved; tongued and grooved joints; on bearers etc.; level fixing to cement and sand base						
Strip flooring; over 300 mm wide						
on 45 x 45 mm blue bat bearers	-	-	-	-	m²	**49.67**
on 10 mm Pro Foam	-	-	-	-	m²	**52.14**
on Uno bat 50 mm bearers	-	-	-	-	m²	**54.92**
on New Era levelling system	-	-	-	-	m²	**56.96**
on Uno bat 62 mm bearers	-	-	-	-	m²	**58.90**
on Duo bat 110 mm bearers	-	-	-	-	m²	**70.87**

Item	PC £	Labour hours	Labour £	Material £	Unit	Total rate £
K20 TIMBER BOARD FLOORING/SHEATHING/ LININGS/CASINGS – cont'd						
Wood strip; 22 mm thick; Junckers pre-treated or other equal and approved flooring systems; tongued and grooved joints; on bearers etc.; level fixing to cement and sand base						
Strip flooring; over 300 mm wide						
Sylva Squash Beech untreated on blue bat bearers	-	-	-	-	m²	**67.56**
Classic Beech clip system	-	-	-	-	m²	**70.05**
Harmoni Oak clip system	-	-	-	-	m²	**70.05**
Classic Beech on blue bat bearers	-	-	-	-	m²	**70.05**
Harmoni Oak on blue bat bearers	-	-	-	-	m²	**70.05**
Unfinished wood strip; 22 mm thick; Havwoods or other equal and approved; tongued and grooved joints; secret fixed; laid on semi-sprung bearers; fixing to cement and sand base; sanded and sealed						
Strip flooring; over 300 mm wide						
Prime Iroko	-	-	-	-	m²	**60.03**
Prime Maple	-	-	-	-	m²	**64.17**
American Oak	-	-	-	-	m²	**65.51**
K30 DEMOUNTABLE PARTITIONS						
Insulated panel and two-hour fire wall system for warehouses etc., comprising white polyester coated galvanised steel frame and 0.55 mm galvanised steel panels either side of rockwool infill						
100 mm thick wall: 31 Rw dB acoustic rating	-	-	-	-	m²	**49.20**
150 mm thick wall: 31 Rw dB acoustic rating	-	-	-	-	m²	**52.89**
intumescent mastic sealant; bedding frames at perimeter of metal fire walls	-	-	-	-	m	**4.30**
Getalit laminated both sides top hung movable acoustic panel wall with concealed uPVC vertical edge profiles, 9 nr 1106 m x 3000 mm panels and type K two point panel support system						
105 mm thick wall: 47 Rw dB acoustic rating	-	-	-	-	m²	**455.10**
105 mm thick wall: 50 Rw dB acoustic rating	-	-	-	-	m²	**492.00**
105 mm thick wall: 53 Rw dB acoustic rating	-	-	-	-	m²	**528.90**
K32 FRAMED PANEL CUBICLE PARTITIONS						
Toilet cubicle partitions; Amwells or other equal and approved; standard colours and ironmongery; assembling and screwing to floor and wall						
"Axis" standard cubicle set; 800 mm x 1500 mm x 1980 mm high per cubicle, with polished aluminium framing; 19 mm melamine-faced chipboard divisions and doors						
One cubicle set; 2 nr panels; 1 nr door	-	3.25	130.00	280.00	nr	**410.00**
range of 3 cubicle sets; 4 nr panels; 3 nr doors	-	9.75	390.00	800.00	nr	**1190.00**
range of 6 cubicle sets; 7 nr panels; 6 nr doors	-	19.50	780.00	1580.00	nr	**2360.00**
Reduction of 1 nr panel for end unit adjoining side wall	-	-	-	-110.00	nr	**-**

K LININGS/SHEATHING/DRY PARTITIONING

Item	PC £	Labour hours	Labour £	Material £	Unit	Total rate £
"Minima" designer cubicle set; 800 mm x 1500 mm x 2100 mm high per cubicle, with satin polished stainless steel framing; 18 mm high pressure laminated (HPL) chipboard divisions and doors						
One cubicle set; 2 nr panels; 1 nr door	-	3.25	130.00	590.00	nr	**720.00**
range of 3 cubicle sets; 4 nr panels; 3 nr doors	-	9.75	390.00	1530.00	nr	**1920.00**
range of 6 cubicle sets; 7 nr panels; 6 nr doors	-	19.50	780.00	2940.00	nr	**3720.00**
Reduction of 1 nr panel for end unit adjoining side wall	-	-	-	-160.00	nr	**-**
"Sylan " corporate cubicle set; 800 mm x 1500 mm x 2400 mm high per cubicle, with sating finished stainless steel ironmongery; 30 mm high pressure laminated (HPL) chipboard divisions and 44 mm solid cored real wood veneered doors and pilasters						
One cubicle set; 2 nr panels; 1 nr door	-	5.00	200.00	1745.00	nr	**1945.00**
range of 3 cubicle sets; 4 nr panels; 3 nr doors	-	15.00	600.00	4715.00	nr	**5315.00**
range of 6 cubicle sets; 7 nr panels; 6 nr doors	-	30.00	1200.00	9165.00	nr	**10365.00**
Reduction of 1 nr panel for end unit adjoining side wall	-	-	-	-355.00	nr	**-**
K33 CONCRETE/TERRAZZO PARTITIONS						
Terrazzo faced partitions; polished on two faces						
Pre-cast reinforced terrazzo faced WC partitions						
38 mm thick; over 300 mm wide	-	-	-	-	m²	**265.38**
50 mm thick; over 300 mm wide	-	-	-	-	m²	**276.13**
Wall post; once rebated						
64 mm x 102 mm	-	-	-	-	m	**122.60**
64 mm x 152 mm	-	-	-	-	m	**134.30**
Centre post; twice rebated						
64 mm x 102 mm	-	-	-	-	m	**127.97**
64 mm x 152 mm	-	-	-	-	m	**139.67**
Lintel; once rebated						
64 mm x 102 mm	-	-	-	-	m	**122.60**
Pair of brass topped plates or sockets cast into posts for fixings (not included)	-	-	-	-	nr	**30.94**
Brass indicator bolt lugs cast into posts for fixings (not included)	-	-	-	-	nr	**14.88**
K40 DEMOUNTABLE SUSPENDED CEILINGS						
Suspended ceilings; Donn Products exposed suspended ceiling system or other equal and approved; hangers plugged and screwed to concrete soffit; 600 mm x 600 mm x 15 mm Cape TAP Ceilings Ltd; "Solitude" tegular fissured tile						
Lining to ceilings; hangers average 400 mm long						
over 300 mm wide	-	0.32	9.09	11.57	m²	**20.66**

K LININGS/SHEATHING/DRY PARTITIONING

Item	PC £	Labour hours	Labour £	Material £	Unit	Total rate £
K40 DEMOUNTABLE SUSPENDED CEILINGS – cont'd						
Suspended ceilings, Gyproc M/F suspended ceiling system or other equal and approved; hangers plugged and screwed to concrete soffit, 900 mm x 1800 mm x 12.50 mm tapered edge wallboard infill; joints filled with joint filler and taped to receive direct direction						
Lining to ceilings; hangers average 400 mm long						
over 300 mm wide	-	-	-	-	m²	**31.48**
not exceeding 300 mm wide in isolated strips	-	-	-	-	m	**27.01**
300 mm - 600 mm wide in isolated strips	-	-	-	-	m	**31.88**
Edge treatments						
20 x 20 mm SAS perimeter shadow gap; screwed to plasterboard	-	-	-	-	m	**5.10**
20 x 20 mm SAS shadow gap around 450 mm dia. column; including 15 x 44 mm batten plugged and screwed to concrete	-	-	-	-	nr	**25.49**
Vertical bulkhead; including additional hangers						
over 300 mm wide	-	-	-	-	m²	**39.68**
not exceeding 300 mm wide in isolated strips	-	-	-	-	m	**38.46**
300 mm - 600 mm wide in isolated strips	-	-	-	-	m	**39.14**
Suspended ceilings; Rockfon, or other equal and approved; "Z" demountable suspended concealed ceiling system; 400 mm long hangers plugged and screwed to concrete soffit.						
Lining to ceilings; 600 mm x 600mm x 20 mm 'Sonar' suspended ceiling tiles						
over 300 mm wide	-	-	-	-	m²	**39.41**
not exceeding 300 mm wide	-	-	-	-	m	**23.52**
Edge trim; shadow-line trim	-	-	-	-	m	**4.69**
Vertical bulkhead, as upstand to rooflight well; including additional hangers; perimeter trim						
300 mm x 600 mm wide	-	-	-	-	m	**43.37**
Suspended ceilings; Ecophon, or other equal and approved; "Z" demountable suspended concealed ceiling system; 400 mm long hangers plugged and screwed to concrete soffit.						
Lining to ceilings; 600 mm x 600mm x 20 mm 'Gedina ET15' suspended ceiling tiles						
over 300 mm wide	-	-	-	-	m²	**34.24**
not exceeding 300 mm wide	-	-	-	-	m	**21.86**
Edge trim; shadow-line trim	-	-	-	-	m	**4.35**
Vertical bulkhead, as upstand to rooflight well; including additional hangers; perimeter trim						
300 mm x 600 mm wide	-	-	-	-	m	**39.95**
Lining to ceilings; 600 mm x 600mm x 20 mm 'Hygiene Performance' washable suspended ceiling tiles						
over 300 mm wide	-	-	-	-	m²	**45.73**
not exceeding 300 mm wide	-	-	-	-	m	**37.66**
Edge trim; shadow-line trim	-	-	-	-	m	**6.13**

K LININGS/SHEATHING/DRY PARTITIONING

Item	PC £	Labour hours	Labour £	Material £	Unit	**Total rate £**
Vertical bulkhead, as upstand to rooflight well; including additional hangers; perimeter trim						
300 mm x 600 mm wide	-	-	-	-	m	**41.46**
Lining to ceilings; 1200 mm x 1200mm x 20 mm 'Focus DG' suspended ceiling tiles						
over 300 mm wide	-	-	-	-	m²	**41.40**
not exceeding 300 mm wide	-	-	-	-	m	**24.83**
Edge trim; shadow-line trim	-	-	-	-	m	**4.35**
Vertical bulkhead, as upstand to rooflight well; including additional hangers; perimeter trim						
300 mm x 600 mm wide	-	-	-	-	m	**42.06**
Suspended ceilings; "Z" demountable suspended ceiling system or other equal and approved; hangers plugged and screwed to concrete soffit, 600 mm x 600 mm x 19 mm "Echostop" glass reinforced fibrous plaster lightweight plain bevelled edge tiles						
Lining to ceilings; hangers average 400 mm long						
over 300 mm wide	-	-	-	-	m²	**84.90**
not exceeding 300 mm wide in isolated strips	-	-	-	-	m	**61.23**
Suspended ceilings; concealed galvanised steel suspension system; hangers plugged and screwed to concrete soffit, "Burgess" white stove enamelled perforated mild steel tiles 600 mm x 600 mm						
Lining to ceilings; hangers average 400 mm long						
over 300 mm wide	-	-	-	-	m²	**40.87**
not exceeding 300 mm wide; in isolated strips	-	-	-	-	m	**36.04**
Suspended ceilings; concealed galvanised steel "Trulok" suspension system or other equal and approved; hangers plugged and screwed to concrete; Armstrong "Ultima Microlok BE Plain" 300 mm x 300 mm x 18 mm mineral ceiling tiles						
Linings to ceilings; hangers average 700 mm long						
over 300 mm wide	-	-	-	-	m²	**25.68**
over 300 mm wide; 3.50 m - 5.00 m high	-	-	-	-	m²	**26.68**
over 300 mm wide; in staircase areas or plant rooms	-	-	-	-	m²	**33.62**
not exceeding 300 mm wide; in isolated strips	-	-	-	-	m	**18.85**
300 mm - 600 mm wide; in isolated strips	-	-	-	-	m	**23.83**
Extra for cutting and fitting around modular downlighter including yoke	-	-	-	-	nr	**15.50**
24 mm x 19 mm white finished angle edge trim	-	-	-	-	m	**3.92**
Vertical bulkhead; including additional hangers						
over 300 mm wide	-	-	-	-	m²	**47.79**
not exceeding 300 mm wide; in isolated strips	-	-	-	-	m	**37.89**
300 mm - 600 mm wide; in isolated strips	-	-	-	-	m	**43.20**

K LININGS/SHEATHING/DRY PARTITIONING

Item	PC £	Labour hours	Labour £	Material £	Unit	Total rate £
K40 DEMOUNTABLE SUSPENDED CEILINGS – cont'd						
Suspended ceilings, metal; SAS system 330; EMAC suspension system; 100 mm Omega C profiles at 1500 mm centres filled in with 1400 mm x 250 mm perforated metal tiles with 18 mm thick x 80 kg/m3 density foil wrapped tissue-faced acoustic pad adhered above; ceiling to achieve 40d Dnwc with 0.7 absorption coefficient						
Linings to ceilings; hangers average 700 mm long						
over 300 mm wide	24.30	0.64	18.18	32.88	m²	**51.06**
Extra for cutting and reinforcing to receive a recessed light maximum 1300 mm x 500 mm.	-	-	-	-	nr	**15.79**
not exceeding 300 mm wide; in isolated strips	-	-	-	-	m	**28.19**
Edge trim; to perimeter	-	-	-	-	m	**14.09**
Edge trim around 450 mm dia. column;	-	-	-	-	nr	**54.12**
Suspended ceilings; galvanised steel suspension system; hangers plugged and screwed to concrete soffit, "Luxalon" stove enamelled aluminium linear panel ceiling, type 80B or other equal and approved, complete with mineral insulation						
Linings to ceilings; hangers average 700 mm long						
over 300 mm wide	-	-	-	-	m²	**77.31**
not exceeding 300 mm wide; in isolated strips	-	-	-	-	m	**37.60**
K41 RAISED ACCESS FLOORS						
Raised flooring system; laid on or fixed to concrete floor						
Full access system; 150 mm high overall; pedestal supports						
PSA light grade; steel finish	-	-	-	-	m²	**38.74**
PSA medium grade; steel finish	-	-	-	-	m²	**40.20**
PSA heavy grade; steel finish	-	-	-	-	m²	**52.31**
Extra for						
factory applied needlepunch carpet	-	-	-	-	m²	**14.53**
factory applied anti-static vinyl	-	-	-	-	m²	**31.36**
factory applied black PVC edge strips	-	-	-	-	m	**4.19**
ramps; 3.00 m x 1.40 m (no finish)	-	-	-	-	nr	**691.88**
steps (no finish)	-	-	-	-	m	**37.82**
forming cut-out for electrical boxes	-	-	-	-	nr	**8.97**

Item	PC £	Labour hours	Labour £	Material £	Unit	Total rate £
L10 WINDOWS/ROOFLIGHTS/SCREENS/LOUVRES						
SUPPLY ONLY PRICES						
NOTE: The following supply only prices are for purpose-made components, to which fixings, sealants etc. labour and overheads and profit need to be added, before they may be used to arrive at a guide price for a complete window. The reader is then referred to the following SUPPLY AND FIX pages for fixing costs based on the overall window size.						
Purpose made window casements; "treated" wrought softwood						
Casements; rebated; moulded						
44 mm thick	-	-	-	60.42	m²	-
57 mm thick	-	-	-	63.44	m²	-
Casements; rebated; moulded; in medium panes						
44 mm thick	-	-	-	96.58	m²	-
57 mm thick	-	-	-	100.63	m²	-
Casements; rebated; moulded; with semi-circular head						
44 mm thick	-	-	-	126.28	m²	-
57 mm thick	-	-	-	130.24	m²	-
Casements; rebated; moulded; to bullseye window						
44 mm thick; 600 mm diameter	-	-	-	200.17	nr	-
44 mm thick; 900 mm diameter	-	-	-	238.47	nr	-
57 mm thick; 600 mm diameter	-	-	-	209.60	nr	-
57 mm thick; 900 mm diameter	-	-	-	28.84	nr	-
Fitting and hanging casements (in factory)						
square or rectangular	-	-	-	14.13	nr	-
semi-circular	-	-	-	22.95	nr	-
bullseye	-	-	-	28.84	nr	-
Purpose made window casements; selected Sapele						
Casements; rebated; moulded						
44 mm thick	-	-	-	68.13	m²	-
57 mm thick	-	-	-	74.63	m²	-
Casements; rebated; moulded; in medium panes						
44 mm thick	-	-	-	110.32	m²	-
57 mm thick	-	-	-	119.02	m²	-
Casements; rebated; moulded with semi-circular head						
44 mm thick	-	-	-	138.96	m²	-
57 mm thick	-	-	-	147.46	m²	-
Casements; rebated; moulded; to bullseye window						
44 mm thick; 600 mm diameter	-	-	-	245.60	nr	-
44 mm thick; 900 mm diameter	-	-	-	295.55	nr	-
57 mm thick; 600 mm diameter	-	-	-	265.86	nr	-
57 mm thick; 900 mm diameter	-	-	-	321.36	nr	-
Fitting and hanging casements (in factory)						
square or rectangular	-	-	-	15.18	nr	-
semi-circular	-	-	-	25.09	nr	-
bullseye	-	-	-	32.10	nr	-

L WINDOWS/DOORS/STAIRS

Item	PC £	Labour hours	Labour £	Material £	Unit	Total rate £
L10 WINDOWS/ROOFLIGHTS/SCREENS/LOUVRES – cont'd						
SUPPLY ONLY PRICES – cont'd						
Purpose made window frames; "treated" wrought softwood						
Frames; rounded; rebated check grooved						
44 mm x 69 mm	-	-	-	16.43	m	-
44 mm x 94 mm	-	-	-	17.22	m	-
44 mm x 119 mm	-	-	-	18.00	m	-
57 mm x 94 mm	-	-	-	18.03	m	-
69 mm x 144 mm	-	-	-	23.78	m	-
90 mm x 140 mm	-	-	-	33.74	m	-
Mullions and transoms; twice rounded, rebated and check grooved						
57 mm x 69 mm	-	-	-	19.13	m	-
57 mm x 94 mm	-	-	-	20.10	m	-
69 mm x 94 mm	-	-	-	22.73	m	-
69 mm x 144 mm	-	-	-	33.38	m	-
Sill; sunk weathered, rebated and grooved						
69 mm x 94 mm	-	-	-	39.90	m	-
69 mm x 144 mm	-	-	-	42.23	m	-
Add 5% to the above material prices for "selected" softwood for staining						
Purpose made window frames; selected Sapele						
Frames; rounded; rebated check grooved						
44 mm x 69 mm	-	-	-	21.65	m	-
44 mm x 94 mm	-	-	-	23.28	m	-
44 mm x 119 mm	-	-	-	24.91	m	-
57 mm x 94 mm	-	-	-	27.25	m	-
69 mm x 144 mm	-	-	-	39.03	m	-
90 mm x 140 mm	-	-	-	55.45	m	-
Mullions and transoms; twice rounded, rebated and check grooved						
57 mm x 69 mm	-	-	-	24.59	m	-
57 mm x 94 mm	-	-	-	28.61	m	-
69 mm x 94 mm	-	-	-	34.31	m	-
69 mm x 144 mm	-	-	-	52.14	m	-
Sill; sunk weathered, rebated and grooved						
69 mm x 94 mm	-	-	-	48.60	m	-
69 mm x 144 mm	-	-	-	53.87	m	-
The following supply only prices are for standard Velfac windows, to which fixings, sealants etc. labour and overheads and profit need to be added, before they may be used to arrive at a guide price for a complete unit.						

L WINDOWS/DOORS/STAIRS

Item	PC £	Labour hours	Labour £	Material £	Unit	Total rate £
Thermally broken composite double glazed aluminium/ timber windows; 'Velfac 200' or other approved; with a maximum glazing U value of 1.5 W/m²K; low e glazing with laminated glass unless otherwise specified; including multi point espagnolette locking mechanisms and other ironmongery						
Outward opening standard fixed sash casement windows						
1200 mm x 1200 mm single fixed pane; low-e glass 4/16/4	-	-	-	260.00	nr	-
2200 mm x 2200 mm single fixed pane; low-e glass 6/12/6	-	-	-	660.00	nr	-
1200 mm x 2200 mm three fixed panes; low-e glass 4/16/4	-	-	-	520.00	nr	-
Outward opening standard sash casement windows						
1600 mm x 1600 mm with two sidehung sashes; low-e glass 4/16/4	-	-	-	530.00	nr	-
1600 mm x 1600 mm with two sidehung projecting sashes; low-e glass 4/16/4	-	-	-	590.00	nr	-
2000 mm x 1600 mm with one sidehung sash next to a tophung projecting sash over a fixed sash; low-e glass 4/16/4	-	-	-	650.00	nr	-
1200 mm x 2200 mm with fixed lower sash and tophung projecting upper sash; lower low-e upper low-e glass 4 toughened/16/6.4; upper low-e glass 4/16/4	-	-	-	515.00	nr	-
1200 mm x 2200 mm with fixed lower sash and fully reversible upper sash; lower low-e upper low-e glass 4 toughened/16/6.4; upper low-e glass 4/16/4	-	-	-	560.00	nr	-
Outward opening standard doors						
2200 mm x 2200 mm French casement patio door; low-e toughened glass 4/16/4	-	-	-	995.00	nr	-
2200 mm x 2200 mm Sliding patio door; low-e glass 4 toughened/16/4 laminated	-	-	-	1240.00	nr	-
Guide price for installation:	-	1.00	60.00	-	m²	**60.00**
SUPPLY AND FIX PRICES						
Standard windows; "treated" wrought softwood; Jeld-Wen or other equal and approved						
Side hung casement windows; factory glazed with low E 24mm double glazing; with 140 mm wide softwood sills; opening casements and ventilators hung on rustproof hinges; fitted with aluminized lacquered finish casement stays and fasteners						
488 mm x 750 mm; ref LEWN07V	100.56	0.65	13.61	108.53	nr	**122.15**
488 mm x 900 mm; ref LEWN09V	102.31	0.74	15.50	110.41	nr	**125.90**
630 mm x 750 mm; ref LEW107C	90.92	0.74	15.50	98.15	nr	**113.65**
630 mm x 750 mm; ref LEW107V	111.50	0.74	15.50	120.31	nr	**135.81**
630 mm x 900 mm; ref LEW109V	115.53	0.83	17.38	124.63	nr	**142.02**
630 mm x 900 mm; ref LEW109CH	98.05	0.74	15.50	105.82	nr	**121.32**
630 mm x 1050 mm; ref LEW110C	101.64	0.93	19.48	109.79	nr	**129.27**
630 mm x 1050 mm; ref LEW110V	120.88	0.74	15.50	130.49	nr	**145.99**
915 mm x 900 mm; ref LEW2NO9W	138.99	1.02	21.36	149.89	nr	**171.25**
915 mm x 1050 mm; ref LEW2N1OW	146.02	1.06	22.20	157.55	nr	**179.76**
915 mm x 1200 mm; ref LEW2N12W	158.00	1.11	23.25	170.45	nr	**193.70**
915 mm x 1350 mm; ref LEW2N13W	165.65	1.25	26.18	178.68	nr	**204.86**
915 mm x 1500 mm; ref LEW2N15W	187.95	1.30	27.23	202.78	nr	**230.01**

L WINDOWS/DOORS/STAIRS

Item	PC £	Labour hours	Labour £	Material £	Unit	Total rate £
L10 WINDOWS/ROOFLIGHTS/SCREENS/LOUVRES – cont'd						
Standard windows; Jeld-Wen Softwood – cont'd						
Side hung casement windows – cont'd						
1200 mm x 750 mm; ref LEW2O7C	142.85	1.06	22.20	154.15	nr	**176.35**
1200 mm x 750 mm; ref LEW2O7CV	180.69	1.06	22.20	194.86	nr	**217.06**
1200 mm x 900 mm; ref LEW2O9C	151.97	1.11	23.25	163.96	nr	**187.21**
1200 mm x 900 mm; ref LEW2O9W	159.92	1.11	23.25	172.51	nr	**195.76**
1200 mm x 900 mm; ref LEW2O9CV	188.81	1.11	23.25	203.60	nr	**226.85**
1200 mm x 1050 mm; ref LEW210C	160.88	1.25	26.18	173.64	nr	**199.82**
1200 mm x 1050 mm; ref LEW210W	169.50	1.25	26.18	182.93	nr	**209.11**
1200 mm x 1050 mm; ref LEW210T	200.00	1.25	26.18	215.75	nr	**241.93**
1200 mm x 1050 mm; ref LEW210CV	196.15	1.25	26.18	211.61	nr	**237.79**
1200 mm x 1200 mm; ref LEW212C	172.97	1.34	28.06	186.77	nr	**214.83**
1200 mm x 1200 mm; ref LEW212W	179.29	1.34	28.06	193.56	nr	**221.63**
1200 mm x 1200 mm; ref LEW212TX	233.00	1.34	28.06	251.36	nr	**279.43**
1200 mm x 1200 mm; ref LEW212CV	206.63	1.34	28.06	222.99	nr	**251.05**
1200 mm x 1350 mm; ref LEW213W	192.47	1.43	29.95	207.75	nr	**237.70**
1200 mm x 1350 mm; ref LEW213CV	228.52	1.43	29.95	246.55	nr	**276.50**
1200 mm x 1500 mm; ref LEW215W	223.21	1.57	32.88	240.93	nr	**273.81**
1770 mm x 750 mm; ref LEW307CC	209.79	1.30	27.23	226.39	nr	**253.61**
1770 mm x 900 mm; ref LEW309CC	223.22	1.57	32.88	240.84	nr	**273.72**
1770 mm x 1050 mm; ref LEW310C	211.56	1.62	33.93	228.29	nr	**262.22**
1770 mm x 1050 mm; ref LEW310T	249.75	1.57	32.88	269.39	nr	**302.28**
1770 mm x 1050 mm; ref LEW310CC	234.25	1.30	27.23	252.71	nr	**279.94**
1770 mm x 1050 mm; ref LEW310CW	246.67	1.30	27.23	266.08	nr	**293.31**
1770 mm x 1200 mm; ref LEW312C	225.94	1.67	34.98	243.86	nr	**278.84**
1770 mm x 1200 mm; ref LEW312T	264.56	1.67	34.98	285.44	nr	**320.42**
1770 mm x 1200 mm; ref LEW312CC	252.46	1.67	34.98	272.41	nr	**307.38**
1770 mm x 1200 mm; ref LEW312CW	261.49	1.67	34.98	282.12	nr	**317.10**
1770 mm x 1200 mm; ref LEW312CVC	294.56	1.67	34.98	317.72	nr	**352.70**
1770 mm x 1350 mm; ref LEW313CC	285.88	1.76	36.86	308.37	nr	**345.24**
1770 mm x 1350 mm; ref LEW313CW	284.73	1.76	36.86	307.14	nr	**344.00**
1770 mm x 1350 mm; ref LEW313CVC	320.72	1.76	36.86	345.88	nr	**382.74**
1770 mm x 1500 mm; ref LEW315T	340.15	1.85	38.75	366.79	nr	**405.53**
2340 mm x 1050 mm; ref LEW410CWC	325.38	1.80	37.70	350.90	nr	**388.60**
2340 mm x 1200 mm; ref LEW412CWC	345.65	1.90	39.79	372.81	nr	**412.61**
2340 mm x 1350 mm; ref LEW413CWC	380.08	2.04	42.73	409.96	nr	**452.69**
Top hung casement windows; factory glazed with low E 24mm double glazing; with 140 mm wide softwood sills; opening casements and ventilators hung on rustproof hinges; fitted with aluminized lacquered finish casement stays						
630 mm x 750 mm; ref LEW107A	98.01	0.74	15.50	105.78	nr	**121.28**
630 mm x 900 mm; ref LEW109A	103.45	0.83	17.38	111.64	nr	**129.03**
630 mm x 1050 mm; ref LEW110A	109.59	0.93	19.48	118.35	nr	**137.83**
915 mm x 750 mm; ref LEW2N07A	119.03	0.97	20.32	128.41	nr	**148.72**
915 mm x 900 mm; ref LEW2N09A	133.56	1.02	21.36	144.04	nr	**165.40**
915 mm x 1050 mm; ref LEW2N10A	142.09	1.06	22.20	153.33	nr	**175.53**
915 mm x 1350 mm; ref LEW2N13AS	179.26	1.25	26.18	193.43	nr	**219.61**
1200 mm x 750 mm; ref LEW207A	140.06	1.06	22.20	151.14	nr	**173.34**
1200 mm x 900 mm; ref LEW209A	152.85	1.11	23.25	164.91	nr	**188.16**
1200 mm x 1050 mm; ref LEW210A	163.42	1.25	26.18	176.38	nr	**202.56**
1200 mm x 1200 mm; ref LEW212A	177.12	1.34	28.06	191.13	nr	**219.19**
1200 mm x 1350 mm; ref LEW213AS	206.81	1.43	29.95	223.18	nr	**253.12**
1200 mm x 1500 mm; ref LEW215AS	225.68	1.57	32.88	243.49	nr	**276.37**
1770 mm x 1050 mm; ref LEW310AE	226.26	1.57	32.88	244.11	nr	**277.00**
1770 mm x 1200 mm; ref LEW312AE	241.13	1.67	34.98	260.11	nr	**295.09**

L WINDOWS/DOORS/STAIRS

Item	PC £	Labour hours	Labour £	Material £	Unit	Total rate £
High performance Hi-Profile top-hung reversible windows; factory glazed with low E 24mm double glazing; weather stripping; opening panes hung on rustproof hinges; fitted with aluminized lacquered espagnolette bolts						
600 mm x 900 mm; ref LECFR609AR	173.24	0.83	17.38	186.74	nr	**204.13**
600 mm x 1050 mm; ref LECFR610AR	182.44	0.93	19.48	196.76	nr	**216.23**
600 mm x 1200 mm; ref LECFR612AR	190.81	1.03	21.57	205.87	nr	**227.44**
600 mm x 1350 mm; ref LECFR613AR	199.03	1.11	23.25	214.70	nr	**237.95**
1200 mm x 900 mm; ref LECFR1209AFR	257.73	1.11	23.25	277.78	nr	**301.03**
1200 mm x 1050 mm; ref LECFR1210AFR	273.77	1.25	26.18	295.15	nr	**321.33**
1200 mm x 1200 mm; ref LECFR1212AFR	287.88	1.34	28.06	310.44	nr	**338.50**
1200 mm x 1350 mm; ref LECFR1213AFR	301.88	1.43	29.95	325.50	nr	**355.45**
1800 mm x 900 mm; ref LECFR1809AFAR	395.01	1.57	32.88	425.73	nr	**458.61**
1800 mm x 1050 mm; ref LECFR1810AFAR	418.37	1.62	33.93	450.87	nr	**484.80**
1800 mm x 1200 mm; ref LECFR1812AFAR	441.56	1.67	34.98	475.92	nr	**510.90**
1800 mm x 1350 mm; ref LECFR1813AFAR	462.94	1.76	36.86	498.94	nr	**535.80**
High performance double hung sash windows with glazing bars; factory glazed with low E 24mm double glazing; solid frames; 63 mm x 175 mm softwood sills; standard flush external linings; spiral spring balances and sash catch						
635 mm x 1050 mm; ref LESV0610B	321.45	1.85	38.75	346.37	nr	**385.11**
635 mm x 1350 mm; ref LESV0613B	355.88	2.04	42.73	383.52	nr	**426.25**
635 mm x 1650 mm; ref LESV0616B	401.86	2.27	47.54	433.10	nr	**480.64**
860 mm x 1050 mm; ref LESV0810B	368.25	2.13	44.61	396.73	nr	**441.34**
860 mm x 1350 mm; ref LESV0813B	410.76	2.41	50.47	442.59	nr	**493.06**
860 mm x 1650 mm; ref LESV0816B	476.83	2.78	58.22	513.79	nr	**572.02**
1085 mm x 1050 mm; ref LESV1010B	420.77	2.41	50.47	453.25	nr	**503.73**
1085 mm x 1350 mm; ref LESV1013B	475.56	2.78	58.22	512.32	nr	**570.54**
1085 mm x 1650 mm; ref LESV1016B	586.80	3.42	71.63	632.14	nr	**703.77**
1725 mm x 1050 mm; ref LESV1710B	616.39	3.42	71.63	663.89	nr	**735.52**
1725 mm x 1350 mm; ref LESV1713B	700.13	4.26	89.22	754.11	nr	**843.33**
1725 mm x 1650 mm; ref LESV1716B	835.64	4.35	91.11	900.06	nr	**991.16**
Add 27% to the above for full factory finish						
Standard windows; Jeld-Wen Hardwood or other equal and approved; factory applied preservative stain base coat						
Side hung casement windows; factory glazed with low E 24mm double glazing; 45 mm x 140 mm hardwood sills; weather stripping; opening sashes on canopy hinges; fitted with fasteners; brown finish ironmongery						
630 mm x 750 mm; ref LEW107CH	189.60	0.88	18.43	204.36	nr	**222.79**
630 mm x 900 mm; ref LEW109CH	200.44	1.11	23.25	216.03	nr	**239.27**
630 mm x 900 mm; ref LEW109VH	221.51	0.88	18.43	238.70	nr	**257.14**
630 mm x 1050 mm; ref LEW2110VH	230.40	1.20	25.13	248.27	nr	**273.40**
915 mm x 900 mm; ref LEWN09WH	272.25	1.39	29.11	293.41	nr	**322.53**
915 mm x 1050 mm; ref LEWN10WH	283.01	1.48	31.00	305.09	nr	**336.09**
915 mm x 1200 mm; ref LEWN12WH	295.48	1.57	32.88	318.51	nr	**351.39**
915 mm x 1350 mm; ref LEWN13WH	307.66	1.67	34.98	331.62	nr	**366.60**
915 mm x 1550 mm; ref LEWN15WH	339.21	1.76	36.86	365.67	nr	**402.53**
1200 mm x 900 mm; ref LEW209CH	294.57	1.57	32.88	317.34	nr	**350.22**
1200 mm x 900 mm; ref LEW209WH	307.83	1.57	32.88	331.61	nr	**364.49**
1200 mm x 1050 mm; ref LEW210CH	306.34	1.67	34.98	330.00	nr	**364.97**
1200 mm x 1050 mm; ref LEW210WH	322.32	1.67	34.98	347.20	nr	**382.18**
1200 mm x 1200 mm; ref LEW212CH	333.76	1.80	37.70	359.61	nr	**397.31**
1200 mm x 1200 mm; ref LEW212WH	337.07	1.80	37.70	363.18	nr	**400.88**
1200 mm x 1350 mm; ref LEW213WH	356.47	1.94	40.63	384.05	nr	**424.68**
1200 mm x 1550 mm; ref LEW215WH	397.35	2.04	42.73	428.04	nr	**470.77**

L WINDOWS/DOORS/STAIRS

Item	PC £	Labour hours	Labour £	Material £	Unit	Total rate £
L10 WINDOWS/ROOFLIGHTS/SCREENS/LOUVRES – cont'd						
Standard windows; Jeld-Wen Hardwood – cont'd						
1770 mm x 1050 mm; ref LEW310CCH	467.96	2.08	43.56	504.14	nr	**547.70**
1770 mm x 1200 mm ; ref LEW312CCH	498.37	2.22	46.50	536.87	nr	**583.36**
2339 mm x 1200 mm; ref LEW412CMCH	643.78	2.41	50.47	693.47	nr	**743.94**
Top hung casement windows; factory glazed with low E 24mm double glazing; 45 mm x 140 mm hardwood sills; weather stripping; opening sashes on canopy hinges; fitted with fasteners; brown finish ironmongery						
630 mm x 900 mm; ref LEW109AH	211.57	0.88	18.43	228.00	nr	**246.43**
630 mm x 1050 mm; ref LEW110AH	221.94	1.20	25.13	239.26	nr	**264.40**
915 mm x 900 mm; ref LEW2N09AH	273.46	1.39	29.11	294.61	nr	**323.72**
915 mm x 1050 mm; ref LEW2N10AH	288.86	1.48	31.00	311.28	nr	**342.28**
915 mm x 1350 mm; ref LEW2N13ASH	332.81	1.67	34.98	358.69	nr	**393.67**
1200 mm x 1050 mm; ref LEW210AH	329.54	1.57	32.88	355.07	nr	**387.95**
1200 mm x 1350 mm; ref LEW213ASH	398.48	1.67	34.98	429.27	nr	**464.24**
1770 mm x 1050 mm; ref LEW310AEH	434.03	1.80	37.70	467.62	nr	**505.32**
Purpose made double hung sash windows; "treated" wrought softwood						
Cased frames of 100 mm x 25 mm grooved inner linings; 114 mm x 25 mm grooved outer linings; 125 mm x 38 mm twice rebated head linings; 125 mm x 32 mm twice rebated grooved pulley stiles; 150 mm x 13 mm linings; 50 mm x 19 mm parting slips; 25 mm x 19 mm inside beads; 150 mm x 75 mm Oak twice sunk weathered throated sill; 50 mm thick rebated and moulded sashes; moulded horns						
over 1.25 m² each; both sashes in medium panes; including spiral spring balances	422.60	2.08	43.56	540.06	m²	**583.63**
As above but with cased mullions	480.35	2.31	48.38	602.22	m²	**650.60**
Purpose made double hung sash windows; selected Sapele						
Cased frames of 100 mm x 25 mm grooved inner linings; 114 mm x 25 mm grooved outer linings; 125 mm x 38 mm twice rebated head linings; 125 mm x 32 mm twice rebated grooved pulley stiles; 150 mm x 13 mm linings; 50 mm x 19 mm parting slips; 25 mm x 19 mm inside beads; 150 mm x 75 mm Oak twice sunk weathered throated sill; 50 mm thick rebated and moulded sashes; moulded horns						
over 1.25 m² each; both sashes in medium panes; including spiral sash balances	466.55	2.78	58.22	587.36	m²	**645.59**
As above but with cased mullions	500.62	3.08	64.51	624.03	m²	**688.54**
Clements 'EB24' range of factory finished steel fixed light; casement, fanlight windows and doors; with a U-value of 2.0 W/m²K (part L compliant); to EN ISO 9001 2000 ; polyester powder coated; factory glazed with low E double glazing; fixed in position; including lugs plugged and screwed						
Basic fixed light including easy-glaze snap-on beads						
508 mm x 292 mm	141.57	2.00	59.41	152.57	nr	**211.97**
508 mm x 457 mm	154.44	2.00	59.41	166.42	nr	**225.83**
508 mm x 628 mm	167.31	2.00	59.41	180.37	nr	**239.78**
508 mm x 923 mm	193.05	2.00	59.41	208.17	nr	**267.58**
508 mm x 1218 mm	218.79	2.50	74.26	235.97	nr	**310.24**

L WINDOWS/DOORS/STAIRS

Item	PC £	Labour hours	Labour £	Material £	Unit	Total rate £
Basic'Tilt and Turn' window; including easy-glaze ali snap-on beads						
508 mm x 292 mm	334.62	2.00	59.41	360.34	nr	**419.74**
508 mm x 457 mm	347.49	2.00	59.41	374.19	nr	**433.60**
508 mm x 628 mm	360.36	2.00	59.41	388.14	nr	**447.55**
508 mm x 923 mm; including fixed light	424.71	2.00	59.41	457.49	nr	**516.90**
508 mm x 1218 mm; including fixed light	450.45	2.50	74.26	485.30	nr	**559.56**
Basic casement; including easy-glaze snap-on beads						
508 mm x 628 mm	398.97	2.00	59.41	429.69	nr	**489.10**
508 mm x 923 mm	424.71	2.00	59.41	457.49	nr	**516.90**
508 mm x 1218 mm	450.45	2.50	74.26	485.30	nr	**559.56**
Double door						
1143 mm x 2057 mm	2522.52	3.50	103.97	2715.56	nr	**2819.53**
Extra over for						
pressed steel sills; to suit above windows	38.61	0.50	10.47	41.65	m	**52.13**
G + bar	77.22	-	-	83.11	m	**-**
simulated leaded light	77.22	-	-	83.11	m	**-**
uPVC windows; 'Profile 22' or other equal and approved; reinforced where appropriate with aluminium alloy; including standard ironmongery; cills and factory glazed with low E 24mm double glazing; fixed in position; including lugs plugged and screwed to brickwork or blockwork						
Casement/fixed light; including e.p.d.m. glazing gaskets and weather seals						
630 mm x 900 mm; ref P109C	62.05	2.50	74.26	67.18	nr	**141.44**
630 mm x 1200 mm; ref P112V	79.05	3.00	89.11	85.58	nr	**174.69**
1200 mm x 1200 mm; ref P212C	98.60	3.50	103.97	106.72	nr	**210.69**
1770 mm x 1200 mm; ref P312CC	153.85	4.00	118.82	166.18	nr	**285.00**
Casement/fixed light; including vents; e.p.d.m. glazing gaskets and weather seals						
630 mm x 900 mm; ref P109V	76.50	2.50	74.26	82.73	nr	**157.00**
630 mm x 1200 mm; ref P112C	70.55	3.00	89.11	76.43	nr	**165.54**
1200 mm x 1200 mm; ref P212W	129.20	3.50	103.97	139.65	nr	**243.62**
1200 mm x 1200 mm; ref P212CV	124.10	3.50	103.97	134.16	nr	**238.13**
1770 mm x 1200 mm; ref P312WW	184.45	4.00	118.82	199.12	nr	**317.93**
1770 mm x 1200 mm; ref P312CV	156.40	4.00	118.82	168.93	nr	**287.74**
uPVC windows; 'Profile 22' or other equal and approved; reinforced where appropriate with aluminium alloy; in refurbishment work, including standard ironmongery; cills & factory glazed with low E 24mm double glazing; removing existing windows and fixing new in position; including lugs plugged and screwed to brickwork or blockwork						
Casement/fixed light; including e.p.d.m. glazing gaskets and weather seals						
630 mm x 900 mm; ref P109C	62.05	5.00	148.52	67.18	nr	**215.70**
630 mm x 1200 mm; ref P112V	79.05	5.50	163.37	85.58	nr	**248.95**
1200 mm x 1200 mm; ref P212C	98.60	6.00	178.23	106.72	nr	**284.95**
1770 mm x 1200 mm; ref P312CC	153.85	6.50	193.08	166.18	nr	**359.26**
Casement/fixed light; including vents; e.p.d.m. glazing gaskets and weather seals						
630 mm x 900 mm; ref P109V	76.50	5.00	148.52	82.73	nr	**231.26**
630 mm x 1200 mm; ref P112C	70.55	5.50	163.37	76.43	nr	**239.81**
1200 mm x 1200 mm; ref P212W	129.20	6.00	178.23	139.65	nr	**317.88**
1200 mm x 1200 mm; ref P212CV	124.10	6.00	178.23	134.16	nr	**312.39**
1770 mm x 1200 mm; ref P312WW	184.45	6.50	193.08	199.12	nr	**392.19**
1770 mm x 1200 mm; ref P312CV	156.40	6.50	193.08	168.93	nr	**362.01**

L WINDOWS/DOORS/STAIRS

Item	PC £	Labour hours	Labour £	Material £	Unit	Total rate £
L10 WINDOWS/ROOFLIGHTS/SCREENS/LOUVRES – cont'd						
Aluminium windows; Schuco AWS 50 proprietary system or equal and approved						
Polyester powder coated solid colour matt finish or natural anodised window system of glass sealed units with 6.4mm low E coated laminated inner pane, air filled cavity and 6mm clear annealed outer pane. Rates to include all brackets, membranes, cills, silicone seals, trade contractor preliminaries, including external access equipment						
Ribbon construction windows 1.5 m high	-	-	-	-	m²	**450.00**
Punched hole windows fixing into prepared apertures by others	-	-	-	-	m²	**500.00**
Extra over for						
1.25 m wide x 1.5 m high opening vents, assuming tilt and turn operation high	-	-	-	-	m²	**150.00**
neutral selective high performance coating in lieu of low E, for assisting in solar control	-	-	-	-	m²	**40.00**
outer glass pane to be toughened and heat soak tested or heat strengthened in lieu of annealed	-	-	-	-	m²	**20.00**
inner laminated glass to be toughened and heat soak tested laminated, or heat strengthened laminated	-	-	-	-	m²	**40.00**
Rooflights, skylights, roof windows and frames; pre-glazed; "treated" Nordic Red Pine and aluminium trimmed "Velux" windows or other equal and approved; type U flashings and soakers (for tiles and pantiles), and sealed double glazing unit (trimming opening not included)						
Roof windows						
550 mm x 780 mm; ref GGL-3073-C02	168.75	1.85	38.75	181.95	nr	**220.70**
550 mm x 980 mm; ref GGL-3073-C04	176.25	2.08	43.56	190.02	nr	**233.59**
660 mm x 1180 mm; ref GGL-3073-F06	206.25	2.31	48.38	222.38	nr	**270.76**
780 mm x 980 mm; ref GGL-3073-M04	195.00	2.31	48.38	210.27	nr	**258.65**
780 mm x 1180 mm; ref GGL-3073-M06	225.19	2.78	58.22	242.96	nr	**301.18**
780 mm x 1400 mm; ref GGL-3073-M08	236.25	2.31	48.38	254.80	nr	**303.18**
940 mm x 1600 mm; ref GGL-3073-P10	281.25	2.78	58.22	303.30	nr	**361.52**
1140 mm x 1180 mm; ref GGL-3073-S06	262.50	2.78	58.22	283.12	nr	**341.34**
1340 mm x 980 mm; ref GGL-3073-U04	262.50	2.78	58.22	283.12	nr	**341.34**
Rooflights, skylights, roof windows and frames; uPVC; plugged and screwed to concrete; or screwed to timber						
Rooflight; Cox "Suntube" range; double skin polycarbonate dome						
230 mm dia.; for flat roof using felt or membrane	242.76	2.50	52.36	261.95	nr	**314.31**
230 mm dia.; for up to 30 degree pitch roof with standard tiles	242.76	3.00	62.83	261.83	nr	**324.66**
230 mm dia.; for up to 30 degree pitch roof with bold roll tiles	261.50	3.00	62.83	282.06	nr	**344.90**
Rooflight; Cox "Galaxy" range; double skin polycarbonate dome						
600 mm x 600 mm	109.78	1.50	31.42	118.52	nr	**149.94**
900 mm x 900 mm	202.60	1.75	36.65	218.54	nr	**255.19**
1200 mm x 1200 mm	310.59	2.00	41.89	334.89	nr	**376.78**

L WINDOWS/DOORS/STAIRS

Item	PC £	Labour hours	Labour £	Material £	Unit	Total rate £
Rooflight; Cox "Trade" range; double skin polycarbonate dome on 150mm PVCU upstand						
630 mm x 630 mm	240.97	2.00	41.89	259.78	nr	**301.67**
930 mm x 930 mm	379.31	2.25	47.12	408.79	nr	**455.92**
1230 mm x 1230 mm	537.28	2.50	52.36	578.94	nr	**631.30**
Rooflight; Cox "2000" range; double skin polycarbonate dome on 235mm PVCU upstand						
900 mm x900 mm	961.22	2.50	52.36	1034.95	nr	**1087.31**
1200 mm x 1200 mm	1320.01	3.00	62.83	1421.16	nr	**1483.99**
Louvres, Brise Soleils and frames; polyester powder coated aluminium; fixing in position including brackets						
Louvres; Levolux or other equal and approved; 5 rows of 400 aerofins set in steel plate frame						
6700 mm x 2200 mm (14.75 m² overall)	-	-	-	-	m²	**300.00**
Brise Soleil; Levolux or other equal and approved; on galvanised steel cantilever beams and runners						
1000 mm deep	-	-	-	-	m	**400.00**
L20 DOORS/SHUTTERS/HATCHES						
Doors; standard matchboarded; wrought softwood						
Matchboarded, framed, ledged and braced doors; 44 mm thick overall; 19 mm thick tongued, grooved and V-jointed boarding; one side vertical boarding						
762 mm x 1981mm	149.58	1.67	34.98	160.99	nr	**195.96**
838 mm x 1981mm	149.58	1.67	34.98	160.99	nr	**195.96**
Doors; standard flush; softwood composition						
Flush door; internal quality; skeleton or cellular core; hardboard faced both sides; lipped on two long edges; Jeld-Wen "Silverwood" or other equal and approved						
457 mm x 1981 mm x 35 mm	25.73	1.16	24.29	27.70	nr	**51.99**
533 mm x 1981 mm x 35 mm	25.73	1.16	24.29	27.70	nr	**51.99**
610 mm x 1981 mm x 35 mm	25.73	1.16	24.29	27.70	nr	**51.99**
686 mm x 1981 mm x 35 mm	25.73	1.16	24.29	27.70	nr	**51.99**
762 mm x 1981 mm x 35 mm	25.73	1.16	24.29	27.70	nr	**51.99**
838 mm x 1981 mm x 35 mm	26.89	1.16	24.29	28.95	nr	**53.24**
626 mm x 2040 mm x 40 mm	26.89	1.16	24.29	28.95	nr	**53.24**
726 mm x 2040 mm x 40 mm	26.89	1.16	24.29	28.95	nr	**53.24**
Flush door; internal quality; skeleton or cellular core; faced both sides; lipped on two long edges; Jeld-Wen "Paint grade veneer" or other equal and approved						
457 mm x 1981 mm x 35 mm	25.27	1.16	24.29	27.19	nr	**51.49**
533 mm x 1981 mm x 35 mm	25.27	1.16	24.29	27.19	nr	**51.49**
610 mm x 1981 mm x 35 mm	25.27	1.16	24.29	27.19	nr	**51.49**
686 mm x 1981 mm x 35 mm	25.27	1.16	24.29	27.19	nr	**51.49**
762 mm x 1981 mm x 35 mm	25.27	1.16	24.29	27.19	nr	**51.49**
838 mm x 1981 mm x 35 mm	27.89	1.16	24.29	30.02	nr	**54.32**
526 mm x 2040 mm x 40 mm	26.32	1.16	24.29	28.32	nr	**52.62**
626 mm x 2040 mm x 40 mm	26.32	1.16	24.29	28.32	nr	**52.62**
726 mm x 2040 mm x 40 mm	26.32	1.16	24.29	28.32	nr	**52.62**
826 mm x 2040 mm x 40 mm	30.00	1.16	24.29	32.29	nr	**56.58**

L WINDOWS/DOORS/STAIRS

Item	PC £	Labour hours	Labour £	Material £	Unit	Total rate £
L20 DOORS/SHUTTERS/HATCHES – cont'd						
Doors; standard flush; softwood composition – cont'd						
Flush door; internal quality; skeleton or cellular core; chipboard veneered; faced both sides; lipped on two long edges; Jeld-Wen "Sapele veneered" or other equal and approved						
457 mm x 1981 mm x 35 mm	40.90	1.25	26.18	44.01	nr	**70.19**
533 mm x 1981 mm x 35 mm	40.90	1.25	26.18	44.01	nr	**70.19**
610 mm x 1981 mm x 35 mm	40.90	1.25	26.18	44.01	nr	**70.19**
686 mm x 1981 mm x 35 mm	40.90	1.25	26.18	44.01	nr	**70.19**
762 mm x 1981 mm x 35 mm	40.90	1.25	26.18	44.01	nr	**70.19**
838 mm x 1981 mm x 35 mm	44.21	1.25	26.18	47.58	nr	**73.76**
526 mm x 2040 mm x 40 mm	42.58	1.25	26.18	45.83	nr	**72.01**
626 mm x 2040 mm x 40 mm	42.58	1.25	26.18	45.83	nr	**72.01**
726 mm x 2040 mm x 40 mm	42.58	1.25	26.18	45.83	nr	**72.01**
826 mm x 2040 mm x 40 mm	45.37	1.25	26.18	48.83	nr	**75.01**
Flush door; half-hour fire resisting (FD30); hardboard faced both sides; Jeld-Wen "Silverwood" or other equal and approved						
762 mm x 1981 mm x 44 mm	43.48	1.62	33.93	46.79	nr	**80.72**
838 mm x 1981 mm x 44 mm	45.21	1.62	33.93	48.66	nr	**82.59**
726 mm x 2040 mm x 44 mm	45.21	1.62	33.93	48.66	nr	**82.59**
826 mm x 2040 mm x 44 mm	45.79	1.62	33.93	49.28	nr	**83.21**
Flush door; half-hour fire resisting (FD30); chipboard veneered; faced both sides; lipped on two long edges; Jeld-Wen "Paint grade veneer" or other equal and approved						
610 mm x 1981 mm x 44 mm	35.79	1.62	33.93	38.52	nr	**72.45**
686 mm x 1981 mm x 44 mm	35.79	1.62	33.93	38.52	nr	**72.45**
762 mm x 1981 mm x 44 mm	35.79	1.62	33.93	38.52	nr	**72.45**
838 mm x 1981 mm x 44 mm	38.42	1.62	33.93	41.35	nr	**75.28**
526 mm x 2040 mm x 44 mm	36.84	1.62	33.93	39.65	nr	**73.58**
626 mm x 2040 mm x 44 mm	36.84	1.62	33.93	39.65	nr	**73.58**
726 mm x 2040 mm x 44 mm	36.84	1.62	33.93	39.65	nr	**73.58**
826 mm x 2040 mm x 44 mm	42.10	1.62	33.93	45.32	nr	**79.24**
Flush door; half-hour fire resisting (FD30); faced both sides; lipped on two long edges; Jeld-Wen "Sapele veneered" or other equal and approved						
610 mm x 1981 mm x 44 mm	57.73	1.71	35.81	62.14	nr	**97.95**
686 mm x 1981 mm x 44 mm	57.73	1.71	35.81	62.14	nr	**97.95**
762 mm x 1981 mm x 44 mm	57.73	1.71	35.81	62.14	nr	**97.95**
838 mm x 1981 mm x 44 mm	62.27	1.71	35.81	67.01	nr	**102.83**
726 mm x 2040 mm x 44 mm	59.42	1.71	35.81	63.95	nr	**99.76**
826 mm x 2040 mm x 44 mm	63.95	1.71	35.81	68.82	nr	**104.63**
Flush door; half-hour fire resisting; chipboard for painting; hardwood lipping two long edges; "Leaderflush" type B30 or other equal and approved						
526 mm x 2040 mm x 44 mm	103.32	1.71	35.81	111.20	nr	**147.01**
626 mm x 2040 mm x 44 mm	104.87	1.71	35.81	112.86	nr	**148.68**
726 mm x 2040 mm x 44 mm	105.92	1.71	35.81	114.00	nr	**149.81**
826 mm x 2040 mm x 44 mm	106.69	1.71	35.81	114.82	nr	**150.63**
Flush door; half-hour fire resisting; American light oak veneer; hardwood lipping all edges; "Leaderflush" type B30 or other equal and approved;						
526 mm x 2040 mm x 44 mm	120.11	1.71	35.81	129.26	nr	**165.08**
626 mm x 2040 mm x 44 mm	123.44	1.71	35.81	132.86	nr	**168.67**
726 mm x 2040 mm x 44 mm	125.76	1.71	35.81	135.35	nr	**171.16**
826 mm x 2040 mm x 44 mm	128.85	1.71	35.81	138.68	nr	**174.49**

L WINDOWS/DOORS/STAIRS

Item	PC £	Labour hours	Labour £	Material £	Unit	Total rate £
Flush door; one-hour fire resisting; Iroko veneer; hardwood lipping all edges; Leaderflush" type B60 or other equal and approved; including groove and "Leaderseal" intumescent strip						
457 mm x 1981 mm x 54 mm	153.25	1.94	40.63	164.94	nr	**205.57**
533 mm x 1981 mm x 54 mm	167.79	1.94	40.63	180.58	nr	**221.21**
610 mm x 1981 mm x 54 mm	170.29	1.94	40.63	183.27	nr	**223.90**
686 mm x 1981 mm x 54 mm	173.41	1.94	40.63	186.63	nr	**227.27**
762 mm x 1981 mm x 54 mm	176.81	1.94	40.63	190.30	nr	**230.93**
838 mm x 1981 mm x 54 mm	180.17	1.94	40.63	193.91	nr	**234.54**
526 mm x 2040 mm x 54 mm	167.95	1.94	40.63	180.76	nr	**221.39**
626 mm x 2040 mm x 54 mm	171.22	1.94	40.63	184.28	nr	**224.91**
726 mmx 2040 mm x 54 mm	176.31	1.94	40.63	189.75	nr	**230.38**
826 mm x 2040 mm x 54 mm	180.43	1.94	40.63	194.19	nr	**234.82**
Flush door; external quality; skeleton or cellular core; plywood faced both sides; lipped on all four edges						
762 mm x 1981 mm x 54 mm	35.70	1.62	33.93	38.42	nr	**72.35**
838 mm x 1981 mm x 54 mm	39.38	1.62	33.93	42.38	nr	**76.31**
Flush door; half-hour fire resisting; external quality with Georgian wired standard glass opening; skeleton or cellular core; plywood faced both sides; lipped on all four edges; including glazing beads						
762 mm x 1981 mm x 54 mm	200.03	1.62	33.93	215.28	nr	**249.21**
838 mm x 1981 mm x 54 mm	204.22	1.62	33.93	219.80	nr	**253.73**
Doors; purpose made panelled; wrought softwood						
Panelled doors; one open panel for glass; including glazing beads						
686 mm x 1981 mm x 44 mm	98.28	1.62	33.93	105.77	nr	**139.70**
762 mm x 1981 mm x 44 mm	99.10	1.62	33.93	106.66	nr	**140.58**
838 mm x 1981 mm x 44 mm	99.92	1.62	33.93	107.54	nr	**141.47**
Panelled doors; two open panel for glass; including glazing beads						
686 mm x 1981 mm x 44 mm	137.41	1.62	33.93	147.89	nr	**181.82**
762 mm x 1981 mm x 44 mm	138.60	1.62	33.93	149.17	nr	**183.10**
838 mm x 1981 mm x 44 mm	139.79	1.62	33.93	150.45	nr	**184.37**
Panelled doors; four 19 mm thick plywood panels; mouldings worked on solid both sides						
686 mm x 1981 mm x 44 mm	208.67	1.62	33.93	224.58	nr	**258.51**
762 mm x 1981 mm x 44 mm	211.37	1.62	33.93	227.48	nr	**261.41**
838 mm x 1981 mm x 44 mm	214.06	1.62	33.93	230.39	nr	**264.31**
Panelled doors; six 25 mm thick panels raised and fielded; mouldings worked on solid both sides						
686 mm x 1981 mm x 44 mm	383.95	1.94	40.63	413.23	nr	**453.86**
762 mm x 1981 mm x 44 mm	387.79	1.94	40.63	417.35	nr	**457.99**
838 mm x 1981 mm x 44 mm	391.62	1.94	40.63	421.48	nr	**462.11**
rebated edges beaded	-	-	-	2.49	m	**-**
rounded edges or heels	-	-	-	0.56	m	**-**
weatherboard fixed to bottom rail	-	0.23	4.82	8.83	m	**13.64**
stopped groove for weatherboard	-	-	-	2.83	m	**-**
Doors; purpose made panelled; selected Sapele						
Panelled doors; one open panel for glass; including glazing beads						
686 mm x 1981 mm x 44 mm	135.27	2.31	48.38	145.59	nr	**193.97**
762 mm x 1981 mm x 44 mm	137.10	2.31	48.38	147.55	nr	**195.93**
838 mm x 1981 mm x 44 mm	138.95	2.31	48.38	149.54	nr	**197.92**
686 mm x 1981 mm x 57 mm	144.59	2.54	53.20	155.62	nr	**208.82**
762 mm x 1981 mm x 57 mm	146.77	2.54	53.20	157.96	nr	**211.16**
838 mm x 1981 mm x 57 mm	148.94	2.54	53.20	160.29	nr	**213.49**

L WINDOWS/DOORS/STAIRS

Item	PC £	Labour hours	Labour £	Material £	Unit	Total rate £
L20 DOORS/SHUTTERS/HATCHES – cont'd						
Doors; purpose made panelled; selected Sapele – cont'd						
Panelled doors; 250 mm wide cross tongued intermediate rail; two open panels for glass; mouldings worked on the solid one side; 19 mm x 13 mm beads one side; fixing with brass cups and screws						
686 mm x 1981 mm x 44 mm	206.81	2.31	48.38	222.58	nr	**270.96**
762 mm x 1981 mm x 44 mm	210.30	2.31	48.38	226.33	nr	**274.71**
838 mm x 1981 mm x 44 mm	220.52	2.31	48.38	237.34	nr	**285.72**
686 mm x 1981 mm x 57 mm	220.52	2.54	53.20	237.34	nr	**290.54**
762 mm x 1981 mm x 57 mm	224.64	2.54	53.20	241.77	nr	**294.97**
838 mm x 1981 mm x 57 mm	228.88	2.54	53.20	246.34	nr	**299.53**
Panelled doors; four panels; (19 mm thick for 44 mm doors, 25 mm thick for 57 mm doors); mouldings worked on solid both sides						
686 mm x 1981 mm x 44 mm	290.59	2.31	48.38	312.75	nr	**361.13**
762 mm x 1981 mm x 44 mm	313.26	2.31	48.38	337.14	nr	**385.52**
838 mm x 1981 mm x 44 mm	328.75	2.31	48.38	353.81	nr	**402.19**
686 mm x 1981 mm x 57 mm	296.48	2.54	53.20	319.08	nr	**372.28**
762 mm x 1981 mm x 57 mm	321.01	2.54	53.20	345.48	nr	**398.68**
838 mm x 1981 mm x 57 mm	302.36	2.54	53.20	325.41	nr	**378.61**
Panelled doors; 150 mm wide stiles in one width; 430 mm wide cross tongued bottom rail; six panels raised and fielded one side; (19 mm thick for 44 mm doors, 25 mm thick for 57 mm doors); mouldings worked on solid both sides						
686 mm x 1981 mm x 44 mm	496.08	2.31	48.38	533.90	nr	**582.28**
762 mm x 1981 mm x 44 mm	547.45	2.31	48.38	589.19	nr	**637.57**
838 mm x 1981 mm x 44 mm	557.84	2.31	48.38	600.38	nr	**648.76**
686 mm x 1981 mm x 57 mm	528.25	2.54	53.20	568.53	nr	**621.73**
762 mm x 1981 mm x 57 mm	582.48	2.54	53.20	626.90	nr	**680.09**
838 mm x 1981 mm x 57 mm	595.74	2.54	53.20	641.16	nr	**694.36**
rebated edges beaded	-	-	-	3.22	m	**-**
rounded edges or heels	-	-	-	0.85	m	**-**
weatherboard fixed to bottom rail	-	0.31	6.49	12.13	m	**18.63**
stopped groove for weatherboard	-	-	-	2.99	m	**-**
Doors; galvanised steel "up and over" type garage doors; Catnic "Horizon 90" or other equal and approved; spring counter balanced; fixed to timber frame (not included)						
Garage door						
2135 mm x 1980 mm	200.85	3.70	77.49	216.57	nr	**294.06**
2135 mm x 2135 mm	230.10	3.70	77.49	248.05	nr	**325.54**
2400 mm x 2135 mm	280.80	3.70	77.49	302.68	nr	**380.17**
3965 mm x 2135 mm	735.15	5.55	116.24	792.14	nr	**908.38**

L WINDOWS/DOORS/STAIRS

Item	PC £	Labour hours	Labour £	Material £	Unit	Total rate £
Doorsets; Anti-Vandal Security door and frame units; Bastion Security Ltd or other equal and approved; to BS 5051; factory primed; fixing with frame anchors to masonry; cutting mortices						
46 mm thick insulated external door with birch grade plywood; sheet steel bonded into door core; 2 mm thick polyester coated laminate finish; hardwood lippings all edges; 95 mm x 65 mm hardwood frame; polyester coated standard ironmongery; weather stripping all round; low projecting aluminium threshold; plugging; screwing						
for 980 mm x 2100 mm structural opening; single door sets; panic bolt	-	-	-	-	nr	**1527.11**
for 1830 mm x 2100 mm structural opening; double door sets; panic bolt	-	-	-	-	nr	**2584.34**
Doorsets; galvanised steel IG "Weatherbeater Original" door and frame units or other equal and approved; treated softwood frame, primed hardwood sill; fixing in position; plugged and screwed to brickwork or blockwork						
Door and frame						
762 mm x 1981 mm; ref IGD01	130.19	2.78	58.22	140.51	nr	**198.74**
Doorsets; steel security door and frame; Hormann or other equal and approved; including ironmongery, weather seals and all necessary fixing accessories						
Horman ref E55-1 door set.						
To suit structural opening 1100 x 2105 mm; fire Rating 30 minutes; acoustic rating 38dB; including stainless steel ironmongery	-	-	-	-	nr	**1928.28**
To suit structural opening 2000 x 2105 mm; fire rating 30 minutes; acoustic rating 38dB; including stainless steel ironmongery	-	-	-	-	nr	**2864.88**
Doorsets; steel door and frame units; Jandor Architectural Ltd or other equal and approved; polyester powder coated; ironmongery						
Single action door set; "Metset MD01" doors and "Metset MF" frames						
900 mm x 2100 mm	-	-	-	-	nr	**2014.18**
pair 1800 mm x 2100 mm	-	-	-	-	nr	**2782.35**
Doorsets; steel bullet-resistant door and frame units; Wormald Doors or other equal and approved; Medite laquered panels; ironmongery						
Door and frame						
1000 mm x 2060mm overall; fixed to masonry	-	-	-	-	nr	**4302.44**
Rolling shutters and collapsible gates; steel counter shutters; Bolton Brady Ltd or other equal and approved; push-up, self-coiling; polyester power coated; fixing by bolting						
Shutters						
3000 mm x 1000 mm	-	-	-	-	nr	**1204.81**
4000 mm x 1000 mm; in two panels	-	-	-	-	nr	**2095.33**

L WINDOWS/DOORS/STAIRS

Item	PC £	Labour hours	Labour £	Material £	Unit	Total rate £
L20 DOORS/SHUTTERS/HATCHES – cont'd						
Rolling shutters and collapsible gates; galvanised steel; Bolton Brady Type 474 or other equal and approved; one hour fire resisting; self-coiling; activated by fusible link; fixing with bolts						
Rolling shutters and collapsible gates						
1000 mm x 2750 mm	-	-	-	-	nr	**1440.54**
1500 mm x 2750 mm	-	-	-	-	nr	**1512.57**
2400 mm x 2750 mm	-	-	-	-	nr	**1787.58**
Rolling shutters and collapsible gates; GRP vertically opening insulated panel shutter doors; electrically operated; Envirodoor Markus Ltd; type HT40 or other equal and approved; manual over-ride, lock interlock, stop and return safety cage, "deadmans" down button, anti-flip device, photo-electric cell and beam deflectors; fixing by bolting						
Shutter doors						
3000 mm x 3000 mm; windows - 2 nr	-	-	-	-	nr	**8278.84**
4000 mm x 4000 mm; windows - 2 nr	-	-	-	-	nr	**10236.67**
5000 mm x 5000 mm; windows - 2 nr	-	-	-	-	nr	**12194.50**
Sliding/folding partitions; aluminium double glazed sliding patio doors; Crittal "Luminaire" or equal and approved; white acrylic finish; with and including 18 thick annealed double glazing; fixed in position; including lugs plugged and screwed to brickwork or blockwork						
Patio doors						
1800 mm x 2100 mm; ref PF1821	1660.45	2.31	48.38	1788.26	nr	**1836.64**
2400 m x 2100 mm; ref PF2421	1992.54	2.78	58.22	2145.68	nr	**2203.90**
2700 mm x 2100 mm; ref PF2721	2213.94	3.24	67.86	2383.95	nr	**2451.81**
Grilles; "Galaxy" nylon rolling counter grille or other equal and approved; Bolton Brady Ltd; colour, off-white; self-coiling; fixing by bolting						
Grilles						
3000 mm x 1000 mm	-	-	-	-	nr	**928.23**
4000 mm x 1000 mm	-	-	-	-	nr	**1397.95**
Sliding/folding partitions; Alco Beldan Ltd or equal and approved						
Sliding/folding partitions						
ref. NW100 Moveable Wall; 5000 mm (wide) x 2495 mm (high)l comrpising 4 nr. 954 mm (wide) standard panels and 1 nr. 954 mm (wide) telescopic panel; sealing; fixing	-	-	-	-	nr	**8932.88**
External softwood door frame composite standard joinery sets						
External door frame composite set; 56 mm x 78mm wide (finished); for external doors						
762 mm x 1981 mm x 44mm	30.30	0.75	15.71	33.01	nr	**48.72**
813 mm x 1981 mm x 44mm	30.52	0.75	15.71	33.25	nr	**48.96**
838 mm x 1981 mm x 44mm	30.52	0.75	15.71	33.25	nr	**48.96**

L WINDOWS/DOORS/STAIRS

Item	PC £	Labour hours	Labour £	Material £	Unit	Total rate £
External door frame composite set; 56 mm x 78mm wide (finished); with 45 mm x 140 mm (finished) hardwood cill; for external doors						
686 mm x 1981 mm x 44mm	44.88	1.00	20.94	48.70	nr	**69.65**
762 mm x 1981 mm x 44mm	46.29	1.00	20.94	50.22	nr	**71.16**
838 mm x 1981 mm x 44mm	48.15	1.00	20.94	52.22	nr	**73.16**
826 mm x 2040 mm x 44mm	49.60	1.00	20.94	53.79	nr	**74.73**
Internal white foiled moisture-resistant MDF door lining composite standard joinery set						
22 mm x 77mm wide (finished) set; with loose stops; for internal doors						
610 mm x 1981 mm x 35mm	13.11	0.70	14.66	14.51	nr	**29.17**
686 mm x 1981 mm x 35mm	13.09	0.70	14.66	14.49	nr	**29.15**
762 mm x 1981 mm x 35mm	13.09	0.70	14.66	14.49	nr	**29.15**
838 mm x 1981 mm x 35mm	17.80	0.70	14.66	19.56	nr	**34.22**
864 mm x 1981 mm x 35mm	13.71	0.70	14.66	15.16	nr	**29.82**
22 mm x 150mm wide (finished) set; with loose stops; for internal doors						
610 mm x 1981 mm x 35mm	16.98	0.70	14.66	18.67	nr	**33.33**
686 mm x 1981 mm x 35mm	16.98	0.70	14.66	18.67	nr	**33.33**
762 mm x 1981 mm x 35mm	16.98	0.70	14.66	18.67	nr	**33.33**
838 mm x 1981 mm x 35mm	17.80	0.70	14.66	19.56	nr	**34.22**
864 mm x 1981 mm x 35mm	17.80	0.70	14.66	19.56	nr	**34.22**
Door frames and door linings, sets; purpose made; wrought softwood						
Jambs and heads; as linings						
32 mm x 63 mm	-	0.16	3.35	6.24	m	**9.59**
32 mm x 100 mm	-	0.16	3.35	7.02	m	**10.37**
32 mm x 140 mm	-	0.16	3.35	7.52	m	**10.87**
Jambs and heads; as frames; rebated, rounded and grooved						
44 mm x 75 mm	-	0.16	3.35	10.00	m	**13.35**
44 mm x 100 mm	-	0.16	3.35	10.76	m	**14.11**
44 mm x 115 mm	-	0.16	3.35	10.86	m	**14.21**
44 mm x 140 mm	-	0.19	3.98	11.38	m	**15.36**
57 mm x 100 mm	-	0.19	3.98	11.54	m	**15.52**
57 mm x 125 mm	-	0.19	3.98	12.18	m	**16.16**
69 mm x 88 mm	-	0.19	3.98	11.70	m	**15.68**
69 mm x 100 mm	-	0.19	3.98	12.57	m	**16.55**
69 mm x 125 mm	-	0.20	4.19	13.34	m	**17.53**
69 mm x 150 mm	-	0.20	4.19	14.13	m	**18.31**
94 mm x 100 mm	-	0.23	4.82	19.02	m	**23.84**
94 mm x 150 mm	-	0.23	4.82	22.16	m	**26.97**
Mullions and transoms; in linings						
32 mm x 63 mm	-	0.11	2.30	8.25	m	**10.55**
32 mm x 100 mm	-	0.11	2.30	9.03	m	**11.33**
32 mm x 140 mm	-	0.11	2.30	9.42	m	**11.72**
Mullions and transoms; in frames; twice rebated, rounded and grooved						
44 mm x 75 mm	-	0.11	2.30	12.50	m	**14.80**
44 mm x 100 mm	-	0.11	2.30	13.02	m	**15.32**
44 mm x 115 mm	-	0.11	2.30	13.02	m	**15.32**
44 mm x 140 mm	-	0.13	2.72	13.53	m	**16.26**
57 mm x 100 mm	-	0.13	2.72	13.69	m	**16.42**
57 mm x 125 mm	-	0.13	2.72	14.34	m	**17.06**
69 mm x 88 mm	-	0.13	2.72	13.53	m	**16.26**
69 mm x 100 mm	-	0.13	2.72	14.32	m	**17.04**
Add 5% to the above material prices for staining						

Item	PC £	Labour hours	Labour £	Material £	Unit	Total rate £
L20 DOORS/SHUTTERS/HATCHES – cont'd						
Door frames and door linings, sets; purpose made; medium density fireboard						
Jambs and heads; as linings						
18 mm x 126 mm	-	0.16	3.35	6.94	m	**10.29**
22 mm x 126 mm	-	0.16	3.35	7.20	m	**10.55**
25mm x 126 mm	-	0.16	3.35	7.38	m	**10.73**
Door frames and door linings, sets; purpose made; selected Sapele						
Jambs and heads; as linings						
32 mm x 63 mm	9.39	0.21	4.40	10.21	m	**14.61**
32 mm x 100 mm	11.60	0.21	4.40	12.59	m	**16.99**
32 mm x 140 mm	12.70	0.21	4.40	13.87	m	**18.27**
Jambs and heads; as frames; rebated, rounded and grooved						
44 mm x 75 mm	15.38	0.21	4.40	16.66	m	**21.06**
44 mm x 100 mm	17.53	0.21	4.40	18.97	m	**23.36**
44 mm x 115 mm	18.12	0.21	4.40	19.70	m	**24.10**
44 mm x 140 mm	18.99	0.25	5.24	20.63	m	**25.87**
57 mm x 100 mm	19.41	0.25	5.24	21.10	m	**26.33**
57 mm x 125 mm	21.24	0.25	5.24	23.06	m	**28.30**
69 mm x 88 mm	19.39	0.25	5.24	20.97	m	**26.21**
69 mm x 100 mm	21.57	0.25	5.24	23.42	m	**28.65**
69 mm x 125 mm	23.75	0.28	5.86	25.76	m	**31.62**
69 mm x 150 mm	25.93	0.28	5.86	28.10	m	**33.97**
94 mm x 100 mm	30.28	0.28	5.86	32.79	m	**38.66**
94 mm x 150 mm	37.15	0.28	5.86	40.18	m	**46.04**
Mullions and transoms; in linings						
32 mm x 63 mm	11.78	0.15	3.14	12.68	m	**15.82**
32 mm x 100 mm	13.99	0.15	3.14	15.05	m	**18.19**
32 mm x 140 mm	15.09	0.15	3.14	16.24	m	**19.38**
Mullions and transoms; in frames; twice rebated, rounded and grooved						
44 mm x 75 mm	18.94	0.15	3.14	20.38	m	**23.53**
44 mm x 100 mm	20.39	0.15	3.14	21.94	m	**25.08**
44 mm x 115 mm	20.97	0.15	3.14	22.57	m	**25.71**
44 mm x 140 mm	21.84	0.17	3.56	23.51	m	**27.07**
57 mm x 100 mm	22.28	0.17	3.56	23.98	m	**27.54**
57 mm x 125 mm	24.11	0.17	3.56	25.95	m	**29.51**
69 mm x 88 mm	21.84	0.17	3.56	23.51	m	**27.07**
69 mm x 100 mm	24.58	0.17	3.56	26.46	m	**30.02**
Sills; once sunk weathered; once rebated, three times grooved						
63 mm x 175 mm	54.41	0.31	6.49	58.55	m	**65.05**
75 mm x 125 mm	52.41	0.31	6.49	56.40	m	**62.90**
75 mm x 150 mm	54.88	0.31	6.49	59.06	m	**65.56**
Door frames and door linings, sets; European Oak						
Sills; once sunk weathered; once rebated, three times grooved						
63 mm x 175 mm	93.60	0.31	6.49	100.74	m	**107.23**
75 mm x 125 mm	92.41	0.31	6.49	99.46	m	**105.95**
75 mm x 150 mm	101.03	0.31	6.49	108.74	m	**115.23**

L WINDOWS/DOORS/STAIRS

Item	PC £	Labour hours	Labour £	Material £	Unit	Total rate £
Bedding and pointing frames						
Pointing wood frames or sills with mastic						
one side	-	0.09	1.38	0.56	m	**1.94**
both sides	-	0.19	2.92	1.12	m	**4.04**
Pointing wood frames or sills with polysulphide sealant						
one side	-	0.09	1.38	1.77	m	**3.15**
both sides	-	0.19	2.92	3.54	m	**6.46**
Bedding wood frames in cement mortar (1:3) and point						
one side	-	0.07	1.83	0.07	m	**1.90**
both sides	-	0.09	2.36	0.09	m	**2.45**
one side in mortar; other side in mastic	-	0.19	4.00	0.63	m	**4.63**
L30 STAIRS/WALKWAYS/BALUSTRADES						
Standard staircases; wrought softwood (parana pine)						
Stairs; 25 mm thick treads with rounded nosings; 9 mm thick plywood risers; 32 mm thick strings; bullnose bottom tread; 50 mm x 75 mm hardwood handrail; 32 mm square plain balusters; 100 mm square plain newel posts						
straight flight; 838 mm wide; 2676 mm going; 2600 mm rise; with two newel posts	-	6.48	135.72	347.57	nr	**483.28**
straight flight with turn; 838 mm wide; 2676 mm going; 2600 mm rise; with two newel posts; three top treads winding	-	6.48	135.72	453.47	nr	**589.18**
dogleg staircase; 838 mm wide; 2676 mm going; 2600 mm rise; with two newel posts; quarter space landing third riser from top	-	6.48	135.72	425.06	nr	**560.77**
dogleg staircase; 838 mm wide; 2676 mm going; 2600 mm rise; with two newel posts; half space landing third riser from top	-	7.40	154.98	523.21	nr	**678.19**
Standard balustrades; wrought softwood						
Landing balustrade; 50 mm x 75 mm hardwood handrail; 32 mm square plain balusters; one end of handrail jointed to newel post; other end built into wall; balusters housed in at bottom (newel post and mortices both not included)						
3.00 m long	-	3.70	77.49	87.82	nr	**165.31**
Hardwood staircases; purpose made; assembled at works						
Fixing only complete staircase including landings, balustrades, etc.						
plugging and screwing to brickwork or blockwork	-	13.88	290.70	2.45	nr	**293.15**
The following are supply only prices for purpose made staircase components in selected Sapele supplied as part of an assembled staircase and may be used to arrive at a guide price for a complete hardwood staircase						
Board landings; cross-tongued joints; 100 mm x 50 mm sawn softwood bearers						
25 mm thick	-	-	-	109.76	m²	-
32 mm thick	-	-	-	123.38	m²	-

L WINDOWS/DOORS/STAIRS

Item	PC £	Labour hours	Labour £	Material £	Unit	Total rate £
L30 STAIRS/WALKWAYS/BALUSTRADES – cont'd						
Supply only prices for purpose made staircase components in selected Sapele – cont'd						
Treads; cross-tongued joints and risers; rounded nosings; tongued, grooved, glued and blocked together; one 175 mm x 50 mm sawn softwood carriage						
25 mm treads; 19 mm risers	-	-	-	220.59	m²	-
ends; quadrant	-	-	-	67.19	nr	-
ends; housed to hardwood	-	-	-	1.24	nr	-
32 mm treads; 25 mm risers	-	-	-	228.52	m²	-
ends; quadrant	-	-	-	86.38	nr	-
ends; housed to hardwood	-	-	-	1.24	nr	-
Winders; cross-tongued joints and risers in one width; rounded nosings; tongued, grooved glued and blocked together; one 175 mm x 50 mm sawn softwood carriage						
25 mm treads; 19 mm risers	-	-	-	306.65	m²	-
32 mm treads; 25 mm risers	-	-	-	313.86	m²	-
wide ends; housed to hardwood	-	-	-	2.47	nr	-
narrow ends; housed to hardwood	-	-	-	1.86	nr	-
Closed strings; in one width; 230 mm wide; rounded twice						
32 mm thick	-	-	-	40.39	m	-
38 mm thick	-	-	-	43.98	m	-
50 mm thick	-	-	-	48.99	m	-
Closed strings; cross-tongued joints; 280 mm wide; once rounded						
32 mm thick	-	-	-	52.43	m	-
extra for short ramp	-	-	-	26.93	nr	-
38 mm thick	-	-	-	57.19	m	-
extra for short ramp	-	-	-	30.61	nr	-
50 mm thick	-	-	-	63.82	m	-
extra for short ramp	-	-	-	37.91	nr	-
The following labours are irrespective of timber width						
ends; fitted	-	-	-	1.60	nr	-
ends; framed	-	-	-	9.38	nr	-
extra for tongued heading joint	-	-	-	4.63	nr	-
Closed strings; ramped; crossed tongued joints 280 mm wide; once rounded						
32 mm thick	-	-	-	52.43	m	-
44 mm thick	-	-	-	57.19	m	-
57 mm thick	-	-	-	63.82	m	-
Apron linings; in one width 230 mm wide						
19 mm thick	-	-	-	13.93	m	-
25 mm thick	-	-	-	16.42	m	-
Handrails; rounded						
40 mm x 50 mm	-	-	-	15.14	m	-
50 mm x 75 mm	-	-	-	18.25	m	-
57 mm x 87 mm	-	-	-	21.42	m	-
69 mm x 100 mm	-	-	-	26.62	m	-
Handrails; moulded						
40 mm x 50 mm	-	-	-	16.84	m	-
50 mm x 75 mm	-	-	-	19.95	m	-
57 mm x 87 mm	-	-	-	23.13	m	-
69 mm x 100 mm	-	-	-	28.31	m	-

L WINDOWS/DOORS/STAIRS

Item	PC £	Labour hours	Labour £	Material £	Unit	**Total rate £**
Add to above for						
grooved once	-	-	-	0.75	m	-
ends; framed	-	-	-	7.08	nr	-
ends; framed on rake	-	-	-	8.69	nr	-
Heading joints to handrail; mitred or raked						
overall size not exceeding 50 mm x 75 mm	-	-	-	34.75	nr	-
overall size not exceeding 69 mm x 100 mm	-	-	-	43.44	nr	-
Knee piece to handrail; mitred or raked						
overall size not exceeding 69 mm x 100 mm	-	-	-	92.67	nr	-
Balusters; stiffeners						
25 mm x 25 mm	-	-	-	3.87	m	-
32 mm x 32 mm	-	-	-	4.41	m	-
44 mm x 44 mm	-	-	-	5.79	m	-
ends; housed	-	-	-	1.74	nr	-
Sub rails						
32 mm x 63 mm	-	-	-	8.93	m	-
ends; framed joint to newel	-	-	-	7.53	nr	-
Knee rails						
32 mm x 140 mm	-	-	-	14.81	m	-
ends; framed joint to newel	-	-	-	7.53	nr	-
Newel posts						
44 mm x 94 mm; half newel	-	-	-	10.29	m	-
69 mm x 69 mm	-	-	-	11.15	m	-
94 mm x 94 mm	-	-	-	22.81	m	-
Newel caps; splayed on four sides						
62.50 mm x 125 mm x 50 mm	-	-	-	10.88	nr	-
100 mm x 100 mm x 50 mm	-	-	-	11.11	nr	-
125 mm x 125 mm x 50 mm	-	-	-	11.66	nr	-
The following are supply only prices for purpose made staircase components in selected American Oak; supplied as part of an assembled staircase						
Board landings; cross-tongued joints; 100 mm x 50 mm sawn softwood bearers						
25 mm thick	-	-	-	173.78	m²	-
32 mm thick	-	-	-	209.84	m²	-
Treads; cross-tongued joints and risers; rounded nosings; tongued, grooved, glued and blocked together; one 175 mm x 50 mm sawn softwood carriage						
25 mm treads; 19 mm risers	-	-	-	288.48	m²	-
ends; quadrant	-	-	-	144.11	nr	-
ends; housed to hardwood	-	-	-	1.75	nr	-
32 mm treads; 25 mm risers	-	-	-	330.56	m²	-
ends; quadrant	-	-	-	177.40	nr	-
ends; housed to hardwood	-	-	-	1.75	nr	-
Winders; cross-tongued joints and risers in one width; rounded nosings; tongued, grooved glued and blocked together; one 175 mm x 50 mm sawn softwood carriage						
25 mm treads; 19 mm risers	-	-	-	365.69	m²	-
32 mm treads; 25 mm risers	-	-	-	399.00	m²	-
wide ends; housed to hardwood	-	-	-	3.56	nr	-
narrow ends; housed to hardwood	-	-	-	2.66	nr	-
Closed strings; in one width; 230 mm wide; rounded twice						
32 mm thick	-	-	-	68.14	m	-
44 mm thick	-	-	-	78.66	m	-
57 mm thick	-	-	-	108.00	m	-

L WINDOWS/DOORS/STAIRS

Item	PC £	Labour hours	Labour £	Material £	Unit	Total rate £
L30 STAIRS/WALKWAYS/BALUSTRADES – cont'd						
Supply only prices for purpose made staircase components in selected American Oak – cont'd						
Closed strings; cross-tongued joints; 280 mm wide; once rounded						
32 mm thick	-	-	-	86.59	m	-
extra for short ramp	-	-	-	49.51	nr	-
38 mm thick	-	-	-	100.35	m	-
extra for short ramp	-	-	-	56.40	nr	-
50 mm thick	-	-	-	137.60	m	-
extra for short ramp	-	-	-	75.02	nr	-
Closed strings; ramped; crossed tongued joints 280 mm wide; once rounded						
32 mm thick	-	-	-	99.58	m	-
44 mm thick	-	-	-	115.42	m	-
57 mm thick	-	-	-	158.23	m	-
Apron linings; in one width 230 mm wide						
19 mm thick	-	-	-	23.21	m	-
25 mm thick	-	-	-	28.44	m	-
Handrails; rounded						
40 mm x 50 mm	-	-	-	18.81	m	-
50 mm x 75 mm	-	-	-	24.10	m	-
57 mm x 87 mm	-	-	-	36.19	m	-
69 mm x 100 mm	-	-	-	48.72	m	-
Handrails; moulded						
40 mm x 50 mm	-	-	-	20.63	m	-
50 mm x 75 mm	-	-	-	25.93	m	-
57 mm x 87 mm	-	-	-	38.01	m	-
69 mm x 100 mm	-	-	-	50.53	m	-
Add to above for						
grooved once	-	-	-	0.93	m	-
ends; framed	-	-	-	9.32	nr	-
ends; framed on rake	-	-	-	11.81	nr	-
Heading joints to handrail; mitred or raked						
overall size not exceeding 50 mm x 75 mm	-	-	-	49.72	nr	-
overall size not exceeding 69 mm x 100 mm	-	-	-	59.05	nr	-
Knee piece to handrail; mitred or raked						
overall size not exceeding 69 mm x 100 mm	-	-	-	105.66	nr	-
Balusters; stiffeners						
25 mm x 25 mm	-	-	-	4.29	m	-
32 mm x 32 mm	-	-	-	5.41	m	-
44 mm x 44 mm	-	-	-	8.54	m	-
ends; housed	-	-	-	2.18	nr	-
Sub rails						
32 mm x 63 mm	-	-	-	11.57	m	-
ends; framed joint to newel	-	-	-	9.32	nr	-
Knee rails						
32 mm x 140 mm	-	-	-	20.11	m	-
ends; framed joint to newel	-	-	-	9.32	nr	-
Newel posts						
44 mm x 94 mm; half newel	-	-	-	15.09	m	-
69 mm x 69 mm	-	-	-	25.71	m	-
94 mm x 94 mm	-	-	-	64.17	m	-
Newel caps; splayed on four sides						
62.50 mm x 125 mm x 50 mm	-	-	-	12.94	nr	-
100 mm x 100 mm x 50 mm	-	-	-	13.67	nr	-
125 mm x 125 mm x 50 mm	-	-	-	15.03	nr	-

L WINDOWS/DOORS/STAIRS

Item	PC £	Labour hours	Labour £	Material £	Unit	Total rate £
Spiral staircases, balustrades and handrails; mild steel; galvanised and polyester powder coated						
Staircase						
2080 mm diameter x 3695 mm high; 18 nr treads; 16 mm diameter intermediate balusters; 1040 mm x 1350 mm landing unit with matching balustrade both sides; fixing with 16 mm diameter resin anchors to masonry at landing and with 12 mm diameter expanding bolts to concrete at base	-	-	-	-	nr	**6038.41**
Aluminium alloy folding loft ladders; "Zig Zag" stairways, model B or other equal and approved; Light Alloy Ltd; on and including plywood backboard; fixing with screws to timber lining (not included)						
Loft ladders						
ceiling height not exceeding 2500 mm; model 888801	-	0.93	19.48	419.51	nr	**438.99**
ceiling height not exceeding 2750 mm; model 888802	-	0.93	19.48	459.97	nr	**479.45**
ceiling height not exceeding 3000 mm; model 888803	-	0.93	19.48	513.36	nr	**532.83**
ceiling height not exceeding 3250 mm; model 888804	-	0.93	19.48	539.19	nr	**558.66**
Access ladders; mild steel						
Ladders						
400 mm wide; 3850 mm long (overall); 12 mm diameter rungs; 65 mm x 15 mm strings; 50 mm x 5 mm safety hoops; fixing with expanded bolts; to masonry; mortices; welded fabrication	-	-	-	-	nr	**1383.75**
Flooring, balustrades and handrails; metalwork						
Chequer plate flooring; galvanised mild steel; over 300 mm wide; bolted to steel supports						
6 mm thick	-	-	-	-	m²	**276.75**
8 mm thick	-	-	-	-	m²	**295.20**
Open mesh flooring; galvanised; over 300 mm wide; bolted to steel supports						
8 mm thick	-	-	-	-	m²	**276.75**
Balustrades; galvanised mild steel CHS posts and top rail, with one infill rail						
1100 mm high	-	-	-	-	m	**230.63**
Balustrades; painted mild steel flat bar posts and CHS top rail, with 3 nr. stainless steel infills						
1100 mm high	-	-	-	-	m	**322.88**
Balustrades; stainless steel flat bar posts and circular handrail, with 3 nr. stainless steel infills						
1100 mm high	-	-	-	-	m	**387.45**
Balustrades; stainless steel 50 mm Ø posts and circular handrail, with 10 mm thick toughened glass infill panels						
1100 mm high	-	-	-	-	m	**825.13**
Balustrades; laminated glass; with stainless steel cap channel to top and including all necessary support fixings						
1100 mm high	-	-	-	-	m	**1355.56**

L WINDOWS/DOORS/STAIRS

Item	PC £	Labour hours	Labour £	Material £	Unit	Total rate £
L30 STAIRS/WALKWAYS/BALUSTRADES – cont'd						
Flooring, balustrades and handrails – cont'd						
Wallrails; painted mild steel CHS wall rail; with wall rose bracket						
42 mm diameter	-	-	-	-	m	**92.25**
Wallrails; stainless steel circular wall rail; with wall rose bracket						
42 mm diameter	-	-	-	-	m	**129.15**
Surface treatment						
At works						
galvanising	-	-	-	-	tonne	**415.13**
shotblasting	-	-	-	-	m²	**3.69**
touch up primer and one coat of two pack epoxy zinc phosphate or chromate primer	-	-	-	-	m²	**7.38**
L40 GENERAL GLAZING						
BASIC GLASS PRICES (£/m²)						
Prices include discounts						
Ordinary transluscent/patterned glass						
4 mm	-	-	-	16.93	m²	-
5 mm	-	-	-	20.59	m²	-
6 mm	-	-	-	22.59	m²	-
Obscured ground sheet glass - patterned						
4 mm white	-	-	-	23.93	m²	-
6 mm white	-	-	-	26.33	m²	-
Rough cast						
6 mm	-	-	-	19.85	m²	-
Ordinary Georgian wired						
7 mm cast	-	-	-	20.17	m²	-
6 mm polish	-	-	-	31.31	m²	-
"Cetuff" toughened; float						
4 mm	-	-	-	18.72	m²	-
5 mm	-	-	-	24.83	m²	-
6 mm	-	-	-	27.34	m²	-
10 mm	-	-	-	44.89	m²	-
Clear laminated; safety						
4.40 mm	-	-	-	26.67	m²	-
6.40 mm	-	-	-	31.84	m²	-
SUPPLY AND FIX PRICES						
NOTE: The following "measured rates" are provided by a glazing sub-contractor and assume in excess of 500 m², within 20 miles of the suppliers branch.						
Standard plain glass; BS EN 14449; clear float; panes area 0.15 m² - 4.00 m²						
3 mm thick; glazed with screwed beads	-	-	-	-	m²	**28.90**
4 mm thick; glazed with screwed beads	-	-	-	-	m²	**30.64**
5 mm thick; glazed with screwed beads	-	-	-	-	m²	**37.32**
6 mm thick; glazed with screwed beads	-	-	-	-	m²	**40.90**

L WINDOWS/DOORS/STAIRS

Item	PC £	Labour hours	Labour £	Material £	Unit	Total rate £
Standard plain glass; BS EN 14449; obscure patterned; panes area 0.15 m² - 4.00 m²						
4 mm thick; glazed with screwed beads	-	-	-	-	m²	**43.34**
6 mm thick; glazed with screwed beads	-	-	-	-	m²	**47.69**
Standard plain glass; BS EN 14449; rough cast; panes area 0.15 m² - 4.00 m²						
6 mm thick; glazed with screwed beads	-	-	-	-	m²	**34.65**
Standard plain glass; BS EN 14449; Georgian wired cast; panes area 0.15 m² - 4.00 m²						
7 mm thick; glazed with screwed beads	-	-	-	-	m²	**35.25**
Extra for lining up wired glass	-	-	-	-	m²	**3.52**
Standard plain glass; BS EN 14449; Georgian wired polished; panes area 0.15 m² - 4.00 m²						
6 mm thick; glazed with screwed beads	-	-	-	-	m²	**54.69**
Extra for lining up wired glass	-	-	-	-	m²	**3.52**
Special glass; BS EN 14449; toughened clear float; panes area 0.15 m² - 4.00 m²						
4 mm thick; glazed with screwed beads	-	-	-	-	m²	**30.73**
5 mm thick; glazed with screwed beads	-	-	-	-	m²	**40.82**
6 mm thick; glazed with screwed beads	-	-	-	-	m²	**44.90**
10 mm thick; glazed with screwed beads	-	-	-	-	m²	**74.51**
Special glass; BS EN 14449; clear laminated safety glass; panes area 0.15 m² - 4.00m²						
4.40 mm thick; glazed with screwed beads	-	-	-	-	m²	**49.23**
6.40 mm thick; glazed with screwed beads	-	-	-	-	m²	**58.90**
Special glass; BS EN 14449; "Pyran" half-hour fire resisting glass or other equal and approved						
6.50 mm thick rectangular panes; glazed with screwed hardwood beads and Sealmaster "Fireglaze" intumescent compound or other equal and approved to rebated frame						
300 mm x 400 mm pane	-	0.37	10.99	46.82	nr	**57.81**
400 mm x 800 mm pane	-	0.46	13.66	117.94	nr	**131.60**
500 mm x 1400 mm pane	-	0.74	21.98	250.83	nr	**272.81**
600 mm x 1800 mm pane	-	0.93	27.63	411.17	nr	**438.80**

L WINDOWS/DOORS/STAIRS

Item	PC £	Labour hours	Labour £	Material £	Unit	Total rate £
L40 GENERAL GLAZING – cont'd						
Special glass; BS EN 14449; "Pyrostop" one-hour fire resisting glass or other equal and approved						
15 mm thick regular panes; glazed with screwed hardwood beads and Sealmaster "Fireglaze" intumescent liner and compound or other equal and approved both sides						
300 mm x 400 mm pane	-	1.11	32.97	96.91	nr	**129.88**
400 mm x 800 mm pane	-	1.39	41.29	192.15	nr	**233.44**
500 mm x 1400 mm pane	-	1.85	54.95	387.14	nr	**442.09**
600 mm x 1800 mm pane	-	2.31	68.62	570.86	nr	**639.48**
Special glass; BS EN 14449; clear laminated security glass						
7.50 mm thick regular panes; glazed with screwed hardwood beads and Intergens intumescent strip						
300 mm x 400 mm pane	-	0.37	10.99	29.23	nr	**40.22**
400 mm x 800 mm pane	-	0.46	13.66	71.75	nr	**85.42**
500 mm x 1400 mm pane	-	0.74	21.98	150.03	nr	**172.01**
600 mm x 1800 mm pane	-	0.93	27.63	254.18	nr	**281.18**
Mirror panels; BS EN 14449; silvered; insulation backing						
4 mm thick float; fixing with adhesive						
1000 mm x 1000 mm	-	-	-	-	nr	**34.84**
1000 mm x 2000 mm	-	-	-	-	nr	**69.74**
1000 mm x 4000 mm	-	-	-	-	nr	**254.18**
Glass louvres; BS EN 14449; with long edges ground or smooth						
6 mm thick float						
150 mm wide	-	-	-	-	m	**15.37**
7 mm thick Georgian wired cast						
150 mm wide	-	-	-	-	m	**21.33**
6 mm thick Georgian wire polished						
150 mm wide	-	-	-	-	m	**30.42**
Factory made double hermetically sealed units; to wood or metal with screwed or clipped beads						
Two panes; BS EN 14449; clear float glass; 4 mm thick; 6 mm air space						
0.35 m² - 2.00 m²	-	-	-	-	m²	**77.49**
Two panes; BS 952; clear float glass; 6 mm thick; 6 mm air space						
0.35 m² - 2.0 m²	-	-	-	-	m²	**90.24**
2.00 m² - 4.00 m²	-	-	-	-	m²	**135.69**
Factory made double hermetically sealed units; with inner pane of Pilkington's K low emissivity coated glass; to wood or metal with screwed or clipped beads						
Two panes; BS EN 14449; clear float glass; 4 mm thick; 6 mm air space						
0.35 m² - 2.00 m²	-	-	-	-	m²	**94.26**
Two panes; BS EN 14449; clear float glass; 6 mm thick; 6 mm air space						
0.35 m² - 2.0 m²	-	-	-	-	m²	**109.74**
2.00 m² - 4.00 m²	-	-	-	-	m²	**165.05**

L WINDOWS/DOORS/STAIRS

Item	PC £	Labour hours	Labour £	Material £	Unit	Total rate £
Factory made triple hermetically sealed units; with inner pane of Pilkington's K low emissivity coated glass; to wood or metal with screwed or clipped beads						
Three panes; BS EN 14449; clear float glass; 4 mm thick; 6 mm air spaces						
0.35 m² - 2.00 m²	-	-	-	-	m²	**151.76**
Three panes; BS EN 14449; clear float glass; 6 mm thick; 6 mm air spaces						
0.35 m² - 2.0 m²	-	-	-	-	m²	**176.72**
2.00 m² - 4.00 m²	-	-	-	-	m²	**265.74**

M SURFACE FINISHES

Item	PC £	Labour hours	Labour £	Material £	Unit	Total rate £
M10 CEMENT:SAND/CONCRETE/GRANOLITHIC SCREEDS/TOPPING						
Cement and sand (1:3) screeds; steel trowelled						
Work to floors; one coat level; to concrete base; screeded; over 300 mm wide						
25 mm thick	-	-	-	-	m²	**9.66**
50 mm thick	-	-	-	-	m²	**11.44**
75 mm thick	-	-	-	-	m²	**15.14**
100 mm thick	-	-	-	-	m²	**18.83**
Add to the above for work to falls and crossfalls and to slopes						
not exceeding 15 degrees from horizontal	-	0.02	0.43	-	m²	**0.43**
over 15 degrees from horizontal	-	0.09	1.94	-	m²	**1.94**
water repellent additive incorporated in the mix	-	0.02	0.43	5.78	m²	**6.21**
oil repellent additive incorporated in the mix	-	0.07	1.51	4.02	m²	**5.53**
Fine concrete (1:4-5) levelling screeds; steel trowlelled						
Work to floors; one coat; level; to concrete base; over 300 mm wide						
50 mm thick	-	-	-	-	m²	**11.44**
75 mm thick	-	-	-	-	m²	**15.14**
Extra over last for isolation joint to perimeter	-	-	-	-	m	**1.53**
Early drying floor screed; RMC Mortars "Readyscreed"; or other equal and approved; steel trowlelled						
Work to floors; one coat; level; to concrete base; over 300 mm wide						
100 mm thick	-	-	-	-	m²	**24.60**
Extra over last for galvanised chicken wire anticrack reinforcement	-	-	-	-	m²	**1.17**
Granolithic paving; cement and granite chippings 5 to dust (1:1:2); steel trowelled						
Work to floors; one coat; level; laid on concrete while green; bonded; over 300 mm wide						
25 mm thick	-	-	-	-	m²	**26.03**
38 mm thick	-	-	-	-	m²	**28.98**
Work to floors; two coat; laid on hacked concrete with slurry; over 300 mm wide						
50 mm thick	-	-	-	-	m²	**32.05**
75 mm thick	-	-	-	-	m²	**39.44**
Work to landings; one coat; level; laid on concrete while green; bonded; over 300 mm wide						
25 mm thick	-	-	-	-	m²	**38.95**
38 mm thick	-	-	-	-	m²	**43.47**
Work to landings; two coat; laid on hacked concrete with slurry; over 300 mm wide						
50 mm thick	-	-	-	-	m²	**48.07**
75 mm thick	-	-	-	-	m²	**59.16**
Add to the above over 300 mm wide for						
liquid hardening additive incorporated in the mix	-	0.04	0.86	0.66	m²	**1.53**
oil-repellent additive incorporated in the mix	-	0.07	1.51	4.02	m²	**5.53**

M SURFACE FINISHES

Item	PC £	Labour hours	Labour £	Material £	Unit	Total rate £
25 mm work to treads; one coat; to concrete base						
225 mm wide	-	0.83	23.05	10.22	m	**33.26**
275 mm wide	-	0.83	23.05	11.44	m	**34.49**
returned end	-	0.17	4.72	-	nr	**4.72**
13 mm skirtings; rounded top edge and coved bottom junction; to brickwork or blockwork base						
75 mm wide on face	-	0.51	14.16	0.49	m	**14.65**
150 mm wide on face	-	0.69	19.16	8.99	m	**28.15**
ends; fair	-	0.04	1.11	-	nr	**1.11**
angles	-	0.06	1.67	-	nr	**1.67**
13 mm outer margin to stairs; to follow profile of and with rounded nosing to treads and risers; fair edge and arris at bottom, to concrete base						
75 mm wide	-	0.83	23.05	4.90	m	**27.95**
angles	-	0.06	1.67	-	nr	**1.67**
13 mm wall string to stairs; fair edge and arris on top; coved bottom junction with treads and risers; to brickwork or blockwork base						
275 mm (extreme) wide	-	0.74	20.55	8.58	m	**29.13**
ends	-	0.04	1.11	-	nr	**1.11**
angles	-	0.06	1.67	-	nr	**1.67**
ramps	-	0.07	1.94	-	nr	**1.94**
ramped and wreathed corners	-	0.09	2.50	-	nr	**2.50**
13 mm outer string to stairs; rounded nosing on top at junction with treads and risers; fair edge and arris at bottom; to concrete base						
300 mm (extreme) wide	-	0.74	20.55	10.62	m	**31.17**
ends	-	0.04	1.11	-	nr	**1.11**
angles	-	0.06	1.67	-	nr	**1.67**
ramps	-	0.07	1.94	-	nr	**1.94**
ramps and wreathed corners	-	0.09	2.50	-	nr	**2.50**
19 mm thick skirtings; rounded top edge and coved bottom junction; to brickwork or blockwork base						
75 mm wide on face	-	0.51	14.16	8.99	m	**23.15**
150 mm wide on face	-	0.69	19.16	13.89	m	**33.05**
ends; fair	-	0.04	0.86	-	nr	**0.86**
angles	-	0.06	1.67	-	nr	**1.67**
19 mm riser; one rounded nosing; to concrete base						
150 mm high; plain	-	0.83	23.05	7.76	m	**30.81**
150 mm high; undercut	-	0.83	23.05	7.76	m	**30.81**
180 mm high; plain	-	0.83	23.05	10.62	m	**33.67**
180 mm high; undercut	-	0.83	23.05	10.62	m	**33.67**

M SURFACE FINISHES

Item	PC £	Labour hours	Labour £	Material £	Unit	Total rate £
M11 MASTIC ASPHALT FLOORING/FLOOR UNDERLAYS						
Mastic asphalt flooring to BS 6925 Type F 1076; black						
20 mm thick; one coat coverings; felt isolating membrane; to concrete base; flat						
over 300 mm wide	-	-	-	-	m²	**17.55**
225 mm - 300 mm wide	-	-	-	-	m²	**32.62**
150 mm - 225 mm wide	-	-	-	-	m²	**35.83**
not exceeding 150 mm wide	-	-	-	-	m²	**43.86**
25 mm thick; one coat coverings; felt isolating membrane; to concrete base; flat						
over 300 mm wide	-	-	-	-	m²	**20.39**
225 mm - 300 mm wide	-	-	-	-	m²	**34.79**
150 mm - 225 mm wide	-	-	-	-	m²	**37.93**
not exceeding 150 mm wide	-	-	-	-	m²	**45.98**
20 mm three coat skirtings to brickwork base						
not exceeding 150 mm girth	-	-	-	-	m	**17.94**
150 mm - 225 mm girth	-	-	-	-	m	**21.93**
225 mm - 300 mm girth	-	-	-	-	m	**25.91**
Mastic asphalt flooring; acid-resisting; black						
20 mm thick; one coat coverings; felt isolating membrane; to concrete base flat						
over 300 mm wide	-	-	-	-	m²	**20.57**
225 mm - 300 mm wide	-	-	-	-	m²	**37.61**
150 mm - 225 mm wide	-	-	-	-	m²	**38.84**
not exceeding 150 mm wide	-	-	-	-	m²	**46.86**
25 mm thick; one coat coverings; felt isolating membrane; to concrete base; flat						
over 300 mm wide	-	-	-	-	m²	**24.29**
225 mm - 300 mm wide	-	-	-	-	m²	**38.66**
150 mm - 225 mm wide	-	-	-	-	m²	**41.84**
not exceeding 150 mm wide	-	-	-	-	m²	**49.89**
20 mm thick; three coat skirtings to brickwork base						
not exceeding 150 mm girth	-	-	-	-	m	**18.12**
150 mm - 225 mm girth	-	-	-	-	m	**21.11**
225 mm - 300 mm girth	-	-	-	-	m	**23.97**
Mastic asphalt flooring to BS 6925 Type F 1451; red						
20 mm thick; one coat coverings; felt isolating membrane; to concrete base; flat						
over 300 mm wide	-	-	-	-	m²	**28.77**
225 mm - 300 mm wide	-	-	-	-	m²	**47.52**
150 mm - 225 mm wide	-	-	-	-	m²	**51.34**
not exceeding 150 mm wide	-	-	-	-	m²	**61.42**
20 mm thick; three coat skirtings to brickwork base						
not exceeding 150 mm girth	-	-	-	-	m	**22.59**
150 mm - 225 mm girth	-	-	-	-	m	**28.77**

M SURFACE FINISHES

Item	PC £	Labour hours	Labour £	Material £	Unit	Total rate £
M12 TROWELLED BITUMEN/RESIN/RUBBER LATEX FLOORING						
Latex cement floor screeds; steel trowelled						
Work to floors; level; to concrete base; over 300 mm wide						
3 mm thick; one coat	-	-	-	-	m²	**4.09**
5 mm thick; two coats	-	-	-	-	m²	**5.76**
Epoxy resin flooring; Altro "Altroflow 3000" or other equal and approved; steel trowelled						
Work to floors; level; to concrete base; over 300 mm wide						
3 mm thick; one coat	-	-	-	-	m²	**28.04**
Isocrete K screeds or other equal and approved; steel trowelled						
Work to floors; level; to concrete base; over 300 mm wide						
35 mm thick; plus polymer bonder coat	-	-	-	-	m²	**14.10**
40 mm thick	-	-	-	-	m²	**13.02**
45 mm thick	-	-	-	-	m²	**13.76**
50 mm thick	-	-	-	-	m²	**14.49**
Work to floors; to falls or cross-falls; to concrete base; over 300 mm wide						
55 mm (average) thick	-	-	-	-	m²	**15.23**
60 mm (average) thick	-	-	-	-	m²	**15.97**
65 mm (average) thick	-	-	-	-	m²	**16.71**
75 mm (average) thick	-	-	-	-	m²	**18.19**
90 mm (average) thick	-	-	-	-	m²	**20.41**
Isocrete K screeds; quick drying; or other equal and approved; steel trowelled						
Work to floors; level or to floors n.e. 15 degrees frojm the horizontal; to concrete base; over 300 mm wide						
55 mm thick	-	-	-	-	m²	**23.57**
75 mm thick	-	-	-	-	m²	**29.47**
Isocrete pumpable"Self Level Plus" screeds; or other equal and approved; protected with "Corex" type polythene; knifed off prior to layin floor finish; flat smooth finish						
Work to floors; level or to floors n.e. 15 degrees frojm the horizontal; to concrete base; over 300 mm wide						
20 mm thick	-	-	-	-	m²	**28.73**
50 mm thick	-	-	-	-	m²	**38.31**
Bituminous lightweight insulating roof screeds						
"Bit-Ag" or similar roof screed or other equal and approved; to falls or cross-falls; bitumen felt vapour barrier; over 300 mm wide						
75 mm (average) thick	-	-	-	-	m²	**46.15**
100 mm (average) thick	-	-	-	-	m²	**58.49**

M SURFACE FINISHES

Item	PC £	Labour hours	Labour £	Material £	Unit	Total rate £
M20 PLASTERED/RENDERED/ROUGHCAST COATINGS						
Cement and sand (1:3) beds and backings						
10 mm thick work to walls; one coat; to brickwork or blockwork base						
over 300 mm wide	-	-	-	-	m²	**16.86**
not exceeding 300 mm wide	-	-	-	-	m	**8.44**
13 mm thick; work to walls; two coats; to brickwork or blockwork base						
over 300 mm wide	-	-	-	-	m²	**20.28**
not exceeding 300 mm wide	-	-	-	-	m	**10.16**
15 mm thick work to walls; two coats; to brickwork or blockwork base						
over 300 mm wide	-	-	-	-	m²	**21.87**
not exceeding 300 mm wide	-	-	-	-	m	**10.95**
Cement and sand (1:3); steel trowelled						
13 mm thick work to walls; two coats; to brickwork or blockwork base						
over 300 mm wide	-	-	-	-	m²	**17.65**
not exceeding 300 mm wide	-	-	-	-	m	**8.83**
16 mm thick work to walls; two coats; to brickwork or blockwork base						
over 300 mm wide	-	-	-	-	m²	**19.77**
not exceeding 300 mm wide	-	-	-	-	m	**9.91**
19 mm thick work to walls; two coats; to brickwork or blockwork base						
over 300 mm wide	-	-	-	-	m²	**22.88**
not exceeding 300 mm wide	-	-	-	-	m	**11.44**
ADD to above						
over 300 mm wide in water repellant cement	-	-	-	-	m²	**4.68**
finishing coat in colour cement	-	-	-	-	m²	**9.98**
Cement-lime-sand (1:2:9); steel trowelled						
19 mm thick work to walls; two coats; to brickwork or blockwork base						
over 300 mm wide	-	-	-	-	m²	**22.19**
not exceeding 300 mm wide	-	-	-	-	m	**11.09**
Cement-lime-sand (1:1:6); steel trowelled						
13 mm thick work to walls; two coats; to brickwork or blockwork base						
over 300 mm wide	-	-	-	-	m²	**18.14**
not exceeding 300 mm wide	-	-	-	-	m	**8.94**
Add to the above over 300 mm wide for waterproof additive	-	-	-	-	m²	**3.06**
19 mm thick work to ceilings; three coats; to metal lathing base						
over 300 mm wide	-	-	-	-	m²	**21.34**
not exceeding 300 mm wide	-	-	-	-	m	**12.47**

M SURFACE FINISHES

Item	PC £	Labour hours	Labour £	Material £	Unit	Total rate £
Sto External render only system; comprising glassfibre mesh reinforcement embedded in 10 mm Sto Levell Cote with Sto Armat Classic Basecoat Render and Stolit K 1.5 Decorative Topcoat Render (white)						
15 mm thick work to walls; two coats; to brickwork or blockwork base						
over 300 mm wide	-	-	-	-	m²	**60.70**
Extra for						
bellcast bead	-	-	-	-	m	**5.77**
external angle with PVC mesh angle bead	-	-	-	-	m	**5.34**
internal angle with Sto Armor angle	-	-	-	-	m	**5.34**
render stop bead	-	-	-	-	m	**5.34**
K-Rend render or similar through-colour render system						
18 mm thick work to walls; two coats; to brickwork or blockwork base; first coat 8 mm standard base coat; second coat 10 mm K-rend silicone WP/FT						
over 300 mm wide	-	-	-	-	m²	**78.98**
Plaster; first 11 mm coat of "Thistle Hardwall" plaster; second 2 mm finishing coat of "Thistle Multi Finish" plaster; steel trowelled						
13 mm thick work to walls; two coats; to brickwork or blockwork base						
over 300 mm wide	-	-	-	-	m²	**14.00**
over 300 mm wide; in staircase areas or plant rooms	-	-	-	-	m²	**16.81**
not exceeding 300 mm wide	-	-	-	-	m	**7.44**
13 mm thick work to isolated brickwork or blockwork columns; two coats						
over 300 mm wide	-	-	-	-	m²	**26.37**
not exceeding 300 mm wide	-	-	-	-	m	**13.18**
Plaster; first 11 mm coat of "Thistle Browning" plaster; second 2 mm finishing coat of "Thistle Multi Finish" plaster; steel trowelled						
13 mm thick; work to walls; two coats; to brickwork or blockwork base						
over 300 mm wide	-	-	-	-	m²	**14.00**
over 300 mm wide; in staircase areas or plant rooms	-	-	-	-	m²	**16.80**
not exceeding 300 mm wide	-	-	-	-	m	**7.44**
13 mm thick work to isolated brickwork or blockwork columns; two coats						
over 300 mm wide	-	-	-	-	m²	**26.38**
not exceeding 300 mm wide	-	-	-	-	m	**11.69**
Plaster; first 8 mm or 11 mm coat of "Thistle Bonding" plaster; second 2 mm finishing coat of "Thistle Multi Finish" plaster;steel trowelled						
13 mm thick work to walls; two coats; to concrete base						
over 300 mm wide	-	-	-	-	m²	**15.68**
over 300 mm wide; in staircase areas or plant rooms	-	-	-	-	m²	**18.56**
not exceeding 300 mm wide	-	-	-	-	m	**7.18**

M SURFACE FINISHES

Item	PC £	Labour hours	Labour £	Material £	Unit	Total rate £
M20 PLASTERED/RENDERED/ROUGHCAST COATINGS – cont'd						
"Thistle Bonding" plaster – cont'd						
13 mm thick work to isolated piers or columns; two coats; to concrete base						
over 300 mm wide	-	-	-	-	m²	**28.00**
not exceeding 300 mm wide	-	-	-	-	m	**13.11**
10 mm thick work to ceilings; two coats; to concrete base						
over 300 mm wide	-	-	-	-	m²	**13.41**
over 300 wide; 3.50 m - 5.00 m high	-	-	-	-	m²	**16.09**
over 300 mm wide; in staircase areas or plant rooms	-	-	-	-	m²	**17.81**
not exceeding 300 mm wide	-	-	-	-	m	**7.49**
10 mm thick work to isolated beams; two coats; to concrete base						
over 300 mm wide	-	-	-	-	m²	**26.82**
over 300 mm wide; 3.50 m - 5.00 m high	-	-	-	-	m²	**28.60**
not exceeding 300 mm wide	-	-	-	-	m	**13.49**
Plaster; one coat "Snowplast" plaster or other equal and approved; steel trowelled						
13 mm thick work to walls; one coat; to brickwork or blockwork base						
over 300 mm wide	-	-	-	-	m²	**15.48**
over 300 mm wide; in staircase areas or plant rooms	-	-	-	-	m²	**18.39**
not exceeding 300 mm wide	-	-	-	-	m	**7.76**
13 thick work to isolated columns; one coat						
over 300 mm wide	-	-	-	-	m²	**18.73**
not exceeding 300 mm wide	-	-	-	-	m	**9.40**
Plaster; first coat of "Limelite" renovating plaster; finishing coat of "Limelite" finishing plaster; or other equal and approved; steel trowelled						
13 mm thick work to walls; two coats; to brickwork or blockwork base						
over 300 mm wide	-	-	-	-	m²	**21.47**
over 300 mm wide; in staircase areas or plant rooms	-	-	-	-	m²	**23.57**
not exceeding 300 mm wide	-	-	-	-	m	**10.73**
Dubbing out existing walls with undercoat plaster; average 6 mm thick						
over 300 mm wide	-	-	-	-	m²	**6.44**
not exceeding 300 mm wide	-	-	-	-	m	**3.25**
Dubbing out existing walls with undercoat plaster; average 12 mm thick						
over 300 mm wide	-	-	-	-	m²	**12.88**
not exceeding 300 mm wide	-	-	-	-	m	**6.44**

M SURFACE FINISHES

Item	PC £	Labour hours	Labour £	Material £	Unit	Total rate £
Plaster; first coat of "Thistle X-ray" plaster or other equal and approved; finishing coat of "Thistle X-ray" finishing plaster or other equal and approved; steel trowelled						
17 mm thick work to walls; two coats; to brickwork or blockwork base						
over 300 mm wide	-	-	-	-	m²	**69.65**
over 300 mm wide; in staircase areas or plant rooms	-	-	-	-	m²	**74.84**
not exceeding 300 mm wide	-	-	-	-	m	**27.86**
17 mm thick work to isolated columns; two coats						
over 300 mm wide	-	-	-	-	m²	**113.00**
not exceeding 300 mm wide	-	-	-	-	m	**45.16**
Plaster; one coat "Thistle" projection plaster or other equal and approved; steel trowelled						
13 mm thick work to walls; one coat; to brickwork or blockwork base						
over 300 mm wide	-	-	-	-	m²	**14.94**
over 300 mm wide; in staircase areas or plant rooms	-	-	-	-	m²	**17.08**
not exceeding 300 mm wide	-	-	-	-	m	**7.45**
10 mm thick work to isolated columns; one coat						
over 300 mm wide	-	-	-	-	m²	**18.17**
not exceeding 300 mm wide	-	-	-	-	m	**9.07**
Plaster; first 11 mm coat of "Thistle Bonding" plaster; second 2 mm finishing coat of "Thistle Multi Finish" plaster;steel trowelled						
13 mm thick work to ceilings; three coats to metal lathing base						
over 300 mm wide	-	-	-	-	m²	**16.27**
over 300 mm wide; in staircase areas or plant rooms	-	-	-	-	m²	**19.52**
not exceeding 300 mm wide	-	-	-	-	m	**8.78**
13 mm thick work to swept soffit of metal lathing arch former						
not exceeding 300 mm wide	-	-	-	-	m	**11.71**
300 mm - 400 mm wide	-	-	-	-	m	**15.66**
13 mm thick work to vertical face of metal lathing arch former						
not exceeding 0.50 m² per side	-	-	-	-	nr	**16.63**
0.50 m² - 1 m² per side	-	-	-	-	nr	**24.95**
Squash court plaster, Prodorite Ltd; first coat "Formula Base" screed or other equal and approved; finishing coat "Formula 90" finishing plaster or other equal and approved; steel trowelled and finished with sponge float						
12 mm thick work to walls; two coats; to brickwork or blockwork base						
over 300 mm wide	-	-	-	-	m²	**32.54**
not exceeding 300 mm wide	-	-	-	-	m	**16.00**
demarcation lines on battens	-	-	-	-	m	**4.61**

M SURFACE FINISHES

Item	PC £	Labour hours	Labour £	Material £	Unit	Total rate £
M20 PLASTERED/RENDERED/ROUGHCAST COATINGS – cont'd						
"Cemrend" self-coloured render or other equal and approved; one coat; to brickwork or blockwork base						
20 mm thick work to walls; to brickwork or blockwork base						
over 300 mm wide	-	-	-	-	m²	**33.21**
not exceeding 300 mm wide	-	-	-	-	m	**19.37**
Tyrolean decorative rendering or similar; 13 mm thick first coat of cement-lime-sand (1:1:6); finishing three coats of "Cullamix"' or other equal and approved; applied with approved hand operated machine external						
To walls; four coats; to brickwork or blockwork base						
over 300 mm wide	-	-	-	-	m²	**35.70**
not exceeding 300 mm wide	-	-	-	-	m	**17.83**
Drydash (pebbledash) finish of Derbyshire Spar chippings or other equal and approved on and including cement-lime-sand (1:2:9) backing						
18 mm thick work to walls; two coats; to brickwork or blockwork base						
over 300 mm wide	-	-	-	-	m²	**30.94**
not exceeding 300 mm wide	-	-	-	-	m	**15.48**
Plaster; one coat "Thistle" board finish or other equal and approved; steel trowelled (prices included within plasterboard rates)						
3 mm thick work to walls or ceilings; one coat; to plasterboard base						
over 300 mm wide	-	-	-	-	m²	**6.31**
over 300 mm wide; in staircase areas or plant rooms	-	-	-	-	m²	**7.57**
not exceeding 300 mm wide	-	-	-	-	m	**2.52**
Plaster; one coat "Thistle" board finish or other and approved; steel trowelled 3 mm work to walls or ceilings; one coat on and including gypsum plasterboard; BS 1230; fixing with nails; 3 mm joints filled with plaster and jute scrim cloth; to softwood base; plain grade baseboard or lath with rounded edges						
9.50 mm thick boards to walls						
over 300 mm wide	-	0.97	13.93	3.57	m²	**17.51**
not exceeding 300 mm wide	-	0.37	5.68	1.03	m	**6.71**
9.50 mm thick boards to walls; in staircase areas or plant rooms						
over 300 mm wide	-	1.06	15.32	3.57	m²	**18.89**
not exceeding 300 mm wide	-	0.46	7.06	1.03	m	**8.09**
9.50 mm thick boards to isolated columns						
over 300 mm wide	-	1.06	15.32	3.57	m²	**18.89**
not exceeding 300 mm wide	-	0.56	8.60	1.03	m	**9.63**
9.50 mm thick boards to ceilings						
over 300 mm wide	-	0.89	12.71	3.57	m²	**16.28**
over 300 mm wide; 3.50 m - 5.00 m high	-	1.03	14.86	3.57	m²	**18.43**
not exceeding 300 mm wide	-	0.43	6.60	1.03	m	**7.63**

M SURFACE FINISHES

Item	PC £	Labour hours	Labour £	Material £	Unit	Total rate £
9.50 mm thick boards to ceilings; in staircase areas or plant rooms						
over 300 mm wide	-	0.98	14.09	3.57	m²	**17.66**
not exceeding 300 mm wide	-	0.47	7.22	1.03	m	**8.25**
9.50 mm thick boards to isolated beams						
over 300 mm wide	-	1.05	15.16	3.57	m²	**18.74**
not exceeding 300 mm wide	-	0.50	7.68	1.03	m	**8.71**
12.50 mm thick boards to walls; in staircase areas or plant rooms						
over 300 mm wide	-	1.12	16.24	3.57	m²	**19.81**
not exceeding 300 mm wide	-	0.50	7.68	1.03	m	**8.71**
12.50 mm thick boards to isolated columns						
over 300 mm wide	-	1.12	16.24	3.57	m²	**19.81**
not exceeding 300 mm wide	-	0.59	9.06	1.03	m	**10.09**
12.50 mm thick boards to ceilings						
over 300 mm wide	-	0.95	13.63	3.57	m²	**17.20**
over 300 mm wide; 3.50 m - 5.00 m high	-	1.06	15.32	3.57	m²	**18.89**
not exceeding 300 mm wide	-	0.45	6.91	1.03	m	**7.94**
12.50 mm thick boards to ceilings; in staircase areas or plant rooms						
over 300 mm wide	-	1.06	15.32	3.57	m²	**18.89**
not exceeding 300 mm wide	-	0.51	7.83	1.03	m	**8.86**
12.50 mm thick boards to isolated beams						
over 300 mm wide	-	1.15	16.70	3.57	m²	**20.27**
not exceeding 300 mm wide	-	0.56	8.60	1.03	m	**9.63**
Accessories						
"Expamet" render beads or other equal and approved; white PVC nosings; to brickwork or blockwork base						
external stop bead; ref 573	-	0.07	1.08	1.38	m	**2.46**
"Expamet" render beads or other equal and approved; stainless steel; to brickwork or blockwork base						
stop bead; ref 546	-	0.07	1.08	3.71	m	**4.78**
stop bead; ref 547	-	0.07	1.08	3.71	m	**4.78**
"Expamet" plaster beads or other equal and approved; galvanised steel; to brickwork or blockwork base						
angle bead; ref 550	-	0.08	1.23	0.79	m	**2.02**
architrave bead; ref 579	-	0.10	1.54	2.12	m	**3.65**
stop bead; ref 562	-	0.07	1.08	0.98	m	**2.05**
stop beads; ref 563	-	0.07	1.08	0.98	m	**2.05**
movement bead; ref 588	-	0.09	1.38	10.30	m	**11.68**
"Expamet" plaster beads or other equal and approved; stainless steel; to brickwork or blockwork base						
angle bead; ref 545	-	0.08	1.23	4.17	m	**5.40**
stop bead; ref 534	-	0.07	1.08	3.71	m	**4.78**
stop bead; ref 533	-	0.07	1.08	3.71	m	**4.78**
"Expamet" thin coat plaster beads or other equal and approved; galvanised steel; to timber base						
angle bead; ref 553	-	0.07	1.08	0.74	m	**1.82**
angle bead; ref 554	-	0.07	1.08	0.62	m	**1.69**
stop bead; ref 560	-	0.06	0.92	0.94	m	**1.86**
stop bead; ref 561	-	0.06	0.92	0.94	m	**1.86**

M SURFACE FINISHES

Item	PC £	Labour hours	Labour £	Material £	Unit	Total rate £
M21 INSULATION WITH RENDERED FINISH						
Sto Therm Classic M-system insulation render						
70 mm EPS insulation fixed with adhesive to SFS structure (measured separately) with horizontal PVC intermediate track and vertical T-spines; with glassfibre mesh reinforcement embedded in Sto Armat Classic Basecoat Render and Stolit K 1.5 Decorative Topcoat Render (white)						
over 300 mm wide	-	-	-	-	m²	**72.06**
70 mm EPS insulation mechanically fixed to SFS structure (measured separately) with horizontal PVC intermediate track and vertical T-spines; with glassfibre mesh reinforcement embedded in Sto Armat Classic Basecoat Render and Stolit K 1.5 Decorative Topcoat Render (white)						
over 300 mm wide	-	-	-	-	m²	**79.28**
rendered heads and reveals not exceeding 100 mm wide; including angle beads	-	-	-	-	m	**19.90**
Extra for						
aluminium starter track at base of insulated render system	-	-	-	-	m	**12.11**
external angle with PVC mesh angle bead	-	-	-	-	m	**5.34**
internal angle with Sto Armor angle	-	-	-	-	m	**5.34**
render stop bead	-	-	-	-	m	**5.34**
Sto seal tape to all vertical abutments	-	-	-	-	m	**4.96**
Sto Armor mat HD mesh reinforcement to areas prone to physical damage (e.g. 1800 mm high adjoining floor level)						
over 300 mm wide	-	-	-	-	m²	**15.94**
M22 SPRAYED MINERAL FIBRE COATINGS						
Prepare and apply by spray "Mandolite CP2" fire protection or other equal and approved on structural steel/metalwork						
16 mm thick (one hour) fire protection						
to walls and columns	-	-	-	-	m²	**9.72**
to ceilings and beams	-	-	-	-	m²	**10.73**
to isolated metalwork	-	-	-	-	m²	**21.39**
22 mm thick (one and a half hour) fire protection						
to walls and columns	-	-	-	-	m²	**11.31**
to ceilings and beams	-	-	-	-	m²	**12.54**
to isolated metalwork	-	-	-	-	m²	**25.08**
28 mm thick (two hour) fire protection						
to walls and columns	-	-	-	-	m²	**13.26**
to ceilings and beams	-	-	-	-	m²	**14.48**
to isolated metalwork	-	-	-	-	m²	**28.95**
52 mm thick (four hour) fire protection						
to walls and columns	-	-	-	-	m²	**20.05**
to ceilings and beams	-	-	-	-	m²	**22.33**
to isolated metalwork	-	-	-	-	m²	**44.43**
Prepare and apply by spray; cementitious "Pyrok WF26" render or other equal and approved; on expanded metal lathing (not included)						
15 mm thick						
to ceilings and beams	-	-	-	-	m²	**30.77**

M SURFACE FINISHES

Item	PC £	Labour hours	Labour £	Material £	Unit	Total rate £
M30 METAL MESH LATHING/ANCHORED REINFORCEMENT FOR PLASTERED COATINGS						
Accessories						
Pre-formed galvanised expanded steel semi-circular arch-frames; "Expamet" or other equal and approved; to suit walls up to 230 mm thick						
for 760 mm opening; ref ESC 30	50.83	0.46	6.18	56.07	nr	**62.25**
for 840 mm opening; ref ESC 32	53.38	0.46	6.18	58.88	nr	**65.06**
for 920 mm opening; ref ESC 36	62.44	0.46	6.18	68.88	nr	**75.06**
for 1220 mm opening; ref ESC 48	77.82	0.46	6.18	85.85	nr	**92.02**
Lathing; Expamet "BB" expanded metal lathing or other equal and approved; BS EN 13658; 50 mm laps						
6 mm thick mesh linings to ceilings; fixing with staples; to softwood base; over 300 mm wide						
ref BB263; 0.500 mm thick	4.43	0.56	7.52	5.01	m²	**12.53**
ref BB264; 0.675 mm thick	6.20	0.56	7.52	7.01	m²	**14.53**
6 mm thick mesh linings to ceilings; fixing with wire; to steelwork; over 300 mm wide						
ref BB263; 0.500 mm thick	-	0.59	7.94	5.01	m²	**12.95**
ref BB264; 0.675 mm thick	-	0.59	7.94	7.01	m²	**14.95**
6 mm thick mesh linings to ceilings; fixing with wire; to steelwork; not exceeding 300 mm wide						
ref BB263; 0.500 mm thick	-	0.37	4.95	5.01	m²	**9.96**
ref BB264; 0.675 mm thick	-	0.37	4.95	7.01	m²	**11.96**
raking cutting	-	0.19	2.92	-	m	**2.92**
cutting and fitting around pipes; not exceeding 0.30 m girth	-	0.28	4.30	-	nr	**4.30**
Lathing; Expamet "Riblath" or "Spraylath" or other equal and approved stiffened expanded metal lathing or similar; 50 mm laps						
10 mm thick mesh lining to walls; fixing with nails; to softwood base; over 300 mm wide						
"Riblath" ref 269; 0.30 mm thick	7.33	0.46	6.18	8.41	m²	**14.59**
"Riblath" ref 271; 0.50 mm thick	8.44	0.46	6.18	9.66	m²	**15.84**
10 mm thick mesh lining to walls; fixing with nails; to softwood base; not exceeding 300 mm wide						
"Riblath" ref 269; 0.30 mm thick	-	0.28	3.76	2.57	m²	**6.33**
"Riblath" ref 271; 0.50 mm thick	-	0.28	3.76	2.95	m²	**6.71**
10 mm thick mesh lining to walls; fixing to brick or blockwork; over 300 mm wide						
"Red-rib" ref 274; 0.50 mm thick	9.31	0.37	4.95	11.09	m²	**16.04**
Stainless steel "Riblath" ref 267; 0.30 mm thick	21.80	0.37	4.95	25.21	m²	**30.16**
10 mm thick mesh lining to ceilings; fixing with wire; to steelwork; over 300 mm wide						
"Riblath" ref 269; 0.30 mm thick	-	0.59	7.94	8.99	m²	**16.93**
"Riblath" ref 271; 0.50 mm thick	-	0.59	7.94	10.24	m²	**18.18**

M SURFACE FINISHES

Item	PC £	Labour hours	Labour £	Material £	Unit	Total rate £
M31 FIBROUS PLASTER						
Fibrous plaster; fixing with screws; plugging; countersinking; stopping; filling and pointing joints with plaster						
16 mm thick plain slab coverings to ceilings						
over 300 mm wide	-	-	-	-	m²	**133.47**
not exceeding 300 mm wide	-	-	-	-	m	**44.89**
Coves; not exceeding 150 mm girth						
per 25 mm girth	-	-	-	-	m	**6.44**
Coves; 150 mm - 300 mm girth						
per 25 mm girth	-	-	-	-	m	**7.89**
Cornices						
per 25 mm girth	-	-	-	-	m	**8.01**
Cornice enrichments						
per 25 mm girth; depending on degree of enrichments	-	-	-	-	m	**9.46**
Fibrous plaster; fixing with plaster wadding filling and pointing joints with plaster; to steel base						
16 mm thick plain slab coverings to ceilings						
over 300 mm wide	-	-	-	-	m²	**133.47**
not exceeding 300 mm wide	-	-	-	-	m	**44.89**
16 mm thick plain casings to stanchions						
per 25 mm girth	-	-	-	-	m	**4.00**
16 mm thick plain casings to beams						
per 25 mm girth	-	-	-	-	m	**4.00**
Gyproc cove or other equal and approved; fixing with adhesive; filling and pointing joints with plaster						
Cove						
125 mm girth	-	0.19	2.92	3.16	m	**6.08**
Angles	-	0.03	0.46	2.31	nr	**2.77**
M40 STONE/CONCRETE/QUARRY/CERAMIC TILING/MOSAIC						
ALTERNATIVE TILE MATERIALS						
Discounts of 10 - 30% available depending on quantity/status						
Dennis Ruabon clay floor quarries (£/1000)						
194 mm x 194 mm x 12.5 mm; square; red	-	-	-	650.00	1000	-
194 mm x 194 mm x 12.5 mm; red; polygon; red	-	-	-	48.18	m²	-
150 mm x 150 mm x 12.5 mm; square; heatherbrown	-	-	-	740.00	1000	-
150 mm x 150 mm x 12.5 mm; studded square; heatherbrown or red	-	-	-	1078.00	1000	-
150 mm x 150 mm x 12.50 mm; polygon; red	-	-	-	68.85	m²	-

M SURFACE FINISHES

Item	PC £	Labour hours	Labour £	Material £	Unit	Total rate £
SUPPLY AND FIX PRICES						
Clay floor quarries; BS EN 10545; class 1; Dennis Ruabon tiles or other equal and approved; level bedding 10 mm thick and jointing in cement and sand (1:3); butt joints; straight both ways; flush pointing with grout; to cement and sand base						
Work to floors; over 300 mm wide						
150 mm x 150 mm x 12.50 mm thick; heatherbrown	-	0.74	15.96	35.32	m²	**51.28**
150 mm x 150 mm x 12.50 mm thick; red	-	0.74	15.96	31.90	m²	**47.86**
194 mm x 194 mm x 12.50 mm thick; heatherbrown	-	0.60	12.94	28.01	m²	**40.95**
Works to floors; in staircase areas or plant rooms						
150 mm x 150 mm x 12.50 mm thick; heatherbrown	-	0.83	17.90	35.32	m²	**53.22**
150 mm x 150 mm x 12.50 mm thick; red	-	0.83	17.90	31.90	m²	**49.80**
194 mm x 194 mm x 12.50 mm thick; heatherbrown	-	0.69	14.88	28.01	m²	**42.89**
Work to floors; not exceeding 300 mm wide						
150 mm x 150 mm x 12.50 mm thick; heatherbrown	-	0.37	7.98	9.35	m	**17.33**
150 mm x 150 mm x 12.50 mm thick; red	-	0.37	7.98	8.30	m	**16.28**
194 mm x 194 mm x 12.50 mm thick; heatherbrown	-	0.31	6.69	6.96	m	**13.65**
fair square cutting against flush edges of existing finishes	-	0.11	1.57	2.30	m	**3.88**
raking cutting	-	0.19	2.78	2.59	m	**5.37**
cutting around pipes; not exceeding 0.30 m girth	-	0.14	2.15	-	nr	**2.15**
extra for cutting and fitting into recessed manhole cover 600 mm x 600 mm	-	0.93	14.28	-	nr	**14.28**
Work to sills; 150 mm wide; rounded edge tiles						
200 mm x 150 mm x 22 mm thick; interior; heatherbrown or red	-	0.31	6.69	8.76	m	**15.45**
150 mm x 173 mm x 58 mm thick; exterior; heatherbrown or red	-	0.32	6.90	36.36	m	**43.26**
fitted end	-	0.14	2.15	-	nr	**2.15**
Coved skirtings; 150 mm high; rounded top edge						
150 mm x 150 mm x 12.50 mm thick; ref. CBTR; heatherbrown or red	-	0.23	4.96	10.08	m	**15.04**
150 mm x 150 mm x 12.50 mm thick; ref. RE; heatherbrown or red	-	0.23	4.96	6.97	m	**11.93**
ends	-	0.04	0.61	-	nr	**0.61**
angles	-	0.14	2.15	2.67	nr	**4.82**
Glazed ceramic wall tiles; BS EN 10545; fixing with adhesive; butt joints; straight both ways; flush pointing with white grout; to plaster base						
Work to walls; over 300 mm wide						
152 mm x 152 mm x 5.50 mm thick; white	11.26	0.56	15.55	18.53	m²	**34.08**
152 mm x 152 mm x 5.50 mm thick; light colours	13.74	0.56	15.55	21.33	m²	**36.88**
152 mm x 152 mm x 5.50 mm thick; dark colours	15.02	0.56	15.55	22.77	m²	**38.32**
extra for RE or REX tile	-	-	-	6.53	m²	**-**
200 mm x 100 mm x 6.50 mm thick; white and light colours	11.26	0.56	15.55	18.53	m²	**34.08**
250 mm x 200 mm x 7 mm thick; white and light colours	12.20	0.56	15.55	19.59	m²	**35.14**
Work to walls; in staircase areas or plant rooms						
152 mm x 152 mm x 5.50 mm thick; white	-	0.62	17.22	18.53	m²	**35.74**
Work to walls; not exceeding 300 mm wide						
152 mm x 152 mm x 5.50 mm thick; white	-	0.28	7.77	5.49	m	**13.26**
152 mm x 152 mm x 5.50 mm thick; light colours	-	0.28	7.77	9.09	m	**16.87**
152 mm x 152 mm x 5.50 mm thick; dark colours	-	0.28	7.77	9.53	m	**17.30**

M SURFACE FINISHES

Item	PC £	Labour hours	Labour £	Material £	Unit	Total rate £
M40 STONE/CONCRETE/QUARRY/CERAMIC TILING/MOSAIC – cont'd						
Glazed ceramic wall tiles – cont'd						
Work to walls; not exceeding 300 mm wide – cont'd						
200 mm x 100 mm x 6.50 mm thick; white and light colours	-	0.28	7.77	5.49	m	**13.26**
250 mm x 200 mm x 7 mm thick; white and light colours	-	0.23	6.39	5.81	m	**12.19**
cutting around pipes; not exceeding 0.30 m girth	-	0.09	1.38	-	nr	**1.38**
Work to sills; 150 mm wide; rounded edge tiles						
152 mm x 152 mm x 5.50 mm thick; white	-	0.23	6.39	2.75	m	**9.13**
fitted end	-	0.09	1.38	-	nr	**1.38**
198 mm x 64.50 mm x 6 mm thick wall tiles; fixing with adhesive; butt joints; straight both ways; flush pointing with white grout; to plaster base						
Work to walls						
over 300 mm wide	23.50	1.67	46.37	32.36	m²	**78.73**
not exceeding 300 mm wide	-	0.65	18.05	9.64	m	**27.69**
20 mm x 20 mm x 5.50 mm thick glazed mosaic wall tiles; fixing with adhesive; butt joints; straight both ways; flush pointing with white grout; to plaster base						
Work to walls						
over 300 mm wide	28.85	1.76	48.87	39.42	m²	**88.29**
not exceeding 300 mm wide	-	0.69	19.16	14.39	m	**33.54**
50 mm x 50 mm x 5.50 mm thick slip resistant mosaic floor tiles, Series 2 or other equal and approved; Langley London Ltd; fixing with adhesive; butt joints; straight both ways; flush pointing with white grout; to cement and sand base						
Work to floors						
over 300 mm wide	28.33	1.76	37.96	40.04	m²	**78.00**
not exceeding 300 mm wide	-	0.69	14.88	14.21	m	**29.09**
Dakota mahogany granite cladding; polished finish; jointed and pointed in coloured mortar (1:2:8)						
20 mm work to floors; level; to cement and sand base						
over 300 mm wide	-	-	-	-	m²	**280.45**
20 mm x 300 mm treads; plain nosings	-	-	-	-	m	**159.59**
raking, cutting	-	-	-	-	m	**28.62**
polished edges	-	-	-	-	m	**35.98**
birdsmouth	-	-	-	-	m	**37.13**
20 mm thick work to walls; to cement and sand base						
over 300 mm wide	-	-	-	-	m²	**286.09**
not exceeding 300 mm wide	-	-	-	-	m	**128.74**
40 mm thick work to walls; to cement and sand base						
over 300 mm wide	-	-	-	-	m²	**475.04**
not exceeding 300 mm wide	-	-	-	-	m	**213.75**

M SURFACE FINISHES

Item	PC £	Labour hours	Labour £	Material £	Unit	Total rate £
Riven Welsh slate floor tiles; level; bedding 10 mm thick and jointing in cement and sand (1:3); butt joints; straight both ways; flush pointing with coloured mortar; to cement and sand base						
Work to floors; over 300 mm wide						
250 mm x 250 mm x 12 mm - 15 mm thick	-	0.56	15.55	37.12	m²	**52.67**
Work to floors; not exceeding 300 mm wide						
250 mm x 250 mm x 12 mm - 15 mm thick	-	0.28	7.77	11.22	m	**18.99**
Roman Travertine marble cladding; polished finish; jointed and pointed in coloured mortar (1:2:8)						
20 mm thick work to floors; level; to cement and sand base						
over 300 mm wide	-	-	-	-	m²	**183.15**
20 mm x 300 mm treads; plain nosings	-	-	-	-	m	**110.33**
raking cutting	-	-	-	-	m	**21.39**
polished edges	-	-	-	-	m	**19.67**
birdsmouth	-	-	-	-	m	**39.43**
20 mm thick work to walls; to cement and sand base						
over 300 mm wide	-	-	-	-	m²	**217.78**
not exceeding 300 mm wide	-	-	-	-	m	**98.52**
40 mm thick work to walls; to cement and sand base						
over 300 mm wide	-	-	-	-	m²	**299.79**
not exceeding 300 mm wide	-	-	-	-	m	**134.90**
M41 TERRAZZO TILING/IN SITU TERRAZZO						
Terrazzo tiles; BS EN 13748; aggregate size random ground grouted and polished to 80's grit finish; standard colour range; 3 mm joints symmetrical layout; bedding in 42 mm cement semi-dry mix (1:4); grouting with neat matching cement						
300 mm x 300 mm x 28 mm (nominal) Terrazzo tile units; hydraulically pressed, mechanically vibrated, steam cured; to floors on concrete base (not included); sealed with penetrating case hardener or other equal and approved; 2 coats applied immediately after final polishing						
plain; laid level	-	-	-	-	m²	**37.53**
plain; to slopes exceeding 15 degrees from horizontal			-	-	m²	**45.75**
to small areas/toilets	-	-	-	-	m²	**85.90**
Accessories						
plastic division strips; 6 mm x 38 mm; set into floor tiling above crack inducing joints, to the nearest full tile module	-	-	-	-	m	**2.59**

M SURFACE FINISHES

Item	PC £	Labour hours	Labour £	Material £	Unit	Total rate £
M41 TERRAZZO TILING/IN SITU TERRAZZO – cont'd						
Specially made terrazzo precast units; BS EN 13748-1; aggregate size random; standard colour range; 3 mm joints; grouting with neat matching cement						
Standard tread and riser square combined terrazzo units (with riser cast down) or other equal and approved; 280 mm wide; 150 mm high; 40 mm thick; machine made; vibrated and fully machine polished; incorporating 1 nr. "Ferodo" anti-slip insert ref. OT40D or other equal and approved cast-in during manufacture; one end polished only						
fixed with cement:sand (1:4) mortar on prepared backgrounds (not included); grouted in neat tinted cement; wiped clean on completion of fixing	-	-	-	-	m	**196.53**
Standard tread square terrazzo units or other equal and approved; 40 mm thick; 280 mm wide; factory polished; incorporating 1 nr. "Ferodo" anti-slip insert ref. OT40D or other equal and approved						
fixed with cement:sand (1:4) mortar on prepared backgrounds (not included); grouted in neat tinted cement; wiped clean on completion of fixing	-	-	-	-	m	**118.25**
extra over for 55 x 55 mm contrasting colour to step nosing	-	-	-	-	m	**45.92**
Standard riser square terrazzo units or other equal and approved; 40 mm thick; 150 mm high; factory polished						
fixed with cemnt:sand (1:4) mortar on prepared backgrounds (not included); grouted in neat tinted cement; wiped clean on completion of fixing	-	-	-	-	m	**73.38**
Standard coved terrazzo skirting units or other equal and approved; 904 mm long; 150 mm high; nominal finish; 23 mm thick; with square top edge						
fixed with cement:sand (1:4) mortar on prepared backgrounds (by others); grouted in neat tinted cement; wiped clean on completion of fixing	-	-	-	-	m	**68.93**
extra over for special internal/external angle pieces to match	-	-	-	-	m	**19.58**
extra over for special polished ends	-	-	-	-	nr	**6.45**
M42 WOOD BLOCK/COMPOSITION BLOCK/ PARQUET FLOORING						
Wood blocks; Havwoods or other equal and approved; 25 mm thick; level; laid to herringbone pattern with 2 block borderl; fixing with adhesive; to cement:sand base; sanded and sealed						
Work to floors; over 300 mm wide						
Merbau	-	-	-	-	m²	**60.86**
Iroko	-	-	-	-	m²	**67.28**
American Oak	-	-	-	-	m²	**75.66**
European Oak	-	-	-	-	m²	**88.90**
Add to wood block flooring over 300 mm wide for						
buff; one coat seal	-	-	-	-	m²	**3.18**
buff; two coats seal	-	-	-	-	m²	**5.04**
sand; three coats for seal or oil	-	-	-	-	m²	**15.91**

M SURFACE FINISHES

Item	PC £	Labour hours	Labour £	Material £	Unit	Total rate £
M50 RUBBER/PLASTICS/CORK/LINO/CARPET TILING/SHEETING						
Linoleum sheet; Forbo-Nairn "Marmoleum Real" or other equal and approved; level; fixing with adhesive; butt joints; to cement and sand base						
Work to floors; over 300 mm wide						
2.50 mm thick	-	0.37	7.98	9.90	m²	**17.88**
3.20 mm thick; marbled	-	0.37	7.98	12.28	m²	**20.26**
Linoleum sheet; Forbo-Nairn "Walton" or other equal and approved; level; with welded seams; fixing with adhesive; to cement and sand base						
Work to floors; over 300 mm wide						
2.50 mm thick; plain	-	0.46	9.92	11.37	m²	**21.29**
Vinyl sheet; Altro "Safety" range or other equal and approved; with welded seams; level; fixing with adhesive; to cement and sand base						
Work to floors; over 300 mm wide						
2.00 mm thick; "Marine "	-	0.56	12.08	16.27	m²	**28.34**
2.50 mm thick; "Classic 25"	-	0.65	14.02	18.10	m²	**32.12**
3.50 mm thick; "Stronghold 30"	-	0.74	15.96	24.43	m²	**40.39**
Slip resistant vinyl sheet; Forbo-Nairn "Surestep" or other equal and approved; level with welded seams; fixing with adhesive; to cement and sand base						
Work to floors; over 300 mm wide						
2.00 mm thick	-	0.46	9.92	12.19	m²	**22.11**
Homogeneous Vinyl sheet; Marleyflor "Plus" or other equal and approved; level; with welded seams; fixing with adhesive; level; to cement and sand base						
Work to floors; over 300 mm wide						
2.00 mm thick	-	0.42	9.06	6.01	m²	**15.07**
2.00 mm thick skirtings						
100 mm high	-	0.11	2.37	1.45	m	**3.82**
Safety sheet; Marleyflor "Granite Multisafe" or other equal and approved; level; with welded seams; fixing with adhesive; level; to cement and sand base						
Work to floors; over 300 mm wide						
2.00 mm thick	-	0.42	9.06	11.63	m²	**20.69**
Vinyl sheet; Marley "Omnisports" or other equal and approved; level; with welded seams; fixing with adhesive; level; to cement and sand base						
Work to floors; over 300 mm wide						
7.65 mm thick; Pro	-	0.90	19.41	23.40	m²	**42.81**
8.75 mm thick; Competition	-	1.00	21.57	27.37	m²	**48.94**

M SURFACE FINISHES

Item	PC £	Labour hours	Labour £	Material £	Unit	Total rate £
M50 RUBBER/PLASTICS/CORK/LINO/CARPET TILING/SHEETING – cont'd						
Vinyl sheet; Gerflor "Gerflex" standard sheet; "Classic" range or other equal and approved; level; with welded seams; fixing with adhesive; to cement and sand base						
Work to floors; over 300 mm wide						
2.00 mm thick	-	0.46	9.92	6.22	m²	**16.14**
Vinyl sheet; Armstrong "Royal" or other equal and approved; level; with welded seams; fixing with adhesive; to cement and sand base						
Work to floors; over 300 mm wide						
2.50 mm thick	-	0.46	9.92	10.50	m²	**20.42**
Vinyl tiles; Armstrong "Royal" or other equal and approved; level; fixing with adhesive; butt joints; straight both ways; to cement and sand base						
Work to floors; over 300 mm wide						
608 mm x 608 mm x 2.00 mm thick	-	0.20	4.31	10.74	m²	**15.05**
Vinyl semi-flexible tiles; Armstrong "Imperial" or other equal and approved; level; fixing with adhesive; butt joints; straight both ways; to cement and sand base						
Work to floors; over 300 mm wide						
250 mm x 250 mm x 2.00 mm thick	-	0.23	4.96	6.33	m²	**11.30**
Vinyl semi-flexible tiles; Marley Homogeneous tiles range or other equal and approved; level; fixing with adhesive; butt joints; straight both ways; to cement and sand base						
Work to floors; over 300 mm wide						
300 mm x 300 mm x 2.00 mm thick; Vylon Plus	-	0.23	4.96	5.14	m²	**10.10**
500 mm x 500 mm x 2.00 mm thick; Marleyflor Plus	-	0.20	4.31	6.25	m²	**10.56**
Vinyl tiles; "Polyflex Plus" or other equal and approved; level; fixing with adhesive; butt joints; straight both ways; to cement and sand base						
Work to floors; over 300 mm wide						
300 mm x 300 mm x 1.50 mm thick	-	0.23	4.96	4.35	m²	**9.31**
300 mm x 300 mm x 2.00 mm thick	-	0.23	4.96	4.80	m²	**9.76**
Vinyl tiles; "Polyflor XL" or other equal and approved; level; fixing with adhesive; butt joints; straight both ways; to cement and sand base						
Work to floors; over 300 mm wide						
300 mm x 300 mm x 2.00 mm thick	-	0.32	6.90	5.46	m²	**12.36**

M SURFACE FINISHES

Item	PC £	Labour hours	Labour £	Material £	Unit	Total rate £
Vinyl tiles; "Polyflor SD"; level; fixing with adhesive; butt joints; straight both ways; to cement and sand base						
Work to floors; over 300 mm wide						
457 mm x 457 mm x 2.00 mm thick	-	0.42	9.06	9.53	m²	**18.58**
Luxury mineral vinyl tiles; Marley "I D Naturelle" or other equal and approved; level; fixing with adhesive; butt joints; straight both ways; to cement and sand base						
Work to floors; over 300 mm wide						
330 mm x 330 mm x 2.00 mm thick	-	0.23	4.96	8.23	m²	**13.19**
Acoustic vinyl tiles; Marley "Tapiflex 243" or other equal and approved; level; fixing with adhesive; butt joints; straight both ways; to cement and sand base						
Work to floors; over 300 mm wide						
500 mm x 500 mm x 2.00 mm thick	-	0.20	4.31	11.00	m²	**15.31**
Linoleum tiles; Marley "Veneto XF" or other equal and approved; level; fixing with adhesive; butt joints; straight both ways; to cement and sand base						
Work to floors; over 300 mm wide						
500 mm x 500 mm x 2.50 mm thick	-	0.20	4.31	12.98	m²	**17.29**
PVC Wall lining; Altro Whiterock; or other equal and approved; fixed directly to plastered brick or blockwork						
Work to walls						
over 300 mm wide	-	-	-	-	m²	**54.12**
not exceeding 300 mm wide	-	-	-	-	m	**27.06**
Linoleum tiles; BS 6826; Forbo-Nairn Floors or other equal and approved; level; fixing with adhesive; butt joints; straight both ways; to cement and sand base						
Work to floors; over 300 mm wide						
2.50 mm thick (marble pattern)	-	0.28	6.04	11.11	m²	**17.15**
Cork tiles Wicanders "Cork-Master" or other equal and approved; level; fixing with adhesive; butt joints; straight both ways; to cement and sand base						
Work to floors; over 300 mm wide						
300 mm x 300 mm x 4.00 mm thick	-	0.37	7.98	21.26	m²	**29.24**
Rubber studded tiles; Altro "Mondopave" or other equal and approved; level; fixing with adhesive; butt joints; straight to cement and sand base						
Work to floors; over 300 mm wide						
500 mm x 500 mm x 2.50 mm thick; type MRB; black	-	0.56	12.08	25.40	m²	**37.48**
500 mm x 500 mm x 4.00 mm thick; type MRB; black	-	0.56	12.08	28.86	m²	**40.94**
Work to landings; over 300 mm wide						
500 mm x 500 mm x 4.00 mm thick; type MRB; black	-	0.74	15.96	28.86	m²	**44.82**
4.00 mm thick to tread						
275 mm wide	-	0.46	9.92	8.58	m	**18.50**
4.00 mm thick to riser						
180 mm wide	-	0.56	12.08	6.09	m	**18.17**

M SURFACE FINISHES

Item	PC £	Labour hours	Labour £	Material £	Unit	Total rate £
M50 RUBBER/PLASTICS/CORK/LINO/CARPET TILING/SHEETING – cont'd						
Sundry floor sheeting underlays						
For floor finishings; over 300 mm wide						
building paper to BS 1521; class A; 75 mm lap (laying only)	-	0.05	0.58	-	m²	**0.58**
3.20 mm thick hardboard	-	0.19	5.28	1.60	m²	**6.87**
6.00 mm thick plywood	-	0.28	7.77	6.65	m²	**14.42**
Skirtings; plastic; Gradus or equivalent						
Set-in skirtings						
100 mm high; ref. SI1002.5P	-	0.11	2.37	1.61	m	**3.99**
150 mm high; ref. SI1502P	-	0.22	4.75	2.43	m	**7.17**
Set-on skirtings						
100 mm high; ref. SO100P	-	0.22	4.75	1.12	m	**5.87**
Stair nosings; aluminium; Gradus or equivalent						
Medium duty hard aluminium alloy stair tread nosings; plugged and screwed in concrete						
56 mm x 32 mm; ref AS11	7.97	0.23	3.53	8.95	m	**12.48**
84 mm x 32 mm; ref AS12	11.87	0.28	4.30	13.24	m	**17.54**
Heavy duty aluminium alloy stair tread nosings; plugged and screwed to concrete						
48 mm x 38 mm; ref HE1	9.40	0.28	4.30	10.53	m	**14.83**
82 mm x 38 mm; ref HE2	14.05	0.32	4.91	15.65	m	**20.56**
Heavy duty carpet tiles; "Heuga 580 Olympic" or other equal and approved; to cement/sand base						
Work to floors						
over 300 mm wide	22.40	0.28	6.04	25.32	m²	**31.35**
M51 EDGE FIXED CARPETING						
Fitted carpeting; Wilton wool/nylon or other equal /approved; 80/20 velvet pile; heavy domestic plain						
Work to floors						
over 300 mm wide	40.33	0.37	5.42	47.74	m²	**53.16**
Work to treads and risers						
over 300 mm wide	-	0.74	10.84	47.74	m²	**58.58**
Underlay to carpeting						
Work to floors						
over 300 mm wide	3.94	0.07	1.03	4.45	m²	**5.47**
raking cutting	-	0.07	0.81	-	m	**0.81**
Sundries						
Carpet gripper fixed to floor; standard edging						
22 mm wide	-	0.04	0.46	0.27	m	**0.73**
M52 DECORATIVE PAPERS/FABRICS						
Lining paper; and hanging						
Plaster walls or columns						
over 300 mm girth (PC £ per roll)	1.98	0.19	2.92	0.36	m²	**3.28**
Plaster ceilings or beams						
over 300 mm girth	-	0.23	3.53	0.36	m²	**3.90**

M SURFACE FINISHES

Item	PC £	Labour hours	Labour £	Material £	Unit	Total rate £
Decorative paper-backed vinyl wallpaper; hanging						
Plaster walls or columns						
over 300 mm girth (PC £ per roll)	10.63	0.23	3.53	2.03	m²	**5.56**
M60 PAINTING/CLEAR FINISHING						
BASIC PAINT PRICES						
Discounts of 10 - 40% available depending on quantity/status						
Paints						
matt emulsion	-	-	-	22.52	5litre	-
gloss	-	-	-	20.04	5litre	-
eggshell gloss	-	-	-	35.74	5litre	-
oil based undercoat	-	-	-	20.04	5litre	-
"Weathershield" gloss	-	-	-	39.59	5litre	-
"Weathershield" undercoat	-	-	-	43.37	5litre	-
"Sandtex" masonry paint						
brilliant white	-	-	-	11.61	5litre	-
coloured	-	-	-	20.63	5litre	-
Primer/undercoats						
acrylic	-	-	-	16.97	5litre	-
red oxide	-	-	-	31.07	5litre	-
water based	-	-	-	20.65	5litre	-
zinc phosphate	-	-	-	38.43	5litre	-
masonry sealer	-	-	-	17.42	5litre	-
mdf primer	-	-	-	29.32	5litre	-
knotting solution	-	-	-	50.36	5litre	-
Special paints						
solar reflective aluminium	-	-	-	42.08	5litre	-
anti-graffiti	-	-	-	133.31	5litre	-
bituminous emulsion	-	-	-	13.73	5litre	-
"Hammerite"	-	-	-	43.18	5litre	-
fire retardant						
undercoat	-	-	-	58.00	5litre	-
top coat	-	-	-	77.25	5litre	-
Stains and Preservatives						
Cuprinol						
"Clear"	-	-	-	24.61	5litre	-
Boiled linseed oil	-	-	-	24.20	5litre	-
Sadolin						
"Extra"	-	-	-	72.85	5litre	-
"New Base"	-	-	-	40.72	5litre	-
Sikkens						
"Cetol HLS"	-	-	-	52.47	5litre	-
"Cetol TS"	-	-	-	75.48	5litre	-
"Cetol Filter 7"	-	-	-	79.82	5litre	-
Protim Solignum						
"Architectural"	-	-	-	72.16	5litre	-
"Green"	-	-	-	30.55	5litre	-
"Cedar"	-	-	-	34.33	5litre	-
Varnishes						
polyurethane	-	-	-	37.30	5litre	-

M SURFACE FINISHES

Item	PC £	Labour hours	Labour £	Material £	Unit	Total rate £
M60 PAINTING/CLEAR FINISHING – cont'd						
SUPPLY AND FIX PRICES						
NOTE: The following prices include for preparing surfaces. Painting woodwork also includes for knotting prior to applying the priming coat and for all stopping of nail holes etc.						
M60 PAINTING/CLEAR FINISHING - INTERNALLY						
One coat primer; on wood surfaces before fixing						
General surfaces						
over 300 mm girth	-	0.08	1.23	0.72	m²	**1.95**
isolated surfaces not exceeding 300 mm girth	-	0.02	0.31	0.26	m	**0.56**
isolated areas not exceeding 0.50 m² irrespective of girth	-	0.06	0.92	0.21	nr	**1.13**
One coat polyurethane sealer; on wood surfaces before fixing						
General surfaces						
over 300 mm girth	-	0.10	1.54	0.71	m²	**2.25**
isolated surfaces not exceeding 300 mm girth	-	0.03	0.46	0.25	m	**0.71**
isolated areas not exceeding 0.50 m²; irrespective of girth	-	0.08	1.23	0.34	nr	**1.57**
One coat of Sikkens "Cetol HLS" stain or other equal and approved; on wood surfaces before fixing						
General surfaces						
over 300 mm girth	-	0.11	1.69	0.82	m²	**2.50**
isolated surfaces not exceeding 300 mm girth	-	0.03	0.46	0.32	m	**0.78**
isolated areas not exceeding 0.50 m²; irrespective of girth	-	0.08	1.23	0.39	nr	**1.62**
One coat of Sikkens "Cetol TS" interior stain or other equal and approved; on wood surfaces before fixing						
General surfaces						
over 300 mm girth	-	0.11	1.69	1.14	m²	**2.83**
isolated surfaces not exceeding 300 mm girth	-	0.03	0.46	0.44	m	**0.90**
isolated areas not exceeding 0.50 m²; irrespective of girth	-	0.08	1.23	0.55	nr	**1.78**
One coat Cuprinol clear wood preservative or other equal and approved; on wood surfaces before fixing						
General surfaces						
over 300 mm girth	-	0.08	1.23	0.53	m²	**1.76**
isolated surfaces not exceeding 300 mm girth	-	0.02	0.31	0.19	m	**0.50**
isolated areas not exceeding 0.50 m²; irrespective of girth	-	0.05	0.77	0.25	nr	**1.02**
One coat HCC Protective Coatings Ltd "Permacor" urethane alkyd gloss finishing coat or other equal and approved; on previously primed steelwork						
Members of roof trusses						
over 300 mm girth	-	0.06	0.92	1.56	m²	**2.48**

M SURFACE FINISHES

Item	PC £	Labour hours	Labour £	Material £	Unit	Total rate £
Two coats emulsion paint						
Brick or block walls						
over 300 mm girth	-	0.21	3.23	1.00	m²	**4.22**
Cement render or concrete						
over 300 mm girth	-	0.20	3.07	0.88	m²	**3.95**
isolated surfaces not exceeding 300 mm girth	-	0.10	1.54	0.27	m	**1.81**
Plaster walls or plaster/plasterboard ceilings						
over 300 mm girth	-	0.18	2.76	0.84	m²	**3.61**
over 300 mm girth; in multi colours	-	0.24	3.69	1.03	m²	**4.71**
over 300 mm girth; in staircase areas	-	0.21	3.23	0.98	m²	**4.20**
cutting in edges on flush surfaces	-	0.08	1.23	-	m	**1.23**
Plaster/plasterboard ceilings						
over 300 mm girth; 3.50 m - 5.00 m high	-	0.21	3.23	0.85	m²	**4.08**
One mist and two coats emulsion paint						
Brick or block walls						
over 300 mm girth	-	0.19	2.92	1.30	m²	**4.21**
Cement render or concrete						
over 300 mm girth	-	0.19	2.92	1.20	m²	**4.11**
Plaster walls or plaster/plasterboard ceilings						
over 300 mm girth	-	0.18	2.76	1.20	m²	**3.96**
over 300 mm girth; in multi colours	-	0.25	3.84	1.22	m²	**5.06**
over 300 mm girth; in staircase areas	-	0.21	3.23	1.20	m²	**4.42**
cutting in edges on flush surfaces	-	0.09	1.38	-	m	**1.38**
Plaster/plasterboard ceilings						
over 300 mm girth; 3.50 m - 5.00 m high	-	0.21	3.23	1.20	m²	**4.42**
One mist Supermatt; one full Supermatt and one full coat of quick drying Acrylic Eggshell						
Brick or block walls						
over 300 mm girth	-	0.19	2.92	1.39	m²	**4.31**
Cement render or concrete						
over 300 mm girth	-	0.19	2.92	1.29	m²	**4.20**
Plaster walls or plaster/plasterboard ceilings						
over 300 mm girth	-	0.18	2.76	1.29	m²	**4.05**
over 300 mm girth; in multi colours	-	0.25	3.84	1.29	m²	**5.12**
over 300 mm girth; in staircase areas	-	0.21	3.23	1.29	m²	**4.51**
cutting in edges on flush surfaces	-	0.09	1.38	-	m	**1.38**
Plaster/plasterboard ceilings						
over 300 mm girth; 3.50 m - 5.00 m high	-	0.21	3.23	1.29	m²	**4.51**
One coat "Tretol No 10 Sealer" or other equal and approved; two coats "Tretol sprayed Supercover Spraytone" emulsion paint or other equal and approved						
Plaster walls or plaster/plasterboard ceilings						
over 300 mm girth	-	-	-	-	m²	**4.92**
Textured plastic; "Artex" or other equal and approved finish						
Plasterboard ceilings						
over 300 mm girth	-	0.19	2.92	1.78	m²	**4.70**
Concrete walls or ceilings						
over 300 mm girth	-	0.23	3.53	1.57	m²	**5.10**

M SURFACE FINISHES

Item	PC £	Labour hours	Labour £	Material £	Unit	Total rate £
M60 PAINTING/CLEAR FINISHING - INTERNALLY – cont'd						
Touch up primer; one undercoat and one finishing coat of gloss oil paint; on wood surfaces						
General surfaces						
over 300 mm girth	-	0.23	3.53	1.66	m²	**5.19**
isolated surfaces not exceeding 300 mm girth	-	0.09	1.38	0.56	m	**1.94**
isolated areas not exceeding 0.50 m²; irrespective of girth	-	0.18	2.76	0.85	nr	**3.62**
Glazed windows and screens						
panes; area not exceeding 0.10 m²	-	0.38	5.84	1.30	m²	**7.14**
panes; area 0.10 m² - 0.50 m²	-	0.31	4.76	0.98	m²	**5.74**
panes; area 0.50 m² - 1.00 m²	-	0.26	3.99	0.80	m²	**4.79**
panes; area over 1.00 m²	-	0.23	3.53	0.68	m²	**4.21**
Knot; one coat primer; stop; one undercoat and one finishing coat of gloss oil paint; on wood surfaces						
General surfaces						
over 300 mm girth	-	0.33	5.07	1.65	m²	**6.72**
isolated surfaces not exceeding 300 mm girth	-	0.13	2.00	0.55	m	**2.55**
isolated areas not exceeding 0.50 m²; irrespective of girth	-	0.25	3.84	1.09	nr	**4.93**
Glazed windows and screens						
panes; area not exceeding 0.10 m²	-	0.56	8.60	1.65	m²	**10.25**
panes; area 0.10 m² - 0.50 m²	-	0.45	6.91	1.38	m²	**8.29**
panes; area 0.50 m² - 1.00 m²	-	0.40	6.14	1.38	m²	**7.52**
panes; area over 1.00 m²	-	0.33	5.07	1.02	m²	**6.08**
One coat primer; one undercoat and one finishing coat of gloss oil paint						
Plaster surfaces						
over 300 mm girth	-	0.30	4.61	2.09	m²	**6.70**
One coat primer; two undercoats and one finishing coat of gloss oil paint						
Plaster surfaces						
over 300 mm girth	-	0.40	6.14	2.82	m²	**8.96**
One coat primer; two undercoats and one finishing coat of eggshell paint						
Plaster surfaces						
over 300 mm girth	-	0.40	6.14	3.10	m²	**9.24**
Touch up primer; one undercoat and one finishing coat of gloss paint; on iron or steel surfaces						
General surfaces						
over 300 mm girth	-	0.23	3.53	1.26	m²	**4.79**
isolated surfaces not exceeding 300 mm girth	-	0.09	1.38	0.43	m	**1.81**
isolated areas not exceeding 0.50 m²; irrespective of girth	-	0.18	2.76	0.68	nr	**3.44**
Glazed windows and screens						
panes; area not exceeding 0.10 m²	-	0.38	5.84	1.29	m²	**7.13**
panes; area 0.10 m² - 0.50 m²	-	0.31	4.76	0.99	m²	**5.75**
panes; area 0.50 m² - 1.00 m²	-	0.26	3.99	0.78	m²	**4.77**
panes; area over 1.00 m²	-	0.23	3.53	0.66	m²	**4.19**
Structural steelwork						
over 300 mm girth	-	0.25	3.84	1.30	m²	**5.14**

M SURFACE FINISHES

Item	PC £	Labour hours	Labour £	Material £	Unit	Total rate £
Members of roof trusses						
over 300 mm girth	-	0.34	5.22	1.49	m²	**6.71**
Ornamental railings and the like; each side measured overall						
over 300 mm girth	-	0.40	6.14	1.62	m²	**7.77**
Iron or steel radiators						
over 300 mm girth	-	0.23	3.53	1.36	m²	**4.90**
Pipes or conduits						
over 300 mm girth	-	0.34	5.22	1.42	m²	**6.64**
not exceeding 300 mm girth	-	0.13	2.00	0.47	m	**2.47**
One coat primer; one undercoat and one finishing coat of gloss oil paint; on iron or steel surfaces						
General surfaces						
over 300 mm girth	-	0.30	4.61	1.26	m²	**5.87**
isolated surfaces not exceeding 300 mm girth	-	0.12	1.84	0.73	m	**2.58**
isolated areas not exceeding 0.50 m²; irrespective of girth	-	0.23	3.53	1.22	nr	**4.75**
Glazed windows and screens						
panes; area not exceeding 0.10 m²	-	0.50	7.68	1.99	m²	**9.67**
panes; area 0.10 m² - 0.50 m²	-	0.40	6.14	1.58	m²	**7.73**
panes; area 0.50 m² - 1.00 m²	-	0.34	5.22	1.37	m²	**6.59**
panes; area over 1.00 m²	-	0.30	4.61	1.22	m²	**5.82**
Structural steelwork						
over 300 mm girth	-	0.33	5.07	1.95	m²	**7.01**
Members of roof trusses						
over 300 mm girth	-	0.45	6.91	2.04	m²	**8.95**
Ornamental railings and the like; each side measured overall						
over 300 mm girth	-	0.51	7.83	2.43	m²	**10.26**
Iron or steel radiators						
over 300 mm girth	-	0.30	4.61	2.04	m²	**6.65**
Pipes or conduits						
over 300 mm girth	-	0.45	6.91	2.04	m²	**8.95**
not exceeding 300 mm girth	-	0.18	2.76	0.67	m	**3.44**
Two coats of bituminous paint; on iron or steel surfaces						
General surfaces						
over 300 mm girth	-	0.23	3.53	0.73	m²	**4.26**
Inside of galvanized steel cistern						
over 300 mm girth	-	0.34	5.22	0.89	m²	**6.11**
Two coats bituminous paint; first coat blinded with clean sand prior to second coat; on concrete surfaces						
General surfaces						
over 300 mm girth	-	0.79	12.13	1.66	m²	**13.79**
Mordant solution; one coat HCC Protective Coatings Ltd "Permacor Alkyd MIO" or other equal and approved; one coat "Permatex Epoxy Gloss" finishing coat or other equal and approved on galvanised steelwork						
Structural steelwork						
over 300 mm girth	-	0.44	6.76	2.81	m²	**9.56**

M SURFACE FINISHES

Item	PC £	Labour hours	Labour £	Material £	Unit	Total rate £
M60 PAINTING/CLEAR FINISHING - INTERNALLY – cont'd						
One coat HCC Protective Coatings Ltd "Epoxy Zinc Primer" or other equal and approved; two coats "Permacor Alkyd MIO" or other equal and approved; one coat "Permacor Epoxy Gloss" finishing coat or other equal and approved on steelwork						
Structural steelwork						
over 300 mm girth	-	0.63	9.68	5.14	m²	**14.81**
Steel protection; HCC Protective Coatings Ltd "Unitherm" or other equal and approved; two coats to steelwork						
Structural steelwork						
over 300 mm girth	-	0.99	15.20	1.75	m²	**16.96**
Two coats of epoxy anti-slip floor paint; on screeded concrete surfaces						
General surfaces						
over 300 mm girth	-	0.25	3.84	10.77	m²	**14.61**
"Nitoflor Lithurin" floor hardener and dust proofer or other equal and approved; Fosroc Expandite Ltd; two coats; on concrete surfaces						
General surfaces						
over 300 mm girth	-	0.24	2.76	0.58	m²	**3.34**
Two coats of boiled linseed oil; on hardwood surfaces						
General surfaces						
over 300 mm girth	-	0.18	2.76	1.65	m²	**4.42**
isolated surfaces not exceeding 300 mm girth	-	0.07	1.08	0.53	m	**1.61**
isolated areas not exceeding 0.50 m²; irrespective of girth	-	0.13	2.00	0.96	nr	**2.96**
Two coats polyurethane varnish; on wood surfaces						
General surfaces						
over 300 mm girth	-	0.18	2.76	1.26	m²	**4.02**
isolated surfaces not exceeding 300 mm girth	-	0.07	1.08	0.46	m	**1.54**
isolated areas not exceeding 0.50 m²; irrespective of girth	-	0.13	2.00	0.15	nr	**2.15**
Three coats polyurethane varnish; on wood surfaces						
General surfaces						
over 300 mm girth	-	0.26	3.99	1.91	m²	**5.91**
isolated surfaces not exceeding 300 mm girth	-	0.10	1.54	0.59	m	**2.12**
isolated areas not exceeding 0.50 m²; irrespective of girth	-	0.19	2.92	1.07	nr	**3.98**

M SURFACE FINISHES

Item	PC £	Labour hours	Labour £	Material £	Unit	Total rate £
One undercoat; and one finishing coat; of "Albi" clear flame retardant surface coating or other equal and approved; on wood surfaces						
General surfaces						
over 300 mm girth	-	0.34	5.22	2.49	m²	**7.71**
isolated surfaces not exceeding 300 mm girth	-	0.14	2.15	0.87	m	**3.02**
isolated areas not exceeding 0.50 m²; irrespective of girth	-	0.19	2.92	1.91	nr	**4.82**
Two undercoats; and one finishing coat; of "Albi" clear flame retardant surface coating or other equal and approved; on wood surfaces						
General surfaces						
over 300 mm girth	-	0.40	6.14	3.51	m²	**9.65**
isolated surfaces not exceeding 300 mm girth	-	0.20	3.07	1.27	m	**4.34**
isolated areas not exceeding 0.50 m²; irrespective of girth	-	0.33	5.07	1.94	nr	**7.01**
Seal and wax polish; dull gloss finish on wood surfaces						
General surfaces						
over 300 mm girth	-	-	-	-	m²	**9.62**
isolated surfaces not exceeding 300 mm girth	-	-	-	-	m	**4.33**
isolated areas not exceeding 0.50m²; irrespective of girth	-	-	-	-	nr	**6.73**
One coat of "Sadolin Extra" or other equal and approved; clear or pigmented; one further coat of "Holdex" clear interior silk matt lacquer or similar						
General surfaces						
over 300 mm girth	-	0.25	3.84	4.77	m²	**8.61**
isolated surfaces not exceeding 300 mm girth	-	0.10	1.54	2.23	m	**3.76**
isolated areas not exceeding 0.50 m²; irrespective of girth	-	0.20	3.07	2.32	nr	**5.39**
Glazed windows and screens						
panes; area not exceeding 0.10 m²	-	0.42	6.45	2.73	m²	**9.18**
panes; area 0.10 m² - 0.50 m²	-	0.33	5.07	2.54	m²	**7.61**
panes; area 0.50 m² - 1.00 m²	-	0.29	4.45	2.35	m²	**6.81**
panes; area over 1.00 m²	-	0.25	3.84	2.23	m²	**6.07**
Two coats of "Sadolin Extra" or other equal and approved; clear or pigmented; two further coats of "Holdex" clear interior silk matt lacquer or similar						
General surfaces						
over 300 mm girth	-	0.40	6.14	8.81	m²	**14.95**
isolated surfaces not exceeding 300 mm girth	-	0.16	2.46	4.40	m	**6.86**
isolated areas not exceeding 0.50 m²; irrespective of girth	-	0.30	4.61	5.03	nr	**9.63**
Glazed windows and screens						
panes; area not exceeding 0.10 m²	-	0.66	10.14	5.40	m²	**15.54**
panes; area 0.10 m² - 0.50 m²	-	0.52	7.99	5.03	m²	**13.01**
panes; area 0.50 m² - 1.00 m²	-	0.45	6.91	4.65	m²	**11.56**
panes; area over 1.00 m²	-	0.40	6.14	4.40	m²	**10.55**

M SURFACE FINISHES

Item	PC £	Labour hours	Labour £	Material £	Unit	Total rate £
M60 PAINTING/CLEAR FINISHING - INTERNALLY – cont'd						
Two coats of Sikkens "Cetol TS" interior stain or other equal and approved; on wood surfaces						
General surfaces						
over 300 mm girth	-	0.19	2.92	2.05	m²	**4.97**
isolated surfaces not exceeding 300 mm girth	-	0.08	1.23	0.72	m	**1.95**
isolated areas not exceeding 0.50 m²; irrespective of girth	-	0.13	2.00	1.12	nr	**3.11**
Body in and wax polish; dull gloss finish; on hardwood surfaces						
General surfaces						
over 300 mm girth	-	-	-	-	m²	**10.82**
isolated surfaces not exceeding 300 mm girth	-	-	-	-	m	**4.88**
isolated areas not exceeding 0.50 m²; irrespective of girth	-	-	-	-	nr	**7.59**
Stain; body in and wax polish; dull gloss finish; on hardwood surfaces						
General surfaces						
over 300 mm girth	-	-	-	-	m²	**14.48**
isolated surfaces not exceeding 300 mm girth	-	-	-	-	m	**6.52**
isolated areas not exceeding 0.50 m²; irrespective of girth	-	-	-	-	nr	**10.15**
Seal; two coats of synthetic resin lacquer; decorative flatted finish; wire down, wax and burnish; on wood surfaces						
General surfaces						
over 300 mm girth	-	-	-	-	m²	**18.25**
isolated surfaces not exceeding 300 mm girth	-	-	-	-	m	**8.53**
isolated areas not exceeding 0.50 m²; irrespective of girth	-	-	-	-	nr	**12.84**
Stain; body in and fully French polish; full gloss finish; on hardwood surfaces						
General surfaces						
over 300 mm girth	-	-	-	-	m²	**21.11**
isolated surfaces not exceeding 300 mm girth	-	-	-	-	m	**9.50**
isolated areas not exceeding 0.50 m²; irrespective of girth	-	-	-	-	nr	**14.78**
Stain; fill grain and fully French polish; full gloss finish; on hardwood surfaces						
General surfaces						
over 300 mm girth	-	-	-	-	m²	**31.39**
isolated surfaces not exceeding 300 mm girth	-	-	-	-	m	**14.12**
isolated areas not exceeding 0.50 m²; irrespective of girth	-	-	-	-	nr	**21.97**
Stain black; body in and fully French polish; ebonized finish; on hardwood surfaces						
General surfaces						
over 300 mm girth	-	-	-	-	m²	**35.80**
isolated surfaces not exceeding 300 mm girth	-	-	-	-	m	**16.11**
isolated areas not exceeding 0.50 m²; irrespective of girth	-	-	-	-	nr	**25.06**

M SURFACE FINISHES

Item	PC £	Labour hours	Labour £	Material £	Unit	Total rate £
M60 PAINTING/CLEAR FINISHING – EXTERNALLY						
Two coats of cement paint, "Sandtex Matt" or other equal and approved						
Brick or block walls						
over 300 mm girth	-	0.26	3.99	1.28	m²	**5.27**
Cement render or concrete walls						
over 300 mm girth	-	0.23	3.53	0.85	m²	**4.38**
Roughcast walls						
over 300 mm girth	-	0.40	6.14	0.85	m²	**6.99**
One coat sealer and two coats of external grade emulsion paint, Dulux "Weathershield" or other equal and approved						
Brick or block walls						
over 300 mm girth	-	0.43	6.60	5.95	m²	**12.56**
Cement render or concrete walls						
over 300 mm girth	-	0.35	5.38	3.97	m²	**9.34**
Concrete soffits						
over 300 mm girth	-	0.40	6.14	3.97	m²	**10.11**
One coat sealer (applied by brush) and two coats of external grade emulsion paint, Dulux "Weathershield" or other equal and approved (spray applied)						
Roughcast						
over 300 mm girth	-	0.29	4.45	8.04	m²	**12.49**
One coat sealer and two coats of anti-graffiti paint (spray applied)						
Brick or block walls						
over 300 mm girth	-	0.23	3.53	2.75	m²	**6.29**
Cement render or concrete walls						
over 300 mm girth	-	0.26	3.99	4.10	m²	**8.09**
2.5 mm of "Vandalene" anti-climb paint (spray applied)						
General surfaces						
over 300 mm girth	-	0.30	4.61	3.90	m²	**8.51**
Two coats solar reflective aluminium paint; on bituminous roofing						
General surfaces						
over 300 mm girth	-	0.44	6.76	11.65	m²	**18.41**
over 300 mm girth	-	0.44	6.76	11.65	m²	**18.41**
Touch up primer; two undercoats and one finishing coat of gloss oil paint; on wood surfaces						
General surfaces						
over 300 mm girth	-	0.35	5.38	1.65	m²	**7.03**
isolated surfaces not exceeding 300 mm girth	-	0.15	2.30	0.44	m	**2.74**
isolated areas not exceeding 0.50 m²; irrespective of girth	-	0.27	4.15	0.88	nr	**5.03**
Glazed windows and screens						
panes; area not exceeding 0.10 m²	-	0.59	9.06	1.46	m²	**10.52**
panes; area 0.10 m² - 0.50 m²	-	0.59	9.06	1.23	m²	**10.29**
panes; area 0.50 m² - 1.00 m²	-	0.47	7.22	1.07	m²	**8.29**
panes; area over 1.00 m²	-	0.35	5.38	0.88	m²	**6.25**

M SURFACE FINISHES

Item	PC £	Labour hours	Labour £	Material £	Unit	Total rate £
M60 PAINTING/CLEAR FINISHING – EXTERNALLY – cont'd						
Touch up primer; two undercoats and one finishing coat of gloss oil paint – cont'd						
Glazed windows and screens; multi-coloured work						
panes; area not exceeding 0.10 m²	-	0.68	10.44	1.46	m²	**11.90**
panes; area 0.10 m² - 0.50 m²	-	0.55	8.45	1.27	m²	**9.71**
panes; area 0.50 m² - 1.00 m²	-	0.47	7.22	1.07	m²	**8.29**
panes; area over 1.00 m²	-	0.41	6.30	0.88	m²	**7.18**
Knot; one coat primer; two undercoats and one finishing coat of gloss oil paint; on wood surfaces						
General surfaces						
over 300 mm girth	-	0.46	7.06	1.79	m²	**8.85**
isolated surfaces not exceeding 300 mm girth	-	0.19	2.92	0.64	m	**3.56**
isolated areas not exceeding 0.50 m²; irrespective of girth	-	0.35	5.38	1.24	nr	**6.62**
Glazed windows and screens						
panes; area not exceeding 0.10 m²	-	0.78	11.98	2.00	m²	**13.98**
panes; area 0.10 m² - 0.50 m²	-	0.62	9.52	1.77	m²	**11.30**
panes; area 0.50 m² - 1.00 m²	-	0.55	8.45	1.36	m²	**9.80**
panes; area over 1.00 m²	-	0.46	7.06	0.94	m²	**8.00**
Glazed windows and screens; multi-coloured work						
panes; area not exceeding 0.10 m²	-	0.89	13.67	2.00	m²	**15.66**
panes; area 0.10 m² - 0.50 m²	-	0.72	11.06	1.79	m²	**12.84**
panes; area 0.50 m² - 1.00 m²	-	0.64	9.83	1.36	m²	**11.18**
panes; area over 1.00 m²	-	0.54	8.29	0.94	m²	**9.23**
Touch up primer; two undercoats and one finishing coat of gloss oil paint; on iron or steel surfaces						
General surfaces						
over 300 mm girth	-	0.35	5.38	1.39	m²	**6.76**
isolated surfaces not exceeding 300 mm girth	-	0.14	2.15	0.38	m	**2.53**
isolated areas not exceeding 0.50 m²; irrespective of girth	-	0.26	3.99	0.79	nr	**4.78**
Glazed windows and screens						
panes; area not exceeding 0.10 m²	-	0.59	9.06	1.44	m²	**10.50**
panes; area 0.10 m² - 0.50 m²	-	0.47	7.22	1.24	m²	**8.46**
panes; area 0.50 m² - 1.00 m²	-	0.41	6.30	1.04	m²	**7.34**
panes; area over 1.00 m²	-	0.35	5.38	0.84	m²	**6.21**
Structural steelwork						
over 300 mm girth	-	0.40	6.14	1.44	m²	**7.58**
Members of roof trusses						
over 300 mm girth	-	0.54	8.29	1.64	m²	**9.93**
Ornamental railings and the like; each side measured overall						
over 300 mm girth	-	0.60	9.21	1.71	m²	**10.92**
Eaves gutters						
over 300 mm girth	-	0.64	9.83	1.81	m²	**11.64**
not exceeding 300 mm girth	-	0.25	3.84	0.75	m	**4.59**
Pipes or conduits						
over 300 mm girth	-	0.54	8.29	1.81	m²	**10.10**
not exceeding 300 mm girth	-	0.21	3.23	0.62	m	**3.84**

M SURFACE FINISHES

Item	PC £	Labour hours	Labour £	Material £	Unit	Total rate £
One coat primer; two undercoats and one finishing coat of gloss oil paint; on iron or steel surfaces						
General surfaces						
over 300 mm girth	-	0.43	6.60	1.66	m²	**8.27**
isolated surfaces not exceeding 300 mm girth	-	0.18	2.76	0.43	m	**3.19**
isolated areas not exceeding 0.50 m²; irrespective of girth	-	0.32	4.91	0.86	nr	**5.77**
Glazed windows and screens						
panes; area not exceeding 0.10 m²	-	0.71	10.90	1.51	m²	**12.42**
panes; area 0.10 m² - 0.50 m²	-	0.56	8.60	1.31	m²	**9.91**
panes; area 0.50 m² - 1.00 m²	-	0.50	7.68	1.11	m²	**8.79**
panes; area over 1.00 m²	-	0.43	6.60	0.86	m²	**7.46**
Structural steelwork						
over 300 mm girth	-	0.48	7.37	1.71	m²	**9.08**
Members of roof trusses						
over 300 mm girth	-	0.64	9.83	1.91	m²	**11.74**
Ornamental railings and the like; each side measured overall						
over 300 mm girth	-	0.72	11.06	1.91	m²	**12.97**
Eaves gutters						
over 300 mm girth	-	0.76	11.67	2.06	m²	**13.73**
not exceeding 300 mm girth	-	0.31	4.76	0.70	m	**5.46**
Pipes or conduits						
over 300 mm girth	-	0.64	9.83	2.06	m²	**11.89**
not exceeding 300 mm girth	-	0.25	3.84	0.68	m	**4.52**
One coat of Andrews "Hammerite" paint or other equal and approved; on iron or steel surfaces						
General surfaces						
over 300 mm girth	-	0.15	2.30	1.15	m²	**3.45**
isolated surfaces not exceeding 300 mm girth	-	0.08	1.23	0.36	m	**1.59**
isolated areas not exceeding 0.50 m²; irrespective of girth	-	0.11	1.69	0.66	nr	**2.35**
Glazed windows and screens						
panes; area not exceeding 0.10 m²	-	0.25	3.84	0.87	m²	**4.71**
panes; area 0.10 m² - 0.50 m²	-	0.19	2.92	0.96	m²	**3.88**
panes; area 0.50 m² - 1.00 m²	-	0.18	2.76	0.87	m²	**3.63**
panes; area over 1.00 m²	-	0.15	2.30	0.87	m²	**3.17**
Structural steelwork						
over 300 mm girth	-	0.17	2.61	1.06	m²	**3.67**
Members of roof trusses						
over 300 mm girth	-	0.23	3.53	1.15	m²	**4.68**
Ornamental railings and the like; each side measured overall						
over 300 mm girth	-	0.26	3.99	1.15	m²	**5.14**
Eaves gutters						
over 300 mm girth	-	0.27	4.15	1.25	m²	**5.39**
not exceeding 300 mm girth	-	0.08	1.23	0.58	m	**1.81**
Pipes or conduits						
over 300 mm girth	-	0.26	3.99	1.06	m²	**5.05**
not exceeding 300 mm girth	-	0.08	1.23	0.48	m	**1.71**

M SURFACE FINISHES

Item	PC £	Labour hours	Labour £	Material £	Unit	Total rate £
M60 PAINTING/CLEAR FINISHING – EXTERNALLY – cont'd						
Two coats of creosote; on wood surfaces						
General surfaces						
over 300 mm girth	-	0.16	2.46	0.33	m²	**2.78**
isolated surfaces not exceeding 300 mm girth	-	0.05	0.77	0.20	m	**0.96**
Two coats of "Solignum" wood preservative or other equal and approved; on wood surfaces						
General surfaces						
over 300 mm girth	-	0.14	2.15	1.08	m²	**3.23**
isolated surfaces not exceeding 300 mm girth	-	0.05	0.77	0.33	m	**1.10**
Three coats of polyurethane; on wood surfaces						
General surfaces						
over 300 mm girth	-	0.29	4.45	2.11	m²	**6.56**
isolated surfaces not exceeding 300 mm girth	-	0.11	1.69	1.04	m	**2.73**
isolated areas not exceeding 0.50 m²; irrespective of girth	-	0.21	3.23	1.20	nr	**4.43**
Two coats of "New Base" primer or other equal and approved; and two coats of "Extra" or other equal and approved; Sadolin Ltd; pigmented; on wood surfaces						
General surfaces						
over 300 mm girth	-	0.43	6.60	4.56	m²	**11.17**
isolated surfaces not exceeding 300 mm girth	-	0.26	3.99	1.56	m	**5.55**
Glazed windows and screens						
panes; area not exceeding 0.10 m²	-	0.71	10.90	3.25	m²	**14.15**
panes; area 0.10 m² - 0.50 m²	-	0.57	8.75	3.06	m²	**11.81**
panes; area 0.50 m² - 1.00 m²	-	0.50	7.68	2.87	m²	**10.55**
panes; area over 1.00 m²	-	0.43	6.60	2.31	m²	**8.91**
Two coats Sikkens "Cetol Filter 7" exterior stain or other equal and approved; on wood surfaces						
General surfaces						
over 300 mm girth	-	0.20	3.07	3.43	m²	**6.50**
isolated surfaces not exceeding 300 mm girth	-	0.09	1.38	1.18	m	**2.56**
isolated areas not exceeding 0.50 m²; irrespective of girth	-	0.14	2.15	1.74	nr	**3.89**

N FURNITURE/EQUIPMENT

Item	PC £	Labour hours	Labour £	Material £	Unit	**Total rate £**
N10/11 GENERAL FIXTURES/KITCHEN FITTINGS						
SUPPLY ONLY PRICES						
NOTE: The fixing of general fixtures will vary considerably dependent upon the size of the fixture and the method of fixing employed. Prices for fixing like sized kitchen fittings may be suitable for certain fixtures, although adjustment to those rates will almost invariably be necessary and the reader is directed to section "G20" for information on bolts, plugging brickwork and blockwork, etc. which should prove useful in building up a suitable rate.						
The following supply only prices are for purpose made fittings components in various materials supplied as part of an assembled fitting and therefore may be used to arrive at a guide price for a complete fitting.						
Fitting components; medium density fibreboard						
Backs, fronts, sides or divisions; over 300 mm wide						
12 mm thick	-	-	-	24.65	m²	-
18 mm thick	-	-	-	26.11	m²	-
25 mm thick	-	-	-	29.08	m²	-
Shelves or worktops; over 300 mm wide						
18 mm thick	-	-	-	26.11	m²	-
25 mm thick	-	-	-	29.08	m²	-
Flush doors; lipped on four edges						
450 mm x 750 mm x 18 mm	-	-	-	37.83	nr	-
450 mm x 750 mm x 25 mm	-	-	-	38.55	nr	-
600 mm x 900 mm x 18 mm	-	-	-	44.65	nr	-
600 mm x 900 mm x 25 mm	-	-	-	45.81	nr	-
Fitting components; moisture-resistant medium density fibreboard						
Backs, fronts, sides or divisions; over 300 mm wide						
12 mm thick	-	-	-	27.59	m²	-
18 mm thick	-	-	-	30.55	m²	-
25 mm thick	-	-	-	33.49	m²	-
Shelves or worktops; over 300 mm wide						
18 mm thick	-	-	-	30.55	m²	-
25 mm thick	-	-	-	33.49	m²	-
Flush doors; lipped on four edges						
450 mm x 750 mm x 18 mm	-	-	-	38.55	nr	-
450 mm x 750 mm x 25 mm	-	-	-	39.64	nr	-
600 mm x 900 mm x 18 mm	-	-	-	45.81	nr	-
600 mm x 900 mm x 25 mm	-	-	-	47.59	nr	-
Fitting components; medium density fibreboard; melamine faced both sides						
Backs, fronts, sides or divisions; over 300 mm wide						
12 mm thick	-	-	-	32.52	m²	-
18 mm thick	-	-	-	36.21	m²	-
Shelves or worktops; over 300 mm wide						
18 mm thick	-	-	-	36.21	m²	-
Flush doors; lipped on four edges						
450 mm x 750 mm x 18 mm	-	-	-	27.05	nr	-
600 mm x 900 mm x 25 mm	-	-	-	34.32	nr	-

N FURNITURE/EQUIPMENT

Item	PC £	Labour hours	Labour £	Material £	Unit	Total rate £
N10/11 GENERAL FIXTURES/KITCHEN FITTINGS - cont'd						
Fitting components; medium density fibreboard; formica faced both sides						
Backs, fronts, sides or divisions; over 300 mm wide						
12 mm thick	-	-	-	99.02	m²	-
18 mm thick	-	-	-	102.97	m²	-
Shelves or worktops; over 300 mm wide						
18 mm thick	-	-	-	102.97	m²	-
Flush doors; lipped on four edges						
450 mm x 750 mm x 18 mm	-	-	-	54.53	nr	-
600 mm x 900 mm x 25 mm	-	-	-	55.71	nr	-
Fitting components; wrought softwood						
Backs, fronts, sides or divisions; cross-tongued joints; over 300 mm wide						
25 mm thick	-	-	-	41.88	m²	-
Shelves or worktops; cross-tongued joints; over 300 mm wide						
25 mm thick	-	-	-	41.88	m²	-
Bearers						
19 mm x 38 mm	-	-	-	2.19	m	-
25 mm x 50 mm	-	-	-	2.43	m	-
44 mm x 44 mm	-	-	-	2.59	m	-
44 mm x 75 mm	-	-	-	2.99	m	-
Bearers; framed; to backs, fronts or sides						
19 mm x 38 mm	-	-	-	5.01	m	-
25 mm x 50 mm	-	-	-	5.42	m	-
50 mm x 50 mm	-	-	-	7.01	m	-
50 mm x 75 mm	-	-	-	8.04	m	-
Add 5% to the above material prices for selected softwood staining						
Fitting components; selected Sapele						
Backs, fronts, sides or divisions; cross-tongued joints; over 300 mm wide						
25 mm thick	-	-	-	78.45	m²	-
Shelves or worktops; cross-tongued joints; over 300 mm wide						
25 mm thick	-	-	-	78.45	m²	-
Bearers						
19 mm x 38 mm	-	-	-	3.93	m	-
25 mm x 50 mm	-	-	-	4.90	m	-
50 mm x 50 mm	-	-	-	5.50	m	-
50 mm x 75 mm	-	-	-	7.16	m	-
Bearers; framed; to backs, fronts or sides						
19 mm x 38 mm	-	-	-	7.60	m	-
25 mm x 50 mm	-	-	-	8.43	m	-
50 mm x 50 mm	-	-	-	11.30	m	-
50 mm x 75 mm	-	-	-	14.16	m	-
Fitting components; Iroko						
Backs, fronts, sides or divisions; cross-tongued joints; over 300 mm wide						
25 mm thick	-	-	-	96.64	m²	-
Shelves or worktops; cross-tongued joints; over 300 mm wide						
25 mm thick	-	-	-	96.64	m²	-

N FURNITURE/EQUIPMENT

Item	PC £	Labour hours	Labour £	Material £	Unit	Total rate £
Draining boards; cross-tongued joints; over 300 mm wide						
25 mm thick	-	-	-	121.08	m²	-
stopped flutes	-	-	-	6.26	m	-
grooves; cross-grain	-	-	-	0.93	m	-
Bearers						
19 mm x 38 mm	-	-	-	4.80	m	-
25 mm x 50 mm	-	-	-	6.14	m	-
50 mm x 50 mm	-	-	-	6.98	m	-
50 mm x 75 mm	-	-	-	9.19	m	-
Bearers; framed; to backs, fronts or sides						
19 mm x 38 mm	-	-	-	8.78	m	-
25 mm x 50 mm	-	-	-	9.78	m	-
50 mm x 50 mm	-	-	-	13.20	m	-
50 mm x 75 mm	-	-	-	17.15	m	-
SUPPLY AND FIX PRICES						
NOTE: Kitchen fittings vary considerably. PC supply prices for reasonable quantities for a moderately priced range of kitchen fittings have been shown.						
Supplying and fixing to backgrounds requiring plugging; including any pre-assembly						
Wall units						
300 mm x 300 mm x 720 mm	56.66	1.11	17.05	61.25	nr	**78.29**
500 mm x 300 mm x 720 mm	66.76	1.16	17.82	72.12	nr	**89.94**
600 mm x 300 mm x 720 mm	74.82	1.30	19.97	80.80	nr	**100.76**
800 mm x 300 mm x 720 mm	113.85	1.48	22.73	122.80	nr	**145.53**
Floor units with drawers						
500 mm x 600 mm x 870 mm	99.43	1.16	17.82	107.28	nr	**125.10**
600 mm x 600 mm x 870 mm	110.68	1.30	19.97	119.39	nr	**139.35**
1000 mm x 600 mm x 870 mm	170.58	1.57	24.11	183.86	nr	**207.97**
Sink units (excluding sink top)						
1000 mm x 600 mm x 870 mm	173.18	1.48	22.73	186.66	nr	**209.39**
Laminated plastics worktops; single rolled edge; prices include for fixing						
38 mm thick; 600 mm wide	26.93	0.37	5.68	30.56	m	**36.25**
extra for forming hole for inset sink	-	0.69	10.60	-	nr	**10.60**
extra for jointing strip at corner intersection of worktops	-	0.14	2.15	6.35	nr	**8.50**
extra for butt and scribe joint at corner intersection of worktops	-	4.16	63.89	-	nr	**63.89**
Lockers and cupboards; Welconstruct Distribution or other equal and approved						
Standard clothes lockers; steel body and door within reinforced 19G frame, powder coated finish, cam locks						
1 compartment; placing in position						
300 mm x 300 mm x 1800 mm	-	0.23	2.65	49.59	nr	**52.24**
380 mm x 380 mm x 1800 mm	-	0.23	2.65	60.70	nr	**63.35**
450 mm x 450 mm x 1800 mm	-	0.28	3.22	69.22	nr	**72.45**
Compartment lockers; steel body and door within reinforced 19G frame, powder coated finish, cam locks						
2 compartments; placing in position						
300 mm x 300 mm x 1800 mm	-	0.23	2.65	54.67	nr	**57.32**
380 mm x 380 mm x 1800 mm	-	0.23	2.65	49.59	nr	**52.24**
450 mm x 450 mm x 1800 mm	-	0.28	3.22	75.25	nr	**78.47**

N FURNITURE/EQUIPMENT

Item	PC £	Labour hours	Labour £	Material £	Unit	Total rate £
N10/11 GENERAL FIXTURES/KITCHEN FITTINGS - cont'd						
Lockers and cupboards; Welconstruct - cont'd						
4 compartments; placing in position						
300 mm x 300 mm x 1800 mm	-	0.23	2.65	64.92	nr	**67.57**
380 mm x 380 mm x 1800 mm	-	0.23	2.65	77.75	nr	**80.39**
450 mm x 450 mm x 1800 mm	-	0.28	3.22	77.75	nr	**80.97**
Timber clothes lockers; veneered MDF finish, routed door, cam locks						
1 compartment; placing in position						
380 mm x 380 mm x 1830 mm	-	0.28	3.22	180.81	nr	**184.03**
4 compartments; placing in position						
380 mm x 380 mm x 1830 mm	-	0.28	3.22	263.47	nr	**266.69**
Cupboards; stainless steel; cam locks						
900 mm x 460 mm x 900 mm; one shelf	-	0.23	2.65	523.49	nr	**526.13**
900 mm x 460 mm x 1200 mm; two shelves	-	0.23	2.65	594.09	nr	**596.74**
900 mm x 460 mm x 1800 mm; three shelves	-	0.23	2.65	736.15	nr	**738.80**
1200 mm x 460 mm x 1800 mm; three shelves	-	0.23	2.65	953.99	nr	**956.63**
Shelving support systems; The Welconstruct Company or other equal and approved						
Shelving support systems; steel body; stove enamelled finish; assembling						
open initial bay; 5 shelves; placing in position						
1000 mm x 300 mm x 1850 mm	-	0.69	10.11	92.90	nr	**103.01**
1000 mm x 600 mm x 1850 mm	-	0.69	10.11	134.32	nr	**144.42**
open extension bay; 5 shelves; placing in position						
1000 mm x 300 mm x 1850 mm	-	0.83	12.16	74.05	nr	**86.20**
1000 mm x 600 mm x 1850 mm	-	0.83	12.16	105.13	nr	**117.29**
closed initial bay; 5 shelves; placing in position						
1000 mm x 300 mm x 1850 mm	-	0.69	10.11	123.90	nr	**134.00**
1000 mm x 600 mm x 1850 mm	-	0.69	10.11	169.62	nr	**179.72**
closed extension bay; 5 shelves; placing in position						
1000 mm x 300 mm x 1850 mm	-	0.83	12.16	105.04	nr	**117.20**
1000 mm x 600 mm x 1850 mm	-	0.83	12.16	140.34	nr	**152.50**
extra for pair of doors; fixing in position						
1000 mm x 1850 mm	-	0.75	10.99	133.89	nr	**144.87**
Cloakroom racks; The Welconstruct Company or other equal and approved						
Cloakroom racks; 40 mm x 40 mm square tube framing, polyester powder coated finish; beech slatted seats and rails to one side only; placing in position						
1675 mm x 325 mm x 1500 mm; 5 nr coat hooks	-	0.30	4.39	317.71	nr	**322.10**
1825 mm x 325 mm x 1500 mm; 15 nr coat hangers	-	0.30	4.39	372.81	nr	**377.21**
Extra for						
shoe baskets	-	-	-	73.27	nr	**-**
mesh bottom shelf	-	-	-	51.23	nr	**-**
Cloakroom racks; 40 mm x 40 mm square tube framing, polyester powder coated finish; beech slatted seats and rails to both sides; placing in position						
1675 mm x 600 mm x 1500 mm; 10 nr coat hooks	-	0.40	5.86	407.25	nr	**413.11**
1825 mm x 600 mm x 1500 mm; 30 nr coat hangers	-	0.40	5.86	464.94	nr	**470.80**
Extra for						
shoe baskets	-	-	-	102.03	nr	**-**
mesh bottom shelf	-	-	-	62.25	nr	**-**

N FURNITURE/EQUIPMENT

Item	PC £	Labour hours	Labour £	Material £	Unit	Total rate £
6 mm thick rectangular glass mirrors; silver backed; fixed with chromium plated domed headed screws; to background requiring plugging						
Mirror with polished edges						
365 mm x 254 mm	6.78	0.74	11.37	8.66	nr	**20.02**
400 mm x 300 mm	8.84	0.74	11.37	11.21	nr	**22.58**
560 mm x 380 mm	15.32	0.83	12.75	19.23	nr	**31.98**
640 mm x 460 mm	20.03	0.93	14.28	25.05	nr	**39.33**
Mirror with bevelled edges						
365 mm x 254 mm	12.08	0.74	11.37	15.22	nr	**26.59**
400 mm x 300 mm	14.15	0.74	11.37	17.77	nr	**29.14**
560 mm x 380 mm	23.57	0.83	12.75	29.44	nr	**42.18**
640 mm x 460 mm	29.46	0.93	14.28	36.73	nr	**51.01**
Door mats						
Entrance mats; "Tuftiguard type C" or other equal and approved; laying in position; 12 mm thick						
900 mm x 550 mm	114.34	0.46	5.29	123.06	nr	**128.36**
1200 mm x 750 mm	205.82	0.46	5.29	221.51	nr	**226.81**
2400 mm x 1200 mm	658.63	0.93	10.70	708.85	nr	**719.55**
Matwells						
Polished aluminium matwell; comprising angle rim with brazed angles and lugs brazed on; to suit mat size						
914 mm x 560 mm; constructed with 25 x 25 x 3 mm angle	25.94	0.93	10.70	27.92	nr	**38.62**
1067 mm x 610 mm; constructed with 34 x 26 x 6 mm angle	35.79	0.93	10.70	38.52	nr	**49.22**
1219 mm x 762 mm; constructed with 50 x 50 x 6 mm angle	87.39	0.93	10.70	94.05	nr	**104.75**
Polished brass matwell; comprising angle rim with brazed angles and lugs brazed on; to suit mat size						
914 mm x 560 mm; constructed with 25 x 25 x 5 mm angle	100.60	0.93	10.70	108.28	nr	**118.97**
1067 mm x 610 mm; constructed with 38 x 38 x 6 mm angle	146.71	0.93	10.70	157.90	nr	**168.60**
Internal blinds; Luxaflex Ltd or other equal and approved						
Roller blinds; Luxaflex EOS type 10 roller; "Compact Fabric"; plain type material; 1219 mm drop; fixing with screws						
1016 mm wide	37.80	0.93	10.70	40.68	nr	**51.38**
2031 mm wide	55.80	1.45	16.68	60.05	nr	**76.74**
2843 mm wide	69.30	1.97	22.66	74.58	nr	**97.25**
Roller blinds; Luxaflex EOS type 10 roller; "Compact Fabric"; fire resisting material; 1219mm drop; fixing with screws						
1016 mm wide	49.50	0.93	10.70	53.27	nr	**63.97**
2031 mm wide	73.80	1.45	16.68	79.43	nr	**96.11**
2843 mm wide	93.60	1.97	22.66	100.74	nr	**123.40**
Roller blinds; Luxaflex EOS type 10 roller; "Light resistant"; blackout material; 1219 mm drop; fixing with screws						
1016 mm wide	63.90	0.93	10.70	68.77	nr	**79.47**
2031 mm wide	107.10	1.45	16.68	115.27	nr	**131.95**
2843 mm wide	144.90	1.97	22.66	155.95	nr	**178.61**

N FURNITURE/EQUIPMENT

Item	PC £	Labour hours	Labour £	Material £	Unit	Total rate £
N10/11 GENERAL FIXTURES/KITCHEN FITTINGS - cont'd						
Internal blinds; Luxaflex - cont'd						
Roller blinds; Luxaflex "Lite-master Crank Op"; 100% blackout; 1219 mm drop; fixing with screws						
1016 mm wide	196.02	1.96	22.55	210.97	nr	**233.52**
2031 mm wide	262.35	2.75	31.64	282.35	nr	**313.99**
2843 mm wide	338.58	3.53	40.61	364.40	nr	**405.01**
Vertical louvre blinds; 89 mm wide louvres; Luxaflex EOS type; "Florida Fabric"; 1219 mm drop; fixing with screws						
1016 mm wide	51.30	0.82	9.43	55.21	nr	**64.65**
2031 mm wide	78.30	1.30	14.96	84.27	nr	**99.23**
3046 mm wide	107.10	1.77	20.36	115.27	nr	**135.63**
Vertical louvre blinds; 127 mm wide louvres; Luxaflex EOS type; "Florida Fabric"; 1219 mm drop; fixing with screws						
1016 mm wide	43.20	0.88	10.12	46.49	nr	**56.62**
2031 mm wide	65.70	1.35	15.53	70.71	nr	**86.24**
3046 mm wide	88.20	1.81	20.82	94.93	nr	**115.75**
N13 SANITARY APPLIANCES/FITTINGS						
Sinks; Armitage Shanks or equal and approved						
Sinks; white glazed fireclay; BS 6465; pointing all round with Dow Corning Hansil silicone sealant						
Belfast sink; 46 cm x 38 cm x 21 cm ref S580001; pair of Nuastyle 21 basin taps with dual indices, chrome handle ref B8262AA; wall mounts ref S8331AA; 38 mm slotted waste, chain and plug, screw stay ref S8766AA; pair of 40.5 cm aluminium alloy build-in brackets with 35.5 cm studs ref S921967; screwing	165.68	2.78	57.72	226.67	nr	**284.39**
Belfast sink; 61 cm x 38 cm x 21 cm ref S580501; pair of Nuastyle 21 basin taps with dual indices, chrome handle ref B8262AA; wall mounts ref S8331AA; 38 mm slotted waste, chain and plug, screw stay ref S8766AA; pair of 40.5 cm aluminium alloy build-in brackets with 35.5 cm studs ref S921967; screwing	196.06	2.78	57.72	259.53	nr	**317.25**
Belfast sink; 76 cm x 38 cm x 21 cm ref S581101; pair of Nuastyle 21 basin taps with dual indices, chrome handle ref B8262AA; wall mounts ref S8331AA; 38 mm slotted waste, chain and plug, screw stay ref S8766AA; pair of 40.5 cm aluminium alloy build-in brackets with 35.5 cm studs ref S921967; screwing	272.96	2.78	57.72	342.45	nr	**400.17**

N FURNITURE/EQUIPMENT

Item	PC £	Labour hours	Labour £	Material £	Unit	Total rate £
Lavatory basins; Armitage Shanks or equal and approved						
Basins; white vitreous china; BS 6465 Part 3; pointing all round with Dow Corning Hansil silcone sealant						
Portman 21 40 cm basin ref S231701; with overflow, chain hole and two tapholes ; pair of Nuastyle 21 basin taps with dual indices ref B8262AA; slotted basin waste with plastic plug, chain waste and plug ref S8800AA; 32 x 75 mm seal plastic standard bottle trap ref S891067; pair of Portman concealed brackets with waste support ref S915067; Isovalve 15 mm plastic servicing valve with outlet for copper ref S900067; screwing	80.12	2.13	44.22	117.59	nr	**161.81**
Portman 21 50 cm basin ref S230901; with overflow, chain hole and two tapholes ; pair of Nuastyle 21 basin taps with dual indices ref B8262AA; slotted basin waste with plastic plug, chain waste and plug ref S8800AA; 32 x 75 mm seal plastic standard bottle trap ref S891067; pair of Portman concealed brackets with waste support ref S915067; Isovalve 15 mm plastic servicing valve with outlet for copper ref S900067; screwing	97.67	2.13	44.22	136.58	nr	**180.81**
Portman 21 60 cm basin ref S225701; with overflow, chain hole and two tapholes ; pair of Nuastyle 21 basin taps with dual indices ref B8262AA; slotted basin waste with plastic plug, chain waste and plug ref S8800AA; 32 x 75 mm seal plastic standard bottle trap ref S891067; pair of Portman concealed brackets with waste support ref S915067; Isovalve 15 mm plastic servicing valve with outlet for copper ref S900067; screwing	128.29	2.13	44.22	169.64	nr	**213.86**
Tiffany 51 cm pedestal basin ref S208001; with two tapholes ; Millenia STD dual control one taphole standard basin mixer with pop-up waste ref S7300AA; pair of Millenia STD handles ref B8000AA; Full pedestal (IS group S2920) ref S292001; Isovalve 15 mm plastic servicing valve with outlet for copper ref S900067; screwing	139.66	2.31	47.96	165.33	nr	**213.29**
Tiffany 56 cm pedestal basin ref S208301; with two tapholes ; Millenia STD dual control one taphole standard basin mixer with pop-up waste ref S7300AA; pair of Millenia STD handles ref B8000AA; Full pedestal (IS group S2920) ref S292001; Isovalve 15 mm plastic servicing valve with outlet for copper ref S900067; screwing	136.77	2.31	47.96	162.22	nr	**210.18**
Tiffany 61 cm pedestal basin ref S208601; with two tapholes ; Millenia STD dual control one taphole standard basin mixer with pop-up waste ref S7300AA; pair of Millenia STD handles ref B8000AA; Full pedestal (IS group S2920) ref S292001; Isovalve 15 mm plastic servicing valve with outlet for copper ref S900067; screwing	142.58	2.31	47.96	170.50	nr	**218.46**
Montana 51 cm pedestal basin ref S210101; with one taphole ; Millenia STD dual control one taphole standard basin mixer with pop-up waste ref S7300AA; pair of Millenia STD handles ref B8000AA; Full pedestal (IS group S2920) ref S292001; Isovalve 15 mm plastic servicing valve with outlet for copper ref S900067; screwing	132.50	2.31	47.96	157.57	nr	**205.53**

N FURNITURE/EQUIPMENT

Item	PC £	Labour hours	Labour £	Material £	Unit	Total rate £
N13 SANITARY APPLIANCES/FITTINGS - cont'd						
Lavatory basins; Armitage Shanks - cont'd						
Montana 58 cm pedestal basin ref S210401; with one taphole ; Millenia STD dual control one taphole standard basin mixer with pop-up waste ref S7300AA; pair of Millenia STD handles ref B8000AA; Full pedestal (IS group S2920) ref S292001; Isovalve 15 mm plastic servicing valve with outlet for copper ref S900067; screwing	135.29	2.31	47.96	160.65	nr	**208.61**
Drinking fountains; Armitage Shanks or equal and approved						
White vitreous china fountains; pointing all round with Dow Corning Hansil silicone selant						
Aqualon wall mounted drinking fountain ref S540101; Aqualon self closing valve with fittings and plastic waste ref S5402AA; 32 x 75 mm seal plastic standard bottle trap ref S891067; screwing	228.20	2.31	47.96	252.82	nr	**300.77**
Polished stainless steel fountains; pointing all round with Dow Corning Hansil silicone selant						
Purita wall mounted drinking fountain ref S5435MY with self closing valve and fittings; 32mm unslotted basin strainer waste ref S8720AA; screwing	219.51	2.31	47.96	236.61	nr	**284.57**
Purita pedestal mounted drinking fountain 90 cm high ref S5440MY with self closing valve and fittings; 32mm unslotted basin strainer waste ref S8720AA; screwing	468.88	2.78	57.72	505.81	nr	**563.52**
Baths; Armitage Shanks or equal and approved						
Sandringham acrylic rectangular bath with chrome plated grips and two tapholes ref S159301; Sandringha,m STD pair of standard bath taps with chrome handles ref S7032AA; bath chain waste with plastic plug and overflow ref S8830AA; cast brass "P" trap with plain outlet and overflow connection; pointing with Dow Corning Hansil silicone selant						
170 cm long x 70 cm wide; white or coloured	111.59	3.50	72.66	120.10	nr	**192.76**
Nisa lowline heavy gauge steel rectangular bath with chrome plated grips and two tapholes ref S176501; Sandringha,m STD pair of standard bath taps with chrome handles ref S7032AA; bath chain waste with plastic plug and overflow ref S8830AA; cast brass "P" trap with plain outlet and overflow connection; pointing with Dow Corning Hansil silicone selant						
170 cm long x 70 cm wide; white or coloured	282.77	3.50	72.66	304.33	nr	**376.99**
Water closets; Armitage Shanks or equal and approved						
White vitreous china pans and cisterns; pointing all round base with Dow Corning Hansil silicone sealant						
Wentworth close coupled washdown closet pan with horizontal outlet ref S316101; Orion 3 plastic toilet seat and cover ref S404501; Panketa pan connector 14° finned ref S430501; Universal close coupled bottom inlet cistern with syphon ref S392001	129.01	3.05	63.32	144.79	nr	**208.12**

N FURNITURE/EQUIPMENT

Item	PC £	Labour hours	Labour £	Material £	Unit	Total rate £
Tiffany back to wall washdown closet pan with horizontal outlet ref S341001; Saturn plastic toilet seat and cover ref S404001; Panketa pan connector 14° finned ref S430501; Conceala 2 6 litre low level side inlet cistern with syphon and lever ref S361767	169.39	3.05	63.32	188.25	nr	**251.57**
Extra over for; Panketa pan connector 90° finned ref S430001	-	-	-	1.40	nr	-
Tiffany close coupled washdown closet pan with horizontal outlet (IS group S3080) ref S308001; Saturn plastic toilet seat and cover ref S404001; Panketa pan connector 14° finned ref S430501; Tiffany 7½ litre close coupled cistern with dual flush valve ref S365001	173.88	3.05	63.32	193.09	nr	**256.41**
Extra over for; Panketa pan connector 90° finned ref S430001	-	-	-	1.40	nr	-
Cameo close coupled washdown closet pan with horizontal outlet (IS group S3080) ref S308001; Accolade/Cameo plastic toilet seat and cover ref S402501; Panketa pan connector 14° finned ref S430501; Cameo 6 litre close coupled cistern with dual flush valve ref S361301	217.16	3.05	63.32	239.67	nr	**302.99**
Extra over for; Panketa pan connector 90° finned ref S430001	-	-	-	1.40	nr	-
Wall urinals; Armitage Shanks or equal and approved						
White vitreous china bowls and cisterns; pointing all round with Dow Corning Hansil silicone sealant						
Single Sanura 40 cm urinal bowl ref S610501; Sanura top inlet spreader ref S6285AA; pair of wall hangers for urinal bowl ref S9725AA; 38 mm plastic domed waste ref S885067; 38 x 75 mm seal plastic standard bottle trap ref S891567; Conceala 4½ litres capacity auto cistern and cover ref S621567; Sanura concealed flushpipe for single urinal bowl ref S6226NU; screwing	152.92	3.70	76.82	182.89	nr	**259.71**
Single Sanura 40 cm urinal bowl ref S610501; Sanura top inlet spreader ref S6285AA; pair of wall hangers for urinal bowl ref S9725AA; 38 mm plastic domed waste ref S885067; 38 x 75 mm seal plastic standard bottle trap ref S891567; Mura 4½ litres capacity auto cistern and cover ref S620001; Sanura/Mura exposed flushpipe for single urinal bowl ref S6220MY; screwing	172.64	3.70	76.82	204.12	nr	**280.93**
Single Sanura 50 cm urinal bowl ref S610001; Sanura top inlet spreader ref S6285AA; pair of wall hangers for urinal bowl ref S9725AA; 38 mm plastic domed waste ref S885067; 38 x 75 mm seal plastic standard bottle trap ref S891567; Conceala 4½ litres capacity auto cistern and cover ref S621567; Sanura concealed flushpipe for single urinal bowl ref S6226NU; screwing	208.02	3.70	76.82	242.27	nr	**319.09**
Single Sanura 50 cm urinal bowl ref S610001; Sanura top inlet spreader ref S6285AA; pair of wall hangers for urinal bowl ref S9725AA; 38 mm plastic domed waste ref S885067; 38 x 75 mm seal plastic standard bottle trap ref S891567; Mura 4½ litres capacity.auto cistern and cover ref S620001; Sanura/Mura exposed flushpipe for single urinal bowl ref S6220MY; screwing	227.74	3.70	76.82	263.50	nr	**340.31**

N FURNITURE/EQUIPMENT

Item	PC £	Labour hours	Labour £	Material £	Unit	Total rate £
N13 SANITARY APPLIANCES/FITTINGS - cont'd						
Wall urinals; Armitage Shanks - cont'd						
Range of 2 nr Sanura 40 cm urinal bowls ref S610501; Sanura top inlet spreader ref S6285AA; pairs of wall hangers for urinal bowls ref S9725AA; 38 mm plastic domed wastes ref S885067; 38 x 75 mm seal plastic standard bottle traps ref S891567; Conceala 9 litres capacity auto cistern and cover ref S621667; Sanura concealed flushpipe for range of 2 nr urinal bowls ref S6227NU; screwing	257.00	6.95	144.29	313.22	nr	**457.51**
Range of 2 nr Sanura 50 cm urinal bowls ref S610001; Sanura top inlet spreader ref S6285AA; pairs of wall hangers for urinal bowls ref S9725AA; 38 mm plastic domed wastes ref S885067; 38 x 75 mm seal plastic standard bottle traps ref S891567; Conceala 9 litres capacity auto cistern and cover ref S621667; Sanura concealed flushpipe for range of 2 nr urinal bowls ref S6227NU; screwing	367.20	6.95	144.29	431.98	nr	**576.27**
Range of 3 nr Sanura 40 cm urinal bowls ref S610501; Sanura top inlet spreader ref S6285AA; pairs of wall hangers for urinal bowls ref S9725AA; 38 mm plastic domed wastes ref S885067; 38 x 75 mm seal plastic standard bottle traps ref S891567; Conceala 9 litres capacity auto cistern and cover ref S621667; Sanura concealed flushpipe for range of 3 nr urinal bowls ref S6228NU; screwing	357.51	10.15	210.72	439.71	nr	**650.43**
Range of 3 nr Sanura 50 cm urinal bowls ref S610001; Sanura top inlet spreader ref S6285AA; pairs of wall hangers for urinal bowls ref S9725AA; 38 mm plastic domed wastes ref S885067; 38 x 75 mm seal plastic standard bottle traps ref S891567; Conceala 9 litres capacity auto cistern and cover ref S621667; Sanura concealed flushpipe for range of 3 nr urinal bowls ref S6228NU; screwing	522.81	10.15	210.72	617.61	nr	**828.34**
Range of 4 nr Sanura 40 cm urinal bowls ref S610501; Sanura top inlet spreader ref S6285AA; pairs of wall hangers for urinal bowls ref S9725AA; 38 mm plastic domed wastes ref S885067; 38 x 75 mm seal plastic standard bottle traps ref S891567; Conceala 9 litres capacity auto cistern and cover ref S621767; Sanura concealed flushpipe for range of 4 nr urinal bowls ref S6229NU; screwing	461.73	13.40	278.20	570.19	nr	**848.38**
Range of 4 nr Sanura 50 cm urinal bowls ref S610001; Sanura top inlet spreader ref S6285AA; pairs of wall hangers for urinal bowls ref S9725AA; 38 mm plastic domed wastes ref S885067; 38 x 75 mm seal plastic standard bottle traps ref S891567; Conceala 9 litres capacity auto cistern and cover ref S621767; Sanura concealed flushpipe for range of 4 nr urinal bowls ref S6229NU; screwing	682.13	13.40	278.20	807.71	nr	**1085.90**
Range of 5 nr Sanura 40 cm urinal bowls ref S610501; Sanura top inlet spreader ref S6285AA; pairs of wall hangers for urinal bowls ref S9725AA; 38 mm plastic domed wastes ref S885067; 38 x 75 mm seal plastic standard bottle traps ref S891567; Conceala 9 litres capacity auto cistern and cover ref S621767; Sanura concealed flushpipe for range of 5 nr urinal bowls ref S6230NU; screwing	562.26	16.65	345.67	696.70	nr	**1042.37**

N FURNITURE/EQUIPMENT

Item	PC £	Labour hours	Labour £	Material £	Unit	Total rate £
Range of 5 nr Sanura 50 cm urinal bowls ref S610001; Sanura top inlet spreader ref S6285AA; pairs of wall hangers for urinal bowls ref S9725AA; 38 mm plastic domed wastes ref S885067; 38 x 75 mm seal plastic standard bottle traps ref S891567; Conceala 9 litres capacity auto cistern and cover ref S621767; Sanura concealed flushpipe for range of 5 nr urinal bowls ref S6230NU; screwing	837.76	16.65	345.67	993.60	nr	**1339.26**
White vitreous china division panels; pointing all round with Dow Corning Hansill silicone sealant						
Urinal division with screw and hanger ref S612001; screwing	47.24	0.70	14.53	51.70	nr	**66.24**
Bidets; Armitage Shanks or equal and approved						
Tiffany back to wall bidet with one taphole ref S491001; vitreous china; chromium plated pop-up waste and mixer tap with hand wheels refs S7500AA and S8000AA						
58 cm x 39 cm; white or coloured	253.07	3.50	72.66	272.37	nr	**345.03**
Shower tray and fittings						
Simplicity shower tray; acrylic; with outlet and grated waste; chain and plug; bedding and pointing in waterproof cement mortar						
760 mm x 760 mm ; white or coloured	124.38	3.00	62.28	133.86	nr	**196.15**
Shower fitting; riser pipe with mixing valve and shower rose; chromium plated; plugging and screwing mixing valve and pipe bracket						
15 mm diameter riser pipe; 127 mm diameter shower rose	243.36	5.00	103.80	261.92	nr	**365.72**
Miscellaneous fittings; Magrini Ltd or equal and approved						
Vertical nappy changing unit						
ref KBCS; screwing	-	0.60	9.21	254.19	nr	**263.41**
Horizontal nappy changing unit						
ref KBHS; screwing	-	0.60	9.21	254.19	nr	**263.41**
Stay Safe baby seat						
ref KBPS; screwing	-	0.55	8.45	63.51	nr	**71.96**
Miscellaneous fittings; Pressalit Ltd or equal and approved						
Grab rails						
300 mm long ref RT100000; screwing	-	0.50	7.68	46.43	nr	**54.11**
450 mm long ref RT101000; screwing	-	0.50	7.68	54.40	nr	**62.08**
600 mm long ref RT102000; screwing	-	0.50	7.68	62.32	nr	**70.00**
800 mm long ref RT103000; screwing	-	0.50	7.68	69.97	nr	**77.65**
1000 mm long ref RT104000; screwing	-	0.50	7.68	80.34	nr	**88.02**
Angled grab rails						
900 mm long, angled 135° ref RT110000; screwing	-	0.50	7.68	100.82	nr	**108.50**
1300 mm long, angled 90° ref RT119000; screwing	-	0.75	11.52	157.92	nr	**169.44**
Hinged grab rails						
600 mm long ref R3016000 ; screwing	-	0.35	5.38	163.33	nr	**168.71**
600 mm long with spring counter balance ref RF016000 ; screwing	-	0.35	5.38	227.50	nr	**232.87**
850 mm long ref R3010000 ; screwing	-	0.35	5.38	198.01	nr	**203.38**
850 mm long with spring counter balance ref RF010000 ; screwing	-	0.35	5.38	243.88	nr	**249.26**

N FURNITURE/EQUIPMENT

Item	PC £	Labour hours	Labour £	Material £	Unit	Total rate £
N15 SIGNS/NOTICES						
Plain script; in gloss oil paint; on painted or varnished surfaces						
Capital letters; lower case letters or numerals						
per coat; per 25 mm high	-	0.09	1.38	-	nr	**1.38**
Stops						
per coat	-	0.02	0.31	-	nr	**0.31**

P BUILDING FABRIC SUNDRIES

Item	PC £	Labour hours	Labour £	Material £	Unit	Total rate £
P10 SUNDRY INSULATION/PROOFING WORK/ FIRE STOPS						
ALTERNATIVE INSULATION PRICES						
Discounts of 10 - 50% available depending on quantity/status						
Insulation (£/m²)						
"Crown FrameTherm Roll 40"						
90 mm	-	-	-	4.37	m²	-
140 mm	-	-	-	6.34	m²	-
"Crown FrameTherm Roll 35"						
90 mm	-	-	-	10.41	m²	-
140 mm	-	-	-	14.90	m²	-
"Crown Factoryclad 40"						
80 mm	-	-	-	3.51	m²	-
100 mm	-	-	-	4.28	m²	-
"Crown Factoryclad 37"						
100 mm	-	-	-	8.10	m²	-
120 mm	-	-	-	9.50	m²	-
"Crown Factoryclad 35"						
100 mm	-	-	-	10.94	m²	-
"Crown Factoryclad 32"						
100 mm	-	-	-	18.90	m²	-
SUPPLY AND FIX PRICES						
"Sisalkraft" building papers/vapour barriers or other equal and approved						
Building paper; 150 mm laps; fixed to softwood						
"Moistop" grade 728 (class A1F)	-	0.08	1.17	0.79	m²	**1.96**
Vapour barrier/reflective insulation 150 mm laps; fixed to softwood						
"Insulex" grade 714; single sided	-	0.08	1.17	0.93	m²	**2.10**
Mat or quilt insulation						
Glass fibre roll; "Crown Loft Roll" or other equal and approved; laid loose between members at 600 mm centres						
100 mm thick	2.67	0.09	1.32	3.02	m²	**4.34**
150 mm thick	3.94	0.10	1.46	4.46	m²	**5.92**
200 mm thick	5.32	0.11	1.61	6.01	m²	**7.62**
Glass fibre quilt; Isowool "Modular roll" or other equal and approved; laid loose between members at 600 mm centres						
60 mm thick	2.48	0.09	1.32	2.67	m²	**3.99**
80 mm thick	3.24	0.10	1.46	3.49	m²	**4.95**
100 mm thick	3.86	0.11	1.61	4.16	m²	**5.77**
150 mm thick	5.90	0.12	1.76	6.35	m²	**8.11**
Mineral fibre quilt; Isowool " APR 1200" or other equal and approved; pinned vertically to softwood						
25 mm thick	1.70	0.08	1.17	1.83	m²	**3.01**
50 mm thick	2.69	0.09	1.32	2.90	m²	**4.22**

P BUILDING FABRIC SUNDRIES

Item	PC £	Labour hours	Labour £	Material £	Unit	Total rate £
P10 SUNDRY INSULATION/PROOFING WORK/ FIRE STOPS – cont'd						
Mat or quilt insulation – cont'd						
"Crown Dritherm Cavity Slab 37" glass fibre batt; as full or partial cavity fill; including cutting and fitting around wall ties and retaining discs						
50 mm thick	2.91	0.12	1.76	3.47	m²	**5.22**
75 mm thick	3.46	0.13	1.90	4.08	m²	**5.98**
100 mm thick	4.50	0.14	2.05	5.25	m²	**7.31**
"Crown Dritherm Cavity Slab 34" glass fibre batt; as full or partial cavity fill; including cutting and fitting around wall ties and retaining discs						
65 mm thick	5.25	0.12	1.76	6.10	m²	**7.86**
75 mm thick	6.07	0.13	1.90	7.04	m²	**8.94**
85 mm thick	7.90	0.13	1.90	9.11	m²	**11.01**
100 mm thick	7.90	0.14	2.05	9.11	m²	**11.16**
"Crown Dritherm Cavity Slab 32" glass fibre batt; as full or partial cavity fill; including cutting and fitting around wall ties and retaining discs						
65 mm thick	6.94	0.12	1.76	8.02	m²	**9.78**
75 mm thick	8.02	0.13	1.90	9.23	m²	**11.14**
85 mm thick	9.03	0.13	1.90	10.38	m²	**12.29**
100 mm thick	10.46	0.14	2.05	11.99	m²	**14.04**
"Crown Frametherm Roll 40" glass fibre semi-rigid or rigid batt; pinned vertically in timber frame construction						
90 mm thick	3.50	0.14	2.05	3.95	m²	**6.00**
140 mm thick	5.07	0.16	2.34	5.73	m²	**8.08**
"Crown Rafter Roll 32" glass fibre flanged building roll; pinned vertically or to slope between timber framing						
50 mm thick	4.85	0.13	1.90	5.48	m²	**7.38**
75 mm thick	6.94	0.14	2.05	7.85	m²	**9.90**
100 mm thick	8.94	0.15	2.20	10.11	m²	**12.30**
Board or slab insulation						
Expanded polystyrene board standard grade SD/N or other equal and approved; fixed with adhesive						
20 mm thick	-	0.14	2.93	4.10	m²	**7.03**
25 mm thick	-	0.14	2.93	4.92	m²	**7.85**
30 mm thick	-	0.14	2.93	5.75	m²	**8.68**
40 mm thick	-	0.15	3.14	6.57	m²	**9.71**
50 mm thick	-	0.16	3.35	9.06	m²	**12.41**
60 mm thick	-	0.17	3.56	10.71	m²	**14.27**
75 mm thick	-	0.18	3.77	13.19	m²	**16.96**
100 mm thick	-	0.19	3.98	17.32	m²	**21.30**
"KIngspan Thermawall TW50" zero ODP rigid urethene insulation board; as full or partial cavity fill; including cutting and fitting around wall ties and retaining discs						
50 mm thick	5.37	0.17	3.56	6.24	m²	**9.80**
75 mm thick	8.00	0.18	3.77	9.21	m²	**12.98**
100 mm thick	10.40	0.19	3.98	11.93	m²	**15.91**
"Styrofoam Floormate 500" extruded polystyrene foam or other equal and approved						
50 mm thick	-	0.46	9.63	5.65	m²	**15.28**
80 mm thick	-	0.46	9.63	9.94	m²	**19.58**
120 mm thick	-	0.46	9.63	14.92	m²	**24.55**

P BUILDING FABRIC SUNDRIES

Item	PC £	Labour hours	Labour £	Material £	Unit	Total rate £
Fire stops						
Cape "Firecheck" channel; intumescent coatings on cut mitres; fixing with brass cups and screws						
19 mm x 44 mm or 19 mm x 50 mm	9.20	0.56	11.73	10.83	m	**22.55**
"Sealmaster" intumescent fire and smoke seals; pinned into groove in timber						
type N30; for single leaf half hour door	6.08	0.28	5.86	6.87	m	**12.74**
type N60; for single leaf one hour door	9.26	0.31	6.49	10.47	m	**16.96**
type IMN or IMP; for meeting or pivot stiles of pair of one hour doors; per stile	9.26	0.31	6.49	10.47	m	**16.96**
intumescent plugs in timber; including boring	-	0.09	1.88	0.31	nr	**2.19**
Rockwool fire stops or other equal and approved; between top of brick/block wall and concrete soffit						
30 mm deep x 100 mm wide	-	0.07	1.47	3.31	m	**4.78**
30 mm deep x 150 mm wide	-	0.09	1.88	5.02	m	**6.90**
30 mm deep x 200 mm wide	-	0.11	2.30	6.71	m	**9.01**
60 mm deep x 100 mm wide	-	0.08	1.68	4.90	m	**6.58**
60 mm deep x 150 mm wide	-	0.10	2.09	7.29	m	**9.38**
60 mm deep x 200 mm wide	-	0.12	2.51	9.82	m	**12.33**
90 mm deep x 100 mm wide	-	0.10	2.09	6.45	m	**8.54**
90 mm deep x 150 mm wide	-	0.12	2.51	9.65	m	**12.16**
90 mm deep x 200 mm wide	-	0.14	2.93	12.88	m	**15.81**
Fire protection compound						
Quelfire QF4, fire protection; filling around pipes, ducts and the like; including all necessary formwork						
300 mm x 300 mm x 250 mm; pipes - 2	-	0.93	16.98	11.65	nr	**28.63**
500 mm x 500 mm x 250 mm; pipes - 2	-	1.16	20.63	34.96	nr	**55.59**
Fire barriers						
Rockwool fire barrier; between top of suspended ceiling and concrete soffit						
one 50 mm layer x 900 mm wide; half hour	-	0.56	11.73	9.81	m²	**21.54**
two 50 mm layers x 900 mm wide; one hour	-	0.83	17.38	19.39	m²	**36.77**
three 50 mm layers x 900 mm wide; two hour	-	1.10	23.04	26.14	m²	**49.18**
Corofil C144 fire barrier to edge of slab; fixed with non-flammable contact adhesive						
to suit void 30 mm wide x 100 mm deep; one hour	-	-	-	-	m	**15.07**
Lamatherm fire barrier or other equal and approved; to void below raised access floors						
75 mm thick x 300 mm high; half hour	-	0.17	3.56	14.22	m	**17.78**
75 mm thick x 600 mm high; half hour	-	0.17	3.56	32.04	m	**35.60**
90 mm thick x 300 mm high; half hour	-	0.17	3.56	16.95	m	**20.51**
90 mm thick x 600 mm high; half hour	-	0.17	3.56	39.62	m	**43.18**
Dow Chemicals "Styrofoam SP"; cold bridging insulation fixed with adhesive to brick, block or concrete base						
Insulation to walls						
50 mm thick	-	0.33	6.91	4.58	m²	**11.49**
75 mm thick	-	0.35	7.33	7.55	m²	**14.88**
Insulation to isolated columns						
50 mm thick	-	0.41	8.59	4.58	m²	**13.17**
75 mm thick	-	0.43	9.01	7.55	m²	**16.55**
Insulation to ceilings						
50 mm thick	-	0.36	7.54	4.58	m²	**12.12**
75 mm thick	-	0.39	8.17	7.55	m²	**15.72**
Insulation to isolated beams						
50 mm thick	-	0.43	9.01	4.58	m²	**13.59**
75 mm thick	-	0.46	9.63	7.55	m²	**17.18**

P BUILDING FABRIC SUNDRIES

Item	PC £	Labour hours	Labour £	Material £	Unit	Total rate £
P11 FOAMED/FIBRE/BEAD CAVITY WALL INSULATION						
Injected insulation						
Mineral Wool cavity wall insulation; for cavity						
75 mm	-	-	-	-	m²	**4.61**
100 mm	-	-	-	-	m²	**4.61**
P20 UNFRAMED ISOLATED TRIMS/SKIRTINGS/ SUNDRY ITEMS						
Medium density fibreboard (Sapele veneered one side); 18 mm thick						
Window boards and the like; rebated; hardwood lipped on one edge						
18 mm x 200 mm	-	0.25	5.24	16.56	m	**21.80**
18 mm x 250 mm	-	0.28	5.86	17.43	m	**23.29**
18 mm x 300 mm	-	0.31	6.49	17.87	m	**24.36**
18 mm x 350 mm	-	0.33	6.91	19.17	m	**26.08**
returned and fitted ends	-	0.20	4.19	3.24	nr	**7.43**
Medium density fibreboard (American White Ash veneered one side); 18 mm thick						
Window boards and the like; rebated; hardwood lipped on one edge						
18 mm x 200 mm	-	0.25	5.24	17.21	m	**22.45**
18 mm x 250 mm	-	0.28	5.86	18.30	m	**24.16**
18 mm x 300 mm	-	0.31	6.49	18.84	m	**25.33**
18 mm x 350 mm	-	0.33	6.91	20.46	m	**27.37**
returned and fitted ends	-	0.20	4.19	3.24	nr	**7.43**
Wrought softwood						
Skirtings, picture rails, dado rails and the like; splayed or moulded						
19 mm x 44 mm; splayed	-	0.09	1.88	3.19	m	**5.07**
19 mm x 44 mm; moulded	-	0.09	1.88	3.40	m	**5.29**
19 mm x 69 mm; splayed	-	0.09	1.88	3.45	m	**5.33**
19 mm x 69 mm; moulded	-	0.09	1.88	3.45	m	**5.33**
19 mm x 94 mm; splayed	-	0.09	1.88	3.87	m	**5.75**
19 mm x 94 mm; moulded	-	0.09	1.88	3.87	m	**5.75**
19 mm x 144 mm; moulded	-	0.11	2.30	4.55	m	**6.85**
19 mm x 169 mm; moulded	-	0.11	2.30	4.81	m	**7.12**
25 mm x 50 mm; moulded	-	0.09	1.88	3.33	m	**5.22**
25 mm x 69 mm; splayed	-	0.09	1.88	3.69	m	**5.57**
25 mm x 94 mm; splayed	-	0.09	1.88	4.13	m	**6.01**
25 mm x 144 mm; splayed	-	0.11	2.30	5.01	m	**7.31**
25 mm x 144 mm; moulded	-	0.11	2.30	5.01	m	**7.31**
25 mm x 169 mm; moulded	-	0.11	2.30	5.57	m	**7.88**
25 mm x 219 mm; moulded	-	0.13	2.72	7.26	m	**9.98**
returned ends	-	0.14	2.93	-	nr	**2.93**
mitres	-	0.09	1.88	-	nr	**1.88**

P BUILDING FABRIC SUNDRIES

Item	PC £	Labour hours	Labour £	Material £	Unit	Total rate £
Architraves, cover fillets and the like; half round; splayed or moulded						
13 mm x 25 mm; half round	-	0.11	2.30	2.99	m	**5.30**
13 mm x 50 mm; moulded	-	0.11	2.30	3.19	m	**5.49**
16 mm x 32 mm; half round	-	0.11	2.30	3.33	m	**5.64**
16 mm x 38 mm; moulded	-	0.11	2.30	3.33	m	**5.64**
16 mm x 50 mm; moulded	-	0.11	2.30	3.33	m	**5.64**
19 mm x 50 mm; splayed	-	0.11	2.30	3.33	m	**5.64**
19 mm x 63 mm; splayed	-	0.11	2.30	3.45	m	**5.75**
19 mm x 69 mm; splayed	-	0.11	2.30	3.66	m	**5.97**
25 mm x 44 mm; splayed	-	0.11	2.30	3.29	m	**5.59**
25 mm x 50 mm; moulded	-	0.11	2.30	3.46	m	**5.76**
25 mm x 63 mm; splayed	-	0.11	2.30	3.61	m	**5.91**
25 mm x 69 mm; splayed	-	0.11	2.30	4.11	m	**6.41**
32 mm x 88 mm; moulded	-	0.11	2.30	4.13	m	**6.43**
38 mm x 38 mm; moulded	-	0.11	2.30	3.65	m	**5.96**
50 mm x 50 mm; moulded	-	0.11	2.30	4.25	m	**6.56**
returned ends	-	0.14	2.93	-	nr	**2.93**
mitres	-	0.09	1.88	-	nr	**1.88**
Stops; screwed on						
16 mm x 38 mm	-	0.09	1.88	1.50	m	**3.39**
16 mm x 50 mm	-	0.09	1.88	1.63	m	**3.51**
19 mm x 38 mm	-	0.09	1.88	1.50	m	**3.39**
25 mm x 38 mm	-	0.09	1.88	1.64	m	**3.52**
25 mm x 50 mm	-	0.09	1.88	1.68	m	**3.57**
Glazing beads and the like						
13 mm x 16 mm	-	0.04	0.84	1.82	m	**2.66**
13 mm x 19 mm	-	0.04	0.84	1.82	m	**2.66**
13 mm x 25 mm	-	0.04	0.84	1.86	m	**2.70**
13 mm x 25 mm; screwed	-	0.04	0.84	3.04	m	**3.87**
13 mm x 25 mm; fixing with brass cups and screws	-	0.04	0.84	3.90	m	**4.74**
16 mm x 25 mm; screwed	-	0.04	0.84	3.04	m	**3.87**
16 mm quadrant	-	0.04	0.84	2.77	m	**3.61**
19 mm quadrant or scotia	-	0.04	0.84	2.77	m	**3.61**
19 mm x 36 mm; screwed	-	0.04	0.84	3.07	m	**3.91**
25 mm x 38 mm; screwed	-	0.04	0.84	3.22	m	**4.06**
25 mm quadrant or scotia	-	0.04	0.84	2.93	m	**3.77**
38 mm scotia	-	0.04	0.84	3.66	m	**4.50**
50 mm scotia	-	0.04	0.84	4.14	m	**4.98**
Isolated shelves, worktops, seats and the like						
19 mm x 150 mm	-	0.15	3.14	3.85	m	**7.00**
19 mm x 200 mm	-	0.20	4.19	5.32	m	**9.51**
25 mm x 150 mm	-	0.15	3.14	4.40	m	**7.54**
25 mm x 200 mm	-	0.20	4.19	6.25	m	**10.44**
32 mm x 150 mm	-	0.15	3.14	5.14	m	**8.28**
32 mm x 200 mm	-	0.20	4.19	7.02	m	**11.20**
Isolated shelves, worktops, seats and the like; cross-tongued joints						
19 mm x 300 mm	-	0.26	5.45	15.86	m	**21.31**
19 mm x 450 mm	-	0.31	6.49	23.90	m	**30.39**
19 mm x 600 mm	-	0.37	7.75	30.93	m	**38.68**
25 mm x 300 mm	-	0.26	5.45	17.01	m	**22.46**
25 mm x 450 mm	-	0.31	6.49	25.78	m	**32.27**
25 mm x 600 mm	-	0.37	7.75	33.52	m	**41.27**
32 mm x 300 mm	-	0.26	5.45	18.01	m	**23.46**
32 mm x 450 mm	-	0.31	6.49	27.39	m	**33.89**
32 mm x 600 mm	-	0.37	7.75	35.74	m	**43.49**

P BUILDING FABRIC SUNDRIES

Item	PC £	Labour hours	Labour £	Material £	Unit	Total rate £
P20 UNFRAMED ISOLATED TRIMS/SKIRTINGS/ SUNDRY ITEMS – cont'd						
Wrought softwood – cont'd						
Isolated shelves, worktops, seats and the like; slatted with 50 wide slats at 75 mm centres						
19 mm thick	-	0.60	12.57	38.64	m	**51.21**
25 mm thick	-	0.60	12.57	39.50	m	**52.07**
32 mm thick	-	0.60	12.57	40.26	m	**52.83**
Window boards, nosings, bed moulds and the like; rebated and rounded						
19 mm x 75 mm	-	0.17	3.56	4.82	m	**8.38**
19 mm x 150 mm	-	0.19	3.98	5.98	m	**9.96**
19 mm x 225 mm; in one width	-	0.24	5.03	7.41	m	**12.43**
19 mm x 300 mm; cross-tongued joints	-	0.28	5.86	17.20	m	**23.06**
25 mm x 75 mm	-	0.17	3.56	5.09	m	**8.65**
25 mm x 150 mm	-	0.19	3.98	6.57	m	**10.55**
25 mm x 225 mm; in one width	-	0.24	5.03	8.32	m	**13.34**
25 mm x 300 mm; cross-tongued joints	-	0.28	5.86	18.61	m	**24.47**
32 mm x 75 mm	-	0.17	3.56	5.35	m	**8.91**
32 mm x 150 mm	-	0.19	3.98	7.09	m	**11.07**
32 mm x 225 mm; in one width	-	0.24	5.03	9.10	m	**14.12**
32 mm x 300 mm; cross-tongued joints	-	0.28	5.86	19.79	m	**25.66**
38 mm x 75 mm	-	0.17	3.56	5.86	m	**9.42**
38 mm x 150 mm	-	0.19	3.98	8.19	m	**12.17**
38 mm x 225 mm; in one width	-	0.24	5.03	10.64	m	**15.67**
38 mm x 300 mm; cross-tongued joints	-	0.28	5.86	22.18	m	**28.05**
returned and fitted ends	-	0.14	2.93	-	nr	**2.93**
Handrails; mopstick						
50 mm diameter	-	0.23	4.82	11.06	m	**15.88**
Handrails; rounded						
44 mm x 50 mm	-	0.23	4.82	10.72	m	**15.54**
50 mm x 75 mm	-	0.25	5.24	11.71	m	**16.95**
63 mm x 87 mm	-	0.28	5.86	12.98	m	**18.85**
75 mm x 100 mm	-	0.32	6.70	15.91	m	**22.62**
Handrails; moulded						
44 mm x 50 mm	-	0.23	4.82	10.72	m	**15.54**
50 mm x 75 mm	-	0.25	5.24	11.71	m	**16.95**
63 mm x 87 mm	-	0.28	5.86	12.98	m	**18.85**
75 mm x 100 mm	-	0.32	6.70	15.91	m	**22.62**
Add 5% to the above material prices for selected softwood for staining						
Medium density fibreboard						
Skirtings, picture rails, dado rails and the like; splayed or moulded						
18 mm x 50 mm; splayed	-	0.09	1.88	3.09	m	**4.97**
18 mm x 50 mm; moulded	-	0.09	1.88	3.09	m	**4.97**
18 mm x 75 mm; splayed	-	0.09	1.88	3.21	m	**5.10**
18 mm x 75 mm; moulded	-	0.09	1.88	3.21	m	**5.10**
18 mm x 100 mm; splayed	-	0.09	1.88	3.34	m	**5.23**
18 mm x 100 mm; moulded	-	0.09	1.88	3.34	m	**5.23**
18 mm x 150 mm; moulded	-	0.11	2.30	3.65	m	**5.95**
18 mm x 175 mm; moulded	-	0.11	2.30	3.78	m	**6.08**
22 mm x 100 mm; splayed	-	0.09	1.88	5.66	m	**7.54**

P BUILDING FABRIC SUNDRIES

Item	PC £	Labour hours	Labour £	Material £	Unit	Total rate £
25 mm x 50 mm; moulded	-	0.09	1.88	3.23	m	**5.12**
25 mm x 75 mm; splayed	-	0.09	1.88	3.42	m	**5.31**
25 mm x 100 mm; splayed	-	0.09	1.88	3.64	m	**5.52**
25 mm x 150 mm; splayed	-	0.11	2.30	4.09	m	**6.40**
25 mm x 150 mm; moulded	-	0.11	2.30	4.09	m	**6.40**
25 mm x 175 mm; moulded	-	0.11	2.30	4.31	m	**6.61**
25 mm x 225 mm; moulded	-	0.13	2.72	4.61	m	**7.33**
returned ends	-	0.14	2.93	-	nr	**2.93**
mitres	-	0.09	1.88	-	nr	**1.88**
Architraves, cover fillets and the like; half round; splayed or moulded						
12 mm x 25 mm; half round	-	0.11	2.30	2.94	m	**5.25**
12 mm x 50 mm; moulded	-	0.11	2.30	3.03	m	**5.34**
15 mm x 32 mm; half round	-	0.11	2.30	2.96	m	**5.27**
15 mm x 38 mm; moulded	-	0.11	2.30	2.99	m	**5.29**
15 mm x 50 mm; moulded	-	0.11	2.30	3.03	m	**5.34**
18 mm x 50 mm; splayed	-	0.11	2.30	3.03	m	**5.34**
18 mm x 63 mm; splayed	-	0.11	2.30	3.17	m	**5.47**
18 mm x 75 mm; splayed	-	0.11	2.30	3.23	m	**5.54**
25 mm x 44 mm; splayed	-	0.11	2.30	3.23	m	**5.54**
25 mm x 50 mm; moulded	-	0.11	2.30	3.23	m	**5.54**
25 mm x 63 mm; splayed	-	0.11	2.30	3.34	m	**5.65**
25 mm x 75 mm; splayed	-	0.11	2.30	3.46	m	**5.76**
30 mm x 88 mm; moulded	-	0.11	2.30	4.47	m	**6.78**
38 mm x 38 mm; moulded	-	0.11	2.30	3.84	m	**6.14**
50 mm x 50 mm; moulded	-	0.11	2.30	4.03	m	**6.33**
returned ends	-	0.14	2.93	-	nr	**2.93**
mitres	-	0.09	1.88	-	nr	**1.88**
Stops; screwed on						
15 mm x 38 mm	-	0.09	1.88	1.68	m	**3.56**
15 mm x 50 mm	-	0.09	1.88	1.73	m	**3.62**
18 mm x 38 mm	-	0.09	1.88	1.72	m	**3.61**
25 mm x 38 mm	-	0.09	1.88	1.81	m	**3.70**
25 mm x 50 mm	-	0.09	1.88	1.90	m	**3.79**
Glazing beads and the like						
12 mm x 16 mm	-	0.04	0.84	1.93	m	**2.77**
12 mm x 19 mm	-	0.04	0.84	1.94	m	**2.78**
12 mm x 25 mm	-	0.04	0.84	1.97	m	**2.80**
12 mm x 25 mm; screwed	-	0.04	0.84	2.80	m	**3.64**
12 mm x 25 mm; fixing with brass cups and screws	-	0.04	0.84	3.23	m	**4.07**
15 mm x 25 mm; screwed	-	0.04	0.84	2.92	m	**3.76**
15 mm quadrant	-	0.04	0.84	2.78	m	**3.62**
18 mm quadrant or scotia	-	0.04	0.84	2.79	m	**3.63**
18 mm x 36 mm; screwed	-	0.04	0.84	2.98	m	**3.81**
25 mm x 38 mm; screwed	-	0.04	0.84	3.11	m	**3.95**
25 mm quadrant or scotia	-	0.04	0.84	2.92	m	**3.75**
38 mm scotia	-	0.04	0.84	2.77	m	**3.61**
50 mm scotia	-	0.04	0.84	3.22	m	**4.06**
Isolated shelves, worktops, seats and the like						
18 mm x 150 mm	-	0.15	3.14	3.65	m	**6.80**
18 mm x 200 mm	-	0.20	4.19	3.84	m	**8.03**
25 mm x 150 mm	-	0.15	3.14	4.19	m	**7.33**
25 mm x 200 mm	-	0.20	4.19	4.48	m	**8.67**
30 mm x 150 mm	-	0.15	3.14	5.88	m	**9.02**
30 mm x 200 mm	-	0.20	4.19	6.51	m	**10.70**

P BUILDING FABRIC SUNDRIES

Item	PC £	Labour hours	Labour £	Material £	Unit	Total rate £
P20 UNFRAMED ISOLATED TRIMS/SKIRTINGS/ SUNDRY ITEMS – cont'd						
Medium density fibreboard – cont'd						
Isolated shelves, worktops, seats and the like; cross-tongued joints						
18 mm x 300 mm	-	0.26	5.45	11.90	m	**17.35**
18 mm x 450 mm	-	0.31	6.49	13.60	m	**20.09**
18 mm x 600 mm	-	0.37	7.75	22.73	m	**30.48**
25 mm x 300 mm	-	0.26	5.45	12.55	m	**17.99**
25 mm x 450 mm	-	0.31	6.49	15.34	m	**21.83**
25 mm x 600 mm	-	0.37	7.75	22.15	m	**29.90**
30 mm x 300 mm	-	0.26	5.45	14.48	m	**19.93**
30 mm x 450 mm	-	0.31	6.49	17.23	m	**23.72**
30 mm x 600 mm	-	0.37	7.75	25.08	m	**32.83**
Isolated shelves, worktops, seats and the like; slatted with 50 wide slats at 75 mm centres						
18 mm thick	-	0.60	12.57	37.28	m	**49.85**
25 mm thick	-	0.60	12.57	39.56	m	**52.12**
30 mm thick	-	0.60	12.57	41.63	m	**54.19**
Window boards, nosings, bed moulds and the like; rebated and rounded						
18 mm x 75 mm	-	0.17	3.56	3.51	m	**7.07**
18 mm x 150 mm	-	0.19	3.98	3.97	m	**7.95**
18 mm x 225 mm	-	0.24	5.03	4.33	m	**9.36**
18 mm x 300 mm	-	0.28	5.86	4.78	m	**10.64**
25 mm x 75 mm	-	0.17	3.56	3.68	m	**7.24**
25 mm x 150 mm	-	0.19	3.98	4.33	m	**8.31**
25 mm x 225 mm	-	0.24	5.03	4.85	m	**9.87**
25 mm x 300 mm	-	0.28	5.86	5.45	m	**11.32**
30 mm x 75 mm	-	0.17	3.56	5.03	m	**8.59**
30 mm x 150 mm	-	0.19	3.98	6.22	m	**10.20**
30 mm x 225 mm	-	0.24	5.03	7.14	m	**12.16**
30 mm x 300 mm	-	0.28	5.86	8.22	m	**14.09**
38 mm x 75 mm	-	0.17	3.56	5.71	m	**9.27**
38 mm x 150 mm	-	0.19	3.98	7.13	m	**11.11**
38 mm x 225 mm	-	0.24	5.03	8.22	m	**13.25**
38 mm x 300 mm	-	0.28	5.86	9.53	m	**15.39**
returned and fitted ends	-	-	-	1.14	nr	**-**
Selected Sapele						
Skirtings, picture rails, dado rails and the like; splayed or moulded						
19 mm x 44 mm; splayed	4.36	0.13	2.72	5.05	m	**7.77**
19 mm x 44 mm; moulded	4.36	0.13	2.72	5.05	m	**7.77**
19 mm x 69 mm; splayed	5.08	0.13	2.72	5.86	m	**8.58**
19 mm x 69 mm; moulded	5.08	0.13	2.72	5.86	m	**8.58**
19 mm x 94 mm; splayed	5.92	0.13	2.72	6.81	m	**9.53**
19 mm x 94 mm; moulded	5.92	0.13	2.72	6.81	m	**9.53**
19 mm x 144 mm; moulded	7.87	0.15	3.14	9.01	m	**12.15**
19 mm x 169 mm; moulded	8.72	0.15	3.14	9.96	m	**13.11**
25 mm x 44 mm; moulded	4.90	0.13	2.72	5.66	m	**8.38**
25 mm x 69 mm; splayed	5.84	0.13	2.72	6.72	m	**9.44**
25 mm x 94 mm; splayed	7.16	0.13	2.72	8.21	m	**10.93**
25 mm x 144 mm; splayed	9.39	0.15	3.14	10.73	m	**13.87**
25 mm x 144 mm; moulded	9.39	0.15	3.14	10.73	m	**13.87**
25 mm x 169 mm; moulded	10.53	0.15	3.14	12.01	m	**15.15**
25 mm x 219 mm; moulded	12.03	0.17	3.56	13.71	m	**17.27**
returned ends	-	0.20	4.19	-	nr	**4.19**
mitres	-	0.14	2.93	-	nr	**2.93**

P BUILDING FABRIC SUNDRIES

Item	PC £	Labour hours	Labour £	Material £	Unit	**Total rate £**
Architraves, cover fillets and the like; half round; splayed or moulded						
13 mm x 25 mm; half round	2.64	0.15	3.14	3.10	m	**6.25**
13 mm x 50 mm; moulded	4.24	0.15	3.14	4.90	m	**8.05**
16 mm x 32 mm; half round	2.69	0.15	3.14	3.15	m	**6.29**
16 mm x 38 mm; moulded	4.08	0.15	3.14	4.73	m	**7.87**
16 mm x 50 mm; moulded	4.36	0.15	3.14	5.05	m	**8.19**
19 mm x 50 mm; splayed	4.36	0.15	3.14	5.05	m	**8.19**
19 mm x 63 mm; splayed	4.74	0.15	3.14	5.47	m	**8.61**
19 mm x 69 mm; splayed	5.08	0.15	3.14	5.86	m	**9.00**
25 mm x 44 mm; splayed	4.72	0.15	3.14	5.45	m	**8.59**
25 mm x 50 mm; moulded	4.90	0.15	3.14	5.66	m	**8.80**
25 mm x 63 mm; splayed	5.48	0.15	3.14	6.30	m	**9.45**
25 mm x 69 mm; splayed	5.84	0.15	3.14	6.72	m	**9.86**
32 mm x 88 mm; moulded	7.11	0.15	3.14	8.15	m	**11.29**
38 mm x 38 mm; moulded	5.59	0.15	3.14	6.43	m	**9.58**
50 mm x 50 mm; moulded	7.31	0.15	3.14	8.38	m	**11.52**
returned ends	-	0.20	4.19	-	nr	**4.19**
mitres	-	0.14	2.93	-	nr	**2.93**
Stops; screwed on						
16 mm x 38 mm	1.80	0.14	2.93	2.04	m	**4.97**
16 mm x 50 mm	1.94	0.14	2.93	2.19	m	**5.12**
19 mm x 38 mm	1.80	0.14	2.93	2.04	m	**4.97**
25 mm x 38 mm	2.27	0.14	2.93	2.57	m	**5.50**
25 mm x 50 mm	2.63	0.14	2.93	2.98	m	**5.91**
Glazing beads and the like						
13 mm x 16 mm	2.35	0.06	1.26	2.66	m	**3.92**
13 mm x 19 mm	2.35	0.06	1.26	2.66	m	**3.92**
13 mm x 25 mm	2.53	0.06	1.26	2.86	m	**4.12**
13 mm x 25 mm; screwed	3.43	0.06	1.26	3.87	m	**5.13**
13 mm x 25 mm; fixing with brass cups and screws	4.22	0.06	1.26	4.77	m	**6.02**
16 mm x 25 mm; screwed	3.43	0.06	1.26	3.87	m	**5.13**
16 mm quadrant	3.30	0.06	1.26	3.73	m	**4.99**
19 mm quadrant or scotia	3.30	0.06	1.26	3.73	m	**4.99**
19 mm x 36 mm; screwed	4.23	0.06	1.26	4.78	m	**6.03**
25 mm x 38 mm; screwed	4.60	0.06	1.26	5.20	m	**6.46**
25 mm quadrant or scotia	3.77	0.06	1.26	4.26	m	**5.52**
38 mm scotia	5.59	0.06	1.26	6.32	m	**7.58**
50 mm scotia	7.31	0.06	1.26	8.26	m	**9.52**
Isolated shelves; worktops, seats and the like						
19 mm x 150 mm	8.02	0.20	4.19	9.06	m	**13.25**
19 mm x 200 mm	9.48	0.28	5.86	10.71	m	**16.57**
25 mm x 150 mm	9.39	0.20	4.19	10.61	m	**14.80**
25 mm x 200 mm	11.28	0.28	5.86	12.74	m	**18.61**
32 mm x 150 mm	10.60	0.20	4.19	11.98	m	**16.17**
32 mm x 200 mm	12.82	0.28	5.86	14.48	m	**20.35**
Isolated shelves, worktops, seats and the like; cross-tongued joints						
19 mm x 300 mm	21.68	0.35	7.33	24.50	m	**31.83**
19 mm x 450 mm	33.94	0.42	8.80	38.36	m	**47.16**
19 mm x 600 mm	45.09	0.51	10.68	50.95	m	**61.63**
25 mm x 300 mm	24.34	0.35	7.33	27.51	m	**34.84**
25 mm x 450 mm	38.25	0.42	8.80	43.22	m	**52.02**
25 mm x 600 mm	50.83	0.51	10.68	57.45	m	**68.13**
32 mm x 300 mm	26.62	0.35	7.33	30.09	m	**37.42**
32 mm x 450 mm	41.93	0.42	8.80	47.38	m	**56.18**
32 mm x 600 mm	55.74	0.51	10.68	62.99	m	**73.67**

P BUILDING FABRIC SUNDRIES

Item	PC £	Labour hours	Labour £	Material £	Unit	Total rate £
P20 UNFRAMED ISOLATED TRIMS/SKIRTINGS/ SUNDRY ITEMS – cont'd						
Selected Sapele – cont'd						
Isolated shelves, worktops, seats and the like; slatted with 50 wide slats at 75 mm centres						
19 mm thick	61.02	0.80	16.75	71.03	m²	**87.78**
25 mm thick	65.32	0.80	16.75	76.00	m²	**92.76**
32 mm thick	69.01	0.80	16.75	80.28	m²	**97.04**
Window boards, nosings, bed moulds and the like; rebated and rounded						
19 mm x 75 mm	6.17	0.22	4.61	7.24	m	**11.85**
19 mm x 150 mm	8.97	0.25	5.24	10.39	m	**15.63**
19 mm x 225 mm; in one width	11.02	0.33	6.91	12.71	m	**19.62**
19 mm x 300 mm; cross-tongued joints	22.42	0.37	7.75	25.59	m	**33.34**
25 mm x 75 mm	6.80	0.22	4.61	7.94	m	**12.55**
25 mm x 150 mm	9.99	0.25	5.24	11.54	m	**16.78**
25 mm x 225 mm; in one width	13.11	0.33	6.91	15.07	m	**21.99**
25 mm x 300 mm; cross-tongued joints	26.05	0.37	7.75	29.70	m	**37.45**
32 mm x 75 mm	7.39	0.22	4.61	8.61	m	**13.22**
32 mm x 150 mm	11.11	0.25	5.24	12.81	m	**18.05**
32 mm x 225 mm; in one width	14.81	0.33	6.91	16.99	m	**23.90**
32 mm x 300 mm; cross-tongued joints	28.66	0.37	7.75	32.64	m	**40.39**
returned and fitted ends	-	0.21	4.40	-	nr	**4.40**
Handrails; rounded						
44 mm x 50 mm	14.07	0.31	6.49	15.90	m	**22.39**
50 mm x 75 mm	16.96	0.33	6.91	19.17	m	**26.08**
63 mm x 87 mm	19.90	0.37	7.75	22.49	m	**30.24**
75 mm x 100 mm	24.73	0.42	8.80	27.95	m	**36.75**
Handrails; moulded						
44 mm x 50 mm	15.65	0.31	6.49	17.68	m	**24.18**
50 mm x 75 mm	18.53	0.33	6.91	20.94	m	**27.86**
63 mm x 87 mm	21.49	0.37	7.75	24.29	m	**32.04**
75 mm x 100 mm	26.30	0.42	8.80	29.72	m	**38.52**
Pin-boards; medium board						
Sundeala "A" pin-board or other equal and approved; fixed with adhesive to backing (not included); over 300 mm wide						
6.40 mm thick	-	0.56	11.73	5.53	m²	**17.26**
Sundries on softwood/hardwood						
Extra over fixing with nails for						
gluing and pinning	-	0.01	0.21	0.07	m	**0.28**
masonry nails	-	-	-	-	m	**0.33**
steel screws	-	-	-	-	m	**0.31**
self-tapping screws	-	-	-	-	m	**0.32**
steel screws; gluing	-	-	-	-	m	**0.54**
steel screws; sinking; filling heads	-	-	-	-	m	**0.69**
steel screws; sinking; pellating over	-	-	-	-	m	**1.49**
brass cups and screws	-	-	-	-	m	**1.85**
Extra over for						
countersinking	-	-	-	-	m	**0.28**
pellating	-	-	-	-	m	**1.31**
Head or nut in softwood						
let in flush	-	-	-	-	nr	**0.69**
Head or nut; in hardwood						
let in flush	-	-	-	-	nr	**1.02**
let in over; pellated	-	-	-	-	nr	**2.39**

P BUILDING FABRIC SUNDRIES

Item	PC £	Labour hours	Labour £	Material £	Unit	Total rate £
Metalwork; mild steel						
Angle section bearers; for building in						
90 mm x 90 mm x 6 mm	-	0.31	6.69	7.44	m	**14.12**
120 mm x 120 mm x 8 mm	-	0.32	6.90	13.92	m	**20.82**
200 mm x 150 mm x 12 mm	-	0.37	7.98	30.85	m	**38.83**
Metalwork; mild steel; galvanized						
Waterbars; groove in timber						
6 mm x 30 mm	-	0.46	9.63	3.96	m	**13.59**
6 mm x 40 mm	-	0.46	9.63	5.20	m	**14.83**
6 mm x 50 mm	-	0.46	9.63	6.52	m	**16.15**
Angle section bearers; for building in						
90 mm x 90 mm x 6 mm	-	0.31	6.69	8.94	m	**15.62**
120 mm x 120 mm x 8 mm	-	0.32	6.90	15.93	m	**22.84**
200 mm x 150 mm x 12 mm	-	0.37	7.98	33.78	m	**41.76**
Dowels; mortice in timber						
8 mm diameter x 100 mm long	-	0.04	0.84	0.19	nr	**1.03**
10 mm diameter x 50 mm long	-	0.04	0.84	0.46	nr	**1.29**
Cramps						
25 mm x 3 mm x 230 mm girth; one end bent, holed and screwed to softwood; other end fishtailed for building in	-	0.06	1.26	1.09	nr	**2.34**
Metalwork; stainless steel						
Angle section bearers; for building in						
90 mm x 90 mm x 6 mm	-	0.31	6.69	48.41	m	**55.10**
120 mm x 120 mm x 8 mm	-	0.32	6.90	71.57	m	**78.48**
200 mm x 150 mm x 12 mm	-	0.37	7.98	186.69	m	**194.67**
P21 IRONMONGERY						
NOTE: Ironmongery is largely a matter of selection and prices vary considerably; indicative prices for reasonable quantities of "good quality" ironmongery are given below.						
Ironmongery; Allgood or other equal and approved; to softwood						
Bolts						
75 x 35 mm Modric anodised aluminium straight barrel bolt	7.95	0.30	6.28	8.55	nr	**14.84**
150 x 35 mm Modric anodised aluminium Modric anodised aluminium straight barrel bolt	9.04	0.30	6.28	9.73	nr	**16.02**
75 x 35 mm Modric anodised aluminium necked barrel bolt	8.89	0.30	6.28	9.57	nr	**15.85**
150 x 35 mm Modric anodised aluminium necked barrel bolt	11.35	0.30	6.28	12.21	nr	**18.50**
11 mm Easiclean socket for wood or stone	1.62	0.10	2.09	1.74	nr	**3.84**
Security hinge bolt chubb WS12	14.83	0.50	10.47	15.96	nr	**26.43**
203 x 19 x 11 mm Complete bolt set, with floor socket and intumescent pack for FD30 and FD60 fire doors	31.36	0.60	12.57	33.76	nr	**46.32**
203/609 x 19 mm Complete bolt set, with floor socket and intumescent pack for FD30 and FD60 fire doors	73.80	0.60	12.57	79.43	nr	**91.99**
Stainless steel indicating bolt complete with outside indicator and emergency release	35.25	0.60	12.57	37.94	nr	**50.51**

P BUILDING FABRIC SUNDRIES

Item	PC £	Labour hours	Labour £	Material £	Unit	Total rate £
P21 IRONMONGERY – cont'd						
Ironmongery; Allgood; to softwood – cont'd						
Catches						
Magnetic catch	0.99	0.20	4.19	1.07	nr	**5.25**
Door closers and furniture						
13 mm Satin chrome rebate component for 7204/08/78/79/86	24.95	0.60	12.57	26.85	nr	**39.42**
90 x 90 mm Modric anodised aluminium electrically powered hold open wall magnet. CE marked to BS EN1155:1997 & A1:2002 3-5-6/3-1-1-3	107.56	0.40	8.38	115.76	nr	**124.14**
Modric anodised aluminium bathroom configuration with quadaxial assembly, turn, release and optional indicator	57.07	0.80	16.75	61.42	nr	**78.18**
Overhead limiting stay; galvanised	12.28	1.00	20.94	13.21	nr	**34.16**
263 x 48 x 48 mm Overhead door closer Fig 6 adjustable power 2-5 with adjustable backcheck and intumescent protected bracket. Certifire listed and CE Marked to BS EN1154 4-8-2/5-1-1-3	61.02	1.00	20.94	65.67	nr	**86.62**
Concealed jamb door closer check action	87.18	1.00	20.94	93.83	nr	**114.77**
75 x 57 x 170 mm Modric anodised aluminium door co-ordinator for pairs of rebated leaves, CE Marked to BS EN1158 3-5-3/5-1-1-0	26.79	0.80	16.75	28.84	nr	**45.59**
263 x 50 x 48 mm Modric anodised aluminium overhead door closer Fig 1 adjustable power 2-5 with adjustable backcheck, intumescent protected bracket. Certifire listed and CE Marked to BS EN1154 4-8/5-1-1-3	59.09	1.00	20.94	63.59	nr	**84.53**
290 x 48 x 50 mm Modric anodised aluminium rectangular overhead door closer with adjustable power and adjustable backcheck intumescent protected arm heavy duty U.L. & certifire listed & CE Marked to BS EN1154 4-8-2/4-1-1-3 and Kitemarked	83.30	1.00	20.94	89.66	nr	**110.60**
Stainless steel overhead door closer Fig 1. projecting armset, Power EN 2-5, CE marked , c/w Backcheck, Latch action and Speed control. Max door width 1100 mm, Max door weight 100kg	80.78	1.00	20.94	86.93	nr	**107.88**
Fully concealed overhead door closer complete with track and arm for single action doors	126.79	0.80	16.75	136.46	nr	**153.21**
92 x 45 mm Stainless steel heavy duty floor pivot set with thrust roller bearing 200kg load capacity. Complete with forged steel intumescent protected double action strap with 10 mm height adjustment, new low profile top centre, and matching cover plate	113.40	2.30	48.17	122.05	nr	**170.22**
92 x 45 mm Stainless steel heavy duty floor pivot set with thrust roller bearing 200kg load capacity, with stainless steel intumescent protected S/A 25 mm offset strap & top centre, matching plate	151.20	2.30	48.17	162.73	nr	**210.90**
Double action pivot set for door maximum width 1100 mm and maximum weight 80kg	69.32	2.30	48.17	74.60	nr	**122.77**
305 x 80 x 50 mm Stainless steel 'Cavalier' floor spring, intumescent protected forged steel D/A strap with 10 mm height adjustment & low profile top centre, matching covers & box. CE Marked to BS EN1154 4-8-*-1-1-3. Adjustable power 2/4	174.36	2.30	48.17	187.65	nr	**235.82**
305 x 80 x 50 mm Stainless steel 'Cavalier' floor spring adjustable power 2/4 stainless steel intumsecent protected S/A 16 mm offset strap & top centre, matching covers & box. Certifire listed and CE marked to BSEN1154 4-8-2/4-1-1-3	193.72	2.30	48.17	208.50	nr	**256.67**

P BUILDING FABRIC SUNDRIES

Item	PC £	Labour hours	Labour £	Material £	Unit	Total rate £
Surface vertical rod push bar panic bolt, reversible, to suit doors 2500x1100 mm maximum, silver finish, CE marked to EN1125 class 3-7-5-1-1-3-2-2-A	237.63	1.50	31.42	255.75	nr	**287.16**
Rim push bar panic latch, reversible, to suit doors 1100 mm wide maximum, silver finish, CE marked to EN1125 class 3-7-5-1-1-3-2-2-A	142.17	1.30	27.23	153.01	nr	**180.24**
76 x 51 x 13 mm Adjustable heavy roller catch satin chrome	6.71	0.60	12.57	7.22	nr	**19.78**
External access device for use with XX10280/2 panic hardware to suit door thickness 45-55 mm, complete with SS3006N lever, SS755 rose, SS796 profile escutcheon and spindle.For use with MA7420A51 or MA7420A55 profile cylinders	28.67	1.30	27.23	30.86	nr	**58.09**
142 x 22 mm Ø Concealed jamb door closer light duty	11.02	0.80	16.75	11.86	nr	**28.61**
80 x 40 x 45 mm Emergency release door stop with holdback facility	69.32	1.00	20.94	74.60	nr	**95.55**
Modric anodised aluminium quadaxial lever assembly tested to BS EN1906 4/7/-/1/1/4/0/U	28.07	0.80	16.75	30.21	pair	**46.97**
Modric anodised aluminium quadaxial lever assembly Tested to BS EN1906 4/7/-/1/1/4/0/U	28.07	0.80	16.75	30.21	pair	**46.97**
Modric anodised aluminium quadaxial lever assembly Tested to BS EN1906 4/7/-/1/1/4/0/U with Biocote® anti-bacterial protection	39.76	0.80	16.75	42.79	pair	**59.55**
Modric stainless steel quadaxial lever assembly Tested to BS EN1906 4/7/-/1/1/4/0/U	40.29	0.80	16.75	43.37	pair	**60.12**
152 x 38 x 13 mm Modric anodised aluminium security door chain leather covered	37.55	0.40	8.38	40.41	nr	**48.79**
50 Ø x 3 mm Modric anodised aluminium circular covered rose for profile cylinder	3.68	0.10	2.09	3.96	nr	**6.06**
50 Ø x 3 mm Modric anodised aluminium circular covered rose with indicator and emergency release	7.61	0.15	3.14	8.19	nr	**11.34**
50 Ø x 3 mm Modric anodised aluminium circular covered rose with heavy turn, 5-8 mm spindle	13.35	0.15	3.14	14.36	nr	**17.51**
Budget lock escutcheon - satin stainless steel 316	4.41	0.10	2.09	4.75	nr	**6.84**
50 Ø x 3 mm Stainless steel circular covered rose for profile cylinder	5.89	0.10	2.09	6.33	nr	**8.43**
50 Ø x 3 mm Stainless steel circular covered rose with indicator and emergency release	7.98	0.15	3.14	8.59	nr	**11.73**
50 Ø x 3 mm Stainless steel circular covered rose with heavy turn, 5-8 mm spindle	16.43	0.15	3.14	17.68	nr	**20.82**
330 x 76 x 1.6 mm Modric anodised aluminium push plate	4.25	0.15	3.14	4.57	nr	**7.71**
330 x 76 x 1.6 mm Stainless steel push plate	9.55	0.15	3.14	10.28	nr	**13.42**
800 x 150 x 1.5 mm Modric anodised aluminium kicking plate, drilled and countersunk with screws.	6.34	0.25	5.24	6.82	nr	**12.05**
900 x 150 x 1.5 mm Modric anodised aluminium kicking plate, drilled and countersunk with screws.	7.13	0.25	5.24	7.67	nr	**12.91**
1000 x 150 x 1.5 mm Modric anodised aluminium kicking plate, drilled & countersunk with screws.	7.92	0.25	5.24	8.52	nr	**13.76**
800 x 150 x 1.5 mm Stainless steel kicking plate, drilled and countersunk with screws.	12.46	0.25	5.24	13.41	nr	**18.64**
900 x 150 x 1.5 mm Stainless steel kicking plate, drilled and countersunk with screws.	14.00	0.25	5.24	15.07	nr	**20.31**
1000 x 150 x 1.5 mm Stainless steel kicking plate, drilled & countersunk with screws.	15.57	0.25	5.24	16.76	nr	**21.99**

P BUILDING FABRIC SUNDRIES

Item	PC £	Labour hours	Labour £	Material £	Unit	Total rate £
P21 IRONMONGERY – cont'd						
Ironmongery; Allgood; to softwood – cont'd						
305 x 70 x 19 mm Ø Modric anodised aluminium grab handle bolt through fixing	16.97	0.40	8.38	18.27	nr	**26.65**
400 x 19 mm Ø Stainless steel D line straight pull handle with M8 threaded holes,fixing centres 300mm	44.57	0.33	6.91	52.03	nr	**58.95**
Hinges						
100 x 75 x 3 mm Stainless steel triple knuckle concealed twin Newtonbearings, button tipped butt hinges, jig drilled for metal doors/frames, complete with M6x12MT 'undercut' machine screws, stainless steel 316 CE marked to EN1935 4-7-7-1-1-4-0-14	20.39	0.25	5.24	21.95	pair	**27.18**
100 x 100 x 3 mm Stainless steel triple knuckle concealed twin Newton bearings, button tipped hinges, jig drilled, stainless steel grade 316 CE marked to EN1935 4-7-7-1-1-4-0-13	27.54	0.25	5.24	29.64	pair	**34.88**
Latches						
Modric anodised aluminium round cylinder for rim night latch, 2 keyed satin nickel plated	21.02	0.40	8.38	22.62	nr	**30.99**
93 x 75 mm Cylinder rim non-deadlocking night latch case only 60 mm backset	14.89	0.40	8.38	16.02	nr	**24.40**
71 series mortice latch, case only, low friction latchbolt, griptight follower, heavy spring for levers. Radius forend and sq strike. CE marked to BS EN12209 3/X/8/1/0G/-/B/02/0	13.80	0.80	16.75	14.85	nr	**31.60**
Modric anodised aluminium latch configuration with quadaxial assembly	43.20	0.80	16.75	46.49	nr	**63.25**
Modric anodised aluminium Nightlatch configuration with quadaxial assembly and single cylinder	81.94	0.80	16.75	88.19	nr	**104.95**
Locks						
44 mm case Bright zinc plated steel mortice budget lock with slotted strike plate 33 mm backset	5.46	0.80	16.75	5.88	nr	**22.63**
76 x 58 mm b/s Stainless steel cubicle mortice deadlock with 8 mm follower	11.65	0.80	16.75	12.53	nr	**29.29**
'A' length European profile double cylinder lock, 2 keyed satin nickel plated	21.18	0.80	16.75	22.79	nr	**39.55**
'A' length European profile cylinder and large turn, 2 keyed satin nickel plated	24.07	0.80	16.75	25.90	nr	**42.66**
'A' length European profile cylinder and large turn, 2 keyed under master key, satin nickel plated	24.80	0.80	16.75	26.70	nr	**43.45**
'A' length European profile single cylinder, 2 keyed satin nickel plated	16.87	0.80	16.75	18.15	nr	**34.91**
'A' length European profile single cylinder, 2 keyed under master key, satin nickel plated	16.87	0.80	16.75	18.15	nr	**34.91**
93 x 60 mm b/s 71 series profile cylinder mortice deadlock, case only. Single throw 22mm deadbolt. radius forend and square strike. CE marked to BS EN12209 3/X/8/1/0/G/4/B/A/0/0	13.80	0.80	16.75	14.85	nr	**31.60**
92 x 60 mm b/s 71 series bathroom lock, case only, low friction latchbolt, griptight follower, heavy spring for levers, twin 8mm followers at 78mm centres. radius forend and square strike. CE marked to BS EN12209 3/X/8/0/0/G-/B/0/2/0	-	0.80	16.75	17.59	nr	**34.35**
93 x 60 mm b/s 71 series profile cylinder mortice lock, case only, low friction latchbolt, griptight follower. Heavy spring for levers, 22mm throw deadbolt, cylinder withdraws bolt bolts. Radius forend and square strike. CE marked to BS EN 12209 3/X/8/1/0G/4/B/A2/0	16.34	0.80	16.75	17.59	nr	**34.35**

P BUILDING FABRIC SUNDRIES

Item	PC £	Labour hours	Labour £	Material £	Unit	Total rate £
92 x 60 mm b/s71 series profile cylinder emergency lock, case only. Low friction latchbolt, griptight follower, heavy spring for lever, single throw 22mm deadbolt, lever can withdraw both bolts. Radius forend and strike	56.33	0.80	16.75	60.63	nr	**77.38**
Modric anodised aluminium lock configuration with quadaxial assembly and cylinder with turn	74.74	0.80	16.75	80.43	nr	**97.19**
Sundries						
76 mm Ø Modric anodised aluminium circular sex symbol male	4.07	0.08	1.68	4.38	nr	**6.05**
76 mm Ø Modric anodised aluminium circular symbol fire door keep locked	4.07	0.08	1.68	4.38	nr	**6.05**
76 mm Ø Modric anodised aluminium circular symbol fire door keep shut	4.07	0.10	2.09	4.38	nr	**6.47**
38 x 47 mm Ø Modric anodised aluminium heavy circular floor door stop with cover	7.51	0.10	2.09	8.09	nr	**10.18**
38 x 47 mm Ø Stainless steel heavy circular floor door stop with cover	10.21	0.10	2.09	10.98	nr	**13.08**
63 x 19 mm Ø Modric ancdised aluminium Circular heavy duty skirting buffer with thief resistant insert	5.37	0.10	2.09	5.78	nr	**7.88**
63 x 19 mm Ø Stainless steel circular heavy duty skirting buffer with thief resistant insert	6.22	0.10	2.09	6.69	nr	**8.79**
152 mm Cabin hook satin chrome on brass	14.13	0.15	3.14	15.21	nr	**18.35**
14 mm Ø x 145 x 94 mm Toilet roll holder, length 145 mm, colour white, satin stainless steel 316	46.54	0.15	3.14	51.44	nr	**54.59**
Towel rail with bushes, fixing centres 450 mm, satin stainless steel 316	53.15	0.25	5.24	61.27	nr	**66.51**
Toilet brush holder with toilet brush, with bushes, satin stainless steel 316	84.16	0.20	4.19	93.29	nr	**97.48**
Set of stainless steel rails, one lift-up rail, 4 straight 600 mm long straight grab rails, and one back rest rail for use in toilets for the disabled, to meet the requirements of Part M of the Building Regulations	479.34	1.50	31.42	515.89	set	**547.31**
Ironmongery; Allgood or other equal and approved; to hardwood						
Bolts						
75 x 35 mm Modric anodised aluminium straight barrel bolt	7.95	0.40	8.38	8.55	nr	**16.93**
150 x 35 mm Modric anodised aluminium Modric anodised aluminium straight barrel bolt	9.04	0.40	8.38	9.73	nr	**18.11**
75 x 35 mm Modric anodised aluminium necked barrel bolt	8.89	0.40	8.38	9.57	nr	**17.95**
150 x 35 mm Modric anodised aluminium necked barrel bolt	11.35	0.40	8.38	12.21	nr	**20.59**
11 mm Easiclean socket for wood or stone	1.62	0.15	3.14	1.74	nr	**4.89**
Security hinge bolt chubb WS12	14.83	0.65	13.61	15.96	nr	**29.58**
203 x 19 x 11 mm Complete bolt set, with floor socket and intumescent pack for FD30 and FD60 fire doors	31.36	0.80	16.75	33.76	nr	**50.51**
203/609 x 19 mm Complete bolt set, with floor socket and intumescent pack for FD30 and FD60 fire doors	73.80	0.80	16.75	79.43	nr	**96.18**
Stainless steel indicating bolt complete with outside indicator and emergency release	35.25	0.80	16.75	37.94	nr	**54.70**

P BUILDING FABRIC SUNDRIES

Item	PC £	Labour hours	Labour £	Material £	Unit	Total rate £
P21 IRONMONGERY – cont'd						
Ironmongery; Allgood; to hardwood – cont'd						
Catches						
Magnetic catch	0.99	0.25	5.24	1.07	nr	**6.30**
Door closers and furniture						
13 mm Satin chrome rebate component for 7204/08/78/79/86	24.95	0.80	16.75	26.85	nr	**43.61**
90 x 90 mm Modric anodised aluminium electrically powered hold open wall magnet. CE marked to BS EN1155:1997 & A1:2002 3-5-6/3-1-1-3	107.56	0.55	11.52	115.76	nr	**127.28**
Modric anodised aluminium bathroom configuration with quadaxial assembly, turn, release and optional indicator	57.07	1.05	21.99	61.42	nr	**83.41**
Overhead limiting stay; galvanised	12.28	1.35	28.27	13.21	nr	**41.49**
263 x 48 x 48 mm Overhead door closer Fig 6 adjustable power 2-5 with adjustable backcheck and intumescent protected bracket. Certifire listed and CE Marked to BS EN1154 4-8-2/5-1-1-3	61.02	1.35	28.27	65.67	nr	**93.95**
Concealed jamb door closer check action	87.18	1.35	28.27	93.83	nr	**122.10**
75 x 57 x 170 mm Modric anodised aluminium door co-ordinator for pairs of rebated leaves, CE Marked to BS EN1158 3-5-3/5-1-1-0	26.79	1.05	21.99	28.84	nr	**50.83**
263 x 50 x 48 mm Modric anodised aluminium overhead door closer Fig 1 adjustable power 2-5 with adjustable backcheck, intumescent protected bracket. Certifire listed and CE Marked to BS EN1154 4-8/5-1-1-3	59.09	1.35	28.27	63.59	nr	**91.86**
290 x 48 x 50 mm Modric anodised aluminium rectangular overhead door closer with adjustable power and adjustable backcheck intumescent protected arm heavy duty U.L. & certifire listed & CE marked to BS EN1154 4-8-2/4-1-1-3 and kitemarked.	83.30	1.35	28.27	89.66	nr	**117.93**
Stainless steel overhead door closer Fig 1. projecting armset, Power EN 2-5, CE marked , c/w backcheck, Latch action and Speed control. Max door width 1100 mm, Max door weight 100kg	80.78	1.35	28.27	86.93	nr	**115.21**
Fully concealed overhead door closer complete with track and arm for single action doors	126.79	1.05	21.99	136.46	nr	**158.45**
92 x 45 mm Stainless steel heavy duty floor pivot set with thrust roller bearing 200kg load capacity. Complete with forged steel intumescent protected double action strap with 10 mm height adjustment, new low profile top centre, and matching cover plate	113.40	3.05	63.88	122.05	nr	**185.93**
92 x 45 mm Stainless steel heavy duty floor pivot set with thrust roller bearing 200kg load capacity, with stainless steel intumescent protected S/A 25 mm offset strap & top centre, matching plate	151.20	3.05	63.88	162.73	nr	**226.61**
Double action pivot set for door maximum width 1100 mm and maximum weight 80kg	69.32	3.05	63.88	74.60	nr	**138.48**
305 x 80 x 50 mm Stainless steel 'Cavalier' floor spring, intumescent protected forged steel D/A strap with 10 mm height adjustment & low profile top centre, matching covers & box. CE Marked to BS EN1154 4-8-*-1-1-3. Adjustable power 2/4	174.36	3.05	63.88	187.65	nr	**251.53**
305 x 80 x 50 mm Stainless steel 'Cavalier' floor spring adjustable power 2/4 stainless steel intumsecent protected S/A 16 mm offset strap & top centre, matching covers & box. Certifire listed and CE marked to BSEN1154 4-8-2/4-1-1-3	193.72	3.05	63.88	208.50	nr	**272.38**

P BUILDING FABRIC SUNDRIES

Item	PC £	Labour hours	Labour £	Material £	Unit	**Total rate £**
Surface vertical rod push bar panic bolt, reversible, tp suit doors 2500x1100 mm maximum, silver finish, CE marked to EN1125 class 3-7-5-1-1-3-2-2-A	237.63	2.00	41.89	255.75	nr	**297.63**
Rim push bar panic latch, reversible, to suit doors 1100 mm wide maximum, silver finish, CE marked to EN1125 class 3-7-5-1-1-3-2-2-A	142.17	1.75	36.65	153.01	nr	**189.67**
76 x 51 x 13 mm Adjustable heavy roller catch satin chrome	6.71	0.80	16.75	7.22	nr	**23.97**
External access device for use with XX10280/2 panic hardware to suit door thickness 45-55 mm, complete with SS3006N lever, SS755 rose, SS796 profile escutcheon and spindle.For use with MA7420A51 or MA7420A55 profile cylinders	28.67	1.75	36.65	30.86	nr	**67.51**
142 x 22 mm Ø Concealed jamb door closer light duty	11.02	1.05	21.99	11.86	nr	**33.85**
80 x 40 x 45 mm Emergency release door stop with holdback facility	69.32	1.35	28.27	74.60	nr	**102.88**
Modric anodised aluminium quadaxial lever assembly tested to BS EN1906 4/7/-/1/1/4/0/U	28.07	1.05	21.99	30.21	pair	**52.20**
Modric anodised aluminium quadaxial lever assembly Tested to BS EN1906 4/7/-/1/1/4/0/U	28.07	1.05	21.99	30.21	pair	**52.20**
Modric anodised aluminium quadaxial lever asembly Tested to BS EN1906 4/7/-/1/1/4/0/U with Biocote® anti-bacterial protection	39.76	1.05	21.99	42.79	pair	**64.78**
Modric stainless steel quadaxial lever assembly Tested to BS EN1906 4/7/-/1/1/4/0/U	40.29	1.05	21.99	43.37	pair	**65.36**
152 x 38 x 13 mm Modric anodised aluminium security door chain leather covered	37.55	0.55	11.52	40.41	nr	**51.93**
50 Ø x 3 mm Modric anodised aluminium circular covered rose for profile cylinder	3.68	0.15	3.14	3.96	nr	**7.10**
50 Ø x 3 mm Modric anodised aluminium circular covered rose with indicator and emergency release	7.61	0.20	4.19	8.19	nr	**12.38**
50 Ø x 3 mm Modric anodised aluminium circular covered rose with heavy turn, 5-8 mm spindle	13.35	0.20	4.19	14.36	nr	**18.55**
Budget lock escutcheon - satin stainless steel 316	4.41	0.15	3.14	4.75	nr	**7.89**
50 Ø x 3 mm Stainless steel circular covered rose for profile cylinder	5.89	0.15	3.14	6.33	nr	**9.48**
50 Ø x 3 mm Stainless steel circular covered rose with indicator and emergency release	7.98	0.20	4.19	8.59	nr	**12.78**
50 Ø x 3 mm Stainless steel circular covered rose with heavy turn, 5-8 mm spindle	16.43	0.20	4.19	17.68	nr	**21.87**
330 x 76 x 1.6 mm Modric anodised aluminium push plate	4.25	0.20	4.19	4.57	nr	**8.76**
330 x 76 x 1.6 mm Stainless steel push plate	9.55	0.20	4.19	10.28	nr	**14.47**
800 x 150 x 1.5 mm Modric anodised aluminium kicking plate, drilled and countersunk with screws.	6.34	0.35	7.33	6.82	nr	**14.15**
900 x 150 x 1.5 mm Modric anodised aluminium kicking plate, drilled and countersunk with screws.	7.13	0.35	7.33	7.67	nr	**15.00**
1000 x 150 x 1.5 mm Modric anodised aluminium kicking plate, drilled & countersunk with screws.	7.92	0.35	7.33	8.52	nr	**15.85**
800 x 150 x 1.5 mm Stainless steel kicking plate, drilled and countersunk with screws.	12.46	0.35	7.33	13.41	nr	**20.74**
900 x 150 x 1.5 mm Stainless steel kicking plate, drilled and countersunk with screws.	14.00	0.35	7.33	15.07	nr	**22.40**
1000 x 150 x 1.5 mm Stainless steel kicking plate, drilled & countersunk with screws.	15.57	0.35	7.33	16.76	nr	**24.09**

P BUILDING FABRIC SUNDRIES

Item	PC £	Labour hours	Labour £	Material £	Unit	Total rate £
P21 IRONMONGERY – cont'd						
Ironmongery; Allgood; to hardwood – cont'd						
305 x 70 x 19 mm Ø Modric anodised aluminium grab handle bolt through fixing	16.97	0.55	11.52	18.27	nr	**29.79**
400 x 19 mm Ø Stainless steel D line straight pull handle with M8 threaded holes, fixing centres 300 mm	44.57	0.45	9.42	52.03	nr	**61.46**
Hinges						
100 x 75 x 3 mm Stainless steel triple knuckle concealed twin Newtonbearings, button tipped butt hinges, jig drilled for metal doors/frames, complete with M6x12MT 'undercut' machine screws, stainless steel 316 CE marked to EN1935 4-7-7-1-1-4-0-14	20.39	0.35	7.33	21.95	pair	**29.28**
100 x 100 x 3 mm Stainless steel triple knuckle concealed twin Newton bearings, button tipped hinges, jig drilled, stainless steel grade 316 CE marked to EN1935 4-7-7-1-1-4-0-13	27.54	0.35	7.33	29.64	pair	**36.97**
Latches						
Modric anodised aluminium round cylinder for rim night latch, 2 keyed satin nickel plated	21.02	0.55	11.52	22.62	nr	**34.14**
93 x 75 mm Cylinder rim non-deadlocking night latch case only 60 mm backset	14.89	0.55	11.52	16.02	nr	**27.54**
71 series mortice latch, case only, low friction latchbolt, griptight follower, heavy spring for levers. Radius forend and sq strike. CE marked to BS EN12209 3/X/8/1/0G/-/B/02/0	13.80	1.05	21.99	14.85	nr	**36.84**
Modric anodised aluminium latch configuration with quadaxial assembly	43.20	1.05	21.99	46.49	nr	**68.48**
Modric anodised aluminium Nightlatch configuration with quadaxial assembly and single cylinder	81.94	1.05	21.99	88.19	nr	**110.18**
Locks						
44 mm case Bright zinc plated steel mortice budget lock with slotted strike plate 33 mm backset	5.46	1.05	21.99	5.88	nr	**27.87**
76 x 58 mm b/s Stainless steel cubicle mortice deadlock with 8 mm follower	11.65	1.05	21.99	12.53	nr	**34.52**
'A' length European profile double cylinder lock, 2 keyed satin nickel plated	21.18	1.05	21.99	22.79	nr	**44.78**
'A' length European profile cylinder and large turn, 2 keyed satin nickel plated	24.07	1.05	21.99	25.90	nr	**47.89**
'A' length European profile cylinder and large turn, 2 keyed under master key, satin nickel plated	24.80	1.05	21.99	26.70	nr	**48.69**
'A' length European profile single cylinder, 2 keyed satin nickel plated	16.87	1.05	21.99	18.15	nr	**40.14**
'A' length European profile single cylinder, 2 keyed under master key, satin nickel plated	16.87	1.05	21.99	18.15	nr	**40.14**
93 x 60 mm b/s 71 series profile cylinder mortice deadlock, case only. Single throw 22mm deadbolt. Radius forend and square strike. CE marked to BS EN12209 3/X/8/1/0/G/4/B/A/0/0	13.80	1.05	21.99	14.85	nr	**36.84**
92 x 60 mm b/s 71 series bathroom lock, case only, low friction latchbolt, griptight follower, heavy spring for levers, twin 8mm followers at 78mm centres. Radius forend and square strike. CE marked to BS EN12209 3/X/8/0/0/G-/B/0/2/0	-	1.05	21.99	17.59	nr	**39.58**

P BUILDING FABRIC SUNDRIES

Item	PC £	Labour hours	Labour £	Material £	Unit	Total rate £
93 x 60 mm b/s 71 series profile cylinder mortice lock, case only, low friction latchbolt, griptight follower. Heavy spring for levers, 22mm throw deadbolt, cylinder withdraws bolt bolts. Radius forend and square strike. CE marked to BS EN 12209 3/X/8/1/0G/4/B/A2/0	16.34	1.05	21.99	17.59	nr	**39.58**
92 x 60 mm b/s71 series profile cylinder emergency lock, case only. Low friction latchbolt, griptight follower, heavy spring for lever, single throw 22mm deadbolt, lever can withdraw both bolts. Radius forend and strike	56.33	1.05	21.99	60.63	nr	**82.62**
Modric anodised aluminium lock configuration with quadaxial assembly and cylinder with turn	74.74	1.05	21.99	80.43	nr	**102.43**
Sundries						
76 mm Ø Modric anodised aluminium circular sex symbol male	4.07	0.10	2.09	4.38	nr	**6.47**
76 mm Ø Modric anodised aluminium circular symbol fire door keep locked	4.07	0.10	2.09	4.38	nr	**6.47**
76 mm Ø Modric anodised aluminium circular symbol fire door keep shut	4.07	0.15	3.14	4.38	nr	**7.52**
38 x 47 mm Ø Modric anodised aluminium heavy circular floor door stop with cover	7.51	0.15	3.14	8.09	nr	**11.23**
38 x 47 mm Ø Stainless steel heavy circular floor door stop with cover	10.21	0.15	3.14	10.98	nr	**14.13**
63 x 19 mm Ø Modric ancdised aluminium Circular heavy duty skirting buffer with thief resistant insert	5.37	0.15	3.14	5.78	nr	**8.92**
63 x 19 mm Ø Stainless steel circular heavy duty skirting buffer with thief resistant insert	6.22	0.15	3.14	6.69	nr	**9.83**
152 mm Cabin hook satin chrome on brass	14.13	0.20	4.19	15.21	nr	**19.40**
14 mm Ø x 145 x 94 mm Toilet roll holder, length 145 mm, colour white, satin stainless steel 316	46.54	0.20	4.19	51.44	nr	**55.63**
Towel rail with bushes, fixing centres 450 mm, satin stainless steel 316	53.15	0.35	7.33	61.27	nr	**68.61**
Toilet brush holder with toilet brush, with bushes, satin stainless steel 316	84.16	0.25	5.24	93.29	nr	**98.52**
Set of stainless steel rails, one lift-up rail, 4 straight 600 mm long straight grab rails, and one back rest rail for use in toilets for the disabled, to meet the requirements of Part M of the Building Regulations	479.34	2.00	41.89	515.89	set	**557.78**
Sliding door gear; Hillaldam Coburn Ltd or other equal and approved; Commercial/Light industrial; for top hung timber/metal doors, weight not exceeding 365 kg						
Sliding door gear						
bottom guide; fixed to concrete in groove	16.26	0.46	9.63	17.50	m	**27.13**
top track	22.59	0.23	4.82	24.31	m	**29.13**
detachable locking bar	20.41	0.31	6.49	21.97	nr	**28.46**
hangers; timber doors	44.60	0.46	9.63	48.00	nr	**57.63**
hangers; metal doors	30.07	0.46	9.63	32.36	nr	**42.00**
head brackets; open, soffit fixing; screwing to timber	6.26	0.32	6.70	6.81	nr	**13.51**
head brackets; open, side fixing; bolting to masonry	6.29	0.46	9.63	9.57	nr	**19.20**
door guide to timber door	5.27	0.23	4.82	5.67	nr	**10.49**
door stop; rubber buffers; to masonry	23.24	0.69	14.45	25.01	nr	**39.46**
drop bolt; screwing to timber	20.86	0.46	9.63	22.45	nr	**32.08**
bow handle; to timber	10.70	0.23	4.82	11.52	nr	**16.34**
Sundries						
rubber door stop; plugged and screwed to concrete	5.43	0.09	1.88	5.84	nr	**7.72**

P BUILDING FABRIC SUNDRIES

Item	PC £	Labour hours	Labour £	Material £	Unit	Total rate £
P30 TRENCHES/PIPEWAYS/PITS FOR BURIED ENGINEERINMG SERVICES						
Excavating trenches; by machine; grading bottoms; earthwork support; filling with excavated material and compacting; disposal of surplus soil on site; spreading on site average 50 m						
Services not exceeding 200 mm nominal size						
average depth of run not exceeding 0.50 m	-	0.28	3.51	1.61	m	**5.12**
average depth of run not exceeding 0.75 m	-	0.37	4.64	2.66	m	**7.30**
average depth of run not exceeding 1.00 m	-	0.79	9.90	5.06	m	**14.96**
average depth of run not exceeding 1.25 m	-	1.16	14.53	6.97	m	**21.50**
average depth of run not exceeding 1.50 m	-	1.48	18.54	9.10	m	**27.64**
average depth of run not exceeding 1.75 m	-	1.85	23.18	11.60	m	**34.78**
average depth of run not exceeding 2.00 m	-	2.13	26.68	13.30	m	**39.98**
Excavating trenches; by hand; grading bottoms; earthwork support; filling with excavated material and compacting; disposal; of surplus soil on site; spreading on site average 50 m						
Services not exceeding 200 mm nominal size						
average depth of run not exceeding 0.50 m	-	0.93	11.65	-	m	**11.65**
average depth of run not exceeding 0.75 m	-	1.39	17.41	-	m	**17.41**
average depth of run not exceeding 1.00 m	-	2.04	25.56	1.72	m	**27.28**
average depth of run not exceeding 1.25 m	-	2.87	35.95	2.37	m	**38.32**
average depth of run not exceeding 1.50 m	-	3.93	49.23	2.88	m	**52.12**
average depth of run not exceeding 1.75 m	-	5.18	64.89	3.49	m	**68.38**
average depth of run not exceeding 2.00 m	-	5.92	74.16	3.83	m	**77.99**
Stop cock pits, valve chambers and the like; excavating; half brick thick walls in common bricks in cement mortar (1:3); on in situ concrete designated mix C20 - 20 mm aggregate bed; 100 mm thick						
Pits						
100 mm x 100 mm x 750 mm deep; internal holes for one small pipe; polypropylene hinged box cover; bedding in cement mortar (1:3)	-	3.89	101.82	54.35	nr	**156.17**
P31 HOLES/CHASES/COVERS/SUPPORTS FOR SERVICES						
Builders' work for electrical installations; cutting away for and making good after electrician; including cutting or leaving all holes, notches, mortices, sinkings and chases, in both the structure and its coverings, for the following electrical points						
Exposed installation						
lighting points	-	0.28	4.23	-	nr	**4.23**
socket outlet points	-	0.46	7.30	-	nr	**7.30**
fitting outlet points	-	0.46	7.30	-	nr	**7.30**
equipment points or control gear points	-	0.65	10.50	-	nr	**10.50**
Concealed installation						
lighting points	-	0.37	5.77	-	nr	**5.77**
socket outlet points	-	0.65	10.50	-	nr	**10.50**
fitting outlet points	-	0.65	10.50	-	nr	**10.50**
equipment points or control gear points	-	0.93	14.72	-	nr	**14.72**

P BUILDING FABRIC SUNDRIES

Item	PC £	Labour hours	Labour £	Material £	Unit	Total rate £
Builders' work for other services installations						
Cutting chases in brickwork						
for one pipe; not exceeding 55 mm nominal size; vertical	-	0.37	4.26	-	m	**4.26**
for one pipe; 55 mm - 110 mm nominal size; vertical	-	0.65	7.48	-	m	**7.48**
Cutting and pinning to brickwork or blockwork; ends of supports						
for pipes not exceeding 55 mm nominal size	-	0.19	4.97	-	nr	**4.97**
for cast iron pipes 55 mm - 110 mm nominal size	-	0.31	8.11	-	nr	**8.11**
Cutting or forming holes for pipes or the like; not exceeding 55 mm nominal size; making good						
reinforced concrete; not exceeding 100 mm deep	-	0.75	12.11	0.68	nr	**12.78**
reinforced concrete; 100 mm - 200 mm deep	-	1.15	18.57	1.03	nr	**19.60**
reinforced concrete; 200 mm - 300 mm deep	-	1.50	24.22	1.35	nr	**25.56**
half brick thick	-	0.31	4.54	-	nr	**4.54**
one brick thick	-	0.51	7.47	-	nr	**7.47**
one and a half brick thick	-	0.83	12.16	-	nr	**12.16**
100 mm blockwork	-	0.28	4.10	-	nr	**4.10**
140 mm blockwork	-	0.37	5.42	-	nr	**5.42**
215 mm blockwork	-	0.46	6.74	-	nr	**6.74**
plasterboard partition or suspended ceiling	-	0.35	5.13	-	nr	**5.13**
Cutting or forming holes for pipes or the like; 55 mm - 110 mm nominal size; making good						
reinforced concrete; not exceeding 100 mm deep	-	1.15	18.57	1.03	nr	**19.60**
reinforced concrete; 100 mm - 200 mm deep	-	1.75	28.25	1.57	nr	**29.82**
reinforced concrete; 200 mm - 300 mm deep	-	2.25	36.32	2.02	nr	**38.34**
half brick thick	-	0.37	5.42	-	nr	**5.42**
one brick thick	-	0.65	9.52	-	nr	**9.52**
one and a half brick thick	-	1.02	14.94	-	nr	**14.94**
100 mm blockwork	-	0.32	4.69	-	nr	**4.69**
140 mm blockwork	-	0.46	6.74	-	nr	**6.74**
215 mm blockwork	-	0.56	8.20	-	nr	**8.20**
plasterboard partition or suspended ceiling	-	0.40	5.86	-	nr	**5.86**
Cutting or forming holes for pipes or the like; over 110 mm nominal size; making good						
reinforced concrete; not exceeding 100 mm deep	-	1.15	18.57	1.03	nr	**19.60**
reinforced concrete; 100 mm - 200 mm deep	-	1.75	28.25	1.57	nr	**29.82**
reinforced concrete; 200 mm - 300 mm deep	-	2.25	36.32	2.02	nr	**38.34**
half brick thick	-	0.46	6.74	-	nr	**6.74**
one brick thick	-	0.79	11.57	-	nr	**11.57**
one and a half brick thick	-	1.25	18.31	-	nr	**18.31**
100 mm blockwork	-	0.42	6.15	-	nr	**6.15**
140 mm blockwork	-	0.56	8.20	-	nr	**8.20**
215 mm blockwork	-	0.69	10.11	-	nr	**10.11**
plasterboard partition or suspended ceiling	-	0.45	6.59	-	nr	**6.59**
Add for making good fair face or facings one side						
pipe; not exceeding 55 mm nominal size	-	0.07	1.83	-	nr	**1.83**
pipe; 55 mm - 110 mm nominal size	-	0.09	2.36	-	nr	**2.36**
pipe; over 110 mm nominal size	-	0.11	2.88	-	nr	**2.88**
Add for fixing sleeve (supply not included)						
for pipe; small	-	0.14	3.66	-	nr	**3.66**
for pipe; large	-	0.19	4.97	-	nr	**4.97**
for pipe; extra large	-	0.28	7.33	-	nr	**7.33**
Add for supplying and fixing one-hour intumescent sleeve						
for pipe; small	-	0.25	4.04	14.62	nr	**18.65**
for pipe; large	-	0.28	4.52	17.87	nr	**22.39**
for pipe; extra large	-	0.30	4.84	36.81	nr	**41.65**

P BUILDING FABRIC SUNDRIES

Item	PC £	Labour hours	Labour £	Material £	Unit	Total rate £
P31 HOLES/CHASES/COVERS/SUPPORTS FOR SERVICES – cont'd						
Builders' work for other services installations – cont'd						
Cutting or forming holes for ducts; girth not exceeding 1.00 m; making good						
half brick thick	-	0.56	8.20	-	nr	**8.20**
one brick thick	-	0.93	13.62	-	nr	**13.62**
one and a half brick thick	-	1.48	21.68	-	nr	**21.68**
100 mm blockwork	-	0.46	6.74	-	nr	**6.74**
140 mm blockwork	-	0.65	9.52	-	nr	**9.52**
215 mm blockwork	-	0.83	12.16	-	nr	**12.16**
plasterboard partition or suspended ceiling	-	0.65	9.52	-	nr	**9.52**
Cutting or forming holes for ducts; girth 1.00 m - 2.00 m; making good						
half brick thick	-	0.65	9.52	-	nr	**9.52**
one brick thick	-	1.11	16.26	-	nr	**16.26**
one and a half brick thick	-	1.76	25.78	-	nr	**25.78**
100 mm blockwork	-	0.56	8.20	-	nr	**8.20**
140 mm blockwork	-	0.74	10.84	-	nr	**10.84**
215 mm blockwork	-	0.93	13.62	-	nr	**13.62**
plasterboard partition or suspended ceiling	-	0.75	10.99	-	nr	**10.99**
Cutting or forming holes for ducts; girth 2.00 m - 3.00 m; making good						
half brick thick	-	1.02	14.94	-	nr	**14.94**
one brick thick	-	1.76	25.78	-	nr	**25.78**
one and a half brick thick	-	2.78	40.72	-	nr	**40.72**
100 mm blockwork	-	0.88	12.89	-	nr	**12.89**
140 mm blockwork	-	1.20	17.58	-	nr	**17.58**
215 mm blockwork	-	1.53	22.41	-	nr	**22.41**
plasterboard partition or suspended ceiling	-	1.00	14.65	-	nr	**14.65**
Cutting or forming holes for ducts; girth 3.00 m - 4.00 m; making good						
half brick thick	-	1.39	20.36	-	nr	**20.36**
one brick thick	-	2.31	33.84	-	nr	**33.84**
one and a half brick thick	-	3.70	54.20	-	nr	**54.20**
100 mm blockwork	-	1.02	14.94	-	nr	**14.94**
140 mm blockwork	-	1.39	20.36	-	nr	**20.36**
215 mm blockwork	-	1.76	25.78	-	nr	**25.78**
plasterboard partition or suspended ceiling	-	1.25	18.31	-	nr	**18.31**
Mortices in brickwork						
for expansion bolt	-	0.19	2.78	-	nr	**2.78**
for 20 mm diameter bolt; 75 mm deep	-	0.14	2.05	-	nr	**2.05**
for 20 mm diameter bolt; 150 mm deep	-	0.23	3.37	-	nr	**3.37**
Mortices in brickwork; grouting with cement mortar (1:1)						
75 mm x 75 mm x 200 mm deep	-	0.28	4.10	0.14	nr	**4.24**
75 mm x 75 mm x 300 mm deep	-	0.37	5.42	0.20	nr	**5.62**
Holes in softwood for pipes, bars, cables and the like						
12 mm thick	-	0.03	0.63	-	nr	**0.63**
25 mm thick	-	0.05	1.05	-	nr	**1.05**
50 mm thick	-	0.09	1.88	-	nr	**1.88**
100 mm thick	-	0.14	2.93	-	nr	**2.93**
Holes in hardwood for pipes, bars, cables and the like						
12 mm thick	-	0.05	1.05	-	nr	**1.05**
25 mm thick	-	0.08	1.68	-	nr	**1.68**
50 mm thick	-	0.14	2.93	-	nr	**2.93**
100 mm thick	-	0.20	4.19	-	nr	**4.19**

P BUILDING FABRIC SUNDRIES

Item	PC £	Labour hours	Labour £	Material £	Unit	Total rate £
NOTE: The following rates for cutting holes and mortices in brickwork or blockwork etc. allow for diamond drilling						
Diamond drilling						
Cutting holes and mortices in brickwork; per 25 mm depth						
25 mm diameter	-	-	-	-	nr	**1.48**
32 mm diameter	-	-	-	-	nr	**1.20**
52 mm diameter	-	-	-	-	nr	**1.43**
78 mm diameter	-	-	-	-	nr	**1.57**
107 mm diameter	-	-	-	-	nr	**1.66**
127 mm diameter	-	-	-	-	nr	**2.03**
152 mm diameter	-	-	-	-	nr	**2.40**
200 mm diameter	-	-	-	-	nr	**3.09**
250 mm diameter	-	-	-	-	nr	**4.66**
300 mm diameter	-	-	-	-	nr	**6.18**
Diamond chasing; per 25 x 25 mm section						
in facing or common brickwork	-	-	-	-	m	**2.81**
in semi-engineering brickwork	-	-	-	-	m	**5.63**
in engineering brickwork	-	-	-	-	m	**7.84**
in lightweight blockwork	-	-	-	-	m	**2.21**
in heavyweight blockwork	-	-	-	-	m	**4.43**
In render/screed	-	-	-	-	m	**8.72**
Forming boxes; 100 x 100 mm; per 25 mm depth						
in facing or common brickwork	-	-	-	-	nr	**1.13**
in semi-engineering brickwork	-	-	-	-	nr	**2.25**
in engineering brickwork	-	-	-	-	nr	**3.14**
in lightweight blockwork	-	-	-	-	nr	**0.89**
in heavyweight blockwork	-	-	-	-	nr	**1.77**
in render/screed	-	-	-	-	nr	**3.49**
Other items						
diamond track mount or ring sawing brickwork	-	-	-	-	m	**5.54**
diamond floor sawing asphalte	-	-	-	-	m	**0.92**
stitch drilling 107 mm diameter hole in brickwork	-	-	-	-	nr	**1.20**
"SFD Screeduct" or other equal and approved; MDT Ducting Ltd; with side flanges; laid within floor screed; galvanised mild steel						
Floor ducting						
100 mm wide x 50 mm deep	10.20	0.19	3.98	11.53	m	**15.51**
extra for						
bend	7.65	0.09	1.88	8.64	nr	**10.53**
tee section	7.65	0.09	1.88	8.64	nr	**10.53**
connector / stop end	0.51	0.09	1.88	0.58	nr	**2.46**
ply cover 15 mm/16 mm thick WBP exterior grade	0.62	0.09	1.88	0.70	m	**2.59**
200 mm wide x 75 mm deep	13.60	0.19	3.98	15.37	m	**19.35**
extra for						
bend	10.20	0.09	1.88	11.53	nr	**13.41**
tee section	10.20	0.09	1.88	11.53	nr	**13.41**
connector / stop end	0.68	0.09	1.88	0.77	nr	**2.65**
ply cover 15 mm/16 mm thick WBP exterior grade	0.97	0.09	1.88	1.10	m	**2.99**

Q PAVING/PLANTING/FENCING/SITE FURNITURE

Item	PC £	Labour hours	Labour £	Material £	Unit	Total rate £
Q10 KERBS/EDGINGS/CHANNELS/PAVING ACCESSORIES						
Excavating; by machine						
Excavating trenches; to receive kerb foundations; average size						
300 mm x 100 mm	-	0.02	0.25	0.40	m	**0.65**
450 mm x 150 mm	-	0.02	0.25	0.81	m	**1.06**
600 mm x 200 mm	-	0.03	0.40	1.13	m	**1.53**
Excavating curved trenches; to receive kerb foundations; average size						
300 mm x 100 mm	-	0.01	0.13	0.65	m	**0.77**
450 mm x 150 mm	-	0.03	0.38	0.97	m	**1.34**
600 mm x 200 mm	-	0.04	0.50	1.21	m	**1.71**
Excavating; by hand						
Excavating trenches; to receive kerb foundations; average size						
150 mm x 50 mm	-	0.02	0.25	-	m	**0.25**
200 mm x 75 mm	-	0.06	0.75	-	m	**0.75**
250 mm x 100 mm	-	0.10	1.25	-	m	**1.25**
300 mm x 100 mm	-	0.13	1.63	-	m	**1.63**
Excavating curved trenches; to receive kerb foundations; average size						
150 mm x 50 mm	-	0.03	0.38	-	m	**0.38**
200 mm x 75 mm	-	0.07	0.88	-	m	**0.88**
250 mm x 100 mm	-	0.11	1.38	-	m	**1.38**
300 mm x 100 mm	-	0.14	1.75	-	m	**1.75**
Plain in situ ready mixed designated concrete; C7.5 - 40 mm aggregate; poured on or against earth or unblinded hardcore						
Foundations	76.84	1.16	18.73	88.91	m^3	**107.63**
Blinding beds						
thickness not exceeding 150 mm	76.84	1.71	27.61	88.91	m^3	**116.51**
Plain in situ ready mixed designated concrete; C10 - 40 mm aggregate; poured on or against earth or unblinded hardcore						
Foundations	78.02	1.16	18.73	90.28	m^3	**109.00**
Blinding beds						
thickness not exceeding 150 mm	78.02	1.71	27.61	90.28	m^3	**117.88**
Plain in situ ready mixed designated concrete; C20 - 20 mm aggregate; poured on or against earth or unblinded hardcore						
Foundations	81.52	1.16	18.73	94.32	m^3	**113.04**
Blinding beds						
thickness not exceeding 150 mm	81.52	1.71	27.61	94.32	m^3	**121.92**
Precast concrete kerbs, channels, edgings, etc.; BS 340; bedded, jointed and pointed in cement mortar (1:3); including haunching up one side with in situ ready mix designated concrete C10 - 40 mm aggregate; to concrete base						
Edgings; straight; square edge, fig 12						
50 mm x 150 mm	-	0.23	4.96	3.11	m	**8.07**
50 mm x 200 mm	-	0.23	4.96	4.01	m	**8.97**
50 mm x 255 mm	-	0.23	4.96	4.25	m	**9.21**

Q PAVING/PLANTING/FENCING/SITE FURNITURE

Item	PC £	Labour hours	Labour £	Material £	Unit	Total rate £
Kerbs; straight						
125 mm x 255 mm; fig 7	-	0.31	6.69	5.40	m	**12.09**
150 mm x 305 mm; fig 6	-	0.31	6.69	9.88	m	**16.56**
Kerbs; curved						
125 mm x 255 mm; fig 7	-	0.46	9.92	7.18	m	**17.10**
150 mm x 305 mm; fig 6	-	0.46	9.92	16.83	m	**26.75**
Channels; 255 x 125 mm; fig 8						
straight	-	0.31	6.69	5.40	m	**12.09**
curved	-	0.46	9.92	7.18	m	**17.10**
Quadrants; fig 14						
305 mm x 305 mm x 150 mm	-	0.32	6.90	9.68	nr	**16.58**
305 mm x 305 mm x 255 mm	-	0.32	6.90	9.68	nr	**16.58**
457 mm x 457 mm x 150 mm	-	0.37	7.98	10.70	nr	**18.68**
457 mm x 457 mm x 255 mm	-	0.37	7.98	10.70	nr	**18.68**
Q20 HARDCORE/GRANULAR/CEMENT BOUND BASES/SUB BASES TO ROADS/PAVINGS						
Filling to make up levels; by machine						
Average thickness not exceeding 0.25 m						
obtained off site; hardcore	-	0.28	3.51	28.10	m³	**31.60**
obtained off site; granular fill type one	-	0.28	3.51	34.50	m³	**38.00**
obtained off site; granular fill type two	-	0.28	3.51	32.55	m³	**36.06**
Average thickness exceeding 0.25 m						
obtained off site; hardcore	-	0.24	3.01	24.04	m³	**27.05**
obtained off site; granular fill type one	-	0.24	3.01	34.22	m³	**37.23**
obtained off site; granular fill type two	-	0.24	3.01	32.27	m³	**35.28**
Filling to make up levels; by hand						
Average thickness not exceeding 0.25 m						
obtained off site; hardcore	-	0.61	7.64	30.02	m³	**37.67**
obtained off site; sand	-	0.71	8.89	43.89	m³	**52.79**
Average thickness exceeding 0.25 m						
obtained off site; hardcore	-	0.51	6.39	25.65	m³	**32.03**
obtained off site; sand	-	0.60	7.52	43.25	m³	**50.77**
Surface treatments						
Compacting						
filling; blinding with sand	-	0.04	0.50	2.02	m²	**2.52**
Q21 IN SITU CONCRETE ROADS/PAVINGS						
Reinforced in situ ready mixed designated concrete; C10 - 40 mm aggregate						
Roads; to hardcore base						
thickness not exceeding 150 mm	74.31	1.85	29.87	85.97	m³	**115.84**
thickness 150 mm - 450 mm	74.31	1.30	20.99	85.97	m³	**106.96**
Reinforced in situ ready mixed designated concrete; C20 - 20 mm aggregate						
Roads; to hardcore base						
thickness not exceeding 150 mm	77.63	1.85	29.87	89.82	m³	**119.69**
thickness 150 mm - 450 mm	77.63	1.30	20.99	89.82	m³	**110.81**
Reinforced in situ ready mixed designated concrete; C25 - 20 mm aggregate						
Roads; to hardcore base						
thickness ot exceeding 150 mm	78.85	1.85	29.87	91.23	m³	**121.09**
thickness 150 mm - 450 mm	78.85	1.30	20.99	91.23	m³	**112.21**

Q PAVING/PLANTING/FENCING/SITE FURNITURE

Item	PC £	Labour hours	Labour £	Material £	Unit	Total rate £
Q21 IN SITU CONCRETE ROADS/PAVINGS – cont'd						
Formwork; sides of foundations; basic finish						
Plain vertical						
height not exceeding 250 mm	-	0.39	7.42	1.55	m	**8.97**
height 250 mm - 500 mm	-	0.57	10.85	2.58	m	**13.43**
height 500 mm - 1.00 m	-	0.83	15.79	4.97	m	**20.77**
add to above for curved radius 6m	-	0.03	0.57	0.23	m	**0.80**
Reinforcement; fabric; BS 4449; lapped; in roads, footpaths or pavings						
Ref A142 (2.22 kg/m²)						
400 mm minimum laps	1.68	0.11	2.09	1.99	m²	**4.08**
Ref A193 (3.02 kg/m²)						
400 mm minimum laps	2.29	0.11	2.09	2.71	m²	**4.80**
Formed joints; Fosroc Expandite "Flexcell" impregnated joint filler or other equal and approved						
Width not exceeding 150 mm						
12.50 mm thick	-	0.14	2.66	1.85	m	**4.51**
25 mm thick	-	0.19	3.62	2.77	m	**6.38**
Width 150 - 300 mm						
12.50 mm thick	-	0.19	3.62	2.93	m	**6.55**
25 mm thick	-	0.19	3.62	4.68	m	**8.30**
Width 300 - 450 mm						
12.50 mm thick	-	0.23	4.38	4.40	m	**8.78**
25 mm thick	-	0.23	4.38	7.02	m	**11.40**
Sealants; Fosroc Expandite "Pliastic N2" hot poured rubberized bituminous compound or other equal and approved						
Width 25 mm						
25 mm depth	-	0.20	3.81	1.86	m	**5.67**
Concrete sundries						
Treating surfaces of unset concrete; grading to cambers; tamping with a 75 mm thick steel shod tamper	-	0.23	3.71	-	m²	**3.71**
Q22 COATED MACADAM/ASPHALT ROADS/ PAVINGS						
NOTE: The prices for all bitumen macadam and hot rolled asphalt materials are for individual courses to roads and footpaths and need combining to arrive at complete specifications and costs for full construction. Intermediate course thicknesses can interpolated so long as BS 594987 allows the material type to be compacted to the required thickness. Costs include for work to falls, crossfalls or slopes not exceeding 15 degrees from horizontal; for laying on prepared bases (prices not included) and for rolling with an appropriate roller. The following rates are based on black bitumen macadam. Red bitumen macadam rates are approximately 50% dearer. PSV is Polished Stone Value.						

Q PAVING/PLANTING/FENCING/SITE FURNITURE

Item	PC £	Labour hours	Labour £	Material £	Unit	Total rate £
Dense bitumen macadam base course; BS 594987 - 1; bitumen penetration 100/125						
Carriageway, hardshoulder and hardstrip						
100 mm thick; one coat; with 0/32mm aggregate size; to clause 5.2	-	-	-	-	m²	**16.76**
200 mm thick; one coat; with 0/32mm aggregate size; to clause 5.2	-	-	-	-	m²	**29.50**
Extra over above items for increase / reduction in 10 mm increments	-	-	-	-	m²	**1.20**
Hot rolled asphalt base course; BS 594987 - 1						
Carriageway, hardshoulder and hardstrip						
150 mm thick; one coat; 60% 0/32 mm aggregate size; to column 2/5	-	-	-	-	m²	**27.87**
200 mm thick; one coat; 60% 0/32 mm aggregate size; to column 2/5	-	-	-	-	m²	**37.06**
Extra over above items for increase / reduction in 10 mm increments	-	-	-	-	m²	**1.54**
Dense bitumen macadam binder course; BS 594987 - 1; bitumen penetration 100/125						
Carriageway, hardshoulder and hardstrip						
60 mm thick; one coat; with 0/32 mm aggregate size; to clause 6.4	-	-	-	-	m²	**10.25**
60 mm thick; one coat; with 0/32 mm aggregate size; to clause 6.5	-	-	-	-	m²	**10.34**
Extra over above items for increase / reduction in 10 mm increments	-	-	-	-	m²	**1.36**
Hot rolled asphalt binder course; BS 594987 - 1						
Carriageway, hardshoulder and hardstrip						
40 mm thick; one coat; 50% 0/14 mm aggregate size; to column 2/2; 55 PSV	-	-	-	-	m²	**9.78**
60 mm thick; one coat; 50% 0/14 mm aggregate size; to column 2/2	-	-	-	-	m²	**11.26**
60 mm thick; one coat; 50% 0/20 mm aggregate size; to column 2/3	-	-	-	-	m²	**11.06**
60 mm thick; one coat; 60% 0/32 mm aggregate size; to column 2/5	-	-	-	-	m²	**10.51**
100 mm thick; one coat; 60% 0/32 mm aggregate size; to column 2/5	-	-	-	-	m²	**16.96**
Extra over above for 10 mm increase/reduction	-	-	-	-	m²	**2.02**
Macadam surface course; BS 594987 - 1; bitumen penetration 100/125						
Carriageway, hardshoulder and hardstrip						
30 mm thick; one coat; medium graded with 0/6 mm nominal aggregate binder; to clause 7.6	-	-	-	-	m²	**7.91**
40 mm thick; one coat; close graded with 0/14 mm nominal aggregate binder; to clause 7.3	-	-	-	-	m²	**7.26**
40 mm thick; one coat; close graded with 0/10 mm nominal aggregate binder; to clause 7.4	-	-	-	-	m²	**7.91**
Extra over above items for increase / reduction in 10 mm increments	-	-	-	-	m²	**1.42**
Extra over above items for coarse aggregate 60-64 PSV	-	-	-	-	m²	**1.44**
Extra over above items for coarse aggregate 65-67 PSV	-	-	-	-	m²	**1.57**
Extra over above items for coarse aggregate 68 PSV	-	-	-	-	m²	**2.09**

Q PAVING/PLANTING/FENCING/SITE FURNITURE

Item	PC £	Labour hours	Labour £	Material £	Unit	Total rate £
Q22 COATED MACADAM/ASPHALT ROADS/ PAVINGS – cont'd						
Hot rolled asphalt surface course; BS 594987 - 1; bitumen penetration 40/60						
Carriageway, hardshoulder and hardstrip						
40 mm thick; one coat; 30% mix 0/10 mm aggregate size; to column 3/2; with 20 mm pre-coated chippings 60-64 PSV	-	-	-	-	m²	**9.77**
40 mm thick; one coat; 30% mix 0/10 mm aggregate size; to column 3/2; with 14 mm pre-coated chippings 60-64 PSV	-	-	-	-	m²	**9.85**
Extra over above items for increase / reduction in 10 mm increments	-	-	-	-	m²	**1.63**
Extra over above items for chippings with 65-67 PSV	-	-	-	-	m²	**0.06**
Extra over above items for chippings with 68 PSV	-	-	-	-	m²	**0.14**
Extra over above items for 6-10KN High Traffic Flows	-	-	-	-	m²	**0.71**
Stone mastic asphalt surface course; BS 594987 - 1						
Carriageway, hardshoulder and hardstrip						
35 mm thick; one coat; with 0/14 mm nominal aggregate size; 55 PSV	-	-	-	-	m²	**8.98**
35 mm thick; one coat; with 0/10 mm nominal aggregate size; 55 PSV	-	-	-	-	m²	**8.98**
Extra over above items for increase / reduction in 10 mm increments	-	-	-	-	m²	**2.01**
Thin surface course with 60 PSV						
Carriageway, hardshoulder and hardstrip						
35 mm thick; one coat; with 0/10 mm nominal aggregate size	-	-	-	-	m²	**8.98**
Extra over above items for increase / reduction in 10 mm increments	-	-	-	-	m²	**1.01**
Extra over above items for coarse aggregate 60-64 PSV	-	-	-	-	m²	**0.25**
Extra over above items for coarse aggregate 65-67 PSV	-	-	-	-	m²	**0.25**
Extra over above items for coarse aggregate 68 PSV	-	-	-	-	m²	**0.46**
Regulating courses						
Carriageway, hardshoulder and hardstrip						
Dense Bitumen Macadam; bitumen penetration 100/125; with 0/20 mm nominal aggregate regulating course (BS 594987 - clause 6.5)	-	-	-	-	tonne	**77.59**
Hot rolled asphalte; 50% 0/20 mm aggregate size (BS 594987 - 1:2003 column 2/3)	-	-	-	-	tonne	**84.56**
Stone mastic asphalte; 0/6 mm aggregate	-	-	-	-	tonne	**108.25**
Bitumen Emulsion tack coats						
Carriageway, hardshoulder and hardstrip						
K1-40; applied 0.35-0.45l/m²	-	-	-	-	m²	**0.14**
K1-70; applied 0.35-0.45l/m²	-	-	-	-	m²	**0.24**

Q PAVING/PLANTING/FENCING/SITE FURNITURE

Item	PC £	Labour hours	Labour £	Material £	Unit	Total rate £
Q23 GRAVEL/HOGGIN/WOODCHIP ROADS/ PAVINGS						
Two coat gravel paving; level and to falls; first layer course clinker aggregate and wearing layer fine gravel aggregate						
Pavings; over 300 mm wide						
50 mm thick	-	0.07	1.51	5.59	m²	**7.10**
63 mm thick	-	0.09	1.94	7.25	m²	**9.19**
Resin bonded gravel paving; level and to falls						
Pavings; over 300 mm wide						
50 mm thick	-	-	-	-	m²	**46.68**
Q25 SLAB/BRICK/BLOCK/SETT/COBBLE PAVINGS						
Artificial stone paving; Charcon's "Moordale Textured" or other equal and approved; to falls or crossfalls; bedding 25 mm thick in cement mortar (1:3); staggered joints; jointing in coloured cement mortar (1:3), brushed in; to sand base						
Pavings; over 300 mm wide						
600 mm x 600 mm x 50 mm thick; natural	12.25	0.39	8.41	16.82	m²	**25.23**
Brick paviors; 215 mm x 103 mm x 65 mm rough stock bricks; to falls or crossfalls; bedding 10 mm thick in cement mortar (1:3); jointing in cement mortar (1:3); as work proceeds; to concrete base						
Pavings; over 300 mm wide; straight joints both ways						
bricks laid flat (PC £ per 1000)	430.00	0.74	19.37	22.07	m²	**41.44**
bricks laid on edge	-	1.04	27.22	32.74	m²	**59.96**
Pavings; over 300 mm wide; laid to herringbone pattern						
bricks laid flat	-	0.93	24.34	22.07	m²	**46.41**
bricks laid on edge	-	1.30	34.03	32.74	m²	**66.77**
Add or deduct for variation of £10.00/1000 in PC of brick paviours						
bricks laid flat	-	-	-	0.49	m²	-
bricks laid on edge	-	-	-	0.73	m²	-
River washed cobble paving; 50 mm -75 mm; to falls or crossfalls; bedding 13 mm thick in cement mortar (1:3); jointing to a height of two thirds of cobbles in dry mortar (1:3); tightly butted, washed and brushed; to concrete						
Pavings; over 300 mm wide						
regular (PC £ per tonne)	113.34	3.70	79.80	26.89	m²	**106.69**
laid to pattern	-	4.63	99.86	26.89	m²	**126.75**

Q PAVING/PLANTING/FENCING/SITE FURNITURE

Item	PC £	Labour hours	Labour £	Material £	Unit	Total rate £
Q25 SLAB/BRICK/BLOCK/SETT/COBBLE PAVINGS – cont'd						
Concrete paving flags; BS EN 1339; to falls or crossfalls; bedding 25 mm thick in cement and sand mortar (1:4); butt joints straight both ways; jointing in cement and sand (1:3); brushed in; to sand base						
Pavings; over 300 mm wide						
450 mm x 600 mm x 50 mm thick; grey	7.26	0.42	9.06	9.65	m²	**18.71**
450 mm x 600 mm x 60 mm thick; coloured	8.06	0.42	9.06	10.55	m²	**19.61**
600 mm x 600 mm x 50 mm thick; grey	5.62	0.39	8.41	7.80	m²	**16.21**
600 mm x 600 mm x 50 mm thick; coloured	6.75	0.39	8.41	9.07	m²	**17.48**
750 mm x 600 mm x 50 mm thick; grey	5.04	0.36	7.76	7.14	m²	**14.90**
750 mm x 600 mm x 50 mm thick; coloured	6.70	0.36	7.76	9.01	m²	**16.78**
900 mm x 600 mm x 50 mm thick; grey	4.50	0.33	7.12	6.53	m²	**13.65**
900 mm x 600 mm x 50 mm thick; coloured	6.17	0.33	7.12	8.41	m²	**15.53**
Concrete rectangular paving blocks; to falls or crossfalls; bedding 50 mm thick in dry sharp sand; filling joints with sharp sand brushed in; on earth base						
Pavings; "Keyblock" or other equal and approved; over 300 mm wide; straight joints both ways						
200 mm x 100 mm x 60 mm thick; grey	7.75	0.69	14.88	11.77	m²	**26.65**
200 mm x 100 mm x 60 mm thick; coloured	8.40	0.69	14.88	12.51	m²	**27.39**
200 mm x 100 mm x 80 mm thick; grey	8.62	0.74	15.96	13.00	m²	**28.96**
200 mm x 100 mm x 80 mm thick; coloured	9.73	0.74	15.96	14.26	m²	**30.22**
Pavings; "Keyblock" or other equal and approved; over 300 mm wide; laid to herringbone pattern						
200 mm x 100 mm x 60 mm thick; grey	-	0.88	18.98	11.77	m²	**30.74**
200 mm x 100 mm x 60 mm thick; coloured	-	0.88	18.98	12.51	m²	**31.49**
200 mm x 100 mm x 80 mm thick; grey	-	0.93	20.06	13.00	m²	**33.06**
200 mm x 100 mm x 80 mm thick; coloured	-	0.93	20.06	14.26	m²	**34.32**
Extra for two row boundary edging to herringbone pavings; 200 mm wide; including a 150 mm high in situ concrete mix C10 - 40 mm aggregate haunching to one side; blocks laid breaking joint						
200 mm x 100 mm x 60 mm; coloured	-	0.28	6.04	2.38	m	**8.42**
200 mm x 100 mm x 80 mm; coloured	-	0.28	6.04	2.49	m	**8.53**
Pavings; "Europa" or other equal and approved; over 300 mm wide; straight joints both ways						
200 mm x 100 mm x 60 mm thick; grey	6.61	0.69	14.88	10.48	m²	**25.36**
200 mm x 100 mm x 60 mm thick; coloured	7.33	0.69	14.88	11.30	m²	**26.18**
200 mm x 100 mm x 80 mm thick; grey	7.88	0.74	15.96	12.16	m²	**28.12**
200 mm x 100 mm x 80 mm thick; coloured	8.64	0.74	15.96	13.03	m²	**28.99**
Pavings; "Pedesta" or other equal and approved; over 300 mm wide; straight joints both ways						
200 mm x 100 mm x 60 mm thick; grey	14.91	0.69	14.88	19.87	m²	**34.75**
200 mm x 100 mm x 60 mm thick; coloured	14.91	0.69	14.88	19.87	m²	**34.75**
200 mm x 100 mm x 80 mm thick; grey	19.02	0.74	15.96	24.75	m²	**40.71**
200 mm x 100 mm x 80 mm thick; coloured	19.02	0.74	15.96	24.75	m²	**40.71**
Pavings; "Intersett" or other equal and approved; over 300 mm wide; straight joints both ways						
200 mm x 100 mm x 60 mm thick; grey	11.50	0.69	14.88	16.01	m²	**30.89**
200 mm x 100 mm x 60 mm thick; coloured	12.77	0.69	14.88	17.44	m²	**32.32**
200 mm x 100 mm x 80 mm thick; grey	13.76	0.74	15.96	18.81	m²	**34.77**
200 mm x 100 mm x 80 mm thick; coloured	15.28	0.74	15.96	20.53	m²	**36.49**

Q PAVING/PLANTING/FENCING/SITE FURNITURE

Item	PC £	Labour hours	Labour £	Material £	Unit	Total rate £
Concrete rectangular paving blocks; to falls or crossfalls; 6 mm wide joints; symmetrical layout; bedding in 15 mm semi-dry cement mortar (1:4); jointing and pointing in cement and sand (1:4); on concrete base						
Pavings; "Trafica" or other equal and approved; over 300 mm wide						
400 mm x 400 mm x 65 mm; Saxon textured; natural	21.99	0.44	9.49	26.16	m²	**35.65**
400 mm x 400 mm x 65 mm; Saxon textured; buff	25.42	0.44	9.49	30.04	m²	**39.53**
400 mm x 400 mm x 65 mm; Perfecta; natural	26.39	0.44	9.49	31.12	m²	**40.61**
400 mm x 400 mm x 65 mm; Perfecta; buff	30.44	0.44	9.49	35.71	m²	**45.20**
450 mm x 450 mm x 70 mm; Saxon textured; natural	22.51	0.43	9.27	26.75	m²	**36.02**
450 mm x 450 mm x 70 mm; Saxon textured; buff	25.89	0.43	9.27	30.56	m²	**39.83**
450 mm x 450 mm x 70 mm; Perfecta; natural	25.61	0.43	9.27	30.25	m²	**39.53**
450 mm x 450 mm x 70 mm; Perfecta; buff	29.64	0.43	9.27	34.80	m²	**44.07**
York stone slab pavings; to falls or crossfalls; bedding 25 mm thick in cement:sand mortar (1:4); 5 mm wide joints; jointing in coloured cement mortar (1:3); brushed in; to sand base						
Pavings; over 300 mm wide						
50 mm thick; random rectangular pattern	88.15	0.69	18.06	99.00	m²	**117.06**
600 mm x 600 mm x 50 mm thick	107.04	0.39	10.21	119.84	m²	**130.04**
600 mm x 900 mm x 50 mm thick	110.19	0.33	8.64	123.31	m²	**131.95**
Granite setts; BS EN 1342; 200 mm x 100 mm x 100 mm; standard "C" dressing; tightly butted to falls or crossfalls; bedding 25 mm thick in cement mortar (1:3); filling joints with dry mortar (1:6); washed and brushed; on concrete base						
Pavings; over 300 mm wide						
straight joints (PC £ per tonne)	188.89	1.48	31.92	4.75	m²	**36.67**
laid to pattern	-	1.85	39.90	59.76	m²	**99.66**
Two rows of granite setts as boundary edging; 200 mm wide; including a 150 mm high ready mixed designated concrete C10 - 40 mm aggregate; haunching to one side; blocks laid breaking joint	-	0.65	14.02	13.65	m	**27.67**
Q26 SPECIAL SURFACINGS/PAVINGS FOR SPORT						
Sundries						
Line marking						
width not exceeding 300 mm	-	0.04	0.61	0.17	m	**0.79**
Q30 SEEDING/TURFING						
Top soil						
Selected from spoil heaps; grading; prepared for turfing or seeding; to general surfaces						
average 75 mm thick	-	0.21	2.63	-	m²	**2.63**
average 100 mm thick	-	0.23	2.88	-	m²	**2.88**
average 125 mm thick	-	0.25	3.13	-	m²	**3.13**
average 150 mm thick	-	0.26	3.26	-	m²	**3.26**
average 175 mm thick	-	0.27	3.38	-	m²	**3.38**
average 200 mm thick	-	0.29	3.63	-	m²	**3.63**

Q PAVING/PLANTING/FENCING/SITE FURNITURE

Item	PC £	Labour hours	Labour £	Material £	Unit	Total rate £
Q30 SEEDING/TURFING – cont'd						
Top soil – cont'd						
Selected from spoil heaps; grading; prepared for turfing or seeding; to cuttings or embankments						
average 75 mm thick	-	0.24	3.01	-	m²	**3.01**
average 100 mm thick	-	0.26	3.26	-	m²	**3.26**
average 125 mm thick	-	0.28	3.51	-	m²	**3.51**
average 150 mm thick	-	0.30	3.76	-	m²	**3.76**
average 175 mm thick	-	0.31	3.88	-	m²	**3.88**
average 200 mm thick	-	0.32	4.01	-	m²	**4.01**
Imported recycled top soil						
Grading; prepared for turfing or seeding; to general surfaces						
average 75 mm thick	-	0.19	2.38	1.94	m²	**4.32**
average 100 mm thick	-	0.20	2.51	2.52	m²	**5.02**
average 125 thick	-	0.22	2.76	3.68	m²	**6.43**
average 150 mm thick	-	0.23	2.88	4.84	m²	**7.72**
average 175 mm thick	-	0.25	3.13	5.42	m²	**8.55**
average 200 mm thick	-	0.26	3.26	6.00	m²	**9.26**
Grading; preparing for turfing or seeding; to cuttings or embankments						
average 75 mm thick	-	0.21	2.63	1.94	m²	**4.57**
average 100 mm thick	-	0.23	2.88	2.52	m²	**5.40**
average 125 mm thick	-	0.25	3.13	3.68	m²	**6.81**
average 150 mm thick	-	0.26	3.26	4.84	m²	**8.10**
average 175 mm thick	-	0.27	3.38	5.42	m²	**8.80**
average 200 mm thick	-	0.29	3.63	6.00	m²	**9.63**
Fertilizer						
Fertilizer 0.07 kg/m²; raking in						
general surfaces (PC £ per 25kg)	15.01	0.03	0.38	0.05	m²	**0.42**
Selected grass seed						
Grass seed; sowing at a rate of 0.042 kg/m² two applications; raking in						
general surfaces (PC £ per 25kg)	97.08	0.06	0.75	0.35	m²	**1.10**
cuttings or embankments	-	0.07	0.88	0.35	m²	**1.23**
Preserved turf from stack on site						
Turfing						
general surfaces	-	0.19	2.38	-	m²	**2.38**
cuttings or embankments; shallow	-	0.20	2.51	-	m²	**2.51**
cuttings or embankments; steep; pegged	-	0.28	3.51	-	m²	**3.51**
Imported turf; cultivated						
Turfing						
general surfaces	2.56	0.19	2.38	2.75	m²	**5.13**
cuttings or embankments; shallow	2.56	0.20	2.51	2.75	m²	**5.26**
cuttings or embankments; steep; pegged	2.56	0.28	3.51	2.75	m²	**6.26**
Q31 PLANTING						
Planting only						
Hedge plants						
height not exceeding 750 mm	-	0.23	2.88	-	nr	**2.88**
height 750 mm - 1.50 m	-	0.56	7.02	-	nr	**7.02**
Saplings						
height not exceeding 3.00 m	-	1.57	19.67	-	nr	**19.67**

Q PAVING/PLANTING/FENCING/SITE FURNITURE

Item	PC £	Labour hours	Labour £	Material £	Unit	Total rate £
Q40 FENCING						
NOTE: The prices for all fencing include for setting posts in position, to a depth of 0.60 m for fences not exceeding 1.40 m high and of 0.76 m for fences over 1.40 m high. The prices allow for excavating post holes; filling to within 150 mm of ground level with concrete and all necessary backfilling.						
Strained wire fencing; BS 1722 Part 3; 4 mm diameter galvanized mild steel plain wire threaded through posts and strained with eye bolts						
Fencing; height 900 mm; three line; concrete posts at 2750 mm centres	-	-	-	-	m	**20.33**
Extra for						
end concrete straining post; one strut	-	-	-	-	nr	**49.45**
angle concrete straining post; two struts	-	-	-	-	nr	**57.44**
Fencing; height 1.07 m; six line; concrete posts at 2750 mm centres	-	-	-	-	m	**21.16**
Extra for						
end concrete straining post; one strut	-	-	-	-	nr	**55.61**
angle concrete straining post; two struts	-	-	-	-	nr	**63.59**
Fencing; height 1.20 m; six line; concrete posts at 2750 mm centres	-	-	-	-	m	**21.28**
Extra for						
end concrete straining post; one strut	-	-	-	-	nr	**57.18**
angle concrete straining post; two struts	-	-	-	-	nr	**65.14**
Fencing; height 1.40 m; eight line; concrete posts at 2750 mm centres	-	-	-	-	m	**21.87**
Extra for						
end concrete straining post; one strut	-	-	-	-	nr	**58.42**
angle concrete straining post; two struts	-	-	-	-	nr	**66.38**
Chain link fencing; BS 1722 Part 1; 3 mm diameter galvanized mild steel wire; 50 mm mesh; galvanized mild steel tying and line wire; three line wires threaded through posts and strained with eye bolts and winding brackets						
Fencing; height 900 mm; galvanized mild steel angle posts at 3.00 m centres	-	-	-	-	m	**28.60**
Extra for						
end steel straining post; one strut	-	-	-	-	nr	**82.44**
angle steel straining post; two struts	-	-	-	-	nr	**95.08**
Fencing; height 900 mm; concrete posts at 3.00 m centres	-	-	-	-	m	**20.73**
Extra for						
end concrete straining post; one strut	-	-	-	-	nr	**44.29**
angle concrete straining post; two struts	-	-	-	-	nr	**52.26**
Fencing; height 1.20 m; galvanized mild steel angle posts at 3.00 m centres	-	-	-	-	m	**21.09**
Extra for						
end steel straining post; one strut	-	-	-	-	nr	**87.97**
angle steel straining post; two struts	-	-	-	-	nr	**112.65**
Fencing; height 1.20 m; concrete posts at 3.00 m centres	-	-	-	-	m	**20.22**
Extra for						
end concrete straining post; one strut	-	-	-	-	nr	**50.69**
angle concrete straining post; two struts	-	-	-	-	nr	**59.88**

Q PAVING/PLANTING/FENCING/SITE FURNITURE

Item	PC £	Labour hours	Labour £	Material £	Unit	Total rate £
Q40 FENCING – cont'd						
Chain link fencing; BS 1722 Part 1 – cont'd						
Fencing; height 1.80 m; galvanized mild steel angle posts at 3.00 m centres	-	-	-	-	m	**23.72**
Extra for						
end steel straining post; one strut	-	-	-	-	nr	**89.47**
angle steel straining post; two struts	-	-	-	-	nr	**111.30**
Fencing; height 1.80 m; concrete posts at 3.00 m centres	-	-	-	-	m	**27.39**
Extra for						
end concrete straining post; one strut	-	-	-	-	nr	**70.88**
angle concrete straining post; two struts	-	-	-	-	nr	**83.62**
Pair of gates and gate posts; gates to match galvanized chain link fencing, with angle framing, braces, etc., complete with hinges, locking bar, lock and bolts; two 100 mm x 100 mm angle section gate posts; each with one strut						
2.44 m x 0.90 m	-	-	-	-	nr	**695.17**
2.44 m x 1.20 m	-	-	-	-	nr	**717.43**
2.44 m x 1.80 m	-	-	-	-	nr	**773.91**
Chain link fencing; BS 1722 Part 1; 3 mm diameter plastic coated mild steel wire; 50 mm mesh; plastic coated mild steel tying and line wire; three line wires threaded through posts and strained with eye bolts and winding brackets						
Fencing; height 900 mm; galvanized mild steel angle posts at 3.00 m centres	-	-	-	-	m	**26.23**
Extra for						
end steel straining post; one strut	-	-	-	-	nr	**72.69**
angle steel straining post; two struts	-	-	-	-	nr	**80.92**
Fencing; height 900 mm; concrete posts at 3.00 m centres	-	-	-	-	m	**19.56**
Extra for						
end concrete straining post; one strut	-	-	-	-	nr	**44.29**
angle concrete straining post; two struts	-	-	-	-	nr	**52.26**
Fencing; height 1.20 m; galvanized mild steel angle posts at 3.00 m centres	-	-	-	-	m	**19.38**
Extra for						
end steel straining post; one strut	-	-	-	-	nr	**76.25**
angle steel straining post; two struts	-	-	-	-	nr	**81.46**
Fencing; height 1.20 m; concrete posts at 3.00 m centres	-	-	-	-	m	**19.83**
Extra for						
end concrete straining post; one strut	-	-	-	-	nr	**50.69**
angle concrete straining post; two struts	-	-	-	-	nr	**59.88**
Fencing; height 1.80 m; galvanized mild steel angle posts at 3.00 m centres	-	-	-	-	m	**21.97**
Extra for						
end steel straining post; one strut	-	-	-	-	nr	**75.47**
angle steel straining post; two struts	-	-	-	-	nr	**90.14**
Fencing; height 1.80 m; concrete posts at 3.00 m centres	-	-	-	-	m	**25.26**
Extra for						
end concrete straining post; one strut	-	-	-	-	nr	**70.88**
angle concrete straining post; two struts	-	-	-	-	nr	**83.62**

Q PAVING/PLANTING/FENCING/SITE FURNITURE

Item	PC £	Labour hours	Labour £	Material £	Unit	Total rate £
Pair of gates and gate posts; gates to match plastic chain link fencing; with angle framing, braces, etc. complete with hinges, locking bar, lock and bolts; two 100 mm x 100 mm angle section gate posts; each with one strut						
2.44 m x 0.90 m	-	-	-	-	nr	**607.86**
2.44 m x 1.20 m	-	-	-	-	nr	**623.71**
2.44 m x 1.80 m	-	-	-	-	nr	**671.76**
Chain link fencing for tennis courts; BS 1722 Part 13; 2.5 diameter galvanised mild wire; 45 mm mesh; line and tying wires threaded through 45 mm x 45 mm x 5 mm galvanised mild steel angle standards, posts and struts; 60 mm x 60 mm x 6 mm straining posts and gate posts; straining posts and struts strained with eye bolts and winding brackets						
Fencing to tennis court 36.00 m x 18.00 m; including gate 1.07 m x 1.98 m; complete with hinges, locking bar, lock and bolts						
height 2745 mm fencing; standards at 3.00 m centres	-	-	-	-	nr	**2677.56**
height 3660 mm fencing; standards at 2.50 m centres	-	-	-	-	nr	**3590.36**
Cleft chestnut pale fencing; BS 1722 Part 4; pales spaced 51 mm apart; on two lines of galvanized wire; 64 mm diameter posts; 76 mm x 51 mm struts						
Fencing; height 900 mm; posts at 2.50 m centres	-	-	-	-	m	**11.34**
Extra for						
straining post; one strut	-	-	-	-	nr	**30.22**
corner straining post; two struts	-	-	-	-	nr	**30.22**
Fencing; height 1.05 m; posts at 2.50 m centres	-	-	-	-	m	**12.84**
Extra for						
straining post; one strut	-	-	-	-	nr	**30.63**
corner straining post; two struts	-	-	-	-	nr	**30.63**
Close boarded fencing; BS 1722 Part 5; 76 mm x 38 mm softwood rails; 89 mm x 19 mm softwood pales lapped 13 mm; 152 mm x 25 mm softwood gravel boards; all softwood "treated"; posts at 3.00 m centres						
Fencing; two rail; concrete posts						
height 1.00 m	-	-	-	-	m	**34.87**
height 1.20 m	-	-	-	-	m	**35.20**
Fencing; three rail; concrete posts						
height 1.40 m	-	-	-	-	m	**38.71**
height 1.60 m	-	-	-	-	m	**38.80**
height 1.80 m	-	-	-	-	m	**40.28**
Precast concrete slab fencing; 305 mm x 38 mm x 1753 mm slabs; fitted into twice grooved concrete posts at 1830 mm centres						
Fencing						
height 1.50 m	-	-	-	-	m	**68.35**
height 1.80 m	-	-	-	-	m	**75.69**

Q PAVING/PLANTING/FENCING/SITE FURNITURE

Item	PC £	Labour hours	Labour £	Material £	Unit	Total rate £
Q40 FENCING – cont'd						
Mild steel unclimbable fencing; in rivetted panels 2440 mm long; 44 mm x 13 mm flat section top and bottom rails; two 44 mm x 19 mm flat section standards; one with foot plate; and 38 mm x 13 mm raking stay with foot plate; 20 mm diameter pointed verticals at 120 mm centres; two 44 mm x 19 mm supports 760 mm long with ragged ends to bottom rail; the whole bolted together; coated with red oxide primer; setting standards and stays in ground at 2440 mm centres and supports at 815 mm centres						
Fencing						
height 1.67 m	-	-	-	-	m	**127.79**
height 2.13 m	-	-	-	-	m	**146.96**
Pair of gates and gate posts, to match mild steel unclimbable fencing; with flat section framing, braces, etc., complete with locking bar, lock, handles, drop bolt, gate stop and holding back catches; two 102 mm x 102 mm hollow section gate posts with cap and foot plates						
2.44 m x 1.67 m	-	-	-	-	nr	**1107.54**
2.44 m x 2.13 m	-	-	-	-	nr	**1277.93**
4.88 m x 1.67 m	-	-	-	-	nr	**1734.33**
4.88 m x 2.13 m	-	-	-	-	nr	**2172.47**
PVC coated, galvanised mild steel high security fencing; "Sentinal Sterling" fencing or other equal and approved; Twil Wire Products Ltd; 50 mm x 50 mm mesh; 3/3.50 mm gauge wire; barbed edge - 1; "Sentinal Bi-steel" colour coated posts or other equal and approved at 2440 mm centres						
Fencing						
1.80 m	-	0.93	11.65	40.23	m	**51.88**
2.10 m	-	1.16	14.53	44.43	m	**58.96**

R DISPOSAL SYSTEMS

Item	PC £	Labour hours	Labour £	Material £	Unit	Total rate £
R10 RAINWATER PIPEWORK/GUTTERS						
Aluminium pipes and fittings; BS EN 612; ears cast on; polyester powder coated finish						
63 mm diameter pipes; plugged and screwed	12.26	0.34	6.16	14.54	m	**20.70**
Extra for						
fittings with one end	-	0.20	3.62	7.54	nr	**11.16**
fittings with two ends	-	0.39	7.06	7.82	nr	**14.89**
fittings with three ends	-	0.56	10.14	10.32	nr	**20.47**
shoe	7.51	0.20	3.62	7.54	nr	**11.16**
bend	8.00	0.39	7.06	7.82	nr	**14.89**
single branch	10.44	0.56	10.14	10.32	nr	**20.47**
offset 228 projection	18.46	0.39	8.17	18.38	nr	**26.55**
offset 304 projection	20.59	0.39	7.06	20.72	nr	**27.79**
access pipe	22.81	0.39	7.06	21.74	nr	**28.80**
connection to clay pipes; cement and sand (1:2) joint	-	0.14	2.54	0.10	nr	**2.63**
76.50 mm diameter pipes; plugged and screwed	14.28	0.37	6.70	16.84	m	**23.54**
Extra for						
shoe	10.30	0.23	4.17	10.45	nr	**14.62**
bend	10.12	0.42	7.61	9.99	nr	**17.60**
single branch	12.57	0.60	10.87	12.43	nr	**23.30**
offset 228 projection	20.41	0.42	7.61	20.15	nr	**27.76**
offset 304 projection	22.58	0.42	7.61	22.55	nr	**30.16**
access pipe	24.92	0.42	7.61	23.51	nr	**31.12**
connection to clay pipes; cement and sand (1:2) joint	-	0.16	2.90	0.10	nr	**3.00**
100 mm diameter pipes; plugged and screwed	24.38	0.42	7.61	28.27	m	**35.87**
Extra for						
shoe	12.41	0.26	4.71	11.88	nr	**16.59**
bend	14.09	0.46	8.33	13.50	nr	**21.83**
single branch	16.84	0.69	12.50	15.83	nr	**28.33**
offset 228 projection	23.62	0.46	8.33	21.45	nr	**29.78**
offset 304 projection	26.22	0.46	8.33	24.32	nr	**32.65**
access pipe	29.54	0.46	8.33	25.78	nr	**34.11**
connection to clay pipes; cement and sand (1:2) joint	-	0.19	3.44	0.10	nr	**3.54**
Roof outlets; circular aluminium; with flat or domed grating; joint to pipe						
50 mm diameter	56.17	0.56	16.63	60.45	nr	**77.09**
75 mm diameter	74.54	0.60	17.82	80.22	nr	**98.04**
100 mm diameter	104.55	0.65	19.31	112.52	nr	**131.83**
150 mm diameter	130.05	0.60	20.50	139.97	nr	**160.46**
Roof outlets; d-shaped; balcony; with flat or domed grating; joint to pipe						
50 mm diameter	66.52	0.56	16.63	71.59	nr	**88.23**
75 mm diameter	74.54	0.60	17.82	80.22	nr	**98.04**
100 mm diameter	93.94	0.65	19.31	101.10	nr	**120.41**
Galvanized wire balloon grating; BS 416 for pipes or outlets						
50 mm diameter	1.10	0.06	1.78	1.19	nr	**2.97**
63 mm diameter	1.12	0.06	1.78	1.21	nr	**2.99**
75 mm diameter	1.19	0.06	1.78	1.28	nr	**3.07**
100 mm diameter	1.31	0.07	2.08	1.41	nr	**3.49**

Item	PC £	Labour hours	Labour £	Material £	Unit	Total rate £
R10 RAINWATER PIPEWORK/GUTTERS – cont'd						
Aluminium gutters and fittings; BS EN 612; polyester powder coated finish						
100 mm half round gutters; on brackets; screwed to timber	11.25	0.32	6.70	15.84	m	**22.54**
Extra for						
stop end	3.05	0.15	3.14	6.31	nr	**9.45**
running outlet	6.76	0.31	6.49	7.27	nr	**13.76**
stop end outlet	6.01	0.15	3.14	8.56	nr	**11.70**
angle	6.25	0.31	6.49	5.60	nr	**12.10**
113 mm half round gutters; on brackets; screwed to timber	11.78	0.32	6.70	16.47	m	**23.18**
Extra for						
stop end	3.20	0.15	3.14	6.51	nr	**9.65**
running outlet	7.37	0.31	6.49	7.90	nr	**14.39**
stop end outlet	6.89	0.15	3.14	9.49	nr	**12.63**
angle	7.03	0.31	6.49	6.33	nr	**12.82**
125 mm half round gutters; on brackets; screwed to timber	13.24	0.37	7.75	19.80	m	**27.54**
Extra for						
stop end	3.91	0.17	3.56	8.81	nr	**12.37**
running outlet	7.97	0.32	6.70	8.44	nr	**15.14**
stop end outlet	7.32	0.17	3.56	11.38	nr	**14.94**
angle	7.81	0.32	6.70	8.41	nr	**15.11**
100 mm ogee gutters; on brackets; screwed to timber	14.03	0.34	7.12	20.38	m	**27.50**
Extra for						
stop end	3.22	0.16	3.35	4.17	nr	**7.52**
running outlet	7.93	0.32	6.70	7.78	nr	**14.48**
stop end outlet	6.14	0.16	3.35	9.68	nr	**13.03**
angle	6.68	0.32	6.70	5.02	nr	**11.72**
112 mm ogee gutters; on brackets; screwed to timber	15.61	0.39	8.17	22.44	m	**30.61**
Extra for						
stop end	3.44	0.16	3.35	4.42	nr	**7.77**
running outlet	8.02	0.32	6.70	7.74	nr	**14.44**
stop end outlet	6.88	0.16	3.35	10.60	nr	**13.95**
angle	7.96	0.32	6.70	6.13	nr	**12.83**
125 mm ogee gutters; on brackets; screwed to timber	17.24	0.39	8.17	24.67	m	**32.84**
Extra for						
stop end	3.76	0.18	3.77	4.78	nr	**8.55**
running outlet	8.77	0.34	7.12	8.43	nr	**15.55**
stop end outlet	7.81	0.18	3.77	11.84	nr	**15.61**
angle	9.28	0.34	7.12	7.29	nr	**14.41**
Cast iron pipes and fittings; BS 416; ears cast on; joints						
65 mm pipes; primed; nailed to masonry	20.24	0.48	8.69	23.23	m	**31.93**
Extra for						
shoe	17.77	0.30	5.43	18.09	nr	**23.52**
bend	10.87	0.53	9.60	10.48	nr	**20.08**
single branch	21.38	0.67	12.14	21.32	nr	**33.46**
offset 225 mm projection	19.38	0.53	9.60	18.50	nr	**28.10**
offset 305 mm projection	22.70	0.53	9.60	21.70	nr	**31.30**
connection to clay pipes; cement and sand (1:2) joint	-	0.14	2.54	0.11	nr	**2.65**

R DISPOSAL SYSTEMS

Item	PC £	Labour hours	Labour £	Material £	Unit	Total rate £
75 mm pipes; primed; nailed to masonry	20.24	0.51	9.24	23.42	m	**32.66**
Extra for						
shoe	17.77	0.32	5.80	18.16	nr	**23.95**
bend	13.20	1.11	21.80	13.12	nr	**34.92**
single branch	23.56	0.69	12.50	23.86	nr	**36.36**
offset 225 mm projection	19.38	0.56	10.14	18.57	nr	**28.71**
offset 305 mm projection	23.82	0.56	10.14	23.00	nr	**33.14**
connection to clay pipes; cement and sand (1:2) joint	-	0.16	2.90	0.11	nr	**3.01**
100 mm pipes; primed; nailed to masonry	27.18	0.56	10.14	31.49	m	**41.63**
Extra for						
shoe	23.58	0.37	6.70	24.07	nr	**30.77**
bend	18.65	0.60	10.87	18.63	nr	**29.49**
single branch	27.46	0.74	13.40	27.48	nr	**40.89**
offset 225 mm projection	38.03	0.60	10.87	38.17	nr	**49.03**
offset 305 mm projection	38.78	0.60	10.87	38.37	nr	**49.24**
connection to clay pipes; cement and sand (1:2) joint	-	0.19	3.44	0.10	nr	**3.54**
100 mm x 75 mm rectangular pipes; primed; nailing to masonry	54.67	0.56	10.14	62.55	m	**72.69**
Extra for						
shoe	66.54	0.37	6.70	68.97	nr	**75.67**
bend	63.35	0.60	10.87	65.46	nr	**76.32**
offset 225 mm projection	89.22	0.37	6.70	90.29	nr	**96.99**
offset 305 mm projection	95.35	0.37	6.70	95.82	nr	**102.52**
connection to clay pipes; cement and sand (1:2) joint	-	0.19	3.44	0.10	nr	**3.54**
Rainwater head; rectangular; for pipes						
65 mm diameter	54.60	0.53	9.60	60.55	nr	**70.15**
75 mm diameter	54.60	0.56	10.14	60.62	nr	**70.76**
100 mm diameter	75.39	0.60	10.87	83.68	nr	**94.55**
Rainwater head; octagonal; for pipes						
65 mm diameter	33.36	0.53	9.60	37.12	nr	**46.72**
75 mm diameter	33.36	0.56	10.14	37.19	nr	**47.33**
100 mm diameter	41.44	0.60	10.87	46.23	nr	**57.09**
Copper wire balloon grating; BS 416 for pipes or outlets						
50 mm diameter	1.34	0.06	1.09	1.45	nr	**2.53**
63 mm diameter	1.74	0.06	1.09	1.87	nr	**2.96**
75 mm diameter	1.36	0.06	1.09	1.46	nr	**2.55**
100 mm diameter	1.53	0.07	1.27	1.64	nr	**2.91**
Cast iron gutters and fittings; BS EN 877						
100 mm half round gutters; primed; on brackets; screwed to timber	10.40	0.37	7.75	14.96	m	**22.71**
Extra for						
stop end	2.62	0.16	3.35	4.37	nr	**7.72**
running outlet	7.60	0.32	6.70	7.42	nr	**14.13**
angle	7.80	0.32	6.70	9.19	nr	**15.89**
115 mm half round gutters; primed; on brackets; screwed to timber	10.84	0.37	7.75	15.48	m	**23.23**
Extra for						
stop end	3.39	0.16	3.35	5.21	nr	**8.56**
running outlet	8.28	0.32	6.70	8.12	nr	**14.82**
angle	8.02	0.32	6.70	9.35	nr	**16.05**
125 mm half round gutters; primed; on brackets; screwed to timber	12.68	0.42	8.80	17.57	m	**26.37**
Extra for						
stop end	3.39	0.19	3.98	5.21	nr	**9.19**
running outlet	9.46	0.37	7.75	9.23	nr	**16.98**
angle	9.46	0.37	7.75	10.55	nr	**18.30**

R DISPOSAL SYSTEMS

Item	PC £	Labour hours	Labour £	Material £	Unit	Total rate £
R10 RAINWATER PIPEWORK/GUTTERS – cont'd						
Cast iron gutters and fittings; BS EN 877 – cont'd						
150 mm half round gutters; primed; on brackets; screwed to timber	21.67	0.46	9.63	27.32	m	**36.95**
Extra for						
stop end	4.71	0.20	4.19	8.54	nr	**12.72**
running outlet	16.39	0.42	8.80	15.97	nr	**24.77**
angle	17.29	0.42	8.80	18.11	nr	**26.90**
100 mm ogee gutters; primed; on brackets; screwed to timber	11.60	0.39	8.17	16.52	m	**24.69**
Extra for						
stop end	2.68	0.17	3.56	5.89	nr	**9.45**
running outlet	8.29	0.34	7.12	8.06	nr	**15.18**
angle	8.14	0.34	7.12	9.54	nr	**16.66**
115 mm ogee gutters; primed; on brackets; screwed to timber	12.76	0.39	8.17	17.86	m	**26.03**
Extra for						
stop end	3.47	0.17	3.56	6.75	nr	**10.31**
running outlet	8.82	0.34	7.12	8.53	nr	**15.66**
angle	8.82	0.34	7.12	10.06	nr	**17.18**
125 mm ogee gutters; primed; on brackets; screwed to timber	13.38	0.43	9.01	18.97	m	**27.97**
Extra for						
stop end	3.47	0.19	3.98	7.11	nr	**11.09**
running outlet	9.63	0.39	8.17	9.35	nr	**17.52**
angle	9.63	0.39	8.17	11.17	nr	**19.34**
3 mm thick galvanised heavy pressed steel gutters and fittings; joggle joints; BS 1091						
200 mm x 100 mm (400 mm girth) box gutter; screwed to timber	-	0.60	10.87	23.09	m	**33.95**
Extra for						
stop end	-	0.32	5.80	12.85	nr	**18.64**
running outlet	-	0.65	11.77	21.29	nr	**33.07**
stop end outlet	-	0.32	5.80	29.79	nr	**35.59**
angle	-	0.65	11.77	23.64	nr	**35.41**
381 mm boundary wall gutters (900 mm girth); bent twice; screwed to timber	-	0.60	10.87	37.98	m	**48.85**
Extra for						
stop end	-	0.37	6.70	21.90	nr	**28.61**
running outlet	-	0.65	11.77	29.31	nr	**41.09**
stop end outlet	-	0.32	5.80	41.16	nr	**46.96**
angle	-	0.65	11.77	34.00	nr	**45.78**
457 mm boundary wall gutters (1200 mm girth); bent twice; screwed to timber	-	0.69	12.50	50.66	m	**63.16**
Extra for						
stop end	-	0.37	6.70	28.11	nr	**34.82**
running outlet	-	0.74	13.40	42.39	nr	**55.79**
stop end outlet	-	0.37	6.70	45.10	nr	**51.80**
angle	-	0.74	13.40	46.01	nr	**59.42**

R DISPOSAL SYSTEMS

Item	PC £	Labour hours	Labour £	Material £	Unit	Total rate £
uPVC external rainwater pipes and fittings; BS EN 12200; slip-in joints						
50 mm pipes; fixing with pipe or socket brackets; plugged and screwed	5.03	0.28	5.07	7.62	m	**12.69**
Extra for						
shoe	2.97	0.19	3.44	4.06	nr	**7.50**
bend	3.47	0.28	5.07	4.61	nr	**9.68**
two bends to form offset 229 mm projection	6.94	0.28	5.07	7.47	nr	**12.55**
connection to clay pipes; cement and sand (1:2) joint	-	0.12	2.17	0.11	nr	**2.29**
68 mm pipes; fixing with pipe or socket brackets; plugged and screwed	3.89	0.31	5.62	6.71	m	**12.32**
Extra for						
shoe	2.97	0.20	3.62	4.41	nr	**8.03**
bend	4.55	0.31	5.62	6.15	nr	**11.76**
single branch	9.15	0.41	7.43	11.22	nr	**18.64**
two bends to form offset 229 mm projection	9.10	0.31	5.62	10.42	nr	**16.04**
loose drain connector; cement and sand (1:2) joint	-	0.14	2.54	11.82	nr	**14.36**
110 mm pipes; fixing with pipe or socket brackets; plugged and screwed	7.80	0.33	5.98	14.99	m	**20.96**
Extra for						
shoe	9.50	0.22	3.98	11.81	nr	**15.79**
bend	14.09	0.33	5.98	16.87	nr	**22.85**
single branch	21.55	0.44	7.97	25.10	nr	**33.07**
two bends to form offset 229 mm projection	28.17	0.33	5.98	31.17	nr	**37.15**
loose drain connector; cement and sand (1:2) joint	-	0.32	5.80	9.79	nr	**15.59**
65 mm square pipes; fixing with pipe or socket brackets; plugged and screwed	3.90	0.31	5.62	6.72	m	**12.34**
Extra for						
shoe	2.97	0.20	3.62	4.41	nr	**8.03**
bend	4.55	0.31	5.62	6.15	nr	**11.76**
single branch	9.15	0.41	7.43	11.22	nr	**18.64**
two bends to form offset 229 mm projection	9.10	0.31	5.62	10.64	nr	**16.25**
drain connector; square to round; cement and sand (1:2) joint	-	0.32	5.80	5.03	nr	**10.83**
Rainwater head; rectangular; for pipes						
50 mm diameter	16.17	0.42	7.61	19.50	nr	**27.11**
68 mm diameter	13.06	0.43	7.79	16.76	nr	**24.55**
110 mm diameter	27.24	0.51	9.24	32.81	nr	**42.05**
65 mm square	13.06	0.43	7.79	16.76	nr	**24.55**
uPVC gutters and fittings; BS EN 12200						
76 mm half round gutters; on brackets screwed to timber	3.82	0.28	5.86	6.28	m	**12.14**
Extra for						
stop end	1.36	0.12	2.51	1.98	nr	**4.49**
running outlet	3.85	0.23	4.82	3.90	nr	**8.72**
stop end outlet	3.83	0.12	2.51	4.36	nr	**6.87**
angle	3.85	0.23	4.82	4.78	nr	**9.60**
112 mm half round gutters; on brackets screwed to timber	3.84	0.31	6.49	7.63	m	**14.12**
Extra for						
stop end	2.13	0.12	2.51	3.09	nr	**5.61**
running outlet	4.19	0.26	5.45	4.28	nr	**9.72**
stop end outlet	4.19	0.12	2.51	5.02	nr	**7.53**
angle	4.68	0.26	5.45	6.30	nr	**11.74**

R DISPOSAL SYSTEMS

Item	PC £	Labour hours	Labour £	Material £	Unit	Total rate £
R10 RAINWATER PIPEWORK/GUTTERS – cont'd						
uPVC gutters and fittings; BS EN 12200 – cont'd						
170 mm half round gutters; on brackets; screwed to timber	8.03	0.31	6.49	14.18	m	**20.67**
Extra for						
stop end	7.65	0.15	3.14	9.85	nr	**12.99**
running outlet	8.05	0.29	6.07	8.15	nr	**14.23**
stop end outlet	7.65	0.15	3.14	9.12	nr	**12.26**
angle	10.48	0.29	6.07	13.65	nr	**19.72**
114 mm rectangular gutters; on brackets; screwed to timber	3.94	0.31	6.49	8.00	m	**14.50**
Extra for						
stop end	2.13	0.12	2.51	3.09	nr	**5.61**
running outlet	4.19	0.29	6.07	4.27	nr	**10.34**
stop end outlet	4.19	0.12	2.51	5.01	nr	**7.52**
angle	5.04	0.26	5.45	6.68	nr	**12.13**
R11 FOUL DRAINAGE ABOVE GROUND						
Cast iron "Timesaver" pipes and fittings or other equal and approved; BS 416						
50 mm pipes; primed; 3 m lengths; fixing with expanding bolts; to masonry	13.35	0.51	9.24	23.87	m	**33.10**
Extra for						
fittings with two ends	-	0.51	9.26	18.79	nr	**28.05**
fittings with three ends	-	0.69	12.50	31.83	nr	**44.33**
bends; short radius	12.18	0.51	9.24	18.79	nr	**28.03**
access bends; short radius	30.00	0.51	9.24	38.46	nr	**47.70**
boss; 38 BSP	25.21	0.51	9.24	32.79	nr	**42.03**
single branch	18.31	0.69	12.50	32.51	nr	**45.01**
isolated "Timesaver" coupling joint	6.91	0.28	5.07	7.63	nr	**12.70**
connection to clay pipes; cement and sand (1:2) joint	-	0.12	2.17	0.10	nr	**2.27**
75 mm pipes; primed; 3 m lengths; fixing with standard brackets; plugged and screwed to masonry	14.93	0.51	9.24	26.78	m	**36.02**
Extra for						
bends; short radius	13.78	0.55	9.96	21.09	nr	**31.05**
access bends; short radius	32.54	0.51	9.24	41.79	nr	**51.03**
boss; 38 BSP	25.21	0.55	9.96	33.70	nr	**43.66**
single branch	20.72	0.79	14.31	36.24	nr	**50.55**
double branch	30.79	1.02	18.48	55.77	nr	**74.24**
offset 115 mm projection	19.76	0.55	9.96	25.75	nr	**35.71**
offset 150 mm projection	23.21	0.55	9.96	29.04	nr	**39.01**
access pipe	29.29	0.55	9.96	36.00	nr	**45.97**
isolated "Timesaver" coupling joint	7.63	0.32	5.80	8.42	nr	**14.22**
connection to clay pipes; cement and sand (1:2) joint	-	0.14	2.54	0.10	nr	**2.63**
100 mm pipes; primed; 3 m lengths; fixing with standard brackets; plugged and screwed to masonry	18.05	0.55	9.96	37.22	m	**47.18**
Extra for						
WC bent connector; 450 mm long tail	26.98	0.55	9.96	33.52	nr	**43.48**
bends; short radius	16.85	0.62	11.23	26.52	nr	**37.75**
access bends; short radius	35.65	0.62	11.23	47.26	nr	**58.49**
boss; 38 BSP	30.10	0.62	11.23	41.15	nr	**52.38**
single branch	26.05	0.93	16.85	45.93	nr	**62.78**
double branch	32.22	1.20	21.74	63.73	nr	**85.47**
offset 225 mm projection	25.37	0.62	11.23	33.27	nr	**44.50**
offset 300 mm projection	27.30	0.62	11.23	34.79	nr	**46.02**
access pipe	30.79	0.62	11.23	38.85	nr	**50.08**
roof connector; for asphalt	29.10	0.62	11.23	39.33	nr	**50.56**

R DISPOSAL SYSTEMS

Item	PC £	Labour hours	Labour £	Material £	Unit	Total rate £
isolated "Timesaver" coupling joint	9.97	0.39	7.06	11.00	nr	**18.06**
transitional clayware socket; cement and sand (1:2) joint	19.82	0.37	6.70	32.96	nr	**39.67**
150 mm pipes; primed; 3 m lengths; fixing with standard brackets; plugged and screwed to masonry	37.68	0.69	12.50	75.25	m	**87.74**
Extra for						
bends; short radius	30.10	0.77	13.95	48.76	nr	**62.71**
access bends; short radius	50.62	0.77	13.95	71.39	nr	**85.34**
boss; 38 BSP	49.12	0.77	13.95	68.67	nr	**82.62**
single branch	64.58	1.11	20.11	103.62	nr	**123.73**
double branch	90.74	1.48	26.81	152.50	nr	**179.31**
access pipe	51.22	0.77	13.95	63.37	nr	**77.31**
isolated "Timesaver" coupling joint	-	0.46	8.33	21.94	nr	**30.27**
transitional clayware socket; cement and sand (1:2) joint	34.71	0.48	8.69	60.33	nr	**69.03**
Cast iron "Ensign" lightweight pipes and fittings or other equal and approved; BS EN 877						
50 mm pipes; primed; 3 m lengths; fixing with standard brackets; plugged and screwed to masonry	-	0.31	5.69	15.67	m	**21.36**
Extra for						
bends; short radius	-	0.27	4.97	13.00	nr	**17.97**
single branch	-	0.33	6.06	22.73	nr	**28.79**
access pipe	-	0.27	4.81	26.69	nr	**31.50**
70 mm pipes; primed; 3 m lengths; fixing with standard brackets; plugged and screwed to masonry	-	0.34	6.26	17.49	m	**23.75**
Extra for						
bends; short radius	-	0.30	5.49	14.50	nr	**19.99**
single branch	-	0.37	6.78	24.41	nr	**31.19**
access pipe	-	0.30	5.49	28.44	nr	**33.93**
100 mm pipes; primed; 3 m lengths; fixing with standard brackets; plugged and screwed to masonry	-	0.37	6.78	20.79	m	**27.57**
Extra for						
bends; short radius	-	0.32	5.90	17.79	nr	**23.69**
single branch	-	0.39	7.14	32.76	nr	**39.90**
double branch	-	0.46	8.44	46.00	nr	**54.44**
access pipe	-	0.32	5.90	32.33	nr	**38.23**
connector	-	0.21	3.83	30.03	nr	**33.86**
reducer	-	0.32	5.90	21.71	nr	**27.62**
Polypropylene (PP) waste pipes and fittings; BS EN 1451; push fit "O" - ring joints						
32 mm pipes; fixing with pipe clips; plugged and screwed	1.53	0.20	3.62	2.65	m	**6.28**
Extra for						
fittings with one end	-	0.15	2.72	1.48	nr	**4.19**
fittings with two ends	-	0.20	3.62	1.50	nr	**5.12**
fittings with three ends	-	0.28	5.07	2.60	nr	**7.67**
access plug	1.27	0.15	2.72	1.48	nr	**4.19**
double socket	0.97	0.14	2.54	1.13	nr	**3.67**
male iron to PP coupling	2.70	0.26	4.71	3.13	nr	**7.84**
sweep bend	1.21	0.20	3.62	1.40	nr	**5.02**
spigot bend	1.77	0.23	4.17	2.05	nr	**6.21**

R DISPOSAL SYSTEMS

Item	PC £	Labour hours	Labour £	Material £	Unit	Total rate £
R11 FOUL DRAINAGE ABOVE GROUND – cont'd						
Polypropylene (PP) waste pipes – cont'd						
40 mm pipes; fixing with pipe clips; plugged and screwed	1.89	0.20	3.62	3.10	m	**6.73**
Extra for						
fittings with one end	-	0.18	3.26	1.54	nr	**4.80**
fittings with two ends	-	0.28	5.07	1.77	nr	**6.84**
fittings with three ends	-	0.37	6.70	2.73	nr	**9.44**
access plug	1.33	0.18	3.26	1.54	nr	**4.80**
double socket	1.00	0.19	3.44	1.16	nr	**4.60**
universal connector	3.06	0.23	4.17	3.54	nr	**7.70**
sweep bend	1.36	0.28	5.07	1.58	nr	**6.65**
spigot bend	1.72	0.28	5.07	1.99	nr	**7.06**
reducer 40 mm - 32 mm	1.21	0.28	5.07	1.40	nr	**6.47**
50 mm pipes; fixing with pipe clips; plugged and screwed	2.42	0.32	5.80	4.51	m	**10.30**
Extra for						
fittings with one end	-	0.19	3.44	2.74	nr	**6.18**
fittings with two ends	-	0.32	5.80	2.94	nr	**8.74**
fittings with three ends	-	0.43	7.79	4.08	nr	**11.87**
access plug	2.37	0.19	3.44	2.74	nr	**6.18**
double socket	2.00	0.21	3.80	2.31	nr	**6.12**
sweep bend	2.61	0.32	5.80	3.02	nr	**8.82**
spigot bend	4.08	0.32	5.80	4.72	nr	**10.52**
reducer 50 mm - 40 mm	1.57	0.32	5.80	1.82	nr	**7.62**
muPVC waste pipes and fittings; BS EN 1329; solvent welded joints						
32 mm pipes; fixing with pipe clips; plugged and screwed	1.66	0.23	4.17	2.80	m	**6.97**
Extra for						
fittings with one end	-	0.16	2.90	1.39	nr	**4.29**
fittings with two ends	-	0.23	4.17	1.49	nr	**5.65**
fittings with three ends	-	0.31	5.62	1.97	nr	**7.59**
access plug	0.96	0.16	2.90	1.39	nr	**4.29**
straight coupling	1.04	0.16	2.90	1.47	nr	**4.37**
expansion coupling	1.83	0.23	4.17	2.34	nr	**6.51**
male iron to muPVC coupling	1.85	0.35	6.34	2.20	nr	**8.54**
sweep bend	1.05	0.23	4.17	1.49	nr	**5.65**
spigot/socket bend	-	0.23	4.17	2.23	nr	**6.40**
sweep tee	1.42	0.31	5.62	1.97	nr	**7.59**
40 mm pipes; fixing with pipe clips; plugged and screwed	2.06	0.28	5.07	3.30	m	**8.37**
Extra for						
fittings with one end	-	0.18	3.26	1.39	nr	**4.65**
fittings with two ends	-	0.28	5.07	1.62	nr	**6.70**
fittings with three ends	-	0.37	6.70	2.39	nr	**9.09**
fittings with four ends	4.36	0.49	8.88	5.38	nr	**14.26**
access plug	0.96	0.18	3.26	1.39	nr	**4.65**
straight coupling	1.03	0.19	3.44	1.47	nr	**4.91**
expansion coupling	2.20	0.28	5.07	2.76	nr	**7.83**
male iron to muPVC coupling	1.85	0.35	6.34	2.20	nr	**8.54**
level invert taper	1.29	0.28	5.07	1.75	nr	**6.82**
sweep bend	1.18	0.28	5.07	1.62	nr	**6.70**
spigot/socket bend	1.98	0.28	5.07	2.51	nr	**7.58**
sweep tee	1.79	0.37	6.70	2.39	nr	**9.09**
sweep cross	4.36	0.49	8.88	5.38	nr	**14.26**

R DISPOSAL SYSTEMS

Item	PC £	Labour hours	Labour £	Material £	Unit	Total rate £
50 mm pipes; fixing with pipe clips; plugged and screwed	3.10	0.32	5.80	5.31	m	**11.10**
Extra for						
fittings with one end	-	0.19	3.44	1.85	nr	**5.29**
fittings with two ends	-	0.32	5.80	2.61	nr	**8.40**
fittings with three ends	-	0.43	7.79	4.27	nr	**12.06**
fittings with four ends	-	0.57	10.32	5.61	nr	**15.94**
access plug	1.38	0.19	3.44	1.85	nr	**5.29**
straight coupling	1.89	0.21	3.80	2.41	nr	**6.22**
expansion coupling	2.98	0.32	5.80	3.62	nr	**9.41**
male iron to muPVC coupling	2.67	0.42	7.61	3.10	nr	**10.71**
level invert taper	1.61	0.32	5.80	2.10	nr	**7.89**
sweep bend	2.07	0.32	5.80	2.61	nr	**8.40**
spigot/socket bend	2.81	0.32	5.80	3.43	nr	**9.23**
sweep tee	1.79	0.37	6.70	2.39	nr	**9.09**
sweep cross	4.57	0.57	10.32	5.61	nr	**15.94**
uPVC overflow pipes and fittings; solvent welded joints						
19 mm pipes; fixing with pipe clips; plugged and screwed	0.98	0.20	3.62	1.86	m	**5.48**
Extra for						
splay cut end	-	0.01	0.18	-	nr	**0.18**
fittings with one end	-	0.16	2.90	1.29	nr	**4.19**
fittings with two ends	-	0.16	2.90	1.49	nr	**4.39**
fittings with three ends	-	0.20	3.62	1.69	nr	**5.32**
straight connector	1.05	0.16	2.90	1.29	nr	**4.19**
female iron to uPVC coupling	-	0.19	3.44	2.01	nr	**5.45**
bend	1.24	0.16	2.90	1.49	nr	**4.39**
bent tank connector	1.93	0.19	3.44	2.16	nr	**5.60**
uPVC pipes and fittings; BS EN 1329; with solvent welded joints (unless otherwise described)						
82 mm pipes; fixing with holderbats; plugged/screwed	7.06	0.37	6.70	11.24	m	**17.95**
Extra for						
socket plug	5.33	0.19	3.44	6.72	nr	**10.16**
slip coupling; push fit	11.64	0.34	6.16	12.84	nr	**19.00**
expansion coupling	5.60	0.37	6.70	7.02	nr	**13.72**
sweep bend	9.41	0.37	6.70	11.21	nr	**17.91**
boss connector	5.14	0.25	4.53	6.51	nr	**11.04**
single branch	13.14	0.49	8.88	15.84	nr	**24.71**
access door	12.52	0.56	10.14	14.23	nr	**24.37**
110 mm pipes; fixing with holderbats; plugged/screwed	7.19	0.41	7.43	11.74	m	**19.16**
Extra for						
socket plug	6.46	0.20	3.62	8.21	nr	**11.84**
slip coupling; push fit	14.58	0.37	6.70	16.08	nr	**22.79**
expansion coupling	5.73	0.41	7.43	7.40	nr	**14.83**
W.C. connector	10.41	0.27	4.89	12.06	nr	**16.96**
sweep bend	11.01	0.41	7.43	13.23	nr	**20.66**
W.C. connecting bend	17.08	0.27	4.89	19.43	nr	**24.32**
access bend	30.53	0.43	7.79	34.77	nr	**42.55**
boss connector	5.14	0.27	4.89	6.76	nr	**11.65**
single branch	14.56	0.54	9.78	17.73	nr	**27.51**
single branch with access	24.93	0.56	10.14	29.17	nr	**39.31**
double branch	35.98	0.68	12.32	41.95	nr	**54.26**
W.C. manifold	14.29	0.27	4.89	17.44	nr	**22.33**
access door	-	0.56	10.14	14.23	nr	**24.37**
access pipe connector	23.39	0.46	8.33	26.89	nr	**35.22**
connection to clay pipes; caulking ring and cement and sand (1:2) joint	-	0.39	7.06	10.38	nr	**17.44**

R DISPOSAL SYSTEMS

Item	PC £	Labour hours	Labour £	Material £	Unit	Total rate £
R11 FOUL DRAINAGE ABOVE GROUND – cont'd						
uPVC pipes and fittings; BS EN 1329 – cont'd						
160 mm pipes; fixing with holderbats; plugged and screwed	18.66	0.46	8.33	30.23	m	**38.57**
Extra for						
socket plug	11.88	0.23	4.17	15.52	nr	**19.69**
slip coupling; push fit	37.32	0.42	7.61	41.17	nr	**48.78**
expansion coupling	17.24	0.46	8.33	21.45	nr	**29.78**
sweep bend	27.41	0.46	8.33	32.66	nr	**40.99**
boss connector	7.28	0.31	5.62	10.45	nr	**16.07**
single branch	30.91	0.61	11.05	37.69	nr	**48.74**
double branch	65.01	0.77	13.95	76.47	nr	**90.42**
access door	22.37	0.56	10.14	25.09	nr	**35.23**
access pipe connector	23.39	0.46	8.33	26.89	nr	**35.22**
Weathering apron; for pipe						
82 mm diameter	2.65	0.31	5.62	3.34	nr	**8.96**
110 mm diameter	3.04	0.35	6.34	3.94	nr	**10.28**
160 mm diameter	9.15	0.39	7.06	11.27	nr	**18.33**
Weathering slate; for pipe						
110 mm diameter	32.32	0.83	15.03	36.24	nr	**51.27**
Vent cowl; for pipe						
82 mm diameter	2.65	0.31	5.62	3.34	nr	**8.96**
110 mm diameter	2.68	0.31	5.62	3.54	nr	**9.15**
160 mm diameter	7.01	0.31	5.62	8.91	nr	**14.52**
Polypropylene ancillaries; screwed joint to waste fitting						
Tubular "S" trap; bath; shallow seal						
40 mm diameter	6.60	0.51	9.24	7.29	nr	**16.52**
Trap; "P"; two piece; 76 mm seal						
32 mm diameter	4.46	0.35	6.34	4.92	nr	**11.26**
40 mm diameter	5.15	0.42	7.61	5.68	nr	**13.29**
Trap; "S"; two piece; 76 mm seal						
32 mm diameter	5.65	0.35	6.34	6.23	nr	**12.57**
40 mm diameter	6.60	0.42	7.61	7.29	nr	**14.89**
Bottle trap; "P"; 76 mm seal						
32 diameter	4.97	0.35	6.34	5.48	nr	**11.82**
40 diameter	5.92	0.42	7.61	6.53	nr	**14.14**
R12 DRAINAGE BELOW GROUND						
NOTE: Prices for drain trenches are for excavation in "firm" soil and it has been assumed that earthwork support will only be required for trenches 1.00 m or more in depth.						
Excavating trenches; by machine; grading bottoms; earthwork support; filling with excavated material and compacting; disposal of surplus soil; spreading on site average 50 m						
Pipes not exceeding 200 mm nominal size						
average depth of trench 0.50 m	-	0.28	3.51	2.18	m	**5.69**
average depth of trench 0.75 m	-	0.37	4.64	3.23	m	**7.86**
average depth of trench 1.00 m	-	0.79	9.90	6.32	m	**16.22**
average depth of trench 1.25 m	-	1.16	14.53	7.21	m	**21.74**
average depth of trench 1.50 m	-	1.48	18.54	8.21	m	**26.75**
average depth of trench 1.75 m	-	1.85	23.18	9.09	m	**32.27**

R DISPOSAL SYSTEMS

Item	PC £	Labour hours	Labour £	Material £	Unit	Total rate £
average depth of trench 2.00 m	-	2.13	26.68	10.33	m	**37.02**
average depth of trench 2.25 m	-	2.64	33.07	13.00	m	**46.07**
average depth of trench 2.50 m	-	3.10	38.84	15.15	m	**53.98**
average depth of trench 2.75 m	-	3.42	42.84	16.92	m	**59.77**
average depth of trench 3.00 m	-	3.75	46.98	18.62	m	**65.60**
average depth of trench 3.25 m	-	4.07	50.99	19.83	m	**70.82**
average depth of trench 3.50 m	-	4.35	54.50	20.96	m	**75.46**
Pipes exceeding 200 mm nominal size; 225 mm nominal size						
average depth of trench 0.50 m	-	0.28	3.51	2.18	m	**5.69**
average depth of trench 0.75 m	-	0.37	4.64	3.23	m	**7.86**
average depth of trench 1.00 m	-	0.79	9.90	6.32	m	**16.22**
average depth of trench 1.25 m	-	1.16	14.53	7.21	m	**21.74**
average depth of trench 1.50 m	-	1.48	18.54	8.21	m	**26.75**
average depth of trench 1.75 m	-	1.85	23.18	9.09	m	**32.27**
average depth of trench 2.00 m	-	2.13	26.68	10.33	m	**37.02**
average depth of trench 2.25 m	-	2.64	33.07	13.00	m	**46.07**
average depth of trench 2.50 m	-	3.10	38.84	15.15	m	**53.98**
average depth of trench 2.75 m	-	3.42	42.84	16.92	m	**59.77**
average depth of trench 3.00 m	-	3.75	46.98	18.62	m	**65.60**
average depth of trench 3.25 m	-	4.07	50.99	19.83	m	**70.82**
average depth of trench 3.50 m	-	4.35	54.50	20.96	m	**75.46**
Pipes exceeding 200 mm nominal size; 300 mm nominal size						
average depth of trencg 0.75 m	-	0.44	5.51	4.04	m	**9.55**
average depth of trench 1.00 m	-	0.93	11.65	6.32	m	**17.97**
average depth of trench 1.25 m	-	1.25	15.66	7.45	m	**23.11**
average depth of trench 1.50 m	-	1.62	20.29	8.45	m	**28.74**
average depth of trench 1.75 m	-	1.85	23.18	9.34	m	**32.51**
average depth of trench 2.00 m	-	2.13	26.68	11.14	m	**37.82**
average depth of trench 2.25 m	-	2.64	33.07	13.48	m	**46.55**
average depth of trench 2.50 m	-	3.10	38.84	15.47	m	**54.31**
average depth of trench 2.75 m	-	3.42	42.84	17.17	m	**60.01**
average depth of trench 3.00 m	-	3.75	46.98	18.86	m	**65.84**
average depth of trench 3.25 m	-	4.07	50.99	20.64	m	**71.62**
average depth of trench 3.50 m	-	4.35	54.50	21.52	m	**76.02**
Pipes exceeding 200 mm nominal size; 375 mm nominal size						
average depth of trench 0.75 m	-	0.46	5.76	4.84	m	**10.61**
average depth of trench 1.00 m	-	0.97	12.15	7.13	m	**19.28**
average depth of trench 1.25 m	-	1.34	16.79	8.83	m	**25.61**
average depth of trench 1.50 m	-	1.71	21.42	9.50	m	**30.92**
average depth of trench 1.75 m	-	1.99	24.93	10.71	m	**35.64**
average depth of trench 2.00 m	-	2.27	28.44	11.38	m	**39.82**
average depth of trench 2.25 m	-	2.82	35.33	14.29	m	**49.62**
average depth of trench 2.50 m	-	3.38	42.34	16.52	m	**58.86**
average depth of trench 2.75 m	-	3.70	46.35	17.97	m	**64.33**
average depth of trench 3.00 m	-	4.02	50.36	19.43	m	**69.79**
average depth of trench 3.25 m	-	4.35	54.50	21.12	m	**75.62**
average depth of trench 3.50 m	-	4.67	58.50	22.57	m	**81.08**
Pipes exceeding 200 mm nominal size; 450 mm nominal size						
average depth of trench 0.75 m	-	0.51	6.39	4.84	m	**11.23**
average depth of trench 1.00 m	-	1.02	12.78	7.61	m	**20.39**
average depth of trench 1.25 m	-	1.48	18.54	9.39	m	**27.93**
average depth of trench 1.50 m	-	1.85	23.18	10.30	m	**33.48**
average depth of trench 1.75 m	-	2.13	26.68	11.27	m	**37.96**
average depth of trench 2.00 m	-	2.45	30.69	12.19	m	**42.88**
average depth of trench 2.25 m	-	3.05	38.21	14.85	m	**53.06**
average depth of trench 2.50 m	-	3.61	45.22	17.33	m	**62.55**

R DISPOSAL SYSTEMS

Item	PC £	Labour hours	Labour £	Material £	Unit	Total rate £
R12 DRAINAGE BELOW GROUND – cont'd						
Excavating trenches; by machine – cont'd						
Pipes exceeding 200 mm nominal size; 450 mm nominal size – cont'd						
average depth of trench 2.75 m	-	3.98	49.86	19.02	m	**68.88**
average depth of trench 3.00 m	-	4.26	53.37	20.80	m	**74.17**
average depth of trench 3.25 m	-	4.63	58.00	22.74	m	**80.74**
average depth of trench 3.50 m	-	5.00	62.64	24.75	m	**87.39**
Pipes exceeding 200 mm nominal size; 600 mm nominal size						
average depth of trench 1.00 m	-	1.11	13.91	8.18	m	**22.09**
average depth of trench 1.25 m	-	1.57	19.67	9.87	m	**29.54**
average depth of trench 1.50 m	-	2.04	25.56	11.44	m	**36.99**
average depth of trench 1.75 m	-	2.31	28.94	12.08	m	**41.02**
average depth of trench 2.00 m	-	2.73	34.20	13.32	m	**47.52**
average depth of trench 2.25 m	-	3.28	41.09	16.47	m	**57.56**
average depth of trench 2.50 m	-	3.89	48.73	19.18	m	**67.92**
average depth of trench 2.75 m	-	4.30	53.87	21.44	m	**75.31**
average depth of trench 3.00 m	-	4.72	59.13	23.46	m	**82.59**
average depth of trench 3.25 m	-	5.09	63.77	25.16	m	**88.92**
average depth of trench 3.50 m	-	5.46	68.40	26.61	m	**95.01**
Pipes exceeding 200 mm nominal size; 900 mm nominal size						
average depth of trench 1.25 m	-	1.90	23.80	11.49	m	**35.29**
average depth of trench 1.50 m	-	2.41	30.19	13.05	m	**43.24**
average depth of trench 1.75 m	-	2.78	34.83	13.94	m	**48.76**
average depth of trench 2.00 m	-	3.10	38.84	15.98	m	**54.82**
average depth of trench 2.25 m	-	3.84	48.11	19.45	m	**67.56**
average depth of trench 2.50 m	-	4.53	56.75	22.41	m	**79.16**
average depth of trench 2.75 m	-	5.00	62.64	24.67	m	**87.31**
average depth of trench 3.00 m	-	5.46	68.40	26.93	m	**95.33**
average depth of trench 3.25 m	-	5.92	74.16	29.19	m	**103.36**
average depth of trench 3.50 m	-	6.38	79.93	31.21	m	**111.14**
Pipes exceeding 200 mm nominal size; 1200 mm nominal size						
average depth of trench 1.50 m	-	2.73	34.20	13.86	m	**48.06**
average depth of trench 1.75 m	-	3.19	39.96	16.12	m	**56.08**
average depth of trench 2.00 m	-	3.56	44.60	18.40	m	**63.00**
average depth of trench 2.25 m	-	4.35	54.50	22.36	m	**76.85**
average depth of trench 2.50 m	-	5.18	64.89	25.64	m	**90.53**
average depth of trench 2.75 m	-	5.69	71.28	28.47	m	**99.75**
average depth of trench 3.00 m	-	6.20	77.67	30.97	m	**108.64**
average depth of trench 3.25 m	-	6.75	84.56	33.55	m	**118.11**
average depth of trench 3.50 m	-	7.26	90.95	36.05	m	**127.00**
Extra over excavating trenches; irrespective of depth; breaking out existing materials						
brick	-	1.80	22.55	9.63	m^3	**32.18**
concrete	-	2.54	31.82	13.24	m^3	**45.06**
reinforced concrete	-	3.61	45.22	19.16	m^3	**64.39**
Extra over excavating trenches; irrespective of depth; breaking out existing hard pavings; 75 mm thick						
tarmacadam	-	0.19	2.38	1.03	m^2	**3.41**
Extra over excavating trenches; irrsepective of depth; breaking out existing hard pavings; 150 mm thick						
concrete	-	0.37	4.64	2.21	m^2	**6.85**
tarmacadam and hardcore	-	0.28	3.51	1.17	m^2	**4.68**

R DISPOSAL SYSTEMS

Item	PC £	Labour hours	Labour £	Material £	Unit	Total rate £
Excavating trenches; by hand; grading bottoms; earthwork support; filling with excavated material and compacting; disposal of surplus soil on site; spreading on site average 50 m						
Pipes not exceeding 200 mm nominal size						
average depth of trench 0.50 m	-	0.93	11.65	-	m	**11.65**
average depth of trench 0.75 m	-	1.39	17.41	-	m	**17.41**
average depth of trench 1.00 m	-	2.04	25.56	1.72	m	**27.28**
average depth of trench 1.25 m	-	2.87	35.95	2.37	m	**38.32**
average depth of trench 1.50 m	-	3.93	49.23	2.88	m	**52.12**
average depth of trench 1.75 m	-	5.18	64.89	3.44	m	**68.34**
average depth of trench 2.00 m	-	5.92	74.16	3.87	m	**78.04**
average depth of trench 2.25 m	-	7.40	92.70	5.17	m	**97.87**
average depth of trench 2.50 m	-	8.88	111.25	6.03	m	**117.27**
average depth of trench 2.75 m	-	9.76	122.27	6.67	m	**128.94**
average depth of trench 3.00 m	-	10.64	133.29	7.32	m	**140.61**
average depth of trench 3.25 m	-	11.52	144.32	7.96	m	**152.28**
average depth of trench 3.50 m	-	12.40	155.34	8.61	m	**163.95**
Pipes exceeding 200 mm nominal size; 225 mm nominal size						
average depth of trench 0.50 m	-	0.93	11.65	-	m	**11.65**
average depth of trench 0.75 m	-	1.39	17.41	-	m	**17.41**
average depth of trench 1.00 m	-	2.04	25.56	1.72	m	**27.28**
average depth of trench 1.25 m	-	2.87	35.95	2.37	m	**38.32**
average depth of trench 1.50 m	-	3.93	49.23	2.88	m	**52.12**
average depth of trench 1.75 m	-	5.18	64.89	3.44	m	**68.34**
average depth of trench 2.00 m	-	5.92	74.16	3.87	m	**78.04**
average depth of trench 2.25 m	-	7.40	92.70	5.17	m	**97.87**
average depth of trench 2.50 m	-	8.88	111.25	6.03	m	**117.27**
average depth of trench 2.75 m	-	9.76	122.27	6.67	m	**128.94**
average depth of trench 3.00 m	-	10.64	133.29	7.32	m	**140.61**
average depth of trench 3.25 m	-	11.52	144.32	7.96	m	**152.28**
average depth of trench 3.50 m	-	12.40	155.34	8.61	m	**163.95**
Pipes exceeding 200 mm nominal size; 300 mm nominal size						
average depth of trench 0.75 m	-	1.62	20.29	-	m	**20.29**
average depth of trench 1.00 m	-	2.36	29.57	1.72	m	**31.29**
average depth of trench 1.25 m	-	3.33	41.72	2.37	m	**44.08**
average depth of trench 1.50 m	-	4.44	55.62	2.88	m	**58.51**
average depth of trench 1.75 m	-	5.18	64.89	3.44	m	**68.34**
average depth of trench 2.00 m	-	5.92	74.16	3.87	m	**78.04**
average depth of trench 2.25 m	-	7.40	92.70	5.17	m	**97.87**
average depth of trench 2.50 m	-	8.88	111.25	6.03	m	**117.27**
average depth of trench 2.75 m	-	9.76	122.27	6.67	m	**128.94**
average depth of trench 3.00 m	-	10.64	133.29	7.32	m	**140.61**
average depth of trench 3.25 m	-	11.52	144.32	7.96	m	**152.28**
average depth of trench 3.50 m	-	12.40	155.34	8.61	m	**163.95**
Pipes exceeding 200 mm nominal size; 375 mm nominal size						
average depth of trench 0.75 m	-	1.80	22.55	-	m	**22.55**
average depth of trench 1.00 m	-	2.64	33.07	1.72	m	**34.79**
average depth of trench 1.25 m	-	3.70	46.35	2.37	m	**48.72**
average depth of trench 1.50 m	-	4.93	61.76	2.88	m	**64.65**
average depth of trench 1.75 m	-	5.74	71.91	3.44	m	**75.35**
average depth of trench 2.00 m	-	6.57	82.31	3.87	m	**86.18**
average depth of trench 2.25 m	-	8.23	103.10	5.17	m	**108.27**
average depth of trench 2.50 m	-	9.90	124.02	6.03	m	**130.05**
average depth of trench 2.75 m	-	10.87	136.18	6.67	m	**142.85**

R DISPOSAL SYSTEMS

Item	PC £	Labour hours	Labour £	Material £	Unit	Total rate £
R12 DRAINAGE BELOW GROUND – cont'd						
Excavating trenches; by hand – cont'd						
Pipes exceeding 200 mm nominal size; 375 mm nominal size – cont'd						
average depth of trench 3.00 m	-	11.84	148.33	7.32	m	**155.65**
average depth of trench 3.25 m	-	12.86	161.10	7.96	m	**169.07**
average depth of trench 3.50 m	-	13.88	173.88	8.61	m	**182.49**
Pipes exceeding 200 mm nominal size; 450 mm nominal size						
average depth of trench 0.75 m	-	2.04	25.56	-	m	**25.56**
average depth of trench 1.00 m	-	2.94	36.83	1.72	m	**38.55**
average depth of trench 1.25 m	-	4.13	51.74	2.37	m	**54.11**
average depth of trench 1.50 m	-	5.41	67.77	2.88	m	**70.66**
average depth of trench 1.75 m	-	6.31	79.05	3.44	m	**82.49**
average depth of trench 2.00 m	-	7.22	90.45	3.87	m	**94.32**
average depth of trench 2.25 m	-	9.05	113.37	5.17	m	**118.54**
average depth of trench 2.50 m	-	10.87	136.18	6.03	m	**142.20**
average depth of trench 2.75 m	-	11.96	149.83	6.67	m	**156.50**
average depth of trench 3.00 m	-	13.04	163.36	7.32	m	**170.68**
average depth of trench 3.25 m	-	14.11	176.76	7.96	m	**184.73**
average depth of trench 3.50 m	-	15.17	190.04	8.61	m	**198.65**
Pipes exceeding 200 mm nominal size; 600 mm nominal size						
average depth of trench 1.00 m	-	3.24	40.59	1.72	m	**42.31**
average depth of trench 1.25 m	-	4.63	58.00	2.37	m	**60.37**
average depth of trench 1.50 m	-	6.20	77.67	2.88	m	**80.56**
average depth of trench 1.75 m	-	7.17	89.82	3.44	m	**93.27**
average depth of trench 2.00 m	-	8.19	102.60	3.87	m	**106.48**
average depth of trench 2.25 m	-	9.20	115.25	5.17	m	**120.42**
average depth of trench 2.50 m	-	11.56	144.82	6.03	m	**150.85**
average depth of trench 2.75 m	-	12.35	154.72	6.67	m	**161.39**
average depth of trench 3.00 m	-	14.80	185.41	7.32	m	**192.73**
average depth of trench 3.25 m	-	16.03	200.82	7.96	m	**208.78**
average depth of trench 3.50 m	-	17.25	216.10	8.61	m	**224.71**
Pipes exceeding 200 mm nominal size; 900 mm nominal size						
average depth of trench 1.25 m	-	5.78	72.41	2.37	m	**74.78**
average depth of trench 1.50 m	-	7.63	95.59	2.88	m	**98.47**
average depth of trench 1.75 m	-	8.88	111.25	3.44	m	**114.69**
average depth of trench 2.00 m	-	10.13	126.90	3.87	m	**130.78**
average depth of trench 2.25 m	-	12.72	159.35	5.17	m	**164.52**
average depth of trench 2.50 m	-	15.31	191.80	6.03	m	**197.82**
average depth of trench 2.75 m	-	16.84	210.96	6.67	m	**217.64**
average depth of trench 3.00 m	-	18.32	229.51	7.32	m	**236.82**
average depth of trench 3.25 m	-	19.84	248.55	7.96	m	**256.51**
average depth of trench 3.50 m	-	21.37	267.71	8.61	m	**276.32**
Pipes exceeding 200 mm nominal size; 1200 mm nominal size						
average depth of trench 1.50 m	-	9.11	114.13	2.88	m	**117.01**
average depth of trench 1.75 m	-	10.59	132.67	3.44	m	**136.11**
average depth of trench 2.00 m	-	12.12	151.83	3.87	m	**155.71**
average depth of trench 2.25 m	-	15.20	190.42	5.17	m	**195.59**
average depth of trench 2.50 m	-	18.27	228.88	6.03	m	**234.91**
average depth of trench 2.75 m	-	20.07	251.43	6.67	m	**258.10**
average depth of trench 3.00 m	-	21.88	274.10	7.32	m	**281.42**
average depth of trench 3.25 m	-	23.66	296.40	7.96	m	**304.37**
average depth of trench 3.50 m	-	25.44	318.70	8.61	m	**327.31**

R DISPOSAL SYSTEMS

Item	PC £	Labour hours	Labour £	Material £	Unit	Total rate £
Extra over excavating trenches irrespective of depth; breaking out existing materials						
brick	-	2.78	34.83	7.00	m³	**41.82**
concrete	-	4.16	52.11	11.65	m³	**63.77**
reinforced concrete	-	5.55	69.53	16.33	m³	**85.86**
concrete; 150 mm thick	-	0.65	8.14	1.64	m²	**9.78**
tarmacadam and hardcore; 150 mm thick	-	0.46	5.76	1.16	m²	**6.92**
Extra over excavating trenches irrespective of depth; breaking out existing hard pavings, 75 mm thick						
tarmacadam	-	0.37	4.64	0.94	m²	**5.57**
Extra over excavating trenches irrespective of depth; breaking out existing hard pavings, 150 mm thick						
concrete	-	0.65	8.14	1.64	m²	**9.78**
tarmacadam and hardcore	-	0.46	5.76	1.16	m²	**6.92**
Sand filling						
Beds; to receive pitch fibre pipes						
600 mm x 50 mm thick	-	0.07	0.88	0.94	m	**1.82**
700 mm x 50 mm thick	-	0.09	1.13	1.10	m	**2.23**
800 mm x 50 mm thick	-	0.11	1.38	1.26	m	**2.63**
Granular (shingle) filling						
Beds; 100 mm thick; to pipes						
100 mm nominal size	-	0.09	1.13	1.95	m	**3.08**
150 mm nominal size	-	0.09	1.13	2.28	m	**3.40**
225 mm nominal size	-	0.11	1.38	2.60	m	**3.98**
300 mm nominal size	-	0.13	1.63	2.93	m	**4.55**
375 mm nominal size	-	0.15	1.88	3.25	m	**5.13**
450 mm nominal size	-	0.17	2.13	3.58	m	**5.71**
600 mm nominal size	-	0.19	2.38	3.90	m	**6.28**
Beds; 150 mm thick; to pipes						
100 mm nominal size	-	0.13	1.63	2.93	m	**4.55**
150 mm nominal size	-	0.15	1.88	3.25	m	**5.13**
225 mm nominal size	-	0.17	2.13	3.58	m	**5.71**
300 mm nominal size	-	0.19	2.38	3.90	m	**6.28**
375 mm nominal size	-	0.22	2.76	4.88	m	**7.63**
450 mm nominal size	-	0.24	3.01	5.20	m	**8.21**
600 mm nominal size	-	0.28	3.51	6.18	m	**9.69**
Beds and benchings; beds 100 mm thick; to pipes						
100 nominal size	-	0.21	2.63	3.58	m	**6.21**
150 nominal size	-	0.23	2.88	3.58	m	**6.46**
225 nominal size	-	0.28	3.51	4.88	m	**8.38**
300 nominal size	-	0.32	4.01	5.53	m	**9.54**
375 nominal size	-	0.42	5.26	7.48	m	**12.74**
450 nominal size	-	0.48	6.01	8.45	m	**14.47**
600 nominal size	-	0.62	7.77	11.06	m	**18.82**
Beds and benchings; beds 150 mm thick; to pipes						
100 nominal size	-	0.23	2.88	3.90	m	**6.78**
150 nominal size	-	0.26	3.26	4.23	m	**7.49**
225 nominal size	-	0.32	4.01	5.86	m	**9.86**
300 nominal size	-	0.42	5.26	7.15	m	**12.42**
375 nominal size	-	0.48	6.01	8.45	m	**14.47**
450 nominal size	-	0.57	7.14	10.08	m	**17.22**
600 nominal size	-	0.68	8.52	13.01	m	**21.53**

R DISPOSAL SYSTEMS

Item	PC £	Labour hours	Labour £	Material £	Unit	Total rate £
R12 DRAINAGE BELOW GROUND – cont'd						
Granular (shingle) filling – cont'd						
Beds and coverings; 100 mm thick; to pipes						
100 nominal size	-	0.33	4.13	4.88	m	**9.01**
150 nominal size	-	0.42	5.26	5.86	m	**11.12**
225 nominal size	-	0.56	7.02	8.13	m	**15.14**
300 nominal size	-	0.67	8.39	9.76	m	**18.15**
375 nominal size	-	0.80	10.02	11.71	m	**21.73**
450 nominal size	-	0.94	11.78	13.98	m	**25.76**
600 nominal size	-	1.22	15.28	17.89	m	**33.17**
Beds and coverings; 150 mm thick; to pipes						
100 nominal size	-	0.50	6.26	7.15	m	**13.42**
150 nominal size	-	0.56	7.02	8.13	m	**15.14**
225 nominal size	-	0.72	9.02	10.41	m	**19.43**
300 nominal size	-	0.86	10.77	12.36	m	**23.13**
375 nominal size	-	1.00	12.53	14.63	m	**27.16**
450 nominal size	-	1.19	14.91	17.56	m	**32.47**
600 nominal size	-	1.44	18.04	21.14	m	**39.18**
Plain in situ ready mixed designated concrete; C10 - 40 mm aggregate						
Beds; 100 mm thick; to pipes						
100 mm nominal size	-	0.17	2.74	4.30	m	**7.05**
150 mm nominal size	-	0.17	2.74	4.30	m	**7.05**
225 mm nominal size	-	0.20	3.23	5.16	m	**8.39**
300 mm nominal size	-	0.23	3.71	6.02	m	**9.74**
375 mm nominal size	-	0.27	4.36	6.88	m	**11.24**
450 mm nominal size	-	0.30	4.84	7.74	m	**12.58**
600 mm nominal size	-	0.33	5.33	8.60	m	**13.92**
900 mm nominal size	-	0.40	6.46	10.32	m	**16.77**
1200 mm nominal size	-	0.54	8.72	13.76	m	**22.47**
Beds; 150 mm thick; to pipes						
100 mm nominal size	-	0.23	3.71	6.02	m	**9.74**
150 mm nominal size	-	0.27	4.36	6.88	m	**11.24**
225 mm nominal size	-	0.30	4.84	7.74	m	**12.58**
300 mm nominal size	-	0.33	5.33	8.60	m	**13.92**
375 mm nominal size	-	0.40	6.46	10.32	m	**16.77**
450 mm nominal size	-	0.43	6.94	11.18	m	**18.12**
600 mm nominal size	-	0.50	8.07	12.90	m	**20.97**
900 mm nominal size	-	0.63	10.17	16.34	m	**26.51**
1200 mm nominal size	-	0.77	12.43	19.78	m	**32.21**
Beds and benchings; beds 100 mm thick; to pipes						
100 mm nominal size	-	0.33	5.33	7.74	m	**13.07**
150 mm nominal size	-	0.38	6.13	8.60	m	**14.73**
225 mm nominal size	-	0.45	7.26	10.32	m	**17.58**
300 mm nominal size	-	0.53	8.56	12.04	m	**20.59**
375 mm nominal size	-	0.68	10.98	15.47	m	**26.45**
450 mm nominal size	-	0.80	12.91	18.05	m	**30.97**
600 mm nominal size	-	1.02	16.47	23.22	m	**39.68**
900 mm nominal size	-	1.65	26.64	37.83	m	**64.47**
1200 mm nominal size	-	2.44	39.39	55.88	m	**95.27**
Beds and benchings; beds 150 mm thick; to pipes						
100 mm nominal size	-	0.38	6.13	8.60	m	**14.73**
150 mm nominal size	-	0.42	6.78	9.45	m	**16.23**
225 mm nominal size	-	0.53	8.56	12.04	m	**20.59**
300 mm nominal size	-	0.68	10.98	15.47	m	**26.45**
375 mm nominal size	-	0.80	12.91	18.05	m	**30.97**
450 mm nominal size	-	0.94	15.18	21.49	m	**36.66**

R DISPOSAL SYSTEMS

Item	PC £	Labour hours	Labour £	Material £	Unit	Total rate £
600 mm nominal size	-	1.20	19.37	27.51	m	**46.88**
900 mm nominal size	-	1.91	30.83	43.85	m	**74.68**
1200 mm nominal size	-	2.70	43.59	61.90	m	**105.49**
Beds and coverings; 100 mm thick; to pipes						
100 mm nominal size	-	0.50	8.07	10.32	m	**18.39**
150 mm nominal size	-	0.58	9.36	12.04	m	**21.40**
225 mm nominal size	-	0.83	13.40	17.19	m	**30.59**
300 mm nominal size	-	1.00	16.14	20.63	m	**36.78**
375 mm nominal size	-	1.21	19.53	24.93	m	**44.46**
450 mm nominal size	-	1.42	22.92	29.23	m	**52.15**
600 mm nominal size	-	1.83	29.54	37.83	m	**67.37**
900 mm nominal size	-	2.79	45.04	57.61	m	**102.65**
1200 mm nominal size	-	3.83	61.83	79.09	m	**140.93**
Beds and coverings; 150 mm thick; to pipes						
100 mm nominal size	-	0.75	12.11	15.47	m	**27.58**
150 mm nominal size	-	0.83	13.40	17.19	m	**30.59**
225 mm nominal size	-	1.08	17.44	22.35	m	**39.79**
300 mm nominal size	-	1.30	20.99	26.65	m	**47.63**
375 mm nominal size	-	1.50	24.22	30.95	m	**55.17**
450 mm nominal size	-	1.79	28.90	36.96	m	**65.86**
600 mm nominal size	-	2.16	34.87	44.71	m	**79.58**
900 mm nominal size	-	3.54	57.15	73.07	m	**130.22**
1200 mm nominal size	-	5.00	80.72	103.17	m	**183.89**
Plain in situ ready mixed designated concrete; C20 - 40 mm aggregate						
Beds; 100 mm thick; to pipes						
100 mm nominal size	-	0.17	2.74	4.50	m	**7.24**
150 mm nominal size	-	0.17	2.74	4.50	m	**7.24**
225 mm nominal size	-	0.20	3.23	5.39	m	**8.62**
300 mm nominal size	-	0.23	3.71	6.29	m	**10.00**
375 mm nominal size	-	0.27	4.36	7.19	m	**11.54**
450 mm nominal size	-	0.30	4.84	8.09	m	**12.93**
600 mm nominal size	-	0.33	5.33	8.98	m	**14.31**
900 mm nominal size	-	0.40	6.46	10.78	m	**17.24**
1200 mm nominal size	-	0.54	8.72	14.37	m	**23.09**
Beds; 150 mm thick; to pipes						
100 mm nominal size	-	0.23	3.71	6.29	m	**10.00**
150 mm nominal size	-	0.27	4.36	7.19	m	**11.54**
225 mm nominal size	-	0.30	4.84	8.09	m	**12.93**
300 mm nominal size	-	0.33	5.33	8.98	m	**14.31**
375 mm nominal size	-	0.40	6.46	10.78	m	**17.24**
450 mm nominal size	-	0.43	6.94	11.68	m	**18.62**
600 mm nominal size	-	0.50	8.07	13.48	m	**21.55**
900 mm nominal size	-	0.63	10.17	17.07	m	**27.24**
1200 mm nominal size	-	0.77	12.43	20.66	m	**33.09**
Beds and benchings; beds 100 mm thick; to pipes						
100 mm nominal size	-	0.33	5.33	8.09	m	**13.42**
150 mm nominal size	-	0.38	6.13	8.98	m	**15.12**
225 mm nominal size	-	0.45	7.26	10.78	m	**18.04**
300 mm nominal size	-	0.53	8.56	12.57	m	**21.13**
375 mm nominal size	-	0.68	10.98	16.17	m	**27.15**
450 mm nominal size	-	0.80	12.91	18.86	m	**31.77**
600 mm nominal size	-	1.02	16.47	24.26	m	**40.72**
900 mm nominal size	-	1.65	26.64	39.52	m	**66.16**
1200 mm nominal size	-	2.44	39.39	58.38	m	**97.77**

R DISPOSAL SYSTEMS

Item	PC £	Labour hours	Labour £	Material £	Unit	Total rate £
R12 DRAINAGE BELOW GROUND – cont'd						
Plain in situ ready mixed designated concrete; C20 - 40 mm aggregate – cont'd						
Beds and benchings; beds 150 mm thick; to pipes						
100 mm nominal size	-	0.38	6.13	8.98	m	**15.12**
150 mm nominal size	-	0.42	6.78	9.88	m	**16.66**
225 mm nominal size	-	0.53	8.56	12.57	m	**21.13**
300 mm nominal size	-	0.68	10.98	16.17	m	**27.15**
375 mm nominal size	-	0.80	12.91	18.86	m	**31.77**
450 mm nominal size	-	0.94	15.18	22.45	m	**37.63**
600 mm nominal size	-	1.20	19.37	28.74	m	**48.12**
900 mm nominal size	-	1.91	30.83	45.81	m	**76.65**
1200 mm nominal size	-	2.70	43.59	64.67	m	**108.26**
Beds and coverings; 100 mm thick; to pipes						
100 mm nominal size	-	0.50	8.07	10.78	m	**18.85**
150 mm nominal size	-	0.58	9.36	12.57	m	**21.94**
225 mm nominal size	-	0.83	13.40	17.96	m	**31.36**
300 mm nominal size	-	1.00	16.14	21.56	m	**37.70**
375 mm nominal size	-	1.21	19.53	26.04	m	**45.58**
450 mm nominal size	-	1.42	22.92	30.54	m	**53.46**
600 mm nominal size	-	1.83	29.54	39.52	m	**69.06**
900 mm nominal size	-	2.79	45.04	60.18	m	**105.23**
1200 mm nominal size	-	3.83	61.83	82.64	m	**144.47**
Beds and coverings; 150 mm thick; to pipes						
100 mm nominal size	-	0.75	12.11	16.17	m	**28.28**
150 mm nominal size	-	0.83	13.40	17.96	m	**31.36**
225 mm nominal size	-	1.08	17.44	23.35	m	**40.79**
300 mm nominal size	-	1.30	20.99	27.84	m	**48.83**
375 mm nominal size	-	1.50	24.22	32.34	m	**56.55**
450 mm nominal size	-	1.79	28.90	38.62	m	**67.52**
600 mm nominal size	-	2.16	34.87	46.71	m	**81.58**
900 mm nominal size	-	3.54	57.15	76.34	m	**133.49**
1200 mm nominal size	-	5.00	80.72	107.78	m	**188.50**
NOTE: The following items unless otherwise described include for all appropriate joints/couplings in the running length. The prices for gullies and rainwater shoes, etc. include for appropriate joints to pipes and for setting on and surrounding accessory with site mixed in situ concrete 10.00 N/mm² - 40 mm aggregate (1:3:6).						
Cast iron "Timesaver" drain pipes and fittings or other equal and approved; BS 437; coated; with mechanical coupling joints						
100 mm pipes; laid straight	23.96	0.46	5.71	32.84	m	**38.55**
100 mm pipes; in runs not exceeding 3 m long	23.96	0.63	7.82	51.24	m	**59.06**
Extra for						
bend; medium radius	28.72	0.56	6.95	44.57	nr	**51.51**
bend; medium radius with access	79.81	0.56	6.95	100.92	nr	**107.87**
bend; long radius	47.46	0.56	6.95	63.89	nr	**70.84**
rest bend	32.94	0.56	6.95	47.87	nr	**54.82**
single branch	38.11	0.69	8.56	69.30	nr	**77.86**
single branch; with access	87.90	0.79	9.80	124.22	nr	**134.02**
double branch	64.78	0.88	10.92	113.50	nr	**124.42**
isolated "Timesaver" joint	15.36	0.32	3.97	16.94	nr	**20.92**
transitional pipe; for WC	22.50	0.46	5.71	41.76	nr	**47.47**

R DISPOSAL SYSTEMS

Item	PC £	Labour hours	Labour £	Material £	Unit	Total rate £
150 mm pipes; laid straight	44.36	0.56	6.95	57.10	m	**64.05**
150 mm pipes; in runs not exceeding 3 m long	-	0.76	9.43	84.01	m	**93.44**
Extra for						
bend; medium radius	66.09	0.65	8.07	85.91	nr	**93.97**
bend; medium radius with access	140.14	0.65	8.07	167.59	nr	**175.66**
bend; long radius	88.50	0.65	8.07	108.13	nr	**116.19**
diminishing pipe	37.44	0.65	8.07	51.80	nr	**59.86**
single branch	82.28	0.79	9.80	92.74	nr	**102.54**
isolated "Timesaver" joint	18.60	0.39	4.84	20.52	nr	**25.36**
Accessories in "Timesaver" cast iron or other equal and approved; with mechanical coupling joints						
Gully fittings; comprising low invert gully trap and round hopper						
100 mm outlet	38.11	0.88	10.92	63.48	nr	**74.40**
150 mm outlet	94.82	1.20	14.89	130.51	nr	**145.40**
Add to above for bellmouth 300 mm high; circular plain grating						
100 mm nominal size; 200 mm grating	39.69	0.42	5.21	66.09	nr	**71.30**
100 mm nominal size; 100 mm horizontal inlet; 200 mm grating	48.52	0.42	5.21	75.84	nr	**81.05**
100 mm nominal size; 100 mm horizontal inlet; 200 mm grating	49.76	0.42	5.21	77.20	nr	**82.42**
Yard gully (Deans); trapped; galvanized sediment pan; 267 mm round heavy grating						
100 mm outlet	257.57	2.68	33.26	326.99	nr	**360.25**
Yard gully (garage); trapless; galvanized sediment pan; 267 mm round heavy grating						
100 mm outlet	262.74	2.50	31.02	312.18	nr	**343.20**
Yard gully (garage); trapped; with rodding eye, galvanised perforated sediment pan; stopper; 267 mm round heavy grating						
100 mm outlet	488.62	2.50	31.02	613.69	nr	**644.71**
Grease trap; internal access; galvanized perforated bucket; lid and frame						
100 mm outlet; 20 gallon capacity	531.69	3.70	45.91	633.90	nr	**679.82**
Cast iron "Ensign" lightweight drain pipes and fittings or other equal and approved; BS EN 877; ductile iron couplings						
100 mm pipes; laid straight	-	0.19	3.47	22.05	m	**25.52**
Extra for						
bend; long radius	-	0.19	3.47	39.54	nr	**43.01**
single branch	-	0.23	4.25	39.93	nr	**44.18**
150 mm pipes; laid straight	-	0.22	4.04	44.09	m	**48.13**
Extra for						
bend; medium radius	-	0.22	4.04	109.29	nr	**113.32**
single branch	-	0.28	5.12	87.88	nr	**93.01**

R DISPOSAL SYSTEMS

Item	PC £	Labour hours	Labour £	Material £	Unit	Total rate £
R12 DRAINAGE BELOW GROUND – cont'd						
Extra strength vitrified clay pipes and fittings; Hepworth "Supersleve" or equivalent; plain ends with push fit polypropylene flexible couplings						
100 mm pipes; laid straight	5.76	0.19	2.36	6.51	m	**8.87**
Extra for						
bend	5.32	0.19	2.36	11.25	nr	**13.60**
access bend	35.00	0.19	2.36	44.78	nr	**47.13**
rest bend	9.42	0.19	2.36	15.87	nr	**18.23**
access pipe	30.41	0.19	2.36	39.15	nr	**41.51**
socket adaptor	5.64	0.16	1.99	9.21	nr	**11.20**
saddle	11.28	0.69	8.56	16.03	nr	**24.60**
single junction	11.49	0.23	2.85	21.05	nr	**23.91**
single access junction	40.48	0.23	2.85	53.81	nr	**56.67**
150 mm pipes; laid straight	10.98	0.23	2.85	12.41	m	**15.27**
Extra for						
bend	10.96	0.22	2.73	21.61	nr	**24.34**
access bend	5.81	0.22	2.73	58.53	nr	**61.26**
rest bend	14.08	0.22	2.73	25.14	nr	**27.87**
taper pipe	13.58	0.22	2.73	21.89	nr	**24.62**
access pipe	41.33	0.22	2.73	55.04	nr	**57.77**
socket adaptor	11.12	0.19	2.36	17.62	nr	**19.98**
adaptor to "HepSeal" pipe	7.89	0.19	2.36	13.98	nr	**16.34**
saddle	16.79	0.83	10.30	24.94	nr	**35.24**
single junction	14.66	0.28	3.47	30.86	nr	**34.34**
single access junction	60.18	0.28	3.47	82.29	nr	**85.77**
Extra strength vitrified clay pipes and fittings; Hepworth "SuperSeal" / "Hepseal" or equivalent; socketted; with push-fit flexible joints						
150 mm SuperSeal pipes; laid straight	13.64	0.30	3.72	15.42	m	**19.14**
Extra for						
bend	26.23	0.23	2.85	25.01	nr	**27.87**
rest bend	14.08	0.20	2.48	11.28	nr	**13.77**
stopper	7.91	0.15	1.86	8.94	nr	**10.80**
taper reducer	13.58	0.23	2.85	10.72	nr	**13.57**
saddle	16.79	0.75	9.31	18.97	nr	**28.28**
single junction	-5.46	0.30	3.72	32.57	nr	**36.30**
225 mm SuperSeal pipes; laid straight	28.31	0.38	4.72	32.00	m	**36.71**
Extra for						
bend	61.46	0.30	3.72	59.86	nr	**63.58**
rest bend	75.08	0.30	3.72	75.25	nr	**78.97**
stopper	13.32	0.19	2.36	15.05	nr	**17.41**
taper reducer	42.33	0.30	3.72	38.24	nr	**41.96**
saddle	62.46	1.00	12.41	70.58	nr	**82.99**
single junction	109.17	0.38	4.72	110.57	nr	**115.28**
300 mm SuperSeal pipes; laid straight	43.43	0.50	6.20	49.07	m	**55.28**
Extra for						
bend	116.73	0.40	4.96	117.19	nr	**122.15**
rest bend	166.34	0.40	4.96	173.25	nr	**178.21**
stopper	28.43	0.25	3.10	32.13	nr	**35.23**
taper reducer	115.55	0.40	4.96	115.86	nr	**120.82**
saddle	108.70	1.33	16.50	122.84	nr	**139.34**
single junction	208.76	0.50	6.20	216.28	nr	**222.49**
400 mm Hepseal pipes; laid straight	101.74	0.67	8.31	114.97	m	**123.28**
Extra for						
bend	382.30	0.54	6.70	397.54	nr	**404.24**
single unequal junction	358.21	0.67	8.31	358.82	nr	**367.13**

R DISPOSAL SYSTEMS

Item	PC £	Labour hours	Labour £	Material £	Unit	Total rate £
450 mm Hepseal pipes; laid straight	132.15	0.83	10.30	149.34	m	**159.64**
Extra for bend	503.43	0.67	8.31	524.11	nr	**532.42**
Extra for single unequal junction	428.53	0.83	10.30	424.52	nr	**434.82**
British Standard quality vitrified clay pipes and fittings; socketted; cement and sand (1:2) joints						
100 mm pipes; laid straight	8.63	0.37	4.59	9.85	m	**14.44**
Extra for						
bend (short/medium/knuckle)	6.04	0.30	3.72	6.92	nr	**10.64**
bend (long/rest/elbow)	14.18	0.30	3.72	13.20	nr	**16.92**
single junction	15.85	0.37	4.59	14.13	nr	**18.72**
double collar	10.41	0.25	3.10	11.86	nr	**14.96**
150 mm pipes; laid straight	13.27	0.42	5.21	15.10	m	**20.31**
Extra for						
bend (short/medium/knuckle)	13.27	0.33	4.09	10.60	nr	**14.69**
bend (long/rest/elbow)	23.97	0.33	4.09	22.68	nr	**26.78**
taper	31.31	0.33	4.09	30.53	nr	**34.63**
single junction	26.22	0.42	5.21	23.75	nr	**28.97**
double collar	17.33	0.28	3.47	19.68	nr	**23.16**
225 mm pipes; laid straight	26.29	0.51	6.33	29.95	m	**36.28**
Extra for						
double collar	40.55	0.33	4.09	45.93	nr	**50.02**
300 mm pipes; laid straight	44.08	0.69	8.56	50.05	m	**58.61**
Accessories in vitrified clay; set in concrete; with polypropylene coupling joints to pipes						
Rodding point; with oval aluminium plate						
100 mm nominal size	31.22	0.46	5.71	40.28	nr	**45.99**
Gully fittings; comprising low back trap and square hopper; 150 mm x 150 mm square gully grid						
100 mm nominal size	26.60	0.79	9.80	38.35	nr	**48.15**
Gully fittings; comprising low back trap and square hopper with back inlet; 150 mm x 150 mm square gully grid						
100 mm nominal size	36.25	0.85	10.55	49.25	nr	**59.80**
Access gully; trapped with rodding eye and integral vertical back inlet; stopper; 150 mm x 150 mm square gully grid						
100 mm nominal size	48.31	0.60	7.45	59.60	nr	**67.05**
Inspection chamber; comprising base; 300 mm or 450 mm raising piece; integral alloy cover and frame; 100 mm inlets						
straight through; 2 nr inlets	165.59	1.85	22.96	193.78	nr	**216.74**
Accessories in polypropylene; cover set in concrete; with coupling joints to pipes						
Inspection chamber; 5 nr 100 mm inlets; cast iron cover and frame						
475 mm diameter x 585 mm deep	197.31	2.13	26.43	230.72	nr	**257.15**
475 mm diameter x 930 mm deep	240.32	2.31	28.67	279.32	nr	**307.99**

R DISPOSAL SYSTEMS

Item	PC £	Labour hours	Labour £	Material £	Unit	Total rate £
R12 DRAINAGE BELOW GROUND – cont'd						
Accessories in vitrified clay; set in concrete; with cement and sand (1:2) joints to pipes						
Yard gully; 225 mm diameter; including domestic duty grating and frame (up to 1 tonne) and combined filter and silk bucket						
100 mm outlet	114.13	2.50	31.02	129.48	nr	**160.51**
100 mm outlet; 100 mm back inlet	159.46	2.70	33.50	180.71	nr	**214.22**
150 mm outlet	114.13	3.50	43.43	129.48	nr	**172.92**
150 mm outlet; 150 mm back inlet	162.73	3.70	45.91	184.41	nr	**230.32**
Yard gully; 225 mm diameter; including medium duty grating and frame (up to 5 tonnes) and combined filter and silk bucket						
100 mm outlet	149.26	2.50	31.02	169.18	nr	**200.20**
100 mm outlet; 100 mm back inlet	197.78	2.70	33.50	224.01	nr	**257.51**
150 mm outle	160.75	3.50	43.43	182.16	nr	**225.59**
150 mm outlet; 150 mm back inlet	201.05	3.70	45.91	227.71	nr	**273.62**
Road gully; trapped with rodding eye and stopper (grate not included)						
300 mm x 600 mm x 100 mm outlet	81.66	3.05	37.85	112.16	nr	**150.00**
300 mm x 600 mm x 150 mm outlet	83.62	3.05	37.85	114.37	nr	**152.22**
400 mm x 750 mm x 150 mm outlet	96.98	3.70	45.91	139.78	nr	**185.69**
450 mm x 900 mm x 150 mm outlet	131.22	4.65	57.70	184.49	nr	**242.19**
Grease trap; with internal access; galvanized perforated bucket; lid and frame						
600 mm x 450 mm x 600 mm deep; 100 mm outlet	668.56	3.89	48.27	785.13	nr	**833.41**
Interceptor; trapped with inspection arm; lever locking stopper; chain and staple; cement and sand (1:2) joints to pipes; building in, and cutting and fitting brickwork around						
100 mm outlet; 100 mm inlet	102.62	3.70	45.91	116.38	nr	**162.30**
150 mm outlet; 150 mm inlet	145.70	4.16	51.62	165.08	nr	**216.70**
225 mm outlet; 225 mm inlet	399.06	4.63	57.45	451.43	nr	**508.89**
Accessories; grates and covers						
Aluminium alloy gully grids; set in position						
120 mm x 120 mm	3.12	0.09	1.12	3.53	nr	**4.64**
150 mm x 150 mm	3.12	0.09	1.12	3.53	nr	**4.64**
225 mm x 225 mm	9.29	0.09	1.12	10.50	nr	**11.61**
100 mm diameter	3.12	0.09	1.12	3.53	nr	**4.64**
150 mm diameter	4.77	0.09	1.12	5.39	nr	**6.51**
225 mm diameter	10.39	0.09	1.12	11.74	nr	**12.85**
Aluminium alloy sealing plates and frames; set in cement and sand (1:3)						
150 mm x 150 mm	11.99	0.23	2.85	13.65	nr	**16.50**
225 mm x 225 mm	21.94	0.23	2.85	24.89	nr	**27.74**
140 mm diameter (for 100 mm)	9.77	0.23	2.85	11.14	nr	**13.99**
197 mm diameter (for 150 mm)	14.06	0.23	2.85	15.98	nr	**18.84**
273 mm diameter (for 225 mm)	22.50	0.23	2.85	25.52	nr	**28.37**
Polypropylene access covers and frames; supplied by Manhole Covers Ltd or other equal and approved; to suit PPIC inspection chambers; bedding and pointing in frame.						
450 mm dia; class A15	15.30	1.30	16.13	18.36	nr	**34.49**
450 mm dia; class B125; kite-marked	39.60	1.30	16.13	45.16	nr	**61.30**

R DISPOSAL SYSTEMS

Item	PC £	Labour hours	Labour £	Material £	Unit	**Total rate £**
Ductile iron heavy duty road gratings and frame; supplied by Manhole Covers Ltd or other equal and approved; bedding and pointing in cement and sand (1:3); one course half brick thick wall in semi-engineering bricks in cement mortar (1:3)						
225 mm x 225 mm x 80 mm hinged and dished road grating and frame; class C250	23.85	2.25	27.92	29.25	nr	**57.17**
300 mm x 300 mm x 80 mm hinged and dished road grating and frame; class C250	35.10	2.25	27.92	41.66	nr	**69.58**
420 mm x 420 mm x 75 mm hinged road grating and frame; class C250; kite-marked	35.55	2.25	27.92	42.16	nr	**70.08**
445 mm x 445 mm x 75 mm double triangular road grating and frame; class C250; kite-marked	37.80	2.25	27.92	44.64	nr	**72.56**
435 mm x 435 mm x 100 mm pedestrian mesh road grating and frame; class D400	66.60	2.25	27.92	76.41	nr	**104.33**
440 mm x 400 mm x 150 mm hinged road grating and frame; class D400; kite-marked	70.20	2.25	27.92	80.38	nr	**108.31**
Vibrated concrete pipes and fittings; with flexible joints; BS 5911 Part 1						
300 mm pipes Class M; laid straight	12.72	0.65	8.07	14.37	m	**22.44**
Extra for						
bend; <= 45 degree	-	0.65	8.07	125.78	nr	**133.84**
bend; > 45 degree	-	0.65	8.07	197.65	nr	**205.71**
junction; 300 mm x 100 mm	-	0.46	5.71	75.47	nr	**81.17**
450 mm pipes Class H; laid straight	18.80	1.02	12.66	21.25	m	**33.90**
Extra for						
bend; <= 45 degree	-	1.02	12.66	185.90	nr	**198.55**
bend; > 45 degree	-	1.02	12.66	292.12	nr	**304.78**
junction; 450 mm x 150 mm	-	0.65	8.07	111.54	nr	**119.60**
600 mm pipes Class H; laid straight	30.56	1.48	18.37	34.53	m	**52.90**
Extra for						
bend; <= 45 degree	-	1.48	18.37	302.18	nr	**320.54**
bend; > 45 degree	-	1.48	18.37	474.85	nr	**493.22**
junction; 600 mm x 150 mm	-	0.83	10.30	181.31	nr	**191.61**
900 mm pipes Class H; laid straight	78.40	2.59	32.14	88.60	m	**120.74**
Extra for						
bend; <= 45 degree	-	2.59	32.14	775.22	nr	**807.36**
bend; > 45 degree	-	2.59	32.14	1218.21	nr	**1250.35**
junction; 900 mm x 150 mm	-	1.02	12.66	243.64	nr	**256.30**
1200 mm pipes Class H; laid straight	135.12	3.70	45.91	152.69	m	**198.61**
Extra for						
bend; <= 45 degree	-	3.70	45.91	1336.07	nr	**1381.99**
bend; > 45 degree	-	3.70	45.91	2099.54	nr	**2145.46**
junction; 1200 mm x 150 mm	-	1.48	18.37	419.91	nr	**438.27**
Accessories in precast concrete; top set in with rodding eye and stopper; cement and sand (1:2) joint to pipe						
Concrete road gully; BS 5911; trapped with rodding eye and stopper; cement and sand (1:2) joint to pipe						
450 mm diameter x 1050 mm deep; 100 mm or 150 mm outlet	37.08	4.39	54.48	62.75	nr	**117.23**

R DISPOSAL SYSTEMS

Item	PC £	Labour hours	Labour £	Material £	Unit	Total rate £
R12 DRAINAGE BELOW GROUND – cont'd						
"Osmadrain" uPVC pipes and fittings or other equal and approved; BS 4660; with ring seal joints						
82 mm pipes; laid straight	10.50	0.15	1.86	11.86	m	**13.72**
Extra for						
bend; short radius	18.44	0.13	1.61	20.34	nr	**21.96**
spigot/socket bend	15.50	0.13	1.61	17.09	nr	**18.71**
adaptor	8.09	0.07	0.87	8.92	nr	**9.79**
single junction	23.98	0.18	2.23	26.46	nr	**28.69**
slip coupler	8.58	0.07	0.87	9.47	nr	**10.33**
100 mm pipes; laid straight	6.60	0.17	2.11	8.60	m	**10.71**
Extra for						
bend; short radius	17.42	0.15	1.86	18.77	nr	**20.63**
bend; long radius	28.21	0.15	1.86	28.88	nr	**30.74**
spigot/socket bend	14.72	0.15	1.86	21.53	nr	**23.39**
socket plug	7.62	0.04	0.50	8.41	nr	**8.91**
adjustable double socket bend	20.85	0.15	1.86	28.37	nr	**30.23**
adaptor to clay	19.64	0.09	1.12	21.31	nr	**22.43**
single junction	20.79	0.21	2.61	20.69	nr	**23.30**
sealed access junction	53.76	0.19	2.36	57.07	nr	**59.43**
slip coupler	8.58	0.09	1.12	9.47	nr	**10.58**
160 mm pipes; laid straight	14.48	0.21	2.61	18.66	m	**21.26**
Extra for						
bend; short radius	41.44	0.18	2.23	44.73	nr	**46.96**
spigot/socket bend	37.57	0.18	2.23	52.95	nr	**55.18**
socket plug	16.37	0.07	0.87	18.06	nr	**18.92**
adaptor to clay	42.70	0.12	1.49	46.23	nr	**47.72**
level invert taper	20.09	0.18	2.23	32.68	nr	**34.92**
single junction	67.86	0.24	2.98	74.86	nr	**77.84**
slip coupler	12.21	0.11	1.36	13.47	nr	**14.83**
uPVC Osma "Ultra-Rib" ribbed pipes and fittings or other equal and approved; WIS approval; with sealed ring push-fit joints						
150 mm pipes; laid straight	-	0.19	2.36	7.68	m	**10.04**
Extra for						
bend; short radius	22.55	0.17	2.11	24.42	nr	**26.53**
adaptor to 160 mm diameter upvc	31.90	0.10	1.24	34.27	nr	**35.51**
adaptor to clay	65.49	0.10	1.24	71.78	nr	**73.02**
level invert taper	9.84	0.18	2.23	9.47	nr	**11.71**
single junction	40.55	0.22	2.73	42.42	nr	**45.15**
225 mm pipes; laid straight	17.02	0.22	2.73	19.24	m	**21.97**
Extra for						
bend; short radius	90.65	0.20	2.48	98.85	nr	**101.33**
adaptor to clay	81.59	0.13	1.61	87.70	nr	**89.31**
level invert taper	15.77	0.20	2.48	13.93	nr	**16.41**
single junction	134.56	0.27	3.35	142.67	nr	**146.02**
300 mm pipes; laid straight	25.29	0.32	3.97	28.57	m	**32.54**
Extra for						
bend; short radius	142.77	0.29	3.60	155.79	nr	**159.38**
adaptor to clay	214.61	0.14	1.74	233.32	nr	**235.06**
level invert taper	51.21	0.29	3.60	51.35	nr	**54.95**
single junction	310.93	0.37	4.59	334.43	nr	**339.02**

R DISPOSAL SYSTEMS

Item	PC £	Labour hours	Labour £	Material £	Unit	**Total rate £**
Interconnecting concrete drainage channel 100 wide; ACO Technologies Ltd ref NK100 or other equal and approved; with Heelguard ductile iron grating suitable for load class 250; bedding and haunching in in situ concrete (not included)						
100 mm wide						
laid level or to falls	-	0.46	5.71	76.80	m	**82.51**
extra for sump unit	-	1.39	17.25	86.72	nr	**103.97**
extra for end caps	-	0.09	1.12	15.30	nr	**16.41**
Interconnecting drainage channel; "Birco-lite" ref 8012 or other equal and approved; Marshalls Plc; galvanised steel grating ref 8041; bedding and haunching in in situ concrete (not included)						
100 mm wide						
laid level or to falls	-	0.46	5.71	36.15	m	**41.86**
extra for 100 mm diameter trapped outlet unit	-	1.39	17.25	79.56	nr	**96.81**
extra for end caps	-	0.09	1.12	4.51	nr	**5.63**
Accessories in uPVC; with ring seal joints to pipes (unless otherwise described)						
Rodding eye						
110 mm diameter	35.61	0.43	5.34	43.59	nr	**48.93**
Universal gulley fitting; comprising gulley trap, plain hopper						
150 mm x 150 mm grate	31.01	0.93	11.54	40.23	nr	**51.77**
Bottle gulley; comprising gulley with bosses closed; sealed access covers						
217 mm x 217 mm grate	61.54	0.78	9.68	73.91	nr	**83.59**
Shallow access pipe; light duty screw down access door assembly						
110 mm diameter	87.50	0.78	9.68	102.54	nr	**112.22**
Shallow access inspection junction; 3 nr 110 mm inlets; light duty screw down access door assembly						
110 mm diameter	135.56	1.11	13.77	152.13	nr	**165.91**
Shallow inspection chamber; 250 mm diameter; 600 mm deep; sealed cover and frame						
4 nr 110 mm outlets/inlets	111.81	1.28	15.88	144.84	nr	**160.72**
Universal inspection chamber; 450 mm diameter; single seal cast iron cover and frame; 4 nr 110 mmoutlets/inlets						
500 mm deep	218.94	1.35	16.75	263.01	nr	**279.77**
730 mm deep	245.23	1.60	19.85	296.32	nr	**316.18**
960 mm deep	271.52	1.85	22.96	329.62	nr	**352.58**
Equal manhole base; 750 mm diameter						
6 nr 160 mm outlets/inlets	320.87	1.21	15.02	366.87	nr	**381.88**
Unequal manhole base; 750 mm diameter						
2 nr 160 mm, 4nr 110 mm outlets/inlets	248.07	1.21	15.02	286.56	nr	**301.57**
Kerb to gullies; class B engineering bricks on edge to three sides in cement mortar (1:3) rendering in cement mortar (1:3) to top and two sides and skirting to brickwork 230 mm high; dishing in cement mortar (1:3) to gully; steel trowelled						
230 mm x 230 mm internally	-	1.39	17.25	1.30	nr	**18.55**

R DISPOSAL SYSTEMS

Item	PC £	Labour hours	Labour £	Material £	Unit	Total rate £
R12 DRAINAGE BELOW GROUND – cont'd						
MANHOLES						
Excavating; by machine						
Manholes						
maximum depth not exceeding 1.00 m	-	0.19	2.38	5.65	m³	**8.03**
maximum depth not exceeding 2.00 m	-	0.21	2.63	6.22	m³	**8.85**
maximum depth not exceeding 4.00 m	-	0.25	3.13	7.26	m³	**10.40**
Excavating; by hand						
Manholes						
maximum depth not exceeding 1.00 m	-	3.05	38.21	-	m³	**38.21**
maximum depth not exceeding 2.00 m	-	3.61	45.22	-	m³	**45.22**
maximum depth not exceeding 4.00 m	-	4.63	58.00	-	m³	**58.00**
Earthwork support (average "risk" prices)						
Maximum depth not exceeding 1.00 m						
distance between opposing faces not exceeding 2.00 m	-	0.14	1.75	2.69	m²	**4.44**
Maximum depth not exceeding 2.00 m						
distance between opposing faces not exceeding 2.00 m	-	0.18	2.25	4.95	m²	**7.20**
Maximum depth not exceeding 4.00 m						
distance between opposing faces not exceeding 2.00 m	-	0.22	2.76	7.21	m²	**9.96**
Disposal; by machine						
Excavated material						
off site; to tip not exceeding 13 km (using lorries) including Landfill Tax based on inactive waste	-	-	-	21.52	m³	**21.52**
on site; depositing on site in spoil heaps; average 50 m distance	-	0.14	1.75	3.79	m³	**5.54**
Disposal; by hand						
Excavated material						
off site; to tip not exceeding 13 km (using lorries) including Landfill Tax based on inactive waste	-	0.74	9.27	29.60	m³	**38.87**
on site; depositing on site in spoil heaps; average 50 m distance	-	1.20	15.03	-	m³	**15.03**
Filling to excavations; by machine						
Average thickness not exceeding 0.25 m						
arising excavations	-	0.14	1.75	2.66	m³	**4.42**
Filling to excavations; by hand						
Average thickness not exceeding 0.25 m						
arising from excavations	-	0.93	11.65	-	m³	**11.65**
Plain in situ ready mixed designated concrete; C10 - 40 mm aggregate						
Beds						
thickness not exceeding 150 mm	78.02	2.78	44.88	90.28	m³	**135.15**
thickness 150 mm - 450 mm	-	2.08	33.58	90.28	m³	**123.85**
thickness exceeding 450 mm	-	1.76	28.41	90.28	m³	**118.69**

R DISPOSAL SYSTEMS

Item	PC £	Labour hours	Labour £	Material £	Unit	Total rate £
Plain in situ ready mixed designated concrete; C20 - 20 mm aggregate						
Beds						
thickness not exceeding 150 mm	81.52	2.78	44.88	94.32	m³	**139.20**
thickness 150 mm - 450 mm	-	2.08	33.58	94.32	m³	**127.90**
thickness exceeding 450 mm	-	1.76	28.41	94.32	m³	**122.73**
Plain in situ ready mixed designated concrete; C25 - 20 mm aggregate; (small quantities)						
Benching in bottoms						
150 mm - 450 mm average thickness	78.85	8.33	154.54	91.23	m³	**245.77**
Reinforced in situ ready mixed designated concrete; C20 - 20 mm aggregate; (small quantities)						
Isolated cover slabs						
thickness not exceeding 150 mm	77.63	6.48	104.61	89.82	m³	**194.43**
Reinforcement; fabric to BS 4449; lapped; in beds or suspended slabs						
Ref A98 (1.54 kg/m²)						
400 mm minimum laps	1.77	0.11	2.09	2.09	m²	**4.18**
Ref A142 (2.22 kg/m²)						
400 mm minimum laps	1.68	0.11	2.09	1.99	m²	**4.08**
Ref A193 (3.02 kg/m²)						
400 mm minimum laps	2.29	0.11	2.09	2.71	m²	**4.80**
Formwork; basic finish						
Soffits of isolated cover slabs						
horizontal	-	2.64	50.23	5.95	m²	**56.19**
Edges of isolated cover slabs						
height not exceeding 250 mm	-	0.78	14.84	1.86	m	**16.70**
Precast concrete circular manhole rings; BS5911 Part 1; bedding, jointing and pointing in cement mortar (1:3) on prepared bed						
Chamber or shaft rings; plain						
900 mm diameter	43.44	5.09	63.16	47.47	m	**110.64**
1050 mm diameter	45.88	6.01	74.58	50.82	m	**125.40**
1200 mm diameter	55.76	6.94	86.12	62.18	m	**148.30**
Chamber or shaft rings; reinforced						
1350 mm diameter	83.28	7.86	97.54	92.52	m	**190.05**
1500 mm diameter	93.24	8.79	109.08	104.68	m	**213.76**
1800 mm diameter	131.08	11.10	137.74	147.57	m	**285.31**
2100 mm diameter	256.64	13.88	172.24	284.87	m	**457.11**
extra for step irons built in	5.60	0.14	1.74	6.03	nr	**7.76**
Reducing slabs						
1200 mm diameter	78.40	5.55	68.87	85.82	nr	**154.69**
1350 mm diameter	118.28	8.79	109.08	130.19	nr	**239.26**
1500 mm diameter	136.44	10.18	126.33	150.45	nr	**276.78**
1800 mm diameter	181.60	12.95	160.70	201.22	nr	**361.92**
Heavy duty cover slabs; to suit rings						
900 mm diameter	46.40	2.78	34.50	50.66	nr	**85.16**
1050 mm diameter	49.80	3.24	40.21	54.46	nr	**94.67**
1200 mm diameter	60.32	3.70	45.91	66.36	nr	**112.28**
1350 mm diameter	91.00	4.16	51.62	100.10	nr	**151.73**
1500 mm diameter	104.88	4.63	57.45	115.77	nr	**173.22**
1800 mm diameter	153.60	5.55	68.87	169.32	nr	**238.19**
2100 mm diameter	325.44	6.48	80.41	355.67	nr	**436.08**

R DISPOSAL SYSTEMS

Item	PC £	Labour hours	Labour £	Material £	Unit	Total rate £
R12 DRAINAGE BELOW GROUND – cont'd						
MANHOLES – cont'd						
Common bricks; in cement mortar (1:3)						
Walls to manholes						
one brick thick	240.00	2.22	58.11	44.32	m²	**102.42**
one and a half brick thick	-	3.24	84.81	66.48	m²	**151.28**
Projections of footings						
two brick thick	-	4.53	118.57	88.63	m²	**207.20**
Class A engineering bricks; in cement mortar (1:3)						
Walls to manholes						
one brick thick (PC £ per 1000)	356.00	2.50	65.44	63.64	m²	**129.07**
one and a half brick thick	-	3.61	94.49	65.80	m²	**160.29**
Projections of footings						
two brick thick	-	5.09	133.23	127.27	m²	**260.50**
Class B engineering bricks; in cement mortar (1:3)						
Walls to manholes						
one brick thick (PC £ per 1000)	288.00	2.50	65.44	52.32	m²	**117.76**
one and a half brick thick	-	3.61	94.49	78.48	m²	**172.97**
Projections of footings						
two brick thick	-	5.09	133.23	104.64	m²	**237.87**
Brickwork sundries						
Extra over for fair face; flush smooth pointing						
manhole walls	-	0.19	4.97	-	m²	**4.97**
Building ends of pipes into brickwork; making good fair face or rendering						
not exceeding 55 mm nominal size	-	0.09	2.36	-	nr	**2.36**
55 mm - 110 mm nominal size	-	0.14	3.66	-	nr	**3.66**
over 110 mm nominal size	-	0.19	4.97	-	nr	**4.97**
Step irons; BS 1247; malleable; galvanized; building into joints						
general purpose pattern	-	0.14	3.66	4.22	nr	**7.88**
Cement and sand (1:3) in situ finishings; steel trowelled						
13 mm work to manhole walls; one coat; to brickwork base over 300 wide	-	0.65	17.01	1.44	m²	**18.46**
Cast iron inspection chambers; with bolted flat covers; BS 437; bedded in cement mortar (1:3); with mechanical coupling joints						
100 mm x 100 mm						
one branch either side	180.37	1.40	17.37	199.70	nr	**217.07**
two branches either side	341.16	2.00	24.82	377.07	nr	**401.89**
150 mm x 100 mm						
one branch either side	223.31	1.55	19.23	247.07	nr	**266.30**
two branches either side	433.42	2.15	26.68	479.57	nr	**506.25**
150 mm x 150 mm						
one branch either side	276.59	1.80	22.34	306.57	nr	**328.91**
two branches either side	533.88	2.60	32.26	590.40	nr	**622.66**

R DISPOSAL SYSTEMS

Item	PC £	Labour hours	Labour £	Material £	Unit	Total rate £
Coated cast or ductile iron access covers and frames; to BS EN124; supplied by Manhole Covers Ltd or other equal and approved; bedding frame in cement and sand (1:3); cover in grease and sand						
Light duty; cast iron; rectangular single seal solid top						
450 mm x 450 mm; class A15	35.10	1.50	18.61	40.20	nr	**58.81**
600 mm x 450 mm; class A15	37.80	1.50	18.61	43.31	nr	**61.92**
600 mm x 600 mm; class A15	59.40	1.50	18.61	67.26	nr	**85.87**
750 mm x 600 mm; class A15	98.50	1.50	18.61	110.40	nr	**129.01**
Light duty; cast iron; rectangular double seal solid top						
600 mm x 450 mm; class A15	71.55	1.50	18.61	80.54	nr	**99.15**
Medium duty; ductile iron; rectangular single seal solid top						
450 mm x 450 mm x 40 mm; class C250; kite-marked	57.60	2.00	24.82	65.27	nr	**90.09**
600 mm x 450 mm x 40 mm; slide-out; class C250; kite-marked	71.55	2.00	24.82	80.66	nr	**105.48**
600 mm x 600 mm x 40 mm; slide-out; class C250; kite-marked	76.50	2.00	24.82	86.12	nr	**110.94**
760 mm x 600 mm x 40 mm; slide-out; class C250; kite-marked	114.30	2.00	24.82	127.82	nr	**152.64**
Heavy duty; ductile iron; solid top						
450 mm x 450 mm x 75 mm; single seal; class C250; kite-marked	80.55	2.50	31.02	90.59	nr	**121.61**
600 mm x 450 mm x 75 mm; single seal; class C250; kite-marked	85.50	2.50	31.02	96.05	nr	**127.08**
600 mm x 600 mm x 75 mm; single seal; class C250; kite-marked	96.30	2.50	31.02	107.97	nr	**138.99**
450 mm x 450 mm x 100 mm; double triangular; class D400; kite-marked	89.10	2.50	31.02	100.02	nr	**131.05**
600 mm x 450 mm x 100 mm; double triangular; class D400; kite-marked	80.10	2.50	31.02	90.09	nr	**121.12**
600 mm x 600 mm x 100 mm; double triangular; class D400; kite-marked	71.10	2.50	31.02	80.17	nr	**111.19**
750 mm x 600 mm x 100 mm; double triangular; class D400; kite-marked	169.78	2.50	31.02	189.02	nr	**220.04**
1220 mm x 675 mm x 100 mm; double triangular; class D400; kite-marked	202.50	3.50	43.43	225.12	nr	**268.55**
British Standard best quality vitrified clay channels; bedding and jointing in cement and sand (1:2)						
Half section straight						
100 mm diameter x 1 m long	4.89	0.74	9.18	5.52	nr	**14.71**
150 mm diameter x 1 m long	8.14	0.93	11.54	9.20	nr	**20.74**
225 mm diameter x 1 m long	18.28	1.20	14.89	20.66	nr	**35.55**
300 mm diameter x 1 m long	37.53	1.48	18.37	42.41	nr	**60.78**
Half section bend						
100 mm diameter	11.24	0.56	6.95	12.71	nr	**19.66**
150 mm diameter	18.43	0.69	8.56	20.83	nr	**29.39**
225 mm diameter	61.28	0.93	11.54	69.25	nr	**80.79**
Taper straight						
150 mm - 100 mm diameter	22.89	0.65	8.07	25.86	nr	**33.93**
225 mm - 150 mm diameter	51.08	0.83	10.30	57.73	nr	**68.03**
Taper bend						
150 mm - 100 mm diameter	34.85	0.83	10.30	39.38	nr	**49.68**
225 mm - 150 mm diameter	99.84	1.06	13.15	112.83	nr	**125.98**
Three quarter section branch bend						
100 mm diameter	12.40	0.46	5.71	14.02	nr	**19.72**
150 mm diameter	20.83	0.69	8.56	23.53	nr	**32.10**
225 mm diameter	75.98	0.93	11.54	85.87	nr	**97.41**

R DISPOSAL SYSTEMS

Item	PC £	Labour hours	Labour £	Material £	Unit	Total rate £
R12 DRAINAGE BELOW GROUND – cont'd						
MANHOLES – cont'd						
uPVC channels; with solvent weld or lip seal coupling joints; bedding in cement and sand						
Half section cut away straight; with coupling either end						
110 mm diameter	46.66	0.28	3.47	67.50	nr	**70.98**
160 mm diameter	87.63	0.37	4.59	126.19	nr	**130.78**
Half section cut away long radius bend; with coupling either end						
110 mm diameter	76.50	0.28	3.47	101.23	nr	**104.71**
160 mm diameter	165.40	0.37	4.59	214.07	nr	**218.67**
Channel adaptor to clay; with one coupling						
110 mm diameter	17.90	0.23	2.85	27.62	nr	**30.47**
160 mm diameter	43.31	0.31	3.85	62.52	nr	**66.37**
Half section bend						
110 mm diameter	29.54	0.31	3.85	33.88	nr	**37.73**
160 mm diameter	50.73	0.46	5.71	58.50	nr	**64.21**
Half section channel connector						
110 mm diameter	8.08	0.07	0.87	10.13	nr	**11.00**
Half section channel junction						
110 mm diameter	22.92	0.46	5.71	26.40	nr	**32.11**
Polypropylene slipper bend						
110 mm diameter	19.88	0.37	4.59	22.96	nr	**27.55**
Glass fibre septic tank; "Klargester" or other equal and approved; fixing lockable manhole cover and frame; placing in position						
3750 litre capacity; 2000 mm diameter; depth to invert						
1000 mm deep; standard grade	732.79	2.27	28.17	870.08	nr	**898.25**
1500 mm deep; heavy duty grade	921.90	2.54	31.52	1073.61	nr	**1105.13**
6000 litre capacity; 2300 mm diameter; depth to invert						
1000 mm deep; standard grade	1181.92	2.45	30.40	1373.81	nr	**1404.21**
1500 mm deep; heavy duty grade	1560.14	2.73	33.88	1780.87	nr	**1814.74**
9000 litre capacity; 2660 mm diameter; depth to invert						
1000 mm deep; standard grade	1796.53	2.64	32.76	2035.27	nr	**2068.04**
1500 mm deep; heavy duty grade	2363.85	2.91	36.11	2645.86	nr	**2681.97**
Glass fibre petrol interceptors; "Klargester" or other equal and approved; placing in position						
2000 litre capacity; 2370 mm x 1300 mm diameter; depth to invert						
1000 mm deep	827.10	2.50	31.02	890.17	nr	**921.19**
4000 litre capacity; 4370 mm x 1300 mm diameter; depth to invert						
1000 mm deep	1418.40	2.68	33.26	1526.55	nr	**1559.81**

R DISPOSAL SYSTEMS

Item	PC £	Labour hours	Labour £	Material £	Unit	Total rate £
R13 LAND DRAINAGE						
Excavating; by hand; grading bottoms; earthwork support; filling to within 150 mm of surface with gravel rejects; remainder filled with excavated material and compacting; disposal of surplus soil on site; spreading on site average 50 m						
Pipes not exceeding 200 nominal size						
average depth of trench 0.75 m	-	1.57	19.67	10.59	m	**30.26**
average depth of trench 1.00 m	-	2.08	26.06	16.79	m	**42.85**
average depth of trench 1.25 m	-	2.91	36.46	21.07	m	**57.53**
average depth of trench 1.50 m	-	5.00	62.64	25.67	m	**88.31**
average depth of trench 1.75 m	-	5.92	74.16	29.94	m	**104.11**
average depth of trench 2.00 m	-	6.85	85.81	34.54	m	**120.36**
Disposal; by machine						
Excavated material						
off site; to tip not exceeding 13 km (using lorries); including Landfill Tax based on inactive waste	-	-	-	21.52	m³	**-**
hand loaded	-	-	-	29.60	m³	**-**
Disposal; by hand						
Excavated material						
off site; to tip not exceeding 13 km (using lorries); including Landfill Tax based on inactive waste	-	0.74	9.27	29.60	m³	**38.87**
Vitrified clay perforated sub-soil pipes; BS 65; Hepworth "Hepline" or other equal and approved						
Pipes; laid straight						
100 mm diameter	6.61	0.20	2.48	7.47	m	**9.95**
150 mm diameter	12.01	0.25	3.10	13.57	m	**16.68**
225 mm diameter	24.21	0.33	4.09	27.36	m	**31.46**

S PIPED SUPPLY SYSTEMS

Item	PC £	Labour hours	Labour £	Material £	Unit	Total rate £
S10/S11 HOT AND COLD WATER						
Copper pipes; BS EN 1057; capillary fittings						
15 mm pipes; fixing with pipe clips and screwed	2.16	0.34	6.16	2.54	m	**8.70**
Extra for						
made bend	-	0.14	2.54	-	nr	**2.54**
stop end	1.32	0.10	1.81	1.46	nr	**3.27**
straight coupling	0.21	0.16	2.90	0.23	nr	**3.13**
union coupling	7.16	0.16	2.90	7.90	nr	**10.79**
reducing coupling	2.44	0.16	2.90	2.69	nr	**5.59**
copper to lead connector	5.58	0.20	3.62	6.16	nr	**9.78**
imperial to metric adaptor	2.87	0.20	3.62	3.17	nr	**6.79**
elbow	0.37	0.16	2.90	0.41	nr	**3.31**
backplate elbow	5.38	0.32	5.80	5.93	nr	**11.73**
return bend	8.05	0.16	2.90	8.88	nr	**11.78**
tee; equal	0.71	0.23	4.17	0.78	nr	**4.95**
tee; reducing	5.86	0.23	4.17	6.46	nr	**10.63**
straight tap connector	1.82	0.47	8.51	2.01	nr	**10.53**
bent tap connector	2.04	0.63	11.41	2.25	nr	**13.66**
tank connector	6.40	0.23	4.17	7.06	nr	**11.23**
22 mm pipes; fixing with pipe clips and screwed	4.33	0.40	7.25	4.99	m	**12.23**
Extra for						
made bend	-	0.19	3.44	-	nr	**3.44**
stop end	2.47	0.12	2.17	2.72	nr	**4.89**
straight coupling	0.55	0.20	3.62	0.61	nr	**4.23**
union coupling	11.47	0.20	3.62	12.65	nr	**16.27**
reducing coupling	2.39	0.20	3.62	2.64	nr	**6.26**
copper to lead connector	7.62	0.29	5.25	8.41	nr	**13.66**
elbow	0.97	0.20	3.62	1.07	nr	**4.70**
backplate elbow	11.54	0.41	7.43	12.73	nr	**20.15**
return bend	15.82	0.20	3.62	17.45	nr	**21.08**
tee; equal	2.25	0.31	5.62	2.48	nr	**8.10**
tee; reducing	1.78	0.31	5.62	1.97	nr	**7.58**
straight tap connector	2.06	0.16	2.90	2.27	nr	**5.17**
28 mm pipes; fixing with pipe clips and screwed	5.45	0.43	7.79	6.26	m	**14.05**
Extra for						
made bend	-	0.23	4.17	-	nr	**4.17**
stop end	4.40	0.14	2.54	4.86	nr	**7.39**
straight coupling	1.22	0.26	4.71	1.35	nr	**6.06**
reducing coupling	3.34	0.26	4.71	3.69	nr	**8.40**
union coupling	11.47	0.26	4.71	12.65	nr	**17.36**
copper to lead connector	14.32	0.36	6.52	15.79	nr	**22.31**
imperial to metric adaptor	7.38	0.36	6.52	8.14	nr	**14.66**
elbow	1.96	0.26	4.71	2.16	nr	**6.87**
return bend	20.21	0.26	4.71	22.30	nr	**27.01**
tee; equal	5.44	0.38	6.88	6.00	nr	**12.88**
tank connector	12.82	0.38	6.88	14.14	nr	**21.03**
35 mm pipes; fixing with pipe clips and screwed	11.03	0.50	9.06	12.56	m	**21.62**
Extra for						
made bend	-	0.28	5.07	-	nr	**5.07**
stop end	9.73	0.16	2.90	10.74	nr	**13.63**
straight coupling	3.98	0.31	5.62	4.39	nr	**10.00**
reducing coupling	7.88	0.31	5.62	8.70	nr	**14.31**
union coupling	21.92	0.31	5.62	24.18	nr	**29.79**
flanged connector	60.45	0.41	7.43	66.69	nr	**74.11**
elbow	8.51	0.31	5.62	9.39	nr	**15.01**
obtuse elbow	12.85	0.31	5.62	14.18	nr	**19.79**
tee; equal	13.87	0.43	7.79	15.30	nr	**23.09**
tank connector	16.44	0.43	7.79	18.14	nr	**25.92**

S PIPED SUPPLY SYSTEMS

Item	PC £	Labour hours	Labour £	Material £	Unit	**Total rate £**
42 mm pipes; fixing with pipe clips; plugged and screwed	13.41	0.56	10.14	15.25	m	**25.40**
Extra for						
made bend	-	0.37	6.70	-	nr	**6.70**
stop end	16.75	0.18	3.26	18.48	nr	**21.74**
straight coupling	6.64	0.36	6.52	7.33	nr	**13.85**
reducing coupling	13.19	0.36	6.52	14.55	nr	**21.07**
union coupling	32.03	0.36	6.52	35.33	nr	**41.85**
flanged connector	72.26	0.46	8.33	79.71	nr	**88.04**
elbow	14.08	0.36	6.52	15.53	nr	**22.05**
obtuse elbow	22.88	0.36	6.52	25.24	nr	**31.76**
tee; equal	22.25	0.48	8.69	24.54	nr	**33.24**
tank connector	21.55	0.48	8.69	23.77	nr	**32.46**
54 mm pipes; fixing with pipe clips; plugged and screwed	17.25	0.62	11.23	19.59	m	**30.82**
Extra for						
made bend	-	0.51	9.24	-	nr	**9.24**
stop end	23.39	0.19	3.44	25.80	nr	**29.24**
straight coupling	12.25	0.41	7.43	13.51	nr	**20.94**
reducing coupling	22.15	0.41	7.43	24.43	nr	**31.86**
union coupling	60.95	0.41	7.43	67.24	nr	**74.67**
flanged connector	109.24	0.46	8.33	120.50	nr	**128.84**
elbow	29.07	0.41	7.43	32.07	nr	**39.50**
obtuse elbow	41.38	0.41	7.43	45.65	nr	**53.08**
tee; equal	44.86	0.53	9.60	49.48	nr	**59.08**
tank connector	32.93	0.53	9.60	36.32	nr	**45.92**
Copper pipes; EN1057:1996; compression fittings						
15 mm pipes; fixing with pipe clips; plugged and screwed	2.16	0.39	7.06	2.54	m	**9.61**
Extra for						
made bend	-	0.14	2.54	-	nr	**2.54**
stop end	2.40	0.09	1.63	2.65	nr	**4.28**
straight coupling	1.93	0.14	2.54	2.12	nr	**4.66**
male coupling	1.72	0.19	3.44	1.89	nr	**5.33**
female coupling	2.06	0.19	3.44	2.28	nr	**5.72**
90 degree bend	2.32	0.14	2.54	2.56	nr	**5.10**
90 degree backplate bend	4.30	0.28	5.07	4.75	nr	**9.82**
tee; equal	3.25	0.20	3.62	3.59	nr	**7.21**
tank coupling	4.89	0.20	3.62	5.39	nr	**9.02**
22 mm pipes; fixing with pipe clips; plugged and screwed	4.33	0.44	7.97	4.99	m	**12.96**
Extra for						
made bend	-	0.19	3.44	-	nr	**3.44**
stop end	3.47	0.11	1.99	3.83	nr	**5.82**
straight coupling	3.14	0.19	3.44	3.46	nr	**6.90**
male coupling	3.69	0.26	4.71	4.07	nr	**8.78**
female coupling	3.02	0.26	4.71	3.33	nr	**8.04**
90 degree bend	3.70	0.19	3.44	4.08	nr	**7.52**
tee; equal	5.37	0.28	5.07	5.92	nr	**11.00**
tee; reducing	8.59	0.28	5.07	9.47	nr	**14.54**
tank coupling	4.82	0.28	5.07	5.32	nr	**10.39**

S PIPED SUPPLY SYSTEMS

Item	PC £	Labour hours	Labour £	Material £	Unit	Total rate £
S10/S11 HOT AND COLD WATER – cont'd						
Copper pipes; compression fittings – cont'd						
28 mm pipes; fixing with pipe clips; plugged and screwed	5.45	0.48	8.69	6.26	m	**14.96**
Extra for						
made bend	-	0.23	4.17	-	nr	**4.17**
stop end	7.44	0.13	2.35	8.21	nr	**10.56**
straight coupling	7.12	0.23	4.17	7.86	nr	**12.02**
male coupling	5.04	0.32	5.80	5.56	nr	**11.36**
female coupling	6.53	0.32	5.80	7.20	nr	**13.00**
90 degree bend	9.19	0.23	4.17	10.13	nr	**14.30**
tee; equal	14.65	0.34	6.16	16.16	nr	**22.32**
tee; reducing	14.14	0.34	6.16	15.60	nr	**21.76**
tank coupling	11.44	0.34	6.16	12.62	nr	**18.77**
35 mm pipes; fixing with pipe clips; plugged and screwed	11.03	0.55	9.96	12.56	m	**22.52**
Extra for						
made bend	-	0.28	5.07	-	nr	**5.07**
stop end	11.67	0.15	2.72	12.87	nr	**15.59**
straight coupling	15.06	0.28	5.07	16.61	nr	**21.69**
male coupling	11.45	0.37	6.70	12.63	nr	**19.33**
female coupling	13.75	0.37	6.70	15.16	nr	**21.87**
tee; equal	26.44	0.39	7.06	29.17	nr	**36.23**
tee; reducing	25.84	0.39	7.06	28.51	nr	**35.57**
tank coupling	20.20	0.39	7.06	22.28	nr	**29.34**
42 mm pipes; fixing with pipe clips; plugged and screwed	13.41	0.61	11.05	15.25	m	**26.30**
Extra for						
made bend	-	0.37	6.70	-	nr	**6.70**
stop end	19.43	0.17	3.08	21.43	nr	**24.51**
straight coupling	19.81	0.32	5.80	21.85	nr	**27.65**
male coupling	17.17	0.42	7.61	18.94	nr	**26.55**
female coupling	18.48	0.42	7.61	20.39	nr	**27.99**
tee; equal	41.57	0.43	7.79	45.86	nr	**53.65**
tee; reducing	39.94	0.43	7.79	44.06	nr	**51.85**
54 mm pipes; fixing with pipe clips; plugged and screwed	17.25	0.67	12.14	19.59	m	**31.73**
Extra for						
made bend	-	0.51	9.24	-	nr	**9.24**
straight coupling	29.62	0.37	6.70	32.67	nr	**39.37**
male coupling	25.36	0.46	8.33	27.97	nr	**36.30**
female coupling	27.11	0.46	8.33	29.91	nr	**38.24**
tee; equal	66.78	0.48	8.69	73.67	nr	**82.36**
tee; reducing	66.78	0.48	8.69	73.67	nr	**82.36**
Copper, brass and gunmetal ancillaries; screwed joints to fittings						
Stopcock; brass/gunmetal capillary joints to copper						
15 mm nominal size	54.60	0.19	3.44	60.23	nr	**63.67**
22 mm nominal size	10.20	0.25	4.53	11.25	nr	**15.78**
28 mm nominal size	29.00	0.31	5.62	32.00	nr	**37.61**
Stopcock; brass/gunmetal compression joints to copper						
15 mm nominal size	17.69	0.17	3.08	19.52	nr	**22.60**
22 mm nominal size	24.89	0.22	3.98	27.46	nr	**31.44**
28 mm nominal size	44.27	0.28	5.07	48.84	nr	**53.91**

S PIPED SUPPLY SYSTEMS

Item	PC £	Labour hours	Labour £	Material £	Unit	**Total rate £**
Stopcock; brass/gunmetal compression joints to polyethylene						
15 mm nominal size	18.32	0.24	4.35	20.21	nr	**24.56**
22 mm nominal size	31.85	0.31	5.62	35.13	nr	**40.75**
28 mm nominal size	33.96	0.37	6.70	37.46	nr	**44.17**
Gunmetal "Fullway" gate valve; capillary joints to copper						
15 mm nominal size	17.11	0.19	3.44	18.87	nr	**22.31**
22 mm nominal size	19.81	0.25	4.53	21.86	nr	**26.38**
28 mm nominal size	27.59	0.31	5.62	30.44	nr	**36.06**
35 mm nominal size	61.54	0.38	6.88	67.88	nr	**74.77**
42 mm nominal size	76.94	0.43	7.79	84.88	nr	**92.67**
54 mm nominal size	111.62	0.49	8.88	123.14	nr	**132.01**
Brass gate valve; compression joints to copper						
15 mm nominal size	20.77	0.28	5.07	22.91	nr	**27.99**
22 mm nominal size	24.48	0.37	6.70	27.01	nr	**33.71**
28 mm nominal size	33.25	0.46	8.33	36.68	nr	**45.01**
Chromium plated; lockshield radiator valve; union outlet						
15 mm nominal size	6.22	0.20	3.62	6.86	nr	**10.48**
PEX/PEM 'JG Speedfit' system; BS 7291 Parts 1, 2 & 3 class S; push-fit fittings						
10 mm PEX barrier pipes; fixing with pipe clips; in wall, floor and roof voids	0.86	0.20	3.62	1.62	m	**5.24**
Extra for						
stop end	1.27	0.05	0.91	1.69	nr	**2.60**
straight connector	1.31	0.10	1.81	2.03	nr	**3.84**
elbow	1.61	0.10	1.81	2.36	nr	**4.17**
stem elbow	2.14	0.10	1.81	2.94	nr	**4.75**
tee; equal	1.86	0.15	2.72	2.92	nr	**5.64**
brass chrome plated service valve	6.61	0.10	1.81	7.87	nr	**9.68**
brass chrome plated ball valve	9.31	0.10	1.81	10.86	nr	**12.67**
15 mm PEX barrier pipes; fixing with pipe clips; in wall, floor and roof voids	0.99	0.22	3.98	1.71	m	**5.70**
15 mm Polybutylene barrier pipes; fixing with pipe clips; in wall, floor and roof voids	1.17	0.22	3.98	1.91	m	**5.89**
Extra for						
stop end	1.31	0.07	1.27	1.74	nr	**3.01**
straight connector	0.98	0.14	2.54	1.66	nr	**4.20**
reducing coupler	2.29	0.14	2.54	3.11	nr	**5.64**
PE-copper coupler	4.54	0.16	2.90	5.60	nr	**8.49**
elbow	1.15	0.14	2.54	1.85	nr	**4.39**
stem elbow	2.26	0.14	2.54	3.07	nr	**5.61**
tee; equal	1.66	0.20	3.62	2.71	nr	**6.33**
tee; reducing	2.79	0.20	3.62	3.95	nr	**7.58**
tank connector	1.63	0.20	3.62	2.09	nr	**5.71**
straight tap connector	1.90	0.28	5.07	2.39	nr	**7.46**
bent tap connector	2.40	0.28	5.07	2.94	nr	**8.01**
angle service valve with tap connector	5.94	0.28	5.07	6.85	nr	**11.92**
stop valve	4.66	0.14	2.54	5.72	nr	**8.25**
brass chrome plated service valve	7.93	0.14	2.54	9.33	nr	**11.86**
brass chrome plated ball valve	10.17	0.14	2.54	11.80	nr	**14.34**
speedfit x union nut flexi hose 500 mm long	5.51	0.28	5.07	6.37	nr	**11.44**

S PIPED SUPPLY SYSTEMS

Item	PC £	Labour hours	Labour £	Material £	Unit	Total rate £
S10/S11 HOT AND COLD WATER – cont'd						
PEX/PEM 'JG Speedfit' system – cont'd						
22 mm PEX barrier pipes; fixing with pipe clips; in wall, floor and roof voids	1.95	0.25	4.53	2.98	m	**7.51**
22 mm Polybutylene barrier pipes; fixing with pipe clips; in wall, floor and roof voids	2.22	0.25	4.53	3.29	m	**7.82**
Extra for						
stop end	1.58	0.09	1.63	2.11	nr	**3.74**
straight connector	1.53	0.18	3.26	2.41	nr	**5.67**
reducing coupler	2.69	0.18	3.26	3.69	nr	**6.95**
PE-copper coupler	5.37	0.20	3.62	6.65	nr	**10.27**
elbow	1.83	0.18	3.26	2.74	nr	**6.01**
stem elbow	3.42	0.18	3.26	4.50	nr	**7.76**
tee; equal	2.47	0.27	4.89	3.81	nr	**8.70**
tee; reducing	2.79	0.27	4.89	4.02	nr	**8.91**
tank connector	2.08	0.27	4.89	2.66	nr	**7.55**
straight tap connector	2.48	0.36	6.52	3.10	nr	**9.62**
stop valve	7.08	0.18	3.26	8.53	nr	**11.79**
brass chrome plated service valve	17.79	0.18	3.26	20.35	nr	**23.61**
brass chrome plated ball valve	20.03	0.18	3.26	22.82	nr	**26.08**
speedfit x union nut flexi hose 500 mm long	6.61	0.36	6.52	7.65	nr	**14.17**
22 x 10 4 Way manifold	5.78	0.36	6.52	7.91	nr	**14.43**
22 x 15 4 Port rail manifold	11.93	0.36	6.52	14.69	nr	**21.21**
22 x 15 4 Zone brass rail manifold	178.58	1.00	18.11	198.52	nr	**216.64**
28 mm PEX barrier pipes; fixing with pipe clips; in wall, floor and roof voids	2.67	0.28	5.07	4.53	m	**9.60**
Extra for						
straight connector	3.86	0.24	4.35	5.10	nr	**9.45**
reducer	3.26	0.24	4.35	4.45	nr	**8.80**
elbow	4.51	0.24	4.35	5.82	nr	**10.17**
tee; equal	6.36	0.36	6.52	8.29	nr	**14.81**
tee; reducing	7.02	0.36	6.52	8.89	nr	**15.41**
Water tanks/cisterns						
Polyethylene cold water feed and expansion cistern; BS 4213; with covers						
ref SC15; 68 litres	36.20	1.16	21.01	38.96	nr	**59.97**
ref SC25; 114 litres	42.60	1.34	24.27	45.85	nr	**70.12**
ref SC40; 182 litres	50.03	1.34	24.27	53.85	nr	**78.12**
ref SC50; 227 litres	69.00	1.80	32.60	74.26	nr	**106.87**
GRP cold water storage cistern; with covers						
ref 899.10; 30 litres	85.53	1.02	18.48	92.06	nr	**110.53**
ref 899.25; 68 litres	107.95	1.16	21.01	116.19	nr	**137.20**
ref 899.40; 114 litres	134.22	1.34	24.27	144.45	nr	**168.72**
ref 899.70; 227 litres	168.30	1.80	32.60	181.13	nr	**213.74**
Storage cylinders/calorifiers						
Copper cylinders; single feed coil indirect; BS 1566 Part 2; grade 3						
ref 2; 96 litres	184.38	1.85	33.51	198.44	nr	**231.94**
ref 3; 114 litres	117.31	2.08	37.68	126.25	nr	**163.93**
ref 7; 117 litres	115.76	2.31	41.84	124.59	nr	**166.43**
ref 8; 140 litres	131.06	2.78	50.35	141.05	nr	**191.41**
ref 9; 162 litres	167.40	3.24	58.69	180.17	nr	**238.85**

S PIPED SUPPLY SYSTEMS

Item	PC £	Labour hours	Labour £	Material £	Unit	Total rate £
Combination copper hot water storage units; coil direct; BS 3198; (hot/cold)						
400 mm x 900 mm; 65/20 litres	136.25	2.59	46.91	146.64	nr	**193.55**
450 mm x 900 mm; 85/25 litres	140.31	3.61	65.39	151.01	nr	**216.40**
450 mm x 1075 mm; 115/25 litres	154.35	4.53	82.05	166.11	nr	**248.17**
450 mm x 1200 mm; 115/45 litres	164.31	5.09	92.20	176.84	nr	**269.04**
Combination copper hot water storage						
450 mm x 900 mm; 85/25 litres	175.96	4.07	73.72	189.38	nr	**263.10**
450 mm x 1200 mm; 115/45 litres	201.21	5.55	100.53	216.56	nr	**317.09**
Thermal insulation						
20mm thick Rockwool "Rocklap" bonded pre-formed mineral glass fibre sectional pipe lagging; aluminum outer foil finish finish; taped to steel or copper pipework; including working over pipe fittings						
around 15/15 pipes	1.79	0.06	1.09	2.03	m	**3.12**
around 20/22 pipes	1.88	0.09	1.63	2.13	m	**3.76**
around 25/28 pipes	2.00	0.10	1.81	2.26	m	**4.07**
around 32/35 pipes	2.17	0.11	1.99	2.46	m	**4.45**
around 40/42 pipes	2.44	0.12	2.17	2.76	m	**4.94**
around 50/54 pipes	2.82	0.14	2.54	3.18	m	**5.72**
19 mm thick rigid mineral glass fibre sectional pipe lagging; canvas or class O lacquered aluminium finish; fixed with aluminium bands to steel or copper pipework; including working over pipe fittings						
around 15/15 pipes	2.66	0.06	1.09	3.01	m	**4.10**
around 20/22 pipes	2.77	0.09	1.63	3.13	m	**4.76**
around 25/28 pipes	2.97	0.10	1.81	3.36	m	**5.17**
around 32/35 pipes	3.12	0.11	1.99	3.53	m	**5.52**
around 40/42 pipes	3.41	0.12	2.17	3.86	m	**6.03**
around 50/54 pipes	3.95	0.14	2.54	4.47	m	**7.01**
60 mm thick glass-fibre filled polyethylene insulating jackets for GRP or polyethylene cold water cisterns; complete with fixing bands; for cisterns size (supply not included)						
450 mm x 300 mm x 300 mm (45 litres)	-	0.37	6.70	-	nr	**6.70**
650 mm x 500 mm x 400 mm (91 litres)	-	0.56	10.14	-	nr	**10.14**
675 mm x 525 mm x 500 mm (136 litres)	-	0.65	11.77	-	nr	**11.77**
675 mm x 575 mm x 525 mm (182 litres)	-	0.74	13.40	-	nr	**13.40**
1000 mm x 625 mm x 525 mm (273 litres)	-	0.79	14.31	-	nr	**14.31**
1125 mm x 650 mm x 575 mm (341 litres)	-	0.79	14.31	-	nr	**14.31**
80 mm thick glass-fibre filled insulating jackets in flame retardant PVC to BS 5615; type 1B; segmental type for hot water cylinders; complete with fixing bands; for cylinders size (supply not included)						
400 mm x 900 mm; ref 2	-	0.31	5.62	-	nr	**5.62**
450 mm x 900 mm; ref 7	-	0.31	5.62	-	nr	**5.62**
450 mm x 1050 mm; ref 8	-	0.37	6.70	-	nr	**6.70**
450 mm x 1200 mm	-	0.46	8.33	-	nr	**8.33**

S PIPED SUPPLY SYSTEMS

Item	PC £	Labour hours	Labour £	Material £	Unit	Total rate £
S13 PRESSURISED WATER						
Blue MDPE pipes; BS EN 12201; mains pipework; no joints in the running length; laid in trenches						
Pipes						
20 mm nominal size	0.78	0.10	1.81	0.87	m	**2.68**
25 mm nominal size	0.91	0.11	1.99	1.01	m	**3.00**
32 mm nominal size	1.54	0.12	2.17	1.70	m	**3.88**
50 mm nominal size	3.67	0.14	2.54	4.07	m	**6.61**
63 mm nominal size	5.82	0.15	2.72	6.46	m	**9.17**
Ductile iron bitumen coated pipes and fittings; BS EN 969; class K9; Stanton's "Tyton" water main pipes or other equal and approved; flexible joints						
100 mm pipes; laid straight	29.55	0.56	6.95	41.26	m	**48.21**
Extra for						
bend; 45 degrees	48.29	0.56	6.95	70.32	nr	**77.27**
branch; 45 degrees; socketed	351.79	0.83	10.30	421.17	nr	**431.47**
tee	76.23	0.83	10.30	109.77	nr	**120.07**
flanged spigot	48.34	0.56	6.95	62.51	nr	**69.46**
flanged socket	45.99	0.56	6.95	59.85	nr	**66.80**
150 mm pipes; laid straight	35.52	0.65	8.07	48.61	m	**56.67**
Extra for						
bend; 45 degrees	75.59	0.65	8.07	102.37	nr	**110.43**
branch; 45 degrees; socketed	448.98	0.97	12.04	532.79	nr	**544.83**
tee	158.41	0.97	12.04	204.42	nr	**216.46**
flanged spigot	56.07	0.65	8.07	71.84	nr	**79.90**
flanged socket	73.19	0.65	8.07	91.18	nr	**99.25**
200 mm pipes; laid straight	48.55	0.93	11.54	67.12	m	**78.66**
Extra for						
bend; 45 degrees	136.42	0.93	11.54	178.67	nr	**190.21**
branch; 45 degrees; socketed	509.93	1.39	17.25	613.00	nr	**630.25**
tee	217.59	1.39	17.25	282.64	nr	**299.89**
flanged spigot	122.11	0.93	11.54	150.24	nr	**161.78**
flanged socket	115.78	0.93	11.54	143.09	nr	**154.63**

T MECHANICAL HEATING/COOLING SYSTEMS ETC

Item	PC £	Labour hours	Labour £	Material £	Unit	Total rate £
T10 GAS/OIL FIRED BOILERS						
Boilers						
Gas fired wall mounted combination boilers; for central heating and hot water supply; Potterton 'Performa' or equivalent; with cream or white; enamelled casing; 32 mm diameter BSPT female flow and return tappings; 102 mm diameter flue socket 13 mm diameter BSPT male draw-off outlet						
24.00 kW output; ref Performa 24	459.84	5.00	103.80	494.90	nr	**598.70**
31.00 kW output; ref Performa 28	567.82	5.00	103.80	611.12	nr	**714.92**
31.00 kW output; ref Performa 28i	591.51	5.00	103.80	636.61	nr	**740.41**
Gas fired wall mounted domestic boilers; for central heating and indirect hot water supply; Potterton 'Profile' or equivalent; with cream or white; enamelled casing; 32 mm diameter BSPT female flow and return tappings; 102 mm diameter flue socket 13 mm diameter BSPT male draw-off outlet						
14.60 kW output (50,000 Btu/Hr); ref Profile 50e L	580.40	5.00	103.80	624.65	nr	**728.46**
23.45 kW output (80,000 Btu/Hr); ref Profile 80e L	789.78	5.00	103.80	850.00	nr	**953.81**
Flues						
Scheidel Rite-Vent ICS Plus flue system; suitable for domestic multifuel appliances; stainless steel; twin wall; insulated; for use internally or externally						
80 mm pipes; including one locking band (fixing brackets measured separately)	-	0.90	16.30	84.30	m	**100.60**
Extra for						
Appliance Connecter	-	0.80	14.49	12.93	nr	**27.42**
30° Bend	-	1.80	32.60	62.89	nr	**95.50**
45° Bend	-	1.80	32.60	59.76	nr	**92.36**
135° Tee; fully welded	-	2.70	48.91	138.41	nr	**187.32**
Inspection Length	-	0.90	16.30	11.01	nr	**27.31**
Drain Plug and Support	-	1.00	18.11	57.94	nr	**76.06**
Damper	-	0.90	16.30	47.10	nr	**63.40**
Angled Flashing including Storm Collar	-	1.25	22.64	68.17	nr	**90.81**
Stub Terminal	-	1.00	18.11	20.66	nr	**38.77**
Tapered Terminal	-	1.00	18.11	43.15	nr	**61.26**
Floor Support (2 piece)	-	1.50	27.17	37.48	nr	**64.65**
Firestop Floor Support (2 piece)	-	1.50	27.17	21.01	nr	**48.18**
Wall Support (Stainless Steel)	-	1.00	18.11	72.21	nr	**90.32**
Wall Sleeve	-	1.20	21.74	31.59	nr	**53.33**
100 mm pipes; including one locking band (fixing brackets measured separately)	-	1.00	18.11	89.47	m	**107.58**
Extra for						
Appliance Connecter	-	0.90	16.30	14.27	nr	**30.57**
30° Bend	-	2.00	36.23	65.78	nr	**102.01**
45° Bend	-	2.00	36.23	62.45	nr	**98.68**
135° Tee; fully welded	-	3.00	54.34	142.34	nr	**196.68**
Inspection Length	-	1.00	18.11	204.82	nr	**222.93**
Drain Plug and Support	-	1.10	19.92	59.99	nr	**79.91**
Damper	-	1.00	18.11	52.05	nr	**70.17**
Angled Flashing including Storm Collar	-	1.40	25.36	68.66	nr	**94.02**
Stub Terminal	-	1.10	19.92	21.00	nr	**40.92**
Tapered Terminal	-	1.10	19.92	45.95	nr	**65.87**
Floor Support (2 piece)	-	1.65	29.89	37.48	nr	**67.37**
Firestop Floor Support (2 piece)	-	1.65	29.89	21.01	nr	**50.90**
Wall Support (Stainless Steel)	-	1.10	19.92	76.31	nr	**96.23**
Wall Sleeve	-	1.35	24.45	31.59	nr	**56.05**

T MECHANICAL HEATING/COOLING SYSTEMS ETC

Item	PC £	Labour hours	Labour £	Material £	Unit	Total rate £
T10 GAS/OIL FIRED BOILERS – cont'd						
Flues – cont'd						
Scheidel Rite-Vent ICS Plus flue system – cont'd						
150 mm pipes; including one locking band (fixing brackets measured separately)	-	1.10	19.92	104.93	m	**124.85**
Extra for						
Appliance Connecter	-	1.00	18.11	18.40	nr	**36.51**
30° Bend	-	2.20	39.85	78.96	nr	**118.81**
45° Bend	-	2.20	39.85	75.08	nr	**114.92**
135° Tee; fully welded	-	3.30	59.77	186.18	nr	**245.95**
Inspection Length	-	1.10	19.92	214.80	nr	**234.72**
Drain Plug and Support	-	1.20	21.74	76.61	nr	**98.34**
Damper	-	1.10	19.92	64.83	nr	**84.75**
Angled Flashing including Storm Collar	-	1.55	28.08	70.57	nr	**98.65**
Stub Terminal	-	1.20	21.74	22.59	nr	**44.32**
Tapered Terminal	-	1.20	21.74	52.81	nr	**74.55**
Floor Support (2 piece)	-	1.80	32.60	38.67	nr	**71.28**
Firestop Floor Support (2 piece)	-	1.80	32.60	21.01	nr	**53.61**
Wall Support (Stainless Steel)	-	1.20	21.74	84.50	nr	**106.24**
Wall Sleeve	-	1.50	27.17	32.53	nr	**59.70**
T31 LOW TEMPERATURE HOT WATER HEATING						
NOTE: The reader is referred to section "S10/S11 Hot and Cold Water" for rates for copper pipework which will equally apply to this section of work. For further and more detailed information the reader is advised to consult *Spon's Mechanical and Electrical Services Price Book.*						
Radiators; Hudevad Heat Emitters or other equal and approved						
Plan Fiona double panel convector; 600 mm high; front, back plates and convector fins with intergrated top grille; wheelhead and lockshield valves						
500 mm long x 68 mm deep; 584 watts output	76.61	1.85	38.41	95.60	nr	**134.01**
1400 mm long x 68 mm deep; 1634 watts output	185.47	2.15	44.64	212.77	nr	**257.40**
1400 mm long x 98 mm deep; 2022 watts output	207.52	2.15	44.64	236.50	nr	**281.14**
P5K horizontal single panel convector; 600 mm high; wheelhead and lockshield valves						
500 mm long; 412 watts output	59.37	1.75	36.33	77.05	nr	**113.38**
1400 mm long; 1154 watts output	121.67	2.15	44.64	144.10	nr	**188.73**
2000 mm long; 1648 watts output	163.32	2.40	49.83	188.92	nr	**238.75**
P5KV vertical single panel convector; 600 mm long; wheelhead and lockshield valves						
1400 mm high; 960 watts output	139.87	2.40	49.83	163.69	nr	**213.52**
2200 mm high; 1492 watts output	195.31	2.60	53.98	223.36	nr	**277.34**

V ELECTRICAL SYSTEMS

Item	PC £	Labour hours	Labour £	Material £	Unit	Total rate £
V21/V22 GENERAL LIGHTING AND LV POWER						
NOTE: The following items indicate approximate prices for wiring of lighting and power points complete, including accessories and socket outlets, but excluding lighting fittings. Consumer control units are shown separately. For a more detailed breakdown of these costs and specialist costs for a complete range of electrical items, reference should be made to *Spon's Mechanical and Electrical Services Price Book.*						
Consumer control units						
8-way 60 amp SP&N surface mounted insulated consumer control units fitted with miniature circuit breakers including 2.00 m long 32 mm screwed welded conduit with three runs of 16 mm2 PVC cables ready for final connections	-	-	-	-	nr	**195.88**
extra for current operated ELCB of 30 mA tripping current	-	-	-	-	nr	**78.35**
As above but 100 amp metal cased consumer unit and 25 mm2 PVC cables	-	-	-	-	nr	**218.26**
extra for current operated ELCB of 30 mA tripping current	-	-	-	-	nr	**179.09**
Final circuits						
Lighting points						
wired in PVC insulated and PVC sheathed cable in flats and houses; insulated in cavities and roof space; protected where buried by heavy gauge PVC conduit	-	-	-	-	nr	**44.77**
as above but in commercial property	-	-	-	-	nr	**61.56**
wired in PVC insulated cable in screwed welded conduit in commercial property	-	-	-	-	nr	**190.28**
as above but in industrial property	-	-	-	-	nr	**207.07**
wired in MICC cable in commercial property	-	-	-	-	nr	**167.90**
as above but in industrial property with PVC sheathed cable	-	-	-	-	nr	**167.90**
Single 13 amp switched socket outlet points						
wired in PVC insulated and PVC sheathed cable in flats and houses on a ring main circuit; protected where buried by heavy gauge PVC conduit	-	-	-	-	nr	**72.75**
as above but in commercial property	-	-	-	-	nr	**83.95**
wired in PVC insulated cable in screwed welded conduit in commercial property	-	-	-	-	nr	**195.88**
as above but in industrial property	-	-	-	-	nr	**218.26**
wired in MICC cable on a ring main ciircuit in commercial property	-	-	-	-	nr	**212.67**
as above but in industrial property with PVC sheathed cable	-	-	-	-	nr	**212.67**
Cooker control units						
45 amp circuit including unit wired in PVC insulated and PVC sheathed cable; protected where buried by heavy gauge PVC conduit	-	-	-	-	nr	**106.33**
as above but wired in PVC insulated cable in screwed welded conduit	-	-	-	-	nr	**246.25**
as above but wired in MICC cable	-	-	-	-	nr	**268.63**

W SECURITY SYSTEMS

Item	PC £	Labour hours	Labour £	Material £	Unit	Total rate £
W20 LIGHTNING PROTECTION						
Lightning protection equipment						
Copper strip roof or down conductors fixed with bracket or saddle clips						
20 mm x 3 mm flat section	-	-	-	-	m	**20.07**
25 mm x 3 mm flat section	-	-	-	-	m	**23.42**
Aluminium strip roof or down conductors fixed with bracket or saddle clips						
20 mm x 3 mm flat section	-	-	-	-	m	**14.73**
25 mm x 3 mm flat section	-	-	-	-	m	**16.07**
Joints in tapes	-	-	-	-	nr	**11.39**
Bonding connections to roof and structural metalwork	-	-	-	-	nr	**66.93**
Testing points	-	-	-	-	nr	**55.00**
Earth electrodes						
16 mm diameter driven copper electrodes in 1220 mm long sectional lengths (minimum 2440 mm long overall)	-	-	-	-	nr	**174.00**
first 2440 mm length driven and tested 25 mm x 3 mm copper strip electrode in 457 mm deep prepared trench	-	-	-	-	m	**13.39**

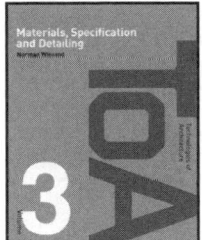

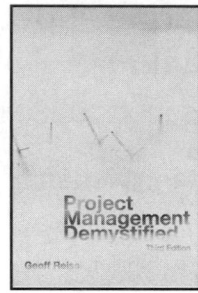

PRICES FOR MEASURED WORKS - MINOR WORKS

INTRODUCTION

The "Prices for Measured Works - Minor Works" are intended to apply to a small project in the outer London area costing about £160,000 (including Preliminaries).

The format of this section follows that of the "Major Works" section with minor variations because of the different nature of the work, and reference should be made to the "Introduction" to that section on page 169.

It has been assumed that reasonable quantities of work are involved, equivalent to quantities for two houses, although clearly this would not apply to all trades and descriptions of work in a project of this value. Where smaller quantities of work are involved it will be necessary to adjust the prices accordingly.

For section "C Demolition/Alteration/Renovation" even smaller quantities have been assumed as can be seen from the stated "PC" of the materials involved.

Where work in an existing building is concerned it has been assumed that the building is vacated and that in all cases there is reasonable access and adequate storage space. Should this not be the case, and if any abnormal circumstances have to be taken into account, an allowance can be made either by a lump sum addition or by suitably modifying the Main Contractor's percentage factor for overheads and profit. Built-up prices include an allowance of 7½% for overheads and 5% for profit, whereas non-analysed sub-contractor prices only include mark-up of 2½% for profit.

Labour rates are based upon typical gang costs divided by the number of primary working operatives for the trade concerned; and for general building work include an allowance for trade supervision, overheads and profit. The "Labour hours" column gives the total hours allocated to a particular item and the "Labour £" the consolidated cost of such labour. "Labour hours" have not always been given for "spot" items because of the inclusion of Sub-Contractor's labour.

The "Material Plant £" column includes the cost of removal of debris by skips or lorries. Alternative materials prices tables can be found in the appropriate "Prices for Measured Works - Major Works" section. As stated earlier, these prices are "list" prices before deduction of quantity discounts, and therefore require 'discount' adjustment before they can be substituted in place of "PC" figures given for Measured Work items. The reader should bear in mind that although large orders are delivered free of charge, smaller orders generally attract a delivery or part load charge and this should be added to the alternative material price prior to substitution in a rate.

No allowance has been made for any Value Added Tax which will probably be payable on the majority of work of this nature.

A PRELIMINARIES/CONTRACT CONDITIONS FOR MINOR WORKS

When pricing Preliminaries all factors affecting the execution of the works must be considered; some of the more obvious have already been mentioned above.

As mentioned in "A Preliminaries" in the "Prices for Measured Work - Major Works" section (page 175), the current trend is for Preliminaries to be priced at between 15% and 19%, but for alterations and additions work in particular, care must be exercised in ensuring that all adverse factors are covered. The reader is advised to identify systematically and separately price all preliminary items with cost/time implications in order to reflect as accurately as possible preliminary costs likely to stem from any particular scheme.

Where the Standard Form of Contract applies two clauses which will affect the pricing of Preliminaries should be noted.

(a) Insurance of the works against Section 6 Perils

(b) Fluctuations
An allowance for any shortfall in recovery of increased costs under whichever clause is contained in the Contract may be covered by the inclusion of a lump sum in the Preliminaries or by increasing the prices by a suitable percentage

ADDITIONS AND NEW WORKS WITHIN EXISTING BUILDINGS

Depending upon the contract size either the prices in "Prices for Measured Work - Major Works" or those prices in "Prices for Measured Work - Minor Works" will best apply.

It is likely, however, that conditions affecting the excavations for foundations might preclude the use of mechanical plant, and that it will be necessary to restrict prices to those applicable to hand excavation.

If, in any circumstances, less than what might be termed "normal quantities" are likely to be involved it is stressed that actual quotations should be invited from specialist Sub-contractors for these works.

JOBBING WORK

Jobbing work is outside the scope of this section and no attempt has been made to include prices for such work.

B NEW ITEMS

Item	PC £	Labour hours	Labour £	Material £	Unit	Total rate £
New items are also included in their appropriate work section						
D41 CRIB WALLS/GABIONS/REINFORCED EARTHWORKS						
Gabion baskets						
Wire mesh gabion baskets; Maccaferri Ltd or other equal and approved; galvanised mesh 80 mm x 100 mm; filling with broken stones 125 mm - 200 mm size						
2.00 x 1.00 x 0.50 m	17.18	1.25	28.28	117.27	nr	**145.55**
2.00 x 1.00 x 0.50 m; pvc coated	21.88	1.25	28.28	122.84	nr	**151.11**
2.00 x 1.00 x 0.50 m	24.08	2.50	56.55	222.37	nr	**278.92**
2.00 x 1.00 x 0.50 m; pvc coated	30.83	2.50	56.55	230.37	nr	**286.92**
Reno mattress gabion baskets or other equal and approved; Maccaferri Ltd; filling with broken stones 125 mm - 200 mm size						
6.00 x 2.00 x 0.17 m	66.33	2.50	56.55	267.37	nr	**323.92**
6.00 x 2.00 x 0.23 m	72.20	3.00	67.86	340.48	nr	**408.34**
6.00 x 2.00 x 0.30 m	83.36	3.50	79.17	425.32	nr	**504.49**
E30 REINFORCEMENT FOR IN SITU CONCRETE						
Bars; stainless steel; LDX2101® (EN 1.4362)						
NOTE: LDX2101® (EN 1.4362) is a new low Ni, Mn bearing stainless steel alloy, which offers greater price stability and cost effectiveness, and is expected to be adopted into the British Standard in the near future.						
32 mm diameter nominal size						
straight	3350.00	18.00	359.01	3857.32	tonne	**4216.33**
bent	3500.00	18.00	359.01	4026.63	tonne	**4385.64**
25 mm diameter nominal size						
straight	3350.00	20.00	398.90	3853.46	tonne	**4252.36**
bent	3500.00	20.00	398.90	4022.77	tonne	**4421.67**
20 mm diameter nominal size						
straight	3350.00	22.00	438.79	3858.71	tonne	**4297.50**
bent	3450.00	22.00	438.79	3971.59	tonne	**4410.38**
16 mm diameter nominal size						
straight	3350.00	24.00	478.68	3868.39	tonne	**4347.07**
bent	3450.00	24.00	478.68	3981.26	tonne	**4459.94**
12 mm diameter nominal size						
straight	3350.00	26.00	518.57	3878.06	tonne	**4396.63**
bent	3400.00	26.00	518.57	3934.50	tonne	**4453.07**
10 mm diameter nominal size						
straight	3350.00	28.00	558.46	3889.13	tonne	**4447.59**
bent	3400.00	28.00	558.46	3945.57	tonne	**4504.03**
8 mm diameter nominal size						
straight	3350.00	30.00	593.77	3898.80	tonne	**4492.57**
bent	3450.00	30.00	593.77	4011.68	tonne	**4605.44**

B NEW ITEMS

Item	PC £	Labour hours	Labour £	Material £	Unit	Total rate £
F30 ACCESSORIES/SUNDRY ITEMS FOR BRICK/BLOCK WALLING						
Damp proof courses						
"Engerseal" polymer elastomeric damp proof course or other equal and approved; 200 mm laps; in gauged morter (1:1:6)						
width exceeding 225 mm; horizontal	3.16	0.27	7.41	3.83	m²	**11.25**
width exceeding 225 mm; forming cavity gutters in hollow walls	-	0.43	11.80	3.83	m²	**15.64**
width not exceeding 225 mm; horizontal	-	0.53	14.55	3.83	m²	**18.38**
width not exceeding 225 mm; horizontal	-	0.80	21.96	3.83	m²	**25.80**
"Zedex CPT" (Co-Polymer Thermoplastic) damp proof course or other equal and approved; 200 mm laps; in gauged mortar (1:1:6)						
width exceeding 225 mm; horizontal	4.34	0.27	7.41	5.27	m²	**12.68**
width exceeding 225 mm; forming cavity gutters in hollow walls	-	0.43	11.80	5.27	m²	**17.07**
width not exceeding 225 mm; horizontal	-	0.53	14.55	5.27	m²	**19.82**
width not exceeding 225 mm; vertical	-	0.80	21.96	5.27	m²	**27.23**
"Alumite" aluminium cored bitumen gas retardent damp proof course or other equal and approved; 200 mm laps; in gauged mortar (1:1;6)						
width exceeding 225 mm; horizontal	6.40	0.35	9.61	7.77	m²	**17.37**
width exceeding 225 mm wide; forming cavity gutters in hollow walls; horizontal	-	0.56	15.37	7.77	m²	**23.14**
width not exceeding 225 mm; horizontal	-	0.68	18.67	7.77	m²	**26.43**
width not exceeding 225 mm; vertical	-	0.92	25.26	7.77	m²	**33.02**
Galvanised steel lintels; "Catnic" or other equal and approved; built into brickwork or blockwork						
90/125 range "CG" open back lintel for cavity wall						
750 mm long	44.74	0.27	7.41	50.55	nr	**57.96**
900 mm long	53.69	0.32	8.78	60.65	nr	**69.44**
1200 mm long	70.51	0.37	10.16	79.63	nr	**89.79**
1500 mm long	87.92	0.43	11.80	99.29	nr	**111.09**
1800 mm long	111.25	0.48	13.18	125.62	nr	**138.79**
2100 mm long	131.84	0.53	14.55	148.86	nr	**163.41**
2400 mm long	186.24	0.64	17.57	210.27	nr	**227.84**
90/125 range "CUB" open back lintel for cavity wall						
2700 mm long	294.86	0.74	20.31	332.87	nr	**353.18**
3000 mm long	327.90	0.85	23.33	370.17	nr	**393.50**
90/125 range "CU" open back lintel for cavity wall						
3300 mm long	355.99	0.95	26.08	401.87	nr	**427.95**
3600 mm long	388.36	1.06	29.10	438.40	nr	**467.50**
3900 mm long	420.42	1.17	32.12	474.59	nr	**506.71**
4200 mm long	441.21	0.53	14.55	498.06	nr	**512.61**
"CN100" single lintel for 75 mm internal wall						
1050 mm long	21.00	0.32	8.78	23.73	nr	**32.52**
1200 mm long	26.10	0.37	10.16	29.49	nr	**39.64**
"CN5XA" single lintel for 100 mm internal wall						
1050 mm long	39.67	0.32	8.78	44.80	nr	**53.58**
1200 mm long	40.70	0.37	10.16	45.96	nr	**56.12**

B NEW ITEMS

Item	PC £	Labour hours	Labour £	Material £	Unit	Total rate £
H10 PATENT GLAZING						
Patent glazing; aluminium alloy bars 2.55 m long at 622 mm centres; fixed to supports						
Roof cladding						
single glazed with 6.4 mm laminated glass	-	-	-	-	m²	**155.00**
thermally broken and double glazed with low-e clear toughened and laminated double glazed units; aluminium finished RAL matt colour	-	-	-	-	m²	**440.00**
Extra for opening roof vents						
600 mm x 900 mm top hung opening roof vent; manually operated	-	-	-	-	nr	**480.00**
600 mm x 900 mm top hung opening roof vent; electrically operated	-	-	-	-	nr	**600.00**
Skylight						
Self-supporting hipped or gable ended lantern/skylight thermally broken and double glazed with low-e clear toughened and laminated double glazed units; aluminium finished RAL matt colour	-	-	-	-	m²	**900.00**
Wall cladding						
single glazed with 6.4 mm laminated glass	-	-	-	-	m²	**160.00**
thermally broken and double glazed with low-e clear toughened and laminated double glazed units; aluminium finished RAL matt colour	-	-	-	-	m²	**460.00**
H20 RIGID SHEET CLADDING						
Prodema ProdEX high density resin-bonded cellulose fibre weatherboarding panels; including secondary supports and fixing						
Walls						
8 mm panels face fixed on to timber batten	-	-	-	-	m²	**164.00**
8 mm panels face fixed on to aluminium rails	-	-	-	-	m²	**184.50**
8 mm panels adhesive fixed on to timber battens or aluminium rails	-	-	-	-	m²	**194.75**
10 mm panels secret fixed on to helping hand aluminium system	-	-	-	-	m²	**225.50**
H60 PLAIN ROOF TILING						
Concrete interlocking tiles; Marley Eternit "Ecologic Ludlow Major" granule finish tiles or other equal and approved; 420 mm x 330 mm; to 75 mm lap; on 25 mm x 38 mm battens and type 1F reinforced underlay						
Roof coverings (PC £ per 1000)	936.00	0.37	11.53	15.09	m²	**26.62**
Extra over coverings for						
fixing every tile	-	0.02	0.62	1.05	m²	**1.67**
eaves; eaves filler	-	0.05	1.56	0.78	m	**2.34**
verges; 150 mm wide asbestos free strip undercloak	-	0.24	7.48	2.20	m	**9.68**
dry verge system; extruded white pvc	-	0.16	4.98	14.04	m	**19.02**
segmental ridge cap to dry verge	-	0.02	0.62	4.48	m	**5.10**
valley trough tiles; cutting both sides	-	0.58	18.07	31.28	m	**49.35**
segmental ridge tiles	-	0.53	16.51	11.77	m	**28.29**
segmental hip tiles; cutting both sides	-	0.69	21.50	15.10	m	**36.60**
dry ridge tiles; segmental including batten sections; unions and filler pieces	-	0.32	9.97	24.77	m	**34.73**
segmental mono-ridge tiles	-	0.53	16.51	21.07	m	**37.58**
gas ridge terminal	-	0.53	16.51	84.34	nr	**100.85**
holes for pipes and the like	-	0.21	6.54	-	nr	**6.54**

B NEW ITEMS

Item	PC £	Labour hours	Labour £	Material £	Unit	**Total rate £**
K32 FRAMED PANEL CUBICLE PARTITIONS						
Toilet cubicle partitions; Amwells or other equal and approved; standard colours and ironmongery; assembling and screwing to floor and wall						
Axis standard cubicle set; 800 mm x 1500 mm x 1980 mm high per cubicle, with polished aluminium framing; 19 mm melamine-faced chipboard divisions and doors						
One cubicle set; 2 nr panels; 1 nr door	-	3.25	143.00	308.00	nr	**451.00**
range of 3 cubicle sets; 4 nr panels; 3 nr doors	-	9.75	429.00	880.00	nr	**1309.00**
range of 6 cubicle sets; 7 nr panels; 6 nr doors	-	19.50	858.00	1738.00	nr	**2596.00**
Reduction of 1 nr panel for end unit adjoining side wall	-	-	-	-121.00	nr	-
Minima designer cubicle set; 800 mm x 1500 mm x 2100 mm high per cubicle, with satin polished stainless steel framing; 18 mm high pressure laminated (HPL) chipboard divisions and doors						
One cubicle set; 2 nr panels; 1 nr door	-	3.25	143.00	649.00	nr	**792.00**
range of 3 cubicle sets; 4 nr panels; 3 nr doors	-	9.75	429.00	1683.00	nr	**2112.00**
range of 6 cubicle sets; 7 nr panels; 6 nr doors	-	19.50	858.00	3234.00	nr	**4092.00**
Reduction of 1 nr panel for end unit adjoining side wall	-	-	-	-176.00	nr	-
Sylan corporate cubicle set; 800 mm x 1500 mm x 2400 mm high per cubicle, with sating finished stainless steel ironmongery; 30 mm high pressure laminated (HPL) chipboard divisions and 44 mm solid cored real wood veneered doors and pilasters						
One cubicle set; 2 nr panels; 1 nr door	-	5.00	220.00	1919.50	nr	**2139.50**
range of 3 cubicle sets; 4 nr panels; 3 nr doors	-	15.00	660.00	5186.50	nr	**5846.50**
range of 6 cubicle sets; 7 nr panels; 6 nr doors	-	30.00	1320.00	10081.50	nr	**11401.50**
Reduction of 1 nr panel for end unit adjoining side wall	-	-	-	-390.50	nr	-
L10 WINDOWS/ROOFLIGHTS/SCREENS/ LOUVRES						
SUPPLY ONLY PRICES						
Thermally broken composite double glazed aluminium/ timber windows; 'Velfac 200' or other approved; with a maximum glazing U value of 1.5 W/m²K; low e glazing with laminated glass unless otherwise specified; including multi point espagnolette locking mechanisms and other ironmongery						
NOTE: The following supply only prices are for standard windows, to which fixings, sealants etc. labour and overheads and profit need to ba added, before they may be used to arrive at a guide price for a complete unit.						
Outward opening standard fixed sash casement windows						
1200 mm x 1200 mm single fixed pane; low-e glass 4/16/4	-	-	-	286.00	nr	-
2200 mm x 2200 mm single fixed pane; low-e glass 6/12/6	-	-	-	726.00	nr	-

B NEW ITEMS

Item	PC £	Labour hours	Labour £	Material £	Unit	Total rate £
1200 mm x 2200 mm three fixed panes; low-e glass 4/16/4	-	-	-	572.00	nr	-
Outward opening standard sash casement windows						
1600 mm x 1600 mm with two sidehung sashes; low-e glass 4/16/4	-	-	-	583.00	nr	-
1600 mm x 1600 mm with two sidehung projecting sashes; low-e glass 4/16/4	-	-	-	649.00	nr	-
2000 mm x 1600 mm with one sidehung sash next to a tophung projecting sash over a fixed sash; low-e glass 4/16/4	-	-	-	715.00	nr	-
2000 mm x 1600 mm with one sidehung sash next to a tophung projecting sash over a fixed sash; low-e glass 4/16/4	-	-	-	566.50	nr	-
1200 mm x 2200 mm with fixed lower sash and fully reversible upper sash; lower low-e upper low-e glass 4 toughened/16/6.4; upper low-e glass 4/16/4	-	-	-	616.00	nr	-
Outward opening standard doors						
2200 mm x 2200 mm French casement patio door; low-e toughened glass 4/16/4	-	-	-	1094.50	nr	-
2200 mm x 2200 mm Sliding patio door; low-e glass 4 toughened/16/4 laminated	-	-	-	1364.00	nr	-
Guide price for installation:	-	1.00	65.00	-	m²	**65.00**
L40 GENERAL GLAZING						
Factory made double hermetically sealed units; with inner pane of Pilkington's "K" low emissivity coated glass; to wood or metal with screwed or clipped beads						
Two panes; BS EN 14449; clear float glass; 4 mm thick; 6 mm air space						
0.35 m² - 2.00 m²	-	-	-	-	m²	**104.63**
Two panes; BS EN 14449; clear float glass; 6 mm thick; 6 mm air space						
0.35 m² - 2.00 m²	-	-	-	-	m²	**121.84**
2.00 m² - 4.00 m²	-	-	-	-	m²	**183.21**
Factory made triple hermetically sealed units; with inner pane of Pilkington's "K" low emissivity coated glass; to wood or metal with screwed or clipped beads						
Three panes; BS EN 14449; clear float glass; 4 mm thick; 6 mm air spaces						
0.35 m² - 2.00 m²	-	-	-	-	m²	**168.36**
Three panes; BS EN 14449; clear float glass; 6 mm thick; 6 mm air spaces						
0.35 m² - 2.00 m²	-	-	-	-	m²	**204.92**
2.00 m² - 4.00 m²	-	-	-	-	m²	**294.97**

B NEW ITEMS

Item	PC £	Labour hours	Labour £	Material £	Unit	Total rate £
R12 DRAINAGE BELOW GROUND						
Accessories; grates and covers						
Polypropylene access covers and frames; supplied by Manhole Covers Ltd or other equal and approved; to suit PPIC inspection chambers; bedding and pointing in frame.						
450 mm dia; class A15	17.00	1.50	19.52	21.47	nr	**41.00**
450 mm dia; class B125; kite-marked	44.00	1.50	19.52	52.71	nr	**72.23**
Ductile iron heavy duty road gratings and frame; supplied by Manhole Covers Ltd or other equal and approved; bedding and pointing in cement and sand (1:3); one course half brick thick wall in semi-engineering bricks in cement mortar (1:3)						
420 mm x 420 x 75 mm hinged road grating and frame; ref C250; kite-marked	39.50	2.60	33.84	49.17	nr	**83.01**
445 mm x 445 x 75 mm double triangular road grating and frame; ref C250; kite-marked	42.00	2.60	33.84	52.07	nr	**85.90**
435 mm x 435 x 100 mm pedestrian mesh road grating and frame; ref D400	74.00	2.60	33.84	89.09	nr	**122.93**
MANHOLES						
Coated cast or ductile iron access covers and frames; to BS EN124; supplied by Manhole Covers Ltd or other equal and approved; bedding frame in cement and sand (1:3); cover in grease and sand						
Light duty; cast iron; rectangular single seal solid top						
750 mm x 600 mm; class A15	109.45	1.75	22.78	128.75	nr	**151.53**
Medium duty; ductile iron; rectangular single seal solid top						
450 mm x 450 mm x 40 mm; class C250; kite-marked	64.00	2.30	29.93	76.17	nr	**106.10**
600 mm x 450 mm x 40 mm; slide-out; class C250; kite-marked	79.50	2.30	29.93	94.10	nr	**124.04**
600 mm xn 600 mm x 40 mm; slide-out; class C250; kite-marked	85.00	2.30	29.93	100.47	nr	**130.40**
760 mm x 600 mm x 40 mm; slide-out; class C250; kite-marked	127.00	2.30	29.93	149.06	nr	**178.99**
Heavy duty; ductile iron; solid top						
450 mm x 450 mm x 75 mm; single seal; class C250; kite-marked	89.50	2.85	37.09	105.67	nr	**142.76**
600 mm x 450 mm x 75 mm; single seal; class C250; kite-marked	95.00	2.85	37.09	112.04	nr	**149.13**
600 mm x 600 mm x 75 mm; single seal; class C250; kite-marked	107.00	2.85	37.09	125.92	nr	**163.01**
450 mm x 450 mm x 100 mm; double triangular; class D400; kite-marked	99.00	2.85	37.09	116.66	nr	**153.75**
600 mm x 450 mm x 100 mm; double triangular; class D400; kite-marked	89.00	2.85	37.09	105.09	nr	**142.19**
750 mm x 600 mm x 100 mm; double triangular; class D400; kite-marked	188.64	2.85	37.09	220.37	nr	**257.47**
1220 mm x 675 x 100 mm; double triangular; D400; kite-marked	225.00	4.00	52.06	262.44	nr	**314.50**

B NEW ITEMS

Item	PC £	Labour hours	Labour £	Material £	Unit	Total rate £
T31 LOW TEMPERATURE HOT WATER HEATING						
Steel radiators and convectors; Hudevad Heat Emitters or other equal and approved						
"Plan Fiona" double panel convector; 600 mm high; front, back plates and convector fins with intergrated top grille; wheelhead and lockshield valves						
500 mm long x 68 mm deep; 584 watts output	98.50	2.00	43.55	129.56	nr	**173.11**
1400 mm long x 68 mm deep; 1634 watts output	238.46	2.50	54.43	287.55	nr	**341.98**
1400 mm long x 98 mm deep; 2022 watts output	266.81	2.50	54.43	319.55	nr	**373.98**
"P5K" horizontal single panel convector; 600 mm high; wheelhead and lockshield valves						
500 mm long; 412 watts output	76.33	2.00	43.55	104.68	nr	**148.22**
1400 mm long; 1154 watts output	156.43	2.50	54.43	195.09	nr	**249.52**
2000 mm long; 1648 watts output	209.98	2.75	59.88	255.53	nr	**315.41**
"P5KV" vertical single panel convector; 600 mm long; wheelhead and lockshield valves						
1400 mm high; 960 watts output	179.84	2.75	59.88	221.51	nr	**281.39**
2200 mm high; 1492 watts output	251.12	3.00	65.32	301.97	nr	**367.29**

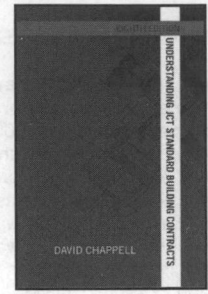

C DEMOLITION/ALTERATION/RENOVATION

Item	PC £	Labour hours	Labour £	Material £	Unit	Total rate £
C20 DEMOLITION						
NOTE: Demolition rates vary considerably from one scheme to another, depending upon access, the type of construction, the method of demolition, whether there are any redundant materials etc. Therefore, it is advisable to obtain specific quotations for each scheme under consideration. However, the following demolition rates, including the cost of removal, but excluding scaffolding costs, may be of some assistance for comparative purposes.						
Demolishing all structures						
Demolishing to ground level; single storey brick out-building; timber flat roofs; volume						
50 m³	-	-	-	-	m³	**13.63**
200 m³	-	-	-	-	m³	**10.02**
500 m³	-	-	-	-	m³	**5.10**
Demolishing to ground level; two storey brick out-building; timber joisted suspended floor and timber flat roofs; volume						
200 m³	-	-	-	-	m³	**7.65**
Demolishing parts of structures						
Breaking up concrete bed						
100 mm thick	-	0.50	8.47	3.80	m²	**12.27**
150 mm thick	-	0.74	12.53	7.11	m²	**19.64**
200 mm thick	-	1.00	16.93	7.61	m²	**24.54**
300 mm thick	-	1.48	25.06	11.16	m²	**36.21**
Breaking up reinforced concrete bed						
100 mm thick	-	0.56	9.48	4.36	m²	**13.84**
150 mm thick	-	0.83	14.05	6.42	m²	**20.47**
200 mm thick	-	1.11	18.79	8.71	m²	**27.51**
300 mm thick	-	1.67	28.28	13.07	m²	**41.35**
Demolishing reinforced concrete column or cutting away casing to steel column	-	11.10	187.94	54.44	m³	**242.38**
Demolishing reinforced concrete beam or cutting away casing to steel beam	-	12.75	215.87	58.21	m³	**274.08**
Demolishing reinforced concrete wall						
100 mm thick	-	1.11	18.79	5.37	m²	**24.17**
150 mm thick	-	1.67	28.28	7.98	m²	**36.26**
225 mm thick	-	2.50	42.33	12.02	m²	**54.35**
300 mm thick	-	3.33	56.38	16.17	m²	**72.56**
Demolishing reinforced concrete suspended slabs						
100 mm thick	-	0.93	15.75	5.08	m²	**20.82**
150 mm thick	-	1.39	23.53	7.44	m²	**30.97**
225 mm thick	-	2.08	35.22	11.15	m²	**46.37**
300 mm thick	-	2.78	47.07	15.10	m²	**62.17**
Breaking up concrete plinth; making good structures	-	4.26	72.13	34.14	m³	**106.26**
Breaking up precast concrete kerb	-	0.46	7.79	1.40	m	**9.18**
Removing precast concrete window sill; materials for re-use	-	1.48	25.06	-	m	**25.06**
Breaking up concrete hearth	-	1.67	28.28	2.33	nr	**30.60**
Demolishing external brick walls; in gauged mortar						
half brick thick	-	0.65	9.99	3.49	m²	**13.47**
two half brick thick skins	-	1.11	17.05	7.44	m²	**24.49**
one brick thick	-	1.11	17.05	7.44	m²	**24.49**
one and a half brick thick	-	1.57	24.12	11.63	m²	**35.75**
two brick thick	-	2.04	31.34	14.88	m²	**46.22**
add for plaster, render or pebbledash per side	-	0.09	1.38	0.70	m²	**2.08**

C DEMOLITION/ALTERATION/RENOVATION

Item	PC £	Labour hours	Labour £	Material £	Unit	Total rate £
C20 DEMOLITION – cont'd						
Demolishing parts of structures – cont'd						
Demolishing external brick walls; in cement mortar						
half brick thick	-	0.97	14.90	3.49	m²	**18.39**
two half brick thick skins	-	1.62	24.89	7.44	m²	**32.33**
one brick thick	-	1.67	25.66	7.44	m²	**33.10**
one and a half brick thick	-	2.27	34.87	11.63	m²	**46.50**
two brick thick	-	2.91	44.70	14.88	m²	**59.59**
add for plaster, render or pebbledash per side	-	0.09	1.38	0.70	m²	**2.08**
Demolishing internal partitions; gauged mortar						
half brick thick	-	0.97	14.90	3.49	m²	**18.39**
one brick thick	-	1.67	25.66	7.44	m²	**33.10**
one and a half brick thick	-	2.36	36.26	11.63	m²	**47.88**
75 mm blockwork	-	0.65	9.99	2.56	m²	**12.54**
90 mm blockwork	-	0.69	10.60	3.02	m²	**13.62**
100 mm blockwork	-	0.74	11.37	3.49	m²	**14.86**
115 mm blockwork	-	0.79	12.14	3.49	m²	**15.62**
125 mm blockwork	-	0.83	12.75	3.72	m²	**16.47**
140 mm blockwork	-	0.88	13.52	3.95	m²	**17.47**
150 mm blockwork	-	0.93	14.29	4.42	m²	**18.70**
190 mm blockwork	-	1.09	16.74	5.58	m²	**22.33**
215 mm blockwork	-	1.20	18.43	6.05	m²	**24.48**
255 mm blockwork	-	1.39	21.35	7.21	m²	**28.56**
add for plaster per side	-	0.09	1.38	0.70	m²	**2.08**
Demolishing internal partitions; cement mortar						
half brick thick	-	1.48	22.74	3.49	m²	**26.22**
one brick thick	-	2.45	37.64	7.44	m²	**45.08**
one and a half brick thick	-	3.42	52.54	11.63	m²	**64.17**
add for plaster per side	-	0.09	1.38	0.70	m²	**2.08**
Breaking up brick plinths	-	3.70	56.84	23.25	m³	**80.09**
Demolishing bund walls or piers in cement mortar						
one brick thick	-	1.30	19.97	7.44	m²	**27.41**
Demolishing walls to roof ventilator housing						
one brick thick	-	1.48	22.74	7.44	m²	**30.18**
Demolishing brick chimney to 300 mm below roof level; sealing off flues with slates						
680 mm x 680 mm x 900 mm high above roof	-	11.56	186.65	36.08	nr	**222.73**
add for each additional 300 height	-	2.31	37.30	6.58	nr	**43.88**
680 mm x 1030 mm x 900 mm high above roof	-	17.40	280.96	52.87	nr	**333.83**
add for each additional 300 height	-	3.46	55.81	10.74	nr	**66.56**
1030 mm x 1030 mm x 900 mm high above roof	-	26.69	431.18	81.44	nr	**512.61**
add for each additional 300 height	-	5.23	84.58	17.00	nr	**101.58**
Demolishing brick chimneys to 300 mm below roof level; sealing off flues with slates; piecing in "treated" sawn softwood rafters and making good roof coverings over to match existing (scaffolding excluded)						
680 mm x 680 mm x 900 mm high above roof	-	-	-	-	nr	**200.22**
add for each additional 300 mm height	-	-	-	-	nr	**33.15**
680 mm x 1030 mm x 900 mm high above roof	-	-	-	-	nr	**294.06**
add for each additional 300 mm height	-	-	-	-	nr	**59.45**
1030 mm x 1030 mm x 900 mm high above roof	-	-	-	-	nr	**437.98**
add for each additional 300 mm height	-	-	-	-	nr	**143.90**

C DEMOLITION/ALTERATION/RENOVATION

Item	PC £	Labour hours	Labour £	Material £	Unit	Total rate £
Removing existing chimney pots; materials for re-use; demolishing defective chimney stack to roof level; re-building using 25% new facing bricks to match existing; providing new lead flashings; parge and core flues, resetting chimney pots including flaunching in cement:mortar (scaffolding excluded)						
680 mm x 680 mm x 900 mm high above roof	-	-	-	-	nr	**450.49**
add for each additional 300 mm height	-	-	-	-	nr	**75.08**
680 mm x 1030 mm x 900 mm high above roof	-	-	-	-	nr	**681.99**
add for each additional 300 mm height	-	-	-	-	nr	**100.11**
1030 mm x 1030 mm x 900 mm high above roof	-	-	-	-	nr	**1001.09**
add for each additional 300 mm height	-	-	-	-	nr	**150.16**
Removing fireplace surround and hearth						
interior tiled	-	1.71	26.27	5.35	nr	**31.62**
cast iron; materials for re-use	-	2.87	44.09	-	nr	**44.09**
stone iron; materials for re-use	-	7.49	115.06	-	nr	**115.06**
Removing fireplace; filling in opening; plastering and extending skirtings; fixing air brick; breaking up hearth and re-screeding						
tiled	-	-	-	-	nr	**168.93**
cast iron; set aside	-	-	-	-	nr	**156.43**
stone; set aside	-	-	-	-	nr	**256.53**
Removing brick-on-edge coping; prepare walls for raising						
one brick thick	-	0.42	11.53	0.47	m	**12.00**
one and a half brick thick	-	0.56	15.37	0.70	m	**16.07**
Demolishing external stone walls in lime mortar						
300 mm thick	-	1.11	17.05	6.98	m²	**24.03**
400 mm thick	-	1.48	22.74	9.30	m²	**32.04**
600 mm thick	-	2.22	34.10	13.95	m²	**48.06**
Demolishing stone walls in lime mortar; clean off; set aside for re-use						
300 mm thick	-	1.67	25.66	2.33	m²	**27.98**
400 mm thick	-	2.22	34.10	3.02	m²	**37.13**
600 mm thick	-	3.33	51.16	4.65	m²	**55.81**
Demolishing metal partitions						
corrugated metal partition	-	0.32	4.92	0.70	m²	**5.61**
lightweight steel mesh security screen	-	0.46	7.07	1.16	m²	**8.23**
solid steel demountable partition	-	0.69	10.60	1.63	m²	**12.23**
glazed sheet demountable partition; including removal of glass	-	0.93	14.29	2.33	m²	**16.61**
Removing metal shutter door and track						
6.20 m x 4.60 m (12.60 m long track)	-	11.10	170.52	34.88	nr	**205.40**
12.40 m x 4.60 m (16.40 m long track)	-	13.88	213.23	69.76	nr	**282.99**
Removing roof timbers complete; including rafters, purlins, ceiling joists, plates, etc., (measured flat on plan)	-	0.31	5.29	2.56	m²	**7.85**
Removing softwood floor construction						
100 mm deep joists at ground level	-	0.23	3.53	0.47	m²	**4.00**
175 mm deep joists at first floor level	-	0.46	7.07	0.93	m²	**8.00**
125 mm deep joists at roof level	-	0.65	9.99	0.70	m²	**10.68**
Removing individual floor or roof members	-	0.25	4.30	0.47	m	**4.77**
Removing infected or decayed floor plates	-	0.34	5.88	0.47	m	**6.35**
Removing boarding; withdrawing nails						
25 mm thick softwood flooring; at ground floor level	-	0.34	5.69	0.70	m²	**6.38**
25 mm thick softwood flooring; at first floor level	-	0.58	9.90	0.70	m²	**10.60**
25 mm thick softwood roof boarding	-	0.68	11.63	0.70	m²	**12.33**
25 mm thick softwood gutter boarding	-	0.74	12.69	0.70	m²	**13.39**
22 mm thick chipboard flooring; at first floor level	-	0.34	5.69	0.70	m²	**6.38**

C DEMOLITION/ALTERATION/RENOVATION

Item	PC £	Labour hours	Labour £	Material £	Unit	Total rate £
C20 DEMOLITION – cont'd						
Demolishing parts of structures – cont'd						
Removing tilting fillet or roll	-	0.14	2.41	0.23	m	**2.65**
Removing fascia or barge boards	-	0.56	9.59	0.23	m	**9.83**
Demolishing softwood stud partitions; including finishings both sides etc						
solid	-	0.42	6.45	2.33	m²	**8.78**
glazed; including removal of glass	-	0.56	8.60	2.33	m²	**10.93**
Removing windows and doors; and set aside or clear away						
single door	-	0.37	10.78	0.70	nr	**11.47**
single door and frame or lining	-	0.74	21.55	1.16	nr	**22.71**
pair of doors	-	0.65	18.93	1.40	nr	**20.32**
pair of doors and frame or lining	-	1.11	32.33	2.33	nr	**34.65**
extra for taking out floor spring box	-	0.70	20.39	0.47	nr	**20.85**
casement window and frame	-	1.11	32.33	1.16	nr	**33.49**
double hung sash window and frame	-	1.57	45.72	2.33	nr	**48.05**
pair of french windows and frame	-	3.70	107.75	3.49	nr	**111.24**
Removing double hung sash window and frame; remove and store for re-use elsewhere	-	2.22	64.65	-	nr	**64.65**
Demolishing staircase; including balustrades						
single straight flight	-	3.24	94.35	23.25	m	**117.61**
dogleg flight	-	4.63	134.83	34.88	m	**169.71**
C30 SHORING/FACADE RETENTION						
NOTE: The requirements for shoring and strutting for the formation of large openings are dependant upon a number of factors, for example, the weight of the superimposed structure to be supported, the number (if any) of windows above, the number of floors and the type of roof to be strutted, whether raking shores are required, the depth to a load-bearing surface, and the duration the support is to be in place. Prices, would therefore, be best built-up by assessing the use and waste of materials and the labour involved, including getting timber from and returning to a yard, cutting away and making good, overhead and profit. This method is considered a more practical way of pricing than endeavouring to price the work on a cubic metre basis of timber used, and has been adopted in preparing the prices of the examples which follow.						
Support of structures not to be demolished						
Strutting to window openings over proposed new openings	-	0.56	12.30	8.74	nr	**21.04**
Plates, struts, braces and hardwood wedges in supports to floors and roof of opening	-	1.11	24.38	25.28	nr	**49.66**
Dead shore and needle using die square timber with sole plates, braces, hardwood wedges and steel dogs	-	27.75	609.54	107.70	nr	**717.24**
Set of two raking shores using die square timber with 50 mm thick wall piece; hardwood wedges and steel dogs; including forming holes for needles and making good	-	33.30	731.45	108.87	nr	**840.32**
Cut holes through one brick wall for die square needle and make good; including facings externally and plaster internally	-	5.56	152.96	1.84	nr	**154.80**

C DEMOLITION/ALTERATION/RENOVATION

Item	PC £	Labour hours	Labour £	Material £	Unit	Total rate £
C41 REPAIRING/RENOVATING/CONSERVING MASONRY						
Repairing/renovating plain/reinforced concrete work						
Reinstating plain concrete bed with site mixed in situ concrete; mix 20.00 N/mm² - 20 mm aggregate (1:2:4), where opening no longer required						
100 mm thick	-	0.44	7.99	10.61	m²	**18.60**
150 mm thick	-	0.72	12.73	15.92	m²	**28.65**
Reinstating reinforced concrete bed with site mixed in situ concrete; mix 20.00 N/mm² - 20 mm aggregate (1:2:4); including mesh reinforcement; where opening no longer required						
100 mm thick	-	0.66	11.72	14.07	m²	**25.78**
150 mm thick	-	0.91	15.95	19.37	m²	**35.32**
Reinstating reinforced concrete suspended floor with site mixed in situ concrete; mix 25.00 N/mm² - 20 mm aggregate (1:1.5:3); including mesh reinforcement and formwork; where opening no longer required						
150 mm thick	-	2.96	53.44	25.18	m²	**78.63**
225 mm thick	-	3.47	55.58	33.08	m²	**88.67**
300 mm thick	-	3.84	61.89	42.11	m²	**104.01**
Reinstating 150 mm x 150 mm x 150 mm perforation through concrete suspended slab; with site mixed in situ concrete; mix 20.00 N/mm² - 20 mm aggregate (1:2:4); including formwork; where opening no longer required	-	0.85	14.12	0.16	nr	**14.28**
Cleaning surfaces of concrete to receive new damp proof membrane	-	0.14	2.15	-	m²	**2.15**
Cleaning out existing minor crack and fill in with cement mortar mixed with bonding agent	-	0.31	4.76	0.78	m	**5.54**
Cleaning out existing crack to form 20 mm x 20 mm groove and fill in with fine cement mixed with bonding agent	-	0.61	9.37	3.57	m	**12.94**
Making good hole where existing pipe removed; 150 mm deep						
50 mm diameter	-	0.39	6.60	0.49	nr	**7.09**
100 mm diameter	-	0.51	8.63	0.60	nr	**9.23**
150 mm diameter	-	0.65	11.01	0.79	nr	**11.79**
Add for each additional 25 mm thick up to 300 mm thick						
50 mm diameter	-	0.08	1.35	0.09	nr	**1.44**
100 mm diameter	-	0.11	1.86	0.13	nr	**1.99**
150 mm diameter	-	0.14	2.37	0.17	nr	**2.54**
Repairing/renovating brick/blockwork						
Cutting out decayed, defective or cracked work and replacing with new common bricks; in gauged mortar (1:1:6)						
half brick thick (PC £ per 1000)	270.00	4.56	125.18	25.14	m²	**150.32**
one brick thick	-	8.88	243.77	51.81	m²	**295.57**
one and a half brick thick	-	12.58	345.34	78.47	m²	**423.81**
two brick thick	-	16.10	441.96	105.14	m²	**547.10**
individual bricks; half brick thick	-	0.28	7.69	0.40	nr	**8.09**

C DEMOLITION/ALTERATION/RENOVATION

Item	PC £	Labour hours	Labour £	Material £	Unit	Total rate £
C41 REPAIRING/RENOVATING/CONSERVING MASONRY – cont'd						
Repairing/renovating brick/blockwork – cont'd						
Cutting out decayed, defective or cracked work and replacing with new facing brickwork in gauged mortar (1:1:6); half brick thick; facing and pointing one side						
small areas; machine made facings (PC £ per1000)	380.00	6.75	185.30	34.53	m²	**219.82**
small areas; hand made facings (PC £ per1000)	600.00	6.75	185.30	51.35	m²	**236.65**
individual bricks; machine made facings (PC £ per1000)	380.00	0.42	11.53	0.54	nr	**12.07**
individual bricks; hand made facings (PC £ per1000)	600.00	0.42	11.53	0.80	nr	**12.33**
ADD or DEDUCT for variation of £10.00/1000 in PC of facing bricks; in flemish bond						
half brick thick	-	-	-	0.73	m²	**-**
Cutting out decayed, defective or cracked soldier arch and replacing with new; repointing to match existing						
machine made facings (PC £ per 1000)	380.00	1.80	49.41	8.07	m	**57.48**
hand made facings (PC £ per 1000)	600.00	1.80	49.41	12.34	m	**61.75**
Cutting out decayed, defective or cracked work in uncoursed stonework; replacing with cement:mortar to match existing						
small areas; 300 mm thick wall	-	5.18	142.20	14.49	m²	**156.68**
small areas; 400 mm thick wall	-	6.48	177.88	19.60	m²	**197.48**
small areas; 600 mm thick wall	-	9.25	253.92	29.83	m²	**283.75**
Cutting out staggered cracks and repointing to match existing along brick joints	-	0.37	10.16	-	m	**10.16**
Cutting out raking cracks in brickwork; stitching in new common bricks and repointing to match existing						
half brick thick	-	2.96	81.26	11.56	m²	**92.81**
one brick thick	-	5.41	148.51	23.74	m²	**172.25**
one and a half brick thick	-	8.09	222.08	35.30	m²	**257.38**
Cutting out raking cracks in brickwork; stitching in new facing bricks; half brick thick; facing and pointing one side to match existing						
machine made facings (PC £ per 1000)	380.00	4.44	121.88	15.07	m²	**136.96**
hand made facings (PC £ per 1000)	600.00	4.44	121.88	22.28	m²	**144.16**
Cutting out raking cracks in cavity brickwork; stitching in new common bricks one side; facing bricks the other side; both skins half brick thick; facing and pointing one side to match existing						
machine made facings (PC £ per 1000)	380.00	7.59	208.35	26.55	m²	**234.90**
hand made facings (PC £ per 1000)	600.00	7.59	208.35	33.76	m²	**242.11**
Cutting away and replacing with new cement mortar (1:3); angle fillets; 50 mm face width	-	0.23	6.31	2.29	m	**8.60**
Cutting out ends of joists and plates from walls; making good in common bricks; in cement mortar (1:3)						
175 mm deep joists; 400 mm centres (bricks PC £ per 1000)	270.00	0.60	16.47	8.06	m	**24.53**
225 mm deep joists; 400 mm centres	-	0.74	20.31	9.37	m	**29.69**
Cutting and pinning to existing brickwork ends of joists	-	0.37	10.16	-	nr	**10.16**

C DEMOLITION/ALTERATION/RENOVATION

Item	PC £	Labour hours	Labour £	Material £	Unit	Total rate £
Making good adjacent work; where intersecting wall removed						
half brick thick	-	0.28	7.69	0.76	m	**8.45**
one brick thick	-	0.37	10.16	1.53	m	**11.68**
100 blockwork	-	0.23	6.31	0.76	m	**7.08**
150 blockwork	-	0.27	7.41	0.76	m	**8.18**
215 blockwork	-	0.32	8.78	1.53	m	**10.31**
255 blockwork	-	0.36	9.88	1.53	m	**11.41**
Removing defective parapet wall; 600 mm high; with two courses of tiles and brick coping over; re-building in new facing bricks, tiles and coping stones						
one brick thick	-	6.16	169.10	62.80	m	**231.89**
Removing defective capping stones and haunching; replacing stones and re-haunching in cement:mortar to match existing						
300 mm thick wall	-	1.25	34.31	4.26	m²	**38.57**
400 mm thick wall	-	1.39	38.16	5.11	m²	**43.27**
600 mm thick wall	-	1.62	44.47	7.67	m²	**52.14**
Cleaning surfaces; moss and lichen from walls	-	0.28	4.30	-	m²	**4.30**
Cleaning surfaces; lime mortar off brickwork; sort and stack for re-use	-	9.25	142.10	-	1000	**142.10**
Repointing in cement mortar (1:1:6); to match existing						
raking out existing decayed joints in brickwork walls	-	0.69	18.94	0.76	m²	**19.70**
raking out existing decayed joints in chimney stacks	-	1.11	30.47	0.76	m²	**31.23**
raking out existing decayed joints in brickwork; re-wedging horizontal flashing	-	0.23	6.31	0.38	m	**6.70**
raking out existing decayed joints in brickwork; re-wedging stepped flashing	-	0.34	9.33	0.38	m	**9.72**
Repointing in cement:mortar (1:3); to match existing						
raking out existing decayed joints in uncoursed stonework	-	1.11	30.47	0.85	m²	**31.32**
Making good hole where small pipe removed						
102 mm brickwork	-	0.19	2.92	0.09	nr	**3.00**
215 mm brickwork	-	0.19	2.92	0.09	nr	**3.00**
327 mm brickwork	-	0.19	2.92	0.09	nr	**3.00**
440 mm brickwork	-	0.19	2.92	0.09	nr	**3.00**
100 mm blockwork	-	0.19	2.92	0.09	nr	**3.00**
150 mm blockwork	-	0.19	2.92	0.09	nr	**3.00**
215 mm blockwork	-	0.19	2.92	0.09	nr	**3.00**
255 mm blockwork	-	0.19	2.92	0.09	nr	**3.00**
Making good hole and facings one side where small pipe removed						
102 mm brickwork	-	0.19	5.22	0.85	nr	**6.07**
215 mm brickwork	-	0.19	5.22	0.85	nr	**6.07**
327 mm brickwork	-	0.19	5.22	0.85	nr	**6.07**
440 mm brickwork	-	0.19	5.22	0.85	nr	**6.07**
Making good hole where large pipe removed						
102 mm brickwork	-	0.28	4.30	0.09	nr	**4.39**
215 mm brickwork	-	0.42	6.45	0.30	nr	**6.75**
327 mm brickwork	-	0.56	8.60	0.51	nr	**9.11**
440 mm brickwork	-	0.69	10.60	0.64	nr	**11.24**
100 mm blockwork	-	0.28	4.30	0.09	nr	**4.39**
150 mm blockwork	-	0.32	4.92	0.17	nr	**5.09**
215 mm blockwork	-	0.37	5.68	0.21	nr	**5.90**
255 mm blockwork	-	0.42	6.45	0.30	nr	**6.75**

C DEMOLITION/ALTERATION/RENOVATION

Item	PC £	Labour hours	Labour £	Material £	Unit	Total rate £
C41 REPAIRING/RENOVATING/CONSERVING MASONRY – cont'd						
Repairing/renovating brick/blockwork – cont'd						
Making good hole and facings one side where large pipe removed						
half brick thick	-	0.25	6.86	0.85	nr	**7.71**
one brick thick	-	0.33	9.06	0.94	nr	**10.00**
one and a half brick thick	-	0.42	11.53	1.02	nr	**12.55**
two brick thick	-	0.50	13.73	1.36	nr	**15.09**
Making good hole where extra large pipe removed						
half brick thick	-	0.37	5.68	0.34	nr	**6.03**
one brick thick	-	0.56	8.60	0.72	nr	**9.33**
one and a half brick thick	-	0.74	11.37	1.28	nr	**12.65**
two brick thick	-	0.93	14.29	1.53	nr	**15.82**
100 mm blockwork	-	0.37	5.68	0.34	nr	**6.03**
150 mm blockwork	-	0.43	6.61	0.51	nr	**7.12**
215 mm blockwork	-	0.46	7.07	0.72	nr	**7.79**
255 mm blockwork	-	0.51	7.83	0.89	nr	**8.72**
Making good hole and facings one side where extra large pipe removed						
half brick thick	-	0.33	9.06	1.32	nr	**10.38**
one brick thick	-	0.44	12.08	1.62	nr	**13.70**
one and a half brick thick	-	0.56	15.37	2.05	nr	**17.42**
two brick thick	-	0.67	18.39	2.77	nr	**21.16**
C50 REPAIRING/RENOVATING/CONSERVING METAL						
Repairing metal						
Overhauling and repairing metal casement windows; adjusting and oiling ironmongery; bringing forward affected parts for redecoration	-	1.39	21.35	6.28	nr	**27.63**
C51 REPAIRING/RENOVATING/CONSERVING TIMBER						
Repairing timber						
Removing or punching in projecting nails; re-fixing softwood or hardwood flooring						
loose boards	-	0.14	3.08	-	m²	**3.08**
floorboards previously set aside	-	0.74	16.25	0.66	m²	**16.92**
Removing damaged softwood flooring; providing and fixing new 25 mm thick plain edge softwood boarding						
small areas	-	1.06	23.28	19.65	m²	**42.93**
individual boards 150 mm wide	-	0.28	6.15	1.93	m	**8.08**
Sanding down and resurfacing existing flooring; preparing, bodying in with shellac and wax polish						
softwood	-	-	-	-	m²	**12.50**
hardwood	-	-	-	-	m²	**15.09**
Fitting existing softwood skirting to new frames or architraves						
75 mm high	-	0.09	1.98	-	m	**1.98**
150 mm high	-	0.12	2.64	-	m	**2.64**
225 mm high	-	0.15	3.29	-	m	**3.29**

C DEMOLITION/ALTERATION/RENOVATION

Item	PC £	Labour hours	Labour £	Material £	Unit	Total rate £
Piecing in new 25 mm x 150 mm moulded softwood skirtings to match existing where old removed; bringing forward for redecoration	-	0.35	6.69	6.65	m	**13.34**
Piecing in new 25 mm x 150 mm moulded softwood skirtings to match existing where socket outlet removed; bringing forward for redecoration	-	0.20	3.81	3.71	nr	**7.52**
Easing and adjusting softwood doors, oiling ironmongery; bringing forward affected parts for redecoration	-	0.71	14.89	1.06	nr	**15.95**
Removing softwood doors, easing and adjusting; re-hanging; oiling ironmongery; bringing forward affected parts for redecoration	-	1.11	23.50	1.41	nr	**24.92**
Removing mortice lock, piecing in softwood doors; bringing forward affected parts for redecoration	-	1.02	21.82	0.83	nr	**22.65**
Fixing only salvaged softwood door	-	1.42	31.19	-	nr	**31.19**
Removing softwood doors; planing 12 mm from bottom edge; re-hanging	-	1.11	24.38	-	nr	**24.38**
Removing softwood doors; altering ironmongery; piecing in and rebating frame and door; re-hanging on opposite stile; bringing forward affected parts for redecoration	-	2.45	52.35	1.65	nr	**54.00**
Removing softwood doors to prepare for fire upgrading; removing ironmongery; replacing existing beads with 25 mm x 38 mm hardwood screwed beads; repairing minor damaged areas; re-hanging on wider butt hinges; adjusting all ironmongery; sealing around frame in cement mortar; bringing forward affected parts for redecoration (replacing glass panes not included)	-	4.85	104.48	18.70	nr	**123.19**
Upgrading and facing up one side of flush doors with 9 mm thick "Supalux"; screwing	-	1.16	25.48	39.77	nr	**65.25**
Upgrading and facing up one side of softwood panelled doors with 9 mm thick "Supalux"; screwing; plasterboard infilling to recesses	-	2.50	54.91	42.16	nr	**97.07**
Taking off existing softwood doorstops; providing and screwing on new 25 mm x 38 mm doorstop; bringing forward for redecoration	-	0.20	3.81	2.72	nr	**6.53**
Cutting away defective 75 mm x 100 mm softwood external door frames; providing and splicing in new piece 300 mm long; bedding in cement mortar (1:3); pointing one side; bringing forward for redecoration	-	1.30	27.38	9.06	nr	**36.44**
Sealing roof trap flush with ceiling	-	0.56	12.30	4.57	nr	**16.87**
Forming opening 762 mm x 762 mm in existing ceiling for new standard roof trap comprising softwood linings, architraves and 6 mm thick plywood trap doors; trimming ceiling joists (making good to ceiling plaster not included)	-	2.50	54.91	76.37	nr	**131.29**
Easing and adjusting softwood casement windows, oiling ironmongery; bringing forward affected parts for redecoration	-	0.48	9.84	0.71	nr	**10.55**
Removing softwood casement windows; easing and adjusting; re-hanging; oiling ironmongery; bringing forward affected parts for redecoration	-	0.71	14.89	0.71	nr	**15.60**
Renewing solid mullion jambs or transoms of softwood casement windows to match existing; bringing forward affected parts for redecoration (taking off and re-hanging adjoining casements not included)	-	2.59	55.13	24.00	nr	**79.14**

C DEMOLITION/ALTERATION/RENOVATION

Item	PC £	Labour hours	Labour £	Material £	Unit	Total rate £
C51 REPAIRING/RENOVATING/CONSERVING TIMBER – cont'd						
Repairing timber – cont'd						
Temporary linings 6 mm thick plywood infill to window while casement under repair	-	0.74	16.25	5.92	nr	**22.17**
Overhauling softwood double hung sash windows; easing, adjusting and oiling pulley wheels; re-hanging sashes on new hemp sash lines; re-assembling; bringing forward affected parts for redecoration	-	2.45	52.94	6.46	nr	**59.40**
Cutting away defective parts of softwood window sills; providing and splicing in new 75 mm x 100 mm weathered and throated pieces 300 mm long; bringing forward affected parts for redecoration	-	1.90	40.86	14.72	nr	**55.58**
Renewing broken stair nosings to treads or landings	-	1.67	36.68	3.50	nr	**40.18**
Cutting out infected or decayed structural members; shoring up adjacent work; providing and fixing new "treated" sawn softwood members pieced in						
Floors or flat roofs						
50 mm x 125 mm	-	0.37	8.13	3.36	m	**11.48**
50 mm x 150 mm	-	0.41	9.01	4.04	m	**13.04**
50 mm x 175 mm	-	0.44	9.66	4.88	m	**14.54**
Pitched roofs						
38 mm x 100 mm	-	0.33	7.25	1.95	m	**9.20**
50 mm x 100 mm	-	0.42	9.23	2.60	m	**11.83**
50 mm x 125 mm	-	0.46	10.10	3.22	m	**13.33**
50 mm x 150 mm	-	0.51	11.20	3.84	m	**15.04**
Kerbs bearers and the like						
50 mm x 75 mm	-	0.42	9.23	2.06	m	**11.29**
50 mm x 100 mm	-	0.52	11.42	2.60	m	**14.02**
75 mm x 100 mm	-	0.63	13.84	3.80	m	**17.64**
Scarfed joint; new to existing; over 450 mm²	-	0.93	20.43	-	nr	**20.43**
Scarfed and bolted joint; new to existing; including bolt let in flush; over 450 mm²	-	1.34	29.43	2.22	nr	**31.66**
C52 FUNGUS/BEETLE ERADICATION						
Treating existing timber						
Removing cobwebs, dust and roof insulation; de-frass; treat exposed joists/rafters with two coats of proprietary insecticide and fungicide; by spray application	-	-	-	-	m²	**12.32**
Treating boarding with two coats of proprietary insecticide and fungicide; by spray application	-	-	-	-	m²	**6.42**
Treating individual timbers with two coats proprietary insecticide and fungicide; by brush application						
boarding	-	-	-	-	m²	**6.42**
structural members	-	-	-	-	m²	**6.42**
skirtings	-	-	-	-	m	**6.42**
Lifting necessary floorboards; treating floors with two coats proprietary insecticide and fungicide; by spray application; re-fixing boards	-	-	-	-	m²	**11.45**
Treating surfaces of adjoining concrete or brickwork with two coats of dry rot fluid; by spray application	-	-	-	-	m²	**6.42**

C DEMOLITION/ALTERATION/RENOVATION

Item	PC £	Labour hours	Labour £	Material £	Unit	Total rate £
C90 ALTERATIONS - SPOT ITEMS						
Composite "spot" items						
NOTE: Few exactly similar composite items of alteration works are encountered on different schemes; for this reason it is considered more accurate for the reader to build up the value of such items from individual prices in the following section. However, for estimating purposes, the following "spot" items have been prepared. Prices include for removal of debris from site but do not include for shoring, scaffolding or re-decoration, except where stated.						
Removing fittings and fixtures						
Removing shelves, window boards and the like	-	0.31	4.76	0.23	m	**5.00**
Removing handrails and balustrades						
tubular handrailing and brackets	-	0.28	4.30	0.23	m	**4.53**
metal balustrades	-	0.46	7.07	0.70	m	**7.76**
Removing handrails and brackets	-	0.09	2.62	0.70	m	**3.32**
Removing sloping timber ramps in corridors; at changes of levels	-	1.85	53.88	3.49	nr	**57.36**
Removing bath panels and bearers	-	0.37	10.78	1.16	nr	**11.94**
Removing kitchen fittings						
wall units	-	0.42	12.23	3.49	nr	**15.72**
floor units	-	0.28	8.15	5.12	nr	**13.27**
larder units	-	0.37	10.78	11.63	nr	**22.40**
built-in cupboards	-	1.39	40.48	23.25	nr	**63.73**
Removing bathroom fittings; making good works disturbed						
toilet roll holder or soap dispenser	-	0.28	4.30	-	nr	**4.30**
towel holder	-	0.56	8.60	-	nr	**8.60**
mirror	-	0.60	9.22	-	nr	**9.22**
Removing pipe casings	-	0.28	8.15	0.93	m	**9.08**
Removing ironmongery; in preparation for re-decoration; and subsequently re-fixing; including providing any new screws necessary	-	0.23	6.70	0.23	nr	**6.93**
Removing, withdrawing nails, etc; making good holes						
carpet fixing strip from floors	-	0.04	0.61		m	**0.61**
curtain track from head of window	-	0.23	3.53	-	m	**3.53**
nameplates or numerals from face of door	-	0.46	7.07	-	nr	**7.07**
fly screen and frame from window	-	0.83	12.75	-	nr	**12.75**
small notice board and frame from walls	-	0.82	12.60	-	nr	**12.60**
fire extinguisher and bracket from walls	-	1.16	17.82	-	nr	**17.82**
Removing plumbing and engineering installations						
Removing sanitary fittings and supports; temporarily capping off services; to receive new (not included)						
sink or lavatory basin	-	0.93	16.10	13.95	nr	**30.06**
bath	-	1.85	32.05	20.93	nr	**52.98**
WC suite	-	1.39	24.08	13.95	nr	**38.03**

C DEMOLITION/ALTERATION/RENOVATION

Item	PC £	Labour hours	Labour £	Material £	Unit	Total rate £
C90 ALTERATIONS - SPOT ITEMS – cont'd						
Removing plumbing and engineering installations – cont'd						
Removing sanitary fittings and supports, complete with associated services, overflows and waste pipes; making good all holes and other works disturbed; bringing forward all surfaces ready for re-decoration						
sink or lavatory basin	-	3.70	64.11	17.86	nr	**81.97**
range of three lavatory basins	-	7.40	128.22	32.53	nr	**160.75**
bath	-	5.55	96.16	24.60	nr	**120.77**
WC suite	-	7.40	128.22	33.48	nr	**161.70**
2 stall urinal	-	14.80	256.44	34.49	nr	**290.92**
3 stall urinal	-	22.20	384.66	70.58	nr	**455.24**
4 stall urinal	-	29.60	512.88	104.48	nr	**617.36**
Removing taps	-	0.09	1.71	-	nr	**1.71**
Clearing blocked wastes without dismantling						
sinks	-	0.46	10.02	-	nr	**10.02**
WC traps	-	0.56	12.19	-	nr	**12.19**
Removing gutterwork and supports						
uPVC or asbestos	-	0.28	4.30	0.23	m	**4.53**
cast iron	-	0.32	4.92	0.47	m	**5.38**
Overhauling sections of rainwater gutterings; cutting out existing joints; adjusting brackets to correct falls; re-making joints						
100 mm diameter uPVC	-	0.23	5.05	0.03	m	**5.08**
100 mm diameter cast iron including bolt	-	0.83	18.23	0.14	m	**18.37**
Removing rainwater heads and supports						
uPVC or asbestos	-	0.27	4.15	0.23	nr	**4.38**
cast iron	-	0.37	5.68	0.47	nr	**6.15**
Removing pipework and supports						
uPVC or asbestos rainwater stack	-	0.28	4.30	0.23	m	**4.53**
cast iron rainwater stack	-	0.32	4.92	0.47	m	**5.38**
cast iron jointed soil stack	-	0.56	8.60	0.47	m	**9.07**
copper or steel water or gas pipework	-	0.14	2.15	0.23	m	**2.38**
cast iron rainwater shoe	-	0.07	1.08	0.23	m	**1.31**
Overhauling and re-making leaking joints in pipework						
100 mm diameter upvc	-	0.19	2.92	0.06	nr	**2.98**
100 mm diameter cast iron including bolt	-	0.74	11.37	0.28	nr	**11.64**
Cleaning out existing rainwater installations						
rainwater gutters	-	0.07	1.08	-	m	**1.08**
rainwater gully	-	0.19	2.92	-	nr	**2.92**
rainwater stack; including head, swan-neck and shoe (not exceeding 10 m long)	-	0.69	10.60	-	nr	**10.60**
Removing the following equipment and ancillaries; capping off services; making good works disturbed (excluding any draining down of system)						
expansion tank; 900 mm x 450 mm x 900 mm	-	1.67	28.93	9.30	nr	**38.23**
hot water cylinder; 450 mm diameter x 1050 mm high	-	1.11	19.23	3.95	nr	**23.19**
cold water tank; 1540 mm x 900 mm x 900 mm	-	2.22	38.47	28.60	nr	**67.07**
cast iron radiator	-	1.85	32.05	7.44	nr	**39.50**
gas water heater	-	3.70	64.11	4.65	nr	**68.76**
gas fire	-	1.85	32.05	6.05	nr	**38.10**
Removing cold water tanks and housing on roof; stripping out and capping off all associated piping; making good works disturbed and roof finishings						
1540 mm x 900 mm x 900 mm	-	11.10	192.33	35.58	nr	**227.91**

C DEMOLITION/ALTERATION/RENOVATION

Item	PC £	Labour hours	Labour £	Material £	Unit	Total rate £
Turning off supplies; dismantling the following fittings; replacing washers; re-assembling and testing						
15 mm diameter tap	-	0.23	5.01		nr	**5.01**
15 mm diameter ball valve	-	0.32	6.97	-	nr	**6.97**
Turning off supplies; removing the following fittings; testing and replacing						
15 mm diameter ball valve	-	0.46	10.02	6.79	nr	**16.80**
Removing lagging from pipès						
up to 42 mm diameter	-	0.09	1.38	0.23	nr	**1.62**
Removing finishings						
Removing plasterboard wall finishings	-	0.37	5.68	-	m²	**5.68**
Removing wall finishings; cutting out and making good cracks						
plasterboard wall finishing	-	0.37	5.68	-	m²	**5.68**
decorative wallpaper and lining	-	0.19	2.92	1.47	m²	**4.39**
heavy wallpaper and lining	-	0.32	4.92	1.47	m²	**6.38**
Hacking off wall finishings						
plaster	-	0.19	2.92	1.16	m²	**4.08**
cement rendering or pebbledash	-	0.37	5.68	1.16	m²	**6.85**
wall tiling and screed	-	0.46	7.07	1.86	m²	**8.93**
Removing wall linings; including battening behind						
plain sheeting	-	0.28	4.30	0.93	m²	**5.23**
matchboarding	-	0.37	5.68	1.40	m²	**7.08**
Removing oak dado wall panel finishings; cleaning off and setting aside for re-use	-	0.60	17.47	-	m²	**17.47**
Removing defective or damaged plaster wall finishings; re-plastering walls with two coats of gypsum plaster; including dubbing out; jointing new to existing						
small areas	-	1.48	33.48	8.16	m²	**41.64**
isolated areas not exceeding 0.50 m²	-	1.06	23.98	4.10	nr	**28.08**
Making good plaster wall finishings with two coats of gypsum plaster where wall or partition removed; dubbing out; trimming back existing and fair jointing to new work						
150 mm wide	-	0.60	13.57	1.23	m	**14.80**
225 mm wide	-	0.74	16.74	1.83	m	**18.57**
300 mm wide	-	0.88	19.91	2.46	m	**22.37**
Removing defective or damaged damp plaster wall finishings, investigating and treating wall; re-plastering walls with two coats of "Thistle Renovating" plaster; including dubbing out; fair jointing to existing work						
small areas	-	1.53	34.61	6.01	m²	**40.62**
isolated areas not exceeding 0.50 m²	-	1.13	25.56	3.01	m²	**28.58**
Dubbing out in cement and sand; average 13 mm thick						
over 300 mm wide	-	0.46	10.41	1.11	m²	**11.51**
Making good plaster wall finishings with plasterboard and skim where wall or partition removed; trimming back existing and fair joint to new work						
150 mm wide	-	0.69	15.61	1.93	m	**17.54**
225 mm wide	-	0.83	18.77	2.24	m	**21.02**
300 mm wide	-	0.93	21.04	2.55	m	**23.59**
Cutting out; making good cracks in plaster wall finishings						
walls	-	0.23	5.20	1.47	m	**6.67**
ceilings	-	0.31	7.01	1.47	m	**8.48**

C DEMOLITION/ALTERATION/RENOVATION

Item	PC £	Labour hours	Labour £	Material £	Unit	Total rate £
C90 ALTERATIONS - SPOT ITEMS – cont'd						
Removing finishings – cont'd						
Making good plaster wall finishings where items removed or holes left						
small pipe or conduit	-	0.06	1.36	0.73	nr	**2.09**
large pipe	-	0.09	2.04	1.47	nr	**3.50**
extra large pipe	-	0.14	3.17	1.16	nr	**4.32**
small recess; eg. electrical switch point	-	0.07	1.58	0.25	nr	**1.83**
Making good plasterboard and skim wall finishings where items removed or holes left						
small pipe or conduit	-	0.06	1.36	0.73	nr	**2.09**
large pipe	-	0.21	4.75	0.66	nr	**5.41**
extra large pipe	-	0.28	6.33	0.86	nr	**7.20**
Removing floor finishings						
carpet and underfelt	-	0.11	1.69	-	m²	**1.69**
linoleum sheet flooring	-	0.09	1.38	-	m²	**1.38**
carpet gripper	-	0.02	0.31	-	m	**0.31**
Removing floor finishings; preparing screed to receive new						
carpet and underfelt	-	0.61	9.37	-	m²	**9.37**
vinyl or thermoplastic tiles	-	0.79	12.14	-	m²	**12.14**
Removing woodblock floor finishings; cleaning off and setting aside for re-use	-	0.69	10.60	-	m²	**10.60**
Breaking up floor finishings						
floor screed	-	0.60	9.22	-	m²	**9.22**
granolithic flooring and screed	-	0.79	12.14	-	m²	**12.14**
terrazzo or ceramic floor tiles and screed	-	0.97	14.90	-	m²	**14.90**
Levelling and repairing floor finishings screed; 5 mm thick						
screed; 5 mm thick; in small areas	-	0.46	10.41	10.15	m²	**20.55**
screed; 5 mm thick; in isolated areas not exceeding 0.50 m²	-	0.32	7.24	5.07	m²	**12.31**
Removing softwood skirtings, picture rails, dado rails, architraves and the like	-	0.09	1.38	-	m	**1.38**
Removing softwood skirtings; cleaning off and setting aside for re-use in making good	-	0.23	3.53	-	m	**3.53**
Breaking up paving						
asphalt	-	0.56	8.60	-	m²	**8.60**
Removing ceiling finishings						
plasterboard and skim; withdrawing nails	-	0.28	4.30	0.70	m²	**5.00**
wood lath and plaster; withdrawing nails	-	0.46	7.07	1.16	m²	**8.23**
suspended ceilings	-	0.69	10.60	1.16	m²	**11.76**
plaster moulded cornice; 25 mm girth	-	0.14	2.15	0.23	m	**2.38**
Removing part of plasterboard ceiling finishings to facilitate insertion of new steel beam	-	1.02	15.67	1.40	m	**17.07**
Removing ceiling linings; including battening behind						
plain sheeting	-	0.42	6.45	0.93	m²	**7.38**
matchboarding	-	0.56	8.60	1.40	m²	**10.00**
Removing defective or damaged ceiling plaster finishings; removing laths or cutting back boarding; preparing and fixing new plasterboard; applying one skim coat of gypsum plaster; fair jointing new to existing						
small areas	-	1.57	35.51	5.89	m²	**41.41**
isolated areas not exceeding 0.50m²	-	1.13	25.56	3.25	m²	**28.81**

C DEMOLITION/ALTERATION/RENOVATION

Item	PC £	Labour hours	Labour £	Material £	Unit	Total rate £
Removing coverings						
Removing roof coverings						
slates	-	0.46	7.07	0.47	m²	**7.53**
slates; set aside for re-use	-	0.56	8.60	-	m²	**8.60**
nibbed tiles	-	0.37	5.68	0.47	m²	**6.15**
nibbed tiles; set aside for re-use	-	0.46	7.07	-	m²	**7.07**
corrugated asbestos sheeting	-	0.37	5.68	0.47	m²	**6.15**
corrugated metal sheeting	-	0.37	5.68	0.47	m²	**6.15**
underfelt and nails	-	0.04	0.61	0.23	m²	**0.85**
three layer felt roofing; cleaning base off for new coverings	-	0.23	3.53	0.47	m²	**4.00**
sheet metal coverings	-	0.46	7.07	0.47	m²	**7.53**
Removing roof coverings; selecting and re-fixing; including providing 25% new; including nails, etc.						
asbestos-free artificial blue/black slates; 500 mm x 250 mm (PC £ per 1000)	859.00	1.02	31.78	8.63	m²	**40.41**
asbestos-free artificial blue/black slates; 600 mm x 300 mm (PC £ per 1000)	1049.00	0.93	28.97	6.70	m²	**35.68**
natural slates; Welsh blue 510 mm x 255 mm (PC £ per 1000)	2750.00	1.11	34.58	16.99	m²	**51.58**
natural slates; Welsh blue 600 mm x 300 mm (PC £ per 1000)	6435.00	0.97	30.22	25.11	m²	**55.33**
clay plain tiles "Dreadnought" machine made; 265 mm x 165 mm (PC £ per 1000)	414.00	1.02	31.78	9.69	m²	**41.46**
concrete interlocking tiles; Marley Eternit "Ludlow Major" or other equal and approved; 413 mm x 330 mm (PC £ per 1000)	872.00	0.65	20.25	3.42	m²	**23.67**
concrete interlocking tiles; Redland "Renown" or other equal and approved; 417 mm x 330 mm; (PC £ per 1000)	908.00	0.65	20.25	3.52	m²	**23.77**
Removing damaged roof coverings in area less than 10 m²; providing and fixing new; including nails, etc.						
asbestos-free artificial blue/black slates; 500 mm x 250 mm	-	1.25	38.94	23.22	m²	**62.16**
asbestos-free artificial blue/black slates; 600 mm x 300 mm	-	1.16	36.14	18.75	m²	**54.89**
natural slates; Welsh blue 510 mm x 255 mm	-	1.34	41.75	61.23	m²	**102.97**
natural slates; Welsh blue 600 mm x 300 mm	-	1.20	37.38	95.27	m²	**132.66**
clay plain tiles "Dreadnought" machine made or other equal and approved; 265 mm x 165 mm	-	1.25	38.94	32.00	m²	**70.04**
concrete interlocking tiles; Marley Eternit "Ludlow Major" or other equal and approved; 413 mm x 330 mm	-	0.83	25.86	10.96	m²	**36.82**
concrete interlocking tiles; Redland "Renown" or other equal and approved; 417 mm x 330 mm	-	0.83	25.86	11.50	m²	**37.36**
Removing individual damaged roof coverings; providing and fixing new; including nails, etc.						
asbestos-free artificial blue/black slates; 500 mm x 250 mm	-	0.23	7.17	1.69	nr	**8.85**
asbestos-free artificial blue/black slates; 600 mm x 300 mm	-	0.23	7.17	1.34	nr	**8.51**
natural slates; Welsh blue 510 mm x 255 mm	-	0.28	8.72	3.36	nr	**12.09**
natural slates; Welsh blue 600 mm x 300 mm	-	0.28	8.72	7.69	nr	**16.41**
clay plain tiles "Dreadnought" machine made or other equal and approved; 265 mm x 165 mm	-	0.14	4.36	0.53	nr	**4.89**
concrete interlocking tiles; Marley Eternit "Ludlow Major" or other equal and approved; 413 mm x 330 mm	-	0.19	5.92	1.07	nr	**6.99**
concrete interlocking tiles; Redland "Renown" or other equal and approved; 417 mm x 330 mm	-	0.19	5.92	1.11	nr	**7.03**

C DEMOLITION/ALTERATION/RENOVATION

Item	PC £	Labour hours	Labour £	Material £	Unit	Total rate £
C90 ALTERATIONS - SPOT ITEMS – cont'd						
Removing coverings – cont'd						
Breaking up roof coverings						
asphalt	-	0.93	14.29	-	m²	**14.29**
Removing half round ridge or hip tile 300 mm long; providing and fixing new	-	0.46	14.33	4.37	nr	**18.70**
Removing defective metal flashings						
horizontal	-	0.19	2.92	0.47	m	**3.38**
stepped	-	0.23	3.53	0.23	m	**3.77**
Turning back bitumen felt and later dressing up face of new brickwork as skirtings; not exceeding 150 mm girth	-	0.93	21.08	0.12	m	**21.20**
Cutting out crack in asphalt roof coverings; making good to match existing						
20 mm thick two coat	-	1.53	24.64	-	m	**24.64**
Removing bitumen felt roof coverings and boarding to allow access for work to top of walls or beams beneath	-	0.74	11.37	-	m	**11.37**
Removing tiling battens; withdrawing nails	-	0.07	1.08	0.23	m²	**1.31**
Examining roof battens; re-nailing where loose; providing and fixing 25% new						
25 mm x 50 mm slating battens at 262 mm centres	-	0.07	2.18	1.14	m²	**3.32**
25 mm x 38 mm tiling battens at 100 mm centres	-	0.19	5.92	2.73	m²	**8.65**
Removing roof battens and nails; providing and fixing new "treated" softwood battens throughout						
25 mm x 50 mm slating battens at 262 mm centres	-	0.11	3.43	4.00	m²	**7.43**
25 mm x 38 mm tiling battens at 100 mm centres	-	0.23	7.17	8.43	m²	**15.60**
Removing underfelt and nails; providing and fixing new						
unreinforced felt	0.54	0.09	2.80	0.88	m²	**3.68**
reinforced felt	1.00	0.09	2.80	1.43	m²	**4.23**
Cutting openings or recesses						
Cutting openings or recesses through reinforced concrete walls						
150 mm thick	-	5.18	94.66	13.27	m²	**107.93**
225 mm thick	-	7.08	128.34	19.84	m²	**148.18**
300 mm thick	-	9.02	162.55	26.77	m²	**189.33**
Cutting openings or recesses through reinforced concrete suspended slabs						
150 mm thick	-	3.93	71.08	20.39	m²	**91.47**
225 mm thick	-	5.83	105.52	20.35	m²	**125.87**
300 mm thick	-	7.26	130.63	25.80	m²	**156.43**
Cutting openings or recesses through slated, boarded and timbered roof; 700 mm x 1100 mm; for new rooflight; including cutting structure and finishings; trimming timbers in rafters and making good roof coverings (kerb and rooflight not included)	-	-	-	-	nr	**375.41**
Cutting openings or recesses through brick or block walls or partitions; for lintels or beams above openings; in gauged mortar						
half brick thick	-	2.45	67.26	3.95	m²	**71.21**
one brick thick	-	4.07	111.73	7.91	m²	**119.63**
one and a half brick thick	-	5.69	156.20	11.86	m²	**168.06**
two brick thick	-	7.31	200.67	15.81	m²	**216.48**

C DEMOLITION/ALTERATION/RENOVATION

Item	PC £	Labour hours	Labour £	Material £	Unit	Total rate £
75 mm blockwork	-	1.48	40.63	2.56	m²	**43.19**
90 mm blockwork	-	1.67	45.84	3.02	m²	**48.87**
100 mm blockwork	-	1.80	49.41	3.49	m²	**52.90**
115 mm blockwork	-	1.84	50.51	3.95	m²	**54.46**
125 mm blockwork	-	2.04	56.00	4.42	m²	**60.42**
140 mm blockwork	-	2.17	59.57	4.88	m²	**64.45**
150 mm blockwork	-	2.27	62.31	5.35	m²	**67.66**
190 mm blockwork	-	2.53	69.45	6.74	m²	**76.19**
215 mm blockwork	-	2.68	73.57	7.44	m²	**81.01**
255 mm blockwork	-	2.94	80.71	8.84	m²	**89.54**
Cutting openings or recesses through brick walls or partitions; for lintels or beams above openings; in cement mortar						
half brick thick	-	3.52	96.63	3.95	m²	**100.58**
one brick thick	-	5.83	160.04	7.91	m²	**167.95**
one and a half brick thick	-	8.14	223.45	11.86	m²	**235.31**
two brick thick	-	10.45	286.87	15.81	m²	**302.68**
Cutting openings or recesses through brick or block walls or partitions; for door or window openings; in gauged mortar						
half brick thick	-	1.25	34.31	3.95	m²	**38.27**
one brick thick	-	2.04	56.00	7.91	m²	**63.91**
one and a half brick thick	-	2.82	77.41	11.86	m²	**89.27**
two brick thick	-	3.65	100.20	15.81	m²	**116.01**
75 mm blockwork	-	0.74	20.31	2.56	m²	**22.87**
90 mm blockwork	-	0.85	23.33	3.02	m²	**26.36**
100 mm blockwork	-	0.93	25.53	3.49	m²	**29.02**
115 mm blockwork	-	0.98	26.90	3.95	m²	**30.86**
125 mm blockwork	-	1.02	28.00	4.42	m²	**32.42**
140 mm blockwork	-	1.07	29.37	4.88	m²	**34.26**
150 mm blockwork	-	1.11	30.47	5.35	m²	**35.82**
190 mm blockwork	-	1.22	33.49	6.74	m²	**40.23**
215 mm blockwork	-	1.34	36.78	7.44	m²	**44.23**
255 mm blockwork	-	1.48	40.63	8.84	m²	**49.46**
Cutting openings or recesses through brick or block walls or partitions; for door or window openings; in cement mortar						
half brick thick	-	1.76	48.31	3.95	m²	**52.27**
one brick thick	-	2.91	79.88	7.91	m²	**87.79**
one and a half brick thick	-	4.02	110.35	11.86	m²	**122.21**
two brick thick	-	5.23	143.57	15.81	m²	**159.38**
Cutting openings or recesses through faced wall 1200 mm x 1200 mm (1.44 m²) for new window; including cutting structure, quoining up jambs, cutting and pinning in suitable precast concrete boot lintel with galvanised steel angle bolted on to support, outer brick soldier course in facing bricks to match existing (new window and frame not included)						
one brick thick wall or two half brick thick skins	-	-	-	-	nr	**506.81**
one and a half brick thick wall	-	-	-	-	nr	**531.84**
two brick thick wall	-	-	-	-	nr	**575.63**
Cutting openings or recesses through 100 mm thick softwood stud partition including framing studwork around, making good boarding and any plaster either side and extending floor finish through opening (new door and frame not included)						
single door and frame	-	-	-	-	nr	**262.79**
pair of doors and frame	-	-	-	-	nr	**344.14**

C DEMOLITION/ALTERATION/RENOVATION

Item	PC £	Labour hours	Labour £	Material £	Unit	Total rate £
C90 ALTERATIONS - SPOT ITEMS – cont'd						
Cutting openings or recesses – cont'd						
Cutting openings or recesses through internal plastered wall for single door and frame; including cutting structure, quoining or making good jambs, cutting and pinning in suitable precast concrete plate lintel(s), making good plasterwork up to new frame both sides and extending floor finish through new opening (new door and frame not included)						
150 mm reinforced concrete wall	-	-	-	-	nr	**300.33**
225 mm reinforced concrete wall	-	-	-	-	nr	**412.95**
half brick thick wall	-	-	-	-	nr	**281.56**
one brick thick wall or two half brick thick skins	-	-	-	-	nr	**369.14**
one and a half brick thick wall	-	-	-	-	nr	**450.49**
two brick thick wall	-	-	-	-	nr	**544.33**
100 mm block wall	-	-	-	-	nr	**262.79**
215 mm block wall	-	-	-	-	nr	**344.14**
Cutting openings or recesses through internal plastered wall for pair of doors and frame; including cutting structure, quoining or making good jambs, cutting and pinning in suitable precast concrete plate lintel(s), making good plasterwork up to new frame both sides and extending floor finish through new opening (new door and frame not included)						
150 mm reinforced concrete wall	-	-	-	-	nr	**431.72**
225 mm reinforced concrete wall	-	-	-	-	nr	**550.60**
half brick thick wall	-	-	-	-	nr	**331.62**
one brick thick wall or two half brick thick skins	-	-	-	-	nr	**456.76**
one and a half brick thick wall	-	-	-	-	nr	**588.15**
two brick thick wall	-	-	-	-	nr	**700.76**
100 mm block wall	-	-	-	-	nr	**312.85**
215 mm block wall	-	-	-	-	nr	**425.46**
Cutting back projections						
Cutting back brick projections flush with adjacent wall						
225 mm x 112 mm	-	0.28	7.69	0.23	m	**7.92**
225 mm x 225 mm	-	0.46	12.63	0.47	m	**13.09**
337 mm x 112 mm	-	0.65	17.84	0.70	m	**18.54**
450 mm x 225 mm	-	0.83	22.78	0.93	m	**23.71**
Cutting back chimney breasts flush with adjacent wall						
half brick thick	-	1.62	44.47	7.44	m²	**51.91**
one brick thick	-	2.17	59.57	11.63	m²	**71.20**
Filling in openings						
Removing doors and frames; making good plaster and skirtings across reveals and heads; leaving as blank openings						
single doors	-	-	-	-	nr	**112.63**
pair of doors	-	-	-	-	nr	**131.39**
Removing doors and frames in 100 mm thick softwood partitions; filling in openings with timber covered on both sides with boarding or lining to match existing; extending skirtings both sides						
single doors	-	-	-	-	nr	**175.19**
pair of doors	-	-	-	-	nr	**231.50**

C DEMOLITION/ALTERATION/RENOVATION

Item	PC £	Labour hours	Labour £	Material £	Unit	Total rate £
Removing single doors and frames in internal walls; filling in openings with brickwork or blockwork; plastering walls and extending skirtings both sides						
half brick thick	-	-	-	-	nr	**193.95**
one brick thick	-	-	-	-	nr	**269.06**
one and a half brick thick	-	-	-	-	nr	**337.87**
two brick thick	-	-	-	-	nr	**425.46**
100 mm blockwork	-	-	-	-	nr	**156.43**
215 mm blockwork	-	-	-	-	nr	**231.50**
Removing pairs of doors and frames in internal walls; filling in openings with brickwork or blockwork; plastering walls and extend skirtings both sides						
half brick thick	-	-	-	-	nr	**312.85**
one brick thick	-	-	-	-	nr	**437.98**
one and a half brick thick	-	-	-	-	nr	**563.12**
two brick thick	-	-	-	-	nr	**681.99**
100 mm blockwork	-	-	-	-	nr	**269.06**
215 mm blockwork	-	-	-	-	nr	**356.64**
Removing 825 mm x 1046 mm (1.16 m²) sliding sash windows and frames in external faced walls; filling in openings with facing brickwork on outside to match existing and common brickwork on inside; plastering internally						
one brick thick or two half brick thick skins	-	-	-	-	nr	**256.53**
one and a half brick thick	-	-	-	-	nr	**287.82**
two brick thick	-	-	-	-	nr	**331.62**
Removing 825 mm x 1406 mm (1.16 m²) curved headed sliding sashed windows in external stuccoed walls; filling in openings with common bricks; stucco on outside and plastering internally						
one brick thick or two half brick thick skins	-	-	-	-	nr	**287.82**
one and a half brick thick	-	-	-	-	nr	**331.62**
two brick thick	-	-	-	-	nr	**412.95**
Removing 825 mm x 1406 mm (1.16 m²) curved headed sliding sash windows in external masonry faced brick walls; filling in openings with facing brickwork on outside and common brickwork on inside; plastering internally						
350 mm wall	-	-	-	-	nr	**750.82**
500 mm wall	-	-	-	-	nr	**819.66**
600 mm wall	-	-	-	-	nr	**900.98**
Quoining up jambs in common bricks; in gauged mortar (1:1:6); as the work proceeds						
half brick thick or skin of hollow wall (PC £ per 1000)	270.00	0.93	25.53	5.35	m	**30.88**
one brick thick	-	1.39	38.16	10.70	m	**48.86**
one and a half brick thick	-	1.80	49.41	16.05	m	**65.46**
two brick thick	-	2.22	60.94	19.88	m	**80.82**
75 mm blockwork	-	0.58	15.92	4.67	m	**20.60**
90 mm blockwork	-	0.62	17.02	5.03	m	**22.05**
100 mm blockwork	-	0.65	17.84	5.67	m	**23.52**
115 mm blockwork	-	0.70	19.22	6.42	m	**25.63**
125 mm blockwork	-	0.74	20.31	7.29	m	**27.61**
140 mm blockwork	-	0.80	21.96	8.41	m	**30.37**
150 mm blockwork	-	0.83	22.78	9.28	m	**32.07**
190 mm blockwork	-	0.93	25.53	11.25	m	**36.78**
215 mm blockwork	-	1.00	27.45	12.48	m	**39.93**
225 mm blockwork	-	1.10	30.20	15.32	m	**45.52**

C DEMOLITION/ALTERATION/RENOVATION

Item	PC £	Labour hours	Labour £	Material £	Unit	Total rate £
C90 ALTERATIONS - SPOT ITEMS – cont'd						
Filling in openings – cont'd						
Closing at jambs with common brickwork half brick thick						
50 mm cavity; including lead-lined hessian based vertical damp proof course	-	0.37	10.16	9.39	m	**19.55**
Quoining up jambs in machine made facings; in gauged mortar (1:1:6); facing and pointing one side to match existing						
half brick thick or skin of hollow wall (PC £ per 1000)	380.00	1.16	31.84	9.22	m	**41.06**
one brick thick	-	1.39	38.16	16.90	m	**55.06**
one and a half brick thick	-	2.13	58.47	25.13	m	**83.60**
two brick thick	-	2.59	71.10	33.20	m	**104.30**
Quoining up jambs in hand made facings; in gauged mortar (1:1:6); facing and pointing one side to match existing						
half brick thick or skin of hollow wall (PC £ per 1000)	600.00	1.16	31.84	13.22	m	**45.07**
one brick thick	-	1.39	38.16	24.91	m	**63.07**
one and a half brick thick	-	2.13	58.47	37.15	m	**95.62**
two brick thick	-	2.59	71.10	49.23	m	**120.33**
Filling existing openings with common brickwork or blockwork in gauged mortar (1:1:6) (cutting and bonding not included)						
half brick thick (PC £ per 1000)	270.00	1.71	46.94	21.19	m²	**68.13**
one brick thick	-	2.82	77.41	43.90	m²	**121.31**
one and a half brick thick	-	3.89	106.79	65.85	m²	**172.63**
two brick thick	-	4.86	133.41	87.80	m²	**221.21**
75 mm blockwork	6.87	0.85	23.33	9.98	m²	**33.32**
90 mm blockwork	-	0.93	25.53	10.49	m²	**36.01**
100 mm blockwork	8.32	0.97	26.63	11.78	m²	**38.40**
115 mm blockwork	9.56	1.05	28.82	13.32	m²	**42.15**
125 mm blockwork	10.39	1.09	29.92	14.36	m²	**44.28**
140 mm blockwork	11.64	1.16	31.84	15.91	m²	**47.75**
150 mm blockwork	12.47	1.20	32.94	16.94	m²	**49.88**
190 mm blockwork	15.80	1.37	37.61	21.83	m²	**59.44**
215 mm blockwork	17.88	1.48	40.63	24.41	m²	**65.04**
255 mm blockwork	22.04	1.65	45.29	29.58	m²	**74.87**
Cutting and bonding ends to existing						
half brick thick	-	0.37	10.16	1.69	m	**11.85**
one brick thick	-	0.54	14.82	3.31	m	**18.13**
one and a half brick thick	-	0.80	21.96	4.04	m	**26.00**
two brick thick	-	1.16	31.84	6.11	m	**37.96**
75 mm blockwork	-	0.16	4.39	0.73	m	**5.12**
90 mm blockwork	-	0.19	5.22	0.80	m	**6.02**
100 mm blockwork	-	0.21	5.76	0.90	m	**6.67**
115 mm blockwork	-	0.23	6.31	1.05	m	**7.36**
125 mm blockwork	-	0.24	6.59	1.13	m	**7.72**
140 mm blockwork	-	0.26	7.14	1.25	m	**8.38**
150 mm blockwork	-	0.27	7.41	1.33	m	**8.74**
190 mm blockwork	-	0.33	9.06	1.76	m	**10.82**
215 mm blockwork	-	0.38	10.43	1.96	m	**12.39**
255 mm blockwork	-	0.44	12.08	2.35	m	**14.43**
half brick thick in facings; to match existing (PC £ per 1000)	380.00	0.56	15.37	4.14	m	**19.52**
Extra over common brickwork for fair face; flush pointing						
walls and the like	-	0.19	5.22	-	m²	**5.22**

C DEMOLITION/ALTERATION/RENOVATION

Item	PC £	Labour hours	Labour £	Material £	Unit	Total rate £
Extra over common bricks for facing bricks in flemish bond; facing and pointing one side						
machine made facings (PC £ per 1000)	380.00	0.97	26.63	10.01	m²	**36.63**
hand made facings (PC £ per 1000)	600.00	0.97	26.63	30.03	m²	**56.66**
ADD or DEDUCT for variation of £10.00/1000 in PC for facing bricks; in flemish bond						
half brick thick	-	-	-	0.73	m²	**-**
Filling in openings to hollow walls with inner skin of common bricks; 50 mm cavity and galvanised steel butterfly ties; outer skin of facings; all in gauged mortar (1:1:6); facing and pointing one side						
two half brick thick skins; outer skin machine made facings (PC £ per 1000)	380.00	4.26	116.94	53.69	m²	**170.64**
two half brick thick skins; outer skin hand made facings (PC £ per 1000)	600.00	4.26	116.94	70.52	m²	**187.46**
Temporary screens						
Providing and erecting; maintaining; temporary dust proof screens; with 50 mm x 75 mm sawn softwood framing; covering one side with 12 mm thick plywood						
over 300 mm wide	-	0.74	16.25	18.48	m²	**34.74**
Providing and erecting; maintaining; temporary screen; with 50 mm x 100 mm sawn softwood framing; covering one side with 13 mm thick insulating board and other side with single layer of polythene sheet						
over 300 mm wide	-	0.93	20.43	9.43	m²	**29.86**
Providing and erecting; maintaining; temporary screen; with 50 mm x 100 mm sawn softwood framing; covering one side with 19 mm thick exterior quality plywood; softwood cappings; including three coats of gloss paint; clearing away						
over 300 mm wide	-	1.85	35.66	25.50	m²	**61.15**

D GROUNDWORK

Item	PC £	Labour hours	Labour £	Material /Plant £	Unit	Total rate £
D20 EXCAVATING AND FILLING						
NOTE: Prices are applicable to excavation in firm soil. Multiplying factors for other soils are as follows:						
Mechanical / Hand						
Clay x 2.00 / x 1.20						
Compact gravel x 3.00 / x 1.50						
Soft chalk x 4.00 / x 2.00						
Hard rock x 5.00 / x 6.00						
Running sand or silt x 6.00 / x 2.00						
Site preparation						
Removing trees						
girth 600 mm - 1.50 m	-	20.35	267.37	-	nr	**267.37**
girth 1.50 - 3.00 m	-	35.61	467.87	-	nr	**467.87**
girth exceeding 3.00 m	-	50.88	668.50	-	nr	**668.50**
Removing tree stumps						
girth 600 mm - 1.50 m	-	1.02	13.40	53.04	nr	**66.44**
girth 1.50 m - 3.00 m	-	1.02	13.40	77.56	nr	**90.96**
girth exceeding 3.00 m	-	1.02	13.40	106.20	nr	**119.60**
Clearing site vegetation						
bushes, scrub, undergrowth, hedges and trees and tree stumps not exceeding 600 mm girth	-	0.03	0.39	-	m²	**0.39**
Lifting turf for preservation						
stacking	-	0.36	4.73	-	m²	**4.73**
Excavating; by machine						
Topsoil for preservation						
average depth 150 mm	-	0.02	0.26	1.33	m²	**1.59**
add or deduct for each 25 mm variation in average depth	-	0.01	0.13	0.31	m²	**0.44**
To reduce levels						
maximum depth not exceeding 0.25 m	-	0.06	0.79	1.74	m³	**2.53**
maximum depth not exceeding 1.00 m	-	0.04	0.53	1.23	m³	**1.75**
maximum depth not exceeding 2.00 m	-	0.06	0.79	1.74	m³	**2.53**
maximum depth not exceeding 4.00 m	-	0.08	1.05	2.15	m³	**3.20**
Basements and the like; commencing level exceeding 0.25 m below exitsing ground level						
maximum depth not exceeding 1.00 m	-	0.09	1.18	1.57	m³	**2.76**
maximum depth not exceeding 2.00 m	-	0.06	0.79	1.21	m³	**2.00**
maximum depth not exceeding 4.00 m	-	0.09	1.18	1.57	m³	**2.76**
maximum depth not exceeding 6.00 m	-	0.10	1.31	2.06	m³	**3.37**
maximum depth not exceeding 8.00 m	-	0.13	1.71	2.42	m³	**4.13**
Pits						
maximum depth not exceeding 0.25m	-	0.36	4.73	6.42	m³	**11.15**
maximum depth not exceeding 1.00 m	-	0.38	4.99	5.69	m³	**10.68**
maximum depth not exceeding 2.00 m	-	0.44	5.78	6.42	m³	**12.20**
maximum depth not exceeding 4.00 m	-	0.53	6.96	7.26	m³	**14.23**
maximum depth not exceeding 6.00 m	-	0.55	7.23	7.63	m³	**14.85**
Extra over pit excavating for commencing level exceeding 0.25 m below existing ground level						
1.00 m below	-	0.03	0.39	0.85	m³	**1.24**
2.00 m below	-	0.06	0.79	1.21	m³	**2.00**
3.00 m below	-	0.07	0.92	1.57	m³	**2.49**
4.00 m below	-	0.10	1.31	2.06	m³	**3.37**

D GROUNDWORK

Item	PC £	Labour hours	Labour £	Material /Plant £	Unit	Total rate £
Trenches, width not exceeding 0.30 m						
maximum depth not exceeding 0.25 m	-	0.30	3.94	5.21	m³	**9.15**
maximum depth not exceeding 1.00 m	-	0.32	4.20	4.48	m³	**8.68**
maximum depth not exceeding 2.00 m	-	0.37	4.86	5.21	m³	**10.07**
maximum depth not exceeding 4.00 m	-	0.45	5.91	6.42	m³	**12.33**
maximum depth not exceeding 6.00 m	-	0.52	6.83	7.63	m³	**14.46**
Trenches, width exceeding 0.30 m						
maximum depth 0.25 m	-	0.27	3.55	4.84	m³	**8.39**
maximum depth 1.00 m	-	0.28	3.68	3.99	m³	**7.67**
maximum depth 2.00 m	-	0.34	4.47	4.84	m³	**9.31**
maximum depth 4.00 m	-	0.40	5.26	5.69	m³	**10.95**
maximum depth 6.00 m	-	0.49	6.44	7.26	m³	**13.70**
Extra over trench excavating for commencing level exceeding 0.25 m below existing ground level						
1.00 m below	-	0.03	0.39	0.85	m³	**1.24**
2.00 m below	-	0.06	0.79	1.21	m³	**2.00**
3.00 m below	-	0.07	0.92	1.57	m³	**2.49**
4.00 m below	-	0.10	1.31	2.06	m³	**3.37**
For pile caps and ground beams between piles						
maximum depth not exceeding 0.25 m	-	0.45	5.91	8.47	m³	**14.39**
maximum depth not exceeding 1.00 m	-	0.40	5.26	7.63	m³	**12.88**
maximum depth not exceeding 2.00 m	-	0.45	5.91	8.47	m³	**14.39**
To bench sloping ground to receive filling						
maximum depth not exceeding 0.25 m	-	0.10	1.31	2.06	m³	**3.37**
maximum depth not exceeding 1.00 m	-	0.07	0.92	2.42	m³	**3.34**
maximum depth not exceeding 2.00 m	-	0.10	1.31	2.06	m³	**3.37**
Extra over any types of excavating irrespective of depth						
excavating below ground water level	-	0.16	2.10	2.78	m³	**4.89**
next existing services	-	2.90	38.10	1.57	m³	**39.68**
around existing services crossing excavation	-	6.70	88.03	4.48	m³	**92.51**
Extra over any types of excavating irrespective of depth for breaking out existing materials						
rock	-	3.42	44.93	17.78	m³	**62.72**
concrete	-	2.90	38.10	12.95	m³	**51.06**
reinforced concrete	-	4.17	54.79	19.22	m³	**74.01**
brickwork; blockwork or stonework	-	2.08	27.33	9.77	m³	**37.10**
Extra over any types of excavating irrespective of depth for breaking out existing hard pavings, 75 mm thick						
coated macadam or asphalt	-	0.22	2.89	0.68	m²	**3.57**
Extra over any types of excavating irrespective of depth for breaking out existing hard pavings; 150 mm thick						
concrete	-	0.45	5.91	2.05	m²	**7.97**
reinforced concrete	-	0.68	8.93	2.76	m²	**11.69**
coated macadam or asphalt and hardcore	-	0.30	3.94	0.74	m²	**4.68**
Working space allowance to excavations						
reduce levels; basements and the like	-	0.09	1.18	1.57	m²	**2.76**
pits	-	0.22	2.89	4.48	m²	**7.37**
trenches	-	0.20	2.63	3.99	m²	**6.62**
pile caps and ground beams between piles	-	0.23	3.02	4.48	m²	**7.50**
Extra over excavating for working space for backfilling with special materials						
hardcore	-	0.16	2.10	17.85	m²	**19.95**
sand	-	0.16	2.10	27.33	m²	**29.43**
40 mm - 20 mm gravel	-	0.16	2.10	30.37	m²	**32.47**
plain in situ ready mixed designated concrete C7.5 - 40 mm aggregate	-	1.07	18.12	58.56	m²	**76.68**

D GROUNDWORK

Item	PC £	Labour hours	Labour £	Material /Plant £	Unit	Total rate £
D20 EXCAVATING AND FILLING – cont'd						
Excavating; by hand						
Topsoil for preservation						
average depth 150 mm	-	0.26	3.42	-	m²	**3.42**
add or deduct for each 25 mm variation in average depth	-	0.03	0.39	-	m²	**0.39**
To reduce levels						
maximum depth not exceeding 0.25 m	-	1.59	20.89	-	m³	**20.89**
maximum depth not exceeding 1.00 m	-	1.80	23.65	-	m³	**23.65**
maximum depth not exceeding 2.00 m	-	1.99	26.15	-	m³	**26.15**
maximum depth not exceeding 4.00 m	-	2.19	28.77	-	m³	**28.77**
Basements and the like; commencing level exceeding 0.25 m below existing ground level						
maximum depth not exceeding 1.00 m	-	2.09	27.46	-	m³	**27.46**
maximum depth not exceeding 2.00 m	-	2.24	29.43	-	m³	**29.43**
maximum depth not exceeding 4.00 m	-	3.01	39.55	-	m³	**39.55**
maximum depth not exceeding 6.00 m	-	3.66	48.09	-	m³	**48.09**
maximum depth not exceeding 8.00 m	-	4.35	57.15	-	m³	**57.15**
Pits						
maximum depth not exceeding 0.25 m	-	2.34	30.74	-	m³	**30.74**
maximum depth not exceeding 1.00 m	-	2.90	38.10	-	m³	**38.10**
maximum depth not exceeding 2.00 m	-	3.60	47.30	-	m³	**47.30**
maximum depth not exceeding 4.00 m	-	4.56	59.91	-	m³	**59.91**
maximum depth not exceeding 6.00 m	-	5.64	74.10	-	m³	**74.10**
Extra over pit excavating for commencing level exceeding 0.25 m below existing ground level						
1.00 m below	-	0.45	5.91	-	m³	**5.91**
2.00 m below	-	0.97	12.74	-	m³	**12.74**
3.00 m below	-	1.43	18.79	-	m³	**18.79**
4.00 m below	-	1.88	24.70	-	m³	**24.70**
Trenches, width not exceeding 0.30 m						
maximum depth not exceeding 0.25 m	-	2.03	26.67	-	m³	**26.67**
maximum depth not exceeding 1.00 m	-	2.99	39.28	-	m³	**39.28**
maximum depth not exceeding 2.00 m	-	3.51	46.12	-	m³	**46.12**
maximum depth not exceeding 4.00 m	-	4.29	56.37	-	m³	**56.37**
maximum depth not exceeding 6.00 m	-	5.53	72.66	-	m³	**72.66**
Trenches, width exceeding 0.30 m						
maximum depth not exceeding 0.25 m	-	1.99	26.15	-	m³	**26.15**
maximum depth not exceeding 1.00 m	-	2.67	35.08	-	m³	**35.08**
maximum depth not exceeding 2.00 m	-	3.12	40.99	-	m³	**40.99**
maximum depth not exceeding 4.00 m	-	3.97	52.16	-	m³	**52.16**
maximum depth not exceeding 6.00 m	-	5.07	66.61	-	m³	**66.61**
Extra over trench excavating for commencing level exceeding 0.25 m below existing ground level						
1.00 m below	-	0.45	5.91	-	m³	**5.91**
2.00 m below	-	0.97	12.74	-	m³	**12.74**
3.00 m below	-	1.46	19.18	-	m³	**19.18**
4.00 m below	-	1.88	24.70	-	m³	**24.70**
For pile caps and ground beams between piles						
maximum depth not exceeding 0.25 m	-	3.05	40.07	-	m³	**40.07**
maximum depth not exceeding 1.00 m	-	3.26	42.83	-	m³	**42.83**
maximum depth not exceeding 2.00 m	-	3.87	50.85	-	m³	**50.85**
To bench sloping ground to receive filling						
maximum depth not exceeding 0.25 m	-	1.43	18.79	-	m³	**18.79**
maximum depth not exceeding 1.00 m	-	1.63	21.42	-	m³	**21.42**
maximum depth not exceeding 2.00 m	-	1.83	24.04	-	m³	**24.04**

D GROUNDWORK

Item	PC £	Labour hours	Labour £	Material /Plant £	Unit	Total rate £
Extra over any types of excavating irrespective of depth						
excavating below ground water level	-	0.36	4.73	-	m³	**4.73**
next existing services	-	1.02	13.40	-	m³	**13.40**
around existing services crossing excavation	-	2.04	26.80	-	m³	**26.80**
Extra over any types of excavating irrespective of depth for breaking out existing materials						
rock	-	5.09	66.88	8.69	m³	**75.57**
concrete	-	4.58	60.18	7.24	m³	**67.42**
reinforced concrete	-	6.11	80.28	10.14	m³	**90.42**
brickwork; blockwork or stonework	-	3.05	40.07	4.35	m³	**44.42**
Extra over any types of excavating irrespective of depth for breaking out existing hard pavings, 60 mm thick						
precast concrete paving slabs	-	0.31	4.07	-	m²	**4.07**
Extra over any types of excavating irrespective of depth for breaking out existing hard pavings, 75 mm thick						
coated macadam or asphalt	-	0.44	5.78	1.19	m²	**6.97**
Extra over any types of excavating irrespective of depth for breaking out existing hard pavings, 150 mm thick						
concrete	-	0.71	9.33	1.02	m²	**10.35**
reinforced concrete	-	0.92	12.09	1.45	m²	**13.54**
coated macadam or asphalt and hardcore	-	0.51	6.70	0.72	m²	**7.42**
Working space allowance to excavations						
reduce levels; basements and the like	-	2.34	30.74	-	m²	**30.74**
pits	-	2.44	32.06	-	m²	**32.06**
trenches	-	2.14	28.12	-	m²	**28.12**
pile caps and ground beams between piles	-	2.54	33.37	-	m²	**33.37**
Extra over excavation for working space for backfilling with special materials						
hardcore	-	0.81	10.64	15.49	m²	**26.13**
sand	-	0.81	10.64	24.97	m²	**35.62**
40 mm - 20 mm gravel	-	0.81	10.64	28.01	m²	**38.65**
plain in situ ready mixed designated concrete; C7.5 - 40 mm aggregate	-	1.12	18.96	56.21	m²	**75.17**
Excavating; by hand; inside existing buildings						
Basements and the like; commencing level exceeding 0.25 m below existing ground level						
maximum depth not exceeding 1.00 m	-	3.14	41.26	-	m³	**41.26**
maximum depth not exceeding 2.00 m	-	3.36	44.15	-	m³	**44.15**
maximum depth not exceeding 4.00 m	-	4.51	59.26	-	m³	**59.26**
maximum depth not exceeding 6.00m	-	5.50	72.26	-	m³	**72.26**
maximum depth not exceeding 8.00 m	-	6.65	87.37	-	m³	**87.37**
Pits						
maximum depth not exceeding 0.25 m	-	3.51	46.12	-	m³	**46.12**
maximum depth not exceeding 1.00 m	-	3.82	50.19	-	m³	**50.19**
maximum depth not exceeding 2.00 m	-	4.58	60.18	-	m³	**60.18**
maximum depth not exceeding 4.00 m	-	5.80	76.20	-	m³	**76.20**
maximum depth not exceeding 6.00 m	-	7.17	94.20	-	m³	**94.20**
Extra over pit excavating for commencing level exceeding 0.25 m below existing ground level						
1.00 m below	-	0.68	8.93	-	m³	**8.93**
2.00 m below	-	1.45	19.05	-	m³	**19.05**
3.00 m below	-	2.14	28.12	-	m³	**28.12**
4.00 m below	-	2.81	36.92	-	m³	**36.92**

D GROUNDWORK

Item	PC £	Labour hours	Labour £	Material /Plant £	Unit	Total rate £
D20 EXCAVATING AND FILLING – cont'd						
Excavating; by hand; inside existing buildings – cont'd						
Trenches, width not exceeding 0.30 m						
maximum depth not exceeding 0.25 m	-	3.05	40.07	-	m³	**40.07**
maximum depth not exceeding 1.00 m	-	3.51	46.12	-	m³	**46.12**
maximum depth not exceeding 2.00 m	-	4.13	54.26	-	m³	**54.26**
maximum depth not exceeding 4.00 m	-	5.04	66.22	-	m³	**66.22**
maximum depth not exceeding 6.00 m	-	6.49	85.27	-	m³	**85.27**
Trenches, width exceeding 0.30 m						
maximum depth not exceeding 0.25 m	-	2.99	39.28	-	m³	**39.28**
maximum depth not exceeding 1.00 m	-	3.14	41.26	-	m³	**41.26**
maximum depth not exceeding 2.00 m	-	3.66	48.09	-	m³	**48.09**
maximum depth not exceeding 4.00 m	-	4.66	61.23	-	m³	**61.23**
maximum depth not exceeding 6.00 m	-	5.96	78.31	-	m³	**78.31**
Extra over trench excavating for commencing level exceeding 0.25 m below existing ground level						
1.00 m below	-	0.68	8.93	-	m³	**8.93**
2.00 m below	-	1.45	19.05	-	m³	**19.05**
3.00 m below	-	2.14	28.12	-	m³	**28.12**
4.00 m below	-	2.81	36.92	-	m³	**36.92**
Extra over any types of excavating irrespective of depth						
excavating below ground water level	-	0.55	7.23	-	m³	**7.23**
Extra over any types of excavating irrespective of depth for breaking out existing materials						
concrete	-	6.86	90.13	7.24	m³	**97.37**
reinforced concrete	-	9.16	120.35	10.14	m³	**130.49**
brickwork; blockwork or stonework	-	4.58	60.18	4.35	m³	**64.52**
Extra over any types of excavating irrespective of depth for breaking out existing hard pavings, 150 mm thick						
concrete	-	1.07	14.06	1.02	m²	**15.08**
reinforced concrete	-	1.38	18.13	1.45	m²	**19.58**
Working space allowance to excavations						
pits	-	3.60	47.30	-	m²	**47.30**
trenches	-	3.21	42.18	-	m²	**42.18**
Earthwork support (average "risk" prices)						
Maximum depth not exceeding 1.00 m						
distance between opposing faces not exceeding 2.00 m	-	0.11	1.45	0.48	m²	**1.92**
distance between opposing faces 2.00 - 4.00 m	-	0.12	1.58	0.56	m²	**2.13**
distance between opposing faces exceeding 4.00 m	-	0.13	1.71	0.70	m²	**2.41**
Maximum depth not exceeding 2.00 m						
distance between opposing faces not exceeding 2.00 m	-	0.13	1.71	0.56	m²	**2.27**
distance between opposing faces 2.00 - 4.00 m	-	0.14	1.84	0.70	m²	**2.54**
distance between opposing faces exceeding 4.00 m	-	0.16	2.10	0.89	m²	**2.99**
Maximum depth not exceeding 4.00 m						
distance between opposing faces not exceeding 2.00 m	-	0.16	2.10	0.70	m²	**2.81**
distance between opposing faces 2.00 - 4.00 m	-	0.18	2.37	0.89	m²	**3.25**
distance between opposing faces exceeding 4.00 m	-	0.19	2.50	1.11	m²	**3.61**
Maximum depth not exceeding 6.00 m						
distance between opposing faces not exceeding 2.00 m	-	0.19	2.50	0.84	m²	**3.33**
distance between opposing faces 2.00 - 4.00 m	-	0.21	2.76	1.11	m²	**3.87**
distance between opposing faces exceeding 4.00 m	-	0.24	3.15	1.39	m²	**4.55**

D GROUNDWORK

Item	PC £	Labour hours	Labour £	Material /Plant £	Unit	Total rate £
Maximum depth not exceeding 8.00 m						
distance between opposing faces not exceeding 2.00 m	-	0.26	3.42	1.11	m²	**4.53**
distance between opposing faces 2.00 - 4.00 m	-	0.31	4.07	1.39	m²	**5.47**
distance between opposing faces exceeding 4.00 m	-	0.37	4.86	1.67	m²	**6.53**
Earthwork support (open boarded)						
Maximum depth not exceeding 1.00 m						
distance between opposing faces not exceeding 2.00 m	-	0.31	4.07	0.98	m²	**5.06**
distance between opposing faces 2.00 - 4.00 m	-	0.34	4.47	1.11	m²	**5.58**
distance between opposing faces exceeding 4.00 m	-	0.39	5.12	1.39	m²	**6.52**
Maximum depth not exceeding 2.00 m						
distance between opposing faces not exceeding 2.00 m	-	0.39	5.12	1.11	m²	**6.24**
distance between opposing faces 2.00 - 4.00 m	-	0.43	5.65	1.34	m²	**6.99**
distance between opposing faces exceeding 4.00 m	-	0.49	6.44	1.67	m²	**8.11**
Maximum depth not exceeding 4.00 m						
distance between opposing faces not exceeding 2.00 m	-	0.49	6.44	1.26	m²	**7.70**
distance between opposing faces 2.00 - 4.00 m	-	0.55	7.23	1.56	m²	**8.78**
distance between opposing faces exceeding 4.00 m	-	0.61	8.01	1.95	m²	**9.96**
Maximum depth not exceeding 6.00 m						
distance between opposing faces not exceeding 2.00 m	-	0.61	8.01	1.39	m²	**9.41**
distance between opposing faces 2.00 - 4.00 m	-	0.68	8.93	1.75	m²	**10.69**
distance between opposing faces exceeding 4.00 m	-	0.78	10.25	2.23	m²	**12.48**
Maximum depth not exceeding 8.00 m						
distance between opposing faces not exceeding 2.00 m	-	0.81	10.64	1.82	m²	**12.46**
distance between opposing faces 2.00 - 4.00 m	-	0.92	12.09	2.10	m²	**14.19**
distance between opposing faces exceeding 4.00 m	-	1.06	13.93	2.79	m²	**16.71**
Earthwork support (close boarded)						
Maximum depth not exceeding 1.00 m						
distance between opposing faces not exceeding 2.00 m	-	0.81	10.64	1.95	m²	**12.59**
distance between opposing faces 2.00 - 4.00 m	-	0.90	11.82	2.23	m²	**14.05**
distance between opposing faces exceeding 4.00 m	-	0.99	13.01	2.79	m²	**15.79**
Maximum depth not exceeding 2.00 m						
distance between opposing faces not exceeding 2.00 m	-	1.02	13.40	2.23	m²	**15.63**
distance between opposing faces 2.00 - 4.00 m	-	1.12	14.72	2.67	m²	**17.39**
distance between opposing faces exceeding 4.00 m	-	1.22	16.03	3.34	m²	**19.37**
Maximum depth not exceeding 4.00 m						
distance between opposing faces not exceeding 2.00 m	-	1.28	16.82	2.51	m²	**19.32**
distance between opposing faces 2.00 - 4.00 m	-	1.43	18.79	3.11	m²	**21.90**
distance between opposing faces exceeding 4.00 m	-	1.58	20.76	3.90	m²	**24.66**
Maximum depth not exceeding 6.00 m						
distance between opposing faces not exceeding 2.00 m	-	1.59	20.89	2.79	m²	**23.68**
distance between opposing faces 2.00 - 4.00 m	-	1.73	22.73	3.51	m²	**26.24**
distance between opposing faces exceeding 4.00 m	-	1.93	25.36	4.46	m²	**29.82**
Maximum depth not exceeding 8.00 m						
distance between opposing faces not exceeding 2.00 m	-	1.93	25.36	3.62	m²	**28.98**
distance between opposing faces 2.00 - 4.00 m	-	2.14	28.12	4.18	m²	**32.30**
distance between opposing faces exceeding 4.00 m	-	2.44	32.06	5.01	m²	**37.07**

D GROUNDWORK

Item	PC £	Labour hours	Labour £	Material /Plant £	Unit	Total rate £
D20 EXCAVATING AND FILLING – cont'd						
Extra over earthwork support for						
Curved	-	0.02	0.26	0.48	m²	**0.74**
Below ground water level	-	0.31	4.07	0.43	m²	**4.50**
Unstable ground	-	0.51	6.70	0.84	m²	**7.54**
Next to roadways	-	0.41	5.39	0.70	m²	**6.09**
Left in	-	0.67	8.80	19.50	m²	**28.31**
Earthwork support (average "risk" prices - inside existing existing buildings)						
Maximum depth not exceeding 1.00 m						
distance between opposing faces not exceeding 2.00 m	-	0.19	2.50	0.70	m²	**3.20**
distance between opposing faces 2.00 - 4.00 m	-	0.21	2.76	0.80	m²	**3.56**
distance between opposing faces exceeding 4.00 m	-	0.24	3.15	0.98	m²	**4.14**
Maximum depth not exceeding 2.00 m						
distance between opposing faces not exceeding 2.00 m	-	0.24	3.15	0.80	m²	**3.96**
distance between opposing faces 2.00 - 4.00 m	-	0.27	3.55	1.07	m²	**4.61**
distance between opposing faces exceeding 4.00 m	-	0.32	4.20	1.21	m²	**5.42**
Maximum depth not exceeding 4.00 m						
distance between opposing faces not exceeding 2.00 m	-	0.31	4.07	1.07	m²	**5.14**
distance between opposing faces 2.00 - 4.00 m	-	0.34	4.47	1.26	m²	**5.73**
distance between opposing faces exceeding 4.00 m	-	0.38	4.99	1.48	m²	**6.47**
Maximum depth not exceeding 6.00 m						
distance between opposing faces not exceeding 2.00 m	-	0.38	4.99	1.20	m²	**6.19**
distance between opposing faces 2.00 - 4.00 m	-	0.42	5.52	1.48	m²	**6.99**
distance between opposing faces exceeding 6.00 m	-	0.48	6.31	1.75	m²	**8.06**
Disposal; by machine						
Excavated material						
inactive waste off site; to tip not exceeding 13 km (using lorries); including Landfill Tax	-	-	-	24.83	m³	**24.83**
active non-hazardous waste off site; to tip not exceeding 13 km (using lorries); including Landfill Tax	-	-	-	104.62	m³	**104.62**
inactive waste on site; depositing in spoil heaps; average 25 m distance	-	-	-	1.02	m³	**1.02**
on site; spreading; average 25 m distance	-	0.23	3.02	0.72	m³	**3.74**
on site; depositing in spoil heaps; average 50 m distance	-	-	-	1.74	m³	**1.74**
on site; spreading; average 50 m distance	-	0.23	3.02	1.33	m³	**4.35**
on site; depositing in spoil heaps; average 100 m distance	-	-	-	3.07	m³	**3.07**
on site; spreading; average 100 m distance	-	0.23	3.02	2.05	m³	**5.07**
on site; depositing in spoil heaps; average 200 m distance	-	-	-	4.40	m³	**4.40**
on site; spreading; average 200 m distance	-	0.23	3.02	2.77	m³	**5.79**
Disposal; by hand						
Excavated material						
inactive waste off site; to tip not exceeding 13 km (using lorries); including Landfill Tax	-	0.81	10.64	32.59	m³	**43.23**
active non-hazardous waste off site; to tip not exceeding 13 km (using lorries); including Landfill Tax	-	1.50	19.71	110.32	m³	**130.03**

D GROUNDWORK

Item	PC £	Labour hours	Labour £	Material /Plant £	Unit	Total rate £
inactive on site; depositing in spoil heaps; average 25 m distance	-	1.12	14.72	-	m³	**14.72**
on site; spreading; average 25 m distance	-	1.48	19.45	-	m³	**19.45**
on site; depositing in spoil heaps; average 50 m distance	-	1.48	19.45	-	m³	**19.45**
on site; spreading; average 50 m distance	-	1.79	23.52	-	m³	**23.52**
on site; depositing in spoil heaps; average 100 m distance	-	2.14	28.12	-	m³	**28.12**
on site; spreading; average 100 m distance	-	2.44	32.06	-	m³	**32.06**
on site; depositing in spoil heaps; average 200 m distance	-	3.15	41.39	-	m³	**41.39**
on site; spreading; average 200 m distance	-	3.46	45.46	-	m³	**45.46**
Filling to excavations; by machine						
Average thickness not exceeding 0.25 m						
arising from the excavations	-	0.19	2.50	2.77	m³	**5.26**
obtained off site; hardcore	25.42	0.21	2.76	33.62	m³	**36.37**
obtained off site; granular fill type one	31.57	0.21	2.76	43.43	m³	**46.19**
obtained off site; granular fill type two	29.70	0.21	2.76	41.16	m³	**43.92**
Average thickness exceeding 0.25 m						
arising from the excavations	-	0.16	2.10	2.05	m³	**4.15**
obtained off site; hardcore	21.79	0.18	2.37	28.49	m³	**30.86**
obtained off site; granular fill type one	31.57	0.18	2.37	42.40	m³	**44.77**
obtained off site; granular fill type two	29.70	0.18	2.37	40.14	m³	**42.50**
Filling to make up levels; by machine						
Average thickness not exceeding 0.25 m						
arising from the excavations	-	0.27	3.55	3.86	m³	**7.40**
obtained off site; imported topsoil	18.90	0.27	3.55	25.91	m³	**29.45**
obtained off site; hardcore	25.42	0.31	4.07	34.64	m³	**38.71**
obatined off site; granular fill type one	31.57	0.31	4.07	44.45	m³	**48.53**
obtained off site; granular fill type two	29.70	0.31	4.07	42.18	m³	**46.26**
obtained off site; sand	35.64	0.31	4.07	49.39	m³	**53.47**
Average thickness exceeding 0.25 m						
arising from the excavations	-	0.22	2.89	2.82	m³	**5.71**
obtained off site; imported topsoil	18.90	0.22	2.89	24.87	m³	**27.76**
obtained off site; hardcore	21.79	0.27	3.55	29.27	m³	**32.82**
obatined off site; granular fill type one	31.57	0.27	3.55	43.19	m³	**46.73**
obtained off site; granular fill type two	29.70	0.27	3.55	40.92	m³	**44.47**
obtained off site; sand	35.64	0.27	3.55	48.13	m³	**51.67**
Filling to excavations; by hand						
Average thickness not exceeding 0.25 m						
arising from the excavations	-	1.25	16.42	-	m³	**16.42**
obtained off site; hardcore	25.42	1.35	17.74	30.85	m³	**48.59**
obtained off site; granular fill type one	31.57	1.60	21.02	38.31	m³	**59.33**
obtained off site; granular fill; type two	29.70	1.60	21.02	36.04	m³	**57.06**
obtained off site; sand	35.64	1.60	21.02	43.25	m³	**64.27**
Average thickness exceeding 0.25 m						
arising from the excavations	-	1.02	13.40	-	m³	**13.40**
obtained off site; hardcore	21.79	1.19	15.64	26.44	m³	**42.08**
obtained off site; granular fill type one	31.57	1.32	17.34	38.31	m³	**55.65**
obtained off site; granular fill; type two	29.70	1.32	17.34	36.04	m³	**53.38**
obtained off site; sand	35.64	1.32	17.34	43.25	m³	**60.59**

D GROUNDWORK

Item	PC £	Labour hours	Labour £	Material /Plant £	Unit	Total rate £
D20 EXCAVATING AND FILLING – cont'd						
Filling to make up levels; by hand						
Average thickness not exceeding 0.25 m						
arising from the excavations	-	1.38	18.13	9.26	m³	**27.39**
obtained off site; imported topsoil	18.90	1.38	18.13	30.59	m³	**48.72**
obtained off site; hardcore	25.42	1.71	22.47	42.34	m³	**64.81**
obtained off site; granular fill type one	31.57	1.82	23.91	50.54	m³	**74.46**
obtained off site; granular fill type two	29.70	1.82	23.91	48.27	m³	**72.19**
obtained off site; sand	35.64	1.82	23.91	55.48	m³	**79.39**
Average thickness exceeding 0.25 m						
arising from the excavations	-	1.19	15.64	7.51	m³	**23.14**
arising from on site spoil heaps; average 25 m distance; multiple handling	-	2.44	32.06	16.39	m³	**48.45**
obtained off site; imported topsoil	18.90	1.19	15.64	28.84	m³	**44.47**
obtained off site; hardcore	21.79	1.57	20.63	37.00	m³	**57.63**
obtained off site; granular fill type one	31.57	1.68	22.07	49.61	m³	**71.69**
obtained off site; granular fill type two	29.70	1.68	22.07	47.34	m³	**69.42**
obtained off site; sand	35.64	1.68	22.07	54.55	m³	**76.62**
Surface packing to filling						
To vertical or battered faces	-	0.19	2.50	0.13	m²	**2.63**
Surface treatments						
Compacting						
filling; blinding with sand	-	0.05	0.66	2.35	m²	**3.01**
bottoms of excavations	-	0.05	0.66	0.02	m²	**0.68**
Trimming						
sloping surfaces	-	0.19	2.50	-	m²	**2.50**
sloping surfaces; in rock	-	1.02	12.31	2.03	m²	**14.33**
Filter membrane; one layer; laid on earth to receive granular material						
"Terram 500" filter membrane; one layer; laid on earth	-	0.05	0.66	0.52	m²	**1.18**
"Terram 700" filter membrane; one layer; laid on earth	-	0.05	0.66	0.56	m²	**1.21**
"Terram 1000" filter membrane; one layer; laid on earth	-	0.05	0.66	0.59	m²	**1.24**
"Terram 2000" filter membrane; one layer; laid on earth	-	0.05	0.66	1.25	m²	**1.90**
D41 CRIB WALLS/GABIONS/REINFORCED EARTHWORKS						
Gabion baskets						
Wire mesh gabion baskets; Maccaferri Ltd or other equal and approved; galvanised mesh 80 mm x 100 mm; filling with broken stones 125 mm - 200 mm size						
2.00 x 1.00 x 0.50 m	17.18	1.25	28.28	117.27	nr	**145.55**
2.00 x 1.00 x 0.50 m; pvc coated	21.88	1.25	28.28	122.84	nr	**151.11**
2.00 x 1.00 x 0.50 m	24.08	2.50	56.55	222.37	nr	**278.92**
2.00 x 1.00 x 0.50 m; pvc coated	30.83	2.50	56.55	230.37	nr	**286.92**
Reno mattress gabion baskets or other equal and approved; Maccaferri Ltd; filling with broken stones 125 mm - 200 mm size						
6.00 x 2.00 x 0.17 m	66.33	2.50	56.55	267.37	nr	**323.92**
6.00 x 2.00 x 0.23 m	72.20	3.00	67.86	340.48	nr	**408.34**
6.00 x 2.00 x 0.30 m	83.36	3.50	79.17	425.32	nr	**504.49**

D GROUNDWORK

Item	PC £	Labour hours	Labour £	Material /Plant £	Unit	Total rate £
D50 UNDERPINNING						
Excavating; by machine						
Preliminary trenches						
maximum depth not exceeding 1.00 m	-	0.23	3.02	8.50	m³	**11.52**
maximum depth not exceeding 2.00 m	-	0.28	3.68	10.24	m³	**13.92**
maximum depth not exceeding 4.00 m	-	0.32	4.20	11.98	m³	**16.19**
Extra over preliminary trench excavating for breaking out existing hard pavings, 150 mm thick						
concrete	-	0.65	8.54	1.02	m²	**9.56**
Excavating; by hand						
Preliminary trenches						
maximum depth not exceeding 1.00 m	-	2.68	35.21	-	m³	**35.21**
maximum depth not exceeding 2.00 m	-	3.05	40.07	-	m³	**40.07**
maximum depth not exceeding 4.00 m	-	3.93	51.64	-	m³	**51.64**
Extra over preliminary trench excavating for breakig out existing hard pavings, 150 mm thick						
concrete	-	0.28	3.68	2.46	m²	**6.14**
Underpinning pits; commencing from 1.00 m below existing ground level						
maximum depth not exceeding 0.25 m	-	4.07	53.47	-	m³	**53.47**
maximum depth not exceeding 1.00 m	-	4.44	58.34	-	m³	**58.34**
maximum depth not exceeding 2.00 m	-	5.32	69.90	-	m³	**69.90**
Underpinning pits; commencing from 2.00 m below existing ground level						
maximum depth not exceeding 0.25 m	-	5.00	65.69	-	m³	**65.69**
maximum depth not exceeding 1.00 m	-	5.37	70.55	-	m³	**70.55**
maximum depth not exceeding 2.00 m	-	6.24	81.99	-	m³	**81.99**
Underpinning pits; commencing from 4.00 m below existing ground level						
maximum depth not exceeding 0.25 m	-	5.92	77.78	-	m³	**77.78**
maximum depth not exceeding 1.00 m	-	6.29	82.64	-	m³	**82.64**
maximum depth not exceeding 2.00 m	-	7.17	94.20	-	m³	**94.20**
Extra over any types of excavating irrespective of depth for excavating below ground water level	-	0.32	4.20	-	m³	**4.20**
Earthwork support to preliminary trenches (open boarded - in 3.00 m lengths)						
Maximum depth not exceeding 1.00 m						
distance between opposing faces not exceeding 2.00 m	-	0.37	4.86	1.82	m²	**6.68**
Maximum depth not exceeding 2.00 m						
distance between opposing faces not exceeding 2.00 m	-	0.46	6.04	2.23	m²	**8.27**
Maximum depth not exceeding 4.00 m						
distance between opposing faces not exceeding 2.00 m	-	0.59	7.75	2.79	m²	**10.54**
Earthwork support to underpinning pits (open boarded - in 3.00 m lengths)						
Maximum depth not exceeding 1.00 m						
distance between opposing faces not exceeding 2.00 m	-	0.41	5.39	1.95	m²	**7.34**
Maximum depth not exceeding 2.00 m						
distance between opposing faces not exceeding 2.00 m	-	0.51	6.70	2.51	m²	**9.21**
Maximum depth not exceeding 4.00 m						
distance between opposing faces not exceeding 2.00 m	-	0.65	8.54	3.06	m²	**11.60**

D GROUNDWORK

Item	PC £	Labour hours	Labour £	Material /Plant £	Unit	Total rate £
D50 UNDERPINNING – cont'd						
Earthwork support to preliminary trenches (closed boarded - in 3.00 m lengths)						
Maximum depth not exceeding 1.00 m						
distance between opposing faces not exceeding 2.00 m	–	0.93	12.22	3.06	m²	15.28
Maximum depth not exceeding 2.00 m						
distance between opposing faces not exceeding 2.00 m	–	1.16	15.24	3.90	m²	19.14
Maximum depth not exceeding 4.00 m						
distance between opposing faces not exceeding 2.00 m	–	1.43	18.79	4.74	m²	23.52
Earthwork support to underpinning pits (closed boarded - in 3.00 m lengths)						
Maximum depth not exceeding 1.00 m						
distance between opposing faces not exceeding 2.00 m	–	1.02	13.40	3.34	m²	16.74
Maximum depth not exceeding 2.00 m						
distance between opposing faces not exceeding 2.00 m	–	1.28	16.82	4.18	m²	21.00
Maximum depth not exceeding 4.00 m						
distance between opposing faces not exceeding 2.00 m	–	1.57	20.63	5.29	m²	25.92
Extra over earthwork support for						
Left in	–	0.69	9.07	19.50	m²	28.57
Cutting away existing projecting foundations						
Concrete						
maximum width 150 mm; maximum depth 150 mm	–	0.15	1.97	0.20	m	2.17
maximum width 150 mm; maximum depth 225 mm	–	0.22	2.89	0.30	m	3.19
maximum width 150 mm; maximum depth 300 mm	–	0.30	3.94	0.40	m	4.34
maximum width 300 mm; maximum depth 300 mm	–	0.58	7.62	0.78	m	8.40
Masonry						
maximum width one brick thick; maximum depth one course high	–	0.04	0.53	0.06	m	0.59
maximum width one brick thick; maximum depth two courses high	–	0.13	1.71	0.18	m	1.88
maximum wodth one brick thick; maximum depth three courses high	–	0.25	3.28	0.34	m	3.62
maximum width one brick thick; maximum depth four courses high	–	0.42	5.52	0.56	m	6.08
Preparing the underside of the existing work to receive the pinning up of the new work						
Width of existing work						
380 mm	–	0.56	7.36	–	m	7.36
600 mm	–	0.74	9.72	–	m	9.72
900 mm	–	0.93	12.22	–	m	12.22
1200 mm	–	1.11	14.58	–	m	14.58
Disposal; by hand						
Excavated material						
off site; to tip not exceeding 13 km (using lorries); including Landfill Tax based on inactive waste	–	0.74	9.72	42.68	m³	52.40

D GROUNDWORK

Item	PC £	Labour hours	Labour £	Material /Plant £	Unit	Total rate £
Filling to excavations; by hand						
Average thickness exceeding 0.25 m						
arising from the excavations	-	0.93	12.22	-	m³	**12.22**
Surface treatments						
Compacting						
bottoms of excavations	-	0.05	0.66	0.02	m²	**0.68**
Plain in situ ready mixed designated concrete C10 - 40 mm aggregate; poured against faces of excavation						
Underpinning						
thickness not exceeding 150 mm	-	3.42	57.90	104.57	m³	**162.48**
thickness 150 - 450 mm	-	2.87	48.59	104.57	m³	**153.17**
thickness exceeding 450 mm	-	2.50	42.33	104.57	m³	**146.90**
Plain in situ ready mixed designated concrete C20 - 20 mm aggregate; poured against faces of excavation						
Underpinning						
thickness not exceeding 150 mm	-	3.42	57.90	109.13	m³	**167.04**
thickness 150 - 450 mm	-	2.87	48.59	109.13	m³	**157.73**
thickness exceeding 450 mm	-	2.50	42.33	109.13	m³	**151.46**
Extra for working around reinforcement	-	0.28	4.74	-	m³	**4.74**
Sawn formwork; sides of foundations in underpinning						
Plain vertical						
height exceeding 1.00 m	-	1.48	29.54	6.97	m²	**36.50**
height not exceeding 250 mm	-	0.51	10.18	2.00	m²	**12.18**
height 250 - 500 mm	-	0.79	15.77	3.72	m²	**19.49**
height 500 mm - 1.00 m	-	1.20	23.95	6.97	m²	**30.91**
Reinforcement bar; BS4449; hot rolled deformed high yield steel bars						
20 mm diameter nominal size						
bent	668.25	27.00	524.77	815.47	tonne	**1340.24**
16 mm diameter nominal size						
bent	668.25	29.00	564.66	818.40	tonne	**1383.06**
12 mm diameter nominal size						
bent	708.75	31.00	604.55	869.34	tonne	**1473.88**
10 mm diameter nominal size						
bent	726.30	33.00	644.44	894.46	tonne	**1538.90**
8 mm diameter nominal size						
bent	726.30	35.00	679.74	897.39	tonne	**1577.14**
Common bricks; in cement mortar (1:3)						
Walls in underpinning						
one brick thick (PC £ per 1000)	270.00	2.22	60.94	43.75	m²	**104.69**
one and a half brick thick	-	3.05	83.73	65.20	m²	**148.93**
two brick thick	-	3.79	104.04	90.15	m²	**194.19**
Class A engineering bricks; In cement mortar (1:3)						
Walls in underpinning						
one brick thick (PC £ per 1000)	390.00	2.22	60.94	59.90	m²	**120.84**
one and a half brick thick	-	3.05	83.73	89.42	m²	**173.14**
two brick thick	-	3.79	104.04	122.44	m²	**226.48**

D GROUNDWORK

Item	PC £	Labour hours	Labour £	Material /Plant £	Unit	Total rate £
D50 UNDERPINNING – cont'd						
Class B engineering bricks; in cement mortar (1:3)						
Walls in underpinning						
one brick thick (PC £ per 1000)	315.00	2.22	60.94	49.24	m²	**110.18**
one and a half brick thick	-	3.05	83.73	73.43	m²	**157.16**
two brick thick	-	3.79	104.04	101.12	m²	**205.16**
Add or deduct for variation of £10.00/1000 in PC of bricks						
one brick thick	-	-	-	1.46	m²	-
one and a half brick thick	-	-	-	2.19	m²	-
two brick thick	-	-	-	2.92	m²	-
Zedex CPT" (Co-Polymer Thermoplastic) damp proof course or other equal and approved; 200 mm laps; in gauged mortar (1:1:6)						
Horizontal						
width exceeding 225 mm	4.34	0.27	7.41	5.27	m²	**12.68**
width not exceeding 225 mm	-	0.53	14.55	5.27	m²	**19.82**
"Hyload" (pitch polymer) damp proof course or equal and approved; 150 mm laps; in cement mortar (1:3)						
Horizontal						
width exceeding 225 mm	5.19	0.23	6.31	6.29	m²	**12.61**
width not exceeding 225 mm	5.31	0.46	12.63	6.44	m²	**19.07**
"Alumite" aluminium cored bitumen gas retardent damp proof course or other equal and approved; 200 mm laps; in gauged mortar (1:1;6)						
Horizontal						
width exceeding 225 mm	6.40	0.35	9.61	7.77	m²	**17.37**
width not exceeding 225 mm	-	0.68	18.67	7.77	m²	**26.43**
Two courses of slates in cement mortar (1:3)						
Horizontal						
width exceeding 225 mm	-	1.39	38.16	32.81	m²	**70.97**
width not exceeding 225 mm	-	2.31	63.41	33.56	m²	**96.97**
Wedging and pinning						
To underside of existing construction with slates in cement mortar (1:3)						
width of wall - half brick thick	-	1.02	28.00	7.63	m	**35.63**
width of wall - one brick thick	-	1.20	32.94	15.26	m	**48.20**
width of wall - one and a half brick thick	-	1.39	38.16	22.89	m	**61.05**

E IN SITU CONCRETE/LARGE PRECAST CONCRETE

Item	PC £	Labour hours	Labour £	Material £	Unit	Total rate £
E10 IN SITU CONCRETE CONSTRUCTION						
Plain in situ ready mixed designated concrete; C7.5 - 40 mm aggregate						
Foundations	77.20	1.39	23.53	96.02	m³	**119.56**
Isolated foundations	-	1.62	27.43	96.02	m³	**123.45**
Beds						
thickness not exceeding 150 mm	-	1.90	32.17	96.02	m³	**128.19**
thickness 150 - 450 mm	-	1.30	22.01	96.02	m³	**118.03**
thickness exceeding 450 mm	-	1.06	17.95	96.02	m³	**113.97**
Screeded beds; protection to compressible formwork						
50 mm thick	-	0.12	2.03	4.69	m²	**6.72**
75 mm thick	-	0.17	2.88	7.02	m²	**9.90**
100 mm thick	-	0.23	3.89	9.37	m²	**13.26**
Filling hollow walls						
thickness not exceeding 150 mm	-	3.61	61.12	96.02	m³	**157.14**
Column casings						
stub columns beneath suspended ground slabs	-	5.20	88.04	96.02	m³	**184.07**
Plain in situ ready mixed designated concrete; C10 - 40 mm aggregate						
Foundations	78.35	1.39	23.53	97.45	m³	**120.98**
Isolated foundations	-	1.62	27.43	97.45	m³	**124.87**
Beds						
thickness not exceeding 150 mm	-	1.90	32.17	97.45	m³	**129.61**
thickness 150 - 450 mm	-	1.30	22.01	97.45	m³	**119.46**
thickness exceeding 450 mm	-	1.06	17.95	97.45	m³	**115.39**
Filling hollow walls						
thickness not exceeding 150 mm	-	3.61	61.12	97.45	m³	**158.57**
Plain in situ ready mixed designated concrete; C10 - 40 mm aggregate; poured on or against earth or unblinded hardcore						
Foundations	78.35	1.43	24.21	99.82	m³	**124.04**
Isolated foundations	-	1.71	28.95	99.82	m³	**128.78**
Beds						
thickness not exceeding 150 mm	-	1.99	33.69	99.82	m³	**133.52**
thickness 150 - 450 mm	-	1.43	24.21	99.82	m³	**124.04**
thickness exceeding 450 mm	-	1.11	18.79	99.82	m³	**118.62**
Plain in situ ready mixed designated concrete; C20 - 20 mm aggregate						
Foundations	81.76	1.39	23.53	101.70	m³	**125.23**
Isolated foundations	-	1.62	27.43	101.70	m³	**129.12**
Beds						
thickness not exceeding 150 mm	-	2.04	34.54	101.70	m³	**136.24**
thickness 150 - 450 mm	-	1.39	23.53	101.70	m³	**125.23**
thickness exceeding 450 mm	-	1.06	17.95	101.70	m³	**119.64**
Filling hollow walls						
thickness not exceeding 150 mm	-	3.61	61.12	101.70	m³	**162.82**
Plain in situ ready mixed concrete; C20 - 20 mm aggregate; poured on or against earth or unblinded hardcore						
Foundations	81.76	1.43	24.21	104.18	m³	**128.39**
Isolated foundations	-	1.71	28.95	104.18	m³	**133.13**
Beds						
thickness not exceeding 150 mm	-	2.13	36.06	104.18	m³	**140.24**
thickness 150 - 450 mm	-	1.48	25.06	104.18	m³	**129.24**
thickness exceeding 450 mm	-	1.11	18.79	104.18	m³	**122.97**

E IN SITU CONCRETE/LARGE PRECAST CONCRETE

Item	PC £	Labour hours	Labour £	Material £	Unit	Total rate £
E10 IN SITU CONCRETE CONSTRUCTION – cont'd						
Reinforced in situ ready mixed designated concrete; C25 - 20 mm aggregate						
Foundations	83.00	1.48	25.06	103.23	m³	**128.29**
Ground beams	-	2.96	50.12	103.23	m³	**153.35**
Isolated foundations	-	1.80	30.48	103.23	m³	**133.71**
Beds						
thickness not exceeding 150 mm	-	2.36	39.96	103.23	m³	**143.19**
thickness 150 - 450 mm	-	1.71	28.95	103.23	m³	**132.19**
thickness exceeding 450 mm	-	1.39	23.53	103.23	m³	**126.77**
Slabs						
thickness not exceeding 150 mm	-	3.75	63.49	103.23	m³	**166.72**
thickness 150 - 450 mm	-	2.96	50.12	103.23	m³	**153.35**
thickness exceeding 450 mm	-	2.68	45.38	103.23	m³	**148.61**
Coffered and troughed slabs						
thickness 150 - 450 mm	-	3.42	57.90	103.23	m³	**161.14**
thickness exceeding 450 mm	-	2.96	50.12	103.23	m³	**153.35**
Extra over for sloping						
not exceeding 15 degrees	-	0.28	4.74	-	m³	**4.74**
over 15 degrees	-	0.56	9.48	-	m³	**9.48**
Walls						
thickness not exceeding 150 mm	-	3.93	66.54	103.23	m³	**169.77**
thickness 150 - 450 mm	-	3.15	53.33	103.23	m³	**156.57**
thickness exceeding 450 mm	-	2.75	46.56	103.23	m³	**149.79**
Beams						
isolated	-	4.26	72.13	103.23	m³	**175.36**
isolated deep	-	4.67	79.07	103.23	m³	**182.30**
attached deep	-	4.26	72.13	103.23	m³	**175.36**
Beam casings						
isolated	-	4.67	79.07	103.23	m³	**182.30**
isolated deep	-	5.09	86.18	103.23	m³	**189.41**
attached deep	-	4.67	79.07	103.23	m³	**182.30**
Columns	-	5.09	86.18	103.23	m³	**189.41**
Column casings	-	5.64	95.49	103.23	m³	**198.72**
Staircases	-	6.38	108.02	103.23	m³	**211.25**
Upstands	-	4.12	69.76	103.23	m³	**172.99**
Reinforced in situ ready mixed designated concrete; C35 - 20 mm aggregate						
Foundations	87.20	1.48	25.06	108.46	m³	**133.51**
Ground beams	-	2.96	50.12	108.46	m³	**158.57**
Isolated foundations	-	1.80	30.48	108.46	m³	**138.93**
Beds						
thickness not exceeding 150 mm	-	2.36	39.96	108.46	m³	**148.41**
thickness 150 - 450 mm	-	1.71	28.95	108.46	m³	**137.41**
thickness exceeding 450 mm	-	1.39	23.53	108.46	m³	**131.99**
Slabs						
thickness not exceeding 150 mm	-	3.75	63.49	108.46	m³	**171.95**
thickness 150 - 450 mm	-	2.96	50.12	108.46	m³	**158.57**
thickness exceeding 450 mm	-	2.68	45.38	108.46	m³	**153.83**
Coffered and troughed slabs						
thickness 150 - 450 mm	-	3.42	57.90	108.46	m³	**166.36**
thickness exceeding 450 mm	-	2.96	50.12	108.46	m³	**158.57**
Extra over for sloping						
not exceeding 15 degrees	-	0.28	4.74	-	m³	**4.74**
over 15 degrees	-	0.56	9.48	-	m³	**9.48**

E IN SITU CONCRETE/LARGE PRECAST CONCRETE

Item	PC £	Labour hours	Labour £	Material £	Unit	Total rate £
Walls						
thickness not exceeding 150 mm	-	3.93	66.54	108.46	m³	**175.00**
thickness 150 - 450 mm	-	3.15	53.33	108.46	m³	**161.79**
thickness exceeding 450 mm	-	2.75	46.56	108.46	m³	**155.02**
Beams						
isolated	-	4.26	72.13	108.46	m³	**180.58**
isolated deep	-	4.67	79.07	108.46	m³	**187.53**
attached deep	-	4.26	72.13	108.46	m³	**180.58**
Beam casings						
isolated	-	4.67	79.07	108.46	m³	**187.53**
isolated deep	-	5.09	86.18	108.46	m³	**194.64**
attached deep	-	4.67	79.07	108.46	m³	**187.53**
Columns	-	5.09	86.18	108.46	m³	**194.64**
Column casings	-	5.64	95.49	108.46	m³	**203.95**
Staircases	-	6.38	108.02	108.46	m³	**216.48**
Upstands	-	4.12	69.76	108.46	m³	**178.21**
Reinforced in situ ready mixed designated concrete; C40 - 20 mm aggregate						
Foundations	87.69	1.48	25.06	109.07	m³	**134.12**
Ground beams	-	2.96	50.12	109.07	m³	**159.18**
Isolated foundations	-	1.80	30.48	109.07	m³	**139.54**
Beds						
thickness not exceeding 150 mm	-	2.36	39.96	109.07	m³	**149.02**
thickness 150 - 450 mm	-	1.71	28.95	109.07	m³	**138.02**
thickness exceeding 450 mm	-	1.39	23.53	109.07	m³	**132.60**
Slabs						
thickness not exceeding 150 mm	-	3.75	63.49	109.07	m³	**172.56**
thickness 150 - 450 mm	-	2.96	50.12	109.07	m³	**159.18**
thickness exceeding 450 mm	-	2.68	45.38	109.07	m³	**154.44**
Coffered and troughed slabs						
thickness 150 - 450 mm	-	3.42	57.90	109.07	m³	**166.97**
thickness exceeding 450 mm	-	2.96	50.12	109.07	m³	**159.18**
Extra over for sloping						
not exceeding 15 degrees	-	0.28	4.74	-	m³	**4.74**
over 15 degrees	-	0.56	9.48	-	m³	**9.48**
Walls						
thickness not exceeding 150 mm	-	3.93	66.54	109.07	m³	**175.61**
thickness 150 - 450 mm	-	3.15	53.33	109.07	m³	**162.40**
thickness exceeding 450 mm	-	2.78	47.07	109.07	m³	**156.13**
Beams						
isolated	-	4.26	72.13	109.07	m³	**181.19**
isolated deep	-	4.67	79.07	109.07	m³	**188.13**
attached deep	-	4.26	72.13	109.07	m³	**181.19**
Beam casings						
isolated	-	4.67	79.07	109.07	m³	**188.13**
isolated deep	-	5.09	86.18	109.07	m³	**195.25**
attached deep	-	4.67	79.07	109.07	m³	**188.13**
Columns	-	5.09	86.18	109.07	m³	**195.25**
Column casings	-	5.64	95.49	109.07	m³	**204.56**
Staircases	-	6.38	108.02	109.07	m³	**217.09**
Upstands	-	4.12	69.76	109.07	m³	**178.82**
Extra over vibrated concrete for						
Reinforcement content over 5%	-	0.58	9.82	-	m³	**9.82**
Grouting with cement mortar (1:1)						
Stanchion bases						
10 mm thick	-	1.06	17.95	0.16	nr	**18.11**
25 mm thick	-	1.33	22.52	0.42	nr	**22.94**

E IN SITU CONCRETE/LARGE PRECAST CONCRETE

Item	PC £	Labour hours	Labour £	Material £	Unit	Total rate £
E10 IN SITU CONCRETE CONSTRUCTION – cont'd						
Grouting with epoxy resin						
Stanchion bases						
10 mm thick	-	1.33	22.52	12.14	nr	**34.66**
25 mm thick	-	1.60	27.09	31.02	nr	**58.11**
Grouting with "Conbextra GP" cementitious grout						
Stanchion bases						
10 mm thick	-	1.33	22.52	1.83	nr	**24.35**
25 mm thick	-	1.60	27.09	4.69	nr	**31.78**
Grouting with "Conbextra HF" flowable cementitious grout						
Stanchion bases						
10 mm thick	-	1.33	22.52	2.26	nr	**24.78**
25 mm thick	-	1.60	27.09	5.78	nr	**32.87**
Filling; plain ready mixed designated concrete; C20 - 20 mm aggregate						
Mortices	-	0.11	1.86	0.62	nr	**2.48**
Holes	-	0.27	4.57	128.12	m³	**132.69**
Chases exceeding 0.01 m²	-	0.21	3.56	128.12	m³	**131.67**
Chases not exceeding 0.01 m²	-	0.16	2.71	1.28	m	**3.99**
Sheeting to prevent moisture loss						
Building paper; lapped joints						
subsoil grade; horizontal on foundations	-	0.02	0.34	0.65	m²	**0.99**
standard grade; horizontal on slabs	-	0.05	0.85	0.98	m²	**1.82**
Polythene sheeting; lapped joints; horizontal on slabs						
250 microns; 0.25 mm thick	-	0.05	0.85	0.67	m²	**1.51**
"Visqueen" sheeting or other equal and approved; lapped joints; horizontal on slabs						
250 microns; 0.25 mm thick	-	0.05	0.85	0.56	m²	**1.40**
300 microns; 0.30 mm thick	-	0.06	1.02	0.64	m²	**1.65**
E20 FORMWORK FOR IN SITU CONCRETE						
NOTE: Generally all formwork based on four uses unless otherwise stated.						
Sides of foundations; basic finish						
Plain vertical						
height exceeding 1.00 m	-	1.70	33.93	10.10	m²	**44.03**
height exceeding 1.00 m; left in	-	1.49	29.73	24.02	m²	**53.76**
height not exceeding 250 mm	-	0.48	9.58	3.90	m	**13.48**
height not exceeding 250 mm; left in	-	0.48	9.58	6.92	m	**16.50**
height 250 - 500 mm	-	0.91	18.16	8.46	m	**26.62**
height 250 - 500 mm; left in	-	0.80	15.96	16.99	m	**32.95**
height 500 mm - 1.00 m	-	1.28	25.54	10.10	m	**35.65**
height 500 mm - 1.00 m ; left in	-	1.22	24.35	24.02	m	**48.37**
Sides of foundations; polystyrene sheet formwork; Cordek "Claymaster"; 50 mm thick						
Plain vertical						
height exceeding 1.00 m; left in	-	0.34	6.79	9.67	m²	**16.46**
height not exceeding 250 mm; left in	-	0.11	2.20	2.42	m	**4.61**
height 250 - 500 mm; left in	-	0.19	3.79	4.84	m	**8.63**
height 500 mm - 1.00 m; left in	-	0.28	5.59	9.67	m	**15.26**

E IN SITU CONCRETE/LARGE PRECAST CONCRETE

Item	PC £	Labour hours	Labour £	Material £	Unit	Total rate £
Sides of foundations; polystyrene sheet formwork; Cordek "Claymaster" or other equal and approved; 75 mm thick						
Plain vertical						
height exceeding 1.00 m; left in	-	0.34	6.79	14.49	m²	**21.28**
height not exceeding 250 mm; left in	-	0.11	2.20	3.62	m	**5.82**
height 250 - 500 mm; left in	-	0.19	3.79	7.25	m	**11.04**
height 500 mm - 1.00 m; left in	-	0.28	5.59	14.49	m	**20.08**
Sides of foundation; polystyrene sheet formwork; Cordek "Claymaster" or other equal and approved; 100 mm thick						
Plain vertical						
height exceeding 1.00 m; left in	-	0.37	7.38	19.30	m²	**26.69**
height not exceeding 250 mm; left in	-	0.12	2.39	4.83	m	**7.22**
height 250 - 500 mm; left in	-	0.20	3.99	9.65	m	**13.64**
height 500 mm - 1.00 m; left in	-	0.31	6.19	19.30	m	**25.49**
Combined heave pressure relief insultaion and compressible board substructure formawork; Cordeck "Cellcore CP" or other equal and approved; butt joints; securely fixed in place						
Plain horizontal						
200 mm thick; beneath slabs; left in	-	0.70	13.97	25.43	m²	**39.40**
250 mm thick; beneath slabs; left in	-	0.75	14.97	28.06	m²	**43.02**
300 mm thick; beneath slabs; left in	-	0.80	15.96	30.50	m²	**46.47**
Sides of ground beams and edges of beds; basic finish						
Plain vertical						
height exceeding 1.00 m	-	1.76	35.12	10.04	m²	**45.16**
height not exceeding 250 mm	-	0.53	10.58	3.84	m	**14.41**
height 250 - 500 mm	-	0.95	18.96	8.39	m	**27.35**
height 500 mm - 1.00 m	-	1.33	26.54	10.04	m	**36.58**
Edges of suspended slabs; basic finish						
Plain vertical						
height not exceeding 250 mm	-	0.80	15.96	3.96	m	**19.92**
height 250 - 500 mm	-	1.17	23.35	6.65	m	**30.00**
height 500 mm - 1.00 m	-	1.86	37.12	10.16	m	**47.28**
Sides of upstands; basic finish						
Plain vertical						
height exceeding 1.00 m	-	2.13	42.51	12.77	m²	**55.28**
height not exceeding 250 mm	-	0.67	13.37	4.08	m	**17.45**
height 250 - 500 mm	-	1.06	21.15	8.64	m	**29.79**
height 500 mm - 1.00 m	-	1.86	37.12	12.77	m	**49.89**
Steps in top surfaces; basic finish						
Plain vertical						
height not exceeding 250 mm	-	0.53	10.58	4.14	m	**14.72**
height 250 - 500 mm	-	0.85	16.96	8.70	m	**25.66**
Steps in soffits; basic finish						
Plain vertical						
height not exceeding 250 mm	-	0.58	11.57	3.29	m	**14.86**
height 250 - 500 mm	-	0.93	18.56	5.97	m	**24.53**

E IN SITU CONCRETE/LARGE PRECAST CONCRETE

Item	PC £	Labour hours	Labour £	Material £	Unit	Total rate £
E20 FORMWORK FOR IN SITU CONCRETE – cont'd						
Machine bases and plinths; basic finish						
Plain vertical						
height exceeding 1.00 m	–	1.70	33.93	10.04	m²	43.97
height not exceeding 250 mm	–	0.53	10.58	3.84	m	14.41
height 250 - 500 mm	–	0.91	18.16	8.39	m	26.55
height 500 mm - 1.00 m	–	1.33	26.54	10.04	m	36.58
Soffits of slabs; basic finish						
Slab thickness not exceeding 200 mm						
horizontal; height to soffit not exceeding 1.50 m	–	1.92	38.32	9.37	m²	47.69
horizontal; height to soffit 1.50 - 3.00 m	–	1.86	37.12	9.49	m²	46.61
horizontal; height to soffit 1.50 - 3.00 m (based on 5 uses)	–	1.76	35.12	7.84	m²	42.96
horizontal; height to soffit 1.50 - 3.00 m (based on 6 uses)	–	1.70	33.93	6.73	m²	40.66
horizontal; height to soffit 3.00 - 4.50 m	–	1.81	36.12	9.80	m²	45.92
horizontal; height to soffit 4.50 - 6.00 m	–	1.92	38.32	10.10	m²	48.42
Slab thickness 200 - 300 mm						
horizontal; height to soffit 1.50 - 3.00 m	–	1.92	38.32	12.59	m²	50.91
Slab thickness 300 - 400 mm						
horizontal; height to soffit 1.50 - 3.00 m	–	1.97	39.31	14.14	m²	53.45
Slab thickness 400 - 500 mm						
horizontal; height to soffit 1.50 - 3.00 m	–	2.07	41.31	15.69	m²	57.00
Slab thickness 500 - 600 mm						
horizontal; height to soffit 1.50 - 3.00 m	–	2.23	44.50	15.69	m²	60.19
Extra over soffits of slabs for						
sloping not exceeding 15 degrees	–	0.21	4.19	–	m²	4.19
sloping exceeding 15 degrees	–	0.43	8.58	–	m²	8.58
Soffits of landings; basic finish						
Slab thickness not exceeding 200 mm						
horizontal; height to soffit 1.50 - 3.00 m	–	1.92	38.32	10.11	m²	48.43
Slab thickness 200 - 300 mm						
horizontal; height to soffit 1.50 - 3.00 m	–	2.02	40.31	13.58	m²	53.89
Slab thickness 300 - 400 mm						
horizontal; height to soffit 1.50 - 3.00 m	–	2.07	41.31	15.23	m²	56.53
Slab thickness 400 - 500 mm						
horizontal; height to soffit 1.50 - 3.00 m	–	2.18	43.50	16.93	m²	60.43
Slab thickness 500 - 600 mm						
horizontal; height to soffit 1.50 - 3.00 m	–	2.34	46.70	16.93	m²	63.63
Extra over soffits of landings for						
sloping not exceeding 15 degrees	–	0.21	4.19	–	m²	4.19
sloping exceeding 15 degrees	–	0.43	8.58	–	m²	8.58
Soffits of coffered or troughed slabs; basic finish						
Cordek "Correx" trough mould or other equal and approved; 300 mm deep; ribs of mould at 600 mm centres and cross ribs at centres of bay; slab thickness 300 - 400 mm						
horizontal; height to soffit 1.50 - 3.00 m	–	2.66	53.08	14.86	m²	67.95
horizontal; height to soffit 3.00 - 4.50 m	–	2.77	55.28	15.17	m²	70.45
horizontal; height to soffit 4.50 - 6.00 m	–	2.88	57.47	15.35	m²	72.83
Top formwork; basic finish						
Sloping exceeding 15 degrees	–	1.60	31.93	6.70	m²	38.63

E IN SITU CONCRETE/LARGE PRECAST CONCRETE

Item	PC £	Labour hours	Labour £	Material £	Unit	Total rate £
Walls; basic finish						
Vertical	-	1.92	38.32	12.59	m²	**50.91**
Vertical; height exceeding 3.00 m above floor level	-	2.34	46.70	12.90	m²	**59.59**
Vertical; interrupted	-	2.23	44.50	12.90	m²	**57.40**
Vertical; to one side only	-	3.73	74.44	16.30	m²	**90.74**
Vertical; exceeding 3.00 m high; inside stairwell	-	2.34	46.70	12.90	m²	**59.59**
Battered	-	2.98	59.47	13.45	m²	**72.92**
Beams; basic finish						
Attached to slabs						
regular shaped; square or rectangular; height to soffit 1.50 - 3.00 m	-	2.34	46.70	12.23	m²	**58.92**
regular shaped; square or rectangular; height to soffit 3.00 - 4.50 m	-	2.44	48.69	12.59	m²	**61.28**
regular shaped; spaure or rectangular; height to soffit 4.50 - 6.00 m	-	2.55	50.89	12.90	m²	**63.78**
Attached to walls						
regular shaped; square or rectangular; height to soffit 1.50 - 3.00 m	-	2.44	48.69	12.23	m²	**60.92**
Isolated						
regular shaped; square or rectangular; height to soffit 1.50 - 3.00 m	-	2.55	50.89	12.23	m²	**63.11**
regular shaped; square or rectangular; height to soffit 3.00 - 4.50 m	-	2.66	53.08	12.59	m²	**65.68**
regular shaped; square or rectangular; height to soffit 4.50 - 6.00 m	-	2.77	55.28	12.90	m²	**68.18**
Extra over for sloping not exceeding 15 degrees	-	0.32	6.39	1.49	m²	**7.87**
Extra over for sloping exceeding 15 degrees	-	0.64	12.77	2.98	m²	**15.75**
Beam casings; basic finish						
Attached to slabs						
regular shaped; square or rectangular; height to soffit 1.50 - 3.00 m	-	2.44	48.69	12.23	m²	**60.92**
regular shaped; square or rectangular; height to soffit 3.00 - 4.50 m	-	2.55	50.89	12.59	m²	**63.48**
Attached to walls						
regular shaped; square or rectangular; height to soffit 1.50 - 3.00 m	-	2.55	50.89	12.23	m²	**63.11**
Isolated						
regular shaped; square or rectangular; height to soffit 1.50 - 3.00 m	-	2.66	53.08	12.23	m²	**65.31**
regular shaped; square or rectangular; height to soffit 3.00 - 4.50 m	-	2.77	55.28	12.59	m²	**67.87**
Extra over beam casings for						
regular shaped; sloping not exceeding 15 degrees	-	0.32	6.39	1.49	m²	**7.87**
regular shaped; sloping exceeding 15 degrees	-	0.64	12.77	2.98	m²	**15.75**
Columns; basic finish						
Attached to walls						
regular shaped; square or rectangular; height to soffit 1.50 - 3.00 m	-	2.34	46.70	10.10	m²	**56.80**
Isolated						
regular shaped; square or rectangular; height to soffit 1.50 - 3.00 m	-	2.44	48.69	10.10	m²	**58.80**
regular shaped; circular; not exceeding 300 mm diameter; height to soffit 1.50 - 3.00 m	-	4.26	85.01	18.18	m²	**103.19**
regular shaped; circular; 300 - 600 mm diameter; height to soffit 1.50 - 3.00 m	-	3.99	79.63	15.69	m²	**95.32**
regular shaped; circular; 600 - 900 mm diameter; height to soffit 1.50 - 3.00 m	-	3.73	74.44	15.39	m²	**89.82**

E IN SITU CONCRETE/LARGE PRECAST CONCRETE

Item	PC £	Labour hours	Labour £	Material £	Unit	Total rate £
E20 FORMWORK FOR IN SITU CONCRETE – cont'd						
Column casings; basic finish						
Attached to walls						
regular shaped; square or rectangular; height to soffit; 1.50 - 3.00 m	-	2.44	48.69	10.10	m²	**58.80**
Isolated						
regular shaped; square or rectangular; height to soffit 1.50 - 3.00 m	-	2.55	50.89	10.10	m²	**60.99**
Recesses or rebates						
12 x 12 mm	-	0.07	1.40	0.30	m	**1.70**
25 x 25 mm	-	0.07	1.40	0.56	m	**1.96**
25 x 50 mm	-	0.07	1.40	0.73	m	**2.13**
50 x 50 mm	-	0.07	1.40	1.02	m	**2.41**
Nibs						
50 x 50 mm	-	0.58	11.57	1.42	m	**12.99**
100 x 100 mm	-	0.83	16.56	1.71	m	**18.27**
100 x 200 mm	-	1.11	22.15	10.44	m	**32.60**
Extra over a basic finish for fine formed finishes						
Slabs	-	0.35	6.98	-	m²	**6.98**
Walls	-	0.35	6.98	-	m²	**6.98**
Beams	-	0.35	6.98	-	m²	**6.98**
Columns	-	0.35	6.98	-	m²	**6.98**
Add to prices for basic formwork for						
Curved radius 6.00 m - 50%						
Curved radius 2.00 m - 100%						
Coating with retardant agent	-	0.01	0.20	0.36	m²	**0.56**
Wall kickers; basic finish						
Height 150 mm	-	0.53	10.58	2.84	m	**13.42**
Height 225 mm	-	0.69	13.77	3.34	m	**17.11**
Suspended wall kickers; basic finish						
Height 150 mm	-	0.67	13.37	2.67	m	**16.04**
Wall ends, soffits and steps in walls; basic finish						
Plain						
width exceeding 1.00 m	-	2.02	40.31	12.59	m²	**52.90**
width not exceeding 250 mm	-	0.64	12.77	2.98	m	**15.76**
width 250 - 500 mm	-	1.01	20.16	6.91	m	**27.06**
width 500 mm - 1.00 m	-	1.60	31.93	12.59	m	**44.52**
Openings in walls						
Plain						
width exceeding 1.00 m	-	2.23	44.50	12.59	m²	**57.09**
width not exceeding 250 mm	-	0.69	13.77	2.98	m	**16.75**
width 250 - 500 mm	-	1.17	23.35	6.91	m	**30.25**
width 500 mm - 1.00 m	-	1.81	36.12	12.59	m	**48.71**
Stairflights						
Width 1.00 m; 150 mm waist; 150 mm undercut risers						
string, width 300 mm	-	5.32	106.17	26.46	m	**132.63**
Width 2.00 m; 300 mm waist; 150 mm undercut risers						
string, width 350 mm	-	9.57	190.98	72.17	m	**263.15**

E IN SITU CONCRETE/LARGE PRECAST CONCRETE

Item	PC £	Labour hours	Labour £	Material £	Unit	Total rate £
Mortices						
Girth not exceeding 500 mm						
depth not exceeding 250 mm; circular	-	0.16	3.19	0.80	nr	**3.99**
Holes						
Girth not exceeding 500 mm						
depth not exceeding 250 mm; circular	-	0.21	4.19	1.22	nr	**5.41**
depth 250 - 500 mm; circular	-	0.33	6.59	3.48	nr	**10.06**
Girth 500 mm - 1.00 m						
depth not exceeding 250 mm; circular	-	0.27	5.39	1.88	nr	**7.27**
depth 250 - 500 mm; circular	-	0.41	8.18	6.04	nr	**14.22**
Girth 1.00 - 2.00 m						
depth not exceeding 250 mm; circular	-	0.48	9.58	6.04	nr	**15.62**
depth 250 - 500 mm; circular	-	0.71	14.17	12.66	nr	**26.83**
Girth 2.00 - 3.00 m						
depth not exceeding 250 mm; circular	-	0.64	12.77	11.75	nr	**24.52**
depth 250 - 500 mm; circular	-	0.95	18.96	70.09	nr	**89.05**
E30 REINFORCEMENT FOR IN SITU CONCRETE						
NOTE: Prices of steel rebar and fabric are particularly volatile at the time of going to press, so Readers are encouraged to contact their suppliers and check prices for currency based on anticipated delivery dates.						
Bars; BS 4449; hot rolled deformed high steel bars grade 500C						
40 mm diameter nominal size						
straight	679.05	17.00	339.06	825.33	tonne	**1164.40**
bent	704.70	17.00	339.06	855.73	tonne	**1194.80**
32 mm diameter nominal size						
straight	673.65	18.00	359.01	820.48	tonne	**1179.49**
bent	699.30	18.00	359.01	850.88	tonne	**1209.89**
25 mm diameter nominal size						
straight	668.25	20.00	398.90	813.58	tonne	**1212.48**
bent	695.25	20.00	398.90	845.58	tonne	**1244.48**
20 mm diameter nominal size						
straight	668.25	22.00	438.79	815.47	tonne	**1254.26**
bent	695.25	22.00	438.79	847.47	tonne	**1286.26**
16 mm diameter nominal size						
straight	668.25	24.00	478.68	818.40	tonne	**1297.08**
bent	695.25	24.00	478.68	850.40	tonne	**1329.08**
12 mm diameter nominal size						
straight	708.75	26.00	518.57	869.34	tonne	**1387.91**
bent	753.30	26.00	518.57	922.14	tonne	**1440.71**
10 mm diameter nominal size						
straight	726.30	28.00	558.46	894.46	tonne	**1452.92**
bent	772.20	28.00	558.46	948.86	tonne	**1507.32**
8 mm diameter nominal size						
straight	762.75	30.00	593.77	940.59	tonne	**1534.36**
links	762.75	33.00	653.60	943.68	tonne	**1597.28**
bent	799.20	30.00	593.77	983.79	tonne	**1577.56**

E IN SITU CONCRETE/LARGE PRECAST CONCRETE

Item	PC £	Labour hours	Labour £	Material £	Unit	Total rate £
E30 REINFORCEMENT FOR IN SITU CONCRETE – cont'd						
Bars; stainless steel; EN 1.4301						
32 mm diameter nominal size						
straight	3400.00	18.00	359.01	3913.76	tonne	**4272.77**
bent	3600.00	18.00	359.01	4139.51	tonne	**4498.52**
25 mm diameter nominal size						
straight	3400.00	20.00	398.90	3909.89	tonne	**4308.79**
bent	3600.00	20.00	398.90	4135.64	tonne	**4534.54**
20 mm diameter nominal size						
straight	3400.00	22.00	438.79	3915.15	tonne	**4353.94**
bent	3600.00	22.00	438.79	4140.90	tonne	**4579.69**
16 mm diameter nominal size						
straight	3400.00	24.00	478.68	3924.82	tonne	**4403.50**
bent	3400.00	24.00	478.68	3924.82	tonne	**4403.50**
12 mm diameter nominal size						
straight	3400.00	26.00	518.57	3934.50	tonne	**4453.07**
bent	3400.00	26.00	518.57	3934.50	tonne	**4453.07**
10 mm diameter nominal size						
straight	3400.00	28.00	558.46	3945.57	tonne	**4504.03**
bent	3400.00	28.00	558.46	3945.57	tonne	**4504.03**
8 mm diameter nominal size						
straight	3400.00	30.00	593.77	3955.24	tonne	**4549.01**
bent	3400.00	30.00	593.77	3955.24	tonne	**4549.01**
Bars; stainless steel; EN 1.4462						
32 mm diameter nominal size						
straight	5400.00	18.00	359.01	6171.26	tonne	**6530.27**
bent	5600.00	18.00	359.01	6397.01	tonne	**6756.02**
25 mm diameter nominal size						
straight	5400.00	20.00	398.90	6167.39	tonne	**6566.29**
bent	5600.00	20.00	398.90	6393.14	tonne	**6792.04**
20 mm diameter nominal size						
straight	5400.00	22.00	438.79	6172.65	tonne	**6611.44**
bent	5600.00	22.00	438.79	6398.40	tonne	**6837.19**
16 mm diameter nominal size						
straight	5400.00	24.00	478.68	6182.32	tonne	**6661.00**
bent	5600.00	24.00	478.68	6408.07	tonne	**6886.75**
12 mm diameter nominal size						
straight	5400.00	26.00	518.57	6192.00	tonne	**6710.57**
bent	5600.00	26.00	518.57	6417.75	tonne	**6936.32**
10 mm diameter nominal size						
straight	5400.00	28.00	558.46	6203.07	tonne	**6761.53**
bent	5600.00	28.00	558.46	6428.82	tonne	**6987.28**
8 mm diameter nominal size						
straight	5400.00	30.00	593.77	6212.74	tonne	**6806.51**
bent	5600.00	30.00	593.77	6438.49	tonne	**7032.26**

E IN SITU CONCRETE/LARGE PRECAST CONCRETE

Item	PC £	Labour hours	Labour £	Material £	Unit	Total rate £
Bars; stainless steel; LDX2101® (EN 1.4362)						
Note: LDX2101® (EN 1.4362) is a new low Ni, Mn bearing stainless steel alloy, which offers greater price stability and cost effectiveness, and is expected to be adopted into the British Standard in the near future.						
32 mm diameter nominal size						
straight	3350.00	18.00	359.01	3857.32	tonne	**4216.33**
bent	3500.00	18.00	359.01	4026.63	tonne	**4385.64**
25 mm diameter nominal size						
straight	3350.00	20.00	398.90	3853.46	tonne	**4252.36**
bent	3500.00	20.00	398.90	4022.77	tonne	**4421.67**
20 mm diameter nominal size						
straight	3350.00	22.00	438.79	3858.71	tonne	**4297.50**
bent	3450.00	22.00	438.79	3971.59	tonne	**4410.38**
16 mm diameter nominal size						
straight	3350.00	24.00	478.68	3868.39	tonne	**4347.07**
bent	3450.00	24.00	478.68	3981.26	tonne	**4459.94**
12 mm diameter nominal size						
straight	3350.00	26.00	518.57	3878.06	tonne	**4396.63**
bent	3400.00	26.00	518.57	3934.50	tonne	**4453.07**
10 mm diameter nominal size						
straight	3350.00	28.00	558.46	3889.13	tonne	**4447.59**
bent	3400.00	28.00	558.46	3945.57	tonne	**4504.03**
8 mm diameter nominal size						
straight	3350.00	30.00	593.77	3898.80	tonne	**4492.57**
bent	3450.00	30.00	593.77	4011.68	tonne	**4605.44**
Fabric; BS 4449						
Ref A98 (1.54 kg/m²)						
400 mm minimum laps	1.86	0.13	2.59	2.31	m²	**4.90**
strips in one width; 600 mm width	1.86	0.16	3.19	2.31	m²	**5.50**
strips in one width; 900 mm width	1.86	0.15	2.99	2.31	m²	**5.30**
strips in one width; 1200 mm width	1.86	0.14	2.79	2.31	m²	**5.10**
Ref A142 (2.22 kg/m²)						
400 mm minimum laps	1.77	0.13	2.59	2.20	m²	**4.79**
strips in one width; 600 mm width	1.77	0.16	3.19	2.20	m²	**5.39**
strips in one width; 900 mm width	1.77	0.15	2.99	2.20	m²	**5.19**
strips in one width; 1200 mm width	1.77	0.14	2.79	2.20	m²	**4.99**
Ref A193 (3.02 kg/m²)						
400 mm minimum laps	2.41	0.13	2.59	2.99	m²	**5.59**
strips in one width; 600 mm width	2.41	0.16	3.19	2.99	m²	**6.18**
strips in one width; 900 mm width	2.41	0.15	2.99	2.99	m²	**5.98**
strips in one width; 1200 mm width	2.41	0.14	2.79	2.99	m²	**5.78**
Ref A252 (3.95 kg/m²)						
400 mm minimum laps	3.16	0.14	2.79	3.92	m²	**6.72**
strips in one width; 600 mm width	3.16	0.17	3.39	3.92	m²	**7.31**
strips in one width; 900 mm width	3.16	0.16	3.19	3.92	m²	**7.11**
strips in one width; 1200 mm width	3.16	0.15	2.99	3.92	m²	**6.92**
Ref A393 (6.16 kg/m²)						
400 mm minimum laps	4.92	0.16	3.19	6.11	m²	**9.30**
strips in one width; 600 mm width	4.92	0.19	3.79	6.11	m²	**9.90**
strips in one width; 900 mm width	4.92	0.18	3.59	6.11	m²	**9.70**
strips in one width; 1200 mm width	4.92	0.17	3.39	6.11	m²	**9.50**

E IN SITU CONCRETE/LARGE PRECAST CONCRETE

Item	PC £	Labour hours	Labour £	Material £	Unit	Total rate £
E30 REINFORCEMENT FOR IN SITU CONCRETE – cont'd						
Fabric; BS 4449 – cont'd						
Ref B196 (3.05 kg/m²)						
400 mm minimum laps	4.49	0.13	2.59	5.57	m²	**8.17**
strips in one width; 600 mm width	4.49	0.16	3.19	5.57	m²	**8.77**
strips in one width; 900 mm width	4.49	0.15	2.99	5.57	m²	**8.57**
strips in one width; 1200 mm width	4.49	0.14	2.79	5.57	m²	**8.37**
Ref B283 (3.73 kg/m²)						
400 mm minimum laps	2.98	0.13	2.59	3.70	m²	**6.29**
strips in one width; 600 mm width	2.98	0.16	3.19	3.70	m²	**6.89**
strips in one width; 900 mm width	2.98	0.15	2.99	3.70	m²	**6.69**
strips in one width; 1200 mm width	2.98	0.14	2.79	3.70	m²	**6.49**
Ref B385 (4.53 kg/m²)						
400 mm minimum laps	3.62	0.14	2.79	4.49	m²	**7.29**
strips in one width; 600 mm width	3.62	0.17	3.39	4.49	m²	**7.89**
strips in one width; 900 mm width	3.62	0.16	3.19	4.49	m²	**7.69**
strips in one width; 1200 mm width	3.62	0.15	2.99	4.49	m²	**7.49**
Ref B503 (5.93 kg/m²)						
400 mm minimum laps	4.74	0.16	3.19	5.89	m²	**9.08**
strips in one width; 600 mm width	4.74	0.19	3.79	5.89	m²	**9.67**
strips in one width; 900 mm width	4.74	0.18	3.59	5.89	m²	**9.48**
strips in one width; 1200 mm width	4.74	0.17	3.39	5.89	m²	**9.28**
Ref B785 (8.14 kg/m²)						
400 mm minimum laps	6.51	0.19	3.79	8.08	m²	**11.87**
strips in one width; 600 mm width	6.51	0.22	4.39	8.08	m²	**12.47**
strips in one width; 900 mm width	6.51	0.21	4.19	8.08	m²	**12.27**
strips in one width; 1200 mm width	6.51	0.20	3.99	8.08	m²	**12.07**
Ref B1131 (10.90 kg/m²)						
400 mm minimum laps	8.72	0.20	3.99	10.83	m²	**14.82**
strips in one width; 600 mm width	8.72	0.26	5.19	10.83	m²	**16.01**
strips in one width; 900 mm width	8.72	0.24	4.79	10.83	m²	**15.61**
strips in one width; 1200 mm width	8.72	0.22	4.39	10.83	m²	**15.21**
Ref D49 (0.77 kg/m²)						
100 mm minimum laps; bent	1.66	0.26	5.19	2.06	m²	**7.25**
E40 DESIGNED JOINTS IN IN SITU CONCRETE						
Formed; Fosroc impregnated fibreboard joint filler or other equal and approved						
Width not exceeding 150 mm						
12.50 mm thick	-	0.16	3.19	1.70	m	**4.90**
20 mm thick	-	0.20	3.99	2.58	m	**6.57**
25 mm thick	-	0.25	4.99	2.69	m	**7.67**
Width 150 - 300 mm						
12.50 mm thick	-	0.26	5.19	2.69	m	**7.88**
20 mm thick	-	0.26	5.19	4.16	m	**9.35**
25 mm thick	-	0.26	5.19	4.91	m	**10.10**
Width 300 - 450 mm						
12.50 mm thick	-	0.31	6.19	3.78	m	**9.96**
20 mm thick	-	0.31	6.19	5.79	m	**11.98**
25 mm thick	-	0.31	6.19	6.83	m	**13.01**

E IN SITU CONCRETE/LARGE PRECAST CONCRETE

Item	PC £	Labour hours	Labour £	Material £	Unit	Total rate £
Formed; Grace Servicised "Kork-pak" waterproof bonded cork joint filler board or other equal and approved						
Width not exceeding 150 mm						
10 mm thick	-	0.16	3.19	4.57	m	**7.76**
13 mm thick	-	0.16	3.19	6.00	m	**9.19**
19 mm thick	-	0.16	3.19	6.89	m	**10.08**
25 mm thick	-	0.16	3.19	4.49	m	**7.69**
Width 150 - 300 mm						
10 mm thick	-	0.20	3.99	8.48	m	**12.47**
13 mm thick	-	0.20	3.99	11.34	m	**15.33**
19 mm thick	-	0.20	3.99	13.12	m	**17.11**
25 mm thick	-	0.20	3.99	8.32	m	**12.32**
Width 300 - 450 mm						
10 mm thick	-	0.26	5.19	12.91	m	**18.10**
13 mm thick	-	0.26	5.19	17.21	m	**22.40**
19 mm thick	-	0.26	5.19	19.88	m	**25.07**
25 mm thick	-	0.26	5.19	12.68	m	**17.87**
Sealants; Fosroc "Pliastic 77" hot poured rubberized bituminous compound or other equal and approved						
Width 10 mm						
25 mm depth	-	0.19	3.79	1.18	m	**4.97**
Width 12.50 mm						
25 mm depth	-	0.20	3.99	1.45	m	**5.44**
Width 20 mm						
25 mm depth	-	0.21	4.19	2.36	m	**6.55**
Width 25 mm						
25 mm depth	-	0.22	4.39	2.90	m	**7.29**
Sealants; Fosroc "Thioflex 600" gun grade two part polysulphide or other equal and approved						
Width 10 mm						
25 mm depth	-	0.06	1.20	4.48	m	**5.68**
Width 12.50 mm						
25 mm depth	-	0.07	1.40	5.61	m	**7.00**
Width 20 mm						
25 mm depth	-	0.08	1.60	8.97	m	**10.56**
Width 25 mm						
25 mm depth	-	0.09	1.80	11.21	m	**13.00**
Sealants; Grace Servicised "Paraseal" polysulphide compound or other equal and approved; priming with Grace Servicised "Servicised P" or other equal and approved						
Width 10 mm						
25 mm depth	-	0.20	3.07	4.29	m	**7.36**
Width 13 mm						
25 mm depth	-	0.20	3.07	5.51	m	**8.59**
Width 19 mm						
25 mm depth	-	0.26	3.99	7.97	m	**11.96**
Width 25 mm						
25 mm depth	-	0.26	3.99	10.41	m	**14.41**

E IN SITU CONCRETE/LARGE PRECAST CONCRETE

Item	PC £	Labour hours	Labour £	Material £	Unit	Total rate £
E40 DESIGNED JOINTS IN IN SITU CONCRETE – cont'd						
Waterstops; Grace Servicised or other equal and approved						
Hydrophilic strip water stop; lapped joints; cast into concrete						
5 x 20 mm "Servistrip AH 205"	-	0.35	5.38	8.08	m	**13.46**
50 x 20 mm "Adcor 500S"	-	0.35	5.38	10.35	m	**15.73**
"Servitite" internal 10 mm thick pvc water stop; flat dumbbell type; heat welded joints; cast into concrete						
Servitite 150; 150 mm wide	9.38	0.26	5.19	11.36	m	**16.54**
flat angle	15.02	0.31	6.18	21.95	nr	**28.13**
vertical angle	14.94	0.31	6.18	21.85	nr	**28.03**
flat three way intersection	21.66	0.41	8.18	31.93	nr	**40.11**
vertical three way intersection	25.10	0.41	8.18	35.91	nr	**44.09**
four way intersection	26.56	0.51	10.17	39.86	nr	**50.04**
Servitite 230; 230 mm wide	13.76	0.26	5.19	16.54	m	**21.73**
flat angle	17.82	0.31	6.18	27.25	nr	**33.44**
vertical angle	21.81	0.31	6.18	31.87	nr	**38.06**
flat three way intersection	25.84	0.41	8.18	39.87	nr	**48.05**
vertical three way intersection	49.30	0.41	8.18	67.02	nr	**75.20**
four way intersection	11.00	0.51	10.17	43.01	nr	**53.18**
Servitite AT200; 200 mm wide	21.22	0.26	5.19	25.39	m	**30.57**
flat angle	20.43	0.31	6.18	33.82	nr	**40.00**
vertical angle	22.56	0.31	6.18	36.28	nr	**42.46**
flat three way intersection	36.55	0.41	8.18	57.57	nr	**65.75**
vertical three way intersection	27.05	0.41	8.18	46.58	nr	**54.75**
four way intersection	42.28	0.51	10.17	69.28	nr	**79.45**
Servitite K305; 305 mm wide	22.71	0.31	6.18	27.16	m	**33.34**
flat angle	29.17	0.36	7.18	47.32	nr	**54.51**
vertical angle	32.08	0.36	7.18	50.69	nr	**57.87**
flat three way intersection	40.98	0.45	8.98	67.80	nr	**76.77**
vertical three way intersection	50.07	0.45	8.98	78.31	nr	**87.29**
four way intersection	55.80	0.56	11.17	91.72	nr	**102.88**
"Serviseal" external pvc water stop; centre bulb type; heat welded joints; cast into concrete						
Serviseal 195; 195 mm wide	6.89	0.26	5.19	8.40	m	**13.59**
flat angle	9.31	0.31	6.18	14.16	nr	**20.34**
vertical angle	16.72	0.31	6.18	22.73	nr	**28.91**
flat three way intersection	15.97	0.41	8.18	23.56	nr	**31.74**
four way intersection	24.09	0.51	10.17	34.64	nr	**44.81**
Serviseal 240; 240 mm wide	8.76	0.26	5.19	10.63	m	**15.81**
flat angle	10.59	0.31	6.18	16.53	nr	**22.71**
vertical angle	17.50	0.31	6.18	24.52	nr	**30.70**
flat three way intersection	17.90	0.41	8.18	27.14	nr	**35.32**
four way intersection	26.66	0.51	10.17	39.40	nr	**49.57**
Serviseal AT240; 240 mm wide	25.04	0.26	5.19	29.92	m	**35.11**
flat angle	19.03	0.31	6.18	34.01	nr	**40.20**
vertical angle	17.94	0.31	6.18	32.74	nr	**38.93**
flat three way intersection	29.30	0.41	8.18	51.91	nr	**60.08**
four way intersection	44.13	0.51	10.17	75.04	nr	**85.21**
Serviseal K320; 320 mm wide	11.76	0.31	6.18	14.18	m	**20.37**
flat angle	23.85	0.36	7.18	34.68	nr	**41.86**
vertical angle	20.30	0.36	7.18	30.58	nr	**37.76**
flat three way intersection	34.91	0.45	8.98	51.04	nr	**60.02**
four way intersection	43.49	0.56	11.17	64.50	nr	**75.67**

E IN SITU CONCRETE/LARGE PRECAST CONCRETE

Item	PC £	Labour hours	Labour £	Material £	Unit	Total rate £
E41 WORKED FINISHES/CUTTING TO IN SITU CONCRETE						
Worked finishes						
Tamping by mechanical means	-	0.02	0.34	0.11	m²	**0.44**
Power floating	-	0.18	3.05	0.35	m²	**3.40**
Trowelling	-	0.33	5.59	-	m²	**5.59**
Hacking						
by mechanical means	0.27	0.33	5.59	0.41	m²	**6.00**
by hand	-	0.71	12.02	-	m²	**12.02**
Lightly shot blasting surface of concrete	-	0.41	6.94	-	m²	**6.94**
Blasting surface of concrete						
to produce textured finish	-	0.71	12.02	0.87	m²	**12.89**
Wood float finish	-	0.13	2.20	-	m²	**2.20**
Tamped finish						
level or to falls	-	0.07	1.19	-	m²	**1.19**
to falls	-	0.10	1.69	-	m²	**1.69**
Spade finish	-	0.16	2.71	-	m²	**2.71**
Cutting chases						
Depth not exceeding 50 mm						
width 10 mm	-	0.33	5.59	0.73	m	**6.32**
width 50 mm	-	0.51	8.63	0.94	m	**9.58**
width 75 mm	-	0.68	11.51	1.14	m	**12.66**
Depth 50 - 100 mm						
width 75 mm	-	0.92	15.58	1.81	m	**17.38**
width 100 mm	-	1.02	17.27	1.93	m	**19.20**
width 100 mm; in reinforced concrete	-	1.53	25.90	3.04	m	**28.95**
Depth 100 - 150 mm						
width 100 mm	-	1.32	22.35	2.25	m	**24.60**
width 100 mm; in reinforced concrete	-	2.04	34.54	3.89	m	**38.43**
width 150 mm	-	1.63	27.60	2.62	m	**30.22**
width 150 mm; in reinforced concrete	-	2.44	41.31	4.36	m	**45.67**
Cutting rebates						
Depth not exceeding 50 mm						
width 50 mm	-	0.51	8.63	0.94	m	**9.58**
Depth 50 - 100 mm						
width 100 mm	-	1.02	17.27	1.93	m	**19.20**
NOTE: The following rates for cutting holes and mortices in concrete allow for diamond drilling.						
Diamond drilling						
Cutting holes and mortices in concrete; per 25 mm depth						
25 mm diameter	-	-	-	-	nr	**2.00**
32 mm diameter	-	-	-	-	nr	**1.55**
52 mm diameter	-	-	-	-	nr	**1.90**
78 mm diameter	-	-	-	-	nr	**2.20**
107 mm diameter	-	-	-	-	nr	**2.40**
127 mm diameter	-	-	-	-	nr	**2.65**
152 mm diameter	-	-	-	-	nr	**3.15**
200 mm diameter	-	-	-	-	nr	**4.20**
250 mm diameter	-	-	-	-	nr	**6.15**
300 mm diameter	-	-	-	-	nr	**8.15**

E IN SITU CONCRETE/LARGE PRECAST CONCRETE

Item	PC £	Labour hours	Labour £	Material £	Unit	Total rate £
E41 WORKED FINISHES/CUTTING TO IN SITU CONCRETE – cont'd						
Diamond drilling – cont'd						
Cutting holes and mortices in reinforced concrete; per 25 mm depth						
25 mm diameter	-	-	-	-	nr	**2.67**
32 mm diameter	-	-	-	-	nr	**2.36**
52 mm diameter	-	-	-	-	nr	**2.25**
78 mm diameter	-	-	-	-	nr	**2.36**
107 mm diameter	-	-	-	-	nr	**2.77**
127 mm diameter	-	-	-	-	nr	**3.23**
152 mm diameter	-	-	-	-	nr	**3.79**
200 mm diameter	-	-	-	-	nr	**5.48**
250 mm diameter	-	-	-	-	nr	**8.30**
300 mm diameter	-	-	-	-	nr	**10.71**
Other items in reinforced concrete						
Diamond chasing; per 25 x 25 mm section	-	-	-	-	m	**12.30**
Forming box; per 25 mm depth (per m of perimeter)	-	-	-	-	m	**4.92**
Diamond floor sawing; per 25 mm depth	-	-	-	-	m	**2.82**
Diamond track mount or ring sawing; per 25 mm depth	-	-	-	-	m	**10.25**
Stitch drilling 107 mm diameter hole; per 25 mm depth	-	-	-	-	nr	**35.88**
E42 ACCESSORIES CAST INTO IN SITU CONCRETE						
Foundation bolt boxes						
Temporary plywood; for group of 4 nr bolts						
75 x 75 x 150 mm	-	0.45	8.98	1.24	nr	**10.22**
75 x 75 x 250 mm	-	0.45	8.98	1.57	nr	**10.55**
Expanded metal; Expamet Building Products Ltd or other equal and approved						
75 mm diameter x 150 mm long	-	0.31	6.19	2.40	nr	**8.58**
75 mm diameter x 300 mm long	-	0.31	6.19	2.47	nr	**8.66**
100 mm diameter x 450 mm long	-	0.31	6.19	2.65	nr	**8.84**
Foundation bolts and nuts						
Black hexagon						
10 mm diameter x 100 mm long	-	0.26	5.19	0.95	nr	**6.14**
12 mm diameter x 120 mm long	-	0.26	5.19	1.45	nr	**6.64**
16 mm diameter x 160 mm long	-	0.31	6.19	4.01	nr	**10.19**
20 mm diameter x 180 mm long	-	0.31	6.19	4.68	nr	**10.87**
Masonry slots						
Galvanised steel; dovetail slots; 1.20 mm thick; 18G						
1000 mm long	-	0.25	4.99	3.67	m	**8.66**
100 mm long	-	0.08	1.60	0.45	nr	**2.05**
Galvanised steel; metal insert slots; Halfen Ltd or other equal and approved; 2.50 mm thick; end caps and foam filling						
41 mm x 41 mm; ref P3270	-	0.41	8.18	10.99	m	**19.17**
41 mm x 41 mm x 100 mm; ref P3250	-	0.10	2.00	1.46	nr	**3.45**
41mm x 41 mm x 150 mm; ref P3251	-	0.10	2.00	1.95	nr	**3.95**

E IN SITU CONCRETE/LARGE PRECAST CONCRETE

Item	PC £	Labour hours	Labour £	Material £	Unit	Total rate £
Cramps						
Mild steel; once bent; one end shot fired into concrete; other end flanged and built into brickwork joint						
200 mm girth	-	0.16	3.51	0.87	nr	**4.39**
Column guards						
White nylon coated steel; "Rigifix"; Huntley and Sparks Ltd or other equal and approved; plugging; screwing to concrete; 1.50 mm thick						
75 x 75 x 1000 mm	-	0.81	16.16	24.18	nr	**40.34**
Galvanised steel; "Rigifix"; Huntley and Sparks Ltd or other equal and approved; 3 mm thick						
75 x 75 x 1000 mm	-	0.61	12.17	17.11	nr	**29.28**
Galvanised steel; "Rigifix"; Huntley and Sparks Ltd or other equal and approved; 4.50 mm thick						
75 x 75 x 1000 mm	-	0.61	12.17	23.00	nr	**35.18**
Stainless steel; "HKW"; Halfen Ltd or other equal and approved; 5 mm thick						
50 x 50 x 1200 mm	-	1.02	20.36	47.93	nr	**68.28**
50 x 50 x 2000 mm	-	1.22	24.35	86.49	nr	**110.84**
E60 PRECAST/COMPOSITE CONCRETE DECKING						
Prestressed precast flooring planks; Bison "Drycast" or other equal and approved; cement:sand (1:3) grout between planks and on prepared bearings						
100 mm thick suspended slabs; horizontal						
600 mm wide planks	-	-	-	-	m²	**42.01**
1200 mm wide planks	-	-	-	-	m²	**39.85**
150 mm thick suspended slabs; horizontal						
1200 mm wide planks	-	-	-	-	m²	**41.54**

F MASONRY

Item	PC £	Labour hours	Labour £	Material £	Unit	Total rate £
F10 BRICK/BLOCK WALLING						
Common bricks; PC £270.00 per 1000; in gauged mortar (1:1:6)						
Walls						
half brick thick	-	1.06	29.10	21.95	m²	**51.05**
half brick thick; building against other work; concrete	-	1.16	31.84	23.48	m²	**55.32**
half brick thick; building overhand	-	1.34	36.78	21.95	m²	**58.73**
half brick thick; curved; 6.00 m radii	-	1.39	38.16	21.95	m²	**60.11**
half brick thick; curved; 1.50 m radii	-	1.80	49.41	25.23	m²	**74.64**
one brick thick	-	1.80	49.41	43.90	m²	**93.31**
one brick thick; curved; 6.00 m radii	-	2.36	64.78	47.18	m²	**111.96**
one brick thick; curved; 1.50 m radii	-	2.91	79.88	47.94	m²	**127.82**
one and a half brick thick	-	2.45	67.26	65.85	m²	**133.11**
one and a half brick thick; battering	-	2.82	77.41	65.85	m²	**143.26**
two brick thick	-	2.96	81.26	87.80	m²	**169.06**
two brick thick; battering	-	3.52	96.63	87.80	m²	**184.43**
337 mm average thick; tapering, one side	-	3.10	85.10	65.85	m²	**150.95**
450 mm average thick; tapering, one side	-	3.98	109.26	87.80	m²	**197.06**
337 mm average thick; tapering both sides	-	3.56	97.73	65.85	m²	**163.58**
450 mm average thick; tapering both sides	-	4.49	123.26	88.56	m²	**211.82**
facework one side, half brick thick	-	1.16	31.84	21.95	m²	**53.79**
facework one side, one brick thick	-	1.90	52.16	43.90	m²	**96.06**
facework one side, one and a half brick thick	-	2.54	69.73	65.85	m²	**135.58**
facework one side, two brick thick	-	3.10	85.10	87.80	m²	**172.90**
facework both sides, half brick thick	-	1.30	35.69	21.95	m²	**57.64**
facework both sides, one brick thick	-	2.04	56.00	43.90	m²	**99.90**
facework both sides, one and a half brick thick	-	2.68	73.57	65.85	m²	**139.42**
facework both sides, two brick thick	-	3.19	87.57	87.80	m²	**175.37**
Isolated piers						
one brick thick	-	2.73	74.94	43.90	m²	**118.84**
two brick thick	-	4.26	116.94	88.56	m²	**205.51**
three brick thick	-	5.37	147.41	133.23	m²	**280.64**
Isolated casings						
half brick thick	-	1.39	38.16	21.95	m²	**60.11**
one brick thick	-	2.36	64.78	43.90	m²	**108.68**
Chimney stacks						
one brick thick	-	2.73	74.94	43.90	m²	**118.84**
two brick thick	-	4.26	116.94	88.56	m²	**205.51**
three brick thick	-	5.37	147.41	133.23	m²	**280.64**
Projections						
225 mm width; 112 mm depth; vertical	-	0.32	8.78	4.64	m	**13.42**
225 mm width; 225 mm depth; vertical	-	0.65	17.84	9.27	m	**27.12**
337 mm width; 225 mm depth; vertical	-	0.97	26.63	13.91	m	**40.54**
440 mm width; 225 mm depth; vertical	-	1.06	29.10	18.54	m	**47.64**
Closing cavities						
width of cavity 50 mm, closing with common brickwork half brick thick; vertical	-	0.32	8.78	1.11	m	**9.90**
width of cavity 50 mm, closing with common brickwork half brick thick; horizontal	-	0.32	8.78	3.44	m	**12.22**
width of cavity 50 mm, closing with common brickwork half brick thick; including damp proof course; vertical	-	0.43	11.80	1.97	m	**13.77**
width of cavity 50 mm, closing with common brickwork half brick thick; including damp proof course; horizontal	-	0.37	10.16	4.06	m	**14.22**
width of cavity 75 mm, closing with common brickwork half brick thick; vertical	-	0.32	8.78	1.63	m	**10.42**
width of cavity 75 mm, closing with common brickwork half brick thick; horizontal	-	0.32	8.78	5.08	m	**13.86**

F MASONRY

Item	PC £	Labour hours	Labour £	Material £	Unit	Total rate £
width of cavity 75 mm, closing with common brickwork half brick thick; including damp proof course; vertical	-	0.43	11.80	2.49	m	**14.30**
width of cavity 50 mm, closing with common brickwork half brick thick; including damp proof course; horizontal	-	0.37	10.16	5.70	m	**15.86**
Bonding to existing						
half brick thick	-	0.32	8.78	1.21	m	**9.99**
one brick thick	-	0.46	12.63	2.41	m	**15.04**
one and a half brick thick	-	0.74	20.31	3.62	m	**23.93**
two brick thick	-	1.02	28.00	4.83	m	**32.83**
Arches						
height on face 102 mm, width of exposed soffit 102 mm, shape of arch - segmental, one ring	-	1.80	39.20	8.03	m	**47.24**
height on face 102 mm, width of exposed soffit 215 mm, shape of arch - segmental, segmental, one ring	-	2.36	55.03	10.62	m	**65.64**
height on face 102 mm, width of exposed soffit 102 mm, shape of arch - semi-circular, one ring	-	2.31	53.66	8.03	m	**61.69**
height on face 102 mm, width of exposed soffit 215 mm, shape of arch - semi-circular, one ring	-	2.87	69.48	10.62	m	**80.10**
height on face 215 mm, width of exposed soffit 102 mm, shape of arch - segmental, two ring	-	2.31	53.66	10.37	m	**64.03**
height on face 215 mm, width of exposed soffit 215 mm, shape of arch - segmental, two ring	-	2.82	68.11	15.29	m	**83.40**
height on face 215 mm, width of exposed soffit 102 mm, shape of arch semi-circular, two ring	-	3.10	76.02	10.37	m	**86.40**
height on face 215 mm, width of exposed soffit 215 mm, shape of arch - semi-circular, two ring	-	3.52	87.89	15.29	m	**103.18**
ADD or DEDUCT to walls for variation of £10.00/1000 in PC of common bricks						
half brick thick	-	-	-	0.73	m²	-
one brick thick	-	-	-	1.46	m²	-
one and a half brick thick	-	-	-	2.19	m²	-
two brick thick	-	-	-	2.92	m²	-
Class B engineering bricks; PC £315.00 per 1000; in cement mortar (1:3)						
Walls						
half brick thick	-	1.16	31.84	24.96	m²	**56.80**
one brick thick	-	1.90	52.16	49.92	m²	**102.08**
one brick thick; building against other work	-	2.27	62.31	52.47	m²	**114.79**
one brick thick; curved; 6.00 m radii	-	2.54	69.73	49.92	m²	**119.64**
one and a half brick thick	-	2.54	69.73	74.88	m²	**144.60**
one and a half brick thick; building against other work	-	3.10	85.10	74.88	m²	**159.98**
two brick thick	-	3.27	89.77	99.84	m²	**189.60**
337 mm average thick; tapering, one side	-	3.28	90.04	74.88	m²	**164.92**
450 mm average thick; tapering, one side	-	4.26	116.94	99.84	m²	**216.78**
337 mm average thick; tapering, both sides	-	3.84	105.41	74.88	m²	**180.29**
450 mm average thick; tapering, both sides	-	4.86	133.41	100.69	m²	**234.10**
facework one side, half brick thick	-	1.30	35.69	24.96	m²	**60.65**
facework one side, one brick thick	-	2.04	56.00	49.92	m²	**105.92**
facework one side, one and a half brick thick	-	2.68	73.57	74.88	m²	**148.45**
facework one side, two brick thick	-	3.28	90.04	99.84	m²	**189.88**
facework both sides, half brick thick	-	1.39	38.16	24.96	m²	**63.12**
facework both sides, one brick thick	-	2.13	58.47	49.92	m²	**108.39**
facework both sides, one and a half brick thick	-	2.78	76.31	74.88	m²	**151.19**
facework both sides, two brick thick	-	3.42	93.88	99.84	m²	**193.72**

F MASONRY

Item	PC £	Labour hours	Labour £	Material £	Unit	Total rate £
F10 BRICK/BLOCK WALLING – cont'd						
Class B engineering bricks; PC £315.00 per 1000 – cont'd						
Isolated piers						
one brick thick	-	2.96	81.26	49.92	m²	**131.17**
two brick thick	-	4.67	128.20	100.69	m²	**228.89**
three brick thick	-	5.74	157.57	151.46	m²	**309.03**
Isolated casings						
half brick thick	-	1.48	40.63	24.96	m²	**65.59**
one brick thick	-	2.54	69.73	49.92	m²	**119.64**
Projections						
225 mm width; 112 mm depth; vertical	-	0.37	10.16	5.28	m	**15.43**
225 mm width; 225 mm depth; vertical	-	0.69	18.94	10.55	m	**29.49**
337 mm width; 225 mm depth; vertical	-	1.02	28.00	15.83	m	**43.83**
440 mm width; 225 mm depth; vertical	-	1.16	31.84	21.11	m	**52.95**
Bonding to existing						
half brick thick	-	0.37	10.16	1.37	m	**11.53**
one brick thick	-	0.56	15.37	2.74	m	**18.12**
one and a half brick thick	-	0.74	20.31	4.12	m	**24.43**
two brick thick	-	1.11	30.47	5.49	m	**35.96**
ADD or DEDUCT to walls for variation of £10.00/1000 in PC of bricks						
half brick thick	-	-	-	0.73	m²	-
one brick thick	-	-	-	1.46	m²	-
one and a half brick thick	-	-	-	2.19	m²	-
two brick thick	-	-	-	2.92	m²	-
Facing bricks; machine made facings; PC £380.00 per 1000; in gauged mortar (1:1:6)						
Walls						
facework one side, half brick thick; stretcher bond	-	1.39	38.16	29.95	m²	**68.11**
facework one side, half brick thick, flemish bond with snapped headers	-	1.62	44.47	29.95	m²	**74.43**
facework one side, half brick thick; stretcher bond; building against other work; concrete	-	1.48	40.63	31.48	m²	**72.11**
facework one side, half brick thick; flemish bond with snapped headers; building against other work; concrete	-	1.71	46.94	31.48	m²	**78.42**
facework one side, half brick thick, stretcher bond; building overhand	-	1.71	46.94	29.95	m²	**76.90**
facework one side, half brick thick; flemish bond with snapped headers; building overhand	-	1.90	52.16	29.95	m²	**82.11**
facework one side, half brick thick; stretcher bond; curved; 6.00 m radii	-	2.04	56.00	29.95	m²	**85.95**
facework one side, half brick thick; flemish bond with snapped headers; curved; 6.00 m radii	-	2.27	62.31	29.95	m²	**92.27**
facework one side, half brick thick; stretcher bond; curved; 1.50 m radii	-	2.54	69.73	34.57	m²	**104.29**
facework one side; half brick thick; stretcher bond; curved; 1.50 m radii	-	2.96	81.26	34.57	m²	**115.82**
facework both sides, one brick thick; two stretcher skins tied together	-	2.41	66.16	62.38	m²	**128.54**
facework both sides, one brick thick; flemish bond	-	3.28	90.04	59.91	m²	**149.95**
facework both sides, one brick thick; two stretcher skins tied together; curved; 6.00 m radii	-	3.42	93.88	66.99	m²	**160.87**
facework both sides, one brick thick; flemish bond; curved; 6.00 m radii	-	3.42	93.88	64.52	m²	**158.40**
facework both sides, one brick thick; two stretcher skins tied together; curved; 1.50 m radii	-	4.12	113.10	72.37	m²	**185.46**

F MASONRY

Item	PC £	Labour hours	Labour £	Material £	Unit	Total rate £
facework both sides, one brick thick; flemish bond; curved; 1.50 m radii	-	2.26	62.04	69.89	m²	**131.93**
Isolated piers						
facework both sides, one brick thick; two stretcher skins tied together	-	2.82	77.41	63.79	m²	**141.21**
facework both sides, one brick thick; flemish bond	-	2.87	78.78	63.79	m²	**142.58**
Isolated casings						
facework one side, half brick thick; stretcher bond	-	2.13	58.47	29.95	m²	**88.43**
facework one side, half brick thick; flemish bond with snapped headers	-	2.36	64.78	29.95	m²	**94.74**
Projections						
225 mm width; 112 mm depth; stretcher bond; vertical	-	0.32	8.78	6.41	m	**15.20**
225 mm width; 112 mm depth; flemish bond with snapped headers; vertical	-	0.42	11.53	6.41	m	**17.94**
225 mm width; 225 mm depth; flemish bond; vertical	-	0.69	18.94	12.83	m	**31.77**
328 mm width; 112 mm depth; stretcher bond; vertical	-	0.65	17.84	9.63	m	**27.47**
328 mm width; 112 mm depth; flemish bond with snapped headers; vertical	-	0.74	20.31	9.63	m	**29.94**
328 mm width; 225 mm depth; flemish bond; vertical	-	1.30	35.69	19.21	m	**54.89**
440 mm width; 112 mm depth; stretcher bond; vertical	-	0.97	26.63	12.83	m	**39.46**
440 mm width; 112 mm depth; flemish bond with snapped headers; vertical	-	1.02	28.00	12.83	m	**40.83**
440 mm width; 225 mm depth; flemish bond; vertical	-	1.85	50.78	25.66	m	**76.44**
Arches						
height on face 215 mm, width of exposed soffit 102 mm, shape of arch - flat	-	1.06	25.01	9.14	m	**34.15**
height on face 215 mm, width of exposed soffit 215 mm, shape of arch - flat	-	1.57	39.47	15.78	m	**55.25**
height on face 215 mm, width of exposed soffit 102 mm, shape of arch - segmental, one ring	-	2.04	44.77	12.15	m	**56.92**
height on face 215 mm, width of exposed soffit 215 mm, shape of arch - segmental, one ring	-	2.45	56.48	18.54	m	**75.02**
height on face 215 mm, width of exposed soffit 102 mm, shape of arch - semi-circular, one ring	-	3.10	74.89	12.15	m	**87.04**
height on face 215 mm, width of exposed soffit 215 mm, shape of arch - semi-circular, one ring	-	4.16	105.01	18.54	m	**123.55**
height on face 215 mm, width of exposed soffit 102 mm, shape of arch - segmental, two ring	-	2.50	57.85	12.15	m	**70.00**
height on face 215 mm, width of exposed soffit 215 mm, shape of arch - segmental, two ring	-	3.24	78.85	18.54	m	**97.39**
height on face 215 mm, width of exposed soffit 102 mm, shape of arch - semi-circular, two ring	-	4.16	105.01	12.15	m	**117.16**
height on face 215 mm, width of exposed soffit 215 mm, shape of arch - semi-circular, two ring	-	5.74	149.74	18.54	m	**168.28**
Arches; cut voussoirs (PC £ per 1000)	3750.00					
height on face 215 mm, width of exposed soffit 102 mm, shape of arch - segmental, one ring	-	2.08	45.98	66.67	m	**112.65**
height on face 215 mm, width of exposed soffit 215 mm, shape of arch - segmental, one ring	-	2.49	58.82	127.59	m	**186.41**
height on face 215 mm, width of exposed soffit 102 mm, shape of arch - semi-circular, one ring	-	2.36	53.89	66.67	m	**120.57**
height on face 215 mm, width of exposed soffit 215 mm, shape of arch - semi-circular, one ring	-	2.96	70.93	127.59	m	**198.52**
height on face 320 mm, width of exposed soffit 102 mm, shape of arch - segmental, one and a half ring	-	2.78	65.76	127.50	m	**193.26**
height on face 320 mm, width of exposed soffit 215 mm, shape of arch - segmental, one and a half ring	-	3.61	89.34	257.78	m	**347.13**

F MASONRY

Item	PC £	Labour hours	Labour £	Material £	Unit	Total rate £
F10 BRICK/BLOCK WALLING – cont'd						
Facing bricks; machine made facings; PC £380.00 per 1000 – cont'd						
Arches; bullnosed specials (PC £ per 1000)	2200.00					
height on face 215 mm, width of exposed soffit 102 mm, shape of arch - flat	-	1.11	26.39	38.58	m	**64.97**
height on face 215 mm, width of exposed soffit 215 mm, shape of arch - flat	-	1.62	40.84	75.41	m	**116.25**
Bullseye windows; 600 mm diameter						
height on face 215 mm, width of exposed soffit 102 mm, two rings	-	5.32	137.87	15.79	nr	**153.66**
height on face 215 mm, width of exposed soffit 215 mm, two rings	-	7.55	199.88	29.62	nr	**229.51**
Bullseye windows; 600 mm diameter; cut voussoirs (PC £ per 1000)	3750.00					
height on face 215 mm, width of exposed soffit 102 mm, one ring	-	4.49	114.29	162.13	nr	**276.43**
height on face 215 mm, width of exposed soffit, 215 mm, one ring	-	6.15	161.45	322.32	nr	**483.77**
Bullseye windows; 1200 mm diameter						
height on face 215 mm, width of exposed soffit 102 mm, two rings	-	8.28	221.85	34.60	nr	**256.45**
height on face 215 mm, width of exposed soffit 215 mm, two rings	-	11.93	325.45	62.95	nr	**388.40**
Bullseye windows; 1200 mm diameter; cut voussoirs (PC £ per 1000)	3750.00					
height on face 215 mm, width of exposed soffit 102 mm, one ring	-	7.03	186.40	282.56	nr	**468.96**
height on face 215 mm, width of exposed soffit 215 mm, one ring	-	9.99	270.38	557.34	nr	**827.72**
ADD or DEDUCT for variation of £10.00/1000 in PC of facing bricks in 102 mm high arches with 215 mm soffit	-	-	-	0.32	m	-
Facework sills						
150 mm x 102 mm; headers on edge; pointing top and one side; set weathering; horizontal	-	0.60	16.47	6.41	m	**22.89**
150 mm x 102 mm; cant headers on edge; pointing top and one side; set weathering; horizontal (PC £ per 1000)	2200.00	0.65	17.84	35.86	m	**53.70**
150 mm x 102 mm; bullnosed specials; headers on flat; pointing top and one side; horizontal (PC £ per 1000)	2200.00	0.56	15.37	35.86	m	**51.23**
Facework copings						
215 mm x 102 mm; headers on edge; pointing top and both sides; horizontal	-	0.46	12.63	6.53	m	**19.16**
260 mm x 102 mm; headers on edge; pointing top and both sides; horizontal	-	0.74	20.31	9.70	m	**30.02**
215 mm x 102 mm; double bullnose specials; headers on edge; pointing top and both sides; horizontal (PC £ per 1000)	2200.00	0.56	15.37	35.97	m	**51.35**
260 mm x 102 mm; single bullnose specials; headers on edge; pointing top and both sides; horizontal (PC £ per 1000)	2200.00	0.74	20.31	77.89	m	**98.21**
ADD or DEDUCT for variation of £10.00/1000 in PC of facing bricks in copings 215 mm wide, 102 mm high	-	-	-	0.16	m	-

F MASONRY

Item	PC £	Labour hours	Labour £	Material £	Unit	Total rate £
Extra over facing bricks for; facework ornamental bands and the like, plain bands						
flush; horizontal; 225 mm width; entirely of stretchers (PC £ per 1000)	430.00	0.23	6.31	0.81	m	**7.12**
Extra over facing bricks for; facework quoins						
flush; mean girth 320 mm (PC £ per 1000)	430.00	0.32	8.78	0.81	m	**9.59**
Bonding to existing						
facework one side, half brick thick; stretcher bond	-	0.56	15.37	1.65	m	**17.02**
facework one side, half brick thick; flemish bond with snapped headers	-	0.56	15.37	1.65	m	**17.02**
facework both sides, one brick thick; two stretcher skins tied together	-	0.74	20.31	3.30	m	**23.62**
facework both sides, one brick thick; flemish bond	-	0.74	20.31	3.30	m	**23.62**
ADD or DEDUCT for variation of £10.00/1000 in PC of facing bricks; in walls built entirely of facings; in stretcher or flemish bond						
half brick thick	-	-	-	0.73	m²	-
one brick thick	-	-	-	1.46	m²	-
Facing bricks; hand made; PC £600.00 per 1000; in gauged mortar (1:1:6)						
Walls						
facework one side, half brick thick; stretcher bond	-	1.39	38.16	45.98	m²	**84.13**
facework one side, half brick thick; flemish bond with snapped headers	-	1.62	44.47	45.98	m²	**90.45**
facework one side, half brick thick; stretcher bond; building against other work; concrete	-	1.48	40.63	47.50	m²	**88.13**
facework one side, half brick thick; flemish bond with snapped headers; building against other work; concrete	-	1.71	46.94	47.50	m²	**94.44**
facework one side, half brick thick; stretcher bond; building overhand	-	1.71	46.94	45.98	m²	**92.92**
facework one side, half brick thick; flemish bond with snapped headers; building overhand	-	1.90	52.16	45.98	m²	**98.13**
facework one side, half brick thick; stretcher bond; curved; 6.00 m radii	-	2.04	56.00	45.98	m²	**101.98**
facework one side, half brick thick; flemish bond with snapped headers; curved; 6.00 m radii	-	2.27	62.31	51.44	m²	**113.75**
facework one side; half brick thick; stretcher bond; curved; 1.50 m radii	-	2.54	69.73	46.08	m²	**116.70**
facework one side; half brick thick; flemish bond with snapped headers; curved; 1.50 m radii	-	2.96	81.26	55.08	m²	**136.33**
facework both sides, one brick thick; two stretcher skins tied together	-	2.41	66.16	94.42	m²	**160.58**
facework both sides; one brick thick; flemish bond	-	2.45	67.26	91.95	m²	**159.21**
facework both sides; one brick thick; two stretcher skins tied together; curved; 6.00 m radii	-	3.28	90.04	101.70	m²	**191.74**
facework both sides; one brick thick; flemish bond; curved; 6.00 m radii	-	3.42	93.88	99.23	m²	**193.12**
facework both sides; one brick thick; two stretcher skins tied together; curved; 1.50 m radii	-	4.12	113.10	109.75	m²	**222.85**
facework both sides; one brick thick; flemish bond; curved; 1.50 m radii	-	4.26	116.94	107.28	m²	**224.22**
Isolated piers						
facework both sides, one brick thick; two stretcher skins tied together	-	2.82	77.41	95.84	m²	**173.25**
facework both sides, one brick thick; flemish bond	-	2.87	78.78	95.84	m²	**174.62**

F MASONRY

Item	PC £	Labour hours	Labour £	Material £	Unit	Total rate £
F10 BRICK/BLOCK WALLING – cont'd						
Facing bricks; hand made; PC £600.00 per 1000 – cont'd						
Isolated casings						
facework one side, half brick thick; stretcher bond	-	2.13	58.47	45.98	m²	**104.45**
facework one side, half brick thick; flemish bond with snapped headers	-	2.36	64.78	45.98	m²	**110.76**
Projections						
225 mm width; 112 mm depth; stretcher bond; vertical	-	0.32	8.78	9.98	m	**18.76**
225 mm width; 112 mm depth; flemish bond with snapped headers; vertical	-	0.42	11.53	9.98	m	**21.50**
225 mm width; 225 mm depth; flemish bond; vertical	-	0.69	18.94	19.95	m	**38.89**
328 mm width; 112 mm depth; stretcher bond; vertical	-	0.65	17.84	14.97	m	**32.82**
328 mm width; 112 mm depth; flemish bond with snapped headers; vertical	-	0.74	20.31	14.97	m	**35.29**
328 mm width; 225 mm depth; flemish bond; vertical	-	1.30	35.69	29.89	m	**65.57**
440 mm width; 112 mm depth; stretcher bond; vertical	-	0.97	26.63	19.95	m	**46.58**
440 mm width; 112 mm depth; flemish bond with snapped headers; vertical	-	1.02	28.00	19.95	m	**47.95**
440 mm width; 225 mm depth; flemish bond; vertical	-	1.85	50.78	39.90	m	**90.69**
Arches						
height on face 215 mm, width of exposed soffit 102 mm, shape of arch - flat	-	1.06	25.01	12.70	m	**37.71**
height on face 215 mm, width of exposed soffit 215 mm, shape of arch - flat	-	1.57	39.47	22.99	m	**62.46**
height on face 215 mm, width of exposed soffit 102 mm, shape of arch - segmental, one ring	-	2.04	44.77	15.71	m	**60.48**
height on face 215 mm, width of exposed soffit 215 mm, shape of arch - segmental, one ring	-	2.45	56.48	25.66	m	**82.14**
height on face 215 mm, width of exposed soffit 102 mm, shape of arch - semi-circular, one ring	-	2.50	57.85	15.71	m	**73.56**
height on face 215 mm, width of exposed soffit 215 mm, shape of arch - semi-circular, one ring	-	3.24	78.85	25.66	m	**104.51**
height on face 215 mm, width of exposed soffit 102 mm, shape of arch - segmental, two ring	-	2.50	57.85	15.71	m	**73.56**
height on face 215 mm, width of exposed soffit 215 mm, shape of arch - segmental, two ring	-	3.24	78.85	25.66	m	**104.51**
height on face 215 mm, width of exposed soffit 102 mm, shape of arch - semi-circular, two ring	-	4.16	105.01	15.71	m	**120.72**
height on face 215 mm, width of exposed soffit 215 mm, shape of arch - semi-circular, two ring	-	5.74	149.74	25.66	m	**175.41**
Arches; cut voussoirs (PC £ per 1000)	4000.00					
height on face 215 mm, width of exposed soffit 102 mm, shape of arch - segmental, one ring	53.33	2.08	45.98	70.72	m	**116.70**
height on face 215 mm, width of exposed soffit 215 mm, shape of arch - segmental, one ring	-	2.59	60.44	135.68	m	**196.11**
height on face 215 mm, width of exposed soffit 102 mm, shape of arch - semi-circular, one ring	-	2.36	53.89	70.72	m	**124.61**
height on face 215 mm, width of exposed soffit 215 mm, shape of arch - semi-circular, one ring	-	2.96	70.93	135.68	m	**206.61**
height on face 320 mm, width of exposed soffit 102 mm, shape of arch - segmental, one and a half ring	-	2.78	65.76	135.59	m	**201.35**
height on face 320 mm, width of exposed soffit 215 mm, shape of arch - segmental, one and a half ring	-	3.61	89.34	273.96	m	**363.31**

F MASONRY

Item	PC £	Labour hours	Labour £	Material £	Unit	Total rate £
Arches; bullnosed specials (PC £ per 1000)	2300.00					
height on face 215 mm, width of exposed soffit 102 mm, shape of arch - flat	-	1.11	26.39	40.20	m	**66.59**
height on face 215 mm, width of exposed soffit 215 mm, shape of arch - flat	-	1.62	40.84	78.69	m	**119.53**
Bullseye windows; 600 mm diameter						
height on face 215 mm, width of exposed soffit 102 mm, two ring	-	5.32	137.87	23.26	nr	**161.13**
height on face 215 mm, width of exposed soffit 215 mm, two ring	-	7.45	198.27	58.32	nr	**256.59**
Bullseye windows; 600 mm diameter; cut voussoirs (PC £ per 1000)	4000.00					
height on face 215 mm, width of exposed soffit 102 mm, one ring	-	4.49	114.29	172.75	nr	**287.05**
height on face 215 mm, width of exposed soffit 215 mm, one ring	-	6.15	161.45	343.56	nr	**505.01**
Bullseye windows; 1200 mm diameter						
height on face 215 mm, width of exposed soffit 102 mm, two ring	-	8.28	221.85	49.55	nr	**271.40**
height on face 215 mm, width of exposed soffit 215 mm, two ring	-	11.93	325.45	92.86	nr	**418.31**
Bullseye windows; 1200 mm diameter; cut voussoirs (PC £ per 1000)	4000.00					
height on face 215 mm, width of exposed soffit 102 mm, one ring	-	7.03	186.40	300.76	nr	**487.16**
height on face 215 mm, width of exposed soffit 215 mm, one ring	-	9.99	270.38	593.75	nr	**864.13**
ADD or DEDUCT for variation of £10.00/1000 in PC of facing bricks in 102 high arches with 215 mm soffit	-	-	-	0.32	m	**-**
Facework sills						
150 mm x 102 mm; headers on edge; pointing top and one side; set weathering; horizontal	-	0.60	16.47	9.98	m	**26.45**
150 mm x 102 mm; cant headers on edge; pointing top and one side; set weathering; horizontal (PC £ per 1000)	2300.00	0.65	17.84	37.48	m	**55.32**
150 mm x 102 mm; bullnosed specials; headers on flat; pointing top and one side; horizontal (PC £ per 1000)	2300.00	0.56	15.37	37.48	m	**52.85**
Facework copings						
215 mm x 102 mm; headers on edge; pointing top and both sides; horizontal	-	0.46	12.63	10.09	m	**22.72**
260 mm x 102 mm; headers on edge; pointing top and both sides; horizontal	-	0.74	20.31	15.04	m	**35.36**
215 mm x 102 mm; double bullnose specials; headers on edge; pointing top and both sides (PC £ per 1000)	2300.00	0.56	15.37	37.59	m	**52.97**
260 mm x 102 mm; single bullnose specials; headers on edge; pointing top and both sides (PC £ per 1000)	2300.00	0.74	20.31	74.90	m	**95.22**
ADD or DEDUCT for variation of £10.00/1000 in PC of facing bricks in copings 215 mm wide, 102 mm high	-	-	-	0.16	m	**-**
Extra over facing bricks for; facework ornamental bands and the like, plain bands						
flush; horizontal; 225 mm width; entirely of stretchers (PC £ per 1000)	650.00	0.23	6.31	0.81	m	**7.12**
Extra over facing bricks for; facework quoins						
flush; mean girth 320 mm (PC £ per 1000)	650.00	0.32	8.78	0.81	m	**9.59**

F MASONRY

Item	PC £	Labour hours	Labour £	Material £	Unit	Total rate £
F10 BRICK/BLOCK WALLING – cont'd						
Facing bricks; hand made; PC £600.00 per 1000 – cont'd						
Bonding ends to existing						
facework one side, half brick thick; stretcher bond	-	0.56	15.37	2.54	m	**17.91**
facework one side, half brick thick; flemish bond with snapped headers	-	0.56	15.37	2.54	m	**17.91**
facework both sides, one brick thick; two stretcher skins tied together	-	0.74	20.31	5.08	m	**25.40**
facework both sides, one brick thick; flemish bond	-	0.74	20.31	5.08	m	**25.40**
ADD or DEDUCT for variation of £10.00/1000 in PC of facing bricks; in walls built entirely of facings; in stretcher or flemish bond						
half brick thick	-	-	-	0.73	m²	**-**
one brick thick	-	-	-	1.46	m²	**-**
Facing bricks slips 50 mm thick; in gauged mortar (1:1:6) built up against concrete including flushing up at back (ties not included)						
Walls (PC £ per 1000)	1250.00	2.13	58.47	93.29	m²	**151.76**
Edges of suspended slabs; 200 mm wide	-	0.65	17.84	18.66	m	**36.50**
Columns; 400 mm wide	-	1.30	35.69	37.32	m	**73.00**
Engineering bricks and facework specials; in cement mortar (1:3)						
Facework steps						
215 mm x 102 mm; all headers-on-edge; edges set with bullnosed specials; pointing top and one side; set weathering; horizontal (specials PC £ per 1000)	2200.00	0.60	16.47	35.89	m	**52.36**
returned ends pointed	-	0.14	3.84	7.06	nr	**10.90**
430 mm x 102 mm; all headers-on-edge; edges set with bullnosed specials; pointing top and one side; set weathering; horizontal (engineering bricks PC £ per 1000)	390.00	0.83	22.78	42.35	m	**65.14**
returned ends pointed	-	0.23	6.31	9.03	nr	**15.34**
Lightweight aerated concrete blocks; Thermalite "Turbo" blocks or other equal and approved; in gauged mortar (1:2:9)						
Walls						
100 mm thick	7.92	0.56	15.37	10.33	m²	**25.70**
115 mm thick	9.11	0.56	15.37	11.87	m²	**27.24**
125 mm thick	9.90	0.56	15.37	12.91	m²	**28.28**
130 mm thick	10.30	0.56	15.37	13.42	m²	**28.79**
140 mm thick	11.09	0.60	16.47	14.45	m²	**30.93**
150 mm thick	11.88	0.60	16.47	15.49	m²	**31.96**
190 mm thick	15.05	0.65	17.84	19.62	m²	**37.46**
200 mm thick	15.84	0.65	17.84	20.65	m²	**38.49**
215 mm thick	17.03	0.65	17.84	22.20	m²	**40.04**
Isolated piers or chimney stacks						
190 mm thick	-	0.97	26.63	19.62	m²	**46.24**
215 mm thick	-	0.97	26.63	22.20	m²	**48.83**
Isolated casings						
100 mm thick	-	0.60	16.47	10.33	m²	**26.80**
115 mm thick	-	0.60	16.47	11.87	m²	**28.34**
125 mm thick	-	0.60	16.47	12.91	m²	**29.38**
140 mm thick	-	0.65	17.84	14.45	m²	**32.30**

F MASONRY

Item	PC £	Labour hours	Labour £	Material £	Unit	Total rate £
Extra over for fair face; flush pointing						
walls; one side	-	0.04	1.10	-	m²	**1.10**
walls; both sides	-	0.09	2.47	-	m²	**2.47**
Closing cavities						
width of cavity 50 mm, closing with lightweight blockwork 100 mm; thick	-	0.27	7.41	0.58	m	**7.99**
width of cavity 50 mm, closing with lightweight blockwork 100 mm; thick; including damp proof course; vertical	-	0.32	8.78	1.44	m	**10.22**
width of cavity 75 mm, closing with lightweight blockwork 100 mm; thick	-	0.27	7.41	0.83	m	**8.25**
width of cavity 75 mm, closing with lightweight blockwork 100 mm; thick; including damp proof course; vertical	-	0.32	8.78	1.69	m	**10.48**
Bonding ends to common brickwork						
100 mm thick	-	0.14	3.84	1.19	m	**5.03**
115 mm thick	-	0.14	3.84	1.37	m	**5.21**
125 mm thick	-	0.28	7.69	1.49	m	**9.18**
130 mm thick	-	0.28	7.69	1.55	m	**9.24**
140 mm thick	-	0.28	7.69	1.68	m	**9.36**
150 mm thick	-	0.28	7.69	1.79	m	**9.47**
190 mm thick	-	0.32	8.78	2.26	m	**11.04**
200 mm thick	-	0.32	8.78	2.38	m	**11.17**
215 mm thick	-	0.37	10.16	2.57	m	**12.73**
Lightweight aerated concrete blocks; Thermalite "Shield" blocks or other equal and approved; in thin joint mortar						
Walls						
75 mm thick	6.54	0.32	8.78	9.34	m²	**18.12**
90 mm thick	6.93	0.34	9.33	10.06	m²	**19.39**
100 mm thick	7.70	0.36	9.88	11.02	m²	**20.90**
140 mm thick	10.78	0.38	10.43	15.08	m²	**25.52**
150 mm thick	11.55	0.36	9.88	16.79	m²	**26.67**
190 mm thick	14.63	0.45	12.35	21.25	m²	**33.60**
200 mm thick	15.40	0.48	13.18	22.37	m²	**35.55**
Isolated piers or chimney stacks						
190 mm thick	-	0.69	18.94	21.25	m²	**40.19**
Isolated casings						
75 mm thick	-	0.40	10.98	9.75	m²	**20.73**
90 mm thick	-	0.40	10.98	10.23	m²	**21.21**
100 mm thick	-	0.40	10.98	11.19	m²	**22.17**
140 mm thick	-	0.43	11.80	15.67	m²	**27.47**
Lightweight aerated concrete blocks; Thermalite "Shield" blocks or other equal and approved; in gauged mortar (1:2:9)						
Walls						
75 mm thick	6.54	0.46	12.63	7.76	m²	**20.38**
90 mm thick	6.93	0.46	12.63	8.27	m²	**20.89**
100 mm thick	7.70	0.56	15.37	9.19	m²	**24.56**
140 mm thick	10.78	0.60	16.47	12.86	m²	**29.33**
150 mm thick	11.55	0.60	16.47	13.78	m²	**30.25**
190 mm thick	14.63	0.60	16.47	17.44	m²	**33.92**
200 mm thick	15.40	0.65	17.84	18.37	m²	**36.21**
Isolated piers or chimney stacks						
190 mm thick	-	0.97	26.63	17.44	m²	**44.07**

F MASONRY

Item	PC £	Labour hours	Labour £	Material £	Unit	Total rate £
F10 BRICK/BLOCK WALLING – cont'd						
Lightweight aerated concrete blocks; Thermalite "Shield" – cont'd						
Isolated casings						
75 mm thick	-	0.60	16.47	8.50	m²	**24.97**
90 mm thick	-	0.60	16.47	9.05	m²	**25.52**
100 mm thick	-	0.60	16.47	9.19	m²	**25.66**
140 mm thick	-	0.65	17.84	12.86	m²	**30.70**
Extra over for fair face; flush pointing						
walls; one side	-	0.04	1.10	-	m²	**1.10**
walls; both sides	-	0.09	2.47	-	m²	**2.47**
Closing cavities						
width of cavity 50 mm, closing with lightweight blockwork 100 mm; thick	-	0.27	7.41	0.52	m	**7.94**
width of cavity 50 mm, closing with lightweight blockwork 100 mm; thick; including damp proof course; vertical	-	0.32	8.78	1.38	m	**10.16**
width of cavity 75 mm, closing with lightweight blockwork 100 mm; thick	-	0.27	7.41	0.75	m	**8.16**
width of cavity 75 mm, closing with lightweight blockwork 100 mm; thick; including damp proof course; vertical	-	0.32	8.78	1.61	m	**10.39**
Bonding ends to common brickwork						
75 mm thick	-	0.09	2.47	0.98	m	**3.45**
90 mm thick	-	0.09	2.47	1.04	m	**3.52**
100 mm thick	-	0.14	3.84	1.06	m	**4.90**
140 mm thick	-	0.28	7.69	1.50	m	**9.19**
150 mm thick	-	0.28	7.69	1.60	m	**9.28**
190 mm thick	-	0.32	8.78	2.02	m	**10.80**
200 mm thick	-	0.32	8.78	2.13	m	**10.92**
Lightweight smooth face aerated concrete blocks; Thermalite "Smooth Face" blocks or other equal and approved; in gauged mortar (1:2:9); flush pointing one side						
Walls						
100 mm thick	10.45	0.65	17.84	13.47	m²	**31.31**
140 mm thick	14.63	0.74	20.31	18.85	m²	**39.17**
150 mm thick	15.68	0.74	20.31	20.20	m²	**40.52**
190 mm thick	19.86	0.83	22.78	25.58	m²	**48.37**
200 mm thick	22.47	0.83	22.78	28.89	m²	**51.67**
215 mm thick	20.90	0.83	22.78	27.01	m²	**49.79**
Isolated piers or chimney stacks						
190 mm thick	-	1.06	29.10	25.58	m²	**54.68**
200 mm thick	-	1.06	29.10	28.89	m²	**57.98**
215 mm thick	-	1.06	29.10	27.01	m²	**56.10**
Isolated casings						
100 mm thick	-	0.79	21.69	13.47	m²	**35.16**
140 mm thick	-	0.83	22.78	18.85	m²	**41.64**
Extra over for flush pointing						
walls; both sides	-	0.04	1.10	-	m²	**1.10**
Bonding ends to common brickwork						
100 mm thick	-	0.28	7.69	1.54	m	**9.22**
140 mm thick	-	0.28	7.69	2.17	m	**9.85**
150 mm thick	-	0.32	8.78	2.31	m	**11.09**
190 mm thick	-	0.37	10.16	2.92	m	**13.08**
200 mm thick	-	0.37	10.16	3.30	m	**13.46**
215 mm thick	-	0.37	10.16	3.10	m	**13.26**

F MASONRY

Item	PC £	Labour hours	Labour £	Material £	Unit	Total rate £
Lightweight smooth face aerated concrete blocks; Thermalite "Party Wall" blocks or other equal and approved; in gauged mortar (1:2:9); flush pointing one side						
Walls						
100 mm thick	7.70	0.65	17.84	9.19	m²	**27.03**
215 mm thick	16.55	0.83	22.78	19.74	m²	**42.53**
Isolated piers or chimney stacks						
215 mm thick	-	1.06	29.10	19.74	m²	**48.84**
Isolated casings						
100 mm thick	-	0.79	21.69	9.19	m²	**30.87**
Extra over for flush pointing						
walls; both sides	-	0.04	1.10	-	m²	**1.10**
Bonding ends to common brickwork						
100 mm thick	-	0.28	7.69	1.06	m	**8.75**
215 mm thick	-	0.37	10.16	2.30	m	**12.45**
Lightweight aerated high strength concrete blocks (7.00 N/mm²); Thermalite "High Strength" blocks or other equal and approved; in cement mortar (1:3)						
Walls						
100 mm thick	9.90	0.56	15.37	12.86	m²	**28.24**
140 mm thick	13.86	0.60	16.47	18.00	m²	**34.47**
150 mm thick	14.85	0.60	16.47	19.29	m²	**35.76**
190 mm thick	18.81	0.65	17.84	24.43	m²	**42.27**
200 mm thick	19.80	0.65	17.84	25.72	m²	**43.56**
215 mm thick	21.29	0.65	17.84	27.65	m²	**45.49**
Isolated piers or chimney stacks						
190 mm thick	-	0.97	26.63	24.43	m²	**51.06**
200 mm thick	-	0.97	26.63	25.72	m²	**52.35**
215 mm thick	-	0.97	26.63	27.65	m²	**54.27**
Isolated casings						
100 mm thick	-	0.60	16.47	12.86	m²	**29.33**
140 mm thick	-	0.65	17.84	18.00	m²	**35.84**
150 mm thick	-	0.65	17.84	19.29	m²	**37.13**
190 mm thick	-	0.79	21.69	24.43	m²	**46.12**
200 mm thick	-	0.79	21.69	25.72	m²	**47.40**
215 mm thick	-	0.79	21.69	27.65	m²	**49.33**
Extra over for flush pointing						
walls; one side	-	0.04	1.10	-	m²	**1.10**
walls; both sides	-	0.09	2.47	-	m²	**2.47**
Bonding ends to common brickwork						
100 mm thick	-	0.28	7.69	1.48	m	**9.16**
140 mm thick	-	0.28	7.69	2.08	m	**9.77**
150 mm thick	-	0.32	8.78	2.22	m	**11.00**
190 mm thick	-	0.37	10.16	2.81	m	**12.96**
200 mm thick	-	0.37	10.16	2.96	m	**13.12**
215 mm thick	-	0.37	10.16	3.19	m	**13.35**
Lightweight aerated concrete blocks; Thermalite "Trenchblock" blocks or other equal and approved; with tongued and grooved joints; in cement mortar (1:4)						
Walls						
255 mm thick	20.61	0.70	19.22	26.78	m²	**46.00**
275 mm thick	22.23	0.75	20.59	28.88	m²	**49.46**
305 mm thick	24.25	0.80	21.96	31.47	m²	**53.43**
355 mm thick	28.70	0.85	23.33	37.08	m²	**60.41**

F MASONRY

Item	PC £	Labour hours	Labour £	Material £	Unit	Total rate £
F10 BRICK/BLOCK WALLING – cont'd						
Concrete blocks; Thermalite "Trenchblock" 7.00N/mm² blocks or other equal and approved; with tongued and grooved joints; in cement mortar (1:4)						
Walls						
255 mm thick	28.05	0.80	21.96	36.02	m²	**57.98**
275 mm thick	30.25	0.85	23.33	38.83	m²	**62.17**
305 mm thick	33.00	0.90	24.71	42.33	m²	**67.04**
355 mm thick	39.05	0.95	26.08	49.93	m²	**76.01**
Lightweight smooth face medium dense concrete blocks; Lignacite standard or paint grade 3.60N/mm² blocks or other equal and approved; in gauged mortar (1:2:9); flush pointing one side						
Walls						
100 mm thick	7.82	0.71	19.49	10.20	m²	**29.70**
140 mm thick	11.45	0.83	22.78	14.90	m²	**37.69**
150 mm thick	12.19	0.85	23.33	15.87	m²	**39.21**
190 mm thick	15.56	1.00	27.45	20.25	m²	**47.70**
215 mm thick	16.68	1.10	30.20	21.77	m²	**51.96**
Isolated piers or chimney stacks						
190 mm thick	-	1.33	36.51	20.25	m²	**56.76**
215 mm thick	-	1.46	40.08	21.77	m²	**61.85**
Isolated casings						
100 mm thick	-	0.88	24.16	10.20	m²	**34.36**
140 mm thick	-	1.04	28.55	14.90	m²	**43.45**
Extra over for flush pointing						
walls; both sides	-	0.04	1.10	-	m²	**1.10**
Bonding ends to common brickwork						
100 mm thick	-	0.28	7.69	1.17	m	**8.86**
140 mm thick	-	0.28	7.69	1.73	m	**9.41**
150 mm thick	-	0.32	8.78	1.83	m	**10.61**
190 mm thick	-	0.37	10.16	2.33	m	**12.49**
215 mm thick	-	0.37	10.16	2.52	m	**12.68**
Lightweight smooth face medium dense concrete blocks; Lignacite standard or paint grade 7.30N/mm² blocks or other equal and approved; in gauged mortar (1:2:9); flush pointing one side						
Walls						
100 mm thick	7.95	0.71	19.49	10.37	m²	**29.86**
140 mm thick	11.55	0.83	22.78	15.03	m²	**37.81**
150 mm thick	12.82	0.85	23.33	16.66	m²	**39.99**
190 mm thick	16.06	1.00	27.45	20.87	m²	**48.32**
215 mm thick	17.79	1.10	30.20	23.15	m²	**53.34**
Isolated piers or chimney stacks						
190 mm thick	-	1.33	36.51	20.87	m²	**57.38**
215 mm thick	-	1.46	40.08	23.15	m²	**63.22**
Isolated casings						
100 mm thick	-	0.88	24.16	10.37	m²	**34.52**
140 mm thick	-	1.03	28.27	15.03	m²	**43.30**
Dense aggregate concrete blocks; Hanson "Conbloc" or other equal and approved; in gauged mortar (1:2:9)						

F MASONRY

Item	PC £	Labour hours	Labour £	Material £	Unit	Total rate £
Walls						
75 mm thick; solid	5.98	0.65	20.51	7.79	m²	**28.30**
100 mm thick; solid	6.61	0.79	24.92	8.71	m²	**33.63**
140 mm thick; solid	13.01	0.97	30.60	16.84	m²	**47.44**
140 mm thick; hollow	12.48	0.83	26.19	16.19	m²	**42.37**
190 mm thick; hollow	14.74	1.06	33.44	19.23	m²	**52.68**
215 mm thick; hollow	15.37	1.16	36.60	20.15	m²	**56.74**
Isolated piers or chimney stacks						
140 mm thick; hollow	-	1.16	36.60	16.19	m²	**52.78**
190 mm thick; hollow	-	1.53	48.27	19.23	m²	**67.50**
215 mm thick; hollow	-	1.76	55.53	20.15	m²	**75.67**
Isolated casings						
75 mm thick; solid	-	0.79	24.92	7.79	m²	**32.72**
100 mm thick; solid	-	0.83	26.19	8.71	m²	**34.89**
140 mm thick; solid	-	1.06	33.44	16.84	m²	**50.28**
Extra over for fair face; flush pointing						
walls; one side	-	0.09	2.84	-	m²	**2.84**
walls; both sides	-	0.14	4.42	-	m²	**4.42**
Bonding ends to common brickwork						
75 mm thick solid	-	0.14	4.42	0.90	m	**5.32**
100 mm thick solid	-	0.28	8.83	1.01	m	**9.84**
140 mm thick solid	-	0.32	10.10	1.94	m	**12.04**
140 mm thick hollow	-	0.32	10.10	1.87	m	**11.97**
190 mm thick hollow	-	0.37	11.67	2.22	m	**13.89**
215 mm thick hollow	-	0.42	13.25	2.34	m	**15.59**
Dense aggregate concrete blocks; (7.00 N/mm²) Forticrete "Shepton Mallet Common" blocks or other equal and approved; in cement mortar (1:3)						
Walls						
75 mm thick; solid	8.81	0.65	20.51	11.36	m²	**31.87**
100 mm thick; hollow	7.19	0.79	24.92	9.50	m²	**34.42**
100 mm thick; solid	6.71	0.79	24.92	8.90	m²	**33.83**
140 mm thick; hollow	10.61	0.83	26.19	13.97	m²	**40.15**
140 mm thick; solid	10.61	0.97	30.60	13.97	m²	**44.57**
190 mm thick; hollow	14.36	1.06	33.44	18.90	m²	**52.34**
190 mm thick; solid	14.34	1.16	36.60	18.88	m²	**55.48**
215 mm thick; hollow	13.02	1.16	36.60	17.38	m²	**53.98**
215 mm thick; solid	15.71	1.34	42.28	20.72	m²	**63.00**
Dwarf support wall						
140 mm thick; solid	-	1.34	42.28	13.97	m²	**56.24**
190 mm thick; solid	-	1.53	48.27	18.88	m²	**67.15**
215 mm thick; solid	-	1.76	55.53	20.72	m²	**76.25**
Isolated piers or chimney stacks						
140 mm thick; hollow	-	1.16	36.60	13.97	m²	**50.56**
190 mm thick; hollow	-	1.53	48.27	18.90	m²	**67.17**
215 mm thick; hollow	-	1.76	55.53	17.38	m²	**72.91**
Isolated casings						
75 mm thick; solid	-	0.79	24.92	11.36	m²	**36.29**
100 mm thick; solid	-	0.83	26.19	8.90	m²	**35.09**
140 mm thick; solid	-	1.06	33.44	13.97	m²	**47.41**
Extra over for fair face; flush pointing						
walls; one side	-	0.09	2.84	-	m²	**2.84**
walls; both sides	-	0.14	4.42	-	m²	**4.42**
Bonding ends to common brickwork						
75 mm thick solid	-	0.14	4.42	1.29	m	**5.71**
100 mm thick solid	-	0.28	8.83	1.02	m	**9.86**
140 mm thick solid	-	0.32	10.10	1.61	m	**11.71**
190 mm thick solid	-	0.37	11.67	2.16	m	**13.84**
215 mm thick solid	-	0.42	13.25	2.39	m	**15.64**

F MASONRY

Item	PC £	Labour hours	Labour £	Material £	Unit	Total rate £
F10 BRICK/BLOCK WALLING – cont'd						
Dense aggregate coloured concrete blocks; Forticrete "Yorkstone" or other equal and approved; in coloured gauged mortar (1:1:6); flush pointing one side						
Walls						
100 mm thick hollow	27.16	0.83	26.19	32.82	m²	**59.01**
100 mm thick solid	27.16	0.83	26.19	32.82	m²	**59.01**
140 mm thick hollow	39.34	0.97	30.60	47.51	m²	**78.11**
140 mm thick solid	39.34	1.06	33.44	47.51	m²	**80.95**
215 mm thick hollow	44.97	1.34	42.28	54.65	m²	**96.93**
Isolated piers or chimney stacks						
140 mm thick solid	-	1.43	45.11	47.51	m²	**92.62**
215 mm thick solid	-	1.80	56.79	51.96	m²	**108.75**
Extra over blocks for						
100 mm thick half lintel blocks ref D14	-	0.28	8.83	25.32	m	**34.16**
140 mm thick half lintel blocks ref H14	-	0.32	10.10	45.30	m	**55.39**
140 mm thick quoin blocks ref H16	-	0.37	11.67	38.30	m	**49.97**
140 mm thick cavity closer blocks ref H17	-	0.37	11.67	41.08	m	**52.75**
140 mm thick cill blocks ref H21	-	0.32	10.10	30.20	m	**40.30**
190 mm thick half lintel blocks ref A14	-	0.37	11.67	28.51	m	**40.19**
Astra-Glaze satin-gloss glazed finish blocks or other equal and approved; Aldwick Design Ltd; standard colours; in gauged mortar (1:1:6); joints raked out; gun applied latex grout to joints						
Walls						
100 mm thick; glazed one side	105.39	1.06	33.44	125.42	m²	**158.86**
extra; glazed square end return	66.12	0.42	13.25	39.83	m	**53.08**
100 mm thick; glazed both sides	138.82	1.30	41.01	165.04	m²	**206.05**
100 mm thick lintel 200 mm high; glazed one side	-	0.97	22.57	33.15	m	**55.72**
F11 GLASS BLOCK WALLING						
NOTE: The following specialist prices for glass block walling; supplied by Roger Wilde Ltd; assume standard blocks in panels of 50 m²; no fire rating; work in straight walls at ground level; and all necessary ancillary fixing; strengthening; easy access; pointing and expansion materials etc.						
Hollow glass block walling; Pittsburgh Corning sealed "Thinline" or other equal and approved; in cement mortar joints; reinforced with 6 mm diameter stainless steel rods; pointed both sides with mastic or other equal and approved						
Walls; facework both sides						
115 mm x 115 mm x 80 mm flemish blocks	-	-	-	-	m²	**597.88**
190 mm x 190 mm x 80 mm flemish blocks; cross reeded or clear blocks	-	-	-	-	m²	**313.55**
240 mm x 240 mm x 80 mm flemish blocks; cross reeded or clear blocks	-	-	-	-	m²	**333.02**
240 mm x 115 mm x 80 mm flemish blocks; cross reeded or clear blocks	-	-	-	-	m²	**445.98**

F MASONRY

Item	PC £	Labour hours	Labour £	Material £	Unit	Total rate £
F20 NATURAL STONE RUBBLE WALLING						
Cotswold Guiting limestone or other equal and approved; laid dry						
Uncoursed random rubble walling						
275 mm thick	-	2.36	64.46	44.69	m²	**109.15**
350 mm thick	-	2.82	76.34	56.87	m²	**133.22**
425 mm thick	-	3.24	86.92	69.06	m²	**155.98**
500 mm thick	-	3.61	96.74	81.25	m²	**177.98**
Cotswold Guiting limestone or other equal and approved; bedded; jointed and pointed in cement; lime mortar (1:2:9)						
Uncoursed random rubble walling; faced and pointed; both sides						
275 mm thick	-	2.27	61.70	49.16	m²	**110.86**
350 mm thick	-	2.50	66.61	62.56	m²	**129.18**
425 mm thick	-	2.78	72.98	75.97	m²	**148.95**
500 mm thick	-	2.96	77.12	89.38	m²	**166.49**
Coursed random rubble walling; rough dressed; faced and pointed one side						
114 mm thick	-	1.71	46.82	63.78	m²	**110.60**
150 mm thick	-	2.16	60.70	64.36	m²	**125.07**
Fair returns on walling						
114 mm wide	-	0.02	0.55	-	m	**0.55**
150 mm wide	-	0.04	1.10	-	m	**1.10**
275 mm wide	-	0.07	1.92	-	m	**1.92**
350 mm wide	-	0.09	2.47	-	m	**2.47**
425 mm wide	-	0.12	3.29	-	m	**3.29**
500 mm wide	-	0.14	3.84	-	m	**3.84**
Fair raking cutting on walling						
114 mm wide	-	0.23	6.35	9.29	m	**15.64**
150 mm wide	-	0.29	8.01	9.29	m	**17.30**
Level uncoursed rubble walling for damp proof courses and the like						
275 mm wide	-	0.22	6.41	3.17	m	**9.58**
350 mm wide	-	0.23	6.70	3.97	m	**10.67**
425 mm wide	-	0.24	6.99	4.85	m	**11.84**
500 mm wide	-	0.26	7.57	5.71	m	**13.29**
Copings formed of rough stones; faced and pointed all round						
275 mm x 200 mm (average) high	-	0.64	17.95	11.49	m	**29.44**
350 mm x 250 mm (average) high	-	0.86	23.81	16.00	m	**39.81**
425 mm x 300 mm (average) high	-	1.12	30.69	22.14	m	**52.83**
500 mm x 300 mm (average) high	-	1.42	38.46	29.91	m	**68.38**
F22 CAST STONE WALLING/DRESSINGS						
Reconstructed limestone walling; "Bradstone 100 bed Weathered Cotswold" or "North Cerney" masonry blocks or other equal and approved; laid to pattern or course recommended; bedded; jointed and pointed in approved coloured cement:lime mortar (1:2:9)						
Walls; facing and pointing one side						
Rebastone Split	-	1.15	31.57	34.11	m²	**65.68**
Rebastone Rustic	-	1.15	31.57	36.48	m²	**68.05**
masonry blocks; random uncoursed	-	1.19	32.67	51.33	m²	**83.99**
extra; returned ends	-	0.43	11.80	28.27	m	**40.07**
extra; plain L shaped quoins	-	0.14	3.84	36.57	m	**40.41**

F MASONRY

Item	PC £	Labour hours	Labour £	Material £	Unit	Total rate £
F22 CAST STONE WALLING/DRESSINGS – cont'd						
Reconstructed limestone walling; "Bradstone 100 – cont'd						
traditional walling; coursed squared	-	1.49	40.90	51.33	m²	**92.23**
squared coursed rubble	-	1.43	39.26	52.67	m²	**91.92**
squared random rubble	-	1.49	40.90	52.48	m²	**93.38**
squared and pitched rock faced walling; coursed	-	1.54	42.27	52.48	m²	**94.75**
rough hewn rock faced walling; random	-	1.60	43.92	52.29	m²	**96.21**
extra; returned ends	-	0.17	4.67	-	m	**4.67**
ashlar; 440 x 215 x 100 mm thick	-	1.25	34.31	51.88	m²	**86.20**
Isolated piers or chimney stacks; facing and pointing one side						
Rebastone Split	-	1.60	43.92	34.11	m²	**78.03**
Rebastone Rustic	-	1.60	43.92	36.48	m²	**80.40**
masonry blocks; random uncoursed	-	1.65	45.29	51.33	m²	**96.62**
traditional walling; coursed squared	-	2.07	56.82	51.33	m²	**108.15**
squared coursed rubble	-	2.02	55.45	52.67	m²	**108.12**
squared random rubble	-	2.07	56.82	52.48	m²	**109.30**
squared and pitched rock faced walling; coursed	-	2.18	59.84	52.48	m²	**112.32**
rough hewn rock faced walling; random	-	2.23	61.22	52.29	m²	**113.50**
ashlar; 440 x 215 x 100 mm thick	-	1.75	48.04	51.88	m²	**99.92**
Isolated casings; facing and pointing one side						
Rebastone Split	-	1.40	38.43	34.11	m²	**72.54**
Rebastone Rustic	-	1.40	38.43	36.48	m²	**74.91**
masonry blocks; random uncoursed	-	1.43	39.26	51.33	m²	**90.58**
traditional walling; coursed squared	-	1.81	49.69	51.33	m²	**101.01**
squared coursed rubble	-	1.76	48.31	52.67	m²	**100.98**
squared random rubble	-	1.81	49.69	52.48	m²	**102.16**
squared and pitched rock faced walling; coursed	-	1.86	51.06	52.48	m²	**103.54**
rough hewn rock faced walling; random	-	1.91	52.43	52.29	m²	**104.72**
ashlar; 440 x 215 x 100 mm thick	-	1.50	41.18	51.88	m²	**93.06**
Fair returns 100 mm wide						
Rebastone Split	-	0.12	3.29	-	m²	**3.29**
Rebastone Rustic	-	0.12	3.29	-	m²	**3.29**
masonry blocks; random uncoursed	-	0.13	3.57	-	m²	**3.57**
traditional walling; coursed squared	-	0.16	4.39	-	m²	**4.39**
squared coursed rubble	-	0.15	4.12	-	m²	**4.12**
squared random rubble	-	0.16	4.39	-	m²	**4.39**
squared and pitched rock faced walling; coursed	-	0.16	4.39	-	m²	**4.39**
rough hewn rock faced walling; random	-	0.17	4.67	-	m²	**4.67**
ashlar; 440 x 215 x 100 mm thick	-	0.16	4.39	-	m²	**4.39**
Fair raking cutting on masonry blocks						
100 mm wide	-	0.19	5.22	-	m	**5.22**
Quoin						
ashlar; 440 x 215 x 215 x 100 mm thick	-	0.85	23.33	68.24	m	**91.57**
Reconstructed limestone dressings; "Bradstone Architectural" dressings in weathered "Cotswold" or "North Cerney" shades or other equal and approved; bedded, jointed and pointed in approved coloured cement:lime mortar (1:2:9)						
Copings; twice weathered and throated						
305 mm x 76 mm; type A	-	0.43	11.80	28.09	m	**39.89**
Extra for						
fair end	-	-	-	13.95	nr	**-**
returned mitred fair end	-	-	-	13.95	nr	**-**

F MASONRY

Item	PC £	Labour hours	Labour £	Material £	Unit	Total rate £
Copings; once weathered and throated						
305 mm x 76 mm	-	0.43	11.80	27.67	m	**39.48**
356 mm x 76 mm	-	0.43	11.80	25.65	m	**37.45**
Extra for						
fair end	-	-	-	13.95	nr	-
returned mitred fair end	-	-	-	13.95	nr	-
Pier caps; four times weathered and throated						
305 mm x 305 mm	-	0.27	7.41	16.57	nr	**23.98**
381 mm x 381 mm	-	0.27	7.41	24.57	nr	**31.98**
457 mm x 457 mm	-	0.32	8.78	33.60	nr	**42.39**
533 mm x 533 mm	-	0.32	8.78	46.65	nr	**55.43**
Splayed corbels						
479 mm x 100 mm x 215 mm	-	0.16	4.39	27.49	nr	**31.88**
665 mm x 100 mm x 215 mm	-	0.21	5.76	38.01	nr	**43.78**
100 mm x 140 mm lintels; rectangular; reinforced with mild steel bars						
all lengths to 2.07 m	-	0.30	8.24	43.86	m	**52.09**
100 mm x 215 mm lintels; rectangular; reinforced with mild steel bars						
all lengths to 2.85 m	-	0.34	9.33	46.84	m	**56.18**
sills to suit standard windows; stooled 100 mm at ends						
150 mm x140 mm; not exceeding 2.00 m long	-	0.32	8.78	59.22	m	**68.00**
197 mm x140 mm; not exceeding 1.97 m long	-	0.32	8.78	66.49	m	**75.27**
Window surround; traditional with label moulding; for single light; sill 146 mm x 133 mm; jambs 146 mm x 146 mm; head 146 mm x 105 mm; including all dowels and anchors						
overall size 508 mm x 1479 mm	179.30	0.95	26.08	204.30	nr	**230.38**
Window surround; traditional with label moulding; three light; for windows 508 mm x 1219 mm; sill 146 mm x 133 mm; jambs 146 mm x 146 mm; head 146 mm x 103 mm; mullions 146 mm x 108 mm; including all dowels and anchors						
overall size 1975 mm x 1479 mm	421.73	2.50	68.63	480.82	nr	**549.44**
Door surround; moulded continuous jambs and head with label moulding; including all dowels and anchors						
door 839 mm x 1981 mm in 102 mm x 64 mm frame	384.80	1.76	48.31	435.31	nr	**483.62**
F30 ACCESSORIES/SUNDRY ITEMS FOR BRICK/ BLOCK/STONE WALLING						
Forming cavities						
In hollow walls						
width of cavity 50 mm; galvanised steel twisted wall ties; three wall ties per m²	-	0.06	1.65	0.98	m²	**2.63**
width of cavity 50 mm; stainless steel butterfly wall ties; three wall ties per m²	-	0.06	1.65	0.32	m²	**1.97**
width of cavity 50 mm; stainless steel twisted wall ties; three wall ties per m²	-	0.06	1.65	1.78	m²	**3.43**
width of cavity 75 mm; galvanised steel twisted wall ties; three wall ties per m²	-	0.06	1.65	1.04	m²	**2.69**
width of cavity 75 mm; stainless steel butterfly wall ties; three wall ties per m²	-	0.06	1.65	0.34	m²	**1.99**
width of cavity 75 mm; stainless steel twisted wall ties; three wall ties per m²	-	0.06	1.65	1.97	m²	**3.62**

F MASONRY

Item	PC £	Labour hours	Labour £	Material £	Unit	Total rate £
F30 ACCESSORIES/SUNDRY ITEMS FOR BRICK/ BLOCK/STONE WALLING – cont'd						
Damp proof courses						
Polythene damp proof course or other equal and approved; 200 mm laps; in gauged mortar (1:1:6)						
width exceeding 225 mm; horizontal	0.75	0.27	7.41	0.91	m²	**8.32**
width exceeding 225 mm; forming cavity gutters in hollow walls	-	0.43	11.80	0.91	m²	**12.72**
width not exceeding 225 mm; horizontal	-	0.53	14.55	0.91	m²	**15.46**
width not exceeding 225 mm; vertical	-	0.80	21.96	0.91	m²	**22.87**
"Engerseal" polymer elastomeric damp proof course or other equal and approved; 200 mm laps; in gauged morter (1:1:6)						
width exceeding 225 mm; horizontal	3.16	0.27	7.41	3.83	m²	**11.25**
width exceeding 225 mm; forming cavity gutters in hollow walls	-	0.43	11.80	3.83	m²	**15.64**
width not exceeding 225 mm; horizontal	-	0.53	14.55	3.83	m²	**18.38**
width not exceeding 225 mm; horizontal	-	0.80	21.96	3.83	m²	**25.80**
"Zedex CPT" (Co-Polymer Thermoplastic) damp proof course or other equal and approved; 200 mm laps; in gauged mortar (1:1:6)						
width exceeding 225 mm; horizontal	4.34	0.27	7.41	5.27	m²	**12.68**
width exceeding 225 mm; forming cavity gutters in hollow walls	-	0.43	11.80	5.27	m²	**17.07**
width not exceeding 225 mm; horizontal	-	0.53	14.55	5.27	m²	**19.82**
width not exceeding 225 mm; vertical	-	0.80	21.96	5.27	m²	**27.23**
"Hyload" (pitch polymer) damp proof course or other equal and approved; 150 mm laps; in gauged mortar (1:1:6)						
width exceeding 225 mm; horizontal	4.82	0.27	7.41	5.85	m²	**13.27**
width exceeding 225 mm; forming cavity gutters in hollow walls	-	0.43	11.80	5.85	m²	**17.66**
width not exceeding 225 mm; horizontal	-	0.53	14.55	5.85	m²	**20.40**
width not exceeding 225 mm; vertical	-	0.80	21.96	5.85	m²	**27.81**
"Nubit" bitumen and polyester based damp proof course or other equal and approved; 200 mm laps; in gauged mortar (1:1:6)						
width exceeding 225 mm; horizontal	-	0.35	9.61	7.23	m²	**16.84**
width exceeding 225 mm wide; forming cavity gutters in hollow walls; horizontal	-	0.56	15.37	7.23	m²	**22.60**
width not exceeding 225 mm; horizontal	-	0.68	18.67	7.23	m²	**25.90**
width not exceeding 225 mm; vertical	-	0.80	21.96	7.23	m²	**29.19**
"Permabit" bitumen polymer damp proof course or other equal and approved; 150 mm laps; in gauged mortar (1:1:6)						
width exceeding over 225 mm	9.80	0.27	7.41	11.89	m²	**19.30**
width exceeding 225 mm; forming cavity gutters in hollow walls	-	0.43	11.80	11.89	m²	**23.70**
width not exceeding 225 mm; horizontal	-	0.53	14.55	11.89	m²	**26.44**
width not exceeding 225 mm; vertical	-	0.80	21.96	11.89	m²	**33.85**
"Alumite" aluminium cored bitumen gas retardent damp proof course or other equal and approved; 200 mm laps; in gauged mortar (1:1;6)						
width exceeding 225 mm; horizontal	6.40	0.35	9.61	7.77	m²	**17.37**
width exceeding 225 mm wide; forming cavity gutters in hollow walls; horizontal	-	0.56	15.37	7.77	m²	**23.14**
width not exceeding 225 mm; horizontal	-	0.68	18.67	7.77	m²	**26.43**
width not exceeding 225 mm; vertical	-	0.92	25.26	7.77	m²	**33.02**

F MASONRY

Item	PC £	Labour hours	Labour £	Material £	Unit	Total rate £
Milled lead damp proof course; BS 1178; 1.80 mm thick (code 4), 175 mm laps, in cement:lime mortar (1:2:9)						
width exceeding 225 mm; horizontal (PC £/kg)	1.73	2.13	58.47	48.16	m²	**106.63**
width not exceeding 225 mm; horizontal	-	3.19	87.57	48.16	m²	**135.73**
Two courses slates in cement mortar (1:3)						
width exceeding 225 mm; horizontal	-	1.60	43.92	14.41	m²	**58.33**
width exceeding 225 mm; vertical	-	2.40	65.88	14.41	m²	**80.29**
"Peter Cox" chemical transfusion damp proof course system or other equal and approved						
half brick thick; horizontal	-	-	-	-	m	**23.88**
one brick thick; horizontal	-	-	-	-	m	**46.31**
one and a half brick thick; horizontal	-	-	-	-	m	**69.32**
Silicone injection damp-proofing; 450 mm centres; making good brickwork						
half brick thick; horizontal	-	-	-	-	m	**12.44**
one brick thick; horizontal	-	-	-	-	m	**21.68**
one and a half brick thick; horizontal	-	-	-	-	m	**34.51**
"Synthprufe" damp proof membrane or other equal and approved; PC £52.15/25 litres; three coats brushed on						
width not exceeding 150 mm; vertical	-	0.36	4.34	6.83	m²	**11.18**
width 150 mm - 225 mm; vertical	-	0.34	4.10	6.83	m²	**10.94**
width 225 mm - 300 mm; vertical	-	0.32	3.86	6.83	m²	**10.70**
width exceeding 300 mm; vertical	-	0.30	3.62	6.83	m²	**10.45**
Joint reinforcement						
"Brickforce" galvanised steel joint reinforcement or other equal and approved						
width 60 mm; ref GBF40W60B25	-	0.06	1.65	0.81	m	**2.46**
width 100 mm; ref GBF40W100B25	-	0.08	2.20	0.94	m	**3.14**
width 150 mm; ref GBF40W150B25	-	0.10	2.75	1.18	m	**3.92**
width 175 mm; ref GBF40W175B25	-	0.11	3.02	1.41	m	**4.43**
"Brickforce" stainless steel joint reinforcement or other equal and approved						
width 60 mm; ref SBF35W60BSC	-	0.06	1.65	1.77	m	**3.42**
width 100 mm; ref SBF35W100BSC	-	0.08	2.20	1.82	m	**4.02**
width 150 mm; ref SBF35W150BSC	-	0.10	2.75	1.86	m	**4.60**
width 175 mm; ref SBF35W175BSC	-	0.11	3.02	2.03	m	**5.05**
width 60 mm; ref SBF40W60BSC	-	0.06	1.65	2.22	m	**3.86**
width 100 mm; ref SBF40W100BSC	-	0.08	2.20	2.29	m	**4.48**
width 150 mm; ref SBF40W150BSC	-	0.10	2.75	2.32	m	**5.07**
width 175 mm; ref SBF40W175BSC	-	0.11	3.02	2.54	m	**5.56**
"Wallforce" stainless steel joint reinforcement or other equal and approved						
width 240 mm; ref SWF35W240	-	0.13	3.57	7.00	m	**10.57**
width 260 mm; ref SWF35W260	-	0.14	3.84	7.31	m	**11.15**
width 275 mm; ref SWF35W275	-	0.15	4.12	-	m	**4.12**
Weather fillets						
Weather fillets in cement mortar (1:3)						
50 mm face width	-	0.13	3.57	0.05	m	**3.62**
100 mm face width	-	0.21	5.76	0.21	m	**5.98**
Bedding wall plates or similar in cement mortar (1:3)						
100 mm wide	-	0.06	1.65	0.09	m	**1.73**
Bedding wood frames in cement mortar (1:3) and point						
one side	-	0.08	2.20	0.09	m	**2.28**
both sides	-	0.11	3.02	0.11	m	**3.13**
one side in mortar; other side in mastic	-	0.22	4.90	0.74	m	**5.64**

F MASONRY

Item	PC £	Labour hours	Labour £	Material £	Unit	Total rate £
F30 ACCESSORIES/SUNDRY ITEMS FOR BRICK/ BLOCK/STONE WALLING – cont'd						
Angle fillets						
Angle fillets in cement mortar (1:3)						
50 mm face width	-	0.13	3.57	0.05	m	**3.62**
100 mm face width	-	0.21	5.76	0.21	m	**5.98**
Pointing in						
Pointing with mastic						
wood frames or sills	-	0.11	1.77	0.65	m	**2.42**
Pointing with polysulphide sealant						
wood frames or sills	-	0.11	1.77	2.06	m	**3.84**
Wedging and pinning						
To underside of existing construction with slates in cement mortar (1:3)						
width of wall - one brick thick	-	0.85	23.33	3.13	m	**26.47**
width of wall - one and a half brick thick	-	1.06	29.10	6.27	m	**35.37**
width of wall - two brick thick	-	1.28	35.14	9.40	m	**44.54**
Joints						
Hacking joints and faces of brickwork or blockwork						
to form key for plaster	-	0.28	3.38	-	m²	**3.38**
Raking out joint in brickwork or blockwork for turned-in edge of flashing						
horizontal	-	0.16	4.39	-	m	**4.39**
stepped	-	0.21	5.76	-	m	**5.76**
Raking out and enlarging joint in brickwork or blockwork for nib of asphalt						
horizontal	-	0.21	5.76	-	m	**5.76**
Cutting grooves in brickwork or blockwork						
for water bars and the like	-	0.27	3.26	0.45	m	**3.71**
for nib of asphalt; horizontal	-	0.27	3.26	0.45	m	**3.71**
Preparing to receive new walls						
top existing 215 mm wall	-	0.21	5.76	-	m	**5.76**
Cleaning and priming both faces; filling with pre-formed closed cell joint filler and pointing one side with polysulphide sealant; 12 mm deep						
expansion joints; 12 mm wide	-	0.27	6.28	4.11	m	**10.38**
expansion joints; 20 mm wide	-	0.31	6.92	6.01	m	**12.93**
expansion joints; 25 mm wide	-	0.37	7.89	7.25	m	**15.14**
Fire resisting horizontal expansion joints; filling with joint filler; fixed with high temperature slip adhesive; between top of wall and soffit						
wall not exceeding 215 mm wide; 10 mm wide joint with 30 mm deep filler (one hour fire seal)	-	0.27	7.41	7.37	m	**14.78**
wall not exceeding 215 mm wide; 10 mm wide joint with 30 mm deep filler (two hour fire seal)	-	0.27	7.41	7.37	m	**14.78**
wall not exceeding 215 mm wide; 20 mm wide joint with 45 mm deep filler (two hour fire seal)	-	0.32	8.78	11.06	m	**19.84**
wall not exceeding 215 mm wide; 30 mm wide joint with 75 mm deep filler (three hour fire seal)	-	0.37	10.16	27.53	m	**37.69**
Fire resisting vertical expansion joints; filling with joint filler; fixed with high temperature slip adhesive; with polysulphide sealant one side; between end of wall of wall and concrete						
wall not exceeding 215 mm wide; 20 mm wide joint with 45 mm deep filler (two hour fire seal)	-	0.43	10.90	15.19	m	**26.08**

F MASONRY

Item	PC £	Labour hours	Labour £	Material £	Unit	Total rate £
Slates and tile sills						
Sills; two courses of machine made plain roofing tiles set weathering; bedded and pointed	-	0.64	17.57	6.32	m	**23.89**
Sundries						
Weep holes						
Perpend units; plastic	-	0.02	0.55	0.23	nr	**0.77**
Chimney pots; red terracotta; plain or cannon-head; setting and flaunching in cement mortar (1:3)						
185 mm diameter x 300 mm long	17.01	1.91	52.43	20.90	nr	**73.34**
185 mm diameter x 600 mm long	30.15	2.13	58.47	35.74	nr	**94.21**
185 mm diameter x 900 mm long	54.81	2.13	58.47	63.57	nr	**122.04**
Air bricks						
Air bricks; red terracotta; building into prepared openings						
215 mm x 65 mm	-	0.08	2.20	1.86	nr	**4.05**
215 mm x 140 mm	-	0.08	2.20	2.57	nr	**4.76**
215 mm x 215 mm	-	0.08	2.20	7.17	nr	**9.37**
Gas flue blocks						
Gas flue system; Schiedel "HP" or other equal and approved; concrete blocks built in; in flue joint mortar mix; cutting brickwork or blockwork around						
recess; ref HP1	-	0.11	3.02	2.96	nr	**5.98**
cover; ref HP2	-	0.11	3.02	6.50	nr	**9.52**
222 mm standard block with nib; ref HP3	-	0.11	3.02	4.77	nr	**7.79**
112 mm standard block with nib; ref HP3112	-	0.11	3.02	3.76	nr	**6.78**
72 mm standard block with nib; ref HP372	-	0.11	3.02	3.76	nr	**6.78**
vent block; ref HP3BH	-	0.11	3.02	7.69	nr	**10.71**
222 mm standard block without nib; ref HP4	-	0.11	3.02	4.77	nr	**7.79**
112 mm standard block without nib; ref HP4112	-	0.11	3.02	3.76	nr	**6.78**
72 mm standard block without nib; ref HP472	-	0.11	3.02	3.76	nr	**6.78**
120 mm side offset block; ref HP5	-	0.11	3.02	5.04	nr	**8.06**
70 mm back offset block; ref HP6	-	0.11	3.02	16.18	nr	**19.20**
vertical exit block; ref HP7	-	0.11	3.02	9.62	nr	**12.64**
angled entry/exit block; ref HP8	-	0.11	3.02	9.57	nr	**12.59**
reverse rebate block; ref HP9	-	0.11	3.02	7.02	nr	**10.04**
corbel block; ref HP10	-	0.11	3.02	9.41	nr	**12.43**
lintel unit; ref HP11	-	0.11	3.02	8.79	nr	**11.81**
Proprietary items						
External door and window cavity closers; "Thermabate" or equivalent; inclusive of flange clips; jointing strips; wall fixing ties and adhesive tape						
closing cavities; width of cavity 50 mm	-	0.16	4.39	4.59	m	**8.98**
closing cavities; width of cavity 75 mm	-	0.16	4.39	5.17	m	**9.56**
closing cavities; width of cavity 100 mm	-	0.16	4.39	5.17	m	**9.56**
"Westbrick" cavity closers or other equal and approved; Manthorpe Building Products Ltd						
closing cavities; width of cavity 50 mm	-	0.16	4.39	6.05	m	**10.44**
"Type H cavicloser" or other equal and approved; uPVC universal cavity closer, insulator and damp-proof course by Cavity Trays Ltd; built into cavity wall as work proceeds, complete with face closer and ties						
closing cavities; width of cavity 50 mm - 100 mm	-	0.08	2.20	6.41	m	**8.60**

F MASONRY

Item	PC £	Labour hours	Labour £	Material £	Unit	Total rate £
F30 ACCESSORIES/SUNDRY ITEMS FOR BRICK/ BLOCK/STONE WALLING – cont'd						
Proprietary items – cont'd						
"Type L" durropolythelene lintel stop ends or other equal and approved; Cavity Trays Ltd; fixing to lintel with butyl anchoring strip; building in as the work proceeds						
adjusted to lintel as required	-	0.05	1.37	0.65	nr	**2.02**
"Type W" polypropylene weeps/vents or other equal and approved; Cavity Trays Ltd; built into cavity wall as work proceeds						
100 mm/115 mm x 65 mm x 10 mm including lock fit wedges	-	0.05	1.37	0.49	nr	**1.86**
extra; extension duct 200 mm/225 mm x 65 mm x 10 mm	-	0.08	2.20	0.83	nr	**3.03**
"Type X" polypropylene abutment cavity tray or other equal and approved; Cavity Trays; built into facing brickwork as the work proceeds; complete with Code 4 lead flashing; intermediate/catchment tray with short leads (requiring soakers); to suit roof of						
17 - 20 degree pitch	-	0.06	1.65	7.77	nr	**9.41**
21 - 25 degree pitch	-	0.06	1.65	7.23	nr	**8.87**
26 - 45 degree pitch	-	0.06	1.65	6.90	nr	**8.55**
"Type X" polypropylene abutment cavity tray or other equal and approved; Cavity Trays; built into facing brickwork as the work proceeds; complete with Code 4 lead flashing; intermediate/catchment tray with long leads (suitable only for corrugated roof tiles); to suit roof of						
17 - 20 degree pitch	-	0.06	1.65	10.50	nr	**12.14**
21 - 25 degree pitch	-	0.06	1.65	9.65	nr	**11.30**
26 - 45 degree pitch	-	0.06	1.65	8.90	nr	**10.55**
"Type X" polypropylene abutment cavity tray or other equal and approved; Cavity Trays; built into facing brickwork as the work proceeds; complete with Code 4 lead flashing; ridge tray with short/long leads; to suit roof of						
17 - 20 degree pitch	-	0.06	1.65	17.67	nr	**19.32**
21 - 25 degree pitch	-	0.06	1.65	16.38	nr	**18.03**
26 - 45 degree pitch	-	0.06	1.65	11.21	nr	**12.86**
Servicised "Bituthene MR" aluminium faced gas resistant cavity flashing or other equal and approved; sealed at joints with ervitape 30mm; in gauged mortar (1:1:6)						
width exceeding 225 mm; horizontal	-	0.91	24.98	21.27	m²	**46.25**
"Expamet" stainless steel wall starters or other equal and approved; plugged and screwed						
to suit walls 60 mm - 75 mm thick	-	0.27	3.26	17.25	m	**20.50**
to suit walls 100 mm - 115 mm thick	-	0.27	3.26	18.97	m	**22.23**
to suit walls 125 mm - 180 mm thick	-	0.43	5.19	25.38	m	**30.57**
to suit walls 190 mm - 260 mm thick	-	0.53	6.40	32.42	m	**38.81**
Stainless steel posts, channels and ties						
Windposts; 130 x 70 x 6 mm; including one piece through ties						
1200 mm long	-	-	-	-	nr	**148.92**
3000 mm long	-	-	-	-	nr	**223.38**
4800 mm long	-	-	-	-	nr	**327.62**

F MASONRY

Item	PC £	Labour hours	Labour £	Material £	Unit	Total rate £
Wall restraint channel ties; vertical channels; welded to steel work; with lateral restraint ties						
channel reference 28/15; tie reference HTS-B9; 200 mm long; one end of tie secured to channel; other end and debonding sleeve built into horizontal joint of masonry at 250 mm centres	-	0.14	3.84	18.84	m	**22.69**
Head restraints; sliding brick anchors reference SBA/L at 900 mm horizontal centres; 450 mm deep; tying into two courses of blockwork; fixed to steelwork						
ties built into horizontal joint of masonry	-	0.06	1.65	27.81	nr	**29.46**
Wall restraint; individually fixed ties; fixed to steelwork						
ties reference HTS-B9; 200 mm long; one end of tie secured to channel; other end and debonding sleeve built into horizontal joint of masonry at 250 mm centres	-	0.02	0.55	0.87	nr	**1.42**
Ties in walls; 200 mm long butterfly type; building into joints of brickwork or blockwork						
galvanised steel or polypropylene	-	0.02	0.55	0.10	nr	**0.65**
stainless steel	-	0.02	0.55	0.13	nr	**0.68**
Ties in walls; 20 mm x 3 mm x 200 mm long twisted wall type; building into joints of brickwork or blockwork						
galvanised steel	-	0.02	0.55	0.40	nr	**0.95**
stainless steel	-	0.02	0.55	0.72	nr	**1.27**
Anchors in walls; 25 mm x 3 mm x 100 mm long; one end dovetailed; other end building into joints of brickwork or blockwork						
galvanised steel	-	0.06	1.65	0.41	nr	**2.06**
stainless steel	-	0.06	1.65	1.02	nr	**2.67**
Fixing cramp 25 mm x 3 mm x 250 mm long; once bent; fixed to back of frame; other end building into joints of brickwork or blockwork						
galvanised steel	-	0.06	1.65	0.39	nr	**2.03**
Galvanised steel lintels; "Catnic" or other equal and approved; built into brickwork or blockwork						
70/125 range "CG" open back lintel for cavity wall						
750 mm long	40.23	0.27	7.41	45.46	nr	**52.87**
900 mm long	48.05	0.32	8.78	54.28	nr	**63.07**
1200 mm long	63.15	0.37	10.16	71.32	nr	**81.48**
1500 mm long	79.47	0.43	11.80	89.74	nr	**101.55**
1800 mm long	109.06	0.48	13.18	123.15	nr	**136.33**
2100 mm long	128.43	0.53	14.55	145.01	nr	**159.56**
2400 mm long	177.32	0.64	17.57	200.19	nr	**217.76**
70/125 range "CUB" open back lintel for cavity wall						
2700 mm long	261.51	0.74	20.31	295.23	nr	**315.54**
3000 mm long	296.96	0.85	23.33	335.24	nr	**358.58**
70/125 range "CU" open back lintel for cavity wall						
3300 mm long	326.73	0.95	26.08	368.84	nr	**394.92**
3600 mm long	356.40	1.06	29.10	402.34	nr	**431.44**
3900 mm long	385.84	1.17	32.12	435.56	nr	**467.68**
4200 mm long	430.87	0.53	14.55	486.39	nr	**500.94**

F MASONRY

Item	PC £	Labour hours	Labour £	Material £	Unit	Total rate £
F30 ACCESSORIES/SUNDRY ITEMS FOR BRICK/ BLOCK/STONE WALLING – cont'd						
Galvanised steel lintels; "Catnic" – cont'd						
90/125 range "CG" open back lintel; for cavity wall						
750 mm long	44.74	0.27	7.41	50.55	nr	**57.96**
900 mm long	53.69	0.32	8.78	60.65	nr	**69.44**
1200 mm long	70.51	0.37	10.16	79.63	nr	**89.79**
1500 mm long	87.92	0.43	11.80	99.29	nr	**111.09**
1800 mm long	111.25	0.48	13.18	125.62	nr	**138.79**
2100 mm long	131.84	0.53	14.55	148.86	nr	**163.41**
2400 mm long	186.24	0.64	17.57	210.27	nr	**227.84**
90/125 range "CUB" open back lintel for cavity wall						
2700 mm long	294.86	0.74	20.31	332.87	nr	**353.18**
3000 mm long	327.90	0.85	23.33	370.17	nr	**393.50**
90/125 range "CU" open back lintel; for cavity wall						
3300 mm long	355.99	0.95	26.08	401.87	nr	**427.95**
3600 mm long	388.36	1.06	29.10	438.40	nr	**467.50**
3900 mm long	420.42	1.17	32.12	474.59	nr	**506.71**
4200 mm long	441.21	0.53	14.55	498.06	nr	**512.61**
"CN92"single lintel for 75 mm internal wall						
1050 mm long	6.80	0.32	8.78	7.70	nr	**16.48**
1200 mm long	7.69	0.37	10.16	8.71	nr	**18.86**
"CN102" single lintel for 100 mm internal wall						
1050 mm long	8.60	0.32	8.78	9.73	nr	**18.52**
1200 mm long	9.49	0.37	10.16	10.74	nr	**20.90**
"CN100" single lintel for 75 mm internal wall						
1050 mm long	21.00	0.32	8.78	23.73	nr	**32.52**
1200 mm long	26.10	0.37	10.16	29.49	nr	**39.64**
"CN5XA" single lintel for 100 mm internal wall						
1050 mm long	39.67	0.32	8.78	44.80	nr	**53.58**
1200 mm long	40.70	0.37	10.16	45.96	nr	**56.12**
F31 PRECAST CONCRETE SILLS/LINTELS/ COPING FEATURES						
Mix 20.00 N/mm² - 20 mm aggregate (1:2:4)						
Lintels; plate; prestressed bedded						
100 mm x 70 mm x 750 mm long	4.02	0.43	11.80	4.56	nr	**16.36**
100 mm x 70 mm x 900 mm long	4.75	0.43	11.80	5.39	nr	**17.19**
100 mm x 70 mm x 1050 mm long	5.64	0.43	11.80	6.40	nr	**18.20**
100 mm x 70 mm x 1200 mm long	6.39	0.43	11.80	7.24	nr	**19.04**
150 mm x 70 mm x 900 mm long	6.52	0.53	14.55	7.40	nr	**21.95**
150 mm x 70 mm x 1050 mm long	7.62	0.53	14.55	8.64	nr	**23.19**
150 mm x 70 mm x 1200 mm long	8.70	0.53	14.55	9.86	nr	**24.41**
220 mm x 70 mm x 900 mm long	10.06	0.64	17.57	11.42	nr	**28.99**
220 mm x 70 mm x 1200 mm long	13.38	0.64	17.57	15.16	nr	**32.73**
220 mm x 70 mm x 1500 mm long	16.77	0.74	20.31	18.99	nr	**39.31**
265 mm x 70 mm x 900 mm long	10.74	0.64	17.57	12.19	nr	**29.76**
265 mm x 70 mm x 1200 mm long	14.34	0.64	17.57	16.25	nr	**33.82**
265 mm x 70 mm x 1500 mm long	17.87	0.74	20.31	20.24	nr	**40.55**
265 mm x 70 mm x 1800 mm long	21.40	0.85	23.33	24.22	nr	**47.55**
Lintels; rectangular; reinforced with mild steel bars; bedded						
100 mm x 145 mm x 900 mm long	9.72	0.64	17.57	10.99	nr	**28.56**
100 mm x 145 mm x 1050 mm long	11.42	0.64	17.57	12.91	nr	**30.48**
100 mm x 145 mm x 1200 mm long	12.44	0.64	17.57	14.07	nr	**31.63**
225 mm x 145 mm x 1200 mm long	16.38	0.85	23.33	18.55	nr	**41.89**
225 mm x 225 mm x 1800 mm long	31.74	1.60	43.92	35.89	nr	**79.81**

F MASONRY

Item	PC £	Labour hours	Labour £	Material £	Unit	Total rate £
Lintels; boot; reinforced with mild steel bars; bedded						
250 mm x 225 mm x 1200 mm long	26.30	1.28	35.14	29.75	nr	**64.89**
275 mm x 225 mm x 1800 mm long	35.26	1.91	52.43	39.86	nr	**92.29**
Padstones						
300 mm x 100 mm x 75 mm	5.37	0.32	8.78	6.09	nr	**14.87**
225 mm x 225 mm x 150 mm	8.15	0.43	11.80	9.24	nr	**21.05**
450 mm x 450 mm x 150 mm	21.54	0.64	17.57	24.48	nr	**42.05**
Mix 30.00 N/mm² - 20 mm aggregate (1:1:2)						
Copings; once weathered; once throated; bedded and pointed						
152 mm x 76 mm	6.52	0.74	20.31	7.82	m	**28.13**
178 mm x 64 mm	7.20	0.74	20.31	8.62	m	**28.93**
305 mm x 76 mm	12.15	0.85	23.33	14.60	m	**37.93**
extra for fair ends	-	-	-	6.29	nr	**-**
extra for angles	-	-	-	7.13	nr	**-**
Copings; twice weathered; twice throated; bedded and pointed						
152 mm x 76 mm	6.52	0.74	20.31	7.82	m	**28.13**
178 mm x 64 mm	7.14	0.74	20.31	8.55	m	**28.86**
305 mm x 76 mm	12.15	0.85	23.33	14.60	m	**37.93**
extra for fair ends	-	-	-	6.29	nr	**-**
extra for angles	-	-	-	7.13	nr	**-**
Sills; splayed top edge; stooled ends; bedded and pointed						
140 mm x 85 mm	18.00	0.85	23.33	21.42	m	**44.75**
140 mm x 85 mm	22.50	0.85	23.33	26.75	m	**50.09**

G STRUCTURAL/CARCASSING METAL/TIMBER

Item	PC £	Labour hours	Labour £	Material £	Unit	Total rate £
G10 STRUCTURAL STEEL FRAMING						
Framing, fabrication; weldable steel; BS EN 10025 Grade S275; hot rolled structural steel sections; welded fabrication						
Columns						
weight not exceeding 40 kg/m	-	-	-	-	tonne	**1849.10**
weight not exceeding 40 kg/m; castellated	-	-	-	-	tonne	**2433.35**
weight not exceeding 40 kg/m; curved	-	-	-	-	tonne	**2428.22**
weight not exceeding 40 kg/m; square hollow section	-	-	-	-	tonne	**2388.25**
weight not exceeding 40 kg/m; circular hollow section	-	-	-	-	tonne	**2519.45**
weight 40 - 100 kg/m	-	-	-	-	tonne	**1610.28**
weight 40 - 100 kg/m; castellated	-	-	-	-	tonne	**2096.13**
weight 40 - 100 kg/m; curved	-	-	-	-	tonne	**2062.30**
weight 40 - 100 kg/m; square hollow section	-	-	-	-	tonne	**2050.00**
weight 40 - 100 kg/m; circular hollow section	-	-	-	-	tonne	**2160.70**
weight exceeding 100 kg/m	-	-	-	-	tonne	**1456.53**
weight exceeding 100 kg/m; castellated	-	-	-	-	tonne	**1852.17**
weight exceeding 100 kg/m; curved	-	-	-	-	tonne	**1843.97**
weight exceeding 100 kg/m; square hollow section	-	-	-	-	tonne	**1978.25**
weight exceeding 100 kg/m; circular hollow section	-	-	-	-	tonne	**2107.40**
Beams						
weight not exceeding 40 kg/m	-	-	-	-	tonne	**1904.45**
weight not exceeding 40 kg/m; castellated	-	-	-	-	tonne	**2493.82**
weight not exceeding 40 kg/m; curved	-	-	-	-	tonne	**2478.45**
weight not exceeding 40 kg/m; square hollow section	-	-	-	-	tonne	**2593.25**
weight not exceeding 40 kg/m; circular hollow section	-	-	-	-	tonne	**3056.55**
weight 40 - 100 kg/m	-	-	-	-	tonne	**1575.42**
weight 40 - 100 kg/m; castellated	-	-	-	-	tonne	**2007.97**
weight 40 - 100 kg/m; curved	-	-	-	-	tonne	**1981.33**
weight 40 - 100 kg/m; square hollow section	-	-	-	-	tonne	**2437.45**
weight 40 - 100 kg/m; circular hollow section	-	-	-	-	tonne	**2837.20**
weight exceeding 100 kg/m	-	-	-	-	tonne	**1452.42**
weight exceeding 100 kg/m; castellated	-	-	-	-	tonne	**1856.28**
weight exceeding 100 kg/m; curved	-	-	-	-	tonne	**1849.10**
weight exceeding 100 kg/m; square hollow section	-	-	-	-	tonne	**2367.75**
weight exceeding 100 kg/m; circular hollow section	-	-	-	-	tonne	**2689.60**
Bracings						
weight not exceeding 40 kg/m	-	-	-	-	tonne	**2263.20**
weight not exceeding 40 kg/m; square hollow section	-	-	-	-	tonne	**2863.85**
weight not exceeding 40 kg/m; circular hollow section	-	-	-	-	tonne	**2863.85**
weight 40 - 100 kg/m	-	-	-	-	tonne	**2105.35**
weight 40 - 100 kg/m; square hollow section	-	-	-	-	tonne	**2693.70**
weight 40 - 100 kg/m; circular hollow section	-	-	-	-	tonne	**2693.70**
weight exceeding 100 kg/m	-	-	-	-	tonne	**1994.65**
weight exceeding 100 kg/m; square hollow section	-	-	-	-	tonne	**2573.78**
weight exceeding 100 kg/m; circular hollow section	-	-	-	-	tonne	**2573.78**

G STRUCTURAL/CARCASSING METAL/TIMBER

Item	PC £	Labour hours	Labour £	Material £	Unit	Total rate £
Purlins and cladding rails						
weight not exceeding 40 kg/m	-	-	-	-	tonne	**1744.55**
weight not exceeding 40 kg/m; square hollow section	-	-	-	-	tonne	**2901.78**
weight not exceeding 40 kg/m; circular hollow section	-	-	-	-	tonne	**2901.78**
weight 40 - 100 kg/m	-	-	-	-	tonne	**1555.95**
weight 40 - 100 kg/m; square hollow section	-	-	-	-	tonne	**2616.82**
weight 40 - 100 kg/m; circular hollow section	-	-	-	-	tonne	**2616.82**
weight exceeding 100 kg/m	-	-	-	-	tonne	**1438.08**
weight exceeding 100 kg/m; square hollow section	-	-	-	-	tonne	**2529.70**
weight exceeding 100 kg/m; circular hollow section	-	-	-	-	tonne	**2529.70**
Grillages						
weight not exceeding 40 kg/m	-	-	-	-	tonne	**1965.95**
weight 40 - 100 kg/m	-	-	-	-	tonne	**1583.63**
weight exceeding 100 kg/m	-	-	-	-	tonne	**1511.88**
Trestles, towers and built up columns						
straight	-	-	-	-	tonne	**2646.55**
Trusses and built up girders						
straight	-	-	-	-	tonne	**2646.55**
curved	-	-	-	-	tonne	**3251.30**
Fittings	-	-	-	-	tonne	**2600.43**
Add to the above forementioned prices for:						
grade 355 steelwork	-	-	-	-	%	**7.69**
Framing, erection						
Trial erection	-	-	-	-	tonne	**472.52**
Permanent erection on site	-	-	-	-	tonne	**472.52**
Surface preparation						
At works						
blast cleaning	-	-	-	-	m²	**3.14**
Surface treatment						
At works						
galvanising	-	-	-	-	m²	**15.04**
shotblasting and priming to SA 2.5	-	-	-	-	m²	**7.72**
touch up primer and one coat of two pack epoxy zinc phosphate primer	-	-	-	-	m²	**5.51**
intumescent paint fire protection (30 minutes); spray applied	-	-	-	-	m²	**12.73**
intumescent paint fire protection (60 minutes); spray applied	-	-	-	-	m²	**19.09**
Extra over for separate decorative sealer top coat	-	-	-	-	m²	**3.82**
On site						
intumescent paint fire protection (30 minutes) to universal beams etc.; spray applied	-	-	-	-	m²	**9.55**
intumescent paint fire protection (30 minutes) to circular columns etc.; spray applied	-	-	-	-	m²	**16.00**
intumescent paint fire protection (60 minutes) to universal beams etc.; spray applied	-	-	-	-	m²	**12.10**
intumescent paint fire protection (60 minutes) to circular columns etc.; spray applied	-	-	-	-	m²	**20.30**
Extra over for separate decorative sealer top coat	-	-	-	-	m²	**3.18**

G STRUCTURAL/CARCASSING METAL/TIMBER

Item	PC £	Labour hours	Labour £	Material £	Unit	Total rate £
G10 STRUCTURAL STEEL FRAMING – cont'd						
Metsec Steel Framing System; or other equal and approved; as inner leaf to external wall; including provision for all openings, abutments, junctions and head details etc.						
Inner leaf; with supports and perimeter sections for external metal cladding (measured separately)						
100 mm thick steel walling	-	-	-	-	m²	**71.68**
150 mm thick steel walling	-	-	-	-	m²	**76.30**
200 mm thick steel walling	-	-	-	-	m²	**87.99**
Inner leaf; with 16 mm Pyroc sheething board						
100 mm thick steel walling	-	-	-	-	m²	**91.17**
150 mm thick steel walling	-	-	-	-	m²	**95.79**
200 mm thick steel walling	-	-	-	-	m²	**107.48**
16 mm Pyroc sheething board fixed to slab perimeter not exceeding 300 mm	-	-	-	-	m	**10.11**
Inner leaf; with 16 mm Pyroc sheething board and 40 mm Thermawall TW55 insulation supported by halfen channels type 28/15 fixed to studs at 450 mm centres.						
100 mm thick steel walling	-	-	-	-	m²	**102.00**
150 mm thick steel walling	-	-	-	-	m²	**106.62**
200 mm thick steel walling	-	-	-	-	m²	**118.31**
16 mm Pyroc sheething board and 40 mm Thermawall TW55 insulation fixed to slab perimeter not exceeding 300 mm	-	-	-	-	m	**11.55**
Cold formed galvanised steel; Kingspan "Multibeam" or other equal and approved						
Cold rolled purlins and cladding rails						
175 x 65 x 1.4 mm gauge purlins or rails; fixed to steelwork	-	0.05	0.95	13.28	m	**14.23**
175 x 65 x 1.6 mm gauge purlins or rails; fixed to steelwork	-	0.05	0.95	14.13	m	**15.09**
175 x 65 x 2.0 mm gauge purlins or rails; fixed to steelwork	-	0.05	0.95	17.08	m	**18.03**
205 x 65 x 1.4 mm gauge purlins or rails; fixed to steelwork	-	0.05	0.95	14.70	m	**15.65**
205 x 65 x 1.6 mm gauge purlins or rails; fixed to steelwork	-	0.05	0.95	16.07	m	**17.03**
205 x 65 x 2.0 mm gauge purlins or rails; fixed to steelwork	-	0.05	0.95	18.18	m	**19.14**
Heavy duty Zed section spacers						
vertically; across cladding rails; fixed to steelwork	-	0.06	1.14	10.06	m	**11.21**
Cleats						
weld-on for 175 mm purlin or rail	-	0.12	2.29	3.80	nr	**6.09**
bolt-on for 175 mm purlin or rail; including fixing bolts	-	0.03	0.57	7.30	nr	**7.88**
weld-on for 205 mm purlin or rail	-	0.12	2.29	4.34	nr	**6.63**
bolt-on for 205 mm purlin or rail; including fixing bolts	-	0.03	0.57	7.93	nr	**8.50**
Tubular ties						
150 mm long; bolted diagonally across purlins or cladding rails	-	0.03	0.57	7.91	nr	**8.48**

G STRUCTURAL/CARCASSING METAL/TIMBER

Item	PC £	Labour hours	Labour £	Material £	Unit	Total rate £
G12 ISOLATED STRUCTURAL METAL MEMBERS						
Isolated structural member; weldable steel; BS EN 10025: 1993 Grade S275; hot rolled structural steel sections						
Plain member; beams						
weight not exceeding 40 kg/m	-	-	-	-	tonne	**1418.60**
weight 40 - 100 kg/m	-	-	-	-	tonne	**1380.67**
weight exceeding 100 kg/m	-	-	-	-	tonne	**1344.80**
Metsec open web steel lattice beams; in single members; raised 3.50 m above ground; ends built in						
Beams; one coat zinc phosphate primer at works						
220 mm deep; to span 6.00 m (11.50 kg/m); ref B22	-	0.20	5.82	43.32	m	**49.14**
270 mm deep; to span 7.00 m (11.50 kg/m); ref B27	-	0.26	7.57	43.32	m	**50.89**
300 mm deep; to span 8.00 m (12.50 kg/m); ref B30	-	0.26	7.57	46.70	m	**54.28**
350 mm deep; to span 9.00 m (14.00 kg/m); ref B35	-	0.26	7.57	52.35	m	**59.92**
350 mm deep; to span 10.00 m (20.00 kg/m); ref D35	-	0.31	9.03	74.92	m	**83.95**
450 mm deep; to span 11.00 m (21.00 kg/m); ref D45	-	0.36	10.48	78.31	m	**88.79**
450 mm deep; to span 12.00 m (32.5 kg/m); ref G45	-	0.51	14.85	121.20	m	**136.05**
Beams; galvanised						
220 mm deep; to span 6.00 m (11.50 kg/m); ref B22	-	0.20	5.82	45.58	m	**51.40**
270 mm deep; to span 7.00 m (11.50 kg/m); ref B27	-	0.26	7.57	45.58	m	**53.15**
300 mm deep; to span 8.00 m (12.50 kg/m); ref B30	-	0.26	7.57	50.09	m	**57.66**
350 mm deep; to span 9.00 m (14.00 kg/m); ref B35	-	0.26	7.57	55.73	m	**63.31**
350 mm deep; to span 10.00 m (20.00 kg/m); ref D35	-	0.31	9.03	79.44	m	**88.47**
450 mm deep; to span 11.00 m (21.00 kg/m); ref D45	-	0.36	10.48	83.95	m	**94.44**
450 mm deep; to span 12.00 m (32.5 kg/m); ref G45	-	0.51	14.85	129.10	m	**143.96**
G20 CARPENTRY/TIMBER FRAMING/FIRST FIXING						
Sawn softwood; untreated						
Floor members						
38 mm x 100 mm	-	0.12	2.64	1.72	m	**4.35**
38 mm x 150 mm	-	0.14	3.08	2.37	m	**5.45**
47 mm x 75 mm	-	0.12	2.64	1.83	m	**4.46**
47 mm x 100 mm	-	0.14	3.08	2.26	m	**5.34**
47 mm x 125 mm	-	0.14	3.08	2.78	m	**5.85**
47 mm x 150 mm	-	0.16	3.51	3.29	m	**6.80**
47 mm x 175 mm	-	0.16	3.51	4.01	m	**7.52**
47 mm x 200 mm	-	0.17	3.73	4.49	m	**8.22**
47 mm x 225 mm	-	0.17	3.73	5.21	m	**8.94**
47 mm x 250 mm	-	0.18	3.95	5.87	m	**9.82**
75 mm x 125 mm	-	0.17	3.73	4.16	m	**7.89**
75 mm x 150 mm	-	0.17	3.73	4.93	m	**8.67**
75 mm x 175 mm	-	0.17	3.73	5.90	m	**9.64**
75 mm x 200 mm	-	0.18	3.95	6.89	m	**10.84**
75 mm x 225 mm	-	0.18	3.95	7.80	m	**11.76**
75 mm x 250 mm	-	0.19	4.17	9.51	m	**13.68**
100 mm x 150 mm	-	0.22	4.83	6.67	m	**11.50**
100 mm x 200 mm	-	0.23	5.05	9.98	m	**15.04**
100 mm x 250 mm	-	0.26	5.71	12.71	m	**18.43**
100 mm x 300 mm	-	0.28	6.15	12.44	m	**18.59**

G STRUCTURAL/CARCASSING METAL/TIMBER

Item	PC £	Labour hours	Labour £	Material £	Unit	Total rate £
G20 CARPENTRY/TIMBER FRAMING/FIRST FIXING – cont'd						
Sawn softwood; untreated – cont'd						
Wall or partition members						
25 mm x 25 mm	-	0.07	1.54	0.68	m	**2.22**
25 mm x 38 mm	-	0.07	1.54	0.77	m	**2.30**
25 mm x 75 mm	-	0.09	1.98	0.95	m	**2.93**
38 mm x 38 mm	-	0.09	1.98	0.90	m	**2.87**
38 mm x 50 mm	-	0.09	1.98	1.13	m	**3.10**
38 mm x 75 mm	-	0.12	2.64	1.39	m	**4.02**
38 mm x 100 mm	-	0.16	3.51	1.72	m	**5.23**
47 mm x 50 mm	-	0.12	2.64	1.21	m	**3.84**
47 mm x 75 mm	-	0.16	3.51	1.86	m	**5.38**
47 mm x 100 mm	-	0.19	4.17	2.30	m	**6.47**
47 mm x 125 mm	-	0.19	4.17	2.81	m	**6.99**
75 mm x 75 mm	-	0.19	4.17	2.53	m	**6.70**
75 mm x 100 mm	-	0.21	4.61	3.54	m	**8.15**
100 mm x 100 mm	-	0.21	4.61	4.61	m	**9.22**
Roof members; flat						
38 mm x 75 mm	-	0.14	3.08	1.39	m	**4.46**
38 mm x 100 mm	-	0.14	3.08	1.72	m	**4.79**
38 mm x 125 mm	-	0.14	3.08	2.04	m	**5.12**
38 mm x 150 mm	-	0.14	3.08	2.37	m	**5.45**
47 mm x 100 mm	-	0.14	3.08	2.26	m	**5.34**
47 mm x 125 mm	-	0.14	3.08	2.78	m	**5.85**
47 mm x 150 mm	-	0.16	3.51	3.29	m	**6.80**
47 mm x 175 mm	-	0.16	3.51	4.01	m	**7.52**
47 mm x 200 mm	-	0.17	3.73	4.49	m	**8.22**
47 mm x 225 mm	-	0.17	3.73	5.21	m	**8.94**
47 mm x 250 mm	-	0.18	3.95	5.87	m	**9.82**
75 mm x 150 mm	-	0.17	3.73	4.93	m	**8.67**
75 mm x 175 mm	-	0.17	3.73	5.90	m	**9.64**
75 mm x 200 mm	-	0.18	3.95	6.89	m	**10.84**
75 mm x 225 mm	-	0.18	3.95	7.80	m	**11.76**
75 mm x 250 mm	-	0.19	4.17	9.51	m	**13.68**
Joist strutting; herringbone strutting						
47 mm x 50 mm; depth of joist 150 mm	-	0.51	11.20	2.82	m	**14.03**
47 mm x 50 mm; depth of joist 175 mm	-	0.51	11.20	2.88	m	**14.08**
47 mm x 50 mm; depth of joist 200 mm	-	0.51	11.20	2.93	m	**14.13**
47 mm x 50 mm; depth of joist 225 mm	-	0.51	11.20	2.99	m	**14.19**
47 mm x 50 mm; depth of joist 250 mm	-	0.51	11.20	3.04	m	**14.24**
Joist strutting; block						
47 mm x 150 mm; depth of joist 150 mm	-	0.31	6.81	3.79	m	**10.60**
47 mm x 175 mm; depth of joist 175 mm	-	0.31	6.81	4.52	m	**11.32**
47 mm x 200 mm; depth of joist 200 mm	-	0.31	6.81	5.00	m	**11.81**
47 mm x 225 mm; depth of joist 225 mm	-	0.31	6.81	5.72	m	**12.53**
47 mm x 250 mm; depth of joist 250 mm	-	0.31	6.81	6.37	m	**13.18**
Cleat; 225 mm x 100 mm x 75 mm	-	0.20	4.39	0.88	nr	**5.27**
Extra for stress grading to above timbers						
general structural (GS) grade	-	-	-	32.76	m³	**-**
special structural (SS) grade	-	-	-	65.52	m³	**-**
Extra for protecting and flameproofing timber with "Celgard CF" protection						
small sections	-	-	-	154.44	m³	**-**
large sections	-	-	-	148.26	m³	**-**
Wrot surfaces						
plain; 50 mm wide	-	0.02	0.44	-	m	**0.44**
plain; 100 mm wide	-	0.03	0.66	-	m	**0.66**
plain; 150 mm wide	-	0.05	1.10	-	m	**1.10**

G STRUCTURAL/CARCASSING METAL/TIMBER

Item	PC £	Labour hours	Labour £	Material £	Unit	Total rate £
Sawn softwood; "Tanalised"						
Floor members						
38 mm x 75 mm	-	0.12	2.64	1.55	m	**4.19**
38 mm x 100 mm	-	0.12	2.64	1.93	m	**4.57**
38 mm x 150 mm	-	0.14	3.08	2.70	m	**5.77**
47 mm x 75 mm	-	0.12	2.64	2.04	m	**4.68**
47 mm x 100 mm	-	0.14	3.08	2.55	m	**5.63**
47 mm x 125 mm	-	0.14	3.08	3.14	m	**6.21**
47 mm x 150 mm	-	0.16	3.51	3.72	m	**7.24**
47 mm x 175 mm	-	0.16	3.51	4.52	m	**8.03**
47 mm x 200 mm	-	0.17	3.73	5.07	m	**8.80**
47 mm x 225 mm	-	0.17	3.73	5.86	m	**9.59**
47 mm x 250 mm	-	0.18	3.95	6.58	m	**10.53**
75 mm x 125 mm	-	0.17	3.73	4.70	m	**8.43**
75 mm x 150 mm	-	0.17	3.73	5.58	m	**9.31**
75 mm x 175 mm	-	0.17	3.73	6.66	m	**10.39**
75 mm x 200 mm	-	0.18	3.95	7.75	m	**11.70**
75 mm x 225 mm	-	0.18	3.95	8.78	m	**12.73**
75 mm x 250 mm	-	0.19	4.17	10.59	m	**14.76**
100 mm x 150 mm	-	0.22	4.83	7.53	m	**12.36**
100 mm x 200 mm	-	0.23	5.05	11.13	m	**16.18**
100 mm x 250 mm	-	0.26	5.71	14.15	m	**19.86**
100 mm x 300 mm	-	0.28	6.15	14.16	m	**20.31**
Wall or partition members						
25 mm x 25 mm	-	0.07	1.54	0.71	m	**2.25**
25 mm x 38 mm	-	0.07	1.54	0.82	m	**2.36**
25 mm x 75 mm	-	0.09	1.98	1.06	m	**3.04**
38 mm x 38 mm	-	0.09	1.98	0.98	m	**2.95**
38 mm x 50 mm	-	0.09	1.98	1.23	m	**3.21**
38 mm x 75 mm	-	0.12	2.64	1.55	m	**4.19**
38 mm x 100 mm	-	0.16	3.51	1.93	m	**5.45**
47 mm x 50 mm	-	0.12	2.64	1.35	m	**3.99**
47 mm x 75 mm	-	0.16	3.51	2.08	m	**5.60**
47 mm x 100 mm	-	0.19	4.17	2.59	m	**6.76**
47 mm x 125 mm	-	0.19	4.17	3.18	m	**7.35**
75 mm x 75 mm	-	0.19	4.17	2.85	m	**7.02**
75 mm x 100 mm	-	0.21	4.61	3.97	m	**8.59**
100 mm x 100 mm	-	0.21	4.61	5.18	m	**9.80**
Roof members; flat						
38 mm x 75 mm	-	0.14	3.08	1.55	m	**4.63**
38 mm x 100 mm	-	0.14	3.08	1.93	m	**5.01**
38 mm x 125 mm	-	0.14	3.08	2.32	m	**5.40**
38 mm x 150 mm	-	0.14	3.08	2.70	m	**5.77**
47 mm x 100 mm	-	0.14	3.08	2.55	m	**5.63**
47 mm x 125 mm	-	0.14	3.08	3.14	m	**6.21**
47 mm x 150 mm	-	0.16	3.51	3.72	m	**7.24**
47 mm x 175 mm	-	0.16	3.51	4.52	m	**8.03**
47 mm x 200 mm	-	0.17	3.73	5.07	m	**8.80**
47 mm x 225 mm	-	0.17	3.73	5.86	m	**9.59**
47 mm x 250 mm	-	0.18	3.95	6.58	m	**10.53**
75 mm x 150 mm	-	0.17	3.73	5.58	m	**9.31**
75 mm x 175 mm	-	0.17	3.73	6.66	m	**10.39**
75 mm x 200 mm	-	0.18	3.95	7.75	m	**11.70**
75 mm x 225 mm	-	0.18	3.95	8.78	m	**12.73**
75 mm x 250 mm	-	0.19	4.17	10.59	m	**14.76**

G STRUCTURAL/CARCASSING METAL/TIMBER

Item	PC £	Labour hours	Labour £	Material £	Unit	Total rate £
G20 CARPENTRY/TIMBER FRAMING/FIRST FIXING – cont'd						
Sawn softwood; "Tanalised" – cont'd						
Roof members; pitched						
25 mm x 100 mm	-	0.12	2.64	1.53	m	**4.17**
25 mm x 125 mm	-	0.12	2.64	2.07	m	**4.70**
25 mm x 150 mm	-	0.16	3.51	2.48	m	**6.00**
25 mm x 175 mm	-	0.18	3.95	2.91	m	**6.86**
25 mm x 200 mm	-	0.19	4.17	3.34	m	**7.51**
38 mm x 100 mm	-	0.16	3.51	1.93	m	**5.45**
38 mm x 125 mm	-	0.16	3.51	2.32	m	**5.84**
38 mm x 150 mm	-	0.16	3.51	2.70	m	**6.21**
38 mm x 175 mm	-	0.18	3.95	3.19	m	**7.15**
38 mm x 200 mm	-	0.19	4.17	3.68	m	**7.86**
47 mm x 50 mm	-	0.12	2.64	1.31	m	**3.95**
47 mm x 75 mm	-	0.16	3.51	2.04	m	**5.56**
47 mm x 100 mm	-	0.19	4.17	2.55	m	**6.72**
47 mm x 125 mm	-	0.19	4.17	3.14	m	**7.31**
47 mm x 150 mm	-	0.21	4.61	3.72	m	**8.33**
47 mm x 175 mm	-	0.21	4.61	4.52	m	**9.13**
47 mm x 200 mm	-	0.21	4.61	5.07	m	**9.68**
47 mm x 225 mm	-	0.21	4.61	5.86	m	**10.47**
75 mm x 100 mm	-	0.26	5.71	3.90	m	**9.61**
75 mm x 125 mm	-	0.26	5.71	4.70	m	**10.41**
75 mm x 150 mm	-	0.26	5.71	5.58	m	**11.29**
100 mm x 150 mm	-	0.31	6.81	7.57	m	**14.38**
100 mm x 175 mm	-	0.31	6.81	8.80	m	**15.61**
100 mm x 200 mm	-	0.31	6.81	11.13	m	**17.94**
100 mm x 225 mm	-	0.33	7.25	12.64	m	**19.89**
100 mm x 250 mm	-	0.33	7.25	14.15	m	**21.40**
Plates						
38 mm x 75 mm	-	0.12	2.64	1.55	m	**4.19**
38 mm x 100 mm	-	0.16	3.51	1.93	m	**5.45**
47 mm x 75 mm	-	0.16	3.51	2.08	m	**5.60**
47 mm x 100 mm	-	0.19	4.17	2.59	m	**6.76**
75 mm x 100 mm	-	0.21	4.61	3.97	m	**8.59**
75 mm x 125 mm	-	0.24	5.27	4.66	m	**9.93**
75 mm x 150 mm	-	0.27	5.93	5.54	m	**11.47**
Plates; fixing by bolting						
38 mm x 75 mm	-	0.21	4.61	1.55	m	**6.17**
38 mm x 100 mm	-	0.25	5.49	1.93	m	**7.43**
47 mm x 75 mm	-	0.25	5.49	2.08	m	**7.57**
47 mm x 100 mm	-	0.28	6.15	2.59	m	**8.74**
75 mm x 100 mm	-	0.31	6.81	3.97	m	**10.78**
75 mm x 125 mm	-	0.33	7.25	4.66	m	**11.91**
75 mm x 150 mm	-	0.36	7.91	5.54	m	**13.45**
Joist strutting; herringbone						
47 mm x 50 mm; depth of joist 150 mm	-	0.51	11.20	3.12	m	**14.32**
47 mm x 50 mm; depth of joist 175 mm	-	0.51	11.20	3.18	m	**14.38**
47 mm x 50 mm; depth of joist 200 mm	-	0.51	11.20	3.24	m	**14.44**
47 mm x 50 mm; depth of joist 225 mm	-	0.51	11.20	3.30	m	**14.50**
47 mm x 50 mm; depth of joist 250 mm	-	0.51	11.20	3.36	m	**14.56**
Joist strutting; block						
47 mm x 150 mm; depth of joist 150 mm	-	0.31	6.81	4.23	m	**11.04**
47 mm x 175 mm; depth of joist 175 mm	-	0.31	6.81	5.02	m	**11.83**
47 mm x 200 mm; depth of joist 200 mm	-	0.31	6.81	5.57	m	**12.38**
47 mm x 225 mm; depth of joist 225 mm	-	0.31	6.81	6.36	m	**13.17**
47 mm x 250 mm; depth of joist 250 mm	-	0.31	6.81	7.09	m	**13.90**

G STRUCTURAL/CARCASSING METAL/TIMBER

Item	PC £	Labour hours	Labour £	Material £	Unit	Total rate £
Cleats						
225 mm x 100 mm x 75 mm	-	0.20	4.39	0.97	nr	**5.37**
Extra for stress grading to above timbers						
general structural (GS) grade	-	-	-	32.76	m³	-
special structural (SS) grade	-	-	-	65.52	m³	-
Extra for protecting and flameproofing timber with "Celgard CF" protection or other equal or approved						
small sections	-	-	-	154.44	m³	-
large sections	-	-	-	148.26	m³	-
Wrot surfaces						
plain; 50 mm wide	-	0.02	0.44	-	m	**0.44**
plain; 100 mm wide	-	0.03	0.66	-	m	**0.66**
plain; 150 mm wide	-	0.05	1.10	-	m	**1.10**
Trussed rafters, stress graded sawn softwood pressure impregnated; raised through two storeys and fixed in position						
"W" type truss (Fink); 22.50 degree pitch; 450 mm eaves overhang						
5.00 m span	-	1.63	35.80	30.21	nr	**66.01**
7.60 m span	-	1.79	39.32	38.00	nr	**77.32**
10.00 m span	-	2.04	44.81	62.24	nr	**107.05**
"W" type truss (Fink); 30 degree pitch; 450 mm eaves overhang						
5.00 m span	-	1.63	35.80	30.49	nr	**66.30**
7.60 m span	-	1.79	39.32	40.14	nr	**79.46**
10.00 m span	-	2.04	44.81	65.62	nr	**110.43**
"W" type truss (Fink); 45 degree pitch; 450 mm eaves overhang						
4.60 m span	-	1.63	35.80	32.17	nr	**67.98**
7.00 m span	-	1.79	39.32	47.73	nr	**87.05**
"Mono" type truss; 17.50 degree pitch; 450 mm eaves overhang						
3.30 m span	-	1.43	31.41	23.93	nr	**55.35**
5.60 m span	-	1.63	35.80	35.81	nr	**71.61**
7.00 m span	-	1.88	41.30	44.74	nr	**86.03**
"Attic" type truss; 45 degree pitch; 450 mm eaves overhang						
5.00 m span	-	3.21	70.51	66.42	nr	**136.93**
7.60 m span	-	3.36	73.80	117.81	nr	**191.62**
9.00 m span	-	3.56	78.20	150.47	nr	**228.66**
"Moelven Toreboda" glulam timber beams or other equal and approved; Moelven Laminated Timber Structures; LB grade whitewood; pressure impregnated; phenbol resorcinal adhesive; clean planed finish; fixed						
Laminated roof beams						
56 mm x 225 mm	-	0.56	12.30	8.01	m	**20.31**
66 mm x 315 mm	-	0.71	15.60	13.23	m	**28.83**
90 mm x 315 mm	-	0.92	20.21	18.04	m	**38.25**
90 mm x 405 mm	-	1.17	25.70	23.20	m	**48.90**
115 mm x 405 mm	-	1.48	32.51	29.65	m	**62.15**
115 mm x 495 mm	-	1.83	40.20	36.22	m	**76.42**
115 mm x 630 mm	-	2.24	49.20	46.10	m	**95.31**

G STRUCTURAL/CARCASSING METAL/TIMBER

Item	PC £	Labour hours	Labour £	Material £	Unit	Total rate £
G20 CARPENTRY/TIMBER FRAMING/FIRST FIXING – cont'd						
"Masterboard" or other equal and approved; 6 mm thick						
Eaves, verge soffit boards, fascia boards and the like						
over 300 mm wide	7.74	0.71	15.60	10.18	m²	**25.77**
75 mm wide	-	0.21	4.61	0.78	m	**5.40**
150 mm wide	-	0.24	5.27	1.53	m	**6.80**
225 mm wide	-	0.29	6.37	2.27	m	**8.64**
300 mm wide	-	0.31	6.81	3.01	m	**9.82**
Plywood; external quality; 12 mm thick						
Eaves, verge soffit boards, fascia boards and the like						
over 300 mm wide	9.01	0.83	18.23	11.71	m²	**29.95**
75 mm wide	-	0.26	5.71	0.90	m	**6.61**
150 mm wide	-	0.30	6.59	1.76	m	**8.35**
225 mm wide	-	0.33	7.25	2.62	m	**9.87**
300 mm wide	-	0.38	8.35	3.48	m	**11.82**
Plywood; external quality; 15 mm thick						
Eaves, verge soffit boards, fascia boards and the like						
over 300 mm wide	11.31	0.83	18.23	14.51	m²	**32.74**
75 mm wide	-	0.26	5.71	1.11	m	**6.82**
150 mm wide	-	0.30	6.59	2.18	m	**8.77**
225 mm wide	-	0.33	7.25	3.24	m	**10.49**
300 mm wide	-	0.38	8.35	4.31	m	**12.66**
Plywood; external quality; 18 mm thick						
Eaves, verge soffit boards, fascia boards and the like						
over 300 mm wide	13.51	0.83	18.23	17.17	m²	**35.40**
75 mm wide	-	0.26	5.71	1.31	m	**7.02**
150 mm wide	-	0.30	6.59	2.58	m	**9.17**
225 mm wide	-	0.33	7.25	3.84	m	**11.09**
300 mm wide	-	0.38	8.35	5.11	m	**13.46**
Plywood; marine quality; 18 mm thick						
Gutter boards; butt joints						
over 300 mm wide	10.47	0.94	20.65	13.48	m²	**34.13**
150 mm wide	-	0.33	7.25	2.02	m	**9.27**
225 mm wide	-	0.38	8.35	3.05	m	**11.40**
300 mm wide	-	0.42	9.23	4.04	m	**13.27**
Eaves, verge soffit boards, fascia boards and the like						
over 300 mm wide	-	0.83	18.23	13.48	m²	**31.71**
75 mm wide	-	0.26	5.71	1.03	m	**6.74**
150 mm wide	-	0.30	6.59	2.02	m	**8.61**
225 mm wide	-	0.33	7.25	3.01	m	**10.26**
300 mm wide	-	0.38	8.35	4.00	m	**12.35**
Plywood; marine quality; 25 mm thick						
Gutter boards; butt joints						
over 300 mm wide	14.54	1.02	22.40	18.42	m²	**40.83**
150 mm wide	-	0.36	7.91	2.76	m	**10.67**
225 mm wide	-	0.41	9.01	4.17	m	**13.17**
300 mm wide	-	0.45	9.88	5.53	m	**15.41**

G STRUCTURAL/CARCASSING METAL/TIMBER

Item	PC £	Labour hours	Labour £	Material £	Unit	Total rate £
Eaves, verge soffit boards, fascia boards and the like						
over 300 mm wide	-	0.90	19.77	18.42	m²	**38.19**
75 mm wide	-	0.27	5.93	1.40	m	**7.33**
150 mm wide	-	0.31	6.81	2.76	m	**9.57**
225 mm wide	-	0.31	6.81	4.13	m	**10.94**
300 mm wide	-	0.41	9.01	5.49	m	**14.49**
Sawn softwood; untreated						
Gutter boards; butt joints						
19 mm thick; sloping	-	1.28	28.12	10.10	m²	**38.21**
19 mm thick; 75 mm wide	-	0.36	7.91	0.77	m	**8.68**
19 mm thick; 150 mm wide	-	0.41	9.01	1.47	m	**10.47**
19 mm thick; 225 mm wide	-	0.45	9.88	2.63	m	**12.51**
25 mm thick; sloping	-	1.28	28.12	15.92	m²	**44.04**
25 mm thick; 75 mm wide	-	0.36	7.91	0.99	m	**8.90**
25 mm thick; 150 mm wide	-	0.41	9.01	2.34	m	**11.35**
25 mm thick; 225 mm wide	-	0.45	9.88	3.69	m	**13.57**
Cesspools with 25 mm thick sides and bottom						
225 mm x 225 mm x 150 mm	-	1.22	26.80	3.09	nr	**29.89**
300 mm x 300 mm x 150 mm	-	1.42	31.19	4.05	nr	**35.24**
Individual supports; firrings						
50 mm wide x 36 mm average depth	-	0.16	3.51	2.08	m	**5.60**
50 mm wide x 50 mm average depth	-	0.16	3.51	3.17	m	**6.69**
50 mm wide x 75 mm average depth	-	0.16	3.51	4.10	m	**7.62**
Individual supports; bearers						
25 mm x 50 mm	-	0.10	2.20	0.94	m	**3.14**
38 mm x 50 mm	-	0.10	2.20	1.20	m	**3.40**
50 mm x 50 mm	-	0.10	2.20	1.25	m	**3.44**
50 mm x 75 mm	-	0.10	2.20	1.90	m	**4.10**
Individual supports; angle fillets						
38 mm x 38 mm	-	0.10	2.20	0.83	m	**3.03**
50 mm x 50 mm	-	0.10	2.20	1.06	m	**3.26**
75 mm x 75 mm	-	0.12	2.64	2.18	m	**4.81**
Individual supports; tilting fillets						
19 mm x 38 mm	-	0.10	2.20	0.52	m	**2.71**
25 mm x 50 mm	-	0.10	2.20	0.81	m	**3.01**
38 mm x 75 mm	-	0.10	2.20	1.25	m	**3.44**
50 mm x 75 mm	-	0.10	2.20	1.60	m	**3.79**
75 mm x 100 mm	-	0.16	3.51	2.90	m	**0.40**
Individual supports; grounds or battens						
13 mm x 19 mm	-	0.05	1.10	0.41	m	**1.50**
13 mm x 32 mm	-	0.05	1.10	0.41	m	**1.50**
25 mm x 50 mm	-	0.05	1.10	0.86	m	**1.96**
Individial supports; grounds or battens; plugged and screwed						
13 mm x 19 mm	-	0.16	3.51	0.47	m	**3.99**
13 mm x 32 mm	-	0.16	3.51	0.47	m	**3.99**
25 mm x 50 mm	-	0.16	3.51	0.93	m	**4.44**
Framed supports; open-spaced grounds or battens; at 300 mm centres one way						
25 mm x 50 mm	-	0.16	3.51	2.85	m²	**6.37**
25 mm x 50 mm; plugged and screwed	-	0.45	9.88	3.10	m²	**12.98**
Framed supports; at 300 mm centres one way and 600 mm centres the other way						
25 mm x 50 mm	-	0.77	16.91	4.28	m²	**21.20**
38 mm x 50 mm	-	0.77	16.91	5.59	m²	**22.51**
50 mm x 50 mm	-	0.77	16.91	5.81	m²	**22.72**
50 mm x 75 mm	-	0.77	16.91	9.09	m²	**26.00**
75 mm x 75 mm	-	0.77	16.91	12.42	m²	**29.33**

G STRUCTURAL/CARCASSING METAL/TIMBER

Item	PC £	Labour hours	Labour £	Material £	Unit	Total rate £
G20 CARPENTRY/TIMBER FRAMING/FIRST FIXING – cont'd						
Sawn softwood; untreated – cont'd						
Framed supports; at 300 mm centres one way and 600 mm centres the other way; plugged and screwed						
25 mm x 50 mm	-	1.28	28.12	4.96	m²	**33.08**
38 mm x 50 mm	-	1.28	28.12	6.27	m²	**34.39**
50 mm x 50 mm	-	1.28	28.12	6.49	m²	**34.61**
50 mm x 75 mm	-	1.28	28.12	9.77	m²	**37.88**
75 mm x 75 mm	-	1.28	28.12	13.10	m²	**41.22**
Framed supports; at 500 mm centres both ways						
25 mm x 50 mm; to bath panels	-	0.92	20.21	5.58	m²	**25.79**
Framed supports; as bracketing and cradling around steelwork						
25 mm x 50 mm	-	1.42	31.19	6.05	m²	**37.24**
50 mm x 50 mm	-	1.53	33.61	8.19	m²	**41.80**
50 mm x 75 mm	-	1.63	35.80	12.78	m²	**48.58**
Sawn softwood; "Tanalised"						
Gutter boards; butt joints						
19 mm thick; sloping	-	1.28	28.12	11.19	m²	**39.30**
19 mm thick; 75 mm wide	-	0.36	7.91	0.85	m	**8.76**
19 mm thick; 150 mm wide	-	0.41	9.01	1.63	m	**10.64**
19 mm thick; 225 mm wide	-	0.45	9.88	2.88	m	**12.76**
25 mm thick; sloping	-	1.28	28.12	17.36	m²	**45.47**
25 mm thick; 75 mm wide	-	0.36	7.91	1.10	m	**9.00**
25 mm thick; 150 mm wide	-	0.41	9.01	2.56	m	**11.56**
25 mm thick; 225 mm wide	-	0.45	9.88	4.01	m	**13.89**
Cesspools with 25 mm thick sides and bottom						
225 mm x 225 mm x 150 mm	-	1.22	26.80	3.38	nr	**30.17**
300 mm x 300 mm x 150 mm	-	1.42	31.19	4.44	nr	**35.63**
Individual supports; firrings						
50 mm wide x 36 mm average depth	-	0.16	3.51	2.19	m	**5.70**
50 mm wide x 50 mm average depth	-	0.16	3.51	3.32	m	**6.83**
50 mm wide x 75 mm average depth	-	0.16	3.51	4.32	m	**7.84**
Individual suports; bearers						
25 mm x 50 mm	-	0.10	2.20	1.02	m	**3.21**
38 mm x 50 mm	-	0.10	2.20	1.31	m	**3.51**
50 mm x 50 mm	-	0.10	2.20	1.39	m	**3.59**
50 mm x 75 mm	-	0.10	2.20	2.12	m	**4.32**
Individual supports; angle fillets						
38 mm x 38 mm	-	0.10	2.20	0.88	m	**3.07**
50 mm x 50 mm	-	0.10	2.20	1.14	m	**3.33**
75 mm x 75 mm	-	0.12	2.64	2.34	m	**4.97**
Individual supports; tilting fillets						
19 mm x 38 mm	-	0.10	2.20	0.54	m	**2.73**
25 mm x 50 mm	-	0.10	2.20	0.84	m	**3.04**
38 mm x 75 mm	-	0.10	2.20	1.33	m	**3.52**
50 mm x 75 mm	-	0.10	2.20	1.70	m	**3.90**
75 mm x 100 mm	-	0.16	3.51	3.18	m	**6.70**
Individual supports; grounds or battens						
13 mm x 19 mm	-	0.05	1.10	0.42	m	**1.51**
13 mm x 32 mm	-	0.05	1.10	0.43	m	**1.53**
25 mm x 50 mm	-	0.05	1.10	0.94	m	**2.04**

G STRUCTURAL/CARCASSING METAL/TIMBER

Item	PC £	Labour hours	Labour £	Material £	Unit	Total rate £
Individual supports; grounds or battens; plugged and screwed						
13 mm x 19 mm	-	0.16	3.51	0.48	m	**4.00**
13 mm x 32 mm	-	0.16	3.51	0.49	m	**4.01**
25 mm x 50 mm	-	0.16	3.51	1.00	m	**4.52**
Framed supports; open-spaced grounds or battens; at 300 mm centres one way						
25 mm x 50 mm	-	0.16	3.51	3.09	m²	**6.61**
25 mm x 50 mm; plugged and screwed	-	0.45	9.88	3.34	m²	**13.22**
Framed supports; at 300 mm centres one way and 600 mm centres the other way						
25 mm x 50 mm	-	0.77	16.91	4.64	m²	**21.56**
38 mm x 50 mm	-	0.77	16.91	6.14	m²	**23.05**
50 mm x 50 mm	-	0.77	16.91	6.53	m²	**23.44**
50 mm x 75 mm	-	0.77	16.91	10.16	m²	**27.08**
75 mm x 75 mm	-	0.77	16.91	14.03	m²	**30.94**
Framed supports; at 300 mm centres one way and 600 mm centres the other way; plugged and screwed						
25 mm x 50 mm	-	1.28	28.12	5.33	m²	**33.44**
38 mm x 50 mm	-	1.28	28.12	6.82	m²	**34.93**
50 mm x 50 mm	-	1.28	28.12	7.21	m²	**35.32**
50 mm x 75 mm	-	1.28	28.12	10.85	m²	**38.96**
75 mm x 75 mm	-	1.28	28.12	14.71	m²	**42.83**
Framed supports; at 500 mm centres both ways						
25 mm x 50 mm; to bath panels	-	0.92	20.21	6.04	m²	**26.25**
Framed supports; as bracketing and cradling around steelwork						
25 mm x 50 mm	-	1.42	31.19	6.56	m²	**37.75**
50 mm x 50 mm	-	1.53	33.61	9.19	m²	**42.80**
50 mm x 75 mm	-	1.63	35.80	14.29	m²	**50.09**
Wrought softwood						
Gutter boards; tongued and grooved joints						
19 mm thick; sloping	-	1.53	33.61	11.85	m²	**45.45**
19 mm thick; 75 mm wide	-	0.41	9.01	1.12	m	**10.13**
19 mm thick; 150 mm wide	-	0.45	9.88	1.73	m	**11.61**
19 mm thick; 225 mm wide	-	0.51	11.20	2.77	m	**13.97**
25 mm thick; sloping	-	1.53	33.61	15.22	m²	**48.83**
25 mm thick; 75 mm wide	-	0.41	9.01	1.25	m	**10.26**
25 mm thick; 150 mm wide	-	0.45	9.88	2.14	m	**12.02**
25 mm thick; 225 mm wide	-	0.51	11.20	3.27	m	**14.48**
Eaves, verge soffit boards, fascia boards and the like						
19 mm thick; over 300 mm wide	-	1.26	27.68	15.50	m²	**43.17**
19 mm thick; 150 mm wide; once grooved	-	0.20	4.39	2.61	m	**7.01**
25 mm thick; 150 mm wide; once grooved	-	0.20	4.39	2.29	m	**6.68**
25 mm thick; 175 mm wide; once grooved	-	0.22	4.83	3.16	m	**7.99**
32 mm thick; 225 mm wide; once grooved	-	0.26	5.71	4.06	m	**9.77**
Wrought softwood; "Tanalised"						
Gutter boards; tongued and grooved joints						
19 mm thick; sloping	-	1.53	33.61	12.93	m²	**46.54**
19 mm thick; 75 mm wide	-	0.41	9.01	1.20	m	**10.21**
19 mm thick; 150 mm wide	-	0.45	9.88	1.89	m	**11.78**
19 mm thick; 225 mm wide	-	0.51	11.20	3.02	m	**14.22**
25 mm thick; sloping	-	1.53	33.61	16.66	m²	**50.26**
25 mm thick; 75 mm wide	-	0.41	9.01	1.36	m	**10.37**
25 mm thick; 150 mm wide	-	0.45	9.88	2.36	m	**12.24**
25 mm thick; 225 mm wide	-	0.51	11.20	3.59	m	**14.80**

G STRUCTURAL/CARCASSING METAL/TIMBER

Item	PC £	Labour hours	Labour £	Material £	Unit	Total rate £
G20 CARPENTRY/TIMBER FRAMING/FIRST FIXING – cont'd						
Wrought softwood; "Tanalised" – cont'd						
Eaves, verge soffit boards, fascia boards and the like						
19 mm thick; over 300 mm wide	-	1.26	27.68	16.58	m²	**44.26**
19 mm thick; 150 mm wide; once grooved	-	0.20	4.39	2.78	m	**7.17**
25 mm thick; 150 mm wide; once grooved	-	0.20	4.39	2.50	m	**6.90**
25 mm thick; 175 mm wide; once grooved	-	0.22	4.83	3.41	m	**8.24**
32 mm thick; 225 mm wide; once grooved	-	0.26	5.71	4.47	m	**10.18**
Straps; mild steel; galvanised						
Straps; standard twisted vertical restraint; fixing to softwood and brick or blockwork						
27.5 mm x 2.50 mm x 400 mm girth	-	0.26	5.71	1.41	nr	**7.12**
27.5 mm x 2.50 mm x 600 mm girth	-	0.27	5.93	1.97	nr	**7.90**
27.5 mm x 2.50 mm x 800 mm girth	-	0.28	6.15	2.79	nr	**8.94**
27.5 mm x 2.50 mm x 1000 mm girth	-	0.31	6.81	3.62	nr	**10.43**
27.5 mm x 2.50 mm x 1200 mm girth	-	0.31	6.81	4.35	nr	**11.15**
Hangers; mild steel; galvanised						
Joist hangers 0.90 mm thick; The Expanded Metal Company Ltd "Speedy" or other equal and approved; for fixing to softwood; joist sizes						
50 mm wide; all sizes to 225 mm deep	1.54	0.12	2.64	2.01	nr	**4.65**
75 mm wide; all sizes to 225 mm deep	1.61	0.16	3.51	2.19	nr	**5.71**
100 mm wide; all sizes to 225 mm deep	1.73	0.19	4.17	2.43	nr	**6.60**
Joist hangers 2.50 mm thick; for building in; joist sizes						
50 mm x 100 mm	2.98	0.08	1.87	3.66	nr	**5.53**
50 mm x 125 mm	2.99	0.08	1.87	3.67	nr	**5.54**
50 mm x 150 mm	2.81	0.10	2.31	3.52	nr	**5.82**
50 mm x 175 mm	2.94	0.10	2.31	3.68	nr	**5.99**
50 mm x 200 mm	3.26	0.12	2.75	4.11	nr	**6.86**
50 mm x 225 mm	3.46	0.12	2.75	4.35	nr	**7.10**
75 mm x 150 mm	4.33	0.10	2.31	5.32	nr	**7.63**
75 mm x 175 mm	4.06	0.10	2.31	5.01	nr	**7.31**
75 mm x 200 mm	4.33	0.12	2.75	5.38	nr	**8.13**
75 mm x 225 mm	4.64	0.12	2.75	5.75	nr	**8.50**
75 mm x 250 mm	4.91	0.14	3.18	6.14	nr	**9.32**
100 mm x 200 mm	5.38	0.12	2.75	6.63	nr	**9.38**
Metal connectors; mild steel; galvanised						
Round toothed plate; for 10 mm or 12 mm diameter bolts						
38 mm diameter; single sided	-	0.02	0.44	0.49	nr	**0.93**
38 mm diameter; double sided	-	0.02	0.44	0.54	nr	**0.98**
50 mm diameter; single sided	-	0.02	0.44	0.52	nr	**0.96**
50 mm diameter; double sided	-	0.02	0.44	0.58	nr	**1.02**
63 mm diameter; single sided	-	0.02	0.44	0.76	nr	**1.20**
63 mm diameter; double sided	-	0.02	0.44	0.85	nr	**1.29**
75 mm diameter; single sided	-	0.02	0.44	1.13	nr	**1.56**
75 mm diameter; double sided	-	0.02	0.44	1.17	nr	**1.61**
framing anchor	-	0.16	3.51	0.92	nr	**4.44**

G STRUCTURAL/CARCASSING METAL/TIMBER

Item	PC £	Labour hours	Labour £	Material £	Unit	Total rate £
Bolts; mild steel; galvanised						
Fixing only bolts; 50 mm - 200 mm long						
6 mm diameter	-	0.03	0.66	-	nr	**0.66**
8 mm diameter	-	0.03	0.66	-	nr	**0.66**
10 mm diameter	-	0.05	1.10	-	nr	**1.10**
12 mm diameter	-	0.05	1.10	-	nr	**1.10**
16 mm diameter	-	0.06	1.32	-	nr	**1.32**
20 mm diameter	-	0.06	1.32	-	nr	**1.32**
Bolts						
Expanding bolts; "Rawlbolt" projecting type or other equal and approved; Rawl Fixings; plated; one nut; one washer						
6 mm diameter; ref M6 10P	-	0.10	2.20	0.69	nr	**2.89**
6 mm diameter; ref M6 25P	-	0.10	2.20	0.79	nr	**2.98**
6 mm diameter; ref M6 60P	-	0.10	2.20	1.00	nr	**3.19**
8 mm diameter; ref M8 25P	-	0.10	2.20	1.17	nr	**3.37**
8 mm diameter; ref M8 60P	-	0.10	2.20	1.20	nr	**3.40**
10 mm diameter; ref M10 15P	-	0.10	2.20	1.48	nr	**3.68**
10 mm diameter; ref M10 30P	-	0.10	2.20	1.55	nr	**3.74**
10 mm diameter; ref M10 60P	-	0.10	2.20	1.62	nr	**3.82**
12 mm diameter; ref M12 15P	-	0.10	2.20	2.41	nr	**4.60**
12 mm diameter; ref M12 30P	-	0.10	2.20	2.52	nr	**4.72**
12 mm diameter; ref M12 75P	-	0.10	2.20	3.12	nr	**5.32**
16 mm diameter; ref M16 35P	-	0.10	2.20	5.78	nr	**7.98**
16 mm diameter; ref M16 75P	-	0.10	2.20	6.40	nr	**8.60**
Expanding bolts; "Rawlbolt" loose bolt type or other equal and approved; Rawl Fixings; plated; one bolt; one washer						
6 mm diameter; ref M6 10L	-	0.10	2.20	0.74	nr	**2.94**
6 mm diameter; ref M6 25L	-	0.10	2.20	0.92	nr	**3.11**
6 mm diameter; ref M6 40L	-	0.10	2.20	0.93	nr	**3.12**
8 mm diameter; ref M8 25L	-	0.10	2.20	1.16	nr	**3.35**
8 mm diameter; ref M8 40L	-	0.10	2.20	1.32	nr	**3.52**
10 mm diameter; ref M10 10L	-	0.10	2.20	1.37	nr	**3.56**
10 mm diameter; ref M10 25L	-	0.10	2.20	1.50	nr	**3.70**
10 mm diameter; ref M10 50L	-	0.10	2.20	1.60	nr	**3.79**
10 mm diameter; ref M10 75L	-	0.10	2.20	1.85	nr	**4.05**
12 mm diameter; ref M12 10L	-	0.10	2.20	2.36	nr	**4.56**
12 mm diameter; ref M12 25L	-	0.10	2.20	2.50	nr	**4.70**
12 mm diameter; ref M12 40L	-	0.10	2.20	2.73	nr	**4.93**
12 mm diameter; ref M12 60L	-	0.10	2.20	2.87	nr	**5.07**
16 mm diameter; ref M16 30L	-	0.10	2.20	5.83	nr	**8.03**
16 mm diameter; ref M16 60L	-	0.10	2.20	6.26	nr	**8.45**
Truss clips; mild steel galvanised						
Truss clips; fixing to softwood; joist size						
38 mm wide	0.30	0.16	3.51	0.74	nr	**4.25**
50 mm wide	0.29	0.16	3.51	0.72	nr	**4.23**
Sole plate angles; mild steel; galvanised						
Sole plate angles; fixing to softwood and concrete						
112 mm x 40 mm x 76 mm	0.69	0.20	4.39	2.13	nr	**6.53**

G STRUCTURAL/CARCASSING METAL/TIMBER

Item	PC £	Labour hours	Labour £	Material £	Unit	Total rate £
G20 CARPENTRY/TIMBER FRAMING/FIRST FIXING – cont'd						
Chemical anchors						
R-CAS Spin-in epoxy acrylate capsules and standard studs or other equal and approved; Rawl Fixings; with nuts and washers; drilling masonry						
capsule ref 60-408; stud ref 60-448	-	0.28	6.15	1.87	nr	**8.02**
capsule ref 60-410; stud ref 60-454	-	0.31	6.81	2.02	nr	**8.83**
capsule ref 60-412; stud ref 60-460	-	0.34	7.47	2.41	nr	**9.87**
capsule ref 60-416; stud ref 60-472	-	0.38	8.35	3.48	nr	**11.83**
capsule ref 60-420; stud ref 60-478	-	0.40	8.79	6.33	nr	**15.12**
capsule ref 60-424; stud ref 60-484	-	0.43	9.45	7.44	nr	**16.89**
R-CAS Spin-in epoxy acrylate capsules and stainless steel studs or other equal and approved; Rawl Fixings; with nuts and washers; drilling masonry						
capsule ref 60-408; stud ref 60-905	-	0.28	6.15	3.18	nr	**9.33**
capsule ref 60-410; stud ref 60-910	-	0.31	6.81	4.15	nr	**10.96**
capsule ref 60-412; stud ref 60-915	-	0.34	7.47	5.60	nr	**13.07**
capsule ref 60-416; stud ref 60-920	-	0.38	8.35	9.22	nr	**17.57**
capsule ref 60-420; stud ref 60-925	-	0.40	8.79	15.48	nr	**24.26**
capsule ref 60-424; stud ref 60-930	-	0.43	9.45	24.93	nr	**34.37**
R-CAS Spin-in epoxy acrylate capsules and standard internal threaded sockets or other equal and approved; Rawl Fixings; drilling masonry						
capsule ref 60-410; socket ref 60-656	-	0.31	6.81	2.24	nr	**9.05**
capsule ref 60-412; socket ref 60-662	-	0.34	7.47	2.70	nr	**10.17**
capsule ref 60-416; socket ref 60-668	-	0.38	8.35	3.39	nr	**11.74**
capsule ref 60-420; socket ref 60-674	-	0.40	8.79	5.23	nr	**14.01**
capsule ref 60-424; socket ref 60-676	-	0.43	9.45	8.40	nr	**17.84**
R-CAS Spin-in epoxy acrylate capsules and stainless steel internal threaded sockets or other equal and approved; Rawl Fixings; drilling masonry						
capsule ref 60-410; socket ref 60-945	-	0.31	6.81	3.89	nr	**10.70**
capsule ref 60-412; socket ref 60-947	-	0.34	7.47	4.42	nr	**11.89**
capsule ref 60-416; socket ref 60-949	-	0.38	8.35	5.95	nr	**14.30**
capsule ref 60-420; socket ref 60-951	-	0.40	8.79	8.32	nr	**17.11**
capsule ref 60-424; socket ref 60-955	-	0.43	9.45	14.99	nr	**24.43**
R-CAS Spin-in epoxy acrylate capsules, perforated sleeves and standard studs or other equal and approved; Rawl Fixings; in low density material; with nuts and washers; drilling masonry						
capsule ref 60-408; sleeve ref 60-538 stud ref 60-448	-	0.28	6.15	4.37	nr	**10.52**
capsule ref 60-410; sleeve ref 60-544; stud ref 60-454	-	0.31	6.81	4.81	nr	**11.62**
capsule ref 60-412; sleeve ref 60-550; stud ref 60-460	-	0.34	7.47	5.48	nr	**12.95**
capsule ref 60-416; sleeve ref 60-562; stud ref 60-472	-	0.38	8.35	6.63	nr	**14.97**

G STRUCTURAL/CARCASSING METAL/TIMBER

Item	PC £	Labour hours	Labour £	Material £	Unit	Total rate £
R-CAS Spin-in epoxy acrylate capsules, perforated sleeves and stainless steel studs or other equal and approved; Rawl Fixings; in low density material; with nuts and washers; drilling masonry						
capsule ref 60-408; sleeve ref 60-538; stud ref 60-905	-	0.28	6.15	5.68	nr	**11.83**
capsule ref 60-410; sleeve ref 60-544; stud ref 60-910	-	0.31	6.81	6.95	nr	**13.76**
capsule ref 60-412; sleeve ref 60-550; stud ref 60-915	-	0.34	7.47	8.68	nr	**16.15**
capsule ref 60-416; sleeve ref 60-562; stud ref 60-920	-	0.38	8.35	12.37	nr	**20.71**
R-CAS Spin-in epoxy acrylate capsules, perforated sleeves and standard internal threaded sockets or other equal and approved; Rawl Fixings; in low density material; with nuts and washers; drilling masonry						
capsule ref 60-408; sleeve ref 60-538; socket ref 60-650	-	0.28	6.15	4.70	nr	**10.85**
capsule ref 60-410; sleeve ref 60-544; socket ref 60-656	-	0.31	6.81	5.04	nr	**11.85**
capsule ref 60-412; sleeve ref 60-550; socket ref 60-662	-	0.34	7.47	5.78	nr	**13.25**
capsule ref 60-416; sleeve ref 60-562; socket ref 60-668	-	0.38	8.35	6.54	nr	**14.88**
R-CAS Spin-in epoxy acrylate capsules, perforated sleeves and stainless steel internal threaded sockets or other equal and approved; Rawl Fixings; in low density material; drilling masonry						
capsule ref 60-408; sleeve ref 60-538; socket ref 60-943	-	0.28	6.15	6.35	nr	**12.50**
capsule ref 60-410; sleeve ref 60-544; socket ref 60-945	-	0.31	6.81	6.69	nr	**13.50**
capsule ref 60-412; sleeve ref 60-550; socket ref 60-947	-	0.34	7.47	7.49	nr	**14.96**
capsule ref 60-416; sleeve ref 60-562; socket ref 60-949	-	0.38	8.35	9.10	nr	**17.44**

H CLADDING/COVERING

Item	PC £	Labour hours	Labour £	Material £	Unit	Total rate £
H10 PATENT GLAZING						
Patent glazing; aluminium alloy bars 2.55 m long at 622 mm centres; fixed to supports						
Roof cladding						
single glazed with 6.4 mm laminated glass	-	-	-	-	m²	**155.00**
single glazed with 7 mm thick Georgian wired cast glass	-	-	-	-	m²	**175.00**
thermally broken and double glazed with low-e clear toughened and laminated double glazed units; aluminium finished RAL matt colour	-	-	-	-	m²	**440.00**
Extra for opening roof vents						
600 mm x 900 mm top hung opening roof vent; manually operated	-	-	-	-	nr	**480.00**
600 mm x 900 mm top hung opening roof vent; electrically operated	-	-	-	-	nr	**600.00**
Skylight						
Self-supporting hipped or gable ended lantern/skylight thermally broken and double glazed with low-e clear toughened and laminated double glazed units; aluminium finished RAL matt colour	-	-	-	-	m²	**900.00**
Associated code 4 lead flashings						
top flashing; 210 mm girth	-	-	-	-	m	**66.00**
bottom flashing; 240 mm girth	-	-	-	-	m	**76.00**
end flashing; 300 mm girth	-	-	-	-	m	**82.00**
Wall cladding						
single glazed with 6.4 mm laminated glass	-	-	-	-	m²	**160.00**
single glazed with 7 mm thick Georgian wired cast glass	-	-	-	-	m²	**180.00**
thermally broken and double glazed with low-e clear toughened and laminated double glazed units; aluminium finished RAL matt colour	-	-	-	-	m²	**460.00**
Extra for aluminium alloy perimeter members						
38 mm x 38 mm x 3 mm angle jamb	-	-	-	-	m	**23.00**
pressed cill member	-	-	-	-	m	**48.00**
pressed channel head and PVC case	-	-	-	-	m	**48.00**
H20 RIGID SHEET CLADDING						
"Resoplan" sheet or other equal and approved; Eternit UK Ltd; flexible neoprene gasket joints; fixing with stainless steel screws and coloured caps						
6 mm thick cladding to walls						
over 300 mm wide	-	2.23	48.98	70.79	m²	**119.77**
not exceeding 300 mm wide	-	0.74	16.25	23.17	m	**39.42**
Eternit 2000 "Glasal" sheet or other equal and approved; Eternit UK Ltd; flexible neoprene gasket joints; fixing with stainless steel screws and coloured caps						
7.50 mm thick cladding to walls						
over 300 mm wide	-	2.23	48.98	59.74	m²	**108.72**
not exceeding 300 mm wide	-	0.74	16.25	19.86	m	**36.11**
external angle trim	-	0.11	2.42	12.06	m	**14.48**
7.50 mm thick cladding to eaves; verges fascias or the like						
100 mm wide	-	0.53	11.64	8.42	m	**20.06**
200 mm wide	-	0.64	14.06	14.14	m	**28.20**
300 mm wide	-	0.74	16.25	19.86	m	**36.11**

H CLADDING/COVERING

Item	PC £	Labour hours	Labour £	Material £	Unit	Total rate £
Prodema ProdEX high density resin-bonded cellulose fibre weatherboarding panels; including secondary supports and fixing						
Walls						
8 mm Panels face fixed on to timber batten	-	-	-	-	m²	**164.00**
8 mm Panels face fixed on to aluminium rails	-	-	-	-	m²	**184.50**
8 mm Panels adhesive fixed on to timber battens or aluminium rails	-	-	-	-	m²	**194.75**
10 mm Panels secret fixed on to helping hand aluminium system	-	-	-	-	m²	**225.50**
H30 FIBRE CEMENT PROFILED SHEET CLADDING						
Asbestos-free corrugated sheets; Eternit "2000" or other equal and approved						
Roof cladding; sloping not exceeding 50 degrees; fixing to steel purlins with hook bolts						
"Profile 3"; natural grey	-	0.27	8.41	20.83	m²	**29.24**
"Profile 3"; coloured	-	0.27	8.41	23.76	m²	**32.18**
"Profile 6"; natural grey	-	0.32	9.97	16.52	m²	**26.49**
"Profile 6"; coloured	-	0.32	9.97	18.63	m²	**28.60**
"Profile 6"; natural grey; insulated 80 mm glass fibre infill; lining panel	-	0.53	16.51	35.48	m²	**51.99**
"Profile 6"; coloured; insulated 80 mm glass fibre infill; lining panel	-	0.53	16.51	40.43	m²	**56.94**
Accessories; to "Profile 3" cladding; natural grey						
eaves filler	-	0.11	3.43	12.65	m	**16.08**
external corner piece	-	0.13	4.05	14.40	m	**18.45**
apron flashing	-	0.13	4.05	12.65	m	**16.70**
plain wing or close fitting two piece adjustable capping to ridge	-	0.19	5.92	11.86	m	**17.78**
ventilating two piece adjustable capping to ridge	-	0.19	5.92	18.25	m	**24.17**
Accessories; to "Profile 6" cladding; natural grey						
eaves filler	-	0.11	3.43	7.63	m	**11.06**
external corner piece	-	0.13	4.05	8.68	m	**12.73**
apron flashing	-	0.13	4.05	8.48	m	**12.53**
underglazing flashing	-	0.13	4.05	11.17	m	**15.22**
plain cranked crown to ridge	-	0.19	5.92	16.77	m	**22.69**
plain wing or close fitting two piece adjustable capping to ridge	-	0.19	5.92	15.25	m	**21.17**
ventilating two piece adjustable capping to ridge	-	0.19	5.92	19.49	m	**25.41**
H31 METAL PROFILED/FLAT SHEET CLADDING/ COVERING/SIDING						
Lightweight galvanised steel roof tiles; Decra Roof Systems "Stratos"; or other equal and approved; coated finish						
Roof coverings	-	0.26	8.10	26.71	m²	**34.81**
Accessories for roof cladding						
pitched "D" ridge	-	0.10	3.12	12.17	m	**15.29**
barge cover (handed)	-	0.10	3.12	13.09	m	**16.21**
in line air vent	-	0.10	3.12	65.67	nr	**68.78**
in line soil vent	-	0.10	3.12	94.73	nr	**97.85**
gas flue terminal	-	0.20	6.23	122.96	nr	**129.19**

H CLADDING/COVERING

Item	PC £	Labour hours	Labour £	Material £	Unit	Total rate £
H31 METAL PROFILED/FLAT SHEET CLADDING/ COVERING/SIDING – cont'd						
Galvanised steel strip troughed sheets; Corus products or other equal and approved						
Roof cladding or decking; sloping not exceeding 50 degrees; fixing to steel purlins with plastic headed self-tapping screws						
0.7 mm thick; 46 profile	-	-	-	-	m²	**14.79**
0.7 mm thick; 60 profile	-	-	-	-	m²	**16.20**
0.7 mm thick; 100 profile	-	-	-	-	m²	**17.61**
Galvanised steel strip troughed sheets; PMF Strip Mills Products or other equal and approved						
Roof cladding; sloping not exceeding 50 degrees; fixing to steel purlins with plastic headed self-tapping screws						
0.7 mm thick type HPS200 13.5/3 corrugated	-	-	-	-	m²	**16.69**
0.7 mm thick type HPS200 R32/1000	-	-	-	-	m²	**15.90**
0.7 mm thick type Acline 40; plasticol finished	-	-	-	-	m²	**22.06**
Extra over last for aluminium roof decking	-	-	-	-	m²	**7.91**
Accessories for roof cladding						
HPS200 Drip flashing; 250 mm girth	-	-	-	-	m	**4.99**
HPS200 Ridge flashing; 375 mm girth	-	-	-	-	m	**6.53**
HPS200 Gable flashing; 500 mm girth	-	-	-	-	m	**8.30**
HPS200 Internal angle; 625 mm girth	-	-	-	-	m	**9.58**
Extra over steel roofing for						
roof penetration; 150 mm dia. opening; with top hat flashing and collar 150 mm high; and silicone joint to roofsheet	-	-	-	-	nr	**66.63**
roof penetration; 250 mm dia. opening; with top hat flashing and collar 150 mm high; and silicone joint to roofsheet	-	-	-	-	nr	**94.81**
Zalutite coated flat steel composite panel cladding; Kingspan or other equal and approved; outer panel 0.7 mm gauge HPS200 colourcoated; HCFC free LPCB FM/FW core and 0.4 mm stucco embossed lining panel with bright white polyester paint finish						
Roof cladding; horizontal fixing to steel purlins (measured elsewhere)						
KS1000RW 80 mm roof panel	-	-	-	-	m²	**44.37**
Wall cladding; vertical fixing to steel cladding rails (measured elsewhere)						
KS1000RW 60 mm wall panel	-	-	-	-	m²	**42.68**
KS1000RW 70 mm wall panel	-	-	-	-	m²	**43.66**
KS1000RW 80 mm wall panel	-	-	-	-	m²	**44.65**
KS1000MR 70 mm wall panel	-	-	-	-	m²	**61.13**
KS1000MR 80 mm wall panel	-	-	-	-	m²	**62.11**
KS900MR 70 mm wall panel	-	-	-	-	m²	**65.07**
KS900MR 80 mm wall panel	-	-	-	-	m²	**66.06**
KS600MR 70 mm wall panel	-	-	-	-	m²	**89.86**
KS600MR 80 mm wall panel	-	-	-	-	m²	**90.70**

H CLADDING/COVERING

Item	PC £	Labour hours	Labour £	Material £	Unit	Total rate £
Extra over for						
raking cutting on 60 mm KS1000RW roof panel; including waste	-	-	-	-	m	**29.94**
raking cutting on 70 mm KS1000RW roof panel; including waste	-	-	-	-	m	**30.33**
raking cutting on 80 mm KS1000RW roof panel; including waste	-	-	-	-	m	**31.65**
raking cutting on 70 mm KS1000MR wall panel; including waste	-	-	-	-	m	**32.09**
raking cutting on 80 mm KS1000MR wall panel; including waste	-	-	-	-	m	**42.20**
raking cutting on 70 mm KS900MR wall panel; including waste	-	-	-	-	m	**42.64**
raking cutting on 80 mm KS900MR wall panel; including waste	-	-	-	-	m	**21.10**
raking cutting on 70 mm KS600MR wall panel; including waste	-	-	-	-	m	**21.54**
raking cutting on 80 mm KS600MR wall panel; including waste	-	-	-	-	m	**21.98**
panel bearers at 1500 mm centres	-	-	-	-	m	**7.43**
vertical tophat joint in HPS200	-	-	-	-	m	**13.09**
vertical tophat joint with cap in HPS200	-	-	-	-	m	**16.86**
cranked panel KS1000MR	-	-	-	-	m	**121.36**
cranked panel KS900MR	-	-	-	-	m	**134.84**
cranked panel KS600MR	-	-	-	-	m	**202.27**
GRP Transluscent rooflights; factory assembled						
Rooflight; vertical fixing to steel purlins (measured elsewhere)						
double skin; class 3 over 1	-	-	-	-	m²	**64.07**
triple skin; class 3 over 1	-	-	-	-	m²	**67.18**
H32 PLASTICS PROFILED SHEET CLADDING/ COVERING/SIDING						
Extended, hard skinned, foamed PVC-UE profiled sections; Swish Celuka or other equal and approved; Class 1 fire rated to BS 476; Part 7; in white finish						
Wall cladding; vertical; fixing to timber						
100 mm shiplap profiles Code 001	-	0.39	8.57	54.84	m²	**63.41**
150 mm shiplap profiles Code 002	-	0.36	7.91	48.69	m²	**56.60**
125 mm feather-edged profiles Code C208	-	0.38	8.35	52.98	m²	**61.33**
Vertical angles	-	0.20	4.39	5.83	m	**10.22**
Raking cutting	-	0.16	3.51	-	m	**3.51**
Holes for pipes and the like	-	0.03	0.66	-	nr	**0.66**
H41 GLASS REINFORCED PLASTICS CLADDING/ FEATURES						
Glass fibre translucent sheeting grade AB class 3						
Roof cladding; sloping not exceeding 50 degrees; fixing to timber purlins with drive screws; to suit						
"Profile 3" or other equal and approved	22.89	0.21	6.54	29.96	m²	**36.50**
"Profile 6" or other equal and approved	21.46	0.27	8.41	28.27	m²	**36.68**
Roof cladding; sloping not exceeding 50 degrees; fixing to timber purlins with hook bolts; to suit						
"Profile 3" or other equal and approved	22.89	0.27	8.41	32.23	m²	**40.64**
Profile 6" or other equal and approved	21.46	0.32	9.97	30.54	m²	**40.51**
"Longrib 1000" or other equal and approved	21.09	0.32	9.97	30.10	m²	**40.07**

H CLADDING/COVERING

Item	PC £	Labour hours	Labour £	Material £	Unit	Total rate £
H60 PLAIN ROOF TILING						
NOTE: The following items of tile roofing unless otherwise described, include for conventional fixing assuming normal exposure with appropriate nails and/or rivets or clips to pressure impregnated softwood battens fixed with galvanised nails; prices also include for all bedding and pointing at verges, beneath ridge tiles etc.						
Clay interlocking pantiles; Sandtoft "20/20" red sand faced or other equal and approved; 370 mm x 223 mm; to 75 mm lap; on 25 mm x 38 mm battens and type 1F reinforced underlay						
Roof coverings (PC £ per 1000)	1093.50	0.48	14.95	23.80	m²	**38.76**
Extra over coverings for						
fixing every tile	-	0.03	0.93	2.51	m²	**3.45**
double course at eaves	-	0.32	9.97	16.10	m	**26.06**
verges	-	0.32	9.97	5.58	m	**15.55**
open valleys; cutting both sides	-	0.19	5.92	5.18	m	**11.10**
dry ridge tiles	-	0.64	19.94	16.25	m	**36.19**
dry hips; cutting both sides	-	0.80	24.92	23.19	m	**48.11**
holes for pipes and the like	-	0.21	6.54	-	nr	**6.54**
Clay pantiles; Sandtoft Goxhill "Old English"; red sand faced or other equal and approved; 342 mm x 241 mm; to 75 mm lap; on 25 mm x 38 mm battens and type 1F reinforced underlay						
Roof coverings (PC £ per 1000)	1107.00	0.48	14.95	27.51	m²	**42.46**
Extra over coverings for						
fixing every tile	-	0.02	0.62	6.29	m²	**6.91**
other colours	-	-	-	1.36	m²	**-**
double course at eaves	-	0.35	10.90	7.34	m	**18.25**
verges; extra single undercloak course of plain tiles	-	0.32	9.97	14.63	m	**24.60**
open valleys; cutting both sides	-	0.19	5.92	5.25	m	**11.17**
ridge tiles; tile slips	-	0.64	19.94	36.72	m	**56.66**
hips; cutting both sides	-	0.80	24.92	41.97	m	**66.89**
holes for pipes and the like	-	0.21	6.54	-	nr	**6.54**
Clay pantiles; William Blyth's "Lincoln" natural or other equal and approved; 343 mm x 280 mm; to 75 mm lap; on 19 mm x 38 mm battens and type 1F reinforced underlay						
Roof coverings (PC £ per 1000)	1200.00	0.48	14.95	28.56	m²	**43.51**
Extra over coverings for						
fixing every tile	-	0.02	0.62	6.29	m²	**6.91**
other colours	-	-	-	1.84	m²	**-**
double course at eaves	-	0.35	10.90	7.78	m	**18.69**
verges; extra single undercloak course of plain tiles	-	0.32	9.97	15.06	m	**25.03**
open valleys; cutting both sides	-	0.19	5.92	5.69	m	**11.61**
ridge tiles; tile slips	-	0.64	19.94	30.36	m	**50.30**
hips; cutting both sides	-	0.80	24.92	36.05	m	**60.98**
holes for pipes and the like	-	0.21	6.54	-	nr	**6.54**

H CLADDING/COVERING

Item	PC £	Labour hours	Labour £	Material £	Unit	Total rate £
Clay plain tiles; Hinton, Perry and Davenhill "Dreadnought" smooth red machine-made or other equal and approved; 265 mm x 165 mm; on 19 mm x 38 mm battens and type 1F reinforced underlay						
Roof coverings; to 64 mm lap (PC £ per 1000)	414.00	1.12	34.89	41.15	m²	**76.04**
Wall coverings; to 38 mm lap	-	1.33	41.43	34.94	m²	**76.38**
Extra over coverings for						
other colours	-	-	-	3.36	m²	**-**
ornamental tiles	-	-	-	22.85	m²	**-**
double course at eaves	-	0.27	8.41	4.81	m	**13.23**
verges	-	0.32	9.97	1.61	m	**11.58**
swept valleys; cutting both sides	-	0.69	21.50	57.34	m	**78.84**
bonnet hips; cutting both sides	-	0.85	26.48	57.34	m	**83.83**
external vertical angle tiles; supplementary nail fixings	-	0.43	13.40	64.58	m	**77.98**
half round ridge tiles	-	0.64	19.94	19.20	m	**39.14**
holes for pipes and the like	-	0.21	6.54	-	nr	**6.54**
Concrete interlocking tiles; Marley Eternit "Anglia" granule finish tiles or other equal and approved; 387 mm x 230 mm; to 75 mm lap; on 25 mm x 38 mm battens and type 1F reinforced underlay						
Roof coverings (PC £ per 1000)	602.00	0.48	14.95	16.19	m²	**31.14**
Extra over coverings for						
fixing every tile	-	0.02	0.62	1.05	m²	**1.67**
eaves; eaves filler	-	0.05	1.56	13.82	m	**15.37**
verges; 150 mm wide asbestos free strip undercloak	-	0.24	7.48	2.20	m	**9.68**
valley trough tiles; cutting both sides	-	0.58	18.07	30.49	m	**48.56**
segmental ridge tiles; tile slips	-	0.58	18.07	17.46	m	**35.53**
segmental hip tiles; tile slips; cutting both sides	-	0.74	23.05	19.60	m	**42.66**
dry ridge tiles; segmental including batten sections; unions and filler pieces	-	0.32	9.97	24.77	m	**34.73**
segmental mono-ridge tiles	-	0.58	18.07	24.39	m	**42.46**
gas ridge terminal	-	0.53	16.51	84.34	nr	**100.85**
holes for pipes and the like	-	0.21	6.54	-	nr	**6.54**
Concrete interlocking tiles; Marley Eternit "Ludlow Major" granule finish tiles or other equal and approved; 420 mm x 330 mm; to 75 mm lap; on 25 mm x 38 mm battens and type 1F reinforced underlay						
Roof coverings (PC £ per 1000)	872.00	0.37	11.53	14.35	m²	**25.87**
Extra over coverings for						
fixing every tile	-	0.02	0.62	1.05	m²	**1.67**
eaves; eaves filler	-	0.05	1.56	0.78	m	**2.34**
verges; 150 mm wide asbestos free strip undercloak	-	0.24	7.48	2.20	m	**9.68**
dry verge system; extruded white pvc	-	0.16	4.98	14.04	m	**19.02**
segmental ridge cap to dry verge	-	0.02	0.62	4.48	m	**5.10**
valley trough tiles; cutting both sides	-	0.58	18.07	31.13	m	**49.20**
segmental ridge tiles	-	0.53	16.51	11.77	m	**28.29**
segmental hip tiles; cutting both sides	-	0.69	21.50	14.87	m	**36.37**
dry ridge tiles; segmental including batten sections; unions and filler pieces	-	0.32	9.97	24.77	m	**34.73**
segmental mono-ridge tiles	-	0.53	16.51	21.07	m	**37.58**
gas ridge terminal	-	0.53	16.51	84.34	nr	**100.85**
holes for pipes and the like	-	0.21	6.54	-	nr	**6.54**

H CLADDING/COVERING

Item	PC £	Labour hours	Labour £	Material £	Unit	Total rate £
H60 PLAIN ROOF TILING – cont'd						
Concrete interlocking tiles; Marley Eternit "Ecologic Ludlow Major" granule finish tiles; 420 mm x 330 mm; to 75 mm lap; on 25 mm x 38 mm battens and type 1F reinforced underlay						
Roof coverings (PC £ per 1000)	936.00	0.37	11.53	15.09	m²	**26.62**
Extra over coverings for						
fixing every tile	-	0.02	0.62	1.05	m²	**1.67**
eaves; eaves filler	-	0.05	1.56	0.78	m	**2.34**
verges; 150 mm wide asbestos free strip undercloak	-	0.24	7.48	2.20	m	**9.68**
dry verge system; extruded white pvc	-	0.16	4.98	14.04	m	**19.02**
segmental ridge cap to dry verge	-	0.02	0.62	4.48	m	**5.10**
valley trough tiles; cutting both sides	-	0.58	18.07	31.28	m	**49.35**
segmental ridge tiles	-	0.53	16.51	11.77	m	**28.29**
segmental hip tiles; cutting both sides	-	0.69	21.50	15.10	m	**36.60**
dry ridge tiles; segmental including batten sections; unions and filler pieces	-	0.32	9.97	24.77	m	**34.73**
segmental mono-ridge tiles	-	0.53	16.51	21.07	m	**37.58**
gas ridge terminal	-	0.53	16.51	84.34	nr	**100.85**
holes for pipes and the like	-	0.21	6.54		nr	**6.54**
Concrete interlocking tiles; Marley Eternit "Mendip" granule finish double pantiles; 420 mm x 330 mm; to 75 mm lap; on 22 mm x 38 mm battens and type 1F reinforced underlay						
Roof coverings (PC £ per 1000)	888.00	0.37	11.53	14.43	m²	**25.95**
Extra over coverings for						
fixing every tile	-	0.02	0.62	1.05	m²	**1.67**
eaves; eaves filler	-	0.02	0.62	13.20	m	**13.83**
verges; 150 mm wide asbestos free strip undercloak	-	0.24	7.48	2.20	m	**9.68**
dry verge system; extruded white pvc	-	0.16	4.98	14.04	m	**19.02**
segmental ridge cap to dry verge	-	0.02	0.62	4.48	m	**5.10**
valley trough tiles; cutting both sides	-	0.58	18.07	31.17	m	**49.24**
segmental ridge tiles	-	0.58	18.07	17.46	m	**35.53**
segmental hip tiles; cutting both sides	-	0.74	23.05	20.62	m	**43.67**
dry ridge tiles; segmental including batten sections; unions and filler pieces	-	0.32	9.97	24.77	m	**34.73**
segmental mono-ridge tiles	-	0.53	16.51	23.92	m	**40.43**
gas ridge terminal	-	0.53	16.51	84.34	nr	**100.85**
holes for pipes and the like	-	0.21	6.54	-	nr	**6.54**
Concrete interlocking tiles; Marley Eternit "Modern" smooth finish tiles; 420 mm x 220 mm; to 75 mm lap; on 25 mm x 38 mm battens and type 1F reinforced underlay						
Roof coverings (PC £ per 1000)	908.00	0.37	11.53	15.10	m²	**26.63**
Extra over coverings for						
fixing every tile	-	0.02	0.62	1.05	m²	**1.67**
verges; 150 mm wide asbestos free strip undercloak	-	0.24	7.48	2.20	m	**9.68**
dry verge system; extruded white pvc	-	0.21	6.54	14.04	m	**20.58**
"Modern" ridge cap to dry verge	-	0.02	0.62	4.48	m	**5.10**
valley trough tiles; cutting both sides	-	0.58	18.07	31.21	m	**49.28**
"Modern" ridge tiles	-	0.53	16.51	12.30	m	**28.81**
"Modern" hip tiles; cutting both sides	-	0.69	21.50	15.53	m	**37.02**
dry ridge tiles; "Modern"; including batten sections; unions and filler pieces	-	0.32	9.97	25.29	m	**35.26**
"Modern" mono-ridge tiles	-	0.53	16.51	21.07	m	**37.58**
gas ridge terminal	-	0.53	16.51	84.34	nr	**100.85**
holes for pipes and the like	-	0.21	6.54	-	nr	**6.54**

H CLADDING/COVERING

Item	PC £	Labour hours	Labour £	Material £	Unit	Total rate £
Concrete interlocking tiles; Marley Eternit "Wessex" smooth finish tiles or other equal and approved; 413 mm x 330 mm; to 75 mm lap; on 25 mm x 38 mm battens and type 1F reinforced underlay						
Roof coverings (PC £ per 1000)	1375.00	0.37	11.53	20.58	m²	**32.11**
Extra over coverings for						
fixing every tile	-	0.02	0.62	1.05	m²	**1.67**
verges; 150 mm wide asbestos free strip undercloak	-	0.24	7.48	2.20	m	**9.68**
dry verge system; extruded white pvc	-	0.21	6.54	14.04	m	**20.58**
"Modern" ridge cap to dry verge	-	0.02	0.62	4.48	m	**5.10**
valley trough tiles; cutting both sides	-	0.58	18.07	32.32	m	**50.39**
"Modern" ridge tiles	-	0.53	16.51	12.30	m	**28.81**
"Modern" hip tiles; cutting both sides	-	0.69	21.50	17.19	m	**38.69**
dry ridge tiles; "Modern"; including batten sections; unions and filler pieces	-	0.32	9.97	25.29	m	**35.26**
"Modern" mono-ridge tiles	-	0.53	16.51	21.07	m	**37.58**
gas ridge terminal	-	0.53	16.51	84.34	nr	**100.85**
holes for pipes and the like	-	0.21	6.54	-	nr	**6.54**
Concrete interlocking slates; Redland "Richmond" smooth finish tiles or other equal and approved; 430 mm x 380 mm; to 75 mm lap; on 25 mm x 38 mm battens and type 1F reinforced underlay						
Roof coverings (PC £ per 1000)	1167.00	0.37	11.53	15.58	m²	**27.11**
Extra over coverings for						
fixing every tile	-	0.02	0.62	1.05	m²	**1.67**
eaves; eaves filler	-	0.02	0.62	6.87	m	**7.49**
verges; extra single undercloak course of plain tiles	-	0.27	8.41	4.48	m	**12.89**
ambi-dry verge system	-	0.21	6.54	13.42	m	**19.96**
ambi-dry verge eave/ridge end piece	-	0.02	0.62	4.80	m	**5.42**
universal valley trough tiles; cutting both sides	-	0.64	19.94	43.45	m	**63.39**
universal hip tiles; cutting both sides	-	0.69	21.50	15.82	m	**37.31**
dry ridge system; universal angle ridge tiles	-	0.27	8.41	31.21	m	**39.62**
universal monopitch angle ridge tiles	-	0.58	18.07	22.46	m	**40.53**
gas ridge terminal	-	0.53	16.51	64.30	nr	**80.81**
ridge vent with 110 mm diameter flexible adaptor	-	0.53	16.51	103.94	nr	**120.45**
holes for pipes and the like	-	0.21	6.54	-	nr	**6.54**
Concrete interlocking slates; Redland "Stonewold II" smooth finish tiles or other equal and approved; 430 mm x 380 mm; to 75 mm lap; on 25 mm x 38 mm battens and type 1F reinforced underlay						
Roof coverings (PC £ per 1000)	1786.00	0.37	11.53	25.14	m²	**36.67**
Extra over coverings for						
fixing every tile	-	0.02	0.62	1.26	m²	**1.88**
verges; extra single undercloak course of plain tiles	-	0.31	9.66	4.48	m	**14.14**
ambi-dry verge system	-	0.21	6.54	13.42	m	**19.96**
ambi-dry verge eave/ridge end piece	-	0.02	0.62	4.80	m	**5.42**
valley trough tiles; cutting both sides	-	0.58	18.07	44.12	m	**62.19**
universal angle ridge tiles	-	0.53	16.51	11.67	m	**28.18**
universal hip tiles; cutting both sides	-	0.69	21.50	18.02	m	**39.51**
dry ridge system; universal angle ridge tiles	-	0.27	8.41	31.21	m	**39.62**
universal monopitch angle ridge tiles	-	0.58	18.07	22.46	m	**40.53**
universal gas flue angle ridge tile	-	0.53	16.51	46.53	nr	**63.04**
universal angle ridge vent tile with 110 mm diameter adaptor	-	0.53	16.51	103.83	nr	**120.34**
holes for pipes and the like	-	0.21	6.54	-	nr	**6.54**

H CLADDING/COVERING

Item	PC £	Labour hours	Labour £	Material £	Unit	Total rate £
H60 PLAIN ROOF TILING – cont'd						
Concrete interlocking tiles; Redland "Norfolk" smooth finish pantiles or other equal and approved; 381 mm x 229 mm; to 75 mm lap; on 25 mm x 38 mm battens and type 1F reinforced underlay						
Roof coverings (PC £ per 1000)	663.00	0.48	14.95	18.97	m²	**33.92**
Extra over coverings for						
fixing every tile	-	0.05	1.56	0.46	m²	**2.02**
eaves; eaves filler	-	0.05	1.56	1.49	m	**3.05**
verges; extra single undercloak course of plain tiles	-	0.32	9.97	8.03	m	**18.00**
valley trough tiles; cutting both sides	-	0.64	19.94	41.70	m	**61.64**
universal ridge tiles	-	0.53	16.51	15.71	m	**32.22**
universal hip tiles; cutting both sides	-	0.69	21.50	19.64	m	**41.14**
universal gas flue ridge tile	-	0.53	16.51	88.71	nr	**105.22**
universal ridge vent tile with 110 mm diameter adaptor	-	0.53	16.51	103.88	nr	**120.39**
holes for pipes and the like	-	0.21	6.54	-	nr	**6.54**
Concrete interlocking tiles; Redland "Regent" granule finish bold roll tiles or other equal and approved; 418 mm x 332 mm; to 75 mm lap; on 25 mm x 38 mm battens and type 1F reinforced underlay						
Roof coverings (PC £ per 1000)	940.00	0.37	11.53	15.33	m²	**26.86**
Extra over coverings for						
fixing every tile	-	0.04	1.25	0.98	m²	**2.22**
eaves; eaves filler	-	0.05	1.56	1.13	m	**2.69**
verges; extra single undercloak course of plain tiles	-	0.27	8.41	3.83	m	**12.25**
cloaked verge system	-	0.16	4.98	9.53	m	**14.52**
valley trough tiles; cutting both sides	-	0.58	18.07	41.12	m	**59.18**
universal ridge tiles	-	0.53	16.51	15.71	m	**32.22**
universal hip tiles; cutting both sides	-	0.69	21.50	19.05	m	**40.55**
dry ridge system; universal ridge tiles	-	0.27	8.41	53.85	m	**62.26**
universal half round mono-pitch ridge tiles	-	0.58	18.07	33.13	m	**51.20**
universal gas flue ridge tile	-	0.53	16.51	88.71	nr	**105.22**
universal ridge vent tile with 110 mm diameter adaptor	-	0.53	16.51	103.88	nr	**120.39**
holes for pipes and the like	-	0.21	6.54	-	nr	**6.54**
Concrete interlocking tiles; Redland "Renown" granule finish tiles or other equal and approved; 418 mm x 330 mm; to 75 mm lap; on 25 mm x 38 mm battens and type 1F reinforced underlay						
Roof coverings (PC £ per 1000)	908.00	0.37	11.53	14.96	m²	**26.49**
Extra over coverings for						
fixing every tile	-	0.02	0.62	0.55	m²	**1.17**
verges; extra single undercloak course of plain tiles	-	0.27	8.41	4.62	m	**13.03**
cloaked verge system	-	0.16	4.98	9.64	m	**14.63**
valley trough tiles; cutting both sides	-	0.58	18.07	41.00	m	**59.07**
universal ridge tiles	-	0.53	16.51	15.71	m	**32.22**
universal hip tiles; cutting both sides	-	0.69	21.50	18.94	m	**40.44**
dry ridge system; universal ridge tiles	-	0.27	8.41	52.53	m	**60.95**
universal half round mono-pitch ridge tiles	-	0.58	18.07	33.13	m	**51.20**
universal gas flue ridge tile	-	0.53	16.51	88.71	nr	**105.22**
universal ridge vent tile with 110 mm diameter adaptor	-	0.53	16.51	103.88	nr	**120.39**
holes for pipes and the like	-	0.21	6.54	-	nr	**6.54**

H CLADDING/COVERING

Item	PC £	Labour hours	Labour £	Material £	Unit	Total rate £
Concrete plain tiles; EN 490 group A; 267 mm x 165 mm; on 25 mm x 38 mm battens and type 1F reinforced underlay						
Roof coverings; to 64 mm lap (PC £ per 1000)	383.00	1.12	34.89	38.95	m²	**73.84**
Wall coverings; to 38 mm lap	-	1.33	41.43	32.99	m²	**74.43**
Extra over coverings for						
ornamental tiles	-	-	-	23.32	m²	**-**
double course at eaves	-	0.27	8.41	4.59	m	**13.01**
verges	-	0.35	10.90	1.76	m	**12.67**
swept valleys; cutting both sides	-	0.69	21.50	39.94	m	**61.43**
bonnet hips; cutting both sides	-	0.85	26.48	39.94	m	**66.42**
external vertical angle tiles; supplementary nail fixings	-	0.43	13.40	28.50	m	**41.90**
half round ridge tiles	-	0.53	16.51	9.72	m	**26.23**
third round hip tiles; cutting both sides	-	0.53	16.51	12.44	m	**28.95**
holes for pipes and the like	-	0.21	6.54	-	nr	**6.54**
Sundries						
Hip irons						
galvanised mild steel; fixing with screws	-	0.11	3.43	2.15	nr	**5.58**
"Rytons Clip strip" or other equal and approved; continuous soffit ventilator						
51 mm wide; plastic; code CS351	-	0.32	9.97	0.97	m	**10.94**
"Rytons over fascia ventilator" or other equal and approved; continuous eaves ventilator						
40 mm wide; plastic; code OFV890	-	0.11	3.43	2.05	m	**5.47**
"Rytons roof ventilator" or other equal and approved; to suit rafters at 600 centres						
250 mm deep x 43 mm high; plastic; code TV600	-	0.11	3.43	1.65	m	**5.08**
"Rytons push and lock ventilators" or other equal and approved; circular						
83 mm diameter; plastic; code PL235	-	0.05	1.10	0.34	nr	**1.44**
Fixing only						
lead soakers (supply cost given elsewhere)	-	0.08	1.74	-	nr	**1.74**
Pressure impregnated softwood counter battens; 25 x 50						
450 mm centres	-	0.07	2.18	2.21	m²	**4.39**
600 mm centres	-	0.05	1.56	1.67	m²	**3.23**
Underlay; BS EN 13707 type 1B; bitumen felt weighing 14kg/10m²; 75 mm laps						
To sloping or vertical surfaces	0.54	0.03	0.93	1.55	m²	**2.49**
Underlay; BS EN 13707 type 1F; reinforced bitumen felt weighing 22.5kg/10 m²; 75 mm laps						
To sloping or vertical surfaces	1.00	0.03	0.93	2.10	m²	**3.04**
Underlay; Visqueen "Tilene 200P" or other equal and approved; micro-perforated sheet; 75 mm laps						
To sloping or vertical surfaces	0.60	0.03	0.93	1.66	m²	**2.60**
Underlay; "Powerlon 250 BM" or other equal and approved; reinforced breather membrane; 75 mm laps						
To sloping or vertical surfaces	0.64	0.03	0.93	1.71	m²	**2.65**
Underlay; "Anticon" sarking membrane or other equal and approved; Euroroof Ltd; polyethylene; 75 mm laps						
To sloping or vertical surfaces	1.08	0.03	0.93	2.20	m²	**3.13**

H CLADDING/COVERING

Item	PC £	Labour hours	Labour £	Material £	Unit	Total rate £
H61 FIBRE CEMENT SLATING						
Asbestos-free artificial slates; Eternit "Garsdale/ E2000T"; to 75 mm lap; on 19 mm x 50 mm battens and type 1F reinforced underlay						
Coverings; 500 mm x 250 mm slates						
roof coverings	-	0.69	21.50	28.43	m²	**49.92**
wall coverings	-	0.85	26.48	28.43	m²	**54.91**
Coverings; 600 mm x 300 mm slates						
roof coverings	-	0.53	16.51	23.16	m²	**39.67**
wall coverings	-	0.69	21.50	23.16	m²	**44.66**
Extra over slate coverings for						
double course at eaves	-	0.27	8.41	5.78	m	**14.19**
verges; extra single undercloak course	-	0.35	10.90	1.18	m	**12.08**
open valleys; cutting both sides	-	0.21	6.54	4.72	m	**11.27**
valley gutters; cutting both sides	-	0.58	18.07	32.49	m	**50.56**
stop end	-	0.11	3.43	11.55	nr	**14.97**
roll top ridge tiles	-	0.64	19.94	38.66	m	**58.60**
stop end	-	0.11	3.43	21.43	nr	**24.85**
mono-pitch ridge tiles	-	0.53	16.51	45.19	m	**61.70**
stop end	-	0.11	3.43	49.14	nr	**52.56**
duo-pitch ridge tiles	-	0.53	16.51	36.60	m	**53.12**
stop end	-	0.11	3.43	36.03	nr	**39.45**
mitred hips; cutting both sides	-	0.21	6.54	4.72	m	**11.27**
holes for pipes and the like	-	0.21	6.54	-	nr	**6.54**
H62 NATURAL SLATING						
NOTE: The following items of slate roofing include for conventional fixing assuming "normal exposure" with appropriate nails and/or rivets or clips to pressure impregnated softwood battens						
Natural slates; BS EN 12326 Part 2; Spanish blue grey; uniform size; to 75 mm lap; on 25 mm x 50 mm battens and type 1F reinforced underlay						
Coverings; 400 mm x 250 mm slates						
roof coverings (PC £ per 1000)	640.00	0.95	29.60	28.64	m²	**58.23**
wall coverings	-	1.22	38.01	28.64	m²	**66.64**
Coverings; 500 mm x 250 mm slates						
roof coverings (PC £ per 1000)	640.00	0.85	26.48	22.18	m²	**48.66**
wall coverings	-	1.02	31.78	22.18	m²	**53.96**
Coverings; 600 mm x 300 mm slates						
roof coverings (PC £ per 1000)	1675.00	0.64	19.94	31.54	m²	**51.47**
wall coverings	-	0.80	24.92	31.54	m²	**56.46**
Extra over coverings for						
double course at eaves	-	0.32	9.97	7.96	m	**17.93**
verges; extra single undercloak course	-	0.44	13.71	4.18	m	**17.89**
open valleys; cutting both sides	-	0.23	7.17	15.88	m	**23.05**
blue/black glass reinforced concrete 152 mm half round ridge tiles	-	0.53	16.51	17.86	m	**34.37**
blue/black glass reinforced concrete 125 mm x 125 mm plain angle ridge tiles	-	0.53	16.51	17.86	m	**34.37**
mitred hips; cutting both sides	-	0.23	7.17	15.88	m	**23.05**
blue/black glass reinforced concrete 152 mm half round hip tiles; cutting both sides	-	0.74	23.05	33.74	m	**56.80**
blue/black glass reinforced concrete 125 mm x 125 mm plain angle hip tiles; cutting both sides	-	0.74	23.05	33.74	m	**56.80**
holes for pipes and the like	-	0.21	6.54	-	nr	**6.54**

H CLADDING/COVERING

Item	PC £	Labour hours	Labour £	Material £	Unit	Total rate £
Natural slates; BS EN 12326 Part 2; Welsh blue grey; uniform size; to 75 mm lap; on 25 mm x 50 mm battens and type 1F reinforced underlay						
Coverings; 400 mm x 250 mm slates						
roof coverings (PC £ per 1000)	1595.00	0.95	29.60	56.37	m²	**85.96**
wall coverings	-	1.22	38.01	56.37	m²	**94.38**
Coverings; 500 mm x 250 mm slates						
roof coverings (PC £ per 1000)	2750.00	0.85	26.48	68.95	m²	**95.43**
wall coverings	-	1.02	31.78	68.95	m²	**100.73**
Coverings; 500 mm x 300 mm slates						
roof coverings (PC £ per 1000)	3080.00	0.80	24.92	64.95	m²	**89.87**
wall coverings	-	0.95	29.60	64.95	m²	**94.54**
Coverings; 600 mm x 300 mm slates						
roof coverings (PC £ per 1000)	6435.00	0.64	19.94	102.62	m²	**122.56**
wall coverings	-	0.80	24.92	102.62	m²	**127.54**
Extra over coverings for						
double course at eaves	-	0.32	9.97	26.58	m	**36.55**
verges; extra single undercloak course	-	0.44	13.71	15.46	m	**29.17**
open valleys; cutting both sides	-	0.23	7.17	61.01	m	**68.18**
blue/black glazed ware 152 mm half round ridge tiles	-	0.53	16.51	12.17	m	**28.69**
blue/black glazed ware 125 mm x 125 mm plain angle ridge tiles	-	0.53	16.51	35.40	m	**51.91**
mitred hips; cutting both sides	-	0.23	7.17	61.01	m	**68.18**
blue/black glazed ware 152 mm half round hip tiles; cutting both sides	-	0.74	23.05	73.19	m	**96.24**
blue/black glazed ware 125 mm x 125 mm plain angle hip tiles; cutting both sides	-	0.74	23.05	96.42	m	**119.47**
holes for pipes and the like	-	0.21	6.54	-	nr	**6.54**
Natural slates; Westmoreland green; random lengths; 457 mm - 229 mm proportionate widths to 75 mm lap; in diminishing courses; on 25 mm x 50 mm battens and type 1F underlay						
Roof coverings (PC £ per tonne)	2100.00	1.22	38.01	143.42	m²	**181.43**
Wall coverings	-	1.54	47.98	143.42	m²	**191.40**
Extra over coverings for						
double course at eaves	-	0.70	21.81	26.60	m	**48.41**
verges; extra single undercloak course slates 152 mm wide	-	0.80	24.92	22.52	m	**47.44**
holes for pipes and the like	-	0.32	9.97	-	nr	**9.97**
H63 RECONSTRUCTED STONE SLATING/TILING						
Reconstructed stone slates; "Hardrow Slates" or other equal and approved; standard colours; or similar; 75 mm lap; on 25 mm x 50 mm battens and type 1F reinforced underlay						
Coverings; 457 mm x 305 mm slates						
roof coverings	23.68	0.85	26.48	37.36	m²	**63.85**
wall coverings	-	1.06	33.02	37.36	m²	**70.39**
Coverings; 457 mm x 457 mm slates						
roof coverings	23.69	0.69	21.50	36.75	m²	**58.24**
wall coverings	-	0.91	28.35	36.75	m²	**65.10**
Extra over 457 mm x 305 mm coverings for						
double course at eaves	-	0.32	9.97	6.91	m	**16.87**
verges; pointed	-	0.44	13.71	0.07	m	**13.78**
open valleys; cutting both sides	-	0.23	7.17	16.31	m	**23.48**
ridge tiles	-	0.53	16.51	42.21	m	**58.73**
hip tiles; cutting both sides	-	0.74	23.05	35.34	m	**58.40**
holes for pipes and the like	-	0.21	6.54	-	nr	**6.54**

H CLADDING/COVERING

Item	PC £	Labour hours	Labour £	Material £	Unit	Total rate £
H63 RECONSTRUCTED STONE SLATING/TILING – cont'd						
Reconstructed stone slates; Bradstone "Cotswold" style or other equal and approved; random lengths 550 mm - 300 mm; proportional widths; to 80 mm lap; in diminishing courses; on 25 mm x 50 mm battens and type 1F reinforced underlay						
Roof coverings (all-in rate inclusive of eaves and verges)	31.35	1.12	34.89	46.46	m²	**81.35**
Extra over coverings for						
open valleys/mitred hips; cutting both sides	-	0.48	14.95	16.72	m²	**31.67**
ridge tiles	-	0.70	21.81	22.19	m	**44.00**
hip tiles; cutting both sides	-	1.12	34.89	37.78	m	**72.67**
holes for pipes and the like	-	0.32	9.97	-	nr	**9.97**
Reconstructed stone slates; Bradstone "Moordale" style or other equal and approved; random lengths 550 mm - 450 mm; proportional widths; to 80 lap; in diminishing course; on 25 mm x 50 mm battens and type 1F reinforced underlay						
Roof coverings (all-in rate inclusive of eaves and verges)	29.50	1.12	34.89	44.26	m²	**79.15**
Extra over coverings for						
open valleys/mitred hips; cutting both sides	-	0.48	14.95	15.73	m²	**30.68**
ridge tiles	-	0.70	21.81	22.19	m	**44.00**
holes for pipes and the like	-	0.32	9.97	-	nr	**9.97**
H64 TIMBER SHINGLING						
Red cedar sawn shingles preservative treated; uniform length 400 mm; to 125 mm gauge; on 25 mm x 38 mm battens and type 1F reinforced underlay						
Roof coverings; 125 mm gauge, 2.28 m²/bundle (PC £ per bundle)	67.50	1.12	34.89	44.09	m²	**78.98**
Wall coverings; 190 mm gauge, 3.47 m²/bundle	-	0.85	26.48	29.48	m²	**55.96**
Extra over coverings for						
double course at eaves	3.71	0.21	6.54	4.50	m	**11.04**
open valleys; cutting both sides	7.42	0.21	6.54	8.59	m	**15.14**
pre-formed ridge capping	-	0.32	9.97	11.89	m	**21.86**
pre-formed hip capping; cutting both sides	-	0.53	16.51	20.49	m	**37.00**
double starter course to cappings	-	0.11	3.43	1.24	m	**4.67**
holes for pipes and the like	-	0.16	4.98	-	nr	**4.98**
H71 LEAD SHEET COVERINGS/FLASHINGS						
Milled Lead; BS EN 12588; on and including Geotec underlay						
The following rates are based upon the measurement allowances and the coverage rules of SMM7 clause M2(a-f)						

H CLADDING/COVERING

Item	PC £	Labour hours	Labour £	Material £	Unit	Total rate £
Roof and dormer coverings						
1.80 mm thick (code 4) roof coverings (20.41 kg per m²)						
flat (in wood roll construction) (PC £ per kg)	2.05	0.90	26.94	72.39	m²	**99.33**
pitched (in welded seam construction)	-	1.00	29.93	72.70	m²	**102.64**
pitched (in welded seam construction)	-	0.90	26.94	72.39	m²	**99.33**
vertical (in welded seam construction)	-	1.00	29.93	69.60	m²	**99.54**
1.80 mm thick (code 4) dormer coverings						
flat (in wood roll construction)	-	0.68	20.20	71.70	m²	**91.90**
pitched (in welded seam construction)	-	0.75	22.45	71.93	m²	**94.38**
pitched (in welded seam construction)	-	0.68	20.20	71.70	m²	**91.90**
vertical (in welded seam construction)	-	1.50	44.90	69.60	m²	**114.50**
2.24 mm thick (code 5 roof coverings (25.40 kg/m²)						
flat (in wood roll construction)	-	0.94	28.29	88.03	m²	**116.32**
pitched (in welded seam construction)	-	1.05	31.43	88.36	m²	**119.79**
pitched (in welded seam construction)	-	0.94	28.29	88.03	m²	**116.32**
vertical (in welded seam construction)	-	1.05	31.43	85.10	m²	**116.53**
2.24 mm thick (code 5) dormer coverings						
flat (in wood roll construction)	-	0.71	21.22	87.30	m²	**108.53**
pitched (in welded seam construction)	-	0.79	23.59	87.55	m²	**111.14**
pitched (in welded seam construction)	-	0.71	21.22	87.30	m²	**108.53**
vertical (in welded seam construction)	-	1.57	47.14	85.10	m²	**132.25**
2.65 mm thick (code 6 roof coverings (30.05 kg/m²)						
flat (in wood roll construction)	-	0.99	29.63	102.61	m²	**132.25**
pitched (in welded seam construction)	-	1.10	32.93	102.95	m²	**135.88**
pitched (in welded seam construction)	-	0.99	29.63	102.61	m²	**132.25**
vertical (in welded seam construction)	-	1.10	32.93	99.54	m²	**132.47**
2.65 mm thick (code 6) dormer coverings						
flat (in wood roll construction)	-	0.74	22.24	101.85	m²	**124.09**
pitched (in welded seam construction)	-	0.82	24.69	102.10	m²	**126.80**
pitched (in welded seam construction)	-	0.74	22.24	101.85	m²	**124.09**
vertical (in welded seam construction)	-	1.65	49.39	99.54	m²	**148.93**
3.15 mm thick (code 7 roof coverings (35.72 kg/m²)						
flat (in wood roll construction)	-	1.06	31.67	120.44	m²	**152.11**
pitched (in welded seam construction)	-	1.18	35.17	120.80	m²	**155.97**
pitched (in welded seam construction)	-	1.06	31.67	120.44	m²	**152.11**
vertical (in welded seam construction)	-	1.18	35.17	117.15	m²	**152.33**
3.15 mm thick (code 7) dormer coverings						
flat (in wood roll construction)	-	0.79	23.74	119.62	m²	**143.36**
pitched (in welded seam construction)	-	0.88	20.37	119.00	m²	**146.26**
pitched (in welded seam construction)	-	0.79	23.74	119.62	m²	**143.36**
vertical (in welded seam construction)	-	1.76	52.77	117.15	m²	**169.93**
3.55 mm thick (code 8 roof coverings (40.26 kg/m²)						
flat (in wood roll construction)	-	1.15	34.36	134.82	m²	**169.18**
pitched (in welded seam construction)	-	1.27	38.16	135.21	m²	**173.37**
pitched (in welded seam construction)	-	1.15	34.36	134.82	m²	**169.18**
vertical (in welded seam construction)	-	1.27	38.16	131.25	m²	**169.42**
3.55 mm thick (code 8) dormer coverings						
flat (in wood roll construction)	-	0.86	25.77	133.92	m²	**159.69**
pitched (in welded seam construction)	-	0.96	28.62	134.22	m²	**162.83**
pitched (in welded seam construction)	-	0.86	25.77	133.92	m²	**159.69**
vertical (in welded seam construction)	-	1.91	57.26	131.25	m²	**188.52**
Sundries						
patination oil to finished work surfaces	-	0.03	0.75	0.30	m²	**1.05**
chalk slurry to underside of panels	-	0.33	9.97	2.73	m²	**12.70**
provision of 45 x 45 mm wood rolls at 600 mm centres (per m)	-	0.10	2.99	1.29	m	**4.28**
dressing over glazing bars and glass	-	0.25	7.48	0.78	m	**8.26**
soldered nail head	-	0.01	0.24	0.08	nr	**0.32**

H CLADDING/COVERING

Item	PC £	Labour hours	Labour £	Material £	Unit	Total rate £
H71 LEAD SHEET COVERINGS/FLASHINGS – cont'd						
1.32 mm thick (code 3) lead flashings, etc.						
Soakers (14.18 kg per m²)						
200 x 200 mm	-	0.02	0.45	1.88	nr	**2.33**
300 x 300 mm	-	0.02	0.45	4.24	nr	**4.69**
1.80 mm thick (code 4) lead flashings, etc.						
Flashings; wedging into grooves						
150 mm girth	-	0.25	7.48	9.51	m	**16.99**
200 mm girth	-	0.25	7.48	12.68	m	**20.16**
240 mm girth	-	0.25	7.48	15.21	m	**22.70**
300 mm girth	-	0.25	7.48	19.02	m	**26.50**
Stepped flashings; wedging into grooves						
180 mm girth	-	0.50	14.97	11.41	m	**26.38**
270 mm girth	-	0.50	14.97	17.12	m	**32.09**
Linings to sloping gutters						
390 mm girth	-	0.40	11.97	24.73	m	**36.70**
450 mm girth	-	0.45	13.47	28.53	m	**42.00**
600 mm girth	-	0.55	16.46	38.03	m	**54.50**
Cappings to hips or ridges						
450 mm girth	-	0.50	14.97	28.53	m	**43.50**
600 mm girth	-	0.60	17.96	38.03	m	**55.99**
Saddle flashings; at intersections of hips and ridges; dressing and bossing						
450 x 450 mm girth	-	0.50	14.97	14.39	m	**29.35**
600 x 600 mm girth	-	0.50	14.97	24.37	m	**39.34**
Slates; with 150 mm high collar						
450 x 450 mm girth; to suit 50 mm diameter pipe	-	0.75	22.45	16.66	m	**39.11**
450 x 450 mm girth; to suit 100 mm diameter pipe	-	0.75	22.45	18.15	m	**40.60**
450 x 450 mm girth; to suit 150 mm diameter pipe	-	0.75	22.45	19.64	m	**42.09**
2.24 mm thick (code 5) lead flashings, etc.						
Flashings; wedging into grooves						
150 mm girth	-	0.25	7.48	11.84	m	**19.32**
200 mm girth	-	0.25	7.48	15.78	m	**23.26**
240 mm girth	-	0.25	7.48	18.93	m	**26.42**
300 mm girth	-	0.25	7.48	23.67	m	**31.15**
Stepped flashings; wedging into grooves						
180 mm girth	-	0.50	14.97	14.20	m	**29.17**
270 mm girth	-	0.50	14.97	21.30	m	**36.27**
Linings to sloping gutters						
390 mm girth	-	0.40	11.97	30.77	m	**42.74**
450 mm girth	-	0.45	13.47	35.50	m	**48.97**
600 mm girth	-	0.55	16.46	47.33	m	**63.80**
Cappings to hips or ridges						
450 mm girth	-	0.50	14.97	35.50	m	**50.47**
600 mm girth	-	0.60	17.96	47.33	m	**65.29**
Saddle flashings; at intersections of hips and ridges; dressing and bossing						
450 x 450 mm girth	-	0.50	14.97	17.53	m	**32.49**
600 x 600 mm girth	-	0.50	14.97	29.95	m	**44.92**
Slates; with 150 mm high collar						
450 x 450 mm girth; to suit 50 mm diameter pipe	-	0.75	22.45	20.17	m	**42.62**
450 x 450 mm girth; to suit 100 mm diameter pipe	-	0.75	22.45	22.02	m	**44.46**
450 x 450 mm girth; to suit 150 mm diameter pipe	-	0.75	22.45	23.88	m	**46.33**

H CLADDING/COVERING

Item	PC £	Labour hours	Labour £	Material £	Unit	Total rate £
H72 ALUMINIUM SHEET COVERINGS/FLASHINGS						
Aluminium roofing; commercial grade; on and including Geotec underlay						
The following rates are based upon nett 'deck' or 'wall' areas, and depart from SMM7 coverage rules						
Roof, dormer and wall coverings						
0.7 mm thick roof coverings; mill finish						
flat (in wood roll construction) (PC per kg)	4.50	1.00	29.93	26.05	m²	**55.98**
eaves detail ED1	-	0.20	5.99	3.33	m²	**9.32**
abutment upstands at perimeters	-	0.33	9.88	1.33	m²	**11.21**
pitched over 3 degrees (in standing seam construction)	-	0.75	22.45	22.05	m²	**44.50**
vertical (in angled or flat seam construction)	-	0.80	23.95	22.05	m²	**46.00**
0.7 mm thick roof coverings; mill finish						
flat (in wood roll construction)	-	1.50	44.90	26.05	m²	**70.95**
eaves detail ED1	-	0.20	5.99	3.33	m²	**9.32**
pitched over 3 degrees (in standing seam construction)	-	1.25	37.42	22.05	m²	**59.47**
vertical (in angled or flat seam construction)	-	1.35	40.41	22.05	m²	**62.46**
0.7 mm thick roof coverings; Pvf2 finish						
flat (in wood roll construction) (PC per kg)	5.60	1.00	29.93	30.93	m²	**60.86**
eaves detail ED1	-	0.20	5.99	4.14	m²	**10.13**
abutment upstands at perimeters	-	0.33	9.88	1.66	m²	**11.54**
pitched over 3 degrees (in standing seam construction)	-	0.75	22.45	25.96	m²	**48.41**
vertical (in angled or flat seam construction)	-	0.80	23.95	25.96	m²	**49.90**
0.7 mm thick roof coverings; Pvf2 finish						
flat (in wood roll construction)	-	1.50	44.90	30.93	m²	**75.83**
eaves detail ED1	-	0.20	5.99	4.14	m²	**10.13**
pitched over 3 degrees (in standing seam construction)	-	1.25	37.42	25.96	m²	**63.37**
vertical (in angled or flat seam construction)	-	1.35	40.41	25.96	m²	**66.37**
0.8 mm thick aluminium flashings, etc.						
Flashings; wedging into grooves; mill finish						
150 mm girth (PC per kg)	4.50	0.25	7.48	2.00	m	**9.48**
240 mm girth	-	0.25	7.48	3.20	m	**10.68**
300 mm girth	-	0.25	7.48	4.00	m	**11.48**
Stepped flashings; wedging into grooves; mill finish						
180 mm girth	-	0.50	14.97	2.40	m	**17.36**
270 mm girth	-	0.50	14.97	3.60	m	**18.56**
Flashings; wedging into grooves; Pvf2 finish						
150 mm girth (PC per kg)	5.60	0.25	7.48	2.49	m	**9.97**
240 mm girth	-	0.25	7.48	3.98	m	**11.46**
300 mm girth	-	0.25	7.48	4.97	m	**12.46**
Stepped flashings; wedging into grooves; Pvf2 finish						
180 mm girth	-	0.50	14.97	2.98	m	**17.95**
270 mm girth	-	0.50	14.97	4.47	m	**19.44**
Sundries						
provision of square batten roll at 500 mm centres (per m)	-	0.10	2.99	1.48	m	**4.47**

H CLADDING/COVERING

Item	PC £	Labour hours	Labour £	Material £	Unit	Total rate £
H72 ALUMINIUM SHEET COVERINGS/FLASHINGS – cont'd						
Aluminium Alumasc "Eavesline" coping system; polyester powder coated						
Coping; fixing straps plugged and screwed to brickwork						
362 mm wide; for parapet wall 271 - 300 mm wide	-	0.58	12.63	32.19	m	**44.82**
Extra for						
90 degree bend	-	0.28	6.10	76.68	nr	**82.78**
90 degree tee junction	-	0.40	8.71	82.27	nr	**90.98**
stop end	-	0.17	3.70	45.61	nr	**49.31**
stop end upstand	-	0.23	5.01	49.19	nr	**54.20**
H73 COPPER STRIP/SHEET COVERINGS/ FLASHINGS						
Copper roofing; BS EN 504; on and including Geotec underlay						
The following rates are based upon nett 'deck' or 'wall' areas, and depart from SMM7 coverage rules						
Roof and dormer coverings						
0.6 mm thick roof coverings; mill finish						
flat (in wood roll construction) (PC per kg)	6.25	1.00	29.93	82.37	m²	**112.31**
eaves detail ED1	-	0.20	5.99	10.17	m	**16.16**
abutment upstands at perimeters	-	0.33	9.88	5.09	m	**14.96**
pitched over 3 degrees (in standing seam construction)	-	0.75	22.45	67.11	m²	**89.56**
vertical (in angled or flat seam construction)	-	0.80	23.95	67.11	m²	**91.06**
0.6 mm thick dormer coverings; mill finish						
flat (in wood roll construction)	-	1.50	44.90	82.37	m²	**127.27**
eaves detail ED1	-	0.20	5.99	10.17	m	**16.16**
pitched over 3 degrees (in standing seam construction)	-	1.25	37.42	67.11	m²	**104.53**
vertical (in angled or flat seam construction)	-	1.35	40.41	67.11	m²	**107.52**
0.6 mm thick roof coverings; oxid finish						
flat (in wood roll construction) (PC per kg)	7.30	1.00	29.93	95.19	m²	**125.12**
eaves detail ED1	-	0.20	5.99	11.88	m	**17.87**
abutment upstands at perimeters	-	0.33	9.88	5.94	m	**15.82**
pitched over 3 degrees (in standing seam construction)	-	0.75	22.45	77.37	m²	**99.82**
vertical (in angled or flat seam construction)	-	0.80	23.95	77.37	m²	**101.31**
0.6 mm thick dormer coverings; oxid finish						
flat (in wood roll construction)	-	1.50	44.90	95.19	m²	**140.09**
eaves detail ED1	-	0.20	5.99	11.88	m	**17.87**
pitched over 3 degrees (in standing seam construction)	-	1.25	37.42	77.37	m²	**114.78**
vertical (in angled or flat seam construction)	-	1.35	40.41	77.37	m²	**117.78**
0.6 mm thick roof coverings; KME pre-patinated finish						
flat (in wood roll construction)	56.50	1.10	32.93	131.48	m²	**164.41**
eaves detail ED1	-	0.20	5.99	16.72	m	**22.71**
abutment upstands at perimeters	-	0.33	9.88	8.36	m	**18.24**
pitched over 3 degrees (in standing seam construction)	-	0.85	25.44	106.40	m²	**131.84**
vertical (in angled or flat seam construction)	-	0.90	26.94	106.40	m²	**133.34**

H CLADDING/COVERING

Item	PC £	Labour hours	Labour £	Material £	Unit	Total rate £
0.6 mm thick dormer coverings; KME pre-patinated finish						
flat (in wood roll construction)	-	1.50	44.90	131.48	m²	**176.38**
eaves detail ED1	-	0.20	5.99	16.72	m	**22.71**
pitched over 3 degrees (in standing seam construction)	-	1.25	37.42	106.40	m²	**143.82**
vertical (in angled or flat seam construction)	-	1.35	40.41	106.40	m²	**146.81**
0.7 mm thick roof coverings; mill finish						
flat (in wood roll construction) (PC per kg)	6.25	1.00	29.93	93.47	m²	**123.40**
eaves detail ED1	-	0.20	5.99	11.65	m	**17.64**
abutment upstands at perimeters	-	0.33	9.88	5.83	m	**15.70**
pitched over 3 degrees (in standing seam construction)	-	0.75	22.45	75.99	m²	**98.44**
vertical (in angled or flat seam construction)	-	0.80	23.95	75.99	m²	**99.94**
0.7 mm thick dormer coverings; mill finish						
flat (in wood roll construction)	-	1.50	44.90	93.47	m²	**138.37**
eaves detail ED1	-	0.20	5.99	11.65	m	**17.64**
pitched over 3 degrees (in standing seam construction)	-	1.25	37.42	75.99	m²	**113.41**
vertical (in angled or flat seam construction)	-	1.35	40.41	75.99	m²	**116.40**
0.7 mm thick roof coverings; oxid finish						
flat (in wood roll construction) (PC per kg)	7.30	1.00	29.93	108.15	m²	**138.09**
eaves detail ED1	-	0.20	5.99	13.61	m	**19.60**
abutment upstands at perimeters	-	0.33	9.88	6.81	m	**16.68**
pitched over 3 degrees (in standing seam construction)	-	0.75	22.45	87.74	m²	**110.19**
vertical (in angled or flat seam construction)	-	0.80	23.95	87.74	m²	**111.68**
0.7 mm thick dormer coverings; oxid finish						
flat (in wood roll construction)	-	1.50	44.90	108.15	m²	**153.05**
eaves detail ED1	-	0.20	5.99	13.61	m	**19.60**
pitched over 3 degrees (in standing seam construction)	-	1.25	37.42	87.74	m²	**125.15**
vertical (in angled or flat seam construction)	-	1.35	40.41	87.74	m²	**128.15**
0.7 mm thick roof coverings; KME pre-patinated finish						
flat (in wood roll construction)	65.00	1.10	32.93	150.35	m²	**183.28**
eaves detail ED1	-	0.20	5.99	19.24	m	**25.22**
abutment upstands at perimeters	-	0.33	9.88	9.62	m	**19.50**
pitched over 3 degrees (in standing seam construction)	-	0.85	25.44	121.50	m²	**146.94**
vertical (in angled or flat seam construction)	-	0.90	26.94	121.50	m²	**148.44**
0.7 mm thick dormer coverings; KME pre-patinated finish						
flat (in wood roll construction)	-	1.50	44.90	150.35	m²	**195.25**
eaves detail ED1	-	0.20	5.99	19.24	m	**25.22**
pitched over 3 degrees (in standing seam construction)	-	1.25	37.42	121.50	m²	**158.91**
vertical (in angled or flat seam construction)	-	1.35	40.41	121.50	m²	**161.90**
0.6 mm thick copper flashings, etc.						
Flashings; wedging into grooves; mill finish						
150 mm girth (PC per kg)	6.25	0.25	7.48	7.63	m	**15.11**
240 mm girth	-	0.25	7.48	12.21	m	**19.69**
300 mm girth	-	0.25	7.48	15.26	m	**22.75**
Stepped flashings; wedging into grooves; mill finish						
180 mm girth	-	0.50	14.97	9.16	m	**24.12**
270 mm girth	-	0.50	14.97	13.74	m	**28.70**

H CLADDING/COVERING

Item	PC £	Labour hours	Labour £	Material £	Unit	Total rate £
H73 COPPER STRIP/SHEET COVERINGS/ FLASHINGS – cont'd						
0.6 mm thick copper flashings, etc. – cont'd						
Flashings; wedging into grooves; oxid finish						
150 mm girth (PC per kg)	7.30	0.25	7.48	8.91	m	**16.40**
240 mm girth	-	0.25	7.48	14.26	m	**21.74**
300 mm girth	-	0.25	7.48	17.82	m	**25.31**
Stepped flashings; wedging into grooves; oxid finish						
180 mm girth	-	0.50	14.97	10.69	m	**25.66**
270 mm girth	-	0.50	14.97	16.04	m	**31.01**
Flashings; wedging into grooves; KME pre-patinated finish						
150 mm girth (PC per m²)	56.50	0.25	7.48	12.54	m	**20.02**
240 mm girth	-	0.25	7.48	20.07	m	**27.55**
300 mm girth	-	0.25	7.48	25.08	m	**32.57**
Stepped flashings; wedging into grooves; KME pre-patinated finish						
180 mm girth	-	0.50	14.97	15.05	m	**30.02**
270 mm girth	-	0.50	14.97	22.57	m	**37.54**
0.7 mm thick copper flashings, etc.						
Flashings; wedging into grooves; mill finish						
150 mm girth (PC per kg)	6.25	0.25	7.48	8.74	m	**16.22**
240 mm girth	-	0.25	7.48	13.99	m	**21.47**
300 mm girth	-	0.25	7.48	17.48	m	**24.97**
Stepped flashings; wedging into grooves; mill finish						
180 mm girth	-	0.50	14.97	10.49	m	**25.46**
270 mm girth	-	0.50	14.97	15.73	m	**30.70**
Flashings; wedging into grooves; oxid finish						
150 mm girth (PC per kg)	7.30	0.25	7.48	10.24	m	**17.73**
240 mm girth	-	0.25	7.48	16.39	m	**23.87**
300 mm girth	-	0.25	7.48	20.48	m	**27.97**
Stepped flashings; wedging into grooves; oxid finish						
180 mm girth	-	0.50	14.97	12.29	m	**27.26**
270 mm girth	-	0.50	14.97	18.44	m	**33.40**
Flashings; wedging into grooves; KME pre-patinated finish						
150 mm girth (PC per m²)	65.00	0.25	7.48	14.43	m	**21.91**
240 mm girth	-	0.25	7.48	23.09	m	**30.57**
300 mm girth	-	0.25	7.48	28.86	m	**36.34**
Stepped flashings; wedging into grooves; KME pre-patinated finish						
180 mm girth	-	0.50	14.97	17.31	m	**32.28**
270 mm girth	-	0.50	14.97	25.97	m	**40.94**
Sundries						
provision of square batten roll at 500 mm centres (per m)	-	0.10	2.99	1.48	m	**4.47**

H CLADDING/COVERING

Item	PC £	Labour hours	Labour £	Material £	Unit	Total rate £
H74 ZINC STRIP/SHEET COVERINGS/FLASHINGS						
Zinc roofing; BS EN 506; on and including Delta Trella underlay						
The following rates are based upon nett 'deck' or 'wall' areas, and depart from SMM7 coverage rules						
Roof, dormer and wall coverings						
0.7 mm thick roof coverings; pre-weathered Rheinzink						
flat (in wood roll construction) (PC per kg)	3.60	1.00	29.93	47.67	m²	**77.61**
eaves detail ED1	-	0.20	5.99	8.04	m²	**14.03**
abutment upstands at perimeters	-	0.33	9.88	4.02	m²	**13.90**
pitched over 3 degrees (in standing seam construction)	-	0.75	22.45	39.63	m²	**62.08**
0.7 mm thick roof coverings; pre-weathered Rheinzink						
flat (in wood roll construction)	-	1.50	44.90	47.67	m²	**92.57**
eaves detail ED1	-	0.20	5.99	8.04	m²	**14.03**
pitched over 3 degrees (in standing seam construction)	-	1.25	37.42	39.63	m²	**77.05**
0.8 mm thick wall coverings; pre-weathered Rheinzink						
vertical (in angled or flat seam construction)	-	0.80	23.95	44.37	m²	**68.32**
0.8 mm thick dormer coverings; pre-weathered Rheinzink						
vertical (in angled or flat seam construction)	-	1.35	40.41	44.37	m²	**84.78**
0.8 mm thick zinc flashings, etc.; pre-weathered Rheinzink						
Flashings; wedging into grooves						
150 mm girth (PC per kg)	3.60	0.25	7.48	4.62	m	**12.10**
240 mm girth	-	0.25	7.48	7.39	m	**14.87**
300 mm girth	-	0.25	7.48	9.23	m	**16.72**
Stepped flashings; wedging into grooves						
180 mm girth	-	0.50	14.97	5.54	m	**20.51**
270 mm girth	-	0.50	14.97	8.31	m	**23.28**
Integral box gutter						
900 mm girth; 2 x bent; 2 x welted	-	1.00	29.93	38.17	m	**68.10**
Valley gutter						
600 mm girth; 2 x bent; 2 x welted	-	0.75	22.45	23.09	m	**45.54**
Hips and ridges						
450 mm girth; 2 x bent; 2 x welted	-	1.00	29.93	13.85	m	**43.78**
Sundries						
provision of square batten roll at 500 mm centres (per m)	-	0.10	2.99	1.48	m	**4.47**

H CLADDING/COVERING

Item	PC £	Labour hours	Labour £	Material £	Unit	Total rate £
H75 STAINLESS STEEL COVERINGS/FLASHINGS						
Terne-coated stainless steel roofing; Associated Lead Mills Ltd; or other equal and approved: on and including Metmatt underlay						
The following rates are based upon nett 'deck' or 'wall' areas, and depart from SMM7 coverage rules						
Roof, dormer and wall coverings in 'Uginox' grade 316; marine						
0.4 mm thick roof coverings						
flat (in wood roll construction) (PC per kg)	9.00	1.00	29.93	66.59	m²	**96.53**
eaves detail ED1	-	0.20	5.99	7.99	m²	**13.98**
abutment upstands at perimeters	-	0.33	9.88	4.00	m²	**13.87**
pitched over 3 degrees (in standing seam construction)	-	0.75	22.45	54.61	m²	**77.06**
0.5mm thick dormer coverings						
flat (in wood roll construction)	-	1.50	44.90	82.58	m²	**127.48**
eaves detail ED1	-	0.20	5.99	7.99	m²	**13.98**
pitched over 3 degrees (in standing seam construction)	-	1.25	37.42	67.39	m²	**104.81**
0.5 mm thick wall coverings						
vertical (in angled or flat seam construction)	-	0.80	23.95	67.39	m²	**91.34**
vertical (with Coulisseau joint construction)	-	1.25	37.42	69.63	m²	**107.04**
0.5 mm thick 'Uginox' grade 316 flashings, etc.						
Flashings; wedging into grooves						
150 mm girth (PC per kg)	9.00	0.25	7.48	7.59	m	**15.07**
240 mm girth	-	0.25	7.48	12.15	m	**19.63**
300 mm girth	-	0.25	7.48	15.18	m	**22.67**
Stepped flashings; wedging into grooves						
180 mm girth	-	0.50	14.97	9.11	m	**24.08**
270 mm girth	-	0.50	14.97	13.66	m	**28.63**
Fan apron						
250 mm girth	-	0.25	7.48	12.65	m	**20.14**
Integral box gutter						
900 mm girth; 2 x bent; 2 x welted	-	1.00	29.93	52.13	m	**82.06**
Valley gutter						
600 mm girth; 2 x bent; 2 x welted	-	0.75	22.45	37.05	m	**59.50**
Hips and ridges						
450 mm girth; 2 x bent; 2 x welted	-	1.00	29.93	22.77	m	**52.71**
Sundries						
provision of square batten roll at 500 mm centres (per m)	-	0.10	2.99	1.48	m	**4.47**

H CLADDING/COVERING

Item	PC £	Labour hours	Labour £	Material £	Unit	Total rate £
H76 FIBRE BITUMEN THERMOPLASTIC SHEET COVERINGS/FLASHINGS						
Glass fibre reinforced bitumen strip slates; "Ruberglas 105" or other equal and approved; 1000 mm x 336 mm mineral finish; to external quality plywood boarding (boarding not included)						
Roof coverings	-	0.27	8.41	13.26	m²	**21.67**
Wall coverings	-	0.43	13.40	13.26	m²	**26.65**
Extra over coverings for						
double course at eaves; felt soaker	-	0.21	6.54	8.80	m	**15.35**
verges; felt soaker	-	0.16	4.98	7.30	m	**12.28**
valley slate; cut to shape; felt soaker and cutting both sides	-	0.48	14.95	11.62	m	**26.58**
ridge slate; cut to shape	-	0.32	9.97	7.30	m	**17.27**
hip slate; cut to shape; felt soaker andcutting both sides	-	0.48	14.95	11.44	m	**26.39**
holes for pipes and the like	-	0.56	17.45	-	nr	**17.45**
Bostik Findley "Flashband Plus" sealing strips and flashings or other equal and approved; special grey finish						
Flashings; wedging at top if required; pressure bonded; to walls						
100 mm girth	-	0.27	5.88	0.74	m	**6.62**
150 mm girth	-	0.35	7.62	1.27	m	**8.89**
225 mm girth	-	0.43	9.36	1.96	m	**11.32**
300 mm girth	-	0.48	10.45	2.38	m	**12.83**
H92 RAINSCREEN CLADDING						
Western Red Cedar tongued and grooved wall cladding on and including treated softwrood battens on breather mambrane, 10 mm Eternit Blueclad board and 50 mm insulation board; the whole fixed to Metsec frame system; including sealing all joints etc.						
26 mm thick cladding to walls; boards laid horizontally	-	-	-	-	m²	**101.47**

J WATERPROOFING

Item	PC £	Labour hours	Labour £	Material £	Unit	Total rate £
J10 SPECIALIST WATERPROOF RENDERING						
"Sika" waterproof rendering or other equal and approved; steel trowelled						
20 mm work to walls; three coat; to concrete base						
over 300 mm wide	-	-			m²	**53.34**
not exceeding 300 mm wide	-	-			m²	**83.30**
25 mm work to walls; three coat; to concrete base						
over 300 mm wide	-	-	-	-	m²	**60.64**
not exceeding 300 mm wide	-	-	-	-	m²	**94.99**
40 mm work to walls; four coat; to concrete base						
over 300 mm wide	-	-	-	-	m²	**90.60**
not exceeding 300 mm wide	-	-	-	-	m²	**146.13**
J20 MASTIC ASPHALT TANKING/DAMP PROOF MEMBRANES						
Mastic asphalt to BS 6925 Type T 1097						
13 mm thick one coat coverings to concrete base; flat; subsequently covered						
over 300 mm wide	-	-	-	-	m²	**14.71**
225 mm - 300 mm wide	-	-	-	-	m²	**49.75**
150 mm - 225 mm wide	-	-	-	-	m²	**55.01**
not exceeding 150 mm wide	-	-	-	-	m²	**69.62**
20 mm thick two coat coverings to concrete base; flat; subsequently covered						
over 300 mm wide	-	-	-	-	m²	**18.33**
225 mm - 300 mm wide	-	-	-	-	m²	**46.05**
150 mm - 225 mm wide	-	-	-	-	m²	**61.29**
not exceeding 150 mm wide	-	-	-	-	m²	**73.62**
30 mm thick three coat coverings to concrete base; flat; subsequently covered						
over 300 mm wide	-	-	-	-	m²	**30.01**
225 mm - 300 mm wide	-	-	-	-	m²	**69.74**
150 mm - 225 mm wide	-	-	-	-	m²	**76.38**
not exceeding 150 mm wide	-	-	-	-	m²	**94.74**
13 mm thick two coat coverings to brickwork base; vertical; subsequently covered						
over 300 mm wide	-	-	-	-	m²	**47.28**
225 mm - 300 mm wide	-	-	-	-	m²	**70.41**
150 mm - 225 mm wide	-	-	-	-	m²	**76.37**
not exceeding 150 mm wide	-	-	-	-	m²	**100.93**
20 mm thick three coat coverings to brickwork base; vertical; subsequently covered						
over 300 mm wide	-	-	-	-	m²	**76.02**
225 mm - 300 mm wide	-	-	-	-	m²	**94.02**
150 mm - 225 mm wide	-	-	-	-	m²	**103.86**
not exceeding 150 mm wide	-	-	-	-	m²	**136.45**
Turning into groove 20 mm deep	-	-	-	-	m	**0.83**
Internal angle fillets; subsequently covered	-	-	-	-	m	**5.74**

J WATERPROOFING

Item	PC £	Labour hours	Labour £	Material £	Unit	Total rate £
J21 MASTIC ASPHALT ROOFING/INSULATION/ FINISHES						
Mastic asphalt to BS 6925 Type R 988						
20 mm thick two coat coverings; felt isolating membrane; to concrete (or timber) base; flat or to falls or slopes not exceeding 10 degrees from horizontal						
over 300 mm wide	-	-	-	-	m²	**19.29**
225 mm - 300 mm wide	-	-	-	-	m²	**32.86**
150 mm - 225 mm wide	-	-	-	-	m²	**39.34**
not exceeding 150 mm wide	-	-	-	-	m²	**52.26**
Add to the above for covering with:						
10 mm thick limestone chippings in hot bitumen	-	-	-	-	m²	**3.61**
coverings with solar reflective paint	-	-	-	-	m²	**3.65**
300 mm x 300 mm x 8 mm g.r.p. tiles in hot bitumen	-	-	-	-	m²	**53.54**
Cutting to line; jointing to old asphalt	-	-	-	-	m	**7.43**
13 mm thick two coat skirtings to brickwork base						
not exceeding 150 mm girth	-	-	-	-	m	**15.97**
150 mm - 225 mm girth	-	-	-	-	m	**18.11**
225 mm - 300 mm girth	-	-	-	-	m	**22.13**
13 mm thick three coat skirtings; expanded metal lathing reinforcement nailed to timber base						
not exceeding 150 mm girth	-	-	-	-	m	**26.48**
150 mm - 225 mm girth	-	-	-	-	m	**31.14**
225 mm - 300 mm girth	-	-	-	-	m	**36.28**
13 mm thick two coat fascias to concrete base						
not exceeding 150 mm girth	-	-	-	-	m	**15.97**
150 mm - 225 mm girth	-	-	-	-	m	**18.11**
20 mm thick two coat linings to channels to concrete base						
not exceeding 150 mm girth	-	-	-	-	m	**34.51**
150 mm - 225 mm girth	-	-	-	-	m	**39.20**
225 mm - 300 mm girth	-	-	-	-	m	**40.07**
20 mm thick two coat lining to cesspools						
250 mm x 150 mm x 150 mm deep	-	-	-	-	nr	**34.58**
Collars around pipes, standards and like members	-	-	-	-	nr	**23.06**
Accessories						
Eaves trim; extruded aluminium alloy; working asphalt into trim						
"Alutrim"; type A roof edging or other equal and approved	-	-	-	-	m	**14.67**
extra; angle	-	-	-	-	nr	**8.19**
Roof screed ventilator - aluminium alloy						
"Extr-aqua-vent" or other equal and approved; set on screed over and including dished sinking; working collar around ventilator	-	-	-	-	nr	**26.94**
J30 LIQUID APPLIED TANKING/DAMP PROOF MEMBRANES						
Tanking and damp proofing						
"Synthaprufe"or other equal and approved; blinding with sand; horizontal on slabs						
two coats	-	0.20	3.07	3.20	m²	**6.27**
three coats	-	0.29	4.46	4.72	m²	**9.18**
"Tretolastex 202T" or other equal and approved; on vertical surfaces of concrete						
two coats	-	0.20	3.07	0.67	m²	**3.74**
three coats	-	0.29	4.46	1.00	m²	**5.46**

J WATERPROOFING

Item	PC £	Labour hours	Labour £	Material £	Unit	Total rate £
J30 LIQUID APPLIED TANKING/DAMP PROOF MEMBRANES – cont'd						
Tanking and damp proofing – cont'd						
One coat Vandex "Super" 0.75 kg/m² slurry or other equal and approved; one consolidating coat of Vandex "Premix" 1 kg/m² slurry or simiar; horizontal on beds						
over 225 mm wide	-	0.36	5.53	4.05	m²	**9.59**
Intergritank MMA (Methyl Methacrylate) resin elastomeric structural waterproffing membrane; in two separate 1mm colour coded coats; or other equal and approved; on a primed substrate						
over 225 mm wide	-	-	-	-	m²	**41.00**
J40 FLEXIBLE SHEET TANKING/DAMP PROOF MEMBRANES						
Tanking and damp proofing						
Visqueen self-adhesive damp proof membrane						
over 300 mm wide; horizontal	-	-	-	-	m²	**8.07**
not exceeding 300 mm wide; horizontal	-	-	-	-	m	**3.10**
Tanking primer for self-adhesive dpm						
over 300 mm wide; horizontal	-	-	-	-	m²	**5.58**
not exceeding 300 mm wide; horizontal	-	-	-	-	m	**2.56**
"Bituthene" sheeting or other equal and approved; lapped joints; horizontal on slabs						
3000 grade	-	0.10	1.54	7.30	m²	**8.84**
8000 grade	-	0.11	1.69	10.19	m²	**11.88**
500HD heavy duty grade	-	0.13	2.00	9.68	m²	**11.68**
"Bituthene" sheeting or other equal and approved; lapped joints; dressed up vertical face of concrete						
8000 grade	-	0.19	2.92	10.19	m²	**13.11**
RIW "Structureseal" tanking and damp proof membrane; or other equal and approved						
over 300 mm wide; horizontal	-	-	-	-	m²	**8.84**
"Structureseal" Fillet						
40 mm x 40mm	-	-	-	-	m	**6.48**
Ruberoid "Plasprufe 2000SA" self-adhesive damp proof membrane						
over 300 mm wide; horizontal	-	-	-	-	m²	**15.38**
not exceeding 300 mm wide; horizontal	-	-	-	-	m	**5.89**
Extra for 50 mm thick sand blinding	-	-	-	-	m²	**2.69**
"Servi-pak" protection board or other equal and approved; butt jointed; taped joints; to horizontal surfaces;						
3 mm thick	-	0.16	2.46	7.58	m²	**10.04**
6 mm thick	-	0.16	2.46	11.48	m²	**13.94**
12 mm thick	-	0.20	3.07	20.49	m²	**23.56**
"Servi-pak" protection board or other equal and approved; butt jointed; taped joints; to vertical surfaces						
3 mm thick	-	0.20	3.07	7.58	m²	**10.65**
6 mm thick	-	0.20	3.07	11.48	m²	**14.55**
12 mm thick	-	0.26	3.99	20.49	m²	**24.48**
"Bituthene" reinforcing strip or other equal and approved; 300 mm wide						
Bitutape 4000	-	0.10	1.54	3.06	m	**4.59**

J WATERPROOFING

Item	PC £	Labour hours	Labour £	Material £	Unit	Total rate £
Expandite "Famflex" hot bitumen bonded waterproof tanking or other equal and approved; 150 mm laps						
horizontal; over 300 mm wide	-	0.41	6.30	18.24	m²	**24.54**
vertical; over 300 mm wide	-	0.67	10.29	18.24	m²	**28.54**
J41 BUILT UP FELT ROOF COVERINGS						
NOTE: The following items of felt roofing, unless otherwise described, include for conventional lapping, laying and bonding between layers and to base; and laying flat or to falls, to crossfalls or to slopes not exceeding 10 degrees - but exclude any insulation etc.						
Felt roofing; BS EN 13707; suitable for flat roofs						
Three layer coverings first layer type 3G; subsequent layers type 3B bitumen glass fibre based felt	-	-	-	-	m²	**16.72**
Extra over felt for covering with and bedding in hot bitumen						
13 mm thick stone chippings	-	-	-	-	m²	**4.84**
300 mm x 300 mm x 8 mm g.r.p. tiles	-	-	-	-	m²	**51.78**
working into outlet pipes and the like	-	-	-	-	m²	**11.67**
Skirtings; three layer; top layer mineral surfaced; dressed over tilting fillet; turned into groove						
not exceeding 200 mm girth	-	-	-	-	m	**12.44**
200 mm - 400 mm girth	-	-	-	-	m	**15.37**
Coverings to kerbs; three layer						
400 mm - 600 mm girth	-	-	-	-	m	**19.91**
Linings to gutters; three layer						
400 mm - 600 mm girth	-	-	-	-	m	**24.19**
Collars around pipes and the like; three layer mineral surface; 150 mm high						
not exceeding 55 mm nominal size	-	-	-	-	nr	**13.21**
55 mm - 110 mm nominal size	-	-	-	-	nr	**13.21**
Three layer coverings; two base layers type 5U bitumen polyester based felt; top layer type 5B polyester based mineral surfaced felt; 10 mm stone chipping covering; bitumen bonded	-	-	-	-	m²	**28.47**
Coverings to kerbs						
not exceeding 200 mm girth	-	-	-	-	m	**12.03**
200 mm - 400 mm girth	-	-	-	-	m	**15.74**
Outlets and dishing to gullies						
300 mm diameter	-	-	-	-	nr	**14.33**
"Andersons" high performance polyester-based roofing system or other equal and approved						
Two layer coverings; first layer HT 125 underlay; second layer HT 350; fully bonded to wood; fibre or cork base	-	-	-	-	m²	**23.60**
Extra over for						
top layer mineral surfaced	-	-	-	-	m²	**2.00**
13 mm thick stone chippings	-	-	-	-	m²	**4.84**
third layer of type 3B as underlay for concrete or screeded base	-	-	-	-	m²	**6.18**
working into outlet pipes and the like	-	-	-	-	nr	**14.31**
Skirtings; two layer; top layer mineral surfaced; dressed over tilting fillet; turned into groove						
not exceeding 200 mm girth	-	-	-	-	m	**12.03**
200 mm - 400 mm girth	-	-	-	-	m	**15.74**

J WATERPROOFING

Item	PC £	Labour hours	Labour £	Material £	Unit	Total rate £
J41 BUILT UP FELT ROOF COVERINGS – cont'd						
"Andersons" high performance polyester-based roofing system – cont'd						
Coverings to kerbs; two layer						
400 mm - 600 mm girth	-	-	-	-	m	**20.39**
Linings to gutters; three layer						
400 mm - 600 mm girth	-	-	-	-	m	**21.92**
Collars around pipes and the like; two layer; 150 high						
not exceeding 55 mm nominal size	-	-	-	-	nr	**14.31**
55 mm - 110 mm nominal size	-	-	-	-	nr	**14.31**
"Ruberoid Challenger SBS" high performance roofing or other equal and approved (10 year guarantee specification)						
Two layer coverings; first and second layers "Ruberglas 120 GP"; fully bonded to wood, fibre or cork base	-	-	-	-	m²	**15.53**
Extra over for						
top layer mineral surfaced	-	-	-	-	m²	**5.49**
13 mm thick stone chippings	-	-	-	-	m²	**4.84**
third layer of "Rubervent 3G" as underlay for concrete or screeded base	-	-	-	-	m²	**6.15**
working into outlet pipes and the like	-	-	-	-	nr	**14.22**
Skirtings; two layer; top layer mineral surfaced; dressed over tilting fillet; turned into groove						
not exceeding 200 mm girth	-	-	-	-	m	**11.85**
200 mm - 400 mm girth	-	-	-	-	m	**15.51**
Coverings to kerbs; two layer						
400 mm - 600 mm girth	-	-	-	-	m	**20.10**
Linings to gutters; three layer						
400 mm - 600 mm girth	-	-	-	-	m	**21.53**
Collars around pipes and the like; two layer, 150 mm high						
not exceeding 55 mm nominal size	-	-	-	-	nr	**14.22**
55 mm - 110 mm nominal size	-	-	-	-	nr	**14.22**
"Ruberfort HP 350" high performance roofing or other equal and approved						
Two layer coverings; first layer Ruberfort HP 180; second layer Ruberfort HP 350; fully bonded; to wood; fibre or cork base	-	-	-	-	m²	**18.31**
Extra over for						
top layer mineral surfaced	-	-	-	-	m²	**7.58**
13 mm thick stone chippings	-	-	-	-	m²	**4.84**
third layer of "Rubervent 3G"; as underlay for concrete or screeded base	-	-	-	-	m²	**6.15**
working into outlet pipes and the like	-	-	-	-	nr	**14.38**
Skirtings; two layer; top layer mineral surface; dressed over tilting fillet; turned into groove						
not exceeding 200 mm girth	-	-	-	-	m	**12.09**
200 mm - 400 mm girth	-	-	-	-	m	**15.82**
Coverings to kerbs; two layer						
400 mm - 600 mm girth	-	-	-	-	m	**20.52**
Linings to gutters; three layer						
400 mm - 600 mm girth	-	-	-	-	m	**26.49**
Collars around pipes and the like; two layer; 150 mm high						
not exceeding 55 mm nominal size	-	-	-	-	nr	**14.38**
55 mm - 110 mm nominal size	-	-	-	-	nr	**14.38**

J WATERPROOFING

Item	PC £	Labour hours	Labour £	Material £	Unit	Total rate £
"Ruberoid Superflex Firebloc" high performance roofing or other equal and approved (15 year guarantee specification)						
Two layer coverings; first layer Superflex 180; second layer Superflex 250; fully bonded to wood; fibre or cork base	-	-	-	-	m²	**22.63**
Extra over for						
top layer mineral surfaced	-	-	-	-	m²	**5.38**
13 mm thick stone chippings	-	-	-	-	m²	**4.84**
third layer of "Rubervent 3G" as underlay for concrete or screeded base	-	-	-	-	m²	**6.15**
working into outlet pipes and the like	-	-	-	-	nr	**16.33**
Skirtings; two layer; top layer mineral surfaced; dressed over tilting fillet; turned into groove						
not exceeding 200 mm girth	-	-	-	-	m	**14.14**
200 mm - 400 mm girth	-	-	-	-	m	**18.66**
Coverings to kerbs; two layer						
400 mm - 600 mm girth	-	-	-	-	m	**24.90**
Linings to gutters; three layer						
400 mm - 600 mm girth	-	-	-	-	m	**26.97**
Collars around pipes and the like; two layer; 150 mm high						
not exceeding 55 mm nominal size	-	-	-	-	nr	**16.33**
55 mm - 110 mm nominal size	-	-	-	-	nr	**16.33**
"Ruberoid Ultra prevENt" high performance roofing or other equal and approved (20 year guarantee specification)						
Two layer coverings; first layer Ultra prevENt underlay; second layer Ultra prevENt mineral surface cap sheet; fully bonded to wood; fibre or cork base	-	-	-	-	m³	**41.64**
Extra over for						
13 mm thick stone chippings	-	-	-	-	m²	**4.84**
third layer of "Rubervent 3G" as underlay for concrete or screeded base	-	-	-	-	m²	**6.15**
working into outlet pipes and the like	-	-	-	-	nr	**19.75**
Skirtings; two layer; dressed over tilting fillet; turned into groove						
not exceeding 200 mm girth	-	-	-	-	m	**17.69**
200 mm - 400 mm girth	-	-	-	-	m	**23.57**
Coverings to kerbs; two layer						
400 mm - 600 mm girth	-	-	-	-	m	**32.53**
Linings to gutters; three layer						
400 mm - 600 mm girth	-	-	-	-	m	**34.00**
Collars around pipes and the like; two layer; 150 high						
not exceeding 55 mm nominal size	-	-	-	-	nr	**19.73**
55 mm - 110 mm nominal size	-	-	-	-	nr	**19.73**
Accessories						
Eaves trim; extruded aluminium alloy; working felt into trim						
"Rubertrim"; type FL/G; 65 mm face	-	-	-	-	m	**14.49**
extra over for external angle	-	-	-	-	nr	**14.59**
Roof screed ventilator - aluminium alloy						
"Extr-aqua-vent" or other equal and approved - set on screed over and including dished sinking and collar	-	-	-	-	nr	**45.84**

J WATERPROOFING

Item	PC £	Labour hours	Labour £	Material £	Unit	Total rate £
J41 BUILT UP FELT ROOF COVERINGS – cont'd						
Insulation board underlays						
Vapour barrier						
reinforced; metal lined	-	-	-	-	m²	**13.86**
Rockwool; Duorock flat insulation board (0.25 U-value)						
140 mm thick	-	-	-	-	m²	**37.97**
Kingspan Thermaroof TR21 zero OPD urethene insulation board						
50 mm thick	-	-	-	-	m²	**23.34**
90 mm thick	-	-	-	-	m²	**40.20**
100 mm thick (0.25 U-value)	-	-	-	-	m²	**44.66**
Wood fibre boards; impregnated; density 220 - 350 kg/m3						
12.70 mm thick	-	-	-	-	m²	**6.28**
Tapered insulation board underlays						
Tapered PIR (Polyisocyanurate) boards; bedded in hot bitumen						
average thickness achieving 0.25W/m²K	27.00	-	-	-	m²	**55.77**
minimum thickness achieving 0.25W/m²K	33.00	-	-	-	m²	**61.92**
Tapered PIR boards; mechanically fastened						
average thickness achieving 0.25W/m²K	-	-	-	-	m²	**58.23**
minimum thickness achieving 0.25W/m²K	-	-	-	-	m²	**64.38**
Tapered Rockwool boards; bedded in hot bitumen						
average thickness achieving 0.25W/m²K	27.00	-	-	-	m²	**84.10**
minimum thickness achieving 0.25W/m²K	33.00	-	-	-	m²	**88.20**
Tapered Rockwool boards; mechanically fastened						
average thickness achieving 0.25W/m²K	-	-	-	-	m²	**86.56**
minimum thickness achieving 0.25W/m²K	-	-	-	-	m²	**92.71**
Insulation board overlays						
Dow "Roofmate SL" extruded polystyrene foam boards or other equal and approved						
75 mm thick	-	-	-	-	m²	**15.44**
140 mm thick (0.25 U-value)	-	-	-	-	m²	**26.96**
Dow "Roofmate LG" extruded polystyrene foam boards or other equal and approved						
80 mm thick	-	-	-	-	m²	**55.01**
100 mm thick	-	-	-	-	m²	**58.97**
120 mm thick	-	-	-	-	m²	**62.85**
J42 SINGLE LAYER PLASTICS ROOF COVERINGS						
"Trocal S" PVC roofing or other equal and approved						
Coverings	-	-	-	-	m²	**23.71**
Skirtings; dressed over metal upstands						
not exceeding 200 mm girth	-	-	-	-	m	**18.41**
200 mm - 400 mm girth	-	-	-	-	m	**22.64**
Coverings to kerbs						
400 mm - 600 mm girth	-	-	-	-	m	**41.44**
Collars around pipes and the like; 150 mm high						
not exceeding 55 mm nominal size	-	-	-	-	nr	**12.66**
55 mm - 110 mm nominal size	-	-	-	-	nr	**12.66**
"Trocal" metal upstands or other equal and approved						
not exceeding 200 mm girth	-	-	-	-	m	**13.42**
200 mm - 400 mm girth	-	-	-	-	m	**17.42**

J WATERPROOFING

Item	PC £	Labour hours	Labour £	Material £	Unit	Total rate £
Sarnafil polymeric waterproofing membrane; ref. S327-12EL; Sarnabar mechanically fastened system; 85mm thick Sarnaform G CFC & HCFC free (0.25 U-value) rigid urethene insulation board mechanially fastened; Sarnavap 1000E vapour control layer loose laid all laps sealed.						
Roof coverings						
Pitch not exceeding 5°; to metal decking or the like	-	-	-	-	m²	**66.14**
Sarnafil polymeric waterproofing membrane; ref. G410 - 12ELF fleece backed membrane; fully adhered system; 90mm thick Sarnaform G CFC & HCFC free (0.25 U-value) insulation board bedded in hot bitumen; BS 747 type 5U felt vapour control layer in hot bitumen; prime concrete with spirit priming solution.						
Roof coverings						
Pitch not exceeding 5°; to concrete base or the like	-	-	-	-	m²	**80.22**
Coverings to kerbs; parapet flashing; Sarnatrim 50 mm deep on face 100 mm fixing arm; standard Sarnafil detail 1.1						
not exceeding 200 mm girth	-	-	-	-	m	**43.43**
200 mm - 400 mm girth	-	-	-	-	m	**45.81**
400 mm - 600 mm girth	-	-	-	-	m	**50.88**
Eaves detail; Sarnatrmetal drip edge to gutter; standard Sarnafil detail 1.3						
not exceeding 200 mm girth	-	-	-	-	m	**23.57**
Skirtings/Upstands; skirting to brickwork with galvanised steel counter flashing to top edge; standard Sarnafil detail 2.3						
not exceeding 200 mm girth	-	-	-	-	m	**35.51**
200 mm - 400 mm girth	-	-	-	-	m	**37.89**
400 mm - 600 mm girth	-	-	-	-	m	**42.95**
Skirtings/Upstands; skirting to brickwork with Sarnametal Raglet to chase; standard Sarnafil detail 2.8						
not exceeding 200 mm girth	-	-	-	-	m	**49.26**
200 mm - 400 mm girth	-	-	-	-	m	**51.64**
400 mm - 600 mm girth	-	-	-	-	m	**56.71**
Collars around pipe standards, and the like						
50 mm diameter x 150 mm high	-	-	-	-	nr	**40.99**
100 mm diameter x 150 mm high	-	-	-	-	nr	**40.99**
Outlets and dishing to gullies						
Fix Sarnadrain pvc rainwater outlet; 110 mm diameter; weld membrane to same; fit plastic leafguard	-	-	-	-	nr	**132.03**

K LININGS/SHEATHING/DRY PARTITIONING

Item	PC £	Labour hours	Labour £	Material £	Unit	Total rate £
K10 PLASTERBOARD DRY LINING/PARTITIONS/ CEILINGS						
Linings; "Gyproc GypLyner" metal framed wall lining system; or other equal and approved; floor and ceiling channels plugged and screwed to concrete						
Tapered edge panels; joints filled with joint filler and joint tape to receive direct decoration; one layer of 12.5 mm thick Gyproc Wallboard; or other equal and approved						
height 2.10 m - 2.40 m	-	1.15	25.48	38.34	m	**63.82**
height 2.40 m - 2.70 m	-	1.26	27.93	42.58	m	**70.50**
height 2.70 m - 3.00 m	-	1.41	31.27	46.92	m	**78.19**
height 3.00 m - 3.30 m	-	1.60	35.50	51.18	m	**86.68**
height 3.30 m - 3.60 m	-	1.79	39.74	44.69	m	**84.43**
height 3.60 m - 3.90 m	-	2.07	45.96	59.70	m	**105.66**
height 3.90 m - 4.20 m	-	2.34	51.95	63.95	m	**115.90**
Linings; "Gyproc GypLyner IWL" independent walling system or other equal and approved; comprising 48 mm wide metal I stud frame; 50 mmm wide metal C stud floor and ceiling channels; plugged and screwed to concrete						
62.5 mm partition; outer skin of 12.50 mm thick tapered edge wallboard one side; joints filled with joint filler and joint tape to receive direct decoration						
height 2.10 m - 2.40 m	-	3.40	75.27	31.05	m	**106.32**
height 2.40 m - 2.70 m	-	3.95	87.65	35.93	m	**123.58**
height 2.70 m - 3.00 m	-	4.35	96.56	39.62	m	**136.18**
height 3.00 m - 3.30 m	-	5.05	112.07	43.36	m	**155.43**
height 3.30 m - 3.60 m	-	5.45	121.02	47.10	m	**168.12**
height 3.60 m - 3.90 m	-	6.00	133.10	50.92	m	**184.02**
height 3.90 m - 4.20 m	-	6.80	150.94	54.70	m	**205.64**
62.5 mm partition; outer skin of 12.50 mm thick tapered edge wallboard one side; filling cavity with "Isowool high performance slab (2405); wallboard joints filled with joint filler and joint tape to receive direct decoration						
height 2.10 m - 2.40 m	-	3.40	75.27	43.09	m	**118.36**
height 2.40 m - 2.70 m	-	3.95	87.65	49.48	m	**137.13**
height 2.70 m - 3.00 m	-	4.35	96.56	54.67	m	**151.23**
height 3.00 m - 3.30 m	-	5.05	112.07	59.91	m	**171.98**
height 3.30 m - 3.60 m	-	5.45	121.02	65.16	m	**186.18**
height 3.60 m - 3.90 m	-	6.00	133.10	70.48	m	**203.59**
height 3.90 m - 4.20 m	-	6.80	150.94	75.77	m	**226.71**

K LININGS/SHEATHING/DRY PARTITIONING

Item	PC £	Labour hours	Labour £	Material £	Unit	Total rate £
Labours and associated additional wrought softwood studwork						
Floor, wall or ceiling battens						
25 mm x 38 mm	-	0.12	2.64	0.73	m	**3.36**
Forming openings in 2400 mm high partition; 25 mm x 38 mm softwood framing						
900 mm x 2100 mm	-	0.51	11.20	4.04	nr	**15.24**
fair ends	-	0.20	4.47	0.75	m	**5.21**
angle	-	0.31	6.95	1.09	m	**8.04**
Cutting and fitting around steel joists, angles, trunking, ducting, ventilators, pipes, tubes, etc.						
over 2.00 m girth	-	0.09	1.98	-	nr	**1.98**
not exceeding 0.30 m girth	-	0.05	1.10	-	nr	**1.10**
0.30 m - 1.00 m girth	-	0.07	1.54	-	nr	**1.54**
1.00 m - 2.00 m girth	-	0.11	2.42	-	nr	**2.42**
"Gyprwall Rapid/db Plus" metal stud housing partitioning system or other equal and approved; floor and ceiling channels plugged and screwed to concrete						
75 mm partition; 43/44 mm studs and channels; one layer of 15 mm thick Gyproc SoundBloc each side; joints filled with joint filler and joint tape to receive direct decoration						
height 2.10 m - 2.40 m; studs at 900 mm centres	-	2.97	66.10	37.96	m	**104.06**
height 2.10 m - 2.40 m; studs at 900 mm centres; with 25 mm Isowool 1200 insulation within the stud cavity	-	2.97	66.10	40.68	m	**106.78**
height 2.10 m - 2.40 m; studs at 450 mm centres	-	4.07	90.26	41.20	m	**131.46**
height 2.10 m - 2.40 m; studs at 450 mm centres; with 25 mm Isowool 1200 insulation within the stud cavity	-	4.07	90.26	43.91	m	**134.18**
height 2.40 m - 2.70 m; studs at 450 mm centres	-	4.48	99.35	45.88	m	**145.23**
height 2.40 m - 2.70 m; studs at 450 mm centres; with 25 mm Isowool 1200 insulation within the stud cavity	-	4.48	99.35	48.59	m	**147.94**
102 mm partition; 70/72 mm studs and channels; one layer of 15 mm thick Gyproc SoundBloc each side; joints filled with joint filler and joint tape to receive direct decoration						
height 2.10 m - 2.40 m; studs at 900 mm centres	-	3.50	77.74	41.61	m	**119.35**
height 2.10 m - 2.40 m; studs at 900 mm centres; with 25 mm Isowool 1200 insulation within the stud cavity	-	3.50	77.74	44.32	m	**122.06**
height 2.10 m - 2.40 m; studs at 450 mm centres	-	4.42	97.95	46.95	m	**144.90**
height 2.10 m - 2.40 m; studs at 450 mm centres; with 25 mm Isowool 1200 insulation within the stud cavity	-	4.42	97.95	49.66	m	**147.61**
height 2.40 m - 2.70 m; studs at 900 mm centres	-	3.65	81.12	45.92	m	**127.04**
height 2.40 m - 2.70 m; studs at 900 mm centres; with 25 mm Isowool 1200 insulation within the stud cavity	-	3.65	81.12	48.63	m	**129.75**
height 2.40 m - 2.70 m; studs at 450 mm centres	-	4.75	105.29	52.23	m	**157.51**
height 2.40 m - 2.70 m; studs at 450 mm centres; with 25 mm Isowool 1200 insulation within the stud cavity	-	4.75	105.29	54.94	m	**160.22**

K LININGS/SHEATHING/DRY PARTITIONING

Item	PC £	Labour hours	Labour £	Material £	Unit	Total rate £
K10 PLASTERBOARD DRY LINING/PARTITIONS/ CEILINGS – cont'd						
"Gyproc" metal stud proprietary partitions or other equal and approved; comprising 48 mm wide metal stud frame; 50 mm wide floor channel plugged and screwed to concrete through 38 mm x 48 mm tanalised softwood sole plate						
Tapered edge panels; joints filled with joint filler and joint tape to receive direct decoration; 80 mm thick partition; one hour; one layer of 15 mm thick "Fireline" board each side						
height 2.10 m - 2.40 m	-	4.30	95.04	27.40	m	**122.44**
height 2.40 m - 2.70 m	-	4.90	108.48	31.69	m	**140.17**
height 2.70 m - 3.00 m	-	5.50	121.82	34.77	m	**156.60**
height 3.00 m - 3.30 m	-	6.34	140.41	38.21	m	**178.62**
height 3.30 m - 3.60 m	-	6.98	154.63	41.09	m	**195.72**
height 3.60 m - 3.90 m	-	8.33	184.41	44.27	m	**228.68**
height 3.90 m - 4.20 m	-	8.93	197.72	47.45	m	**245.18**
angles	-	0.20	4.47	1.88	m	**6.35**
T-junctions	-	0.10	2.20	-	m	**2.20**
fair ends	-	0.20	4.47	0.64	m	**5.10**
Tapered edge panels; joints filled with joint filler and joint tape to receive direct decoration; 100 mm thick partition; two hour; two layers of 12.50 mm thick "Fireline" board both sides						
height 2.10 m - 2.40 m	-	5.32	117.45	40.91	m	**158.36**
height 2.40 m - 2.70 m	-	6.06	133.96	46.90	m	**180.86**
height 2.70 m - 3.00 m	-	6.80	150.38	51.67	m	**202.05**
height 3.00 m - 3.30 m	-	6.75	149.41	56.80	m	**206.21**
height 3.30 m - 3.60 m	-	8.51	188.24	61.36	m	**249.59**
height 3.60 m - 3.90 m	-	8.33	184.41	66.27	m	**250.68**
height 3.90 m - 4.20 m	-	10.73	237.26	71.10	m	**308.36**
angles	-	0.31	6.88	2.04	m	**8.92**
T-junctions	-	0.10	2.20	-	m	**2.20**
fair ends	-	0.31	6.88	0.79	m	**7.67**
Gypsum plasterboard; BS EN 520; plain grade tapered edge wallboard; fixing on dabs or with nails; joints left open to receive "Artex" finish or other equal and approved; to softwood base						
9.50 mm board to ceilings						
over 300 mm wide	-	0.26	5.71	2.52	m²	**8.23**
9.50 mm board to beams						
girth not exceeding 600 mm	-	0.31	6.81	1.53	m²	**8.34**
girth 600 mm - 1200 mm	-	0.41	9.01	3.05	m²	**12.05**
12.50 mm board to ceilings						
over 300 mm wide	-	0.33	7.25	2.62	m²	**9.86**
12.50 mm board to beams						
girth not exceeding 600 mm	-	0.31	6.81	1.61	m²	**8.42**
girth 600 mm - 1200 mm	-	0.41	9.01	3.14	m²	**12.15**

K LININGS/SHEATHING/DRY PARTITIONING

Item	PC £	Labour hours	Labour £	Material £	Unit	Total rate £
Gypsum plasterboard to BS EN 520; fixing on dabs or with nails; joints filled with joint filler and joint tape to receive direct decoration; to softwood base						
Plain grade tapered edge wallboard						
9.50 mm board to walls						
wall height 2.40 m - 2.70 m	-	1.02	22.83	24.64	m	**47.47**
wall height 2.70 m - 3.00 m	-	1.16	25.97	27.39	m	**53.36**
wall height 3.00 m - 3.30 m	-	1.34	30.02	30.13	m	**60.16**
wall height 3.30 m - 3.60 m	-	1.53	34.26	32.94	m	**67.20**
9.50 mm board to reveals and soffits of openings and recesses						
not exceeding 300 mm wide	-	0.20	4.47	3.53	m	**7.99**
300 mm - 600 mm wide	-	0.41	9.15	6.06	m	**15.21**
9.50 mm board to faces of columns - 4 nr						
not exceeding 600 mm total girth	-	0.52	11.61	7.11	m	**18.71**
600 mm - 1200 mm total girth	-	1.02	22.80	12.18	m	**34.98**
1200 mm - 1800 mm total girth	-	1.32	29.57	17.25	m	**46.82**
9.50 mm board to ceilings						
over 300 mm wide	-	0.43	9.59	9.13	m²	**18.72**
9.50 mm board to faces of beams - 3 nr						
not exceeding 600 mm total girth	-	0.61	13.61	7.06	m	**20.67**
600 mm - 1200 mm total girth	-	1.12	25.03	12.13	m	**37.16**
1200 mm - 1800 mm total girth	-	1.43	32.03	17.20	m	**49.23**
12.50 mm board to walls						
wall height 2.40 m - 2.70 m	-	1.20	26.85	26.40	m	**53.25**
wall height 2.70 m - 3.00 m	-	1.34	29.99	29.33	m	**59.32**
wall height 3.00 m - 3.30 m	-	1.48	33.13	32.27	m	**65.41**
wall height 3.30 m - 3.60 m	-	1.76	39.38	35.27	m	**74.65**
12.50 mm board to reveals and soffits of openings and recesses						
not exceeding 300 mm wide	-	0.20	4.47	3.75	m	**8.22**
300 mm - 600 mm wide	-	0.41	9.15	6.47	m	**15.62**
12.50 mm board to faces of columns - 4 nr						
not exceeding 600 mm total girth	-	0.52	11.61	7.56	m	**19.16**
600 mm - 1200 mm total girth	-	1.02	22.80	13.00	m	**35.81**
1200 mm - 1800 mm total girth	-	1.32	29.57	18.45	m	**48.02**
12.50 mm board to ceilings						
over 300 mm wide	-	0.44	9.81	9.78	m²	**19.59**
12.50 mm board to faces of beams - 3 nr						
not exceeding 600 mm total girth	-	0.61	13.61	7.48	m	**21.10**
600 mm - 1200 mm total girth	-	1.12	25.03	12.93	m	**37.96**
1200 mm - 1800 mm total girth	-	1.43	32.03	18.38	m	**50.40**
external angle; with joint tape bedded and covered with "Jointex" or other equal and approved	-	0.12	2.71	0.47	m	**3.19**
Tapered edge wallboard TEN						
12.50 mm board to walls						
wall height 2.40 m - 2.70 m	-	1.20	26.85	9.99	m	**36.84**
wall height 2.70 m - 3.00 m	-	1.34	29.99	11.10	m	**41.09**
wall height 3.00 m - 3.30 m	-	1.48	33.13	12.22	m	**45.35**
wall height 3.30 m - 3.60 m	-	1.76	39.38	13.39	m	**52.77**
12.50 mm board to reveals and soffits of openings and recesses						
not exceeding 300 mm wide	-	0.20	4.47	1.93	m	**6.40**
300 mm - 600 mm wide	-	0.41	9.15	2.82	m	**11.97**
12.50 mm board to faces of columns - 4 nr						
not exceeding 600 mm total girth	-	0.52	11.61	3.91	m	**15.52**
600 mm - 1200 mm total girth	-	1.02	22.80	5.71	m	**28.52**
1200 mm - 1800 mm total girth	-	1.32	29.57	7.51	m	**37.08**

K LININGS/SHEATHING/DRY PARTITIONING

Item	PC £	Labour hours	Labour £	Material £	Unit	Total rate £
K10 PLASTERBOARD DRY LINING/PARTITIONS/ CEILINGS – cont'd						
Tapered edge wallboard TEN – cont'd						
12.50 mm board to ceilings						
over 300 mm wide	-	0.44	9.81	3.70	m²	**13.51**
12.50 mm board to faces of beams - 3 nr						
not exceeding 600 mm total girth	-	0.61	13.61	3.84	m	**17.45**
600 mm - 1200 mm total girth	-	1.12	25.03	5.64	m	**30.67**
1200 mm - 1800 mm total girth	-	1.43	32.03	7.44	m	**39.47**
external angle; with joint tape bedded and covered with "Jointex" or other equal and approved	-	0.12	2.71	0.47	m	**3.19**
Tapered edge plank						
19 mm plank to walls						
wall height 2.40 m - 2.70 m	-	1.11	24.81	15.87	m	**40.68**
wall height 2.70 m - 3.00 m	-	1.30	29.05	17.64	m	**46.68**
wall height 3.00 m - 3.30 m	-	1.43	32.00	19.41	m	**51.41**
wall height 3.30 m - 3.60 m	-	1.71	38.22	21.23	m	**59.45**
19 mm plank to reveals and soffits of openings and recesses						
not exceeding 300 mm wide	-	0.22	4.90	2.58	m	**7.49**
300 mm - 600 mm wide	-	0.46	10.25	4.13	m	**14.37**
19 mm plank to faces of columns - 4 nr						
not exceeding 600 mm total girth	-	0.56	12.48	5.22	m	**17.70**
600 mm - 1200 mm total girth	-	1.07	23.90	8.33	m	**32.23**
1200 mm - 1800 mm total girth	-	1.38	30.89	11.43	m	**42.32**
19 mm plank to ceilings						
over 300 mm wide	-	0.47	10.47	5.88	m²	**16.35**
19 mm plank to faces of beams - 3 nr						
not exceeding 600 mm total girth	-	0.67	14.93	5.14	m	**20.08**
600 mm - 1200 mm total girth	-	1.17	26.13	8.25	m	**34.38**
1200 mm - 1800 mm total girth	-	1.48	33.12	11.36	m	**44.48**
Thermal Board						
27 mm board to walls						
wall height 2.40 m - 2.70 m	-	1.16	25.91	11.74	m	**37.65**
wall height 2.70 m - 3.00 m	-	1.34	29.92	13.05	m	**42.98**
wall height 3.00 m - 3.30 m	-	1.48	33.07	14.36	m	**47.43**
wall height 3.30 m - 3.60 m	-	1.80	40.19	15.73	m	**55.92**
27 mm board to reveals and soffits of openings and recesses						
not exceeding 300 mm wide	-	0.23	5.12	2.13	m	**7.25**
300 mm - 600 mm wide	-	0.47	10.47	3.21	m	**13.68**
27 mm board to faces of columns - 4 nr						
not exceeding 600 mm total girth	-	0.57	12.70	4.30	m	**17.00**
600 mm - 1200 mm total girth	-	1.13	25.22	6.49	m	**31.71**
1200 mm - 1800 mm total girth	-	1.42	31.77	8.68	m	**40.45**
27 mm board to ceilings						
over 300 mm wide	-	0.51	11.35	4.35	m²	**15.70**
27 mm board to faces of beams - 3 nr						
not exceeding 600 mm total girth	-	0.62	13.80	4.23	m	**18.03**
600 mm - 1200 mm total girth	-	1.17	26.10	6.42	m	**32.52**
1200 mm - 1800 mm total girth	-	1.58	35.32	8.61	m	**43.93**
50 mm board to walls						
wall height 2.40 m - 2.70 m	-	1.16	25.91	13.78	m	**39.69**
wall height 2.70 m - 3.00 m	-	1.39	31.02	15.34	m	**46.36**
wall height 3.00 m - 3.30 m	-	1.62	36.14	16.87	m	**53.01**
wall height 3.30 m - 3.60 m	-	1.90	42.39	18.47	m	**60.86**

K LININGS/SHEATHING/DRY PARTITIONING

Item	PC £	Labour hours	Labour £	Material £	Unit	Total rate £
50 mm board to reveals and soffits of openings and recesses						
not exceeding 300 mm wide	-	0.26	5.78	2.36	m	**8.14**
300 mm - 600 mm wide	-	0.51	11.35	3.68	m	**15.02**
50 mm board to faces of columns - 4 nr						
not exceeding 600 mm total girth	-	0.62	13.80	4.82	m	**18.62**
600 mm - 1200 mm total girth	-	1.23	27.42	7.45	m	**34.87**
1200 mm - 1800 mm total girth	-	1.58	35.28	10.09	m	**45.37**
50 mm board to ceilings						
over 300 mm wide	-	0.54	12.01	5.11	m²	**17.12**
50 mm board to faces of beams - 3 nr						
not exceeding 600 mm total girth	-	0.65	14.46	4.87	m	**19.33**
600 mm - 1200 mm total girth	-	1.30	28.95	7.55	m	**36.51**
1200 mm - 1800 mm total girth	-	1.74	38.84	10.24	m	**49.07**
White plastic faced gypsum plasterboard to BS EN 520; industrial grade square edge wallboard; fixing on dabs or with screws; butt joints; to softwood base						
12.50 mm board to walls						
wall height 2.40 m - 2.70 m	-	0.79	17.35	9.78	m	**27.13**
wall height 2.70 m - 3.00 m	-	0.93	20.43	10.86	m	**31.29**
wall height 3.00 m - 3.30 m	-	1.06	23.28	11.94	m	**35.22**
wall height 3.30 m - 3.60 m	-	1.20	26.36	13.02	m	**39.38**
12.50 mm board to reveals and soffits of openings and recesses						
not exceeding 300 mm wide	-	0.17	3.73	1.10	m	**4.84**
300 mm - 600 mm wide	-	0.32	7.03	2.17	m	**9.20**
12.50 mm board to faces of columns - 4 nr						
not exceeding 600 mm total girth	-	0.43	9.45	2.28	m	**11.72**
600 mm - 1200 mm total girth	-	0.85	18.67	4.49	m	**23.16**
1200 mm - 1800 mm total girth	-	1.12	24.60	6.67	m	**31.27**
Plasterboard jointing system; filling joint with jointing compounds						
To ceilings						
to suit 9.50 mm or 12.50 mm thick boards	-	0.10	2.20	2.66	m	**4.86**
Angle trim; plasterboard edge support system						
To ceilings						
to suit 9.50 mm or 12.50 mm thick boards	-	0.10	2.20	2.48	m	**4.68**
Gypsum SoundBloc plasterboard with higehr density core; fixing on dabs or with nails; joints filled with joint filler and joint tape to receive direct decoration; to softwood base						
Tapered edge board						
12.50 mm board to walls						
wall height 2.40 m - 2.70 m	-	1.06	23.72	13.28	m	**36.99**
wall height 2.70 m - 3.00 m	-	1.22	27.30	14.77	m	**42.07**
wall height 3.00 m - 3.30 m	-	1.36	30.45	16.24	m	**46.69**
wall height 3.30 m - 3.60 m	-	1.57	35.13	17.72	m	**52.86**
12.50 mm board to ceilings						
over 300 mm wide	-	0.45	10.03	4.92	m²	**14.95**
15.00 mm board to walls						
wall height 2.40 m - 2.70 m	-	1.10	24.59	15.14	m	**39.73**
wall height 2.70 m - 3.00 m	-	1.25	27.96	16.83	m	**44.79**
wall height 3.00 m - 3.30 m	-	1.40	31.33	18.51	m	**49.84**
wall height 3.30 m - 3.60 m	-	1.61	36.01	20.20	m	**56.21**

K LININGS/SHEATHING/DRY PARTITIONING

Item	PC £	Labour hours	Labour £	Material £	Unit	Total rate £
K10 PLASTERBOARD DRY LINING/PARTITIONS/ CEILINGS – cont'd						
Gypsum SoundBloc plasterboard – cont'd						
15.00 mm board to reveals and soffits of openings and recesses						
not exceeding 300 mm wide	-	0.22	4.90	1.81	m	**6.72**
300 mm - 600 mm wide	-	0.42	9.37	3.07	m	**12.44**
15.00 mm board to ceilings						
over 300 mm wide	-	0.47	10.47	5.67	m²	**16.14**
Two layers of gypsum plasterboard to BS 1230; plain grade square and tapered edge wallboard; fixing on dabs or with nails; joints filled with joint filler and joint tape; top layer to receive direct decoration; to softwood base						
19 mm two layer board to walls						
wall height 2.40 m - 2.70 m	-	1.43	31.84	31.60	m	**63.43**
wall height 2.70 m - 3.00 m	-	1.62	36.08	35.13	m	**71.20**
wall height 3.00 m - 3.30 m	-	1.85	41.23	38.65	m	**79.87**
wall height 3.30 m - 3.60 m	-	2.13	47.44	42.22	m	**89.66**
19 mm two layer board to reveals and soffits of openings and recesses						
not exceeding 300 mm wide	-	0.31	6.88	4.37	m	**11.25**
300 mm - 600 mm wide	-	0.61	13.54	7.65	m	**21.19**
19 mm two layer board to faces of columns - 4 nr						
not exceeding 600 mm total girth	-	0.77	17.10	8.84	m	**25.93**
600 mm - 1200 mm total girth	-	1.38	30.71	15.42	m	**46.13**
1200 mm - 1800 mm total girth	-	1.83	40.77	22.00	m	**62.77**
25 mm two layer board to walls						
wall height 2.40 m - 2.70 m	-	1.53	34.03	33.35	m	**67.38**
wall height 2.70 m - 3.00 m	-	1.71	38.05	37.07	m	**75.12**
wall height 3.00 m - 3.30 m	-	1.94	43.20	40.79	m	**83.99**
wall height 3.30 m - 3.60 m	-	2.27	50.52	44.55	m	**95.06**
25 mm two layer board to reveals and soffits of openings and recesses						
not exceeding 300 mm wide	-	0.31	6.88	4.59	m	**11.48**
300 mm - 600 mm wide	-	0.61	13.54	8.05	m	**21.59**
25 mm two layer board to faces of columns - 4 nr						
not exceeding 600 mm total girth	-	0.77	17.10	9.29	m	**26.38**
600 mm - 1200 mm total girth	-	1.38	30.71	16.24	m	**46.96**
1200 mm - 1800 mm total girth	-	1.83	40.77	23.20	m	**63.98**
Gyproc Dri-Wall dry lining system or other equal and approved; plain grade tapered edge wallboard; fixed to walls with adhesive; joints filled with joint filler and joint tape; to receive direct decoration						
9.50 mm board to walls						
wall height 2.40 m - 2.70 m	-	1.20	26.78	27.66	m	**54.45**
wall height 2.70 m - 3.00 m	-	1.39	31.02	30.72	m	**61.74**
wall height 3.00 m - 3.30 m	-	1.62	36.17	33.77	m	**69.95**
wall height 3.30 m - 3.60 m	-	1.85	41.29	36.89	m	**78.18**
9.50 mm board to reveals and soffits of openings and recesses						
not exceeding 300 mm wide	-	0.26	5.78	3.81	m	**9.59**
300 mm - 600 mm wide	-	0.51	11.35	6.67	m	**18.02**

K LININGS/SHEATHING/DRY PARTITIONING

Item	PC £	Labour hours	Labour £	Material £	Unit	Total rate £
9.50 mm board to faces of columns - 4 nr						
not exceeding 600 mm total girth	-	0.65	14.46	7.62	m	**22.08**
600 mm - 1200 mm total girth	-	1.26	28.08	13.58	m	**41.65**
1200 mm - 1800 mm total girth	-	1.58	35.28	19.07	m	**54.35**
Angle; with joint tape bedded and covered with "Jointex" or other equal and approved						
internal	-	0.06	1.36	0.47	m	**1.83**
external	-	0.12	2.71	0.47	m	**3.19**
Gyproc Dri-Wall M/F dry lining system; mild steel furrings fixed to walls with adhesive; tapered edge wallboard screwed to furrings; joints filled with joint filler and joint tape						
12.50 mm board to walls						
wall height 2.40 m - 2.70 m	-	1.62	36.01	33.31	m	**69.32**
wall height 2.70 m - 3.00 m	-	1.85	41.13	36.99	m	**78.12**
wall height 3.00 m - 3.30 m	-	2.13	47.38	40.70	m	**88.07**
wall height 3.30 m - 3.60 m	-	2.45	54.47	44.46	m	**98.93**
12.50 mm board to reveals and soffits of openings and recesses						
not exceeding 300 mm wide	-	0.26	5.78	3.54	m	**9.33**
300 mm - 600 mm wide	-	0.51	11.35	6.14	m	**17.49**
Gypsum cladding; Glasroc "Firecase s" board; fixed with adhesive; joints pointed in adhesive						
25 mm thick column linings, faces - 4; 2 hour fire protection rating						
not exceeding 600 mm girth	-	0.33	7.25	21.49	m	**28.74**
600 mm - 1200 mm girth	-	0.50	10.98	32.86	m	**43.85**
1200 mm - 1800 mm girth	-	0.66	14.50	44.24	m	**58.74**
30 mm thick beam linings, faces - 3; 2 hour fire protection rating						
not exceeding 600 mm girth	-	0.65	14.28	20.75	m	**35.03**
600 mm - 1200 mm girth	-	1.00	21.97	33.60	m	**55.57**
1200 mm - 1800 mm girth	-	1.30	28.56	46.45	m	**75.01**
Vermiculite gypsum cladding; "Vermiculux" board or other equal and approved; fixed with adhesive; joints pointed in adhesive						
25 mm thick column linings, faces - 4; 2 hour fire protection rating						
not exceeding 600 mm girth	-	0.33	7.25	18.43	m	**25.68**
600 mm - 1200 mm girth	-	0.50	10.98	36.48	m	**47.46**
1200 mm - 1800 mm girth	-	0.66	14.50	54.52	m	**69.02**
30 mm thick beam linings, faces - 3; 2 hour fire protection rating						
not exceeding 600 mm girth	-	0.65	14.28	24.20	m	**38.48**
600 mm - 1200 mm girth	-	1.00	21.97	48.02	m	**69.98**
1200 mm - 1800 mm girth	-	1.30	28.56	71.83	m	**100.39**
55 mm thick column linings, faces - 4 ; 4 hour fire protection rating						
not exceeding 600 mm girth	-	0.92	20.21	51.41	m	**71.62**
600 mm - 1200 mm girth	-	1.12	24.60	102.42	m	**127.03**
1200 mm - 1800 mm girth	-	1.53	33.61	153.44	m	**187.05**
60 mm thick beam linings, faces - 3; 4 hour fire protection rating						
not exceeding 600 mm girth	-	0.77	16.91	55.54	m	**72.46**
600 mm - 1200 mm girth	-	0.92	20.21	110.70	m	**130.91**
1200 mm - 1800 mm girth	-	1.12	24.60	165.85	m	**190.46**

K LININGS/SHEATHING/DRY PARTITIONING

Item	PC £	Labour hours	Labour £	Material £	Unit	Total rate £
K10 PLASTERBOARD DRY LINING/PARTITIONS/ CEILINGS – cont'd						
Vermiculite gypsum cladding – cont'd						
Add to the above for						
plus 3% for work 3.50 m - 5.00 m high						
plus 6% for work 5.00 m - 6.50 m high						
plus 12% for work 6.50 m - 8.00 m high						
plus 18% for work over 8.00 m high						
Cutting and fitting around steel joints, angles, trunking, ducting, ventilators, pipes, tubes, etc.						
over 2.00 m girth	-	0.45	9.88	-	m	**9.88**
not exceeding 0.30 m girth	-	0.31	6.81	-	nr	**6.81**
0.30 m - 1.00 m girth	-	0.41	9.01	-	nr	**9.01**
1.00 m - 2.00 m girth	-	0.56	12.30	-	nr	**12.30**
K11 RIGID SHEET FLOORING/SHEATHING/ LININGS/CASINGS						
Blockboard (Birch faced)						
Lining to walls 18 mm thick						
over 300 mm wide	5.78	0.51	11.20	7.32	m²	**18.52**
not exceeding 300 mm wide	-	0.32	7.03	2.22	m	**9.25**
holes for pipes and the like	-	0.05	1.10	-	nr	**1.10**
Chipboard (plain)						
Lining to walls 12 mm thick						
over 300 mm wide	2.29	0.39	8.57	3.09	m²	**11.66**
not exceeding 300 mm wide	-	0.22	4.83	0.95	m	**5.78**
holes for pipes and the like	-	0.02	0.44	-	nr	**0.44**
Lining to walls 15 mm thick						
over 300 mm wide	2.69	0.41	9.01	3.57	m²	**12.58**
not exceeding 300 mm wide	-	0.24	5.27	1.09	m	**6.37**
holes for pipes and the like	-	0.03	0.66	-	nr	**0.66**
Two-sided 15 mm thick pipe casing; to softwood framing (not included)						
300 mm girth	-	0.60	13.18	1.21	m	**14.39**
600 mm girth	-	0.69	15.16	2.19	m	**17.35**
Three-sided 15 mm thick pipe casing; to softwood framing (not included)						
450 mm girth	-	1.30	28.56	1.82	m	**30.37**
900 mm girth	-	1.53	33.61	3.32	m	**36.93**
Extra for 400 x 400 removable access panel; brass cups and screws; additional framing	-	1.02	22.40	1.48	nr	**23.88**
Lining to walls 18 mm thick						
over 300 mm wide	3.22	0.43	9.45	4.30	m²	**13.74**
not exceeding 300 mm wide	-	0.28	6.15	1.29	m	**7.44**
holes for pipes and the like	-	0.05	1.10	-	nr	**1.10**
Fire-retardant chipboard; Antivlam or other equal and approved; Class 1 spread of flame						
Lining to walls 12 mm thick						
over 300 mm wide	-	0.39	8.57	10.11	m²	**18.68**
not exceeding 300 mm wide	-	0.22	4.83	3.06	m	**7.89**
holes for pipes and the like	-	0.02	0.44	-	nr	**0.44**
Lining to walls 18 mm thick						
over 300 mm wide	-	0.43	9.45	13.11	m²	**22.55**
not exceeding 300 mm wide	-	0.28	6.15	3.96	m	**10.11**
holes for pipes and the like	-	0.05	1.10	-	nr	**1.10**

K LININGS/SHEATHING/DRY PARTITIONING

Item	PC £	Labour hours	Labour £	Material £	Unit	Total rate £
Lining to walls 22 mm thick						
over 300 mm wide	-	0.44	9.66	16.92	m²	**26.59**
not exceeding 300 mm wide	-	0.31	6.81	5.10	m	**11.91**
holes for pipes and the like	-	0.06	1.32	-	nr	**1.32**
Chipboard Melamine faced; white matt finish; laminated masking strips						
Lining to walls 15 mm thick						
over 300 mm wide	3.37	1.06	23.28	4.70	m²	**27.98**
not exceeding 300 mm wide	-	0.69	15.16	1.54	m	**16.70**
holes for pipes and the like	-	0.07	1.54	-	nr	**1.54**
Chipboard boarding and flooring						
Boarding to floors; butt joints						
18 mm thick	4.20	0.31	6.81	5.49	m²	**12.30**
Boarding to floors; tongued and grooved joints						
18 mm thick	4.10	0.32	7.03	5.36	m²	**12.39**
22 mm thick	5.27	0.36	7.91	6.79	m²	**14.70**
Acoustic Chipboard flooring						
Boarding to floors; tongued and grooved joints						
chipboard on blue bat bearers	-	-	-	-	m²	**21.85**
chipboard on New Era levelling system	-	-	-	-	m²	**29.94**
Laminated engineered board flooring; 180 or 240 mm face widths; with 6 mm wear surface down to tongue; pre-finished laquered, oiled or untreated.						
Boarding to floors; micro bevel or square edge						
Country laquered; on 10 mm Pro Foam	-	-	-	-	m²	**51.92**
Rustic laquered; on 10 mm Pro Foam	-	-	-	-	m²	**54.34**
Plywood flooring						
Boarding to floors; tongued and grooved joints						
18 mm thick	9.43	0.44	9.66	11.83	m²	**21.50**
22 mm thick	7.57	0.48	10.54	9.57	m²	**20.12**
Plywood; external quality; 18 mm thick						
Boarding to roofs; butt joints						
flat to falls	13.51	0.41	9.01	16.78	m²	**25.79**
sloping	13.51	0.43	9.45	16.78	m²	**26.23**
vertical	13.51	0.58	12.74	16.78	m²	**29.52**
Plywood; external quality; 12 mm thick						
Boarding to roofs; butt joints						
flat to falls	9.01	0.41	9.01	11.33	m²	**20.33**
sloping	9.01	0.43	9.45	11.33	m²	**20.77**
vertical	9.01	0.58	12.74	11.33	m²	**24.07**
Glazed hardboard to BS EN 622; on and including 38 mm x 38 mm sawn softwood framing						
3.20 mm thick panel						
to side of bath	-	1.82	39.98	7.31	nr	**47.29**
to end of bath	-	0.71	15.60	2.27	nr	**17.87**
Insulation board to BS EN 622						
Lining to walls 12 mm thick						
over 300 mm wide	1.81	0.24	5.27	2.51	m²	**7.78**
not exceeding 300 mm wide	-	0.14	3.08	0.78	m	**3.85**
holes for pipes and the like	-	0.01	0.22	-	nr	**0.22**

K LININGS/SHEATHING/DRY PARTITIONING

Item	PC £	Labour hours	Labour £	Material £	Unit	Total rate £
K11 RIGID SHEET FLOORING/SHEATHING/ LININGS/CASINGS – cont'd						
Non-asbestos board; "Masterboard" or other equal and approved; sanded finish						
Lining to walls 6 mm thick						
over 300 mm wide	7.74	0.33	7.25	9.63	m²	**16.88**
not exceeding 300 mm wide	-	0.20	4.39	2.90	m	**7.29**
Lining to ceilings 6 mm thick						
over 300 mm wide	-	0.44	9.66	9.63	m²	**19.29**
not exceeding 300 mm wide	-	0.28	6.15	2.90	m	**9.05**
holes for pipes and the like	-	0.02	0.44	-	nr	**0.44**
Lining to walls 9 mm thick						
over 300 mm wide	16.37	0.37	8.13	20.09	m²	**28.22**
not exceeding 300 mm wide	-	0.20	4.39	6.04	m	**10.43**
Lining to ceilings 9 mm thick						
over 300 mm wide	-	0.45	9.88	20.09	m²	**29.98**
not exceeding 300 mm wide	-	0.30	6.59	6.04	m	**12.63**
holes for pipes and the like	-	0.03	0.66	-	nr	**0.66**
Non-asbestos board; "Supalux" or other equal and approved; sanded finish						
Lining to walls 6 mm thick						
over 300 mm wide	12.82	0.33	7.25	15.79	m²	**23.04**
not exceeding 300 mm wide	-	0.20	4.39	4.75	m	**9.14**
Lining to ceilings 6 mm thick						
over 300 mm wide	-	0.44	9.66	15.79	m²	**25.46**
not exceeding 300 mm wide	-	0.28	6.15	4.75	m	**10.90**
holes for pipes and the like	-	0.02	0.44	-	nr	**0.44**
Lining to walls 9 mm thick						
over 300 mm wide	19.07	0.37	8.13	23.38	m²	**31.50**
not exceeding 300 mm wide	-	0.20	4.39	7.02	m	**11.41**
Lining to ceilings 9 mm thick						
over 300 mm wide	-	0.45	9.88	23.38	m²	**33.26**
not exceeding 300 mm wide	-	0.30	6.59	7.02	m	**13.61**
holes for pipes and the like	-	0.03	0.66	-	nr	**0.66**
Lining to walls 12 mm thick						
over 300 mm wide	25.26	0.41	9.01	30.89	m²	**39.89**
not exceeding 300 mm wide	-	0.24	5.27	9.27	m	**14.55**
Lining to ceilings 12 mm thick						
over 300 mm wide	-	0.54	11.86	30.89	m²	**42.75**
not exceeding 300 mm wide	-	0.32	7.03	9.27	m	**16.30**
holes for pipes and the like	-	0.05	1.10	-	nr	**1.10**
Non-asbestos board; "Monolux 40" or other equal and approved; 6 mm x 50 mm "Supalux" cover fillets or other equal and approved one side						
Lining to walls 19 mm thick						
over 300 mm wide	51.17	0.71	15.60	64.74	m²	**80.34**
not exceeding 300 mm wide	-	0.51	11.20	21.72	m	**32.92**
Lining to walls 25 mm thick						
over 300 mm wide	61.34	0.77	16.91	77.09	m²	**94.00**
not exceeding 300 mm wide	-	0.54	11.86	25.42	m	**37.28**

K LININGS/SHEATHING/DRY PARTITIONING

Item	PC £	Labour hours	Labour £	Material £	Unit	Total rate £
Plywood (Eastern European); internal quality						
Lining to walls 4 mm thick						
over 300 mm wide	3.06	0.38	8.35	4.02	m²	**12.37**
not exceeding 300 mm wide	-	0.24	5.27	1.23	m	**6.50**
Lining to ceilings 4 mm thick						
over 300 mm wide	-	0.51	11.20	4.02	m²	**15.22**
not exceeding 300 mm wide	-	0.32	7.03	1.23	m	**8.26**
holes for pipes and the like	-	0.02	0.44	-	nr	**0.44**
Lining to walls 6 mm thick						
over 300 mm wide	3.99	0.41	9.01	5.15	m²	**14.16**
not exceeding 300 mm wide	-	0.27	5.93	1.57	m	**7.50**
Lining to ceilings 6 mm thick						
over 300 mm wide	-	0.54	11.86	5.15	m²	**17.01**
not exceeding 300 mm wide	-	0.36	7.91	1.57	m	**9.48**
holes for pipes and the like	-	0.02	0.44	-	nr	**0.44**
Two-sided 6 mm thick pipe casings; to softwood framing (not included)						
300 mm girth	-	0.83	18.23	1.69	m	**19.92**
600 mm girth	-	1.02	22.40	3.14	m	**25.54**
Three-sided 6 mm thick pipe casing; to softwood framing (not included)						
450 mm girth	-	1.16	25.48	2.53	m	**28.01**
900 mm girth	-	1.39	30.53	4.75	m	**35.28**
Lining to walls 12 mm thick						
over 300 mm wide	6.59	0.47	10.32	8.31	m²	**18.63**
not exceeding 300 mm wide	-	0.31	6.81	2.52	m	**9.33**
Lining to ceilings 12 mm thick						
over 300 mm wide	-	0.62	13.62	8.31	m²	**21.93**
not exceeding 300 mm wide	-	0.41	9.01	2.52	m	**11.52**
holes for pipes and the like	-	0.03	0.66	-	nr	**0.66**
Lining to walls 18 mm thick						
over 300 mm wide	9.64	0.51	11.20	12.01	m²	**23.21**
not exceeding 300 mm wide	-	0.33	7.25	3.63	m	**10.87**
Lining to ceilings 18 mm thick						
over 300 mm wide	-	0.66	14.50	12.01	m²	**26.51**
not exceeding 300 mm wide	-	0.43	9.45	3.63	m	**13.07**
holes for pipes and the like	-	0.03	0.66	-	nr	**0.66**
Plywood (Eastern European); external quality						
Lining to walls 4 mm thick						
over 300 mm wide	4.40	0.38	8.35	5.65	m²	**14.00**
not exceeding 300 mm wide	-	0.24	5.27	1.72	m	**6.99**
Lining to ceilings 4 mm thick						
over 300 mm wide	-	0.51	11.20	5.65	m²	**16.85**
not exceeding 300 mm wide	-	0.32	7.03	1.72	m	**8.75**
holes for pipes and the like	-	0.02	0.44	-	nr	**0.44**
Lining to walls 6.5 mm thick						
over 300 mm wide	4.40	0.41	9.01	5.65	m²	**14.65**
not exceeding 300 mm wide	-	0.27	5.93	1.72	m	**7.65**
Lining to ceilings 6.5 mm thick						
over 300 mm wide	-	0.54	11.86	5.65	m²	**17.51**
not exceeding 300 mm wide	-	0.36	7.91	1.72	m	**9.62**
holes for pipes and the like	-	0.02	0.44	-	nr	**0.44**

K LININGS/SHEATHING/DRY PARTITIONING

Item	PC £	Labour hours	Labour £	Material £	Unit	Total rate £
K11 RIGID SHEET FLOORING/SHEATHING/ LININGS/CASINGS – cont'd						
Plywood (Eastern European); external quality – cont'd						
Two-sided 6.5 mm thick pipe casings; to softwood framing (not included)						
300 mm girth	-	0.83	18.23	1.83	m	**20.07**
600 mm girth	-	1.02	22.40	3.43	m	**25.84**
Three-sided 6.5 mm thick pipe casing; to softwood framing (not included)						
450 mm girth	-	1.16	25.48	2.75	m	**28.23**
900 mm girth	-	1.39	30.53	5.19	m	**35.72**
Lining to walls 9 mm thick						
over 300 mm wide	5.83	0.43	9.45	7.38	m²	**16.83**
not exceeding 300 mm wide	-	0.29	6.37	2.24	m	**8.61**
Lining to ceilings 9 mm thick						
over 300 mm wide	-	0.58	12.74	7.38	m²	**20.12**
not exceeding 300 mm wide	-	0.38	8.35	2.24	m	**10.59**
holes for pipes and the like	-	0.03	0.66	-	nr	**0.66**
Lining to walls 12 mm thick						
over 300 mm wide	6.75	0.47	10.32	8.51	m²	**18.83**
not exceeding 300 mm wide	-	0.31	6.81	2.58	m	**9.38**
Two-sided 12 mm thick pipe casing; to softwood framing (not included)						
300 mm girth	-	0.79	17.35	3.74	m	**21.10**
600 mm girth	-	0.93	20.43	7.25	m	**27.68**
Three-sided 12 mm thick pipe casing; to softwood framing (not included)						
450 mm girth	-	1.02	22.40	5.61	m	**28.02**
900 mm girth	-	1.20	26.36	10.92	m	**37.28**
Extra for 400 x 400 removable access panel; brass cups and screws; additional framing	-	1.02	22.40	1.48	nr	**23.88**
Lining to ceilings 12 mm thick						
over 300 mm wide	-	0.62	13.62	8.51	m²	**22.13**
not exceeding 300 mm wide	-	0.41	9.01	2.58	m	**11.58**
holes for pipes and the like	-	0.03	0.66	-	nr	**0.66**
Extra over wall linings fixed with nails for screwing	-	-	-	-	m²	**2.02**
Preformed white melamine faced plywood casings; Pendock Profiles Ltd or other equal and approved; to softwood battens (not included)						
Skirting trunking profile; plain butt joints in the running length						
45 mm x 150 mm; ref TK150	-	0.12	2.64	54.68	m	**57.31**
extra for stop end	-	0.05	1.10	6.69	nr	**7.79**
extra for external corner	-	0.10	2.20	49.88	nr	**52.08**
extra for internal corner	-	0.10	2.20	49.88	nr	**52.08**
Casing profiles						
150 mm x 150 mm ref MX150/150; 5 mm thick	-	0.12	2.64	74.88	m	**77.51**
extra for stop end	-	0.05	1.10	13.33	nr	**14.43**
extra for external corner	-	0.10	2.20	66.48	nr	**68.67**
extra for internal corner	-	0.10	2.20	49.88	nr	**52.08**
Internal quality American Cherry veneered plywood; 6 mm thick						
Lining to walls						
over 300 mm wide	5.75	0.44	9.66	7.21	m²	**16.87**
not exceeding 300 mm wide	-	0.30	6.59	2.21	m	**8.80**

K LININGS/SHEATHING/DRY PARTITIONING

Item	PC £	Labour hours	Labour £	Material £	Unit	Total rate £
"Tacboard" or other equal and approved; Eternit UK Ltd; fire resisting boards; butt joints; to softwood base						
Lining to walls; 6 mm thick						
over 300 mm wide	-	0.33	7.25	10.37	m²	**17.62**
not exceeding 300 mm wide	-	0.20	4.39	3.16	m	**7.55**
Lining to walls; 9 mm thick						
over 300 mm wide	-	0.37	8.13	18.91	m²	**27.04**
not exceeding 300 mm wide	-	0.22	4.83	5.72	m	**10.55**
Lining to walls; 12 mm thick						
over 300 mm wide	-	0.41	9.01	24.55	m²	**33.56**
not exceeding 300 mm wide	-	0.24	5.27	7.41	m	**12.68**
"Tacfire" or other equal and approved; Eternit UK Ltd; fire resisting boards						
Lining to walls; 6 mm thick						
over 300 mm wide	-	0.33	7.25	13.85	m²	**21.10**
not exceeding 300 mm wide	-	0.20	4.39	4.20	m	**8.60**
Lining to walls; 9 mm thick						
over 300 mm wide	-	0.37	8.13	21.09	m²	**29.22**
not exceeding 300 mm wide	-	0.22	4.83	6.37	m	**11.21**
Lining to walls; 12 mm thick						
over 300 mm wide	-	0.41	9.01	27.75	m²	**36.76**
not exceeding 300 mm wide	-	0.24	5.27	8.37	m	**13.64**
K14 GLASS REINFORCED GYPSUM LININGS/ PANELLING/CASINGS/MOULDINGS						
Glass reinforced gypsum Glasroc Multi-board or other equal and approved; fixing with nails; joints filled with joint filler and joint tape; finishing with "Jointex" or other equal and approved to receive decoration; to softwood base						
10 mm board to walls						
wall height 2.40 m - 2.70 m	-	1.02	22.83	61.42	m	**84.25**
wall height 2.70 m - 3.00 m	-	1.16	25.97	68.25	m	**94.22**
wall height 3.00 m - 3.30 m	-	1.34	30.02	75.08	m	**105.10**
wall height 3.30 m - 3.60 m	-	1.53	34.26	81.97	m	**116.23**
12.50 mm board to walls						
wall height 2.40 m - 2.70 m	-	1.06	23.71	80.30	m	**104.01**
wall height 2.70 m - 3.00 m	-	1.20	26.85	89.23	m	**116.08**
wall height 3.00 m - 3.30 m	-	1.39	31.12	98.16	m	**129.28**
wall height 3.30 m - 3.60 m	-	1.57	35.14	107.14	m	**142.28**
K20 TIMBER BOARD FLOORING/SHEATHING/ LININGS/CASINGS						
Sawn softwood; untreated						
Boarding to roofs; 150 mm wide boards; butt joints						
19 mm thick; flat; over 300 mm wide	7.20	0.45	9.88	9.51	m²	**19.40**
19 mm thick; flat; not exceeding 300 mm wide	-	0.31	6.81	2.89	m	**9.70**
19 mm thick; sloping; over 300 mm wide	-	0.51	11.20	9.51	m²	**20.72**
19 mm thick; sloping; not exceeding 300 mm wide	-	0.33	7.25	2.89	m	**10.14**
19 mm thick; sloping; laid diagonally; over 300 mm wide	-	0.64	14.06	9.51	m²	**23.57**
19 mm thick; sloping; laid diagonally; not exceeding 300 mm wide	-	0.41	9.01	2.89	m	**11.90**

K LININGS/SHEATHING/DRY PARTITIONING

Item	PC £	Labour hours	Labour £	Material £	Unit	Total rate £
K20 TIMBER BOARD FLOORING/SHEATHING/ LININGS/CASINGS – cont'd						
Sawn softwood; untreated – cont'd						
Boarding to roofs; 150 mm boards; butt joints – cont'd						
25 mm thick; flat; over 300 mm wide	-	0.45	9.88	15.34	m²	**25.22**
25 mm thick; flat; not exceeding 300 mm wide	-	0.31	6.81	4.64	m	**11.45**
25 mm thick; sloping; over 300 mm wide	-	0.51	11.20	15.34	m²	**26.54**
25 mm thick; sloping; not exceeding 300 mm wide	-	0.33	7.25	4.64	m	**11.89**
25 mm thick; sloping; laid diagonally; over 300 mm wide	-	0.64	14.06	15.34	m²	**29.40**
25 mm thick; sloping; laid diagonally; not exceeding 300 mm wide	-	0.41	9.01	4.64	m	**13.65**
Boarding to tops or cheeks of dormers; 150 mm wide boards; butt joints						
19 mm thick; laid diagonally; over 300 mm wide	-	0.81	17.79	9.51	m²	**27.31**
19 mm thick; laid diagonally; not exceeding 300 mm wide	-	0.51	11.20	2.89	m	**14.10**
19 mm thick; laid diagonally; area not exceeding 1.00 m² irrespective of width	-	1.02	22.40	9.01	nr	**31.41**
Sawn softwood; "Tanalised"						
Boarding to roofs; 150 mm wide boards; butt joints						
19 mm thick; flat; over 300 mm wide	-	0.45	9.88	10.60	m²	**20.49**
19 mm thick; flat; not exceeding 300 mm wide	-	0.31	6.81	3.22	m	**10.03**
19 mm thick; sloping; over 300 mm wide	-	0.51	11.20	10.60	m²	**21.81**
19 mm thick; sloping; not exceeding 300 mm wide	-	0.33	7.25	3.22	m	**10.47**
19 mm thick; sloping; laid diagonally; over 300 mm wide	-	0.64	14.06	10.60	m²	**24.66**
19 mm thick; sloping; laid diagonally; not exceeding 300 mm wide	-	0.41	9.01	3.22	m	**12.22**
25 mm thick; flat; over 300 mm wide	-	0.45	9.88	16.77	m²	**26.66**
25 mm thick; flat; not exceeding 300 mm wide	-	0.31	6.81	5.07	m	**11.88**
25 mm thick; sloping; over 300 mm wide	-	0.51	11.20	16.77	m²	**27.98**
25 mm thick; sloping; not exceeding 300 mm wide	-	0.33	7.25	5.07	m	**12.32**
25 mm thick; sloping; laid diagonally; over 300 mm wide	-	0.64	14.06	16.77	m²	**30.83**
25 mm thick; sloping; laid diagonally; not exceeding 300 mm wide	-	0.41	9.01	5.07	m	**14.08**
Boarding to tops or cheeks of dormers; 150 mm wide boards; butt joints						
19 mm thick; laid diagonally; over 300 mm wide	-	0.81	17.79	10.60	m²	**28.39**
19 mm thick; laid diagonally; not exceeding 300 mm wide	-	0.51	11.20	3.22	m	**14.42**
19 mm thick; laid diagonally; area not exceeding 1.00 m² irrespective of width	-	1.02	22.40	10.10	nr	**32.50**
Wrought softwood						
Boarding to floors; butt joints						
19 mm x 75 mm boards	-	0.61	13.40	12.88	m²	**26.28**
19 mm x 125 mm boards	-	0.56	12.30	9.87	m²	**22.17**
22 mm x 150 mm boards	-	0.52	11.42	10.94	m²	**22.36**
25 mm x 100 mm boards	-	0.56	12.30	11.95	m²	**24.25**
25 mm x 150 mm boards	-	0.52	11.42	12.16	m²	**23.58**
25 mm boarding and bearers to floors; butt joints; in making good where partitions removed or openings formed (boards running in direction of partition)						
150 mm wide	-	0.28	6.15	2.16	m	**8.31**
225 mm wide	-	0.42	9.23	3.53	m	**12.76**
300 mm wide	-	0.56	12.30	4.24	m	**16.54**

K LININGS/SHEATHING/DRY PARTITIONING

Item	PC £	Labour hours	Labour £	Material £	Unit	Total rate £
25 mm boarding and bearers to floors; butt joints; in making good where partitions removed or openings formed (boards running at right angles to partition)						
150 mm wide	-	0.42	9.23	5.79	m	**15.02**
225 mm wide	-	0.65	14.28	7.58	m	**21.86**
300 mm wide	-	0.83	18.23	8.59	m	**26.82**
450 mm wide	-	1.25	27.46	11.11	m	**38.56**
Boarding to floors; tongued and grooved joints						
19 mm x 75 mm boards	-	0.71	15.60	13.87	m²	**29.46**
19 mm x 125 mm boards	-	0.67	14.72	11.09	m²	**25.81**
22 mm x 150 mm boards	-	0.62	13.62	11.26	m²	**24.88**
25 mm x 100 mm boards	-	0.67	14.72	14.19	m²	**28.90**
25 mm x 150 mm boards	-	0.62	13.62	13.29	m²	**26.91**
Boarding to internal walls; tongued and grooved and V-jointed						
12 mm x 100 mm boards	-	0.81	17.79	10.14	m²	**27.93**
16 mm x 100 mm boards	-	0.81	17.79	11.45	m²	**29.24**
19 mm x 100 mm boards	-	0.81	17.79	12.10	m²	**29.90**
19 mm x 125 mm boards	-	0.77	16.91	13.17	m²	**30.08**
19 mm x 125 mm boards; chevron pattern	-	1.22	26.80	13.17	m²	**39.96**
25 mm x 125 mm boards	-	0.77	16.91	14.56	m²	**31.48**
12 mm x 100 mm boards; knotty pine	-	0.81	17.79	8.57	m²	**26.36**
Boarding to internal ceilings						
12 mm x 100 mm boards	-	1.02	22.40	10.14	m²	**32.54**
16 mm x 100 mm boards	-	1.02	22.40	11.45	m²	**33.85**
19 mm x 100 mm boards	-	1.02	22.40	12.10	m²	**34.51**
19 mm x 125 mm boards	-	0.97	21.31	13.17	m²	**34.47**
19 mm x 125 mm boards; chevron pattern	-	1.42	31.19	13.17	m²	**44.36**
25 mm x 125 mm boards	-	0.97	21.31	14.56	m²	**35.87**
12 mm x 100 mm boards; knotty pine	-	1.02	22.40	8.57	m²	**30.97**
Boarding to roofs; tongued and grooved joints						
19 mm thick; flat to falls	-	0.56	12.30	11.26	m²	**23.56**
19 mm thick; sloping	-	0.61	13.40	11.26	m²	**24.66**
19 mm thick; sloping; laid diagonally	-	0.79	17.35	11.26	m²	**28.62**
25 mm thick; flat to falls	-	0.56	12.30	14.25	m²	**26.55**
25 mm thick; sloping	-	0.61	13.40	14.25	m²	**27.65**
Boarding to tops or cheeks of dormers; tongued and grooved joints						
19 mm thick; laid diagonally	-	1.02	22.40	11.26	m²	**33.67**
Wrought softwood; "Tanalised"						
Boarding to roofs; tongued and grooved joints						
19 mm thick; flat to falls	-	0.56	12.30	12.35	m²	**24.65**
19 mm thick; sloping	-	0.61	13.40	12.35	m²	**25.75**
19 mm thick; sloping; laid diagonally	-	0.79	17.35	12.35	m²	**29.70**
25 mm thick; flat to falls	-	0.56	12.30	15.68	m²	**27.98**
25 mm thick; sloping	-	0.61	13.40	15.68	m²	**29.08**
Boarding to tops or cheeks of dormers; tongued and grooved joints						
19 mm thick; laid diagonally	-	1.02	22.40	12.35	m²	**34.76**

K LININGS/SHEATHING/DRY PARTITIONING

Item	PC £	Labour hours	Labour £	Material £	Unit	Total rate £
K20 TIMBER BOARD FLOORING/SHEATHING/ LININGS/CASINGS – cont'd						
Wood strip; 22 mm thick; Junckers All in Beech "Sylva Sport Premium" pre-treated or other equal and approved; tongued and grooved joints; on bearers etc.; level fixing to cement and sand base						
Strip flooring; over 300 mm wide						
on 45 x 45 mm blue bat bearers	-	-	-	-	m²	**55.19**
on 10 mm Pro Foam	-	-	-	-	m²	**57.94**
on Uno bat 50 mm bearers	-	-	-	-	m²	**61.02**
on New Era levelling system	-	-	-	-	m²	**63.29**
on Uno bat 62 mm bearers	-	-	-	-	m²	**65.44**
on Duo bat 110 mm bearers	-	-	-	-	m²	**78.74**
Wood strip; 22 mm thick; Junckers pre-treated or other equal and approved flooring systems; tongued and grooved joints; on bearers etc.; level fixing to cement and sand base						
Strip flooring; over 300 mm wide						
"Sylva Squash Beech untreated" on blue bat bearers	-	-	-	-	m²	**76.24**
"Classic Beech" clip system	-	-	-	-	m²	**77.83**
"Harmoni Oak" clip system	-	-	-	-	m²	**77.83**
"Classic Beech" on blue bat bearers	-	-	-	-	m²	**77.83**
"Harmoni Oak" on blue bat bearers	-	-	-	-	m²	**77.83**
Unfinished wood strip; 22 mm thick; Havwoods or other equal and approved; tongued and grooved joints; secret fixed; laid on semi-sprung bearers; fixing to cement and sand base; sanded and sealed						
Strip flooring; over 300 mm wide						
Prime Iroko	-	-	-	-	m²	**66.72**
Prime Maple	-	-	-	-	m²	**71.31**
American Oak	-	-	-	-	m²	**72.79**
K30 DEMOUNTABLE PARTITIONS						
Insulated panel and two-hour fire wall system for warehouses etc., comprising white polyester coated galvanised steel frame and 0.55 mm galvanised steel panels either side of rockwool infill						
100 mm thick wall; 31 Rw dB acoustic rating	-	-	-	-	m²	**51.25**
150 mm thick wall; 31 Rw dB acoustic rating	-	-	-	-	m²	**55.09**
intumescent mastic sealant; bedding frames at perimeters of metal fire walls	-	-	-	-	m	**4.48**
Getalit laminated both sides top hung movable acoustic panel wall with concealed uPVC vertical edge profiles, 9 nr 1106 mm x 3000 mm panels and type K two point panel support system						
105 mm thick wall; 47 Rw dB acoustic rating	-	-	-	-	m²	**474.06**
105 mm thick wall; 47 Rw dB acoustic rating	-	-	-	-	m²	**512.50**
105 mm thick wall; 47 Rw dB acoustic rating	-	-	-	-	m²	**550.94**

K LININGS/SHEATHING/DRY PARTITIONING

Item	PC £	Labour hours	Labour £	Material £	Unit	Total rate £
K32 FRAMED PANEL CUBICLE PARTITIONS						
Toilet cubicle partitions; Amwells or other equal and approved; standard colours and ironmongery; assembling and screwing to floor and wall						
Axis standard cubicle set; 800 mm x 1500 mm x 1980 mm high per cubicle, with polished aluminium framing; 19 mm melamine-faced chipboard divisions and doors						
One cubicle set; 2 nr panels; 1 nr door	-	3.25	143.00	308.00	nr	**451.00**
range of 3 cubicle sets; 4 nr panels; 3 nr doors	-	9.75	429.00	880.00	nr	**1309.00**
range of 6 cubicle sets; 7 nr panels; 6 nr doors	-	19.50	858.00	1738.00	nr	**2596.00**
Reduction of 1 nr panel for end unit adjoining side wall	-	-	-	-121.00	nr	-
Minima designer cubicle set; 800 mm x 1500 mm x 2100 mm high per cubicle, with satin polished stainless steel framing; 18 mm high pressure laminated (HPL) chipboard divisions and doors						
One cubicle set; 2 nr panels; 1 nr door	-	3.25	143.00	649.00	nr	**792.00**
range of 3 cubicle sets; 4 nr panels; 3 nr doors	-	9.75	429.00	1683.00	nr	**2112.00**
range of 6 cubicle sets; 7 nr panels; 6 nr doors	-	19.50	858.00	3234.00	nr	**4092.00**
Reduction of 1 nr panel for end unit adjoining side wall	-	-	-	-176.00	nr	-
Sylan corporate cubicle set; 800 mm x 1500 mm x 2400 mm high per cubicle, with sating finished stainless steel ironmongery; 30 mm high pressure laminated (HPL) chipboard divisions and 44 mm solid cored real wood veneered doors and pilasters						
One cubicle set; 2 nr panels; 1 nr door	-	5.00	220.00	1919.50	nr	**2139.50**
range of 3 cubicle sets; 4 nr panels; 3 nr doors	-	15.00	660.00	5186.50	nr	**5846.50**
range of 6 cubicle sets; 7 nr panels; 6 nr doors	-	30.00	1320.00	10081.50	nr	**11401.50**
Reduction of 1 nr panel for end unit adjoining side wall	-	-	-	-390.50	nr	-
K40 DEMOUNTABLE SUSPENDED CEILINGS						
Suspended ceilings; Donn Products exposed suspended ceiling system or other equal and approved; hangers plugged and screwed to concrete soffit; 600 mm x 600 mm x 15 mm Cape TAP Ceilings Ltd; Solitude tegular fissured tile						
Lining to ceilings; hangers average 400 mm long over 300 mm wide	-	0.40	9.48	11.48	m²	**20.96**
Suspended ceilings; Gyproc M/F suspended ceiling system or other equal and approved; hangers plugged and screwed to concrete soffit, 900 mm x 1800 mm x 12.5 mm tapered edge wallboard infill; joints filled with joint filler and taped to receive direct decoration						
Lining to ceilings; hangers average 400 mm long over 300 mm wide	-	-	-	-	m²	**37.88**
not exceeding 300 mm wide in isolated strips	-	-	-	-	m	**31.00**
300 mm - 600 mm wide in isolated strips	-	-	-	-	m	**38.84**

K LININGS/SHEATHING/DRY PARTITIONING

Item	PC £	Labour hours	Labour £	Material £	Unit	Total rate £
K40 DEMOUNTABLE SUSPENDED CEILINGS – cont'd						
Suspended ceilings; Gyproc M/F – cont'd						
Edge treatments						
20 x 20 mm SAS perimeter shadow gap; screwed to plasterboard	-	-	-	-	m	**5.88**
20 x 20 mm SAS shadow gap around 450 mm dia. column; including 15 x 44 mm batten plugged and screwed to concrete	-	-	-	-	m	**29.32**
Vertical bulkhead; including additional hangers						
over 300 mm wide	-	-	-	-	m²	**47.72**
not exceeding 300 mm wide in isolated strips	-	-	-	-	m	**46.36**
300 mm - 600 mm wide in isolated strips	-	-	-	-	m	**47.12**
Suspended ceilings; Rockfon, or other equal and approved; Z demountable suspended concealed ceiling system; 400 mm long hangers plugged and screwed to concrete soffit.						
Lining to ceilings; 600 mm x 600mm x 20 mm 'Sonar' suspended ceiling tiles						
over 300 mm wide	-	-	-	-	m²	**53.99**
not exceeding 300 mm wide	-	-	-	-	m	**31.48**
Edge trim; shadow-line trim	-	-	-	-	m	**6.51**
Vertical bulkhead, as upstand to rooflight well; including additional hangers; perimeter trim						
300 mm x 600 mm wide	-	-	-	-	m	**58.71**
Suspended ceilings; Ecophon, or other equal and approved; Z demountable suspended concealed ceiling system; 400 mm long hangers plugged and screwed to concrete soffit.						
Lining to ceilings; 600 mm x 600mm x 20 mm 'Gedina ET15' suspended ceiling tiles						
over 300 mm wide	-	-	-	-	m²	**46.73**
not exceeding 300 mm wide	-	-	-	-	m	**29.01**
Edge trim; shadow-line trim	-	-	-	-	m	**6.03**
Vertical bulkhead, as upstand to rooflight well; including additional hangers; perimeter trim						
300 mm x 600 mm wide	-	-	-	-	m	**53.90**
Lining to ceilings; 600 mm x 600mm x 20 mm 'Hygiene Performance' washable suspended ceiling tiles						
over 300 mm wide	-	-	-	-	m²	**62.41**
not exceeding 300 mm wide	-	-	-	-	m	**50.86**
Edge trim; shadow-line trim	-	-	-	-	m	**8.32**
Vertical bulkhead, as upstand to rooflight well; including additional hangers; perimeter trim						
300 mm x 600 mm wide	-	-	-	-	m	**56.01**
Lining to ceilings; 1200 mm x 1200mm x 20 mm 'Focus DG' suspended ceiling tiles						
over 300 mm wide	-	-	-	-	m²	**56.75**
not exceeding 300 mm wide	-	-	-	-	m	**33.16**
Edge trim; shadow-line trim	-	-	-	-	m	**6.03**
Vertical bulkhead, as upstand to rooflight well; including additional hangers; perimeter trim						
300 mm x 600 mm wide	-	-	-	-	m	**56.88**

K LININGS/SHEATHING/DRY PARTITIONING

Item	PC £	Labour hours	Labour £	Material £	Unit	Total rate £
Suspended ceilings; Z demountable suspended ceiling system or other equal and approved; hangers plugged and screwed to concrete soffit, 600 mm x 600 mm x 19 mm Echostop glass reinforced fibrous plaster lightweight plain bevelled edge tiles						
Lining to ceilings; hangers average 400 mm long						
over 300 mm wide	-	-	-	-	m²	**115.34**
not exceeding 300 mm wide	-	-	-	-	m	**80.95**
Suspended ceilings; concealed galvanised steel suspension system; hangers plugged and screwed to concrete soffit, Burgess white stove enamelled perforated mild steel tiles 600 mm x 600 mm						
Lining to ceilings; hangers average 400 mm long						
over 300 mm wide	-	-	-	-	m²	**52.50**
not exceeding 300 mm wide	-	-	-	-	m	**46.10**
Suspended ceilings; concealed galvanised steel suspension system; hangers plugged and screwed to concrete soffit, Burgess white stove enamelled perforated mild steel tiles 600 mm x 600 mm						
Linings to ceilings; hangers average 700 mm long						
over 300 mm wide	-	-	-	-	m²	**32.08**
over 300 mm wide; 3.50 m - 5.00 m high	-	-	-	-	m²	**33.35**
over 300 mm wide; in staircase areas or plant rooms	-	-	-	-	m²	**42.03**
not exceeding 300 mm wide; in isolated strips	-	-	-	-	m	**23.54**
300 mm - 600 mm wide; in isolated strips	-	-	-	-	m	**29.75**
Extra for cutting and fitting around modular downlighter including yoke	-	-	-	-	nr	**21.76**
24 mm x 19 mm white finished angle edge trim	-	-	-	-	m	**5.05**
Vertical bulkhead; including additional hangers						
over 300 mm wide	-	-	-	-	m²	**59.73**
not exceeding 300 mm wide; in isolated strips	-	-	-	-	m	**47.24**
300 mm - 600 mm wide; in isolated strips	-	-	-	-	m	**54.01**
Suspended ceilings, metal; SAS system 330; EMAC suspension system; 100 mm Omega C profiles at 1500 mm centres filled in with 1400 mm x 250 mm perforated metal tiles with 18 mm thick x 80 kg/m3 density foill wrapped tissue-faced acoustic pad adhered above; ceiling to achieve 40d Dnwc with 0.7 absorption coefficient						
Linings to ceilings; hangers average 700 mm long						
over 300 mm wide	-	-	-	-	m²	**57.76**
Extra for cutting and reinforcing to receive a recessed light maximum 1300 mm x 500 mm.	-	-	-	-	nr	**16.50**
not exceeding 300 mm wide; in isolated strips	-	-	-	-	m	**29.47**
Edge trim; to perimeter	-	-	-	-	m	**14.73**
Edge trim around 450 mm dia. column	-	-	-	-	nr	**56.58**

K LININGS/SHEATHING/DRY PARTITIONING

Item	PC £	Labour hours	Labour £	Material £	Unit	Total rate £
K40 DEMOUNTABLE SUSPENDED CEILINGS – cont'd						
Suspended ceilings; galvanised steel suspension system; hangers plugged and screwed to concrete soffit, Luxalon stove enamelled aluminium linear panel ceiling, type 80B or other equal and approved, complete with mineral insulation						
Linings to ceilings; hangers average 700 mm long						
over 300 mm wide	-	-	-	-	m²	**102.79**
not exceeding 300 mm wide; in isolated strips	-	-	-	-	m	**49.71**

L WINDOWS/DOORS/STAIRS

Item	PC £	Labour hours	Labour £	Material £	Unit	Total rate £
L10 WINDOWS/ROOFLIGHTS/SCREENS/LOUVRES						
SUPPLY ONLY PRICES						
NOTE: The following supply only prices are for purpose-made components, to which fixings, sealants etc. labour and overheads and profit need to be added, before they may be used to arrive at a guide price for a complete window. The reader is then referred to the following SUPPLY AND FIX pages for fixing costs based on the overall window size.						
Purpose made window casements; "treated" wrought softwood						
Casements; rebated; moulded						
44 mm thick	-	-	-	64.16	m²	-
57 mm thick	-	-	-	67.38	m²	-
Casements; rebated; moulded; in medium panes						
44 mm thick	-	-	-	102.56	m²	-
57 mm thick	-	-	-	106.86	m²	-
Casements; rebated; moulded; with semi-circular head						
44 mm thick	-	-	-	134.10	m²	-
57 mm thick	-	-	-	138.30	m²	-
Casements; rebated; moulded; to bullseye window						
44 mm thick; 600 mm diameter	-	-	-	212.56	nr	-
44 mm thick; 900 mm diameter	-	-	-	253.22	nr	-
57 mm thick; 600 mm diameter	-	-	-	222.57	nr	-
57 mm thick; 900 mm diameter	-	-	-	30.63	nr	-
Fitting and hanging casements (in factory)						
square or rectangular	-	-	-	15.00	nr	-
semi-circular	-	-	-	24.38	nr	-
bullseye	-	-	-	30.63	nr	-
Purpose made window casements; selected Sapele						
Casements; rebated; moulded						
44 mm thick	-	-	-	72.34	m²	-
57 mm thick	-	-	-	79.25	m²	-
Casements; rebated; moulded; in medium panes						
44 mm thick	-	-	-	117.16	m²	-
57 mm thick	-	-	-	126.38	m²	-
Casements; rebated; moulded with semi-circular head - supply only						
44 mm thick	-	-	-	147.56	m²	-
57 mm thick	-	-	-	156.59	m²	-
Casements; rebated; moulded; to bullseye window						
44 mm thick; 600 mm diameter	-	-	-	260.80	nr	-
44 mm thick; 900 mm diameter	-	-	-	313.84	nr	-
57 mm thick; 600 mm diameter	-	-	-	282.31	nr	-
57 mm thick; 900 mm diameter	-	-	-	341.25	nr	-
Fitting and hanging casements (in factory)						
square or rectangular	-	-	-	16.12	nr	-
semi-circular	-	-	-	26.65	nr	-
bullseye	-	-	-	34.09	nr	-

L WINDOWS/DOORS/STAIRS

Item	PC £	Labour hours	Labour £	Material £	Unit	Total rate £
L10 WINDOWS/ROOFLIGHTS/SCREENS/LOUVRES – cont'd						
SUPPLY ONLY PRICES – cont'd						
Purpose made window frames; "treated" wrought softwood						
Frames; rounded; rebated check grooved						
44 mm x 69 mm	-	-	-	16.61	m	-
44 mm x 94 mm	-	-	-	17.41	m	-
44 mm x 119 mm	-	-	-	18.20	m	-
57 mm x 94 mm	-	-	-	18.24	m	-
69 mm x 144 mm	-	-	-	24.05	m	-
90 mm x 140 mm	-	-	-	34.13	m	-
Mullions and transoms; twice rounded, rebated and check grooved						
57 mm x 69 mm	-	-	-	19.35	m	-
57 mm x 94 mm	-	-	-	20.32	m	-
69 mm x 94 mm	-	-	-	22.99	m	-
69 mm x 144 mm	-	-	-	33.76	m	-
Sill; sunk weathered, rebated and grooved						
69 mm x 94 mm	-	-	-	40.35	m	-
69 mm x 144 mm	-	-	-	42.71	m	-
Add 5% to the above material prices for "selected" softwood for staining						
Purpose made window frames; selected Sapele						
Frames; rounded; rebated check grooved						
44 mm x 69 mm	-	-	-	21.89	m	-
44 mm x 94 mm	-	-	-	23.54	m	-
44 mm x 119 mm	-	-	-	25.19	m	-
57 mm x 94 mm	-	-	-	27.55	m	-
69 mm x 144 mm	-	-	-	39.47	m	-
90 mm x 140 mm	-	-	-	56.08	m	-
Mullions and transoms; twice rounded, rebated and check grooved						
57 mm x 69 mm	-	-	-	24.87	m	-
57 mm x 94 mm	-	-	-	28.93	m	-
69 mm x 94 mm	-	-	-	34.70	m	-
69 mm x 144 mm	-	-	-	52.74	m	-
Sill; sunk weathered, rebated and grooved						
69 mm x 94 mm	-	-	-	49.15	m	-
69 mm x 144 mm	-	-	-	54.48	m	-
Thermally broken composite double glazed aluminium/ timber windows; 'Velfac 200' or other approved; with a maximum glazing U value of 1.5 W/m²K; low e glazing with laminated glass unless otherwise specified; including multi point espagnolette locking mechanisms and other ironmongery						
NOTE: The following supply only prices are for standard windows, to which fixings, sealants etc. labour and overheads and profit need to ba added, before they may be used to arrive at a guide price for a complete unit.						

L WINDOWS/DOORS/STAIRS

Item	PC £	Labour hours	Labour £	Material £	Unit	**Total rate £**
Outward opening standard fixed sash casement windows						
1200 mm x 1200 mm single fixed pane; low-e glass 4/16/4	-	-	-	286.00	nr	-
2200 mm x 2200 mm single fixed pane; low-e glass 6/12/6	-	-	-	726.00	nr	-
1200 mm x 2200 mm three fixed panes; low-e glass 4/16/4	-	-	-	572.00	nr	-
Outward opening standard sash casement windows						
1600 mm x 1600 mm with two sidehung sashes; low-e glass 4/16/4	-	-	-	583.00	nr	-
1600 mm x 1600 mm with two sidehung projecting sashes; low-e glass 4/16/4	-	-	-	649.00	nr	-
2000 mm x 1600 mm with one sidehung sash next to a tophung projecting sash over a fixed sash; low-e glass 4/16/4	-	-	-	715.00	nr	-
2000 mm x 1600 mm with one sidehung sash next to a tophung projecting sash over a fixed sash low-e glass 4/16/4	-	-	-	566.50	nr	-
1200 mm x 2200 mm with fixed lower sash and fully reversible upper sash; lower low-e upper low-e glass 4 toughened/16/6.4; upper low-e glass 4/16/4	-	-	-	616.00	nr	-
Outward opening standard doors						
2200 mm x 2200 mm French casement patio door; low-e toughened glass 4/16/4	-	-	-	1094.50	nr	-
2200 mm x 2200 mm Sliding patio door; low-e glass 4 toughened/16/4 laminated	-	-	-	1364.00	nr	-
Guide price for installation:	-	1.00	65.00	-	m²	**65.00**
SUPPLY AND FIX PRICES						
Standard windows; "treated" wrought softwood; Jeld-Wen or other equal and approved						
Side hung casement windows without glazing bars; factory glazed with low E 24 mm double glazing; with 140 mm wide softwood sills; opening casements and ventilators hung on rustproof hinges; fitted with aluminized lacquered finish casement stays and fasteners						
488 mm x 750 mm; ref LEWN07V	140.86	0.69	15.16	159.40	nr	**174.56**
488 mm x 900 mm; ref LEWN09V	143.23	0.83	18.23	162.07	nr	**180.30**
630 mm x 750 mm; ref LEW107V	156.11	0.83	18.23	176.61	nr	**194.84**
630 mm x 750 mm; ref LEW107C	127.28	0.83	18.23	144.07	nr	**162.31**
630 mm x 900 mm; ref LEW109CH	137.26	0.83	18.23	155.34	nr	**173.57**
630 mm x 900 mm; ref LEW109C	161.74	0.93	20.43	182.96	nr	**203.39**
630 mm x 1050 mm; ref LEW110C	142.30	1.02	22.40	161.16	nr	**183.56**
630 mm x 1050 mm; ref LEW110V	169.22	0.83	18.23	191.55	nr	**209.78**
915 mm x 900 mm; ref LEW2NO9W	194.59	1.11	24.38	220.04	nr	**244.43**
915 mm x 1050 mm; ref LEW2N1OW	204.43	1.16	25.48	231.29	nr	**256.77**
915 mm x 1200 mm; ref LEW2N12W	221.20	1.20	26.36	250.22	nr	**276.58**
915 mm x 1350 mm; ref LEW2N13W	231.90	1.39	30.53	262.30	nr	**292.83**
915 mm x 1500 mm; ref LEW2N15W	263.13	1.43	31.41	297.68	nr	**329.09**
1200 mm x 750 mm; ref LEW2O7C	200.00	1.16	25.48	226.29	nr	**251.77**
1200 mm x 750 mm; ref LEW2O7CV	252.96	1.16	25.48	286.07	nr	**311.55**
1200 mm x 900 mm; ref LEW2O9C	212.76	1.20	26.36	240.69	nr	**267.05**
1200 mm x 900 mm; ref LEW2O9W	223.89	1.20	26.36	253.25	nr	**279.61**
1200 mm x 900 mm; ref LEW2O9CV	264.33	1.20	26.36	298.90	nr	**325.26**

Item	PC £	Labour hours	Labour £	Material £	Unit	Total rate £
L10 WINDOWS/ROOFLIGHTS/SCREENS/LOUVRES – cont'd						
Standard softwood windows; Jeld-Wen – cont'd						
Side hung casement windows – cont'd						
1200 mm x 1050 mm; ref LEW210C	225.22	1.39	30.53	254.90	nr	**285.43**
1200 mm x 1050 mm; ref LEW210W	237.31	1.39	30.53	268.54	nr	**299.07**
1200 mm x 1050 mm; ref LEW210T	280.00	1.39	30.53	316.73	nr	**347.26**
1200 mm x 1050 mm; ref LEW210CV	274.61	1.39	30.53	310.64	nr	**341.17**
1200 mm x 1200 mm; ref LEW212C	242.16	1.48	32.51	274.15	nr	**306.66**
1200 mm x 1200 mm; ref LEW212W	250.87	1.48	32.51	283.98	nr	**316.49**
1200 mm x 1200 mm; ref LEW212TX	326.19	1.48	32.51	369.00	nr	**401.51**
1200 mm x 1200 mm; ref LEW212CV	289.28	1.48	32.51	327.34	nr	**359.85**
1200 mm x 1350 mm; ref LEW213W	269.46	1.57	34.49	304.97	nr	**339.46**
1200 mm x 1350 mm; ref LEW213CV	319.93	1.57	34.49	361.93	nr	**396.42**
1200 mm x 1500 mm; ref LEW215W	312.49	1.71	37.56	353.67	nr	**391.23**
1770 mm x 750 mm; ref LEW307CC	293.71	1.43	31.41	332.33	nr	**363.74**
1770 mm x 900 mm; ref LEW309CC	312.50	1.71	37.56	353.55	nr	**391.11**
1770 mm x 1050 mm; ref LEW310C	296.18	1.80	39.54	335.13	nr	**374.67**
1770 mm x 1050 mm; ref LEW310T	349.65	1.71	37.56	395.48	nr	**433.04**
1770 mm x 1050 mm; ref LEW310CC	327.95	1.43	31.41	370.98	nr	**402.40**
1770 mm x 1050 mm; ref LEW310CW	345.34	1.43	31.41	390.61	nr	**422.02**
1770 mm x 1200 mm; ref LEW312C	316.31	1.85	40.64	357.98	nr	**398.62**
1770 mm x 1200 mm; ref LEW312T	370.39	1.85	40.64	419.03	nr	**459.66**
1770 mm x 1200 mm; ref LEW312CC	353.44	1.85	40.64	399.89	nr	**440.52**
1770 mm x 1200 mm; ref LEW312CW	366.08	1.85	40.64	414.16	nr	**454.79**
1770 mm x 1200 mm; ref LEW312CVC	412.38	1.85	40.64	466.42	nr	**507.06**
1770 mm x 1350 mm; ref LEW313CC	400.23	1.94	42.61	452.70	nr	**495.31**
1770 mm x 1350 mm; ref LEW313CW	398.62	1.94	42.61	450.89	nr	**493.50**
1770 mm x 1350 mm; ref LEW313CVC	449.01	1.94	42.61	507.76	nr	**550.38**
1770 mm x 1500 mm; ref LEW315T	476.21	2.04	44.81	538.47	nr	**583.28**
2340 mm x 1050 mm; ref LEW410CWC	455.54	1.99	43.71	515.14	nr	**558.85**
2340 mm x 1200 mm; ref LEW412CWC	483.92	2.08	45.69	547.30	nr	**592.99**
2340 mm x 1350 mm; ref LEW413CWC	532.11	2.22	48.76	601.84	nr	**650.60**
Top hung casement windows; factory glazed with low E 24 mm double glazing; with 140 mm wide softwood sills; opening casements and ventilators hung on rustproof hinges; fitted with aluminized lacquered finish casement stays						
630 mm x 750 mm; ref LEW107A	137.21	0.83	18.23	155.29	nr	**173.52**
630 mm x 900 mm; ref LEW109A	144.84	0.93	20.43	163.89	nr	**184.32**
630 mm x 1050 mm; ref LEW110A	153.43	1.02	22.40	173.73	nr	**196.13**
915 mm x 750 mm; ref LEW2N07A	166.64	1.06	23.28	188.50	nr	**211.79**
915 mm x 900 mm; ref LEW2N09A	186.98	1.11	24.38	211.46	nr	**235.84**
915 mm x 1050 mm; ref LEW2N10A	198.93	1.16	25.48	225.08	nr	**250.56**
915 mm x 1350 mm; ref LEW2N13AS	250.96	1.39	30.53	283.95	nr	**314.48**
1200 mm x 750 mm; ref LEW207A	196.08	1.16	25.48	221.87	nr	**247.35**
1200 mm x 900 mm; ref LEW209A	214.00	1.20	26.36	242.09	nr	**268.45**
1200 mm x 1050 mm; ref LEW210A	228.79	1.39	30.53	258.92	nr	**289.45**
1200 mm x 1200 mm; ref LEW212A	247.97	1.48	32.51	280.57	nr	**313.08**
1200 mm x 1350 mm; ref LEW213AS	289.53	1.57	34.49	327.61	nr	**362.10**
1200 mm x 1500 mm; ref LEW215AS	315.95	1.71	37.56	357.44	nr	**395.00**
1770 mm x 1050 mm; ref LEW310AE	316.76	1.71	37.56	358.36	nr	**395.92**
1770 mm x 1200 mm; ref LEW312AE	337.57	1.85	40.64	381.85	nr	**422.49**

L WINDOWS/DOORS/STAIRS

Item	PC £	Labour hours	Labour £	Material £	Unit	Total rate £
High performance Hi-Profile top hung reversible windows; factory glazed with low E 24 mm double glazing; weather stripping; opening panes hung on rustproof hinges; fitted with aluminized lacquered espagnolette bolts						
600 mm x 900 mm; ref LECFR609AR	242.53	0.93	20.43	274.16	nr	**294.59**
600 mm x 1050 mm; ref LECFR610AR	255.42	1.02	22.40	288.85	nr	**311.25**
600 mm x 1200 mm; ref LECFR612AR	267.14	1.13	24.82	302.21	nr	**327.03**
600 mm x 1350 mm; ref LECFR613AR	278.63	1.24	27.24	315.19	nr	**342.42**
1200 mm x 900 mm; ref LECFR1209AFR	360.82	1.22	26.80	407.82	nr	**434.62**
1200 mm x 1050 mm; ref LECFR1210AFR	383.29	1.38	30.31	433.31	nr	**463.62**
1200 mm x 1200 mm; ref LECFR1212AFR	403.04	1.47	32.29	455.74	nr	**488.03**
1200 mm x 1350 mm; ref LECFR1213AFR	422.64	1.57	34.49	477.87	nr	**512.35**
1800 mm x 900 mm; ref LECFR1809AFAR	553.01	1.73	38.00	625.03	nr	**663.03**
1800 mm x 1050 mm; ref LECFR1810AFAR	585.71	1.78	39.10	661.93	nr	**701.03**
1800 mm x 1200 mm; ref LECFR1812AFAR	618.18	1.84	40.42	698.71	nr	**739.13**
1800 mm x 1350 mm; ref LECFR1813AFAR	648.12	1.94	42.61	732.51	nr	**775.12**
High performance double hung sash windows with glazing bars; factory glazed with low E 24 mm double glazing; solid frames; 63 mm x 175 mm softwood sills; standard flush external linings; spiral spring balances and sash catch						
635 mm x 1050 mm; ref LESV0610B	450.04	2.04	44.81	508.52	nr	**553.33**
635 mm x 1350 mm; ref LESV0613B	498.24	2.22	48.76	563.06	nr	**611.83**
635 mm x 1650 mm; ref LESV0616B	562.60	2.50	54.91	635.84	nr	**690.76**
860 mm x 1050 mm; ref LESV0810B	515.55	2.36	51.84	582.47	nr	**634.31**
860 mm x 1350 mm; ref LESV0813B	575.07	2.64	57.99	649.79	nr	**707.78**
860 mm x 1650 mm; ref LESV0816B	667.57	3.05	66.99	754.33	nr	**821.32**
1085 mm x 1050 mm; ref LESV1010B	589.08	2.64	57.99	665.46	nr	**723.45**
1085 mm x 1350 mm; ref LESV1013B	665.78	3.05	66.99	752.17	nr	**819.17**
1085 mm x 1650 mm; ref LESV1016B	821.52	3.75	82.37	928.10	nr	**1010.47**
1715 mm x 1050 mm; ref LESV1710B	862.95	3.75	82.37	974.73	nr	**1057.10**
1715 mm x 1350 mm; ref LESV1713B	980.17	4.67	102.58	1107.18	nr	**1209.76**
1715 mm x 1650 mm; ref LESV1716B	1169.90	4.76	104.56	1321.47	nr	**1426.02**
Add to the above material prices for full factory finish	-	-	-	27.00	%	-
Standard windows; Jeld-Wen Hardwood or other equal and approved; factory applied preservative stain base coat						
Side hung casement windows; factory glazed with low E 24 mm double glazing; 45 mm x 140 mm hardwood sills; weather stripping; opening sashes on canopy hinges; fitted with fasteners; brown finish ironmongery						
630 mm x 750 mm; ref LEW107CH	265.44	0.97	21.31	300.02	nr	**321.33**
630 mm x 900 mm; ref LEW109CH	280.62	1.20	26.36	317.15	nr	**343.51**
630 mm x 900 mm; ref LEW109VH	310.12	0.97	21.31	350.45	nr	**371.76**
630 mm x 1050 mm; ref LEW110VH	322.56	1.34	29.43	364.49	nr	**393.93**
915 mm x 900 mm; ref LEW2N09WH	381.15	1.53	33.61	430.77	nr	**464.38**
915 mm x 1050 mm; ref LEWN10WH	396.21	1.62	35.58	447.90	nr	**483.48**
915 mm x 1200 mm; ref LEWN12WH	413.67	1.71	37.56	467.60	nr	**505.16**
915 mm x 1350 mm; ref LEWN13WH	430.73	1.85	40.64	486.86	nr	**527.50**
915 mm x 1550 mm; ref LEWN15WH	474.89	1.94	42.61	536.84	nr	**579.46**
1200 mm x 900 mm; ref LEW209CH	412.40	1.71	37.56	465.91	nr	**503.47**
1200 mm x 900 mm; ref LEW209WH	430.97	1.71	37.56	486.86	nr	**524.42**
1200 mm x 1050 mm; ref LEW210CH	428.87	1.85	40.64	484.50	nr	**525.13**
1200 mm x 1050 mm; ref LEW210WH	451.25	1.85	40.64	509.75	nr	**550.39**
1200 mm x 1200 mm; ref LEW212CH	467.27	1.99	43.71	527.97	nr	**571.68**
1200 mm x 1200 mm; ref LEW212WH	471.90	1.99	43.71	533.20	nr	**576.91**

Item	PC £	Labour hours	Labour £	Material £	Unit	Total rate £
L10 WINDOWS/ROOFLIGHTS/SCREENS/LOUVRES – cont'd						
Standard hardwood windows; Jeld-Wen – cont'd						
Side hung casement windows – cont'd						
1200 mm x 1350 mm; ref LEW213WH	499.06	2.13	46.79	563.85	nr	**610.64**
1200 mm x 1550 mm; ref LEW215WH	556.28	2.22	48.76	628.45	nr	**677.21**
1770 mm x 1050 mm; ref LEW310CCH	655.14	2.31	50.74	740.17	nr	**790.91**
1770 mm x 1200 mm; ref LEW312CCH	697.71	2.45	53.82	788.22	nr	**842.04**
2339 mm x 1200 mm; ref LEW412CMCH	901.29	2.64	57.99	1018.14	nr	**1076.13**
Top hung casement windows; factory glazed with low E 24 mm double glazing; 45 mm x 140 mm hardwood sills; weather stripping; opening sashes on canopy hinges; fitted with fasteners; brown finish ironmongery						
630 mm x 900 mm; ref LEW109AH	296.20	0.97	21.31	334.74	nr	**356.04**
630 mm x 1050 mm; ref LEW110AH	310.72	1.34	29.43	351.26	nr	**380.70**
915 mm x 900 mm; ref LEW2N09AH	382.84	1.53	33.61	432.54	nr	**466.15**
915 mm x 1050 mm; ref LEW2N10AH	404.40	1.62	35.58	457.01	nr	**492.59**
915 mm x 1350 mm; ref LEW2N13ASH	465.94	1.85	40.64	526.60	nr	**567.24**
1200 mm x 1050 mm; ref LEW210AH	461.36	1.71	37.56	521.30	nr	**558.86**
1200 mm x 1350 mm; ref LEW212ASH	557.87	1.85	40.64	630.24	nr	**670.88**
1770 mm x 1050 mm; ref LEW310AEH	607.64	1.94	42.61	686.55	nr	**729.16**
Purpose made double hung sash windows; "treated" wrought softwood						
Cased frames of 100 mm x 25 mm grooved inner linings; 114 mm x 25 mm grooved outer linings; 125 mm x 44 mm twice rebated head linings; 125 mm x 32 mm twice rebated grooved pulley stiles; 150 mm x 13 mm linings; 50 mm x 19 mm partings slips; 25 mm x 19 mm inside beads; 150 mm x 69 mm Oak twice sunk weathered throated sill; 57 mm thick rebated and moulded sashes; moulded horns						
over 1.25 m² each; both sashes in medium panes; including spiral spring balances	482.98	2.31	50.74	635.73	m²	**686.47**
As above but with cased mullions	548.98	2.54	55.79	710.23	m²	**766.02**
Purpose made double hung sash windows; selected Sapele						
Cased frames of 100 mm x 25 mm grooved inner linings; 114 mm x 25 mm grooved outer linings; 125 mm x 38 mm twice reabated head linings; 125 mm x 32 mm twice rebated grooved pulley stiles; 150 mm x 13 mm linings; 50 mm x 19 mm parting slips; 25 mm x 19 mm inside beads; 150 mm x 75 mm Oak twice sunk weathered throated sill; 50 mm thick rebated and moulded sashes; moulded horns						
over 1.25 m² each; both sashes in medium panes; including spiral sash balances	533.20	3.05	66.99	692.42	m²	**759.41**
As above but with cased mullions	572.14	3.38	74.24	736.37	m²	**810.61**

L WINDOWS/DOORS/STAIRS

Item	PC £	Labour hours	Labour £	Material £	Unit	Total rate £
Clements 'EB24' range of factory finished steel fixed light; casement and fanlight windows and doors; with a U-value of 2.0 W/m²K (part L compliant); to EN ISO 9001 2000 ; polyester powder coated; factory glazed with low E double glazing; fixed in position; including lugs plugged and screwed to brickwork or blockwork						
Basic fixed light including easy-glaze ali snap-on beads						
508 mm x 292 mm	157.30	2.20	68.54	177.82	nr	**246.36**
508 mm x 457 mm	171.60	2.20	68.54	193.96	nr	**262.50**
508 mm x 628 mm	185.90	2.20	68.54	210.24	nr	**278.78**
508 mm x 923 mm	214.50	2.20	68.54	242.66	nr	**311.20**
508 mm x 1218 mm	243.10	2.20	68.54	275.08	nr	**343.61**
Basic'Tilt and Turn' window; including easy-glaze ali snap-on beads						
508 mm x 292 mm	371.80	2.20	68.54	419.94	nr	**488.48**
508 mm x 457 mm	386.10	2.20	68.54	436.08	nr	**504.62**
508 mm x 628 mm	400.40	2.20	68.54	452.36	nr	**520.89**
508 mm x 923 mm; including fixed light	471.90	2.20	68.54	533.20	nr	**601.74**
508 mm x 1218 mm; including fixed light	500.50	2.20	68.54	565.62	nr	**634.15**
Basic casement; including easy-glaze ali snap-on beads						
508 mm x 628 mm	443.30	2.20	68.54	500.78	nr	**569.32**
508 mm x 923 mm	471.90	2.20	68.54	533.20	nr	**601.74**
508 mm x 1218 mm	500.50	2.20	68.54	565.62	nr	**634.15**
Double door						
1143 mm x 2057 mm	2802.80	3.30	102.81	3164.61	nr	**3267.41**
Extra over for						
pressed steel sills; to suit above windows	42.90	0.60	13.18	48.56	m	**61.74**
G + bar	85.80	-	-	96.85	m	**-**
simulated leaded lights	85.80	-	-	96.85	m	**-**
uPVC windows; 'Profile 22' or other equal and approved; reinforced where appropriate with aluminium alloy; including standard ironmongery; cills and factory glazed with low E 24mm double glazing; fixed in position; including lugs plugged and screwed to brickwork or blockwork						
Casement/fixed light; including e.p.d.m.glazing gaskets and weather seals						
630 mm x 900 mm; ref P109C	73.00	2.75	85.67	82.94	nr	**168.61**
630 mm x 1200 mm; ref P112C	83.00	2.75	85.67	94.36	nr	**180.03**
1200 mm x 1200 mm; ref P212C	116.00	2.75	85.67	131.75	nr	**217.42**
1770 mm x 1200 mm; ref P312CC	181.00	2.75	85.67	205.11	nr	**290.79**
Casement/fixed light; including vents, e.p.d.m.glazing gaskets and weather seals						
630 mm x 900 mm; ref P109C	90.00	2.75	85.67	102.13	nr	**187.80**
630 mm x 1200 mm; ref P112V	93.00	2.75	85.67	105.65	nr	**191.32**
1200 mm x 1200 mm; ref P212W	152.00	2.75	85.67	172.38	nr	**258.05**
1200 mm x 1200 mm; ref P212CV	146.00	2.75	85.67	165.61	nr	**251.28**
1770 mm x 1200 mm; ref P312WW	217.00	2.75	85.67	245.75	nr	**331.42**
1770 mm x 1200 mm; ref P312CV	184.00	2.75	85.67	208.50	nr	**294.17**

L WINDOWS/DOORS/STAIRS

Item	PC £	Labour hours	Labour £	Material £	Unit	Total rate £
L10 WINDOWS/ROOFLIGHTS/SCREENS/LOUVRES – cont'd						
uPVC windows; 'Profile 22'; reinforced where appropriate with aluminium alloy; in refurbishment work; including standard ironmongery; cills and factory glazed with low E 24mm double glazing; removing existing window and fixing in position; including lugs plugged and screwed to brickwork or blockwork						
Casement/fixed light; including e.p.d.m.glazing gaskets and weather seals						
630 mm x 900 mm; ref P109C	73.00	5.50	171.34	82.94	nr	**254.28**
630 mm x 1200 mm; ref P112C	83.00	5.50	171.34	94.36	nr	**265.71**
1200 mm x 1200 mm; ref P212C	116.00	5.50	171.34	131.75	nr	**303.09**
1770 mm x 1200 mm; ref P312CC	181.00	5.50	171.34	205.11	nr	**376.46**
Casement/fixed light; including vents, e.p.d.m.glazing gaskets and weather seals						
630 mm x 900 mm; ref P109C	90.00	5.50	171.34	102.13	nr	**273.47**
630 mm x 1200 mm; ref P112V	93.00	5.50	171.34	105.65	nr	**276.99**
1200 mm x 1200 mm; ref P212W	152.00	5.50	171.34	172.38	nr	**343.73**
1200 mm x 1200 mm; ref P212CV	146.00	5.50	171.34	165.61	nr	**336.95**
1770 mm x 1200 mm; ref P312WW	217.00	5.50	171.34	245.75	nr	**417.09**
1770 mm x 1200 mm; ref P312CV	184.00	5.50	171.34	208.50	nr	**379.85**
"Kawneer" aluminium window frame system or other equal and approved; polyester powder coated glazing bars; glazed with double hermetically sealed units in toughened safety glass; one 6 mm thick air space; overall 18 mm						
Vertical surfaces						
single tier; aluminium glazing bars at 850 mm centres x 890 mm long; timber supports at 890 mm centres	-	-	-	-	m²	**241.28**
"Kawneer" aluminium window frame system or other equal and approved; polyester powder coated glazing bars; glazed with clear toughened safety glass; 10.70 mm thick						
Vertical surfaces						
single tier; aluminium glazing bars at 850 mm centres x 890 mm long; timber supports at 890 mm centres	-	-	-	-	m²	**229.22**
Rooflights, skylights, roof windows and frames; pre-glazed; "treated" Nordic Red Pine and aluminium trimmed "Velux" windows or other equal and approved; type U flashings and soakers (for tiles and pantiles), and sealed double glazing unit (trimming opening not included)						
Roof windows						
550 mm x 780 mm; ref GGL-3073-C02	202.50	2.05	45.03	229.02	nr	**274.05**
550 mm x 980 mm; ref GGL-3073-C04	211.50	2.31	50.74	239.18	nr	**289.92**
660 mm x 1180 mm; ref GGL-3073-F06	247.50	2.54	55.79	279.91	nr	**335.70**
780 mm x 980 mm; ref GGL-3073-M04	234.00	2.54	55.79	264.67	nr	**320.46**
780 mm x 1180 mm; ref GGL-3073-M06	270.00	3.05	66.99	305.57	nr	**372.57**
780 mm x 1400 mm; ref GGL-3073-M08	283.50	2.54	55.79	320.72	nr	**376.51**
940 mm x 1600 mm; ref GGL-3073-P10	337.50	3.05	66.99	381.76	nr	**448.76**
1140 mm x 1180 mm; ref GGL-3073-S06	315.00	3.05	66.99	356.37	nr	**423.36**
1340 mm x 980 mm; ref GGL-3073-U04	315.00	3.05	66.99	356.37	nr	**423.36**

L WINDOWS/DOORS/STAIRS

Item	PC £	Labour hours	Labour £	Material £	Unit	Total rate £
Rooflights, skylights, roof windows and frames; uPVC; plugged and screwed to concrete; or screwed to timber						
Rooflight; Cox "Suntube" range or other equal and approved; double skin polycarbonate dome						
230 mm dia.; for flat roof using felt or membrane	285.60	3.10	68.09	323.29	nr	**391.39**
230 mm dia.; for up to 30 degrees pitch roof with standard tiles	285.60	3.75	82.37	323.13	nr	**405.50**
230 mm dia.; for up to 30 degrees pitch roof with bold roll tiles	307.65	3.75	82.37	348.10	nr	**430.47**
Rooflight; Cox "Galaxy" range or other equal and approved; double skin polycarbonate dome						
630 mm x 630 mm	129.15	1.90	41.73	145.78	nr	**187.51**
930 mm x 930 mm	238.35	2.20	48.32	269.04	nr	**317.36**
1230 mm x 1230 mm	365.40	2.50	54.91	412.45	nr	**467.36**
Rooflight; Cox " Trade" range or other equal and approved; double skin polycarbonate dome on 150mm PVCU upstand						
600 mm x 600 mm	283.50	2.50	54.91	320.00	nr	**374.91**
900 mm x 900 mm	446.25	2.80	61.50	503.70	nr	**565.21**
1200 mm x 1200 mm	632.10	3.10	68.09	713.48	nr	**781.58**
Rooflight; Cox "2000" range or other equal and approved; double skin polycarbonate dome on 235mm PVCU upstand						
900 mm x 900 mm	1130.85	3.10	68.09	1277.03	nr	**1345.13**
1200 mm x 1200 mm	1552.95	3.75	82.37	1753.56	nr	**1835.93**
Louvres and frames; polyester powder coated aluminium; fixing in position including brackets						
Louvre; Colt Double Bank Universal 2UL; or other equal and approved; including insect mesh						
1500mm x 675 mm (approx. 1.00 m²)				-	nr	**766.19**
L20 DOORS/SHUTTERS/HATCHES						
Doors; standard matchboarded; wrought softwood						
Matchboarded, framed, ledged and braced doors; 44 mm thick overall; 19 mm thick tongued, grooved and V-jointed boarding; one side vertical boarding						
762 mm x 1981 mm	209.41	1.85	40.64	236.37	nr	**277.01**
838 mm x 1981 mm	209.41	1.85	40.64	236.37	nr	**277.01**
Doors; standard flush; softwood composition						
Flush door; internal quality; skeleton or cellular core; hardboard faced both sides; Jeld-Wen "Silverwood" or other equal and approved						
457 mm x 1981 mm x 35 mm	36.03	1.30	28.56	40.67	nr	**69.22**
533 mm x 1981 mm x 35 mm	36.03	1.30	28.56	40.67	nr	**69.22**
610 mm x 1981 mm x 35 mm	36.03	1.30	28.56	40.67	nr	**69.22**
686 mm x 1981 mm x 35 mm	36.03	1.30	28.56	40.67	nr	**69.22**
762 mm x 1981 mm x 35 mm	36.03	1.30	28.56	40.67	nr	**69.22**
838 mm x 1981 mm x 35 mm	37.65	1.30	28.56	42.50	nr	**71.06**
626 mm x 2040 mm x 40 mm	37.65	1.30	28.56	42.50	nr	**71.06**
726 mm x 2040 mm x 40 mm	37.65	1.30	28.56	42.50	nr	**71.06**

L WINDOWS/DOORS/STAIRS

Item	PC £	Labour hours	Labour £	Material £	Unit	Total rate £
L20 DOORS/SHUTTERS/HATCHES – cont'd						
Doors; standard flush; softwood composition – cont'd						
Flush door; internal quality; skeleton or cellular core; faced both sides; lipped on two long edges; Jeld-Wen "Paint grade veneer" or other equal and approved						
457 mm x 1981 mm x 35 mm	35.37	1.30	28.56	39.92	nr	**68.48**
533 mm x 1981 mm x 35 mm	35.37	1.30	28.56	39.92	nr	**68.48**
610 mm x 1981 mm x 35 mm	35.37	1.30	28.56	39.92	nr	**68.48**
686 mm x 1981 mm x 35 mm	35.37	1.30	28.56	39.92	nr	**68.48**
762 mm x 1981 mm x 35 mm	35.37	1.30	28.56	39.92	nr	**68.48**
838 mm x 1981 mm x 35 mm	39.05	1.30	28.56	44.08	nr	**72.64**
526 mm x 2040 mm x 40 mm	36.84	1.30	28.56	41.58	nr	**70.14**
626 mm x 2040 mm x 40 mm	36.84	1.30	28.56	41.58	nr	**70.14**
726 mm x 2040 mm x 40 mm	36.84	1.30	28.56	41.58	nr	**70.14**
826 mm x 2040 mm x 40 mm	42.00	1.30	28.56	47.41	nr	**75.96**
Flush door; internal quality; skeleton or cellular core; veneer faced both sides; lipped on two long edges; Jeld-Wen "Sapele veneered" or other equal and approved						
457 mm x 1981 mm x 35 mm	57.25	1.39	30.53	64.62	nr	**95.16**
533 mm x 1981 mm x 35 mm	57.25	1.39	30.53	64.62	nr	**95.16**
610 mm x 1981 mm x 35 mm	57.25	1.39	30.53	64.62	nr	**95.16**
686 mm x 1981 mm x 35 mm	57.25	1.39	30.53	64.62	nr	**95.16**
762 mm x 1981 mm x 35 mm	57.25	1.39	30.53	64.62	nr	**95.16**
838 mm x 1981 mm x 35 mm	61.89	1.39	30.53	69.86	nr	**100.39**
526 mm x 2040 mm x 40 mm	59.61	1.39	30.53	67.29	nr	**97.82**
626 mm x 2040 mm x 40 mm	59.61	1.39	30.53	67.29	nr	**97.82**
726 mm x 2040 mm x 40 mm	59.61	1.39	30.53	67.29	nr	**97.82**
826 mm x 2040 mm x 40 mm	63.52	1.39	30.53	71.70	nr	**102.23**
Flush door; half-hour fire resisting (FD30); hardboard faced both sides; Jeld-Wen "Silverwood" or other equal and approved						
762 mm x 1981 mm x 44 mm	62.97	1.80	39.54	71.07	nr	**110.61**
838 mm x 1981 mm x 44 mm	63.29	1.80	39.54	71.44	nr	**110.98**
726 mm x 2040 mm x 44 mm	63.29	1.80	39.54	71.44	nr	**110.98**
826 mm x 2040 mm x 44 mm	64.11	1.80	39.54	72.36	nr	**111.90**
Flush door; half-hour fire resisting (FD30); chipboard veneered; faced both sides; lipped on two long edges; Jeld-Wen "Paint grade veneer" or other equal and approved						
610 mm x 1981 mm x 44 mm	50.11	1.80	39.54	56.56	nr	**96.09**
686 mm x 1981 mm x 44 mm	50.11	1.80	39.54	56.56	nr	**96.09**
762 mm x 1981 mm x 44 mm	50.11	1.80	39.54	56.56	nr	**96.09**
838 mm x 1981 mm x 44 mm	53.79	1.80	39.54	60.71	nr	**100.25**
526 mm x 2040 mm x 44 mm	51.58	1.80	39.54	58.22	nr	**97.75**
626 mm x 2040 mm x 44 mm	51.58	1.80	39.54	58.22	nr	**97.75**
726 mm x 2040 mm x 44 mm	51.58	1.80	39.54	58.22	nr	**97.75**
826 mm x 2040 mm x 44 mm	58.95	1.80	39.54	66.54	nr	**106.07**
Flush door; half-hour fire resisting (FD30); veneer faced both sides; lipped on two long edges; Jeld-Wen "Sapele veneered" or other equal and approved						
610 mm x 1981 mm x 44 mm	80.83	1.90	41.73	91.24	nr	**132.97**
686 mm x 1981 mm x 44 mm	80.83	1.90	41.73	91.24	nr	**132.97**
762 mm x 1981 mm x 44 mm	80.83	1.90	41.73	91.24	nr	**132.97**
838 mm x 1981 mm x 44 mm	87.17	1.90	41.73	98.39	nr	**140.13**
726 mm x 2040 mm x 44 mm	83.19	1.90	41.73	93.90	nr	**135.63**
826 mm x 2040 mm x 44 mm	89.52	1.90	41.73	101.05	nr	**142.78**

L WINDOWS/DOORS/STAIRS

Item	PC £	Labour hours	Labour £	Material £	Unit	Total rate £
Flush door; half hour fire resisting; chipboard for painting; hardwood lipping two long edges; "Leaderflush" type B30 or other equal and approved						
526 mm x 2040 mm x 44 mm	114.80	1.90	41.73	129.58	nr	**171.31**
626 mm x 2040 mm x 44 mm	116.52	1.90	41.73	131.52	nr	**173.26**
726 mm x 2040 mm x 44 mm	117.69	1.90	41.73	132.84	nr	**174.58**
826 mm x 2040 mm x 44 mm	118.54	1.90	41.73	133.80	nr	**175.54**
Flush door; half hour fire resisting; American light oak veneer; hardwood lipping all edges; "Leaderflush" type B30 or other equal and approved						
526 mm x 2040 mm x 44 mm	133.45	1.90	41.73	150.63	nr	**192.37**
626 mm x 2040 mm x 44 mm	137.16	1.90	41.73	154.82	nr	**196.55**
726 mm x 2040 mm x 44 mm	139.73	1.90	41.73	157.72	nr	**199.45**
826 mm x 2040 mm x 44 mm	143.17	1.90	41.73	161.60	nr	**203.34**
Flush door; one hour fire resisting; Iroko veneer; hardwood lipping all edges; "Leaderflush" type B60 or other equal and approved; including groove and "Leaderseal" intumescent strip						
457 mm x 1981 mm x 54 mm	170.28	2.13	46.79	192.20	nr	**238.99**
533 mm x 1981 mm x 54 mm	186.43	2.13	46.79	210.43	nr	**257.22**
610 mm x 1981 mm x 54 mm	189.21	2.13	46.79	213.57	nr	**260.36**
686 mm x 1981 mm x 54 mm	192.68	2.13	46.79	217.49	nr	**264.27**
762 mm x 1981 mm x 54 mm	196.46	2.13	46.79	221.75	nr	**268.54**
838 mm x 1981 mm x 54 mm	200.19	2.13	46.79	225.96	nr	**272.75**
526 mm x 2040 mm x 54 mm	186.61	2.13	46.79	210.64	nr	**257.42**
626 mm x 2040 mm x 54 mm	190.25	2.13	46.79	214.74	nr	**261.53**
726 mm x 2040 mm x 54 mm	195.90	2.13	46.79	221.12	nr	**267.91**
826 mm x 2040 mm x 54 mm	200.48	2.13	46.79	226.29	nr	**273.08**
Flush door; external quality; skeleton or cellular core; plywood faced both sides; lipped on all four edges						
762 mm x 1981 mm x 54 mm	53.55	1.80	39.54	60.44	nr	**99.98**
838 mm x 1981 mm x 54 mm	59.06	1.80	39.54	66.67	nr	**106.20**
Flush door; half hour fire resisting; external quality with Georgian wired standard glass opening; skeleton or cellular core; plywood faced both sides; lipped on all four edges; including glazing beads						
762 mm x 1981 mm x 54 mm	300.04	1.80	39.54	338.67	nr	**378.21**
838 mm x 1981 mm x 54 mm	306.34	1.80	39.54	345.78	nr	**385.32**
Doors; purpose made panelled; wrought softwood						
Panelled doors; one open panel for glass; including glazing beads						
686 mm x 1981 mm x 44 mm	112.33	1.80	39.54	126.79	nr	**166.33**
762 mm x 1981 mm x 44 mm	113.27	1.80	39.54	127.85	nr	**167.39**
838 mm x 1981 mm x 44 mm	114.20	1.80	39.54	128.91	nr	**168.45**
Panelled doors; two open panel for glass; including glazing beads						
686 mm x 1981 mm x 44 mm	157.07	1.80	39.54	177.29	nr	**216.83**
762 mm x 1981 mm x 44 mm	158.42	1.80	39.54	178.82	nr	**218.36**
838 mm x 1981 mm x 44 mm	159.78	1.80	39.54	180.35	nr	**219.89**
Panelled doors; four 19 mm thick plywood panels; mouldings worked on solid both sides						
686 mm x 1981 mm x 44 mm	238.50	1.80	39.54	269.21	nr	**308.74**
762 mm x 1981 mm x 44 mm	241.58	1.80	39.54	272.69	nr	**312.23**
838 mm x 1981 mm x 44 mm	244.67	1.80	39.54	276.17	nr	**315.71**

L WINDOWS/DOORS/STAIRS

Item	PC £	Labour hours	Labour £	Material £	Unit	Total rate £
L20 DOORS/SHUTTERS/HATCHES – cont'd						
Doors; purpose made panelled; wrought softwood – cont'd						
Panelled doors; six 25 mm thick panels raised and fielded; mouldings worked on solid both sides						
686 mm x 1981 mm x 44 mm	438.84	2.13	46.79	495.34	nr	**542.13**
762 mm x 1981 mm x 44 mm	443.23	2.13	46.79	500.30	nr	**547.08**
838 mm x 1981 mm x 44 mm	447.61	2.13	46.79	505.24	nr	**552.03**
rebated edges beaded	-	-	-	2.98	m	**-**
rounded edges or heels	-	-	-	0.68	m	**-**
weatherboard fixed to bottom rail	-	0.28	6.15	10.58	m	**16.73**
stopped groove for weatherboard	-	-	-	3.39	m	**-**
Doors; purpose made panelled; selected Sapele						
Panelled doors; one open panel for glass; including glazing beads						
686 mm x 1981 mm x 44 mm	154.59	2.54	55.79	174.49	nr	**230.28**
762 mm x 1981 mm x 44 mm	156.67	2.54	55.79	176.84	nr	**232.64**
838 mm x 1981 mm x 44 mm	158.78	2.54	55.79	179.23	nr	**235.02**
686 mm x 1981 mm x 57 mm	165.24	2.82	61.94	186.51	nr	**248.45**
762 mm x 1981 mm x 57 mm	167.73	2.82	61.94	189.32	nr	**251.26**
838 mm x 1981 mm x 57 mm	170.20	2.82	61.94	192.12	nr	**254.06**
Panelled doors; 250 mm wide cross tongued intermediate rail; two open panels for glass; mouldings worked on the solid one side; 19 mm x 13 mm beads one side; fixing with brass cups and screws						
686 mm x 1981 mm x 44 mm	236.34	2.54	55.79	266.77	nr	**322.56**
762 mm x 1981 mm x 44 mm	240.34	2.54	55.79	271.28	nr	**327.08**
838 mm x 1981 mm x 44 mm	252.00	2.54	55.79	284.45	nr	**340.24**
686 mm x 1981 mm x 57 mm	252.00	2.82	61.94	284.45	nr	**346.39**
762 mm x 1981 mm x 57 mm	256.72	2.82	61.94	289.78	nr	**351.72**
838 mm x 1981 mm x 57 mm	261.57	2.82	61.94	295.24	nr	**357.19**
Panelled doors; four panels; (19 mm thick for 44 mm doors, 25 mm thick for 57 mm doors); mouldings worked on solid both sides						
686 mm x 1981 mm x 44 mm	332.10	2.54	55.79	374.85	nr	**430.65**
762 mm x 1981 mm x 44 mm	357.98	2.54	55.79	404.08	nr	**459.87**
838 mm x 1981 mm x 44 mm	375.69	2.54	55.79	424.06	nr	**479.85**
686 mm x 1981 mm x 57 mm	338.81	2.82	61.94	382.43	nr	**444.37**
762 mm x 1981 mm x 57 mm	366.84	2.82	61.94	414.07	nr	**476.02**
838 mm x 1981 mm x 57 mm	345.53	2.82	61.94	390.02	nr	**451.96**
Panelled doors; 150 mm wide stiles in one width; 430 mm wide cross tongued bottom rail; six panels raised and fielded one side; (19 mm thick for 44 mm doors, 25 mm thick for 57 mm doors); mouldings worked on solid both sides						
686 mm x 1981 mm x 44 mm	566.91	2.54	55.79	639.90	nr	**695.69**
762 mm x 1981 mm x 44 mm	625.62	2.54	55.79	706.16	nr	**761.96**
838 mm x 1981 mm x 44 mm	637.50	2.54	55.79	719.58	nr	**775.37**
686 mm x 1981 mm x 57 mm	603.68	2.82	61.94	681.40	nr	**743.35**
762 mm x 1981 mm x 57 mm	665.64	2.82	61.94	751.35	nr	**813.29**
838 mm x 1981 mm x 57 mm	717.40	2.82	61.94	809.76	nr	**871.70**
rebated edges beaded	-	-	-	3.86	m	**-**
rounded edges or heels	-	-	-	1.02	m	**-**
weatherboard fixed to bottom rail	-	0.32	7.03	14.54	m	**21.57**
stopped groove for weatherboard	-	-	-	3.58	m	**-**

L WINDOWS/DOORS/STAIRS

Item	PC £	Labour hours	Labour £	Material £	Unit	Total rate £
Doors; galvanised steel "up and over" type garage doors; Catnic "Horizon 90" or other equal and approved; spring counterbalanced; fixed to timber frame (not included)						
Garage door						
2135 mm x 1980 mm	262.65	4.07	89.40	297.01	nr	**386.41**
2135 mm x 2135 mm	300.90	4.07	89.40	340.18	nr	**429.58**
2400 mm x 2135 mm	367.20	4.07	89.40	415.11	nr	**504.51**
3965 mm x 2135 mm	961.35	6.11	134.21	1086.38	nr	**1220.59**
Doorsets; galvanised steel IG "Weatherbeater Original" door and frame units or other equal and approved; treated softwood frame, primed hardwood sill; fixing in position; plugged and screwed to brickwork or blockwork						
Door and frame						
762 mm x 1981 mm; ref IGD1	174.11	3.05	66.99	197.07	nr	**264.06**
Doorsets; steel door and frame units; Jandor Architectural Ltd or other equal and approved; polyester powder coated; ironmongery						
Single action door set; "Metset MD01" doors and "Metset MF" frames						
900 mm x 2100 mm	-	-	-	-	nr	**2237.98**
pair 1800 mm x 2100 mm	-	-	-	-	nr	**3091.50**
Doorsets; steel bullet-resistant door and frame units; Wormald Doors or other equal and approved; Medite laquered panels; ironmongery						
Door and frame						
1000 mm x 2060mm overall; fixed to masonry	-	-	-	-	nr	**4302.44**
Rolling shutters and collapsible gates; steel counter shutters; Bolton Brady Ltd or other equal and approved; push-up, self-coiling; polyester power coated; fixing by bolting						
Shutters						
3000 mm x 1000 mm	-	-	-	-	nr	**1382.34**
4000 mm x 1000 mm; in two panels	-	-	-	-	nr	**2400.90**
Rolling shutters and collapsible gates; galvanised steel; Bolton Brady Type 474 or other equal and approved; one hour fibre resisting; self-coiling; activated by fusible link; fixing with bolts						
Rolling shutters and collapsible gates						
1000 mm x 2750 mm	-	-	-	-	nr	**1673.35**
1500 mm x 2750 mm	-	-	-	-	nr	**1760.66**
2400 mm x 2750 mm	-	-	-	-	nr	**2037.13**

L WINDOWS/DOORS/STAIRS

Item	PC £	Labour hours	Labour £	Material £	Unit	Total rate £
L20 DOORS/SHUTTERS/HATCHES – cont'd						
Sliding/folding partitions; aluminium double glazed sliding patio doors; Crittal "Luminaire" or other equal and approved; white acrylic finish; with and including 18 thick annealed double glazing; fixed in position; including lugs plugged and screwed to brickwork or blockwork						
Patio doors						
1800 mm x 2100 mm; ref PF1821	1844.94	2.54	55.79	2084.10	nr	**2139.89**
2400 mm x 2100 mm; ref PF2421	2213.94	3.05	66.99	2500.60	nr	**2567.60**
2700 mm x 2100 mm; ref PF2721	2459.93	3.56	78.20	2778.27	nr	**2856.47**
Grilles; "Galaxy" nylon rolling counter grille or other equal and approved; Bolton Brady Ltd; colour, off-white; self-coiling; fixing by bolting						
Grilles						
3000 mm x 1000 mm	-	-	-	-	nr	**1056.22**
4000 mm x 1000 mm	-	-	-	-	nr	**1615.40**
External softwood door frame composite standard joinery sets						
External door frame composite set; 56 mm x 78mm wide (finished); for external doors						
762 mm x 1981 mm x 44 mm	48.48	0.83	18.23	55.26	nr	**73.49**
813 mm x 1981 mm x 44 mm	48.84	0.83	18.23	55.67	nr	**73.90**
838 mm x 1981 mm x 44 mm	48.84	0.83	18.23	55.67	nr	**73.90**
External door frame composite set; 56 mm x 78mm wide (finished); with 45 mm x 140 mm (finished) hardwood cill; for external doors						
686 mm x 1981 mm x 44 mm	71.81	1.10	24.16	81.59	nr	**105.76**
762 mm x 1981 mm x 44 mm	74.06	1.10	24.16	84.14	nr	**108.30**
838 mm x 1981 mm x 44 mm	77.03	1.10	24.16	87.49	nr	**111.65**
826 mm x 2040 mm x 44 mm	79.37	1.10	24.16	90.13	nr	**114.29**
Internal white foiled moisture-resistant MDF door lining composite standard joinery set						
22 mm x 77mm wide (finished) set; with loose stops; for internal doors						
610 mm x 1981 mm x 35 mm	19.99	0.78	17.13	23.11	nr	**40.24**
686 mm x 1981 mm x 35 mm	19.97	0.78	17.13	23.08	nr	**40.21**
762 mm x 1981 mm x 35 mm	19.97	0.78	17.13	23.08	nr	**40.21**
838 mm x 1981 mm x 35 mm	20.91	0.78	17.13	24.14	nr	**41.27**
864 mm x 1981 mm x 35 mm	20.91	0.78	17.13	24.14	nr	**41.27**
22 mm x 150mm wide (finished) set; with loose stops; for internal doors						
610 mm x 1981 mm x 35 mm	25.89	0.78	17.13	29.77	nr	**46.90**
686 mm x 1981 mm x 35 mm	25.66	0.78	17.13	29.51	nr	**46.64**
762 mm x 1981 mm x 35 mm	25.89	0.78	17.13	29.77	nr	**46.90**
838 mm x 1981 mm x 35 mm	27.15	0.78	17.13	31.18	nr	**48.32**
864 mm x 1981 mm x 35 mm	27.15	0.78	17.13	31.18	nr	**48.32**

L WINDOWS/DOORS/STAIRS

Item	PC £	Labour hours	Labour £	Material £	Unit	Total rate £
Door frames and door linings, sets; purpose made; wrought softwood						
Jambs and heads; as linings						
32 mm x 63 mm	-	0.19	4.17	7.50	m	**11.68**
32 mm x 100 mm	-	0.19	4.17	8.43	m	**12.60**
32 mm x 140 mm	-	0.19	4.17	9.04	m	**13.22**
Jambs and heads; as frames; rebated, rounded and grooved						
44 mm x 75 mm	-	0.19	4.17	12.02	m	**16.19**
44 mm x 100 mm	-	0.19	4.17	12.91	m	**17.09**
44 mm x 115 mm	-	0.19	4.17	13.05	m	**17.22**
44 mm x 140 mm	-	0.19	4.17	13.67	m	**17.84**
57 mm x 100 mm	-	0.19	4.17	13.86	m	**18.04**
57 mm x 125 mm	-	0.19	4.17	14.63	m	**18.81**
69 mm x 88 mm	-	0.19	4.17	14.04	m	**18.22**
69 mm x 100 mm	-	0.19	4.17	15.10	m	**19.27**
69 mm x 125 mm	-	0.23	5.05	16.02	m	**21.08**
69 mm x 150 mm	-	0.23	5.05	16.96	m	**22.01**
94 mm x 100 mm	-	0.28	6.15	22.83	m	**28.98**
94 mm x 150 mm	-	0.28	6.15	26.59	m	**32.74**
Mullions and transoms; in linings						
32 mm x 63 mm	-	0.14	3.08	9.87	m	**12.95**
32 mm x 100 mm	-	0.14	3.08	10.82	m	**13.90**
32 mm x 140 mm	-	0.14	3.08	11.31	m	**14.38**
Mullions and transoms; in frames; twice rebated, rounded and grooved						
44 mm x 75 mm	-	0.14	3.08	14.98	m	**18.06**
44 mm x 100 mm	-	0.14	3.08	15.62	m	**18.69**
44 mm x 115 mm	-	0.14	3.08	15.62	m	**18.69**
44 mm x 140 mm	-	0.14	3.08	16.22	m	**19.30**
57 mm x 100 mm	-	0.14	3.08	16.41	m	**19.49**
57 mm x 125 mm	-	0.14	3.08	17.20	m	**20.27**
69 mm x 88 mm	-	0.14	3.08	16.22	m	**19.30**
69 mm x 100 mm	-	0.14	3.08	17.16	m	**20.23**
Add 5% to the above material prices for selected softwood for staining						
Door frames and door linings, sets; purpose made; medium density fibreboard						
Jambs and heads; as linings						
18 mm x 126 mm	-	0.19	4.17	8.34	m	**12.51**
22 mm x 126 mm	-	0.19	4.17	8.64	m	**12.82**
25 mm x 126 mm	-	0.19	4.17	8.88	m	**13.05**
Door frames and door linings, sets; purpose made; selected Sapele						
Jambs and heads; as linings						
32 mm x 63 mm	8.15	0.23	5.05	9.34	m	**14.39**
32 mm x 100 mm	13.26	0.23	5.05	15.10	m	**20.16**
32 mm x 140 mm	14.53	0.23	5.05	16.67	m	**21.72**
Jambs and heads; as frames; rebated, rounded and grooved						
44 mm x 75 mm	17.58	0.23	5.05	19.98	m	**25.03**
44 mm x 100 mm	20.03	0.23	5.05	22.75	m	**27.80**
44 mm x 115 mm	20.72	0.23	5.05	23.66	m	**28.71**
44 mm x 140 mm	21.70	0.28	6.15	24.76	m	**30.91**
57 mm x 100 mm	22.19	0.28	6.15	25.32	m	**31.47**
57 mm x 125 mm	24.28	0.28	6.15	27.67	m	**33.82**

Item	PC £	Labour hours	Labour £	Material £	Unit	Total rate £
L20 DOORS/SHUTTERS/HATCHES – cont'd						
Door frames and door linings, sets; purpose made; selected Sapele – cont'd						
Jambs and heads; as frames – cont'd						
69 mm x 88 mm	22.16	0.28	6.15	25.15	m	**31.30**
69 mm x 100 mm	24.65	0.28	6.15	28.10	m	**34.25**
69 mm x 125 mm	27.13	0.32	7.03	30.89	m	**37.92**
69 mm x 150 mm	29.63	0.32	7.03	33.72	m	**40.75**
94 mm x 100 mm	34.60	0.32	7.03	39.32	m	**46.35**
94 mm x 150 mm	42.45	0.32	7.03	48.19	m	**55.22**
Mullions and transoms; in linings						
32 mm x 63 mm	13.47	0.19	4.17	15.20	m	**19.38**
32 mm x 100 mm	15.98	0.19	4.17	18.04	m	**22.21**
32 mm x 140 mm	17.25	0.19	4.17	19.47	m	**23.64**
Mullions and transoms; in frames; twice rebated, rounded and grooved						
44 mm x 75 mm	21.65	0.19	4.17	24.43	m	**28.61**
44 mm x 100 mm	23.31	0.19	4.17	26.31	m	**30.48**
44 mm x 115 mm	23.97	0.19	4.17	27.05	m	**31.23**
44 mm x 140 mm	24.96	0.19	4.17	28.18	m	**32.35**
57 mm x 100 mm	25.45	0.19	4.17	28.73	m	**32.90**
57 mm x 125 mm	27.55	0.19	4.17	31.10	m	**35.28**
69 mm x 88 mm	24.96	0.19	4.17	28.18	m	**32.35**
69 mm x 100 mm	28.10	0.19	4.17	31.71	m	**35.89**
Sills; once sunk weathered; once rebated, three times grooved						
57 mm x 175 mm	62.18	0.32	7.03	70.18	m	**77.21**
69 mm x 125 mm	59.89	0.32	7.03	67.61	m	**74.64**
69 mm x 150 mm	62.72	0.32	7.03	70.80	m	**77.82**
Door frames and door linings, sets; European Oak						
Sills; once sunk weathered; once rebated, three times grooved						
57 mm x 175 mm	106.97	0.32	7.03	120.75	m	**127.78**
69 mm x 125 mm	105.61	0.32	7.03	119.21	m	**126.24**
69 mm x 150 mm	115.47	0.32	7.03	130.33	m	**137.36**
Bedding and pointing frames						
Pointing wood frames or sills with mastic						
one side	-	0.09	1.45	0.65	m	**2.10**
both sides	-	0.19	3.06	1.30	m	**4.36**
Pointing wood frames or sills with polysulphide sealant						
one side	-	0.09	1.45	2.06	m	**3.51**
both sides	-	0.19	3.06	4.13	m	**7.19**
Bedding wood frames in cement mortar (1:3) and point						
one side	-	0.09	2.47	0.09	m	**2.56**
both sides	-	0.09	2.47	0.11	m	**2.58**
one side in mortar; other side in mastic	-	0.19	4.19	0.74	m	**4.93**

L WINDOWS/DOORS/STAIRS

Item	PC £	Labour hours	Labour £	Material £	Unit	Total rate £
L30 STAIRS/WALKWAYS/BALUSTRADES						
Standard staircases; wrought softwood (parana pine)						
Stairs; 25 mm thick treads with rounded nosings; 9 mm thick plywood risers; 32 mm thick strings; bullnose bottom tread; 50 mm x 75 mm hardwood handrail; 32 mm square plain balusters; 100 mm square plain newel posts						
straight flight; 838 mm wide; 2676 mm going; 2600 mm rise; with two newel posts	-	7.12	156.39	418.25	nr	**574.64**
straight flight with turn; 838 wide; 2676 going; 2600 rise; with two newel posts; three top treads winding	-	7.12	156.39	433.24	nr	**589.64**
dogleg staircase; 838 mm wide; 2676 mm going; 2600 mm rise; with two newel posts; quarter space landing third riser from top	-	7.12	156.39	511.74	nr	**668.14**
dogleg staircase; 838 mm wide; 2676 mm going; 2600 mm rise; with two newel posts; half space landing third riser from top	-	8.14	178.80	630.81	nr	**809.61**
Standard balustrades; wrought softwood						
Landing balustrade; 50 mm x 75 mm hardwood handrail; 32 mm square plain balusters; one end of handrail jointed to newel post; other end built into wall; balusters housed in at bottom (newel post and mortices both not included)						
3.00 m long	-	4.07	89.40	105.84	nr	**195.24**
Hardwood staircases; purpose made; assembled at works						
Fixing only complete staircase including landings, balustrades, etc.						
plugging and screwing to brickwork or blockwork	-	15.26	335.19	3.30	nr	**338.49**
The following are supply only prices for purpose made staircase components in selected Sapele supplied as part of an assembled staircase and may be used to arrive at a guide price for a complete hardwood staircase						
Board landings; cross-tongued joints; 100 mm x 50 mm sawn softwood bearers						
25 mm thick	-	-	-	131.56	m²	-
32 mm thick	-	-	-	147.89	m²	-
Treads; cross-tongued joints and risers; rounded nosings; tongued, grooved, glued and blocked together; one 175 mm x 50 mm sawn softwood carriage						
25 mm treads; 19 mm risers	-	-	-	264.39	m²	-
ends; quadrant	-	-	-	80.54	nr	-
ends; housed to hardwood	-	-	-	1.49	nr	-
32 mm treads; 25 mm risers	-	-	-	273.91	m²	-
ends; quadrant	-	-	-	103.53	nr	-
ends; housed to hardwood	-	-	-	1.49	nr	-

L WINDOWS/DOORS/STAIRS

Item	PC £	Labour hours	Labour £	Material £	Unit	Total rate £
L30 STAIRS/WALKWAYS/BALUSTRADES – cont'd						
Supply only prices for purpose made selected Sapele staircase components – cont'd						
Winders; cross-tongued joints and risers in one width; rounded nosings; tongued, grooved glued and blocked together; one 175 mm x 50 mm sawn softwood carriage						
25 mm treads; 19 mm risers	-	-	-	367.56	m²	-
32 mm treads; 25 mm risers	-	-	-	376.19	m²	-
wide ends; housed to hardwood	-	-	-	2.96	nr	-
narrow ends; housed to hardwood	-	-	-	2.24	nr	-
Closed strings; in one width; 230 mm wide; rounded twice						
32 mm thick	-	-	-	48.41	m	-
44 mm thick	-	-	-	52.72	m	-
57 mm thick	-	-	-	58.71	m	-
Closed strings; cross-tongued joints; 280 mm wide; once rounded						
32 mm thick	-	-	-	62.84	m	-
extra for short ramp	-	-	-	32.28	nr	-
44 mm thick	-	-	-	68.54	m	-
extra for short ramp	-	-	-	36.69	nr	-
57 mm thick	-	-	-	76.50	m	-
extra for short ramp	-	-	-	45.44	nr	-
Closed strings; ramped; crossed tongued joints 280 mm wide; once rounded						
32 mm thick	-	-	-	62.84	m	-
44 mm thick	-	-	-	68.54	m	-
57 mm thick	-	-	-	76.50	m	-
Apron linings; in one width 230 mm wide						
19 mm thick	-	-	-	16.70	m	-
25 mm thick	-	-	-	19.69	m	-
Handrails; rounded						
40 mm x 50 mm	-	-	-	18.15	m	-
50 mm x 75 mm	-	-	-	21.88	m	-
57 mm x 87 mm	-	-	-	25.67	m	-
69 mm x 100 mm	-	-	-	31.90	m	-
Handrails; moulded						
40 mm x 50 mm	-	-	-	20.19	m	-
50 mm x 75 mm	-	-	-	23.91	m	-
57 mm x 87 mm	-	-	-	27.73	m	-
69 mm x 100 mm	-	-	-	33.93	m	-
Add to above for						
grooved once	-	-	-	0.90	m	-
ends; framed	-	-	-	8.50	nr	-
ends; framed on rake	-	-	-	10.41	nr	-
Heading joints to handrail; mitred or raked						
overall size not exceeding 50 mm x 75 mm	-	-	-	41.65	nr	-
overall size not exceeding 69 mm x 100 mm	-	-	-	52.06	nr	-
Knee piece to handrail; mitred or raked						
overall size not exceeding 69 mm x 100 mm	-	-	-	111.07	nr	-
Balusters; stiffeners						
25 mm x 25 mm	-	-	-	4.64	m	-
32 mm x 32 mm	-	-	-	5.29	m	-
44 mm x 44 mm	-	-	-	6.94	m	-
ends; housed	-	-	-	2.08	nr	-
Sub rails						
32 mm x 63 mm	-	-	-	10.70	m	-
ends; framed joint to newel	-	-	-	9.02	nr	-

L WINDOWS/DOORS/STAIRS

Item	PC £	Labour hours	Labour £	Material £	Unit	Total rate £
Knee rails						
32 mm x 140 mm	-	-	-	17.76	m	-
ends; framed joint to newel	-	-	-	9.02	nr	-
Newel posts						
44 mm x 94 mm; half newel	-	-	-	12.33	m	-
69 mm x 69 mm	-	-	-	13.37	m	-
94 mm x 94 mm	-	-	-	27.34	m	-
Newel caps; splayed on four sides						
62.50 mm x 125 mm x 50 mm	-	-	-	13.04	nr	-
100 mm x 100 mm x 50 mm	-	-	-	13.31	nr	-
125 mm x 125 mm x 50 mm	-	-	-	13.98	nr	-
The following are supply only prices for purpose made staircase components in selected American Oak; supplied as part of an assembled staircase						
Board landings; cross-tongued joints; 100 mm x 50 mm sawn softwood bearers						
25 mm thick	-	-	-	208.29	m²	-
32 mm thick	-	-	-	251.53	m²	-
Treads; cross-tongued joints and risers; rounded nosings; tongued, grooved, glued and blocked together; one 175 mm x 50 mm sawn softwood carriage						
25 mm treads; 19 mm risers	-	-	-	345.79	m²	-
ends; quadrant	-	-	-	172.73	nr	-
ends; housed to hardwood	-	-	-	2.12	nr	-
32 mm treads; 25 mm risers	-	-	-	396.22	m²	-
ends; quadrant	-	-	-	212.63	nr	-
ends; housed to hardwood	-	-	-	2.12	nr	-
Winders; cross-tongued joints and risers in one width; rounded nosings; tongued, grooved glued and blocked together; one 175 mm x 50 mm sawn softwood carriage						
25 mm treads; 19 mm risers	-	-	-	438.33	m²	-
32 mm treads; 25 mm risers	-	-	-	478.26	m²	-
wide ends; housed to hardwood	-	-	-	4.26	nr	-
narrow ends; housed to hardwood	-	-	-	3.19	nr	-
Closed strings; in one width; 230 mm wide; rounded twice						
32 mm thick	-	-	-	81.69	m	-
44 mm thick	-	-	-	94.28	m	-
57 mm thick	-	-	-	129.45	m	-
Closed strings; cross-tongued joints; 280 mm wide; once rounded						
32 mm thick	-	-	-	103.79	m	-
extra for short ramp	-	-	-	59.34	nr	-
44 mm thick	-	-	-	120.30	m	-
extra for short ramp	-	-	-	67.60	nr	-
57 mm thick	-	-	-	164.92	m	-
extra for short ramp	-	-	-	89.92	nr	-
Closed strings; ramped; crossed tongued joints 280 mm wide; once rounded						
32 mm thick	-	-	-	119.36	m	-
44 mm thick	-	-	-	138.34	m	-
57 mm thick	-	-	-	189.66	m	-
Apron linings; in one width 230 mm wide						
19 mm thick	-	-	-	27.82	m	-
25 mm thick	-	-	-	34.09	m	-

L WINDOWS/DOORS/STAIRS

Item	PC £	Labour hours	Labour £	Material £	Unit	Total rate £
L30 STAIRS/WALKWAYS/BALUSTRADES – cont'd						
Supply only prices for selected American Oak purpose made staircase components – cont'd						
Handrails; rounded						
40 mm x 50 mm	-	-	-	22.54	m	-
50 mm x 75 mm	-	-	-	28.89	m	-
57 mm x 87 mm	-	-	-	43.36	m	-
69 mm x 100 mm	-	-	-	58.38	m	-
Handrails; moulded						
40 mm x 50 mm	-	-	-	24.73	m	-
50 mm x 75 mm	-	-	-	31.08	m	-
57 mm x 87 mm	-	-	-	45.55	m	-
69 mm x 100 mm	-	-	-	60.57	m	-
Add to above for						
grooved once	-	-	-	1.12	m	-
ends; framed	-	-	-	11.17	nr	-
ends; framed on rake	-	-	-	14.15	nr	-
Heading joints to handrail; mitred or raked						
overall size not exceeding 50 mm x 75 mm	-	-	-	59.60	nr	-
overall size not exceeding 69 mm x 100 mm	-	-	-	70.77	nr	-
Knee piece to handrail; mitred or raked						
overall size not exceeding 69 mm x 100 mm	-	-	-	126.65	nr	-
Balusters; stiffeners						
25 mm x 25 mm	-	-	-	5.13	m	-
32 mm x 32 mm	-	-	-	6.48	m	-
44 mm x 44 mm	-	-	-	10.24	m	-
ends; housed	-	-	-	2.61	nr	-
Sub rails						
32 mm x 63 mm	-	-	-	13.87	m	-
ends; framed joint to newel	-	-	-	11.17	nr	-
Knee rails						
32 mm x 140 mm	-	-	-	24.11	m	-
ends; framed joint to newel	-	-	-	11.17	nr	-
Newel posts						
44 mm x 94 mm; half newel	-	-	-	18.09	m	-
69 mm x 69 mm	-	-	-	30.81	m	-
94 mm x 94 mm	-	-	-	76.91	m	-
Newel caps; splayed on four sides						
62.50 mm x 125 mm x 50 mm	-	-	-	15.51	nr	-
100 mm x 100 mm x 50 mm	-	-	-	16.39	nr	-
125 mm x 125 mm x 50 mm	-	-	-	18.03	nr	-
Aluminium alloy folding loft ladders; "Zig Zag" stairways, model B or other equal and approved; Light Alloy Ltd; on and including plywood backboard; fixing with screws to timber lining (not included)						
Loft ladders						
ceiling height not exceeding 2500 mm; model 888801	-	1.02	22.40	569.21	nr	**591.62**
ceiling height not exceeding 2800 mm; model 888802	-	1.02	22.40	624.12	nr	**646.52**
ceiling height not exceeding 3100 mm; model 888803	-	1.02	22.40	696.55	nr	**718.96**

L WINDOWS/DOORS/STAIRS

Item	PC £	Labour hours	Labour £	Material £	Unit	Total rate £
Flooring, balustrades and handrails; metalwork						
Chequer plate flooring; galvanised mild steel; over 300 mm wide; bolted to steel supports						
6 mm thick	-	-	-	-	m²	**307.50**
8 mm thick	-	-	-	-	m²	**328.00**
Open mesh flooring; galvanised; over 300 mm wide; bolted to steel supports						
8 mm thick	-	-	-	-	m²	**307.50**
Balustrades; galvanised mild steel CHS posts and top rail, with one infill rail						
1100 mm high	-	-	-	-	m	**256.25**
Balustrades; painted mild steel flat bar posts and CHS top rail, with 3 nr. stainless steel infills						
1100 mm high	-	-	-	-	m	**358.75**
Balustrades; stainless steel flat bar posts and circular handrail, with 3 nr. stainless steel infills						
1100 mm high	-	-	-	-	m	**430.50**
Balustrades; stainless steel 50 mm Ø posts and circular handrail, with 10 mm thick toughened glass infill panels						
1100 mm high	-	-	-	-	m	**789.25**
Balustrades; laminated glass; with stainless steel cap channel to top and including all necessary support fixings						
1100 mm high	-	-	-	-	m	**1296.63**
Wallrails; painted mild steel CHS wall rail; with wall rose bracket						
42 mm diameter	-	-	-	-	m	**102.50**
Wallrails; stainless steel circular wall rail; with wall rose bracket						
42 mm diameter	-	-	-	-	m	**143.50**
Surface treatment						
At works						
galvanising	-	-	-	-	tonne	**461.25**
shotblasting	-	-	-	-	m²	**4.10**
touch up primer and one coat of two pack epoxy zinc phosphate or chromate primer	-	-	-	-	m²	**8.20**
L40 GENERAL GLAZING						
Standard plain glass; BS EN 14449; clear float; panes area 0.15 m² - 4.00 m²						
3 mm thick; glazed with screwed beads	-	-	-	-	m²	**32.09**
4 mm thick; glazed with screwed beads	-	-	-	-	m²	**34.01**
5 mm thick; glazed with screwed beads	-	-	-	-	m²	**41.43**
6 mm thick; glazed with screwed beads	-	-	-	-	m²	**45.39**
Standard plain glass; BS EN 14449; obscure patterned; panes area 0.15 m² - 4.00 m²						
4 mm thick; glazed with screwed beads	-	-	-	-	m²	**48.11**
6 mm thick; glazed with screwed beads	-	-	-	-	m²	**52.94**

L WINDOWS/DOORS/STAIRS

Item	PC £	Labour hours	Labour £	Material £	Unit	Total rate £
L40 GENERAL GLAZING – cont'd						
Standard plain glass; BS EN 14449; rough cast; panes area 0.15 m² - 4.00 m²						
6 mm thick; glazed with screwed beads	-	-	-	-	m²	**38.45**
Standard plain glass; BS EN 14449; Georgian wired cast; panes area 0.15 m² - 4.00 m²						
7 mm thick; glazed with screwed beads	-	-	-	-	m²	**39.13**
Extra for lining up wired glass	-	-	-	-	m²	**3.91**
Standard plain glass; BS EN 14449; Georgian wired polished; panes area 0.15 m² - 4.00 m²						
6 mm thick; glazed with screwed beads	-	-	-	-	m²	**60.70**
Extra for lining up wired glass	-	-	-	-	m²	**3.91**
Special glass; BS EN 14449; toughened clear float; panes area 0.15 m² - 4.00 m²						
4 mm thick; glazed with screwed beads	-	-	-	-	m²	**34.12**
5 mm thick; glazed with screwed beads	-	-	-	-	m²	**45.31**
6 mm thick; glazed with screwed beads	-	-	-	-	m²	**49.85**
10 mm thick; glazed with screwed beads	-	-	-	-	m²	**82.71**
Special glass; BS EN 14449; clear laminated safety glass; panes area 0.15 m² - 4.00 m²						
4.40 mm thick; glazed with screwed beads	-	-	-	-	m²	**54.66**
6.40 mm thick; glazed with screwed beads	-	-	-	-	m²	**65.38**
Special glass; BS EN 14449; "Pyran" half-hour fire resisting glass or other equal and approved						
6.50 mm thick rectangular panes; glazed with screwed hardwood beads and Sealmaster "Fireglaze" intumescent compound or other equal and approved to rebated frame						
300 mm x 400 mm pane	-	0.42	13.08	54.98	nr	**68.07**
400 mm x 800 mm pane	-	0.51	15.89	138.05	nr	**153.94**
500 mm x 1400 mm pane	-	0.83	25.86	293.24	nr	**319.09**
600 mm x 1800 mm pane	-	1.02	31.78	485.40	nr	**517.17**
Special glass; BS EN 14449; "Pyrostop" one-hour fire resisting glass or other equal and approved						
15 mm thick regular panes; glazed with screwed hardwood beads and Sealmaster "Fireglaze" intumescent liner and compound or other equal and approved both sides						
300 mm x 400 mm pane	-	1.20	37.38	116.62	nr	**154.00**
400 mm x 800 mm pane	-	1.53	47.66	230.10	nr	**277.77**
500 mm x 1400 mm pane	-	2.04	63.55	460.89	nr	**524.45**
600 mm x 1800 mm pane	-	2.54	79.13	677.38	nr	**756.51**

L WINDOWS/DOORS/STAIRS

Item	PC £	Labour hours	Labour £	Material £	Unit	Total rate £
Special glass; BS EN 14449; clear laminated security glass						
7.50 mm thick regular panes; glazed with screwed hardwood beads and "Intergens" intumescent strip or other equal and approved						
300 mm x 400 mm pane	-	0.42	13.08	34.50	nr	**47.58**
400 mm x 800 mm pane	-	0.51	15.89	84.29	nr	**100.18**
500 mm x 1400 mm pane	-	0.83	25.86	175.88	nr	**201.74**
600 mm x 1800 mm pane	-	1.02	31.78	302.63	nr	**334.41**
Mirror panels; BS EN 14449; silvered backing						
4 mm thick float; fixing with adhesive						
1000 mm x 1000 mm	-	-	-	-	nr	**38.68**
1000 mm x 2000 mm	-	-	-	-	nr	**77.41**
1000 mm x 4000 mm	-	-	-	-	nr	**282.13**
Glass louvres; BS EN 14449; with long edges ground or smooth						
6 mm thick float						
150 mm wide	-	-	-	-	m	**17.06**
7 mm thick Georgian wired cast						
150 mm wide	-	-	-	-	m	**23.67**
6 mm thick Georgian wire polished						
150 mm wide	-	-	-	-	m	**33.76**
Factory made double hermetically sealed units; to wood or metal with screwed or clipped beads						
Two panes; BS EN 14449; clear float glass; 4 mm thick; 6 mm air space						
0.35 m² - 2.00 m²	-	-	-	-	m²	**86.01**
Two panes; BS EN 14449; clear float glass; 6 mm thick; 6 mm air space						
0.35 m² - 2.00 m²	-	-	-	-	m²	**100.17**
2.00 m² - 4.00 m²	-	-	-	-	m²	**150.62**
Factory made double hermetically sealed units; with inner pane of Pilkington's "K" low emissivity coated glass; to wood or metal with screwed or clipped beads						
Two panes; BS EN 14449; clear float glass; 4 mm thick; 6 mm air space						
0.35 m² - 2.00 m²	-	-	-	-	m²	**104.63**
Two panes; BS EN 14449; clear float glass; 6 mm thick; 6 mm air space						
0.35 m² - 2.00 m²	-	-	-	-	m²	**121.84**
2.00 m² - 4.00 m²	-	-	-	-	m²	**183.21**
Factory made triple hermetically sealed units; with inner pane of Pilkington's "K" low emissivity coated glass; to wood or metal with screwed or clipped beads						
Three panes; BS EN 14449; clear float glass; 4 mm thick; 6 mm air spaces						
0.35 m² - 2.00 m²	-	-	-	-	m²	**168.36**
Three panes; BS EN 14449; clear float glass; 6 mm thick; 6 mm air spaces						
0.35 m² - 2.00 m²	-	-	-	-	m²	**204.92**
2.00 m² - 4.00 m²	-	-	-	-	m²	**294.97**

M SURFACE FINISHES

Item	PC £	Labour hours	Labour £	Material £	Unit	Total rate £
M10 CEMENT:SAND/CONCRETE/GRANOLITHIC SCREEDS/TOPPINGS						
Cement:sand (1:3) screeds ; steel trowelled						
Work to floors; one coat level; to concrete base; screeded; over 300 mm wide						
25 mm thick	-	-	-	-	m²	**11.23**
50 mm thick	-	-	-	-	m²	**13.29**
75 mm thick	-	-	-	-	m²	**17.59**
100 mm thick	-	-	-	-	m²	**21.89**
Add to the above for work to falls and crossfalls and to slopes						
not exceeding 15 degrees from horizontal	-	0.02	0.45	-	m²	**0.45**
over 15 degrees from horizontal	-	0.10	2.26	-	m²	**2.26**
water repellent additive incorporated in the mix	-	0.02	0.45	6.07	m²	**6.52**
oil repellent additive incorporated in the mix	-	0.08	1.81	5.62	m²	**7.43**
Fine concrete (1:4-5) levelling screed; steel trowelled						
Work to floors; one coat; level; to concrete base; over 300 mm wide						
50 mm thick	-	-	-	-	m²	**13.29**
75 mm thick	-	-	-	-	m²	**17.59**
Extra over last for isolation joint to perimeter	-	-	-	-	m	**1.59**
Early drying floor screed; RMC Mortars "Readyscreed"; or other equal and approved; steel trowelled						
Work to floors; one coat; level; to concrete base; over 300 mm wide						
100 mm thick	-	-	-	-	m²	**25.63**
Extra over last for galvanised chicken wire anticrack reinforcement	-	-	-	-	m²	**1.22**
Granolithic paving; cement and granite chippings 5 mm to dust (1:1:2); steel trowelled						
Work to floors; one coat; level; laid on concrete while green; bonded; over 300 mm wide						
25 mm thick	-	-	-	-	m²	**30.25**
38 mm thick	-	-	-	-	m²	**33.68**
Work to floors; two coat; laid on hacked concrete with slurry; over 300 mm wide						
50 mm thick	-	-	-	-	m²	**37.24**
75 mm thick	-	-	-	-	m²	**45.84**
Work to landings; one coat; level; laid on concrete while green; bonded; over 300 mm wide						
25 mm thick	-	-	-	-	m²	**45.27**
38 mm thick	-	-	-	-	m²	**50.51**
Work to landings; two coat; laid on hacked concrete with slurry; over 300 mm wide						
50 mm thick	-	-	-	-	m²	**55.87**
75 mm thick	-	-	-	-	m²	**68.76**
Add to the above over 300 mm wide for						
liquid hardening additive incorporated in the mix	-	0.05	1.13	0.70	m²	**1.83**
oil-repellent additive incorporated in the mix	-	0.08	1.81	5.62	m²	**7.43**
25 mm work to treads; one coat; to concrete base						
225 mm wide	-	0.92	26.79	10.86	m	**37.66**
275 mm wide	-	0.92	26.79	12.17	m	**38.96**
returned end	-	0.19	5.53	-	nr	**5.53**

M SURFACE FINISHES

Item	PC £	Labour hours	Labour £	Material £	Unit	Total rate £
13 mm skirtings; rounded top edge and coved bottom junction; to brickwork or blockwork base						
75 mm wide on face	-	0.56	16.31	0.52	m	**16.83**
150 mm wide on face	-	0.77	22.42	9.56	m	**31.98**
ends; fair	-	0.05	1.46	-	nr	**1.46**
angles	-	0.07	2.04	-	nr	**2.04**
13 mm outer margin to stairs; to follow profile of and with rounded nosing to treads and risers; fair edge and arris at bottom, to concrete base						
75 mm wide	-	0.92	26.79	5.22	m	**32.01**
angles	-	0.07	2.04	-	nr	**2.04**
13 mm wall string to stairs; fair edge and arris on top; coved bottom junction with treads and risers; to brickwork or blockwork base						
275 mm (extreme) wide	-	0.81	23.59	9.13	m	**32.72**
ends	-	0.05	1.46	-	nr	**1.46**
angles	-	0.07	2.04	-	nr	**2.04**
ramps	-	0.08	2.33	-	nr	**2.33**
ramped and wreathed corners	-	0.10	2.91	-	nr	**2.91**
13 mm outer string to stairs; rounded nosing on top at junction with treads and risers; fair edge and arris at bottom; to concrete base						
300 mm (extreme) wide	-	0.81	23.59	11.30	m	**34.89**
ends	-	0.05	1.46	-	nr	**1.46**
angles	-	0.07	2.04	-	nr	**2.04**
ramps	-	0.08	2.33	-	nr	**2.33**
ramps and wreathed corners	-	0.10	2.91	-	nr	**2.91**
19 mm thick skirtings; rounded top edge and coved bottom junction; to brickwork or blockwork base						
75 mm wide on face	-	0.56	16.31	9.56	m	**25.87**
150 mm wide on face	-	0.77	22.42	14.78	m	**37.20**
ends; fair	-	0.05	1.13	-	nr	**1.13**
angles	-	0.07	2.04	-	nr	**2.04**
19 mm risers; one rounded nosing; to concrete base						
150 mm high; plain	-	0.92	26.79	8.26	m	**35.05**
150 mm high; undercut	-	0.92	26.79	8.26	m	**35.05**
180 mm high; plain	-	0.92	26.79	11.30	m	**38.09**
180 mm high; undercut	-	0.92	26.79	11.30	m	**38.09**
M11 MASTIC ASPHALT FLOORING/FLOOR UNDERLAYS						
Mastic asphalt flooring to BS 6925 Type F 1076; black						
20 mm thick; one coat coverings; felt isolating membrane; to concrete base; flat						
over 300 mm wide	-	-	-	-	m²	**20.65**
225 mm - 300 mm wide	-	-	-	-	m²	**38.38**
150 mm - 225 mm wide	-	-	-	-	m²	**42.15**
not exceeding 150 mm wide	-	-	-	-	m²	**51.59**
25 mm thick; one coat coverings; felt isolating membrane; to concrete base; flat						
over 300 mm wide	-	-	-	-	m²	**23.98**
225 mm - 300 mm wide	-	-	-	-	m²	**40.93**
150 mm - 225 mm wide	-	-	-	-	m²	**44.62**
not exceeding 150 mm wide	-	-	-	-	m²	**54.09**
20 mm thick; three coat skirtings to brickwork base						
not exceeding 150 mm girth	-	-	-	-	m	**21.11**
150 mm - 225 mm girth	-	-	-	-	m	**25.80**
225 mm - 300 mm girth	-	-	-	-	m	**30.49**

M SURFACE FINISHES

Item	PC £	Labour hours	Labour £	Material £	Unit	Total rate £
M11 MASTIC ASPHALT FLOORING/FLOOR UNDERLAYS – cont'd						
Mastic asphalt flooring; acid-resisting; black						
20 mm thick; one coat coverings; felt isolating membrane; to concrete base flat						
over 300 mm wide	-	-	-	-	m²	**24.20**
225 mm - 300 mm wide	-	-	-	-	m²	**44.25**
150 mm - 225 mm wide	-	-	-	-	m²	**45.70**
not exceeding 150 mm wide	-	-	-	-	m²	**55.13**
25 mm thick; one coat coverings; felt isolating membrane; to concrete base; flat						
over 300 mm wide	-	-	-	-	m²	**28.58**
225 mm - 300 mm wide	-	-	-	-	m²	**45.49**
150 mm - 225 mm wide	-	-	-	-	m²	**49.23**
not exceeding 150 mm wide	-	-	-	-	m²	**58.69**
20 mm thick; three coat skirtings to brickwork base						
not exceeding 150 mm girth	-	-	-	-	m	**21.32**
150 mm - 225 mm girth	-	-	-	-	m	**24.84**
225 mm - 300 mm girth	-	-	-	-	m	**28.20**
Mastic asphalt flooring to BS 6925 Type F 1451; red						
20 mm thick; one coat coverings; felt isolating membrane; to concrete base; flat						
over 300 mm wide	-	-	-	-	m²	**33.85**
225 mm - 300 mm wide	-	-	-	-	m²	**55.91**
150 mm - 225 mm wide	-	-	-	-	m²	**60.40**
not exceeding 150 mm wide	-	-	-	-	m²	**72.25**
20 mm thick; three coat skirtings to brickwork base						
not exceeding 150 mm girth	-	-	-	-	m	**26.58**
150 mm - 225 mm girth	-	-	-	-	m	**33.85**
M12 TROWELLED BITUMEN/RESIN/RUBBER LATEX FLOORING						
Latex cement floor screeds; steel trowelled						
Work to floors; level; to concrete base; over 300 mm wide						
3 mm thick; one coat	-	-	-	-	m²	**4.54**
5 mm thick; two coats	-	-	-	-	m²	**6.40**
Isocrete K screeds or other equal and approved; steel trowelled						
Work to floors; level; to concrete base; over 300 mm wide						
35 mm thick; plus polymer bonder coat	-	-	-	-	m²	**16.39**
40 mm thick	-	-	-	-	m²	**15.13**
45 mm thick	-	-	-	-	m²	**15.99**
50 mm thick	-	-	-	-	m²	**16.85**
Work to floors; to falls or crossfalls; to concrete base; over 300 mm wide						
55 mm (average) thick	-	-	-	-	m²	**17.70**
60 mm (average) thick	-	-	-	-	m²	**18.56**
65 mm (average) thick	-	-	-	-	m²	**19.42**
75 mm (average) thick	-	-	-	-	m²	**21.14**
90 mm (average) thick	-	-	-	-	m²	**23.72**

M SURFACE FINISHES

Item	PC £	Labour hours	Labour £	Material £	Unit	Total rate £
Isocrete K screeds; quick drying; or other equal and approved; steel trowelled						
Work to floors; level or to floors n.e. 15 degrees frojm the horizontal; to concrete base; over 300 mm wide						
55 mm thick	-	-	-	-	m²	**22.55**
75 mm thick	-	-	-	-	m²	**28.19**
Isocrete pumpableSelf Level Plus screeds; or other equal and approved; protected with Corex type polythene; knifed off prior to layin floor finish; flat smooth finish						
Work to floors; level or to floors n.e. 15 degrees frojm the horizontal; to concrete base; over 300 mm wide						
20 mm thick	-	-	-	-	m²	**27.48**
50 mm thick	-	-	-	-	m²	**36.64**
Bituminous lightweight insulating roof screeds						
"Bit-Ag" or similar roof screed or other equal and approved; to falls or crossfalls; bitumen felt vapour barrier; over 300 mm wide						
75 mm (average) thick	-	-	-	-	m²	**54.29**
100 mm (average) thick	-	-	-	-	m²	**68.81**
M20 PLASTERED/RENDERED/ROUGHCAST COATINGS						
Prepare and brush down 2 coats of "Unibond" bonding agent or other equal and approved						
Brick or block walls						
over 300 mm wide	-	0.16	3.62	1.05	m²	**4.67**
Concrete walls or ceilings						
over 300 mm wide	-	0.12	2.71	0.84	m²	**3.56**
Cement:sand (1:3) beds and backings						
10 mm thick work to walls; one coat; to brickwork or blockwork base						
over 300 mm wide	-	-	-	-	m²	**19.60**
not exceeding 300 mm wide	-	-	-	-	m	**9.81**
13 mm thick; work to walls; two coats; to brickwork or blockwork base						
over 300 mm wide	-	-	-	-	m²	**23.57**
not exceeding 300 mm wide	-	-	-	-	m	**11.80**
15 mm thick work to walls; two coats; to brickwork or blockwork base						
over 300 mm wide	-	-	-	-	m²	**25.42**
not exceeding 300 mm wide	-	-	-	-	m	**12.72**

M SURFACE FINISHES

Item	PC £	Labour hours	Labour £	Material £	Unit	Total rate £
M20 PLASTERED/RENDERED/ROUGHCAST COATINGS – cont'd						
Cement:sand (1:3); steel trowelled						
13 mm thick work to walls; two coats; to brickwork or blockwork base						
over 300 mm wide	-	-	-	-	m²	**20.51**
not exceeding 300 mm wide	-	-	-	-	m	**10.26**
16 mm thick work to walls; two coats; to brickwork or blockwork base						
over 300 mm wide	-	-	-	-	m²	**22.98**
not exceeding 300 mm wide	-	-	-	-	m	**11.52**
19 mm thick work to walls; two coats; to brickwork or blockwork base						
over 300 mm wide	-	-	-	-	m²	**26.59**
not exceeding 300 mm wide	-	-	-	-	m	**13.29**
ADD to above						
over 300 mm wide in water repellant cement	-	-	-	-	m²	**5.44**
finishing coat in colour cement	-	-	-	-	m²	**11.60**
Cement:lime:sand (1:2:9); steel trowelled						
19 mm thick work to walls; two coats; to brickwork or blockwork base						
over 300 mm wide	-	-	-	-	m²	**25.78**
not exceeding 300 mm wide	-	-	-	-	m	**12.89**
Cement:lime:sand (1:1:6); steel trowelled						
13 mm thick work to walls; two coats; to brickwork or blockwork base						
over 300 mm wide	-	-	-	-	m²	**21.09**
not exceeding 300 mm wide	-	-	-	-	m	**10.39**
Add to the above over 300 wide for						
waterproof additive	-	-	-	-	m²	**3.55**
19 mm thick work to ceilings; three coats; to metal lathing base						
over 300 mm wide	-	-	-	-	m²	**24.80**
not exceeding 300 mm wide	-	-	-	-	m	**14.50**
Sto External render only system; comprising glassfibre mesh reinforcement embedded in 10 mm Sto Levell Cote with Sto Armat Classic Basecoat Render and Stolit K 1.5 Decorative Topcoat Render (white)						
15 mm thick work to walls; two coats; to brickwork or blockwork base						
over 300 mm wide	-	-	-	-	m²	**69.80**
Extra for						
bellcast bead	-	-	-	-	m	**6.64**
external angle with PVC mesh angle bead	-	-	-	-	m	**6.14**
internal angle with Sto Armor angle	-	-	-	-	m	**6.14**
render stop bead	-	-	-	-	m	**6.14**
K-Rend render or similar through-colour render system						
18 mm thick work to walls; two coats; to brickwork or blockwork base; first coat 8 mm standard base coat; second coat 10 mm K-rend silicone WP/FT						
over 300 mm wide	-	-	-	-	m²	**75.54**

M SURFACE FINISHES

Item	PC £	Labour hours	Labour £	Material £	Unit	Total rate £
Plaster; first 11 mm coat of "Thistle Hardwall" plaster; second 2 mm finishing coat of "Thistle Multi Finish" plaster; steel trowelled						
13 mm thick work to walls; two coats; to brickwork or blockwork base						
over 300 mm wide	-	-	-	-	m²	**16.27**
over 300 mm wide; in staircase areas or plant rooms	-	-	-	-	m²	**19.54**
not exceeding 300 mm wide	-	-	-	-	m	**8.65**
13 mm thick work to isolated brickwork or blockwork columns; two coats						
over 300 mm wide	-	-	-	-	m²	**30.64**
not exceeding 300 mm wide	-	-	-	-	m	**15.32**
Plaster; first 11 mm coat of "Thistle Browning" plaster; second 2 mm finishing coat of "Thistle Multi Finish" plaster; steel trowelled						
13 mm thick; work to walls; two coats; to brickwork or blockwork base						
over 300 mm wide	-	-	-	-	m²	**16.27**
over 300 mm wide; in staircase areas or plant rooms	-	-	-	-	m²	**19.53**
not exceeding 300 mm wide	-	-	-	-	m	**8.65**
13 mm thick work to isolated columns; two coats						
over 300 mm wide	-	-	-	-	m²	**30.65**
not exceeding 300 mm wide	-	-	-	-	m	**13.59**
Plaster; first 8 mm or 11 mm coat of "Thistle Bonding" plaster; second 2 mm finishing coat of "Thistle Multi Finish" plaster; steel trowelled						
13 mm thick work to walls; two coats; to concrete base						
over 300 mm wide	-	-	-	-	m²	**18.22**
over 300 mm wide; in staircase areas or plant rooms	-	-	-	-	m²	**21.57**
not exceeding 300 mm wide	-	-	-	-	m	**8.34**
13 mm thick work to isolated piers or columns; two coats; to concrete base						
over 300 mm wide	-	-	-	-	m²	**32.55**
not exceeding 300 mm wide	-	-	-	-	m	**15.24**
10 mm thick work to ceilings; two coats; to concrete base						
over 300 mm wide	-	-	-	-	m²	**15.58**
over 300 mm wide; 3.50 m - 5.00 m high	-	-	-	-	m²	**18.70**
over 300 mm wide; in staircase areas or plant rooms	-	-	-	-	m²	**20.70**
not exceeding 300 mm wide	-	-	-	-	m	**8.71**
10 mm thick work to isolated beams; two coats; to concrete base						
over 300 mm wide	-	-	-	-	m²	**31.17**
over 300 mm wide; 3.50 m - 5.00 m high	-	-	-	-	m²	**33.23**
not exceeding 300 mm wide	-	-	-	-	m	**14.99**
Plaster; one coat "Snowplast" plaster or other equal and approved; steel trowelled						
13 mm thick work to walls; one coat; to brickwork or blockwork base						
over 300 mm wide	-	-	-	-	m²	**17.99**
over 300 mm wide; in staircase areas or plant rooms	-	-	-	-	m²	**21.37**
not exceeding 300 mm wide	-	-	-	-	m	**9.02**
13 mm thick work to isolated columns; one coat						
over 300 mm wide	-	-	-	-	m²	**21.77**
not exceeding 300 mm wide	-	-	-	-	m	**10.92**

M SURFACE FINISHES

Item	PC £	Labour hours	Labour £	Material £	Unit	Total rate £
M20 PLASTERED/RENDERED/ROUGHCAST COATINGS – cont'd						
Plaster; first coat of "Limelite" renovating plaster; finishing coat of "Limelite" finishing plaster; or other equal and approved; steel trowelled						
13 mm thick work to walls; two coats; to brickwork or blockwork base						
over 300 mm wide	-	-	-	-	m²	**24.95**
over 300 mm wide; in staircase areas or plant rooms	-	-	-	-	m²	**27.39**
not exceeding 300 mm wide	-	-	-	-	m	**12.47**
Dubbing out existing walls with undercoat plaster; average 6 mm thick						
over 300 mm wide	-	-	-	-	m²	**7.48**
not exceeding 300 mm wide	-	-	-	-	m	**3.78**
Dubbing out existing walls with undercoat plaster; average 12 mm thick						
over 300 mm wide	-	-	-	-	m²	**14.97**
not exceeding 300 mm wide	-	-	-	-	m	**7.48**
Plaster; first coat of "Thistle X-ray" plaster or other equal and approved; finishing coat of "Thistle X-ray" finishing plaster or other equal and approved; steel trowelled						
17 mm thick work to walls; two coats; to brickwork or blockwork base						
over 300 mm wide	-	-	-	-	m²	**80.95**
over 300 mm wide; in staircase areas or plant rooms	-	-	-	-	m²	**86.98**
not exceeding 300 mm wide	-	-	-	-	m	**32.37**
17 mm thick work to isolated columns; two coats						
over 300 mm wide	-	-	-	-	m²	**16.73**
not exceeding 300 mm wide	-	-	-	-	m	**52.48**
Plaster, one coat "Thistle" projection plaster or other equal and approved; steel trowelled						
13 mm thick work to walls; one coat; to brickwork or blockwork base						
over 300 mm wide	-	-	-	-	m²	**17.36**
over 300 mm wide; in staircase areas or plant rooms	-	-	-	-	m²	**19.85**
not exceeding 300 mm wide	-	-	-	-	m	**8.66**
10 mm thick work to isolated columns; one coat						
over 300 mm wide	-	-	-	-	m²	**21.12**
not exceeding 300 mm wide	-	-	-	-	m	**10.54**
Plaster; first 11 mm coat of "Thistle Bonding" plaster; second 2 mm finishing coat of "Thistle Multi Finish" plaster; steel trowelled						
13 mm thick work to ceilings; three coats to metal lathing base						
over 300 mm wide	-	-	-	-	m²	**18.91**
over 300 mm wide; in staircase areas or plant rooms	-	-	-	-	m²	**22.69**
not exceeding 300 mm wide	-	-	-	-	m	**10.20**
13 mm thick work to swept soffit of metal lathing arch former						
not exceeding 300 mm wide	-	-	-	-	m	**13.61**
300 mm - 400 mm wide	-	-	-	-	m	**18.20**
13 mm thick work to vertical face of metal lathing arch former						
not exceeding 0.50 m² per side	-	-	-	-	nr	**19.33**
0.50 m² - 1.00 m² per side	-	-	-	-	nr	**28.99**

M SURFACE FINISHES

Item	PC £	Labour hours	Labour £	Material £	Unit	Total rate £
"Tyrolean" decorative rendering or other equal and approved; 13 mm thick first coat of cement:lime:sand (1:1:6); finishing three coats of "Cullamix" or other equal and approved applied with approved hand operated machine external						
To walls; four coats; to brickwork or blockwork base						
over 300 mm wide	-	-	-	-	m²	**41.48**
not exceeding 300 mm wide	-	-	-	-	m	**20.72**
Drydash (pebbledash) finish of Derbyshire Spar chippings or other equal and approved on and including cement:lime:sand (1:2:9) backing						
18 mm thick work to walls; two coats; to brickwork or blockwork base						
over 300 mm wide	-	-	-	-	m²	**35.96**
not exceeding 300 mm wide	-	-	-	-	m	**17.99**
Plaster; one coat "Thistle" board finish or other equal and approved; steel trowelled (prices included within plasterboard rates)						
3 mm thick work to walls or ceilings; one coat; to plasterboard base						
over 300 mm wide	-	-	-	-	m²	**7.33**
over 300 mm wide; in staircase areas or plant rooms	-	-	-	-	m²	**8.80**
not exceeding 300 mm wide	-	-	-	-	m	**2.93**
Plaster; one coat "Thistle" board finish or other equal and approved; steel trowelled 3 mm work to walls or ceilings; one coat on and including gypsum plasterboard; BS 1230; fixing with nails; 3 mm joints filled with plaster and jute scrim cloth; to softwood base; plain grade baseboard or lath with rounded edges						
9.50 mm thick boards to walls						
over 300 mm wide	-	1.07	16.10	4.35	m²	**20.46**
not exceeding 300 mm wide	-	0.42	6.76	2.93	m	**9.70**
9.50 mm thick boards to walls; in staircase areas or plant rooms						
over 300 mm wide	-	1.17	17.71	4.35	m²	**22.07**
not exceeding 300 mm wide	-	0.52	8.38	2.93	m	**11.31**
9.50 mm thick boards to isolated columns						
over 300 mm wide	-	1.17	17.71	4.35	m²	**22.07**
not exceeding 300 mm wide	-	0.62	9.99	2.93	m	**12.92**
9.50 mm thick boards to ceilings						
over 300 mm wide	-	0.99	14.81	4.35	m²	**19.17**
over 300 mm wide; 3.50 m - 5.00 m high	-	1.14	17.23	4.35	m²	**21.58**
not exceeding 300 mm wide	-	0.47	7.57	2.93	m	**10.50**
9.50 mm thick boards to ceilings; in staircase areas or plant rooms						
over 300 mm wide	-	1.09	16.43	4.35	m²	**20.78**
not exceeding 300 mm wide	-	0.52	8.38	2.93	m	**11.31**
9.50 mm thick boards to isolated beams						
over 300 mm wide	-	1.16	17.55	4.35	m²	**21.91**
not exceeding 300 mm wide	-	0.56	9.02	2.93	m	**11.95**

M SURFACE FINISHES

Item	PC £	Labour hours	Labour £	Material £	Unit	Total rate £
M20 PLASTERED/RENDERED/ROUGHCAST COATINGS – cont'd						
Plaster; one coat "Thistle" board finish on and including gypsum plasterboard – cont'd						
12.50 mm thick boards to walls; in staircase areas or plant rooms						
over 300 mm wide	-	1.12	16.91	4.35	m²	**21.26**
not exceeding 300 mm wide	-	0.53	8.54	2.93	m	**11.47**
12.50 mm thick boards to isolated columns						
over 300 mm wide	-	1.16	17.55	4.35	m²	**21.91**
not exceeding 300 mm wide	-	0.56	9.02	2.93	m	**11.95**
12.50 mm thick boards to ceilings						
over 300 mm wide	-	1.05	15.78	4.35	m²	**20.13**
over 300 mm wide; 3.50 m - 5.00 m high	-	1.17	17.71	4.35	m²	**22.07**
not exceeding 300 mm wide	-	0.50	8.05	2.93	m	**10.99**
12.50 mm thick boards to ceilings; in staircase areas or plant rooms						
over 300 mm wide	-	1.17	17.71	4.35	m²	**22.07**
not exceeding 300 mm wide	-	0.56	9.02	2.93	m	**11.95**
12.50 mm thick boards to isolated beams						
over 300 wide	-	1.27	19.32	4.35	m²	**23.68**
not exceeding 300 mm wide	-	0.62	9.99	2.93	m	**12.92**
Accessories						
"Expamet" render beads or other equal and approved; white PVC nosings; to brickwork or blockwork base						
external stop bead; ref 573	-	0.08	1.29	1.78	m	**3.07**
"Expamet" render beads or other equal and approved; stainless steel; to brickwork or blockwork base						
stop bead; ref 546	-	0.08	1.29	4.78	m	**6.07**
stop bead; ref 547	-	0.08	1.29	4.78	m	**6.07**
"Expamet" plaster beads or other equal and approved; galvanised steel; to brickwork or blockwork base						
angle bead; ref 550	-	0.09	1.45	1.02	m	**2.47**
architrave bead; ref 579	-	0.11	1.77	2.73	m	**4.51**
stop bead; ref 562	-	0.08	1.29	1.26	m	**2.55**
stop beads; ref 563	-	0.08	1.29	1.26	m	**2.55**
movement bead; ref 588	-	0.10	1.61	13.29	m	**14.90**
"Expamet" plaster beads or other equal and approved; stainless steel; to brickwork or blockwork base						
angle bead; ref 545	-	0.09	1.45	5.38	m	**6.83**
stop bead; ref 534	-	0.08	1.29	4.78	m	**6.07**
stop bead; ref 533	-	0.08	1.29	4.78	m	**6.07**
"Expamet" thin coat plaster beads or other equal and approved; galvanised steel; to timber base						
angle bead; ref 553	-	0.08	1.29	0.96	m	**2.25**
angle bead; ref 554	-	0.08	1.29	0.80	m	**2.09**
stop bead; ref 560	-	0.07	1.13	1.21	m	**2.34**
stop bead; ref 561	-	0.07	1.13	1.21	m	**2.34**

M SURFACE FINISHES

Item	PC £	Labour hours	Labour £	Material £	Unit	Total rate £
M21 INSULATION WITH RENDERED FINISH						
Sto Therm Classic M-system insulation render						
70 mm EPS insulation fixed with adhesive to SFS structure (measured separately) with horizontal PVC intermediate track and vertical T-spines; with glassfibre mesh reinforcement embedded in Sto Armat Classic Basecoat Render and Stolit K 1.5 Decorative Topcoat Render (white)						
over 300 mm wide	-	-	-	-	m²	82.87
70 mm EPS insulation mechanically fixed to SFS structure (measured separately) with horizontal PVC intermediate track and vertical T-spines; with glassfibre mesh reinforcement embedded in Sto Armat Classic Basecoat Render and Stolit K 1.5 Decorative Topcoat Render (white)						
over 300 mm wide	-	-	-	-	m²	91.17
rendered heads and reveals not exceeding 100 mm wide; including angle beadsover 300 mm wide	-	-	-	-	m	22.88
Extra for						
aluminium starter track at base of insulated render system	-	-	-	-	m	13.93
external angle with PVC mesh angle bead	-	-	-	-	m	6.14
internal angle with Sto Armor angle	-	-	-	-	m	6.14
render stop bead	-	-	-	-	m	6.14
Sto seal tape to all vertical abutments	-	-	-	-	m	5.70
Sto Armor mat HD mesh reinforcement to areas prone to physical damage (e.g. 1800 mm high adjoining floor level)						
over 300 mm wide	-	-	-	-	m²	18.33
M22 SPRAYED MINERAL FIBRE COATINGS						
Prepare and apply by spray "Mandolite CP2" fire protection or other equal and approved on structural steel/metalwork						
16 mm thick (one hour) fire protection						
to walls and columns	-	-	-	-	m²	11.27
to ceilings and beams	-	-	-	-	m²	12.52
to isolated metalwork	-	-	-	-	m²	24.67
22 mm thick (one and a half hour) fire protection						
to walls and columns	-	-	-	-	m²	12.97
to ceilings and beams	-	-	-	-	m²	14.40
to isolated metalwork	-	-	-	-	m²	28.94
28 mm thick (two hour) fire protection						
to walls and columns	-	-	-	-	m²	15.22
to ceilings and beams	-	-	-	-	m²	16.92
to isolated metalwork	-	-	-	-	m²	33.18
52 mm thick (four hour) fire protection						
to walls and columns	-	-	-	-	m²	23.30
to ceilings and beams	-	-	-	-	m²	25.68
to isolated metalwork	-	-	-	-	m²	51.36
Prepare and apply by spray; cementitious "Pyrok WF26" render or other equal and approved; on expanded metal lathing (not included)						
15 mm thick						
to ceilings and beams	-	-	-	-	m²	35.57

M SURFACE FINISHES

Item	PC £	Labour hours	Labour £	Material £	Unit	Total rate £
M30 METAL MESH LATHING/ANCHORED REINFORCEMENT FOR PLASTERED COATING						
Accessories						
Pre-formed galvanised expanded steel semi-circular arch-frames; "Expamet" or other equal and approved; to suit walls up to 230 mm thick						
for 760 mm opening; ref ESC 30	62.56	0.52	7.24	72.38	nr	**79.62**
for 840 mm opening; ref ESC 32	65.70	0.52	7.24	76.01	nr	**83.25**
for 920 mm opening; ref ESC 36	76.85	0.52	7.24	88.91	nr	**96.16**
for 1220 mm opening; ref ESC 48	95.78	0.52	7.24	110.81	nr	**118.05**
"Newlath 2000" damp free lathing or other equal and approved; plugging and screwing to background at 250 mm centres each way						
Linings to walls						
over 300 mm wide	-	1.28	28.12	11.31	m²	**39.43**
not exceeding 300 mm wide	-	0.55	12.08	3.54	m²	**15.62**
bottom erdge strip	-	0.20	4.39	5.09	m	**9.48**
Lathing; Expamet "BB" expanded metal lathing or other and approved; BS EN 13658; 50 mm laps						
6 mm thick mesh linings to ceilings; fixing with staples; to softwood base; over 300 mm wide						
ref BB263; 0.500 mm thick	5.45	0.61	8.49	6.46	m²	**14.96**
ref BB264; 0.675 mm thick	7.63	0.61	8.49	9.05	m²	**17.54**
6 mm thick mesh linings to ceilings; fixing with wire; to steelwork; over 300 mm wide						
ref BB263; 0.500 mm thick	-	0.65	9.06	6.46	m²	**15.52**
ref BB264; 0.675 mm thick	-	0.65	9.06	9.05	m²	**18.10**
6 mm thick mesh linings to ceilings; fixing with wire; to steelwork; not exceeding 300 mm wide						
ref BB263; 0.500 mm thick	-	0.41	5.71	6.46	m²	**12.18**
ref BB264; 0.675 mm thick	-	0.41	5.71	9.05	m²	**14.76**
raking cutting	-	0.20	3.22	-	m	**3.22**
cutting and fitting around pipes; not exceeding 0.30 m girth	-	0.31	4.99	-	nr	**4.99**
Lathing; Expamet "Riblath" or "Spraylath" stiffened expanded metal lathing or other equal and approved; 50 mm laps						
10 mm thick mesh lining to walls; fixing with nails; to softwood base; over 300 mm wide						
"Riblath" ref 269; 0.30 mm thick	7.73	0.52	7.24	9.33	m²	**16.58**
"Riblath" ref 271; 0.50 mm thick	8.90	0.52	7.24	10.72	m²	**17.96**
10 mm thick mesh lining to walls; fixing with nails; to softwood base; not exceeding 300 mm wide						
"Riblath" ref 269; 0.30 mm thick	-	0.31	4.31	2.86	m²	**7.17**
"Riblath" ref 271; 0.50 mm thick	-	0.31	4.31	3.28	m²	**7.59**
10 mm thick mesh lining to walls; fixing to brick or blockwork; over 300 mm wide						
"Red-rib" ref 274; 0.50 mm thick	9.82	0.41	5.71	12.41	m²	**18.13**
stainless steel "Riblath" ref 267; 0.30 mm thick	22.99	0.41	5.71	28.03	m²	**33.74**
10 mm thick mesh lining to ceilings; fixing with wire; to steelwork; over 300 mm wide						
"Riblath" ref 269; 0.30 mm thick	-	0.65	9.06	10.11	m²	**19.17**
"Riblath" ref 271; 0.50 mm thick	-	0.65	9.06	11.50	m²	**20.55**

M SURFACE FINISHES

Item	PC £	Labour hours	Labour £	Material £	Unit	Total rate £
M31 FIBROUS PLASTER						
Fibrous plaster; fixing with screws; plugging; countersinking; stopping; filling and pointing joints with plaster						
16 mm thick plain slab coverings to ceilings						
over 300 mm wide	-	-	-	-	m²	**148.30**
not exceeding 300 mm wide	-	-	-	-	m	**49.88**
Coves; not exceeding 150 mm girth						
per 25 mm girth	-	-	-	-	m	**7.15**
Coves; 150 mm - 300 mm girth						
per 25 mm girth	-	-	-	-	m	**8.76**
Cornices						
per 25 mm girth	-	-	-	-	m	**8.90**
Cornice enrichments						
per 25 mm girth; depending on degree of enrichments	-	-	-	-	m	**10.51**
Fibrous plaster; fixing with plaster wadding filling and pointing joints with plaster; to steel base						
16 mm thick plain slab coverings to ceilings						
over 300 mm wide	-	-	-	-	m²	**148.30**
not exceeding 300 mm wide	-	-	-	-	m	**49.88**
16 mm thick plain casings to stanchions						
per 25 mm girth	-	-	-	-	m	**4.45**
16 mm thick plain casings to beams						
per 25 mm girth	-	-	-	-	m	**4.45**
Gyproc cove or other equal and approved; fixing with adhesive; filling and pointing joints with plaster						
Coves						
125 mm girth	-	0.20	3.22	3.85	m	**7.08**
angles	-	0.03	0.48	2.81	nr	**3.30**
M40 STONE/CONCRETE/QUARRY/CERAMIC TILING/MOSAIC						
Clay floor quarries; BS EN 19545; class 1; Dennis Ruabon tiles or other equal and approved; level bedding 10 mm thick and jointing in cement and sand (1:3); butt joints; straight both ways; flush pointing with grout; to cement and sand base						
Work to floors; over 300 mm wide						
150 mm x 150 mm x 12.50 mm thick; heatherbrown	-	0.81	18.32	41.67	m²	**59.99**
150 mm x 150 mm x 12.50 mm thick; red	-	0.81	18.32	37.64	m²	**55.96**
194 mm x 194 mm x 12.50 mm thick; brown	-	0.67	15.16	33.05	m²	**48.21**
Works to floors; in staircase areas or plant rooms						
150 mm x 150 mm x 12.50 mm thick; heatherbrown	-	0.92	20.81	41.67	m²	**62.48**
150 mm x 150 mm x 12.50 mm thick; red	-	0.92	20.81	37.64	m²	**58.45**
194 mm x 194 mm x 12.50 mm thick; brown	-	0.77	17.42	33.05	m²	**50.47**
Work to floors; not exceeding 300 mm wide						
150 mm x 150 mm x 12.50 mm thick; heatherbrown	-	0.41	9.27	11.03	m²	**20.31**
150 mm x 150 mm x 12.50 mm thick; red	-	0.41	9.27	9.80	m²	**19.07**
194 mm x 194 mm x 12.50 mm thick; brown	-	0.33	7.46	8.31	m²	**15.77**

M SURFACE FINISHES

Item	PC £	Labour hours	Labour £	Material £	Unit	Total rate £
M40 STONE/CONCRETE/QUARRY/CERAMIC TILING/MOSAIC – cont'd						
Clay floor quarries; BS EN 19545 – cont'd						
fair square cutting against flush edges of existing finishes	-	0.12	1.81	2.72	m	**4.53**
raking cutting	-	0.21	3.22	3.05	m	**6.28**
cutting around pipes; not exceeding 0.30 m girth	-	0.16	2.58	-	nr	**2.58**
extra for cutting and fitting into recessed manhole cover 600 mm x 600 mm	-	1.02	16.43	-	nr	**16.43**
Work to sills; 150 mm wide; rounded edge tiles						
203 mm x 150 mm x 22 mm thick; interior; heatherbrown or red	-	0.33	7.46	8.76	m	**16.23**
150 mm x 173 mm x 58 mm thick; exterior; heatherbrown or red	-	0.33	7.46	36.23	m	**43.70**
fitted end	-	0.16	2.58	-	m	**2.58**
Coved skirtings; 150 mm high; rounded top edge						
150 mm x 150 mm x 12.50 mm thick; ref. CBTR; heatherbrown or red	-	0.26	5.88	10.30	m	**16.19**
150 mm x 150 mm x 12.50 mm thick; ref. RE; heatherbrown or red	-	0.26	5.88	7.20	m	**13.08**
ends	-	0.05	0.81	-	nr	**0.81**
angles	-	0.16	2.58	2.65	nr	**5.23**
Glazed ceramic wall tiles; BS EN 19545; fixing with adhesive; butt joints; straight both ways; flush pointing with white grout; to plaster base						
Work to walls; over 300 mm wide						
152 mm x 152 mm x 5.50 mm thick; white	12.78	0.61	17.76	22.97	m²	**40.73**
152 mm x 152 mm x 5.50 mm thick; light colours	15.59	0.61	17.76	26.30	m²	**44.06**
152 mm x 152 mm x 5.50 mm thick; dark colours	17.04	0.61	17.76	28.02	m²	**45.78**
extra for RE or REX tile	-	-	-	7.77	m²	**-**
200 mm x 100 mm x 6.50 mm thick; white and light colours	12.78	0.61	17.76	22.97	m²	**40.73**
250 mm x 200 mm x 7 mm thick; white and light colours	13.84	0.61	17.76	24.23	m²	**41.99**
Work to walls; in staircase areas or plant rooms						
152 mm x 152 mm x 5.50 mm thick; white	-	0.68	19.80	22.97	m²	**42.77**
Work to walls; not exceeding 300 mm wide						
152 mm x 152 mm x 5.50 mm thick; white	-	0.31	9.03	6.80	m	**15.83**
152 mm x 152 mm x 5.50 mm thick; light colours	-	0.31	9.03	11.52	m	**20.55**
152 mm x 152 mm x 5.50 mm thick; dark colours	-	0.31	9.03	12.04	m	**21.07**
200 mm x 100 mm x 6.50 mm thick; white and light colours	-	0.31	9.03	6.80	m	**15.83**
250 mm x 200 mm x 7 mm thick; white and light colours	-	0.26	7.57	7.18	m	**14.75**
cutting around pipes; not exceeding 0.30 m girth	-	0.10	1.61	-	nr	**1.61**
Work to sills; 150 mm wide; rounded edge tiles						
152 mm x 152 mm x 5.50 mm thick; white	-	0.26	7.57	3.40	m	**10.97**
fitted end	-	0.10	1.61	-	nr	**1.61**
198 mm x 64.50 mm x 6 mm thick wall tiles; fixing with adhesive; butt joints; straight both ways; flush pointing with white grout; to plaster base						
Work to walls						
over 300 mm wide	27.05	1.83	53.29	39.87	m²	**93.17**
not exceeding 300 mm wide	-	0.71	20.68	11.87	m	**32.55**

M SURFACE FINISHES

Item	PC £	Labour hours	Labour £	Material £	Unit	Total rate £
20 mm x 20 mm x 5.50 mm thick glazed mosaic wall tiles; fixing with adhesive; butt joints; straight both ways; flush pointing with white grout; to plaster base						
Work to walls						
over 300 mm wide	33.19	1.93	56.20	48.54	m²	**104.75**
not exceeding 300 mm wide	-	0.77	22.42	18.01	m	**40.44**
50 mm x 50 mm x 5 mm thick slip resistant mosaic floor tiles, Series 2 or other equal and approved; fixing with adhesive; butt joints; straight both ways; flush pointing with white grout; to cement:sand base						
Work to floors						
over 300 mm wide	33.13	1.93	43.66	50.07	m²	**93.73**
not exceeding 300 mm wide	-	0.77	17.42	17.99	m	**35.41**
Riven Welsh slate floor tiles; level; bedding 10 mm thick and jointing in cement and sand (1:3); butt joints; straight both ways; flush pointing with coloured mortar; to cement:sand base						
Work to floors; over 300 mm wide						
250 mm x 250 mm x 12 mm - 15 mm thick	-	0.61	17.76	46.49	m²	**64.25**
Work to floors; not exceeding 300 mm wide						
250 mm x 250 mm x 12 mm - 15 mm thick	-	0.31	9.03	14.04	m	**23.07**
M41 TERRAZZO TILING/ IN SITU TERRAZZO						
Terrazzo tiles BS EN 13748; aggregate size random ground grouted and polished to 80's grit finish; standard colour range; 3 mm jonts symmetrical layout; bedding in 42 mm cement semi-dry mix (1:4); grouting with neat matching cement						
300 mm x 300 mm x 28 mm (nominal) Terrazzo tile units; hydraulically pressed, mechanically vibrated, steam cured; to floors on concrete base (not included); sealed with penetrating case hardener or other equal and approved; 2 coats applied immediately after final polishing						
plain; laid level	-	-	-	-	m²	**45.92**
plain; to slopes exceeding 15 degrees from horizontal	-	-	-	-	m²	**55.97**
to small areas/toilets	-	-	-	-	m²	**100.10**
Accessories						
plastic division strips 6 mm x 38 mm; set into floor tiling above crack inducing joints; to the nearest full tile module	-	-	-	-	m	**3.02**

M SURFACE FINISHES

Item	PC £	Labour hours	Labour £	Material £	Unit	Total rate £
M42 WOOD BLOCK/COMPOSITION BLOCK/ PARQUET FLOORING						
Wood blocks; Havwoods or other equal and approved; 25 mm thick; level; laid to herringbone pattern with 2 block border; fixing with adhesive; to cement:sand base; sanded and sealed						
Work to floors; over 300 mm wide						
Merbau	-	-	-	-	m²	**67.62**
Iroko	-	-	-	-	m²	**74.76**
American Oak	-	-	-	-	m²	**84.07**
European Oak	-	-	-	-	m²	**98.78**
Add to wood block flooring over 300 mm wide for						
buff; one coat seal	-	-	-	-	m²	**3.54**
buff; two coats seal	-	-	-	-	m²	**5.60**
sand; three coats for seal or oil	-	-	-	-	m²	**17.68**
M50 RUBBER/PLASTICS/CORK/LINO/CARPET TILING/SHEETING						
Linoleum sheet; Forbo-Nairn "Marmoleum Real" or other equal and approved; level; fixing with adhesive; butt joints; to cement:sand base						
Work to floors; over 300 mm wide						
2.50 mm thick; plain	-	0.41	9.27	13.35	m²	**22.62**
3.20 mm thick; marbled	-	0.41	9.27	16.56	m²	**25.83**
Linoleum sheet; Forbo-Nairn "Walton" or other equal and approved; level; with welded seams; fixing with adhesive; to cement:sand base						
Work to floors; over 300 mm wide						
2.50 mm thick	-	0.51	11.54	15.33	m²	**26.86**
Vinyl sheet; Altro "Safety" range or other equal and approved; with welded seams; level; fixing with adhesive; to cement:sand base						
Work to floors; over 300 mm wide						
2.00 mm thick; "Marine"	-	0.61	13.80	18.96	m²	**32.75**
2.50 mm thick; "Classic 25"	-	0.71	16.06	21.09	m²	**37.15**
3.50 mm thick; "Stronghold 30"	-	0.81	18.32	28.47	m²	**46.79**
Slip resistant vinyl sheet; Forbo-Nairn "Surestep" or other equal and approved; level with welded seams; fixing with adhesive; to cement:sand base						
Work to floors; over 300 mm wide						
2.00 mm thick	-	0.51	11.54	16.43	m²	**27.97**
Homogeneous vinyl sheet; Marleyflor Plus or other equal and approved; level; with welded seams; fixing with adhesive; level; to cement and sand base						
Work to floors; over 300 mm wide						
2.00 mm thick	-	0.45	10.18	8.11	m²	**18.29**
2.00 mm thick skirtings						
100 high	-	0.12	2.71	2.00	m	**4.71**

M SURFACE FINISHES

Item	PC £	Labour hours	Labour £	Material £	Unit	Total rate £
Safety sheet; Marleyflor Granite Multisafe or other equal and approved; level; with welded seams; fixing with adhesive; level; to cement and sand base						
Work to floors; over 300 mm wide						
2.00 mm thick	-	0.45	10.18	15.68	m²	**25.86**
Vinyl sheet; Marley "Omnisports" or other equal and approved; level; with welded seams; fixing with adhesive; level; to cement:sand base						
Work to floors; over 300 mm wide						
7.65 mm thick; Pro	-	1.00	22.62	30.83	m²	**53.45**
8.75 mm thick; Competition	-	1.10	24.88	36.06	m²	**60.94**
Vinyl sheet; "Gerflex" standard sheet; "Classic" range or other equal and approved; level; with welded seams; fixing with adhesive; to cement:sand base						
Work to floors; over 300 mm wide						
2.00 mm thick	-	0.51	11.54	8.31	m²	**19.85**
Vinyl sheet; Armstrong "Royal" or other equal and approved; level; with welded seams; fixing with adhesive; to cement:sand base						
Work to floors; over 300 mm wide						
2.50 mm thick	-	0.51	11.54	13.16	m²	**24.70**
Vinyl tiles; Armstrong "Royal" or other equal and approved; level; fixing with adhesive; butt joints; straight both ways; to cement:sand base						
Work to floors; over 300 mm wide						
300 mm x 300 mm x 2.00 mm thick	-	0.23	5.20	13.46	m²	**18.66**
Vinyl semi-flexible tiles; Armstrong "Imperial" or other equal and approved; level; fixing with adhesive; butt joints; straight both ways; to cement:sand base						
Work to floors; over 300 mm wide						
250 mm x 250 mm x 2.00 mm thick	-	0.26	5.88	7.95	m²	**13.83**
Vinyl semi-flexible tiles; Marley Homogeneous range or other equal and approved; level; fixing with adhesive; butt joints; straight both ways; to cement:sand base						
Work to floors; over 300 mm wide						
300 mm x 300 mm x 2.00 mm thick; Vylon Plus	-	0.26	5.88	7.13	m²	**13.01**
500 mm x 500 mm x 2.00 mm thick; Marleyflor Plus	-	0.26	5.88	8.43	m²	**14.31**
Vinyl tiles; "Polyflor SD" or other equal and approved; level; fixing with adhesive; butt joints; straight both ways; to cement:sand base						
Work to floors; over 300 mm wide						
457 mm x 457 mm x 2.00 mm thick	-	0.45	10.18	12.85	m²	**23.02**

M SURFACE FINISHES

Item	PC £	Labour hours	Labour £	Material £	Unit	Total rate £
M50 RUBBER/PLASTICS/CORK/LINO/CARPET TILING/SHEETING – cont'd						
Vinyl tiles; "Polyflex" or other equal and approved; level; fixing with adhesive; butt joints; straight both ways; to cement:sand base						
Work to floors; over 300 mm wide						
300 mm x 300 mm x 1.50 mm thick	-	0.26	5.88	5.87	m²	**11.75**
300 mm x 300 mm x 2.00 mm thick	-	0.26	5.88	6.48	m²	**12.36**
Vinyl tiles; "Polyflor XL" or other equal and approved; level; fixing with adhesive; butt joints; straight both ways; to cement:sand base						
Work to floors; over 300 mm wide						
300 mm x 300 mm x 2.00 mm thick	-	0.36	8.14	7.36	m²	**15.51**
Luxury mineral vinyl tiles; Marley I D Naturelle or other equal and approved; level; fixing with adhesive; butt joints; straight both ways; to cement and sand base						
Work to floors; over 300 mm wide						
330 mm x 330 mm x 2.00 mm thick	-	0.26	5.88	11.10	m²	**16.98**
Acoustic vinyl tiles; Marley Tapiflex 243 or other equal and approved; level; fixing with adhesive; butt joints; straight both ways; to cement and sand base						
Work to floors; over 300 mm wide						
500 mm x 500 mm x 2.00 mm thick	-	0.23	5.20	14.83	m²	**20.03**
Linoleum tiles; Marley Veneto XF or other equal and approved; level; fixing with adhesive; butt joints; straight both ways; to cement and sand base						
Work to floors; over 300 mm wide						
500 mm x 500 mm x 2.50 mm thick	-	0.23	5.20	17.50	m²	**22.70**
PVC Wall lining; Altro Whiterock; or other equal and approved; fixed directly to plastered brick or blockwork						
Work to walls						
over 300 mm wide	-	-	-	-	m²	**56.58**
not exceeding 300 mm wide	-	-	-	-	m²	**28.29**
Linoleum tiles; BS 6826; Forbo-Nairn Floors or other equal and approved; level; fixing with adhesive; butt joints; straight both ways; to cement:sand base						
Work to floors; over 300 mm wide						
2.50 mm thick (marble pattern)	-	0.31	7.01	14.97	m²	**21.99**
Cork tiles Wicanders "Cork-Master" or other equal and approved; level; fixing with adhesive; butt joints; straight both ways; to cement:sand base						
Work to floors; over 300 mm wide						
300 mm x 300 mm x 4.00 mm thick	-	0.41	9.27	27.12	m²	**36.39**

M SURFACE FINISHES

Item	PC £	Labour hours	Labour £	Material £	Unit	Total rate £
Rubber studded tiles; Altro "Mondopave" or other equal and approved; level; fixing with adhesive; butt joints; straight to cement:sand base						
Work to floors; over 300 mm wide						
500 mm x 500 mm x 2.50 mm thick; type MRB; black	-	0.61	13.80	29.60	m²	**43.40**
500 mm x 500 mm x 4.00 mm thick; type MRB; black	-	0.61	13.80	33.63	m²	**47.43**
Work to landings; over 300 mm wide						
500 mm x 500 mm x 4.00 mm thick; type MRB; black	-	0.81	18.32	33.63	m²	**51.95**
4.00 mm thick to treads						
275 mm wide	-	0.51	11.54	10.00	m	**21.53**
4.00 mm thick to risers						
180 mm wide	-	0.61	13.80	7.10	m	**20.90**
Sundry floor sheeting underlays						
For floor finishings; over 300 mm wide						
building paper to BS 1521; class A; 75 mm lap (laying only)	-	0.06	0.72	-	m²	**0.72**
3.20 mm thick hardboard	-	0.20	5.82	1.98	m²	**7.81**
6.00 mm thick plywood	-	0.31	9.03	7.99	m²	**17.02**
Skirtings; plastic; Gradus or equivalent						
Set-in skirtings; type P						
100 mm high; ref SI 100 2.5P	-	0.12	2.71	2.04	m	**4.76**
150 mm high; ref SI 150 2P	-	0.24	5.43	3.09	m	**8.51**
Set-on skirtings; type P						
100 mm high; ref SO 100 P	-	0.24	5.43	1.33	m	**6.76**
Stair nosings; aluminium; Gradus or equivalent						
Medium duty hard aluminium alloy stair tread nosings; plugged and screwed in concrete						
56 mm x 32 mm; ref AS11	9.57	0.26	4.19	11.27	m	**15.46**
84 mm x 32 mm; ref AS12	14.24	0.31	4.99	16.68	m	**21.67**
Heavy duty aluminium alloy stair tread nosings; plugged and screwed to concrete						
48 mm x 38 mm; ref HE1	11.29	0.31	4.99	13.26	m	**18.25**
82 mm x 38 mm; ref HE2	16.86	0.36	5.80	19.71	m	**25.50**
Heavy duty carpet tiles; Heuga "580 Olympic" or other and approved; to cement:sand base						
Work to floors						
over 300 mm wide	22.40	0.31	7.01	26.55	m²	**33.56**
M51 EDGE FIXED CARPETING						
Fitted carpeting; Wilton wool/nylon or other equal and approved; 80/20 velvet pile; heavy domestic plain						
Work to floors						
over 300 mm wide	45.07	0.42	6.45	55.96	m²	**62.41**
Work to treads and risers						
over 300 mm wide	-	0.83	12.75	55.96	m²	**68.71**
Underlay to carpeting						
Work to floors						
over 300 mm wide	4.40	0.08	1.23	5.21	m²	**6.44**
Sundries						
Carpet gripper fixed to floor; standard edging						
22 mm wide	-	0.05	0.60	0.32	m	**0.92**

M SURFACE FINISHES

Item	PC £	Labour hours	Labour £	Material £	Unit	Total rate £
M52 DECORATIVE PAPERS/FABRICS						
Lining paper; and hanging						
Plaster walls or columns						
over 300 mm girth (PC £ per roll)	2.22	0.20	3.22	0.54	m²	**3.76**
Plaster ceilings or beams						
over 300 mm girth	-	0.24	3.87	0.54	m²	**4.41**
Decorative vinyl wallpaper; and hanging						
Plaster walls or columns						
over 300 mm girth (PC £ per roll)	11.11	0.26	4.19	2.46	m²	**6.65**
M60 PAINTING/CLEAR FINISHING						
M60 PREPARATION OF EXISTING SURFACES - INTERNALLY						
NOTE: The prices for preparation given hereunder assume that existing surfaces are in fair condition and should be increased for badly dilapidated surfaces.						
Wash down walls; cut out and make good cracks						
Emulsion painted surfaces; including bringing forward bare patches	-	0.07	1.13	-	m²	**1.13**
Gloss painted surfaces	-	0.05	0.81	-	m²	**0.81**
Wash down ceilings; cut out and make good cracks						
Distempered surfaces	-	0.09	1.45	-	m²	**1.45**
Emulsion painted surfaces; including bringing forward bare patches	-	0.10	1.61	-	m²	**1.61**
Gloss painted surfaces	-	0.09	1.45	-	m²	**1.45**
Wash down plaster cornices; cut out and make good cracks						
Distempered surfaces	-	0.13	2.09	-	m²	**2.09**
Emulsion painted surfaces; including bringing forward bare patches	-	0.16	2.58	-	m²	**2.58**
Wash and rub down iron and steel surfaces; bringing forward						
General surfaces						
over 300 mm girth	-	0.12	1.93	-	m²	**1.93**
isolated surfaces not exceeding 300 mm girth	-	0.05	0.81	-	m²	**0.81**
isolated areas not exceeding 0.50 m² irrespective of girth	-	0.09	1.45	-	nr	**1.45**
Glazed windows and screens						
panes; area not exceeding 0.10 m²	-	0.20	3.22	-	m²	**3.22**
panes; area 0.10 - 0.50 m²	-	0.16	2.58	-	m²	**2.58**
panes; area 0.50 - 1.00 m²	-	0.13	2.09	-	m²	**2.09**
panes; area over 1.00 m²	-	0.12	1.93	-	m²	**1.93**

M SURFACE FINISHES

Item	PC £	Labour hours	Labour £	Material £	Unit	Total rate £
Wash and rub down wood surfaces; prime bare patches; bringing forward						
General surfaces						
over 300 mm girth	-	0.19	3.06	-	m²	**3.06**
isolated surfaces not exceeding 300 mm girth	-	0.07	1.13	-	m²	**1.13**
isolated areas not exceeding 0.50 m² irrespective of girth	-	0.14	2.25	-	nr	**2.25**
Glazed windows and screens						
panes; area not exceeding 0.10 m²	-	0.31	4.99	-	m²	**4.99**
panes; area 0.10 - 0.50 m²	-	0.24	3.87	-	m²	**3.87**
panes; area 0.50 - 1.00 m²	-	0.20	3.22	-	m²	**3.22**
panes; area over 1.00 m²	-	0.19	3.06	-	m²	**3.06**
Wash down and remove paint with chemical stripper from iron, steel or wood surfaces						
General surfaces						
over 300 mm girth	-	0.56	9.02	-	m²	**9.02**
isolated surfaces not exceeding 300 mm girth	-	0.24	3.87	-	m²	**3.87**
isolated areas not exceeding 0.50 m² irrespective of girth	-	0.42	6.77	-	nr	**6.77**
Glazed windows and screens						
panes; area not exceeding 0.10 m²	-	1.18	19.01	-	m²	**19.01**
panes; area 0.10 - 0.50 m²	-	0.94	15.14	-	m²	**15.14**
panes; area 0.50 - 1.00 m²	-	0.81	13.05	-	m²	**13.05**
panes; area over 1.00 m²	-	0.71	11.44	-	m²	**11.44**
Burn off and rub down to remove paint from iron, steel or wood surfaces						
General surfaces						
over 300 mm girth	-	0.68	10.95	-	m²	**10.95**
isolated surfaces not exceeding 300 mm girth	-	0.31	4.99	-	m²	**4.99**
isolated areas not exceeding 0.50 m² irrespective of girth	-	0.52	8.38	-	nr	**8.38**
Glazed windows and screens						
panes; area not exceeding 0.10 m²	-	1.48	23.84	-	m²	**23.84**
panes; area 0.10 - 0.50 m²	-	1.18	19.01	-	m²	**19.01**
panes; area 0.50 - 1.00 m²	-	1.02	16.43	-	m²	**16.43**
panes; area over 1.00 m²	-	0.89	14.34	-	m²	**14.34**
M60 PAINTING/CLEAR FINISHING - INTERNALLY						
NOTE: The following prices include for preparing surfaces. Painting woodwork also includes for knotting prior to applying the priming coat and for the stopping of all nail holes etc.						
One coat primer; on wood surfaces before fixing						
General surfaces						
over 300 mm girth	-	0.10	1.61	0.86	m²	**2.47**
isolated surfaces not exceeding 300 mm girth	-	0.03	0.48	0.30	m	**0.79**
isolated areas not exceeding 0.50 m² irrespective of girth	-	0.07	1.13	0.25	nr	**1.38**
One coat polyurethane sealer; on wood surfaces before fixing						
General surfaces						
over 300 mm girth	-	0.12	1.93	0.84	m²	**2.77**
isolated surfaces not exceeding 300 mm girth	-	0.04	0.64	0.30	m	**0.94**
isolated areas not exceeding 0.50 m² irrespective of girth	-	0.09	1.45	0.40	nr	**1.85**

M SURFACE FINISHES

Item	PC £	Labour hours	Labour £	Material £	Unit	Total rate £
M60 PAINTING/CLEAR FINISHING - INTERNALLY – cont'd						
One coat of Sikkens "Cetol HLS" stain on wood surfaces before fixing						
General surfaces						
over 300 mm girth	-	0.13	2.09	0.96	m²	**3.06**
isolated surfaces not exceeding 300 mm girth	-	0.04	0.64	0.37	m	**1.02**
isolated areas not exceeding 0.50 m² irrespective of girth	-	0.09	1.45	0.46	nr	**1.91**
One coat of Sikkens "Cetol TS" interior stain; on wood surfaces before fixing						
General surfaces						
over 300 mm girth	-	0.13	2.09	1.35	m²	**3.44**
isolated surfaces not exceeding 300 mm girth	-	0.04	0.64	0.52	m	**1.16**
isolated areas not exceeding 0.50 m² irrespective of girth	-	0.09	1.45	0.65	nr	**2.10**
One coat Cuprinol clear wood preservative; on wood surfaces before fixing						
General surfaces						
over 300 mm girth	-	0.09	1.45	0.63	m²	**2.07**
isolated surfaces not exceeding 300 mm girth	-	0.03	0.48	0.22	m	**0.71**
isolated areas not exceeding 0.50 m² irrespective of girth	-	0.06	0.97	0.29	nr	**1.26**
One coat HCC Protective Coatings Ltd "Permacor" urethane alkyd gloss finishing coat or other equal and approved; on previously primed steelwork						
Members of roof trusses						
over 300 mm girth	-	0.08	1.29	1.94	m²	**3.23**
Two coats emulsion paint						
Brick or block walls						
over 300 mm girth	-	0.25	4.03	1.08	m²	**5.11**
Cement render or concrete						
over 300 mm girth	-	0.24	3.87	1.04	m²	**4.90**
isolated surfaces not exceeding 300 mm girth	-	0.12	1.93	0.32	m	**2.25**
Plaster walls or plaster/plasterboard ceilings						
over 300 mm girth	-	0.22	3.54	0.99	m²	**4.54**
over 300 mm girth; in multi colours	-	0.30	4.83	1.21	m²	**6.04**
over 300 mm girth; in staircase areas	-	0.25	4.03	1.15	m²	**5.18**
cutting in edges on flush surfaces	-	0.09	1.45	-	m	**1.45**
Plaster/plasterboard ceilings						
over 300 mm girth; 3.50 m - 5.00 m high	-	0.25	4.03	1.01	m²	**5.03**
One mist and two coats emulsion paint						
Brick or block walls						
over 300 mm girth	-	0.23	3.70	1.53	m²	**5.23**
Cement render or concrete						
over 300 mm girth	-	0.23	3.70	1.41	m²	**5.12**
Plaster walls or plaster/plasterboard ceilings						
over 300 mm girth	-	0.22	3.54	1.41	m²	**4.96**
over 300 mm girth; in multi colours	-	0.31	4.99	1.44	m²	**6.43**
over 300 mm girth; in staircase areas	-	0.25	4.03	1.41	m²	**5.44**
cutting in edges on flush surfaces	-	0.10	1.61	-	m	**1.61**
Plaster/plasterboard ceilings						
over 300 mm girth; 3.50 m - 5.00 m high	-	0.21	3.38	1.41	m²	**4.79**

M SURFACE FINISHES

Item	PC £	Labour hours	Labour £	Material £	Unit	Total rate £
One mist Supermatt; one full Supermatt and one full coat of quick drying Acrylic Eggshell						
Brick or block walls						
over 300 mm girth	-	0.23	3.70	1.65	m²	**5.36**
Cement render or concrete						
over 300 mm girth	-	0.23	3.70	1.52	m²	**5.23**
Plaster walls or plaster/plasterboard ceilings						
over 300 mm girth	-	0.22	3.54	1.52	m²	**5.07**
over 300 mm girth; in multi colours	-	0.31	4.99	1.52	m²	**6.52**
over 300 mm girth; in staircase areas	-	0.25	4.03	1.52	m²	**5.55**
cutting in edges on flush surfaces	-	0.10	1.61	-	m	**1.61**
Plaster/plasterboard ceilings						
over 300 mm girth; 3.50 m - 5.00 m high	-	0.21	3.38	1.52	m²	**4.91**
One coat "Tretol No 10" sealer or other equal and approved; two coats Tretol sprayed "Supercover Spraytone" emulsion paint or other equal and approved						
Plaster walls or plaster/plasterboard ceilings						
over 300 mm girth	-	-	-	-	m²	**5.46**
Textured plastic; "Artex" finish or other equal and approved						
Plasterboard ceilings						
over 300 mm girth	-	0.23	3.70	2.10	m²	**5.80**
Concrete walls or ceilings						
over 300 mm girth	-	0.25	4.03	1.85	m²	**5.88**
Touch up primer; one undercoat and one finishing coat of gloss oil paint; on wood surfaces						
General surfaces						
over 300 mm girth	-	0.27	4.35	1.96	m²	**6.30**
isolated surfaces not exceeding 300 mm girth	-	0.11	1.77	0.66	m	**2.43**
isolated areas not exceeding 0.50 m² irrespective of girth	-	0.20	3.22	1.01	nr	**4.23**
Glazed windows and screens						
panes; area not exceeding 0.10 m²	-	0.44	7.09	1.54	m²	**8.63**
panes; area 0.10 - 0.50 m²	-	0.36	5.80	1.16	m²	**6.96**
panes; area 0.50 - 1.00 m²	-	0.31	4.99	0.94	m²	**5.93**
panes; area over 1.00 m²	-	0.27	4.35	0.80	m²	**5.15**
Knot; one coat primer; stop; one undercoat and one finishing coat of gloss oil paint; on wood surfaces						
General surfaces						
over 300 mm girth	-	0.39	6.28	1.95	m²	**8.23**
isolated surfaces not exceeding 300 mm girth	-	0.16	2.58	0.65	m	**3.23**
isolated areas not exceeding 0.50 m² irrespective of girth	-	0.30	4.83	1.28	nr	**6.12**
Glazed windows and screens						
panes; area not exceeding 0.10 m²	-	0.67	10.79	1.95	m²	**12.74**
panes; area 0.10 - 0.50 m²	-	0.54	8.70	1.62	m²	**10.32**
panes; area 0.50 - 1.00 m²	-	0.46	7.41	1.62	m²	**9.03**
panes; area over 1.00 m²	-	0.39	6.28	1.20	m²	**7.48**
One coat primer; one undercoat and one finishing coat of gloss oil paint						
Plaster surfaces						
over 300 mm girth	-	0.35	5.64	2.47	m²	**8.11**

M SURFACE FINISHES

Item	PC £	Labour hours	Labour £	Material £	Unit	Total rate £
M60 PAINTING/CLEAR FINISHING - INTERNALLY – cont'd						
One coat primer; two undercoats and one finishing coat of gloss oil paint						
Plaster surfaces						
over 300 mm girth	-	0.46	7.41	3.32	m²	**10.73**
One coat primer; two undercoats and one finishing coat of eggshell paint						
Plaster surfaces						
over 300 mm girth	-	0.46	7.41	3.65	m²	**11.06**
Touch up primer; one undercoat and one finishing coat of gloss paint; on iron or steel surfaces						
General surfaces						
over 300 mm girth	-	0.27	4.35	1.49	m²	**5.84**
isolated surfaces not exceeding 300 mm girth	-	0.11	1.77	0.50	m	**2.28**
isolated areas ne 0.50 m² irrespective of girth	-	0.20	3.22	0.80	nr	**4.02**
Glazed windows and screens						
panes; area not exceeding 0.10 m²	-	0.44	7.09	1.53	m²	**8.61**
panes; area 0.10 - 0.50 m²	-	0.36	5.80	1.16	m²	**6.96**
panes; area 0.50 - 1.00 m²	-	0.31	4.99	0.92	m²	**5.91**
panes; area over 1.00 m²	-	0.27	4.35	0.78	m²	**5.12**
Structural steelwork						
over 300 mm girth	-	0.30	4.83	1.54	m²	**6.37**
Members of roof trusses						
over 300 mm girth	-	0.40	6.44	1.76	m²	**8.20**
Ornamental railings and the like; each side measured overall						
over 300 mm girth	-	0.46	7.41	1.92	m²	**9.32**
Iron or steel radiators						
over 300 mm girth	-	0.27	4.35	1.61	m²	**5.96**
Pipes or conduits						
over 300 mm girth	-	0.40	6.44	1.67	m²	**8.11**
not exceeding 300 mm girth	-	0.16	2.58	0.55	m	**3.13**
One coat primer; one undercoat and one finishing coat of gloss oil paint; on iron or steel surfaces						
General surfaces						
over 300 mm girth	-	0.27	4.35	1.49	m²	**5.84**
isolated surfaces not exceeding 300 mm girth	-	0.14	2.25	0.87	m	**3.12**
isolated areas ne 0.50 m² irrespective of girth	-	0.27	4.35	1.43	nr	**5.78**
Glazed windows and screens						
panes; area not exceeding 0.10 m²	-	0.58	9.34	2.35	m²	**11.69**
panes; area 0.10 - 0.50 m²	-	0.46	7.41	1.87	m²	**9.28**
panes; area 0.50 - 1.00 m²	-	0.40	6.44	1.61	m²	**8.05**
panes; area over 1.00 m²	-	0.35	5.64	1.43	m²	**7.07**
Structural steelwork						
over 300 mm girth	-	0.39	6.28	2.30	m²	**8.58**
Members of roof trusses						
over 300 mm girth	-	0.53	8.54	2.41	m²	**10.95**
Ornamental railings and the like; each side measured overall						
over 300 mm girth	-	0.60	9.66	2.87	m²	**12.53**
Iron or steel radiators						
over 300 mm girth	-	0.35	5.64	2.41	m²	**8.05**
Pipes or conduits						
over 300 mm girth	-	0.53	8.54	2.41	m²	**10.95**
not exceeding 300 mm girth	-	0.20	3.22	0.79	m	**4.01**

M SURFACE FINISHES

Item	PC £	Labour hours	Labour £	Material £	Unit	Total rate £
Two coats of bituminous paint; on iron or steel surfaces						
General surfaces						
over 300 mm girth	-	0.27	4.35	0.86	m²	**5.21**
Inside of galvanized steel cistern						
over 300 mm girth	-	0.40	6.44	1.04	m²	**7.49**
Two coats bituminous paint; first coat blinded with clean sand prior to second coat; on concrete surfaces						
General surfaces						
over 300 mm girth	-	0.93	14.98	1.96	m²	**16.94**
Mordant solution; one coat HCC Protective Coatings Ltd "Permacor alkyd MIO" or other equal and approved; one coat "Permacor Epoxy Gloss" finishing coat or other equal and approved on galvanised steelwork						
Structural steelwork						
over 300 mm girth	-	0.52	8.38	3.31	m²	**11.69**
One coat HCC Protective Coatings Ltd "Permacor Epoxy Zinc Primer" or other equal and approved; two coats "Permacor alkyd MIO" or other equal and approved; one coat "Permatex Epoxy Gloss" finishing coat or other equal and approved on steelwork						
Structural steelwork						
over 300 mm girth	-	0.74	11.92	6.06	m²	**17.98**
Steel protection; HCC Protective Coatings Ltd "Unitherm" or other equal and approved; two coats to steelwork						
Structural steelwork						
over 300 mm girth	-	1.16	18.68	2.07	m²	**20.75**
Two coats of epoxy anti-slip floor paint; on screeded concrete surfaces						
General surfaces						
over 300 mm girth	-	0.28	4.51	12.71	m²	**17.22**
"Nitoflor Lithurin" floor hardener and dust proofer or other equal and approved; Fosroc Expandite Ltd or other equal and approved; two coats; on concrete surfaces						
General surfaces						
over 300 mm girth	-	0.28	3.38	0.68	m²	**4.05**
Two coats of boiled linseed oil; on hardwood surfaces						
General surfaces						
over 300 mm girth	-	0.20	3.22	1.58	m²	**4.81**
isolated surfaces not exceeding 300 mm girth	-	0.08	1.29	0.51	m	**1.80**
isolated areas not exceeding 0.50 m² irrespective of girth	-	0.16	2.58	0.92	nr	**3.49**

M SURFACE FINISHES

Item	PC £	Labour hours	Labour £	Material £	Unit	Total rate £
M60 PAINTING/CLEAR FINISHING - INTERNALLY – cont'd						
Two coats polyurethane varnish; on wood surfaces						
General surfaces						
over 300 mm girth	-	0.20	3.22	1.48	m²	**4.70**
isolated surfaces not exceeding 300 mm girth	-	0.08	1.29	0.55	m	**1.84**
isolated areas not exceeding 0.50 m² irrespective of girth	-	0.16	2.58	0.18	nr	**2.75**
Three coats polyurethane varnish; on wood surfaces						
General surfaces						
over 300 mm girth	-	0.31	4.99	2.26	m²	**7.25**
isolated surfaces not exceeding 300 mm girth	-	0.12	1.93	0.69	m	**2.63**
isolated areas not exceeding 0.50 m² irrespective of girth	-	0.23	3.70	1.26	nr	**4.96**
One undercoat; and one finishing coat; of "Albi" clear flame retardant surface coating or other equal and approved; on wood surfaces						
General surfaces						
over 300 mm girth	-	0.40	6.44	4.10	m²	**10.54**
isolated surfaces not exceeding 300 mm girth	-	0.17	2.74	1.42	m	**4.15**
isolated areas not exceeding 0.50 m² irrespective of girth	-	0.23	3.70	3.11	nr	**6.81**
Two undercoats; and one finishing coat; of "Albi" clear flame retardant surface coating or other equal and approved; on wood surfaces						
General surfaces						
over 300 mm girth	-	0.47	7.57	6.01	m²	**13.58**
isolated surfaces not exceeding 300 mm girth	-	0.24	3.87	2.08	m	**5.94**
isolated areas not exceeding 0.50 m² irrespective of girth	-	0.39	6.28	3.33	nr	**9.61**
Seal and wax polish; dull gloss finish on wood surfaces						
General surfaces						
over 300 mm girth	-	-	-	-	m²	**10.68**
isolated surfaces not exceeding 300 mm girth	-	-	-	-	m	**4.82**
isolated areas not exceeding 0.50 m² irrespective of girth	-	-	-	-	nr	**7.48**
One coat of "Sadolin Extra" or other equal and approved; clear or pigmented; one further coat of "Holdex" clear interior silk matt lacquer						
General surfaces						
over 300 mm girth	-	0.30	4.83	5.99	m²	**10.82**
isolated surfaces not exceeding 300 mm girth	-	0.12	1.93	2.80	m	**4.73**
isolated areas not exceeding 0.50 m² irrespective of girth	-	0.22	3.54	2.91	nr	**6.46**
Glazed windows and screens						
panes; area not exceeding 0.10 m²	-	0.49	7.89	3.43	m²	**11.32**
panes; area 0.10 - 0.50 m²	-	0.39	6.28	3.19	m²	**9.47**
panes; area 0.50 - 1.00 m²	-	0.34	5.48	2.96	m²	**8.43**
panes; area over 1.00 m²	-	0.30	4.83	2.80	m²	**7.63**

M SURFACE FINISHES

Item	PC £	Labour hours	Labour £	Material £	Unit	Total rate £
Two coats of "Sadolins Extra" or other equal and approved; clear or pigmented; two further coats of "Holdex" clear interior silk matt lacquer or other equal and approved						
General surfaces						
over 300 mm girth	-	0.46	7.41	11.07	m²	**18.48**
isolated surfaces not exceeding 300 mm girth	-	0.19	3.06	5.54	m	**8.60**
isolated areas not exceeding 0.50 m² irrespective of girth	-	0.35	5.64	6.32	nr	**11.96**
Glazed windows and screens						
panes; area not exceeding 0.10 m²	-	0.77	12.40	6.79	m²	**19.19**
panes; area 0.10 - 0.50 m²	-	0.61	9.83	6.32	m²	**16.14**
panes; area 0.50 - 1.00 m²	-	0.54	8.70	5.85	m²	**14.55**
panes; area over 1.00 m²	-	0.46	7.41	5.54	m²	**12.95**
Two coats of Sikkens "Cetol TS" interior stain or other equal and approved; on wood surfaces						
General surfaces						
over 300 mm girth	-	0.21	3.38	2.42	m²	**5.80**
isolated surfaces not exceeding 300 mm girth	-	0.09	1.45	0.85	m	**2.30**
isolated areas not exceeding 0.50 m² irrespective of girth	-	0.16	2.58	1.32	nr	**3.90**
Body in and wax polish; dull gloss finish; on hardwood surfaces						
General surfaces						
over 300 mm girth	-	-	-	-	m²	**12.02**
isolated surfaces not exceeding 300 mm girth	-	-	-	-	m	**5.42**
isolated areas not exceeding 0.50 m² irrespective of girth	-	-	-	-	nr	**8.43**
Stain; body in and wax polish; dull gloss finish; on hardwood surfaces						
General surfaces						
over 300 mm girth	-	-	-	-	m²	**16.09**
isolated surfaces not exceeding 300 mm girth	-	-	-	-	m	**7.25**
isolated areas not exceeding 0.50 m² irrespective of girth	-	-	-	-	nr	**11.28**
Seal; two coats of synthetic resin lacquer, decorative flatted finish; wire down, wax and burnish; on wood surfaces						
General surfaces						
over 300 mm girth	-	-	-	-	m²	**20.27**
isolated surfaces not exceeding 300 mm girth	-	-	-	-	m	**9.48**
isolated areas not exceeding 0.50 m² irrespective of girth	-	-	-	-	nr	**14.26**
Stain; body in and fully French polish; full gloss finish; on hardwood surfaces						
General surfaces						
over 300 mm girth	-	-	-	-	m²	**23.46**
isolated surfaces not exceeding 300 mm girth	-	-	-	-	m	**10.55**
isolated areas not exceeding 0.50 m² irrespective of area	-	-	-	-	nr	**16.42**

M SURFACE FINISHES

Item	PC £	Labour hours	Labour £	Material £	Unit	Total rate £
M60 PAINTING/CLEAR FINISHING - INTERNALLY – cont'd						
Stain; fill grain and fully French polish; full gloss finish; on hardwood surfaces						
General surfaces						
over 300 mm girth	-	-	-	-	m²	**34.88**
isolated surfaces not exceeding 300 mm girth	-	-	-	-	m	**15.69**
isolated areas not exceeding 0.50 m² irrespective of girth	-	-	-	-	nr	**24.41**
Stain black; body in and fully French polish; ebonized finish; on hardwood surfaces						
General surfaces						
over 300 mm girth	-	-	-	-	m²	**39.78**
isolated surfaces not exceeding 300 mm girth	-	-	-	-	m	**17.90**
isolated areas not exceeding 0.50 m² irrespective of girth	-	-	-	-	nr	**27.85**
M60 PREPARATION OF EXISTING SURFACES - EXTERNALLY						
Wash and rub down iron and steel surfaces; bringing forward						
General surfaces						
over 300 mm girth	-	0.15	2.42	-	m²	**2.42**
isolated surfaces not exceeding 300 mm girth	-	0.05	0.81	-	m	**0.81**
isolated areas not exceeding 0.50 m² irrespective of girth	-	0.11	1.77	-	nr	**1.77**
Glazed windows and screens						
panes; area not exceeding 0.10 m²	-	0.24	3.87	-	m²	**3.87**
panes; area 0.10 - 0.50 m²	-	0.20	3.22	-	m²	**3.22**
panes; area 0.50 - 1.00 m²	-	0.17	2.74	-	m²	**2.74**
panes; area over 1.00 m²	-	0.15	2.42	-	m²	**2.42**
Wash and rub down wood surfaces; prime bare patches; bringing forward						
General surfaces						
over 300 mm girth	-	0.24	3.87	-	m²	**3.87**
isolated surfaces not exceeding 300 mm girth	-	0.09	1.45	-	m	**1.45**
isolated areas not exceeding 0.50 m² irrespective of girth	-	0.18	2.90	-	nr	**2.90**
Glazed windows and screens						
panes; area not exceeding 0.10 m²	-	0.41	6.60	-	m²	**6.60**
panes; area 0.10 - 0.50 m²	-	0.32	5.15	-	m²	**5.15**
panes; area 0.50 - 1.00 m²	-	0.28	4.51	-	m²	**4.51**
panes; area over 1.00 m²	-	0.24	3.87	-	m²	**3.87**
Wash down and remove paint with chemical stripper from iron, steel or wood surfaces						
General surfaces						
over 300 mm girth	-	0.74	11.92	-	m²	**11.92**
isolated surfaces not exceeding 300 mm girth	-	0.33	5.32	-	m	**5.32**
isolated areas not exceeding 0.50 m² irrespective of girth	-	0.56	9.02	-	nr	**9.02**
Glazed windows and screens						
panes; area not exceeding 0.10 m²	-	1.58	25.45	-	m²	**25.45**
panes; area 0.10 - 0.50 m²	-	1.26	20.30	-	m²	**20.30**
panes; area 0.50 - 1.00 m²	-	1.09	17.56	-	m²	**17.56**
panes; area over 1.00 m²	-	0.94	15.14	-	m²	**15.14**

M SURFACE FINISHES

Item	PC £	Labour hours	Labour £	Material £	Unit	Total rate £
Burn off and rub down to remove paint from iron, steel or wood surfaces						
General surfaces						
over 300 mm girth	-	0.91	14.66	-	m²	**14.66**
isolated surfaces not exceeding 300 mm girth	-	0.42	6.77	-	m	**6.77**
isolated surfaces not exceeding 0.50 m²	-	0.68	10.95	-	nr	**10.95**
Glazed windows and screens						
panes over 1m²	-	1.18	19.01	-	m²	**19.01**
panes 0.50 - 1.00m²	-	1.35	21.74	-	m²	**21.74**
panes 0.10 - 0.50m²	-	1.57	25.29	-	m²	**25.29**
panes not exceeding 0.10m²	-	1.97	31.73	-	m²	**31.73**
M60 PAINTING/CLEAR FINISHING - EXTERNALLY						
Two coats of cement paint, "Sandtex Matt" or other equal and approved						
Brick or block walls						
over 300 mm girth	-	0.31	4.99	1.51	m²	**6.50**
Cement render or concrete walls						
over 300 mm girth	-	0.27	4.35	1.00	m²	**5.35**
Roughcast walls						
over 300 mm girth	-	0.46	7.41	1.00	m²	**8.41**
One coat sealer and two coats of external grade emulsion paint, Dulux "Weathershield" or other equal and approved						
Brick or block walls						
over 300 mm girth	-	0.50	8.05	7.26	m²	**15.31**
Cement render or concrete walls						
over 300 mm girth	-	0.39	6.28	4.84	m²	**11.12**
Concrete soffits						
over 300 mm girth	-	0.46	7.41	4.84	m²	**12.25**
One coat sealer (applied by brush) and two coats of external grade emulsion paint, Dulux "Weathershield" or other equal and approved (spray applied)						
Roughcast						
over 300 mm girth	-	0.33	5.32	9.80	m²	**15.12**
One coat sealer and two coats of anti-graffiti paint (spray applied)						
Brick or block walls						
over 300 mm girth	-	0.26	4.19	3.26	m²	**7.45**
Cement render or concrete walls						
over 300 mm girth	-	0.30	4.83	4.85	m²	**9.68**
2.5 mm of "Vandalene" anti-climb paint (spray applied)						
General surfaces						
over 300 mm girth	-	0.35	5.64	4.62	m²	**10.25**
Two coats solar reflective aluminium paint; on bituminous roofing						
General surfaces						
over 300 mm girth	-	0.52	8.38	13.74	m²	**22.12**

M SURFACE FINISHES

Item	PC £	Labour hours	Labour £	Material £	Unit	Total rate £
M60 PAINTING/CLEAR FINISHING - EXTERNALLY – cont'd						
Touch up primer; two undercoats and one finishing coat of gloss oil paint; on wood surfaces						
General surfaces						
over 300 mm girth	-	0.42	6.77	1.95	m²	**8.72**
isolated surfaces not exceeding 300 mm girth	-	0.17	2.74	0.52	m	**3.26**
isolated areas not exceeding 0.50 m² irrespective of girth	-	0.31	4.99	1.04	nr	**6.03**
Glazed windows and screens						
panes; area not exceeding 0.10 m²	-	0.69	11.11	1.72	m²	**12.84**
panes; area 0.10 - 0.50 m²	-	0.69	11.11	1.45	m²	**12.57**
panes; area 0.50 - 1.00 m²	-	0.56	9.02	1.27	m²	**10.29**
panes; area over 1.00 m²	-	0.42	6.77	1.04	m²	**7.80**
Glazed windows and screens; multi- coloured work						
panes; area not exceeding 0.10 m²	-	0.80	12.89	1.72	m²	**14.61**
panes; area 0.10 - 0.50 m²	-	0.64	10.31	1.50	m²	**11.80**
panes; area 0.50 - 1.00 m²	-	0.56	9.02	1.27	m²	**10.29**
panes; area over 1.00 m²	-	0.48	7.73	1.04	m²	**8.77**
Knot; one coat primer; two undercoats and one finishing coat of gloss oil paint; on wood surfaces						
General surfaces						
over 300 mm girth	-	0.55	8.86	2.11	m²	**10.97**
isolated surfaces not exceeding 300 mm girth	-	0.22	3.54	0.75	m	**4.30**
isolated areas not exceeding 0.50 m² irrespective of girth	-	0.42	6.77	1.46	nr	**8.23**
Glazed windows and screens						
panes; area not exceeding 0.10 m²	-	0.92	14.82	2.35	m²	**17.17**
panes; area 0.10 - 0.50 m²	-	0.73	11.76	2.09	m²	**13.85**
panes; area 0.50 - 1.00 m²	-	0.64	10.31	1.60	m²	**11.91**
panes; area over 1.00 m²	-	0.55	8.86	1.11	m²	**9.97**
Glazed windows and screens; multi-coloured work						
panes; area not exceeding 0.10 m²	-	1.05	16.91	2.35	m²	**19.27**
panes; area 0.10 - 0.50 m²	-	0.85	13.69	2.11	m²	**15.80**
panes; area 0.50 - 1.00 m²	-	0.75	12.08	1.60	m²	**13.68**
panes; area over 1.00 m²	-	0.63	10.15	1.11	m²	**11.26**
Touch up primer; two undercoats and one finishing coat of gloss oil paint; on iron or steel surfaces						
General surfaces						
over 300 mm girth	-	0.42	6.77	1.64	m²	**8.40**
isolated surfaces not exceeding 300 mm girth	-	0.17	2.74	0.45	m	**3.19**
isolated areas not exceeding 0.50 m² irrespective of girth	-	0.31	4.99	0.93	nr	**5.92**
Glazed windows and screens						
panes; area not exceeding 0.10 m²	-	0.69	11.11	1.70	m²	**12.82**
panes; area 0.10 - 0.50 m²	-	0.56	9.02	1.47	m²	**10.49**
panes; area 0.50 - 1.00 m²	-	0.48	7.73	1.23	m²	**8.96**
panes; area over 1.00 m²	-	0.42	6.77	0.99	m²	**7.76**
Structural steelwork						
over 300 mm girth	-	0.47	7.57	1.70	m²	**9.27**
Members of roof trusses						
over 300 mm girth	-	0.63	10.15	1.93	m²	**12.08**
Ornamental railings and the like; each side measured overall						
over 300 mm girth	-	0.71	11.44	2.02	m²	**13.45**

M SURFACE FINISHES

Item	PC £	Labour hours	Labour £	Material £	Unit	Total rate £
Eaves gutters						
over 300 mm girth	-	0.75	12.08	2.13	m²	**14.21**
not exceeding 300 mm girth	-	0.30	4.83	0.89	m	**5.72**
Pipes or conduits						
over 300 mm girth	-	0.63	10.15	2.13	m²	**12.28**
not exceeding 300 mm girth	-	0.25	4.03	0.73	m	**4.76**
One coat primer; two undercoats and one finishing coat of gloss oil paint; on iron or steel surfaces						
General surfaces						
over 300 mm girth	-	0.50	8.05	1.96	m²	**10.01**
isolated surfaces not exceeding 300 mm girth	-	0.20	3.22	0.51	m	**3.73**
isolated areas not exceeding 0.50 m² irrespective of girth	-	0.38	6.12	1.01	nr	**7.13**
Glazed windows and screens						
panes; area not exceeding 0.10 m²	-	0.83	13.37	1.78	m²	**15.15**
panes; area 0.10 - 0.50 m²	-	0.67	10.79	1.55	m²	**12.34**
panes; area 0.50 - 1.00 m²	-	0.58	9.34	1.31	m²	**10.65**
panes; area over 1.00 m²	-	0.50	8.05	1.01	m²	**9.06**
Structural steelwork						
over 300 mm girth	-	0.56	9.02	2.02	m²	**11.04**
Members of roof trusses						
over 300 mm girth	-	0.75	12.08	2.26	m²	**14.34**
Ornamental railings and the like; each side measured overall						
over 300 mm girth	-	0.85	13.69	2.26	m²	**15.95**
Eaves gutters						
over 300 mm girth	-	0.90	14.50	2.43	m²	**16.93**
not exceeding 300 mm girth	-	0.36	5.80	0.83	m	**6.63**
Pipes or conduits						
over 300 mm girth	-	0.75	12.08	2.43	m²	**14.51**
not exceeding 300 mm girth	-	0.30	4.83	0.81	m	**5.64**
One coat of Andrews "Hammerite" paint or other equal and approved; on iron or steel surfaces						
General surfaces						
over 300 mm girth	-	0.18	2.90	1.36	m²	**4.26**
isolated surfaces not exceeding 300 mm girth	-	0.09	1.45	0.42	m	**1.07**
isolated areas not exceeding 0.50 m² irrespective of girth	-	0.13	2.09	0.78	nr	**2.87**
Glazed windows and screens						
panes; area not exceeding 0.10 m²	-	0.30	4.83	1.02	m²	**5.86**
panes; area 0.10 - 0.50 m²	-	0.23	3.70	1.13	m²	**4.84**
panes; area 0.50 - 1.00 m²	-	0.21	3.38	1.02	m²	**4.40**
panes; area over 1.00 m²	-	0.18	2.90	1.02	m²	**3.92**
Structural steelwork						
over 300 mm girth	-	0.20	3.22	1.25	m²	**4.47**
Members of roof trusses						
over 300 mm girth	-	0.27	4.35	1.36	m²	**5.71**
Ornamental railings and the like; each side measured overall						
over 300 mm girth	-	0.30	4.83	1.36	m²	**6.19**
Eaves gutters						
over 300 mm girth	-	0.31	4.99	1.47	m²	**6.46**
not exceeding 300 mm girth	-	0.10	1.61	0.68	m	**2.29**
Pipes or conduits						
over 300 mm girth	-	0.30	4.83	1.25	m²	**6.08**
not exceeding 300 mm girth	-	0.10	1.61	0.57	m	**2.18**

M SURFACE FINISHES

Item	PC £	Labour hours	Labour £	Material £	Unit	Total rate £
M60 PAINTING/CLEAR FINISHING - EXTERNALLY – cont'd						
Two coats of creosote; on wood surfaces						
General surfaces						
over 300 mm girth	-	0.19	3.06	0.38	m²	**3.44**
isolated surfaces not exceeding 300 mm girth	-	0.06	0.97	0.23	m	**1.20**
Two coats of "Solignum" wood preservative or other equal and approved; on wood surfaces						
General surfaces						
over 300 mm girth	-	0.17	2.74	1.86	m²	**4.59**
isolated surfaces not exceeding 300 mm girth	-	0.06	0.97	0.55	m	**1.51**
Three coats of polyurethane; on wood surfaces						
General surfaces						
over 300 mm girth	-	0.33	5.32	2.49	m²	**7.80**
isolated surfaces not exceeding 300 mm girth	-	0.13	2.09	1.23	m	**3.32**
isolated areas not exceeding 0.50 m² irrespective of girth	-	0.25	4.03	1.42	nr	**5.45**
Two coats of "New Base" primer or other and approved; and two coats of "Extra"; Sadolin Ltd or other equal and approved; pigmented; on wood surfaces						
General surfaces						
over 300 mm girth	-	0.50	8.05	5.74	m²	**13.79**
isolated surfaces not exceeding 300 mm girth	-	0.31	4.99	1.95	m	**6.95**
Glazed windows and screens						
panes; area not exceeding 0.10 m²	-	0.84	13.53	4.08	m²	**17.61**
panes; area 0.10 - 0.50 m²	-	0.68	10.95	3.84	m²	**14.80**
panes; area 0.50 - 1.00 m²	-	0.58	9.34	3.61	m²	**12.95**
panes; area over 1.00 m²	-	0.50	8.05	2.90	m²	**10.95**
Two coats Sikkens "Cetol Filter 7" exterior stain or other equal and approved; on wood surfaces						
General surfaces						
over 300 mm girth	-	0.21	3.38	4.05	m²	**7.43**
isolated surfaces not exceeding 300 mm girth	-	0.09	1.45	1.39	m	**2.84**
isolated areas not exceeding 0.50 m² irrespective of girth	-	0.16	2.58	2.06	nr	**4.63**

N FURNITURE/EQUIPMENT

Item	PC £	Labour hours	Labour £	Material £	Unit	Total rate £
N10/11 GENERAL FIXTURES/KITCHEN FITTINGS						
SUPPLY ONLY PRICES						
NOTE: The fixing of general fixtures will vary considerably depending upon the size of the fixture and the method of fixing employed. Prices for fixing like sized kitchen fittings may be suitable for certain fixtures, although adjustment to those rates will almost invariably be necessary and the reader is directed to section "G20" for information on bolts, plugging brickwork and blockwork etc. which should prove useful in building up a suitable rate.						
The following supply only prices are for purpose made fittings components in various materials supplied as part of an assembled fitting and therefore may be used to arrive at a guide price for a complete fitting.						
Fitting components; medium density fibreboard						
Backs, fronts, sides or divisions; over 300 mm wide						
12 mm thick	-	-	-	29.54	m²	-
18 mm thick	-	-	-	31.31	m²	-
25 mm thick	-	-	-	34.86	m²	-
Shelves or worktops; over 300 mm wide						
18 mm thick	-	-	-	31.31	m²	-
25 mm thick	-	-	-	34.86	m²	-
Flush doors; lipped on four edges						
450 mm x 750 mm x 18 mm	-	-	-	45.34	nr	-
450 mm x 750 mm x 25 mm	-	-	-	46.21	nr	-
600 mm x 900 mm x 18 mm	-	-	-	53.52	nr	-
600 mm x 900 mm x 25 mm	-	-	-	54.92	nr	-
Fitting components; moisture-resistant medium density fibreboard						
Backs, fronts, sides or divisions; over 300 mm wide						
12 mm thick	-	-	-	33.07	m²	-
18 mm thick	-	-	-	36.62	m²	-
25 mm thick	-	-	-	40.15	m²	-
Shelves or worktops; over 300 mm wide						
18 mm thick	-	-	-	36.62	m²	-
25 mm thick	-	-	-	40.15	m²	-
Flush doors; lipped on four edges						
450 mm x 750 mm x 18 mm	-	-	-	46.21	nr	-
450 mm x 750 mm x 25 mm	-	-	-	47.51	nr	-
600 mm x 900 mm x 18 mm	-	-	-	54.92	nr	-
600 mm x 900 mm x 25 mm	-	-	-	57.07	nr	-
Fitting components; medium density board; melamine faced both sides						
Backs, fronts, sides or divisions; over 300 mm wide						
12 mm thick	-	-	-	38.98	m²	-
18 mm thick	-	-	-	43.40	m²	-
Shelves or worktops; over 300 mm wide						
18 mm thick	-	-	-	43.40	m²	-
Flush doors; lipped on four edges						
450 mm x 750 mm x 18 mm	-	-	-	32.42	nr	-
600 mm x 900 mm x 25 mm	-	-	-	41.15	nr	-

N FURNITURE/EQUIPMENT

Item	PC £	Labour hours	Labour £	Material £	Unit	Total rate £
N10/11 GENERAL FIXTURES/KITCHEN FITTINGS – cont'd						
SUPPLY ONLY PRICES – cont'd						
Fitting components; medium density board; formica faced both sides						
Backs, fronts, sides or divisions; over 300 mm wide						
12 mm thick	-			118.70	m²	
18 mm thick	-			123.43	m²	
Shelves or worktops; over 300 mm wide						
18 mm thick	-			123.43	m²	
Flush doors; lipped on four edges						
450 mm x 750 mm x 18 mm	-	-	-	65.37	nr	-
600 mm x 900 mm x 25 mm	-			66.78	nr	
Fitting components; wrought softwood						
Backs, fronts, sides or divisions; cross-tongued joints; over 300 mm wide						
25 mm thick	-	-		50.20	m²	
Shelves or worktops; cross-tongued joints; over 300 mm wide						
25 mm thick	-	-	-	50.20	m²	-
Bearers						
19 mm x 38 mm	-	-	-	2.63	m	-
25 mm x 50 mm	-	-	-	2.91	m	-
44 mm x 44 mm	-	-	-	3.10	m	-
44 mm x 69 mm	-	-	-	3.58	m	-
Bearers; framed; to backs, fronts or sides						
19 mm x 38 mm	-	-	-	6.00	m	-
25 mm x 50 mm	-	-	-	6.50	m	-
44 mm x 44 mm	-	-	-	8.40	m	-
44 mm x 69 mm	-	-	-	9.64	m	-
Add 5% to the above materials prices for selected softwood for staining						
Fitting components; selected Sapele						
Backs, fronts, sides or divisions; cross-tongued joints; over 300 mm wide						
25 mm thick	-	-	-	94.02	m²	-
Shelves or worktops; cross-tongued joints; over 300 mm wide						
25 mm thick	-	-	-	94.02	m²	-
Bearers						
19 mm x 38 mm	-	-	-	4.71	m	-
25 mm x 50 mm	-	-	-	5.87	m	-
50 mm x 50 mm	-	-	-	6.60	m	-
50 mm x 75 mm	-	-	-	8.58	m	-
Bearers; framed; to backs, fronts or sides						
19 mm x 38 mm	-	-	-	9.11	m	-
25 mm x 50 mm	-	-	-	10.10	m	-
50 mm x 50 mm	-	-	-	13.54	m	-
50 mm x 75 mm	-	-	-	16.98	m	-

N FURNITURE/EQUIPMENT

Item	PC £	Labour hours	Labour £	Material £	Unit	Total rate £
Fitting components; Iroko						
Backs, fronts, sides or divisions; cross-tongued joints; over 300 mm wide						
25 mm thick	-	-	-	115.84	m²	-
Shelves or worktops; cross-tongued joints; over 300 mm wide						
25 mm thick	-	-	-	115.84	m²	-
Draining boards; cross-tongued joints; over 300 mm wide						
25 mm thick	-	-	-	145.12	m²	-
stopped flutes	-	-	-	7.51	m	-
grooves; cross-grain	-	-	-	1.11	m	-
Bearers						
19 mm x 38 mm	-	-	-	5.75	m	-
25 mm x 50 mm	-	-	-	7.36	m	-
50 mm x 50 mm	-	-	-	8.36	m	-
50 mm x 75 mm	-	-	-	11.02	m	-
Bearers; framed; to backs, fronts or sides						
19 mm x 38 mm	-	-	-	10.52	m	-
25 mm x 50 mm	-	-	-	11.72	m	-
50 mm x 50 mm	-	-	-	15.82	m	-
50 mm x 75 mm	-	-	-	20.55	m	-
SUPPLY AND FIX PRICES						
NOTE: Kitchen fittings prices vary considerably. PC supply prices for reasonable quantities for a moderately priced range of kitchen fittings have been shown.						
Supplying and fixing to backgrounds requiring plugging; including any pre-assembly						
Wall units						
300 mm x 300 mm x 720 mm	79.32	1.20	19.33	89.89	nr	**109.22**
500 mm x 300 mm x 720 mm	93.47	1.30	20.94	105.87	nr	**126.80**
600 mm x 300 mm x 720 mm	104.75	1.43	23.03	118.60	nr	**141.64**
800 mm x 300 mm x 720 mm	159.39	1.62	26.09	180.28	nr	**206.37**
Floor units with drawers						
500 mm x 600 mm x 870 mm	139.20	1.30	20.94	157.49	nr	**178.43**
600 mm x 600 mm x 870 mm	153.36	1.43	23.03	173.47	nr	**196.50**
1000 mm x 600 mm x 870 mm	238.82	1.71	27.54	269.93	nr	**297.47**
Sink units (excluding sink top)						
1000 mm x 600 mm x 870 mm	242.46	1.62	26.09	274.04	nr	**300.13**
Larder units						
500 mm x 600 mm x 870 mm	139.20	1.50	24.16	157.86	nr	**182.02**
Laminated plastics worktops; single rolled edge; prices include for fixing						
38 mm thick; 600 mm wide	37.70	0.41	6.60	44.86	m	**51.46**
extra for forming hole for inset sink	-	0.77	12.40	-	nr	**12.40**
extra for jointing strip at corner intersection of worktops	-	0.16	2.58	9.33	nr	**11.91**
extra for butt and scribe joint at corner intersection of worktops	-	4.58	73.77	-	nr	**73.77**

N FURNITURE/EQUIPMENT

Item	PC £	Labour hours	Labour £	Material £	Unit	Total rate £
N10/11 GENERAL FIXTURES/KITCHEN FITTINGS – cont'd						
Lockers and cupboards; Welconstruct Distribution or other equal and approved						
Standard clothes lockers; steel body and door within reinforced 19G frame, powder coated finish, cam locks						
1 compartment; placing in position						
300 mm x 300 mm x 1800 mm	-	0.25	3.02	65.02	nr	**68.03**
380 mm x 380 mm x 1800 mm	-	0.25	3.02	79.58	nr	**82.59**
450 mm x 450 mm x 1800 mm	-	0.30	3.62	90.75	nr	**94.37**
Compartment lockers; steel body and door with reinforced 19G frame, powder coated finish, cam locks						
2 compartments; placing in position						
300 mm x 300 mm x 1800 mm	-	0.25	3.02	71.68	nr	**74.69**
380 mm x 380 mm x 1800 mm	-	0.25	3.02	65.02	nr	**68.03**
450 mm x 450 mm x 1800 mm	-	0.30	3.62	98.65	nr	**102.27**
4 compartments; placing in position						
300 mm x 300 mm x 1800 mm	-	0.25	3.02	85.11	nr	**88.12**
380 mm x 380 mm x 1800 mm	-	0.25	3.02	101.93	nr	**104.94**
450 mm x 450 mm x 1800 mm	-	0.30	3.62	101.93	nr	**105.55**
Timber clothes lockers; veneered MDF finish, routed door, cam locks						
1 compartment; placing in position						
380 mm x 380 mm x 1830 mm	-	0.30	3.62	237.04	nr	**240.66**
4 compartments; placing in position						
380 mm x 380 mm x 1830 mm	-	0.30	3.62	345.40	nr	**349.02**
Cupboards; stainless steel; cam locks						
900 mm x 460 mm x 900 mm; with one shelf	-	0.25	3.02	686.28	nr	**689.30**
900 mm x 460 mm x 1200 mm; with two shelves	-	0.25	3.02	778.84	nr	**781.85**
900 mm x 460 mm x 1800 mm; with three shelves	-	0.25	3.02	965.08	nr	**968.10**
1200 mm x 460 mm x 1800 mm; with three shelves	-	0.25	3.02	1250.65	nr	**1253.67**
Shelving support systems; Welconstruct Distribution or other equal and approved						
Shelving support systems; steel body; stove enamelled finish; assembling						
open initial bay; 5 shelves; placing in position						
1000 mm x 300 mm x 1850 mm	-	0.75	11.52	121.79	nr	**133.31**
1000 mm x 600 mm x 1850 mm	-	0.75	11.52	176.09	nr	**187.61**
open extension bay; 5 shelves; placing in position						
1000 mm x 300 mm x 1850 mm	-	0.90	13.83	97.07	nr	**110.90**
1000 mm x 600 mm x 1850 mm	-	0.90	13.83	137.82	nr	**151.65**
closed initial bay; 5 shelves; placing in position						
1000 mm x 300 mm x 1850 mm	-	0.75	11.52	162.43	nr	**173.95**
1000 mm x 600 mm x 1850 mm	-	0.75	11.52	222.36	nr	**233.89**
closed extension bay; 5 shelves; placing in position						
1000 mm x 300 mm x 1850 mm	-	0.90	13.83	137.71	nr	**151.53**
1000 mm x 600 mm x 1850 mm	-	0.90	13.83	183.99	nr	**197.81**
extra for pair of doors; fixing in position						
1000 mm x 1850 mm	-	0.81	12.44	175.52	nr	**187.96**

N FURNITURE/EQUIPMENT

Item	PC £	Labour hours	Labour £	Material £	Unit	Total rate £
Cloakroom racks; Welconstruct Distribution or other equal and approved						
Cloackroom racks; 40 mm x 40 mm square tube framing; polyester coated finish; beech slatted seats and rails to one side only; placing in position						
1675 mm x 325 mm x 1500 mm; 5 nr coat hooks	-	0.33	5.07	416.51	nr	**421.58**
1825 mm x 325 mm x 1500 mm; 15 nr coat hangers	-	0.33	5.07	488.75	nr	**493.82**
Extra for						
shoe baskets	-	-	-	96.06	nr	-
mesh bottom shelf	-	-	-	67.16	nr	-
Cloackroom racks; 40 mm x 40 mm square tube framing; polyester coated finish; beech slatted seats and rails to both sides; placing in position						
1675 mm x 600 mm x 1500 mm; 10 nr coat hooks	-	0.44	6.76	533.90	nr	**540.66**
1825 mm x 600 mm x 1500 mm; 30 nr coat hangers	-	0.44	6.76	609.52	nr	**616.28**
Extra for						
shoe baskets	-	-	-	133.76	nr	-
mesh bottom shelf	-	-	-	72.30	nr	-
6 mm thick rectangular glass mirrors; silver backed; fixed with chromium plated domed headed screws; to background requiring plugging						
Mirror with polished edges						
365 mm x 254 mm	7.53	0.81	13.05	8.86	nr	**21.91**
400 mm x 300 mm	-	0.81	13.05	11.44	nr	**24.49**
560 mm x 380 mm	17.00	0.92	14.82	19.55	nr	**34.37**
640 mm x 460 mm	22.23	1.02	16.43	25.45	nr	**41.88**
Mirror with bevelled edges						
365 mm x 254 mm	13.40	0.81	13.05	15.49	nr	**28.53**
400 mm x 300 mm	15.70	0.81	13.05	18.08	nr	**31.13**
560 mm x 380 mm	26.16	0.92	14.82	29.89	nr	**44.71**
640 mm x 460 mm	32.70	1.02	16.43	37.27	nr	**53.70**
Door mats						
Entrance mats; "Tuftiguard type C" or other equal and approved; laying in position; 12 mm thick						
900 mm x 550 mm	127.05	0.51	6.15	143.41	nr	**149.56**
1200 mm x 750 mm	228.69	0.51	6.15	258.13	nr	**264.29**
2400 mm x 1200 mm	731.81	1.02	12.31	826.03	nr	**838.34**
Matwells						
Polished aluminium matwell; comprising angle rim with brazed angles and lugs brazed on; to suit mat size						
914 mm x 560 mm; constructed with 25 x 25 x 3 mm angle	28.82	1.02	12.31	32.53	nr	**44.84**
1067 mm x 610 mm; constructed with 34 x 26 x 6 mm angle	39.76	1.02	12.31	44.88	nr	**57.19**
1219 mm x 762 mm; constructed with 50 x 50 x 6 mm angle	97.10	1.02	12.31	109.60	nr	**121.91**
Polished brass matwell; comprising angle rim with brazed angles and lugs brazed on; to suit mat size						
914 mm x 560 mm; constructed with 25 x 25 x 5 mm angle	111.78	1.02	12.31	126.17	nr	**138.48**
1067 mm x 610 mm; constructed with 38 x 38 x 6 mm angle	163.01	1.02	12.31	184.00	nr	**196.31**

N FURNITURE/EQUIPMENT

Item	PC £	Labour hours	Labour £	Material £	Unit	Total rate £
N10/11 GENERAL FIXTURES/KITCHEN FITTINGS – cont'd						
Internal blinds; Luxaflex Ltd or other equal and approved						
Roller blinds; Luxaflex EOS type 10 roller; "Compact Fabric"; plain type material; 1219 mm drop; fixing with screws						
1016 mm wide	42.00	1.02	12.31	47.41	nr	**59.72**
2031 mm wide	62.00	1.60	19.31	69.98	nr	**89.29**
2843 mm wide	80.00	2.16	26.06	90.30	nr	**116.36**
Roller blinds; Luxaflex EOS type 10 roller; "Compact Fabric"; fire resisting material; 1219 mm drop; fixing with screws						
1016 mm wide	55.00	1.02	12.31	62.08	nr	**74.39**
2031 mm wide	82.00	1.60	19.31	92.56	nr	**111.86**
2843 mm wide	108.00	2.16	26.06	121.91	nr	**147.97**
Roller blinds; Luxaflex EOS type 10 roller; "Light-resistant"; blackout material; 1219 mm drop; fixing with screws						
1016 mm wide	42.00	1.02	12.31	47.41	nr	**59.72**
2031 mm wide	62.00	1.60	19.31	69.98	nr	**89.29**
2843 mm wide	80.00	2.16	26.06	90.30	nr	**116.36**
Roller blinds; Luxaflex "Lite-master Crank Op"; 100% blackout; 1219 mm drop; fixing with screws						
1016 mm wide	219.00	2.15	25.94	247.20	nr	**273.14**
2031 mm wide	294.00	3.02	36.44	331.85	nr	**368.29**
2843 mm wide	380.00	3.89	46.94	428.93	nr	**475.86**
Vertical louvre blinds; 89 mm wide louvres; Luxaflex EOS type; "Florida Fabric"; 1219 mm drop; fixing with screws						
1016 mm wide	55.00	0.91	10.98	62.08	nr	**73.06**
2031 mm wide	83.00	1.42	17.13	93.69	nr	**110.82**
3046 mm wide	112.00	1.94	23.41	126.42	nr	**149.83**
Vertical louvre blinds; 127 mm wide louvres; Luxaflex EOS type; "Florida Fabric"; 1219 mm drop; fixing with screws						
1016 mm wide	48.00	0.97	11.70	54.18	nr	**65.88**
2031 mm wide	73.00	1.50	18.10	82.40	nr	**100.50**
3046 mm wide	98.00	2.00	24.13	110.62	nr	**134.75**
N13 SANITARY APPLIANCES/FITTINGS						
Sinks; Armitage Shanks or equal and approved						
Sinks; white glazed fireclay; BS 6465; pointing all round with Dow Corning Hansil silicone sealant						
Belfast sink; 46 cm x 38 cm x 21 cm ref S580001; pair of Nuastyle 21 basin taps with dual indices, chrome handle ref B8262AA; wall mounts ref S8331AA; 38 mm slotted waste, chain and plug, screw stay ref S8766AA; pair of 40.5 cm aluminium alloy build-in brackets with 35.5 cm studs ref S921967; screwing	182.25	3.05	66.41	261.50	nr	**327.91**
Belfast sink; 61 cm x 38 cm x 21 cm ref S580501; pair of Nuastyle 21 basin taps with dual indices, chrome handle ref B8262AA; wall mounts ref S8331AA; 38 mm slotted waste, chain and plug, screw stay ref S8766AA; pair of 40.5 cm aluminium alloy build-in brackets with 35.5 cm studs ref S921967; screwing	215.67	3.05	66.41	299.41	nr	**365.82**

N FURNITURE/EQUIPMENT

Item	PC £	Labour hours	Labour £	Material £	Unit	Total rate £
Belfast sink; 76 cm x 38 cm x 21 cm ref S581101; pair of Nuastyle 21 basin taps with dual indices, chrome handle ref B8262AA; wall mounts ref S8331AA; 38 mm slotted waste, chain and plug, screw stay ref S8766AA; pair of 40.5 cm aluminium alloy build-in brackets with 35.5 cm studs ref S921967; screwing	300.26	3.05	66.41	395.07	nr	**461.48**
Lavatory basins; Armitage Shanks or equal and approved						
Basins; white vitreous china; BS 6465 Part 3; pointing all round with Dow Corning Hansil silcone sealant						
Portman 21 40 cm basin ref S231701; with overflow, chain hole and two tapholes ; pair of Nuastyle 21 basin taps with dual indices ref B8262AA; slotted basin waste with plastic plug, chain waste and plug ref S8800AA; 32 x 75 mm seal plastic standard bottle trap ref S891067; pair of Portman concealed brackets with waste support ref S915067; Isovalve 15 mm plastic servicing valve with outlet for copper ref S900067; screwing	88.13	2.35	51.17	135.66	nr	**186.83**
Portman 21 50 cm basin ref S230901; with overflow, chain hole and two tapholes ; pair of Nuastyle 21 basin taps with dual indices ref B8262AA; slotted basin waste with plastic plug, chain waste and plug ref S8800AA; 32 x 75 mm seal plastic standard bottle trap ref S891067; pair of Portman concealed brackets with waste support ref S915067; Isovalve 15 mm plastic servicing valve with outlet for copper ref S900067; screwing	107.44	2.35	51.17	157.57	nr	**208.74**
Portman 21 60 cm basin ref S225701; with overflow, chain hole and two tapholes ; pair of Nuastyle 21 basin taps with dual indices ref B8262AA; slotted basin waste with plastic plug, chain waste and plug ref S8800AA; 32 x 75 mm seal plastic standard bottle trap ref S891067; pair of Portman concealed brackets with waste support ref S915067; Isovalve 15 mm plastic servicing valve with outlet for copper ref S900067; screwing	141.12	2.35	51.17	195.71	nr	**246.88**
Tiffany 51 cm pedestal basin ref S208001; with two tapholes ; Millenia STD dual control one taphole standard basin mixer with pop-up waste ref S7300AA; pair of Millenia STD handles ref B8000AA; Full pedestal (IS group S2920) ref S292001; Isovalve 15 mm plastic servicing valve with outlet for copper ref S900067; screwing	153.63	2.55	55.52	190.73	nr	**246.26**
Tiffany 56 cm pedestal basin ref S208301; with two tapholes ; Millenia STD dual control one taphole standard basin mixer with pop-up waste ref S7300AA; pair of Millenia STD handles ref B8000AA; Full pedestal (IS group S2920) ref S292001; Isovalve 15 mm plastic servicing valve with outlet for copper ref S900067; screwing	150.45	2.55	55.52	187.14	nr	**242.67**

N FURNITURE/EQUIPMENT

Item	PC £	Labour hours	Labour £	Material £	Unit	Total rate £
N13 SANITARY APPLIANCES/FITTINGS – cont'd						
Lavatory basins; Armitage Shanks – cont'd						
Tiffany 61 cm pedestal basin ref S208601; with two tapholes ; Millenia STD dual control one taphole standard basin mixer with pop-up waste ref S7300AA; pair of Millenia STD handles ref B8000AA; Full pedestal (IS group S2920) ref S292001; Isovalve 15 mm plastic servicing valve with outlet for copper ref S900067; screwing	156.84	2.55	55.52	194.36	nr	**249.88**
Montana 51 cm pedestal basin ref S210101; with one taphole ; Millenia STD dual control one taphole standard basin mixer with pop-up waste ref S7300AA; pair of Millenia STD handles ref B8000AA; Full pedestal (IS group S2920) ref S292001; Isovalve 15 mm plastic servicing valve with outlet for copper ref S900067; screwing	145.75	2.55	55.52	181.79	nr	**237.31**
Montana 58 cm pedestal basin ref S210401; with one taphole ; Millenia STD dual control one taphole standard basin mixer with pop-up waste ref S7300AA; pair of Millenia STD handles ref B8000AA; Full pedestal (IS group S2920) ref S292001; Isovalve 15 mm plastic servicing valve with outlet for copper ref S900067; screwing	148.82	2.55	55.52	185.33	nr	**240.86**
Drinking fountains; Armitage Shanks or similar						
White vitreous china fountains; pointing all round with Dow Corning Hansil silicone selant						
Aqualon wall mounted drinking fountain ref S540101; Aqualon self closing valve with fittings and plastic waste ref S5402AA; 32 x 75 mm seal plastic standard bottle trap ref S891067; screwing	251.02	2.55	55.52	291.66	nr	**347.19**
Polished stainless steel fountains; pointing all round with Dow Corning Hansil silicone selant						
Purita wall mounted drinking fountain ref S5435MY with self closing valve and fittings; 32mm unslotted basin strainer waste ref S8720AA; screwing	241.46	2.55	55.52	272.97	nr	**328.49**
Purita pedestal mounted drinking fountain 90 cm high ref S5440MY with self closing valve and fittings; 32mm unslotted basin strainer waste ref S8720AA; screwing	502.90	3.05	66.41	583.53	nr	**649.94**
Baths; Armitage Shanks or equal and approved						
Sandringham acrylic rectangular bath with chrome plated grips and two tapholes ref S159301; Sandringha,m STD pair of standard bath taps with chrome handles ref S7032AA; bath chain waste with plastic plug and overflow ref S8830AA; cast brass P trap with plain outlet and overflow connection; pointing with Dow Corning Hansil silicone selant						
170 cm long x 70 cm wide; white or coloured	122.75	3.85	83.83	138.55	nr	**222.38**
Nisa lowline heavy gauge steel rectangular bath with chrome plated grips and two tapholes ref S176501; Sandringha,m STD pair of standard bath taps with chrome handles ref S7032AA; bath chain waste with plastic plug and overflow ref S8830AA; cast brass P trap with plain outlet and overflow connection; pointing with Dow Corning Hansil silicone selant						
170 cm long x 70 cm wide; white or coloured	311.05	3.85	83.83	351.09	nr	**434.92**

N FURNITURE/EQUIPMENT

Item	PC £	Labour hours	Labour £	Material £	Unit	Total rate £
Water closets; Armitage Shanks or equal and approved						
White vitreous china pans and cisterns; pointing all round base with Dow Corning Hansil silicone sealant						
Wentworth close coupled washdown closet pan with horizontal outlet ref S316101; Orion 3 plastic toilet seat and cover ref S404501; Panketa pan connector 14º finned ref S430501; Universal close coupled bottom inlet cistern with syphon ref S392001	141.91	3.35	72.94	167.04	nr	**239.99**
Tiffany back to wall washdown closet pan with horizontal outlet ref S341001; Saturn plastic toilet seat and cover ref S404001; Panketa pan connector 14º finned ref S430501; Conceala 2 6 litre low level side inlet cistern with syphon and lever ref S361767	186.33	3.35	72.94	217.18	nr	**290.12**
Extra over for; Panketa pan connector 90º finned ref S430001	-	-	-	1.61	nr	**-**
Tiffany close coupled washdown closet pan with horizontal outlet (IS group S3080) ref S308001; Saturn plastic toilet seat and cover ref S404001; Panketa pan connector 14º finned ref S430501; Tiffany 7½ litre close coupled cistern with dual flush valve ref S365001	191.55	3.35	72.94	223.08	nr	**296.02**
Extra over for; Panketa pan connector 90º finned ref S430001	-	-	-	1.61	nr	**-**
Cameo close coupled washdown closet pan with horizontal outlet (IS group S3080) ref S308001; Accolade/Cameo plastic toilet seat and cover ref S402501; Panketa pan connector 14º finned ref S430501; Cameo 6 litre close coupled cistern with dual flush valve ref S361301	238.88	3.35	72.94	276.49	nr	**349.43**
Extra over for; Panketa pan connector 90º finned ref S430001	-	-	-	1.61	nr	**-**
Wall urinals; Armitage Shanks or equal and approved						
White vitreous china bowls and cisterns; pointing all round with Dow Corning Hansil silicone sealant						
Single Sanura 40 cm urinal bowl ref S610501, Sanura top inlet spreader ref S6285AA; pair of wall hangers for urinal bowl ref S9725AA; 38 mm plastic domed waste ref S885067; 38 x 75 mm seal plastic standard bottle trap ref S891567; Conceala 4½ litres capacity auto cistern and cover ref S621567; Sanura concealed flushpipe for single urinal bowl ref S6226NU; screwing	168.21	4.05	88.18	211.00	nr	**299.18**
Single Sanura 40 cm urinal bowl ref S610501; Sanura top inlet spreader ref S6285AA; pair of wall hangers for urinal bowl ref S9725AA; 38 mm plastic domed waste ref S885067; 38 x 75 mm seal plastic standard bottle trap ref S891567; Mura 4½ litres capacity auto cistern and cover ref S620001; Sanura/Mura exposed flushpipe for single urinal bowl ref S6220MY; screwing	189.90	4.05	88.18	235.48	nr	**323.66**

N FURNITURE/EQUIPMENT

Item	PC £	Labour hours	Labour £	Material £	Unit	Total rate £
N13 SANITARY APPLIANCES/FITTINGS – cont'd						
Wall urinals; Armitage Shanks – cont'd						
Single Sanura 50 cm urinal bowl ref S610001; Sanura top inlet spreader ref S6285AA; pair of wall hangers for urinal bowl ref S9725AA; 38 mm plastic domed waste ref S885067; 38 x 75 mm seal plastic standard bottle trap ref S891567; Conceala 4½ litres capacity auto cistern and cover ref S621567; Sanura concealed flushpipe for single urinal bowl ref S6226NU; screwing	228.82	4.05	88.18	279.50	nr	**367.68**
Single Sanura 50 cm urinal bowl ref S610001; Sanura top inlet spreader ref S6285AA; pair of wall hangers for urinal bowl ref S9725AA; 38 mm plastic domed waste ref S885067; 38 x 75 mm seal plastic standard bottle trap ref S891567; Mura 4½ litres capacity auto cistern and cover ref S620001; Sanura/Mura exposed flushpipe for single urinal bowl ref S6220MY; screwing	250.51	4.05	88.18	303.98	nr	**392.17**
Range of 2 nr Sanura 40 cm urinal bowls ref S610501; Sanura top inlet spreader ref S6285AA; pairs of wall hangers for urinal bowls ref S9725AA; 38 mm plastic domed wastes ref S885067; 38 x 75 mm seal plastic standard bottle traps ref S891567; Conceala 9 litres capacity auto cistern and cover ref S621667; Sanura concealed flushpipe for range of 2 nr urinal bowls ref S6227NU; screwing	282.70	7.65	166.57	361.35	nr	**527.92**
Range of 2 nr Sanura 50 cm urinal bowls ref S610001; Sanura top inlet spreader ref S6285AA; pairs of wall hangers for urinal bowls ref S9725AA; 38 mm plastic domed wastes ref S885067; 38 x 75 mm seal plastic standard bottle traps ref S891567; Conceala 9 litres capacity auto cistern and cover ref S621667; Sanura concealed flushpipe for range of 2 nr urinal bowls ref S6227NU; screwing	403.92	7.65	166.57	498.36	nr	**664.93**
Range of 3 nr Sanura 40 cm urinal bowls ref S610501; Sanura top inlet spreader ref S6285AA; pairs of wall hangers for urinal bowls ref S9725AA; 38 mm plastic domed wastes ref S885067; 38 x 75 mm seal plastic standard bottle traps ref S891567; Conceala 9 litres capacity auto cistern and cover ref S621667; Sanura concealed flushpipe for range of 3 nr urinal bowls ref S6228NU; screwing	393.26	11.20	243.86	507.27	nr	**751.14**
Range of 3 nr Sanura 50 cm urinal bowls ref S610001; Sanura top inlet spreader ref S6285AA; pairs of wall hangers for urinal bowls ref S9725AA; 38 mm plastic domed wastes ref S885067; 38 x 75 mm seal plastic standard bottle traps ref S891567; Conceala 9 litres capacity auto cistern and cover ref S621667; Sanura concealed flushpipe for range of 3 nr urinal bowls ref S6228NU; screwing	575.09	11.20	243.86	712.51	nr	**956.38**
Range of 4 nr Sanura 40 cm urinal bowls ref S610501; Sanura top inlet spreader ref S6285AA; pairs of wall hangers for urinal bowls ref S9725AA; 38 mm plastic domed wastes ref S885067; 38 x 75 mm seal plastic standard bottle traps ref S891567; Conceala 9 litres capacity auto cistern and cover ref S621767; Sanura concealed flushpipe for range of 4 nr urinal bowls ref S6229NU; screwing	507.90	14.75	321.16	657.80	nr	**978.96**

N FURNITURE/EQUIPMENT

Item	PC £	Labour hours	Labour £	Material £	Unit	Total rate £
Range of 4 nr Sanura 50 cm urinal bowls ref S610001; Sanura top inlet spreader ref S6285AA; pairs of wall hangers for urinal bowls ref S9725AA; 38 mm plastic domed wastes ref S885067; 38 x 75 mm seal plastic standard bottle traps ref S891567; Conceala 9 litres capacity auto cistern and cover ref S621767; Sanura concealed flushpipe for range of 4 nr urinal bowls ref S6229NU; screwing	750.34	14.75	321.16	931.46	nr	**1252.62**
Range of 5 nr Sanura 40 cm urinal bowls ref S610501; Sanura top inlet spreader ref S6285AA; pairs of wall hangers for urinal bowls ref S9725AA; 38 mm plastic domed wastes ref S885067; 38 x 75 mm seal plastic standard bottle traps ref S891567; Conceala 9 litres capacity auto cistern and cover ref S621767; Sanura concealed flushpipe for range of 5 nr urinal bowls ref S6230NU; screwing	618.49	18.30	398.46	803.75	nr	**1202.21**
Range of 5 nr Sanura 50 cm urinal bowls ref S610001; Sanura top inlet spreader ref S6285AA; pairs of wall hangers for urinal bowls ref S9725AA; 38 mm plastic domed wastes ref S885067; 38 x 75 mm seal plastic standard bottle traps ref S891567; Conceala 9 litres capacity auto cistern and cover ref S621767; Sanura concealed flushpipe for range of 5 nr urinal bowls ref S6230NU; screwing	921.54	18.30	398.46	1146.27	nr	**1544.73**
White vitreous china division panels; pointing all round with Dow Corning Hansill silicone sealant						
Urinal division with screw and hanger ref S612001; screwing	51.96	0.75	16.33	59.65	nr	**75.98**
Bidets; Armitage Shanks or equal and approved						
Tiffany back to wall bidet with one taphole ref S491001; vitreous china; chromium plated pop-up waste and mixer tap with hand wheels refs S7500AA and S8000AA						
58 cm x 39 cm; white or coloured	278.38	3.85	83.83	314.22	nr	**398.05**
Shower tray and fittings						
Simplicity shower tray; acrylic; with outlet and grated waste; chain and plug; bedding and pointing in waterproof cement mortar						
760 mm x 760 mm ; white or coloured	136.82	3.30	71.85	154.43	nr	**226.29**
Shower fitting; riser pipe with mixing valve and shower rose; chromium plated; plugging and screwing mixing valve and pipe bracket						
15 mm diameter riser pipe; 127 mm diameter shower rose	267.70	5.50	119.75	302.16	nr	**421.92**
Miscellaneous fittings; Magrini Ltd or equal and approved						
Vertical nappy changing unit						
ref KBCS; screwing	-	0.65	10.47	295.71	nr	**306.18**
Horizontal nappy changing unit						
ref KBHS; screwing	-	0.65	10.47	295.71	nr	**306.18**
Stay Safe baby seat						
ref KBPS; screwing	-	0.60	9.66	88.02	nr	**97.68**

N FURNITURE/EQUIPMENT

Item	PC £	Labour hours	Labour £	Material £	Unit	Total rate £
N13 SANITARY APPLIANCES/FITTINGS – cont'd						
Miscellaneous fittings; Pressalit Ltd or equal and approved						
Grab rails						
300 mm long ref RT100000; screwing	-	0.55	8.86	59.15	nr	**68.00**
450 mm long ref RT101000; screwing	-	0.55	8.86	63.28	nr	**72.14**
600 mm long ref RT102000; screwing	-	0.55	8.86	72.51	nr	**81.37**
800 mm long ref RT103000; screwing	-	0.55	8.86	81.42	nr	**90.28**
1000 mm long ref RT104000; screwing	-	0.55	8.86	93.51	nr	**102.37**
Angled grab rails						
900 mm long, angled 135° ref RT110000; screwing	-	0.55	8.86	161.92	nr	**170.78**
1300 mm long, angled 90° ref RT119000; screwing	-	0.80	12.89	183.86	nr	**196.74**
Hinged grab rails						
600 mm long ref R3016000 ; screwing	-	0.40	6.44	190.25	nr	**196.70**
600 mm long with spring counter balance ref RF016000 ; screwing	-	0.40	6.44	265.03	nr	**271.47**
800 mm long ref R3010000 ; screwing	3.36	0.40	6.44	230.67	nr	**237.11**
800 mm long with spring counter balance ref RF010000 ; screwing	-	0.40	6.44	284.12	nr	**290.57**
N15 SIGNS/NOTICES						
Plain script; in gloss oil paint; on painted or varnished surfaces						
Capital letters; lower case letters or numerals						
per coat; per 25 mm high	-	0.10	1.61	-	nr	**1.61**
Stops						
per coat	-	0.02	0.32	-	nr	**0.32**

P BUILDING FABRIC SUNDRIES

Item	PC £	Labour hours	Labour £	Material £	Unit	Total rate £
P10 SUNDRY INSULATION/PROOFING WORK/ FIRE STOPS						
"Sisalkraft" building papers/vapour barriers or other equal and approved						
Building paper; 150 mm laps; fixed to softwood						
"Moistop" grade 728 (class A1F)	-	0.09	1.38	1.16	m²	**2.54**
Vapour barrier/reflective insulation 150 mm laps; fixed to softwood						
"Insulex" grade 714; single sided	-	0.09	1.38	1.37	m²	**2.75**
Mat or quilt insulation						
Glass fibre roll; "Crown Loft Roll" or other equal and approved; laid loose between members at 600 mm centres						
100 mm thick	3.34	0.11	1.69	3.96	m²	**5.65**
150 mm thick	4.93	0.12	1.84	5.84	m²	**7.69**
200 mm thick	6.65	0.13	2.00	7.88	m²	**9.88**
Glass fibre quilt; Isowool "Modular roll" or other equal and approved; laid loose between members at 600 mm centres						
60 mm thick	3.33	0.10	1.54	3.76	m²	**5.30**
80 mm thick	4.35	0.11	1.69	4.91	m²	**6.60**
100 mm thick	5.19	0.12	1.84	5.85	m²	**7.70**
150 mm thick	7.92	0.13	2.00	8.94	m²	**10.94**
Mineral fibre quilt; Isowool "APR 1200" or other equal and approved; pinned vertically to softwood						
25 mm thick	2.29	0.09	1.38	2.58	m²	**3.97**
50 mm thick	3.62	0.10	1.54	4.08	m²	**5.62**
"Crown Dritherm Cavity Slab 37" glass fibre batt or other equal and approved; as full or partial cavity fill; including retaining discs						
50 mm thick	3.62	0.13	2.00	4.60	m²	**6.59**
75 mm thick	4.32	0.14	2.15	5.42	m²	**7.58**
100 mm thick	5.62	0.15	2.30	6.97	m²	**9.27**
"Crown Dritherm Cavity Slab 34" glass fibre batt or other equal and approved; as full or partial cavity fill; including retaining discs						
65 mm thick	6.56	0.13	2.00	8.08	m²	**10.08**
75 mm thick	7.59	0.14	2.15	9.30	m²	**11.45**
85 mm thick	8.53	0.14	2.15	10.41	m²	**12.57**
100 mm thick	9.88	0.15	2.30	12.01	m²	**14.32**
"Crown Dritherm Cavity Slab 32" glass fibre batt or other equal and approved; as full or partial cavity fill; including retaining discs						
65 mm thick	8.68	0.13	2.00	10.59	m²	**12.59**
75 mm thick	10.02	0.14	2.15	12.18	m²	**14.33**
85 mm thick	11.29	0.14	2.15	13.69	m²	**15.84**
100 mm thick	13.07	0.15	2.30	15.80	m²	**18.10**
"Crown Frametherm Roll 40" glass fibre semi-rigid or rigid batt or other equal and approved; pinned vertically in timber frame construction						
90 mm thick	4.37	0.15	2.30	5.18	m²	**7.48**
140 mm thick	6.34	0.17	2.61	7.51	m²	**10.13**
"Crown Rafter Roll 32" glass fibre flanged building roll; pinned vertically or to slope between timber framing						
50 mm thick	6.06	0.19	4.17	7.18	m²	**11.36**
75 mm thick	8.68	0.20	4.39	10.29	m²	**14.68**
100 mm thick	11.18	0.21	4.61	13.25	m²	**17.86**

P BUILDING FABRIC SUNDRIES

Item	PC £	Labour hours	Labour £	Material £	Unit	Total rate £
P10 SUNDRY INSULATION/PROOFING WORK/ FIRE STOPS – cont'd						
Board or slab insulation						
"Kingspan Thermawall TW50" zero ODP rigid urethene insulation board or other equal and approved; as full or partial cavity fill; including retaining discs and cutting around ties						
50 mm thick	10.06	0.19	4.17	12.22	m²	**16.40**
75 mm thick	14.99	0.20	4.39	18.08	m²	**22.47**
100 mm thick	19.49	0.21	4.61	23.40	m²	**28.02**
Expanded polystyrene board standard grade SD/N or other equal and approved; fixed with adhesive						
20 mm thick	-	0.15	3.29	5.03	m²	**8.32**
25 mm thick	-	0.15	3.29	6.01	m²	**9.30**
30 mm thick	-	0.16	3.51	7.00	m²	**10.52**
40 mm thick	-	0.45	9.88	7.28	m²	**17.17**
50 mm thick	-	0.18	3.95	10.97	m²	**14.92**
60 mm thick	-	0.19	4.17	12.94	m²	**17.12**
75 mm thick	-	0.20	4.39	15.91	m²	**20.31**
100 mm thick	-	0.21	4.61	20.86	m²	**25.47**
"Styrofoam Floormate 500" extruded polystyrene foam or other equal and approved						
50 mm thick	-	0.51	11.20	8.30	m²	**19.50**
80 mm thick	-	0.51	11.20	14.60	m²	**25.80**
120 mm thick	-	0.51	11.20	21.90	m²	**33.10**
Fire stops						
Cape "Firecheck" channel or other equal and approved; intumescent coatings on cut mitres; fixing with brass cups and screws						
19 mm x 44 mm or 19 mm x 50 mm	10.80	0.61	13.40	13.35	m	**26.74**
"Sealmaster" intumescent fire and smoke seals or other equal and approved; pinned into groove in timber						
type N30; for single leaf half hour door	7.82	0.31	6.81	9.27	m	**16.08**
type N60; for single leaf one hour door	11.91	0.33	7.25	14.11	m	**21.36**
type IMN or IMP; for meeting or pivot stiles of pair of one hour doors; per stile	11.91	0.33	7.25	14.11	m	**21.36**
intumescent plugs in timber; including boring	-	0.10	2.20	0.03	nr	**2.22**
Rockwool fire stops or other equal and approved; between top of brick/block wall and concrete soffit						
30 mm deep x 100 mm wide	-	0.08	1.76	4.22	m	**5.98**
30 mm deep x 150 mm wide	-	0.10	2.20	6.39	m	**8.58**
30 mm deep x 200 mm wide	-	0.12	2.64	8.54	m	**11.18**
60 mm deep x 100 mm wide	-	0.09	1.98	6.25	m	**8.22**
60 mm deep x 150 mm wide	-	0.10	2.20	9.28	m	**11.47**
60 mm deep x 200 mm wide	-	0.12	2.64	12.50	m	**15.14**
90 mm deep x 100 mm wide	-	0.11	2.42	8.21	m	**10.63**
90 mm deep x 150 mm wide	-	0.13	2.86	12.29	m	**15.15**
90 mm deep x 200 mm wide	-	0.15	3.29	16.40	m	**19.70**
Fire protection compound						
Quelfire "QF4", fire protection compound or other equal and approved; filling around pipes, ducts and the like; including all necessary formwork						
300 mm x 300 mm x 250 mm; pipes - 2	-	1.25	23.28	47.14	nr	**70.42**

P BUILDING FABRIC SUNDRIES

Item	PC £	Labour hours	Labour £	Material £	Unit	Total rate £
Fire barriers						
Corofil C144 fire barrier to edge of slab; fixed with non-flammable contact adhesive						
to suit void 30 mm wide x 100 mm deep; one hour	-	-	-	-	m	**15.70**
Rockwool fire barrier or other and approved between top of suspended ceiling and concrete soffit						
one 50 mm layer x 900 mm wide; half hour	-	0.61	13.40	5.98	m²	**19.38**
two 50 mm layers x 900 mm wide; one hour	-	0.92	20.21	11.67	m²	**31.88**
three 50 mm layers x 900 mm wide; two hour	-	1.22	26.80	17.36	m²	**44.16**
Lamatherm fire barrier or other equal and approved; to void below raised access floors						
75 mm thick x 300 mm wide; half hour	-	0.17	3.73	18.01	m	**21.75**
75 mm thick x 600 mm wide; half hour	-	0.17	3.73	40.80	m	**44.54**
90 mm thick x 300 mm wide; half hour	-	0.17	3.73	21.59	m	**25.32**
90 mm thick x 600 mm wide; half hour	-	0.17	3.73	50.45	m	**54.19**
Dow Chemicals "Styrofoam SP" or other equal and approved; cold bridging insulation fixed with adhesive to brick, block or concrete base						
Insulation to walls						
50 mm thick	-	0.37	8.13	6.64	m²	**14.77**
75 mm thick	-	0.39	8.57	11.00	m²	**19.56**
Insulation to isolated columns						
50 mm thick	-	0.44	9.66	6.64	m²	**16.31**
75 mm thick	-	0.48	10.54	11.00	m²	**21.54**
Insulation to ceilings						
50 mm thick	-	0.40	8.79	6.64	m²	**15.43**
75 mm thick	-	0.43	9.45	11.00	m²	**20.44**
Insulation to isolated beams						
50 mm thick	-	0.48	10.54	6.64	m²	**17.18**
75 mm thick	-	0.51	11.20	11.00	m²	**22.20**
P11 FOAMED/FIBRE/BEAD CAVITY WALL INSULATION						
Injected insulation						
Cavity wall insulation; injecting cavity with mineral wool						
75 mm cavity	-	-	-	-	m²	**5.13**
100 mm cavity	-	-	-	-	m²	**5.13**
P20 UNFRAMED ISOLATED TRIMS/SKIRTINGS/ SUNDRY ITEMS						
Medium density fibreboard (Sapele veneered one side); 18 mm thick						
Window boards and the like; rebated; hardwood lipped on one edge						
18 mm x 200 mm	-	0.28	6.15	19.43	m	**25.59**
18 mm x 250 mm	-	0.31	6.81	20.45	m	**27.26**
18 mm x 300 mm	-	0.33	7.25	20.96	m	**28.21**
18 mm x 350 mm	-	0.37	8.13	22.49	m	**30.62**
returned and fitted ends	-	0.22	4.83	3.61	nr	**8.44**

P BUILDING FABRIC SUNDRIES

Item	PC £	Labour hours	Labour £	Material £	Unit	Total rate £
P20 UNFRAMED ISOLATED TRIMS/SKIRTINGS/ SUNDRY ITEMS – cont'd						
Medium density fibreboard (American White Ash veneered one side); 18 mm thick						
Window boards and the like; rebated; hardwood lipped on one edge						
18 mm x 200 mm	-	0.28	6.15	20.20	m	**26.35**
18 mm x 250 mm	-	0.31	6.81	21.47	m	**28.28**
18 mm x 300 mm	-	0.33	7.25	22.10	m	**29.35**
18 mm x 350 mm	-	0.37	8.13	24.00	m	**32.13**
returned and fitted ends	-	0.22	4.83	3.61	nr	**8.44**
Wrought softwood						
Skirtings, picture rails, dado rails and the like; splayed or moulded						
19 mm x 44 mm; splayed	-	0.10	2.20	3.84	m	**6.03**
19 mm x 44 mm; moulded	-	0.10	2.20	4.09	m	**6.29**
19 mm x 69 mm; splayed	-	0.10	2.20	4.15	m	**6.35**
19 mm x 69 mm; moulded	-	0.10	2.20	4.15	m	**6.35**
19 mm x 94 mm; splayed	-	0.10	2.20	4.65	m	**6.85**
19 mm x 94 mm; moulded	-	0.10	2.20	4.65	m	**6.85**
19 mm x 144 mm; moulded	-	0.12	2.64	5.47	m	**8.11**
19 mm x 169 mm; moulded	-	0.12	2.64	5.78	m	**8.42**
25 mm x 50 mm; moulded	-	0.10	2.20	4.16	m	**6.36**
25 mm x 69 mm; splayed	-	0.10	2.20	4.44	m	**6.63**
25 mm x 94 mm; splayed	-	0.10	2.20	4.97	m	**7.16**
25 mm x 144 mm; splayed	-	0.12	2.64	6.02	m	**8.65**
25 mm x 144 mm; moulded	-	0.12	2.64	6.02	m	**8.65**
25 mm x 169 mm; moulded	-	0.12	2.64	6.70	m	**9.33**
25 mm x 219 mm; moulded	-	0.14	3.08	8.72	m	**11.79**
returned ends	-	0.16	3.51	-	nr	**3.51**
mitres	-	0.10	2.20	-	nr	**2.20**
Architraves, cover fillets and the like; half round; splayed or moulded						
13 mm x 25 mm; half round	-	0.12	2.64	3.60	m	**6.24**
13 mm x 50 mm; moulded	-	0.12	2.64	3.82	m	**6.46**
16 mm x 32 mm; half round	-	0.12	2.64	4.01	m	**6.65**
16 mm x 38 mm; moulded	-	0.12	2.64	4.01	m	**6.65**
16 mm x 50 mm; moulded	-	0.12	2.64	4.01	m	**6.65**
19 mm x 50 mm; splayed	-	0.12	2.64	4.01	m	**6.65**
19 mm x 63 mm; splayed	-	0.11	2.42	4.15	m	**6.57**
19 mm x 69 mm; splayed	-	0.12	2.64	4.41	m	**7.04**
25 mm x 44 mm; splayed	-	0.12	2.64	3.96	m	**6.59**
25 mm x 50 mm; moulded	-	0.12	2.64	4.16	m	**6.80**
25 mm x 63 mm; splayed	-	0.12	2.64	4.35	m	**6.99**
25 mm x 69 mm; splayed	-	0.12	2.64	4.94	m	**7.58**
32 mm x 88 mm; moulded	-	0.12	2.64	4.97	m	**7.60**
38 mm x 38 mm; moulded	-	0.12	2.64	4.39	m	**7.03**
50 mm x 50 mm; moulded	-	0.12	2.64	5.12	m	**7.75**
returned ends	-	0.16	3.51	-	nr	**3.51**
mitres	-	0.10	2.20	-	nr	**2.20**
Stops; screwed on						
16 mm x 38 mm	-	0.10	2.20	1.81	m	**4.01**
16 mm x 50 mm	-	0.10	2.20	1.96	m	**4.16**
19 mm x 38 mm	-	0.10	2.20	1.81	m	**4.01**
25 mm x 38 mm	-	0.10	2.20	1.97	m	**4.17**
25 mm x 50 mm	-	0.10	2.20	2.03	m	**4.22**

P BUILDING FABRIC SUNDRIES

Item	PC £	Labour hours	Labour £	Material £	Unit	Total rate £
Glazing beads and the like						
13 mm x 16 mm	-	0.05	1.10	2.18	m	**3.28**
13 mm x 19 mm	-	0.05	1.10	2.18	m	**3.28**
13 mm x 25 mm	-	0.05	1.10	2.24	m	**3.33**
13 mm x 25 mm; screwed	-	0.05	1.10	2.36	m	**3.46**
13 mm x 25 mm; fixing with brass cups and screws	-	0.05	1.10	2.24	m	**3.33**
16 mm x 25 mm; screwed	-	0.05	1.10	3.76	m	**4.86**
16 mm quadrant	-	0.05	1.10	3.33	m	**4.42**
19 mm quadrant or scotia	-	0.05	1.10	3.33	m	**4.42**
19 mm x 36 mm; screwed	-	0.05	1.10	3.81	m	**4.90**
25 mm x 38 mm; screwed	-	0.05	1.10	3.98	m	**5.08**
25 mm quadrant or scotia	-	0.05	1.10	3.52	m	**4.61**
38 mm scotia	-	0.05	1.10	4.24	m	**5.34**
50 mm scotia	-	0.05	1.10	4.96	m	**6.06**
Isolated shelves, worktops, seats and the like						
19 mm x 150 mm	-	0.17	3.73	4.62	m	**8.35**
19 mm x 200 mm	-	0.22	4.83	6.38	m	**11.21**
25 mm x 150 mm	-	0.17	3.73	5.27	m	**9.01**
25 mm x 200 mm	-	0.22	4.83	7.50	m	**12.33**
32 mm x 150 mm	-	0.17	3.73	6.16	m	**9.89**
32 mm x 200 mm	-	0.22	4.83	8.41	m	**13.24**
Isolated shelves, worktops, seats and the like; cross-tongued joints						
19 mm x 300 mm	-	0.29	6.37	19.01	m	**25.38**
19 mm x 450 mm	-	0.34	7.47	28.65	m	**36.12**
19 mm x 600 mm	-	0.41	9.01	37.07	m	**46.08**
25 mm x 300 mm	-	0.29	6.37	20.39	m	**26.76**
25 mm x 450 mm	-	0.34	7.47	30.90	m	**38.37**
25 mm x 600 mm	-	0.41	9.01	40.18	m	**49.19**
32 mm x 300 mm	-	0.29	6.37	21.59	m	**27.96**
32 mm x 450 mm	-	0.34	7.47	32.83	m	**40.30**
32 mm x 600 mm	-	0.41	9.01	42.84	m	**51.84**
Isolated shelves, worktops, seats and the like; slatted with 50 mm wide slats at 75 mm centres						
19 mm thick	-	0.66	14.50	45.32	m	**59.81**
25 mm thick	-	0.66	14.50	46.33	m	**60.82**
32 mm thick	-	0.66	14.50	47.21	m	**61.71**
Window boards, nosings, bed moulds and the like; rebated and rounded						
19 mm x 75 mm	-	0.19	4.17	5.67	m	**9.04**
19 mm x 150 mm	-	0.20	4.39	7.02	m	**11.41**
19 mm x 219 mm; in one width	-	0.27	5.93	8.72	m	**14.65**
19 mm x 300 mm; cross-tongued joints	-	0.31	6.81	20.18	m	**26.99**
25 mm x 75 mm	-	0.19	4.17	6.01	m	**10.18**
25 mm x 150 mm	-	0.20	4.39	7.74	m	**12.13**
25 mm x 219 mm; in one width	-	0.27	5.93	9.78	m	**15.71**
25 mm x 300 mm; cross-tongued joints	-	0.31	6.81	21.83	m	**28.64**
32 mm x 75 mm	-	0.19	4.17	6.31	m	**10.48**
32 mm x 150 mm	-	0.20	4.39	8.35	m	**12.74**
32 mm x 219 mm; in one width	-	0.27	5.93	10.70	m	**16.63**
32 mm x 300 mm; cross-tongued joints	-	0.31	6.81	23.22	m	**30.03**
38 mm x 75 mm	-	0.19	4.17	6.91	m	**11.08**
38 mm x 150 mm	-	0.20	4.39	9.59	m	**13.98**
38 mm x 219 mm; in one width	-	0.27	5.93	12.51	m	**18.44**
38 mm x 300 mm; cross-tongued joints	-	0.31	6.81	26.01	m	**32.82**
returned and fitted ends	-	0.16	3.51	-	nr	**3.51**
Handrails; mopstick						
50 mm diameter	-	0.26	5.71	12.95	m	**18.66**

P BUILDING FABRIC SUNDRIES

Item	PC £	Labour hours	Labour £	Material £	Unit	Total rate £
P20 UNFRAMED ISOLATED TRIMS/SKIRTINGS/ SUNDRY ITEMS – cont'd						
Wrought softwood – cont'd						
Handrails; rounded						
44 mm x 50 mm	-	0.26	5.71	12.55	m	**18.26**
50 mm x 75 mm	-	0.28	6.15	13.71	m	**19.86**
57 mm x 87 mm	-	0.31	6.81	15.20	m	**22.01**
69 mm x 100 mm	-	0.36	7.91	18.63	m	**26.54**
Handrails; moulded						
44 mm x 50 mm	-	0.23	5.05	12.55	m	**17.61**
50 mm x 75 mm	-	0.25	5.49	13.71	m	**19.20**
57 mm x 87 mm	-	0.28	6.15	15.20	m	**21.35**
69 mm x 100 mm	-	0.32	7.03	18.63	m	**25.66**
Add 5% to the above material prices for selected softwood for staining						
Medium Density Fibreboard						
Skirtings, picture rails, dado rails and the like; splayed or moulded						
18 mm x 50 mm; splayed	3.01	0.14	3.08	3.72	m	**6.79**
18 mm x 50 mm; moulded	3.01	0.14	3.08	3.72	m	**6.79**
18 mm x 75 mm; splayed	3.13	0.14	3.08	3.87	m	**6.94**
18 mm x 75 mm; moulded	3.13	0.14	3.08	3.87	m	**6.94**
18 mm x 100 mm; splayed	3.27	0.14	3.08	4.03	m	**7.10**
18 mm x 100 mm; moulded	3.27	0.14	3.08	4.03	m	**7.10**
18 mm x 150 mm; moulded	3.57	0.17	3.73	4.39	m	**8.12**
18 mm x 175 mm; moulded	3.71	0.17	3.73	4.55	m	**8.28**
22 mm x 100 mm; splayed	3.44	0.14	3.08	4.23	m	**7.30**
25 mm x 50 mm; moulded	3.15	0.14	3.08	3.89	m	**6.97**
25 mm x 75 mm; splayed	3.34	0.14	3.08	4.12	m	**7.20**
25 mm x 100 mm; splayed	3.56	0.14	3.08	4.37	m	**7.45**
25 mm x 150 mm; splayed	4.02	0.17	3.73	4.92	m	**8.66**
25 mm x 150 mm; moulded	4.02	0.17	3.73	4.92	m	**8.66**
25 mm x 175 mm; moulded	4.24	0.17	3.73	5.18	m	**8.91**
25 mm x 225 mm; moulded	4.54	0.19	4.17	5.54	m	**9.71**
returned ends	-	0.22	4.83	-	nr	**4.83**
mitres	-	0.16	3.51	-	nr	**3.51**
Architraves, cover fillets and the like; half round; splayed or moulded						
12 mm x 25 mm; half round	2.86	0.17	3.73	3.54	m	**7.28**
12 mm x 50 mm; moulded	2.95	0.17	3.73	3.65	m	**7.39**
15 mm x 32 mm; half round	2.88	0.17	3.73	3.57	m	**7.30**
15 mm x 38 mm; moulded	2.90	0.17	3.73	3.60	m	**7.33**
15 mm x 50 mm; moulded	2.95	0.17	3.73	3.65	m	**7.39**
15 mm x 50 mm; splayed	2.95	0.17	3.73	3.65	m	**7.39**
15 mm x 63 mm; splayed	3.08	0.17	3.73	3.81	m	**7.55**
15 mm x 75 mm; splayed	3.15	0.17	3.73	3.89	m	**7.63**
25 mm x 44 mm; splayed	3.15	0.17	3.73	3.89	m	**7.63**
25 mm x 50 mm; moulded	3.15	0.17	3.73	3.89	m	**7.63**
25 mm x 63 mm; splayed	3.27	0.17	3.73	4.03	m	**7.76**
25 mm x 75 mm; splayed	3.38	0.17	3.73	4.16	m	**7.89**
30 mm x 88 mm; moulded	4.41	0.17	3.73	5.38	m	**9.11**
38 mm x 88 mm; moulded	3.76	0.17	3.73	4.62	m	**8.35**
50 mm x 50 mm; moulded	3.96	0.17	3.73	4.84	m	**8.58**
returned ends	-	0.22	4.83	-	nr	**4.83**
mitres	-	0.16	3.51	-	nr	**3.51**
Stops; screwed on						
15 mm x 38 mm	1.63	0.16	3.51	1.93	m	**5.44**
15 mm x 50 mm	1.68	0.16	3.51	2.00	m	**5.51**

P BUILDING FABRIC SUNDRIES

Item	PC £	Labour hours	Labour £	Material £	Unit	Total rate £
18 mm x 38 mm	1.67	0.16	3.51	1.98	m	**5.50**
25 mm x 38 mm	1.76	0.16	3.51	2.09	m	**5.60**
25 mm x 50 mm	1.85	0.16	3.51	2.20	m	**5.71**
Glazing beads and the like						
12 mm x 16 mm	1.95	0.05	1.10	2.32	m	**3.42**
12 mm x 19 mm	1.97	0.05	1.10	2.33	m	**3.43**
12 mm x 25 mm	1.99	0.05	1.10	2.36	m	**3.46**
12 mm x 25 mm; screwed	2.84	0.08	1.76	3.64	m	**5.39**
12 mm x 25 mm; fixing with brass cups and screws	3.27	0.16	3.51	3.87	m	**7.38**
15 mm x 25 mm; screwed	2.86	0.08	1.76	3.66	m	**5.42**
15 mm quadrant	2.81	0.07	1.54	3.33	m	**4.87**
18 mm quadrant or scotia	2.83	0.07	1.54	3.35	m	**4.89**
18 mm x 36 mm; screwed	2.92	0.07	1.54	3.73	m	**5.27**
25 mm x 38 mm; screwed	3.05	0.07	1.54	3.89	m	**5.43**
25 mm quadrant or scotia	2.95	0.07	1.54	3.50	m	**5.03**
38 mm scotia	2.80	0.07	1.54	3.32	m	**4.86**
50 mm scotia	3.25	0.07	1.54	3.86	m	**5.39**
Isolated shelves, worktops, seats and the like						
18 mm x 150 mm	3.70	0.22	4.83	4.38	m	**9.21**
18 mm x 200 mm	3.89	0.31	6.81	4.61	m	**11.42**
25 mm x 150 mm	4.24	0.22	4.83	5.02	m	**9.85**
25 mm x 200 mm	4.53	0.31	6.81	5.37	m	**12.18**
30 mm x 150 mm	5.94	0.22	4.83	7.04	m	**11.88**
30 mm x 200 mm	6.59	0.31	6.81	7.81	m	**14.62**
Isolated shelves, worktops, seats and the like; cross-tongued joints						
18 mm x 300 mm	12.03	0.39	8.57	14.26	m	**22.83**
18 mm x 450 mm	13.75	0.45	9.88	16.30	m	**26.18**
18 mm x 600 mm	22.93	0.56	12.30	27.17	m	**39.47**
25 mm x 300 mm	12.69	0.39	8.57	15.04	m	**23.61**
25 mm x 450 mm	15.51	0.45	9.88	18.39	m	**28.27**
25 mm x 600 mm	22.40	0.56	12.30	26.54	m	**38.84**
30 mm x 300 mm	14.64	0.39	8.57	17.36	m	**25.92**
30 mm x 450 mm	17.42	0.45	9.88	20.65	m	**30.54**
30 mm x 600 mm	25.37	0.56	12.30	30.07	m	**42.37**
Isolated shelves, worktops, seats and the like; slatted with 50 mm wide slats at 75 mm centres						
18 mm thick	36.40	0.66	14.50	43.72	m²	**58.22**
25 mm thick	38.65	0.66	14.50	46.39	m²	**60.88**
30 mm thick	40.69	0.66	14.50	48.81	m²	**63.31**
Window boards, nosings, bed moulds and the like; rebated and rounded						
18 mm x 75 mm	3.20	0.24	5.27	4.14	m	**9.41**
18 mm x 150 mm	3.65	0.28	6.15	4.68	m	**10.83**
18 mm x 225 mm	4.02	0.37	8.13	5.12	m	**13.25**
18 mm x 300 mm	4.46	0.41	9.01	5.64	m	**14.65**
25 mm x 75 mm	3.38	0.24	5.27	4.35	m	**9.63**
25 mm x 150 mm	4.02	0.28	6.15	5.12	m	**11.27**
25 mm x 225 mm	4.53	0.37	8.13	5.72	m	**13.85**
25 mm x 300 mm	5.13	0.41	9.01	6.43	m	**15.44**
30 mm x 75 mm	4.71	0.24	5.27	5.93	m	**11.21**
30 mm x 150 mm	5.89	0.28	6.15	7.33	m	**13.48**
30 mm x 225 mm	-	0.37	8.13	8.40	m	**16.53**
30 mm x 300 mm	7.86	0.41	9.01	9.67	m	**18.68**
38 mm x 75 mm	5.38	0.24	5.27	6.73	m	**12.00**
38 mm x 150 mm	6.76	0.28	6.15	8.36	m	**14.51**
38 mm x 225 mm	7.86	0.37	8.13	9.67	m	**17.80**
38 mm x 300 mm	9.15	0.41	9.01	11.20	m	**20.20**
returned and fitted ends	-	0.23	5.05	-	nr	**5.05**

P BUILDING FABRIC SUNDRIES

Item	PC £	Labour hours	Labour £	Material £	Unit	Total rate £
P20 UNFRAMED ISOLATED TRIMS/SKIRTINGS/ SUNDRY ITEMS – cont'd						
Selected Sapele						
Skirtings, picture rails, dado rails and the like; splayed or moulded						
19 mm x 44 mm; splayed	4.99	0.14	3.08	6.07	m	**9.14**
19 mm x 44 mm; moulded	4.99	0.14	3.08	6.07	m	**9.14**
19 mm x 69 mm; splayed	5.81	0.14	3.08	7.04	m	**10.11**
19 mm x 69 mm; moulded	5.81	0.14	3.08	7.04	m	**10.11**
19 mm x 94 mm; splayed	6.77	0.14	3.08	8.18	m	**11.26**
19 mm x 94 mm; moulded	6.77	0.14	3.08	8.18	m	**11.26**
19 mm x 144 mm; moulded	9.00	0.17	3.73	10.82	m	**14.55**
19 mm x 169 mm; moulded	9.96	0.17	3.73	11.96	m	**15.69**
25 mm x 44 mm; moulded	5.60	0.14	3.08	6.80	m	**9.87**
25 mm x 75 mm; splayed	6.68	0.14	3.08	8.07	m	**11.14**
25 mm x 94 mm; splayed	8.19	0.14	3.08	9.86	m	**12.93**
25 mm x 144 mm; splayed	10.72	0.17	3.73	12.86	m	**16.60**
25 mm x 144 mm; moulded	10.72	0.17	3.73	12.86	m	**16.60**
25 mm x 169 mm; moulded	12.02	0.17	3.73	14.40	m	**18.13**
25 mm x 219 mm; moulded	13.73	0.19	4.17	16.43	m	**20.60**
returned end	-	0.22	4.83	-	nr	**4.83**
mitres	-	0.16	3.51	-	nr	**3.51**
Architraves, cover fillets and the like; half round; splayed or moulded						
13 mm x 25 mm; half round	3.02	0.17	3.73	3.74	m	**7.47**
13 mm x 50 mm; moulded	4.84	0.17	3.73	5.90	m	**9.63**
16 mm x 32 mm; half round	3.07	0.17	3.73	3.79	m	**7.53**
16 mm x 38 mm; moulded	4.66	0.17	3.73	5.68	m	**9.42**
16 mm x 50 mm; moulded	4.99	0.17	3.73	6.07	m	**9.80**
19 mm x 50 mm; splayed	4.99	0.17	3.73	6.07	m	**9.80**
19 mm x 63 mm; splayed	5.41	0.17	3.73	6.57	m	**10.31**
19 mm x 69 mm; splayed	5.81	0.17	3.73	7.04	m	**10.77**
25 mm x 44 mm; splayed	5.39	0.17	3.73	6.54	m	**10.28**
25 mm x 50 mm; moulded	5.60	0.17	3.73	6.80	m	**10.53**
25 mm x 63 mm; splayed	6.26	0.17	3.73	7.57	m	**11.31**
25 mm x 69 mm; splayed	6.68	0.17	3.73	8.07	m	**11.80**
32 mm x 88 mm; moulded	8.12	0.17	3.73	9.77	m	**13.51**
38 mm x 88 mm; moulded	6.39	0.17	3.73	7.73	m	**11.46**
50 mm x 50 mm; moulded	8.35	0.17	3.73	10.06	m	**13.79**
returned end	-	0.22	4.83	-	nr	**4.83**
mitres	-	0.16	3.51	-	nr	**3.51**
Stops; screwed on						
16 mm x 38 mm	2.06	0.16	3.51	2.44	m	**5.95**
16 mm x 50 mm	2.21	0.16	3.51	2.62	m	**6.14**
19 mm x 38 mm	2.06	0.16	3.51	2.44	m	**5.95**
25 mm x 38 mm	2.59	0.16	3.51	3.08	m	**6.59**
25 mm x 50 mm	3.01	0.16	3.51	3.57	m	**7.08**
Glazing beads and the like						
13 mm x 16 mm	2.69	0.07	1.54	3.19	m	**4.73**
13 mm x 19 mm	2.69	0.07	1.54	3.19	m	**4.73**
13 mm x 25 mm	2.89	0.07	1.54	3.43	m	**4.96**
13 mm x 25 mm; screwed	3.92	0.07	1.54	4.92	m	**6.45**
13 mm x 25 mm; fixing with brass cups and screws	4.82	0.07	1.54	5.71	m	**7.25**
16 mm x 25 mm; screwed	3.92	0.08	1.76	4.92	m	**6.67**
16 mm quadrant	3.77	0.07	1.54	4.47	m	**6.01**
19 mm quadrant or scotia	3.77	0.07	1.54	4.47	m	**6.01**
19 mm x 36 mm; screwed	4.84	0.07	1.54	6.02	m	**7.55**

P BUILDING FABRIC SUNDRIES

Item	PC £	Labour hours	Labour £	Material £	Unit	Total rate £
25 mm x 38 mm; screwed	5.26	0.07	1.54	6.51	m	**8.05**
25 mm quadrant or scotia	4.31	0.07	1.54	5.11	m	**6.64**
38 mm scotia	6.39	0.07	1.54	7.57	m	**9.11**
50 mm scotia	8.35	0.07	1.54	9.90	m	**11.44**
Isolated shelves; worktops, seats and the like						
19 mm x 150 mm	9.16	0.22	4.83	10.86	m	**15.69**
19 mm x 200 mm	10.83	0.31	6.81	12.83	m	**19.64**
25 mm x 150 mm	10.72	0.22	4.83	12.71	m	**17.54**
25 mm x 200 mm	12.89	0.31	6.81	15.27	m	**22.08**
32 mm x 150 mm	12.11	0.22	4.83	14.36	m	**19.19**
32 mm x 200 mm	14.65	0.31	6.81	17.36	m	**24.17**
Isolated shelves, worktops, seats and the like; cross-tongued joints						
19 mm x 300 mm	24.78	0.39	8.57	29.36	m	**37.93**
19 mm x 450 mm	38.79	0.45	9.88	45.98	m	**55.86**
19 mm x 600 mm	51.52	0.56	12.30	61.06	m	**73.36**
25 mm x 300 mm	27.82	0.39	8.57	32.97	m	**41.54**
25 mm x 450 mm	43.70	0.45	9.88	51.79	m	**61.67**
25 mm x 600 mm	58.08	0.56	12.30	68.84	m	**81.14**
32 mm x 300 mm	30.43	0.39	8.57	36.06	m	**44.63**
32 mm x 450 mm	47.92	0.45	9.88	56.80	m	**66.68**
32 mm x 600 mm	63.70	0.56	12.30	75.50	m	**87.80**
Isolated shelves, worktops, seats and the like; slatted with 50 mm wide slats at 75 mm centres						
19 mm thick	69.73	0.88	19.33	83.23	m²	**102.56**
25 mm thick	74.65	0.88	19.33	89.06	m²	**108.39**
32 mm thick	78.86	0.88	19.33	94.05	m²	**113.38**
Window boards, nosings, bed moulds and the like; rebated and rounded						
19 mm x 75 mm	7.06	0.24	5.27	8.71	m	**13.99**
19 mm x 150 mm	10.26	0.28	6.15	12.51	m	**18.66**
19 mm x 219 mm; in one width	12.59	0.37	8.13	15.27	m	**23.40**
19 mm x 300 mm; cross-tongued joints	25.62	0.41	9.01	30.72	m	**39.72**
25 mm x 75 mm	7.77	0.24	5.27	9.56	m	**14.83**
25 mm x 150 mm	11.42	0.28	6.15	13.89	m	**20.04**
25 mm x 219 mm; in one width	14.97	0.37	8.13	18.09	m	**26.22**
25 mm x 300 mm; cross-tongued joints	29.77	0.41	9.01	35.64	m	**44.64**
32 mm x 75 mm	8.45	0.24	5.27	10.36	m	**15.64**
32 mm x 150 mm	12.70	0.28	6.15	15.40	m	**21.55**
32 mm x 219 mm; in one width	16.92	0.37	8.13	20.41	m	**28.53**
32 mm x 300 mm; cross-tongued joints	32.75	0.41	9.01	39.16	m	**48.17**
returned and fitted ends	-	0.23	5.05	-	nr	**5.05**
Handrails; rounded						
44 mm x 50 mm	16.08	0.33	7.25	19.05	m	**26.30**
50 mm x 75 mm	19.37	0.37	8.13	22.96	m	**31.09**
57 mm x 87 mm	22.74	0.41	9.01	26.95	m	**35.96**
69 mm x 100 mm	28.27	0.45	9.88	33.50	m	**43.38**
Handrails; moulded						
44 mm x 50 mm	17.88	0.33	7.25	21.20	m	**28.44**
50 mm x 75 mm	21.18	0.37	8.13	25.10	m	**33.23**
57 mm x 87 mm	24.55	0.41	9.01	29.10	m	**38.10**
69 mm x 100 mm	30.06	0.45	9.88	35.63	m	**45.51**
Pin-boards; medium board						
Sundeala "A" pin-board or other equal and approved; fixed with adhesive to backing (not included); over 300 mm wide						
6.40 mm thick	-	0.61	13.40	6.52	m²	**19.92**

P BUILDING FABRIC SUNDRIES

Item	PC £	Labour hours	Labour £	Material £	Unit	Total rate £
P20 UNFRAMED ISOLATED TRIMS/SKIRTINGS/ SUNDRY ITEMS – cont'd						
Sundries on softwood/hardwood						
Extra over fixing with nails for						
gluing and pinning	-	0.02	0.44	0.10	m	**0.54**
masonry nails	-	-	-	-	m	**0.34**
steel screws	-	-	-	-	m	**0.32**
self-tapping screws	-	-	-	-	m	**0.33**
steel screws; gluing	-	-	-	-	m	**0.57**
steel screws; sinking; filling heads	-	-	-	-	m	**0.73**
steel screws; sinking; pellating over	-	-	-	-	m	**1.57**
brass cups and screws	-	-	-	-	m	**1.93**
Extra over for						
countersinking	-	-	-	-	m	**0.30**
pellating	-	-	-	-	m	**1.37**
Head or nut in softwood						
let in flush	-	-	-	-	nr	**0.73**
Head or nut; in hardwood						
let in flush	-	-	-	-	nr	**1.07**
let in over; pellated	-	-	-	-	nr	**2.51**
Metalwork; mild steel						
Angle section bearers; for building in						
90 mm x 90 mm x 6 mm	-	0.33	7.46	9.44	m	**16.90**
120 mm x 120 mm x 8 mm	-	0.36	8.14	17.67	m	**25.81**
200 mm x 150 mm x 12 mm	-	0.41	9.27	39.15	m	**48.42**
Metalwork; mild steel; galvanized						
Water bars; groove in timber						
6 mm x 30 mm	-	0.51	11.20	5.02	m	**16.22**
6 mm x 40 mm	-	0.51	11.20	6.60	m	**17.80**
6 mm x 50 mm	-	0.51	11.20	8.27	m	**19.48**
Angle section bearers; for building in						
90 mm x 90 mm x 6 mm	-	0.33	7.46	11.34	m	**18.81**
120 mm x 120 mm x 8 mm	-	0.36	8.14	20.22	m	**28.36**
200 mm x 150 mm x 12 mm	-	0.41	9.27	42.86	m	**52.14**
Dowels; mortice in timber						
8 mm diameter x 100 mm long	-	0.05	1.10	0.22	nr	**1.32**
10 mm diameter x 50 mm long	-	0.05	1.10	0.53	nr	**1.62**
Cramps						
25 mm x 3 mm x 230 mm girth; one end bent, holed and screwed to softwood; other end fishtailed for building in	-	0.07	1.54	1.25	nr	**2.79**
Metalwork; stainless steel						
Angle section bearers; for building in						
90 mm x 90 mm x 6 mm	-	0.33	7.46	55.85	m	**63.32**
120 mm x 120 mm x 8 mm	-	0.36	8.14	82.57	m	**90.71**
200 mm x 150 mm x 12 mm	-	0.41	9.27	215.37	m	**224.65**
P21 IRONMONGERY						
NOTE: Ironmongery is largely a matter of selection and prices vary considerably, indicative prices for reasonable quantities of "good quality" ironmongery are given hereafter.						

P BUILDING FABRIC SUNDRIES

Item	PC £	Labour hours	Labour £	Material £	Unit	Total rate £
Iromongery; Allgood or other equal and approved; to softwood						
Bolts						
75 x 35 mm Modric anodised aluminium straight barrel bolt	8.83	0.33	7.25	9.97	nr	**17.22**
150 x 35 mm Modric anodised aluminium straight barrel bolt	10.05	0.33	7.25	11.34	nr	**18.59**
75 x 35 mm Modric anodised aluminium necked barrel bolt	9.88	0.33	7.25	11.15	nr	**18.40**
150 x 35 mm Modric anodised aluminium necked barrel bolt	12.61	0.33	7.25	14.23	nr	**21.48**
11 mm Easiclean socket for wood or stone	1.80	0.11	2.42	2.03	nr	**4.45**
Security hinge bolt chubb WS12	16.48	0.55	12.08	18.60	nr	**30.68**
203 x 19 x 11 mm Complete bolt set, with floor socket and intumescent pack for FD30 and FD60 fire doors	34.85	0.66	14.50	39.34	nr	**53.83**
203/609 x 19 mm Complete bolt set, with floor socket and intumescent pack for FD30 and FD60 fire doors	82.00	0.66	14.50	92.56	nr	**107.05**
Stainless steel indicating bolt complete with outside indicator and emergency release	39.17	0.66	14.50	44.21	nr	**58.71**
Catches						
Magnetic catch	1.10	0.22	4.83	1.24	nr	**6.07**
Door closers and furniture						
13 mm Satin chrome rebate component for 7204/08/78/79/86	27.72	0.66	14.50	31.29	nr	**45.79**
90 x 90 mm Modric anodised aluminium electrically powered hold open wall magnet. CE marked to BS EN1155:1997 & A1:2002 3-5-6/3-1-1-3	119.51	0.44	9.66	134.90	nr	**144.56**
Modric anodised aluminium bathroom configuration with quadaxial assembly, turn, release and optional indicator	63.41	0.88	19.33	71.57	nr	**90.90**
Overhead limiting stay; galvanised	13.64	1.10	24.16	15.40	nr	**39.56**
263 x 48 x 48 mm Overhead door closer Fig 6 adjustable power 2-5 with adjustable backcheck and intumescent protected bracket. Certifire listed and CE Marked to BS EN1154 4-8-2/5-1-1-3	67.80	1.10	24.16	76.53	nr	**100.69**
Concealed jamb door closer check action	96.87	1.10	24.16	109.34	nr	**133.50**
75 x 57 x 170 mm Modric anodised aluminium door co-ordinator for pairs of rebated leaves, CE Marked to BS EN1158 3-5-3/5-1-1-0	29.77	0.88	19.33	33.60	nr	**52.93**
263 x 50 x 48 mm Modric anodised aluminium overhead door closer Fig 1 adjustable power 2-5 with adjustable backcheck, intumescent protected bracket. Certifire listed and CE Marked to BS EN1154 4-8/5-1-1-3	65.65	1.10	24.16	74.10	nr	**98.26**
290 x 48 x 50 mm Modric anodised aluminium rectangular overhead door closer with adjustable power and adjustable backcheck intumescent protected arm heavy duty U.L. & certifire listed & CE Marked to BS EN1154 4-8-2/4-1-1-3 and Kitemarked.	92.56	1.10	24.16	104.48	nr	**128.64**
Stainless steel overhead door closer Fig 1 projecting armset, Power EN 2-5, CE marked , c/w Backcheck, Latch action and Speed control. Max door width 1100 mm, Max door weight 100kg	89.75	1.10	24.16	101.31	nr	**125.47**
Fully concealed overhead door closer complete with track and arm for single action doors	140.88	0.88	19.33	159.02	nr	**178.35**

P BUILDING FABRIC SUNDRIES

Item	PC £	Labour hours	Labour £	Material £	Unit	Total rate £
P21 IRONMONGERY – cont'd						
Iromongery; Allgood; to softwood – cont'd						
Door closers and furniture – cont'd						
92 x 45 mm Stainless steel heavy duty floor pivot set with thrust roller bearing 200kg load capacity. Complete with forged steel intumescent protected double action strap with 10 mm height adjustment, new low profile top centre, and matching cover plate	126.00	2.50	54.91	142.22	nr	**197.14**
92 x 45 mm Stainless steel heavy duty floor pivot set with thrust roller bearing 200kg load capacity, with stainless steel intumescent protected S/A 25 mm offset strap & top centre, matching plate	168.00	2.50	54.91	189.63	nr	**244.54**
Double action pivot set for door maximum width 1100 mm and maximum weight 80kg	77.02	2.50	54.91	86.94	nr	**141.85**
305 x 80 x 50 mm Stainless steel 'Cavalier' floor spring, intumescent protected forged steel D/A strap with 10 mm height adjustment & low profile top centre, matching covers & box. CE Marked to BS EN1154 4-8-*-1-1-3. Adjustable power 2/4	193.73	2.50	54.91	218.67	nr	**273.59**
305 x 80 x 50 mm Stainless steel 'Cavalier' floor spring adjustable power 2/4 stainless steel intumsecent protected S/A 16 mm offset strap & top centre, matching covers & box. Certifire listed and CE marked to BSEN1154 4-8-2/4-1-1-3	215.25	2.50	54.91	242.96	nr	**297.88**
Surface vertical rod push bar panic bolt, reversible, tp suit doors 2500x1100 mm maximum, silver finish, CE marked to EN1125 class 3-7-5-1-1-3-2-2-A	264.03	1.65	36.24	298.02	nr	**334.27**
Rim push bar panic latch, reversible, to suit doors 1100 mm wide maximum, silver finish, CE marked to EN1125 class 3-7-5-1-1-3-2-2-A	157.97	1.43	31.41	178.31	nr	**209.72**
76 x 51 x 13 mm Adjustable heavy roller catch satin chrome	7.45	0.66	14.50	8.41	nr	**22.91**
External access device for use with XX10280/2 panic hardware to suit door thickness 45-55 mm, complete with SS3006N lever, SS755 rose, SS796 profile escutcheon and spindle.For use with MA7420A51 or MA7420A55 profile cylinders	31.86	1.43	31.41	35.96	nr	**67.37**
142 x 22 mm Ø Concealed jamb door closer light duty	12.24	0.88	19.33	13.82	nr	**33.15**
80 x 40 x 45 mm Emergency release door stop with holdback facility	77.02	1.10	24.16	86.94	nr	**111.10**
Modric anodised aluminium quadaxial lever assembly tested to BS EN1906 4/7/-/1/1/4/0/U	31.19	0.88	19.33	35.21	pair	**54.54**
Modric anodised aluminium quadaxial lever assembly tested to BS EN1906 4/7/-/1/1/4/0/U	31.19	0.88	19.33	35.21	pair	**54.54**
Modric anodised aluminium quadaxial lever assembly tested to BS EN1906 4/7/-/1/1/4/0/U with Biocote® anti-bacterial protection	44.18	0.88	19.33	49.87	pair	**69.20**
Modric stainless steel quadaxial lever assembly Tested to BS EN1906 4/7/-/1/1/4/0/U	44.77	0.88	19.33	50.53	pair	**69.86**
152 x 38 x 13 mm Modric anodised aluminium security door chain leather covered	41.72	0.44	9.66	47.09	nr	**56.76**
50 Ø x 3 mm Circular covered rose for profile cylinder	4.09	0.11	2.42	4.62	nr	**7.03**
50 Ø x 3 mm Modric anodised aluminium circular covered rose with indicator and emergency release	8.46	0.15	3.29	9.55	nr	**12.84**
50 Ø x 3 mm Modric anodised aluminium circular covered rose with heavy turn, 5-8 mm spindle	14.83	0.15	3.29	16.74	nr	**20.03**
Budget lock escutcheon - satin stainless steel 316	4.90	0.11	2.42	5.53	nr	**7.95**

P BUILDING FABRIC SUNDRIES

Item	PC £	Labour hours	Labour £	Material £	Unit	Total rate £
50 Ø x 3 mm Stainless steel circular covered rose for profile cylinder	6.54	0.11	2.42	7.38	nr	**9.80**
50 Ø x 3 mm Stainless steel circular covered rose with indicator and emergency release	8.87	0.17	3.73	10.01	nr	**13.75**
50 Ø x 3 mm Stainless steel circular covered rose with heavy turn, 5-8 mm spindle	18.25	0.17	3.73	20.60	nr	**24.33**
330 x 76 x 1.6 mm Modric anodised aluminium push plate	4.72	0.17	3.73	5.33	nr	**9.06**
330 x 76 x 1.6 mm Stainless steel push plate	10.61	0.17	3.73	11.98	nr	**15.71**
800 x 150 x 1.5 mm Modric anodised aluminium kicking plate, drilled and countersunk with screws.	7.04	0.28	6.15	7.95	nr	**14.10**
900 x 150 x 1.5 mm Modric anodised aluminium kicking plate, drilled and countersunk with screws.	7.92	0.28	6.15	8.94	nr	**15.09**
1000 x 150 x 1.5 mm Modric anodised aluminium kicking plate, drilled & countersunk with screws.	8.80	0.28	6.15	9.93	nr	**16.08**
800 x 150 x 1.5 mm Stainless steel kicking plate, drilled and countersunk with screws.	13.84	0.28	6.15	15.62	nr	**21.77**
900 x 150 x 1.5 mm Stainless steel kicking plate, drilled and countersunk with screws.	15.56	0.28	6.15	17.56	nr	**23.71**
1000 x 150 x 1.5 mm Stainless steel kicking plate, drilled & countersunk with screws.	17.30	0.28	6.15	19.53	nr	**25.68**
305 x 70 x 19 mm Ø Modric anodised aluminium grab handle bolt through fixing	18.86	0.44	9.66	21.29	nr	**30.95**
400 x 19 mm Ø Stainless steel D line straight pull handle with M8 threaded holes, fixing centres 300 mm	49.52	0.36	7.91	60.64	nr	**68.54**
Hinges						
100 x 75 x 3 mm Stainless steel triple knuckle concealed twin Newtonbearings, button tipped butt hinges, jig drilled for metal doors/frames, complete with M6x12MT 'undercut' machine screws, stainless steel 316 CE marked to EN1935 4-7-7-1-1-4-0-13	22.66	0.28	6.15	25.58	pair	**31.73**
100 x 100 x 3 mm Stainless steel triple knuckle concealed twin Newton bearings, button tipped hinges, jig drilled, stainless steel grade 316 CE marked to EN1935 4-7-7-1-1-4-0-13	30.60	0.28	6.15	34.54	pair	**40.69**
Latches						
Modric anodised aluminium round cylinder for rim night latch, 2 keyed satin nickel plated	23.35	0.44	9.66	26.36	nr	**36.02**
93 x 75 mm Cylinder rim non-deadlocking night latch case only 60 mm backset	16.54	0.44	9.66	18.67	nr	**28.33**
71 series mortice latch, case only, low friction latchbolt, griptight follower, heavy spring for levers. Radius forend and sq strike. CE marked to BS EN12209 3/X/8/1/0G/-/B/02/0	15.33	0.88	19.33	17.30	nr	**36.63**
Modric anodised aluminium latch configuration with quadaxial assembly	48.00	0.88	19.33	54.18	nr	**73.51**
Modric anodised aluminium nightlatch configuration with quadaxial assembly and single cylinder	91.05	0.88	19.33	102.77	nr	**122.10**
Locks						
44 mm case Bright zinc plated steel mortice budget lock with slotted strike plate 33 mm backset	6.07	0.88	19.33	6.85	nr	**26.18**
76 x 58 mm b/s Stainless steel cubicle mortice deadlock with 8 mm follower	12.94	0.88	19.33	14.61	nr	**33.94**
'A' length European profile double cylinder lock, 2 keyed satin nickel plated	23.53	0.88	19.33	26.56	nr	**45.89**
'A' length European profile cylinder and large turn, 2 keyed satin nickel plated	26.74	0.88	19.33	30.18	nr	**49.51**
'A' length European profile cylinder and large turn, 2 keyed under master key, satin nickel plated	27.56	0.88	19.33	31.11	nr	**50.44**

P BUILDING FABRIC SUNDRIES

Item	PC £	Labour hours	Labour £	Material £	Unit	Total rate £
P21 IRONMONGERY – cont'd						
Iromongery; Allgood; to softwood – cont'd						
Locks – cont'd						
'A' length European profile single cylinder, 2 keyed satin nickel plated	18.74	0.88	19.33	21.15	nr	**40.48**
'A' length European profile single cylinder, 2 keyed under master key, satin nickel plated	18.74	0.88	19.33	21.15	nr	**40.48**
93 x 60 mm b/s 71 series profile cylinder mortice deadlock, case only. Single throw 22mm deadbolt. Radius forend and square strike. CE marked to BS EN12209 3/X/8/1/0/G/4/B/A/0/0	15.33	0.88	19.33	17.30	nr	**36.63**
92 x 60 mm b/s 71 series bathroom lock, case only, low friction latchbolt, griptight follower, heavy spring for levers, twin 8mm followers at 78mm centres. Radius forend and square strike. CE marked to BS EN12209 3/X/8/0/0/G-/B/0/2/0	18.16	0.88	19.33	20.50	nr	**39.83**
93 x 60 mm b/s 71 series profile cylinder mortice lock, case only, low friction latchbolt, griptight follower. Heavy spring for levers, 22mm throw deadbolt, cylinder withdraws bolt bolts. Radius forend and square strike. CE marked to BS EN12209 3/X/8/1/0G/4/B/A2/0	19.07	0.88	19.33	21.53	nr	**40.85**
92 x 60 mm b/s71 series profile cylinder emergency lock, case only. Low friction latchbolt, griptight follower, heavy spring for lever, single throw 22mm deadbolt, lever can withdraw both bolts. Radius forend and strike	62.59	0.88	19.33	70.65	nr	**89.98**
Modric anodised aluminium lock configuration with quadaxial assembly and cylinder with turn	83.04	0.88	19.33	93.73	nr	**113.06**
Sundries						
76 mm Ø Modric anodised aluminium circular sex symbol male	4.52	0.09	1.98	5.10	nr	**7.08**
76 mm Ø Modric anodised aluminium circular symbol fire door keep locked	4.52	0.09	1.98	5.10	nr	**7.08**
76 mm Ø Modric anodised aluminium circular symbol fire door keep shut	4.52	0.11	2.42	5.10	nr	**7.52**
38 x 47 mm Ø Modric anodised aluminium heavy circular floor door stop with cover	8.35	0.11	2.42	9.43	nr	**11.84**
38 x 47 mm Ø Stainless steel heavy circular floor door stop with cover	11.34	0.10	2.20	12.80	nr	**15.00**
63 x 19 mm Ø Modric anodised aluminium circular heavy duty skirting buffer with thief resistant insert	5.97	0.11	2.42	6.74	nr	**9.15**
63 x 19 mm Ø Stainless steel circular heavy duty skirting buffer with thief resistant insert	6.91	0.11	2.42	7.80	nr	**10.22**
152 mm Cabin hook satin chrome on brass	15.70	0.17	3.73	17.72	nr	**21.46**
Stainless steel toilet roll holder, length 145 mm, colour white, satin stainless steel 316	51.71	0.17	3.73	59.95	nr	**63.68**
Stainless steel towel rail with bushes, fixing centres 450 mm, satin stainless steel 316	59.06	0.28	6.15	71.40	nr	**77.55**
Stainless steel toilet brush holder with toilet brush, with bushes, satin stainless steel 316	93.51	0.22	4.83	108.71	nr	**113.54**
Set of stainless steel rails, one lift-up, 3 straight, and one backrest for use in toilets for the disabled, to meet the requirements of Part M of the Building Regulations	532.60	1.65	36.24	601.17	nr	**637.42**

P BUILDING FABRIC SUNDRIES

Item	PC £	Labour hours	Labour £	Material £	Unit	Total rate £
Ironmongery; Allgood or other equal and approved; to hardwood						
Bolts						
75 x 35 mm Modric anodised aluminium straight barrel bolt	8.83	0.45	9.88	9.97	nr	**19.85**
150 x 35 mm Modric anodised aluminium straight barrel bolt	10.05	0.45	9.88	11.34	nr	**21.23**
75 x 35 mm Modric anodised aluminium necked barrel bolt	9.88	0.45	9.88	11.15	nr	**21.04**
150 x 35 mm Modric anodised aluminium necked barrel bolt	12.61	0.45	9.88	14.23	nr	**24.12**
11 mm Easiclean socket for wood or stone	1.80	0.15	3.29	2.03	nr	**5.33**
Security hinge bolt chubb WS12	16.48	0.75	16.47	18.60	nr	**35.08**
203 x 19 x 11 mm Complete bolt set, with floor socket and intumescent pack for FD30 and FD60 fire doors	34.85	0.90	19.77	39.34	nr	**59.11**
203/609 x 19 mm Complete bolt set, with floor socket and intumescent pack for FD30 and FD60 fire doors	82.00	0.90	19.77	92.56	nr	**112.33**
Stainless steel indicating bolt complete with outside indicator and emergency release	39.17	0.90	19.77	44.21	nr	**63.98**
Catches						
Magnetic catch	1.10	0.30	6.59	1.24	nr	**7.83**
Door closers and furniture						
13 mm Satin chrome rebate component for 7204/08/78/79/86	27.72	0.90	19.77	31.29	nr	**51.06**
90 x 90 mm Modric anodised aluminium electrically powered hold open wall magnet. CE marked to BS EN1155:1997 & A1:2002 3-5-6/3-1-1-3	119.51	0.60	13.18	134.90	nr	**148.08**
Modric anodised aluminium bathroom configuration with quadaxial assembly, turn, release and optional indicator	63.41	1.15	25.26	71.57	nr	**96.83**
Overhead limiting stay; galvanised	13.64	1.45	31.85	15.40	nr	**47.25**
263 x 48 x 48 mm Overhead door closer Fig 6 adjustable power 2-5 with adjustable backcheck and intumescent protected bracket. Certifire listed and CE Marked to BS EN1154 4-8-2/5-1-1-3	67.80	1.45	31.85	76.53	nr	**108.38**
Concealed jamb door closer check action	96.87	1.45	31.85	109.34	nr	**141.19**
75 x 57 x 170 mm Modric anodised aluminium door co-ordinator for pairs of rebated leaves, CE Marked to BS EN1158 3-5-3/5-1-1-0	29.77	1.15	25.26	33.60	nr	**58.86**
263 x 50 x 48 mm Modric anodised aluminium overhead door closer Fig 1 adjustable power 2-5 with adjustable backcheck, intumescent protected bracket. Certifire listed and CE Marked to BS EN1154 4-8/5-1-1-3	65.65	1.45	31.85	74.10	nr	**105.95**
290 x 48 x 50 mm Modric anodised aluminium rectangular overhead door closer with adjustable power and adjustable backcheck intumescent protected arm heavy duty U.L. & certifire listed & CE marked to BS EN1154 4-8-2/4-1-1-3 and kitemarked.	92.56	1.45	31.85	104.48	nr	**136.33**
Stainless steel overhead door closer Fig 1. projecting armset, Power EN 2-5, CE marked , c/w Backcheck, Latch action and Speed control. Max door width 1100 mm, Max door weight 100kg	89.75	1.45	31.85	101.31	nr	**133.16**
Fully concealed overhead door closer complete with track and arm for single action doors	140.88	1.15	25.26	159.02	nr	**184.28**

P BUILDING FABRIC SUNDRIES

Item	PC £	Labour hours	Labour £	Material £	Unit	Total rate £
P21 IRONMONGERY – cont'd						
Iromongery; Allgood; to hardwood – cont'd						
Door closers and furniture – cont'd						
92 x 45 mm Stainless steel heavy duty floor pivot set with thrust roller bearing 200kg load capacity. Complete with forged steel intumescent protected double action strap with 10 mm height adjustment, new low profile top centre, and matching cover plate	126.00	3.30	72.49	142.22	nr	**214.71**
92 x 45 mm Stainless steel heavy duty floor pivot set with thrust roller bearing 200kg load capacity, with stainless steel intumescent protected S/A 25 mm offset strap & top centre, matching plate	168.00	3.30	72.49	189.63	nr	**262.12**
Double action pivot set for door maximum width 1100 mm and maximum weight 80kg	77.02	3.30	72.49	86.94	nr	**159.42**
305 x 80 x 50 mm Stainless steel 'Cavalier' floor spring, intumescent protected forged steel D/A strap with 10 mm height adjustment & low profile top centre, matching covers & box. CE Marked to BS EN1154 4-8-*-1-1-3. Adjustable power 2/4	193.73	3.30	72.49	218.67	nr	**291.16**
305 x 80 x 50 mm Stainless steel 'Cavalier' floor spring adjustable power 2/4 stainless steel intumsecent protected S/A 16 mm offset strap & top centre, matching covers & box. Certifire listed and CE marked to BSEN1154 4-8-2/4-1-1-3	215.25	3.30	72.49	242.96	nr	**315.45**
Surface vertical rod push bar panic bolt, reversible, tp suit doors 2500x1100 mm maximum, silver finish, CE marked to EN1125 class 3-7-5-1-1-3-2-2-A	264.03	2.20	48.32	298.02	nr	**346.35**
Rim push bar panic latch, reversible, to suit doors 1100 mm wide maximum, silver finish, CE marked to EN1125 class 3-7-5-1-1-3-2-2-A	157.97	1.90	41.73	178.31	nr	**220.04**
76 x 51 x 13 mm Adjustable heavy roller catch satin chrome	7.45	0.90	19.77	8.41	nr	**28.18**
External access device for use with XX10280/2 panic hardware to suit door thickness 45-55 mm, complete with SS3006N lever, SS755 rose, SS796 profile escutcheon and spindle.For use with MA7420A51 or MA7420A55 profile cylinders	31.86	1.90	41.73	35.96	nr	**77.70**
142 x 22 mm Ø Concealed jamb door closer light duty	12.24	1.15	25.26	13.82	nr	**39.08**
80 x 40 x 45 mm Emergency release door stop with holdback facility	77.02	1.45	31.85	86.94	nr	**118.79**
Modric anodised aluminium quadaxial lever assembly tested to BS EN1906 4/7/-/1/1/4/0/U	31.19	1.15	25.26	35.21	pair	**60.47**
Modric anodised aluminium quadaxial lever assembly tested to BS EN1906 4/7/-/1/1/4/0/U	31.19	1.15	25.26	35.21	pair	**60.47**
Modric anodised aluminium quadaxial lever assembly tested to BS EN1906 4/7/-/1/1/4/0/U with Biocote® anti-bacterial protection	44.18	1.15	25.26	49.87	pair	**75.13**
Modric stainless steel quadaxial lever assembly Tested to BS EN1906 4/7/-/1/1/4/0/U	44.77	1.15	25.26	50.53	pair	**75.79**
152 x 38 x 13 mm Modric anodised aluminium security door chain leather covered	41.72	0.60	13.18	47.09	nr	**60.27**
50 Ø x 3 mm Circular covered rose for profile cylinder	4.09	0.15	3.29	4.62	nr	**7.91**
50 Ø x 3 mm Modric anodised aluminium circular covered rose with indicator and emergency release	8.46	0.20	4.39	9.55	nr	**13.94**
50 Ø x 3 mm Modric anodised aluminium circular covered rose with heavy turn, 5-8 mm spindle	14.83	0.20	4.39	16.74	nr	**21.13**
Budget lock escutcheon - satin stainless steel 316	4.90	0.15	3.29	5.53	nr	**8.83**

P BUILDING FABRIC SUNDRIES

Item	PC £	Labour hours	Labour £	Material £	Unit	Total rate £
50 Ø x 3 mm Stainless steel circular covered rose for profile cylinder	6.54	0.15	3.29	7.38	nr	**10.68**
50 Ø x 3 mm Stainless steel circular covered rose with indicator and emergency release	8.87	0.25	5.49	10.01	nr	**15.50**
50 Ø x 3 mm Stainless steel circular covered rose with heavy turn, 5-8 mm spindle	18.25	0.25	5.49	20.60	nr	**26.09**
330 x 76 x 1.6 mm Modric anodised aluminium push plate	4.72	0.25	5.49	5.33	nr	**10.82**
330 x 76 x 1.6 mm Stainless steel push plate	10.61	0.25	5.49	11.98	nr	**17.47**
800 x 150 x 1.5 mm Modric anodised aluminium kicking plate, drilled and countersunk with screws.	7.04	0.35	7.69	7.95	nr	**15.63**
900 x 150 x 1.5 mm Modric anodised aluminium kicking plate, drilled and countersunk with screws.	7.92	0.35	7.69	8.94	nr	**16.63**
1000 x 150 x 1.5 mm Modric anodised aluminium kicking plate, drilled & countersunk with screws.	8.80	0.35	7.69	9.93	nr	**17.62**
800 x 150 x 1.5 mm Stainless steel kicking plate, drilled and countersunk with screws.	13.84	0.35	7.69	15.62	nr	**23.31**
900 x 150 x 1.5 mm Stainless steel kicking plate, drilled and countersunk with screws.	15.56	0.35	7.69	17.56	nr	**25.25**
1000 x 150 x 1.5 mm Stainless steel kicking plate, drilled & countersunk with screws.	17.30	0.35	7.69	19.53	nr	**27.22**
305 x 70 x 19 mm Ø Modric anodised aluminium grab handle bolt through fixing	18.86	0.60	13.18	21.29	nr	**34.47**
400 x 19 mm Ø Stainless steel D line straight pull handle with M8 threaded holes, fixing centres 300 mm	49.52	0.50	10.98	60.64	nr	**71.62**
Hinges						
100 x 75 x 3 mm Stainless steel triple knuckle concealed twin Newtonbearings, button tipped butt hinges, jig drilled for metal doors/frames, complete with M6x12MT 'undercut' machine screws, stainless steel 316 CE marked to EN1935 4-7-7-1-1-4-0-13	22.66	0.35	7.69	25.58	pair	**33.27**
100 x 100 x 3 mm Stainless steel triple knuckle concealed twin Newton bearings, button tipped hinges, jig drilled, stainless steel grade 316 CE marked to EN1935 4-7-7-1-1-4-0-13	30.60	0.35	7.69	34.54	pair	**42.23**
Latches						
Modric anodised aluminium round cylinder for rim night latch, 2 keyed satin nickel plated	23.35	0.60	13.18	26.36	nr	**39.54**
93 x 75 mm Cylinder rim non-deadlocking night latch case only 60 mm backset	16.54	0.60	13.18	18.67	nr	**31.85**
71 series mortice latch, case only, low friction latchbolt, griptight follower, heavy spring for levers. Radius forend and sq strike. CE marked to BS EN12209 3/X/8/1/0G/-/B/02/0	15.33	1.15	25.26	17.30	nr	**42.56**
Modric anodised aluminium latch configuration with quadaxial assembly	48.00	1.15	25.26	54.18	nr	**79.44**
Modric anodised aluminium nightlatch configuration with quadaxial assembly and single cylinder	91.05	1.15	25.26	102.77	nr	**128.03**
Locks						
44 mm case Bright zinc plated steel mortice budget lock with slotted strike plate 33 mm backset	6.07	1.15	25.26	6.85	nr	**32.11**
76 x 58 mm b/s Stainless steel cubicle mortice deadlock with 8 mm follower	12.94	1.15	25.26	14.61	nr	**39.87**
'A' length European profile double cylinder lock, 2 keyed satin nickel plated	23.53	1.15	25.26	26.56	nr	**51.82**
'A' length European profile cylinder and large turn, 2 keyed satin nickel plated	26.74	1.15	25.26	30.18	nr	**55.44**
'A' length European profile cylinder and large turn, 2 keyed under master key, satin nickel plated	27.56	1.15	25.26	31.11	nr	**56.37**

P BUILDING FABRIC SUNDRIES

Item	PC £	Labour hours	Labour £	Material £	Unit	Total rate £
P21 IRONMONGERY – cont'd						
Iromongery; Allgood; to hardwood – cont'd						
Locks – cont'd						
'A' length European profile single cylinder, 2 keyed satin nickel plated	18.74	1.15	25.26	21.15	nr	**46.41**
'A' length European profile single cylinder, 2 keyed under master key, satin nickel plated	18.74	1.15	25.26	21.15	nr	**46.41**
93 x 60 mm b/s 71 series profile cylinder mortice deadlock, case only. Single throw 22mm deadbolt. Radius forend and square strike. CE marked to BS EN12209 3/X/8/1/0/G/4/B/A/0/0	15.33	1.15	25.26	17.30	nr	**42.56**
92 x 60 mm b/s 71 series bathroom lock, case only, low friction latchbolt, griptight follower, heavy spring for levers, twin 8mm followers at 78mm centres. Radius forend and square strike. CE marked to BS EN12209 3/X/8/0/0/G-/B/0/2/0	18.16	1.15	25.26	20.50	nr	**45.76**
93 x 60 mm b/s 71 series profile cylinder mortice lock, case only, low friction latchbolt, griptight follower. Heavy spring for levers, 22mm throw deadbolt, cylinder withdraws bolt bolts. Radius forend and square strike. CE marked to BS EN12209 3/X/8/1/0G/4/B/A2/0	19.07	1.15	25.26	21.53	nr	**46.79**
92 x 60 mm b/s71 series profile cylinder emergency lock, case only. Low friction latchbolt, griptight follower, heavy spring for lever, single throw 22mm deadbolt, lever can withdraw both bolts. Radius forend and strike	62.59	1.15	25.26	70.65	nr	**95.91**
Modric anodised aluminium lock configuration with quadaxial assembly and cylinder with turn	83.04	1.15	25.26	93.73	nr	**118.99**
Sundries						
76 mm Ø Modric anodised aluminium circular sex symbol male	4.52	0.10	2.20	5.10	nr	**7.30**
76 mm Ø Modric anodised aluminium circular symbol fire door keep locked	4.52	0.10	2.20	5.10	nr	**7.30**
76 mm Ø Modric anodised aluminium circular symbol fire door keep shut	4.52	0.15	3.29	5.10	nr	**8.40**
38 x 47 mm Ø Modric anodised aluminium heavy circular floor door stop with cover	8.35	0.15	3.29	9.43	nr	**12.72**
38 x 47 mm Ø Stainless steel heavy circular floor door stop with cover	11.34	0.15	3.29	12.80	nr	**16.09**
63 x 19 mm Ø Modric anodised aluminium circular heavy duty skirting buffer with thief resistant insert	5.97	0.15	3.29	6.74	nr	**10.03**
63 x 19 mm Ø Stainless steel circular heavy duty skirting buffer with thief resistant insert	6.91	0.15	3.29	7.80	nr	**11.09**
152 mm Cabin hook satin chrome on brass	15.70	0.25	5.49	17.72	nr	**23.21**
Stainless steel toilet roll holder, length 145 mm, colour white, satin stainless steel 316	51.71	0.25	5.49	59.95	nr	**65.44**
Stainless steel towel rail with bushes, fixing centres 450 mm, satin stainless steel 316	59.06	0.35	7.69	71.40	nr	**79.09**
Stainless steel toilet brush holder with toilet brush, with bushes, satin stainless steel 316	93.51	0.30	6.59	108.71	nr	**115.30**
Set of stainless steel rails, one lift-up, 3 straight, and one backrest for use in toilets for the disabled, to meet the requirements of Part M of the Building Regulations	532.60	2.20	48.32	601.17	nr	**649.50**

P BUILDING FABRIC SUNDRIES

Item	PC £	Labour hours	Labour £	Material £	Unit	Total rate £
Sliding door gear; Hillaldam Coburn Ltd or other equal and approved; Commercial/Light industrial; for top hung timber/metal doors, weight not exceeding 365 kg						
Sliding door gear						
bottom guide; fixed to concrete in groove	20.91	0.46	10.10	23.60	m	**33.70**
top track	29.04	0.23	5.05	32.78	m	**37.83**
detachable locking bar	26.24	0.31	6.81	29.62	nr	**36.43**
hangers; timber doors	57.34	0.46	10.10	64.72	nr	**74.83**
hangers; metal doors	38.66	0.46	10.10	43.64	nr	**53.74**
head brackets; open, side fixing; bolting to masonry	8.08	0.46	10.10	12.86	nr	**22.97**
head brackets; open, soffit fixing; screwing to timber	7.88	0.32	7.03	8.98	nr	**16.01**
door guide to timber door	13.55	0.23	5.05	15.30	nr	**20.35**
door stop; rubber buffers; to masonry	29.88	0.69	15.16	33.73	nr	**48.88**
drop bolt; screwing to timber	26.82	0.46	10.10	30.27	nr	**40.38**
bow handle; to timber	13.76	0.23	5.05	15.53	nr	**20.58**
Sundries						
rubber door stop; plugged and screwed to concrete	6.78	0.09	1.98	7.65	nr	**9.63**
P30 TRENCHES/PIPEWAYS/PITS FOR BURIED ENGINEERING SERVICES						
Excavating trenches; by machine; grading bottoms; earthwork support; filling with excavated material and compacting; disposal of surplus soil; spreading on site average 50 m						
Services not exceeding 200 mm nominal size						
average depth of run 0.50 m	-	0.30	3.94	1.86	m	**5.80**
average depth of run 0.75 m	-	0.45	5.91	3.07	m	**8.99**
average depth of run 1.00 m	-	0.90	11.82	6.09	m	**17.92**
average depth of run 1.25 m	-	1.33	17.47	8.41	m	**25.89**
average depth of run 1.50 m	-	1.71	22.47	10.95	m	**33.42**
average depth of run 1.75 m	-	2.08	27.33	13.94	m	**41.27**
average depth of run 2.00 m	-	2.46	32.32	15.95	m	**48.27**
Excavating trenches; by hand; grading bottoms; earthwork support; filling with excavated material and compacting; disposal; of surplus soil; spreading on site average 50 m						
Services not exceeding 200 mm nominal size						
average depth of run 0.50 m	-	1.06	13.93	-	m	**13.93**
average depth of run 0.75 m	-	1.60	21.02	-	m	**21.02**
average depth of run 1.00 m	-	2.34	30.74	2.26	m	**33.00**
average depth of run 1.25 m	-	3.29	43.23	3.10	m	**46.33**
average depth of run1.50 m	-	4.52	59.39	3.78	m	**63.17**
average depth of run 1.75 m	-	5.96	78.31	4.57	m	**82.88**
average depth of run 2.00 m	-	6.81	89.47	5.02	m	**94.50**
Stop cock pits, valves chambers and the like; excavating; half brick thick walls in common bricks in cement mortar (1:3); on in situ concrete designated C20 - 20 mm aggregate bed; 100 mm thick						
Pits						
100 mm x 100 mm x 750 mm deep; internal holes for one small pipe; polypropylene hinged box cover; bedding in cement mortar (1:3)	-	4.49	123.26	67.92	nr	**191.18**

P BUILDING FABRIC SUNDRIES

Item	PC £	Labour hours	Labour £	Material £	Unit	Total rate £
P31 HOLES/CHASES/COVERS/SUPPORTS FOR SERVICES						
Builders' work for electrical installations; cutting away for and making good after electrician; including cutting or leaving all holes, notches, mortices, sinkings and chases, in both the structure and its coverings, for the following electrical points						
Exposed installation						
lighting points	-	0.32	5.13	-	nr	**5.13**
socket outlet points	-	0.54	8.94	-	nr	**8.94**
fitting outlet points	-	0.54	8.94	-	nr	**8.94**
equipment points or control gear points	-	0.75	12.74	-	nr	**12.74**
Concealed installation						
lighting points	-	0.43	6.98	-	nr	**6.98**
socket outlet points	-	0.75	12.74	-	nr	**12.74**
fitting outlet points	-	0.75	12.74	-	nr	**12.74**
equipment points or control gear points	-	1.06	17.64	-	nr	**17.64**
Builders' work for other services installations						
Cutting chases in brickwork						
for one pipe; not exceeding 55 mm nominal size; vertical	-	0.43	5.19	-	m	**5.19**
for one pipe; 55 mm - 110 mm nominal size; vertical	-	0.74	8.93	-	m	**8.93**
Cutting and pinning to brickwork or blockwork; ends of supports						
for pipes not exceeding 55 mm nominal size	-	0.21	5.76	-	nr	**5.76**
for cast iron pipes 55 mm - 110 mm nominal size	-	0.35	9.61	-	nr	**9.61**
Cutting or forming holes for pipes or the like; not exceeding 55 mm nominal size; making good						
reinforced concrete; not exceeding 100 mm deep	-	0.82	13.88	0.85	nr	**14.73**
reinforced concrete; 100 mm - 200 mm deep	-	1.25	21.16	1.24	nr	**22.41**
reinforced concrete; 200 mm - 300 mm deep	-	1.65	27.94	1.71	nr	**29.65**
half brick thick	-	0.35	5.38	-	nr	**5.38**
one brick thick	-	0.58	8.91	-	nr	**8.91**
one and a half brick thick	-	0.95	14.59	-	nr	**14.59**
100 mm blockwork	-	0.32	4.92	-	nr	**4.92**
140 mm blockwork	-	0.43	6.61	-	nr	**6.61**
215 mm blockwork	-	0.53	8.14	-	nr	**8.14**
plasterboard partition or suspended ceiling	-	0.38	5.84	-	nr	**5.84**
Cutting or forming holes for pipes or the like; 55 mm - 110 mm nominal size; making good						
reinforced concrete; not exceeding 100 mm deep	-	1.05	17.78	1.09	nr	**18.87**
reinforced concrete; 100 mm - 200 mm deep	-	1.60	27.09	1.65	nr	**28.74**
reinforced concrete; 200 mm - 300 mm deep	-	2.10	35.56	2.17	nr	**37.73**
half brick thick	-	0.43	6.61	-	nr	**6.61**
one brick thick	-	0.74	11.37	-	nr	**11.37**
one and a half brick thick	-	1.17	17.97	-	nr	**17.97**
100 mm blockwork	-	0.37	5.68	-	nr	**5.68**
140 mm blockwork	-	0.53	8.14	-	nr	**8.14**
215 mm blockwork	-	0.64	9.83	-	nr	**9.83**
plasterboard partition or suspended ceiling	-	0.44	6.76	-	nr	**6.76**
Cutting or forming holes for pipes or the like; over 110 mm nominal size; making good						
reinforced concrete; not exceeding 100 mm deep	-	1.25	21.16	1.29	nr	**22.46**
reinforced concrete; 100 mm - 200 mm deep	-	1.90	32.17	1.96	nr	**34.13**
reinforced concrete; 200 mm - 300 mm deep	-	2.45	41.48	2.54	nr	**44.02**
half brick thick	-	0.53	8.14	-	nr	**8.14**
one brick thick	-	0.91	13.98	-	nr	**13.98**

P BUILDING FABRIC SUNDRIES

Item	PC £	Labour hours	Labour £	Material £	Unit	Total rate £
one and a half brick thick	-	1.43	21.97	-	nr	**21.97**
100 mm blockwork	-	0.48	7.37	-	nr	**7.37**
140 mm blockwork	-	0.64	9.83	-	nr	**9.83**
215 mm blockwork	-	0.80	12.29	-	nr	**12.29**
plasterboard partition or suspended ceiling	-	0.49	7.53	-	nr	**7.53**
Add for making good fair face or facings one side						
pipe; not exceeding 55 mm nominal size	-	0.08	2.20	-	nr	**2.20**
pipe; 55 mm - 110 mm nominal size	-	0.11	3.02	-	nr	**3.02**
pipe; over 110 mm nominal size	-	0.13	3.57	-	nr	**3.57**
Add for fixing sleeve (supply not included)						
for pipe; small	-	0.16	4.39	-	nr	**4.39**
for pipe; large	-	0.21	5.76	-	nr	**5.76**
for pipe; extra large	-	0.32	8.78	-	nr	**8.78**
Add for supplying and fixing one-hour intumescent sleeve						
for pipe; small	-	0.28	4.74	18.77	nr	**23.51**
for pipe; large	-	0.31	5.25	22.95	nr	**28.20**
for pipe; extra large	-	0.34	5.76	47.27	nr	**53.03**
Cutting or forming holes for ducts; girth not exceeding 1.00 m; making good						
half brick thick	-	0.64	9.83	-	nr	**9.83**
one brick thick	-	1.06	16.28	-	nr	**16.28**
one and a half brick thick	-	1.70	26.12	-	nr	**26.12**
100 mm blockwork	-	0.53	8.14	-	nr	**8.14**
140 mm blockwork	-	0.74	11.37	-	nr	**11.37**
215 mm blockwork	-	0.95	14.59	-	nr	**14.59**
plasterboard partition or suspended ceiling	-	0.75	11.52	-	nr	**11.52**
Cutting or forming holes for ducts; girth 1.00 m - 2.00 m; making good						
half brick thick	-	0.74	11.37	-	nr	**11.37**
one brick thick	-	1.28	19.66	-	nr	**19.66**
one and a half brick thick	-	2.02	31.03	-	nr	**31.03**
100 mm blockwork	-	0.64	9.83	-	nr	**9.83**
140 mm blockwork	-	0.85	13.06	-	nr	**13.06**
215 mm blockwork	-	1.06	16.28	-	nr	**16.28**
plasterboard partition or suspended ceiling	-	0.85	13.06	-	nr	**13.06**
Cutting or forming holes for ducts; girth 2.00 m - 3.00 m; making good						
half brick thick	-	1.17	17.97	-	nr	**17.97**
one brick thick	-	2.02	31.03	-	nr	**31.03**
one and a half brick thick	-	3.19	49.01	-	nr	**49.01**
100 mm blockwork	-	1.01	15.52	-	nr	**15.52**
140 mm blockwork	-	1.38	21.20	-	nr	**21.20**
215 mm blockwork	-	1.76	27.04	-	nr	**27.04**
plasterboard partition or suspended ceiling	-	1.10	16.90	-	nr	**16.90**
Cutting or forming holes for ducts; girth 3.00 m - 4.00 m; making good						
half brick thick	-	1.60	24.58	-	nr	**24.58**
one brick thick	-	2.66	40.86	-	nr	**40.86**
one and a half brick thick	-	4.26	65.44	-	nr	**65.44**
100 mm blockwork	-	1.17	17.97	-	nr	**17.97**
140 mm blockwork	-	1.60	24.58	-	nr	**24.58**
215 mm blockwork	-	2.02	31.03	-	nr	**31.03**
plasterboard partition or suspended ceiling	-	1.35	20.74	-	nr	**20.74**
Mortices in brickwork						
for expansion bolt	-	0.21	3.23	-	nr	**3.23**
for 20 mm diameter bolt; 75 mm deep	-	0.16	2.46	-	nr	**2.46**
for 20 mm diameter bolt; 150 mm deep	-	0.27	4.15	-	nr	**4.15**

P BUILDING FABRIC SUNDRIES

Item	PC £	Labour hours	Labour £	Material £	Unit	Total rate £
P31 HOLES/CHASES/COVERS/SUPPORTS FOR SERVICES – cont'd						
Builders' work for other services installations – cont'd						
Mortices in brickwork; grouting with cement mortar (1:1)						
75 mm x 75 mm x 200 mm deep	-	0.32	4.92	-	nr	**4.92**
75 mm x 75 mm x 300 mm deep	-	0.43	6.61	-	nr	**6.61**
Holes in softwood for pipes, bars, cables and the like						
12 mm thick	-	0.04	0.88	-	nr	**0.88**
25 mm thick	-	0.06	1.32	-	nr	**1.32**
50 mm thick	-	0.11	2.42	-	nr	**2.42**
100 mm thick	-	0.16	3.51	-	nr	**3.51**
Holes in hardwood for pipes, bars, cables and the like						
12 mm thick	-	0.06	1.32	-	nr	**1.32**
25 mm thick	-	0.09	1.98	-	nr	**1.98**
50 mm thick	-	0.16	3.51	-	nr	**3.51**
100 mm thick	-	0.23	5.05	-	nr	**5.05**
NOTE: The following rates for cutting holes and mortices in brickwork or blockwork etc. allow for diamond drilling.						
Diamond drilling						
Cutting holes and mortices in brickwork; per 25 mm depth						
25 mm diameter	-	-	-	-	nr	**1.64**
32 mm diameter	-	-	-	-	nr	**1.33**
52 mm diameter	-	-	-	-	nr	**1.59**
78 mm diameter	-	-	-	-	nr	**1.74**
107 mm diameter	-	-	-	-	nr	**1.84**
127 mm diameter	-	-	-	-	nr	**2.25**
152 mm diameter	-	-	-	-	nr	**2.67**
200 mm diameter	-	-	-	-	nr	**3.43**
250 mm diameter	-	-	-	-	nr	**5.18**
300 mm diameter	-	-	-	-	nr	**6.87**
Diamond chasing; per 25 x 25 mm section						
in facing or common brickwork	-	-	-	-	m	**3.13**
in semi-engineering brickwork	-	-	-	-	m	**6.25**
in engineering brickwork	-	-	-	-	m	**8.71**
in lightweight blockwork	-	-	-	-	m	**2.46**
in heavtweight blockwork	-	-	-	-	m	**4.92**
in render/screed	-	-	-	-	m	**9.69**
Forming boxes; 100 x 100 mm; per 25 mm depth						
in facing or common brickwork	-	-	-	-	nr	**1.25**
in semi-engineering brickwork	-	-	-	-	nr	**2.50**
in engineering brickwork	-	-	-	-	nr	**3.48**
in lightweight blockwork	-	-	-	-	nr	**0.98**
in heavtweight blockwork	-	-	-	-	nr	**1.97**
in render/screed	-	-	-	-	nr	**3.87**
Other items						
track mounter or ring sawing in brickwork	-	-	-	-	m	**6.15**
floor sawing in apshalte	-	-	-	-	m	**1.02**
stitch drilling 107 mm diameter hole in brickwork	-	-	-	-	nr	**1.33**

P BUILDING FABRIC SUNDRIES

Item	PC £	Labour hours	Labour £	Material £	Unit	Total rate £
"SFD Screeduct" or other equal and approved; MDT Ducting Ltd; with side flanges; laid within floor screed; galvanised mild steel						
Floor ducting						
100 mm wide x 50 mm deep	13.68	0.21	4.61	16.21	m	**20.83**
Extra for						
bend	10.26	0.11	2.42	12.16	nr	**14.58**
tee section	10.26	0.11	2.42	12.16	nr	**14.58**
connector / stop end	0.68	0.11	2.42	0.81	nr	**3.23**
ply cover 15 mm/16 mm thick WBP exterior grade	2.00	0.11	2.42	2.36	m	**4.78**
200 mm wide x 75 mm deep	18.24	0.21	4.61	21.62	m	**26.23**
Extra for						
bend	13.68	0.11	2.42	16.21	nr	**18.63**
tee section	13.68	0.11	2.42	16.21	nr	**18.63**
connector / stop end	0.91	0.12	2.64	1.08	nr	**3.72**
ply cover 15 mm/16 mm thick WBP exterior grade	6.55	0.11	2.42	7.77	m	**10.19**

Q PAVING/PLANTING/FENCING/SITE FURNITURE

Item	PC £	Labour hours	Labour £	Material £	Unit	Total rate £
Q10 KERBS/EDGINGS/CHANNELS/PAVING ACCESSORIES						
Excavating; by machine						
Excavating trenches; to receive kerb foundations; average size						
300 mm x 100 mm	-	0.02	0.26	0.47	m	**0.73**
450 mm x 150 mm	-	0.02	0.26	0.93	m	**1.19**
600 mm x 200 mm	-	0.04	0.53	1.30	m	**1.83**
Excavating curved trenches; to receive kerb foundations; average size						
300 mm x 100 mm	-	0.01	0.13	0.75	m	**0.88**
450 mm x 150 mm	-	0.03	0.39	1.12	m	**1.51**
600 mm x 200 mm	-	0.05	0.66	1.40	m	**2.05**
Excavating; by hand						
Excavating trenches; to receive kerb foundations; average size						
150 mm x 50 mm	-	0.02	0.26	-	m	**0.26**
200 mm x 75 mm	-	0.07	0.92	-	m	**0.92**
250 mm x 100 mm	-	0.12	1.58	-	m	**1.58**
300 mm x 100 mm	-	0.15	1.97	-	m	**1.97**
Excavating curved trenches; to receive kerb foundations; average size						
150 mm x 50 mm	-	0.04	0.53	-	m	**0.53**
200 mm x 75 mm	-	0.08	1.05	-	m	**1.05**
250 mm x 100 mm	-	0.13	1.71	-	m	**1.71**
300 mm x 100 mm	-	0.16	2.10	-	m	**2.10**
Plain in situ ready mixed designated concrete; C7.5 - 40 mm aggregate; poured on or against earth or unblinded hardcore						
Foundations	81.06	1.33	22.52	98.37	m³	**120.89**
Blinding beds						
not exceeding 150 mm thick	81.06	1.97	33.35	98.37	m³	**131.72**
Plain in situ ready mixed designated concrete; C10 - 40 mm aggregate; poured on or against earth or unblinded hardcore						
Foundations	82.26	1.33	22.52	99.82	m³	**122.34**
Blinding beds						
not exceeding 150 mm thick	82.26	1.97	33.35	99.82	m³	**133.18**
Plain in situ ready mixed designated concrete; C20 - 20 mm aggregate; poured on or against earth or unblinded hardcore						
Foundations	85.85	1.33	22.52	104.18	m³	**126.70**
Blinding beds						
not exceeding 150 mm thick	85.85	1.97	33.35	104.18	m³	**137.53**
Precast concrete kerbs, channels, edgings, etc.; BS 340; bedded, jointed and pointed in cement mortar (1:3); including haunching up one side with in situ ready mixed designated concrete C10 - 40 mm aggregate; to concrete base						
Edgings; straight; square edge, fig 12						
50 mm x 150 mm	-	0.27	6.11	3.53	m	**9.63**
50 mm x 200 mm	-	0.27	6.11	4.57	m	**10.68**
50 mm x 255 mm	-	0.27	6.11	4.85	m	**10.96**

Q PAVING/PLANTING/FENCING/SITE FURNITURE

Item	PC £	Labour hours	Labour £	Material £	Unit	Total rate £
Kerbs; straight						
125 mm x 255 mm; fig 7	-	0.35	7.92	6.14	m	**14.06**
150 mm x 305 mm; fig 6	-	0.35	7.92	11.36	m	**19.27**
Kerbs; curved						
125 mm x 255 mm; fig 7	-	0.53	11.99	8.22	m	**20.20**
150 mm x 305 mm; fig 6	-	0.53	11.99	19.46	m	**31.45**
Channels; 255 mm x 125 mm; fig 8						
straight	-	0.35	7.92	6.14	m	**14.06**
curved	-	0.53	11.99	8.22	m	**20.20**
Quadrants; fig 14						
305 mm x 305 mm x 150 mm	-	0.37	8.37	11.23	nr	**19.60**
305 mm x 305 mm x 255 mm	-	0.37	8.37	11.23	nr	**19.60**
457 mm x 457 mm x 150 mm	-	0.43	9.73	12.42	nr	**22.14**
457 mm x 457 mm x 255 mm	-	0.43	9.73	12.42	nr	**22.14**
Q20 HARDCORE/GRANULAR/CEMENT BOUND BASES/SUB-BASES TO ROADS/PAVINGS						
Filling to make up levels; by machine						
Average thickness not exceeding 0.25 m						
obtained off site; hardcore	-	0.32	4.20	32.72	m³	**36.93**
obtained off site; granular fill type one	-	0.32	4.20	40.18	m³	**44.38**
obtained off site; granular fill type two	-	0.32	4.20	37.91	m³	**42.11**
Average thickness exceeding 0.25 m						
obtained off site; hardcore	-	0.28	3.68	28.00	m³	**31.68**
obtained off site; granular fill type one	-	0.28	3.68	39.86	m³	**43.54**
obtained off site; granular fill type two	-	0.28	3.68	37.59	m³	**41.27**
Filling to make up levels; by hand						
Average thickness not exceeding 0.25 m						
obtained off site; hardcore	-	0.70	9.20	34.95	m³	**44.14**
obtained off site; sand	-	0.82	10.77	48.03	m³	**58.81**
Average thickness exceeding 0.25 m						
obtained off site; hardcore	-	0.58	7.62	29.85	m³	**37.47**
obtained off site; sand	-	0.69	9.07	47.29	m³	**56.35**
Surface treatments						
Compacting						
filling; blinding with sand	-	0.05	0.66	2.35	m²	**3.01**
Q21 IN SITU CONCRETE ROADS/PAVINGS						
Reinforced in situ ready mixed designated concrete; C10 - 40 mm aggregate						
Roads; to hardcore base						
thickness not exceeding 150 mm	78.35	2.17	36.74	95.07	m³	**131.81**
thickness 150 mm - 450 mm	78.35	1.53	25.90	95.07	m³	**120.97**
Reinforced in situ ready mixed designated concrete; C20 - 20 mm aggregate						
Roads; to hardcore base						
thickness not exceeding 150 mm	81.76	2.17	36.74	99.21	m³	**135.95**
thicness 150 mm - 450 mm	81.76	1.53	25.90	99.21	m³	**125.12**
Reinforced in situ ready mixed designated concrete; C25 - 20 mm aggregate						
Roads; to hardcore base						
thickness not exceeding 150 mm	83.00	2.17	36.74	100.71	m³	**137.45**
thickness 150 mm - 450 mm	83.00	1.53	25.90	100.71	m³	**126.62**

Q PAVING/PLANTING/FENCING/SITE FURNITURE

Item	PC £	Labour hours	Labour £	Material £	Unit	Total rate £
Q21 IN SITU CONCRETE ROADS/PAVINGS – cont'd						
Formwork; sides of foundations; basic finish						
Plain vertical						
height not exceeding 250 mm	-	0.44	8.78	1.98	m	**10.76**
height 250 mm - 500 mm	-	0.66	13.17	3.29	m	**16.46**
height 500 mm - 1.00 m	-	0.95	18.96	6.31	m	**25.27**
add to above for curved radius 6.00 m	-	0.04	0.80	0.30	m	**1.10**
Reinforcement; fabric; BS 4449; lapped; in roads, footpaths or pavings						
Ref A142 (2.22 kg/m²)						
400 mm minimum laps	1.77	0.13	2.59	2.20	m²	**4.79**
Ref A193 (3.02 kg/m²)						
400 mm minimum laps	2.41	0.13	2.59	2.99	m²	**5.59**
Formed joints; Fosroc Expandite "Flexcell" impregnated joint filler or other equal and approved						
Width not exceeding 150 mm						
12.50 mm thick	-	0.16	3.19	2.18	m	**5.38**
25 mm thick	-	0.21	4.19	3.25	m	**7.44**
Width 150 - 300 mm						
12.50 mm thick	-	0.21	4.19	3.48	m	**7.67**
25 mm thick	-	0.21	4.19	5.51	m	**9.70**
Width 300 - 450 mm						
12.50 mm thick	-	0.27	5.39	5.22	m	**10.61**
25 mm thick	-	0.27	5.39	8.27	m	**13.66**
Sealants; Fosroc Expandite "Pliastic N2" hot poured rubberized bituminous compound or other equal and approved						
Width 25 mm						
25 mm depth	-	0.22	4.39	2.17	m	**6.56**
Concrete sundries						
Treating surfaces of unset concrete; grading to cambers; tamping with a 75 mm thick steel shod tamper	-	0.27	4.57	-	m²	**4.57**
Q22 COATED MACADAM/ASPHALT ROADS/ PAVINGS						
NOTE: The prices for all bitumen macadam and hot rolled asphalt materials are for individual courses to roads and footpaths and need combining to arrive at complete specifications and costs for full construction. Intermediate course thicknesses can be interpolated so long as BS 594987 allows the material type to be compacted to the required thickness.						
Costs include for work to falls, crossfalls or slopes not exceeding 15 degrees from horizontal; for laying on prepared bases (prices not included) and for rolling with an appropriate roller. The following rates are based on black bitumen macadam. Red bitumen macadam rates are approximately 50% dearer. PSV is Polished Stone Value.						

Q PAVING/PLANTING/FENCING/SITE FURNITURE

Item	PC £	Labour hours	Labour £	Material £	Unit	Total rate £
Dense Bitumen macadam base course; BS 594987 - 1; bitumen penetration 100/125						
Carriageway, hardshoulder and hardstop						
100 mm thick; one coat; with 0/32 mm aggregate size; to clause 5.2	-	-	-	-	m²	**18.05**
200 mm thick; one coat; with 0/32 mm aggregate size; to clause 5.2	-	-	-	-	m²	**30.05**
Extra over above items for increase / reduction in 10 mm increments	-	-	-	-	m²	**1.20**
Hot rolled asphalt base course; BS 594987 - 1						
Carriageway, hardshoulder and hardstop						
150 mm thick; one coat; 60% 0/32 mm aggregate size; to column 2/5	-	-	-	-	m²	**28.25**
200 mm thick; one coat; 60% 0/32 mm aggregate size; to column 2/5	-	-	-	-	m²	**37.66**
Extra over above items for increase / reduction in 10 mm increments	-	-	-	-	m²	**1.54**
Dense Bitumen macadam binder course; BS 594987 - 1; bitumen penetration 100/125						
Carriageway, hardshoulder and hardstop						
60 mm thick; one coat; with 0/20 mm aggregate size; to clause 6.4	-	-	-	-	m²	**11.55**
60 mm thick; one coat; with 0/32 mm aggregate size; to clause 6.5	-	-	-	-	m²	**11.65**
Extra over above items for increase / reduction in 10 mm increments	-	-	-	-	m²	**1.36**
Hot rolled asphalt binder course; BS 594987 - 1						
Carriageway, hardshoulder and hardstop						
40 mm thick; one coat; 50% 0/14 aggregate size; to column 2/2; 55 PSV	-	-	-	-	m²	**11.42**
60 mm thick; one coat; 50% 0/14 aggregate size; to column 2/2	-	-	-	-	m²	**12.57**
60 mm thick; one coat; 50% 0/20 aggregate size; to column 2/2	-	-	-	-	m²	**12.36**
60 mm thick; one coat; 60% 0/32 aggregate size; to column 2/2	-	-	-	-	m²	**11.81**
100 mm thick; one coat; 60% 0/32 aggregate size; to column 2/2	-	-	-	-	m²	**18.32**
Extra over above items for increase / reduction in 10 mm increments	-	-	-	-	m²	**2.02**
Macadam surface course; BS 594987 - 1; bitumen penetration 100/125						
Carriageway, hardshoulder and hardstop						
30 mm thick; one coat; medium graded with 0/6 mm nominal aggregate binder; to clause 7.6	-	-	-	-	m²	**9.55**
40 mm thick; one coat; close graded with 0/14 mm nominal aggregate binder; to clause 7.3	-	-	-	-	m²	**8.75**
40 mm thick; one coat; close graded with 0/10 mm nominal aggregate binder; to clause 7.4	-	-	-	-	m²	**9.55**
Extra over above items for						
increase / reduction in 10 mm increments	-	-	-	-	m²	**1.42**
coarse aggregate 60-64 PSV	-	-	-	-	m²	**1.44**
coarse aggregate 65-67 PSV	-	-	-	-	m²	**1.57**
coarse aggregate 68 PSV	-	-	-	-	m²	**2.09**

Q PAVING/PLANTING/FENCING/SITE FURNITURE

Item	PC £	Labour hours	Labour £	Material £	Unit	Total rate £
Q22 COATED MACADAM/ASPHALT ROADS/ PAVINGS – cont'd						
Hot rolled asphalt surface course; BS 594987 - 1; bitumen penetration 40/60						
Carriageway, hardshoulder and hardstop						
40 mm thick; one coat; 30% mix 0/10 aggregate size; to column 3/2; with 20 mm pre-coated chippings 60-64 PSV	-	-	-	-	m²	**11.73**
40 mm thick; one coat; 30% mix 0/10 aggregate size; to column 3/2; with 14 mm pre-coated chippings 60-64 PSV	-	-	-	-	m²	**11.83**
Extra over above items for increase / reduction in 10 mm increments	-	-	-	-	m²	**1.63**
Extra over above items for coarse aggregate 65-67 PSV	-	-	-	-	m²	**0.06**
Extra over above items for coarse aggregate 68 PSV	-	-	-	-	m²	**0.14**
Extra over above items for 6-10KN High Traffic Flows	-	-	-	-	m²	**0.71**
Stone mastic asphalt surface course; BS 594987 - 1						
Carriageway, hardshoulder and hardstop						
35 mm thick; one coat; with 0/14 mm nominal aggregate size; 55 PSV	-	-	-	-	m²	**10.25**
35 mm thick; one coat; with 0/10 mm nominal aggregate size; 55 PSV	-	-	-	-	m²	**10.25**
Extra over above items for increase / reduction in 10 mm increments	-	-	-	-	m²	**2.01**
Thin surface course with 60 PSV						
Carriageway, hardshoulder and hardstop						
35 mm thick; one coat; with 0/10 mm nominal aggregate size	-	-	-	-	m²	**10.25**
Extra over above items for increase / reduction in 10 mm increments	-	-	-	-	m²	**1.01**
Extra over above items for coarse aggregate 60-64 PSV	-	-	-	-	m²	**0.25**
Extra over above items for coarse aggregate 65-67 PSV	-	-	-	-	m²	**0.25**
Extra over above items for coarse aggregate 68 PSV	-	-	-	-	m²	**0.46**
Regulating courses						
Carriageway, hardshoulder and hardstop						
Dense Bitumen Macadam; bitumen penetration 100/125; with 0/20 mm nominal aggregate regulating course (BS 594987 - clause 6.5)	-	-	-	-	tonne	**95.22**
Hot rolled asphalte; 50% 0/20 aggregate size (BS 594987 - 1:2003 column 2/3)	-	-	-	-	tonne	**102.19**
Stone mastic asphalte; 0/6 mm aggregate	-	-	-	-	tonne	**129.79**
Bitumen Emulsion tack coats						
Carriageway, hardshoulder and hardstop						
K1-40; applied 0.35-0.45l/m²	-	-	-	-	m²	**0.15**
K1-70; applied 0.35-0.45l/m²	-	-	-	-	m²	**0.24**

Q PAVING/PLANTING/FENCING/SITE FURNITURE

Item	PC £	Labour hours	Labour £	Material £	Unit	Total rate £
Q23 GRAVEL/HOGGIN ROADS/PAVINGS						
Two coat gravel paving; level and to falls; first layer course clinker aggregate and wearing layer fine gravel aggregate						
Pavings; over 300 mm wide						
50 mm thick	-	0.08	1.81	6.37	m²	**8.18**
63 mm thick	-	0.10	2.26	8.25	m²	**10.51**
Resin bonded gravel paving; level and to falls						
Pavings; over 300 mm wide						
50 mm thick	-	-	-	-	m²	**48.80**
Q25 SLAB/BRICK/BLOCK/SETT/COBBLE PAVINGS						
Artificial stone paving; Charcon 'Moordale Textured'; or other equal and approved; to falls or crossfalls; bedding 25 mm thick in cement mortar (1:3); staggered joints; jointing in coloured cement mortar (1:3), brushed in; to sand base						
Pavings; over 300 mm wide						
600 mm x 600 mm x 50 mm thick; natural	13.61	0.44	9.95	19.64	m²	**29.59**
Brick paviors; 215 mm x 103 mm x 65 mm rough stock bricks; to falls or crossfalls; bedding 10 mm thick in cement mortar (1:3); jointing in cement mortar (1:3); as work proceeds; to concrete base						
Pavings; over 300 mm wide; straight joints both ways						
bricks laid flat (PC £ per 1000)	465.00	0.85	23.33	25.13	m²	**48.46**
bricks laid on edge	-	1.19	32.67	37.26	m²	**69.93**
Pavings; over 300 mm wide; laid to herringbone pattern						
bricks laid flat	-	1.06	29.10	25.13	m²	**54.23**
bricks laid on edge	-	1.49	40.90	37.26	m²	**78.17**
Add or deduct for variation of £10.00/1000 in PC of brick paviours						
bricks laid flat	-	-	-	0.51	m²	-
bricks laid on edge	-	-	-	0.77	m²	-
River washed cobble paving; 50 mm - 75 mm; to falls or crossfalls; bedding 13 mm thick in cement mortar (1:3); jointing to a height of two thirds of cobbles in dry mortar (1:3); tightly butted, washed and brushed; to concrete						
Pavings; over 300 mm wide						
regular (PC £ per tonne)	139.92	4.26	96.36	34.31	m²	**130.67**
laid to pattern	-	5.32	120.34	34.31	m²	**154.65**

Q PAVING/PLANTING/FENCING/SITE FURNITURE

Item	PC £	Labour hours	Labour £	Material £	Unit	Total rate £
Q25 SLAB/BRICK/BLOCK/SETT/COBBLE PAVINGS – cont'd						
Concrete paving flags; BS EN 1339; to falls or crossfalls; bedding 25 mm thick in cement and sand mortar (1:4); butt joints straight both ways; jointing in cement and sand (1:3); brushed in; to sand base						
Pavings; over 300 mm wide						
450 mm x 600 mm x 50 mm thick; grey	8.07	0.48	10.86	11.26	m²	**22.12**
450 mm x 600 mm x 60 mm thick; coloured	8.95	0.48	10.86	12.32	m²	**23.17**
600 mm x 600 mm x 50 mm thick; grey	6.25	0.44	9.95	9.11	m²	**19.06**
600 mm x 600 mm x 50 mm thick; coloured	7.50	0.44	9.95	10.59	m²	**20.54**
750 mm x 600 mm x 50 mm thick; grey	5.60	0.42	9.50	8.34	m²	**17.84**
750 mm x 600 mm x 50 mm thick; coloured	7.44	0.42	9.50	10.53	m²	**20.03**
900 mm x 600 mm x 50 mm thick; grey	5.00	0.38	8.60	7.63	m²	**16.23**
900 mm x 600 mm x 50 mm thick; coloured	6.85	0.38	8.60	9.83	m²	**18.42**
Concrete rectangular paving blocks; to falls or crossfalls; bedding 50 mm thick in dry sharp sand; filling joints with sharp sand brushed in; on earth base						
Pavings; 'Keyblock' or other equal and approved; over 300 mm wide; straight joints both ways						
200 mm x 100 mm x 65 mm thick; grey	8.39	0.80	18.10	14.00	m²	**32.10**
200 mm x 100 mm x 65 mm thick; coloured	9.12	0.80	18.10	14.86	m²	**32.96**
200 mm x 100 mm x 80 mm thick; grey	9.35	0.85	19.23	15.47	m²	**34.70**
200 mm x 100 mm x 80 mm thick; coloured	10.56	0.85	19.23	16.91	m²	**36.13**
Pavings; 'Keyblock' or other equal and approved; over 300 mm wide; laid to herringbone pattern						
200 mm x 100 mm x 60 mm thick; grey	-	1.00	22.62	14.00	m²	**36.62**
200 mm x 100 mm x 60 mm thick; coloured	-	1.00	22.62	14.86	m²	**37.48**
200 mm x 100 mm x 80 mm thick; grey	-	1.06	23.98	15.47	m²	**39.45**
200 mm x 100 mm x 80 mm thick; coloured	-	1.06	23.98	16.91	m²	**40.88**
Extra for two row boundary edging to herringbone paved areas; 200 wide; including a 150 mm high ready mixed designated concrete C10 - 40 mm aggregate haunching to one side; blocks laid breaking joint						
200 mm x 100 mm x 65 mm; coloured	-	0.32	7.24	2.66	m	**9.90**
200 mm x 100 mm x 80 mm; coloured	-	0.32	7.24	2.78	m	**10.02**
Pavings; 'Europa' or other equal and approved; over 300 mm wide; straight joints both ways						
200 mm x 100 mm x 60 mm thick; grey	7.34	0.80	18.10	12.75	m²	**30.85**
200 mm x 100 mm x 60 mm thick; coloured	8.15	0.80	18.10	13.71	m²	**31.81**
200 mm x 100 mm x 80 mm thick; grey	8.75	0.85	19.23	14.76	m²	**33.99**
200 mm x 100 mm x 80 mm thick; coloured	9.60	0.85	19.23	15.77	m²	**35.00**
Pavings; 'Pedesta' or other equal and approved; over 300 mm wide; straight joints both ways						
200 mm x 100 mm x 60 mm thick; grey	16.57	0.80	18.10	23.70	m²	**41.79**
200 mm x 100 mm x 60 mm thick; coloured	16.57	0.80	18.10	23.70	m²	**41.79**
200 mm x 100 mm x 80 mm thick; grey	21.13	0.85	19.23	29.44	m²	**48.66**
200 mm x 100 mm x 80 mm thick; coloured	21.13	0.85	19.23	29.44	m²	**48.66**
Pavings; 'Intersett' or other equal and approved; over 300 mm wide; straight joints both ways						
200 mm x 100 mm x 60 mm thick; grey	12.14	0.80	18.10	18.44	m²	**36.54**
200 mm x 100 mm x 60 mm thick; coloured	13.48	0.80	18.10	20.03	m²	**38.13**
200 mm x 100 mm x 80 mm thick; grey	14.52	0.85	19.23	21.60	m²	**40.83**
200 mm x 100 mm x 80 mm thick; coloured	16.12	0.85	19.23	23.50	m²	**42.73**

Q PAVING/PLANTING/FENCING/SITE FURNITURE

Item	PC £	Labour hours	Labour £	Material £	Unit	Total rate £
Concrete rectangular paving blocks; to falls or crossfalls; 6 mm wide joints; symmetrical layout; bedding in 15 mm semi-dry cement mortar (1:4); jointing and pointing in cement:sand (1:4); on concrete base						
Pavings; 'Trafica' or other equal and approved; over 300 mm wide						
400 mm x 400 mm x 65 mm; Saxon textured; natural	23.86	0.51	11.54	29.82	m²	**41.36**
400 mm x 400 mm x 65 mm; Saxon textured; buff	27.54	0.51	11.54	34.19	m²	**45.72**
400 mm x 400 mm x 65 mm; Perfecta; natural	28.61	0.51	11.54	35.45	m²	**46.98**
400 mm x 400 mm x 65 mm; Perfecta; buff	33.07	0.51	11.54	40.73	m²	**52.27**
450 mm x 450 mm x 70 mm; Saxon textured; natural	24.43	0.49	11.08	30.50	m²	**41.58**
450 mm x 450 mm x 70 mm; Saxon textured; buff	28.08	0.49	11.08	34.82	m²	**45.90**
450 mm x 450 mm x 70 mm; Perfecta; natural	27.74	0.49	11.08	34.42	m²	**45.50**
450 mm x 450 mm x 70 mm; Perfecta; buff	32.17	0.49	11.08	39.67	m²	**50.75**
York stone slab pavings; to falls or crossfalls; bedding 25 mm thick in cement and sand mortar (1:4); 5 wide joints; jointing in coloured cement mortar (1:3); brushed in; to sand base						
Pavings; over 300 mm wide						
50 mm thick; random rectangular pattern	115.43	0.80	21.96	135.63	m²	**157.59**
600 mm x 600 mm x 50 mm thick	131.52	0.44	12.08	154.24	m²	**166.32**
600 mm x 900 mm x 50 mm thick	139.92	0.38	10.43	163.96	m²	**174.39**
Granite setts; BS EN 1342; 200 mm x 100 mm x 100 mm; standard "C" dressing; tightly butted to falls or crossfalls; bedding 25 mm thick in cement mortar (1:3); filling joints with dry mortar (1:6); washed and brushed; on concrete base						
Pavings; over 300 mm wide						
straight joints (PC £ per tonne)	251.86	1.70	38.45	82.34	m²	**120.79**
laid to pattern	-	2.13	48.18	82.34	m²	**130.52**
Two rows of granite setts as boundary edging; 200 mm wide; including a 150 mm high in situ concrete mix 10.00 N/mm² - 40 mm aggregate (1:3.6); haunching to one side; blocks laid breaking joint	-	0.74	16.74	18.32	m	**35.06**
Q26 SPECIAL SURFACINGS/PAVINGS FOR SPORT						
Sundries						
Line marking						
width not exceeding 300 mm	-	0.05	0.81	0.20	m	**1.01**
Q30 SEEDING/TURFING						
Top soil						
Selected from spoil heaps; grading; prepared for turfing or seeding; to general surfaces						
average 75 mm thick	-	0.23	3.02	-	m²	**3.02**
average 100 mm thick	-	0.25	3.28	-	m²	**3.28**
average 125 mm thick	-	0.27	3.55	-	m²	**3.55**
average 150 mm thick	-	0.29	3.81	-	m²	**3.81**
average 175 mm thick	-	0.30	3.94	-	m²	**3.94**
average 200 mm thick	-	0.31	4.07	-	m²	**4.07**

Q PAVING/PLANTING/FENCING/SITE FURNITURE

Item	PC £	Labour hours	Labour £	Material £	Unit	Total rate £
Q30 SEEDING/TURFING – cont'd						
Top soil – cont'd						
Selected from spoil heaps; grading; prepared for turfing or seeding; to cuttings or embankments						
average 75 mm thick	-	0.27	3.55	-	m²	**3.55**
average 100 mm thick	-	0.29	3.81	-	m²	**3.81**
average 125 mm thick	-	0.30	3.94	-	m²	**3.94**
average 150 mm thick	-	0.31	4.07	-	m²	**4.07**
average 175 mm thick	-	0.33	4.34	-	m²	**4.34**
average 200 mm thick	-	0.36	4.73	-	m²	**4.73**
Imported recycled top soil						
Grading; prepared for turfing or seeding; to general surfaces						
average 75 mm thick	-	0.21	2.76	2.26	m²	**5.01**
average 100 mm thick	-	0.23	3.02	2.93	m²	**5.95**
average 125 mm thick	-	0.24	3.15	4.29	m²	**7.44**
average 150 mm thick	-	0.26	3.42	5.64	m²	**9.05**
average 175 mm thick	-	0.27	3.55	6.31	m²	**9.86**
average 200 mm thick	-	0.29	3.81	6.99	m²	**10.80**
Grading; preparing for turfing or seeding; to cuttings or embankments						
average 75 mm thick	-	0.23	3.02	2.26	m²	**5.28**
average 100 mm thick	-	0.26	3.42	2.93	m²	**6.35**
average 125 mm thick	-	0.27	3.55	4.29	m²	**7.83**
average 150 mm thick	-	0.29	3.81	5.64	m²	**9.45**
average 175 mm thick	-	0.30	3.94	6.31	m²	**10.26**
average 200 mm thick	-	0.31	4.07	6.99	m²	**11.06**
Fertilizer						
Fertilizer 0.07 kg/m²; raking in						
general surfaces (PC £ per 25kg)	16.67	0.03	0.39	0.05	m²	**0.45**
Selected grass seed						
Grass seed; sowing at a rate of 0.042 kg/m² two applications; raking in						
general surfaces (PC £ per 25kg)	107.86	0.07	0.92	0.41	m²	**1.33**
cuttings or embankments	-	0.08	1.05	0.41	m²	**1.46**
Preserved turf from stack on site						
Turfing						
general surfaces	-	0.20	2.63	-	m²	**2.63**
cuttings or embankments; shallow	-	0.22	2.89	-	m²	**2.89**
cuttings or embankments; steep; pegged	-	0.31	4.07	-	m²	**4.07**
Imported turf; cultivated						
Turfing						
general surfaces	2.84	0.20	2.63	3.21	m²	**5.84**
cuttings or embankments; shallow	2.84	0.22	2.89	3.21	m²	**6.10**
cuttings or embankments; steep; pegged	2.84	0.31	4.07	3.21	m²	**7.28**
Q31 PLANTING						
Planting only						
Hedge plants						
height not exceeding 750 mm	-	0.26	3.42	-	nr	**3.42**
height 750 mm - 1.50 m	-	0.61	8.01	-	nr	**8.01**
Saplings						
height not exceeding 3.00 m	-	1.73	22.73	-	nr	**22.73**

Q PAVING/PLANTING/FENCING/SITE FURNITURE

Item	PC £	Labour hours	Labour £	Material £	Unit	Total rate £
Q40 FENCING						
NOTE: The prices for all fencing include for setting posts in position, to a depth of 0.60 m for fences not exceeding 1.40 m high and of 0.76 m for fences over 1.40 m high. The prices allow for excavating post holes; filling to within 150 mm of ground level with concrete and all necessary backfilling.						
Strained wire fences; BS 1722 Part 3; 4 mm diameter galvanized mild steel plain wire threaded through posts and strained with eye bolts						
Fencing; height 900 mm; three line; concrete posts at 2750 mm centres	-	-	-	-	m	**22.59**
Extra for						
end concrete straining post; one strut	-	-	-	-	nr	**54.94**
angle concrete straining post; two struts	-	-	-	-	nr	**63.82**
Fencing; height 1.07 m; six line; concrete posts at 2750 mm centres	-	-	-	-	m	**23.51**
Extra for						
end concrete straining post; one strut	-	-	-	-	nr	**61.79**
angle concrete straining post; two struts	-	-	-	-	nr	**70.66**
Fencing; height 1.20 m; six line; concrete posts at 2750 mm centres	-	-	-	-	m	**23.64**
Extra for						
end concrete straining post; one strut	-	-	-	-	nr	**63.53**
angle concrete straining post; two struts	-	-	-	-	nr	**72.38**
Fencing; height 1.40 m; eight line; concrete posts at 2750 mm centres	-	-	-	-	m	**24.30**
Extra for						
end concrete straining post; one strut	-	-	-	-	nr	**64.91**
angle concrete straining post; two struts	-	-	-	-	nr	**73.76**
Chain link fencing; BS 1722 Part 1; 3 mm diameter galvanized mild steel wire; 50 mm mesh; galvanized mild steel tying and line wire; three line wires threaded through posts and strained with eye bolts and winding brackets						
Fencing; height 900 mm; galvanized mild steel angle posts at 3.00 m centres	-	-	-	-	m	**31.78**
Extra for						
end steel straining post; one strut	-	-	-	-	nr	**91.60**
angle steel straining post; two struts	-	-	-	-	nr	**105.64**
Fencing; height 900 mm; concrete posts at 3.00 m centres	-	-	-	-	m	**23.03**
Extra for						
end concrete straining post; one strut	-	-	-	-	nr	**49.21**
angle concrete straining post; two struts	-	-	-	-	nr	**58.07**
Fencing; height 1.20 m; galvanized mild steel angle posts at 3.00 m centres	-	-	-	-	m	**23.44**
Extra for						
end steel straining post; one strut	-	-	-	-	nr	**97.74**
angle steel straining post; two struts	-	-	-	-	nr	**125.17**
Fencing; height 1.20 m; concrete posts at 3.00 m centres	-	-	-	-	m	**22.46**
Extra for						
end concrete straining post; one strut	-	-	-	-	nr	**56.32**
angle concrete straining post; two struts	-	-	-	-	nr	**66.53**

Q PAVING/PLANTING/FENCING/SITE FURNITURE

Item	PC £	Labour hours	Labour £	Material £	Unit	Total rate £
Q40 FENCING – cont'd						
Chain link fencing; BS 1722 Part 1; 3 mm diameter galvanized mild steel wire – cont'd						
Fencing; height 1.80 m; galvanized mild steel angle posts at 3.00 m centres	-	-	-	-	m	**26.36**
Extra for						
end steel straining post; one strut	-	-	-	-	nr	**99.41**
angle steel straining post; two struts	-	-	-	-	nr	**123.67**
Fencing; height 1.80 m; concrete posts at 3.00 m centres	-	-	-	-	m	**30.43**
Extra for						
end concrete straining post; one strut	-	-	-	-	nr	**78.75**
angle concrete straining post; two struts	-	-	-	-	nr	**92.91**
Pair of gates and gate posts; gates to match galvanized chain link fencing, with angle framing, braces, etc., complete with hinges, locking bar, lock and bolts; two 100 mm x 100 mm angle section gate posts; each with one strut						
2.44 m x 0.90 m	-	-	-	-	nr	**772.41**
2.44 m x 1.20 m	-	-	-	-	nr	**797.15**
2.44 m x 1.80 m	-	-	-	-	nr	**859.90**
Chain link fencing; BS 1722 Part 1; 3 mm diameter plastic coated mild steel wire; 50 mm mesh; plastic coated mild steel tying and line wire; three line wires threaded through posts and strained with eye bolts and winding brackets						
Fencing; height 900 mm; galvanized mild steel angle posts at 3.00 m centres	-	-	-	-	m	**29.15**
Extra for						
end steel straining post; one strut	-	-	-	-	nr	**80.77**
angle steel straining post; two struts	-	-	-	-	nr	**89.91**
Fencing; height 900 mm; concrete posts at 3.00 m centres	-	-	-	-	m	**21.74**
Extra for						
end concrete straining post; one strut	-	-	-	-	nr	**49.21**
angle concrete straining post; two struts	-	-	-	-	nr	**58.07**
Fencing; height 1.20 m; galvanized mild steel angle posts at 3.00 m centres	-	-	-	-	m	**21.53**
Extra for						
end steel straining post; one strut	-	-	-	-	nr	**84.72**
angle steel straining post; two struts	-	-	-	-	nr	**90.51**
Fencing; height 1.20 m; concrete posts at 3.00 m centres	-	-	-	-	m	**22.03**
Extra for						
end concrete straining post; one strut	-	-	-	-	nr	**56.32**
angle concrete straining post; two struts	-	-	-	-	nr	**66.53**
Fencing; height 1.80 m; galvanized mild steel angle posts at 3.00 m centres	-	-	-	-	m	**24.41**
Extra for						
end steel straining post; one strut	-	-	-	-	nr	**83.85**
angle steel straining post; two struts	-	-	-	-	nr	**100.15**
Fencing; height 1.80 m; concrete posts at 3.00 m centres	-	-	-	-	m	**28.07**
Extra for						
end concrete straining post; one strut	-	-	-	-	nr	**78.75**
angle concrete straining post; two struts	-	-	-	-	nr	**92.91**

Q PAVING/PLANTING/FENCING/SITE FURNITURE

Item	PC £	Labour hours	Labour £	Material £	Unit	Total rate £
Pair of gates and gate posts; gates to match plastic chain link fencing; with angle framing, braces, etc. complete with hinges, locking bar, lock and bolts; two 100 mm x 100 mm angle section gate posts; each with one strut						
2.44 m x 0.90 m	-	-	-	-	nr	**675.40**
2.44 m x 1.20 m	-	-	-	-	nr	**693.01**
2.44 m x 1.80 m	-	-	-	-	nr	**746.40**
Chain link fencing for tennis courts; BS 1722 Part 13; 2.50 mm diameter galvanised mild wire; 45 mm mesh; line and tying wires threaded through 45 mm x 45 mm x 5 mm galvanised mild steel angle standards, posts and struts; 60 mm x 60 mm x 6 mm straining posts and gate posts; straining posts and struts strained with eye bolts and winding brackets						
Fencing to tennis court 36.00 m x 18.00 m; including gate 1.07 mm x 1.98 m complete with hinges, locking bar, lock and bolts						
height 2745 mm; standards at 3.00 m centres	-	-	-	-	nr	**3326.97**
height 3660 mm; standards at 2.50 m centres	-	-	-	-	nr	**4501.60**
Cleft chestnut pale fencing; BS 1722 Part4; pales spaced 51 mm apart; on two lines of galvanized wire; 64 mm diameter posts; 76 mm x 51 mm struts						
Fencing; height 900 mm; posts at 2.50 m centres	-	-	-	-	m	**13.11**
Extra for						
straining post; one strut	-	-	-	-	nr	**34.93**
corner straining post; two struts	-	-	-	-	nr	**34.93**
Fencing; height 1.05 m; posts at 2.50 m centres	-	-	-	-	m	**14.85**
Extra for						
straining post; one strut	-	-	-	-	nr	**35.40**
corner straining post; two struts	-	-	-	-	nr	**35.40**
Close boarded fencing; BS 1722 Part 5; 76 mm x 38 mm softwood rails; 89 mm x 19 mm softwood pales lapped 13 mm; 152 mm x 25 mm softwood gravel boards; all softwood "treated"; posts at 3.00 m centres						
Fencing; two rail; concrete posts						
1.00 m	-	-	-	-	m	**40.30**
1.20 m	-	-	-	-	m	**40.68**
Fencing; three rail; concrete posts						
1.40 m	-	-	-	-	m	**44.74**
1.60 m	-	-	-	-	m	**44.85**
1.80 m	-	-	-	-	m	**46.56**
Precast concrete slab fencing; 305 mm x 38 mm x 1753 mm slabs; fitted into twice grooved concrete posts at 1830 mm centres						
Fencing						
height 1.50 m	-	-	-	-	m	**79.00**
height 1.80 m	-	-	-	-	m	**87.49**

Q PAVING/PLANTING/FENCING/SITE FURNITURE

Item	PC £	Labour hours	Labour £	Material £	Unit	Total rate £
Q40 FENCING – cont'd						
Mild steel unclimbable fencing; in rivetted panels 2440 mm long; 44 mm x 13 mm flat section top and bottom rails; two 44 mm x 19 mm flat section standards; one with foot plate, and 38 mm x 13 mm raking stay with foot plate; 20 mm diameter pointed verticals at 120 mm centres; two 44 mm x 19 mm supports 760 mm long with ragged ends to bottom rail; the whole bolted together; coated with red oxide primer; setting standards and stays in ground at 2440 mm centres and supports at 815 mm centres						
Fencing						
height 1.67 m	-	-	-	-	m	**155.10**
height 2.13 m	-	-	-	-	m	**177.25**
Pair of gates and gate posts, to match mild steel unclimbable fencing; with flat section framing, braces, etc., complete with locking bar, lock, handles, drop bolt, gate stop and holding back catches; two 102 mm x 102 mm hollow section gate posts with cap and foot plates						
2.44 m x 1.67 m	-	-	-	-	nr	**1364.54**
2.44 m x 2.13 m	-	-	-	-	nr	**1596.67**
4.88 m x 1.67 m	-	-	-	-	nr	**2110.12**
4.88 m x 2.13 m	-	-	-	-	nr	**2672.82**
PVC coated, galvanised mild steel high security fencing; "Sentinel Sterling fencing" or other equal and approved; Twil Wire Products Ltd; 50 mm x 50 mm mesh; 3 mm/3.50 mm gauge wire; barbed edge - 1; "Sentinal Bi-steel" colour coated posts at 2440 mm centres						
Fencing						
height 1.80 m	-	1.02	13.40	50.63	m	**64.03**
height 2.10 m	-	1.02	13.40	55.92	m	**69.32**

R DISPOSAL SYSTEMS

Item	PC £	Labour hours	Labour £	Material £	Unit	**Total rate £**
R10 RAINWATER PIPEWORK/GUTTERS						
Aluminium pipes and fittings; BS EN 612; ears cast on; polyester powder coated finish						
63 mm diameter pipes; plugged and screwed	15.33	0.41	7.79	19.07	m	**26.85**
Extra for						
fittings with one end	-	0.24	4.56	9.88	nr	**14.44**
fittings with two ends	-	0.46	8.74	10.25	nr	**18.99**
fittings with three ends	-	0.67	12.73	13.52	nr	**26.25**
shoe	9.38	0.24	4.56	9.88	nr	**14.44**
bend	10.01	0.46	8.74	10.25	nr	**18.99**
single branch	13.05	0.67	12.73	13.52	nr	**26.25**
offset 228 mm projection	23.08	0.46	10.10	24.09	nr	**34.20**
offset 304 mm projection	25.73	0.46	8.74	27.16	nr	**35.90**
access pipe	28.51	0.46	8.74	28.50	nr	**37.24**
connection to clay pipes; cement and sand (1:2) joint	-	0.17	3.23	0.12	nr	**3.35**
76.50 mm diameter pipes; plugged and screwed	17.85	0.44	8.36	22.07	m	**30.43**
Extra for						
shoe	12.88	0.28	5.32	13.70	nr	**19.02**
bend	12.64	0.50	9.50	13.09	nr	**22.59**
single branch	15.71	0.72	13.68	16.28	nr	**29.96**
offset 228 mm projection	25.52	0.50	9.50	26.42	nr	**35.91**
offset 304 mm projection	28.23	0.50	9.50	29.56	nr	**39.05**
access pipe	31.15	0.50	9.50	30.82	nr	**40.32**
connection to clay pipes; cement and sand (1:2) joint	-	0.19	3.61	0.12	nr	**3.73**
100 mm diameter pipes; plugged and screwed	30.47	0.50	9.50	37.05	m	**46.55**
Extra for						
shoe	15.51	0.32	6.08	15.57	nr	**21.65**
bend	17.61	0.56	10.64	17.69	nr	**28.33**
single branch	21.05	0.83	15.77	20.74	nr	**36.51**
offset 228 mm projection	29.52	0.56	10.64	28.12	nr	**38.75**
offset 304 mm projection	32.77	0.56	10.64	31.88	nr	**42.52**
access pipe	36.92	0.56	10.64	33.79	nr	**44.43**
connection to clay pipes; cement and sand (1:2) joint	-	0.22	4.18	0.12	nr	**4.30**
Roof outlets; circular aluminium; with flat or domed grating; joint to pipe						
50 mm diameter	70.21	0.67	20.87	79.25	nr	**100.13**
75 mm diameter	93.17	0.72	22.43	105.17	nr	**127.60**
100 mm diameter	130.69	0.78	24.30	147.52	nr	**171.82**
150 mm diameter	162.56	0.83	25.86	183.49	nr	**209.35**
Roof outlets; d-shaped; balcony; with flat or domed grating; joint to pipe						
50 mm diameter	83.15	0.67	20.87	93.86	nr	**114.73**
75 mm diameter	95.60	0.72	22.43	107.91	nr	**130.34**
100 mm diameter	117.42	0.78	24.30	132.54	nr	**156.84**
Galvanized wire balloon grating; BS 416 for pipes or outlets						
50 mm diameter	1.38	0.07	2.18	1.56	nr	**3.74**
63 mm diameter	1.40	0.07	2.18	1.58	nr	**3.76**
75 mm diameter	1.49	0.07	2.18	1.68	nr	**3.86**
100 mm diameter	1.64	0.09	2.80	1.85	nr	**4.65**

R DISPOSAL SYSTEMS

Item	PC £	Labour hours	Labour £	Material £	Unit	Total rate £
R10 RAINWATER PIPEWORK/GUTTERS – cont'd						
Aluminium gutters and fittings; BS EN 612; polyester powder coated finish						
100 mm half round gutters; on brackets; screwed to timber	14.06	0.39	8.57	20.76	m	**29.33**
Extra for						
stop end	3.81	0.18	3.95	8.27	nr	**12.22**
running outlet	8.45	0.37	8.13	9.53	nr	**17.66**
stop end outlet	7.51	0.18	3.95	11.22	nr	**15.17**
angle	7.81	0.37	8.13	7.34	nr	**15.47**
113 mm half round gutters; on brackets; screwed to timber	14.73	0.39	8.57	21.60	m	**30.16**
Extra for						
stop end	4.00	0.18	3.95	8.53	nr	**12.49**
running outlet	9.21	0.37	8.13	10.36	nr	**18.48**
stop end outlet	8.62	0.18	3.95	12.44	nr	**16.39**
angle	8.78	0.37	8.13	8.29	nr	**16.42**
125 mm half round gutters; on brackets; screwed to timber	16.55	0.44	9.66	25.95	m	**35.61**
Extra for						
stop end	4.88	0.20	4.39	11.55	nr	**15.94**
running outlet	9.96	0.39	8.57	11.06	nr	**19.62**
stop end outlet	9.15	0.20	4.39	14.92	nr	**19.31**
angle	9.76	0.39	8.57	11.02	nr	**19.59**
100 mm ogee gutters; on brackets; screwed to timber	17.54	0.41	9.01	26.72	m	**35.73**
Extra for						
stop end	4.02	0.19	4.17	5.46	nr	**9.64**
running outlet	9.91	0.39	8.57	10.20	nr	**18.76**
stop end outlet	7.68	0.19	4.17	12.69	nr	**16.86**
angle	8.35	0.39	8.57	6.57	nr	**15.14**
112 mm ogee gutters; on brackets; screwed to timber	19.51	0.46	10.10	29.41	m	**39.52**
Extra for						
stop end	4.30	0.19	4.17	5.79	nr	**9.96**
running outlet	10.02	0.39	8.57	10.15	nr	**18.71**
stop end outlet	8.60	0.19	4.17	13.89	nr	**18.06**
angle	9.95	0.39	8.57	8.03	nr	**16.60**
125 mm ogee gutters; on brackets; screwed to timber	21.55	0.46	10.10	32.35	m	**42.45**
Extra for						
stop end	4.70	0.21	4.61	6.27	nr	**10.88**
running outlet	10.96	0.41	9.01	11.04	nr	**20.05**
stop end outlet	9.76	0.21	4.61	15.51	nr	**20.13**
angle	11.60	0.41	9.01	9.55	nr	**18.56**
Cast iron pipes and fittings; BS 416; ears cast on; joints						
65 mm pipes; primed; nailed to masonry	22.77	0.57	10.83	27.45	m	**38.27**
Extra for						
shoe	19.99	0.35	6.65	21.38	nr	**28.03**
bend	12.23	0.63	11.97	12.41	nr	**24.38**
single branch	24.05	0.80	15.20	25.21	nr	**40.41**
offset 225 mm projection	21.81	0.63	11.97	21.87	nr	**33.84**
offset 305 mm projection	25.53	0.63	11.97	25.64	nr	**37.61**
connection to clay pipes; cement and sand (1:2) joint	-	0.17	3.23	0.14	nr	**3.37**

R DISPOSAL SYSTEMS

Item	PC £	Labour hours	Labour £	Material £	Unit	Total rate £
75 mm pipes; primed; nailed to masonry	22.77	0.61	11.59	27.69	m	**39.28**
Extra for						
shoe	19.99	0.39	7.41	21.47	nr	**28.87**
bend	14.85	1.33	27.08	15.52	nr	**42.60**
single branch	26.50	0.83	15.77	28.22	nr	**43.98**
offset 225 mm projection	21.81	0.67	12.73	21.95	nr	**34.68**
offset 305 mm projection	26.79	0.67	12.73	27.18	nr	**39.91**
connection to clay pipes; cement and sand (1:2) joint	-	0.19	3.61	0.14	nr	**3.75**
100 mm pipes; primed; nailed to masonry	30.58	0.67	12.73	37.22	m	**49.94**
Extra for						
shoe	26.53	0.44	8.36	28.45	nr	**36.81**
bend	20.98	0.72	13.68	22.03	nr	**35.71**
single branch	30.90	0.89	16.91	32.51	nr	**49.41**
offset 225 mm projection	42.79	0.72	13.68	45.09	nr	**58.76**
offset 305 mm projection	43.62	0.72	13.68	45.33	nr	**59.01**
connection to clay pipes; cement and sand (1:2) joint	-	0.22	4.18	0.12	nr	**4.30**
100 mm x 75 mm rectangular pipes; primed; nailing to masonry	61.50	0.67	12.73	73.86	m	**86.59**
Extra for						
shoe	74.85	0.44	8.36	81.43	nr	**89.79**
bend	71.27	0.72	13.68	77.28	nr	**90.96**
offset 225 mm projection	100.38	0.44	8.36	106.59	nr	**114.94**
offset 305 mm projection	107.27	0.44	8.36	113.10	nr	**121.46**
connection to clay pipes; cement and sand (1:2) joint	-	0.22	4.18	0.12	nr	**4.30**
Rainwater head; rectangular; for pipes						
65 mm diameter	61.42	0.63	11.97	71.48	nr	**83.45**
75 mm diameter	61.42	0.67	12.73	71.57	nr	**84.29**
100 mm diameter	84.82	0.72	13.68	98.79	nr	**112.46**
Rainwater head; octagonal; for pipes						
65 mm diameter	37.53	0.63	11.97	43.84	nr	**55.81**
75 mm diameter	37.53	0.67	12.73	43.92	nr	**56.65**
100 mm diameter	46.62	0.72	13.68	54.60	nr	**68.27**
Copper wire balloon grating; BS 416 for pipes or outlets						
50 mm diameter	1.68	0.07	1.33	1.90	nr	**3.23**
63 mm diameter	2.17	0.07	1.33	2.45	nr	**3.78**
75 mm diameter	1.70	0.07	1.33	1.92	nr	**3.25**
100 mm diameter	1.91	0.09	1.71	2.16	nr	**3.87**
Cast iron gutters and fittings; BS EN 877						
100 mm half round gutters; primed; on brackets; screwed to timber	11.70	0.44	9.66	17.71	m	**27.38**
Extra for						
stop end	2.94	0.19	4.17	5.21	nr	**9.38**
running outlet	8.55	0.39	8.57	8.79	nr	**17.35**
angle	8.78	0.39	8.57	10.90	nr	**19.46**
115 mm half round gutters; primed; on brackets; screwed to timber	12.19	0.44	9.66	18.33	m	**28.00**
Extra for						
stop end	3.82	0.19	4.17	6.20	nr	**10.37**
running outlet	9.31	0.39	8.57	9.61	nr	**18.18**
angle	9.03	0.39	8.57	11.09	nr	**19.65**
125 mm half round gutters; primed; on brackets; screwed to timber	14.27	0.50	10.98	20.80	m	**31.79**
Extra for						
stop end	3.82	0.22	4.83	6.20	nr	**11.04**
running outlet	10.65	0.44	9.66	10.92	nr	**20.58**
angle	10.65	0.44	9.66	12.50	nr	**22.16**

R DISPOSAL SYSTEMS

Item	PC £	Labour hours	Labour £	Material £	Unit	Total rate £
R10 RAINWATER PIPEWORK/GUTTERS – cont'd						
Cast iron gutters and fittings; BS EN 877 – cont'd						
150 mm half round gutters; primed; on brackets; screwed to timber	24.38	0.56	12.30	32.31	m	**44.61**
Extra for						
stop end	5.30	0.24	5.27	10.12	nr	**15.39**
running outlet	18.44	0.50	10.98	18.87	nr	**29.85**
angle	19.45	0.50	10.98	21.41	nr	**32.40**
100 mm ogee gutters; primed; on brackets; screwed to timber	13.04	0.46	10.10	19.56	m	**29.67**
Extra for						
stop end	3.02	0.20	4.39	7.01	nr	**11.40**
running outlet	9.32	0.41	9.01	9.54	nr	**18.55**
angle	9.15	0.41	9.01	11.31	nr	**20.31**
115 mm ogee gutters; primed; on brackets; screwed to timber	14.35	0.46	10.10	21.14	m	**31.25**
Extra for						
stop end	3.91	0.20	4.39	8.01	nr	**12.41**
running outlet	9.93	0.41	9.01	10.10	nr	**19.10**
angle	9.93	0.41	9.01	11.92	nr	**20.92**
125 mm ogee gutters; primed; on brackets; screwed to timber	15.05	0.52	11.42	22.44	m	**33.87**
Extra for						
stop end	3.91	0.23	5.05	8.45	nr	**13.50**
running outlet	10.84	0.46	10.10	11.07	nr	**21.17**
angle	10.84	0.46	10.10	13.24	nr	**23.34**
3 mm thick galvanised heavy pressed steel gutters and fittings; joggle joints; BS 1091						
200 mm x 100 mm (400 mm girth) box gutter; screwed to timber	-	0.72	13.68	31.10	m	**44.78**
Extra for						
stop end	-	0.39	7.41	17.31	nr	**24.72**
running outlet	-	0.78	14.82	28.69	nr	**43.51**
stop end outlet	-	0.39	7.41	40.17	nr	**47.58**
angle	-	0.78	14.82	31.86	nr	**46.67**
381 mm boundary wall gutters; 900 mm girth; screwed to timber	-	0.72	13.68	51.18	m	**64.86**
Extra for						
stop end	-	0.44	8.36	29.52	nr	**37.88**
running outlet	-	0.78	14.82	39.51	nr	**54.33**
stop end outlet	-	0.39	7.41	55.42	nr	**62.83**
angle	-	0.78	14.82	45.68	nr	**60.50**
457 mm boundary wall gutters; 1200 mm girth; screwed to timber	-	0.83	15.77	68.29	m	**84.05**
Extra for						
stop end	-	0.44	8.36	37.90	nr	**46.26**
running outlet	-	0.89	16.91	57.14	nr	**74.04**
stop end outlet	-	0.44	8.36	60.81	nr	**69.16**
angle	-	0.89	16.91	61.84	nr	**78.75**

R DISPOSAL SYSTEMS

Item	PC £	Labour hours	Labour £	Material £	Unit	Total rate £
uPVC external rainwater pipes and fittings; BS EN 12200; slip-in joints						
50 mm pipes; fixing with pipe or socket brackets; plugged and screwed	6.19	0.33	6.27	9.84	m	**16.11**
Extra for						
shoe	3.66	0.22	4.18	5.24	nr	**9.42**
bend	4.27	0.33	6.27	5.95	nr	**12.22**
two bends to form offset 229 mm projection	8.54	0.33	6.27	9.65	nr	**15.92**
connection to clay pipes; cement and sand (1:2) joint	-	0.15	2.85	0.14	nr	**2.99**
68 mm pipes; fixing with pipe or socket brackets; plugged and screwed	4.78	0.37	7.03	8.66	m	**15.69**
Extra for						
shoe	3.66	0.24	4.56	5.69	nr	**10.25**
bend	5.60	0.37	7.03	7.94	nr	**14.97**
single branch	11.26	0.49	9.31	14.48	nr	**23.79**
two bends to form offset 229 mm projection	11.20	0.37	7.03	13.45	nr	**20.48**
loose drain connector; cement and sand (1:2) joint	-	0.17	3.23	15.26	nr	**18.49**
110 mm pipes; fixing with pipe or socket brackets; plugged and screwed	9.60	0.40	7.60	19.35	m	**26.95**
Extra for						
shoe	11.69	0.27	5.13	15.24	nr	**20.37**
bend	17.34	0.40	7.60	21.78	nr	**29.38**
single branch	26.52	0.54	10.26	32.40	nr	**42.66**
two bends to form offset 229 mm projection	34.67	0.40	7.60	40.24	nr	**47.84**
loose drain connector; cement and sand (1:2) joint	-	0.39	7.41	12.64	nr	**20.05**
65 mm square pipes; fixing with pipe or socket brackets; plugged and screwed	4.80	0.37	7.03	8.68	m	**15.71**
Extra for						
shoe	3.66	0.24	4.56	5.69	nr	**10.25**
bend	5.60	0.37	7.03	7.94	nr	**14.97**
single branch	11.26	0.49	9.31	14.48	nr	**23.79**
two bends to form offset 229 mm projection	11.20	0.37	7.03	13.74	nr	**20.76**
drain connector; square to round; cement and sand (1:2) joint	-	0.39	7.41	6.49	nr	**13.90**
Rainwater head; rectangular; for pipes						
50 mm diameter	19.90	0.50	9.50	25.19	nr	**34.69**
68 mm diameter	16.07	0.52	9.88	21.65	nr	**31.53**
110 mm diameter	33.53	0.61	11.59	42.37	nr	**53.95**
65 mm square	16.07	0.52	9.88	21.65	nr	**31.53**
uPVC gutters and fittings; BS EN 12200						
76 mm half round gutters; on brackets screwed to timber	4.70	0.33	7.25	8.11	m	**15.36**
Extra for						
stop end	1.68	0.15	3.29	2.56	nr	**5.86**
running outlet	4.74	0.28	6.15	5.03	nr	**11.18**
stop end outlet	4.72	0.15	3.29	5.63	nr	**8.93**
angle	4.74	0.28	6.15	6.18	nr	**12.33**
112 mm half round gutters; on brackets screwed to timber	4.72	0.37	8.13	9.31	m	**17.43**
Extra for						
stop end	2.62	0.15	3.29	3.99	nr	**7.29**
running outlet	5.16	0.32	7.03	5.52	nr	**12.55**
stop end outlet	5.16	0.15	3.29	6.48	nr	**9.77**
angle	5.76	0.32	7.03	8.13	nr	**15.16**

R DISPOSAL SYSTEMS

Item	PC £	Labour hours	Labour £	Material £	Unit	Total rate £
R10 RAINWATER PIPEWORK/GUTTERS – cont'd						
uPVC gutters and fittings; BS EN 12200 – cont'd						
170 mm half round gutters; on brackets; screwed to timber	9.88	0.37	8.13	18.30	m	**26.43**
Extra for						
stop end	4.44	0.18	3.95	6.95	nr	**10.91**
running outlet	9.90	0.34	7.47	10.52	nr	**17.99**
stop end outlet	9.42	0.18	3.95	11.77	nr	**15.73**
angle	12.90	0.34	7.47	17.62	nr	**25.09**
114 mm rectangular gutters; on brackets; screwed to timber	4.84	0.37	8.13	10.20	m	**18.32**
Extra for						
stop end	2.62	0.15	3.29	3.99	nr	**7.29**
running outlet	5.16	0.34	7.47	5.52	nr	**12.99**
stop end outlet	5.16	0.15	3.29	6.48	nr	**9.77**
angle	6.20	0.32	7.03	8.64	nr	**15.67**
R11 FOUL DRAINAGE ABOVE GROUND						
Cast iron "Timesaver" pipes and fittings or other equal and approved; BS 416						
50 mm pipes; primed; 2.00 m lengths; fixing with expanding bolts; to masonry	15.02	0.61	11.59	28.43	m	**40.02**
Extra for						
fittings with two ends	-	0.61	11.59	22.18	nr	**33.76**
fittings with three ends	-	0.83	15.77	37.56	nr	**53.32**
bends; short radius	13.70	0.61	11.59	22.18	nr	**33.76**
access bends; short radius	33.75	0.61	11.59	45.38	nr	**56.96**
boss; 38 BSP	28.36	0.66	12.54	38.69	nr	**51.23**
single branch	20.60	0.83	15.77	38.36	nr	**54.12**
isolated "Timesaver" coupling joint	7.78	0.33	6.27	9.00	nr	**15.27**
connection to clay pipes; cement and sand (1:2) joint	-	0.15	2.85	0.12	nr	**2.97**
75 mm pipes; primed; 3.00 m lengths; fixing with standard brackets; plugged and screwed to masonry	16.80	0.61	11.59	31.61	m	**43.20**
Extra for						
bends; short radius	15.50	0.67	12.73	24.88	nr	**37.61**
access bends; short radius	36.61	0.61	11.59	49.31	nr	**60.89**
boss; 38 BSP	28.36	0.67	12.73	39.76	nr	**52.49**
single branch	23.33	0.94	17.86	42.78	nr	**60.63**
double branch	34.64	1.22	23.18	65.80	nr	**88.97**
offset 115 mm projection	22.23	0.67	12.73	30.38	nr	**43.10**
offset 150 mm projection	26.11	0.67	12.73	34.27	nr	**47.00**
access pipe	32.95	0.67	12.73	42.48	nr	**55.21**
isolated "Timesaver" coupling joint	8.59	0.39	7.41	9.93	nr	**17.34**
connection to clay pipes; cement and sand (1:2) joint	-	0.17	3.23	0.12	nr	**3.35**
100 mm pipes; primed; 3.00 m lengths; fixing with standard brackets; plugged and screwed to masonry	20.30	0.67	12.73	43.93	m	**56.66**
Extra for						
WC bent connector; 450 mm long tail	30.35	0.67	12.73	39.55	nr	**52.27**
bends; short radius	18.95	0.74	14.06	31.29	nr	**45.35**
access bends; short radius	40.10	0.74	14.06	55.76	nr	**69.82**
boss; 38 BSP	33.87	0.74	14.06	48.55	nr	**62.61**
single branch	29.30	1.11	21.09	54.20	nr	**75.28**
double branch	36.24	1.44	27.36	75.20	nr	**102.55**
offset 225 mm projection	28.54	0.74	14.06	39.26	nr	**53.31**
offset 300 mm projection	30.71	0.74	14.06	41.04	nr	**55.10**

R DISPOSAL SYSTEMS

Item	PC £	Labour hours	Labour £	Material £	Unit	Total rate £
access pipe	34.64	0.74	14.06	45.83	nr	**59.89**
roof connector; for asphalt	32.74	0.74	14.06	46.40	nr	**60.46**
isolated "Timesaver" coupling joint	11.21	0.46	8.74	12.97	nr	**21.71**
transitional clayware socket; cement and sand (1:2) joint	22.30	0.44	8.36	38.90	nr	**47.26**
150 mm pipes; primed; 3.00 m lengths; fixing with standard brackets; plugged and screwed to masonry	42.39	0.83	15.77	88.80	m	**104.57**
Extra for						
bends; short radius	33.87	0.93	17.67	57.53	nr	**75.20**
access bends; short radius	56.94	0.93	17.67	84.23	nr	**101.90**
boss; 38 BSP	55.26	0.93	17.67	81.03	nr	**98.69**
single branch	72.65	1.33	25.27	122.26	nr	**147.52**
double branch	102.08	1.78	33.81	179.94	nr	**213.75**
access pipe	57.62	0.93	17.67	74.76	nr	**92.43**
isolated "Timesaver" coupling joint	-	0.56	10.64	25.89	nr	**36.52**
transitional clayware socket; cement and sand (1:2) joint	39.05	0.57	10.83	71.19	nr	**82.02**
Cast iron "Ensign" lightweight pipes and fittings or other equal and approved; BS EN 877						
50 mm pipes; primed 3.00 m lengths; fixing with standard brackets; plugged and screwed to masonry	-	0.31	5.97	18.49	m	**24.46**
Extra for						
bends; short radius	-	0.27	5.21	15.34	nr	**20.55**
single branch	-	0.33	6.35	26.82	nr	**33.17**
access pipe	-	0.27	5.21	31.49	nr	**36.70**
70 mm pipes; primed 3.00 m lengths; fixing with standard brackets; plugged and screwed to masonry	-	0.34	6.51	20.63	m	**27.15**
Extra for						
bends; short radius	-	0.30	5.75	17.11	nr	**22.86**
single branch	-	0.37	7.11	28.80	nr	**35.91**
access pipe	-	0.30	5.75	33.55	nr	**39.31**
100 mm pipes; primed 3.00 m lengths; fixing with standard brackets; plugged and screwed to masonry	-	0.37	7.11	24.53	m	**31.64**
Extra for						
bends; short radius	-	0.32	6.19	20.99	nr	**27.18**
single branch	-	0.39	7.49	38.65	nr	**46.15**
double branch	-	0.46	8.85	54.28	nr	**63.13**
access pipe	-	0.32	6.19	38.15	nr	**44.34**
connector	-	0.21	4.02	35.43	nr	**39.45**
reducer	-	0.32	6.19	25.62	nr	**31.81**
Polypropylene (PP) waste pipes and fittings; BS EN 1451; push fit "O" - ring joints						
32 mm pipes; fixing with pipe clips; plugged and screwed	1.88	0.24	4.56	3.28	m	**7.83**
Extra for						
fittings with one end	-	0.18	3.42	1.81	nr	**5.23**
fittings with two ends	-	0.24	4.56	1.84	nr	**6.40**
fittings with three ends	-	0.33	6.27	3.19	nr	**9.46**
access plug	1.57	0.18	3.42	1.81	nr	**5.23**
double socket	1.20	0.17	3.23	1.39	nr	**4.62**
male iron to PP coupling	3.33	0.32	6.08	3.85	nr	**9.93**
sweep bend	1.49	0.24	4.56	1.72	nr	**6.28**
spigot bend	2.18	0.28	5.32	2.52	nr	**7.84**

R DISPOSAL SYSTEMS

Item	PC £	Labour hours	Labour £	Material £	Unit	Total rate £
R11 FOUL DRAINAGE ABOVE GROUND – cont'd						
Polypropylene (PP) waste pipes and fittings; BS EN 1451 – cont'd						
40 mm pipes; fixing with pipe clips; plugged and screwed	2.32	0.24	4.56	3.83	m	**8.39**
Extra for						
fittings with one end	-	0.21	3.99	1.90	nr	**5.89**
fittings with two ends	-	0.33	6.27	2.17	nr	**8.44**
fittings with three ends	-	0.44	8.36	3.36	nr	**11.72**
access plug	1.64	0.21	3.99	1.90	nr	**5.89**
double socket	1.23	0.22	4.18	1.43	nr	**5.60**
universal connector	3.76	0.28	5.32	4.35	nr	**9.67**
sweep bend	1.68	0.33	6.27	1.94	nr	**8.21**
spigot bend	2.11	0.33	6.27	2.44	nr	**8.71**
reducer 40 mm - 32 mm	1.49	0.33	6.27	1.72	nr	**7.99**
50 mm pipes; fixing with pipe clips; plugged and screwed	2.98	0.39	7.41	5.55	m	**12.96**
Extra for						
fittings with one end	-	0.23	4.37	3.37	nr	**7.74**
fittings with two ends	-	0.39	7.41	3.62	nr	**11.03**
fittings with three ends	-	0.52	9.88	5.02	nr	**14.90**
access plug	2.91	0.23	4.37	3.37	nr	**7.74**
double socket	2.46	0.26	4.94	2.84	nr	**7.78**
sweep bend	3.21	0.39	7.41	3.71	nr	**11.12**
spigot bend	5.02	0.39	7.41	5.80	nr	**13.21**
reducer 50 mm - 40 mm	1.94	0.39	7.41	2.24	nr	**9.65**
muPVC waste pipes and fittings; BS EN 1329; solvent welded joints						
32 mm pipes; fixing with pipe clips; plugged and screwed	2.05	0.28	5.32	3.62	m	**8.94**
Extra for						
fittings with one end	-	0.19	3.61	1.79	nr	**5.40**
fittings with two ends	-	0.28	5.32	1.92	nr	**7.24**
fittings with three ends	-	0.37	7.03	2.54	nr	**9.57**
access plug	1.18	0.19	3.61	1.79	nr	**5.40**
straight coupling	1.28	0.19	3.61	1.90	nr	**5.51**
expansion coupling	2.25	0.28	5.32	3.02	nr	**8.34**
male iron to muPVC coupling	2.27	0.43	8.17	2.84	nr	**11.01**
sweep bend	1.30	0.28	5.32	1.92	nr	**7.24**
spigot/socket bend	-	0.28	5.32	2.88	nr	**8.20**
sweep tee	1.74	0.37	7.03	2.54	nr	**9.57**
40 mm pipes; fixing with pipe clips; plugged and screwed	2.53	0.33	6.27	4.27	m	**10.54**
Extra for						
fittings with one end	-	0.21	3.99	1.79	nr	**5.78**
fittings with two ends	-	0.33	6.27	2.10	nr	**8.37**
fittings with three ends	-	0.44	8.36	3.08	nr	**11.44**
fittings with four ends	5.37	0.59	11.21	6.95	nr	**18.16**
access plug	1.18	0.21	3.99	1.79	nr	**5.78**
straight coupling	1.27	0.22	4.18	1.89	nr	**6.07**
expansion coupling	2.71	0.33	6.27	3.56	nr	**9.83**
male iron to muPVC coupling	2.27	0.43	8.17	2.84	nr	**11.01**
level invert taper	1.59	0.33	6.27	2.26	nr	**8.53**
sweep bend	1.45	0.33	6.27	2.10	nr	**8.37**
spigot/socket bend	2.44	0.33	6.27	3.24	nr	**9.51**
sweep tee	2.21	0.44	8.36	3.08	nr	**11.44**
sweep cross	5.37	0.59	11.21	6.95	nr	**18.16**

R DISPOSAL SYSTEMS

Item	PC £	Labour hours	Labour £	Material £	Unit	Total rate £
50 mm pipes; fixing with pipe clips; plugged and screwed	3.82	0.39	7.41	6.86	m	**14.26**
Extra for						
fittings with one end	-	0.23	4.37	2.39	nr	**6.76**
fittings with two ends	-	0.39	7.41	3.36	nr	**10.77**
fittings with three ends	-	0.52	9.88	5.52	nr	**15.39**
fittings with four ends	-	0.69	13.11	7.24	nr	**20.35**
access plug	1.70	0.23	4.37	2.39	nr	**6.76**
straight coupling	2.33	0.26	4.94	3.11	nr	**8.05**
expansion coupling	3.67	0.39	7.41	4.67	nr	**12.08**
male iron to muPVC coupling	3.28	0.50	9.50	4.01	nr	**13.50**
level invert taper	1.98	0.39	7.41	2.71	nr	**10.12**
sweep bend	2.54	0.39	7.41	3.36	nr	**10.77**
spigot/socket bend	3.46	0.39	7.41	4.43	nr	**11.84**
sweep tee	2.21	0.44	8.36	3.08	nr	**11.44**
sweep cross	5.62	0.69	13.11	7.24	nr	**20.35**
uPVC overflow pipes and fittings; solvent welded joints						
19 mm pipes; fixing with pipe clips; plug and screwed	1.21	0.24	4.56	2.42	m	**6.98**
Extra for						
splay cut end	-	0.02	0.38	-	nr	**0.38**
fittings with one end	-	0.19	3.61	1.70	nr	**5.31**
fittings with two ends	-	0.19	3.61	1.97	nr	**5.58**
fittings with three ends	-	0.24	4.56	2.23	nr	**6.79**
straight connector	1.29	0.19	3.61	1.70	nr	**5.31**
female iron to uPVC coupling	-	0.22	4.18	2.65	nr	**6.83**
bend	1.52	0.19	3.61	1.97	nr	**5.58**
bent tank connector	2.38	0.22	4.18	2.85	nr	**7.03**
uPVC pipes and fittings; BS EN 1329; with solvent welded joints (unless otherwise described)						
82 mm pipes; fixing with holderbats; plug and screwed	8.69	0.44	8.36	14.52	m	**22.88**
Extra for						
socket plug	6.56	0.22	4.18	8.64	nr	**12.82**
slip coupling; push fit	14.33	0.41	7.79	16.58	nr	**24.37**
expansion coupling	6.90	0.44	8.36	9.03	nr	**17.39**
sweep bend	11.58	0.44	8.36	14.45	nr	**22.80**
boss connector	6.33	0.30	5.70	8.37	nr	**14.07**
single branch	16.18	0.59	11.21	20.40	nr	**31.61**
access door	15.41	0.67	12.73	18.35	nr	**31.08**
110 mm pipes; fixing with holderbats; plugged and screwed	8.85	0.49	9.31	15.16	m	**24.46**
Extra for						
socket plug	7.95	0.24	4.56	10.57	nr	**15.13**
slip coupling; push fit	17.94	0.44	8.36	20.76	nr	**29.12**
expansion coupling	7.05	0.49	9.31	9.52	nr	**18.83**
W.C. connector	12.81	0.32	6.08	15.56	nr	**21.63**
sweep bend	13.55	0.49	9.31	17.05	nr	**26.36**
W.C. connecting bend	21.02	0.32	6.08	25.06	nr	**31.14**
access bend	37.58	0.51	9.69	44.84	nr	**54.53**
boss connector	6.33	0.32	6.08	8.69	nr	**14.77**
single branch	17.92	0.65	12.35	22.84	nr	**35.19**
single branch with access	30.68	0.67	12.73	37.60	nr	**50.33**
double branch	44.29	0.81	15.39	54.08	nr	**69.47**
W.C. manifold	17.59	0.32	6.08	22.46	nr	**28.54**
access door	-	0.67	12.73	18.35	nr	**31.08**
access pipe connector	28.78	0.56	10.64	34.67	nr	**45.31**
connection to clay pipes; caulking ring and cement and sand (1:2) joint	-	0.46	8.74	13.37	nr	**22.11**

R DISPOSAL SYSTEMS

Item	PC £	Labour hours	Labour £	Material £	Unit	Total rate £
R11 FOUL DRAINAGE ABOVE GROUND – cont'd						
uPVC pipes and fittings; BS EN 1329 – cont'd						
160 mm pipes; fixing with holderbats; plugged and screwed	22.96	0.56	10.64	39.03	m	**49.67**
Extra for						
socket plug	14.62	0.28	5.32	19.96	nr	**25.28**
slip coupling; push fit	45.94	0.50	9.50	53.15	nr	**62.65**
expansion coupling	21.22	0.56	10.64	27.61	nr	**38.25**
sweep bend	33.74	0.56	10.64	42.08	nr	**52.72**
boss connector	8.96	0.37	7.03	13.42	nr	**20.45**
single branch	38.04	0.79	15.01	47.90	nr	**62.91**
double branch	80.01	0.93	17.67	98.56	nr	**116.23**
access door	27.53	0.67	12.73	32.38	nr	**45.10**
access pipe connector	28.78	0.56	10.64	34.67	nr	**45.31**
Weathering apron; for pipe						
82 mmm diameter	3.26	0.38	7.22	4.30	nr	**11.52**
110 mm diameter	3.74	0.43	8.17	5.07	nr	**13.24**
160 mm diameter	11.26	0.46	8.74	14.51	nr	**23.24**
Weathering slate; for pipe						
110 mm diameter	39.78	1.00	19.00	46.76	nr	**65.75**
Vent cowl; for pipe						
82 mm diameter	3.26	0.37	7.03	4.30	nr	**11.33**
110 mm diameter	3.30	0.37	7.03	4.55	nr	**11.58**
160 mm diameter	8.63	0.37	7.03	11.46	nr	**18.49**
Polypropylene ancillaries; screwed joint to waste fitting						
Tubular "S" trap; bath; shallow seal						
40 mm diameter	8.13	0.61	11.59	9.40	nr	**20.99**
Trap; "P"; two piece; 76 mm seal						
32 mm diameter	5.49	0.43	8.17	6.35	nr	**14.52**
40 mm diameter	6.34	0.50	9.50	7.33	nr	**16.83**
Trap; "S"; two piece; 76 mm seal						
32 mm diameter	6.95	0.43	8.17	8.04	nr	**16.21**
40 mm diameter	8.13	0.50	9.50	9.40	nr	**18.90**
Bottle trap; "P"; 76 mm seal						
32 mm diameter	6.11	0.43	8.17	7.07	nr	**15.24**
40 mm diameter	7.29	0.45	8.55	8.43	nr	**16.98**
R12 DRAINAGE BELOW GROUND						
NOTE: Prices for drain trenches are for excavation in "firm" soil and it has been assumed that earthwork support will only be required for trenches 1.00 m or more in depth.						

R DISPOSAL SYSTEMS

Item	PC £	Labour hours	Labour £	Material £	Unit	Total rate £
Excavating trenches; by machine; grading bottoms; earthwork support; filling with excavated material and compacting; disposal of surplus soil on site; spreading on site average 50 m						
Pipes not exceeding 200 mm nominal size						
average depth of trench 0.50 m	-	0.30	3.94	2.51	m	**6.46**
average depth of trench 0.75 m	-	0.45	5.91	3.72	m	**9.64**
average depth of trench 1.00 m	-	0.89	11.69	7.57	m	**19.26**
average depth of trench 1.25 m	-	1.33	17.47	8.69	m	**26.17**
average depth of trench 1.50 m	-	1.71	22.47	9.91	m	**32.38**
average depth of trench 1.75 m	-	2.08	27.33	11.03	m	**38.36**
average depth of trench 2.00 m	-	2.46	32.32	12.53	m	**44.85**
average depth of trench 2.25 m	-	3.02	39.68	15.81	m	**55.48**
average depth of trench 2.50 m	-	3.57	46.91	18.42	m	**65.33**
average depth of trench 2.75 m	-	3.95	51.90	20.57	m	**72.47**
average depth of trench 3.00 m	-	4.31	56.63	22.63	m	**79.26**
average depth of trench 3.25 m	-	4.66	61.23	24.13	m	**85.36**
average depth of trench 3.50 m	-	4.99	65.56	25.54	m	**91.10**
Pipes exceeding 200 mm nominal size; 225 mm nominal size						
average depth of trench 0.50 m	-	0.30	3.94	2.51	m	**6.46**
average depth of trench 0.75 m	-	0.45	5.91	3.72	m	**9.64**
average depth of trench 1.00 m	-	0.89	11.69	7.57	m	**19.26**
average depth of trench 1.25 m	-	1.33	17.47	8.69	m	**26.17**
average depth of trench 1.50 m	-	1.71	22.47	9.91	m	**32.38**
average depth of trench 1.75 m	-	2.08	27.33	11.03	m	**38.36**
average depth of trench 2.00 m	-	2.46	32.32	12.53	m	**44.85**
average depth of trench 2.25 m	-	3.02	39.68	15.81	m	**55.48**
average depth of trench 2.50 m	-	3.57	46.91	18.42	m	**65.33**
average depth of trench 2.75 m	-	3.95	51.90	20.57	m	**72.47**
average depth of trench 3.00 m	-	4.31	56.63	22.63	m	**79.26**
average depth of trench 3.25 m	-	4.66	61.23	24.13	m	**85.36**
average depth of trench 3.50 m	-	4.99	65.56	25.54	m	**91.10**
Pipes exceeding 200 mm nominal size; 300 mm nominal size						
average depth of trench 0.75 m	-	0.51	6.70	4.66	m	**11.36**
average depth of trench 1.00 m	-	1.04	13.66	7.57	m	**21.23**
average depth of trench 1.25 m	-	1.42	18.66	8.97	m	**27.63**
average depth of trench 1.50 m	-	1.86	24.44	10.19	m	**34.63**
average depth of trench 1.75 m	-	2.16	28.38	11.31	m	**39.69**
average depth of trench 2.00 m	-	2.46	32.32	13.46	m	**45.78**
average depth of trench 2.25 m	-	3.02	39.68	16.36	m	**56.04**
average depth of trench 2.50 m	-	3.57	46.91	18.80	m	**65.70**
average depth of trench 2.75 m	-	3.95	51.90	20.85	m	**72.75**
average depth of trench 3.00 m	-	4.31	56.63	22.91	m	**79.54**
average depth of trench 3.25 m	-	4.66	61.23	25.06	m	**86.29**
average depth of trench 3.50 m	-	4.99	65.56	26.19	m	**91.75**
Extra over excavating trenches; irrespective of depth; breaking out materials						
brick	-	2.08	27.33	11.11	m³	**38.44**
concrete	-	2.91	38.23	15.28	m³	**53.51**
reinforced concrete	-	4.17	54.79	22.11	m³	**76.89**
Extra over excavating trenches; irrespective of depth; breaking out existing hard pavings; 75 mm thick						
tarmacadam	-	0.22	2.89	1.18	m²	**4.08**
Extra over excavating trenches; irrespective of depth; breaking out existing hard pavings; 150 mm thick						
concrete	-	0.45	5.91	2.55	m²	**8.46**
tarmacadam and hardcore	-	0.30	3.94	1.35	m²	**5.29**

R DISPOSAL SYSTEMS

Item	PC £	Labour hours	Labour £	Material £	Unit	Total rate £
R12 DRAINAGE BELOW GROUND – cont'd						
Excavating trenches; by hand; grading bottoms; earthwork support; filling with excavated material and compacting; disposal of surplus soil on site; spreading on site average 50 m						
Pipes not exceeding 200 mm nominal size; average depth						
average depth of trench 0.50 m	-	1.02	13.40	-	m	**13.40**
average depth of trench 0.75 m	-	1.53	20.10	-	m	**20.10**
average depth of trench 1.00 m	-	2.24	29.43	2.26	m	**31.69**
average depth of trench 1.25 m	-	3.15	41.39	3.10	m	**44.49**
average depth of trench 1.50 m	-	4.33	56.89	3.78	m	**60.67**
average depth of trench 1.75 m	-	5.70	74.89	4.52	m	**79.41**
average depth of trench 2.00 m	-	6.51	85.53	5.08	m	**90.61**
average depth of trench 2.25 m	-	8.14	106.95	6.77	m	**113.72**
average depth of trench 2.50 m	-	9.77	128.37	7.90	m	**136.27**
average depth of trench 2.75 m	-	10.74	141.11	8.75	m	**149.86**
average depth of trench 3.00 m	-	11.70	153.72	9.59	m	**163.32**
average depth of trench 3.25 m	-	12.67	166.47	10.44	m	**176.91**
average depth of trench 3.50 m	-	13.64	179.21	11.29	m	**190.50**
Pipes exceeding 200 mm nominal size; 225 mm nominal size						
average depth of trench 0.50 m	-	1.02	13.40	-	m	**13.40**
average depth of trench 0.75 m	-	1.53	20.10	-	m	**20.10**
average depth of trench 1.00 m	-	2.24	29.43	2.26	m	**31.69**
average depth of trench 1.25 m	-	3.15	41.39	3.10	m	**44.49**
average depth of trench 1.50 m	-	4.33	56.89	3.78	m	**60.67**
average depth of trench 1.75 m	-	5.70	74.89	4.52	m	**79.41**
average depth of trench 2.00 m	-	6.51	85.53	5.08	m	**90.61**
average depth of trench 2.25 m	-	8.14	106.95	6.77	m	**113.72**
average depth of trench 2.50 m	-	9.77	128.37	7.90	m	**136.27**
average depth of trench 2.75 m	-	10.74	141.11	8.75	m	**149.86**
average depth of trench 3.00 m	-	11.70	153.72	9.59	m	**163.32**
average depth of trench 3.25 m	-	12.67	166.47	10.44	m	**176.91**
average depth of trench 3.50 m	-	13.64	179.21	11.29	m	**190.50**
Pipes exceeding 200 mm nominal size; 300 mm nominal size						
average depth of trench 0.75 m	-	1.79	23.52	-	m	**23.52**
average depth of trench 1.00 m	-	2.60	34.16	2.26	m	**36.42**
average depth of trench 1.25 m	-	3.66	48.09	3.10	m	**51.19**
average depth of trench 1.50 m	-	4.88	64.12	3.78	m	**67.90**
average depth of trench 1.75 m	-	5.70	74.89	4.52	m	**79.41**
average depth of trench 2.00 m	-	6.51	85.53	5.08	m	**90.61**
average depth of trench 2.25 m	-	8.14	106.95	6.77	m	**113.72**
average depth of trench 2.50 m	-	9.77	128.37	7.90	m	**136.27**
average depth of trench 2.75 m	-	10.74	141.11	8.75	m	**149.86**
average depth of trench 3.00 m	-	11.74	154.25	9.59	m	**163.84**
average depth of trench 3.25 m	-	12.67	166.47	10.44	m	**176.91**
average depth of trench 3.50 m	-	13.64	179.21	11.29	m	**190.50**
Extra over excavating trenches; irrespective of depth; breaking out existing materials						
brick	-	3.05	40.07	8.07	m³	**48.14**
concrete	-	4.58	60.18	13.45	m³	**73.62**
reinforced concrete	-	6.11	80.28	18.84	m³	**99.11**
Extra over excavating trenches; irrespective of depth; breaking out existing hard pavings; 75 mm thick						
tarmacadam	-	0.41	5.39	1.08	m²	**6.47**

R DISPOSAL SYSTEMS

Item	PC £	Labour hours	Labour £	Material £	Unit	Total rate £
Extra over excavating trenches; irrespective of depth; breaking out existing hard pavings; 150 mm thick						
concrete	-	0.71	9.33	1.89	m²	**11.22**
tarmacadam and hardcore	-	0.51	6.70	1.34	m²	**8.04**
Extra for taking up						
precast concrete paving slabs	-	0.33	4.34	-	m²	**4.34**
Sand filling						
Beds; to receive pitch fibre pipes						
600 mm x 50 mm thick	-	0.08	1.05	1.27	m	**2.32**
700 mm x 50 mm thick	-	0.10	1.31	1.48	m	**2.79**
800 mm x 50 mm thick	-	0.12	1.58	1.69	m	**3.27**
Granular (shingle) filling						
Beds; 100 mm thick; to pipes						
100 mm nominal size	-	0.10	1.31	2.27	m	**3.59**
150 mm nominal size	-	0.10	1.31	2.65	m	**3.97**
225 mm nominal size	-	0.12	1.58	3.03	m	**4.61**
300 mm nominal size	-	0.14	1.84	3.41	m	**5.25**
Beds; 150 mm thick; to pipes						
100 mm nominal size	-	0.14	1.84	3.41	m	**5.25**
150 mm nominal size	-	0.17	2.23	3.79	m	**6.02**
225 mm nominal size	-	0.19	2.50	4.17	m	**6.67**
300 mm nominal size	-	0.20	2.63	4.55	m	**7.18**
Beds and benchings; beds 100 mm thick; to pipes						
100 mm nominal size	-	0.23	3.02	4.17	m	**7.19**
150 mm nominal size	-	0.26	3.42	4.17	m	**7.59**
225 mm nominal size	-	0.31	4.07	5.68	m	**9.76**
300 mm nominal size	-	0.36	4.73	6.44	m	**11.17**
Beds and benchings; beds 150 mm thick; to pipes						
100 mm nominal size	-	0.26	3.42	4.55	m	**7.96**
150 mm nominal size	-	0.29	3.81	4.93	m	**8.74**
225 mm nominal size	-	0.36	4.73	6.82	m	**11.55**
300 mm nominal size	-	0.45	5.91	8.34	m	**14.25**
Beds and coverings; 100 mm thick; to pipes						
100 mm nominal size	-	0.37	4.86	5.68	m	**10.54**
150 mm nominal size	-	0.45	5.91	6.82	m	**12.74**
225 mm nominal size	-	0.61	8.01	9.47	m	**17.49**
300 mm nominal size	-	0.73	9.59	11.37	m	**20.96**
Beds and coverings; 150 mm thick; to pipes						
100 mm nominal size	-	0.55	7.23	8.34	m	**15.56**
150 mm nominal size	-	0.61	8.01	9.47	m	**17.49**
225 mm nominal size	-	0.80	10.51	12.13	m	**22.64**
300 mm nominal size	-	0.94	12.35	14.40	m	**26.75**
Plain in situ ready mixed designated concrete; C10 - 40 mm aggregate						
Beds; 100 mm thick; to pipes						
100 mm nominal size	-	0.21	3.56	4.76	m	**8.31**
150 mm nominal size	-	0.21	3.56	4.76	m	**8.31**
225 mm nominal size	-	0.26	4.40	5.70	m	**10.11**
300 mm nominal size	-	0.29	4.91	6.66	m	**11.57**
Beds; 150 mm thick; to pipes						
100 mm nominal size	-	0.29	4.91	6.66	m	**11.57**
150 mm nominal size	-	0.33	5.59	7.61	m	**13.19**
225 mm nominal size	-	0.37	6.26	8.56	m	**14.82**
300 mm nominal size	-	0.42	7.11	9.51	m	**16.62**

R DISPOSAL SYSTEMS

Item	PC £	Labour hours	Labour £	Material £	Unit	Total rate £
R12 DRAINAGE BELOW GROUND – cont'd						
Plain in situ ready mixed designated concrete; C10 - 40 mm aggregate – cont'd						
Beds and benchings; beds 100 mm thick; to pipes						
100 mm nominal size	-	0.42	7.11	8.56	m	**15.67**
150 mm nominal size	-	0.47	7.96	9.51	m	**17.46**
225 mm nominal size	-	0.56	9.48	11.41	m	**20.89**
300 mm nominal size	-	0.66	11.17	13.31	m	**24.48**
Beds and benchings; beds 150 mm thick; to pipes						
100 mm nominal size	-	0.47	7.96	9.51	m	**17.46**
150 mm nominal size	-	0.52	8.80	10.45	m	**19.26**
225 mm nominal size	-	0.66	11.17	13.31	m	**24.48**
300 mm nominal size	-	0.84	14.22	17.11	m	**31.33**
Beds and coverings; 100 mm thick; to pipes						
100 mm nominal size	-	0.63	10.67	11.41	m	**22.07**
150 mm nominal size	-	0.73	12.36	13.31	m	**25.67**
225 mm nominal size	-	1.05	17.78	19.01	m	**36.79**
300 mm nominal size	-	1.25	21.16	22.82	m	**43.98**
Beds and coverings; 150 mm thick; to pipes						
100 mm nominal size	-	0.93	15.75	17.11	m	**32.86**
150 mm nominal size	-	1.05	17.78	19.01	m	**36.79**
225 mm nominal size	-	1.35	22.86	24.72	m	**47.57**
300 mm nominal size	-	1.62	27.43	29.47	m	**56.89**
Plain in situ ready mixed designated concrete; C20 - 40 mm aggregate						
Beds 100 mm thick; to pipes						
100 mm nominal size	-	0.21	3.56	4.97	m	**8.52**
150 mm nominal size	-	0.21	3.56	4.97	m	**8.52**
225 mm nominal size	-	0.26	4.40	5.95	m	**10.35**
300 mm nominal size	-	0.29	4.91	6.95	m	**11.86**
Beds; 150 mm thick; to pipes						
100 mm nominal size	-	0.29	4.91	6.95	m	**11.86**
150 mm nominal size	-	0.33	5.59	7.94	m	**13.52**
225 mm nominal size	-	0.37	6.26	8.93	m	**15.20**
300 mm nominal size	-	0.42	7.11	9.92	m	**17.03**
Beds and benchings; beds 100 mm thick; to pipes						
100 mm nominal size	-	0.42	7.11	8.93	m	**16.04**
150 mm nominal size	-	0.47	7.96	9.92	m	**17.88**
225 mm nominal size	-	0.56	9.48	11.91	m	**21.39**
300 mm nominal size	-	0.66	11.17	13.89	m	**25.06**
Beds and benchings; beds 150 mm thick; to pipes						
100 mm nominal size	-	0.47	7.96	9.92	m	**17.88**
150 mm nominal size	-	0.52	8.80	10.91	m	**19.71**
225 mm nominal size	-	0.66	11.17	13.89	m	**25.06**
300 mm nominal size	-	0.84	14.22	17.86	m	**32.08**
Beds and coverings; 100 mm thick; to pipes						
100 mm nominal size	-	0.63	10.67	11.91	m	**22.57**
150 mm nominal size	-	0.73	12.36	13.89	m	**26.25**
225 mm nominal size	-	1.05	17.78	19.84	m	**37.62**
300 mm nominal size	-	1.25	21.16	23.81	m	**44.98**
Beds and coverings; 150 mm thick; to pipes						
100 mm nominal size	-	0.93	15.75	17.86	m	**33.60**
150 mm nominal size	-	1.05	17.78	19.84	m	**37.62**
225 mm nominal size	-	1.35	22.86	25.80	m	**48.65**
300 mm nominal size	-	1.62	27.43	30.75	m	**58.18**

R DISPOSAL SYSTEMS

Item	PC £	Labour hours	Labour £	Material £	Unit	Total rate £
NOTE: The following items unless otherwise described include for all appropriate joints/couplings in the running length. The prices for gullies and rainwater shoes, etc. include for appropriate joints to pipes and for setting on and surrounding accessory with site mixed in situ concrete 10.00 N/mm² - 40 mm aggregate (1:3:6)						
Cast iron "Timesaver" drain pipes and fittings or other equal and approved; BS 437; coated; with mechanical coupling joints						
100 mm pipes; laid straight	26.96	0.53	6.90	38.75	m	**45.65**
100 mm pipes; in runs not exceeding 3.00 m long	26.96	0.72	9.37	51.94	m	**61.31**
Extra for						
bend; medium radius	32.31	0.64	8.33	52.58	nr	**60.91**
bend; medium radius with access	89.78	0.64	8.33	119.08	nr	**127.41**
bend; long radius	53.40	0.64	8.33	75.38	nr	**83.71**
rest bend	37.06	0.64	8.33	56.48	nr	**64.81**
single branch	42.88	0.80	10.41	81.76	nr	**92.18**
single branch; with access	98.88	0.91	11.84	146.56	nr	**158.41**
double branch	72.88	1.01	13.14	133.91	nr	**147.06**
isolated "Timesaver" joint	17.28	0.37	4.82	19.99	nr	**24.81**
transitional pipe; for WC	25.31	0.53	6.90	49.27	nr	**56.17**
150 mm pipes; laid straight	49.90	0.64	8.33	67.37	m	**75.70**
150 mm pipes; in runs not exceeding 3.00 m long	49.90	0.87	11.32	83.35	m	**94.68**
Extra for						
bend; medium radius	74.35	0.74	9.63	101.36	nr	**110.99**
bend; medium radius with access	157.65	0.74	9.63	197.74	nr	**207.37**
bend; long radius	99.57	0.74	9.63	127.58	nr	**137.21**
diminishing pipe	42.12	0.74	9.63	61.11	nr	**70.74**
single branch	92.56	0.91	11.84	109.42	nr	**121.26**
isolated "Timesaver" joint	20.93	0.44	5.73	24.21	nr	**29.94**
Accessories in "Timesaver" cast iron or other equal and approved; with mechanical coupling joints						
Gully fittings; comprising low invert gully trap and round hopper						
100 mm outlet	45.26	1.01	13.14	77.32	nr	**90.46**
150 mm outlet	112.60	1.38	17.96	160.44	nr	**178.40**
Add to above for; bellmouth 300 mm high; circular plain grating						
100 mm nominal size; 200 mm grating	47.13	0.48	6.25	81.20	nr	**87.45**
100 mm nominal size; 100 mm horizontal inlet; 200 mm grating	57.62	0.48	6.25	93.34	nr	**99.58**
100 mm nominal size; 100 mm horizontal inlet; 200 mm grating	59.09	0.48	6.25	95.04	nr	**101.29**
Yard gully (Deans); trapped; galvanized sediment pan; 267 mm round heavy grating						
100 mm outlet	305.86	3.08	40.08	405.37	nr	**445.46**
Yard gully (garage); trapless; galvanized sediment pan; 267 mm round heavy grating						
100 mm outlet	312.01	2.88	37.48	386.93	nr	**424.41**
Yard gully (garage); trapped; with rodding eye, galvanised perforated sediment pan; stopper; 267 mm round heavy grating						
100 mm outlet	580.23	2.88	37.48	762.43	nr	**799.91**
Grease trap; internal access; galvanized perforated bucket; lid and frame						
100 mm outlet; 20 gallon capacity	631.38	4.26	55.44	785.36	nr	**840.80**

R DISPOSAL SYSTEMS

Item	PC £	Labour hours	Labour £	Material £	Unit	Total rate £
R12 DRAINAGE BELOW GROUND – cont'd						
Cast iron "Ensign" lightweight pipes and fittings or other equal and approved; BS EN 877; ductile iron couplings						
100 mm pipes; laid straight	-	0.19	3.58	26.02	m	**29.60**
Extra for						
bend; long radius	-	0.19	3.58	46.65	nr	**50.23**
single branch	-	0.23	4.35	47.12	nr	**51.47**
150 mm pipes; laid straight	-	0.22	4.12	52.02	m	**56.14**
Extra for						
bend; long radius	-	0.22	4.12	128.94	nr	**133.07**
single branch	-	0.28	5.26	103.69	nr	**108.95**
Extra strength vitrified clay pipes and fittings; Hepworth "Supersleve" or other equal and approved; plain ends with push fit polypropylene flexible couplings						
100 mm pipes; laid straight	7.09	0.21	2.73	8.40	m	**11.13**
Extra for						
bend	6.55	0.21	2.73	14.52	nr	**17.25**
access bend	43.07	0.21	2.73	57.80	nr	**60.53**
rest bend	11.59	0.21	2.73	20.49	nr	**23.22**
access pipe	37.43	0.21	2.73	50.54	nr	**53.27**
socket adaptor	6.94	0.19	2.47	11.89	nr	**14.37**
saddle	13.89	0.80	10.41	20.70	nr	**31.11**
single junction	14.14	0.27	3.51	27.18	nr	**30.69**
single access junction	49.82	0.27	3.51	69.46	nr	**72.98**
150 mm pipes; laid straight	13.52	0.27	3.51	16.02	m	**19.54**
Extra for						
bend	13.49	0.26	3.38	27.90	nr	**31.28**
access bend	7.15	0.26	3.38	75.55	nr	**78.93**
rest bend	17.33	0.26	3.38	32.45	nr	**35.83**
taper pipe	16.71	0.26	3.38	28.26	nr	**31.64**
access pipe	50.87	0.26	3.38	71.04	nr	**74.42**
socket adaptor	13.68	0.21	2.73	22.75	nr	**25.48**
adaptor to "HepSeal" pipe	9.71	0.21	2.73	18.05	nr	**20.78**
saddle	20.66	0.95	12.36	32.19	nr	**44.55**
single junction	18.05	0.32	4.16	39.84	nr	**44.00**
single access junction	74.06	0.32	4.16	106.23	nr	**110.39**
Extra strength vitrified clay pipes and fittings; Hepworth "SuperSeal" or other equal and approved; socketted; with push-fit flexible joints						
150 mm pipes; laid straight	16.79	0.34	4.42	19.90	m	**24.33**
Extra for						
bend	32.28	0.27	3.51	32.29	nr	**35.80**
rest bend	17.33	0.23	2.99	14.57	nr	**17.56**
stopper	9.74	0.17	2.21	11.54	nr	**13.75**
taper reducer	16.71	0.27	3.51	13.84	nr	**17.35**
saddle	20.66	0.86	11.19	24.49	nr	**35.68**
single junction	42.19	0.34	4.42	42.04	nr	**46.47**
225 mm pipes; laid straight	34.85	0.44	5.73	41.30	m	**47.03**
Extra for						
bend	75.65	0.34	4.42	77.27	nr	**81.69**
rest bend	92.41	0.34	4.42	97.13	nr	**101.56**
stopper	16.39	0.22	2.86	19.43	nr	**22.29**
taper reducer	52.10	0.34	4.42	49.36	nr	**53.79**
saddle	76.87	1.15	14.97	91.11	nr	**106.07**
single junction	134.36	0.44	5.73	142.72	nr	**148.45**

R DISPOSAL SYSTEMS

Item	PC £	Labour hours	Labour £	Material £	Unit	Total rate £
300 mm pipes; laid straight	53.45	0.57	7.42	63.35	m	**70.76**
Extra for						
bend	143.66	0.45	5.86	151.26	nr	**157.12**
rest bend	204.72	0.45	5.86	223.63	nr	**229.48**
stopper	34.99	0.29	3.77	41.47	nr	**45.25**
taper reducer	143.82	0.45	5.86	151.45	nr	**157.30**
saddle	133.78	1.54	20.04	158.56	nr	**178.60**
400 mm pipes; laid straight	125.22	0.77	10.02	148.40	m	**158.43**
Extra for						
bend	470.53	0.62	8.07	513.14	nr	**521.21**
single unequal junction	440.88	0.77	10.02	463.16	nr	**473.18**
450 mm pipes; laid straight	162.65	0.95	12.36	192.77	m	**205.13**
Extra for						
bend	619.61	0.77	10.02	676.52	nr	**686.54**
single unequal junction	527.42	0.95	12.36	547.98	nr	**560.34**
British Standard quality vitrified clay pipes and fittings; socketted; cement:sand (1:2) joints						
100 mm pipes; laid straight	10.62	0.43	5.60	12.71	m	**18.30**
Extra for						
bend (short/medium/knuckle)	7.43	0.34	4.42	8.93	nr	**13.36**
bend (long/rest/elbow)	17.46	0.34	4.42	17.04	nr	**21.46**
single junction	19.50	0.43	5.60	18.24	nr	**23.83**
double collar	12.81	0.29	3.77	15.30	nr	**19.08**
150 mm pipes; laid straight	16.34	0.48	6.25	19.49	m	**25.73**
Extra for						
bend (short/medium/knuckle)	16.34	0.38	4.95	13.68	nr	**18.62**
bend (long/rest/elbow)	29.50	0.38	4.95	29.27	nr	**34.22**
taper	38.54	0.38	4.95	39.41	nr	**44.35**
single junction	32.27	0.48	6.25	30.66	nr	**36.90**
double collar	21.33	0.32	4.16	25.40	nr	**29.57**
225 mm pipes; laid straight	32.36	0.58	7.55	38.65	m	**46.20**
Extra for						
double collar	49.91	0.38	4.95	59.28	nr	**64.22**
300 mm pipes; laid straight	54.26	0.80	10.41	64.60	m	**75.01**
Accessories in vitrified clay; set in concrete; with polypropylene coupling joints to pipes						
Rodding point; with oval aluminium plate						
100 mm nominal size	38.42	0.53	6.90	51.68	nr	**58.58**
Gully fittings; comprising low back trap and square hopper; 150 mm x 150 mm square gully grid						
100 mm nominal size	32.74	0.91	11.84	49.19	nr	**61.03**
Access gully; trapped with rodding eye and integral vertical back inlet; stopper; 150 mm x 150 mm square gully grid						
100 mm nominal size	59.46	0.69	8.98	76.62	nr	**85.60**
Inspection chamber; comprising base; 300 mm or 450 mm raising piece; integral alloy cover and frame; 100 mm inlets						
straight through; 2 nr inlets	203.80	2.13	27.72	250.11	nr	**277.84**
Accessories in polypropylene; cover set in concrete; with coupling joints to pipes						
Inspection chamber; 5 nr 100 mm inlets; cast iron cover and frame						
475 mm diameter x 585 mm deep	242.85	2.44	31.76	296.38	nr	**328.14**
475 mm diameter x 930 mm deep	295.78	2.66	34.62	359.12	nr	**393.74**

R DISPOSAL SYSTEMS

Item	PC £	Labour hours	Labour £	Material £	Unit	Total rate £
R12 DRAINAGE BELOW GROUND – cont'd						
Accessories in vitrified clay; set in concrete; with cement:sand (1:2) joints to pipes						
Yard gully; 225 mm diameter; including domestic duty grating and frame (up to 1 tonne) and combined filter and silk bucket						
100 mm outlet	140.47	2.90	37.74	167.06	nr	**204.80**
100 mm outlet; 100 mm back inlet	196.26	3.10	40.34	233.18	nr	**273.53**
150 mm outlet	140.47	4.00	52.06	167.06	nr	**219.12**
150 mm outlet; 150 mm back inlet	200.29	4.25	55.31	237.95	nr	**293.26**
Yard gully; 225 mm diameter; including medium duty grating and frame (up to 5 tonnes) and combined filter and silk bucket						
100 mm outlet	183.70	2.90	37.74	218.30	nr	**256.04**
100 mm outlet; 100 mm back inlet	243.42	3.10	40.34	289.07	nr	**329.41**
150 mm outlet	197.84	4.00	52.06	235.05	nr	**287.11**
150 mm outlet; 150 mm back inlet	247.45	4.25	55.31	293.84	nr	**349.16**
Road gully; trapped with rodding eye and stopper (grate not included)						
300 mm x 600 mm x 100 mm outlet	100.50	3.51	45.68	141.11	nr	**186.79**
300 mm x 600 mm x 150 mm outlet	102.92	3.51	45.68	143.97	nr	**189.65**
400 mm x 750 mm x 150 mm outlet	119.36	4.26	55.44	174.86	nr	**230.30**
450 mm x 900 mm x 150 mm outlet	168.70	5.32	69.24	239.99	nr	**309.23**
Grease trap; with internal access; galvanized perforated bucket; lid and frame						
600 mm x 450 mm x 600 mm deep; 100 mm outlet	822.85	4.47	58.17	1008.04	nr	**1066.21**
Interceptor; trapped with inspection arm; lever locking stopper; chain and staple; cement and sand (1:2) joints to pipes; building in, and cutting and fitting brickwork around						
100 mm outlet; 100 mm inlet	126.30	4.26	55.44	150.22	nr	**205.66**
150 mm outlet; 150 mm inlet	179.34	4.78	62.21	213.09	nr	**275.30**
225 mm outlet; 225 mm inlet	491.15	5.32	69.24	582.70	nr	**651.93**
Accessories; grates and covers						
Aluminium alloy gully grids; set in position						
120 mm x 120 mm	3.84	0.11	1.43	4.55	nr	**5.98**
150 mm x 150 mm	3.84	0.11	1.43	4.55	nr	**5.98**
225 mm x 225 mm	11.43	0.11	1.43	13.55	nr	**14.98**
100 mm diameter	3.84	0.11	1.43	4.55	nr	**5.98**
150 mm diameter	5.87	0.11	1.43	6.96	nr	**8.39**
225 mm diameter	12.78	0.11	1.43	15.15	nr	**16.58**
Aluminium alloy sealing plates and frames; set in cement and sand (1:3)						
150 x 150	14.76	0.27	3.51	17.61	nr	**21.12**
225 x 225	27.00	0.27	3.51	32.11	nr	**35.63**
300 x 150	28.30	0.27	3.51	33.66	nr	**37.17**
140 diameter (for 100 mm)	12.02	0.27	3.51	14.37	nr	**17.88**
197 diameter (for 150 mm)	17.30	0.27	3.51	20.62	nr	**24.14**
273 diameter (for 225 mm)	27.69	0.27	3.51	32.93	nr	**36.44**
Polypropylene access covers and frames; supplied by Manhole Covers Ltd or other equal and approved; to suit PPIC inspection chambers; bedding and pointing in frame.						
450 mm dia; class A15	17.00	1.50	19.52	21.47	nr	**41.00**
450 mm dia; class B125; kite-marked	44.00	1.50	19.52	52.71	nr	**72.23**

R DISPOSAL SYSTEMS

Item	PC £	Labour hours	Labour £	Material £	Unit	Total rate £
Ductile iron heavy duty road gratings and frame; supplied by Manhole Covers Ltd or other equal and approved; bedding and pointing in cement and sand (1:3); one course half brick thick wall in semi-engineering bricks in cement mortar (1:3)						
225 mm x 225 mm x 80 mm hinged and dished road grating and frame; class C250	26.50	2.60	33.84	34.13	nr	**67.97**
300 mm x 300 mm x 80 mm hinged and dished road grating and frame; class C250	39.00	2.60	33.84	48.60	nr	**82.43**
420 mm x 420 x 75 mm hinged road grating and frame; ref C250; kite-marked	39.50	2.60	33.84	49.17	nr	**83.01**
445 mm x 445 x 75 mm double triangular road grating and frame; ref C250; kite-marked	42.00	2.60	33.84	52.07	nr	**85.90**
435 mm x 435 x 100 mm pedestrian mesh road grating and frame; ref D400	74.00	2.60	33.84	89.09	nr	**122.93**
440 mm x 440 mm x 150 mm hinged road grating and frame; class D400; kite-marked	78.00	2.60	33.84	93.72	nr	**127.55**
Accessories in precast concrete; top set in with rodding eye and stopper; cement and sand (1:2) joint to pipe						
Concrete road gully; BS 5911; trapped with rodding eye and stopper; cement and sand (1:2) joint to pipe						
450 mm diameter x 1050 mm deep; 100 mm or 150 mm outlet	46.35	5.05	65.72	77.65	nr	**143.38**
"Osmadrain" uPVC pipes and fittings or other equal and approved; BS 4660; with ring seal joints						
82 mm pipes; laid straight	13.00	0.17	2.21	15.41	m	**17.62**
Extra for						
bend; short radius	22.70	0.14	1.82	26.26	nr	**28.08**
spigot/socket bend	19.07	0.14	1.82	22.07	nr	**23.89**
adaptor	9.95	0.08	1.04	11.51	nr	**12.56**
single junction	29.52	0.19	2.47	34.15	nr	**36.63**
slip coupler	10.56	0.08	1.04	12.22	nr	**13.26**
100 mm pipes; laid straight	8.12	0.19	2.47	11.10	m	**13.57**
Extra for						
bend; short radius	21.44	0.17	2.21	24.23	nr	**26.44**
bend; long radius	34.72	0.17	2.21	37.28	nr	**39.50**
spigot/socket bend	18.12	0.17	2.21	27.78	nr	**29.99**
socket plug	9.38	0.05	0.65	10.86	nr	**11.51**
adjustable double socket bend	25.66	0.17	2.21	36.61	nr	**38.83**
adaptor to clay	24.17	0.10	1.30	27.51	nr	**28.81**
single junction	25.58	0.23	2.99	26.71	nr	**29.71**
sealed access junction	66.17	0.20	2.60	73.67	nr	**76.27**
slip coupler	10.56	0.10	1.30	12.22	nr	**13.52**
160 mm pipes; laid straight	17.82	0.23	2.99	24.08	m	**27.07**
Extra for						
bend; short radius	51.00	0.19	2.47	57.74	nr	**60.21**
spigot/socket bend	46.24	0.19	2.47	68.35	nr	**70.82**
socket plug	20.14	0.08	1.04	23.31	nr	**24.35**
adaptor to clay	52.56	0.13	1.69	59.67	nr	**61.36**
level invert taper	24.73	0.19	2.47	42.19	nr	**44.66**
single junction	83.52	0.27	3.51	96.63	nr	**100.14**
slip coupler	15.02	0.12	1.56	17.38	nr	**18.94**

R DISPOSAL SYSTEMS

Item	PC £	Labour hours	Labour £	Material £	Unit	Total rate £
R12 DRAINAGE BELOW GROUND – cont'd						
uPVC Osma "Ultra-Rib" ribbed pipes and fittings or other equal and approved; WIS approval; with sealed ring push-fit joints						
150 mm pipes; laid straight	-	0.21	2.73	9.92	m	**12.65**
Extra for						
bend; short radius	27.76	0.19	2.47	31.52	nr	**34.00**
adaptor to 160 diameter upvc	39.26	0.11	1.43	44.24	nr	**45.67**
adaptor to clay	80.60	0.11	1.43	92.66	nr	**94.09**
level invert taper	12.11	0.19	2.47	12.23	nr	**14.70**
single junction	49.90	0.24	3.12	54.76	nr	**57.89**
225 mm pipes; laid straight	20.95	0.24	3.12	24.83	m	**27.96**
Extra for						
bend; short radius	111.57	0.22	2.86	127.59	nr	**130.45**
adaptor to clay	100.42	0.14	1.82	113.20	nr	**115.02**
level invert taper	19.41	0.22	2.86	17.98	nr	**20.85**
single junction	165.62	0.30	3.90	184.16	nr	**188.07**
300 mm pipes; laid straight	31.12	0.36	4.69	36.88	m	**41.57**
Extra for						
bend; short radius	175.72	0.32	4.16	201.09	nr	**205.25**
adaptor to clay	264.14	0.16	2.08	301.17	nr	**303.25**
level invert taper	63.03	0.32	4.16	66.29	nr	**70.45**
single junction	382.68	0.41	5.34	431.68	nr	**437.02**
Interconnecting concrete drainage channel 100 mm wide; ACO Techologies Ltd ref. NK100; or other equal and approved; with Heelguard ductile iron grating suitale for load class 250; bedding and haunching in in situ concrete (not included)						
100 mm wide						
laid level or to falls	-	0.51	6.64	106.61	m	**113.25**
extra for sump unit	-	1.53	19.91	124.83	nr	**144.74**
extra for end caps	-	0.10	1.30	7.81	nr	**9.11**
Interconnecting drainage channel; "Birco-lite" ref 8012 or other equal and approved; Marshalls Plc; galvanised steel grating ref 8041; bedding and haunching in in situ concrete (not included)						
100 mm wide						
laid level or to falls	-	0.51	6.64	60.17	m	**66.81**
extra for 100 mm diameter trapped outlet unit	-	1.53	19.91	116.20	nr	**136.11**
extra for end caps	-	0.10	1.30	6.76	nr	**8.06**
Accessories in uPVC; with ring seal joints to pipes (unless otherwise described)						
Rodding eye						
110 mm diameter	43.83	0.47	6.12	55.47	nr	**61.59**
Universal gulley fitting; comprising gulley trap, plain hopper						
150 mm x 150 mm grate	38.17	1.02	13.27	50.82	nr	**64.09**
Bottle gulley; comprising gulley with bosses closed; sealed access covers						
217 mm x 217 mm grate	76.06	0.85	11.06	94.66	nr	**105.73**
Shallow access pipe; light duty screw down access door assembly						
110 mm diameter	107.69	0.85	11.06	131.25	nr	**142.31**

R DISPOSAL SYSTEMS

Item	PC £	Labour hours	Labour £	Material £	Unit	Total rate £
Shallow access junction; 3 nr 110 mm inlets; light duty screw down access door assembly						
110 mm diameter	166.85	1.22	15.88	195.89	nr	**211.77**
Shallow inspection chamber; 250 mm diameter; 600 mm deep; sealed cover and frame						
4 nr 110 mm outlets/inlets	137.61	1.41	18.35	182.97	nr	**201.32**
Universal inspection chamber; 450 mm diameter; single seal cast iron cover and frame; 4 nr 110 mm outlets/inlets						
500 mm deep	269.46	1.49	19.39	335.52	nr	**354.92**
730 mm deep	301.82	1.76	22.91	377.72	nr	**400.63**
960 mm deep	334.18	2.04	26.55	419.91	nr	**446.46**
Equal manhole base; 750 mm diameter						
6 nr 160 mm outlets/inlets	394.91	1.33	17.31	471.17	nr	**488.47**
Unequal manhole base; 750 mm diameter						
2 nr 160 mm, 4 nr 110 mm outlets/inlets	304.90	1.33	17.31	367.03	nr	**384.34**
Kerb to gullies; class B engineering bricks on edge to three sides in cement mortar (1:3) rendering in cement mortar (1:3) to top and two sides and skirting to brickwork 230 mm high; dishing in cement mortar (1:3) to gully; steel trowelled						
230 mm x 230 mm internally	-	1.53	19.91	1.53	nr	**21.45**
MANHOLES						
Excavating; by machine						
Manholes						
maximum depth not exceeding 1.00 m	-	0.22	2.89	6.52	m^3	**9.41**
maximum depth not exceeding 2.00 m	-	0.24	3.15	7.17	m^3	**10.32**
maximum depth not exceeding 4.00 m	-	0.29	3.81	8.38	m^3	**12.19**
Excavating; by hand						
Manholes						
maximum depth not exceeding 1.00 m	-	3.52	46.25	-	m^3	**46.25**
maximum depth not exceeding 2.00 m	-	4.16	54.66	-	m^3	**54.66**
maximum depth not exceeding 4.00 m	-	5.32	69.90	-	m^3	**69.90**
Earthwork support (average "risk" prices)						
Maximum depth not exceeding 1.00 m						
distance between opposing faces not exceeding 2.00 m	-	0.16	2.10	3.20	m^2	**5.30**
Maximum depth not exceeding 2.00 m						
distance between opposing faces not exceeding 2.00 m	-	0.19	2.50	5.83	m^2	**8.32**
Maximum depth not exceeding 4.00 m						
distance between opposing faces not exceeding 2.00 m	-	0.24	3.15	8.46	m^2	**11.61**
Disposal; by machine						
Excavated material						
off site; to tip not exceeding 13 km (using lorries) including Landfill Tax based on inactive waste	-	-	-	24.83	m^3	**-**
on site depositing; in spoil heaps; average 50 m distance	-	0.16	2.10	4.37	m^3	**6.47**

R DISPOSAL SYSTEMS

Item	PC £	Labour hours	Labour £	Material £	Unit	Total rate £
R12 DRAINAGE BELOW GROUND – cont'd						
MANHOLES – cont'd						
Disposal; by hand						
Excavated material						
off site; to tip not exceeding 13 km (using lorries) including Landfill Tax based on inactive waste	-	0.81	10.64	34.14	m³	**44.79**
on site depositing; in spoil heaps; average 50 m distance	-	1.32	17.34	-	m³	**17.34**
Filling to excavations; by machine						
Average thickness not exceeding 0.25 m						
arising from the exacavations	-	0.16	2.10	3.07	m³	**5.18**
Filling to excavations; by hand						
Average thickness not exceeding 0.25 m						
arising from the exacavations	-	1.02	13.40	-	m³	**13.40**
Plain in situ ready mixed designated concrete; C10 - 40 mm aggregate						
Beds						
thickness not exceeding 150 mm	82.26	3.20	54.18	99.82	m³	**154.00**
thickness 150 mm - 450 mm	-	2.40	40.63	99.82	m³	**140.46**
thickness exceeding 450 mm	-	2.02	34.20	99.82	m³	**134.03**
Plain in situ ready mixed designated concrete; C20 - 20 mm aggregate						
Beds						
thickness not exceeding 150 mm	85.85	3.20	54.18	104.18	m³	**158.36**
thickness 150 mm - 450 mm	-	2.40	40.63	104.18	m³	**144.81**
thickness exceeding 450 mm	-	2.02	34.20	104.18	m³	**138.38**
Plain in situ ready mixed designated concrete; C25 - 20 mm aggregate; (small quantities)						
Benching in bottoms						
150 mm - 450 mm average thickness	83.00	9.57	186.23	100.71	m³	**286.94**
Reinforced in situ ready mixed designated concrete; C20 - 20 mm aggregate; (small quantities)						
Isolated cover slabs						
thickness not exceeding 150 mm	81.76	7.45	126.14	99.21	m³	**225.35**
Reinforcement; fabric to BS 4449; lapped; in beds or suspended slabs						
Ref A98 (1.54 kg/m²)						
400 mm minimum laps	1.86	0.13	2.59	2.31	m²	**4.90**
Ref A142 (2.22 kg/m²)						
400 mm minimum laps	1.77	0.13	2.59	2.20	m²	**4.79**
Ref A193 (3.02 kg/m²)						
400 mm minimum laps	2.41	0.13	2.59	2.99	m²	**5.59**
Formwork; basic finish						
Soffits of isolated cover slabs						
horizontal	-	3.03	60.47	7.15	m²	**67.62**
Edges of isolated cover slabs						
height not exceeding 250 mm	-	0.90	17.96	2.30	m	**20.26**

R DISPOSAL SYSTEMS

Item	PC £	Labour hours	Labour £	Material £	Unit	Total rate £
Common bricks; in cement mortar (1:3)						
Walls to manholes						
one brick thick (PC £ per 1000)	270.00	2.55	70.00	52.30	m²	**122.30**
one and a half brick thick	-	3.73	102.39	78.44	m²	**180.84**
Projections of footings						
two brick thick	-	5.22	143.30	104.59	m²	**247.89**
Class A engineering bricks; in cement mortar (1:3)						
Walls to manholes						
one brick thick (PC £ per 1000)	390.00	2.88	79.06	71.67	m²	**150.73**
one and a half brick thick	-	4.15	113.92	74.23	m²	**188.15**
Projections of footings						
two brick thick	-	5.85	160.59	143.34	m²	**303.93**
Class B engineering bricks; in cement mortar (1:3)						
Walls to manholes						
one brick thick (PC £ per 1000)	315.00	2.88	79.06	58.88	m²	**137.94**
one and a half brick thick	-	4.15	113.92	88.32	m²	**202.24**
Projections of footings						
two brick thick	-	5.85	160.59	117.76	m²	**278.35**
Brickwork sundries						
Extra over for fair face; flush smooth pointing						
manhole walls	-	0.21	5.76	-	m²	**5.76**
Building ends of pipes into brickwork; making good fair face or rendering						
not exceeding 55 mm nominal size	-	0.11	3.02	-	nr	**3.02**
55 mm - 110 mm nominal size	-	0.16	4.39	-	nr	**4.39**
over 110 mm nominal size	-	0.21	5.76	-	nr	**5.76**
Step irons; BS 1247; malleable; galvanized; building into joints						
general purpose pattern	-	0.16	4.39	5.31	nr	**9.70**
Cement:sand (1:3) in situ finishings; steel trowelled						
13 mm work to manhole walls; one coat; to brickwork base over 300 mm wide	-	0.74	20.31	1.70	m²	**22.02**
Cast iron inspection chambers; with bolted flat covers; BS 437; bedded in cement mortar (1:3); with mechanical coupling joints						
100 mm x 100 mm						
one branch	194.03	1.11	14.45	225.34	nr	**239.78**
one branch either side	365.77	1.67	21.73	424.03	nr	**445.77**
150 mm x 100 mm						
one branch	244.82	1.34	17.44	284.10	nr	**301.54**
one branch either side	475.12	1.90	24.73	551.40	nr	**576.13**
150 mm x 150 mm						
one branch	304.04	1.43	18.61	353.47	nr	**372.08**
one branch either side	586.32	2.04	26.55	680.06	nr	**706.61**

R DISPOSAL SYSTEMS

Item	PC £	Labour hours	Labour £	Material £	Unit	Total rate £
R12 DRAINAGE BELOW GROUND – cont'd						
MANHOLES – cont'd						
Coated cast or ductile iron access covers and frames; to BS EN 124; supplied by Manhole Covers Ltd or other equal and approved; bedding frame in cement and sand (1:3); cover in grease and sand						
Light duty; cast iron; rectangular single seal solid top						
450 mm x 450 mm; class A15	39.00	1.75	22.78	46.93	nr	**69.70**
600 mm x 450 mm; class A15	42.00	1.75	22.78	50.56	nr	**73.33**
600 mm x 600 mm; class A15	66.00	1.75	22.78	78.48	nr	**101.26**
750 mm x 600 mm; class A15	109.45	1.75	22.78	128.75	nr	**151.53**
Light duty; cast iron; rectangular double seal solid top						
600 mm x 450 mm; class A15	79.50	1.75	22.78	93.94	nr	**116.72**
Medium duty; ductile iron; rectangular single seal solid top						
450 mm x 450 mm x 40 mm; class C250; kite-marked	64.00	2.30	29.93	76.17	nr	**106.10**
600 mm x 450 mm x 40 mm; slide-out; class C250; kite-marked	79.50	2.30	29.93	94.10	nr	**124.04**
600 mm xn 600 mm x 40 mm; slide-out; class C250; kite-marked	85.00	2.30	29.93	100.47	nr	**130.40**
760 mm x 600 mm x 40 mm; slide-out; class C250; kite-marked	127.00	2.30	29.93	149.06	nr	**178.99**
Heavy duty; ductile iron; solid top						
450 mm x 450 mm x 75 mm; single seal; class C250; kite-marked	89.50	2.85	37.09	105.67	nr	**142.76**
600 mm x 450 mm x 75 mm; single seal; class C250; kite-marked	95.00	2.85	37.09	112.04	nr	**149.13**
600 mm x 600 mm x 75 mm; single seal; class C250; kite-marked	107.00	2.85	37.09	125.92	nr	**163.01**
450 mm x 450 mm x 100 mm; double triangular; class D400; kite-marked	99.00	2.85	37.09	116.66	nr	**153.75**
600 mm x 450 mm x 100 mm; double triangular; class D400; kite-marked	89.00	2.85	37.09	105.09	nr	**142.19**
600 mm x 600 mm x 100 mm; double triangular; class D400; kite-marked	79.00	2.85	37.09	93.52	nr	**130.62**
750 mm x 600 mm x 100 mm; double triangular; class D400; kite-marked	188.64	2.85	37.09	220.37	nr	**257.47**
1220 mm x 675 x 100 mm; double triangular; class D400; kite-marked	225.00	4.00	52.06	262.44	nr	**314.50**
British Standard best quality vitrified clay channels; bedding and jointing in cement:sand (1:2)						
Half section straight						
100 mm diameter x 1.00 m long	6.02	0.85	11.06	7.13	nr	**18.19**
150 mm diameter x 1.00 m long	10.02	1.06	13.80	11.87	nr	**25.67**
225 mm diameter x 1.00 m long	22.50	1.38	17.96	26.67	nr	**44.63**
300 mm diameter x 1.00 m long	46.19	1.70	22.12	54.75	nr	**76.87**
Half section bend						
100 mm diameter	13.84	0.64	8.33	16.40	nr	**24.73**
150 mm diameter	22.69	0.80	10.41	26.89	nr	**37.30**
225 mm diameter	75.42	1.06	13.80	89.39	nr	**103.19**
Half section taper straight						
150 mm - 100 mm diameter	28.17	0.74	9.63	33.38	nr	**43.02**
225 mm - 150 mm diameter	62.87	0.95	12.36	74.52	nr	**86.88**

R DISPOSAL SYSTEMS

Item	PC £	Labour hours	Labour £	Material £	Unit	Total rate £
Taper bend						
150 mm - 100 mm diameter	42.89	0.95	12.36	50.83	nr	**63.19**
225 mm - 150 mm diameter	122.88	1.22	15.88	145.64	nr	**161.51**
Three quarter section branch bend						
100 mm diameter	15.26	0.53	6.90	18.09	nr	**24.99**
150 mm diameter	25.63	0.80	10.41	30.38	nr	**40.79**
225 mm diameter	93.52	1.06	13.80	110.84	nr	**124.63**
uPVC channels; with solvent weld or lip seal coupling joints; bedding in cement:sand						
Half section cut away straight; with coupling either end						
110 mm diameter	57.42	0.32	4.16	87.13	nr	**91.30**
160 mm diameter	107.85	0.43	5.60	162.88	nr	**168.48**
Half section cut away long radius bend; with coupling either end						
110 mm diameter	94.16	0.32	4.16	130.67	nr	**134.84**
160 mm diameter	203.57	0.43	5.60	276.33	nr	**281.93**
Channel adaptor to clay; with one coupling						
110 mm diameter	22.03	0.27	3.51	35.65	nr	**39.16**
160 mm diameter	53.30	0.35	4.56	80.71	nr	**85.26**
Half section bend						
110 mm diameter	36.35	0.35	4.56	43.72	nr	**48.27**
160 mm diameter	62.44	0.53	6.90	75.48	nr	**82.37**
Half section channel connector						
110 mm diameter	9.94	0.08	1.04	13.05	nr	**14.09**
Half section channel junction						
110 mm diameter	28.21	0.53	6.90	34.06	nr	**40.96**
Polypropylene slipper bend						
110 mm diameter	24.46	0.43	5.60	29.63	nr	**35.22**
Glass fibre septic tank; "Klargester" or other equal and approved; fixing lockable manhole cover and frame; placing in position						
3750 litre capacity; 2000 mm diameter; depth to invert						
1000 mm deep; standard grade	814.22	2.50	32.54	1013.91	nr	**1046.45**
1500 mm deep; heavy duty grade	1024.34	2.82	36.70	1251.09	nr	**1287.79**
6000 litre capacity; 2300 mm diameter; depth to invert						
1000 mm deep; standard grade	1313.25	2.68	34.88	1600.92	nr	**1635.80**
1500 mm deep; heavy duty grade	1733.49	3.01	39.17	2075.26	nr	**2114.44**
9000 litre capacity; 2660 mm diameter; depth to invert						
1000 mm deep; standard grade	1996.14	2.91	37.87	2371.73	nr	**2409.60**
1500 mm deep; heavy duty grade	2626.50	3.19	41.52	3083.25	nr	**3124.76**
Glass fibre petrol interceptors; "Klargester" or other equal and approved; placing in position						
2000 litre capacity; 2370 mm x 1300 mm diameter; depth to invert						
1000 mm deep	919.00	2.73	35.53	1037.32	nr	**1072.85**
4000 litre capacity; 4370 mm x 1300 mm diameter; depth to invert						
1000 mm deep	1576.00	2.96	38.52	1778.91	nr	**1817.43**

R DISPOSAL SYSTEMS

Item	PC £	Labour hours	Labour £	Material £	Unit	Total rate £
R13 LAND DRAINAGE						
Excavating; by hand; grading bottoms; earthwork support; filling to within 150 mm of surface with gravel rejects; remainder filled with excavated material and compacting; disposal of surplus soil on site; spreading on site average 50 m						
Pipes not exceeding 200 mm nominal size						
average depth of trench 0.75 m	-	1.71	22.47	12.72	m	**35.19**
average depth of trench 1.00 m	-	2.31	30.35	20.01	m	**50.36**
average depth of trench 1.25 m	-	3.19	41.91	25.06	m	**66.97**
average depth of trench 1.50 m	-	5.50	72.26	30.48	m	**102.74**
average depth of trench 1.75 m	-	6.52	85.66	35.52	m	**121.19**
average depth of trench 2.00 m	-	7.54	99.07	40.94	m	**140.01**
Disposal; by machine						
Excavated material						
off site; to tip not exceeding 13 km (using lorries); including Landfill Tax based on inactive waste	-	-	-	24.83	m³	-
Disposal; by hand						
Excavated material						
off site; to tip not exceeding 13 km (using lorries); including Landfill Tax based on inactive waste	-	0.88	11.56	34.14	m³	**45.71**
Vitrified clay perforated sub-soil pipes; BS 65; Hepworth "Hepline" or other equal and approved						
Pipes; laid straight						
100 mm diameter	8.14	0.23	2.99	9.64	m	**12.64**
150 mm diameter	14.78	0.29	3.77	17.52	m	**21.30**
225 mm diameter	29.80	0.38	4.95	35.32	m	**40.26**

S PIPED SUPPLY SYSTEMS

Item	PC £	Labour hours	Labour £	Material £	Unit	Total rate £
S12 HOT AND COLD WATER (SMALL SCALE)						
Copper pipes; EN1057:1996; capillary fittings						
15 mm pipes; fixing with pipe clips and screwed	3.71	0.37	7.03	4.53	m	**11.56**
Extra for						
made bend	-	0.17	3.23	-	nr	**3.23**
stop end	1.76	0.12	2.28	2.04	nr	**4.32**
straight coupling	0.28	0.19	3.61	0.32	nr	**3.93**
union coupling	9.54	0.19	3.61	11.04	nr	**14.65**
reducing coupling	3.26	0.19	3.61	3.77	nr	**7.38**
copper to lead connector	7.44	0.24	4.56	8.61	nr	**13.17**
imperial to metric adaptor	3.83	0.24	4.56	4.43	nr	**8.99**
elbow	0.50	0.19	3.61	0.57	nr	**4.18**
backplate elbow	7.17	0.39	7.41	8.29	nr	**15.70**
return bend	10.74	0.19	3.61	12.42	nr	**16.03**
tee; equal	0.94	0.28	5.32	1.09	nr	**6.41**
tee; reducing	7.81	0.28	5.32	9.03	nr	**14.35**
straight tap connector	2.43	0.56	10.64	2.81	nr	**13.45**
bent tap connector	2.72	0.76	14.44	3.15	nr	**17.58**
tank connector	8.54	0.28	5.32	9.88	nr	**15.20**
22 mm pipes; fixing with pipe clips and screwed	7.42	0.43	8.17	8.92	m	**17.09**
Extra for						
made bend	-	0.22	4.18	-	nr	**4.18**
stop end	3.29	0.15	2.85	3.80	nr	**6.65**
straight coupling	0.74	0.24	4.56	0.85	nr	**5.41**
union coupling	15.29	0.24	4.56	17.69	nr	**22.25**
reducing coupling	3.19	0.24	4.56	3.69	nr	**8.25**
copper to lead connector	10.16	0.34	6.46	11.75	nr	**18.21**
elbow	1.30	0.24	4.56	1.50	nr	**6.06**
backplate elbow	15.38	0.49	9.31	17.80	nr	**27.11**
return bend	21.10	0.24	4.56	24.41	nr	**28.97**
tee; equal	3.00	0.37	7.03	3.47	nr	**10.50**
tee; reducing	2.38	0.37	7.03	2.75	nr	**9.78**
straight tap connector	2.74	0.19	3.61	3.17	nr	**6.78**
tank connector	7.42	0.37	7.03	8.79	nr	**15.82**
28 mm pipes; fixing with pipe clips and screwed	9.35	0.46	8.74	11.21	m	**19.95**
Extra for						
made bend	-	0.28	5.32	-	nr	**5.32**
stop end	5.87	0.17	3.23	6.79	nr	**10.02**
straight coupling	1.63	0.31	5.89	1.89	nr	**7.78**
reducing coupling	4.46	0.31	5.89	5.16	nr	**11.04**
union coupling	15.29	0.31	5.89	17.69	nr	**23.58**
copper to lead connector	19.09	0.43	8.17	22.08	nr	**30.25**
imperial to metric adaptor	9.84	0.43	8.17	11.38	nr	**19.55**
elbow	2.61	0.31	5.89	3.02	nr	**8.91**
return bend	26.95	0.31	5.89	31.18	nr	**37.07**
tee; equal	7.25	0.45	8.55	8.39	nr	**16.93**
tank connector	9.35	0.45	8.55	11.08	nr	**19.63**
35 mm pipes; fixing with pipe clips and screwed	18.90	0.53	10.07	22.54	m	**32.60**
Extra for						
made bend	-	0.33	6.27	-	nr	**6.27**
stop end	12.98	0.19	3.61	15.01	nr	**18.62**
straight coupling	5.30	0.37	7.03	6.14	nr	**13.17**
reducing coupling	10.51	0.37	7.03	12.16	nr	**19.19**
union coupling	29.22	0.37	7.03	33.81	nr	**40.84**
flanged connector	88.60	0.49	9.31	102.51	nr	**111.82**
elbow	11.35	0.37	7.03	13.13	nr	**20.16**
obtuse elbow	17.14	0.37	7.03	19.83	nr	**26.85**
tee; equal	18.49	0.51	9.69	21.39	nr	**31.08**
tank connector	21.92	0.51	9.69	25.36	nr	**35.05**

S PIPED SUPPLY SYSTEMS

Item	PC £	Labour hours	Labour £	Material £	Unit	Total rate £
S12 HOT AND COLD WATER (SMALL SCALE) – cont'd						
Copper pipes; EN1057:1996; capillary fittings – cont'd						
42 mm pipes; fixing with pipe clips; plugged and screwed	22.99	0.59	11.21	27.38	m	**38.59**
Extra for						
made bend	-	0.44	8.36	-	nr	**8.36**
stop end	22.34	0.21	3.99	25.84	nr	**29.83**
straight coupling	8.86	0.43	8.17	10.25	nr	**18.41**
reducing coupling	17.58	0.43	8.17	20.34	nr	**28.51**
union coupling	42.70	0.43	8.17	49.41	nr	**57.58**
flanged connector	96.34	0.56	10.64	111.47	nr	**122.11**
elbow	18.77	0.43	8.17	21.71	nr	**29.88**
obtuse elbow	30.50	0.43	8.17	35.29	nr	**43.46**
tee; equal	29.66	0.57	10.83	34.32	nr	**45.15**
tank connector	28.73	0.57	10.83	33.24	nr	**44.07**
54 mm pipes; fixing with pipe clips; plugged and screwed	29.57	0.65	12.35	35.18	m	**47.53**
Extra for						
made bend	-	0.61	11.59	-	nr	**11.59**
stop end	31.18	0.23	4.37	36.08	nr	**40.45**
straight coupling	16.33	0.49	9.31	18.89	nr	**28.20**
reducing coupling	29.53	0.49	9.31	34.16	nr	**43.47**
union coupling	81.27	0.49	9.31	94.03	nr	**103.34**
flanged connector	145.65	0.56	10.64	168.51	nr	**179.15**
elbow	38.76	0.49	9.31	44.84	nr	**54.15**
obtuse elbow	55.18	0.49	9.31	63.84	nr	**73.15**
tee; equal	59.81	0.63	11.97	69.20	nr	**81.16**
tank connector	43.90	0.63	11.97	50.80	nr	**62.76**
Copper pipes; EN1057:1996; compression fittings						
15 mm pipes; fixing with pipe clips; plugged and screwed	3.71	0.42	7.98	4.53	m	**12.51**
Extra for						
made bend	-	0.17	3.23	-	nr	**3.23**
stop end	3.20	0.11	2.09	3.70	nr	**5.79**
straight coupling	2.57	0.17	3.23	2.97	nr	**6.20**
reducing set	2.58	0.19	3.61	2.98	nr	**6.59**
male coupling	2.29	0.22	4.18	2.65	nr	**6.83**
female coupling	2.75	0.22	4.18	3.18	nr	**7.36**
90 degree bend	3.10	0.17	3.23	3.58	nr	**6.81**
90 degree backplate bend	5.74	0.33	6.27	6.64	nr	**12.91**
tee; equal	4.34	0.24	4.56	5.02	nr	**9.58**
tank coupling	6.52	0.24	4.56	7.54	nr	**12.10**
22 mm pipes; fixing with pipe clips; plugged and screwed	7.42	0.47	8.93	8.92	m	**17.85**
Extra for						
made bend	-	0.22	4.18	-	nr	**4.18**
stop end	4.63	0.13	2.47	5.36	nr	**7.83**
straight coupling	4.18	0.22	4.18	4.84	nr	**9.02**
reducing set	3.23	0.06	1.14	3.74	nr	**4.88**
male coupling	4.92	0.31	5.89	5.69	nr	**11.58**
female coupling	4.02	0.31	5.89	4.66	nr	**10.54**
90 degree bend	4.93	0.22	4.18	5.70	nr	**9.88**
tee; equal	7.16	0.33	6.27	8.28	nr	**14.55**
tee; reducing	11.45	0.33	6.27	13.24	nr	**19.51**
tank coupling	7.23	0.33	6.27	8.37	nr	**14.64**

S PIPED SUPPLY SYSTEMS

Item	PC £	Labour hours	Labour £	Material £	Unit	Total rate £
28 mm pipes; fixing with pipe clips; plugged and screwed	9.35	0.51	9.69	11.21	m	**20.90**
Extra for						
made bend	-	0.28	5.32	-	nr	**5.32**
stop end	9.92	0.16	3.04	11.48	nr	**14.52**
straight coupling	9.50	0.28	5.32	10.99	nr	**16.31**
male coupling	6.72	0.39	7.41	7.77	nr	**15.18**
female coupling	8.70	0.39	7.41	10.07	nr	**17.48**
90 degree bend	12.25	0.28	5.32	14.17	nr	**19.49**
tee; equal	19.53	0.41	7.79	22.59	nr	**30.38**
tee; reducing	18.86	0.41	7.79	21.82	nr	**29.60**
tank coupling	15.25	0.41	7.79	17.64	nr	**25.43**
35 mm pipes; fixing with pipe clips; plugged and screwed	18.90	0.57	10.83	22.54	m	**33.36**
Extra for						
made bend	-	0.33	6.27	-	nr	**6.27**
stop end	15.56	0.18	3.42	18.00	nr	**21.42**
straight coupling	20.08	0.33	6.27	23.23	nr	**29.50**
male coupling	15.26	0.44	8.36	17.66	nr	**26.02**
female coupling	18.33	0.44	8.36	21.20	nr	**29.56**
tee; equal	35.26	0.46	8.74	40.79	nr	**49.53**
tee; reducing	34.46	0.46	8.74	39.86	nr	**48.60**
tank coupling	26.93	0.46	8.74	31.15	nr	**39.89**
42 mm pipes; fixing with pipe clips; plugged and screwed	22.99	0.64	12.16	27.38	m	**39.54**
Extra for						
made bend	-	0.44	8.36	-	nr	**8.36**
stop end	25.90	0.20	3.80	29.97	nr	**33.77**
straight coupling	26.41	0.39	7.41	30.55	nr	**37.96**
male coupling	22.90	0.50	9.50	26.49	nr	**35.99**
female coupling	24.64	0.50	9.50	28.51	nr	**38.01**
tee; equal	55.43	0.52	9.88	64.13	nr	**74.01**
tee; reducing	53.26	0.52	9.88	61.62	nr	**71.49**
54 mm pipes; fixing with pipe clips; plugged and screwed	29.57	0.69	13.11	35.18	m	**48.29**
Extra for						
made bend	-	0.61	11.59	-	nr	**11.59**
straight coupling	39.49	0.44	8.36	45.69	nr	**54.05**
male coupling	33.81	0.56	10.64	39.11	nr	**49.75**
female coupling	36.15	0.56	10.64	41.83	nr	**52.47**
tee; equal	89.04	0.57	10.83	103.02	nr	**113.84**
tee; reducing	89.04	0.57	10.83	103.02	nr	**113.84**
Copper, brass and gunmetal ancillaries; screwed joints to fittings						
Stopcock; brass/gunmetal capillary joints to copper						
15 mm nominal size	7.28	0.22	4.18	8.42	nr	**12.60**
22 mm nominal size	13.60	0.30	5.70	15.73	nr	**21.43**
28 mm nominal size	38.67	0.38	7.22	44.74	nr	**51.96**
Stopcock; brass/gunmetal compression joints to copper						
15 mm nominal size	23.59	0.20	3.80	27.30	nr	**31.09**
22 mm nominal size	33.18	0.27	5.13	38.39	nr	**43.52**
28 mm nominal size	59.03	0.33	6.27	68.30	nr	**74.57**
Stopcock; brass/gunmetal compression joints to polyethylene						
15 mm nominal size	24.43	0.29	5.51	28.27	nr	**33.78**
22 mm nominal size	42.46	0.37	7.03	49.13	nr	**56.16**
28 mm nominal size	45.28	0.44	8.36	52.39	nr	**60.75**

S PIPED SUPPLY SYSTEMS

Item	PC £	Labour hours	Labour £	Material £	Unit	Total rate £
S12 HOT AND COLD WATER (SMALL SCALE) – cont'd						
Copper, brass and gunmetal ancillaries – cont'd						
Gunmetal "Fullway" gate valve; capillary joints to copper						
15 mm nominal size	22.81	0.22	4.18	26.39	nr	**30.57**
22 mm nominal size	26.42	0.30	5.70	30.56	nr	**36.26**
28 mm nominal size	36.79	0.38	7.22	42.57	nr	**49.79**
35 mm nominal size	82.05	0.45	8.55	94.93	nr	**103.48**
42 mm nominal size	102.59	0.52	9.88	118.70	nr	**128.57**
54 mm nominal size	148.83	0.59	11.21	172.19	nr	**183.40**
Brass gate valve; compression joints to copper						
15 mm nominal size	27.70	0.33	6.27	32.04	nr	**38.31**
22 mm nominal size	32.64	0.44	8.36	37.76	nr	**46.12**
28 mm nominal size	44.34	0.56	10.64	51.30	nr	**61.93**
Chromium plated; lockshield radiator valve; union outlet						
15 mm nominal size	8.29	0.24	4.56	9.59	nr	**14.15**
PEX/PEM 'JG Speedfit' system; BS 7921 Parts 1, 2 & 3 class S; push-fit fittings						
10 mm PEX barrier pipes; fixing with pipe clips; in wall, floor and roof voids	1.07	0.22	4.18	2.12	m	**6.30**
Extra for						
stop end	1.59	0.05	0.95	2.22	nr	**3.17**
straight connector	1.64	0.11	2.09	2.66	nr	**4.75**
elbow	2.01	0.11	2.09	3.09	nr	**5.18**
stem elbow	2.67	0.11	2.09	3.85	nr	**5.94**
tee; equal	2.32	0.16	3.04	3.83	nr	**6.87**
brass chrome plated service valve	8.26	0.11	2.09	10.32	nr	**12.41**
brass chrome plated ball valve	11.64	0.11	2.09	14.23	nr	**16.32**
15 mm PEX barrier pipes; fixing with pipe clips; in wall, floor and roof voids	1.24	0.24	4.56	2.24	m	**6.80**
15 mm Polybutylene barrier pipes; fixing with pipe clips; in wall, floor and roof voids	1.46	0.24	4.56	2.50	m	**7.06**
Extra for						
stop end	1.64	0.08	1.52	2.28	nr	**3.80**
straight connector	1.22	0.15	2.85	2.18	nr	**5.02**
reducing coupler	2.86	0.15	2.85	4.07	nr	**6.92**
PE-copper coupler	5.68	0.17	3.23	7.34	nr	**10.56**
elbow	1.44	0.15	2.85	2.43	nr	**5.28**
stem elbow	2.82	0.15	2.85	4.03	nr	**6.88**
tee; equal	2.08	0.22	4.18	3.55	nr	**7.73**
tee; reducing	3.49	0.22	4.18	5.18	nr	**9.36**
tank connector	2.04	0.22	4.18	2.74	nr	**6.92**
straight tap connector	2.38	0.30	5.70	3.14	nr	**8.83**
bent tap connector	3.00	0.30	5.70	3.85	nr	**9.55**
angle service valve with tap connector	7.43	0.30	5.70	8.98	nr	**14.68**
stop valve	5.82	0.15	2.85	7.50	nr	**10.35**
brass chrome plated service valve	9.91	0.15	2.85	12.23	nr	**15.08**
brass chrome plated ball valve	12.71	0.15	2.85	15.47	nr	**18.32**
speedfit x union nut flexi hose 500 mm long	6.89	0.30	5.70	8.35	nr	**14.05**
22 mm PEX barrier pipes; fixing with pipe clips; in wall, floor and roof voids	2.43	0.27	5.13	3.90	m	**9.03**

S PIPED SUPPLY SYSTEMS

Item	PC £	Labour hours	Labour £	Material £	Unit	Total rate £
22 mm Polybutylene barrier pipes; fixing with pipe clips; in wall, floor and roof voids	2.78	0.27	5.13	4.31	m	**9.44**
Extra for						
stop end	1.98	0.10	1.90	2.77	nr	**4.66**
straight connector	1.91	0.19	3.61	3.16	nr	**6.77**
reducing coupler	3.36	0.19	3.61	4.84	nr	**8.45**
PE-copper coupler	6.71	0.22	4.18	8.71	nr	**12.89**
elbow	2.29	0.19	3.61	3.60	nr	**7.21**
stem elbow	4.28	0.19	3.61	5.90	nr	**9.51**
tee; equal	3.09	0.29	5.51	5.00	nr	**10.51**
tee; reducing	3.49	0.29	5.51	5.28	nr	**10.78**
tank connector	2.60	0.29	5.51	3.48	nr	**8.99**
straight tap connector	3.10	0.39	7.41	4.06	nr	**11.47**
stop valve	8.85	0.19	3.61	11.19	nr	**14.80**
brass chrome plated service valve	22.24	0.19	3.61	26.68	nr	**30.29**
brass chrome plated ball valve	25.40	0.19	3.61	30.34	nr	**33.95**
speedfit x union nut flexi hose 500 mm long	8.26	0.39	7.41	10.03	nr	**17.44**
22 x 10 4 Way manifold	7.23	0.39	7.41	10.37	nr	**17.78**
22 x 15 4 Port rail manifold	14.91	0.39	7.41	19.25	nr	**26.66**
22 x 15 4 Zone brass rail manifold	223.22	1.08	20.52	260.26	nr	**280.78**
28 mm PEX barrier pipes; fixing with pipe clips; in wall, floor and roof voids	3.34	0.30	5.70	5.94	m	**11.64**
Extra for						
straight connector	4.82	0.26	4.94	6.69	nr	**11.63**
reducer	4.08	0.26	4.94	5.83	nr	**10.77**
elbow	5.64	0.26	4.94	7.64	nr	**12.58**
tee; equal	7.95	0.39	7.41	10.86	nr	**18.27**
tee; reducing	8.77	0.39	7.41	11.65	nr	**19.06**
Water tanks/cisterns						
Polyethylene cold water feed and expansion cistern; BS 4213; with covers						
ref SC15; 68 litres	57.92	1.39	26.41	65.38	nr	**91.78**
ref SC25; 114 litres	68.16	1.61	30.59	76.93	nr	**107.52**
ref SC40; 182 litres	80.05	1.61	30.59	90.36	nr	**120.95**
ref SC50; 227 litres	110.40	2.16	41.03	124.61	nr	**165.65**
GRP cold water storage cistern; with covers						
ref 899.10; 30 litres	136.85	1.22	23.18	154.47	nr	**177.65**
ref 899.25; 68 litres	172.73	1.39	26.41	194.96	nr	**221.37**
ref 899.40; 114 litres	214.75	1.61	30.59	242.40	nr	**272.98**
ref 899.70; 227 litres	269.28	2.16	41.03	303.95	nr	**344.98**
Storage cylinders/calorifiers						
Copper cylinders; single feed coil indirect; BS 1566 Part 2; grade 3						
ref 2; 96 litres	-	2.22	42.17	234.13	nr	**276.30**
ref 3; 114 litres	131.97	2.50	47.49	148.96	nr	**196.46**
ref 7; 117 litres	130.24	2.78	52.81	147.00	nr	**199.81**
ref 8; 140 litres	147.44	3.33	63.26	166.42	nr	**229.68**
ref 9; 162 litres	188.33	3.89	73.90	212.57	nr	**286.47**
Combination copper hot water storage units; coil direct; BS 3198; (hot/cold)						
400 mm x 900 mm; 65/20 litres	153.28	3.11	59.08	173.01	nr	**232.09**
450 mm x 900 mm; 85/25 litres	157.85	4.33	82.26	178.17	nr	**260.43**
450 mm x 1075 mm; 115/25 litres	173.64	5.44	103.34	196.00	nr	**299.34**
450 mm x 1200 mm; 115/45 litres	184.85	6.11	116.07	208.65	nr	**324.72**
Combination copper hot water storage						
450 mm x 900 mm; 85/25 litres	197.96	4.88	92.70	223.45	nr	**316.15**
450 mm x 1200 mm; 115/45 litres	226.37	6.66	126.52	255.51	nr	**382.03**

S PIPED SUPPLY SYSTEMS

Item	PC £	Labour hours	Labour £	Material £	Unit	Total rate £
S12 HOT AND COLD WATER (SMALL SCALE) – cont'd						
Thermal insulation						
20 mm thick Rockwool "Rocklap" bonded pre-formed mineral glass fibre sectional pipe lagging; with aluminium foil finish; fixed to steel or copper pipework; including working over pipe fittings						
around 15/15 pipes	2.18	0.07	1.33	2.58	m	**3.91**
around 20/22 pipes	2.29	0.11	2.09	2.71	m	**4.80**
around 25/28 pipes	2.43	0.12	2.28	2.88	m	**5.16**
around 32/35 pipes	2.64	0.13	2.47	3.13	m	**5.60**
around 40/42 pipes	2.97	0.15	2.85	3.52	m	**6.37**
around 50/54 pipes	3.42	0.17	3.23	4.05	m	**7.28**
30 mm thick Rockwool "Rocklap" bonded pre-formed mineral glass fibre sectional pipe lagging; with aluminium foil finish; fixed to steel or copper pipework; including working over pipe fittings						
around 15/15 pipes	3.23	0.07	1.33	3.83	m	**5.16**
around 20/22 pipes	3.36	0.11	2.09	3.98	m	**6.07**
around 25/28 pipes	3.61	0.12	2.28	4.27	m	**6.55**
around 32/35 pipes	3.79	0.13	2.47	4.49	m	**6.96**
around 40/42 pipes	4.15	0.15	2.85	4.91	m	**7.76**
around 50/54 pipes	4.80	0.17	3.23	5.69	m	**8.92**
60 mm thick glass-fibre filled polyethylene insulating jackets for GRP or polyethylene cold water cisterns; complete with fixing bands; for cisterns size						
450 mm x 300 mm x 300 mm (45 litres)	-	0.44	8.36	-	nr	**8.36**
650 mm x 500 mm x 400 mm (91 litres)	-	0.67	12.73	-	nr	**12.73**
675 mm x 525 mm x 500 mm (136 litres)	-	0.78	14.82	-	nr	**14.82**
675 mm x 575 mm x 525 mm (182 litres)	-	0.89	16.91	-	nr	**16.91**
1000 mm x 625 mm x 525 mm (273 litres)	-	0.94	17.86	-	nr	**17.86**
1125 mm x 650 mm x 575 mm (341 litres)	-	0.94	17.86	-	nr	**17.86**
80 mm thick glass-fibre filled insulating jackets in flame retardant PVC to BS 5615 type 1B; segmental type for hot water cylinders; complete with fixing bands; for cylinders size						
400 mm x 900 mm; ref 2	-	0.37	7.03	-	nr	**7.03**
450 mm x 900 mm; ref 7	-	0.37	7.03	-	nr	**7.03**
450 mm x 1050 mm; ref 8	-	0.44	8.36	-	nr	**8.36**
450 mm x 1200 mm	-	0.56	10.64	-	nr	**10.64**
S13 PRESSURISED WATER						
Blue MDPE pipes; BS 6572; mains pipework; no joints in the running length; laid in trenches						
Pipes						
20 mm nominal size	0.98	0.12	2.28	1.14	m	**3.42**
25 mm nominal size	1.14	0.13	2.47	1.33	m	**3.79**
32 mm nominal size	1.92	0.15	2.85	2.23	m	**5.08**
50 mm nominal size	4.59	0.17	3.23	5.34	m	**8.57**
63 mm nominal size	7.28	0.18	3.42	8.46	m	**11.88**

S PIPED SUPPLY SYSTEMS

Item	PC £	Labour hours	Labour £	Material £	Unit	Total rate £
Ductile iron bitumen coated pipes and fittings; EN598; class K9; Stanton's "Tyton" water main pipes or other equal and approved; flexible joints						
100 mm pipes; laid straight	33.24	0.67	8.72	48.69	m	**57.41**
Extra for						
bend; 45 degrees	54.32	0.67	8.72	82.97	nr	**91.69**
branch; 45 degrees; socketted	395.77	1.00	13.01	496.93	nr	**509.94**
tee	85.76	1.00	13.01	129.51	nr	**142.53**
flanged spigot	54.39	0.67	8.72	73.75	nr	**82.47**
flanged socket	51.74	0.67	8.72	70.61	nr	**79.33**
150 mm pipes; laid straight	39.96	0.78	10.15	57.35	m	**67.50**
Extra for						
bend; 45 degrees	85.04	0.78	10.15	120.78	nr	**130.93**
branch; 45 degrees; socketted	505.11	1.17	15.23	628.63	nr	**643.86**
tee	178.21	1.17	15.23	241.19	nr	**256.42**
flanged spigot	63.08	0.78	10.15	84.76	nr	**94.91**
flanged socket	82.34	0.78	10.15	107.58	nr	**117.74**
200 mm pipes; laid straight	54.62	1.11	14.45	79.19	m	**93.64**
Extra for						
bend; 45 degrees	153.48	1.11	14.45	210.81	nr	**225.25**
branch; 45 degrees; socketted	573.67	1.67	21.73	723.27	nr	**745.00**
tee	244.79	1.67	21.73	333.48	nr	**355.22**
flanged spigot	137.38	1.11	14.45	177.27	nr	**191.72**
flanged socket	130.26	1.11	14.45	168.83	nr	**183.28**

T MECHANICAL HEATING/COOLING SYSTEMS ETC

Item	PC £	Labour hours	Labour £	Material £	Unit	Total rate £
T10 GAS/OIL FIRED BOILERS						
Boilers						
Gas fired wall mounted combination domestic boilers; for central heating and hot water supply; Potterton 'Performa' or equivalent; with cream or white enamelled casing; 32 mm diameter BSPT female flow and return tappings; 102 mm diameter flue socket 13 mm diameter BSPT male draw-off outlet						
24.00 kW output; ref Performa 24	591.22	6.00	130.64	667.34	nr	**797.98**
31.00 kW output; ref Performa 28	591.22	6.00	130.64	667.34	nr	**797.98**
31.00 kW output; ref Performa 28i	591.22	6.00	130.64	667.34	nr	**797.98**
Gas fired wall mounted domestic boilers; for central heating and indirect hot water supply; Potterton 'Profile' or equivalent; with cream or white enamelled casing; 32 mm diameter BSPT female flow and return tappings; 102 mm diameter flue socket 13 mm diameter BSPT male draw-off outlet						
14.60 kW output (50,000 Btu/Hr); ref Profile 50e L	746.23	6.00	130.64	842.30	nr	**972.94**
23.45 kW output (80,000 Btu/Hr); ref Profile 80e L	1063.74	6.00	130.64	1200.69	nr	**1331.33**
Flues						
Scheidel Rite-Vent ICS Plus flue system; suitable for domestic multifuel appliances; stainless steel; twin wall; insulated; for use internally or externally						
80 mm pipes; including one locking band (fixing brackets measured separately)	-	1.00	19.00	98.23	m	**117.23**
Extra for						
Appliance Connecter	-	1.00	19.00	15.06	nr	**34.06**
30° Bend	-	2.00	37.99	56.31	nr	**94.30**
45° Bend	-	2.00	37.99	69.63	nr	**107.63**
135° Tee; fully welded	-	3.00	56.99	53.94	nr	**110.93**
Inspection Length	-	1.00	19.00	232.65	nr	**251.65**
Drain Plug and Support	-	1.00	19.00	91.41	nr	**110.41**
Damper	-	1.00	19.00	54.88	nr	**73.88**
Angled Flashing including Storm Collar	-	1.50	28.50	79.44	nr	**107.93**
Stub Terminal	-	1.00	19.00	24.08	nr	**43.07**
Tapered Terminal	-	1.00	19.00	50.17	nr	**69.16**
Floor Support (2 piece)	-	1.50	28.50	43.68	nr	**72.17**
Firestop Floor Support (2 piece)	-	1.50	28.50	24.48	nr	**52.98**
Wall Support (Stainless Steel)	-	1.00	19.00	184.59	nr	**203.59**
Wall Sleeve	-	1.50	28.50	36.81	nr	**65.31**
100 mm pipes; including one locking band (fixing brackets measured separately)	-	1.00	19.00	104.26	m	**123.26**
Extra for						
Appliance Connecter	-	1.00	19.00	16.63	nr	**35.62**
30° Bend	-	2.50	47.49	58.02	nr	**105.51**
45° Bend	-	2.50	47.49	72.78	nr	**120.27**
135° Tee; fully welded	-	3.50	66.49	56.14	nr	**122.63**
Inspection Length	-	1.00	19.00	238.68	nr	**257.68**
Drain Plug and Support	-	1.50	28.50	96.77	nr	**125.26**
Damper	-	1.00	19.00	60.66	nr	**79.65**
Angled Flashing including Storm Collar	-	1.50	28.50	80.02	nr	**108.51**
Stub Terminal	-	1.50	28.50	24.47	nr	**52.97**
Tapered Terminal	-	1.50	28.50	53.54	nr	**82.04**
Floor Support (2 piece)	-	2.00	37.99	43.68	nr	**81.67**
Firestop Floor Support (2 piece)	-	2.00	37.99	24.48	nr	**62.48**
Wall Support (Stainless Steel)	-	1.50	28.50	190.43	nr	**218.92**
Wall Sleeve	-	1.50	28.50	36.81	nr	**65.31**

T MECHANICAL HEATING/COOLING SYSTEMS ETC

Item	PC £	Labour hours	Labour £	Material £	Unit	Total rate £
150 mm pipes; including one locking band (fixing brackets measured separately)	-	1.50	28.50	122.27	m	**150.77**
Extra for						
Appliance Connecter	-	1.00	19.00	21.44	nr	**40.44**
30° Bend	-	2.50	47.49	75.39	nr	**122.88**
45° Bend	-	2.50	47.49	87.49	nr	**134.98**
135° Tee; fully welded	-	4.00	75.99	68.05	nr	**144.04**
Inspection Length	-	1.50	28.50	250.30	nr	**278.80**
Drain Plug and Support	-	1.50	28.50	107.51	nr	**136.00**
Damper	-	1.50	28.50	75.54	nr	**104.04**
Angled Flashing including Storm Collar	-	2.00	37.99	82.24	nr	**120.23**
Stub Terminal	-	1.50	28.50	26.32	nr	**54.82**
Tapered Terminal	-	1.50	28.50	61.54	nr	**90.03**
Floor Support (2 piece)	-	2.00	37.99	45.06	nr	**83.06**
Firestop Floor Support (2 piece)	-	2.00	37.99	24.48	nr	**62.48**
Wall Support (Stainless Steel)	-	1.50	28.50	245.76	nr	**274.26**
Wall Sleeve	-	1.50	28.50	37.90	nr	**66.40**
T31 LOW TEMPERATURE HOT WATER HEATING						
NOTE: The reader is referred to section "S12 Hot and Cold Water (Small Scale)" for rates for copper pipework which will equally applly to this section of work. For further and more detailed information the reader is advised to consult *Spon's Mechanical and Electrical Services Price Book*.						
Steel radiators and convectors; Hudevad Heat Emitters or other equal and approved						
"Plan Fiona" double panel convector; 600 mm high; front, back plates and convector fins with intergrated top grille; wheelhead and lockshield valves						
500 mm long x 68 mm deep; 584 watts output	98.50	2.00	43.55	129.56	nr	**173.11**
1400 mm long x 68 mm deep; 1634 watts output	238.46	2.50	54.43	287.55	nr	**341.98**
1400 mm long x 98 mm deep; 2022 watts output	266.81	2.50	54.43	319.55	nr	**373.98**
"P5K" horizontal single panel convector; 600 mm high; wheelhead and lockshield valves						
500 mm long; 412 watts output	76.33	2.00	43.55	104.68	nr	**148.22**
1400 mm long; 1154 watts output	156.43	2.50	54.43	195.09	nr	**249.52**
2000 mm long; 1648 watts output	209.98	2.75	59.88	255.53	nr	**315.41**
"P5KV" vertical single panel convector; 600 mm long; wheelhead and lockshield valves						
1400 mm high; 960 watts output	179.84	2.75	59.88	221.51	nr	**281.39**
2200 mm high; 1492 watts output	251.12	3.00	65.32	301.97	nr	**367.29**

V ELECTRICAL SYSTEMS

Item	PC £	Labour hours	Labour £	Material £	Unit	Total rate £
V21/V22 GENERAL LIGHTING AND LV POWER						
NOTE: The following items indicate approximate prices for wiring of lighting and power points complete, including accessories and socket outlets, but excluding lighting fittings. Consumer control units are shown separately. For a more detailed breakdown of these costs and specialist costs for a complete range of electrical items, reference should be made to *Spon's Mechanical and Electrical Services Price Book.*						
Consumer control units						
8-way 60 amp SP&N surface mounted insulated consumer control units fitted with miniature circuit breakers including 2 m long 32 mm screwed welded conduit with three runs of 16 mm2 PVC cables ready for final connections	-	-	-	-	nr	**215.47**
extra for current operated ELCB of 30 mA tripping current	-	-	-	-	nr	**86.19**
As above but 100 amp metal cased consumer unit and 25 mm2 PVC cables	-	-	-	-	nr	**240.09**
extra for current operated ELCB of 30 mA tripping current	-	-	-	-	nr	**197.00**
Final circuits						
Lighting points						
wired in PVC insulated and PVC sheathed cable in flats and houses; insulated in cavities and roof space; protected where buried by heavy gauge PVC conduit	-	-	-	-	nr	**49.25**
as above but in commercial property	-	-	-	-	nr	**67.72**
wired in PVC insulated cable in screwed welded conduit in commercial property	-	-	-	-	nr	**209.31**
as above but in industrial property	-	-	-	-	nr	**227.78**
wired in MICC cable in commercial property	-	-	-	-	nr	**184.68**
as above but in industrial property with PVC sheathed cable	-	-	-	-	nr	**184.68**
Single 13 amp switched socket outlet points						
wired in PVC insulated and PVC sheathed cable in flats and houses on a ring main circuit; protected where buried by heavy gauge PVC conduit	-	-	-	-	nr	**80.03**
as above but in commercial property	-	-	-	-	nr	**92.34**
wired in PVC insulated cable in screwed welded conduit throughout on a ring main in commercial property	-	-	-	-	nr	**215.47**
as above but in industrial property	-	-	-	-	nr	**240.09**
wired in MICC cable on a ring main circuit in commercial property	-	-	-	-	nr	**233.93**
as above but in industrial property with PVC sheathed cable	-	-	-	-	nr	**233.93**
Cooker control units						
45 amp circuit including unit wired in PVC insulated and PVC sheathed cable; protected where buried by heavy gauge PVC conduit	-	-	-	-	nr	**116.97**
as above but wired in PVC insulated cable in screwed welded conduit	-	-	-	-	nr	**270.87**
as above but wired in MICC cable	-	-	-	-	nr	**295.50**

W SECURITY SYSTEMS

Item	PC £	Labour hours	Labour £	Material £	Unit	Total rate £
W20 LIGHTNING PROTECTION						
Lightning protection equipment						
Copper strip roof or down conductors fixed with bracket or saddle clips						
20 mm x 3 mm flat section	-	-	-	-	m	**21.99**
25 mm x 3 mm flat section	-	-	-	-	m	**25.65**
Aluminium strip roof or down conductors fixed with bracket or saddle clips						
20 mm x 3 mm flat section	-	-	-	-	m	**16.13**
25 mm x 3 mm flat section	-	-	-	-	m	**17.60**
Joints in tapes	-	-	-	-	nr	**12.47**
Bonding connections to roof and structural metalwork	-	-	-	-	nr	**73.31**
Testing points	-	-	-	-	nr	**60.24**
Earth electrodes						
16 mm diameter driven copper electrodes in 1220 mm long sectional lengths (minimum 2440 mm long overall)	-	-	-	-	nr	**190.58**
first 2440 mm length driven and tested 25 mm x 3 mm copper strip electrode in 457 mm deep prepared trench	-	-	-	-	m	**14.67**

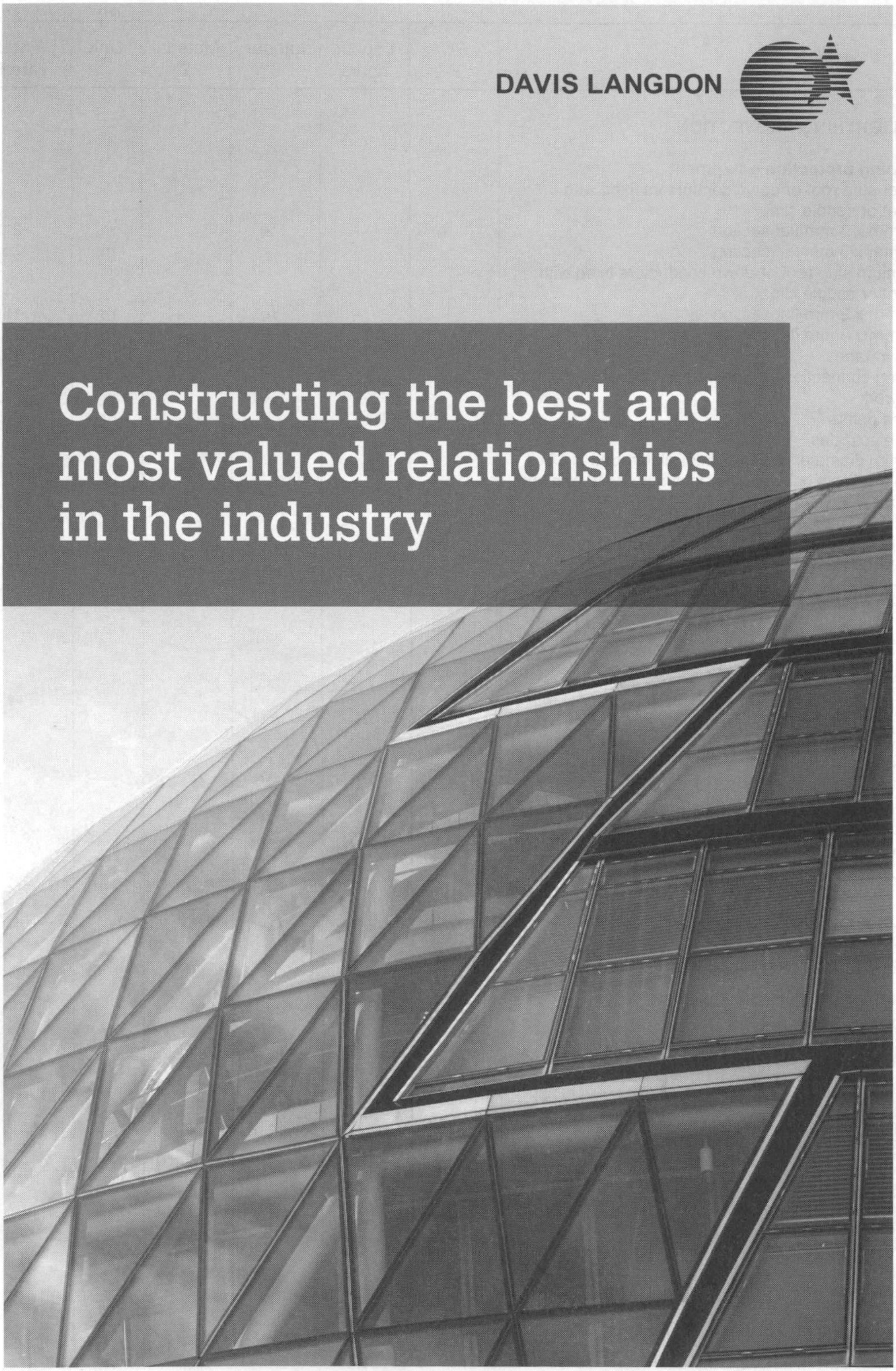

PART 5

Fees for Professional Services

Extracts from the scales of fees for architects, quantity surveyors and consulting engineers are given together with extracts from the Town and Country Planning Regulations 2008 and Building Regulation Charges. These extracts are reproduced by kind permission of the bodies concerned, in the case of Building Regulation Charges, by kind permission of the London Borough of Ealing. Attention is drawn to the fact that the full scales are not reproduced here and that the extracts are given for guidance only. The full authority scales should be studied before concluding any agreement and the reader should ensure that the fees quoted here are still current at the time of reference.

ARCHITECTS' FEES

The format of the RIBA Agreements 2007 is different from previous RIBA Standard Forms of Appointment to suit both paper and electronic usage, allowing users to customise the content of the components to suit their project.

Standard Agreement for the appointment of an Architect (S-Con-07), which replaces SFA/99 and CE/99
Standard Agreement for the appointment of a Consultant (S-Con-07), which replaces PM/99, PS/99 and DB1/99, which is discontinued
Concise Agreement for the appointment of an Architect (C-Con-07), which replaces SW/99
Concise Agreement for the appointment of a Consultant (C-Con-07)
Domestic Project Agreement for the appointment of an Architect (D-Con-07), which replaces the Domestic Project Pack
Domestic Project Agreement for the appointment of a Consultant (D-Con-07
Agreement for the appointment of a Sub-Consultant (SubCon-07), which replaces SC/99
Supplementary Schedule for a Contractor's Design Services (SS-CD-07), which replaces DB2/99

For brief précis on the above agreements, refer to page 798. For a current 'Guide to Architect's fees refer to page 799.

QUANTITY SURVEYORS' FEES

CONSULTING ENGINEERS' FEES

TOWN AND COUNTRY PLANNING REGULATION FEES 2008

THE BUILDING (LOCAL AUTHORITY CHARGES) REGULATIONS 1998

ARCHITECTS' FEES

The RIBA 2007 Agreements have been redesigned to be:

- in line with current working practices, legislative changes and procurement methods;
- attractive to clients, architects and other consultants, with robust but fair terms;
- a flexible system of components that can be assembled and customised to create tailored and bespoke contracts;
- suitable for a wide range of projects and services;
- based upon the updated RIBA Outline Plan of Work 2007;
- available in paper and electronic formats.

Each agreement comprises the selected Conditions of Appointment (i.e. Standard, Concise or Domestic), related components, and a schedule or schedules of Services.

Notes on use and completion and model letters for business clients and domestic clients are included with each pack. The new format provides 'pick and mix' options, perhaps in combination with project-specific schedules. The agreements are also suitable for architects or consultants performing roles other than their traditional ones.

All forms require the Architect to agree with the Client the amount of professional indemnity insurance cover for the project.

Standard Agreement for the appointment of an Architect (AS-Con-07) or a Consultant (CS-Con-07)
The 'core' Conditions of the RIBA Standard Conditions of Appointment set out in explicit terms the obligations of the parties including the rules for the application of particular clauses. They are designed to apportion risk fairly between the architect/consultant and the client, whether or not the client has any experience of building projects.

Concise Agreement for the appointment of an Architect (AC-Con-07) or a Consultant (CC-Con-07)
The obligations are similar to those under the RIBA Standard Conditions, and they include the relevant statutory obligations. However, some of the rules or procedural requirements in the Standard Conditions do not appear. It is, of course, implicit that 'normal standards' are consistent with the requirements of the architect/consultant's professional code of practice.

In deciding to use these Conditions, the parties should carefully consider whether they are compatible with the complexity of the Project and the proposed procurement route, and whether the 'missing' provisions will increase the individual risks of the parties.

Domestic Project Agreement for the appointment of an Architect (AD-Con-07) or a Consultant (CD-Con-07)
D-Con-07 is designed for use where the client requires work on his or her home.

Agreement for the appointment of a Sub-Consultant (SubCon-07)
Suitable for use where a Consultant wishes (another Consultant (Sub-Consultant) or Specialist) to perform a part of his responsibility but not for use where the intention is for the Client to appoint Consultants or Specialists directly. Used with Articles of Agreement. Includes draft form of Warranty to the Client.

Supplementary Schedule for a Contractor's Design Services (SS-CD-07)
A supplement to amend (S-Con-07) where an Architect or Consultant is appointed by the Contractor Client to prepare Contractor's Proposals under a Design and Build contract. Includes replacement Services Supplement and notes on completion for initial appointment and for "consultant switch".

Of the above documents, S-Con-07 is the core document, which is used as the basis for all the documents in the RIBA 2007 suite. It should be suitable for any Project to be procured in the 'traditional' manner. Supplements are available for use with Design and Build procurement. It is used with Articles of Agreement and formal attestation underhand or as a deed.

For further information, Readers are advised to log onto the RIBA Publications web-site at www.ribabookshops.com/agreements

ARCHITECTS' FEES

Guides

A guide, 'A Client's Guide to Engaging an Architect', is available from RIBA Bookshops (www.ribabookshops.com, +44 (0)20 7256 7222)

and at the time of going to press a new edition is expected in October 2008.

This guide includes an introduction to the services an Architect can be expected to provide, advice on the forms to use, linking the RIBA Plan of Work Stages with fees (which are a matter of negotiation) and classifying buildings according to three levels of complexity.

Generally, the more complex the building, the higher the level of fee.

Example categories include:

Simple: for buildings such as car parks, warehouses, factories and speculative retail schemes

Average: for buildings such as offices, most retail outlets, general housing, schools etc., and

Complex: for multi-purpose developments, specialist buildings e.g. hospitals, research laboratories etc.

QUANTITY SURVEYORS' FEES

Author's Note:

The Royal Institution of Chartered Surveyors formally abolished the standard Quantity Surveyors' fee scales with effect from 31 December 1998. However, in the absence of any alternative guideline and for the benefit of readers, the following fee scales have been reproduced with the permission of the Royal Institution of Chartered Surveyors, which owns the copyright.

SCALE 36 INCLUSIVE OF PROFESSIONAL CHARGES FOR QUANTITY SURVEYING SERVICES FOR BUILDING WORKS ISSUED BY THE ROYAL INSTITUTION OF CHARTERED SURVEYORS. This scale has been abolished. See Author's Note above.

EFFECTIVE FROM JULY 1988

1.0. GENERALLY

1.1 This scale is for the use when an inclusive scale of professional charges is considered to appropriate by mutual agreement between the employer and the quantity surveyor.

1.2. This scale does not apply to civil engineering works, housing schemes financed by local authorities and the Housing Corporation and housing improvement work for which separate scales of fees have been published.

1.3. The fees cover quantity surveying services as may be required in connection with a building project irrespective of the type of contract from initial appointment to final certification of the contractor's account such as:

(a) Budget estimating; cost planning and advice on tendering procedures and contract arrangements.

(b) Preparing tendering documents for main contract and specialist sub-contracts; examining tenders received and reporting thereon or negotiating tenders and pricing with a selected contractor and/or sub-contractors.

(c) Preparing recommendations for interim payments on account to the contractor; preparing periodic assessments of anticipated final cost and reporting thereon; measuring work and adjusting variations in accordance with the terms of the contract and preparing final account, pricing same and agreeing totals with the contractor.

(d) Providing a reasonable number of copies of bills of quantities and other documents; normal travelling and other expenses. Additional copies of documents, abnormal travelling and other expenses (e.g. in remote areas or overseas) and the provision of checkers on site shall be charged in addition by prior arrangement with the employer.

1.4. If any of the materials used in the works are supplied by the employer or charged at a preferential rate, then the actual or estimated market value thereof shall be included in the amounts upon which fees are to be calculated.

1.5. If the quantity surveyor incurs additional costs due to exceptional delays in building operations or any other cause beyond the control of the quantity surveyor then the fees may be adjusted by agreement between the employer and the quantity surveyor to cover the reimbursement of these additional costs.

1.6. The fees and charges are in all cases exclusive of value added tax which will be applied in accordance with legislation.

1.7. Copyright in bills of quantities and other documents prepared by the quantity surveyor is reserved to the quantity surveyor.

QUANTITY SURVEYORS' FEES

2.0. INCLUSIVE SCALE

2.1. The fees for the services outlined in para.1.3, subject to the provision of para. 2.2, shall be as follows:

(a) Category A: Relatively complex works and/or works with little or no repetition.

Examples:
Ambulance and fire stations; banks; cinemas; clubs; computer buildings; council offices; crematoria; fitting out of existing buildings; homes for the elderly; hospitals and nursing homes; laboratories; law courts; libraries; "one off" houses; petrol stations; places of religious worship; police stations; public houses, licensed premises; restaurants; sheltered housing; sports pavilions; theatres; town halls; universities, polytechnics and colleges of further education (other than halls of residence and hostels); and the like.

Value of work	£	Category A fee			
		£			£
Up to	150 000	380	+ 6.0%	(Minimum fee £3 380)	
150 000 -	300 000	9 380	+ 5.0%	on balance over	150 000
300 000 -	600 000	16 880	+ 4.3%	on balance over	300 000
600 000 -	1 500 000	29 780	+ 3.4%	on balance over	600 000
1 500 000 -	3 000 000	60 380	+ 3.0%	on balance over	1 500 000
3 000 000 -	6 000 000	105 380	+ 2.8%	on balance over	3 000 000
Over	6 000 000	189 380	+ 2.4%	on balance over	6 000 000

(b) Category B: Less complex works and/or works with some element of repetition.

Examples:
Adult education facilities; canteens; church halls; community centres; departmental stores; enclosed sports stadia and swimming baths; halls of residence; hostels; motels; offices other than those included in Categories A and C; railway stations; recreation and leisure centres; residential hotels; schools; self-contained flats and maisonettes; shops and shopping centres; supermarkets and hypermarkets; telephone exchanges; and the like.

Value of work	£	Category B fee			
		£			£
Up to	150 000	360	+ 5.8%	(Minimum fee £3 260)	
150 000 -	300 000	9 060	+ 4.7%	on balance over	150 000
300 000 -	600 000	16 110	+ 3.9%	on balance over	300 000
600 000 -	1 500 000	27 810	+ 2.8%	on balance over	600 000
1 500 000 -	3 000 000	53 010	+ 2.6%	on balance over	1 500 000
3 000 000 -	6 000 000	92 010	+ 2.4%	on balance over	3 000 000
Over	6 000 000	164 101	+ 2.0%	on balance over	6 000 000

(c) Category C: Simple works and/or works with a substantial element of repetition.

Examples:
Factories; garages; multi-storey car parks; open-air sports stadia; structural shell offices not
Fitted out; warehouses; workshops; and the like.

QUANTITY SURVEYORS' FEES

Value of work	£	Category C fee		
		£		£
Up to	150 000	300 + 4.9%	(Minimum fee £2 750)	
150 000 -	300 000	7 650 + 4.1%	on balance over	150 000
300 000 -	600 000	13 800 + 3.3%	on balance over	300 000
600 000 -	1 500 000	23 700 + 2.5%	on balance over	600 000
1 500 000 -	3 000 000	46 200 + 2.2%	on balance over	1 500 000
3 000 000 -	6 000 000	79 200 + 2.0%	on balance over	3 000 000
Over	6 000 000	139 200 + 1.6%	on balance over	6 000 000

(d) Fees shall be calculated upon the total of the final account for the whole of the work including all nominated sub-contractors' and nominated supplier's accounts. When work normally included in a building contract is the subject of a separate contract for which the quantity surveyor has not been paid fees under any other clause hereof, the value of such work shall be included in the amount upon which fees are charged.

(e) When a contract comprises buildings which fall into more than one category, the fee shall be calculated as follows:

(i) The amount upon which fees are chargeable shall be allocated to the categories of work applicable and the amounts so allocated expressed as percentages of the total amount upon which fees are chargeable.

(ii) Fees shall then be calculated for each category on the total amount upon which fees are chargeable.

(iii) The fee chargeable shall then be calculated by applying the percentages of work in each category to the appropriate total fee and adding the resultant amounts.

(iv) A consolidated percentage fee applicable to the total value of the work may be charged by prior agreement between the employer and the quantity surveyor. Such a percentage shall be based on this scale and on the estimated cost of the various categories of work and calculated in accordance with the principles stated above.

(f) When a project is subject to a number of contracts then, for the purpose of calculating fees, the values of such contracts shall not be aggregated but each contract shall be taken separately and the scale of charges (paras. 2.1 (a) to (e)) applied as appropriate.

2.2. Air conditioning, heating, ventilating and electrical services

(a) When the services outlined in para. 1.3 are provided by the quantity surveyor for the air conditioning, heating, ventilating and electrical services there shall be a fee for these services in addition to the fee calculated in accordance with para. 2.1 as follows:

Value of work	£	Additional fee		
		£		£
Up to	120 000	5.0%		
120 000 -	240 000	6 000 + 4.7%	on balance over	120 000
240 000 -	480 000	11 640 + 4.0%	on balance over	240 000
480 000 -	750 000	21 240 + 3.6%	on balance over	480 000
750 000 -	1 000 000	30 960 + 3.0%	on balance over	750 000
1 000 000 -	4 000 000	38 460 + 2.7%	on balance over	1 000 000
Over	4 000 000	119 200 + 2.4%	on balance over	4 000 000

QUANTITY SURVEYORS' FEES

(b) The value of such services, whether the subject of separate tenders or not, shall be aggregated and the total value of work so obtained used for the purpose of calculating the additional fee chargeable in accordance with para. (a). (Except that when more than one firm of consulting engineers is engaged on the design of these services, the separate values for which each such firm is responsible shall be aggregated and the additional fees charged shall be calculated independently on each such total value so obtained.)

(c) Fees shall be calculated upon the basis of the account for the whole of the air conditioning, heating, ventilating and electrical services for which bills of quantities and final accounts have been prepared by the quantity surveyor.

2.3 Works of alteration
On works of alteration or repair, or on those sections of the work which are mainly works of alteration or repair, there shall be a fee of 1.0% in addition to the fee calculated in accordance with paras. 2.1 and 2.2.

2.4. Works of redecoration and associated minor repairs
On works of redecoration and associated minor repairs, there shall be a fee of 1.5% in addition to the fee calculated in accordance with paras. 2.1 and 2.2.

2.5. Generally
If the works are substantially varied at any stage or if the quantity surveyor is involved in an excessive amount of abortive work, then the fees shall be adjusted by agreement between the employer and the quantity surveyor.

3.0. ADDITIONAL SERVICES

3.1. For additional services not normally necessary, such as those arising as a result of the termination of a contract before completion, liquidation, fire damage to the buildings, services in connection with arbitration, litigation and investigation of the validity of contractors' claims, services in connection with taxation matters, and all similar services where the employer specifically instructs the quantity surveyor, the charges shall be in accordance with para. 4.0.

4.0. TIME CHARGES

4.1. (a) For consultancy and other services performed by a principal, a fee by arrangement according to the circumstances including the professional status and qualifications of the quantity surveyor.

(b) When a principal does work which would normally be done by a member of staff, the charge shall be calculated as para. 4.2 below.

4.2. (a) For services by a member of staff, the charges for which are to be based on the time involved, such charges shall be calculated on the hourly cost of the individual involved plus 145%.

(b) A member of staff shall include a principal doing work normally done by an employee (as para. 4.1 (b) above), technical and supporting staff, but shall exclude secretarial staff or staff engaged upon general administration.

(c) For the purpose of para. 4.2 (b) above, a principal's time shall be taken at the rate applicable to a senior assistant in the firm.

(d) The supervisory duties of a principal shall be deemed to be included in the addition of 145% as para. 4.2 (a) above and shall not be charged separately.

(e) The hourly cost to the employer shall be calculated by taking the sum of the annual cost of the member of staff of:

(i) Salary and bonus but excluding expenses;

(ii) Employer's contributions payable under any Pension and Life Assurance Schemes;

QUANTITY SURVEYORS' FEES

(iii) Employer's contributions made under the National Insurance Acts, the Redundancy Payments Act and any other payments made in respect of the employee by virtue of any statutory requirements; and

(iv) Any other payments or benefits made or granted by the employer in pursuance of the terms of employment of the member of staff;

and dividing by 1,650.

5.0. INSTALMENT PAYMENTS

5.1 In the absence of agreement to the contrary, fees shall be paid by instalments as follows:

(a) Upon acceptance by the employer of a tender for the works, one half of the fee calculated on the amount of the accepted tender.

(b) The balance by instalments at intervals to be agreed between the date of the first certificate and one month after final certification of the contractor's account.

5.2. (a) In the event of no tender being accepted, one half of the fee shall be paid within three months of completion of the tender documents. The fee shall be calculated upon the basis of the lowest original bona fide tender received. In the event of no tender being received, the fee shall be calculated upon a reasonable valuation of the works based upon the tender documents.

(b) In the event of the project being abandoned at any stage other than those covered by the foregoing, the proportion of fee payable shall be by agreement between the employer and the quantity surveyor.

NOTE: In the foregoing context "bona fide tender" shall be deemed to mean a tender submitted in good faith without major errors of computation and not subsequently withdrawn by the tenderer.

SCALE 37 ITEMISED SCALE OF PROFESSIONAL CHARGES FOR QUANTITY SURVEYING SERVICES FOR BUILDING WORK ISSUED BY THE ROYAL INSTITUTION OF CHARTERED SURVEYORS.
This scale has been abolished. See Author's Note on page 800.

EFFECTIVE FROM JULY 1988

1.0. GENERALLY

1.1. The fees are in all cases exclusive of travelling and other expenses (for which the actual disbursement is recoverable unless there is some prior arrangement for such charges) and of the cost of reproduction of bills of quantities and other documents, which are chargeable in addition at net cost.

1.2. The fees are in all cases exclusive of services in connection with the allocation of the cost of the works for purposes of calculating value added tax for which there shall be an additional fee based on the time involved (see paras. 19.1 and 19.2).

1.3. If any of the materials used in the works are supplied by the employer or charged at a preferential rate, then the actual or estimated market value thereof shall be included in the amounts upon which fees are to be calculated.

1.4. The fees are in all cases exclusive of preparing a specification of the materials to be used and the works to be done, but the fees for preparing bills of quantities and similar documents do include for incorporating preamble clauses describing the materials and workmanship (from instructions given by the architect and/or consulting engineer).

QUANTITY SURVEYORS' FEES

1.5. If the quantity surveyor incurs additional costs due to exceptional delays in building operations or any other cause beyond the control of the quantity surveyor then the fees may be adjusted by agreement between the employer and the quantity surveyor to cover the reimbursement of these additional costs.

1.6. The fees and charges are in all cases exclusive of value added tax which will be applied in accordance with legislation.

1.7. Copyright in bills of quantities and other documents prepared by the quantity surveyor is reserved to the quantity surveyor.

CONTRACTS BASED ON BILLS OF QUANTITIES: PRE-CONTRACT SERVICES

2.0. BILLS OF QUANTITIES

2.1. Basic scale

For preparing bills of quantities and examining tenders received and reporting thereon.

(a) Category A: Relatively complex works and/or works with little or no repetition.

Examples:
Ambulance and fire stations; banks; cinemas; clubs; computer buildings; council offices; crematoria; fitting out of existing buildings; homes for the elderly; hospitals and nursing homes; laboratories; law courts; libraries; "one off" houses; petrol stations; places of religious worship; police stations; public houses; licensed premises; restaurants; sheltered housing; sports pavilions; theatres; town halls; universities, polytechnics and colleges of further education (other than halls of residence and hostels); and the like.

Value of work £		Category A fee				
					£	£
Up to	150 000	230	+	3.0%	(Minimum fee £1730)	
150 000 -	300 000	4 730	+	2.3%	on balance over	150 000
300 000 -	600 000	8 180	+	1.8%	on balance over	300 000
600 000 -	1 500 000	13 580	+	1.5%	on balance over	600 000
1 500 000 -	3 000 000	27 080	+	1.2%	on balance over	1 500 000
3 000 000 -	6 000 000	45 080	+	1.1%	on balance over	3 000 000
Over	6 000 000	78 080	+	1.0%	on balance over	6 000 000

(b) Category B: Less complex works and/or works with some element of repetition.

Examples:
Adult education facilities; canteens; church halls; community centres; departmental stores; enclosed sports stadia and swimming baths; halls of residence; hostels; motels; offices other than those included in Categories A and C; railway stations; recreation and leisure centres; residential hotels; schools; self-contained flats and maisonettes; shops and shopping centres; supermarkets and hypermarkets; telephone exchanges; and the like.

QUANTITY SURVEYORS' FEES

Value of work	£	Category B fee		
		£		£
Up to	150 000	210 + 2.8%	(Minimum fee £1 680)	
150 000 -	300 000	4 410 + 2.0%	on balance over	150 000
300 000 -	600 000	7 410 + 1.5%	on balance over	300 000
600 000 -	1 500 000	11 910 + 1.1%	on balance over	600 000
1 500 000 -	3 000 000	21 810 + 1.0%	on balance over	1 500 000
3 000 000 -	6 000 000	36 810 + 0.9%	on balance over	3 000 000
Over	6 000 000	63 810 + 0.8%	on balance over	6 000 000

(c) Category C: Simple works and/or works with a substantial element of repetition

Examples:
Factories; garages; multi-storey car parks; open-air sports stadia; structural shell offices not fitted out; warehouses; workshops and the like.

Value of work	£	Category C fee		
		£		£
Up to	150 000	180 + 2.5%	(Minimum fee £1 430)	
150 000 -	300 000	3 930 + 1.8%	on balance over	150 000
300 000 -	600 000	6 630 + 1.2%	on balance over	300 000
600 000 -	1 500 000	10 230 + 0.9%	on balance over	600 000
1 500 000 -	3 000 000	18 330 + 0.8%	on balance over	1 500 000
3 000 000 -	6 000 000	30 330 + 0.7%	on balance over	3 000 000
Over	6 000 000	51 330 + 0.6%	on balance over	6 000 000

(d) The scales of fees for preparing bills of quantities (paras. 2.1 (a) to (c)) are overall scales based upon the inclusion of all provisional and prime cost items, subject to the provision of para. 2.1 (g). When work normally included in a building contract is the subject of a separate contract for which the quantity surveyor has not been paid fees under any other clause hereof, the value of such work shall be included in the amount upon which fees are charged.

(e) Fees shall be calculated upon the accepted tender for the whole of he work subject to the provisions of para. 2.6. In the event of no tender being accepted, fees shall be calculated upon the basis of the lowest original bona fide tender received. In the event of no such tender being received, the fees shall be calculated upon a reasonable valuation of the works based upon the original bills of quantities.

NOTE: In the foregoing context "bona fide tender" shall be deemed to mean a tender submitted in good faith without major errors of computation and not subsequently withdrawn by the tenderer.

(f) In calculating the amount upon which fees are charged the total of any credits and the totals of any alternative bills shall be aggregated and added to the amount described above. The value of any omission or addition forming part of an alternative bill shall not be added unless measurement or abstraction from the original dimension sheets was necessary.

(g) Where the value of the air conditioning, heating, ventilating and electrical services included in the tender documents together exceeds 25% of the amount calculated as described in paras. 2.1 (d) and (e), then, subject to the provisions of para. 2.2, no fee is chargeable on the amount by which the value of these services exceeds the said 25%. In this context the term "value" excludes general contractor's profit, attendance, builder's work in connection with the services, preliminaries and any similar additions.

QUANTITY SURVEYORS' FEES

(h) When a contract comprises buildings which fall into more than one category, the fee shall be calculated as follows:

(i) The amount upon which fees are chargeable shall be allocated to the categories of work applicable and the amounts so allocated expressed as percentages of the total amount upon which fees are chargeable.
(ii) Fees shall then be calculated for each category on the total amount upon which fees are chargeable.
(iii) The fee chargeable shall then be calculated by applying the percentages of work in each category to the appropriate total fee and adding the resultant amounts.

(j) When a project is the subject of a number of contracts then, for the purpose of calculating fees, the values of such contracts shall not be aggregated but each contract shall be taken separately and the scale of charges (paras. 2.1 (a) to (h)) applied as appropriate.
(k) Where the quantity surveyor is specifically instructed to provide cost planning services the fee calculated in accordance with paras. 2.1 (a) to (j) shall be increased by a sum calculated in accordance with the following table and based upon the same value of work as that upon which the aforementioned fee has been calculated:

Categories A & B: (as defined in paras. 2.1 (a) and (b)).

Value of work	£	Fee		
		£		£
Up to	600 000	0.70%		
600 000 -	3 000 000	4 200 + 0.40%	on balance over	600 000
3 000 000 -	6 000 000	13 800 + 0.35%	on balance over	3 000 000
Over	6 000 000	24 300 + 0.30%	on balance over	6 000 000

Category C: (as defined in paras. 2.1 (c))

Value of work	£	Fee		
		£		£
Up to	600 000	0.50%		
600 000	3 000 000	3 000 + 0.30%	on balance over	600 000
3 000 000	6 000 000	10 200 + 0.25%	on balance over	3 000 000
Over	6 000 000	17 700 + 0.20%	on balance over	6 000 000

2.2. Air conditioning, heating, ventilating and electrical services

(a) Where bills of quantities are prepared by the quantity surveyor for the air conditioning, heating, ventilating and electrical services there shall be a fee for these services (which shall include examining tenders received and reporting thereon), in addition to the fee calculated in accordance with para. 2.1, as follows:

Value of work	£	Additional fee		
		£		£
Up to	120 000	2.50%		
120 000 -	240 000	3 000 + 2.25%	on balance over	120 000
240 000 -	480 000	5 700 + 2.00%	on balance over	240 000
480 000 -	750 000	10 500 + 1.75%	on balance over	480 000
750 000 -	1 000 000	15 225 + 1.25%	on balance over	750 000
Over	1 000 000	18 350 + 1.15%	on balance over	1 000 000

QUANTITY SURVEYORS' FEES

(b) The values of such services, whether the subject of separate tenders or not, shall be aggregated and the total value of work so obtained used for the purpose of calculating the additional fee chargeable in accordance with para. (a). (Except that when more than one firm of consulting engineers is engaged on the design of these services, the separate values for which each such firm is responsible shall be aggregated and the additional fees charged shall be calculated independently on each such total value so obtained.)

(c) Fees shall be calculated upon the accepted tender for the whole of the air conditioning, heating, ventilating and electrical services for which bills of quantities have been prepared by the quantity surveyor. In the event of no tender being accepted, fees shall be calculated upon the basis of the lowest original bona fide tender received. In the event of no such tender being received, the fees shall be calculated upon a reasonable valuation of the services based upon the original bills of quantities.

NOTE In the foregoing context "bona fide tender"' shall be deemed to mean a tender submitted in good faith without major errors of computation and not subsequently withdrawn by the tenderer.

(d) When cost planning services are provided by the quantity surveyor for air conditioning, heating, ventilating and electrical services (or for any part of such services) there shall be an additional fee based on the time involved (see paras. 19.1 and 19.2). Alternatively the fee may be on a lump sum or percentage basis agreed between the employer and the quantity surveyor.

NOTE The incorporation of figures for air conditioning, heating, ventilating and electrical services provided by the consulting engineer is deemed to be included in the quantity surveyor's services under para. 2.1.

2.3. Works of alteration

On works of alteration or repair, or on those sections of the works which are mainly works of alteration or repair, there shall be a fee of 1.0% in addition to the fee calculated in accordance with paras. 2.1 and 2.2.

2.4. Works of redecoration and associated minor repairs,

On works of redecoration and associated minor repairs, there shall be a fee of 1.5% in addition to the fee calculated in accordance with paras. 2.1 and 2.2.

2.5. Bills of quantities prepared in special forms

Fees calculated in accordance with paras. 2.1, 2.2, 2.3 and 2.4 include for the preparation of bills of quantities on a normal trade basis. If the employer requires additional information to be provided in the bills of quantities or the bills to be prepared in an elemental, operational or similar form, then the fee may be adjusted by agreement between the employer and the quantity surveyor.

2.6. Reduction of tenders

(a) When cost planning services have been provided by the quantity surveyor and a tender, when received, is reduced before acceptance, and if the reductions are not necessitated by amended instructions of the employer or by the inclusion in the bills of quantities of items which the quantity surveyor has indicated could not be contained within the approved estimate, then in such a case no charge shall be made by the quantity surveyor for the preparation of bills of reductions and the fee for the preparation of the bills of quantities shall be based on the amount of the reduced tender.

(b) When cost planning services have not been provided by the quantity surveyor and if a tender, when received, is reduced before acceptance, fees are to be calculated upon the amount of the unreduced tender. When the preparation of bills of reductions is required, a fee is chargeable for preparing such bills of reductions as follows:

QUANTITY SURVEYORS' FEES

(i) 2.0% upon the gross amount of all omissions requiring measurement or abstraction from original dimensional sheets.
(ii) 3.0% upon the gross amount of all additions requiring measurement.
(iii) 0.5% upon the gross amount of all remaining additions.

NOTE: The above scale for the preparation of bills of reductions applies to work in all categories.

2.7 Generally
If the works are substantially varied at any stage or if the quantity surveyor is involved in an excessive amount of abortive work, then the fees shall be adjusted by agreement between the employer and the quantity surveyor.

3.0. NEGOTIATING TENDERS

3.1. (a) For negotiating and agreeing prices with a contractor:

Value of work	£	Fee		
		£		£
Up to	150 000	0.5%		
150 000	600 000	750 + 0.3%	on balance over	150 000
600 000	1 200 000	2 100 + 0.2%	on balance over	600 000
Over	1 200 000	3 300 + 0.1%	on balance over	1 200 000

(b) The fee shall be calculated on the total value of the works as defined in paras. 2.1 (d), (e), (f), (g) and (j).

(c) For negotiating and agreeing prices with a contractor for air conditioning, heating, ventilating and electrical services there shall be an additional fee as para. 3.1 (a) calculated on the total value of such services as defined in para. 2.2 (b).

4.0. CONSULTATIVE SERVICES AND PRICING BILLS OF QUANTITIES

4.1. Consultative services
Where the quantity surveyor is appointed to prepare approximate estimates, feasibility studies or submissions for the approval of financial grants or similar services, then the fee shall be based on the time involved (see paras. 19.1 and 19.2) or alternatively, on a lump sum or percentage basis agreed between the employer and the quantity surveyor.

4.2. Pricing bills of quantities

(a) For pricing bills of quantities, if instructed, to provide an estimate comparable with tenders, the fee shall be one-third (33.33%) of the fee for negotiating and agreeing prices with a contractor, calculated in accordance with paras. 3.1 (a) and (b).

(b) For pricing bills of quantities, if instructed, to provide an estimate comparable with tenders for air conditioning, heating, ventilating and electrical services the fee shall be one-third (33.33%) of the fee calculated in accordance with para. 3.1. (c).

CONTRACTS BASED ON BILLS OF QUANTITIES: POST-CONTRACT SERVICES

Alternative scales (I and II) for post-contract services are set out below to be used at the quantity surveyor's discretion by prior agreement with the employer.

5.0. ALTERNATIVE I: OVERALL SCALE OF CHARGES FOR POST-CONTRACT SERVICES

5.1. If the quantity surveyor appointed to carry out the post-contract services did not prepare the bills of quantities then the fees in paras. 5.2 and 5.3 shall be increased to cover the additional services undertaken by the quantity surveyor.

QUANTITY SURVEYORS' FEES

5.2. Basic scale

For taking particulars and reporting valuations for interim certificates for payments on account to the contractor, preparing periodic assessments of anticipated final cost and reporting thereon, measuring and making up bills of variations including pricing and agreeing totals with the contractor, and adjusting fluctuations in the cost of labour and materials if required by the contract.

(a) Category A: Relatively complex works and/or works with little or no repetition.

Examples:
Ambulance and fire stations; banks; cinemas; clubs; computer buildings; council offices; crematoria; fitting out existing buildings; homes for the elderly; hospitals and nursing homes; laboratories; law courts; libraries; "one-off" houses; petrol stations; places of religious worship; police stations; public houses; licensed premises; restaurants; sheltered housing; sports pavilions; theatres; town halls; universities, polytechnics and colleges of further education (other than halls of residence and hostels); and the like.

Value of work	£	Category A fee		
		£		£
Up to	150 000	150 + 2.0%	(Minimum fee £1 150)	
150 000 -	300 000	3 150 + 1.7%	on balance over	150 000
300 000 -	600 000	5 700 + 1.6%	on balance over	300 000
600 000 -	1 500 000	10 500 + 1.3%	on balance over	600 000
1 500 000 -	3 000 000	22 200 + 1.2%	on balance over	1 500 000
3 000 000 -	6 000 000	40 200 + 1.1%	on balance over	3 000 000
Over	6 000 000	73 200 + 1.0%	on balance over	6 000 000

(b) Category B: Less complex works and/or works with some element of repetition.

Examples:
Adult education facilities; canteens; church halls; community centres; departmental stores; enclosed sports stadia and swimming baths; halls of residence; hostels; motels; offices other than those included in Categories A and C; railway stations; recreation and leisure centres; residential hotels; schools; self-contained flats and maisonettes; shops and shopping centres; supermarkets and hypermarkets; telephone exchanges; and the like.

Value of work	£	Category B fee		
		£		£
Up to	150 000	150 + 2.0%	(Minimum fee £1 150)	
150 000 -	300 000	3 150 + 1.7%	on balance over	150 000
300 000 -	600 000	5 700 + 1.5%	on balance over	300 000
600 000 -	1 500 000	10 200 + 1.1%	on balance over	600 000
1 500 000 -	3 000 000	20 100 + 1.0%	on balance over	1 500 000
3 000 000 -	6 000 000	35 100 + 0.9%	on balance over	3 000 000
Over	6 000 000	62 100 + 0.8%	on balance over	6 000 000

QUANTITY SURVEYORS' FEES

(c) Category C: Simple works and/or works with a substantial element of repetition.

Examples:
Factories; garages; multi-storey car parks; open-air sports stadia; structural shell offices not fitted out; warehouses; workshops; and the like.

Value of work	£	Category C fee		
		£		£
Up to	150 000	120 + 1.6%	(Minimum fee £920)	
150 000 -	300 000	2 520 + 1.5%	on balance over	150 000
300 000 -	600 000	4 770 + 1.4%	on balance over	300 000
600 000 -	1 500 000	8 970 + 1.1%	on balance over	600 000
1 500 000 -	3 000 000	18 870 + 0.9%	on balance over	1 500 000
3 000 000 -	6 000 000	32 370 + 0.8%	on balance over	3 000 000
Over	6 000 000	56 370 + 0.7%	on balance over	6 000 000

(d) The scales of fees for post-contract services (paras. 5.2 (a) to (c)) are overall scales based upon the inclusion of all nominated sub-contractors' and nominated suppliers' accounts, subject to the provision of para. 5.2 (g). When work normally included in a building contract is the subject of a separate contract for which the quantity surveyor has not been paid fees under any other clause hereof, the value of such work shall be included in the amount on which fees are charged.

(e) Fees shall be calculated upon the basis of the account for the whole of the work, subject to the provisions of para. 5.3.

(f) In calculating the amount on which fees are charged the total of any credits is to be added to the amount described above.

(g) Where the value of air conditioning, heating, ventilating and electrical services included in the tender documents together exceeds 25% of the amount calculated as described in paras. 5.2. (d) and (e) above, then, subject to provisions of para. 5.3, no fee is chargeable on the amount by which the value of these services exceeds the said 25%. In this context the term "value" excludes general contractors' profit, attendance, builders work in connection with the services, preliminaries and other similar additions.

(h) When a contract comprises buildings which fall into more than one category, the fee shall be calculated as follows:

(i) The amount upon which fees are chargeable shall be allocated to the categories of work applicable and the amounts so allocated expressed as percentages of the total amount upon which fees are chargeable.

(ii) Fees shall then be calculated for each category on the total amount upon which fees are chargeable.

(iii) The fee chargeable shall then be calculated by applying the percentages of work in each category to the appropriate total fee and adding the resultant amounts.

(j) When a project is the subject of a number of contracts then, for the purposes of calculating fees, the values of such contracts shall not be aggregated but each contract shall be taken separately and the scale of charges (paras. 5.2 (a) to (h)), applied as appropriate.

(k) When the quantity surveyor is required to prepare valuations of materials or goods off site, an additional fee shall be charged based on the time involved (see paras. 19.1 and 19.2).

(l) The basic scale for post-contract services includes for a simple routine of periodically estimating final costs. When the employer specifically requests a cost monitoring service which involves the quantity surveyor in additional or abortive measurement an additional fee shall be charged based on the time involved (see paras. 19.1 and 19.2),

QUANTITY SURVEYORS' FEES

or alternatively on a lump sum or percentage basis agreed between the employer and the quantity surveyor.

(m) The above overall scales of charges for post-contract services assume normal conditions when the bills of quantities are based on drawings accurately depicting the building work the employer requires. If the works are materially varied to the extent that substantial re-measurement is necessary then the fee for post contract services shall be adjusted by agreement between the employer and the quantity surveyor.

5.3. Air conditioning, heating, ventilating and electrical services

(a) Where final accounts are prepared by the quantity surveyor for the air conditioning, heating, ventilating and electrical services there shall be a fee for these services, in addition to the fee calculated in accordance with para. 5.2, as follows:

Value of work	£	Additional fee		
		£		£
Up to	120 000	2.00%		
120 000 -	240 000	2 400 + 1.60%	on balance over	120 000
240 000 -	1 000 000	4 320 + 1.25%	on balance over	240 000
1 000 000 -	4 000 000	13 820 + 1.00%	on balance over	1 000 000
Over	4 000 000	43 820 + 0.90%	on balance over	4 000 000

(b) The values of such services, whether the subject of separate tenders or not, shall be aggregated and the total value of work so obtained used for the purpose of calculating the additional fee chargeable in accordance with para. (a). (Except that when more than one firm of consulting engineers is engaged on the design of these services the separate values for which each such firm is responsible shall be aggregated and the additional fee charged shall be calculated independently on each such total value so obtained.)

(c) The scope of the services to be provided by the quantity surveyor under para. (a) above shall be deemed to be equivalent to those described for the basic scale for post-contract services.

(d) When the quantity surveyor is required to prepare periodic valuations of materials or goods off site, an additional fee shall be charged based on the time involved (see paras. 19.1 and 19.2).

(e) The basic scale for post-contract services includes for a simple routine of periodically estimating final costs. When the employer specifically requests a cost monitoring service which involves the quantity surveyor in additional or abortive measurement an additional fee shall be based on the time involved (see paras. 19.1 and 19.2), or alternatively on a lump sum or percentage basis agreed between the employer and the quantity surveyor.

(f) Fees shall be calculated upon the basis of the account for the whole of the air conditioning, heating, ventilating and electrical services for which final accounts have been prepared by the quantity surveyor.

6.0. ALTERNATIVE II: SCALE OF CHARGES FOR SEPARATE STAGES OF POST-CONTRACT SERVICES

6.1. If the quantity surveyor appointed to carry out the post-contract services did not prepare the bills of quantities then the fees in paras. 6.2 and 6.3 shall be increased to cover the additional services undertaken by the quantity surveyor.

NOTE: The scales of fees in paras. 6.2 and 6.3 apply to work in all categories (including air conditioning, heating, ventilating and electrical services).

QUANTITY SURVEYORS' FEES

6.2. Valuations for interim certificates

(a) For taking particulars and reporting valuations for interim certificates for payments on account to the contractor.

Total of valuations £		Fee		
		£		£
Up to	300 000	0.5%		
300 000 -	1 000 000	1 500 + 0.4%	on balance over	300 000
1 000 000 -	6 000 000	4 300 + 0.3%	on balance over	1 000 000
Over	6 000 000	19 300 + 0.2%	on balance over	6 000 000

NOTES:

1. Subject to note 2 below, the fees are to be calculated on the total of all interim valuations (i.e. the amount of the final account less only the net amount of the final valuation).
2. When consulting engineers are engaged in supervising the installation of air conditioning, heating, ventilating and electrical services and their duties include reporting valuations for inclusion in interim certificates for payments on account in respect of such services, then valuations so reported shall be excluded from any total amount of valuations used for calculating fees.

(b) When the quantity surveyor is required to prepare valuations of materials or goods off site, an additional fee shall be charged based on the time involved (see paras. 19.1 and 19.2).

6.3. Preparing accounts of variation upon contracts
For measuring and making up bills of variations including pricing and agreeing totals with the contractor:

(a) An initial lump sum of £600 shall be payable on each contract.
(b) 2.0% upon the gross amount of omissions requiring measurement or abstraction from the original dimension sheets.
(c) 3.0% upon the gross amount of additions requiring measurement and upon dayworks.
(d) 0.5% upon the gross amount of remaining additions which shall be deemed to include all nominated sub-contractors' and nominated suppliers' accounts which do not involve measurement or checking of quantities but only checking against lump sum estimates.
(e) 3.0% upon the aggregate of the amounts of the increases and/or decreases in the cost of labour and materials in accordance with any fluctuations clause in the conditions of contract, except where a price adjustment formula applies.
(f) On contracts where fluctuations are calculated by the use of a price adjustment formula method the following scale shall be applied to the account for the whole of the work:

Value of work £		Fee		
		£		£
Up to	300 000	300 + 0.5%		
300 000 -	1 000 000	1 800 + 0.3%	on balance over	300 000
Over	1 000 000	3 900 + 0.1%	on balance over	1 000 000

(g) When consulting engineers are engaged in supervising the installation of air conditioning, heating, ventilating and electrical services and their duties include for the adjustment of accounts and pricing and agreeing totals with the sub-contractors for inclusion in the measured account, then any totals so agreed shall be excluded from any amounts used for calculating fees.

QUANTITY SURVEYORS' FEES

6.4. Cost monitoring services
The fee for providing all approximate estimates of final cost and/or a cost monitoring service shall be based on the time involved (see paras. 19.1 and 19.2), or alternatively on a lump sum or percentage basis agreed between the employer and the quantity surveyor.

7.0. BILLS OF APPROXIMATE QUANTITIES, INTERIM CERTIFICATES AND FINAL ACCOUNTS

7.1. Basic scale
For preparing bills of approximate quantities suitable for obtaining competitive tenders which will provide a schedule of prices and a reasonably close forecast of the cost of the works, but subject to complete re-measurement, examining tenders and reporting thereon, taking particulars and reporting valuations for interim certificates for payments on account to the contractor, preparing periodic assessments of anticipated final cost and reporting thereon, measuring and preparing final account, including pricing and agreeing totals with the contractor and adjusting fluctuations in the cost of labour and materials if required by the contract:

(a) Category A: Relatively complex works and/or works with little or no repetition.
Examples:
Ambulance and fire stations; banks; cinemas; clubs; computer buildings; council offices; crematoria; fitting out existing buildings; homes for the elderly; hospitals and nursing homes; laboratories; law courts; libraries; "one-off" houses; petrol stations; places of religious worship; police stations; public houses; licensed premises; restaurants; sheltered housing; sports pavilions; theatres; town halls; universities, polytechnics and colleges of further education (other than halls of residence and hostels); and the like.

Value of work	£	Category A fee		
		£		£
Up to	150 000	380 + 5.0%	(Minimum fee £2 880)	
150 000 -	300 000	7 880 + 4.0%	on balance over	150 000
300 000 -	600 000	13 880 + 3.4%	on balance over	300 000
600 000 -	1 500 000	24 080 + 2.8%	on balance over	600 000
1 500 000 -	3 000 000	49 280 + 2.4%	on balance over	1 500 000
3 000 000 -	6 000 000	85 280 + 2.2%	on balance over	3 000 000
Over	6 000 000	151 280 + 2.0%	on balance over	6 000 000

(b) Category B: Less complex works and/or works with some element of repetition
Examples:
Adult education facilities; canteens; church halls; community centres; departmental stores; enclosed sports stadia and swimming baths; halls of residence; hostels; motels; offices other than those included in Categories A and C; railway stations; recreation and leisure centres; residential hotels; schools; self-contained flats and maisonettes; shops and shopping centres; supermarkets and hypermarkets; telephone exchanges; and the like.

Value of work	£	Category B fee		
		£		£
Up to	150 000	360 + 4.8%	(Minimum fee £2 760)	
150 000 -	300 000	7 560 + 3.7%	on balance over	150 000
300 000 -	600 000	13 110 + 3.0%	on balance over	300 000
600 000 -	1 500 000	22 110 + 2.2%	on balance over	600 000
1 500 000 -	3 000 000	41 910 + 2.0%	on balance over	1 500 000
3 000 000 -	6 000 000	71 910 + 1.8%	on balance over	3 000 000
Over	6 000 000	125 910 + 1.6%	on balance over	6 000 000

QUANTITY SURVEYORS' FEES

(c) Category C: Simple works and/or works with a substantial element of repetition.
Examples:
Factories; garages; multi-storey car parks; open air sports stadia; structural shell offices not fitted out; warehouses; workshops; and the like.

Value of work	£	Category C fee		
		£		£
Up to	150 000	300 + 4.1%	(Minimum fee £2 350)	
150 000 -	300 000	6 450 + 3.3%	on balance over	150 000
300 000 -	600 000	11 400 + 2.6%	on balance over	300 000
600 000 -	1 500 000	19 200 + 2.0%	on balance over	600 000
1 500 000 -	3 000 000	37 200 + 1.7%	on balance over	1 500 000
3 000 000 -	6 000 000	62 700 + 1.5%	on balance over	3 000 000
Over	6 000 000	107 700 + 1.3%	on balance over	6 000 000

(d) The scales of fees for pre-contract and post-contract services (paras. 7.1 (a) to (c)) are overall scales based upon the inclusion of all nominated sub-contractors' and nominated suppliers' accounts, subject to the provision of para. 7.1. (g). When work normally included in a building contract is the subject of a separate contract for which the quantity surveyor has not been paid fees under any other clause hereof, the value of such work shall be included in the amount on which fees are charged.

(e) Fees shall be calculated upon the basis of the account for the whole of the work, subject to the provisions of para. 7.2.

(f) In calculating the amount on which fees are charged the total of any credits is to be added to the amount described above.

(g) Where the value of air conditioning, heating, ventilating and electrical services included in tender documents together exceeds 25% of the amount calculated as described in paras. 7.1. (d) and (e), then, subject to the provisions of para. 7.2 no fee is chargeable on the amount by which the value of these services exceeds the said 25%. In this context the term "value" excludes general contractors' profit, attendance, builders' work in connection with the services, preliminaries and any other similar additions.

(h) When a contract comprises buildings which fall into more than one category, the fee shall be calculated as follows.

(i) The amount upon which fees are chargeable shall be allocated to the categories of work applicable and the amount so allocated expressed as percentages of the total amount upon which fees are chargeable.

(ii) Fees shall then be calculated for each category on the total amount upon which fees are chargeable.

(iii) The fee chargeable shall then be calculated by applying the percentages of work in each category to the appropriate total fee adding the resultant amounts.

(j) When a project is the subject of a number of contracts then, for the purpose of calculating fees, the values of such contracts shall not be aggregated but each contract shall be taken separately and the scale of charges (paras. 7.1(a) to (h)) applied as appropriate.

QUANTITY SURVEYORS' FEES

(k) Where the quantity surveyor is specifically instructed to provide cost planning services, the fee calculated in accordance with paras. 7.1 (a) to (j) shall be increased by a sum calculated in accordance with the following table and based upon the same value of work as that upon which the aforementioned fee has been calculated:

Categories A & B: (as defined in paras. 7.1 (a) and (b))

Value of work	£	Fee		
		£		£
Up to	600 000	0.70%		
600 000 -	3 000 000	4 200 + 0.40%	on balance over	600 000
3 000 000 -	6 000 000	13 800 + 0.35%	on balance over	3 000 000
Over	6 000 000	24 300 + 0.30%	on balance over	6 000 000

Category C: (as defined in para. 7.1 (c))

Value of work	£	Fee		
		£		£
Up to	600 000	0.50%		
600 000 -	3 000 000	3 000 + 0.30%	on balance over	600 000
3 000 000 -	6 000 000	10 200 + 0.25%	on balance over	3 000 000
Over	6 000 000	17 700 + 0.20%	on balance over	6 000 000

(l) When the quantity surveyor is required to prepare valuations of materials or goods off site, an additional fee shall be charged based on the time involved (see paras. 19.1 and 19.2).

(m) The basic scale for post-contract services includes for a simple routine of periodically estimating final costs. When the employer specifically requests a cost monitoring service which involves the quantity surveyor in additional or abortive measurement an additional fee shall be charged based on the time involved (see paras. 19.1 and 19.2), or alternatively on a lump sum or percentage basis agreed between the employer and the quantity surveyor.

7.2. Air conditioning, heating, ventilating and electrical services

(a) Where bills of approximate quantities and final accounts are prepared by the quantity surveyor for the air conditioning, heating, ventilating and electrical services there shall be a fee for these services in addition to the fee calculated in accordance with para. 7.1 as follows:

Value of work	£	Category A fee		
		£		£
Up to	120 000	4.50%		
120 000 -	240 000	5 400 + 1.85%	on balance over	120 000
240 000 -	480 000	10 020 + 3.25%	on balance over	240 000
480 000 -	750 000	17 820 + 3.00%	on balance over	480 000
750 000 -	1 000 000	25 920 + 2.50%	on balance over	750 000
1 000 000 -	4 000 000	32 170 + 2.15%	on balance over	1 000 000
Over	4 000 000	96 670 + 2.05%	on balance over	4 000 000

QUANTITY SURVEYORS' FEES

(b) The value of such services, whether the subject of separate tenders or not, shall be aggregated and the value of work so obtained used for the purpose of calculating the additional fee chargeable in accordance with para. (a). (Except that when more than one firm of consulting engineers is engaged on the design of these services, the separate values for which each such firm is responsible shall be aggregated and the additional fees charged shall be calculated independently on each such total value so obtained.)

(c) The scope of the services to be provided by the quantity surveyor under para. (a) above shall be deemed to be equivalent to those described for the basic scale for pre-contract and post-contract services.

(d) When the quantity surveyor is required to prepare valuations of materials or goods off site, an additional fee shall be charged based on the time involved (see paras. 19.1 and 19.2).

(e) The basic scale for post-contract services includes for a simple routine of periodically estimating final costs. When the employer specifically requests a cost monitoring service, which involves the quantity surveyor in additional or abortive measurement, an additional fee shall be charged based on the time involved (see paras. 19.1 and 19.2), or alternatively on a lump sum or percentage basis agreed between the employer and the quantity surveyor.

(f) Fees shall be calculated upon the basis of the account for the whole of the air conditioning, heating, ventilating and electrical services for which final accounts have been prepared by the quantity surveyor.

(g) When cost planning services are provided by the quantity surveyor for air conditioning, heating, ventilating and electrical services (or for any part of such services) there shall be an additional fee based on the time involved (see paras. 19.1 and 19.2) or alternatively on a lump sum or percentage basis agreed between the employer and quantity surveyor.

NOTE: The incorporation of figures for air conditioning, heating, ventilating and electrical services provided by the consulting engineer is deemed to be included in the quantity surveyor's services under para 7.1.

7.3. Works of alteration

On works of alteration or repair, or on those sections of the work which are mainly works of alteration or repair, there shall be a fee of 1.0% in addition to the fee calculated in accordance with paras. 7.1 and 7.2

7.4. Works of redecoration and associated minor repairs

On works of redecoration and associated minor repairs, there shall be a fee of 1.5% in addition to the fee calculated in accordance with paras. 7.1 and 7.2.

7.5. Bills of quantities and/or final accounts prepared in special forms

Fees calculated in accordance with paras. 7.1, 7.2, 7.3 and 7.4 include for the preparation of bills of quantities and/or final accounts on a normal trade basis. If the employer requires additional information to be provided in the bills of quantities and/or final accounts or the bills and/or final accounts to be prepared in an elemental, operational or similar form, then the fee may be adjusted by agreement between the employer and the quantity surveyor.

7.6. Reduction of tenders

(a) When cost planning services have been provided by the quantity surveyor and a tender, when received, is reduced before acceptance and if the reductions are not necessitated by amended instructions of the employer or by the inclusion in the bills of approximate quantities of items which the quantity surveyor has indicated could not be contained within the approved estimate, then in such a case no charge shall be made by the quantity surveyor for the preparation of bills of reductions and the fee for the preparation of bills of approximate quantities shall be based on the amount of the reduced tender.

QUANTITY SURVEYORS' FEES

(b) When cost planning services have not been provided by the quantity surveyor and if a tender, when received, is reduced before acceptance, fees are to be calculated upon the amount of the unreduced tender. When the preparation of bills of reductions is required, a fee is chargeable for preparing such bills of reductions as follows:

(i) 2.0% upon the gross amount of all omissions requiring measurement or abstraction from original dimension sheets.
(ii) 3.0% upon the gross amount of all additions requiring measurement.
(iii) 0.5% upon the gross amount of all remaining additions.

NOTE: The above scale for the preparation of bills of reductions applies to work in all categories.

7.7. Generally
If the works are substantially varied at any stage or if the quantity surveyor is involved in an excessive amount of abortive work, then the fees shall be adjusted by agreement between the employer and the quantity surveyor.

8.0. NEGOTIATING TENDERS

8.1 (a) For negotiating and agreeing prices with a contractor:

Value of work	£	Fee £		£
Up to	150 000	0.5%		
150 000 -	600 000	750 + 0.3%	on balance over	150 000
600 000 -	1 200 000	2 100 + 0.2%	on balance over	600 000
Over	1 200 000	3 300 + 0.1%	on balance over	1 200 000

(b) The fee shall be calculated on the total value of the works as defined in paras. 7.1 (d), (e), (f), (g) and (j).

(c) For negotiating and agreeing prices with a contractor for air conditioning, heating, ventilating and electrical services there shall be an additional fee as para. 8.1 (a) calculated on the total value of such services as defined in para. 7.2 (b).

9.0. CONSULTATIVE SERVICES AND PRICING BILLS OF APPROXIMATE QUANTITIES

9.1. Consultative services
Where the quantity surveyor is appointed to prepare approximate estimates, feasibility studies or submissions for the approval of financial grants or similar services, then the fee shall be based on the time involved (see paras. 19.1 and 19.2) or alternatively, on a lump sum or percentage basis agreed between the employer and the quantity surveyor.

9.2. Pricing bills of approximate quantities
For pricing bills of approximate quantities, if instructed, to provide an estimate comparable with tenders, the fees shall be the same as for the corresponding services in paras. 4.2 (a) and (b).

10.0. INSTALMENT PAYMENTS

10.1. For the purpose of instalment payments the fee for preparation of bills of approximate quantities only shall be the equivalent of forty per cent (40%) of the fees calculated in accordance with the appropriate sections of paras. 7.1 to 7.5, and the fee for providing cost planning services shall be in accordance with the appropriate sections of para. 7.1 (k); both fees shall be based on the total value of the bills of approximate quantities ascertained in accordance with the provisions of para. 2.1 (e).

QUANTITY SURVEYORS' FEES

10.2. In the absence of agreement to the contrary, fees shall be paid by instalments as follows:

(a) Upon acceptance by the employer of a tender for the works the above defined fees for the preparation of bills of approximate quantities and for providing cost planning services.
(b) In the event of no tender being accepted, the aforementioned fees shall be paid within three months of completion of the bills of approximate quantities.
(c) The balance by instalments at intervals to be agreed between the date of the first certificate and one month after certification of the contractor's account.

10.3. In the event of the project being abandoned at any stage other than those covered by the foregoing, the proportion of fee payable shall be by agreement between the employer and the quantity surveyor.

11.0. SCHEDULES OF PRICES

11.1. The fee for preparing, pricing and agreeing schedules of prices shall be based on the time involved (see paras. 19.1 and 19.2). Alternatively, the fee may be on a lump sum or percentage basis agreed between the employer and the quantity surveyor.

12.0. COST PLANNING AND APPROXIMATE ESTIMATES

12.1. The fee for providing cost planning services or for preparing approximate estimates shall be based on the time involved (see paras. 19.1 and 19.2). Alternatively, the fee may be on a lump sum or percentage basis agreed between the employer and the quantity surveyor.

CONTRACTS BASED ON SCHEDULES OF PRICES: POST-CONTRACT SERVICES

13.0. FINAL ACCOUNTS

13.1. Basic Scale

(a) For taking particulars and reporting valuations for interim certificates for payments on account to the contractor, preparing periodic assessments of anticipated final cost and reporting thereon, measuring and preparing final account including pricing and agreeing totals with the contractor, and adjusting fluctuations in the cost of labour and materials if required by the contract, the fee shall be equivalent to sixty per cent (60%) of the fee calculated in accordance with paras. 7.1 (a) to (j).
(b) When the quantity surveyor is required to prepare valuations of materials or goods off site, an additional fee shall be charged on the basis of the time involved (see paras. 19.1 and 19.2).
(c) The basic scale for post-contract services includes for a simple routine of periodically estimating final costs. When the employer specifically requests a cost monitoring service which involves the quantity surveyor in additional or abortive measurement an additional fee shall be charged based on the time involved (see paras. 19.1 and 19.2), or alternatively on a lump sum or percentage basis agreed between the employer and the quantity surveyor.

13.2. Air conditioning, heating, ventilating and electrical services
Where final accounts are prepared by the quantity surveyor for the air conditioning, heating, ventilating and electrical services there shall be a fee for these services, in addition to the fee calculated in accordance with para. 13.1, equivalent to sixty per cent (60%) of the fee calculated in accordance with paras. 7.2 (a) to (f).

13.3. Works of alterations
On works of alteration or repair, or on those sections of the work which are mainly works of alteration or repair, there shall be a fee of 1.0% in addition to the fee calculated in accordance with paras. 13.1 and 13.2.

QUANTITY SURVEYORS' FEES

13.4. Works of redecoration and associated minor repairs
On works of redecoration and associated minor repairs, there shall be a fee of 1.5% in addition to the fee calculated in accordance with paras. 13.1 and 13.2.

13.5. Final accounts prepared in special forms
Fees calculated in accordance with paras. 13.1,13.2,13.3 and 13.4 include for the preparation of final accounts on a normal trade basis. If the employer requires additional information to be provided in the final accounts or the accounts to be prepared in an elemental, operational or similar form, then the fee may be adjusted by agreement between the employer and the quantity surveyor.

PRIME COST CONTRACTS: PRE-CONTRACT AND POST-CONTRACT SERVICES

14.0. COST PLANNING

14.1. The fee for providing a cost planning service shall be based on the time involved (see paras. 19.1 and 19.2). Alternatively, the fee may be on a lump sum or percentage basis agreed between the employer and the quantity surveyor.

15.0. ESTIMATES OF COST

15.1. (a) For preparing an approximate estimate, calculated by measurement, of the cost of work, and, if required under the terms of the contract, negotiating, adjusting and agreeing the estimate:

Value of work	£	Fee		
		£		£
Up to	30 000	1.25%		
30 000 -	150 000	375 + 1.00%	on balance over	30 000
150 000 -	600 000	1 575 + 0.75%	on balance over	150 000
Over	600 000	4 950 + 0.50%	on balance over	600 000

(b) The fee shall be calculated upon the total of the approved estimates.

16.0. FINAL ACCOUNTS

16.1. (a) For checking prime costs, reporting for interim certificates for payments on account to the contractor and preparing final accounts:

Value of work	£	Fee		
		£		£
Up to	30 000	2.50%		
150 000 -	150 000	750 + 2.00%	on balance over	30 000
150 000 -	600 000	3 150 + 1.50%	on balance over	150 000
Over	600 000	9 900 + 1.25%	on balance over	600 000

(b) The fee shall be calculated upon the total of the final account with the addition of the value of credits received for old materials removed and less the value of any work charged for in accordance with para. 16.1 (c).

(c) On the value of any work to be paid for on a measured basis, the fee shall be 3%.

(d) When the quantity surveyor is required to prepare valuations of materials or goods off site, an additional fee shall be charged based on the time involved (see paras. 19.1 and 19.2).

(e) The above charges do not include the provision of checkers on the site. If the quantity surveyor is required to provide such checkers an additional charge shall be made by arrangement.

QUANTITY SURVEYORS' FEES

17.0. COST REPORTING AND MONITORING SERVICES

17.1. The fee for providing cost reporting and/or monitoring services (e.g. preparing periodic assessments of anticipated final costs and reporting thereon) shall be based on the time involved (see paras. 19.1 and 19.2) or alternatively, on a lump sum or percentage basis agreed between the employer and the quantity surveyor.

18.0. ADDITIONAL SERVICES

18.1. For additional services not normally necessary, such as those arising as a result of the termination of a contract before completion, liquidation, fire damage to the buildings, services in connection with arbitration, litigation and investigation of the validity of contractors' claims, services in connection with taxation matters and all similar services where the employer specifically instructs the quantity surveyor, the charges shall be in accordance with paras. 19.1 and 19.2.

19.0. TIME CHARGES

19.1. (a) For consultancy and other services performed by a principal, a fee by arrangement according to the circumstances including the professional status and qualifications of the quantity surveyor.

(b) When a principal does work which would normally be done by a member of staff, the charge shall be calculated as para. 19.2 below.

19.2. (a) For services by a member of staff, the charges for which are to be based on the time involved, such charges shall be calculated on the hourly cost of the individual involved plus 145%.

(b) A member of staff shall include a principal doing work normally done by an employee (as para. 19.1 (b) above), technical and supporting staff, but shall exclude secretarial staff or staff engaged upon general administration.

(c) For the purpose of para. 19.2 (b) above, a principal's time shall be taken at the rate applicable to a senior assistant in the firm.

(d) The supervisory duties of a principal shall be deemed to be included in the addition of 145% as para. 19.2 (a) above and shall not be charged separately.

(e) The hourly cost to the employer shall be calculated by taking the sum of the annual cost of the member of staff of:

(i) Salary and bonus but excluding expenses;

(ii) Employer's contributions payable under any Pension and Life Assurance Schemes;

(iii) Employer's contributions made under the National Insurance Acts, the Redundancy Payments Act and any other payments made in respect of the employee by virtue of any statutory requirements; and

(iv) Any other payments or benefits made or granted by the employer in pursuance of the terms of employment of the member of staff;

and dividing by 1,650.

19.3. The foregoing Time Charges under paras. 19.1 and 19.2 are intended for use where other paras. of the Scale (not related to Time Charges) form a significant proportion of the overall fee. In all other cases an increased time charge may be agreed.

20.0. INSTALMENT PAYMENTS

20.1. In the absence of agreement to the contrary, payments to the quantity surveyor shall be made by instalments by arrangement between the employer and the quantity surveyor.

QUANTITY SURVEYORS' FEES

SCALE 40 PROFESSIONAL CHARGES FOR QUANTITY SURVEYING SERVICES IN CONNECTION WITH HOUSING SCHEMES FOR LOCAL AUTHORITIES

This scale has been abolished. See Author's Note on page 800.

EFFECTIVE FROM FEBRUARY 1983

1.0 GENERALLY

1.1 The scale is applicable to housing schemes of self-contained dwellings regardless of type (e.g. houses, maisonettes, bungalows or flats) and irrespective of the amount of repetition of identical types or blocks within an individual housing scheme and shall also apply to all external works forming part of the contract for the housing scheme. This scale does not apply to improvement to existing dwellings.

1.2 The fees set out below cover the following quantity surveying services as may be required:

(a) Preparing bills of quantities or other tender documents; checking tenders received or negotiating tenders and pricing with a selected contractor; reporting thereon.

(b) Preparing recommendations for interim payments on account to the contractor; measuring work and adjusting variations in accordance with the terms of the contract and preparing the final account; pricing same and agreeing totals with the contractor; adjusting fluctuations in the cost of labour and materials if required by the contract.

(c) Preparing periodic financial statements showing the anticipated final cost by means of a simple routine of estimating final costs and reporting thereon, but excluding cost monitoring (see para. 1.4).

1.3 Where the quantity surveyor is appointed to prepare approximate estimates to establish and substantiate the economic viability of the scheme and to obtain the necessary approvals and consents, or to enable the scheme to be designed and constructed within approved cost criteria an additional fee shall be charged based on the time involved (see para. 7.0) or, alternatively, on a lump sum or percentage basis agreed between the employer and the quantity surveyor. (Cost planning services, see para. 3.0).

1.4 When the employer specifically requests a post-contract cost monitoring service which involves the quantity surveyor in additional or abortive work an additional fee shall be charged based on the time involved (see para. 7.0) or, alternatively, on a lump sum or percentage basis agreed between the employer and the quantity surveyor.

1.5 The fees are in all cases exclusive of travelling and other expenses (for which the actual disbursement is recoverable unless there is some prior arrangement for such charges) and of the cost of reproduction of bills of quantities and other documents, which are chargeable in addition at net cost.

1.6 The fees are in all cases exclusive of services in connection with the allocation of the cost of the works for purposes of calculating value added tax for which there shall be an additional fee based on the time involved (see para. 7.0).

1.7 When work normally included in a building contract is the subject of a separate contract for which the quantity surveyor has not been paid fees under any other clause thereof, the value of such work shall be included in the amount upon which fees are charged.

1.8 If any of the materials used in the works are supplied by the employer or charged at a preferential rate, then the estimated or actual value thereof shall be included in the amount upon which fees are to be calculated.

QUANTITY SURVEYORS' FEES

1.9 The fees are in all cases exclusive of preparing a specification of the materials to be used and the works to be done, but the fees for preparing bills of quantities and similar documents do include for incorporating preamble clauses describing the materials and workmanship (from information given by the architect and/or consulting engineer).

1.10 If the quantity surveyor incurs additional costs due to exceptional delays in building operations or any other cause beyond the control of the quantity surveyor, then the fees shall be adjusted by agreement between the employer and the quantity surveyor to cover the reimbursement of these additional costs.

1.11 When a project is the subject of a number of contracts then for the purposes of calculating fees, the values of such contracts shall not be aggregated but each contract shall be taken separately and the scale of charges applied as appropriate.

1.12 The fees and charges are in all cases exclusive of value added tax which will be applied in accordance with legislation.

1.13 Copyright in bills of quantities and other documents prepared by the quantity surveyor is reserved to the quantity surveyor.

2.0 BASIC SCALE

2.1 The basic fee for the services outlined in para. 1.2 shall be as follows:-

Value of work		Fee		
		£		£
Up to	75 000	250 + 4.6%		
75 000 -	150 000	3 700 + 3.6%	on balance over	75 000
150 000 -	750 000	6 400 + 2.3%	on balance over	150 000
750 000 -	1 500 000	20 200 + 1.7%	on balance over	750 000
Over	1 500 000	32 950 + 1.5%	on balance over	1 500 000

2.2 Fees shall be calculated upon the total of the final account for the whole of the work including all nominated sub-contractors' and nominated suppliers' accounts.

2.3 For services in connection with accommodation designed for the elderly or the disabled or other special category occupants for whom special facilities are required an addition of 10% shall be made to the fee calculated in accordance with para. 2.1.

2.4 When additional fees under para. 2.3 are chargeable on a part or parts of a scheme, the value of basic fee to which the additional percentages shall be applied shall be determined by the proportion that the values of the various types of accommodation bear to the total of those values.

2.5 When the quantity surveyor is required to prepare an interim valuation of materials or goods off site, an additional fee shall be charged based on the time involved (see para. 7.0).

2.6 If the works are substantially varied at any stage and if the quantity surveyor is involved in an excessive amount of abortive work, then the fee shall be adjusted by agreement between the employer and the quantity surveyor.

2.7 The fees payable under paras. 2.1 and 2.3 include for the preparation of bills of quantities or other tender documents on a normal trade basis. If the employer requires additional information to be provided in bills of quantities, or bills of quantities to be prepared in an elemental, operational or similar form, then the fee may be adjusted by agreement between the employer and the quantity surveyor.

QUANTITY SURVEYORS' FEES

3.0 COST PLANNING

3.1 When the quantity surveyor is specifically instructed to provide cost planning services, the fee calculated in accordance with paras. 2.1 and 2.3 shall be increased by a sum calculated in accordance with the following table and based upon the amount of the accepted tender.

Value of work	£	Fee		
		£		£
Up to	150 000	0.45%		
150 000 -	750 000	675 + 0.35%	on balance over	150 000
Over	750 000	2 775 + 0.25%	on balance over	750 000

3.2 Cost planning is defined as the process of ascertaining a cost limit, where necessary, within the guidelines set by any appropriate Authority, and thereafter checking the cost of the project within that limit throughout the design process. It includes the preparation of a cost plan (based upon elemental analysis or other suitable criterion) checking and revising it where required and effecting the necessary liaison with other consultants employed.

3.3 (a) When cost planning services have been provided by the quantity surveyor and bills of reductions are required, then no charge shall be made by the quantity surveyor for the bills of reductions unless the reductions are necessitated by amended instructions of the employer or by the inclusion in the bills of quantities of items which the quantity surveyor has indicated could not be contained within the approved estimate.

(b) When cost planning services have not been provided by the quantity surveyor and bills of reductions are required, a fee is chargeable for preparing such bills of reductions:

(i) 2.0% upon the gross amount of all omissions requiring measurement or abstraction from original dimension sheets.

(ii) 3.0% upon the gross amount of all additions requiring measurement.

(iii) 0.5% upon the gross amount of all remaining additions.

4.0 HEATING, VENTILATING AND ELECTRICAL SERVICES

(a) When bills of quantities and the final account are prepared by the quantity surveyor for the heating, ventilating and electrical services, there shall be a fee for these services in addition to the fee calculated in accordance with paras. 2.1 and 2.3 as follows:

Value of work	£	Fee		
		£		£
Up to	60 000	4.50%		
60 000 -	120 000	2 700 + 3.85%	on balance over	60 000
120 000 -	240 000	5 010 + 3.25%	on balance over	120 000
240 000 -	375 000	8 910 + 3.00%	on balance over	240 000
375 000 -	500 000	12 960 + 2.50%	on balance over	375 000
Over	500 000	16 085 + 2.15%	on balance over	500 000

(b) The value of such services, whether the subject of separate tenders or not shall be aggregated and the total value of work so obtained used for the purpose of calculating the additional fee chargeable in accordance with para. (a). (Except that when more than one firm of consulting engineers is engaged on the design of these services, the separate values for which each such firm is responsible shall be aggregated and the additional fees charged shall be calculated independently on each such total value so obtained).

QUANTITY SURVEYORS' FEES

(c) The scope of the services to be provided by the quantity surveyor under para. (a) above shall be deemed to be equivalent to those outlined in para. 1.2.

(d) Fee shall be calculated upon the basis of the account for the whole of the heating, ventilating and electrical services for which final accounts have been prepared by the quantity surveyor.

5.0 INSTALMENT PAYMENTS

5.1 In the absence of agreement to the contrary, fees shall be paid by instalments as follows:

(a) Upon receipt by the employer of a tender for the works sixty per cent (60%) of the fees calculated in accordance with paras. 2.0 and 4.0 in the amount of the accepted tender plus the appropriate recoverable expenses and the full amount of the fee for cost planning services if such services have been instructed by the employer.

(b) The balance of fees and expenses by instalments at intervals to be agreed between the date of the first certificate and one month after final certification of the contractor's account.

5.2 In the event of no tender being accepted, sixty per cent (60%) of the fees, plus the appropriate recoverable expenses, and the full amount of the fee for cost planning services if such services have been instructed by the employer, shall be paid within three months of the completion of the tender documents. The fee shall be calculated on the amount of the lowest original bona fide tender received. In the event of no tender being received, the fee shall be calculated on a reasonable valuation of the work based upon the tender documents.

NOTE: In the foregoing context "bona fide tender" shall be deemed to mean a tender submitted in good faith without major errors of computation and not subsequently withdrawn by the tenderer.

5.3 In the event of the project being abandoned at any stage other than those covered by the foregoing, the proportion of fee payable shall be by agreement between the employer and the quantity surveyor.

5.4 When the quantity surveyor is appointed to carry out post-contract services only and has not prepared the bills of quantities then the fees shall be agreed between the employer and the quantity surveyor as a proportion of the scale set out in paras. 2.0 and 4.0 with an allowance for the necessary familiarisation and any additional services undertaken by the quantity surveyor. The percentages stated in paras. 5.1 and 5.2 are not intended to be used as a means of calculating the fees payable for post-contract services only.

6.0 ADDITIONAL SERVICES

6.1 For additional services not normally necessary such as those arising as a result of the termination of a contract before completion, liquidation, fire damage to the buildings, services in connection with arbitration, litigation and investigation of the validity of contractors' claims, services in connection with taxation matters, and all similar services where the employer specifically instructs the quantity surveyor, the charge shall be in accordance with para. 7.0.

7.0 TIME CHARGES

7.1 (a) For consultancy and other services performed by a principal, a fee by arrangement according to the circumstances, including the professional status and qualifications of the quantity surveyor.

(b) When a principal does work which would normally be done by a member of staff, the charge shall be calculated as para. 7.2.

QUANTITY SURVEYORS' FEES

7.2 (a) For services by a member of staff, the charges for which are to be based on the time involved, such hourly charges shall be calculated on the basis of annual salary (including bonus and any other payments or benefits previously agreed with the employer) multiplied by a factor of 2.5, plus reimbursement of payroll costs, all divided by 1600. Payroll costs shall include inter alia employer's contributions payable under any Pension and Life Assurance Schemes, employer's contributions made under the National Insurance Acts, the Redundancy Payments Act and any other payments made in respect of the employee by virtue of any statutory requirements. In this connection it would not be unreasonable in individual cases to take account of the cost of providing a car as part of the "salary" of staff engaged on time charge work when considering whether the salaries paid to staff engaged on such work are reasonable.

(b) A member of staff shall include a principal doing work normally done by an employee (as para. 7.1 (b) above), technical and supporting staff, but shall exclude secretarial staff or staff engaged upon general administration.

(c) For the purpose of para. 7.2 (b) above a principal's time shall be taken at the rate applicable to a senior assistant in the firm.

(d) The supervisory duties of a principal shall be deemed to be included in the multiplication factor as para. 7.2 (a) above and shall not be charged separately.

7.3 The foregoing Time Charges under paras. 7.1 and 7.2 are intended for use where other paras. of the scale (not related to Time Charges) form a significant proportion of the overall fee. In all other cases an increased Time Charge may be agreed.

SCALE 44 PROFESSIONAL CHARGES FOR QUANTITY SURVEYING SERVICES IN CONNECTION WITH IMPROVEMENTS TO EXISTING HOUSING AND ENVIRONMENTAL IMPROVEMENT WORKS

This scale has been abolished. See Author's Note on page 800.

EFFECTIVE FROM FEBRUARY 1973

1. This scale of charges is applicable to all works of improvement to existing housing for local authorities, development corporations, housing associations and the like and to environmental improvement works associated therewith or of a similar nature.

2. The fees set out below cover such quantity surveying services as may be required in connection with an improvement project irrespective of the type of contract or contract documentation from initial appointment to final certification of the contractor's account such as:

(a) Preliminary cost exercises and advice on tendering procedures and contract arrangements.

(b) Providing cost advice to assist the design and construction of the project within approved cost limits.

(c) Preliminary inspection of a typical dwelling of each type.

(d) Preparation of tender documents; checking tenders received and reporting thereon or negotiating tenders and agreeing prices with a selected contractor.

(e) Making recommendations for and, where necessary, preparing bills of reductions except in cases where the reductions are necessitated by amended instructions of the employer or by the inclusion in the bills of quantities of items which the quantity surveyor has indicated could not be contained within the approved estimate.

(f) Analysing tenders and preparing details for submission to a Ministry or Government Department and attending upon the employer in any negotiations with such Ministry or Government Department.

(g) Recording the extent of work required to every dwelling before work commences.

(h) Preparing recommendations for interim payments on account to the contractor; preparing periodic assessments of the anticipated final cost of the works and reporting thereon

(j) Measurement of work and adjustment of variations and fluctuations in the cost of labour and materials in accordance with the terms of the contract and preparing final account, pricing same and agreeing totals with the contractor.

QUANTITY SURVEYORS' FEES

3. The services listed in para. 2 do not include the carrying out of structural surveys.

4. The fees set out below have been calculated on the basis of experience that all of the services described above will not normally be required and in consequence these scales shall not be abated if, by agreement, any of the services are not required to be provided by the quantity surveyor.

IMPROVEMENT WORKS TO HOUSING

5. The fee for quantity surveying services in connection with improvement works to existing housing and external works in connection therewith shall be calculated from a sliding scale based upon the total number of houses or flats in a project divided by the total number of types substantially the same in design and plan as follows:

Total number of houses or flats divided by total number of types substantially the same in design and plan	Fee
not exceeding 1	see note below
exceeding 1 but not exceeding 2	7.0%
exceeding 2 but not exceeding 3	5.0%
exceeding 3 but not exceeding 4	4.5%
exceeding 4 but not exceeding 20	4.0%
exceeding 20 but not exceeding 50	3.6%
exceeding 50 but not exceeding 100	3.2%
exceeding 100	3.0%

and to the result of the computation shall be added 12.5%

NOTE: For schemes of only one house or flat per type an appropriate fee is to be agreed between the employer and the quantity surveyor on a percentage, lump sum or time basis.

ENVIRONMENTAL IMPROVEMENT WORKS

6. The fee for quantity surveying services in connection with environmental improvement works associated with improvements to existing housing or environmental improvement works of a similar nature shall be as follows:

Value of work £		Fee		
		£		£
Up to	50 000	4.5%		
50 000 -	200 000	2 250 + 3.0%	on balance over	50 000
200 000 -	500 000	6 750 + 2.1%	on balance over	200 000
Over	500 000	13 050 + 2.0%	on balance over	500 000

and to the result of that computation shall be added 12.5%

GENERALLY

7. When tender documents prepared by a quantity surveyor for an earlier scheme are re-used without amendment by the quantity surveyor for a subsequent scheme or part thereof for the same employer, the percentage fee in respect of such subsequent scheme or the part covered by such reused documents shall be reduced by 20%.

8. The foregoing fees shall be calculated upon the separate totals of the final account for improvement works to housing and environmental Government works respectively including all nominated sub-contractors' and nominated suppliers' accounts and (subject to para. 5 above) regardless of the amount of repetition within the scheme. When environmental improvement works are the subject of a number of contracts then for the purpose of calculating fees, the values of such contracts shall not be aggregated but each contract shall be taken separately and the scale of charges in para. 6 above applied as appropriate.

QUANTITY SURVEYORS' FEES

9. In cases where any of the materials used in the works are supplied by the employer, the estimated or actual value thereof is to be included in the total on which the fee is calculated.

10. In the absence of agreement to the contrary, fees shall be paid by instalments as follows:

(a) Upon acceptance by the employer of a tender for the works, one half of the fee calculated on the amount of the accepted tender.
(b) The balance by instalments at intervals to be agreed between the date of the first certificate and one month after final certification of the contractor's account.

11. (a) In the event of no tender being accepted, one half of the fee shall be paid within three months of completion of the tender documents. The fee shall be calculated on the amount of the lowest original bona fide tender received. If no such tender has been received, the fee shall be calculated upon a reasonable valuation of the work based upon the tender documents.
(b) In the event of the project being abandoned at any stage other than those covered by the foregoing, the proportion of fee payable shall be by agreement between the employer and the quantity surveyor.

12. If the works are substantially varied at any stage or if the quantity surveyor is involved in an excessive amount of abortive work, then the fee shall be adjusted by agreement between the employer and the quantity surveyor.

13. When the quantity surveyor is required to perform additional services in connection with the allocation of the costs of the works for purposes of calculating value added tax there shall be an additional fee based on the time involved.

14. For additional services not normally necessary such as those arising as a result of the termination of the contract before completion, liquidation, fire damage to the buildings, services in connection with arbitration, litigation and claims on which the employer specifically instructs the surveyor to investigate and report, there shall be an additional fee to be agreed between the employer and the quantity surveyor.

15. Copyright in the bills of quantities and other documents prepared by the quantity surveyor is reserved to the quantity surveyor.

16. The foregoing fees are in all cases exclusive of travelling expenses and lithography or other charges for copies of documents, the net amount of such expenses and charges to be paid for in addition. Subsistence expenses, if any, to be charged by arrangement with the employer.

17. The foregoing fees and charges are in all cases exclusive of value added tax which shall be applied in accordance with legislation current at the time the account is rendered.

SCALE 45 PROFESSIONAL CHARGES FOR QUANTITY SURVEYING SERVICES IN CONNECTION WITH HOUSING SCHEMES FINANCED BY THE HOUSING CORPORATION
EFFECTIVE FROM JANUARY 1982 - reprinted 1989
This scale has been abolished. See Author's Note on page 800.

1. (a) This scale of charges has been agreed between The Royal Institution of Chartered Surveyors and the Housing Corporation and shall apply to housing schemes of self-contained dwellings financed by the Housing Corporation regardless of type (e.g. houses, maisonettes, bungalows or flats) and irrespective of the amount of repetition of identical types or blocks within a scheme.
(b) This scale does not apply to services in connection with improvements to existing dwellings.

2. The fees set out below cover the following quantity surveying services as may be required in connection with the particular project:

(a) Preparing such estimates of cost as are required by the employer to establish and substantiate the economic viability of the scheme and to obtain the necessary approvals and consents from the Housing Corporation but excluding cost planning services (see para. 10)
(b) Providing pre-contract cost advice (e.g. approximate estimates on a floor area or similar basis) to enable the scheme to be designed and constructed within the approved cost criteria but excluding cost planning services (see para. 10).

QUANTITY SURVEYORS' FEES

(c) Preparing bills of quantities or other tender documents; checking tenders received or negotiating tenders and pricing with a selected contractor; reporting thereon.

(d) Preparing an elemental analysis of the accepted tender (RICS/BCIS Detailed Form of Cost Analysis excluding the specification notes or equivalent).

(e) Preparing recommendations for interim payments on account to the contractor; measuring the work and adjusting variations in accordance with the terms of the contract and preparing the final account, pricing same and agreeing totals with the contractor; adjusting fluctuations in the cost of labour and materials if required by the contract.

(f) Preparing periodic post-contract assessments of the anticipated final cost by means of a simple routine of periodically estimating final costs and reporting thereon, but excluding a cost monitoring service specifically required by the employer.

3. The fees set out below are exclusive of travelling and of other expenses (for which the actual disbursement is recoverable unless there is some special prior arrangement for such charges) and the cost of reproduction of bills of quantities and other documents, which are chargeable in addition at net cost.

4. Copyright in the bills of quantities and other documents prepared by the quantity surveyor is reserved to the quantity surveyor.

5. (a) The basic fee for the services outlined in para. 2 (regardless of the extent of services described in para. 2) shall be as follows:

Value of work £		Fee		
		£		£
Up to	75 000	210 + 3.8%		
75 000 -	150 000	3 060 + 3.0%	on balance over	75 000
150 000 -	750 000	5 310 + 2.0%	on balance over	150 000
750 000 -	1 500 000	17 310 + 1.5%	on balance over	750 000
Over	1 500 000	28 560 + 1.3%	on balance over	1 500 000

(b) (i) For services in connection with Categories 1 and 2 Accommodation designed for Old People in accordance with the standards described in Ministry of Housing and Local Government Circulars 82/69 and 27/70 (Welsh Office Circulars 84/69 & 30/70), there shall be a fee in addition to that in accordance with para. 5 (a), calculated as follows:

Category 1	An addition of five per cent (5%) to the basic fee calculated in accordance with para. 5 (a)
Category 2	An addition of twelve and a half per cent (12.5%) to the basic fee calculated in accordance with para. 5 (a).

(ii) For services in connection with Accommodation designed for the Elderly in Scotland in accordance with the standards described in Scottish Housing Handbook Part 5, Housing for the Elderly, the fee shall be calculated as follows:

Mainstream and Amenity Housing	Basic fee in accordance with para. 5 (a)
Basic Sheltered Housing (i.e. Amenity Housing plus Warden's accommodation and alarm system)	An addition of five per cent (5%) to the basic fee calculated in accordance with para. 5 (a)
Sheltered Housing, including optional facilities	An addition of twelve and a half per cent (12.5%) of the basic fee calculated in accordance with para. 5 (a)

QUANTITY SURVEYORS' FEES

(c) (i) For services in connection with Accommodation designed for Disabled People in accordance with the standards described in Department of Environment Circular 92/75 (Welsh Office Circular 163/75), there shall be an addition of fifteen per cent (15%) to the fee calculated in accordance with paragraph 5 (a).

(ii) For services in connection with Accommodation designed for the Disabled in Scotland in accordance with the standards described in Scottish Housing Handbook Part 6, Housing for the Disabled, there shall be an addition of fifteen per cent (15%) to the fee calculated in accordance with para. 5 (a).

(d) For services in connection with Accommodation designed for Disabled Old People, the fee shall be calculated in accordance with para. 5 (c).

(e) For services in connection with Subsidised Fair Rent New Build Housing, there shall be a fee, in addition to that in accordance with paras. 5 (a) to (d), calculated as follows:

Value of work	£	Category A fee		
		£		£
Up to	75 000	20 + 0.40%		
75 000 -	150 000	320 + 0.20%	on balance over	75 000
150 000 -	500 000	470 + 0.07%	on balance over	150 000
Over	500 000	715		

6. (a) Where additional fees under paras. 5 (b) to (d) are chargeable on a part or parts of a scheme, the value of basic fee to which the additional percentages shall be applied shall be determined by the proportion that the values of the various types of accommodation bear to the total of those values.

(b) Fees shall be calculated upon the total of the final account for the whole of the work including all nominated sub-contractors' and nominated suppliers' accounts.

(c) If any of the materials used in the works are supplied free of charge to the contractor, the estimated or actual value thereof shall be included in the amount upon which fees are to be calculated.

(d) When a project is the subject of a number of contracts then, for the purpose of calculating fees, the values of such contracts shall not be aggregated but each contract shall be taken separately and the scale of charges applied as appropriate.

7. If bills of quantities and final accounts are prepared by the quantity surveyor for the heating, ventilating or electrical services, there shall be an additional fee by agreement between the employer and the quantity surveyor subject to the approval of the Housing Corporation.

8. In the absence of agreement to the contrary, fees shall be paid by instalments as follows:

(a) Upon receipt by the employer of a tender for the works, or when the employer certifies to the Housing Corporation that the tender documents have been completed, a sum on account representing ninety per cent (90%) of the anticipated sum under para. 8 (b) below.

(b) Upon acceptance by the employer of a tender for the works, sixty per cent (60%) of the fee calculated on the amount of the accepted tender, plus the appropriate recoverable expenses.

(c) The balance of fees and expenses by instalments at intervals to be agreed between the date of the first certificate and one month after final certification of the contractor's account.

9. (a) In the event of no tender being accepted, sixty per cent (60%) of the fee and the appropriate recoverable expenses shall be paid within six months of completion of the tender documents. The fee shall be calculated on the amount of the lowest original bona fide tender received. In the event of no tender being received, the fee shall be calculated upon a reasonable valuation of the work based upon the tender documents.

NOTE: In the foregoing context "bona fide tender" shall be deemed to mean a tender submitted in good faith without major errors of computation and not subsequently withdrawn by the tenderer.

QUANTITY SURVEYORS' FEES

(b) In the event of part of the project being postponed or abandoned after the preparation of the bills of quantities or other tender documents, sixty per cent (60%) of the fee on this part shall be paid within three months of the date of postponement or abandonment.

(c) In the event of the project being postponed or abandoned at any stage other than those covered by the foregoing, the proportion of fee payable shall be by agreement between the employer and the quantity surveyor.

10. (a) Where with the approval of the Housing Corporation the employer instructs the quantity surveyor to carry out cost planning services there shall be a fee additional to that charged under para. 5 as follows:

Value of work £		Category A fee		
		£		£
Up to	150 000	0.45%		
150 000 -	750 000	675 + 0.35%	on balance over	150 000
Over	750 000	2 775 + 0.25%	on balance over	750 000

(b) Cost planning is defined as the process of ascertaining a cost limit where necessary, within guidelines set by any appropriate Authority, and thereafter checking the cost of the project within that limit throughout the design process. It includes the preparation of a cost plan (based upon elemental analysis or other suitable criterion) checking and revising it where required and effecting the necessary liaison with the other consultants employed.

11. If the quantity surveyor incurs additional costs due to exceptional delays in building operations or any other cause beyond the control of the quantity surveyor, then the fees shall be adjusted by agreement between the employer and the quantity surveyor to cover reimbursement of costs.

12. When the quantity surveyor is required to prepare an interim valuation of materials or goods off site, an additional fee shall be charged based on the time involved (see paras. 15 and 16) in respect of each such valuation.

13. If the Works are materially varied to the extent that substantial re-measurement is necessary, then the fee may be adjusted by agreement between the employer and the quantity surveyor.

14. For additional services not normally necessary, such as those arising as a result of the termination of a contract before completion, fire damage to the buildings, cost monitoring (see para. 2 (f)), services in connection with arbitration, litigation and investigation of the validity of contractors' claims, services in connection with taxation matters and similar all services where the employer specifically instructs the quantity surveyor, the charges shall be in accordance with paras. 15 & 16.

15. (a) For consultancy and other services performed by a principal, a fee by arrangement according to the circumstances, including the professional status and qualifications of the quantity surveyor.

(b) When a principal does work which would normally be done by a member of staff, the charge shall be calculated as para. 16.

16. (a) For services by a member of staff, the charges for which are to be based on the time involved, such hourly charges shall be calculated on the basis of annual salary (including bonus and any other payments or benefits previously agreed with the employer) multiplied by a factor of 2.5, plus reimbursement of payroll costs, all divided by 1600. Payroll costs shall include inter alia employer's contributions payable under any Pension and Life Assurance Schemes, employer's contributions made under the National Insurance Acts, the Redundancy Payments Act and any other payments made in respect of the employee by virtue of any statutory requirements in this connection it would not be unreasonable in individual cases to take account of the cost of providing a car as part of the "salary" of staff engaged on time charge work when considering whether the salaries paid to staff engaged on such work are reasonable.

(b) A member of staff shall include a principal doing work normally done by an employee (as para. 15 (b) above), technical and supporting staff, but shall exclude secretarial staff or staff engaged upon general administration.

QUANTITY SURVEYORS' FEES

(c) For the purpose of para. 16 (b) above a principal's time shall be taken at the rate applicable to a senior assistant in the firm.

(d) The supervisory duties of a principal shall be deemed to be included in the multiplication factor as para. 16 (a) above and shall not be charged separately.

17. The foregoing Time Charges under paras. 15 and 16 are intended for use where other paras. of the scale (not related to Time Charges) form a significant proportion of the overall fee. In all other cases an increased time charge may be agreed.

18. (a) In the event of the employment of the contractor being determined due to bankruptcy or liquidation, the fee for the services outlined in para. 2, and for the additional services required, shall be recalculated to the aggregate of the following:

(i) Fifty per cent (50%) of the fee in accordance with paragraphs 5 and 6 calculated upon the total of the Notional Final Account in accordance with the terms of the original contracts.

(ii) Fifty per cent (50%) of the fee in accordance with paragraphs 5 and 6 calculated upon the aggregate of the total value (which may differ from the total of interim valuations) of work up to the date of determination in accordance with the terms of the original contract plus the total of the final account for the completion contract;

(iii) A charge based upon time involved (in accordance with paragraphs 15 and 16) in respect of dealing with those matters specifically generated by the liquidation (other than normal post-contract services related to the completion contract), which may include (inter alia):

Site inspection and (where required) security (initial and until the replacement contractor takes possession);
Taking instructions from and/or advising the employer;
Representing the employer at meeting(s) of creditors;
Making arrangements for the continued employment of sub-contractors and similar related matters;
Preparing bills of quantities or other appropriate documents for the completion contract, obtaining tenders, checking and reporting thereon;
The additional cost (over and above the preparation of the final account for the completion contract) of pre-paring the Notional Final Account; pricing the same;
Negotiations with the liquidator (trustee or receiver).

(b) In calculating fees under para. 18 (a) (iii) above, regard shall be taken of any services carried out by the quantity surveyor for which a fee will ultimately be chargeable under para. 18 (a) (i) and (ii) above in respect of which a suitable abatement shall be made from the fee charged (e.g. measurement of variations for purposes of the completion contract where such would contribute towards the preparation of the contract final account).

(c) Any interim instalments of fees paid under para. 8 in respect of services outlined in para. 2 shall be deducted from the overall fee computed as outlined herein.

(d) In the absence of agreement to the contrary fees and expenses in respect of those services outlined in para. 18 (a) (iii) above up to acceptance of a completion tender shall be paid upon such acceptance; the balance of fees and expenses shall be paid in accordance with para. 8 (c).

(e) For the purpose of this Scale the term "Notional Final Account" shall be deemed to mean an account indicating that which would have been payable to the original contractor had he completed the whole of the works and before deduction of interim payments to him.

19. The fees and charges are in all cases exclusive of Value Added Tax which will be applied in accordance with legislation.

QUANTITY SURVEYORS' FEES

EXPLANATORY NOTE:
(Source: Chartered Quantity Surveyor, August 1986)

For rehabilitation projects the basic fee set out in paragraph 5 (a) of the scale will apply with the addition of a further 1% fee calculated upon the total of the final account for rehabilitation works including ali nominated sub-contractors' and nominated suppliers' accounts.
In the case of special housing categories (e.g., elderly people) the additional percentage should be applied before the application of the additional percentage set out in paragraph 5 (b). The provisions of paragraph 6 (a) of the scale will also apply.
There is no longer any distinction between "hostel" and "cluster dwellings" which now have a single category of shared housing.
For shared housing new build projects other than those specified below the fee should be calculated in accordance with paragraph 5 (a) plus an enhancement of 10%.
For shared housing rehabilitation projects other than those specified below the fee should be calculated in accordance with paragraph 5 (a) of the scale plus 1% plus an enhancement of 10%.
For shared housing projects comprising wheelchair accommodation (as described in the Housing Corporation's Design and Contract Criteria) or frail elderly accommodation (as described in Housing Corporation circular HCO1/85) the fee should be calculated in accordance with paragraph 5 (a), (plus 1% for rehabilitation schemes where applicable) plus an enhancement of 15%.
The additional percentage set out in paragraph 5 (b) does not apply to shared housing projects, but the provisions of paragraph 6 (a) are applicable.

SCALE 46 PROFESSIONAL CHARGES FOR QUANTITY SURVEYING SERVICES IN CONNECTION WITH LOSS ASSESSMENT OF DAMAGE TO BUILDINGS FROM FIRE, ETC ISSUED BY THE ROYAL INSTITUTION OF CHARTERED SURVEYORS.
This scale has been abolished. See Author's Note on page 800.

EFFECTIVE FROM JULY 1988

1. This scale of professional charges is for use in assessing loss resulting from damage to buildings by fire etc., under the "building" section of an insurance policy and is applicable to all categories of buildings.

2. The fees set out below cover the following quantity surveying services as may be required in connection with the particular loss assessment:
   (a) Examining the insurance policy.
   (b) Visiting the building and taking all necessary site notes.
   (c) Measuring at site and/or from drawings and preparing itemised statement of claim and pricing same.
   (d) Negotiating and agreeing claim with the loss adjuster.

3. The fees set out below are exclusive of the following:
   (a) Travelling and other expenses (for which the actual disbursement is recoverable unless there is some special prior arrangement for such charge.)
   (b) Cost of reproduction of all documents, which are chargeable in addition at net cost.

4. Copyright in all documents prepared by the quantity surveyor is reserved.
5. (a) The fees for the services outlined in paragraph 2 shall be as follows:

Agreed Amount of £		Fee		
		£		£
Up to	60 000		see note 5(c) below	
60 000 -	180 000	2.5%		
180 000 -	360 000	4 500 + 2.3%	on balance over	180 000
360 000 -	720 000	8 640 + 2.0%	on balance over	360 000
Over	720 000	15 840 + 1.5%	on balance over	720 000

and to the result of that computation shall be added 12.5%

QUANTITY SURVEYORS' FEES

(b) The sum on which the fees above shall be calculated shall be arrived at after having given effect to the following:

(i) The sum shall be based on the amount of damage, including such amounts in respect of architects', surveyors and other consultants' fees for reinstatement, as admitted by the loss adjuster.

(ii) When a policy is subject to an average clause, the sum shall be the agreed amount before the adjustment for "average".

(iii) When, in order to apply the average clause, the reinstatement value of the whole subject is calculated and negotiated an additional fee shall be charged commensurate with the work involved.

(c) Subject to 5 (b) above, when the amount of the sum on which fees shall be calculated is under £60,000 the fee shall be based on time involved as defined in Scale 37 (July 1988) para. 19 or on a lump sum or percentage basis agreed between the building owner and the quantity surveyor.

6. The foregoing scale of charges is exclusive of any services in connection with litigation and arbitration.

7. The fees and charges are in all cases exclusive of value added tax which shall be applied in accordance with legislation.

QUANTITY SURVEYORS' FEES

SCALE 47 PROFESSIONAL CHARGES FOR THE ASSESSMENT OF REPLACEMENT COSTS BUILDINGS FOR INSURANCE, CURRENT COST ACCOUNTING AND OTHER PURPOSES ISSUED BY THE ROYAL INSTITUTION OF CHARTERED SURVEYORS

This scale has been abolished. See Author's Note on page 800.

EFFECTIVE FROM JULY 1988

1.0 GENERALLY

1.1. The fees are in all cases exclusive of travelling and other expenses (for which the actual disbursement is recoverable unless there is some prior arrangement for such charges).

1.2. The fees and charges are in all cases exclusive of value added tax which will be applied in accordance with legislation.

2.0 ASSESSMENT OF REPLACEMENT COSTS OF BUILDINGS FOR INSURANCE PURPOSES

2.1. Assessing the current replacement cost of buildings where adequate drawings for the purpose are available.

Assessed current costs		Fee			
		£			£
Up to	700 000		0.2%		
140 000 -	700 000	280 +	0.075%	on balance over	140 000
700 000 -	4 200 000	700 +	0.025%	on balance over	700 000
Over	4 200 000	1 575 +	0.01%	on balance over	4 200 000

2.2. Fees to be calculated on the assessed cost, i.e. base value, for replacement purposes including allowances for demolition and the clearance but excluding inflation allowances and professional fees.

2.3. Where drawings adequate for the assessment of costs are not available or where other circumstances require that measurements of the whole or part of the buildings are taken, an additional fee shall be charged based on the time involved or alternatively on a lump sum basis agreed between the employer and the surveyor.

2.4 when the assessment is for buildings of different character or on more than one site, the costs shall not be aggregated for the purpose of calculating fees.

2.5 For current cost accounting purposes this scale refers only to the assessment of replacement cost of buildings.

2.6 The scale is appropriate for initial assessments but for annual review or a regular reassessment the fee should be by arrangement having regard to the scale and to the amount of work involved and the time taken.

2.7. The fees are exclusive of services in connection with negotiations with brokers, accountants or insurance companies for which there shall be an additional fee based upon the time involved.

CONSULTING ENGINEERS' FEES

INTRODUCTION

A scale of professional charges for consulting engineering services is published by the Association for Consultancy and Engineering (ACE)

Copies of the document can be obtained direct from : -

The Association of Consultancy and Engineering
Alliance House
Caxton Street
London SW1H OQL
Tel 0207 222 6557
Fax 0207 222 0750

Comparisons

Instead of the previous arrangement of having different agreements designed for each major discipline of engineering, these new agreements have been developed primarily to suit the different roles that Consulting Engineers may be required to perform, with variants of some of them for different disciplines. The agreements have been standardised as far as possible whilst retaining essential differences.

Greater attention is required than with previous agreements to ensure the documents are completed properly. This is because of the perceived need to allow for a wider choice of arrangements, particularly of methods of payment.

The agreements are not intended to be used as unsigned reference material with the details of an engagement being covered in an exchange of letters, although much of their content could be used as a basis for drafting such correspondence.

Forms of Agreement

The initial agreements are for use where a Consulting Engineer is engaged as follows :

Agreement as a Lead Consultant

Agreement directly by the Client, but not as Lead Consultant

Agreement to provide design services for a design and construct Contractor

Short Form Agreement (Report and Advisory Services)

Agreement as a Project Manager

The ACE/APS Agreement for Planning Supervisor (2002), is now invalid, following the new CDM regulations, which came into force on 6th April 2007

Each of Agreements A, B and C are published in two variants

Variant 1 Civil and Structural Engineering

Variant 2 The Engineering of Electrical and Mechanical Services in Buildings

Each agreement comprises the following :-

Memorandum of Agreement

Conditions of Engagement

Appendix I - Services of the Consulting Engineer

Appendix II - Remuneration of the Consulting Engineer

For the latest information, Readers are advised to log onto the ACE web-site at www.acenet.co.uk

CONSULTING ENGINEERS' FEES

Memorandum of Agreement

There is a different memorandum for each agreement, reflecting in each instance the particular relationships between the parties. It is essential that the memorandum be fully completed. Spaces are provided for entry of important and specific details relevant to each commission, such as nominated individuals, limits of liability, requirements for professional indemnity insurance, the frequency of site visits and meetings, and requirements for collateral warranties. All the memoranda are arranged for execution under hand; some also have provision for execution as deeds.

Conditions of Engagement

These have been standardised as far as possible and thus contain much that is common between the agreements, but parts differ and are peculiar to individual agreements to reflect the responsibilities applying. The conditions can normally stand as drafted but clauses may be deleted and others be added should the circumstances so require for a particular commission.

Appendix I - Services

This appendix, which has significant differences between the agreements and variants, describes the services to be performed. These services include both standard Normal Services, the majority of which will usually be required, and standard Additional Services of which only some will be required. Standard Normal Services may be deleted if not required or not relevant to a particular commission; further Services, both Normal and Additional may be added in spaces provided. It may be agreed in advance, when known that certain of the Additional Services will clearly be required, that these will be treated and paid for as Normal Services for a particular commission.

Appendix II - Remuneration / Fees and Disbursements

This appendix provides alternate means of assessing the consulting engineer's fees and disbursements. It identifies, when completed, which of those services listed in Appendix I are to be performed within the overall fee applicable for Normal Services. Figures need to be entered on such details as time charge rates, fee percentages and interest rates on delayed payments. Alternatives which do not apply require deletion and those remaining completion, so that the appendix when incorporated within an engagement contract describes the exact arrangements applicable to that commission.

Collateral Warranties

The association is convinced that collateral warranties are generally unnecessary and should only be used in exceptional circumstances. The interests of clients, employers and others are better protected by taking out project or BUILD type latent defects insurance. Nevertheless, in response to observations raised when the pilot editions excluded any mention of warranties, references and arrangements have been included in the Memorandum and elsewhere by which Consulting Engineers may agree to enter into collateral warranty agreements; these should however only be given when the format and requirements thereof have been properly defined and recorded in advance of undertaking the commission.

Requirements for the provision of collateral warranties will be justified even less with commissions under Agreement D than with those under the other ACE agreements. Occasional calls may be made for them, such as when a client intends to dispose of property and needs evidence of a duty of care being owed to specific third parties, but these will be few and far between.

Remuneration

Guidance on appropriate levels of fees to be charged is given at the end of each agreement. Firms and their clients may use this or other sources, including their own records, to determine suitable fee arrangements.

CONSULTING ENGINEERS' FEES

Need for formal documentation

The Association of Consulting Engineers recommends that formal written documentation should be executed to record the details of each commission awarded to a Consulting Engineer. These Conditions are published as model forms of agreement suitable for the purpose. However, even if these particular Conditions are not used, it is strongly recommended that, whenever a Consulting Engineer is appointed, there should be at least an exchange of letters defining the duties to be performed and the terms of payment.

Appointments outside the United Kingdom

These conditions of Engagement are designed for use within the UK. For work overseas it is impracticable to give definite recommendations; circumstances differ too widely between countries. There are added complications in documentation relating to local legislation, import customs, conditions of payment, insurance, freight, etc. Furthermore, it is often necessary to arrange for visits to be made by principals and senior staff whose absence abroad during such periods represents a serious reduction of their earning power. The additional duties, responsibilities and non-recoverable costs involved, and the extra work on general co-ordination, should be reflected in the levels of fees. Special arrangements are also necessary to cover travelling and other out-of-pocket expenses in excess of those normally incurred on similar work in the UK, including such matters as local cost-of-living allowances and the cost of providing home-leave facilities for expatriate staff.

CONDITIONS OF ENGAGEMENT

Obligations

The following is a brief summary of the conditions of engagement. It is recommended that reference should be made to the full document of the Association of Consulting Engineers Conditions of Engagement, 1995 before making an engagement.

Obligations of the Consulting Engineer

The responsibilities of the Consultant Engineer for the works are as set out in the actual agreement The various standard clauses in the Conditions relate to such matters as differentiating between Normal and Additional services, the duty to exercise skill and care, the need for Client's written consent to the assignment or transfer of any benefit or obligation of the agreement, the rendering of advice if requested on the appointment of other consultants and specialist sub- consultants, any recommendations for design of any part of the Works by Contractors or Sub-contractors (with the proviso that the Consulting Engineer is not responsible for detailed design of contractors or for defects or omissions in such design), the designation of a Project Leader, the need for timeliness in requests to the Client for information etc., freezing the design once it has been given Client approval and the specific exclusion of any duty to advise on the actual or possible presence of pollution or contamination or its consequences.

Obligations of the Client

The Consultant Engineer shall be supplied with all necessary data and information in good time. The Client shall designate a Representative authorised to make decisions on his behalf and ensure that all decisions, instructions, and approvals are given in time so as not to delay or disrupt the Consultant Engineer.

Site Staff

The Consulting Engineer may employ site staff he feels are required to perform the task, subject to the prior written agreement of the Client. The Client shall bear the cost of local office accommodation, equipment and running costs.

CONSULTING ENGINEERS' FEES

Commencement, Determination, Postponement, Disruption and Delay

The Consulting Engineer's appointment commences at the date of the execution of the Memorandum of Agreement or such earlier date when the Consulting Engineer first commenced the performance of the Services, subject to the right of the Client to determine or postpone all or any of the Services at any time by Notice.

The Client or the Consulting Engineer may determine the appointment in the event of a breach of the Agreement by the other party after two weeks notice. In addition, the Consulting Engineer may determine his appointment after two weeks notice in the event of the Client failing to make proper payment.

The Consulting Engineer may suspend the performance of all or any of the Services for up to twenty-six weeks if he is prevented or significantly impeded from performance by circumstances outside his control. The appointment may be determined by either party in the event of insolvency subject to the issue of notice of determination.

Payments

The Client shall pay fees for the performance of the agreed service(s) together with all fees and charges to the local or other authorities for seeking and obtaining statutory permissions, for all site staff on a time basis, together with additional payments for any variation or the disruption of the Consulting Engineer's work due to the Client varying the task list or brief or to delay caused by the Client, others or unforeseeable events.

If any part of any invoice submitted by the Consulting Engineer is contested, payment shall be made in full of all that is not contested.

Payments shall be made within 28 days of the date of the Consulting Engineer's invoice; interest shall be added to all amounts remaining unpaid thereafter.

Ownership of Documents and Copyright

The Consulting Engineer retains the copyright in all drawings, reports, specifications, calculations etc. prepared in connection with the Task; with the agreement of the Consulting Engineer and subject to certain conditions, the Client may have a licence to copy and use such intellectual property solely for his own purpose on the Task in hand, subject to reservations.

The Consulting Engineer must obtain the client's permission before he publishes any articles, photographs or other illustrations relating to the Task, nor shall he disclose to any person any information provided by the Client as private and confidential unless so authorised by the Client.

Liability, Insurance and Warranties

The liability of the Consulting Engineer is defined, together with the duty of the Client to indemnify the Consulting Engineer against all claims etc. in excess of the agreed liability limit.

The Consulting Engineer shall maintain Professional Indemnity Insurance for an agreed amount and period at commercially reasonable rates, together with Public Liability Insurance and shall produce the brokers' certificates for inspection to show that the required cover is being maintained as and when requested by the Client.

The Consulting Engineer shall enter into and provide collateral warranties for the benefit of other parties if so agreed.

Disputes and Differences

Provision is made for mediation to solve disputes, subject to a time limit of six weeks of the appointment of the mediator at which point it should be referred to an independent adjudicator. Further action could be by referring the dispute to an arbitrator.

THE TOWN AND COUNTRY PLANNING (FEES FOR APPLICATIONS AND DEEMED APPLICATIONS) (AMENDMENT) (ENGLAND) REGULATIONS 2008

S.I. 2008/843 – operative from 6th April 2008

The following extracts from the Town and Country Planning Fees Regulations, available from HMSO, relate only to those applications which meet the "deemed to qualify clauses" laid down in regulations 1 to 12 of S.I. No. 1989/193, amended by S.I. 1990/2473, 1991/2735, 1992/1817, 1992/3052, 1993/3170, 1997/37, 2001/2719, 2002/768, 2005/843 and 2006/994. These regulations apply in relation to England only.

General increase in fees

2.—(1) The Town and Country Planning (Fees for Applications and Deemed Applications) Regulations 1989(**b**) are amended as follows.

(2) After regulation 1(2)(a) (application), insert —

"(aa) to requests for confirmation that a condition or conditions attached to a grant of planning permission has been complied with where the request is made on or after 6th April 2008;".

(3) In regulation 10A (fees for applications for certificates of lawful use or development)—

(a) in paragraph (5)(b), for "£135" substitute "£170";

(b) in paragraph (6)(a), for "£265" substitute "£335"; and

(c) in paragraph (6)(b)—

(i) for "£13,250" substitute "£16,565", and

(ii) for "£50,000" substitute "£250,000".

(4) In regulation 11A (fees for certain applications under the General Permitted Development Order)—

(a) in paragraph (1)(a), for "£50" substitute "£70"; and

(b) in paragraph (1)(b), for "£265" substitute "£335".

(5) After regulation 11B (fees in respect of the monitoring of mining and landfill sites), insert —

"Fee for confirmation of compliance with condition attached to planning permission: England

11D.—(1) Where a request is made to a local planning authority for written confirmation of compliance with a condition or conditions attached to a grant of planning permission, a fee shall be paid to that authority as follows—

(a) where the request relates to a permission for development which falls within category 6 or 7(a) of Part 2 to Schedule 1, £25 for each request;

(b) where the request relates to a permission for development which falls within any other category of that Schedule, £85 for each request.

(2) Any fee paid under this regulation shall be refunded if the local planning authority fails to give the written confirmation requested within a period of twelve weeks from the date on which the authority received the request.".

(6) In Part 1 of Schedule 1 (general provisions)—

(a) in paragraphs 4(1) and 6(2), for "£265" substitute "£335";

(b) in paragraphs 7 and 7A, for "£135" substitute "£170";

THE TOWN AND COUNTRY PLANNING (FEES FOR APPLICATIONS AND DEEMED APPLICATIONS) (AMENDMENT) (ENGLAND) REGULATIONS 2008

(c) omit paragraph 7B;

(d) in paragraph 15(2)(a), for "£265", substitute "£335";

(e) in paragraph 15(2)(b)–

(i) for "£6,625" substitute "£8,285"; and

(ii) for "£25,000" substitute "£125,000".

(7) For Part 2 of Schedule 1 (scale of fees), substitute the Part set out in Schedule 1 to these Regulations.

(8) For Schedule 2 (scale of fees for advertisement applications), substitute the Schedule set out in Schedule 2 to these Regulations.

PART II: SCALE OF FEES

Category of development	Fee payable
I. Operations **1**. The erection of dwellinghouses (other than development within category 6 below).	(a) Where the application is for outline planning permission and - (i) the site area does not exceed 2.5 hectares, £335 for each 0.1 hectare of the site area; (ii) the site area exceeds 2.5 hectares, £8,285 and an additional £100 for each 0.1 hectare in excess of 2.5 hectares, subject to a maximum in total of £125,000. (b) in other cases - (i) where the number of dwellinghouses to be created by the development is 70 or fewer, £335 for each dwellinghouse; (ii) where the number of dwellinghouses to be created by the development exceeds 70, £16,565, and an additional £100 for each dwellinghouse in excess of 70 dwellinghouses, subject to a maximum in total of £250,000.
2. The erection of buildings (other than buildings in categories 1,3,4,5 or 7).	a) Where the application is for outline planning permission and - (i) the site area does not exceed 2.5 hectares, £335 for each 0.1 hectare of the site area; (ii) the site area exceeds 2.5 hectares, £8,285, and an additional £100 for each 0.1 hectare in excess of 2.5 hectares, subject to a maximum in total of £125,000. (b) in other cases - (i) where no floor space is to be created by the development, £170; (ii) where the area of gross floor space to be created by the development does not exceed 40 square metres, £170; (iii) where the area of the gross floor space to be created by the development exceeds 40 square metres, but does not exceed 95 square metres, £335;

THE TOWN AND COUNTRY PLANNING (FEES FOR APPLICATIONS AND DEEMED APPLICATIONS) (AMENDMENT) (ENGLAND) REGULATIONS 2008

Category of development	Fee payable
2. The erection of buildings (other than buildings in categories 1,3,4,5 or 7). – cont'd	(iv) where the area of the gross floor space to be created by the development exceeds 95 square metres, but does not exceed 3750 square metres, £335 for each 95 square metres of that area; (v) where the area of gross floor space to be created by the development exceeds 3750 square metres, £16,565, and an additional £100 for each 95 square metres in excess of 3750 square metres, subject to a maximum in total of £250,000.
3. The erection, on land used for the purposes of agriculture, of buildings to be used for agricultural purposes (other than buildings in category 4).	(a) Where the application is for outline planning permission and - (i) the site area does not exceed 2.5 hectares, £335 for each 0.1 hectare of the site area; (ii) the site area exceeds 2.5 hectares, £8,285, and an additional £100 for each additional 0.1 hectare in excess of 2.5 hectares, subject to a maximum in total of £125,000. (b) in other cases - (i) where the area of gross floor space to be created by the development does not exceed 465 square metres, £70; (ii) where the area of gross floor space to be created by the development exceeds 465 square metres but does not exceed 540 square metres, £335; (iii) where the area of the gross floor space to be created by the development exceeds 540 square metres but does not exceed 4215 square metres, £335 for the first 540 square metres, and an additional £335 for each 95 square metres in excess of 540 square metres; and (iv) where the area of gross floor space to be created by the development exceeds 4215 square metres, £16,565, and an additional £100 for each 95 square metres in excess of 4215 square metres, subject to a maximum in total of £250,000.
4. The erection of glasshouses on land used for the purposes of agriculture.	(a) Where the gross floor space to be created by the development does not exceed 465 square metres, £70; (b) where the gross floor space to be created by the development exceeds 465 square metres, £1,870.
5. The erection, alteration or replacement of plant or machinery.	(a) Where the site area does not exceed 5 hectares, £335 for each 0.1 hectare of the site area; (b) where the site area exceeds 5 hectares, £16,565, and an additional £100 for each 0.1 hectare in excess of 5 hectares, subject to a maximum in total of £250,000.

THE TOWN AND COUNTRY PLANNING (FEES FOR APPLICATIONS AND DEEMED APPLICATIONS) (AMENDMENT) (ENGLAND) REGULATIONS 2008

Category of development	Fee payable
6. The enlargement, improvement or other alteration of existing dwellinghouses.	(a) Where the application relates to one dwellinghouse, £150; (b) where the application relates to 2 or more dwellinghouses, £295.
7. (a) The carrying out of operations (including the erection of a building) within the curtilage of an existing dwellinghouse, for purposes ancillary to the enjoyment of the dwellinghouse as such, or the erection or construction of gates, fences, walls or other means of enclosure along a boundary of the curtilage of an existing dwellinghouse; or	£150.
(b) the construction of carparks, service roads and other means of access on land used for the purposes of a single undertaking, where the development is required for a purpose incidental to the existing use of the land.	£170
8. The carrying out of any operations connected with exploratory drilling for oil or natural gas.	(a) Where the site area does not exceed 7.5 hectares, £335 for each 0.1 hectares of the site area; (b) where the site area exceeds 7.5 hectares, £25,000, and an additional £100 for each 0.1 hectare in excess of 7.5 hectares, subject to a maximum in total of £250,000.
9. The carrying out of any operations not coming within any of the above categories.	(a) In the case of operations for the winning and working of minerals - (i) where the site area does not exceed 15 hectares, £170 for each 0.1 hectare of the site area; (ii) where the site area exceeds 15 hectares, £25,315, and an additional £100 for each 0.1 hectare in excess of 15 hectares, subject to a maximum in total of £65,000; (b) in any other case, £170 for each 0.1 hectare of the site area, subject to a maximum of £250,000.

II. Uses of land

Category of development	Fee payable
10. The change of use of a building to use as one or more separate dwellinghouses.	(a) Where the change of use is from a previous use as a single dwellinghouse to use as two or more single dwellinghouses- (i) where the change of use is to use as 70 or fewer dwellinghouses, £335 for each additional dwellinghouse; (ii) where the change of use is to use as more than 70 dwellinghouses £16,565, and an additional £100 for each dwellinghouse in excess of 70 dwellinghouses, subject to a maximum in total of £250,000;

THE TOWN AND COUNTRY PLANNING (FEES FOR APPLICATIONS AND DEEMED APPLICATIONS) (AMENDMENT) (ENGLAND) REGULATIONS 2008

II. Uses of land – cont'd

10. The change of use of a building to use as one or more separate dwellinghouses – cont'd	(b) in all other cases- (i) where the change of use is to use as 70 or fewer dwellinghouses, £335 for each dwellinghouse; (ii) where the change of use is to use as more than 70 dwellinghouses £16,565, and an additional £100 for each dwellinghouse in excess of 70 dwellinghouses, subject to a maximum in total of £250,000.
11. (a) The use of land for the disposal of refuse or waste materials or for the deposit of material remaining after minerals have been extracted from land; or (b) for use of land for the storage of minerals in the open.	(a) Where the site area does not exceed 15 hectares, £170 for each 0.1 hectare of the site area; (b) where the site area exceeds 15 hectares £25,315, and an additional £100 for each 0.1 hectare in excess of 15 hectares, subject to a maximum in total of £65,000.
12. The making of a material change in the use of a building or land (other than a material change of use coming within any of the above categories).	£335.

SCHEDULE 2

SCALE OF FEES IN RESPECT OF APPLICATIONS FOR CONSENT TO DISPLAY ADVERTISEMENTS

	Category of advertisement	Fee payable
1.	Advertisements displayed on business premises, on the forecourt of business premises or on other land within the curtilage of business premises, wholly with reference to all or any of the following matters: (a) the nature of the business or other activity carried out on the premises; (b) the goods sold or the services provided on the premises; or (c) the name and qualifications of the person carrying on such business or activity or supplying such goods or services.	£95.
2.	Advertisements for the purpose of directing members of the public to, or otherwise drawing attention to the existence of, business premises which are in the same locality as the site on which the advertisement is to be displayed but which are not visible from that site.	£95.
3.	All other advertisements.	£335.

THE BUILDING (LOCAL AUTHORITY CHARGES) REGULATIONS 1998

Author's Note:

On the 31st July 1998 the Minister for Construction, announced his intention of improving the flexibility with which local authorities responsible for building control in England and Wales could respond to competition from the private sector by devolving to individual authorities the setting of charges for building control functions carried out in respect of the Building Regulations 1991.

The Building (Local Authority Charges) Regulations 1998 (the Charges Regulations) require each local authority to prepare a Scheme within which they are to fix their charges. They came into effect on the 1st April 1999. In a number of major cities, uniform levels of fees have been adopted. In some local authorities charges have fallen a third in comparison to those prescribed within the 1991 Regulations. A number of authorities have adopted the Local Government Association (LGA) Model Fee Scheme 2000, which is for local authority distribution only.

Consultation should be made to each local authority for their Charges, however as guidance we have kindly been given permission by the London Borough of Ealing to publish the Charges for their district, which includes Acton W3.

CHARGE SCHEDULES

With effect from 1st April 2002, there are three main charge Schedule Tables:

Table 1, For erection of one or more small new domestic buildings and connected work, ie. houses and flats up to 3 storeys in height with an internal floor area not exceeding 300 m²;
Table 2, For erection of certain small domestic building, and extensions, ie detached garages and carports not exceeding 40 m² and not exempt, and extensions including all new loft conversions up to a total of 60 m² at the same time;
Table 3, For building work other than where Tables 1 and 2 apply. Charges relate to estimated cost of the works.

EXEMPTIONS

No fees are charged, where we are satisfied **work** is **solely** for the purpose of providing means of access **for disabled persons** or within a building, or for providing facilities designed to secure greater health, safety, welfare or convenience and is carried out in relation to a building to which members of the public are admitted or is a dwelling occupied by a disabled person.

CHARGE BANDS (1st January 2008 to 31st December 2008)

Charges, based upon the Total Estimated Costs of the Works, vary from one local authority to another, based on the following bands. Readers should approach the relevant local authority to ascertain their charge rates for projects under consideration. Under table 3, we have shown the published fees for projects in Ealing up to £200,000.

No.	Band	Fee up to bottom of band value	Indicative calculations for extra value
i.	Up to £200,000	£150.46	(see table)
ii.	£200,001 to £1,000,000	£1,380	£3.60 per £1,000
iii.	£1,000,001 to £10,000,000	£4,260	£2.65 per £1,000
iv.	Over £10,000,000	£28,110	£1.70 per £1,000

But fees vary from one authority to another, so always check.

THE BUILDING (LOCAL AUTHORITY CHARGES) REGULATIONS 1998

TABLE 1

CHARGES FOR ERECTION OF ONE OF MORE SMALL NEW DOMESTICBUILDINGS AND CONNECTED WORK

Building Notice submissions are mainly for residential/small domestic where the full **Charge** must be paid at time of notification. These are mainly used for small domestic alterations and not house extensions. Where structural work is involved calculations need to be provided. Upon satisfactory completion of works on site a Completion Certificate may be issued.

NOTES

Table 1 is no longer a separate table and now refers users to Tables 2 and 3.
Dwellings in excess of 300 m² in floor area (excluding garage or carport) are to be calculated on estimated cost on accordance with Table 3.
Buildings in excess of 3 storeys (including any basements) to be calculated on estimated cost in accordance with Table 3.

The Charges in this table includes for works of drainage in connection with erection of a building(s), even where those drainage works are commenced in advance of the plans for the building being deposited.
The charges include for an integral garage and where a garage or carport shares at least one wall of the domestic building. Detached garages are not included in this Table (see Notes Table 2).
Where a Plan or Inspection Charge exceeds £5,000.00 the Council may agree payment in instalments.

Where all dwellings on a site or an estate are substantially the same, it may be possible to offer a discount of 30% reduction on PLAN Charge OR equivalent reduction on the BUILDING NOTICE Charge.

TABLE 2

CHARGES FOR ERECTION OF CERTAIN SMALL DOMESTIC BUILDINGS, GARAGES, CARPORTS AND EXTENSIONS

Table 2: Flat Rate combined Plan and Inspection charge	Charge £	VAT £	Total £
Detached garages or carport with floor area under 40m²	188.08	32.91	220.99
Detached garages or carport between 40m² & 60m² floor area	338.54	59.25	397.79
Extension's of a dwelling with floor area under 10m²	338.54	59.25	397.79
Extension's of a dwelling with floor area between 10m² & 40m²	494.00	86.45	580.45
Extension's of a dwelling with floor area between 40m² & 60m²	644.47	112.78	757.25

NOTES

Detached garages and carports having an internal floor area not exceeding 30 m² are 'exempt buildings', providing that in the case of a garage it is sited at least 1.0 m away from the boundary or is constructed substantially of non-combustible materials.
A carport extension having an internal floor area not exceeding 30 m² would be exempt if it is fully open on at least 2 sides.

THE BUILDING (LOCAL AUTHORITY CHARGES) REGULATIONS 1998

Use Table 3 for domestic entensions and garages over 60 m², four or more storey buildings, alterations and commercial work.
A new Dormer Windows which does not increase the usable floor area would be an alteration so use Table 3.
If the total floor area of all extensions being done at the same time exceeds 60 m² use Table 3.
Loft conversions with new internal useable floor area in roof space are to be treated as an extension in this Table 2.
Chargeable installations of Cavity Fill Insulation, and Unvented Hot Water Systems should use Table 3 (see Table 3 notes).
Extensions to a building that is **NOT wholly domestic** should use **Table 3.**

Where on an estate erections of garages or extensions are substantially the same, it may be possible to offer a discount of 30% reduction on PLAN Charge OR a 7½% reduction on the BUILDING NOTICE Charge.

TABLE 3

CHARGES FOR BUILDING WORK OTHER THAN TO WHICH TABLES 1 AND 2 APPLY. CHARGES RELATE TO ESTIMATED COST.

Full Plan applications apply to designated use projects, where the **Plan Charge,** equivalent to 25% of the **Full Charge** must be paid in the deposit of plans. A subsequent re-submission, further to a Rejection of Plans, will NOT attract an additional fee for essentially the same work. If the inspection charge is not paid upon the deposit of plans, an invoice will be raised after the first inspection on site as this is when **Inspection Charge** becomes payable. Where work is to be done to Shops, Factories, Offices, Railway Premises, Hotels and Boarding Houses, and Non-domestic Workplaces a Full Application should be made. It is also appropriate for Domestic Loft Conversions and other Extensions and Erections of Domestic Buildings. It is important to start work before 3 years have expired or the Plan Approval may be withdrawn. Upon satisfactory completion of the works on site a Completion Certificate may be issued.

See Table 3 example, for applications up to £200,000, on following page

NOTES

If some building work is covered by Table 2 and some by Table 3, both fees are payable.
Estimated cost of work should not include any professional fees (e.g. Architect, Quantity Surveyor, etc) nor any VAT.
Installation of cavity fill insulation in accordance with Part D of Schedule 1 to the Principle Regulations where installation is not certified to an approved standard or is not installed by an approved installer, or is part of a larger project this Table Building Notice Charge is payable.
Installation of an unvented hot water system in accordance with Part G3 of Schedule 1 to the Principle Regulations where the installation is not part of a larger project and where the authority carry out an inspection, this Table Building Notice Charge is payable.

If application is for erection of work substantially the same type under current regulations, it may be possible to offer a discount of PLAN CHARGE by 30% OR BUILDING NOTICE Charge by 7½%.

THE BUILDING (LOCAL AUTHORITY CHARGES) REGULATIONS 1998

TABLE 3 EXAMPLE: Total Estimated Cost of Works – up to £200,000

	Charge	VAT	Total		Charge	VAT	Total
500 or less	150.46	26.33	176.79	100,001 to 102,000	1034.64	181.06	1215.70
501 - 5,000	232.23	40.64	272.87	102,001 to 104,000	1041.68	182.29	1223.97
5,001 - 8,000	253.97	44.44	298.41	104,001 to 106,000	1048.72	183.53	1232.25
8,001 - 10,000	274.47	48.03	322.51	106,001 to 108,000	1055.76	184.76	1240.52
10,001 - 12,000	294.98	51.62	346.60	108,001 to 110,000	1062.80	185.99	1248.79
12,001 - 14,000	315.48	55.21	370.69	110,001 to 112,000	1069.84	187.22	1257.06
14,001 - 16,000	335.99	58.80	394.78	112,001 to 114,000	1076.88	188.45	1335.33
16,001 - 18,000	356.49	62.39	418.88	114,001 to 116,000	1083.92	189.69	1273.61
18,001 - 20,000	376.99	65.97	442.97	116,001 to 118,000	1090.96	190.92	1281.88
20,001 - 22,000	396.09	69.32	465.41	118,001 to 120,000	1098.00	192.15	1290.15
22,001 - 24,000	415.04	72.63	487.67	120,001 to 122,000	1105.04	193.38	1298.42
24,001 - 26,000	433.85	75.92	509.77	122,001 to 124,000	1112.08	194.61	1306.69
26,001 - 28,000	452.51	79.19	531.70	124,001 to 126,000	1119.12	195.85	1314.97
28,001 - 30,000	471.02	82.43	553.45	126,001 to 128,000	1126.16	197.08	1323.24
30,001 - 32,000	489.40	85.64	575.04	128,001 to 130,000	1133.20	198.31	1331.51
32,001 - 34,000	507.62	88.83	596.45	130,001 to 132,000	1140.24	199.54	1339.78
34,001 - 36,000	525.70	92.00	617.70	132,001 to 134,000	1147.28	200.77	1348.05
36,001 - 38,000	543.64	95.14	638.77	134,001 to 136,000	1154.32	202.01	1706.33
38,001 - 40,000	561.43	98.25	659.67	136,001 to 138,000	1161.36	203.24	1364.60
40,001 - 42,000	579.07	101.34	680.41	138,001 to 140,000	1168.40	204.47	1372.87
42,001 - 44,000	596.57	104.40	700.97	140,001 to 142,000	1175.44	205.70	1381.14
44,001 - 46,000	613.92	107.44	721.36	142,001 to 144,000	1182.48	206.93	1389.41
46,001 - 48,000	631.13	110.45	741.58	144,001 to 146,000	1189.52	208.17	1397.69
48,001 - 50,000	648.20	113.43	761.63	146,001 to 148,000	1196.56	209.40	1405.96
50,001 - 52,000	665.11	116.40	781.51	148,001 to 150,000	1203.60	210.63	1414.23
52,001 - 54,000	681.89	119.33	801.22	150,001 to 152,000	1210.64	211.86	1422.50
54,001 - 56,000	698.52	122.24	820.76	152,001 to 154,000	1217.68	213.09	1430.77
56,001 - 58,000	715.00	125.12	840.12	154,001 to 156,000	1224.72	214.33	1439.05
58,001 - 60,000	731.34	127.98	859.32	156,001 to 158,000	1231.76	215.56	1447.32
60,001 - 62,000	747.53	130.82	878.35	158,001 to 160,000	1238.80	216.79	1455.59
62,001 - 64,000	763.58	133.63	897.20	160,001 to 162,000	1245.84	218.02	1463.86
64,001 - 66,000	779.48	136.41	915.89	162,001 to 164,000	1252.88	219.25	1472.13
66,001 - 68,000	795.24	139.17	934.40	164,001 to 166,000	1259.92	220.49	1480.41
68,001 - 70,000	810.85	141.90	952.75	166,001 to 168,000	1266.96	221.72	1488.68
70,001 - 72,000	826.31	144.61	970.92	168,001 to 170,000	1274.00	222.95	1496.95
72,001 - 74,000	841.64	147.29	988.92	170,001 to 172,000	1281.04	224.18	1505.22
74,001 - 76,000	856.81	149.94	1006.75	172,001 to 174,000	1288.08	225.41	1513.49
76,001 - 78,000	871.84	152.57	1024.42	174,001 to 176,000	1295.12	226.65	1521.77
78,001 - 80,000	886.73	155.18	1041.91	176,001 to 178,000	1302.16	227.88	1530.04
80,001 - 82,000	901.47	157.76	1059.23	178,001 to 180,000	1309.20	229.11	1538.31
82,001 - 84,000	916.06	160.31	1076.38	180,001 to 182,000	1316.24	230.34	1546.58
84,001 - 86,000	930.51	162.84	1093.35	182,001 to 184,000	1323.28	231.57	1554.85
86,001 - 88,000	944.82	165.34	1110.16	184,001 to 186,000	1330.32	232.81	1563.13
88,001 - 90,000	958.98	167.82	1126.80	186,001 to 188,000	1337.36	234.04	1571.40
90,001 - 92,000	972.99	170.27	1143.27	188,001 to 190,000	1344.40	235.27	1579.67
92,001 - 94,000	986.86	172.70	1159.56	190,001 to 192,000	1701.44	236.50	1587.94
94,001 - 96,000	1000.59	175.10	1175.69	192,001 to 194,000	1708.48	237.73	1596.21
96,001 - 98,000	1014.17	177.48	1191.65	194,001 to 196,000	1365.52	238.97	1604.49
98,001 - 100,000	1027.60	179.83	1207.43	196,001 to 198,000	1372.56	240.20	1612.76
				198,001 to 200,000	1379.60	241.43	1621.03

PART 6

Rates of Wages

BUILDING INDUSTRY - ENGLAND, WALES AND SCOTLAND

In May 2006, the Building and Civil Engineering Joint Negotiating Committee agreed a new three year agreement on pay and conditions for building and civil engineering operatives.

The Working Rule Agreement includes a pay structure with a general operative and additional skilled rates of pay as well as craft rate. Plus rates and additional payments will be consolidated into basic pay to provide the following rates (for a normal 39 hour week) which will come into effect from the following dates:

Effective from 30 June 2008

The following basic rates of pay will apply:

	Rate per 39-hour week (£)	Rate per hour (£)
Craft Rate	401.70	10.30
Skill Rate 1	382.98	9.82
Skill Rate 2	368.94	9.46
Skill Rate 3	345.15	8.85
Skill Rate 4	325.65	8.35
General operative	302.25	7.75

Holidays with Pay and Benefits Schemes

The Building and Civil Engineering benefits scheme has unveiled a new holiday pay plan following the introduction of the Working Time Directive. From 2 August 1999 there are no fixed holiday credits, instead employers will calculate appropriate sums to fund operatives' holiday pay entitlement and make regular monthly payments into the B&CE scheme.

For full details contact B&CE on 01293 526911.

Employers contribution towards retirement benefit is paid at £5.00 per week, effective from June 2007.

Death and accident cover is provided free.

Young Operatives

Effective from 26 June 2000

The rates of wages for young labourers shall be the following proportions of the General Operatives basic rates:

At 16 years of age 50% of the relevant rate
At 17 years of age 70% of the relevant rate
At 18 years of age or over 100% of the relevant rate

BUILDING INDUSTRY - ENGLAND, WALES AND SCOTLAND

Apprentices/Trainees

The Construction Apprenticeship Scheme (CAS) operates throughout Great Britain from 1 August 1998 it is open to all young people from the age of 16 years. For further information telephone CAS helpline - 01485 578 333.

Apprentice rates - effective from 30 June 2008

Please note that these rates are for guidance only:	Rate per 39-hour week (£)	Rate per hour (£)
Year 1	167.31	4.29
Year 2	216.06	5.54*
Year 3 without NVQ2	252.72	6.48
Year 3 With NVQ2	321.36	8.24
Year 3 With NVQ3	401.70	10.30
On Completion of Apprenticeship With NVQ2	401.70	10.30

* Note: If an apprentice is 22 years and over, and in his/her second year of training, then the National Minimum Wage of £5.52 per hour currently applies, and will increase to £5.73 per hour, as from 1 October 2008.

BUILDING AND ALLIED TRADES JOINT INDUSTRIAL COUNCIL

Authorised rates of wages in the building industry in England and Wales agreed by the Building and Allied Trades Joint Industrial Council.

Effective from 9 June 2008

Subject to the conditions in the Working Rule Agreement the standard weekly rates of wages shall be as follows:

	Rate per 39-hour week (£)	Rate per hour (£)
Craft operative (NQV3)	405.99	10.41
Craft operative (NQV2)	349.05	8.95
Adult general operative	301.47	7.73

For the latest wage/conditions information, go to www.fmb.org.uk/publications/batjic

ROAD HAULAGE WORKERS EMPLOYED IN THE BUILDING INDUSTRY - effective from 30 June 2008

Authorised rates of pay for road haulage workers in the building industry recommended by the Builders Employers Confederation.

Employers

Construction Confederation
Union
56 - 64 Leonard Street
London
EC2A 4JX
Tel: 0207 608 5039
Fax: 0207 608 5001
E-mail: enquiries@constructionconfederation.co.uk

Operatives

The Transport and General Workers
Transport House
128 Theobold's Road
London
WC1X 8TN
Tel: 0207 611 2500
Fax: 0207 611 2555
E-mail: pgwu@tgwu.org.uk

	£
Lorry Drivers (irrespective of the gross weight of the vehicle driven)	401.70

PLUMBING AND MECHANICAL ENGINEERING SERVICES INDUSTRY

PLUMBING AND MECHANICAL ENGINEERING SERVICES INDUSTRY

Authorised rates of wages agreed by the Joint Industry Board for the Plumbing and Mechanical Engineering Services Industry in England and Wales

First part effective from 7 January 2008

The Joint Industry Board for Plumbing and Mechanical Engineering Services in England and Wales
Brook House
Brook Street
St Neots
Huntingdon
Cambridgeshire
PE19 2HW

Tel: 01480 476 925
Fax: 01480 403 081
E-mail: info@jib-pmes.org.uk

	Rate per hour 07/01/2008* £
Operatives	
Technical Plumber and Gas Service Technician	13.46
Advanced Plumber and Gas Service Engineer	12.12
Trained plumber and Gas Service Fitter	10.39
Apprentices (see Note below)**	
1st year of Training	5.04
2nd year of Training	5.77
3rd year of Training	6.52
3rd year of Training with NVQ Level 2	7.92
4th year of Training	8.03
4th year of Training with NVQ Level 2	9.10
4th year of Training with NVQ Level 3	10.05
Adult Trainees	
1st 6 months of Employment	8.11
2nd 6 months of Employment	8.70
3rd 6 months of Employment	9.07

Notes:

* As from 7th January 2008, overtime is payable after 41 hours work per week.

** Where Apprentices have achieved NVQs, the appropriate rate is payable from the date of attainment except that it shall not be any earlier than the commencement of the promulgated year of Training in which it applies.

PLUMBING AND MECHANICAL ENGINEERING SERVICES INDUSTRY

Authorised rates of wages agreed by the Joint Industry Board for the Plumbing Industry in Scotland and Northern Ireland.

Effective from 2nd June 2008

The Joint Industry Board for the Plumbing Industry in Scotland and Northern Ireland
2 Walker Street
Edinburgh
EH3 7LB

Tel: 0131 225 2255
Fax: 0131 226 7638

	Rate per hour 2/06/2008 £
Operatives Plumbers & Gas Service Operatives	
Plumber and Gas Service Fitter	10.61
Advanced Plumber and Gas Service Engineer	12.08
Technician Plumber and Gas Service Technician	13.38
Plumbing Labourer	9.46
Apprentice Plumbers and Fitters	
1st Year Apprentice	3.08
2nd Year Apprentice	4.60
3rd Year Apprentice	5.56
4th Year Apprentice	7.19
Adult Trainees	
Year 1	5.62
Year 2	6.48
Year 3	8.08

PART 7

Daywork and Prime Cost

When work is carried out which cannot be valued in any other way it is customary to assess the value on a cost basis with an allowance to cover overheads and profit. The basis of costing is a matter for agreement between the parties concerned, but definitions of prime cost for the building industry have been prepared and published jointly by the Royal Institution of Chartered Surveyors and the National Federation of Building Trades Employers (now the Construction Confederation) for the convenience of those who wish to use them. These documents are reproduced with the permission of the Royal Institution of Chartered Surveyors, which owns the copyright.

The daywork schedule published by the Civil Engineering Contractors Association is included in the A & B's companion title, *"Spons Civil Engineering and Highway Works Price Book"*.

For larger Prime Cost contracts the reader is referred to the form of contract issued by the Royal Institute of British Architects.

BUILDING INDUSTRY

DEFINITION OF PRIME COST OF DAYWORK CARRIED OUT UNDER A BUILDING CONTRACT (JUNE 2007 - THIRD EDITION)

This definition of Prime Cost is published by the Royal Institution of Chartered Surveyors and the Construction Confederation, for convenience and for use by people who choose to use it. Members of the Construction Confederation are not in any way debarred from defining Prime Cost and rendering their accounts for work carried out on that basis in any way they choose. Building owners are advised to reach agreement with contractors on the Definition of Prime Cost to be used prior to issuing instructions.

INTRODUCTION

This new edition of the Definition includes two options for dealing with the prime cost of labour:

Option 'A' – Percentage Addition, is based upon the traditional method of pricing labour in daywork, and allows for a percentage addition to be made for incidental costs, overheads and profit, to the prime cost of labour applicable at the time the daywork is carried out.

Option 'B' – All inclusive Rates, includes not only the prime cost of labour but also includes an allowance for incidental costs, overheads and profit. The all-inclusive rates are deemed to be fixed for the period of the contract. However, where a fluctuating price contract is used, or where the rates in the contract are to be index-linked, the all-inclusive rates shall be adjusted by a suitable index in accordance with the contract conditions.

Model documentation, intended for inclusion in a building contract, is included in Appendix A, which illustrates how the Definition of Prime Cost may be applied in practice.

Example calculations of the Prime Cost of Labour in Daywork are given in Appendix B

BUILDING INDUSTRY

SECTION 1 - APPLICATION

1.1 This Definition provides a basis for the valuation of daywork executed under such building contracts as provide for its use.

1.2. It is not applicable in the case of daywork executed after the date of practical completion.

1.3. It is applicable to works carried out incidental to contract work but may not be deemed appropriate for use in 'daywork only' work or work carried out on an 'hourly' basis only, for which the 'Definition of Prime Cost of Building Works of a Jobbing or Maintenance Character' may be more suitable.

1.4. The terms 'contract' and 'contractor' herein shall be read as 'sub-contract' and 'sub-contractor' as applicable.

1.5. Dayworks are to be calculated by reference to the rate(s) current and prevailing on the day the work is carried out, except where Option 'B' for labour is used which may be adjusted by a suitable index in accordance with the contract conditions.

SECTION 2 - COMPOSITION OF TOTAL CHARGES

2.1 The prime cost of daywork comprises the sum of the following costs:

2.1.1 Labour as defined in Section 3.

2.1.2 Material and goods as defined in Section 4.

2.1.3 Plant as defined in Section 5.

2.2 Incidental costs, overheads and profit as defined in Section 6, as provided in the building contract and expressed therein as percentage adjustments are applicable to each of 2.1.1 (Option A for Labour – Section 3) – 2.1.3 NB: If using Option 'B' for the labour element of prime cost in Section 3, incidental costs, overheads and profit are deemed included.

SECTION 3 - LABOUR

Option A – Percentage Addition

3.1. The prime cost of labour is defined in 3.5.Incidental costs, overheads and profit should be added as defined in Section 6.

3.2. The standard wage rates, payments and expenses referred to below and the standard working hours referred to in 3.3 are those laid down for the time being in the rules or decisions of the Construction Industry Joint Council (CIJC) and the terms of the Building and Civil Engineering Benefits Scheme (managed by the Building and Civil Engineering Holidays Scheme Management Ltd) applicable to the works, or the rules or decisions or agreements of such body, other than the CIJC, as may be applicable relating to the grade and type of operative concerned at the time when and in the area where the daywork is executed.

3.3. Hourly base rates for labour are computed by dividing the annual prime cost of labour, based upon standard working hours and as defined in 3.5, by the number of standard working hours per annum (see Example 1 on page 862).

3.4. The hourly rates computed in accordance with 3.3 shall be applied in respect of the time spent by operatives directly engaged on daywork, including those operating mechanical plant and transport and erecting and dismantling other plant (unless otherwise expressly provided in the building contract) and handling and distributing the materials and goods used in the daywork.

BUILDING INDUSTRY

3.5. The annual prime cost of labour comprises the following:

(a) Standard or guaranteed minimum weekly earnings.*
(b) All other guaranteed minimum payments (unless included in Section 6). *
(c) Differentials or extra payments in respect of skill, responsibility, discomfort, inconvenience or risk (excluding those in respect of supervisory responsibility - see 3.6). *
(d) Payments in respect of public holidays.
(e) Any amounts which may become payable by the Contractor to or in respect of operatives arising from the operation of the rules or decisions referred to in 3.2 which are not provided for in 3.5 (a)-(d) or in Section 6. *
(f) Employer's contributions to industry's annual holiday with pay scheme or payment in lieu thereof.
(g) Employer's contributions to industry's welfare benefits scheme or payment in lieu thereof.
(h) Employer's National Insurance contributions applicable to 3.5 (a) - (g).
(i) Any contribution, levy or tax imposed by statute, payable by the contractor in his capacity as an employer, or compliance with any legislation which has a direct effect on the cost of labour. *

3.6 Differentials or extra payments in respect of supervisory responsibility are excluded from the annual prime cost (see Section 6). The time of supervisory staff such as principals, foremen, gangers, leading hands and the like, when working manually, is admissible under this Section only at the appropriate standard/normal rates for the grade of operative suitable for the operation concerned.

3.7 An example calculation of a typical standard hourly base rate is provided in Example 1 on page 862.

Non-Productive Overtime

3.8 * The prime cost for non-productive overtime should be based only on the hourly payments for items marked with an asterisk in 3.5 #

3.9 An example calculation of a typical non-productive overtime rate is provided in Example 2 on page 863.

Option B – All-Inclusive Rates

3.10 The prime cost of labour is based on the all-inclusive rates for labour provided for in the building contract. The all-inclusive rates are to include all costs associated with employing the labour including all items listed in 3.5.

3.11 The all-inclusive hourly rates are also to include all costs, fixed and time-related charges, overheads and profit (as defined in Section 6) in connection with labour.

3.12 The all-inclusive hourly rates shall be applied in respect of the time actually spent by the operatives directly engaged on daywork, including those operating mechanical plant and transport and erecting and dismantling other plant (unless otherwise expressly provided in the building contract) and handling and distributing the materials and goods used in the daywork.

3.13 The time of supervisory staff, such as principals, foremen, gangers, leading hands and the like, when working manually, is admissible under this Section only at the appropriate all-inclusive hourly rates for the grade of operative suitable for the operations concerned. Any extra payment in respect of supervisory responsibility is not allowable.

3.14 The all-inclusive rates are deemed to be fixed for the period of the contract. However, where a fluctuating price contract is used, or where the rates in the contract are to be index-linked, the all-inclusive rates shall be adjusted by a suitable index in accordance with the contract conditions.

Non-Productive Overtime

3.15 Allowance for non-productive overtime should be made in accordance with the Model Documentation included in Appendix A. #

BUILDING INDUSTRY

SECTION 4 - MATERIALS AND GOODS

4.1. The prime cost of materials and goods obtained specifically for the daywork is the invoice cost after deducting all trade discounts and any portion of cash discounts in excess of 5%, plus any appropriate handling and delivery charges.

4.2. The prime cost of materials and goods supplied from the Contractor's stock is based upon the current market prices after deducting all trade discounts and any portion of cash discounts in excess of 5%, plus any appropriate handling charges.

4.3. Any Value Added Tax which is treated, or is capable of being treated, as input tax (as defined in the Finance Act, 1972, or any re-enactment or amendment thereof or substitution therefor) by the Contractor is excluded, for the purpose of calculations.

SECTION 5 - PLANT

5.1. Unless otherwise stated in the building contract, the prime cost of plant comprises the cost of the following:

- (a) Use or hire of mechanical operated plant and transport for the time employed/engaged for the daywork.
- (b) Use of non-mechanical plant (excluding non-mechanical hand tools) for the time employed/engaged for the daywork.
- (c) Transport/delivery to and from site and erection and dismantling where applicable.
- (d) Qualified professional operators (e.g. crane drivers) not employed by the contractor (see 5.5 below).

5.2. Where plant is hired, the prime cost of plant shall be the invoice cost after deducting all trade discounts and any portion of cash discount in excess of 5%.

5.3. Where plant is not hired, the prime cost of plant shall be calculated in accordance with the latest edition of the Royal Institution of Chartered Surveyor's (RICS) Schedule of Basic Plant Charges for Use in Connection with Daywork Under a Building Contract.

5.4. The use of non-mechanical hand tools and of erected scaffolding, staging, trestles or the like is excluded (see Section 6).

5.5. Where hired or other plant is operated by the Contractor's operatives, the operative's time is to be included under Section 3 unless otherwise provided in the contract.

5.6. Any Value Added Tax which is treated, or is capable of being treated, as input tax (as defined by the Finance Act, 1972, or any re-enactment or amendment thereof or substitution therefor) by the Contractor is excluded, for the purposes of calculation.

SECTION 6 - INCIDENTAL COSTS, OVERHEADS AND PROFIT

6.1 The percentage adjustments provided in the building contract, which are applicable to each of the totals of Sections 3 (Option A), 4 and 5, include the following: **#**

- (a) Head Office charges.
- (b) Site staff, including site supervision.
- (c) The additional cost of overtime (other than that referred to in **#**).
- (d) Time lost due to inclement weather.
- (e) The additional cost of bonuses and all other incentive payments in excess of any guaranteed minimum included in 3.5 (a).
- (f) Apprentices study time.
- (g) Subsistence, lodging and periodic allowances.
- (h) Fares and travelling allowances.
- (i) Sick pay or insurance in respect thereof.

BUILDING INDUSTRY

(j) Third-party and employers' liability insurance.
(k) Liability in respect of redundancy payments to employees.
(l) Employers' National Insurance contributions not included in Section 3.5.
(m) Tool allowances.
(n) Use and maintenance of non-mechanical hand tools.
(o) Use of erected scaffolding, staging, trestles or the like.
(p) Use of tarpaulins, plastic sheeting or the like, all necessary protective clothing, artificial lighting, safety and welfare facilities, storage and the like that may be available on the site.
(q) Any variation to basic rates required by the Contractor in cases where the building contract provides for the use of a specified schedule of basic plant charges (to the extent that no other provision is made for such variation – see Section 5).
(r) All other liabilities and obligations whatsoever not specifically referred to in this Section nor chargeable under any other Section.
(s) Any variation in welfare/pension payments from industry standard.
(t) Profit, (including main contractor's profit as appropriate).

Non-Productive Overtime

6.2 When calculating the percentage adjustment for incidental costs, overheads and profit, if the Option A calculation of price cost of labour is prescribed in the contract, it should be borne in mind that not all items listed in 6.1 are necessarily applicable to non-productive overtime. When Option B is prescribed, non-productive overtime should be shown separately in the contract documents as detailed in the Model Documentation in Appendix A

The additional cost of non-productive overtime, where specifically ordered by the Architect/Supervising Officer/Contract Administrator/Employer's Agent, shall only be chargeable on the terms of prior written agreement between the parties to the building contract.

APPENDIX A

Model Documentation for Inclusion in a Building Contract

This model document is included to illustrate how the Definition of Prime Cost may be applied in practice. It does not form part of the Definition. It is, however, in a form agreed between the RICS and the Construction Confederation and its use in this form amended only as required to suit the specific building contract is encouraged.

Where using Option A for Labour

Dayworks

The Contractor will be paid as defined below for the cost of works carried out as daywork in accordance with the building contract.

For building works, the prime cost of daywork will be calculated in accordance with the latest *Definition of Prime Cost of Daywork carried out under a Building Contract, (State edition______)*, published by the Royal Institution of Chartered Surveyors and the Construction Confederation.

For electrical works, the prime cost of daywork will be calculated in accordance with the latest *Definition of Prime Cost of Daywork carried out under an Electrical Contract, (State edition______)*, published by the Royal Institution of Chartered Surveyors, the Electrical Contractors' Association and 'SELECT' the Electrical Contractors' Association of Scotland.

For heating and ventilating work etc, the prime cost of daywork will be calculated in accordance with the latest *Definition of Prime Cost of Daywork carried out under a Heating, Ventilating, Air-Conditioning, Refrigeration, Pipework and/or Domestic Engineering Contract, (State edition______)*, published by the Royal Institution of Chartered Surveyors and the Heating and Ventilating Contractors' Association

BUILDING INDUSTRY

APPENDIX A (continued)

Where using Option A for Labour (continued)

Dayworks (continued)

For plumbing work, the prime cost of daywork will be calculated in accordance with the latest *Definition of Prime Cost of Daywork carried out under a Plumbing Contract, (State edition______)*, published by the Royal Institution of Chartered Surveyors, the Association of Plumbing and Heating Contractors and the Scottish and Northern Ireland Plumbing Employers' Confederation.*

** It is anticipated that the 1st Edition of this Definition will be published in 2007. Until such time, reference should be made to the April 1985 formula agreed between the Royal Institution of Chartered Surveyors, the National Association of Plumbing, Heating and Mechanical Services Contractors and the Scottish and Northern Ireland Plumbing Employers' Federation.*

Labour

Building Operatives	Provisional Sum	£
Add for Incidental Costs, Overheads and Profit	...........%	£
Electrical Operatives	Provisional Sum	£
Add for Incidental Costs, Overheads and Profit	...........%	£
Heating and Ventilating Operatives	Provisional Sum	£
Add for Incidental Costs, Overheads and Profit	...........%	£
Plumbing Operatives	Provisional Sum	£
Add for Incidental Costs, Overheads and Profit	...........%	£

Non-productive Overtime

Building Operatives	Provisional Sum	£
Add for Incidental Costs, Overheads and Profit	...........%	£
Electrical Operatives	Provisional Sum	£
Add for Incidental Costs, Overheads and Profit	...........%	£
Heating and Ventilating Operatives	Provisional Sum	£
Add for Incidental Costs, Overheads and Profit	...........%	£
Plumbing Operatives	Provisional Sum	£
Add for Incidental Costs, Overheads and Profit	...........%	£

BUILDING INDUSTRY

Where using Option B for Labour

Dayworks

The Contractor will be paid as defined below for the cost of works carried out as daywork in accordance with the building contract.

For building works, the prime cost of daywork will be calculated in accordance with the latest *Definition of Prime Cost of Daywork carried out under a Building Contract, (State edition______)*, published by the Royal Institution of Chartered Surveyors and the Construction Confederation.

For electrical works, the prime cost of daywork will be calculated in accordance with the latest *Definition of Prime Cost of Daywork carried out under an Electrical Contract, (State edition______)*, published by the Royal Institution of Chartered Surveyors, the Electrical Contractors' Association and 'SELECT' the Electrical Contractors' Association of Scotland.

For heating and ventilating work etc, the prime cost of daywork will be calculated in accordance with the latest *Definition of Prime Cost of Daywork carried out under a Heating, Ventilating, Air-Conditioning, Refrigeration, Pipework and/or Domestic Engineering Contract, (State edition______)*, published by the Royal Institution of Chartered Surveyors and the Heating and Ventilating Contractors' Association

For plumbing work, the prime cost of daywork will be calculated in accordance with the latest *Definition of Prime Cost of Daywork carried out under a Plumbing Contract, (State edition______)*, published by the Royal Institution of Chartered Surveyors, the Association of Plumbing and Heating Contractors and the Scottish and Northern Ireland Plumbing Employers' Confederation.**

*** It is anticipated that the 1st Edition of this Definition will be published in 2007. Until such time, reference should be made to the April 1985 formula agreed between the Royal Institution of Chartered Surveyors, the National Association of Plumbing, Heating and Mechanical Services Contractors and the Scottish and Northern Ireland Plumbing employers' Federation.*

Labour

The Contractor must state below the all-inclusive prime cost hourly rates required for labour as defined in Section 3 (Option B) and the core working ours to which they apply.

Core Hours

General Operatives	£........... per hour
Skilled Operatives (all grades)	£........... per hour
Craft Operatives	£........... per hour
Other Grades/Trades:	
....	£........... per hour
....	£........... per hour
....	£........... per hour
....	£........... per hour
....	£........... per hour

Core hours are ___am to ___pm Monday to Friday (excluding statutory holidays)

BUILDING INDUSTRY

Labour (continued)

Overtime specifically ordered by the Architect/Supervising Officer/Contract Administrator/Employers Agent

The non-productive element of overtime should be as defined in the relevant Working Rule Agreement. However, if different, please state below.

Trade	Day	Time	Non-Productive Element (hours)
....	....	..........to.........	....
....	....	..........to.........	....
....	....	..........to.........	....
....	....	..........to.........	....
....	....	..........to.........	....
....	....	..........to.........	....
....	....	..........to.........	....

Provide the all-inclusive prime cost of labour as defined in Section 3 (Option B)

Productive Hours

[ ] hours (Provisional) General Operatives @ £.........per hour £

[ ] hours (Provisional) General Operatives @ £.........per hour £

[ ] hours (Provisional) General Operatives @ £.........per hour £

Other Grades/Trades:

[ ] hours (Provisional) General Operatives @ £.........per hour £

[ ] hours (Provisional) General Operatives @ £.........per hour £

[ ] hours (Provisional) General Operatives @ £.........per hour £

Non-Productive Hours

[ ] hours (Provisional) General Operatives @ £.........per hour £

[ ] hours (Provisional) General Operatives @ £.........per hour £

[ ] hours (Provisional) General Operatives @ £.........per hour £

BUILDING INDUSTRY

Other Grades/Trades:

[ ] hours (Provisional) General Operatives @ £.........per hour £

[ ] hours (Provisional) General Operatives @ £.........per hour £

[ ] hours (Provisional) General Operatives @ £.........per hour £

Materials and Goods

Provide for the prime cost of materials and goods as defined in Section 4 (Provisional) £[ ]

Add the percentage addition for incidental costs, overheads and profit as defined in Section 6 _______%

Plant

Provide for the prime cost of plant hired by the Contractor as defined in Section 5 (Provisional) £[ ]

Add the percentage addition for incidental costs, overheads and profit as defined in Section 6 _______%

Rates for plant not hired by the Contractor shall be as set out in *The Schedule of Basic Plant Charges for Use in Connection with Daywork Under a Building Contract* published by the Royal Institution of Chartered Surveyors [______ Edition dated ________]

Provide for the prime cost of plant not hired by the Contractor, as defined in Section 5 (Provisional) £[ ]

Add the percentage addition for incidental costs, overheads and profit as defined in Section 6 _______%

BUILDING INDUSTRY

APPENDIX B

Example Calculations of Prime Cost of Labour in Daywork

Example 1

Option A

Example of calculation of typical standard hourly base rate (as defined in Section 3) for CIJC Building Craft operative and General Operative based upon rates applicable 6th April 2007.

		Rate (£)	Craft Operative	Rate (£)	General Operative
Basic Wages:	46.2 weeks	363.48	£16,792.78	273.39	£12,630.62
Extra Payments:	Where applicable	-	-		-
Sub Total:			£16,792.78		£12,630.62
National Insurance:	12.80% above ET (46.2 wks @£100.00pw)		£1,558.12		£1,025.36
Holidays with Pay:	226 hours	9.32	£2,106.32	7.01	£1,584.26
Welfare Benefit:	52 stamps	10.90	£566.80	10.90	£566.80
CITB Levy:	0.5% of	18,899.10	£94.50	14,214.88	£71.07
Annual labour cost:			**£21,118.52**		**£15,878.11**
Hourly Base Rate:	Divide by 1802 hours		**£11.72**		**£8.81**

For the convenience of readers, the example which appears on the previous page has been updated by the Editors for rates applicable 30th June 2008.

		Rate (£)	Craft Operative	Rate (£)	General Operative
Basic Wages:	46.2 weeks	401.70	£18,558.54	302.25	£13,963.95
Extra Payments:	Where applicable	-	-		-
Sub Total:			£18,558.54		£13,963.95
National Insurance:	12.80% above ET (46.2 wks @£105.00pw)		£1,784.07		£1,195.97
Holidays with Pay:	226 hours	10.30	£2,327.80	7.75	£1,751.50
Welfare Benefit:	52 stamps	11.00	£572.00	11.00	£572.00
CITB Levy:	0.5% of	20,886.34	£104.43	15,715.45	£78.58
Annual labour cost:			**£23,346.84**		**£17,562.00**
Hourly Base Rate:	Divide by 1802 hours		**£12.96**		**£9.75**

Note:

(1) Standard working hours per annum calculated as follows:

52 weeks @ 39 hours	2028
Less \	
hours annual holiday	163
hours public holiday	63
Standard working hours per year	1802

(2) It has been assumed that employers who follow the CIJC Working Rules Agreement will match the employee pension contributions (part of welfare benefit) between £3.00 and £10.00 per week. Furthermore it has been assumed that employees have contributed £10.00 per week to the pension scheme and £1.00 per week for life insurance.

BUILDING INDUSTRY

(3) It should be noted that all labour costs incurred by the Contractor in his capacity as an employer other than those contained in the hourly base rate, are to be taken into account under Section 6.

(4) The above example is for the convenience of users only and does not form part of the Definition; all the basic costs are subject to re-examination according to the time when and in the area where the daywork is executed.

Example 2

Non Productive Overtime

Option A

Example of calculation of typical non productive overtime rate (as defined in section 3) for CIJC Building Craft Operative and General Operative based upon rates applicable 6th April 2007.

		Rate (£)	Craft Operative	Rate (£)	General Operative
Basic Wages:	46.2 weeks	363.48	£16,792.78	273.39	£12,630.62
Extra Payments:	Where applicable	-	-		-
Sub Total:			£16,792.78		£12,630.62
National Insurance:	12.80% above ET (46.2 wks @£100.00pw)		£1,558.12		£1,025.36
CITB Levy:	0.5% of	16,792.78	£83.96	12,630.62	£63.15
Annual labour cost:			**£18,434.86**		**£13,719.13**
Hourly Base Rate:	Divide by 1802 hours		**£10.23**		**£7.61**

For the convenience of readers, the example which appears on the previous page has been updated by the Editors for rates applicable 30th June 2008.

		Rate (£)	Craft Operative	Rate (£)	General Operative
Basic Wages:	46.2 weeks	401.70	£18,558.54	302.25	£13,963.95
Extra Payments:	Where applicable	-	-		-
Sub Total:			£18,558.54		£13,963.95
National Insurance:	12.80% above ET (46.2 wks @£105.00pw)		£1,784.07		£1,195.97
CITB Levy:	0.5% of	18,558.54	£92.79	13,963.95	£69.82
Annual labour cost:			**£20,435.40**		**£15,229.74**
Hourly Base Rate:	Divide by 1802 hours		**£11.34**		**£8.45**

Note:

(1) Standard working hours per annum calculated as follows:

52 weeks @ 39 hours	2028
Less \	
hours annual holiday	163
hours public holiday	63
Standard working hours per year	1802

BUILDING INDUSTRY

(2) It should be noted that all labour costs incurred by the Contractor in his capacity as an employer other than those contained in the hourly base rate, are to be taken into account under Section 6.

(3) The above example is for the convenience of users only and does not form part of the Definition; all the basic costs are subject to re-examination according to the time when and in the area where the daywork is executed.

BUILDING INDUSTRY

DEFINITION OF PRIME COST OF BUILDING WORKS OF A JOBBING OR MAINTENANCE CHARACTER (1980 EDITION)

This definition of Prime Cost is published by the Royal Institution of Chartered Surveyors and the National Federation of Building Trades Employers, for convenience and for use by people who choose to use it. Members of the National Federation of Building Trades Employers are not in any way debarred from defining Prime Cost and rendering their accounts for work carried out on that basis in any way they choose. Building owners are advised to reach agreement with contractors on the Definition of Prime Cost to be used prior to issuing instructions.

SECTION 1 - APPLICATION

1.1. This definition provides a basis for the valuation of work of a jobbing or maintenance character executed under such building contracts as provide for its use.

1.2. It is not applicable in any other circumstances, such as daywork executed under or incidental to a building contract.

SECTION 2 - COMPOSITION OF TOTAL CHARGES

2.1. The prime cost of jobbing work comprises the sum of the following costs:

(a) Labour as defined in Section 3.
(b) Materials and goods as defined in Section 4.
(c) Plant, consumable stores and services as defined in Section 5.
(d) Sub-contracts as defined in Section 6.

2.2. Incidental costs, overhead and profit as defined in Section 7 and expressed as percentage adjustments are applicable to each of 2.1 (a)-(d).

SECTION 3 - LABOUR

3.1. Labour costs comprise all payments made to or in respect of all persons directly engaged upon the work, whether on or off the site, except those included in Section 7.

3.2. Such payments are based upon the standard wage rates, emoluments and expenses as laid down for the time being in the rules or decisions of the National Joint Council for the Building Industry and the terms of the Building and Civil Engineering Annual and Public Holiday Agreements applying to the works, or the rules of decisions or agreements of such other body as may relate to the class of labour concerned, at the time when and in the area where the work is executed, together with the Contractor's statutory obligations, including:

(a) Guaranteed minimum weekly earnings (e.g. Standard Basic Rate of Wages and Guaranteed Minimum Bonus Payment in the case of NJCBI rules).
(b) All other guaranteed minimum payments (unless included in Section 7).
(c) Payments in respect of incentive schemes or productivity agreements applicable to the works.
(d) Payments in respect of overtime normally worked; or necessitated by the particular circumstances of the work; or as otherwise agreed between the parties.
(e) Differential or extra payments in respect of skill, responsibility, discomfort or inconvenience.
(f) Tool allowance.
(g) Subsistence and periodic allowances.
(h) Fares, travelling and lodging allowances.
(j) Employer's contributions to annual holiday credits.
(k) Employer's contributions to death benefit schemes.
(l) Any amounts which may become payable by the Contractor to or in respect of operatives arising from the operation of the rules referred to in 3.2 which are not provided for in 3.2 (a)-(k) or in Section 7.

BUILDING INDUSTRY

(m) Employer's National Insurance contributions and any contribution, levy or tax imposed by statute, payable by the Contractor in his capacity as employer.

Note:

Any payments normally made by the Contractor which are of a similar character to those described in 3.2 (a)-(c) but which are not within the terms of the rules and decisions referred to above are applicable subject to the prior agreement of the parties, as an alternative to 3.2 (a)-(c).

3.3. The wages or salaries of supervisory staff, timekeepers, storekeepers, and the like, employed on or regularly visiting site, where the standard wage rates, etc., are not applicable, are those normally paid by the Contractor together with any incidental payments of a similar character to 3.2 (c) - (k).

3.4. Where principals are working manually their time is chargeable, in respect of the trades practised, in accordance with 3.2.

SECTION 4 - MATERIALS AND GOODS

4.1. The prime cost of materials and goods obtained by the Contractor from stockists or manufacturers is the invoice cost after deduction of all trade discounts but including cash discounts not exceeding 5 per cent, and includes the cost of delivery to site.

4.2. The prime cost of materials and goods supplied from the Contractor's stock is based upon the current market prices plus any appropriate handling charges.

4.3. The prime cost under 4.1 and 4.2 also includes any costs of:

(a) non-returnable crates or other packaging.
(b) returning crates and other packaging less any credit obtainable.

4.4. Any Value Added Tax which is treated, or is capable of being treated, as input tax (as defined in the Finance Act, 1972 or any re-enactment thereof) by the Contractor is excluded.

SECTION 5 - PLANT, CONSUMABLE STORES AND SERVICES

5.1. The prime cost of plant and consumable stores as listed below is the cost at hire rates agreed between the parties or in the absence of prior agreement at rates not exceeding those normally applied in the locality at the time when the works are carried out, or on a use and waste basis where applicable:

(a) Machinery in workshops.
(b) Mechanical plant and power-operated tools.
(c) Scaffolding and scaffold boards.
(d) Non-mechanical plant excluding hand tools.
(e) Transport including collection and disposal of rubbish.
(f) Tarpaulins and dust sheets.
(g) Temporary roadways, shoring, planking and strutting, hoarding, centering, formwork, temporary fans, partitions or the like.
(h) Fuel and consumable stores for plant and power-operated tools unless included in 5.1 (a), (b), (d) or (e) above.
(j) Fuel and equipment for drying out the works and fuel for testing mechanical services.

BUILDING INDUSTRY

5.2. The prime cost also includes the net cost incurred by the Contractor of the following services, excluding any such cost included under Sections 3, 4 or 7:

(a) Charges for temporary water supply including the use of temporary plumbing and storage.
(b) Charges for temporary electricity or other power and lighting including the use of temporary installations.
(c) Charges arising from work carried out by local authorities or public undertakings.
(d) Fees, royalties and similar charges.
(e) Testing of materials.
(f) The use of temporary buildings including rates and telephone and including heating and lighting not charged under (b) above.
(g) The use of canteens, sanitary accommodation, protective clothing and other provision for the welfare of persons engaged in the work in accordance with the current Working Rule Agreement and any Act of Parliament, statutory instrument, rule, order, regulation or bye-law.
(h) The provision of safety measures necessary to comply with any Act of Parliament.
(j) Premiums or charges for any performance bonds or insurances which are required by the Building Owner and which are not referred to elsewhere in this Definition.

SECTION 6 - SUB-CONTRACTS

6.1. The prime cost of work executed by sub-contractors, whether nominated by the Building Owner or appointed by the Contractor, is the amount which is due from the Contractor to the sub-contractors in accordance with the terms of the sub-contracts after deduction of all discounts except any cash discount offered by any sub-contractor to the Contractor not exceeding 2.5%.

SECTION 7 - INCIDENTAL COSTS, OVERHEADS AND PROFIT

7.1. The percentage adjustments provided in the building contract, which are applicable to each of the totals of Sections 3-6, provide for the following:

(a) Head Office charges.
(b) Off-site staff including supervisory and other administrative staff in the Contractor's workshops and yard.
(c) Payments in respect of public holidays.
(d) Payments in respect of apprentices' study time.
(e) Sick pay or insurance in respect thereof.
(f) Third party employer's liability insurance.
(g) Liability in respect of redundancy payments made to employees.
(h) Use, repair and sharpening of non-mechanical hand tools.
(j) Any variations to basic rates required by the Contractor in cases where the building contract provides for the use of a specified schedule of basic plant charges (to the extent that no other provision is made for such variation).
(k) All other liabilities and obligations whatsoever not specifically referred to in this Section nor chargeable under any other section.
(l) Profit.

BUILDING INDUSTRY

SPECIMEN ACCOUNT FORMAT

If this Definition of Prime Cost is followed the Contractor's account could be in the following format:

	£
Labour (as defined in Section 3)	
Add ___ % (see Section 7)	
Materials and goods (as defined in Section 4)	
Add ___ % (see Section 7)	
Plant, consumable stores and services (as defined in Section 5)	
Add ___ % (see Section 7)	
Sub-contracts (as defined in Section 6)	
Add ___ % (see Section 7)	________
	£ ________

VAT to be added if applicable.

SCHEDULE OF BASIC PLANT CHARGES (1st MAY 2001 ISSUE)

This Schedule is published by the Royal Institution of Chartered and is for use in connection with Dayworks under a Building Contract.

EXPLANATORY NOTES

1. The rates in the Schedule are intended to apply solely to daywork carried out under and incidental to a Building Contract. They are NOT intended to apply to:
   (i) Jobbing or any other work carried out as a main or separate contract; or
   (ii) Work carried out after the date of commencement of the Defects Liability Period.

2. The rates apply only to plant and machinery already on site, whether hired or owned by the Contractor.

3. The rates, unless otherwise stated, include the cost of fuel and power of every description, lubricating oils, grease, maintenance, sharpening of tools, replacement of spare parts, all consumable stores and for licences and insurances applicable to items of plant.

4. The rates, unless otherwise stated, do not include the costs of drivers and attendants.

5. The rates are base costs and may be subject to the overall adjustment for price movement, overheads and profit, quoted by the Contractor prior to the placing of the Contract.

6. The rates should be applied to the time during which the plant is actually engaged in daywork.

7. Whether or not plant is chargeable on daywork depends on the daywork agreement in use and the inclusion of an item of plant in this schedule does not necessarily indicate that the item is chargeable.

8. Rates for plant not included in the Schedule or which is not already on site and is specifically provided or hired for daywork shall be settled at prices which are reasonably related to the rates in the Schedule having regard to any overall adjustment quoted by the Contractor in the Conditions of Contract.

BUILDING INDUSTRY

MECHANICAL PLANT AND TOOLS Item of plant	Size/Rating	Unit	Rate per Hour £
PUMPS			
Mobile Pumps			
Including pump hoses, values and strainers etc.			
Diaphragm	50 mm diameter	Each	0.87
Diaphragm	76 mm diameter	Each	1.29
Submersible	50 mm diameter	Each	1.18
Induced Flow	50 mm diameter	Each	1.54
Induced Flow	76 mm diameter	Each	2.05
Centrifugal self priming	50 mm diameter	Each	1.96
Centrifugal self priming	102 mm diameter	Each	2.52
Centrifugal self priming	152 mm diameter	Each	3.87
SCAFFOLDING, SHORING, FENCING			
Complete Scaffolding			
Mobile working towers, single width	1.80 m x 0.80 m x 7.00 m high	Each	2.00
Mobile working towers, single width	1.80 m x 0.80 m x 9.00 m high	Each	2.80
Mobile working towers, double width	1.80 m x 1.40 m x 7.00 m high	Each	2.15
Mobile working towers, double width	1.80 m x 1.40 m x 15.00 m high	Each	5.10
Chimney scaffold, single unit		Each	1.79
Chimney scaffold, twin unit		Each	2.05
Chimney scaffold, four unit		Each	3.59
Trestles			
Trestle, adjustable	Any height	Pair	0.10
Trestle, painters	1.80 m high	Pair	0.21
Trestle, painters	2.40 m high	Pair	0.26
Shoring, Planking and Strutting			
'Acrow' adjustable prop	Sizes up to 4.90 m (open)	Each	0.10
'Strong boy' support attachment		Each	0.15
Adjustable trench struts	Sizes up to 1.67m (open)	Each	0.10
Trench sheet		Metre	0.01
Backhole trench box		Each	1.00
Temporary Fencing			
Including block and coupler			
Site fencing steel grid panel	3.50 m x 2.00 m	Each	0.08
Anti-climb site steel grid fence panel	3.50 m x 2.00 m	Each	0.08
LIFTING APPLIANCES AND CONVEYORS			
Cranes			
Mobile Cranes			
Rates are inclusive of drivers			
Lorry mounted, telescopic jib			
Two wheel drive	6 tonnes	Each	24.40
Two wheel drive	7 tonnes	Each	25.00
Two wheel drive	8 tonnes	Each	25.62
Two wheel drive	10 tonnes	Each	26.90
Two wheel drive	12 tonnes	Each	28.25
Two wheel drive	15 tonnes	Each	29.66
Two wheel drive	18 tonnes	Each	31.14
Two wheel drive	20 tonnes	Each	32.70
Two wheel drive	25 tonnes	Each	34.33

BUILDING INDUSTRY

LIFTING APPLIANCES AND CONVEYORS – cont'd

Item of plant	**Size/Rating**		**Unit**	**Rate per Hour £**
Mobile cranes – cont'd				
Rates are inclusive of divers				
Lorry mounted telescopic jib – cont'd				
Four wheel drive	10 tonnes		Each	27.44
Four wheel drive	12 tonnes		Each	28.81
Four wheel drive	15 tonnes		Each	30.25
Four wheel drive	20 tonnes		Each	33.35
Four wheel drive	25 tonnes		Each	35.19
Four wheel drive	30 tonnes		Each	37.12
Four wheel drive	45 tonnes		Each	39.16
Four wheel drive	50 tonnes		Each	41.32
Track-mounted tower crane				
Rates are inclusive of divers				
Note: Capacity equals maximum lift in				
Tonnes times maximum radius at which it Can be lifted	Capacity (Metre/tonnes)	Height under hook above Ground (m)		
	Up to	Up to		
Tower crane	10	17	Each	7.99
Tower crane	15	17	Each	8.59
Tower crane	20	18	Each	9.18
Tower crane	25	20	Each	11.56
Tower crane	30	22	Each	13.78
Tower crane	40	22	Each	18.09
Tower crane	50	22	Each	22.20
Tower crane	60	22	Each	24.32
Tower crane	70	22	Each	23.00
Tower crane	80	22	Each	25.91
Tower crane	110	22	Each	26.45
Tower crane	125	30	Each	29.38
Tower crane	125	30	Each	32.35
Static tower cranes				
Rates inclusive of driver				
To be charged at 90% of the above rates for tower mounted tower cranes				
Crane Equipment				
Mucking tipping skip	Up to 0.25 m³		Each	0.56
Muck tipping skip	0.5 m³		Each	0.67
Muck tipping skip	0.75 m³		Each	0.82
Muck tipping skip	1.00 m³		Each	1.03
Muck tipping skip	1.50 m³		Each	1.18
Muck tipping skip	2.00 m³		Each	1.38
Mortar skips	Up to 0.38 m³		Each	0.41
Boat skips	1.00 m³		Each	1.08
Boat skips	1.50 m³		Each	1.33
Boat skips	2.00 m³		Each	1.59
Concrete skips, hand levered	0.50 m³		Each	1.00
Concrete skips, hand levered	0.75 m³		Each	1.10
Concrete skips, hand levered	1.00 m³		Each	1.25
Concrete skips, hand levered	1.50 m³		Each	1.50

BUILDING INDUSTRY

Item of plant	Size/Rating		Unit	Rate per Hour £
Concrete skips, hand levered	2.00 m^3		Each	1.65
Concrete skips, geared	0.50 m^3		Each	1.30
Concrete skips, geared	0.75 m^3		Each	1.40
Concrete skips, geared	1.00 m^3		Each	1.55
Concrete skips, geared	1.50 m^3		Each	1.80
Concrete skips, geared	2.00 m^3		Each	2.05
Hoists				
Scaffold hoists	200 kg		Each	1.92
Rack and pinion (goods only)	500 kg		Each	3.31
Rack and pinion (goods only)	1100 kg		Each	4.28
Rack and pinion goods and passenger	15 person, 1200 kg		Each	5.62
Wheelbarrow chain sling			Each	0.31
Conveyors				
Belt conveyors				
Conveyor	7.50 m long x 400 mm wide		Each	6.41
Miniveyor, control box and loading hopper	3.00 m unit		Each	3.59
Other Conveying Equipment				
Wheelbarrow			Each	0.21
Hydraulic superlift			Each	2.95
Pavac slab lifter			Each	1.03
Hand pad and hose attachment			Each	0.26
Lifting Trucks				
Fork lift	Payload	Maximum Lift		
Fork lift, two wheel drive	1100 kg	up to 3.00 m	Each	4.87
Fork lift, two wheel drive	2540 kg	up to 3.70 m	Each	5.12
Fork lift, two wheel drive	1524 kg	up to 6.00 m	Each	6.04
Fork lift, two wheel drive	2600 kg	up to 5.40 m	Each	7.69
Lifting Platforms				
Hydraulic platform (Cherry picker)	7.50 m		Each	4.23
Hydraulic platform (Cherry picker)	13.00 m		Each	9.23
Scissors lift	7.80 m		Each	7.56
Telescopic handlers	7.00 m, 2 tonne		Each	7.18
Telescopic handlers	13.00 m, 3 tonne		Each	8.72
Lifting and Jacking Gear				
Pipe winch including gantry	1.00 tonne		Sets	1.92
Pipe winch including gantry	3.00 tonne		Sets	3.21
Chain block	1.00 tonne		Each	0.45
Chain block	2.00 tonne		Each	0.71
Chain block	5.00 tonne		Each	1.22
Pull lift (Tirfor winch)	1.00 tonne		Each	0.64
Pull lift (Tirfor winch)	1.60 tonne		Each	0.90
Pull lift (Tirfor winch)	3.20 tonne		Each	1.15
Brother or chain slings, two legs	not exceeding 4.20 tonnes		Set	0.35
Brother or chain slings, two legs	not exceeding 5.50 tonnes		Set	0.45
Brother or chain slings, four legs	not exceeding 3.10 tonnes		Set	0.41
Brother or chain slings, four legs	not exceeding 11.20 tonnes		Set	1.28

BUILDING INDUSTRY

CONSTRUCTION VEHICLES

Item of plant	Size/Rating	Unit	Rate per Hour £
Lorries			
Plated lorries			
Rates are inclusive of driver			
Platform lorries	7.50 tonnes	Each	19.00
Platform lorries	17.00 tonnes	Each	21.00
Platform lorries	24.00 tonnes	Each	26.00
Platform lorries with winch and skids	7.50 tonnes	Each	21.40
Platform lorries with crane	17.00 tonnes	Each	27.50
Platform lorries with crane	24.00 tonnes	Each	32.10
Tipper Lorries			
Rates are inclusive of driver			
Tipper lorries	15.00/17.00 tonnes	Each	19.50
Tipper lorries	24.00 tonnes	Each	21.40
Tipper lorries	30.00 tonnes	Each	27.10
Dumpers			
Site use only (excluding tax, insurance and extra cost of DEFV etc. when operating on highway)	Makers capacity		
Two wheel drive	0.80 tonnes	Each	1.20
Two wheel drive	1.00 tonnes	Each	1.30
Two wheel drive	1.20 tonnes	Each	1.60
Four wheel drive	2.00 tonnes	Each	2.50
Four wheel drive	3.00 tonnes	Each	3.00
Four wheel drive	4.00 tonnes	Each	3.50
Four wheel drive	5.00 tonnes	Each	4.00
Four wheel drive	6.00 tonnes	Each	4.50
Dumper Trucks			
Rates are inclusive of drivers			
Dumper trucks	10.00/13.00 tonnes	Each	20.00
Dumper trucks	18.00/20.00 tonnes	Each	20.40
Dumper trucks	22.00/25.00 tonnes	Each	26.30
Dumper trucks	35.00/40.00 tonnes	Each	36.60
Tractors			
Agricultural Type			
Wheeled, rubber-clad tyred			
Light	48 h.p.	Each	4.65
Heavy	65 h.p.	Each	5.15
Crawler Tractors			
With bull or angle dozer	80/90 h.p.	Each	21.40
With bull or angle dozer	115/130 h.p.	Each	25.10
With bull or angle dozer	130/150 h.p.	Each	26.00
With bull or angle dozer	155/175 h.p.	Each	27.74
With bull or angle dozer	210/230 h.p.	Each	28.00
With bull or angle dozer	300/340 h.p.	Each	31.10
With bull or angle dozer	400/440 h.p.	Each	46.90
With loading shovel	0.80 m^3	Each	25.00
With loading shovel	1.00 m^3	Each	28.00
With loading shovel	1.20 m^3	Each	32.00
With loading shovel	1.40 m^3	Each	36.00
With loading shovel	1.80 m^3	Each	45.00

BUILDING INDUSTRY

Item of plant	Size/Rating	Unit	Rate per Hour £
Light vans			
Ford escort or the like		Each	4.74
Ford transit or the like	1.00 tonnes	Each	6.79
Luton Box Van or the like	1.80 tonnes	Each	8.33
Water/Fuel Storage			
Mobile water container	110 litres	Each	0.28
Water bowser	1100 litres	Each	0.55
Water bowser	3000 litres	Each	0.74
Mobile fuel container	110 litres	Each	0.28
Fuel bowser	1100 litres	Each	0.65
Fuel bowser	3000 litres	Each	1.02
EXCAVATIONS AND LOADERS			
Excavators			
Wheeled, hydraulic	7.00/10.00 tonnes	Each	12.00
Wheeled, hydraulic	11.00/13.00 tonnes	Each	12.70
Wheeled, hydraulic	15.00/16.00 tonnes	Each	14.80
Wheeled, hydraulic	17.00/18.00 tonnes	Each	16.70
Wheeled, hydraulic	20.00/23.00 tonnes	Each	14.70
Crawler, hydraulic	12.00/14.00 tonnes	Each	12.00
Crawler, hydraulic	15.00/17.50 tonnes	Each	14.00
Crawler, hydraulic	20.00/23.00 tonnes	Each	16.00
Crawler, hydraulic	25.00/30.00 tonnes	Each	21.00
Crawler, hydraulic	30.00/35.00 tonnes	Each	30.00
Mini excavators	1000/1500 kg	Each	4.50
Mini excavators	2150/2400 kg	Each	5.50
Mini excavators	2700/3500 kg	Each	6.50
Mini excavators	3500/4500 kg	Each	8.50
Mini excavators	4500/6000 kg	Each	9.50
Loaders			
Wheeled skip loader		Each	4.50
Shovel loaders, four wheel drive	1.60 kg	Each	12.00
Shovel loaders, four wheel drive	2.40 kg	Each	19.00
Shovel loaders, four wheel drive	3.60 kg	Each	22.00
Shovel loaders, four wheel drive	4.40 kg	Each	23.00
Shovel loaders, crawlers	0.80 kg	Each	11.00
Shovel loaders, crawlers	1.20 kg	Each	14.00
Shovel loaders, crawlers	1.60 kg	Each	16.00
Shovel loaders, crawlers	2.00 kg	Each	17.00
Skid steer loaders wheeled	300/400 kg payload	Each	6.00
Excavator Loaders			
Wheeled tractor type with black-hoe Excavator			
Four wheel drive	2.50/3.50 tonnes	Each	7.00
Four wheel drive, 2 wheel steer	7.00/8.00 tonnes	Each	9.00
Four wheel drive, 4 wheel steer	7.00/8.00 tonnes	Each	10.00
Crawler, hydraulic	12 tonnes	Each	20.00
Crawler, hydraulic	20 tonnes	Each	16.00
Crawler, hydraulic	30 tonnes	Each	35.00
Crawler, hydraulic	40 tonnes	Each	38.00

BUILDING INDUSTRY

COMPACTION EQUIPMENT **Item of plant**	**Size/Rating**	**Unit**	**Rate per Hour £**
Attachments			
Breakers for excavators		Each	7.50
Breakers for mini excavators		Each	3.60
Breakers for back-hoe excavator/loaders		Each	6.00
Rollers			
Vibrating roller	368 – 430 kg	Each	1.68
Single roller	533 kg	Each	1.92
Single roller	750 kg	Each	2.41
Twin roller	698 kg	Each	1.93
Twin roller	851 kg	Each	2.41
Twin roller with seat and steering wheel	1067 kg	Each	3.03
Twin roller with seat and steering wheel	1397 kg	Each	3.17
Pavement rollers	3.00 – 4.00 tonnes dead weight	Each	3.18
Pavement rollers	4.00 – 6.00 tonnes	Each	4.13
Pavement rollers	6.00 – 10.00 tonnes	Each	4.84
Rammers			
Tamper rammer 2 stoke-petrol	225 mm – 275 mm	Each	1.59
Soil Compactors			
Plate compactor	375 mm – 400 mm	Each	1.20
Plate compactor rubber pad	375 mm – 1400 mm	Each	0.33
Plate compactor reversible plate-petrol	400 mm	Each	2.20
CONCRETE EQUIPEMENT			
Concrete/Mortar Mixers			
Open drum without hopper	0.90/0.06 m³	Each	0.62
Open drum without hopper	0.12/0.09 m³	Each	0.68
Open drum without hopper	0.15/0.10 m³	Each	0.72
Open drum with hopper	0.20/0/15 m³	Each	0.80
Concrete/Mortar Transport Equipment			
Concrete pump including hose, valve and couplers			
Lorry mounted concrete pump	23 m maximum distance	Each	36.00
Lorry mounted concrete pump	50 m maximum distance	Each	46.00
Concrete Equipment			
Vibrator, poker, petrol type	Up to 75 mm diameter	Each	1.62
Air vibrator (excluding compressor and hose)	Up to 75 mm diameter	Each	0.79
Extra poker heads	25/36/60 mm diameter	Each	0.77
Vibrating Screed unit with beam	5.00 m	Each	1.77
Vibrating Screed unit with adjusting beam	3.00 – 5.00 m	Each	2.18
Power float	725 mm – 900 mm	Each	1.72
Power grouter		Each	0.92
TESTING EQUIPMENT			
Pipe Testing Equipment			
Pressure testing pump, electric		Sets	1.87
Pipe pressure testing equipment, hydraulic		Sets	2.46
Pressure test pump		Sets	0.64

BUILDING INDUSTRY

SITE ACCOMODATION AND TEMPORARY SERVICES			**Rate per**
Item of plant	**Size/Rating**	**Unit**	**Hour £**
Heating equipment			
Space heaters – propane	80,000 Btu/hr	Each	0.77
Space heaters – propane/electric	125,000 Btu/hr	Each	1.56
Space heaters – propane/electric	250,000 Btu/hr	Each	1.79
Space heaters – propane	125,000 Btu/hr	Each	1.33
Space heaters – propane	260,000 Btu/hr	Each	1.64
Cabinet headers		Each	0.41
Cabinet heater catalytic		Each	0.46
Electric halogen heaters		Each	1.28
Ceramic heaters	3kW	Each	0.79
Fan heaters	3kW	Each	0.41
Cooling fan		Each	1.15
Mobile cooling unit – small		Each	1.38
Mobile cooling unit – large		Each	1.54
Air conditioning unit		Each	2.62
Site Lighting and Equipment			
Tripod floodlight	500W	Each	0.36
Tripod floodlight	1000W	Each	0.34
Towable floodlight	4 x 1000W	Each	2.00
Hand held floodlight	500W	Each	0.22
Rechargeable light		Each	0.62
Inspection light		Each	0.15
Plasterers light		Each	0.56
Lighting mast		Each	0.92
Festoon light string	33.00 m	Each	0.31
Site Electrical Equipment			
Extension leads	240V/14.00 m	Each	0.20
Extension leads	110V/14.00 m	Each	0.20
Cable reel	25.00 m 110V/240V	Each	0.28
Cable reel	50.00 m 110V240V	Each	0.33
4 way junction box	110V	Each	0.17
Power Generating Units			
Generator – petrol	2kVA	Each	1.08
Generator – silenced petrol	2kVA	Each	1.54
Generator – petrol	3kVA	Each	1.38
Generator – diesel	5kVA	Each	1.92
Generator – silenced diesel	8kVA	Each	3.59
Generator – silenced diesel	15kVA	Each	7.69
Trail adaptor	240V	Each	0.20
Transformers			
Transformer	3kVA	Each	0.36
Transformer	5kVA	Each	0.51
Transformer	7.50kVA	Each	0.82
Transformer	10kVA	Each	0.87
Rubbish Collection and Disposal Equipment			
<u>Rubbish Chutes</u>			
Standard plastic module	1.00 m section	Each	0.18
Steel liner insert		Each	0.26
Steel top hopper		Each	0.20
Plastic side entry hopper/line		Each	0.20

BUILDING INDUSTRY

SITE ACCOMODATION AND TEMPORARY SERVICES – cont'd

Item of plant	Size/Rating	Unit	Rate per Hour £
Dust Extraction Plant			
Dust extraction unit, light duty		Each	1.03
Dust extraction unit, heavy duty		Each	1.64
SITE EQUIPMENT - Welding Equipment			
Arc-(Electric) Complete With Leads			
Welder generator – petrol	200 amp	Each	2.26
Welder generator – diesel	300/350 amp	Each	3.33
Welder generator – diesel	400 amp	Each	4.74
Extra welding lead sets		Each	0.29
Gas-Oxy Welder			
Welding and cutting set (including oxygen And acetylene, excluding underwater Equipment and thermic boring)			
Small		Each	1.41
Large		Each	2.00
Mig welder		Each	1.00
Fume extractor		Each	0.92
Road Works Equipment			
Traffic lights, main/generator	2-way	Set	4.01
Traffic lights, main/generator	3-way	Set	7.92
Traffic lights, main/generator	4-way	Set	9.81
Traffic lights, main/generator – trailer Mounted	2-way	Set	3.98
Flashing light		Each	0.20
Road safety cone	450 mm	10	0.26
Safety cone	750 mm	10	0.38
Safety barrier plank	1.25 m	Each	0.03
Safety barrier plank	2.00 m	Each	0.04
Road sign		Each	0.26
DPC Equipment			
Damp proofing injection machine		Each	1.49
Cleaning Equipment			
Vacuum cleaner (industrial wet) single motor		Each	0.62
Vacuum cleaner (industrial wet) twin motor		Each	1.23
Vacuum cleaner (industrial wet) triple motor		Each	1.44
Vacuum cleaner (industrial wet) back Pack		Each	0.97
Pressure washer, light duty, electric	1450 PSI	Each	0.97
Pressure washer, heavy duty, diesel	2500 PSI	Each	2.69
Cold pressure washer, electric		Each	1.79
Hot pressure washer, petrol		Each	2.92
Cold pressure washer, petrol		Each	2.00
Sandblast attachment to last washer		Each	0.54
Drain cleaning attachment to last washer		Each	0.31
Surface Preparation Equipment			
Rotavators	5 h.p.	Each	1.67

BUILDING INDUSTRY

Item of plant	**Size/Rating**	**Unit**	**Rate per Hour £**
Scrabbler, up to three heads		Each	1.15
Scrabbler, pole		Each	1.50
Scrabbler, multi-headed floor		Each	4.00
Floor preparation machine		Each	2.82
Compressors and Equipment			
Portable Compressors			
Compressors – electric	0.23 m³/min	Each	1..59
Compressors – petrol	0.28 m³/min	Each	1.74
Compressors – petrol	0.71 m³/min	Each	2.00
Compressors – diesel	up to 2.83 m³/min	Each	1.24
Compressors – diesel	up to 3.68 m³/min	Each	1.49
Compressors – diesel	up to 4.25 m³/min	Each	1.60
Compressors – diesel	up to 4.81 m³/min	Each	1.92
Compressors – diesel	up to 7.64 m³/min	Each	3.08
Compressors – diesel	up to 11.32 m³/min	Each	4.23
Compressors – diesel	up to 18.40 m³/min	Each	5.73
Mobile Compressors			
Lorry mounted compressors *(machine plus lorry only)*	2.86 – 4.24 m³/min	Each	12.50
Tractor mounted compressors *(machine plus rubber tyred tractor*	2.86 – 3.40 m³/min	Each	13.50
Accessories (Pneumatic Tools) *(with and including up to 15.00 m of air hose)*			
Demolition pick		Each	1.03
Breakers (with six steels) light	up to 150 kg	Each	0.79
Breakers (with six steels) medium	295 kg	Each	1.08
Breakers (with six steels) heavy	386 kg	Each	1.44
Rock drill (for use with compressor)			
Hand held		Each	0.90
Additional hoses	15.00 m	Each	0.16
Muffer, tool silencer		Each	0.14
Breakers			
Demolition hammer drill, heavy duty, Electric		Each	1.00
Road breaker, electric		Each	1.65
Road breaker, 2 stroke, petrol		Each	2.05
Hydraulic breaker unit, light duty, petrol		Each	2.05
Hydraulic breaker unit, heavy duty, petrol		Each	2.60
Hydraulic breaker unit, heavy duty, diesel		Each	2.95
Quarrying and Tooling Equipment			
Block and stone splitter, hydraulic	600 mm x 600 mm	Each	1.35
Block and stone splitter, manual		Each	1.10
Steel Reinforcement Equipment			
Bar bending machine – manual	up to 13 mm diameter rods	Each	0.90
Bar bending machine – manual	up to 20 mm diameter rods	Each	1.28
Bar bending machine – electric	up to 38 mm diameter rods	Each	2.82
Bar bending machine – electric	up to 40 mm diameter rods	Each	3.85
Bar bending machine – electric	up to 13 mm diameter rods	Each	1.54
Bar bending machine – electric	up to 20 mm diameter rods	Each	2.05
Bar bending machine – electric	up to 40 mm diameter rods	Each	2.82
Bar bending machine – 3 phase	up to 40 mm diameter rods	Each	3.85

BUILDING INDUSTRY

SITE EQUIPMENT – cont'd **Item of plant**	**Size/Rating**	**Unit**	**Rate per Hour £**
Dehumidifiers			
110/240v Water	68 litres extraction per 24 hours	Each	1.28
110/240v Water	90 litres extraction per 24 hours	Each	1.85
SMALL TOOLS			
Saws			
Masonry saw bench	350 mm – 500 mm diameter	Each	2.80
Floor saw	350 mm diameter, 125 mm max. cut	Each	1.90
Floor saw	450 mm diameter, 150 mm max. cut	Each	2.60
Floor saw, reversible	Max. Cut 300 mm	Each	13.00
Chop/cut saw, electric	350 mm diameter	Each	1.33
Circular saw, electric	230 mm diameter	Each	0.60
Tyrannosaw		Each	1.20
Reciprocating saw		Each	0.60
Door trimmer		Each	0.90
Chainsaw, petrol	500 mm	Each	2.13
Full chainsaw safety kit		Each	0.50
Working jig		Each	0.60
Pipework Equipment			
Pipe bender	15 mm – 22 mm	Each	0.33
Pipe bender, hydraulic	50 mm	Each	0.60
Pipe bender, electric	50 mm – 150 mm diameter	Each	1.35
Pipe cutter, hydraulic		Each	1.84
Tripod pipe vice		Set	0.40
Ratchet threader	12 mm – 32 mm	Each	0.55
Pipe threading machine, electric	12 mm – 75 mm	Each	2.40
Pipe threading machine, electric	12 mm – 100 mm	Each	3.00
Impact wrench, electric		Each	0.54
Impact wrench, two stroke, petrol		Each	4.49
Impact wrench, heavy duty, electric		Each	1.13
Plumber's furnace, calor gas or similar		Each	2.16
Hand-held Drills and equipment			
Impact or hammer drill	Up to 25 mm diameter	Each	0.50
Impact or hammer drill	35 mm diameter	Each	0.90
Angle heads drill		Each	0.70
Stirrer, mixed drills		Each	0.70
Paint, Insulation Application Equipment			
Airless spray unit		Each	4.20
Portaspray unit		Each	1.65
HPVL turbine spray unit		Each	1.65
Compressor and spray gun		Each	2.20
Other Handtools			
Screwing machine	13 mm – 50 mm diameter	Each	0.77
Screwing machine	25 mm – 100 mm diameter	Each	1.57
Staple gun		Each	0.33
Air nail gun	110V	Each	3.33
Cartridge hammer		Each	1.00
Tongue and groove nailer complete			
With mallet		Each	0.93
Chasing machine	152 mm	Each	1.72
Chasing machine	76 mm – 203 mm	Each	5.99

BUILDING INDUSTRY

Item of plant	Size/Rating	Unit	Rate per Hour £
Floor grinder		Each	3.00
Floor plane		Each	3.67
Diamond concrete planer		Each	2.05
Autofeed screwdriver, electric		Each	1.13
Laminate trimmer		Each	0.64
Biscuit jointer		Each	0.87
Random orbital sander		Each	0.73
Floor sander		Each	1.33
Palm, delta, flap or belt sander		Each	0.38
Saw cutter, two strokes, petrol	300 mm	Each	1.26
Grinder, angle or cutter	Up to 225 mm	Each	0.60
Grinder, angle or cutter	300 mm	Each	1.10
Mortar raking tool attachment		Each	0.15
Floor/polish scrubber	325 mm	Each	1.03
Floor tile stripper		Each	1.74
Wallpaper stripper, electric		Each	0.56
Electric scraper		Each	0.51
Hot air paint stripper		Each	0.38
Electric diamond tile cutter	All sizes	Each	1.38
Hand tile cutter		Each	0.36
Electric needle gun		Each	1.08
Needle chipping gun		Each	0.72
Pedestrian floor sweeper	1.2 m wide	Each	0.87

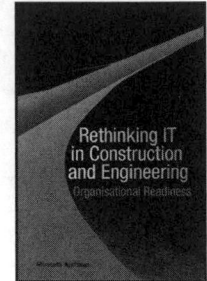

PART 8

Tables and Memoranda

This part of the book contains the following sections:

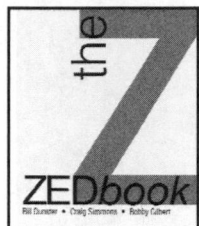

Tables and Memoranda

CONVERSION TABLES

	Unit	Conversion factors			
Length					
Millimetre	mm	1 in	= 25.4 mm	1 mm	= 0.0394 in
Centimetre	cm	1 in	= 2.54 cm	1 cm	= 0.3937 in
Metre	m	1 ft	= 0.3048 m	1 m	= 3.2808 ft
		1 yd	= 0.9144 m		= 1.0936 yd
Kilometre	km	1 mile	= 1.6093 km	1km	=0.6214 mile

Note:

1 cm	= 10 mm	1 ft	= 12 in
1 m	= 1 000 mm	1 yd	= 3 ft
1 km	= 1 000 m	1 mile	= 1 760 yd

	Unit	Conversion factors			
Area					
Square Millimetre	mm^2	1 in^2	= 645.2 mm^2	1 mm^2	= 0.0016 in^2
Square Centimetre	cm^2	1 in^2	= 6.4516 cm^2	1 cm^2	= 1.1550 in^2
Square Metre	m^2	1 ft^2	= 0.0929 m^2	1 m^2	= 10.764 ft^2
		1 yd^2	= 0.8361 m^2	1 m^2	= 1.1960 yd^2
Square Kilometre	km^2	1 $mile^2$	= 2.590 km^2	1 km^2	=0.3861 $mile^2$

Note:

1 cm^2	= 100 mm^2	1 ft^2	= 144 in^2
1 m^2	= 10 000 cm^2	1 yd^2	= 9 ft^2
1 km^2	= 100 hectares	1 acre	= 4 840 yd^2
		1 $mile^2$	= 640 acres

	Unit	Conversion factors			
Volume					
Cubic Centimetre	cm^3	1 cm^3	= 0.0610 in^3	1 in^3	= 16.387 cm^3
Cubic Decimetre	dm^3	1 dm^3	= 0.0353 ft^3	1 ft^3	= 28.329 dm^3
Cubic Metre	m^3	1 m^3	= 35.3147 ft^3	1 ft^3	= 0.0283 m^3
		1 m^3	= 1.3080 yd^3	1 yd^3	= 0.7646 m^3
Litre	l	1 l	= 1.76 pint	1 pint	= 0.5683 l
			= 2.113 US pt		=0.4733 US l

Note:

1 dm^3	= 1 000 cm^3	1 ft^3	= 1 728 in^3	1 pint	= 20 fl oz
1 m^3	= 1 000 dm^3	1 yd^3	= 27 ft^3	1 gal	= 8 pints
1 l	= 1 dm^3				

Neither the Centimetre nor Decimetre are SI units, and as such their use, particularly that of the Decimetre, is not widespread outside educational circles.

	Unit	Conversion factors			
Mass					
Milligram	mg	1 mg	= 0.0154 grain	1 grain	=64.935 mg
Gram	g	1 g	= 0.0353 oz	1 oz	= 28.35 g
Kilogram	kg	1 kg	= 2.2046 lb	1 lb	=0.4536 kg
Tonne	t	1 t	= 0.9842 ton	1 ton	= 1.016 t

Note:

1 g	= 1000 mg		1 oz	= 437.5 grains	1 cwt	= 112 lb
1 kg	= 1000 g	1 lb	= 16 oz	1 ton	= 20 cwt	
1 t	= 1000 kg	1 stone	= 14 lb			

CONVERSION TABLES

Force	Unit	Conversion factors			
Newton	N	1 lbf	= 4.448 N	1 kgf	=9.807N
Kilonewton	kN	1 lbf	= 0.004448 kN	1 ton f	=9.964kN
Meganewton	MN	100 tonf	= 0.9964 MN		

Pressure and stress

Kilonewton per		1 lbf/in^2	= 6.895 kN/m^2
square metre	kN/m^2	1 bar	= 100 kN/m^2
Meganewton per		1 $tonf/ft^2$	= 107.3 kN/m^2 = 0.1073
MN/m^2			
square metre	MN/m^2	1 kgf/cm^2	= 98.07 kN/m^2
		1 lbf/ft^2	= 0.04788 kN/m^2

Coefficient of consolidation (Cv) or swelling

Square metre per year	m^2/year	1 cm^2/s	= 3 154 m^2/year
		1 ft^2/year	= 0.0929 m^2/year

Coefficient of permeability

Metre per second	m/s	1 cm/s	= 0.01 m/s
Metre per year	m/year	1 ft/year	= 0.3048 m/year
			= 0.9651 x $(10)^8$m/s

Temperature

Degree Celsius ^{O}C $\quad ^{O}C = \frac{5}{9} \times (^{O}F - 32)$ $\qquad ^{O}F = \frac{(9 \times ^{O}C)}{5} + 32$

FORMULAE

Two dimensional figures

Figure	Area
Square	$(side)^2$
Rectangle	Length x breadth
Triangle	½ (base x height) or √(s(s - a)(s-b)(s-c)) where a, b and c are the lengths of the three sides, and $s = \frac{a + b + c}{2}$ or $a^2 = b^2 + c^2 - (2b\ c\cos A)$ where A is the angle opposite side a
Hexagon	2.6 x $(side)^2$
Octagon	4.83 x $(side)^2$
Trapezoid	height x ½ (base + top)
Circle	3.142 x $radius^2$ or 0.7854 x $diameter^2$ (circumference = 2 x 3.142 x radius or 3.142 x diameter)

FORMULAE

Two dimensional figures

Figure	Area
Sector of a circle	½ x length of arc x radius
Segment of a circle	area of sector - area of triangle
Ellipse	3.142 x AB (where A = ½ x height and B = ½ x length)
Bellmouth	$\frac{3}{14}$ x radius^2.

Three dimensional figures

Figure	Volume	Surface Area
Prism	Area of base x height	circumference of base x height
Cube	(side)^3	6 x (side)^2
Cylinder	3.142 x radius^2 x height	2 x 3.142 x radius x (height - radius)
Sphere	$\frac{4}{3}$ x 3.142 x radius^3	4 x 3.142 x radius^2
Segment of a sphere	$\frac{(3.142 \times h) \times (3 \times r^2 + h^2)}{6}$	2 x 3.142 x r x h
Pyramid	1/3 of area of base x height	½ x circumference of base x slant height
Cone	1/3 x 3.142 x radius^2 x h	3.142 x radius x slant height
Frustrum of a pyramid	1/3 x height [A + B + √(AB)] where A is the area of the large end and B is the area of the small end	½ x mean circumference x slant height
Frustrum of a cone	(1/3 x 3.142 x height ($R^2 + r^2$ + R x r)) where R is the radius of the large end and r is the radius of the small end	3.142 x slant height x (R + r)

Other formulae

Formula	Description
Pythagoras' two theorum	$A^2 = B^2 + C^2$ where A is the hypotenuse of a right-angled triangle and B and C are the adjacent sides
Simpson's Rule	Volume = $\frac{x}{3}$ [($y_1 + y_n$) + 2($y_3 + y_5$) + 4($y_2 + y_4$)] The volume to be measured must be represented by an odd number of cross-sections (y_1 -y_n) taken at fixed intervals (x), the sum of the areas at even numbered intermediate cross-sections (y_2, y_4, etc.) is multiplied by 4 and the sum of the areas at odd numbered intermediate cross-sections (y_3, y_5, etc.) is multiplied by 2, and the end cross-sections (y_1 and y_n) taken once only. The resulting *weighted average* of these areas is multiplied by 1/3 of the distance between the cross-sections (x) to give the total volume.

FORMULAE

Other formulae

Formula	**Description**
Trapezoidal Rule	(0.16 x [Total length of trench] x [area of first section x 4 times area of middle section + area of last section])

Note: Both Simpson's Rule and Trapezoidal Rule are useful in accurately calculating the volume of an irregular trench, or similar longitudinal earthworks movement, e.g. road construction.

DESIGN LOADINGS FOR BUILDINGS

Note: Refer to BS 6399: Part 1: 1996 Code of Practice for Dead and Imposed Loads min. loading examples.

Definitions

Dead load: The load due to the weight of all walls, permanent partitions, floors, roofs and finishes, including services and all other permanent construction.

Imposed load: The load assumed to be produced by the intended occupancy or use, including the weight of moveable partitions, distributed, concentrated, impact, inertia and snow loads, but excluding wind loads.

Distributed load: The uniformly distributed static loads per square metre of plan area which provide for the effects of normal use. Where no values are given for concentrated load it may be assumed that the tabulated distributed load is adequate for design purposes.

Note: The general recommendations are not applicable to certain atypical usages particularly where mechanical stacking, plant or machinery are to be installed and in these cases the designer should determine the loads from a knowledge of the equipment and processes likely to be employed.

The additional imposed load to provide for partitions, where their positions are not shown on the plans, on beams and floors, where these are capable of effective lateral distributional of the load, is a uniformly distributed load per square metre of not less than one-third of the weight per metre run by the partitions but not less than 1 kN/m^2.

Floor area usage	**Distributed load KN/m²**	**Concentrated load kN/300 m²**
Industrial occupancy class (workshops, factories)		
Foundries	20.0	-
Cold storage of storage height	5.0 for each metre with a minimum of 15.0	9.0
Paper storage, for printing plants	4.0 for each metre of storage height	9.0
Storage, other than types listed separately	2.4 for each metre of storage height	7.0
Type storage and other areas in printing plants	12.5	9.0

DESIGN LOADINGS FOR BUILDINGS

Floor area usage load	Distributed load kN/m^2	Concentrated load kN/300m^2
Industrial occupancy class (workshops, factories) - cont'd		
Boiler rooms, motor rooms, fan rooms and the like, including the weight of machinery	7.5	4.5
Factories, workshops and similar buildings	5.0	4.5
Corridors, hallways, foot bridges, etc. subject to loads greater than for crowds, such as wheeled vehicles, trolleys and the like	5.0	4.5
Corridors, hallways, stairs, landings, footbridges, etc.	4.0	4.5
Machinery halls, circulation spaces therein	4.0	4.5
Laboratories (including equipment), kitchens, laundries	3.0	4.5
Workrooms, light without storage	2.5	1.8
Toilet rooms	2.0	-
Cat walks	-	1.0 at 1 m centres
Institutional and educational occupancy class (prisons, hospitals, schools, colleges)		
Dense mobile stacking (books) on mobile trolleys	4.8 for each metre of stack height but with a minimum of 9.6	7.0
Stack rooms (books)	2.4 for each metre of stack height but with a minimum of 6.5	7.0
Stationery stores	4.0 for each metre of storage height	9.0
Boiler rooms, motor rooms, fan rooms and the like, including the weight of machinery	7.5	4.5
Corridors, hallways, etc. subject to loads greater than from crowds, such as wheeled vehicles, trolleys and the like	5.0	4.5

DESIGN LOADINGS FOR BUILDINGS

Floor area usage load	Distributed load kN/m^2	Concentrated load $kN/300m^2$
Institutional and educational occupancy class (prisons, hospitals, schools, colleges) - cont'd		
Drill rooms and drill halls	5.0	9.0
Assembly areas without fixed seating, stages gymnasia	5.0	3.6
Bars	5.0	-
Projection rooms	5.0	-
Corridors, hallways, aisles, stairs, landings, foot-bridges, etc.	4.0	4.5
Reading rooms with book storage, e.g. libraries	4.0	4.5
Assembly areas with fixed seating	4.0	-
Laboratories (including equipment), kitchens, laundries	3.0	4.5
Corridors, hallways, aisles, landings, stairs, etc. not subject to crowd loading	3.0	2.7
Classrooms, chapels	3.0	2.7
Reading rooms without book storage	2.5	4.5
Areas for equipment	2.0	1.8
X-ray rooms, operating rooms, utility rooms	2.0	4.5
Dining rooms, lounges, billiard rooms	2.0	2.7
Dressing rooms, hospital bedrooms and wards	2.0	1.8
Toilet rooms	2.0	-
Bedrooms, dormitories	1.5	1.8
Balconies	same as rooms to which they give access but with a minimum of 4.0	1.5 per metre run concentrated at the outer edge
Fly galleries	4.5 kN per metre run distributed uniformly over the width	-

DESIGN LOADINGS FOR BUILDINGS

Floor area usage load	Distributed load kN/m^2	Concentrated load kN/300 m^2
Offices occupancy class (offices, banks)		
Stationery stores	4.0 for each metre of storage height	9.0
Boiler rooms, motor rooms, fan rooms and the like, including the weight of machinery	7.5	4.5
Corridors, hallways, etc. subject to loads greater than from crowds, such as wheeled vehicles, trolleys and the like	5.0	4.5
File rooms, filing and storage space	5.0	4.5
Corridors, hallways, stairs, landings, footbridges, etc.	4.0	4.5
Offices with fixed computers or similar equipment	3.5	4.5
Laboratories (including equipment), kitchens, laundries	3.0	-
Banking halls	3.0	4.5
Offices for general use	2.5	2.7
Toilet rooms	2.0	-
Balconies	Same as rooms to which they give access but with a minimum of 4.0	1.5 per metre run concentrated at the outer edge
Cat walks	-	1.0 at 1 m centre
Public assembly occupancy class (halls, auditoria, restaurants, museums, libraries, non-residential clubs, theatres, broadcasting studios, grandstands)		
Dense mobile stacking (books) on mobile trucks	4.8 for each metre of stack height but with a minimum of 9.6	7.0
Stack rooms (books)	2.4 for each metre of stack height but with a minimum of 6.5	7.0
Boiler rooms, motor rooms fan rooms and the like, including the weight of machinery	7.5	4.5
Stages	7.5	4.5

DESIGN LOADINGS FOR BUILDINGS

Floor area usage load	**Distributed load kN/m^2**	**Concentrated load kN/300m^2**
Public assembly occupancy class (halls, auditoria, restaurants, museums, libraries, non-residential clubs, theatres, broadcasting studios, grandstands) - cont'd		
Corridors, hallways, etc. subject to loads greater than from crowds, such as wheeled vehicles, trolleys and the like. Corridors, stairs, and passage ways in grandstands	5.0	4.5
Drill rooms and drill halls	5.0	9.0
Assembly areas without fixed seating dance halls, gymnasia, grancstands	5.0	3.6
Projection rooms, bars	5.0	-
Museum floors and art galleries for exhibition purposes	4.0	4.5
Corridors, hallways, stairs, landings, footbridges, etc.	4.0	4.5
Reading rooms with book storage, e.g. libraries	4.0	4.5
Assembly areas with fixed seating	4.0	-
Kitchens, laundries	3.0	4.5
Chapels, churches	3.0	2.7
Reading rooms without book storage	2.5	4.5
Grids	2.5	-
Areas for equipment	2.0	1.8
Dining rooms, lounges, billiard rooms	2.0	2.7
Dressing rooms	2.0	1.8
Toilet rooms	2.0	-
Balconies	Same as rooms to which they give access but with a minimum of 4.0	1.5 per metre run concentrated at the outer edge
Fly galleries	4.5 kN per metre run distributed uniformly over the width	
Cat walks	-	1.0 at 1 m centres

DESIGN LOADINGS FOR BUILDINGS

Floor area usage load	Distributed load kN/m^2	Concentrated load kN/300 m^2
Residential occupancy class		
Self contained dwelling units and communal areas in blocks of flats not more than three storeys in height and with not more than four self-contained dwelling units per floor accessible from one staircase		
All usages	1.5	1.4
Boarding houses, lodging houses, guest houses, hostels, residential clubs and communal areas in blocks of flats other than type 1		
Boiler rooms, motor rooms, fan rooms and the like including the weight of machinery	7.5	4.5
Communal kitchens, laundries	3.0	4.5
Corridors, hallways, stairs, landings, footbridges etc.	3.0	4.5
Dining rooms, lounges, billiard rooms	2.0	2.7
Toilet rooms	2.0	-
Bedrooms, dormitories	1.5	1.8
Balconies	Same as rooms to which they give access but with a minimum of 3.0	1.5 per metre run concentrated at the outer edge
Cat walks	-	1.0 at 1 m centres
Hotels and Motels		
Boiler rooms, motor rooms, fan rooms and the like, including the weight of machinery	7.5	4.5
Assembly areas without fixed seating, dance halls	5.0	3.6
Bars	5.0	-
Assembly areas with fixed seating	4.0	-
Corridors, hallways, stairs, landings, footbridges, etc.	4.0	4.5
Kitchens, laundries	3.0	4.5
Dining rooms, lounges, billiard rooms	2.0	2.7
Bedrooms	2.0	1.8

DESIGN LOADINGS FOR BUILDINGS

Floor area usage load	Distributed load kN/m^2	Concentrated load kN/300m^2
Hotels and Motels – cont'd		
Toilet rooms	2.0	-
Balconies	Same as rooms to which they give access but with a minimum of 4.0	1.5 per metre run concentrated at the outer edge
Cat Walks	-	1.0 at 1 m centres
Retail occupancy class (shops, departmental stores, supermarkets)		
Cold storage	5.0 for each metre of storage height with a minimum of 15.0	9.0
Stationery stores	4.0 for each metre of storage height	9.0
Storage, other than types separately	2.4 for each metre of storage height	7.0
Boiler rooms, motor rooms, fan rooms and the like, including the weight of machinery	7.5	4.5
Corridors, hallways, etc. subject to loads greater than from crowds, such as wheeled vehicles, trolleys and the like	5.0	4.5
Corridors, hallways, stairs, landings, footbridges, etc.	4.0	4.5
Shop floors for the display and sale of merchandise	4.0	3.6
Kitchens, laundries	3.0	4.5
Toilet rooms	2.0	-
Balconies	Same as rooms to which they give access but with a minimum of 4.0	1.5 per metre run concentrated at the outer edge
Cat walks	-	1.0 at 1 m centres
Storage occupancy class (warehouses)		
Cold storage	5.0 for each metre of storage height with a minimum of 15.0	9.0
Dense mobile stacking (books) on mobile trucks	4.8 for each metre of storage height with a minimum of 15.0	7.0
Paper storage, for printing plants	4.0 for each metre of storage height	9.0

DESIGN LOADINGS FOR BUILDINGS

Floor area usage load	Distributed load kN/m^2	Concentrated load $kN/300\ m^2$
Stationery stores	4.0 for each metre of storage height	9.0
Storage, other than types listed separately, warehouses	2.4 for each metre of storage height	7.0
Motor rooms, fan rooms and the like, including the weight of machinery	7.5	4.5
Corridors, hallways, foot-bridges, etc. subject to loads greater than for crowds, such as wheeled vehicles, trolleys and the like	5.0	4.5
Cat walks	-	1.0 at 1 m centres
Vehicular occupancy class (garages, car parks, vehicle access ramps)		
Motor rooms, fan rooms and the like, including the weight of machinery	7.5	4.5
Driveways and vehicle ramps, other than in garages for the parking only of passenger vehicles and light vans not exceeding 2500 kg gross mass	5.0	9.0
Repair workshops for all types of vehicles, parking for vehicles exceeding 2500 kg gross mass including driveways and ramps	5.0	9.0
Footpaths, terraces and plazas leading from ground level with no obstruction to vehicular traffic, pavement lights	5.0	9.0
Corridors, hallways, stairs, landings, footbridges, etc. subject to crowd loading	4.0	4.5
Footpaths, terraces and plazas leading from ground level but restricted to pedestrian traffic only	4.0	4.5
Car parking only, for passenger vehicles and light vans not exceeding 2500 kg gross mass including garages, driveways and ramps	2.5	9.0
Cat walks	-	1.0 at 1 m centres

PLANNING PARAMETERS

PLANNING PARAMETERS

Definitions

* For precise definitions consult the Code of Measuring Practice published by the Royal Institution of Chartered Surveyors and the Incorporated Society of valuers and Auctioneers.

General definitions

Plot ratio *
Ratio of GEA to site area where the site area is expressed as one.

Gross external area (GEA) *
Gross area on each floor including the external walls of all spaces except open balconies and fire escapes, upper levels of atria and areas less than 1.5 m (5ft) such as under roof slopes, open covered ways or minor canopies, open vehicle parking areas, terraces and party walls beyond the centre line. Measured over structural elements and services space such as partitions and plant rooms. Roof level plant rooms may be excluded from the planning area

Site area *
Total area of the site within the site title boundaries measured on the horizontal plane.

Gross site area *
The site area, plus any area of adjoining roads enclosed by extending the side boundaries of the site up to the centre of the road, or to 6 m (20 ft) out from the frontage, whatever is the less.

Gross internal floor area (GIFA)/ Gross internal area (GIA) *
Gross area measured on the same basis as GEA, but excluding external wall thickness, and for rating GIA, excluding areas with a headroom of less than 1.5 m, except under stairs.

Net internal floor area (NIFA) *
Net usable area measured to the internal finish of the external walls excluding all auxiliary and ancillary spaces such as WC's and lobbies, ducts, lift, tank and plant space etc, staircases, lift wells and major access circulation, fire escape corridors and lobbies, major switchroom space and areas used by external authorities, internal structural walls and columns, car parking and areas with less than 1.5 m headroom, such as under roof slopes, corridors used in common with other occupiers or of a permanent essential nature such as fire corridors, smoke lobbies, space occupied by permanent air-conditioning, heating or cooling apparatus and surface mounted ducting causing space to be unusable.

Cubic content *
The GEA multiplied by the vertical height from the lowest basement floor or average ground to the average height of the roof.

Internal cube
The GIFA of each floor multiplied by its storey height.

Ceiling height *
The height between the floor surface and the underside of the ceiling.

Building frontage *
The measurement along the front of the building from the outside of the external walls or the centre line of party walls.

External wall area
The wall area of all the enclosed spaces fulfilling the functional requirements of the buildings measured on the outer face of the external walls and overall windows and doors etc.

Wall to floor ratio
The factor produced by dividing the external wall area by the GIFA.

PLANNING PARAMETERS

Window to external wall ratio
The factor produced by dividing the external windows and door area by the external wall area.
Circulation (C)
Circulation and ancillary area measured on plan on each floor for staircases, lift lobbies, lift wells, lavatories, cleaners' cupboards usually represented as the allowances for circulation and ancillary space as a percentage of NIFA.

Plant area
Plant rooms and vertical duct space.

Retail definitions

Sales area *
NIFA usable for retailing excluding store rooms unless formed by non-structural partitions.

Storage area *
NIFA not forming part of the sales area and usable only for storage.

Shop frontage *
Overall external frontage to shop premises including entrance and return shop frontage, but excluding recesses, doorways and the like of other accommodation.

Overall frontage *
Overall measurement in a straight line across the front of the building and any return frontage, from the outside of external walls and / or the entire line or party walls.

Shop width *
Internal measurement between inside faces of external walls at shop front or other points of reference.

Shop depth *
Overall measurement from back of pavement or forecourt to back of sales area measured over any non-structural partitions.

Built depth *
Overall external ground level measurement from front to rear walls of building.

Zone A
Front zone of 6 m in standard retail units 6 m x 24 m.

Housing definitions

Number of persons housed
The total number for whom actual bed spaces are provided in the dwellings as designed.

Average number of persons per dwelling
The total number of persons housed divided by the total number of dwellings.

Density
The total number of persons housed divided by the site in hectares or acres.
The total number of units divided by the site area in hectares or acres.

PLANNING PARAMETERS

Functional units

As a "rule of thumb" guide to establish a cost per functional unit, or as a check on economy of design in terms of floor area, the following indicative functional unit areas have been derived from historical data. For indicative unit costs see "Building Prices per Functional Units" (Part 2: Approximate Estimating) on page 49.

Car parking	- surface	20 - 22 m^2/car
	- multi storey	23 - 27 m^2/car
	- basement	28 - 37 m^2/car
Concert Halls		8 m^2/seat
Halls of residence	- college/polytechnic	25 - 35 m^2/bedroom
	- university	30 - 50 m^2/bedroom
Hospitals	- district general	65 - 85 m^2/bed
	- teaching	120 + m^2/bed
	- private	75 - 100 m^2/bed
Hotels	- budget	28 - 35 m^2/bedroom
	- luxury city centre	70 - 130 m^2/bedroom
Housing		**Gross internal floor area**
Private developer:	1 Bedroom Flat	45 - 50 m^2
	2 Bedroom Flat	55 - 65 m^2
	2 Bedroom House	55 - 65 m^2
	3 Bedroom House	70 - 90 m^2
	4 Bedroom House	90 - 100 m^2
Offices	- high density open plan	20 m^2/person
	- low density cellular	15 m^2/person
Schools	- nursery	3 - 5 m^2/child
	- secondary	6 - 10 m^2/child
	- boarding	10 -12 m^2/child
Theatres	- small, local	3 m^2/seat to
	- large, prestige	7 m^2/seat

Typical planning parameters

The following are indicative planning design and functional criteria derived from historical data for a number of major building types.

Gross internal floor areas (GIFA)

Offices

Feasibility assessment of GIFA for:

Curtain wall office	GEA x 0.97
Solid wall office	GEA x 0.95

These measures apply except for thick stone façades - take measurements on site.

PLANNING PARAMETERS

Typical dimensions measured on plan between the internal finishes of the external walls for:

Speculative offices	13.75 m
Open plan offices	15.25 m
Open plan / cellular offices	18.3 m

Retail

Typical gross internal floor areas:

Food courts, comprising	232 to 372 m^2
Kiosks	37 m^2
Services - per seat	1.1 to 1.5 m^2
Seating area in mall - per seat	1.2 to 1.7 m^2
Retail Kiosks	56 to 75 m^2
Small specialist shops	465 to 930 m^2
Electrical goods	930 to 1 395 m^2
DIY	930 to 4 645 m^2
Furniture / carpets	1858 to 5 575 m^2
Toys	3715 to 4 645 m^2
Superstores	3715 to 5 575 m^2
Department stores within shopping centres	5575 to 27 870 m^2
Specialist shopping centres	5574 to 9 290 m^2

Leisure

Standard sizes:

Large sports halls	Medium sports halls	Small sports halls
36.5 x 32 x 9.1 m	29 x 26 x 7.6 - 9.1 m	29.5x16.5x6.7-7.6m
32 x 26 x 7.6 - 9.1 m	32 x 23 x 7.6 - 9.1 m	26x16.5x6.7-7.6m
	32 x 17 x 6.7 - 7.6 m	22.5x16.5x6.7-7.6m

Community halls
17.2 x 15.6 x 6.7 m
17 x 8.5 x 6.7 m

Court sizes:

badminton	13.4 x 6.1 m	volleyball	18 x 9 m
basketball	26+2 x 14+1 m	tug of war	35 (min) x 5 m (min)
handball	30 - 40 x 17 - 20 m	bowls	4.5 x 32 m (min) per rink
hockey	36 - 44 x 18 - 22 m	cricket nets	3.05 (min) x 33.5 m per net
women's lacrosse	27 - 36 x 15 - 21m	snooker	3.7 x 1.9 m table size
men's lacrosse	46 - 48 x 18 - 24m	ice hockey	56.61 x 26 - 30.5 m
netball	30.50 x 15.25 m	racquets	18.288 x 9.144 m
tennis	23.77 x 10.97 m	squash	9.754 x 6.4 x 5.64 m

PLANNING PARAMETERS

Leisure - cont'd

Typical swimming pool dimensions:

Olympic standard	50 m x 21 m (8 lanes) water depth 1.8 m (constant)
ASA, national and county championship standard	25 or 33.3 m long with width multiple of 2.1 m wide lanes minimum water depth 900 mm 1 m springboard needs minimum 3 m water depth
Learner pool	width 7.0 - 7.5 m depth 600 - 900 mm
Toddlers pool	450 mm depth
Leisure pool	informal shape: will sometimes encompass 25 m in one direction to accommodate roping-off for swimming lanes; water area from 400 - 750 m^2
Splash pool	minimum depth 1.05 m
Changing cubicles	minimum dimensions: 914 x 1057 mm

Note: For 25 m pool developments the ratio of water area to gross floor area may average 1:3. For free form leisure pool developments, a typical ratio is 1:5.5.

Multiplex space planning data:

Ideal number of screens	10 (minimum six)
Average area per screen	325 m^2
Typical dimensions:	71 x 45 m (10 screens) 66 x 43 m (8 screens) plus 20 m^2 food area

Housing

Typical densities	**Persons per hectare**	**Units per hectare**
Urban	200	90
Suburban	150	55
Rural	110	35

Typical gross internal floor areas for housing associations / local authorities schemes:

	(m^2)
Bungalows	
one-bed	48
two-bed	55 - 65
Houses	
one-bed	44
two-bed	62 - 80
three-bed	75 - 95
four-bed	111 - 145
Flats	
bedsitters	23
one-bed	35 - 63
two-bed	55 - 80
three-bed	75 - 100

PLANNING PARAMETERS

Gross internal floor areas for private developments are much more variable and may be smaller or larger than the indicative areas shown above, depending on the target market. Standards for private housing are set out in the NHBC's Registered House Builders Handbook. There are no floor space minima, but heating , kitchen layout, kitchens and linen storage, WC provisions, and the number of electrical socket outlets are included.

Average housing room sizes - net internal floor areas:

	Living room (m^2)	Kitchen (m^2)	Bathroom (m^2)	Main bedroom (m^2)	Average bedroom size (m^2)
Bungalows					
one-bed	15.0	6.0	3.5	11.0	-
two-bed	17.0	9.0	3.5	12.5	10.0
Houses					
one-bed	14.5	6.5	3.5	11.0	-
two-bed	17.5	9.5	4.5	10.0	9.0
three-bed	17.5	13.5	7.0	13.0	10.5
four-bed	22.5	12.5	8.0	17.5	12.5
Flats					
bedsitters	18.0	-	3.0	-	-
one-bed	13.5	7.5	4.5	10.0	-
two-bed	17.0	10.0	5.5	13.5	11.5
three-bed	23.0	3.5	5.5	14.0	14.0

Storage accommodation for housing

NHBC requirements are that in every dwelling, enclosed domestic storage accommodation shall be provided as follows:

Area of dwelling (m^2)	Minimum volume of storage (m^3)
less than 60	1.3
60 - 80	1.7
over 80	2.3

Hotels

Typical gross internal floor areas per bedroom:	m^2
Five star, city centre hotel	60+
Four star, city centre / provincial centre hotel	45 to 55
Three star, city / provincial hotel	40 to 45
Three / two star, provincial hotel	33 to 40
Three / two star bedroom extension	26 to 30

Indicative space standards (unit):

Suites including bedroom, living room bathroom and hall (nr)	55 to 65
Double bedrooms including bathroom and lobby (nr)	
large	30 to 35
average	25 to 30
small	20 to 25
disabled	3 to 5 m^2 extra

PLANNING PARAMETERS

Hotels - cont'd

Restaurant (seat)	
first class	1.85
speciality/grill	1.80
Coffee shop (seat)	1.80
Bar (customer standing)	0.40 to 0.45
Food preparation/main kitchen/storage	40% to 50% of restaurant and bar areas
Banquet (seat)	1.40
Catering to banquets	10 % to 25 % of banquet area
Function/meeting rooms (person)	1.50
Staff areas (person)	0.40 to 0.60
Staff restaurant and kitchen (seat)	0.70 to 0.90
Service rooms (floor)	30 to 50
General storage and housekeeping	1.5 to 2% of bedroom and circulation areas
Front hall, entrance areas, lounge	2 to 3% average (up to 5%) of total hotel area
Administrative areas	Allowances based on number of accounts staff. Additional area if self accounting 15 to 25 per cent for bedroom floors depending on number of storeys, layout and operating principles, 20 to 25 per cent for public areas
Plant rooms and ducts	4 to 5 % of total hotel area for non-air-conditioned areas, 7 to 8% for air conditioned areas

Typical internal bedroom dimensions:

Bedroom including bathroom	
five star	8.0 m x 4.0 m
four star	7.5 m x 3.75 m
three/two star	7.0 m x 3.5 m
Typical corridor width	1.4 m to 1.6 m

PLANNING PARAMETERS

Circulation (C)

Figures represent net area which is gross area less space to be set aside for staircases, lift lobbies, lift wells, lavatories, cleaners' cupboards, service risers, plant space, etc.

Typical NIFA to GIFA areas:	**Percentage of GIFA**
Offices	
2 to 4 storey	82 - 87
5 to 9 storey	76 - 82
10 to 14 storey	72 - 76
15 to 19 storey	68 - 72
20 + storeys	65 - 68
Adjustments	
for fancoil air-conditioned offices	deduct 2 - 3
for VAV air-conditioned offices	deduct 6 - 7
Flats	
Staircase access	85
Enclosed balcony	83
Internal corridor and lobby	80

Typical sales to gross internal areas

Retail	
Superstores	45 - 55
Department stores	50 - 60
Retail warehouses	75 - 85

Wall and Window to floor ratios

Typical ratios based on historic data:

Legend:
- (1) W/F - External wall to gross floor area (GIFA) ratio
- (2) W/W - External window to external wall ratio
- (3) IW/F - Internal wall to gross floor area (GIFA) ratio

Building types	(1) W/F	(2) W/W	(3) IW/F
Industrial			
warehouse	0.45	0.04	-
factory	0.60	0.14	-
nursery	0.70	0.14	-
Offices			
open	0.80	0.35	0.30
cellular	0.80	0.35	1.10

PLANNING PARAMETERS

Plant area	**Percentage of GIFA**
Industrial 3 - 5	
Offices	4 - 11
	Percentage of treated floor area
Leisure	
all air, low velocity	4.0 - 6.0
induction	2.0 - 3.0
fan coil	1.5 - 2.5
VAV	3.0 - 4.5
versatemp	1.5 - 2.0
boiler plant (excluding hws cylinders)	0.8 - 1.8
oil tank room	1.0 - 2.0
refrigeration plant (excluding cooling towers)	1.0 - 2.0
supply and extract ventilation	3.0 - 5.0
electrical (excluding input substation or standby generation)	0.5 - 1.5
lift rooms	0.2 - 0.5
toilet ventilation	0.3 - 1.0

Other key dimensions

Structural grid and cladding rail spacing for industrial buildings

Typical economic dimensions	**m**
spans	18
column spacing	6 - 7.5
purlin spacing	1.8

Wall to core for offices

Typical dimensions measured on plan between the internal finish of external wall to finish of core	7.3

Floor to floor heights

Typical dimensions, measured on section

Industrial	
top of ground slab to top of first floor slab	3.9 - 4.5
top of first floor slab to underside of beams / eaves	3.4 - 3.7

Minimum dimensions; floor finish to floor finish

Offices	
speculative centrally heated	3.3
speculative air-conditioned	3.8
trading floors air-conditioned	4.7
Hotels	
bedrooms	2.7 - 3
public areas	3.5 - 3.6

PLANNING PARAMETERS

Floor to underside of structure heights

Industrial	
Minimum internal clear height	
minimum cost stacking warehouse/light industrial	5 - 5.5
minimum height for storage racking	7.5
turret trucks used for stacking	9
automatic warehouse with stacker cranes	15 - 30

Clearance for structural members, sprinklers and lighting in addition to the above

Retail	
Clear height from floor to underside of beams / eaves:	
shop sales area	3.3 - 3.8
shop non-sales area	3.2 - 3.6
retail warehouse	4.75 - 5.5
Leisure	
Specified by each sport's governing body	
badminton/tennis to county standard	7.6
badminton/tennis/ trampolining to international standard	9.1
pool hall from pool surround	8.4 - 8.9

Floor to underside of structure heights

Industrial floor to eaves height	
Typical dimensions measured on section:	
low bay warehouse	6
high bay warehouse	9 - 18

Floor to ceiling height

Typical dimensions measured on section:	
Industrial	
top of ground slab to underside of first floor slab	3.7 - 4.3
top of first floor finish to ceiling finish	2.75 - 3
Minimum dimensions measured on section from floor finish to ceiling finish:	
Offices	
Speculative offices	2.6
Trading floors	3
Leisure	
Multiple cinemas	6
Fitness/dance studios	5 - 6
Snooker room	3
Projectile room	3
Changing rooms	3.5
Houses	
Ground floor	2.1 - 2.55
First floor	2.35 - 2.55
Flats	2.25 - 2.65
Bungalows	2.4
Hotels	
Bedrooms	2.5
Lounges	2.7
Meeting rooms	2.8
Restaurant / coffee shop / bar	3

PLANNING PARAMETERS

Raised floor areas	**mm**
Minimum clear void for:	
Speculative offices	100 - 200
Trading floors	300

Note: one floor box per 9 m^2

Suspended ceilings	**mm**
Minimum clear voids (beneath beams)	
Mechanically ventilated offices	300
Fan coil air-conditioned offices	450
VAV air-conditioned offices	550
Trading floors	760

Typical floor loadings

For more precise floor loadings according to usage refer to section on **DESIGN LOADINGS FOR BUILDINGS** earlier in this section.

Typical loadings (based on minimum uniformly distributed loads plus 25% for partition loads) are:

	KN/m^2
Industrial 24 - 37	
Offices	5 - 7
Retail warehouse / storage	24 - 29
Shop sales areas	6
Shop storage	12
Public assembly areas	6
Residential dwelling units	2 - 2.5
Residential corridor areas	4
Hotel bedrooms	3
Hotel corridor areas	4
Plant rooms	9
Car parks and access ramps	3 - 4

PLANNING PARAMETERS

Fire protection and means of escape

BS 5588: Fire Precautions in the Design and Construction of Building: includes details of:
- angle between escape routes
- disposition of fire resisting construction
- permitted travel distances

The Building Regulations fire safety approved document B 1992 provides advice on interpretation of the Building Regulations and is still the relevant controlling legislation for fire regulations, although the Loss Prevention Council have recently produced an advisory note, the *Code of Practice for the Construction of Buildings* which argues for a higher performance than the mandatory regulations.

Some minimum periods of fire resistance in minutes for elements of a structure are reproduced hereafter, based on Appendix A Table A2 of the Building Regulations fire safety approved document B, but refer to the relevant documentation to ensure that the information is current.

Building group		**Minimum fire resistance in minutes**					
		Basement storey		**Ground and Upper storey**			
		<10m deep	**>10m deep**	**<5m high**	**>20m high**	**<30m high**	**>30m high**
Industrial							
	not sprinklered	120	90	60	90	120	not allowed
	sprinklered	90	60	30*	60	60	120#
Offices							
	not sprinklered	90	60	30*	60	90	not allowed
	sprinklered	60	60	30*	30*	60	120#
Shop, commercial and leisure							
	not sprinklered	90	60	60	60	90	not allowed
	sprinklered	60	60	30*	60	60	120#
Residential dwelling houses							
		-	30*	30*	60	-	-

* Increase to a minimum of 60 minutes for compartment walls separating buildings

Reduce to 90 minutes for elements not forming part of the structural frame

Section 20

Applies to buildings in the Greater London area - refer to *London Building Acts (Amendment) Act 1939: Section 20, Code of Practice*. Major cost considerations include 2 hour fire resistance to reinforced concrete columns, possible requirement for sprinkler installation in offices and / or basement car parks, automatic controls and smoke detection in certain ventilation trucking systems, 4 hour fire resistance to fire fighting lift/stair/lobby enclosures and requirements for ventilated lobbies with a minimum floor area of 5.5 m^2 to fire fighting staircases.

Sprinkler installations

Sprinkler installations should be considered where any of the following are likely to occur:

- rapid fire spread likely, for example warehouses with combustible goods/packaging
- large un-compartmented areas
- high financial or consequential loss arising from fire damage

Refer to BS 5306: Part 2: 1990 for specification of sprinkler systems and associated Technical Bulletins from the Fire Officers Committee.

PLANNING PARAMETERS

Sanitary provisions

For the provisions of sanitary appliances refer to BS 6465: Part 1: 1994, which suggests the following minimum requirements (refer to the relevant documentation to ensure information is correct).

Factories (table 5)	**Males**	**Females**
WC's	1 per 25 persons or part thereof	1 per 25 persons or part thereof
Urinals	As required	Not applicable
Baths or showers	As required	As required
	Male and Female	
Wash basins	1 per 20 persons; for clean processes 1 per 10 persons; for dirty processes 1 per 5 persons; for injurious processes	

Housing (table 1)	**2 - 4 person**	**5 person**	**6 person and over**
One level, e.g. bungalows and flats			
WC's	1	1	2
Bath	1	1	1
Wash basin *	1	1	1
Sink and drainer	1	1	1
On two or more levels, e.g. houses and maisonettes			
WC's	1	2	2
Bath	1	1	1
Wash basin *	1	1	1
Sink and drainer	1	1	1

* in addition, allow one extra wash basin in every separate WC compartment which does not adjoin a bathroom.

Tables 2 and 3 deal with sanitary provisions for elderly people

Office building and shops (table 4)	**Number per male and per female staff**
WC's (no urinals) and wash hand basins	1 for 1 to 15 persons 2 for 16 to 30 persons 3 for 31 to 50 persons 4 for 51 to 75 persons 5 for 76 to 100 persons add 1 for every additional 25 persons or part thereof
Cleaners' sink	At least 1 per floor

For WC's (urinals provided), urinals, incinerators, etc. refer to BS 6465: Part 1: 1984. One unisex type WC and one smaller compartment for each sex on each floor where male and female toilets are provided - refer to BS 5810: 1979 and Building Regulations 1985 Schedule 2 (shortly to be replaced by part M).

PLANNING PARAMETERS

Swimming pools (table 11)

	For spectators Males	Females	For bathers Males	Females
WC's	1 for 1 - 200 persons 2 for 201 - 500 persons 3 for 501 - 1000 persons Over 1000 persons, 3 plus 1 for every additional 500 persons or part thereof	1 for 1 - 100 persons 2 for 101 - 250 persons 3 for 251 - 500 persons Over 500 persons, 3 plus 1for every additional 400 persons or part thereof	1 per 20 changing places	1 per 10 changing places
Urinals	1 per 50 persons	n/a	1 per 20 changing places	n/a
Wash basins	1 per 60 persons	1 per 60 persons	1 per 15 changing places	1 per 15 changing places
Showers	n/a	n/a	1 per 8 changing places	1 per 8 changing places

Refer also to BS 6465: Part 1: 1994 for sanitary provisions for schools, leisure, hotels and restaurants, etc.

Minimum cooling and ventilation requirements

General offices	40 W/m^2
Trading floors	60 W/m^2
Fresh air supply	
offices/dance halls	8 - 12 litres/person/second
bars	12 - 18 litres/person/second

Recommended design values for internal environmental temperatures and empirical values for air infiltration and natural ventilation allowances

	Temperature °C (winter	Air infiltration rate (changes per hour)	Ventilation allowance (W/m^3 degrees C)
Warehouses			
working and packing spaces	16	0.5	0.17
storage space	13	0.25	0.08
Industrial			
production	16	0.5	0.17
offices	20	1.0	0.33
Offices	20	1.0	0.33
Shops			
small	18	1.0	0.33
large	18	0.5	0.17
department store	18	0.25	0.08
fitting rooms	21	1.5	0.50
store rooms	15	0.5	0.17

PLANNING PARAMETERS

Recommended design values for internal environmental temperatures and empirical values for air infiltration and natural ventilation allowances - cont'd

	Temperature °C (winter	Air infiltration rate (changes per hour)	Ventilation allowance (W/m^3 degrees C)
Housing			
living rooms	21	1.0	0.33
bedrooms	18	0.5	0.17
bed sitting rooms	21	1.0	0.33
bathrooms	22	2.0	0.67
lavatory, cloakrooms	18	1.5	0.50
entrance halls, staircases, corridors	16	1.5	0.50
Hotels			
bedrooms (standard)	22	1.0	0.33
bedrooms (luxury)	24	1.0	0.33
public rooms	21	1.0	0.33
corridors	18	1.5	0.50
foyers	18	1.5	0.50

Typical design temperatures and mechanical ventilation allowances for leisure buildings

	Air temperature °C	**Mechanical airchange rates (changes per hour)**
Leisure buildings		
ice rink	below 25(heating temperature in winter:-8)	6
sports hall	16 - 21	3
squash courts	16 - 21	3
bowls halls	16 - 21	3
activity rooms	16 - 21	3
function room/bar	21 $\pm$ 2	2 - 4
fitness / dance studio	16 - 21	3 - 6
snooker room	16 - 21	3 - 6
projectile room	16 - 21	3 - 6
changing rooms	22	10
swimming pools	28	4 - 6
bar and cafe areas	23	2 - 4
administration areas	21	2 - 4

	Pool water temperature °C
Swimming pools	
main pool	27
splash pool	27
learners pool	28 – 30
diving pool	27
leisure pool	29
jacuzzi pool	35

Ventilation rates must be related to the control of condensation. The criteria is the water area and the recommended basis is 20 litres/per m^2 of water surface, plus a margin (say 20 per cent) to allow for the effect of wet surrounds.

PLANNING PARAMETERS

Typical lighting levels

Lighting levels for a number of common building types are given below. For more precise minimum requirements refer to the IES Code.

	Lux
Industrial building - production/assembly areas	100 - 1000(varies)
Offices	500
Conventional shops with counters or wall displays and self-service shops	500
Supermarkets	500
Covered shopping precincts and arcades	
main circulation paces	100 - 200
lift, stairs, escalators	150
external covered walkways	30
Sports buildings	
multi use sports halls	500
squash courts	500
dance / fitness studio	300
snooker room	500 on table
projectile room	300 generally 1000 on target
Homes	
living rooms	
general	50
casual reading	150
bedrooms	
general	50
bedhead	150
studios	
desk and prolonged reading	300
kitchens	
working areas	300
bathrooms	100
halls and landings	150
stairs	100
Hotels	
internal corridors	200
guest room sleep area; stair wells	300
guest room activity area; housekeeping areas	500
meeting / banquet facilities	800

Electrical socket outlets (NHBC)

	Desirable provision	Minimum provision
Homes		
working area of kitchen	4	4
dining area	2	1
living area	5	3
first or only double bedroom	3	2
other double bedrooms	2	2
single bedrooms	2	2
hall and landing	1	1
store/workshop/garage	1	-
single study bedrooms	2	2
single bed sitting rooms in family dwellings	3	3
single bed sitting rooms in self contained bed sitting room dwellings	5	5

PLANNING PARAMETERS

Lifts

Performance standard to be not less than BS 5655: Lifts and service Lifts.

Industrial
Typical goods lift - 1000 kg

Offices
Dependant on number of storeys and planning layout, usually based on:

	Number of lifts
< 4 storeys	1
$> = 4$ storeys and $< 10\,000$m^2 GIA	2
$> = 4$ storeys and $> 10\,000$m^2 GIA	3

Hotels
Dependant on number of bedrooms, number of storeys and planning layout.
Typical examples

120 bed hotel on 3 floors	two 6 - 8 person lifts and service lift
200 bed hotel on 10 floors	four 13 person lifts and fireman's lift and service lift

Car park

Typical car space requirements	**One car space per**
Industrial 45 - 55m^2 GIA	
Offices	
medium tech	28 - 37 m^2 GIFA
high tech 19 - 25 m^2 GIFA	
Retail	
superstores	8 - 10 m^2 GIFA
shopping centres/out of town retailing	18 - 23 m^2 GIFA
furniture/DIY stores	20 - 30 m^2 GIFA
Leisure	
swimming pools	
patrons	10 m^2 pool area
staff	2 nr staff
leisure centres	
patrons	10 m^2 activity area
Residential	1 - 2 dwellings (depending on garage space, standard of dwelling, etc)

Goods and reception and service vehicles

Typical goods reception bay suitable for two 15 m articulated lorries with 1.5 m clearance either side. Loading bays must be level and have a clear height of 4.73 m. Approach routes should have a clear minimum height of 5.03 m. Minimum articulated lorry turning circle 13 m.

Typical design load for service yard 20 KN/m^2.

PLANNING PARAMETERS

Recommended sizes of various sports facilities

Archery (Clout) 7.3 m firing area	Range 109.728 (Women), 146.304 (Men) 182.88 (Normal range)
Baseball	Overall 60 m x 70 m
Basketball	14 m x 26 m
Camogie	91 - 110 m x 54 - 68 m
Discus and Hammer	Safety cage 2.74 m square Landing area 45 arc (65° safety) 70 m radius
Football, American	Pitch 109.80 m x 48.80 m overall 118.94 m x 57.94 m
Football, Association	NPFA rules Senior pitches 96 - 100 m x 60 - 64 m Junior pitches 90 m x 46 - 55 m International 100 - 110 m x 64 - 75 m
Football, Australian Rules	Overall 135 - 185 m x 110 - 155 m
Football, Canadian	Overall 145.74 m x 59.47 m
Football, Gaelic	128 - 146.40 m x 76.80 - 91.50 m
Football, Rugby League	111 - 122 m x 68 m
Football, Rugby Union	144 m max x 69 m
Handball	91 - 110 m x 55 - 65 m
Hockey	91.50 m x 54.90 m
Hurling	137 m x 82 m
Javelin	Runway 36.50 m x 4.27 m Landing area 80 - 95 m long, 48 m wide
Jump, High	Running area 38.80 m x 19 m Landing area 5 m x 4 m
Jump, Long	Runway 45 m x 1.22 m Landing area 9 m x 2.75 m
Jump, Triple	Runway 45 m x 1.22 m Landing area 7.30 m x 2.75 m
Korfball	90 m x 40 m
Lacrosse	(Mens) 100 m x 55 m (Womens) 110 m x 73 m
Netball	15.25 m x 30.48 m
Pole Vault	Runway 45 m x 1.22 m Landing area 5 m x 5 m

PLANNING PARAMETERS

Recommended sizes of various sports facilities - cont'd

Polo	275 m x 183 m
Rounders	Overall 19 m x 17 m
400m Running Track	115.61 m bend length x 2 84.39 m straight length x 2 Overall 176.91 m long x 92.52 m wide
Shot Putt	Base 2.135 m diameter Landing area 65° arc, 25 m radius from base
Shinty	128 - 183 m x 64 - 91.50 m
Tennis	Court 23.77 m x 10.97 m Overall minimum 36.27 m x 18.29 m
Tug-of-war	46 m x 5 m

SOUND INSULATION

Sound reduction requirements as Building Regulations (E1/2/3)

The Building Regulations on airborne and impact sound (E1/2/3) state simply that both airborne and impact sound must be reasonably reduced in floors and walls. No minimum reduction is given but the following tables give example sound reductions for various types of constructions.

Sound reductions of typical walls	**Average sound reduction (dB)**
13 mm Fibreboard	20
16 mm Plasterboard	25
6 mm Float glass	30
16 mm Plasterboard, plastered both sides	35
75 mm Plastered concrete blockwork (100 mm)	44
110 mm half brick wall, half brick thick, plastered both sides	43
240 mm Brick wall one brick thick, plastered both sides	48
Timber stud partitioning with plastered metal lathing both sides	35
Cupboards used as partitions	30
Cavity block wall, plastered both sides	42
75 mm Breeze block cavity wall, plastered both sides	50
100 mm Breeze block cavity wall, plastered both sides including 50 mm air-gap and plasterboard suspended ceiling	55
As above with 150 mm Breeze blocks	65
19 mm T & G boarding on timber joists including plasterboard ceiling and plaster skim coat	32
As above including metal lash and plaster ceiling	37
As above with solid sound proofing material between joists approx 98 kg per sq metre	55
As above with floating floor of T & G boarding on batten and soundproofing quilt	75

SOUND INSULATION

Impact noise is particularly difficult to reduce satisfactorily. The following are the most efficient methods of reducing such sound.

1) Carpet on underlay of rubber or felt;
2) Pugging between joists (e.g. Slag Wool); and
3) A good suspended ceiling system.

Sound requirements

Housing

NHBC requirements are that any partition between a compartment containing a WC and a living-room or bedroom shell have an average sound insulation index of not less than 35 dB over the frequency range of 100 - 3150 Hz when tested in accordance with BS2750.

Hotels

Bedroom to bedroom or bedroom to corridor 48dB

THERMAL INSULATION

Thermal properties of various building elements

Thickness (mm)	Material	(m^2k/W) R	(W/m^2K) U - Value
n/a	Internal and external surface resistance	0.18	-
	Air-gap cavity	0.18	-
103	Brick skin	0.12	-
	Dense concrete block		
100	ARC conbloc	0.09	11.11
140	ARC conbloc	0.13	7.69
190	ARC conbloc	0.18	5.56
	Lightweight aggregate block		
100	Celcon standard	0.59	1.69
125	Celcon standard	0.74	1.35
150	Celcon standard	0.88	1.14
200	Celcon standard	1.18	0.85
	Lightweight aggregate thermal block		
125	Celcon solar	1.14	0.88
150	Celcon solar	1.36	0.74
200	Celcon solar	1.82	0.55
	Insulating board		
25	Dritherm	0.69	1.45
50	Dritherm	1.39	0.72
75	Dritherm	2.08	0.48
13	Lightweight plaster “Carlite”	0.07	14.29
13	Dense plaster “Thistle”	0.02	50.00
	Plasterboard		
9.5	British gypsum	0.06	16.67
12.7	British gypsum	0.08	12.50
40	Screed 0.10	10.00	
150	Reinforced concrete	0.12	8.33
100	Dow roofmate insulation	3.57	0.28

THERMAL INSULATION

Resistance to the passage of heat

Provisions meeting the requirement set out in the Building Regulations (L2/3):

		Minimum U - Value
a)	**Dwellings**	
	Roof	0.35
	Exposed wall	0.60
	Exposed floor	0.60
b)	**Residential, Offices, Shops and Assembly Buildings**	
	Roof	0.06
	Exposed wall	0.60
	Exposed floor	0.60
c)	**Industrial, Storage and Other Buildings**	
	Roof	0.70
	Exposed wall	0.70
	Exposed floor	0.70

TYPICAL CONSTRUCTIONS MEETING THERMAL REQUIREMENTS

External wall, masonry construction:

Concrete blockwork	**U - Value**
200 mm lightweight concrete block, 25 mm air-gap, 10 mm plasterboard	0.68
200 mm lightweight concrete block, 20 mm EPS slab, 10 mm plasterboard	0.54
200 mm lightweight concrete block, 25 mm air-gap, 25 mm EPS slab,10 mm plasterboard	0.46
Brick/Cavity/Brick	
105 mm brickwork, 50 mm UF foam, 105 mm brickwork, 3 mm lightweight plaster	0.55
Brick/Cavity/Block	
105 mm brickwork, 50 mm cavity, 125 mm Thermalite block, 3 mm lightweight plaster	0.59
105 mm brickwork, 50 mm cavity, 130 mm Thermalite block, 3 mm lightweight plaster	0.57
105 mm brickwork, 50 mm cavity, 130 mm Thermalite block, 3 mm dense plaster	0.59
105 mm brickwork, 50 mm cavity, 100 mm Thermalite block, foilbacked plasterboard	0.55
105 mm brickwork, 50 mm cavity, 115 mm Thermalite block, 9.5 mm plasterboard	0.58
105 mm brickwork, 50 mm cavity, 115 mm Thermalite block, foilbacked plasterboard	0.52
105 mm brickwork, 50 mm cavity, 125 mm Theramlite block, 9.5 mm plasterboard	0.55
105 mm brickwork, 50 mm cavity, 100 mm Thermalite block, 25 mm insulating plasterboard	0.53
105 mm brickwork, 50 mm cavity, 125 mm Thermalite block, 25 mm insulating plasterboard	0.47

TYPICAL CONSTRUCTIONS MEETING THERMAL REQUIREMENTS

Brick/Cavity/Block - cont'd		**U-Value**
105 mm brickwork, 25 mm cavity, 25 mm insulation, 115 mm Thermalite block, lightweight plaster		0.44
Render, 100 mm "Shield" block, 50 mm cavity, 100 mm Thermalite block, lightweight plaster		0.50
Render, 100 mm "Shield" block, 50 mm cavity, 115 mm Thermalite block, lightweight plaster		0.47
Render, 100 mm "Shield" block, 50 mm cavity, 125 mm Thermalite block, lightweight plaster		0.45
Tile hanging		
10 mm tile on battens and felt, 150 mm Thermalite block, lightweight plaster		0.57
25 mm insulating plasterboard		0.46
10 mm tile on battens and felt, 190 mm Thermalite block, lightweight plaster		0.47
25 mm insulating plasterboard		0.40
10 mm tile on battens and felt, 200 mm Thermalite block, lightweight plaster		0.45
25 mm insulated plasterboard		0.38
10 mm tile on battens, breather paper, 25 mm air-gap, 50 mm glass fibre quilts, 10 mm plasterboard		0.56
10 mm tile on battens, breather paper, 25 mm air-gap, 75 mm glass fibre quilts, 10 mm plasterboard		0.41
10 mm tile on battens, breather paper, 25 mm air-gap, 100 mm glass fibre quilts, 10 mm plasterboard		0.33
Pitched roofs		
Slate or concrete tiles, felt, airspace, Rockwool flexible slabs laid between rafters, plasterboard		
Slab	40 mm thick	0.62
	50 mm thick	0.52
	60 mm thick	0.45
	75 mm thick	0.38
	100 mm thick	0.29
Concrete tiles, sarking felt, rollbatts between joists, plasterboard		
Insulation	100 mm thick	0.31
	120 mm thick	0.26
	140 mm thick	0.23
	160 mm thick	0.21

TYPICAL CONSTRUCTIONS MEETING THERMAL REQUIREMENTS

Pitched roofs - cont'd **U-Value**

Steel frame Rockwool insulation sandwiched between steel exterior profiled sheeting and interior sheet lining

		U-Value
Insulation	60 mm thick	0.53
	80 mm thick	0.41
	100 mm thick	0.34

Steel frame, steel profiled sheeting, Rockwool insulation over purlins and plasterboard lining

		U-Value
Insulation	60 mm thick	0.51
	80 mm thick	0.38
	100 mm thick	0.32
	120 mm thick	0.27
	140 mm thick	0.24
	160 mm thick	0.21

Flat roofs

Asphalt, Rockwool roof slabs, 25 mm timber boarding, timber joists and 9.5 mm plasterboard

		U-Value
Insulation	30 mm thick	0.68
	40 mm thick	0.57
	50 mm thick	0.49
	60 mm thick	0.44
	70 mm thick	0.39
	80 mm thick	0.35
	90 mm thick	0.32
	100 mm thick	0.29

Asphalt, Rockwool roof slabs on 150 mm dense concrete deck and screed with 16 mm plaster finish

		U-Value
Insulation	40 mm thick	0.68
	50 mm thick	0.57
	60 mm thick	0.49
	70 mm thick	0.43
	80 mm thick	0.39
	90 mm thick	0.35
	100 mm thick	0.32

Asphalt, Rockwool roof slabs on 150 mm dense concrete deck and screed with suspended plasterboard ceiling

		U-Value
Insulation	40 mm thick	0.60
	50 mm thick	0.52
	60 mm thick	0.45
	70 mm thick	0.40
	80 mm thick	0.36
	90 mm thick	0.33
	100 mm thick	0.30

TYPICAL CONSTRUCTIONS MEETING THERMAL REQUIREMENTS

Flat roofs - cont'd		**U-Value**
Steel frame, asphalt on insulation slabs on troughed steel decking		
Insulation	50 mm thick	0.59
	60 mm thick	0.51
	70 mm thick	0.45
	80 mm thick	0.39
	90 mm thick	0.35
	100 mm thick	0.33
Steel frame, asphalt on insulation slabs on troughed steel decking including suspended plasterboard ceiling		
Insulation	40 mm thick	0.67
	50 mm thick	0.57
	60 mm thick	0.49
	70 mm thick	0.43
	80 mm thick	0.38
	90 mm thick	0.34
	100 mm thick	0.32

WEIGHTS OF VARIOUS MATERIALS

Material		**kg/m^3**	**Material**		**kg/m^3**
Aggregates					
Ashes		610	Lime:	Chalk (lump)	704
Cement	(Portland)	1600		Ground	961
Chalk		2406		Quick	880
Chippings	(stone)	1762	Sand:	Dry	1707
Clinker	(furnace)	800		Wet	1831
	(concrete)	1441	Water		1000
Ballast or stone		2241	Shale/Whinstone		2637
Pumice		640	Broken stone		1709
Gravel		1790	Pitch		1152
Metals					
Aluminium		2559	Lead		11260
Brass		8129	Tin		7448
Bronze		8113	Zinc		7464
Gunmetal		8475			
Iron:	Cast	7207			
	Wrought	7687			
Stone and brickwork					
Blockwork:			Brickwork:		
	Aerated	650		Common Fletton	1822
	Dense concrete	1800		Glazed brick	2080
	Lightweight concrete	1200		Staffordshire Blue	2162
	Pumice concrete	1080		Red Engineering	2240
				Concrete	1841

WEIGHTS OF VARIOUS MATERIALS

Material	kg/m³	Material	kg/m³
Stone and brickwork - cont'd			
Stone:			
Artificial	2242	Granite	2642
Bath	2242	Marble	2742
Blue Pennant	2682	Portland	2170
Cragleith	2322	Slate	2882
Darley Dale	2370	York	2402
Forest of Dean	2386	Terra-cotta	2116
Wood			
Blockboard	500 - 700	Jarrah	816
Cork Bark	80	Maple	752
Hardboard:		Mahogany:	
Standard	940 - 1000	Honduras	576
Tempered	940 - 1060	Spanish	1057
Wood chipboard:		Oak:	
Type I	650 - 750	English	848
Type II	680 - 800	American	720
Type III	650 - 800	Austrian & Turkish	704
Type II/III	680 - 800	Pine:	
Laminboard	500 - 700	Pitchpine	800
Timber:		Red Deal	576
Ash	800	Yellow Deal	528
Baltic spruce	480	Spruce	496
Beech	816	Sycamore	530
Birch	720	Teak:	
Box	961	African	961
Cedar	480	Indian	656
Chestnut	640	Moulmein	736
Ebony	1217	Walnut:	
Elm	624	English	496
Greenheart	961	Black	720

MEMORANDA FOR EACH TRADE

EXCAVATION AND EARTHWORK

Transport capacities

Type of vehicle	Capacity of vehicle m³ (solid)
Standard wheelbarrow	0.08
2 ton truck (2.03 t)	1.15
3 ton truck (3.05 t)	1.72
4 ton truck (4.06 t)	2.22
5 ton truck (5.08 t)	2.68
6 ton truck (6.10 t)	3.44
2 cubic yard dumper (1.53 m³)	1.15
3 cubic yard dumper (2.29 m³)	1.72
6 cubic yard dumper (4.59 m³)	3.44
10 cubic yard dumper (7.65 m3)	5.73

MEMORANDA FOR EACH TRADE

Planking and strutting

Maximum depth of excavation in various soils without the use of earthwork support

Ground conditions	Metres (m)
Compact soil	3.65
Drained loam	1.85
Dry sand	0.30
Gravelly earth	0.60
Ordinary earth	0.90
Stiff clay	3.00

It is important to note that the above table should only be used as a guide. Each case must be taken on its merits and, as the limited distances given above are approached, careful watch must be kept for the slightest signs of caving in.

Baulkage of soils after excavation

Soil type	Approximate bulk of 1m^3 after excavation
Vegetable soil and loam	25 - 30%
Soft clay	30 - 40%
Stiff clay	10 - 15%
Gravel	20 - 25%
Sand	40 - 50%
Chalk	40 - 50%
Rock, weathered	30 - 40%
Rock, unweathered	50 - 60%

CONCRETE WORK

Approximate average weights of materials

Materials Percentage of	Weight per m^3 voids (%)	(kg)
Sand	39	1660
Gravel 10 - 20 mm	45	1440
Gravel 35 - 75 mm	42	1555
Crushed stone	50	1330
Crushed granite		
(over 15 mm)	50	1345
(n.e. 15 mm)	47	1440
"All-in" ballast	32	1800

MEMORANDA FOR EACH TRADE

CONCRETE WORK - cont'd

Common mixes for various types of work per m^3

Recommended mix	Class of work suitable for:	Cement (kg)	Sand (kg)	Coarse Aggregate (kg)	No. of 50 kg bags of of cement per m^3 of combined aggregate
1:3:6	Roughest type of mass concrete such as footings, road haunchings 300 mm thick	208	905	1509	4.00
1:2.5:5	Mass concrete of better class than 1:3:6such as bases for machinery, walls below ground.	249	881	1474	5.00
1:2:4	Most ordinary uses of concrete such as mass walls above ground, road slabs etc. and general reinforced concrete work	304	889	1431	6.00
1:1.5:3	Watertight floors, pavements, and walls tanks, pits, steps, paths, surface of two course roads, reinforced concrete where extra strength is required	371	801	1336	7.50
1:1:2	Work of thin section such as fence posts and small precast work	511	720	1206	10.50

Bar reinforcement

Cross-sectional area and mass

Nominal sizes (m)	Cross-sectional area (mm^2)	Mass per metre run (kg)
6*	28.3	0.222
8	50.3	0.395
10	78.5	0.616
12	113.1	0.888
16	201.1	1.579
20	314.2	2.466
25	490.9	3.854
32	804.2	6.313
40	1256.6	9.864
50*	1963.5	15.413

* Where a bar larger than 40 mm is to be used the recommended size is 50 mm. Where a bar smaller than 8 mm is to be used the recommended size is 6 mm.

MEMORANDA FOR EACH TRADE

Fabric reinforcement

Preferred range of designated fabric types and stock sheet sizes

Fabric reference	**Longitudinal wires**			**Cross wires**			
	Nominal wire size (mm)	**Pitch (mm)**	**Area (mm^2/m)**	**Nominal wire size (mm)**	**Pitch (mm)**	**Area (mm^2/m)**	**Mass (kg/m^2)**
Square mesh							
A393	10	200	393	10	200	393	6.16
A252	8	200	252	8	200	252	3.95
A193	7	200	193	7	200	193	3.02
A142	6	200	142	6	200	142	2.22
A98	5	200	98	5	200	98	1.54
Structural mesh							
B1131	12	100	1131	8	200	252	10.90
B785	10	100	785	8	200	252	8.14
B503	8	100	503	8	200	252	5.93
B385	7	100	385	7	200	193	4.53
B283	6	100	283	7	200	193	3.73
B196	5	100	196	7	200	193	3.05
Long mesh							
C785	10	100	785	6	400	70.8	6.72
C636	9	100	636	6	400	70.8	5.55
C503	8	100	503	5	400	49.0	4.34
C385	7	100	385	5	400	49.0	3.41
C283	6	100	283	5	400	49.0	2.61
Wrapping mesh							
D98	5	200	98	5	200	98	1.54
D49	2.5	100	49	2.5	100	49	0.77

Stock sheet size 4.8 m x 2.4 m, Area 11.52 m^2

Average weight kg/m^3 of steelwork reinforcement in concrete for various building elements

Substructure	**kg/m^3 concrete**		
Pile caps	110 - 150	Plate slab	150 - 220
Tie beams	130 - 170	Cant slab	145 - 210
Ground beams	230 - 330	Ribbed floors	130 - 200
Bases	125 - 180	Topping to block floor	30 - 40
Footings	100 - 150	Columns	210 - 310
Retaining walls	150 - 210	Beams	250 - 350
Raft	60 - 70	Stairs	130 - 170
Slabs - one way	120 - 200	Walls - normal	40 - 100
Slabs - two way	110 - 220	Walls - wind	70 - 125

Note: For exposed elements add the following % :

Walls 50%, Beams 100%, Columns 15%

MEMORANDA FOR EACH TRADE

BRICKWORK AND BLOCKWORK

Number of bricks required for various types of work per m^2 of walling

Description	Brick size 215 x 102.5 x 50 mm	Brick size 215 x 102.5 x 65 mm
Half brick thick		
Stretcher bond	74	59
English bond	108	86
English garden wall bond	90	72
Flemish bond	96	79
Flemish garden wall bond	83	66
One brick thick and cavity wall of two half brick skins		
Stretcher bond	148	119

Quantities of bricks and mortar required per m^2 of walling

	Unit	No of bricks required	Mortar required (cubic metres)		
Standard bricks			No frogs	Single frogs	Double frogs
Brick size 215 x 102.5 x 50 mm					
half brick wall (103 mm)	m^2	72	0.022	0.027	0.032
2 x half brick cavity wall (270 mm)	m^2	144	0.044	0.054	0.064
one brick wall (215 mm)	m^2	144	0.052	0.064	0.076
one and a half brick wall (322 mm)	m^2	216	0.073	0.091	0.108
Mass brickwork	m^3	576	0.347	0.413	0.480
Brick size 215 x 102.5 x 65 mm					
half brick wall (103 mm)	m^2	58	0.019	0.022	0.026
2 x half brick cavity wall (270 mm)	m^2	116	0.038	0.045	0.055
one brick wall (215 mm)	m^2	116	0.046	0.055	0.064
one and a half brick wall (322 mm)	m^2	174	0.063	0.074	0.088
Mass brickwork	m^3	464	0.307	0.360	0.413
Metric modular bricks				Perforated	
Brick size 200 x 100 x 75 mm					
90 mm thick	m^2	67	0.016	0.019	
190 mm thick	m^2	133	0.042	0.048	
290 mm thick	m^2	200	0.068	0.078	
Brick size 200 x 100 x 100 mm					
90 mm thick	m^2	50	0.013	0.016	
190 mm thick	m^2	100	0.036	0.041	
290 mm thick	m^2	150	0.059	0.067	
Brick size 300 x 100 x 75 mm					
90 mm thick	m^2	33	-	0.015	
300 x 100 x 100 mm					
90 mm thick	m^2	44	0.015	0.018	

Note: Assuming 10 mm thick joints.

MEMORANDA FOR EACH TRADE

Mortar required per m² blockwork (9.88 blocks/m²)

Wall thickness	75	90	100	125	140	190	215
Mortar m^3/m^2	0.005	0.006	0.007	0.008	0.009	0.013	0.014

Standard available block sizes

Block	**Length x height Co-ordinating size**	**Work size**	**Thicknesses**
A	400 x 100	390 x 90	(75, 90, 100,140 & 190 mm
	400 x 200	440 x 190	(
	450 x 225	440 x 215	(75, 90, 100140, 190, & 215 mm
B	400 x 100	390 x 90	(75, 90, 100
	400 x 200	390 x 190	(140 & 190 mm
	450 x 200	440 x 190	(
	450 x 225	440 x 215	(75, 90, 100
	450 x 300	440 x 290	(140, 190, & 215 mm
	600 x 200	590 x 190	(
	600 x 225	590 x 215	(
C	400 x 200	390 x 190	(
	450 x 200	440 x 190	(
	450 x 225	440 x 215	(60 & 75 mm
	450 x 300	440 x 290	(
	600 x 200	590 x 190	(
	600 x 225	590 x 215	(

ROOFING

Total roof loadings for various types of tiles/slates

		Roof load (slope) kg/m²		
		Slate/Tile	**Roofing underlay and battens**	**Total dead load kg/m²**
Asbestos cement slate (600 x 300)		21.50	3.14	24.64
Clay tile	interlocking	67.00	5.50	72.50
	plain	43.50	2.87	46.37
Concrete tile	interlocking	47.20	2.69	49.89
	plain	78.20	5.50	83.70
Natural slate (18" x 10")		35.40	3.40	38.80
		Roof load (plan) kg/m²		
Asbestos cement slate (600 x 300)		28.45	76.50	104.95
Clay tile	interlocking	53.54	76.50	130.04
	plain	83.71	76.50	60.21
Concrete tile	interlocking	57.60	76.50	134.10
	plain	96.64	76.50	173.14

MEMORANDA FOR EACH TRADE

ROOFING - cont'd

Tiling data

Product		**Lap (mm)**	**Gauge of battens**	**No. slates per m^2**	**Battens (m/m^2)**	**Weight as laid (kg/m^2)**
CEMENT SLATES						
Eternit slates	600 x 300 mm	100	250	13.4	4.00	19.50
(Duracem)		90	255	13.1	3.92	19.20
		80	260	12.9	3.85	19.00
		70	265	12.7	3.77	18.60
	600 x 350 mm	100	250	11.5	4.00	19.50
		90	255	11.2	3.92	19.20
	500 x 250 mm	100	200	20.0	5.00	20.00
		90	205	19.5	4.88	19.50
		80	210	19.1	4.76	19.00
		70	215	18.6	4.65	18.60
	400 x 200 mm	90	155	32.3	6.45	20.80
		80	160	31.3	6.25	20.20
		70	165	30.3	6.06	19.60
CONCRETE TILES/SLATES						
Redland Roofing						
Stonewold slate	430 x 380 mm	75	355	8.2	2.82	51.20
Double Roman tile	418 x 330 mm	75	355	8.2	2.91	45.50
Grovebury pantile	418 x 332 mm	75	343	9.7	2.91	47.90
Norfolk pantile	381 x 227 mm	75	306	16.3	3.26	44.01
		100	281	17.8	3.56	48.06
Renown inter-locking tile	418 x 330 mm	75	343	9.7	2.91	46.40
"49" tile	381 x 227 mm	75	306	16.3	3.26	44.80
		100	281	17.8	3.56	48.95
Plain, vertical tiling	265 x 165 mm	35	115	52.7	8.70	62.20
Marley Roofing						
Bold roll tile	420 x 330 mm	75	344	9.7	2.90	47.00
		100	-	10.5	3.20	51.00
Modern roof tile	420 x 330 mm	75	338	10.2	3.00	54.00
		100	-	11.0	3.20	58.00
Ludlow major	420 x 330 mm	75	338	10.2	3.00	45.00
		100	-	11.0	3.20	49.00
Ludlow plus	387 x 229 mm	75	305	16.1	3.30	47.00
		100	-	17.5	3.60	51.00
Mendip tile	420 x 330 mm	75	338	10.2	3.00	47.00
		100	-	11.0	3.20	51.00
Wessex	413 x 330 mm	75	338	10.2	3.00	54.00
		100	-	11.0	3.20	58.00
Plain tile	267 x 165 mm	65	100	60.0	10.00	76.00
		75	95	64.0	10.50	81.00
		85	90	68.0	11.30	86.00
Plain vertical	267 x 165 mm	35	110	53.0	8.70	67.00
tiles (feature)		34	115	56.0	9.10	71.00

MEMORANDA FOR EACH TRADE

Slate nails, quantity per kilogram

Length	Type			
	Plain wire	**Galvanised wire**	**Copper nail**	**Zinc nail**
28.5 mm	325	305	325	415
34.4 mm	286	256	254	292
50.8 mm	242	224	194	200

Metal sheet coverings

Thicknesses and weights of sheet metal coverings

Lead to BS 1178

BS Code No	3	4	5	6	7	8
Colour Code	Green	Blue	Red	Black	White	Orange
Thickness (mm)	1.25	1.80	2.24	2.50	3.15	3.55
kg/m^2	14.18	20.41	25.40	30.05	35.72	40.26

Copper to BS 2870

Thickness (mm)		0.60	0.70
Bay width			
Roll (mm)		500	650
Seam (mm)		525	600
Standard width to form bay	600	750	
Normal length of sheet	1.80	1.80	

Zinc to BS 849

Zinc Gauge (Nr)	9	10	11	12	13	14	15	16
Thickness (mm)	0.43	0.48	0.56	0.64	0.71	0.79	0.91	1.04
Density kg/m^2	3.1	3.2	3.8	4.3	4.8	5.3	6.2	7.0

Aluminium to BS 4868

Thickness (mm)	0.5	0.6	0.7	0.8	0.9	1.0	1.2
Density kg/m^2	12.8	15.4	17.9	20.5	23.0	25.6	30.7

MEMORANDA FOR EACH TRADE

ROOFING - cont'd

Type of felt	Nominal mass per unit area (kg/10m)	Nominal mass per unit area of fibre base (g/m^2)	Nominal length of roll (m)
Class 1			
1B fine granule surfaced bitumen	14 18 25	220 330 470	10 or 20 10 or 20 10
1E mineral surfaced bitumen	38	470	10
1F reinforced bitumen	15	160 (fibre) 110 (hessian)	15
1F reinforced bitumen, aluminium faced	13	160 (fibre) 110 (hessian)	15
Class 2			
2B fine granule surfaced bitumen asbestos	18	500	10 or 20
2E mineral surfaced bitumen asbestos	38	600	10
Class 3			
3B fine granule surfaced bitumen glass fibre	18	60	20
3E mineral surfaced bitumen glass fibre	28	60	10
3E venting base layer bitumen glass fibre	32	60*	10
3H venting base layer bitumen glass fibre	17	60*	20

* Excluding effect of perforations

MEMORANDA FOR EACH TRADE

WOODWORK

Conversion tables (for timber only)

Inches	Millimetres	Feet	Metres
1	25	1	0.300
2	50	2	0.600
3	75	3	0.900
4	100	4	1.200
5	125	5	1.500
6	150	6	1.800
7	175	7	2.100
8	200	8	2.400
9	225	9	2.700
10	250	10	3.000
11	275	11	3.300
12	300	12	3.600
13	325	13	3.900
14	350	14	4.200
15	375	15	4.500
16	400	16	4.800
17	425	17	5.100
18	450	18	5.400
19	475	19	5.700
20	500	20	6.000
21	525	21	6.300
22	550	22	6.600
23	575	23	6.900
24	600	24	7.200

Planed softwood
The finished end section size of planed timber is usually 3/16" less than the original size from which it is produced. This however varies slightly depending upon availability of material and origin of the species used.

Standards (timber) to cubic metres and cubic metres to standards (timber)

Cubic metres	**Cubic metres** standards	**Standards**
4.672	1	0.214
9.344	2	0.428
14.017	3	0.642
18.689	4	0.856
23.361	5	1.070
28.033	6	1.284
32.706	7	1.498
37.378	8	1.712
42.050	9	1.926
46.722	10	2.140
93.445	20	4.281
140.167	30	6.421
186.890	40	8.561
233.612	50	10.702
280.335	60	12.842
327.057	70	14.982
373.779	80	17.122

MEMORANDA FOR EACH TRADE

WOODWORK - cont'd

1 cu metre = 35.3148 cu ft = 0.21403 std

1 cu ft = 0.028317 cu metres

1 std = 4.67227 cu metres

Basic sizes of sawn softwood available (cross sectional areas)

Thickness (mm)	**Width (mm)**								
	75	100	125	150	175	200	225	250	300
16	X	X	X	X					
19	X	X	X	X					
22	X	X	X	X					
25	X	X	X	X	X	X	X	X	X
32	X	X	X	X	X	X	X	X	X
36	X	X	X	X					
38	X	X	X	X	X	X	X		
44	X	X	X	X	X	X	X	X	X
47*	X	X	X	X	X	X	X	X	X
50	X	X	X	X	X	X	X	X	X
63	X	X	X	X	X	X	X		
75		X	X	X	X	X	X	X	X
100		X		X		X		X	X
150				X		X			X
200						X			
250								X	
300									X

* This range of widths for 47 mm thickness will usually be found to be available in construction quality only.

Note: The smaller sizes below 100 mm thick and 250 mm width are normally but not exclusively of European origin. Sizes beyond this are usually of North and South American origin.

MEMORANDA FOR EACH TRADE

Basic lengths of sawn softwood available (metres)

1.80	2.10	3.00	4.20	5.10	6.00	7.20
	2.40	3.30	4.50	5.40	6.30	
	2.70	3.60	4.80	5.70	6.60	
		3.90			6.90	

Note: Lengths of 6.00 m and over will generally only be available from North American species and may have to be recut from larger sizes.

Reductions from basic size to finished size by planning of two opposed faces

Purpose		Reductions from basic sizes for timber			
		15 - 35 mm	**36 - 100 mm**	**101 - 150 mm**	**over 150 mm**
a)	Constructional timber	3 mm	3 mm	5 mm	6 mm
b)	Matching interlocking boards	4 mm	4 mm	6 mm	6 mm
c)	Wood trim not specified in BS 584	5 mm	7 mm	7 mm	9 mm
d)	Joinery and cabinet work	7 mm	9 mm	11 mm	13 mm

Note: The reduction of width or depth is overall the extreme size and is exclusive of any reduction of the face by the machining of a tongue or lap joints.

Maximum spans for various roof trusses

Maximum permissible spans for rafters for Fink trussed rafters

Basic size (mm)	Actual size (mm)	Pitch (degrees)								
		15 (m)	17.5 (m)	20 (m)	22.5 (m)	25 (m)	27.5 (m)	30 (m)	32.5 (m)	35 (m)
38 x 75	35 x 72	6.03	6.16	6.29	6.41	6.51	6.60	6.70	6.80	6.90
38 x 100	35 x 97	7.48	7.67	7.83	7.97	8.10	8.22	8.34	8.47	8.61
38 x 125	35 x 120	8.80	9.00	9.20	9.37	9.54	9.68	9.82	9.98	10.16
44 x 75	41 x 72	6.45	6.59	6.71	6.83	6.93	7.03	7.14	7.24	7.35
44 x 100	41 x 97	8.05	8.23	8.40	8.55	8.68	8.81	8.93	9.09	9.22
44 x 125	41 x 120	9.38	9.60	9.81	9.99	10.15	10.31	10.45	10.64	10.81
50 x 75	47 x 72	6.87	7.01	7.13	7.25	7.35	7.45	7.53	7.67	7.78
50 x 100	47 x 97	8.62	8.80	8.97	9.12	9.25	9.38	9.50	9.66	9.80
50 x 125	47 x 120	10.01	10.24	10.44	10.62	10.77	10.94	11.00	11.00	11.00

MEMORANDA FOR EACH TRADE

WOODWORK - cont'd

Sizes of internal and external doorsets

Description	**Internal**		**External**	
	Size (mm)	**Permissible deviation**	**Size (mm)**	**Permissible deviation**
Co-ordinating dimension: height of door leaf height sets	2100		2100	
Co-ordinating dimension: height of ceiling height set	2300 2350 2400 2700 3000		2300 2350 2400 2700 3000	
Co-ordinating dimension: width of all door sets S = Single leaf set D = Double leaf set	600 S 700 S 800 S&D 900 S&D 1000 S&D 1200 D 1500 D 1800 D 2100 D		900 S 1000 S 1200 D 1500 D 1800 D 2100 D	
Work size: height of door leaf height set	2090	± 2.0	2095	± 2.0
Work size: height of ceiling height set	2285) 2335) 2385) 2685) 2985)	± 2.0	2295) 2345) 2395) 2695) 2995)	± 2.0
Work size: width of all door sets S = Single leaf set D = Double leaf set	590 S) 690 S) 790 S&D) 890 S&D) 990 S&D) 1190 D) 1490 D) 1790 D) 2090 D)	± 2.0	895 S) 995 S) 1195 D) 1495 D) 1795 D) 2095 D)	± 2.0
Width of door leaf in single leaf sets F = Flush leaf P = Panel leaf	526 F) 626 F) 726 F&P) 826 F&P) 926 F&P)	± 1.5	806 F&P) 906 F&P)	± 1.5

MEMORANDA FOR EACH TRADE

Description	Internal		External	
	Size (mm)	**Permissible deviation**	**Size (mm)**	**Permissible deviation**
Width of door leaf	362 F	)	552 F&P)	
in double leaf sets	412 F	)	702 F&P)	± 1.5
F = Flush leaf	426 F	)	852 F&P)	
P = Panel leaf	562 F&P	) ± 1.5	1002 F&P)	
	712 F&P	)		
	826 F&P	)		
	1012 F&P	)		
Door leaf height for all door sets	2040	± 1.5	1994	± 1.5

STRUCTURAL STEELWORK

Tables showing the mass and surface area per metre run for various steel members

Size (mm)	Mass (kg/m)	Surface area per m run (m^2)
Universal beams		
914 x 419	388	3.404
	343	3.382
914 x 305	289	2.988
	253	2.967
	224	2.948
	201	2.932
838 x 292	226	2.791
	194	2.767
	176	2.754
762 x 267	197	2.530
	173	2.512
	147	2.493
686 x 254	170	2.333
	152	2.320
	140	2.310
	125	2.298
610 x 305	238	2.421
	179	2.381
	149	2.361
610 x 229	140	2.088
	125	2.075
	113	2.064
	101	2.053
533 x 210	122	1.872
	109	1.860
	101	1.853
	92	1.844
	82	1.833

MEMORANDA FOR EACH TRADE

STRUCTURAL STEELWORK - cont'd

Tables showing the mass and surface area per metre run for various steel members - cont'd

Size (mm)	Mass (kg/m)	Surface area per m run (m^2)
Universal beams - cont'd		
457 x 191	98	1.650
	89	1.641
	82	1.633
	74	1.625
	67	1.617
457 x 152	82	1.493
	74	1.484
	67	1.474
	60	1.487
	52	1.476
406 x 178	74	1.493
	67	1.484
	60	1.476
	54	1.468
406 x 140	46	1.332
	39	1.320
356 x 171	67	1.371
	57	1.358
	51	1.351
	45	1.343
356 x 127	39	1.169
	33	1.160
305 x 165	54	1.245
	46	1.235
	40	1.227
305 x 127	48	1.079
	42	1.069
	37	1.062
305 x 102	33	1.006
	28	0.997
	25	0.988
254 x 146	43	1.069
	37	1.060
	31	1.050
254 x 102	28	0.900
	25	0.893
	22	0.887
203 x 133	30	0.912
	25	0.904

MEMORANDA FOR EACH TRADE

Tables showing the mass and surface area per metre run for various steel members - cont'd

Size (mm)	Mass (kg/m)	Surface area per m run (m^2)
Universal columns		
356 x 406	634	2.525
	551	2.475
	467	2.425
	393	2.379
	340	2.346
	287	2.312
	235	2.279
356 x 368	202	2.187
	177	2.170
	153	2.154
	129	2.137
305 x 305	283	1.938
	240	1.905
	198	1.872
	158	1.839
	137	1.822
	118	1.806
	97	1.789
254 x 254	167	1.576
	132	1.543
	107	1.519
	89	1.502
	73	1.485
203 x 203	86	1.236
	71	1.218
	60	1.204
	52	1.194
	46	1.187
152 x 152	37	0.912
	30	0.900
	23	0.889
Joists		
254 x 203	81.85	1.193
254 x 114	37.20	0.882
203 x 152	52.09	0.911
152 x 127	37.20	0.722
127 x 114	29.76	0.620
127 x 114	26.79	0.635
114 x 114	26.79	0.600
102 x 102	23.07	0.528

MEMORANDA FOR EACH TRADE

STRUCTURAL STEELWORK - cont'd

Tables showing the mass and surface area per metre run for various steel members - cont'd

Size (mm)	Mass (kg/m)	Surface area per m run (m^2)
Joists - cont'd		
89 x 89	19.35	0.460
76 x 76	12.65	0.403

Circular hollow sections - outside dia (mm)	Mass (kg/m)	Surface area per m run (m^2)	Thickness (mm)
21.30	1.43	0.067	3.20
26.90	1.87	0.085	3.20
33.70	1.99	0.106	2.60
	2.41	0.106	3.20
	2.93	0.106	4.00
42.40	2.55	0.133	2.60
	3.09	0.133	3.20
	3.79	0.133	4.00
48.30	3.56	0.152	3.20
	4.37	0.152	4.00
	5.34	0.152	5.00
60.30	4.51	0.189	3.20
	5.55	0.189	4.00
	6.82	0.189	5.00
76.10	5.75	0.239	3.20
	7.11	0.239	4.00
	8.77	0.239	5.00
88.90	6.76	0.279	3.20
	8.38	0.279	4.00
	10.30	0.279	5.00
114.30	9.83	0.359	3.60
	13.50	0.359	5.00
	16.80	0.359	6.30
139.70	16.60	0.439	5.00
	20.70	0.439	6.30
	26.00	0.439	8.00
	32.00	0.439	10.00
168.30	20.10	0.529	5.00
	25.20	0.529	6.30
	31.60	0.529	8.00
	39.00	0.529	10.00

MEMORANDA FOR EACH TRADE

Tables showing the mass and surface area per metre run for various steel members - cont'd

Size (mm)	Mass (kg/m)	Surface area per m run (m^2)	Thickness (mm)
Circular hollow sections- outside diameter (mm) - cont'd			
193.70	23.30	0.609	5.00
	29.10	0.609	6.30
	36.60	0.609	8.00
	45.30	0.609	10.00
	55.90	0.609	12.50
	70.10	0.609	16.00
219.10	33.10	0.688	6.30
	41.60	0.688	8.00
	51.60	0.688	10.00
	63.70	0.688	12.50
	80.10	0.688	16.00
	98.20	0.688	20.00
273.00	41.40	0.858	6.30
	52.30	0.858	8.00
	64.90	0.858	10.00
	80.30	0.858	12.50
	101.00	0.858	16.00
	125.00	0.858	20.00
	153.00	0.858	25.00
323.90	62.30	1.020	8.00
	77.40	1.020	10.00
	96.00	1.020	12.50
	121.00	1.020	16.00
	150.00	1.020	20.00
	184.00	1.020	25.00
406.40	97.80	1.280	10.00
	121.00	1.280	12.50
	154.00	1.280	16.00
	191.00	1.280	20.00
	235.00	1.280	25.00
	295.00	1.280	32.00
457.00	110.00	1.440	10.00
	137.00	1.440	12.50
	174.00	1.440	16.00
	216.00	1.440	20.00
	266.00	1.440	25.00
	335.00	1.440	32.00
	411.00	1.440	40.00

MEMORANDA FOR EACH TRADE

STRUCTURAL STEELWORK - cont'd

Tables showing the mass and surface area per metre run for various steel members - cont'd

Size (mm)	Mass (kg/m)	Surface area per m run (m^2)	Thickness (mm)
Square hollow sections			
20 x 20	1.12	0.076	2.00
	1.35	0.074	2.50
30 x 30	2.14	0.114	2.50
	2.51	0.113	3.00
40 x 40	2.92	0.155	2.50
	3.45	0.154	3.00
	4.46	0.151	4.00
50 x 50	4.66	0.193	3.20
	5.72	0.191	4.00
	6.97	0.189	5.00
60 x 60	5.67	0.233	3.20
	6.97	0.231	4.00
	8.54	0.229	5.00
70 x 70	7.46	0.272	3.60
	10.10	0.269	5.00
80 x 80	8.59	0.312	3.60
	11.70	0.309	5.00
	14.40	0.306	6.30
90 x 90	9.72	0.352	3.60
	13.30	0.349	5.00
	16.40	0.346	6.30
100 x 100	12.00	0.391	4.00
	14.80	0.389	5.00
	18.40	0.386	6.30
	22.90	0.383	8.00
	27.90	0.379	10.00
120 x 120	18.00	0.469	5.00
	22.30	0.466	6.30
	27.90	0.463	8.00
	34.20	0.459	10.00
150 x 150	22.70	0.589	5.00
	28.30	0.586	6.30
	35.40	0.583	8.00
	43.60	0.579	10.00
	53.40	0.573	12.50
	66.40	0.566	16.00

MEMORANDA FOR EACH TRADE

Tables showing the mass and surface area per metre run for various steel members - cont'd

Size (mm)	Mass (kg/m)	Surface area per m run (m^2)	Thickness (mm)
Square hollow sections - cont'd			
180 x 180	34.20	0.706	6.30
	43.00	0.703	8.00
	53.00	0.699	10.00
	65.20	0.693	12.50
	81.40	0.686	16.00
200 x 200	38.20	0.786	6.30
	48.00	0.783	8.00
	59.30	0.779	10.00
	73.00	0.773	12.50
	91.50	0.766	16.00
250 x 250	48.10	0.986	6.30
	60.50	0.983	8.00
	75.00	0.979	10.00
	92.60	0.973	12.50
	117.00	0.966	16.00
300 x 300	90.70	1.180	10.00
	112.00	1.170	12.50
	142.00	1.170	16.00
350 x 350	106.00	1.380	10.00
	132.00	1.370	12.50
	167.00	1.370	16.00
400 x 400	122.00	1.580	10.00
	152.00	1.570	12.50
Rectangular hollow sections			
50 x 30	2.92	0.155	2.50
	3.66	0.153	3.20
60 x 40	4.66	0.193	3.20
	5.72	0.191	4.00
80 x 40	5.67	0.232	3.20
	6.97	0.231	4.00
90 x 50	7.46	0.272	3.60
	10.10	0.269	5.00
100 x 50	6.75	0.294	3.00
	7.18	0.293	3.20
	8.86	0.291	4.00
100 x 60	8.59	0.312	3.60
	11.70	0.309	5.00
	14.40	0.306	6.30

MEMORANDA FOR EACH TRADE

STRUCTURAL STEELWORK - cont'd

Tables showing the mass and surface area per metre run for various steel members - cont'd

Size (mm)	**Mass (kg/m)**	**Surface area per m run (m^2)**	**Thickness (mm)**
Rectangular hollow sections - cont'd			
120 x 60	9.72	0.352	3.60
	13.30	0.349	5.00
	16.40	0.346	6.30
120 x 80	14.80	0.389	5.00
	18.40	0.386	6.30
	22.90	0.383	8.00
	27.90	0.379	10.00
150 x 100	18.70	0.489	5.00
	23.30	0.486	6.30
	29.10	0.483	8.00
	35.70	0.479	10.00
160 x 80	18.00	0.469	5.00
	22.30	0.466	6.30
	27.90	0.463	8.00
	34.20	0.459	10.00
200 x 100	22.70	0.589	5.00
	28.30	0.586	6.30
	35.40	0.583	8.00
	43.60	0.579	10.00
250 x 150	38.20	0.786	6.30
	48.00	0.783	8.00
	59.30	0.779	10.00
	73.00	0.773	12.50
	91.50	0.766	16.00
300 x 200	48.10	0.986	6.30
	60.50	0.983	8.00
	75.00	0.979	10.00
	92.60	0.973	12.50
	117.00	0.966	16.00
400 x 200	90.70	1.180	10.00
	112.00	1.170	12.50
	142.00	1.170	16.00
450 x 250	106.00	1.380	10.00
	132.00	1.370	12.50
	167.00	1.370	16.00

MEMORANDA FOR EACH TRADE

Tables showing the mass and surface area per metre run for various steel members - cont'd

Size (mm)	**Mass (kg/m)**	**Surface area per m run (m^2)**
Channels		
432 x 102	65.54	1.217
381 x 102	55.10	1.118
305 x 102	46.18	0.966
305 x 89	41.69	0.920
254 x 89	35.74	0.820
254 x 76	28.29	0.774
229 x 89	32.76	0.770
229 x 76	26.06	0.725
203 x 89	29.78	0.720
203 x 76	23.82	0.675
178 x 89	26.81	0.671
178 x 76	20.84	0.625
152 x 89	23.84	0.621
152 x 76	17.88	0.575
127 x 64	14.90	0.476

Angles - sum of leg lengths	**Thickness (mm)**	**Mass (kg/m)**	**Surface area per m run (m^2)**
50	3	1.11	0.10
	4	1.45	0.10
	5	1.77	0.10
80	4	2.42	0.16
	5	2.97	0.16
	6	3.52	0.16
90	4	2.74	0.18
	5	3.38	0.18
	6	4.00	0.18
100	5	3.77	0.20
	6	4.47	0.20
	8	5.82	0.20
115	5	4.35	0.23
	6	5.16	0.23
	8	6.75	0.23

MEMORANDA FOR EACH TRADE

STRUCTURAL STEELWORK - cont'd

Tables showing the mass and surface area per metre run for various steel members - cont'd

Angles - sum of leg lengths	Thickness (mm)	Mass (kg/m)	Surface area per m run (m^2)
120	5	4.57	0.24
	6	5.42	0.24
	8	7.09	0.24
	10	8.69	0.24
125	6	5.65	0.25
	8	7.39	0.25
200	8	12.20	0.40
	10	15.00	0.40
	12	17.80	0.40
	15	21.90	0.40
225	10	17.00	0.45
	12	20.20	0.45
	15	24.80	0.45
240	8	14.70	0.48
	10	18.20	0.48
	12	21.60	0.48
	15	26.60	0.48
300	10	23.00	0.60
	12	27.30	0.60
	15	33.80	0.60
	18	40.10	0.60
350	12	32.00	0.70
	15	39.60	0.70
	18	47.10	0.70
400	16	48.50	0.80
	18	54.20	0.80
	20	59.90	0.80
	24	71.10	0.80

MEMORANDA FOR EACH TRADE

PLUMBING AND MECHANICAL INSTALLATIONS

Dimensions and weights of tubes

Outside diameter (mm)	Internal dia (mm)	Weight per m (kg)	Internal dia (mm)	Weight per m (kg)	Internal dia (mm)	Weight per m (kg)
Copper to EN 1057:1996						
	Table X		**Table Y**		**Table Z**	
6	4.80	0.0911	4.40	0.1170	5.00	0.0774
8	6.80	0.1246	6.40	0.1617	7.00	0.1054
10	8.80	0.1580	8.40	0.2064	9.00	0.1334
12	10.80	0.1914	10.40	0.2511	11.00	0.1612
15	13.60	0.2796	13.00	0.3923	14.00	0.2031
18	16.40	0.3852	16.00	0.4760	16.80	0.2918
22	20.22	0.5308	19.62	0.6974	20.82	0.3589
28	26.22	0.6814	25.62	0.8985	26.82	0.4594
35	32.63	1.1334	32.03	1.4085	33.63	0.6701
42	39.63	1.3675	39.03	1.6996	40.43	0.9216
54	51.63	1.7691	50.03	2.9052	52.23	1.3343
76.1	73.22	3.1287	72.22	4.1437	73.82	2.5131
108	105.12	4.4666	103.12	7.3745	105.72	3.5834
133	130.38	5.5151	-	-	130.38	5.5151
159	155.38	8.7795	-	-	156.38	6.6056

MEMORANDA FOR EACH TRADE

PLUMBING AND MECHANICAL INSTALLATIONS - cont'd

Dimensions and weights of tubes - cont'd

Nominal size (mm)	Outside diameter max (mm)	min (mm)	Wall thickness (mm)	Weight (kg/m)	Weight screwed and socketted (kg/m)
Steel pipes to BS 1387					
Light gauge					
6	10.1	9.7	1.80	0.361	0.364
8	13.6	13.2	1.80	0.517	0.521
10	17.1	16.7	1.80	0.674	0.680
15	21.4	21.0	2.00	0.952	0.961
20	26.9	26.4	2.35	1.410	1.420
25	33.8	33.2	2.65	2.010	2.030
32	42.5	41.9	2.65	2.580	2.610
40	48.4	47.8	2.90	3.250	3.290
50	60.2	59.6	2.90	4.110	4.180
65	76.0	75.2	3.25	5.800	5.920
80	88.7	87.9	3.25	6.810	6.980
100	113.9	113.0	3.65	9.890	10.200
Medium gauge					
6	10.4	9.8	2.00	0.407	0.410
8	13.9	13.3	2.35	0.650	0.654
10	17.4	16.8	2.35	0.852	0.858
15	21.7	21.1	2.65	1.220	1.230
20	27.2	26.6	2.65	1.580	1.590
25	34.2	33.4	3.25	2.440	2.460
32	42.9	42.1	3.25	3.140	3.170
40	48.8	48.0	3.25	3.610	3.650
50	60.8	59.8	3.65	5.100	5.170
65	76.6	75.4	3.65	6.510	6.630
80	89.5	88.1	4.05	8.470	8.640

MEMORANDA FOR EACH TRADE

Nominal size (mm)	Outside diameter max (mm)	min (mm)	Wall thickness (mm)	Weight (kg/m)	Weight screwed and socketted (kg/m)
Medium gauge - cont'd					
100	114.9	113.3	4.50	12.100	12.400
125	140.6	138.7	4.85	16.200	16.700
150	166.1	164.1	4.85	19.200	19.800
Heavy gauge					
6	10.4	9.8	2.65	0.493	0.496
8	13.9	13.3	2.90	0.769	0.773
10	17.4	16.8	2.90	1.020	1.030
15	21.7	21.1	3.25	1.450	1.460
20	27.2	26.6	3.25	1.900	1.910
25	34.2	33.4	4.05	2.970	2.990
32	42.9	42.1	4.05	3.840	3.870
40	48.8	48.0	4.05	4.430	4.470
50	60.8	59.8	4.50	6.170	6.240
65	76.6	75.4	4.50	7.900	8.020
80	89.5	88.1	4.85	10.100	10.300
100	114.9	113.3	5.40	14.400	14.700
125	140.6	138.7	5.40	17.800	18.300
150	166.1	164.1	5.40	21.200	21.800
Stainless steel pipes to BS 4127					
8	8.045	7.940	0.60	0.1120	
10	10.045	9.940	0.60	0.1419	
12	12.045	11.940	0.60	0.1718	
15	15.045	14.940	0.60	0.2174	
18	18.045	17.940	0.70	0.3046	
22	22.055	21.950	0.70	0.3748	
28	28.055	27.950	0.80	0.5469	

MEMORANDA FOR EACH TRADE

PLUMBING AND MECHANICAL INSTALLATIONS – cont'd

Maximum distances between pipe supports

Pipe material onto distances	BS nominal pipe size inch	mm	Pipes fitted vertically support distances in metres	Pipes fitted horizontally low gradients support in metres
Copper	0.50	15.0	1.90	1.3
	0.75	22.0	2.50	1.9
	1.00	28.0	2.50	1.9
	1.25	35.0	2.80	2.5
	1.50	42.0	2.80	2.5
	2.00	54.0	3.90	2.5
	2.50	67.0	3.90	2.8
	3.00	76.1	3.90	2.8
	4.00	108.0	3.90	2.8
	5.00	133.0	3.90	2.8
	6.00	159.0	3.90	2.8
muPVC	1.25	32.0	1.20	0.5
	1.50	40.0	1.20	0.5
	2.00	50.0	1.20	0.6
Polypropylene	1.25	32.0	1.20	0.5
	1.50	40.0	1.20	0.5
PVC	-	82.4	1.20	0.5
	-	110.0	1.80	0.9
	-	160.0	1.80	1.2

Litres of water storage required per person in various types of building

Type of building	Storage per person (litres)
Houses and flats	90
Hostels	90
Hotels	135
Nurse's home and medical quarters	115
Offices with canteens	45
Offices without canteens	35
Restaurants, per meal served	7
Boarding school	90
Day schools	30

MEMORANDA FOR EACH TRADE

Cold water plumbing - thickness of insulation required against frost

Bore of tube	Pipework within buildings declared thermal conductivity (W/m degrees C)		
	Up to 0.040	0.041 to 0.055	0.056 to 0.070
(mm)	Minimum thickness of insulation (mm)		
15	32	50	75
20	32	50	75
25	32	50	75
32	32	50	75
40	32	50	75
50	25	32	50
65	25	32	50
80	25	32	50
100	19	25	38

Cisterns

Capacities and dimensions of galvanised mild steel cisterns from BS 417

Capacity (litres)	BS type	Dimensions (mm) length	width	depth
18	SCM 45	457	305	305
36	SCM 70	610	305	371
54	SCM 90	610	406	371
68	SCM 110	610	432	432
86	SCM 135	610	457	482
114	SCM 180	686	508	508
159	SCM 230	736	559	559
191	SCM 270	762	584	610
227	SCM 320	914	610	584
264	SCM 360	914	660	610
327	SCM 450/1	1220	610	610
336	SCM 450/2	965	686	686
423	SCM 570	965	762	787
491	SCM 680	1090	864	736
709	SCM 910	1170	889	889

Capacities of cold water polypropylene storage cisterns from BS 4213

Capacity (litres)	BS type	Maximum height (mm)
18	PC 4	310
36	PC 8	380
68	PC 15	430
91	PC 20	510
114	PC 25	530
182	PC 40	610
227	PC 50	660
273	PC 60	660
318	PC 70	660
455	PC 100	760

MEMORANDA FOR EACH TRADE

HEATING AND HOT WATER INSTALLATIONS

Storage capacity and recommended power of hot water storage boilers

Type of building	Storage at 65°C (litres per person)	Boiler power to 65°C (kW per person)
Flats and dwellings		
(a) Low rent properties	25	0.5
(b) Medium rent properties	30	0.7
(c) High rent properties	45	1.2
Nurses homes	45	0.9
Hostels	30	0.7
Hotels		
(a) Top quality - upmarket	45	1.2
(b) Average quality - low market	35	0.9
Colleges and schools		
(a) Live-in accommodation	25	0.7
(b) Public comprehensive	5	0.1
Factories	5	0.1
Hospitals		
(a) General	30	1.5
(b) Infectious	45	1.5
(c) Infirmaries	25	0.6
(d) Infirmaries (inc. laundry facilities)	30	0.9
(e) Maternity	30	2.1
(f) Mental	25	0.7
Offices	5	0.1
Sports pavilions	35	0.3

Thickness of thermal insulation for heating installations

Size of tube (mm)	Declared thermal conductivity			
	Up to 0.025	0.026 to 0.040	0.041 to 0.055	0.056 to 0.070
	Minimum thickness of insulation			
LTHW Systems				
15	25	25	38	38
20	25	32	38	38
25	25	38	38	38
32	32	38	38	50
40	32	38	38	50
50	38	38	50	50
65	38	50	50	50
80	38	50	50	50

MEMORANDA FOR EACH TRADE

Size of tube (mm)	Minimum thickness of insulation Up to 0.025	0.026 to 0.040	0.041 to 0.055	0.056 to 0.070
LTHW Systems - cont'd				
100	38	50	50	63
125	38	50	50	63
150	50	50	63	63
200	50	50	63	63
250	50	63	63	63
300	50	63	63	63
Flat surfaces	50	63	63	63
MTHW Systems and condensate				
		Declared thermal conductivity		
15	25	38	38	38
20	32	38	38	50
25	38	38	38	50
32	38	50	50	50
40	38	50	50	50
50	38	50	50	50
65	38	50	50	50
80	50	50	50	63
100	50	63	63	63
125	50	63	63	63
150	50	63	63	63
200	50	63	63	63
250	50	63	63	75
300	63	63	63	75
Flat surfaces	63	63	63	75
HTHW Systems and steam				
15	38	50	50	50
20	38	50	50	50
25	38	50	50	50
32	50	50	50	63
40	50	50	50	63
50	50	50	75	75
65	50	63	75	75
80	50	63	75	75
100	63	63	75	100
125	63	63	100	100
150	63	63	100	100
200	63	63	100	100
250	63	75	100	100
300	63	75	100	100
Flat surfaces	63	75	100	100

MEMORANDA FOR EACH TRADE

HEATING AND HOT WATER INSTALLATIONS - cont'd

Capacities and dimensions of copper indirect cylinders (coil type) from BS 1566

Capacity (litres)	BS Type	External diameter (mm)	External height over dome (mm)
96	0	300	1600
72	1	350	900
96	2	400	900
114	3	400	1050
84	4	450	675
95	5	450	750
106	6	450	825
117	7	450	900
140	8	450	1050
162	9	450	1200
206	9 E	450	1500
190	10	500	1200
245	11	500	1500
280	12	600	1200
360	13	600	1500
440	14	600	1800

		Internal diameter (mm)	Height (mm)
109	BSG 1M	457	762
136	BSG 2M	457	914
159	BSG 3M	457	1067
227	BSG 4M	508	1270
273	BSG 5M	508	1473
364	BSG 6M	610	1372
455	BSG 7M	610	1753
123	BSG 8M	457	838

Energy costs (July 2001)

GAS SUPPLIES

The last year has seen the wholesale gas market remain somewhat volatile. Suppliers source their gas from this market unless they have a related company producing gas when they can purchase using the transfer pricing mechanism, which is again market based and is equally volatile. The volatility has invoked a continuing increase in the wholesale price of gas and such increases continue to be reflected in the price that the end-user has to pay.

The reasons suggested for the present situation include heavy buying by suppliers in European markets and the link of such markets with oil prices. Another reason is that a new system has been introduced to allocate and price the capacity suppliers require to put their gas into the network from the producers. This capacity has been restricted due to maintenance of the system which has forced entry capacity prices to rise. The suppliers are indicating that rates have reached their peak and will gradually recover in the next few months but not to levels experienced in the last year.

Domestic Markets

This sector refers to individual supply points, which do not consume more than 73,250 kw/hrs (2,500 therms) per annum. Suppliers must supply at their published rates, although there are exceptions to this for bulk purchasing schemes, which can reduce prices by approximately 3%. By contracting with an independent supplier, savings can still be achieved over British Gas Tariffs. Care must be exercised when selecting a supplier, look beyond the savings as some supply contracts contain onerous risk clauses. A typical "all-in" supply rate would be 1.30 p/kw hr for a domestic property. This shows an increase on last year.

MEMORANDA FOR EACH TRADE

Energy costs (July 2001) - cont'd

GAS SUPPLIES – cont'd

Commercial
This sector refers to all other gas supplies. During the last 12 months the price of gas in this sector has continued to increase. For a typical supply consuming 1,000,000 kw/annum, rates approaching 1.2 p/kw are not uncommon.

ELECTRICITY SUPPLIES

In contrast with the significant downtrend last year the market has somewhat levelled out.

Over 100 KVA Supplies
For supplies in this sector of the market there are many options to choose from regarding the charging structure. A typical contract for a supply site with an annual expenditure of £50,000 can expect an "all-in" rate in the region of 4.5 p/kw hr dependent on the load factor. Supplies in this sector require half hourly meters with the associated telephone line in order to collect the half hourly consumption data.

Under 100 KVA Supplies (Non Domestic)
This sector of the market completed its deregulation in 1999. All consumers can purchase their electricity from any authorised supplier, generally a Regional Electricity Company (REC) or Generator, although there are other independent companies in the market place. As suppliers in this market place have established themselves, their pricing structures have matured, many no longer just offering discounts off the local REC's tariff, but offering a pricing structure to meet the consumer's needs. However, the process of changing supplier has not in some cases been the smooth process that was intended with some suppliers experiencing extreme difficulties in managing the transfer process. In extreme cases the industry regulator OFGEM has suspended some suppliers from taking on further business until they, OFGEM, are satisfied the companies in question have the ability to manage the process.

Supply rates achieved vary from region to region, but a typical average rate for a day night supply with an annual expenditure of £500 remains at around 6 p/kw hr.

Domestic Tariff Supplies
Again this sector of the market completed its deregulation process in 1999. Generally the principle is the same as the "Under 100 KVA" market except that the typical discounts are lower, a typical discount being 15% off the host REC tariff.

GENERALLY

For users who are able to group purchase their fuels (eg. schools, health trusts, local authorities, housing associations and any other organisation with multiple supplies) further savings can be achieved. Advice on how to go about this or energy purchasing in general can be obtained from the editor's, Davis Langdon & Everest's, Cambridge office, Tel: 01223 351 258, Fax: 01223 321 002 who have considerable experience in both purchasing energy and the establishment of bulk purchasing schemes.

CLIMATE CHANGE LEVY

This levy, which is a tax on industrial and commercial use of energy is designed to encourage businesses to use less energy and so reduce carbon dioxide emissions. It came into effect on 1st April 2001 and applies to electricity, natural gas, coal, coke and liquid petroleum gas (LPG) but is not levied on standard charges.

The rates for 2001 – 2002 are as follows:

Electricity	0.43 p/kWh
Natural Gas, Coal, Coke	0.15 p/kWh
Liquid Petroleum Gas (LPG)	0.07 p/kWh

which could add 8% - 15% to the energy bills of most businesses. VAT is charged on the levy. Energy supplies are responsible for collecting this levy from customers.

MEMORANDA FOR EACH TRADE

HEATING AND HOT WATER INSTALLATIONS - cont'd

Energy costs (July 2000) - cont'd

CLIMATE CHANGE LEVY – cont'd

National insurance contributions

The Government has reduced the level of employer's National Insurance contributions by the same amount it expects the levy to raise – so, supposedly, there will be no increase in taxation, but the impact is likely to vary company to company, or even sector by sector. The reduction in employers' National Insurance contributions is 0.30%.

Exemptions

Where a taxable commodity (electricity, gas, coal etc.) is used for non energy purpose, e.g. coal is used as a raw material to make carbon filters, the levy is not due. Additionally, where an organisation uses a taxable commodity to produce another taxable commodity, this is also exempt from the levy, e.g. burning gas in a power station to produce electricity. Further, in certain circumstances combined Heat and Power Plants (CHP) are exempt from the levy, and if VAT is paid at the reduced level, i.e. 5% (domestic rate) on any supplies these are not levied.

For further information you may care to access the Customs and Excise website http//www.hmce.gov.uk. A Climate Change Levy Helpdesk also exists on Tel: 0161 827 0332, Fax: 0161 827 0356. Again the Cambridge Office of Davis Langdon & Everest are happy to advise.

MEMORANDA FOR EACH TRADE

VENTILATION AND AIR-CONDITIONING

Typical fresh air supply factors in typical situations

Building type	**Litres of fresh air per second per person**	**Litres of fresh air per second per m² floor area**
General offices	5 - 8	1.30
Board rooms	18 - 25	6.00
Private offices	5 - 12	1.20 - 2.00
Dept. stores	5 - 8	3.00
Factories	20 - 30	0.80
Garages	-	8.00
Bars	12 - 18	-
Dance halls	8 - 12	-
Hotel rooms	8 - 12	1.70
Schools	14	-
Assembly halls	14	-
Drawing offices	16	-

Note: As a global figure for fresh air allow per 1000 m^2 1.20 m^3/second.

Typical air-changes per hour in typical situations

Building type	**Air changes per hour**
Residences	1 - 2
Churches	1 - 2
Storage buildings	1 - 2
Libraries	3 - 4
Book stacks	1 - 2
Banks	5 - 6
Offices	4 - 6
Assembly halls	5 - 10
Laboratories	4 - 6
Internal bathrooms	5 - 6
Laboratories - internal	6 - 8
Restaurants/cafes	10 - 15
Canteens	8 - 12
Small kitchens	20 - 40
Large kitchens	10 - 20
Boiler houses	15 - 30

MEMORANDA FOR EACH TRADE

GLAZING

Float and polished plate glass

Nominal thickness (mm)	Tolerance on thickness (mm)	Approximate weight (kg/m^2)	Normal maximum size (mm)
3	+ 0.2	7.50	2140 x 1220
4	+ 0.2	10.00	2760 x 1220
5	+ 0.2	12.50	3180 x 2100
6	+ 0.2	15.00	4600 x 3180
10	+ 0.3	25.00)	
12	+ 0.3	30.00)	6000 x 3300
15	+ 0.5	37.50	3050 x 3000
19	+ 1.0	47.50)	
25	+ 1.0	63.50)	3000 x 2900

Clear sheet glass

Nominal thickness (mm)	Tolerance on thickness (mm)	Approximate weight (kg/m^2)	Normal maximum size (mm)
2 *	+ 0.2	5.00	1920 x 1220
3	+ 0.3	7.50	2130 x 1320
4	+ 0.3	10.00	2760 x 1220
5 *	+ 0.3	12.50)	
6 *	+ 0.3	15.00)	2130 x 2400

Cast glass

Nominal thickness (mm)	Tolerance on thickness (mm)	Approximate weight (kg/m^2)	Normal maximum size (mm)
3	+ 0.4		
	- 0.2	6.00)	
4	+ 0.5	7.50)	2140 x 1280
5	+ 0.5	9.50	2140 x 1320
6	+ 0.5	11.50)	
10	+ 0.8	21.50)	3700 x 1280

Wired glass

(Cast wired glass)

Nominal thickness (mm)	Tolerance on thickness (mm)	Approximate weight (kg/m^2)	Normal maximum size (mm)
6	+ 0.3	-)	
	- 0.7	)	3700 x 1840
7	+ 0.7	-)	

(Polished wire glass)

Nominal thickness (mm)	Tolerance on thickness (mm)	Approximate weight (kg/m^2)	Normal maximum size (mm)
6	+ 1.0	-	330 x 1830

* The 5 mm and 6 mm thickness are known as *thick drawn sheet*. Although 2 mm sheet glass is available it is not recommended for general glazing purposes.

MEMORANDA FOR EACH TRADE

DRAINAGE

Width required for trenches for various diameters of pipes

Pipe diameter (mm)	**Trench n.e. 1.50 m deep**	**Trench over 1.50 m deep**
n.e. 100 mm	450 mm	600 mm
100 - 150 mm	500 mm	650 mm
150 - 225 mm	600 mm	750 mm
225 - 300 mm	650 mm	800 mm
300 - 400 mm	750 mm	900 mm
400 - 450 mm	900 mm	1050 mm
450 - 600 mm	1100 mm	1300 mm

Weights and dimensions of typically sized uPVC pipes

Nominal size	**Mean outside diameter (mm)**		**Wall thickness**	**Weight kg per metre**
Standard pipes	**min**	**max**		
82.40	82.40	82.70	3.20	1.20
110.00	110.00	110.40	3.20	1.60
160.00	160.00	160.60	4.10	3.00
200.00	200.00	200.60	4.90	4.60
250.00	250.00	250.70	6.10	7.20
Perforated pipes				
Heavy grade as above				
Thin wall				
82.40	82.40	82.70	1.70	-
110.00	110.00	110.40	2.20	-
160.00	160.00	160.60	3.20	-

Vitrified clay pipes

Product	**Nominal diameter (mm)**	**Effective pipe length (mm)**	**Limits of bore load**		**Crushing strength per metre length (kN/m)**	**Weight kg/pipe (/m)**
			min	**max**		
Supersleve	100	1600	96	105	35.00	15.63 (9.77)
Hepsleve	150	1600	146	158	22.00 (normal)	36.50 (22.81)
Hepseal	150	1500	146	158	22.00	37.04 (24.69)
	225	1750	221	235	28.00	95.24 (54.42)
	300	2500	295	313	34.00	196.08 (78.43)
	400	2500	394	414	44.00	357.14 (142.86)

MEMORANDA FOR EACH TRADE

DRAINAGE – cont'd

Vitrified clay pipes – cont'd

Product	Nominal diameter (mm)	Effective pipe length (mm)	Limits of bore load min	max	Crushing strength per metre length (kN/m)	Weight kg/pipe (/m)
Supersleve	100	1600	96	105	35.00	15.63 (9.77)
Hepseal	450	2500	444	464	44.00	500.00 (200.00)
	500	2500	494	514	48.00	555.56 (222.22)
	600	3000	591	615	70.00	847.46 (282.47)
	700	3000	689	719	81.00	1111.11 (370.37)
	800	3000	788	822	86.00	1351.35 (450.35)
	1000	3000	985	1027	120.00	2000.00 (666.67)
Hepline	100	1250	95	107	22.00	15.15 (12.12)
	150	1500	145	160	22.00	32.79 (21.86)
	225	1850	219	239	28.00	74.07 (40.04)
	300	1850	292	317	34.00	105.28 (56.90)
Hepduct	90	1500	-	-	28.00	12.05 (8.03)
(Conduit) 150	100	1600	-	-	28.00	14.29 (8.93)
	125	1250	-	-	22.00	21.28 (17.02)
	150	1250	-	-	22.00	28.57 (22.86)
	225	1850	-	-	28.00	64.52 (34.88)
	300	1850	-	-	34.00	111.11 (60.06)

USEFUL ADDRESSES FOR FURTHER INFORMATION

ACOUSTICAL INVESTIGATION & RESEARCH ORGANISATION LTD (AIRO)
Duxon's Turn
Maylands Avenue
Hemel Hempstead
Hertfordshire
HP2 4SB
Tel: 01442 247 146
Fax: 01442 256 749
E-mail: airo@bcs.org.uk
Web: www.airo.co.uk

ALUMINIUM FEDERATION LTD (ALFED)
Broadway House
Calthorpe Road
Five Ways
Birmingham
West Midlands
B15 1TN
Tel: 0121 456 1103
Fax: 0870 138 9714
E-mail: alfed@alfed.org.uk
Web: www.alfed.org.uk

ALUMINIUM FINISHING ASSOCIATION
Broadway House
Calthorpe Road
Five Ways
Birmingham
West Midlands
B15 1TN
Tel: 0121 456 1103
Fax: 0870 138 9714
E-mail: alfed@alfed.org.uk
Web: www.alfed.org.uk

ALUMINIUM ROLLED PRODUCTS MANUFACTURERS ASSOCIATION
Broadway House
Calthorpe Road
Five Ways
Birmingham
West Midlands
B15 1TN
Tel: 0121 456 1103
Fax: 0870 138 9714
E-mail: info@alfed.org.uk
Web: www.alfed.org.uk

AMERICAN HARDWOOD EXPORT COUNCIL (AHEC)
3 St. Michaels Alley
London
EC3V 9DS
Tel: 0207 626 4111
Fax: 0207 626 4222
E-mail: info@ahec.co.uk
Web: www.ahec-europe.org

ANCIENT MONUMENTS SOCIETY (AMS)
Saint Ann's Vestry Hall
2 Church Entry
London
EC4V 5HB
Tel: 0207 236 3934
Fax: 0207 329 3677
E-mail: office@ancientmonumentssociety.org.uk
Web: www.ancientmonumentssociety.org.uk

APA - THE ENGINEERED WOOD ASSOCIATION
Claridge House
29 Barnes, High Street
London
SW13 9LW
Tel: 0845 123 3721
Fax: 0208 282 1660
E-mail: cdp@apa-europe.org
Web: apa-europe.org

ARCHITECTURAL ADVISORY SERVICE CENTRE (POWDER/ANODIC METAL FINISHES)
Barn One
Barn Road
Conswick
Buckinhamshire
HP27 9RW
Tel: 01844 342 425
Fax: 01844 274 781
E-mail: daveparsons@aasc.org.uk

ARCHITECTURAL ASSOCIATION (AA)
34 - 36 Bedford Square
London
WC1B 3ES
Tel: 0207 887 4000
Fax: 0207 414 0782

ARCHITECTURAL CLADDING ASSOCIATION (ACA)
60 Charles Street
Leicester
Leicestershire
LE1 1FB
Tel: 0116 253 6161
Fax: 0116 251 4568
Email: aca@britishprecast.org
Web: www.britishprecast.org/aca

ASBESTOS INFORMATION CENTRE (AIC)
PO Box 69
Widnes
Cheshire
WA8 9GW
Tel: 0151 420 5866

USEFUL ADDRESSES FOR FURTHER INFORMATION

ASBESTOS REMOVAL CONTRACTORS' ASSOCIATION (ARCA)
Friars House
6 Parkway
Chelmsford,
Essex
CM2 ONF
Tel: 01245 259 744
Fax: 01245 490 722

ASSOCIATION OF INTERIOR SPECIALISTS
Olton Bridge
245 Warwick Road
Solihull
West Midlands
B92 7AH
Tel: 0121 707 0077
Fax: 0121 706 1949
E-mail: info@ais-interiors.org.uk
Web: www.ais-interiors.org.uk

BOX CULVERT ASSOCIATION (BCA)
60 Charles Street
Leicester
Leicestershire
LE1 1FB
Tel: 0116 253 6161
Fax: 0116 251 4568
E-mail: cjb@britishprecast.org
Web: www.britishprecast.org

BRITISH ADHESIVES & SEALANTS ASSOCIATION(BASA)
33 Fellowes Way
Stevenage
Hertfordshire
SG2 8BW
Tel: 01438 358 514

BRITISH AGGREGATE CONSTRUCTION MATERIALS INDUSTRIES LTD (BACMI)
156 Buckingham Palace Road
London
SW1W 9TR
Tel: 0207 730 8194

BRITISH APPROVALS FOR FIRE EQUIPMENT (BAFE)
Thames House
29 Thames Sreet
Kingston upon Thames
Surrey
KT1 1PH
Tel: 0208 541 1950
Fax: 0208 547 1564
E-mail: bafe@abft.org.uk
Web: www.bafe.org.uk

BRITISH APPROVALS SERVICE FOR CABLES (BASEC)
23 Presley Way
Crownhill
Milton Keynes
Buckinghamshire
MK8 0ES
Tel: 01908 267 300
Fax: 01908 267 255
E-mail: mail@basec.org.uk
Web: www.basec.org.uk

BRITISH ARCHITECTURAL LIBRARY (BAL)
Royal Institute of British Architects
66 Portland Place
London
W1N 4AD
Tel: 0906 302 0400
Fax: 0207 307 3812

BITISH ASSOCIATION OF LANDSCAPE INDUSTRIES (BALI)
Landscape House
National Agricultural Centre
Stoneleigh Park
Warwickshire
CV8 2LG
Tel: 0870 770 4971
Fax: 0870 770 4972
E-mail: contact@bali.co.uk
Web: www.bali.co.uk

BRITISH BATHROOM COUNCIL (BATHROOM MANUFACTURERS ASSOCIATION)
Federation House
Station Road
Stoke-on-Trent
Staffordshire
ST4 2RT
Tel: 01782 747 123
Fax: 01782 747 161
E-mail: info@bathroom-association.org.uk
Web: www.bathroom-assciation.org

BRITISH BOARD OF AGREMENT (BBA)
PO Box 195
Bucknalls Lane
Garston
Watford
Hertfordshire
WD25 9BA
Tel: 01923 665 300
Fax: 01923 665 301
E-mail: info.bba.stal.co.uk
Web: www.bbacerts.co.uk

USEFUL ADDRESSES FOR FURTHER INFORMATION

BRITISH CABLES ASSOCIATION (BCA)
37a Walton Road
East Molesey
Surrey
KT8 0DH
Tel: 0208 941 4079
Fax: 0208 783 0104
E-mail: admin@bcauk.org
Web: www.bcauk.org

BRITISH CARPET MANUFACTURERS ASSOCIATION LTD (BCMA)
PO Box 1155
MCF Complex
60 New Road
Kidderminster
Worcestershire
DY10 1AQ
Tel: 01562 755 568
Fax: 01562 865 4055
E-mail: info@carpetfederation.com
Web: www.carpetfoundation.com

BRITISH CEMENT ASSOCIATION (BCA), CENTRE FOR CONCRETE INFORMATION
Century House
Telford Avenue
Crowthorne
Berkshire
RG45 6YS
Tel: 01344 762 676
Fax: 01344 761 214

BRITISH CERAMIC CONFEDERATION (BCC)
Federation House
Station Road
Stoke-on-Trent
Staffordshire
ST4 2SA
Tel: 01782 744 631
Fax: 01782 744 102
E-mail: bcc@ceramfed.co.uk
Web: www.ceramfed.co.uk

BRITISH CERAMIC RESEARCH LTD (BCR)
Queens Road
Penkhull
Stoke-on-Trent
Staffordshire
ST4 7LQ
Tel: 01782 845 431
Fax: 01782 412 331
Web: www.ceram.co.uk

BRITISH CERAMIC TILE COUNCIL (BCTC TILE ASSOCIATION)
Federation house
Station Road
Stoke On Trent
ST4 2RT
Tel: 01782 747 147
Fax: 01782 747 161
E-mail: tiles@netcentral.co.uk

BRITISH COMBUSTION EQUIPMENT MANUFACTURERS ASSOCIATION (BCEMA)
58 London Road
Leicester
LE2 0QD
Tel: 0116 275 7111
Fax: 0116 275 7222
E-mail: bcema@btconnect.com
Web: bcema.co.uk

BRITISH CONCRETE MASONRY ASSOCIATION (BCMA)
Grove Crescent House
18 Grove Place
Bedford
MK40 3JJ
Tel/fax: 01234 353 745

BRITISH CONSTRUCTIONAL STEELWORK ASSOCIATION LTD (BCSA)
4 Whitehall Court
Westminster
London
SW1A 2ES
Tel: 0207 839 8566
Fax: 0207 976 1634
Web: www.steelconstruction.org

BRITISH CONTRACT FURNISHING ASSOCIATION (BCFA)
Suite 2/4
The Business Design Centre
52 Upper Street
Islington Green
London
N1 0QH
Tel: 0207 226 6641
Fax: 0207 288 6190

BRITISH DECORATORS ASSOCIATION (BDA)
32 Coton Road
Nuneaton
Warwickshire
CV11 5TW
Tel: 01203 353 776
Fax: 01203 354 4513

USEFUL ADDRESSES FOR FURTHER INFORMATION

BRITISH ELECTROTECHNICAL APPROVALS BOARD (BEAB)
1 Station View
Guildford
Surrey
GU1 4JY
Tel: 01483 455 466
Fax: 01483 455 477
E-mail: info@beab.com
Web: www.beab.co.uk

BRITISH FIRE PROTECTION SYSTEMS ASSOCIATION LTD (BFPSA)
Thames House
29 Thames Street
Kingston-upon-Thames
Surrey
KT1 1PH
Tel: 0208 549 5855
Fax: 0208 547 1564
Web: www.bfpsa.org.uk

BRITISH FURNITURE MANUFACTURERS FEDERATION
LTD (BFM Ltd)
30 Harcourt Street
London
W1H 2AA
Tel: 0207 724 0851
Fax: 0207 723 0622

BRITISH GEOLOGICAL SURVEY (BGS)
Keyworth Headquarters
Kingsley Drive
Dunham Centre
Nottingham
Nottinghamshire
NG12 5GG
Tel: 0115 936 3100
Fax: 0115 936 3200
E-mail: enquiries@bgs.ac.ukc.uk
Web: www.thebgs.co.uk

BRITISH INSTITUTE OF ARCHITECTURAL TECHNOLOGISTS (BIAT)
397 City Road
London
EC1V 1NH
Tel: 0207 278 2206
Fax: 0207 837 3194
Web: www.biat.org.uk

BRITISH LAMINATED FABRICATORS ASSOCIATION
6 Bath Place
Rivington Street
London
EC2A 3JE
Tel: 0870 444 1500
Fax: 0207 457 5000
Web: www.bpf.co.uk

BRITISH LIBRARY BIBLIOGRAPHIC SERVICE AND DOCUMENT SUPPLY
Boston Spa
Wetherby
West Yorkshire
LS23 7BQ
Tel: 0870 444 1500
Fax: 0207 457 5000
E-mail: nbs-info@bl.uk
Web: www.bl.uk

BRITISH LIBRARY ENVIRONMENTAL INFORMATION SERVICE
96 Euston Road
London
NW1 2DB
Tel: 0870 444 1500
Fax: 0207 457 5000
E-mail: stms@bl.uk
Web: www.bl.uk/environment

BRITISH NON-FERROUS METALS FEDERATION
Broadway House
60 Calthorpe Road
Edgbaston
Birmingham
West Midlands
B15 1TN
Tel: 0121 456 6110
Fax: 0121 456 2274
Email: copperuk@compuserve.com

BRITISH PLASTICS FEDERATION (BPF)
Plastics & Rubber Advisory Service
6 Bath Place
Rivington Street
London
EC2A 3JE
Tel: 0207 457 5000
Fax: 0207 457 5045

USEFUL ADDRESSES FOR FURTHER INFORMATION

BRITISH PRECAST CONCRETE FEDERATION LTD
60 Charles Street
Leicester
Leicestershire
LE1 1FB
Tel: 0116 253 6161
Fax: 0116 251 4568
E-mail: birpre@aol.com
Web: www.britishprecast.org.uk

BRITISH PROPERTY FEDERATION (BPF)
35 Catherine Place
London
SW1E 6DY
Tel: 0207 828 0111
Fax: 0207 824 3442

BRITISH RUBBER MANUFACTURERS' ASSOCIATION LTD (BRMA)
6 Bath Place
Rivington Street
London
EC2A 3JE
Tel: 0207 457 5040
Fax: 0207 972 9008
E-mail: mail@brma.co.uk
Web: www.brma.co.uk

BRITISH STANDARDS INSTITUTION (BSI)
389 Chiswick High Road
London
W4 4AL
Tel: 0208 996 9000
Fax: 0208 996 7400

BRITISH WATER
1 Queen Anne's Gate
London
SW1H 9BT
Tel: 0207 957 4554
Fax: 0207 957 4565
E-mail: info@britishwater.co.uk
Web: www.britishwater.co.uk

BRITISH WOOD PRESERVING & DAMP PROOFING ASSOCIATION (BWPDA)
6 Office Village
Romford Road
London
E15 4ED
Tel: 0208 519 2588
Fax: 0208 519 3444

BRITISH WOODWORKING FEDERATION
55 Tufton Street
London
SW1 3QL
Tel: 0870 458 6939
Fax: 0870 458 6949
Email: bwf@bwf.org.uk
Web: www.bwf.org.uk

BUILDING CENTRE
The Building Centre
26 Store Street
London
WC1E 7BT
Tel: 0207 692 4000
Fax: 0207 580 9641
E-mail: information@buildingcentre.co.uk
Web: www.buildingcentre.co.uk

BUILDING COST INFORMATION SERVICE LTD (BCIS)
Royal Institution of Chartered Surveyors
3 Cadogan Gate
London
SW1X 0AS
Tel: 0207 695 1500
Fax: 0207 695 1501
E-mail: bcis@bcis.co.uk
Web: www.bcis.co.uk

BUILDING EMPLOYERS CONFEDERATION (BEC)
55 Tufton Street
Westminster
London
SW1P 3QL
Tel: 0870 89 89 090
Fax: 0870 89 89 095
E-mail: enquiries@thecc.org.uk
Web: www.thecc.org.uk

BUILDING MAINTENANCE INFORMATION (BMI)
Royal Institution of Chartered Surveyors
3 Cadogan Gate
London
SW1X OAS
Tel: 0207 695 1500
Fax: 0207 695 1501
E-mail: acowan@bcis.co.uk
Web: www.bcis.co.uk

USEFUL ADDRESSES FOR FURTHER INFORMATION

BUILDING RESEARCH ESTABLISHMENT (BRE)
BRE Garston
Watford
WD5 9XX
Tel: 01923 664 000
E-mail: enquiries@bre.co.uk
Web: www.bre.co.uk

BUILDING RESEARCH ESTABLISHMENT: SCOTLAND (BRE)
Kelvin Road
East Kilbride
Glasgow
G75 0RZ
Tel: 01355 233 001
Fax: 01355 241 895
Web: www.bre.co.uk

BUILDING SERVICES RESEARCH AND INFORMATION ASSOCIATION Ltd
Old Bracknell Lane West
Bracknell
Berkshire
RG12 7AH
Tel: 01344 465 600
Fax: 01344 465 626
E-mail: bsria@bsria.co.uk
Web: www.bsria.co.uk

BUILT ENVIRONMMENT RESEARCH GROUP
c/o Disabled Living Foundation
380 - 384 Harrow Road
London
W9 2HU
Tel: 0207 289 6111
Fax: 0207 273 4340
Email: advise@dlf.org.uk
Web: www.dlf.org.uk

CATERING EQUIPMENT MANUFACTURERS ASSOCIATION (CEMA)
Carlyle House
235 Vauxhall Bridge Road
London
SW1V 1EJ
Tel: 0207 233 7724
Fax: 0207 828 0667

CHARTERED INSTITUTE OF BUILDING (CIOB)
Englemere
Kings Ride
Ascot
Berkshire
SL5 8BJ
Tel: 01344 630 700
Fax: 01344 630 777

CHARTERED INSTITUTION OF BUILDING SERVICES ENGINEERS (CIBSE)
Delta House
222 Balham High Road
London
SW12 9BS
Tel: 0208 675 5211
Fax: 0208 675 5449
Web: www.cibse.org

CLAY PIPE DEVELOPMENT ASSOCIATION (CPDA)
Treetops House
Billingdon
Chesham
HP5 2XL
Tel: 01494 791 456
Fax: 01494 792 378
E-mail: cpda@aol.com

CLAY ROOF TILE COUNCIL
Federation House
Station Road
Stoke-on-Trent
Staffordshire
ST4 2SA
Tel: 01782 744 631
Fax: 01782 744 102
E-mail: crtc@ceramfed.co.uk

COLD ROLLED SECTIONS ASSOCIATION (CRSA)
National Metal Forming Centre
47 Birmingham Road
West Bromwich
West Midlands
B70 6PY
Tel: 0121 601 6350
Fax: 0121 601 6373
E-mail: crsa@crsauk.com
Web: www.crsauk.com

COMMONWEALTH ASSOCIATION OF ARCHITECTS (CAA)
PO BOX 508
Edgware
HA8 9XZ
Tel: 44 20895 10550
E-mail: inof@comarchitect.org
Web: www.comarchitect.org

USEFUL ADDRESSES FOR FURTHER INFORMATION

CONCRETE BRIDGE DEVELOPMENT GROUP
Riverside House
4 Meadows Business Park
Station Approach
Blackwater, Camberley
Surrey
GU17 9AB
Tel: 01276 607 140
Fax: 01276 607 141
Web: www.cconcrete.org.uk

CONCRETE LINTEL ASSOCIATION
60 Charles Street
Leicester
Leicestershire
LE1 1FB
Tel: 0116 253 6161
Fax: 0116 251 4568
E-mail: birpre@aol.com
Web: www.britishprecast.org.uk

CONCRETE PIPE ASSOCIATION (CPA)
60 Charles Street
Leicester
Leicestershire
LE1 1FB
Tel: 0116 253 6161
Fax: 0116 251 4568
E-mail: birpre@aol.com
Web: www.britishprecast.org.uk

CONCRETE REPAIR ASSOCIATION (CRA)
Association House
235 Ash Road
Aldershot
Hampshire
GU12 4DD
Tel: 01252 321 302
Fax: 01252 333 901
E-mail: info@associationhouse.org.uk
Web: www.concreterepair.org.uk

CONCRETE SOCIETY ADVISORY SERVICE
Riverside House
4 Meadows Business Park
Station Approach
Blackwater, Camberley
Surrey
GU17 9AB
Tel: 01276 607 140
Fax: 01276 607 141
Web: www.concrete.org.uk

CONCRETE SOCIETY Ltd
3 Eatongateuse
112 Windsor Road
Slough
Berks
SL1 2JA
Tel: 01753 693 313
Fax: 01753 692 333

CONFEDERATION OF BRITISH INDUSTRY (CBI)
Centre Point
103 New Oxford Street
London
WC1A 1DU
Tel: 0207 379 7400
Fax: 0207 240 1578

CONSTRUCT - CONCRETE STRUCTURES GROUP LTD
Riverside House
4 Meadows Business Park
Station Approach
Blackwater, Camberley
Surrey
GU17 9AB
Tel: 01276 38444
Fax: 01276 38899
E-mail: enquiries@construct.org.uk
Web: www.construct.org.uk

CONSTRUCTION EMPLOYERS FEDERATION LTD (CEF)
143 Malone Road
Belfast
Northern Ireland
BT9 6SU
Tel: 01232 877 143
Fax: 01232 877 155

CONSTRUCTION INDUSTRY RESEARCH & INFORMATION ASSOCIATION (CIRIA)
Classic House
174-180 Old Street
London
EC1V 9BP
Tel: 0207 549 3300
Fax: 0207 253 0523
E-mail: enquiries@ciria.org.uk
Web: www.ciria.org.uk

CONSTRUCTION PLANT-HIRE ASSOCIATION (CPA)
28 Eccleston Street
London
SW1P 3AU
Tel: 0207 730 7117

USEFUL ADDRESSES FOR FURTHER INFORMATION

CONTRACT FLOORING ASSOCIATION (CFA)
4c Saint Mary's Place
The Lace Market
Nottingham
Nottinghamshire
NG1 1PH
Tel: 0115 941 1126
Fax: 0115 941 2238
E-mail: info@cfa.org.uk
Web: www.cfa.org.uk

CONTRACTORS MECHANICAL PLANT ENGINEERS (CMPE)
3 Hillview
Hornbeam
Waterlooville
Hampshire
PO8 9EY
Tel: 023 9236 5829
Fax: 023 9236 5829

COPPER DEVELOPMENT ASSOCIATION
Verulam Industrial Estate
224 London Road
Saint Albans
Hertfordshire
AL1 1AQ
Tel: 01727 731 205
Fax: 01727 731 216
E-mail: copperdev@compuserve.com
Web: www.cda.org.uk

CORUS RESEARCH DEVELOPMENT AND TECHNOLOGY
Swinden Technology Centre
Moorgate
Rotherham
South Yorkshire
S60 3AR
Tel: 01709 820 166
Fax: 01709 825 337

COUNCIL FOR ALUMINIUM IN BUILDING (CAB)
191 Cirencester Road
Charlton Kings
Cheltenham
Gloucestershire
GL53 8DF
Tel: 01242 578 278
Fax: 01242 578 283

DRY STONE WALLING ASSOCIATION OF GREAT BRITAIN (DSWA)
Westmorland County Showground
Lane Fram
Crooklands, Milnthorpe
Cumbria
LA7 7NH
Tel: 01539 567 953
E-mail: information@dswa.org.uk
Web: www.dswa.org.uk

ELECTRICAL CONTRACTORS ASSOCIATION (ECA)
ESCA House
34 Palace Court
Bayswater
London
W2 4HY
Tel: 0207 313 4800
Fax: 0207 221 7344
E-mail: electricalcontractors@eca.co.uk
Web: www.eca.uk

ELECTRICAL CONTRACTORS ASSOCIATION OF SCOTLAND (SELECT)
The walled Gardens
Bush Estate
Midlothian
Scotland
EH26 0SB
Tel: 0131 445 5577
Fax: 0131 445 5548
E-mail: admin@select.org.uk
Web: www.select.org.uk

ELECTRICAL INSTALLATION EQUIPMENT MANUFACTURERS ASSOCIATION LTD (EIEMA)
Beama Installation Ltd
Westminster Tower
3 Albert Embankment
London
SE1 7SL
Tel: 0207 793 3013
Fax: 0207 793 3003
E-mail: cac@beama.org.uk
Web: www.eiema.org.uk

EUROPEAN LIQUID ROOFING ASSOCIATION (ELRA)
Fields House
Gower Road
Haywards Heath
West Sussex
RH16 4PL
Tel: 01444 417 458
Fax: 01444 415 616

USEFUL ADDRESSES FOR FURTHER INFORMATION

FACULTY OF BUILDING
Central Office
35 Hayworth Road
Sandiacre
Nottingham
Nottinghamshire
NG10 5LL
Tel: 0115 949 0641
Fax: 0115 949 1664
E-mail: mail@faculty-of-building.co.uk
Web: www.faculty-of-building.co.uk

FEDERATION OF MANUFACTURERS OF CONSTRUCTION EQUIPMENT & CRANES
Ambassador House
Brigstock Road
Thornton Heath
Surrey
CR7 7JG
Tel: 0208 665 5727
Fax: 0208 665 6447
E-mail: cea@admin.co.uk
Web: www.coneq.org.uk

FEDERATION OF MASTER BUILDERS
Gordon Fisher House
14 - 15 Great James Street
London
WC1N 3DP
Tel: 0207 242 7583
Fax: 0207 404 0296

FEDERATION OF PILING SPECIALISTS
Forum Court
83 Coppers Cope Road
Beckenham
Kent
BR3 1NR
Tel: 0208 663 0947
Fax: 0208 663 0949
E-mail: fps@fps.org.ik
Web: www.fps.org.uk

FEDERATION OF PLASTERING & DRYWALL CONTRACTORS
Construction House
56 - 64 Leonard Street
London
EC2A 4JX
Tel: 0207 608 5092
Fax: 0207 608 5081
Web: www.fpdc.org

FENCING CONTRACTORS ASSOCIATION
Warren Road
Trellech
Monmouthshire
NP5 4PQ
Tel: 07000 560 722
Fax: 01600 860 614

FINNISH PLYWOOD INTERNATIONAL
PO BOX 99
Welwyn Garden City
Herts
AL6 0HS
Tel: 01438 798 746
Fax: 01438 798 305

FLAT ROOFING ALLIANCE
Fields House
Gower Road
Haywards Heath
West Sussex
RH16 4PL
Tel: 01444 440 027
Fax: 01444 415 616

FURNITURE INDUSTRY RESEARCH ASSOCIATION (FIRA INTERNATIONAL LTD)
Maxwell Road
Stevenage
Hertfordshire
SG1 2EW
Tel: 01438 777 700
Fax: 01438 777 800
E-mail: info@fira.co.uk
Web: www.fira.co.uk

GLASS & GLAZING FEDERATION (GGF)
44 - 48 Borough High Street
London
SE1 1XB
Tel: 0870 024 4255
Fax: 0870 024 4266
E-mail: info@ggf.org.uk
Web: www.ggf.org.uk

HEATING & VENTILATING CONTRACTORS' ASSOCIATION
ESCA House
34 Palace Court
Bayswater
London
W2 4JG
Tel: 0207 313 4900
Fax: 0207 727 9268

USEFUL ADDRESSES FOR FURTHER INFORMATION

HOUSING CORPORATION HEADQUARTERS
Maple House
149 Tottenham Court Road
London
W1N 7BN
Tel: 0845 230 7000
Fax: 0207 393 2111
E-mail: enquiries@housingcorp.gsx.gov.uk
Web: www.housingcorp.gov.uk

INSTITUTE OF ACOUSTICS
77A Saint Peter' Street
Saint Albans
Hertfordshire
AL1 3BN
Tel: 01727 848 195
Fax: 01727 850 553
E-mail: ioa@ioa.org.uk
Web: www.ioa.org.uk

INSTITUTE OF ASPHALT TECHNOLOGY
Paper Mews Place
290 High Street
Dorking
Surrey
RH4 1QT
Tel: 01306 742792
Fax: 01306 888902
Web: www.instofasphalt.org
Email: secretary@instoasphalt.demon.co.uk

INSTITUTE OF MAINTENANCE AND BUILDING MANAGEMENT
Keets House
30 East Street
Farnham
Surrey
GU9 7SW
Tel: 01252 710 994
Fax: 01252 737 741
Email: imbm.@btconnect.com
Web: www.imbm.org.uk

INSTITUTE OF MATERIALS
Headquarters
1 Carlton House Terrace
London
SW1Y 5DB
Tel: 0207 451 7300
Fax: 0207 839 1702
Web: www.materials.org.uk

INSTITUTE OF PLUMBING
64 Station Lane
Hornchurch
Essex
RM12 6NB
Tel: 01708 472 791
Fax: 01708 448 987
E-mail: info@iphe.org.ukk
Web: www.plumbers.org.uk

INSTITUTE OF WASTES MANAGEMENT
9 Saxon Court
St Peter's Gardens
Northampton
NN1 1SX
Tel: 01604 620 426
Fax: 01604 621 339
E-mail: technical@iwm.co.uk
Web: www.iwm.co.uk

INSTITUTE OF WOOD SCIENCE
Stocking Lane
Hughenden Valley
High Wycombe
Buckinghamshire
HP14 4NU
Tel: 01494 565 374
Fax: 01494 565 395
E-mail: info@iwsc.org.uk
Web: www.iwsc.org.uk

INSTITUTION OF CIVIL ENGINEERS (ICE)
1 Great George Street
London
SW1P 3AA
Tel: 0207 222 7722
Fax: 0207 222 7500

INSTITUTION OF INCORPORATED ENGINEERS
Savoy Hill House
Savoy Hill
London
WC2R 0BS
Tel: 0207 836 3357
Fax: 0207 497 9006

INSTITUTION OF STRUCTURAL ENGINEERS (ISE)
11 Upper Belgrave Street
London
SW1X 8BH
Tel: 0207 235 4535
Fax: 0207 235 4294

USEFUL ADDRESSES FOR FURTHER INFORMATION

INTERPAVE (THE PRECAST CONCRETE PAVING & KERB ASSOCIATION)
60 Charles Street
Leicester
Leicestershire
LE1 1FB
Tel: 0116 253 6161
Fax: 0116 251 4568

JOINT CONTRACTS TRIBUNAL LTD
9 Cavendish Place
London
W1G 0GD
Tel: 0207 630 8650
Fax: 0207630 8670
Email: spanform@jctltd.co.uk
Web: www.jctltd.co.uk

KITCHEN SPECIALISTS ASSOCIATION
12 TopBarn Business Centre
Holt Heath
Worcester
Worcestershire
WR6 6NH
Tel: 01905 621 787
Fax: 01905 621 887
Web: www.kbsa.co.uk

LIGHTING ASSOCIATION LTD
Stafford Park 7
Telford
Shropshire
TF3 3BQ
Tel: 01952 290 905
Fax: 01952 290 906
E-mail: enquiries@lightingasscociation.com
Web: http://www.lightingassociation.com/

MASTIC ASPHALT COUNCIL LTD
PO BOX e House
Hastings
Kent
TN35 4WL
Tel: 01424 814 400
Fax: 01424 814 446

METAL CLADDING & ROOFING MANUFACTURERS ASSOCIATION
18 Mere Farm Road
Prenton
Birkenhead
Merseyside
CH43 9TT
Tel: 0151 652 3846
Fax: 0151 653 4080
E-mail: mcrma@compuserve.com
Web: www.mcrma.co.uk

NATIONAL HOUSE-BUILDING COUNCIL (NHBC)
Ash house
Linford Wood
Milton Keynes
MK14 6ET
Tel: 01494 434 477
Fax: 01494 728 521
E-mail: sts@nhbc.co.uk
Web: www.nhbc.co.uk

NATURAL SLATE QUARRIES ASSOCIATION
26 Store Street
London
WC1E 7BT
Tel: 0207 323 3770
Fax: 0207 323 0307
Email: enquiries@constprod.org.uk
Web: www.constprod.org.uk

NHS ESTATES
Departments of Health
1 Trevelyan Square
Boar Lane
Leeds
West Yorkshire
LS1 6AE
Tel: 0113 254 7000
Fax: 0113 254 7299

ORDNANCE SURVEY
Romsey Road
Southampton
SO16 4GU
Tel: 08456 050 504
Fax: 02380 792 615
E-mail: custiomerservices@ordnancesurvey.co.uk
Web: www.ordnancesurvey.co.uk

PIPELINE INDUSTRIES GUILD
14 - 15 Belgrave Square
London
SW1X 8PS
Tel: 0207 235 7938
Fax: 0207 235 0074
E-mail: hqsec@pipeguild.co.uk
Web: www.pipeguild.co.uk

PLASTIC PIPE MANUFACTURERS SOCIETY
89 Cornwall Street
Birmingham
West Midlands
B3 3BY
Tel: 0121 236 1866
Fax: 0121 262 3370

USEFUL ADDRESSES FOR FURTHER INFORMATION

PRECAST FLOORING FEDERATION
60 Charles Street
Leicester
Leicestershire
LE1 1FB
Tel: 0116 253 6161
Fax: 0116 251 4568
E-mail: info@pff.org.uk
Web: www.pff.org.uk

PRESTRESSED CONCRETE ASSOCIATION
60 Charles Street
Leicester
Leicestershire
LE1 1FB
Tel: 0116 253 6161
Fax: 0116 251 4568
E-mail: pca@britishprecast.org
Web: www.britishprecast.org

PROPERTY CONSULTANTS SOCIETY LTD
107a Tarrant Street
Arundel
West Sussex
BN18 9DP
Tel: 01903 883 787
Fax: 01903 889 590
E-mail: pcs@propco.freeserve.co.uk

QUARRY PRODUCTS ASSOCIATION
Gillingham House
38-44 Gillingham Street
London
SW1 V1HU
Tel: 0207 963 8000
Fax: 0207 963 8001
E-mail: info@qpa.org
Web: www.qpa.org

READY-MIXED CONCRETE BUREAU
Century House
Telford Avenue
Crowthorne
Berkshire
RG45 6YS
Tel: 01344 725 732
Fax: 01344 774 976
E-mail: speed@qpa.org
Web: www.rcb.org.uk

REINFORCED CONCRETE COUNCIL
Riverside House
4 Meadows Business Park
Station Approach
Camberley
Berkshire
GU17 9AB
Tel: 01276 607140
Fax: 01276 607141
Web: www.rcc-info.org.uk
Email: rcc@bca.org.uk

ROYAL INCORPORATION OF ARCHITECTS IN SCOTLAND (RIAS)
15 Rutland Square
Edinburgh
Scotland
EH1 2BE
Tel: 0131 229 7545
Fax: 0131 228 2188
E-mail: info@rias.org.uk
Web: www.rias.org.uk

ROYAL INSTITUTE OF BRITISH ARCHITECTS (RIBA)
66 Portland Place
London
W1N 4AD
Tel: 0207 580 5533
Fax: 0207 255 1541

ROYAL INSTITUTION OF CHARTERED SURVEYORS (RICS)
12 Great George Street
Parliament Square
London
SW1P 3AD
Tel: 0207 222 7000
Fax: 0207 222 9430
Web: www.rics.org.uk

ROYAL TOWN PLANNING INSTITUTE (RTPI)
26 Portland Place
London
W1N 4BEL
Tel: 0207 636 9107
Fax: 0207 323 1582
E-mail: online@rtpi.org.uk

USEFUL ADDRESSES FOR FURTHER INFORMATION

RURAL DESIGN AND BUILDING ASSOCIATION
ATSS House
Station Road East
Stowmarket
Suffolk
IP12 41RQ
Tel: 01449 676 049
Fax: 01449 770 028
E-mail: secretary@rdba.org.uk
Web: www.rdba.org.uk

SCOTTISH BUILDING EMPLOYERS FEDERATION
Carron Grange
Carron Grange Avenue
Stenhousemuir
Scotland
FK5 3BQ
Tel: 01324 555 550
Fax: 01324 555 551
E-mail: info@scottish-building.co.uk
Web: www.scottish-building.co.uk

SCOTTISH HOMES- Community Scotland
Thistle House
91 Haymarket Terrace
Edinburgh
Scotland
EH12 5HE
Tel: 0131 313 0044
Fax: 0131 313 2680
Web: www.communitiescotland.gsi.gov.uk

SCOTTISH NATURAL HERITAGE
Communications Directorate
12 Hope Terrace
Edinburgh
EH9 2AS
Tel: 0131 447 4784
Fax: 0131 446 2277
Web: www.snh.org.uk

SINGLE PLY ROOFING ASSOCIATION
177 Bagnall Road
Basford
Nottinghamshire
NG6 8SJ
Tel: 0115 703 332

SMOKE CONTROL ASSOCIATION
2 Waltham Court
Milley Lane, Hare Hatch
Reading
Berkshire
RG10 9TH
Tel: 0118 940 3416
Fax: 0118 940 6258
Web: www.feta.co.uk
Email: info@feta.co.uk

SOCIETY FOR THE PROTECTION OF ANCIENT BUILDINGS (SPAB)
37 Spital Square
London E1 6DY
Tel: 0207 377 1644
Fax: 0207 247 5296
E-mail: info@spab.org.uk
Web: www.spab.org.uk

SOCIETY OF GLASS TECHNOLOGY
Don Valley House
Saville Street East
Sheffield
South Yorkshire
S4 7UQ
Tel: 0114 263 4455
Fax: 0114 263 4411
E-mail: info@sgt.org
Web: www.sgt.org

SOIL SURVEY AND LAND RESEARCH INSTITUTE
Cranfield University
Silsoe Campus
Bedford
Bedfordshire
MK45 4DT
Tel: 01525 863 000
Fax: 01525 863 253
E-mail: nsri@cranfield.ac.uk
Web: www.cranfield.ac.uk/sslrc

SOLAR ENERGY SOCIETY
c/o School of Engineering
Oxford Brookes University
Gipsy Lane
Headington, Oxford
OX3 0BP

SPON'S A&B PRICE BOOK EDITORS
David Wood
'Woodlands'
Shrubbery Avenue
Weston-Super-Mare
North Somerset
BS23 2JT
Tel: 01934 643171
Jay Kotecha, David Holmes and Max Wilkes (Davis Langdon)

SPORT ENGLAND
3rd Floor, Victoria House
Bloomsbury Square
London
WC1B 4SE
Tel: 0845 850 8508
Fax: 020 7383 5740
Web: www.sportengland.org
Email: info@sportengland.org

USEFUL ADDRESSES FOR FURTHER INFORMATION

SPORT SCOTLAND
Caledonia House
South Gyle
Edinburgh
Scotland
EH12 9DQ
Tel: 0131 317 7200
Fax: 0131 317 7202
E-mail: info@sportscotland.org.uk
Web: www.sportscotland.org.uk

SPORTS COUNCIL FOR WALES
Welsh Institute of Sport
Sophia Gardens
Cardiff
CF11 9SW
Tel: 02920 300 500
Fax: 02920 300 600
Web: www.sports-council-wales.co.uk
Email: scw@scw.co.uk

SPORTS TURF RESEARCH INSTITUTE (STRI)
Saint Ives Estate
Bingley
West Yorkshire
BD16 1AU
Tel: 01274 565 131
Fax: 01274 561 891
E-mail: info@stri.co.uk
Web: www.stri.co.uk

SPRAYED CONCRETE ASSOCIATION
Association House
235 Ash Road
Aldershot
Hampshire
GU12 4DD
Tel: 01252 321 302
Fax: 01252 333 901

STAINLESS STEEL ADVISORY SERVICE
Room 2.04
The Innovation Centre
217 Portabella
Sheffield
South Yorkshire
S1 4DP
Tel: 0114 224 2240
Fax: 0114 273 0444
Email: ssas@materials.org.uk

STEEL CONSTRUCTION INSTITUTE
Silwood Park
Ascot
Berkshire
SL5 7QN
Tel: 01344 623 345
Fax: 01344 622 944
E-mail: reception@steel-sci.com
Web: www.steel-sci.org

STEEL WINDOW ASSOCIATION
The Building Centre
26 Store Street
London
WC1E 7BT
Tel: 0207 637 3571
Fax: 0207 637 3572
E-mail: info@steel-window-association.co.uk
Web: www.steel-window-association.co.uk

STONE FEDERATION GREAT BRITAIN
18 Mansfield Street
London
W1M 9FGX
Tel: 0207 580 5404
Fax: 0207 636 5984

SUSPENDED ACCESS EQUIPMENT
MANUFACTURERS ASSOCIATION
18 Mansfield Street
London
W1M 9FGX
Tel: 0207 580 5404
Fax: 0207 636 5984

SWIMMING POOL & ALLIED TRADES ASSOCIATION
(SPATA)
Spata House
1a Junction Road
Andover
Hampshire
SP10 3QT
Tel: 01264 356210
Fax: 01264 332628
Web: www.spata.co.uk
Email: admin@spata.co.uk

USEFUL ADDRESSES FOR FURTHER INFORMATION

THERMAL INSULATION CONTRACTORS ASSOCIATION
Tica House
Allington Way
Yarm Road Business Park
Darlington
County Durham
DL1 4QB
Tel: 01325 466 704
Fax: 01325 487 691
E-mail: enquiries@tica-acad.co.uk
Web: www.tica-acad.co.uk

TIMBER RESEARCH & DEVELOPMENT ASSOCIATION (TRADA)
Stocking Lane
Hughenden Valley
High Wycombe
Buckinghamshire
HP14 4ND
Tel: 01494 569 600
Fax: 01494 565 487
E-mail: information@trada.co.uk
Web: www.trada.co.uk

TIMBER TRADE FEDERATION
4th Floor
Clareville House
26-27 Oxenden Street
London
SW1Y 4EL
Tel: 0207 839 1891
Fax: 0207 930 0094
E-mail: ttf@ttf.co.uk
Web: www.ttf.co.uk

TOWN & COUNTRY PLANNING ASSOCIATION (TCPA)
17 Carlton House Terrace
London
SW1Y 5AS
Tel: 0207 930 8903
Fax: 0207 930 3280
E-mail: tcpa@tcpa.org.uk
Web: tcpa.org.uk

TREE COUNCIL
71 Newcomen Street
London
SE1 1WT
Tel: 0207 7407 9992
Fax: 0207 7407 9908
Email: info@treecouncil.org.uk
Web: www.treecouncil.org.uk

TRUSSED RAFTER ASSOCIATION
31 Station Road
Sutton
Retford
Nottinghamshire
DN22 8PZ
Tel: 01777 869281
Fax: 01777 869281
Web: www.tra.org.uk
Email: info@tra.org.uk

TWI (FORMERLY THE WELDING INSTITUTE)
Granta Park
Great Abington
Cambridge
Cambridgeshire
CB1 6AL
Tel: 01223 891 162
Fax: 01223 892 588
E-mail: twi@twi.co.uk
Web: www.twi.co.uk

UK STEEL ASSOCIATION: REINFORCEMENT MANUFACTURING PRODUCT GROUP
Broadway House
Tothill Street
London
SW1H 9NQ
Tel: 0207 222 7777
Fax: 0207 222 2782

UNDERFLOOR HEATING MANUFACTURERS' ASSOCIATION
Belhaven House
67 Walton Road
East Moseley
Surrey
KT8 0DB
Tel: 0208 941 7177
Fax: 0208 941 815
Web: www.uhma.org.uk

VERMICULITE INFORMATION SERVICE
1A Guildford Business Park
Guildford
Surrey
GU2 8XG
Tel: 01483 242 100
Fax: 01483 242 101
E-mail: info@palabora.co.uk
Web: www.palabora.co.uk

USEFUL ADDRESSES FOR FURTHER INFORMATION

WALLCOVERING MANUFACTURERS ASSOCIATION
James House
Bridge Street
Leatherhead
Surrey
KT22 7EP
Tel: 01372 360 660
Fax: 01372 376 069
E-mail: Alison.brown@bcf.co.uk

WATERHEATER MANUFACTURERS ASSOCIATION
15 Edge Lanedge
Stetford
Manchester
M32 8HN
Tel: 0161 865 8915
Fax: 0161 866 8242

WATER RESEARCH CENTRE
Henley Road
Medmenham
Marlow
Buckinghamshire
SL7 2HD
Tel: 01491 636 500
Fax: 01491 636 501
E-mail: solutions@wrcplc.co.uk
Web: www.wrcplc.co.uk

WATER SERVICES ASSOCIATION
1 Queen Anne's Gate
London
SW1H 9BT
Tel: 0207 957 4567
Fax: 0207 344 1866
Email: info@water.org.uk

WELDING MANUFACTURERS' ASSOCIATION
Westminster Tower
3 Albert Embankment
London
SE1 7SL
Tel: 0207 793 3041
Fax: 0207 582 8020
Email: wma@beama.org.uk
Web: www.WMA.uk.com

WOOD PANEL INDUSTRIES FEDERATION
Grantham
LincolnshireHertfordshire
NG31 6LR
Tel: 01476 563 707
Fax: 01476 579 314
E-mail: enquiries@wpif.org.uk

ZINC DEVELOPMENT ASSOCIATION
42 Weymouth Street
London
W1N 3LQ
Tel: 0207 499 6636
Fax: 0207 493 1355

USEFUL WEB-SITES WITH COMPARATIVE PRICES

The following Web addresses/links are to a number of companies which publish construction material prices on the world-wide web. These are often basis/catalogue prices and phone-calls to the companies on individual projects will often result in additional discounts due to project size, a contractor's buying power etc.. Whenever using these sites, readers are advised to note whether prices exclude or include VAT. Examination of these sites can provide useful comparative prices for a wide range of construction materials

Company Name	Web url (http:// link)	Materials	Type of organisation	View-on-line or Download	Comments
Bernards Door Furniture Direct	www.doorfurnituredirect.co.uk	Ironmongery	Ironmongers	View-on-line	
Blanchford	www.blanchford.com	A range of building materials	General Builders Merchants	View-on-line	
British Hardwoods	www.britishhardwoods.co.uk	Hardwood joinery, etc.	Manufacturers	View-on-line	
Buildbase	www.buildbase.co.uk	A range of materials	General Builders Merchants	View-on-line	Identifies recent price rises
Buttles Products	www.buttles.com	A range of materials	General Builders Merchants	View-on-line	
Colour Centre	www.colourcentre.com	Painting materials	Decorating Suppliers	View and Download	
Construction Fixings	www.constructionfixings.com	Anchors, bolts, fixings, etc.	Manufacturers	Download	
Coppard Plant Hire Ltd	www.coppard.co.uk	Tool and plant hire	Plant hire	View-on-line	Includes site establishment items
Fairalls	www.fairalls.co.uk	A range of building materials	General Builders Merchants	View-on-line	
Fulham Timber Merchants Ltd	www.fulhamtimber.co.uk	Timber materials	Timber Merchants	View-on-line	
Harrison Hire Ltd	www.harrisonhire.co.uk	Tool and plant hire	Plant hire	View-on-line	
Joseph Parr (Middlesbrough) Ltd	www.jparrboro.co.uk	A range of building materials	Timber and Builders Merchants	View-on-line	
Long and Somerville	www.longandsomerville.co.uk	A range of building materials	Heavyside Builders Merchants	View-on-line	
Old House Store	www.oldhousestore.co.uk	A range of traditional building materials	Builders Merchants	View-on-line	For lime mortars and other traditional materials
Polypipe Terrain	www.polypipe.com	uPVC pipes etc.	Manufacturers	Download	
Richard Potter	www.fortimber.demon.co.uk	Timber materials	Timber Merchants	View-on-line	
Saint-Gobain Pipelines	www.saint-gobain-pipelines.co.uk	Pipes/ ittings, cast iron	Manufacturers	Download	
Screwfix	www.screwfix.com	A range of building materials	Builders Merchants	View-on-line	
Sheffield Insulations	www.sheffins.co.uk	Insulation materials	Builders Merchants	Download	

USEFUL WEB-SITES WITH COMPARATIVE PRICES

Company Name	Web url (http:// link)	Materials	Type of organisation	View-on-line or Download	Comments
Travis Perkins	www.trademate.com	A range of building materials	General Builders Merchants	View-on-line	
Wolseley UK Ltd	www.wolseley.co.uk	Heating and plumbing goods	Builders Merchants	View-on-line	Access to other companies in the group
Yorkshire Fittings	www.yorkshirefittings.co.uk	Pipe fittings, copper	Manufacturers	View and Download	

Index

References in brackets after page numbers refer to SMM7 and the Common Arrangement of Work sections.

Software and eBook Single-User Licence Agreement

We welcome you as a user of this Taylor & Francis Software and eBook and hope that you find it a useful and valuable tool. Please read this document carefully. **This is a legal agreement** between you (hereinafter referred to as the "Licensee") and Taylor and Francis Books Ltd. (the "Publisher"), which defines the terms under which you may use the Product. **By breaking the seal and opening the document inside the back cover of the book containing the access code you agree to these terms and conditions outlined herein. If you do not agree to these terms you must return the Product to your supplier intact, with the seal on the document unbroken.**

1. Definition of the Product

The product which is the subject of this Agreement, *Spon's Architects' and Builders' Price Book 2009* Software and eBook (the "Product") consists of:

1.1 Underlying data comprised in the product (the "Data")
1.2 A compilation of the Data (the "Database")
1.3 Software (the "Software") for accessing and using the Database
1.4 An electronic book containing the data in the price book (the "eBook")

2. Commencement and licence

2.1 This Agreement commences upon the breaking open of the document containing the access code by the Licensee (the "Commencement Date").

2.2 This is a licence agreement (the "Agreement") for the use of the Product by the Licensee, and not an agreement for sale.

2.3 The Publisher licenses the Licensee on a non-exclusive and non-transferable basis to use the Product on condition that the Licensee complies with this Agreement. The Licensee acknowledges that it is only permitted to use the Product in accordance with this Agreement.

3. Multiple use

For more than one user or for a wide area network or consortium, use is only permissible with the purchase from the Publisher of a multiple-user licence and adherence to the terms and conditions of that licence.

4. Installation and Use

4.1 The Licensee may provide access to the Product for individual study in the following manner: The Licensee may install the Product on a secure local area network on a single site for use by one user.

4.2 The Licensee shall be responsible for installing the Product and for the effectiveness of such installation.

4.3 Text from the Product may be incorporated in a coursepack. Such use is only permissible with the express permission of the Publisher in writing and requires the payment of the appropriate fee as specified by the Publisher and signature of a separate licence agreement.

4.4 The Product is a free addition to the book and no technical support will be provided.

5. Permitted Activities

5.1 The Licensee shall be entitled:

- 5.1.1 to use the Product for its own internal purposes;
- 5.1.2 to download onto electronic, magnetic, optical or similar storage medium reasonable portions of the Database provided that the purpose of the Licensee is to undertake internal research or study and provided that such storage is temporary;

5.2 The Licensee acknowledges that its rights to use the Product are strictly set out in this Agreement, and all other uses (whether expressly mentioned in Clause 6 below or not) are prohibited.

6. Prohibited Activities

The following are prohibited without the express permission of the Publisher:

6.1 The commercial exploitation of any part of the Product.

6.2 The rental, loan, (free or for money or money's worth) or hire purchase of this product, save with the express consent of the Publisher.

6.3 Any activity which raises the reasonable prospect of impeding the Publisher's ability or opportunities to market the Product.

6.4 Any networking, physical or electronic distribution or dissemination of the product save as expressly permitted by this Agreement.

6.5 Any reverse engineering, decompilation, disassembly or other alteration of the Product save in accordance with applicable national laws.

6.6 The right to create any derivative product or service from the Product save as expressly provided for in this Agreement.

6.7 Any alteration, amendment, modification or deletion from the Product, whether for the purposes of error correction or otherwise.

7. General Responsibilities of the License

7.1 The Licensee will take all reasonable steps to ensure that the Product is used in accordance with the terms and conditions of this Agreement.

7.2 The Licensee acknowledges that damages may not be a sufficient remedy for the Publisher in the event of breach of this Agreement by the Licensee, and that an injunction may be appropriate.

7.3 The Licensee undertakes to keep the Product safe and to use its best endeavours to ensure that the product does not fall into the hands of third parties, whether as a result of theft or otherwise.

7.4 Where information of a confidential nature relating to the product of the business affairs of the Publisher comes into the possession of the Licensee pursuant to this Agreement (or otherwise), the Licensee agrees to use such information solely for the purposes of this Agreement, and under no circumstances to disclose any element of the information to any third party save strictly as permitted under this Agreement. For the avoidance of doubt, the Licensee's obligations under this sub-clause 7.4 shall survive the termination of this Agreement.

8. Warrant and Liability

8.1 The Publisher warrants that it has the authority to enter into this agreement and that it has secured all rights and permissions necessary to enable the Licensee to use the Product in accordance with this Agreement.

8.2 The Publisher warrants that the Product as supplied on the Commencement Date shall be free of defects in materials and workmanship, and undertakes to replace any defective Product within 28 days of notice of such defect being received provided such notice is received within 30 days of such supply. As an alternative to replacement, the Publisher agrees fully to refund the Licensee in such circumstances, if the Licensee so requests, provided that the Licensee returns this copy of *Spon's Architects' and Builders' Price Book 2009* to the Publisher. The provisions of this sub-clause 8.2 do not apply where the defect results from an accident or from misuse of the product by the Licensee.

8.3 Sub-clause 8.2 sets out the sole and exclusive remedy of the Licensee in relation to defects in the Product.

8.4 The Publisher and the Licensee acknowledge that the Publisher supplies the Product on an "as is" basis. The Publisher gives no warranties:

8.4.1 that the Product satisfies the individual requirements of the Licensee; or

8.4.2 that the Product is otherwise fit for the Licensee's purpose; or

8.4.3 that the Data are accurate or complete or free of errors or omissions; or

8.4.4 that the Product is compatible with the Licensee's hardware equipment and software operating environment.

8.5 The Publisher hereby disclaims all warranties and conditions, express or implied, which are not stated above.

8.6 Nothing in this Clause 8 limits the Publisher's liability to the Licensee in the event of death or personal injury resulting from the Publisher's negligence.

8.7 The Publisher hereby excludes liability for loss of revenue, reputation, business, profits, or for indirect or consequential losses, irrespective of whether the Publisher was advised by the Licensee of the potential of such losses.

8.8 The Licensee acknowledges the merit of independently verifying Data prior to taking any decisions of material significance (commercial or otherwise) based on such data. It is agreed that the Publisher shall not be liable for any losses which result from the Licensee placing reliance on the Data or on the Database, under any circumstances.

8.9 Subject to sub-clause 8.6 above, the Publisher's liability under this Agreement shall be limited to the purchase price.

9. Intellectual Property Rights

9.1 Nothing in this Agreement affects the ownership of copyright or other intellectual property rights in the Data, the Database of the Software.

9.2 The Licensee agrees to display the Publishers' copyright notice in the manner described in the Product.

9.3 The Licensee hereby agrees to abide by copyright and similar notice requirements required by the Publisher, details of which are as follows:

"© 2009 Taylor & Francis. All rights reserved. All materials in *Spon's Architects' and Builders' Price Book 2009* are copyright protected. All rights reserved. No such materials may be used, displayed, modified, adapted, distributed, transmitted, transferred, published or otherwise reproduced in any form or by any means now or hereafter developed other than strictly in accordance with the terms of the licence agreement enclosed with *Spon's Architects' and Builders' Price Book 2009*. However, text and images may be printed and copied for research and private study within the preset program limitations. Please note the copyright notice above, and that any text or images printed or copied must credit the source."

9.4 This Product contains material proprietary to and copyedited by the Publisher and others. Except for the licence granted herein, all rights, title and interest in the Product, in all languages, formats and media throughout the world, including copyrights therein, are and remain the property of the Publisher or other copyright holders identified in the Product.

10. Non-assignment

This Agreement and the licence contained within it may not be assigned to any other person or entity without the written consent of the Publisher.

11. Termination and Consequences of Termination.

11.1 The Publisher shall have the right to terminate this Agreement if:

11.1.1 the Licensee is in material breach of this Agreement and fails to remedy such breach (where capable of remedy) within 14 days of a written notice from the Publisher requiring it to do so; or

11.1.2 the Licensee becomes insolvent, becomes subject to receivership, liquidation or similar external administration; or

11.1.3 the Licensee ceases to operate in business.

11.2 The Licensee shall have the right to terminate this Agreement for any reason upon two month's written notice. The Licensee shall not be entitled to any refund for payments made under this Agreement prior to termination under this sub-clause 11.2.

11.3 Termination by either of the parties is without prejudice to any other rights or remedies under the general law to which they may be entitled, or which survive such termination (including rights of the Publisher under sub-clause 7.4 above).

11.4 Upon termination of this Agreement, or expiry of its terms, the Licensee must destroy all copies and any back up copies of the product or part thereof.

12. General

12.1 **Compliance with export provisions**

The Publisher hereby agrees to comply fully with all relevant export laws and regulations of the United Kingdom to ensure that the Product is not exported, directly or indirectly, in violation of English law.

12.2 **Force majeure**

The parties accept no responsibility for breaches of this Agreement occurring as a result of circumstances beyond their control.

12.3 **No waiver**

Any failure or delay by either party to exercise or enforce any right conferred by this Agreement shall not be deemed to be a waiver of such right.

12.4 **Entire agreement**

This Agreement represents the entire agreement between the Publisher and the Licensee concerning the Product. The terms of this Agreement supersede all prior purchase orders, written terms and conditions, written or verbal representations, advertising or statements relating in any way to the Product.

12.5 **Severability**

If any provision of this Agreement is found to be invalid or unenforceable by a court of law of competent jurisdiction, such a finding shall not affect the other provisions of this Agreement and all provisions of this Agreement unaffected by such a finding shall remain in full force and effect.

12.6 **Variations**

This agreement may only be varied in writing by means of variation signed in writing by both parties.

12.7 **Notices**

All notices to be delivered to: Spon's Price Books, Taylor & Francis Books Ltd., 2 Park Square, Milton Park, Abingdon, Oxfordshire, OX14 4RN, UK.

12.8 **Governing law**

This Agreement is governed by English law and the parties hereby agree that any dispute arising under this Agreement shall be subject to the jurisdiction of the English courts.

If you have any queries about the terms of this licence, please contact:

Spon's Price Books
Taylor & Francis Books Ltd.
2 Park Square, Milton Park, Abingdon, Oxfordshire, OX14 4RN
Tel: +44 (0) 20 7017 6672
Fax: +44 (0) 20 7017 6702
www.tandfbuiltenvironment.com/

Software Installation and Use Instructions

System requirements

Minimum

- Pentium processor
- 256 MB of RAM
- 20 MB available hard disk space
- Microsoft Windows 98/2000/NT/ME/XP/Vista
- SVGA screen
- Internet connection

Recommended

- Intel 466 MHz processor
- 512 MB of RAM (1,024MB for Vista)
- 100 MB available hard disk space
- Microsoft Windows XP/Vista
- XVGA screen
- Broadband Internet connection

Microsoft® is a registered trademark and Windows™ is a trademark of the Microsoft Corporation.

Installation

Spon's Architects' and Builders' Price Book 2009 Electronic Version is supplied solely by internet download. No CD-ROM is supplied.

In your internet browser type in www.ebookstore.tandf.co.uk/supplements/pricebook and follow the instructions on screen. Then type in the unique access code which is sealed inside the back cover of this book.

If the access code is successfully validated, a web page with details of the available download content will be displayed.

Please note: you will only be allowed one download of these files and onto one computer.

Click on the download links to download the file. A folder called *PriceBook* will be added to your desktop, which will need to be unzipped. Then click on the application called *install*.

Use

- The installation process will create a folder containing the price book program links as well as a program icon on your desktop.
- Double click the icon (from the folder or desktop) installed by the Setup program.
- Follow the instructions on screen.

Technical Support

Support for the installation is provided on http://www.ebookstore.tandf.co.uk/html/helpdesk.asp

The *Electronic Version* is a free addition to the book. For help with the running of the software please visit www.pricebooks.co.uk

---

Multiple-user use of the Spon Press Software and eBook

To buy a licence to install your Spon Press Price Book Software and eBook on a secure local area network or a wide area network, and for the supply of network key files, for an agreed number of users please contact:

Spon's Price Books
Taylor & Francis Books Ltd.
2 Park Square, Milton Park, Abingdon, Oxfordshire, OX14 4RN
Tel: +44 (0) 207 017 6672
Fax: +44 (0) 207 017 6072
www.pricebooks.co.uk

Number of users	Licence cost
2–5	£450
6–10	£915
11–20	£1400
21–30	£2150
31–50	£4200
51–75	£5900
76–100	£7100
Over 100	Please contact Spon for details